STATISTICAL FIELD THEORY

Statistical Field Theory

An Introduction to Exactly Solved Models in Statistical Physics

Second Edition

Giuseppe Mussardo

SISSA (Scuola Internazionale Superiore di Studi Avanzati), Trieste

OXFORD
UNIVERSITY PRESS

OXFORD
UNIVERSITY PRESS

Great Clarendon Street, Oxford, OX2 6DP,
United Kingdom

Oxford University Press is a department of the University of Oxford. It furthers the University's objective of excellence in research, scholarship, and education by publishing worldwide. Oxford is a registered trade mark of Oxford University Press in the UK and in certain other countries

First Edition published in 2010
Second Edition published in 2020

Published in the United States of America by Oxford University Press
198 Madison Avenue, New York, NY 10016, United States of America

British Library Cataloguing in Publication Data

Data available

Library of Congress Control Number: 2019954655

ISBN 978–0–19–878810–2

DOI: 10.1093/oso/9780198788102.001.0001

Ulrich thought that the general and the particular are nothing but two faces of the same coin.

Robert Musil, *The Man Without Qualities*

Preface to the first edition

This book is an introduction to statistical field theory, an important subject of theoretical physics that has undergone formidable progress in recent years. Most of the attractiveness of this field comes from its profound interdisciplinary nature and its mathematical elegance; it sets outstanding challenges in several scientific areas, such as statistical mechanics, quantum field theory (QFT) and mathematical physics.

Statistical field theory deals, in short, with the behaviour of classical or quantum systems consisting of an enormous number of degrees of freedom. Those systems have different phases, and the rich spectrum of the phenomena they give rise introduces several questions: What is their ground state in each phase? What is the nature of the phase transitions? What is the spectrum of the excitations? Can we compute the correlation functions of their order parameters? Can we estimate their finite size effects? An ideal guide to the fascinating area of phase transitions is provided by the remarkable Ising model.

There are several reasons to choose the Ising model as a pathfinder in the field of critical phenomena. The first one is its simplicity—an essential quality to illustrate the key physical features of the phase transitions, without masking their derivation with worthless technical details. In the Ising model, the degrees of freedom are simple boolean variables $\sigma_{\vec{i}}$, whose values are $\sigma_{\vec{i}} = \pm 1$, defined on the sites $\vec{i}$ of a d-dimensional lattice. For these essential features, the Ising model has always played an important role in statistical physics, both at pedagogical and methodological levels.

However, this is not the only reason of our choice. The simplicity of the Ising model is, in fact, quite deceptive. Despite its apparent innocent look, the Ising model has shown an extraordinary ability in describing several physical situations and a remarkable theoretical richness. For instance, the detailed analysis of its properties involves several branches of mathematics, quite distinguished for their elegance: here we mention only combinatoric analysis, functions of complex variables, elliptic functions, the theory of non-linear differential and integral equations, the theory of the Fredholm determinant, and, finally, the subject of infinite dimensional algebras. Although this is only a partial list, it is sufficient to prove that the Ising model is an ideal playground for several areas of pure and applied mathematics.

Equally rich is its range of physical aspects. Therefore, its study offers the possibility to acquire a rather general comprehension of the phase transitions. Phase transitions are remarkable collective phenomena, characterized by sharp and discontinous changes of the physical properties of a statistical system. Such discontinuities typically occur at particular values of the external parameters (e.g. temperature or pressure); close to these critical values, there is a divergence of the mean values of many thermodynamical quantities, accompanied by anomalous fluctuations and power law behaviour of correlation functions. From an experimental point of view, phase transitions have an extremely rich phenomenology, ranging from the superfluidity of certain materials to the superconductivity of others, from the mesomorphic transformations of liquid crystals to the magnetic properties of iron. For example, liquid helium He^4 shows exceptional superfluid properties at temperatures lower than $T_c = 2.19K$, while several alloys show phase transitions equally remarkable, with an abrupt vanishing of the electrical resistance for very low values of the temperature.

The aim of the theory of phase transitions is to reach a general understanding of all the phenomena mentioned here on the basis of a few physical principles. Such a theoretical synthesis is made possible by a fundamental aspect of critical phenomena: their universality. This is a crucial property that depends on two basic features: the internal symmetry of the order parameters and the dimensionality of the lattice. In short, this means that, despite the differences that two systems may have at their microscopic level, as far as they share the two features mentioned, their critical behaviours are surprisingly identical.[1] It is for these universal aspects that the theory of the phase transitions is one of the pillars of statistical mechanics and, simultaneously, of theoretical physics. As a matter of fact, it embraces concepts and ideas that proved to be the building blocks of the modern understanding of the fundamental interactions in nature. The universal behaviour, for instance, has its natural demonstration within the general ideas of the renormalization group, while the existence itself of a phase transition can be interpreted as a spontaneously symmetry breaking of the Hamiltonian of the system. As well known, both are common concepts in QFT, another important area of theoretical physics, i.e. the theory that deals with the fundamental interactions of the last constituents of the matter—the elementary particles.

The relationship between two theories that describes such different phenomena may appear, at first sight, quite surprising. However, it becomes more comprehensible if we take into account two aspects: first, that both theories deal with systems of infinite degrees of freedom, and second, that, close to the phase transitions, the excitations of the systems have the same dispersion relations of the elementary particles.[2] Due to the essential identity of the two theories, we should not be surprised to discover that the two-dimensional Ising model, at temperature T slightly away from T_c and in absence of an external magnetic field, is equivalent to a fermionic neutral particle (a Majorana fermion), which satisfies a Dirac equation. Similarly, at $T = T_c$ but in the presence of an external magnetic field B, the two-dimensional Ising model may be regarded as a QFT with eight scalar particles of different masses.

The use of QFT, i.e. those formalisms and methods that led to brilliant results in the study of the fundamental interactions of photons, electrons and all other elementary particles, has produced remarkable progress both in the understanding of phase transitions and in the computation of their universal quantities. As explained in this book, our

[1] This becomes evident by choosing an appropriate combination of the thermodynamical variables of the two systems.

[2] The explicit identification between the two theories can be proved by adopting for both the path of integral formalism.

study significantly benefits from such a possibility: since phase transitions are phenomena that involve the long distance scales of the systems—the infrared scales—the adoption of the continuum formalism of field theory is not only extremely advantageous from a mathematical point of view, but also is perfectly justified from a physical point of view. By adopting the QFT approach, the discrete structure of the original statistical models shows itself only through an ultraviolet microscopic scale, related to the lattice spacing. However, it is worth pointing out that this scale is absolutely necessary to regularize the ultraviolet divergencies of QFT and to implement its renormalization.

The main advantage of QFT is that it embodies a strong set of constraints coming from the compatibility of quantum mechanics with special relativity. This turns into general relations, such as the completeness of the multi-particle states or the unitarity of their scattering processes. Thanks to these general properties, QFT makes it possible to understand, in a very simple and direct way, underlying aspects of phase transitions that may appear mysterious, or at least not evident, in the discrete formulation of the corresponding statistical model.

There is one subject that has particularly improved thanks to this continuum formulation: the set of two-dimensional statistical models, for which we can achieve a classification of the fixed points and a detailed characterization of their classes of universality. Let us briefly discuss the nature of the two-dimensional QFTs.

Right at the critical points, the QFTs are massless. Such theories are invariant under the conformal group, i.e. the set of geometrical transformations that implement a scaling of the length of the vectors while preserving their relative angle. But, in two dimensions conformal transformations coincide with mappings by analytic functions of a complex variable, characterized by an infinite-dimensional algebra known as Virasoro algebra. This enables us to identify first the operator content of the models (in terms of the irreducible representations of the Virasoro algebra) and then to determine the exact expressions of the correlators (by solving certain linear differential equations). In recent years, thanks to the methods of conformal field theory CFT, physicists have reached the exact solutions of a huge number of interacting quantum theories, with the determination of all their physical quantities, such as anomalous dimensions, critical exponents, structure constants of the operator product expansions, correlation functions, partition functions, etc.

Away from criticality, QFTs are, instead, generally massive. Their analysis can be often carried out only by perturbative approaches. However, there are some favourable cases that give rise to integrable models of great physical relevance. The integrable models are characterized by the existence of an infinite number of conserved charges. In such fortunate circumstances, the exact solution of the off-critical models can be achieved by means of S-matrix theory. This approach allows us to compute the exact spectrum of the excitations and the matrix elements of the operators on the set of these asymptotic states. Both these data can be thus employed to compute the correlation functions by the spectral series. These expressions enjoy remarkable convergence properties that turn out to be particularly useful for the control of their behaviours both at large and short distances. Finally, in the integrable cases, it is also possible to study the exact thermodynamical properties and the finite size effects of the QFTs. Exact predictions for many universal quantities can also be obtained. For the two-dimensional Ising model, for instance, there are two distinct integrable theories, one corresponding to its thermal perturbation (i.e. $T \neq T_c, B = 0$), the other to the magnetic deformation ($B \neq 0, T = T_c$). In the last case, a universal quantity is given, for instance, by the ratio of the masses of the lowest excitations, expressed by the famous golden ratio $m_2/m_1 = 2\cos\frac{\pi}{5} = \frac{\sqrt{5}+1}{2}$.

In addition to their notable properties, the exact solution provided by the integrable theories is an important step towards the general study of the scaling region close to the critical points. In fact, they permit an efficient perturbative scheme to study non-integrable effects, in particular, to follow how the mass spectrum changes by varying the coupling constants. Thanks to this approach, new progress has been made in understanding several statistical models, in particular the class of universality of the Ising model by varying the temperature T and the magnetic field B. Non-integrable field theories present an extremely interesting set of new physical phenomena, such as confinement of topological excitations, decay processes of the heavier particles, the presence of resonances in scattering processes or false vacuum decay, etc. The analytic control of such phenomena is one of the most interesting results of QFT in the realm of statistical physics.

This book is a long and detailed journey through several fields of physics and mathematics. It is based on the elaboration of the lecture notes for a PhD course, held by the author at the International School for Advances Studies (Trieste). During this elaboration process, particular attention has been paid to achieving a coherent and complete picture of all surveyed topics. The effort done to emphasize the deep relations among several areas of physics and mathematics reflects the profound belief of the author in the substantial unity of scientific knowledge.

This book is designed for students in physics or mathematics (at the graduate level or in the last years of their undergraduate courses). For this reason, its style is greatly pedagogical; it assumes only some basis of mathematics, statistical physics and quantum mechanics. Nevertheless, we count on the intellectual curiosity of the reader.

Preface to the second edition

Almost nine years have passed since the first edition of this volume. During this period many of the subjects that were discussed in the book grew in importance, deeply influencing the development of many other areas of physics and mathematics. Topics and concepts as CFT, quantum integrability, S-matrix, braiding group, Bethe ansatz, Majorana fermions—just to mention very few—have become indeed more and more decisive tools in areas as diverse as quantum systems out of equilibrium, cold atoms physics, topological phases of matter, quantum computations, non-abelian anyon statistics, strongly correlated systems, quantum entanglement, 2D quantum gravity, holography and higher dimensional conformal models. In short, they are now an essential part of the modern background of a theoretical physicist.

The intriguing ideas mentioned here turn out to be not only bold speculations of purely theoretical nature but also guiding lines of a series of significant experimental advances on a scale never experienced before.

There are more things in heaven and earth, Horatio,
than are dreamt of in your philosophy.

If necessary, the famous words of Hamlet remind us that the natural world has always been the place of very fascinating novelties and, no doubt, this is also the case of low-dimensional QFTs, thanks in particular to new material design and cold atom simulations. As a significant representative of many other beautiful experiments and a key example of the ideal marriage between theory and experiment, it is enough to mention here the explicit realization[3] of the class of universality of the 2D Ising model in a magnetic field, a model not only important for its physical and historical relevance but also for its deep relation with beautiful mathematical object such as E_8 Lie algebra, exact S-matrix theory and form factors.

The great vitality of the subject provides reasons to publish this second edition. In addition to a substantial editing of the text, misprints and references of the previous version, this new edition also features three new chapters that deal with some important subjects missing from the first edition: (a) boundary field theories; (b) semi-classical methods and (c) truncated conformal space approach. The addition of these extra chapters motivates a different arrangement of the previous chapters, now organised into six parts, and a few comments may help to underline the relevance of these new topics.

Boundary field theory has proved to be a very rich subject from its very beginning, and in practice it is the correct theoretical framework to address many important and diverse

[3] R. Coldea et al. Quantum Criticality in an Ising Chain: Experimental Evidence for Emergent E_8 Symmetry, *Science* 327 (2010), 177–80.

subjects, including genuine aspects of statistical physics (the study and classification of surface critical behaviour, the computation of surface critical exponents, etc.), quantum impurity problems (e.g. the Kondo problem), or the dynamics of extended quantum systems suddenly brought out of equilibrium through a protocol known as *quantum quench*. In all these cases there are very distinguished sets of equations and constraints that shape the dynamics of these systems both at the critical point and away from it. A crucial object that emerges from these studies is the concept of boundary states, and the characterization of such states – as discussed in detail in the text – assumes a particular elegant form both in CFTs and integrable models. So, boundary field theory nicely integrates the other parts of the volume devoted to conformal invariance and integrability and considerably enlarges their scope.

To appreciate the value and the role of the other new subjects we need to spend few words on the topic of non-integrable model has recently captured more and more the attention of many researchers. In fact, given all we learned from the exact analysis of integrable models, it was natural to try to extend the frontiers of our knowledge also to those models that cannot be solved exactly. Regardless, they make up the majority of models that a scientist encounters, and therefore any advance in their understanding, no matter how approximate, may be regarded as highly valuable. In this respect, there have been significant advances in elucidating many of the properties of these models, the level of precision depending of course on the degree of their '*non-integrability*', so to speak. The first edition of this book ended with a chapter devoted to the study of a particular class of non-integrable models, i.e. those that can be regarded as close to integrable ones and therefore treated with a particular form of perturbation theory based on the form factors of the integrable theories. However, this class does not exhaust all possible cases. To fill partially this gap, in the second edition we have included three other chapters that deal, in one way or another, with different aspects of non-integrable models.

The first two chapters address semi-classical methods, a subject that has a long tradition in QFT. This set of techniques has its own goals and formalisms but as far as we are concerned its main advantages is to rely neither on integrability nor the opposite. In general, the application of these methods may be quite an elaborated procedure, although it is a great deal simpler if applied to the analysis of QFTs with vacua degeneracy and topological kink excitations. These are our starting points for a thorough study of an important subject of QFTs: the spectrum of the bound states of bosonic or fermionic particles.

The third chapter presents the so-called truncated conformal space approach is a purely numerical algorithm, although it is quite different from the more familiar Monte Carlo approaches or similar. Its implementation does not require the integrability of the theory (but it does not exclude either); rather it is entirely based on the underlying CFT that describes the ultraviolet fixed point of a QFT. Such a CFT provides the Hilbert space for setting up, on a cylinder geometry of radius R, a suitable truncated Hamiltonian built up on the conformal states and their structure constants. While all matrix elements of this Hamiltonian can be determined in an exact form, its eigenvalues as functions of R have to be computed numerically. In this way we can have access to important information regarding the spectrum of the excitations, their number below threshold, the presence of false vacua and resonances, finite size corrections of the various masses, vacuum expectation values of various operators, etc. In short, this algorithm provides concrete and detailed answers to a series of questions related to the scaling region around the critical points. Interestingly enough, it also can be extended to higher dimensional theories.

Although these new chapters increase the extent of the first edition, we consider them necessary instrumental to update, balance and complete the new edition. As before, we count on the intellectual curiosity of our readers.

Structure of the book

In this book many topics are discussed at a fairly advanced level but using a pedagogical approach. I believe that a student could highly profit from some exposure to such treatments. The book is divided in six parts.

Part 1: **Preliminary Notions** (Chapters 1, 2 and 3)

Part I deals with the fundamental aspects of phase transitions and uses explicit examples coming from the Ising model or similar systems.

Chapter 1: The first chapter provides a straighforward introduction of essential ideas on second-order phase transitions and their theoretical challenge. We focus on some important issues, e.g. order parameters, correlation length, correlation functions, scaling behaviour, critical exponents, etc. We also discuss the Ising model and its most significant developments during the years of its study. The chapter also contains two appendices that summarize all relevant results of classical and statistical mechanics.

Chapter 2: This chapter deals with one-dimensional statistical models, e.g. the Ising model and its generalizations (Potts model, systems with $O(n)$ or Z_n symmetry, etc.). It discusses several methods of solution: the recursive method, the transfer matrix approach or series expansion techniques. It also covers general properties of these methods—valid on higher dimensional lattices. The contents of this chapter are quite simple and pedagogical but extremely useful for understanding the following sections of the book. One of the appendices at the end of the chapter is devoted to a famous problem of topology, i.e. the four-colour problem, and its relation with the two-dimensional Potts model.

Chapter 3: This chapter discusses the approximation schemes to approach lattice statistical models that are not exactly solvable. In addition to the mean field approximation, we also consider the Bethe–Peierls approach to the Ising model. Moreover, there is a thorough discussion of the Gaussian model and its spherical version—two important systems with several points of interest. One of the appendices features a detailed analysis of the random walk on different lattices: apart from the importance of the subject on its own, it is shown that the random walk is responsible for the critical properties of the spherical model.

Part 2: **Bi-dimensional Lattice Models** (Chapters 4, 5 and 6)

Part II provides a general introduction to the key ideas of equilibrium statistical mechanics of discrete systems.

Chapter 4: The chapter begins by discussing the Peierls argument (it permits proof of the existence of a phase transition in the two-dimensional Ising model). The rest of the chapter deals with duality transformations that link the low- and the high-temperature phases of several statistical models. Particularly important is the proof of the so-called *star-triangle identity*. This identity is crucial in the later discussion of the transfer matrix of the Ising model (Chapter 6).

Chapter 5: The key topics of the chapter are the two exact combinatorial solutions of the two-dimensional Ising model. Although no subsequent topic depends on them, both the mathematical and the physical aspects of these solutions are so elegant that they deserve a special attention.

Chapter 6: This chapter discusses the exact solution of the two-dimensional Ising model achieved through the transfer matrix formalism. A crucial role is played by the commutativity properties of the transfer matrices, which lead to a functional equation for their eigenvalues. The exact free-energy of the model and its critical point can be identified by means of the lowest eigenvalue.

Part 3: Quantum Field Theory and Conformal Invariance (Chapters 7–14)

Part III makes up the central part of the book and discusses the aims of QFT and some of its fundamental results. A central point is the bootstrap method of CFTs. The main goal of this part is to show the extraordinary efficiency of these techniques for the analysis of critical phenomena.

Chapter 7: This chapter emphasizes the main reasons for adopting the methods of QFT to study the critical phenomena. It presents both the canonical quantization and the path integral formulation of the field theories together with the analysis of the perturbation theory. Everything in this chapter will be needed sooner or later, as it highlights most of the relevant aspects of QFT.

Chapter 8: This chapter introduces the key ideas of the renormalization group. They involve the scaling transformations of a system and their implementations in the space of the coupling constants. This analysis leads to the important notion of relevant, irrelevant and marginal operators and then to the universality of the critical phenomena.

Chapter 9: This chapter introduces a crucial aspect of the Ising model—its fermionic nature—and the property of the model. In the continuum limit, a Dirac equation for neutral Majorana fermions emerges. The details of the derivation are much less important than understanding why it is possible. This chapter also covers the simplicity and the exactness of the result.

Chapter 10: This chapter introduces the notion of conformal transformations and the important topic of the massless QFTs associated to the critical points of the statistical models. It establishes the important conceptual result that the classification of all possible critical phenomena in two dimensions consists of finding out all possible irreducible representations of the Virasoro algebra.

Chapter 11: This chapter discusses the so-called minimal conformal models, characterized by a finite number of representations. It shows that all correlation functions of these models satisfy linear differential equations and their explicit solutions are given by using the Coulomb gas method. Their exact partition functions can be obtained by enforcing the modular invariance of the theory.

Chapter 12: Free theories are usually regarded as trivial examples of quantum systems. This chapter proves that this is not the case of the CFTs associated to the free bosonic and fermionic fields. The subject is not only full of beautiful mathematical identities but is also the source of deep physical concepts with far-reaching applications.

Chapter 13: The conformal transformations may be part of a larger group of symmetry and this chapter discusses several of its extensions: supersymmetry, Z_n transformations, and current algebras. The Appendix provides the reader with a self-contained discussion on the Lie algebras.

Chapter 14: This chapter discusses the identification of a class of universality—one of the central questions in statistical physics. This chapter discusses in detail the class of universality of several models, such as the Ising model, the tricritical Ising model, or the Potts model.

Part 4: **Away from Criticality** (Chapters 15–19)

This part of the book develops the analysis of the statistical models away from criticality.

Chapter 15: Chapter 15 introduces the notion of *the scaling region* near the critical points, identified by the deformations of the critical action by means of the relevant operators. The renormalization group flows that originate from these deformations are subjected to important constraints that can be expressed in terms of sum-rules. This chapter also discusses the nature of the perturbative series based on the conformal theories.

Chapter 16: This chapter covers the general properties of the integrable quantum field theories. They are illustrated by means of significant examples, such as the Sine–Gordon model or the Toda field theories based on the simple roots of a Lie algebra. For the deformations of a conformal theory, it shows how to set up an efficient counting algorithm to prove the integrability of the corresponding model.

Chapter 17: This chapter deals with the analytic theory of the S-matrix of the integrable models. It focuses on the dynamical principle of *bootstrap*, which gives rise to a recursive structure of the amplitudes. Several dynamical quantities, such as mass ratios or three-coupling constants, have an elegant mathematic formulation, which is also of easy geometrical interpretation.

Chapter 18: This chapter discusses the Ising model in a magnetic field, which is one of the most beautiful example of an integrable model. It presents its exact S-matrix and the exact spectrum of its excitations, which consist of eight particles of different masses. Similarly, it discusses the exact scattering theory behind the thermal deformation of the tricritical Ising model and the unusual features of the exact S-matrix of the non-unitary Yang–Lee model. Other important examples are provided by $O(n)$ invariant models: when $n = 2$, one obtains the important case of the Sine–Gordon model.

Chapter 19: Chapter 19 discusses how the correlation functions of the various fields are at the heart of a quantum field theory. In the case of integrable models, the correlators can be expressed in terms of the spectral series based on the matrix elements on the asymptotic states. These matrix elements, also known as form factors, satisfy a set of functional and recursive equations that can exactly solved in many cases of physical interest.

Part 5: Finite Size Effects (Chapters 20 and 21)

This part of the book discuss various formalisms for addressing the physical phenomena that emerge in a finite geometry or at finite temperature.

Chapter 20: This chapter covers the thermodynamic Bethe ansatz permits to study finite size and finite temperature effects of an integrable model. Here we derive the integral equations that determine the free energy and we give their physical interpretation.

Chapter 21: This chapter covers QFT with boundaries—a very rich subject both from purely theoretical point of view or applicative aims. It presents the basis results relative to critical systems and massive integrable field theories.

Part 6: Non-integrable Aspects (Chapters 22 and 25)

This part of the book deals with non-integrable QFTs. It presents the rich phenomenology that emerges in these cases and discusses two formalisms that allow us to reach a certain control on these new phenomena.

Chapter 22: This chapter introduces a perturbative technique based on the form factors to study non-integrable models. Such a technique permits the computation of the corrections to the mass spectrum, the vacuum energy, the scattering amplitudes, and so on.

Chapter 23: This chapter covers the setting of a semi-classical method to address the computation of the spectrum of bound states in QFT with a set of degenerate vacua connected by kink excitations.

Chapter 24: This chapter addresses the semi-classical formalism relative to fermion field in a bosonic background. It also discusses the conditions under which the overall theory presents supersymmetry and the consequences thereof.

Chapter 25: The truncated conformal space approach provides a very efficient numerical algorithm to study many properties of a perturbed CFT defined on a finite geometry, typically an infinite cylinder of radius R. These include: the masses of the various excitations, their number below threshold, the presence of false vacua and resonances, on-shell three-particle coupling, etc. The implementation of this approach does not depend on the integrability of the off-critical model and therefore it is a very useful tool to extend our study to the entire scaling region around the critical point. This chapter discusses the basis of such an algorithm and shows some interesting applications thereof.

Problems Each chapter of this book includes a series of problems. They have different levels of difficulty: some of them relate directly to the essential material of the chapters, while others are instead designed to introduce new applications or even new topics. The problems are an integral part of the course and their solution is a crucial step for the understanding of the whole subject.

Mathematical Aspects Several chapters have one or more appendices devoted to some mathematical aspects encountered in the text. Far from being a collection of formulae, these appendices aim to show the profound relationship between mathematics and physics. Quite often, they also give the opportunity to achieve a comprehension of mathematical results by means of a physical intuition. Some appendices are also devoted to providing a historical perspective.

References An annotated bibliography appears at the end of each chapter. The list of references, either books or articles, is by no means meant to be a comprehensive survey of the present literature. Rather, they are meant to guide the reader a bit deeper if he/she wishes to go on. They also refer to the list of material consulted in preparing the chapters. There are no quotation to references in the text, except for few technical points.

Acknowledgements

During the years I have had the pleasure to collaborate and discuss many of the themes of this book with several colleagues and friends.

First of all, I thank John Cardy: his extraordinary scientific vision has been along the years a very precious guide. Similarly, I have also a special debt to Adam Schwimmer and Vladimir Rittenberg for their constant encouragement and for having the enthusiam to face any scientific themes with a great perspicacity and intelligence. I am also particularly grateful to my friend Aliosha Zamolodchikov, with whom I had the privilege to discuss many important topics. I also thank Gesualdo Delfino for a long collaboration on the two-dimensional QFT.

I have been fortunate in having the benefit of collaboration and discussion with numerous colleagues who have generously shared their insights, in particular Olivier Babelon, Denis Bernard, Pasquale Calabrese, Andrea Cappelli, Jean-Sebastian Caux, Ed Corrigan, Boris Dubrovin, Patrick Dorey, Fabian Essler, Paul Fendley, Robert Konik, Vladimir Kravstov, Andrea Gambassi, André LeClair, Serguey Lukyanov, Alexander Nersesyan, Paul Pearce, Hubert Saleur, Giuseppe Santoro, Kareljan Schoutens, Fedor Smirnov, Erik Tonni, Alexei Tsvelik, Sasha Zamolodchikov, and Jean-Bernard Zuber. I also thank my collaborators Carlo Acerbi, Alvise Bastianello, Michele Burrello, Daniel Cabra, Filippo Colomo, Axel Cubero, Andrea De Luca, Luca Lepori, Alessandro De Martino, Davide Fioravanti, Davide Fioretto, Marton Kormos, Anne Koubek, Marco Moriconi, Paola Mosconi, Alessandro Mossa, Milosz Panfil, Silvia Penati, Valentina Riva, Alessandro Silva, Prospero Simonetti, Galen Sotkov, Spyros Sotiriadis, Gabor Takacs, Roberto Tateo, Andrea Trombettoni, and Jacopo Viti.

Contents

Part 1 Preliminary Notions

1 Introduction 3
- 1.1 Phase Transitions 4
- 1.2 The Ising Model 19
- 1.3 Ernst Ising 21
- Appendix 1.A. Ensembles in Classical Statistical Mechanics 22
- Appendix 1.B. Ensembles in Quantum Statistical Mechanics 27
- Problems 41

2 One-dimensional Systems 48
- 2.1 Recursive Approach 48
- 2.2 Transfer Matrix 55
- 2.3 Series Expansions 64
- 2.4 Critical Exponents and Scaling Laws 66
- 2.5 The Potts Model 67
- 2.6 Models with $O(n)$ Symmetry 73
- 2.7 Models with Z_n Symmetry 81
- 2.8 Feynman Gas 84
- Appendix 2.A. Special Functions 86
- Appendix 2.B. n-dimensional Solid Angle 93
- Appendix 2.C. The Four-colour Problem 95
- Problems 102

3 Approximate Solutions 106
- 3.1 Mean Field Theory of the Ising Model 106
- 3.2 Mean Field Theory of the Potts Model 112
- 3.3 Bethe–Peierls Approximation 116
- 3.4 The Gaussian Model 120
- 3.5 The spherical model 130
- Appendix 3.A. The Saddle Point Method 137
- Appendix 3.B. Brownian Motion on a Lattice 141
- Problems 154

Part 2 Bi-dimensional Lattice Models

4 Duality of the Two-dimensional Ising Model 161
4.1 Peierls Argument 162
4.2 Duality Relation in Square Lattices 163
4.3 Duality Relation: Hexagonal and Triangular Lattices 170
4.4 Star-triangle Identity 172
4.5 Ising Model Critical Temperature: Triangle and Hexagonal Lattices 174
4.6 Duality in Two Dimensions 176
Appendix 4.A. Numerical Series 183
Appendix 4.B. Poisson Sum Formula 184
Problems 186

5 Combinatorial Solutions of the Ising Model 189
5.1 Combinatorial Approach 189
5.2 Dimer Method 200
Problems 209

6 Transfer Matrix of the Two-dimensional Ising Model 211
6.1 Baxter's Approach 212
6.2 Eigenvalue Spectrum at the Critical Point 223
6.3 Away from the Critical Point 227
6.4 Yang–Baxter Equation and R-matrix 227
Problems 233

Part 3 Quantum Field Theory and Conformal Invariance

7 Quantum Field Theory 237
7.1 Motivations 237
7.2 Order Parameters and Lagrangian 239
7.3 Field Theory of the Ising Model 243
7.4 Correlation Functions and Propagator 246
7.5 Perturbation Theory and Feynman Diagrams 250
7.6 Legendre Transformation and Vertex Functions 256
7.7 Spontaneous Symmetry Breaking and Multi-criticality 259
7.8 Renormalization 264
7.9 Field Theory in Minkowski Space 267
7.10 Particles 272
7.11 Correlation Functions and Scattering Processes 276
Appendix 7.A. Feynman Path Integral Formulation 278
Appendix 7.B. Relativistic Invariance 280
Appendix 7.C. Noether Theorem 283
Problems 285

8 **Renormalization Group** 289
8.1 Introduction 289
8.2 Reducing the Degrees of Freedom 291
8.3 Transformation Laws and Effective Hamiltonians 292
8.4 Fixed Points 296
8.5 The Ising Model 300
8.6 The Gaussian Model 303
8.7 Operators and Quantum Field Theory 304
8.8 Functional Form of the Free Energy 307
8.9 Critical Exponents and Universal Ratios 309
8.10 β-Functions 313
Problems 317

9 **Fermionic Formulation of the Ising Model** 319
9.1 Introduction 319
9.2 Transfer Matrix and Hamiltonian Limit 320
9.3 Order and Disorder Operators 325
9.4 Perturbation Theory 327
9.5 Expectation Values of Order and Disorder Operators 329
9.6 Diagonalization of the Hamiltonian 331
9.7 Dirac Equation 336
Problems 339

10 **Conformal Field Theory** 341
10.1 Introduction 341
10.2 The Algebra of Local Fields 342
10.3 Conformal Invariance 347
10.4 Quasi-primary Fields 351
10.5 Two-dimensional Conformal Transformations 353
10.6 Ward Identity and Primary Fields 358
10.7 Central Charge and Virasoro Algebra 362
10.8 Representation Theory 368
10.9 Hamiltonian on a Cylinder Geometry and Casimir Effect 378
10.10 Entanglement Entropy 381
Appendix 10.A. Moebius Transformations 386
Problems 395

11 **Minimal Conformal Models** 399
11.1 Introduction 399
11.2 Null Vectors and Kac Determinant 400
11.3 Unitary Representations 403

11.4 Minimal Models 405
11.5 Coulomb Gas 412
11.6 Landau–Ginzburg Formulation 425
11.7 Modular Invariance 429
Appendix 11.A. Hypergeometric functions 437
Problems 440

12 Conformal Field Theory of Free Bosonic and Fermionic Fields 443

12.1 Introduction 443
12.2 Conformal Field Theory of Free Bosonic Fields 443
12.3 Conformal Field Theory of a Free Fermionic Field 455
12.4 Bosonization 467
Problems 471

13 Conformal Field Theories with Extended Symmetries 476

13.1 Introduction 476
13.2 Superconformal Models 476
13.3 Parafermion Models 482
13.4 Kac–Moody Algebra 489
13.5 Conformal Models as Cosets 502
Appendix 13.A. Lie Algebra 505
Problems 516

14 The Arena of Conformal Models 518

14.1 Introduction 518
14.2 The Ising Model 518
14.3 The Universality Class of the Tricritical Ising Model 530
14.4 3-state Potts Model 533
14.5 The Yang–Lee Model 537
14.6 Conformal Models with $O(n)$ Symmetry 539
Problems 542

Part 4 Away from Criticality

15 In the Vicinity of the Critical Points 545

15.1 Introduction 545
15.2 Conformal Perturbation Theory 548
15.3 Example: The Two-point Function of the Yang–Lee model 554
15.4 Renormalization Group and β-functions 556
15.5 c-theorem 562
15.6 Applications of the c theorem 565
15.7 Δ theorem 572

16 Integrable Quantum Field Theories 575

16.1 Introduction 575
16.2 The Sinh–Gordon Model 576
16.3 The Sine–Gordon Model 582
16.4 The Bullogh–Dodd Model 588
16.5 Integrability versus Non-integrability 590
16.6 The Toda Field Theories 593
16.7 Toda Field Theories with Imaginary Coupling Constant 604
16.8 Deformation of Conformal Conservation Laws 606
16.9 Multiple Deformations of Conformal Field Theories 615
Problems 619

17 *S*-matrix Theory 622

17.1 Analytic Scattering Theory 623
17.2 General Properties of Purely Elastic Scattering Matrices 634
17.3 Unitarity and Crossing Invariance Equations 641
17.4 Analytic Structure and Bootstrap Equations 646
17.5 Conserved Charges and Consistency Equations 651
Appendix 17.A. Historical Developments of the *S*-matrix Theory 655
Appendix 17.B. Scattering Processes in Quantum Mechanics 659
Appendix 17.C. *n*-particle Phase Space 664
Problems 672

18 Exact S-matrices 676

18.1 Yang–Lee and Bullogh–Dodd Models 676
18.2 $\Phi_{1,3}$ Integrable Deformation of the Conformal Minimal Models $\mathcal{M}_{2,2n+3}$ 680
18.3 Multiple Poles 682
18.4 *S*-matrices of the Ising Model 684
18.5 The Tricritical Ising Model at $T \neq T_c$ 692
18.6 Thermal Deformation of the 3-state Potts Model 696
18.7 General Expression Toda Field Theories 700
18.8 Non-relativistic Limit of Toda Field Theories 701
18.9 Models with Internal $O(n)$ Invariance 704
18.10 *S*-matrix of the Sine-Gordon Model 710
18.11 S-matrices for $\Phi_{1,3}$, $\Phi_{1,2}$, $\Phi_{2,1}$ Deformation of Minimal Models 715
18.12 Elastic SUSY *S*-matrix 730
Problems 739

19 Form Factors and Correlation Functions 744

19.1 General Properties of the Form Factors 745
19.2 Watson's Equations 748

19.3 Recursive Equations 751
19.4 The Operator Space 752
19.5 Correlation Functions 754
19.6 Form Factors of the Stress-energy Tensor 757
19.7 Vacuum Expectation Values 759
19.8 Ultraviolet Limit 763
19.9 The Ising Model at $T \neq T_c$ 766
19.10 Form Factors of the Sinh–Gordon Model 772
19.11 The Ising Model in a Magnetic Field 780
Problems 785

Part 5 Finite Size Effects

20 Thermodynamic Bethe Ansatz 791

20.1 Introduction 791
20.2 Casimir Energy 791
20.3 Bethe Relativistic Wave Function 794
20.4 Derivation of Thermodynamics 796
20.5 The Meaning of Pseudo-energy 803
20.6 Infrared and Ultraviolet Limits 806
20.7 The Coefficient of Bulk Energy 808
20.8 The General Form of the TBA Equations 811
20.9 The Exact Relation $\lambda(m)$ 814
20.10 Examples 817
20.11 Thermodynamics of the Free Field Theories 821
20.12 L-channel Quantization 822
20.13 LeClair–Mussardo Formula 828
Problems 834

21 Boundary Field Theory 836

21.1 Introduction 836
21.2 Stress-energy Tensor in Boundary CFT 837
21.3 Conformal Boundary Operators 839
21.4 Conformal Boundary States 842
21.5 Operator Product Expansion Involving a Boundary Operator 847
21.6 Massive Integrable Boundary Field Theory 848
21.7 Boundary States 850
21.8 Massive Boundary Ising Model 851
21.9 Correlation Functions 854
Problems 858

Part 6 Non-Integrable Aspects

22 Form Factor Perturbation Theory 863

22.1 Breaking Integrability 863
22.2 Multiple Deformations of the Conformal Field Theories 865
22.3 Form Factor Perturbation Theory 867
22.4 First-order Perturbation Theory 871
22.5 Non-locality and Confinement of the Excitations 874
22.6 Multi-frequency Sine–Gordon Model 876
Problems 880

23 Particle Spectrum by Semi-classical Methods 882

23.1 Introduction 882
23.2 Kinks 883
23.3 A Semi-classical Formula for the Kink Matrix Elements 888
23.4 Universal Mass Formula 890
23.5 Symmetric Wells 893
23.6 Asymmetric Wells 897
23.7 Double Sine–Gordon Model 901
Problems 910

24 Interacting Fermions and Supersymmetric Models 913

24.1 Introduction 913
24.2 Fermion in a Bosonic Background 914
24.3 The Fermionic Bound States in $\mathcal{T} = 0$ Sector 918
24.4 Symmetric Wells 921
24.5 Supersymmetric Theory 923
24.6 General Results in SUSY Theories 925
24.7 Integrable SUSY Models 927
24.8 Non-integrable Multi-frequency Super Sine-Gordon Models 930
24.9 Phase Transition and Meta-stable States 933
24.10 Summary 937
Problems 939

25 Truncated Hilbert Space Approach 943

25.1 Truncated Hamiltonians of Quantum Mechanics 944
25.2 Truncated Hamiltonian of the Deformed Conformal Models 951
25.3 Finite-size Mass Corrections 962
25.4 The Scaling Region of the Ising Model 964
Problems 973

Index 975

Part 1

Preliminary Notions

1
Introduction

La sapienza è figliola della sperienza.
Leonardo da Vinci, *Codice Forster III, 14 recto*

This chapter introduces some general concepts of Statistical Mechanics and Phase Transitions in order to give a rapid overview of the different topics of the subject and their physical relevance. For the sake of clarity and simplicity, we focus on the magnetic systems, but we stress that the concepts discussed here are of a more general nature and can be applied to other systems as well. In particular, the significant role played by the *correlation length* in the phase transitions and the important properties of *universality* observed in those phenomena. We show that near a phase transition the thermodynamic quantities of a system present an anomalous power law behaviour, parameterized by a set of *critical exponents*. The universal properties shown by phase transitions are manifested by the exact coincidence of the critical exponents of systems that share the same symmetry of their Hamiltonian and the dimensionality of their lattice, but may be, nevertheless, quite different at a microscopic level. From this point of view, the study of phase transitions consists of the classification of all possible *universality classes*. This important property finds its full theoretical justification in the context of the renormalization group ideas, a subject discussed in one of the following chapters.

This chapter also introduces the Ising model and recalls the most significant progresses in the understanding of its features: (i) the duality transformation found by Kramers and Wannier for the partition function of the bi-dimensional case in the absence of a magnetic field, (ii) the exact solution of the lattice model given by Onsager, and (iii) the exact solution provided by Zamolodchikov (with methods borrowed from QFT) of the bi-dimensional Ising model in a magnetic field at the critical value T_c of the temperature.

The appendices at the end of the chapter provide the basic notions of the various ensembles used in statistical mechanics, both at the classical and quantum level, with a discussion of their physical properties.

Statistical Field Theory: An Introduction to Exactly Solved Models in Statistical Physics. Second Edition. Giuseppe Mussardo, Oxford University Press (2020). © Giuseppe Mussardo. DOI: 10.1093/oso/9780198788102.001.0001

1.1 Phase Transitions

1.1.1 Competitive Principles

The atoms of certain materials have a magnetic dipole due either to the spin of the orbital electrons or to the motion of the electrons around the nucleus, or both. In many materials, the magnetic dipoles of the atoms are randomly oriented and the total magnetic field produced by them is then zero (Figure 1.1).

However, in certain compounds or in substances like iron or cobalt, for the effect of the interactions between the atomic dipoles is observed a macroscopic magnetic field different from zero (Figure 1.2). In those *ferromagnetic* materials, this phenomenon is observed for values of the temperature less than a critical value T_c, known as the *Curie temperature*, whose value depends on the material in question. At $T = T_c$ these materials undergo a *phase transition*, i.e. there is a change of the physical properties of the system: in our example, this consists of a spontaneous magnetization on the macroscopic scales created by the alignment of the microscopic dipoles.

The occurrence of a phase transition is the result of two competitive instances: the first tends to minimize the energy, while the second tends to maximize the entropy.

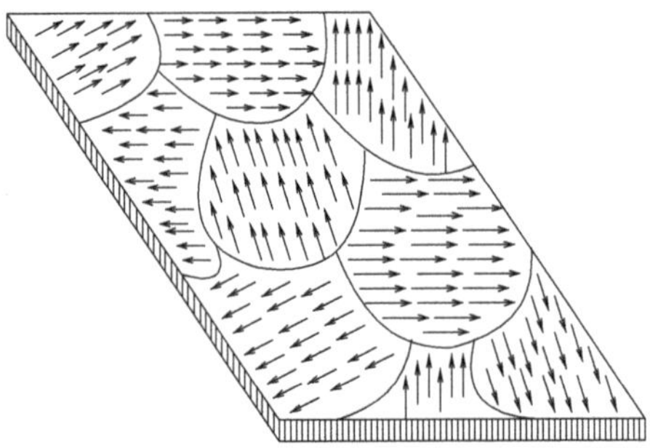

Fig. 1.1 *Magnetic domains for $T > T_c$.*

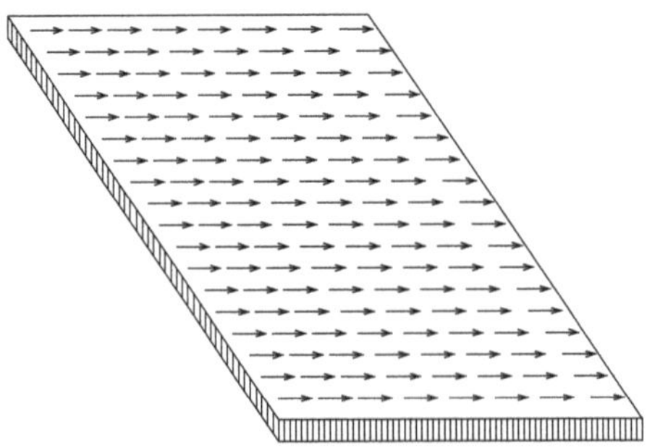

Fig. 1.2 *Alignment of the spins for $T < T_c$.*

- **Principle of Energy Minimization**
In ferromagnetic materials, the configuration of the magnetic dipoles of each atom (that we denote simply as *spins*) tends to minimize the total energy of the system. This minimization is achieved when all spins are aligned. The origin of the atomic dipoles, as well as their interaction, are due to quantum effects. In the following, however, we focus our attention on the classical aspects of this problem, i.e. we consider as given the interaction among the dipoles and the spins of the system as classical degrees of freedom. In this framework, the physical problem can be expressed in a mathematical form as follows: first of all, each spin, placed at the site i of a d-dimensional lattice, is associated a vector $\vec{S}_i$; secondly, their interaction is described by a Hamiltonian $\mathcal{H}$. The simplest version of these Hamiltonians is given by

$$\mathcal{H} = -\frac{\mathcal{J}}{2} \sum_{\langle ij \rangle} \vec{S}_i \cdot \vec{S}_j, \tag{1.1.1}$$

where $\mathcal{J} > 0$ is the coupling constant and the notation $\langle ij \rangle$ stays for a sum to the neighbour spins. The lowest energy configurations are clearly those in which all spins are aligned along one direction.

If the minimization of the energy was the only principle that the spins should follow, we would inevitably observe a giant magnetic field in many substances. The reason why this does not happen is due to another competitive principle.

- **Principle of Entropy Maximization**
Among the extraordinary large number of configurations of the system, those in which the spins align with each other along a common direction are quite special. Hence, unless a great amount of energy is needed to orient in a different direction spins that are at neighbour sites, the number of configurations in which the spins are randomly oriented is much larger than the number of the configurations in which they are completely aligned. It is well known that the measure of the disorder in a system is expressed by the *entropy* S: if we denote by $\omega(E)$ the number of states of the system at energy E, its definition is given by the Boltzmann's formula

$$S(E) = k \log \omega(E), \tag{1.1.2}$$

where k is one of the fundamental constants in Physics, known as *the Boltzmann constant*.

If the tendency to reach the status of maximum disorder was the only physical principle at work, clearly we could never observe any system with a spontaneous magnetization.

Classification scheme of phase transitions
In the modern classification scheme, phase transitions are divided into two broad categories: first-order and second-order phase transitions. First-order phase transitions

are those that involve a latent heat. At the transition point, a system either absorbs or releases a fixed amount of energy while its temperature stays constant. First-order phase transitions are characterized by a finite value of the correlation length. In turns, this implies the presence of a mixed-phase regime, in which some parts of the system have completed the transition and others have not. This is what happens, for instance, when we decrease the temperature of water to its freezing value T_f: the water does not instantly freeze, but forms a *mixture* of water and ice. The presence of a latent heat signals that the structure of the material is drastically changing at $T = T_f$: above T_f, there is no crystal lattice and the water molecules can wander around in a disordered path, while below T_f there is the lattice of ice crystal, where the molecules are packed into a face-centred cubic lattice. In addition to the phase transition of water, many other important phase transitions fall in this category, including Bose–Einstein condensation.

The second class of phase transitions are the continuous phase transitions, also called second-order phase transitions. These have no associated latent heat and they are also characterized by the divergence of the correlation length at the critical point. Examples of second-order phase transitions are the ferromagnetic transition, superconductor, and the superfluid transition. Landau was the first to set up a phenomenological theory of second-order phase transitions. Several transitions are also known as infinite-order phase transitions. They are continuous but break no symmetries. The most famous example is the Kosterlitz–Thouless transition in the two-dimensional XY model. Many quantum phase transitions in two-dimensional electron gases also belong to this class.

As the example of the magnetic dipoles shows, the macroscopic physical systems in which there is a very large number of degrees of freedom are subjected to two different instances: one that tends to order them to minimize the energy, the other that tends instead to disorder them to maximize the entropy. However, to have a real competition between these two different tendencies, we must take into account another important physical quantity: the *temperature* of the system. Its role is determined by the law of statistical mechanics.

1.1.2 Partition Function

One of the most important advances in nineteenth century physics has been the discovery of the exact probabilistic function that rules the microscopic configurations of a system at equilibrium. This is a fundamental law of the statistical mechanics.[1] To express such a law, let us denote by $\mathcal{C}$ a generic state of the system (in our example, a state is specified

[1] In the following we are mainly concerned with the laws of classical statical mechanics. Moreover, we use the formulation of the statistical mechanics given by the *canonical ensemble*. The different ensembles used in statistical mechanics, both in classical and quantum physics, are found in the appendix of this chapter.

once it is known the orientation of each magnetic dipole). Assume that the total number N of the spin is sufficiently large (we will see that a phase transition may occur only when $N \to \infty$). Moreover, assume that the system is at thermal equilibrium, namely that the spins and the surrounding environment exchange energy at a common value T of the temperature. Within these assumptions, the probability that a given configuration $\mathcal{C}$ of the system is realized is given by the Boltzmann law

$$P[\mathcal{C}] = \frac{e^{-E(\mathcal{C})/kT}}{Z}, \tag{1.1.3}$$

where $E(\mathcal{C})$ is the energy of the configuration $\mathcal{C}$ while T is the absolute temperature. A common notation is $\beta = 1/kT$. The expectation value of any physical observable $\mathcal{O}$ is then expressed by the statistical average on all configurations, with weights given by the Boltzmann law

$$\langle \mathcal{O} \rangle = Z^{-1} \sum_{\mathcal{C}} \mathcal{O}(\mathcal{C})\, e^{-\beta E(\mathcal{C})}. \tag{1.1.4}$$

The quantity Z in the denominator is the *partition function* of the system, defined by

$$Z(N,\beta) = \sum_{\mathcal{C}} e^{-\beta E(\mathcal{C})}. \tag{1.1.5}$$

It ensures the proper normalization of the probabilities, $\sum_{\mathcal{C}} P[\mathcal{C}] = 1$. For its own definition, this quantity contains all relevant physical quantities of the statistical system at equilibrium. By making a change of variable, it can be expressed as

$$\begin{aligned} Z(N,\beta) &= \sum_{\mathcal{C}} e^{-\beta E(\mathcal{C})} = \sum_{E} \omega(E)\, e^{-\beta E} = \sum_{E} e^{-\beta E + \log \omega(E)} = \\ &= \sum_{E} e^{\beta[TS-E]} \equiv e^{-\beta F(N,\beta)}, \end{aligned} \tag{1.1.6}$$

where $F(N,\beta)$ is the *free energy* of the system. This is an extensive quantity related to the *internal energy* $U = \langle H \rangle$ and the *entropy* $S = -\langle (\frac{\partial F}{\partial T})_N \rangle$ by the thermodynamical relation

$$F = U - TS. \tag{1.1.7}$$

Namely, we have

$$\begin{aligned} \langle U \rangle &= \frac{\partial}{\partial \beta}(\beta F) \\ \langle S \rangle &= \beta^2 \frac{\partial F}{\partial \beta}. \end{aligned} \tag{1.1.8}$$

The extensive property of F comes from the definition of $Z(N,\beta)$, because if the system is made of two weakly interacting sub-systems, $Z(N,\beta)$ is given by the product of their partition functions.[2] The proof of eqn. (1.1.7) is obtained starting by the identity

$$\sum_{\mathcal{C}} e^{\beta[F(N,\beta)-E(\mathcal{C})]} = 1.$$

Taking a derivative with respect to β of both terms we have

$$\sum_{\mathcal{C}} e^{\beta[F(N,\beta)-E(\mathcal{C})]} \left[F(N,\beta) - E(\mathcal{C}) + \beta\left(\frac{\partial F}{\partial \beta}\right)_N\right] = 0,$$

i.e. precisely formula (1.1.7). This equation enables us to easily understand the occurrence of different phases in the system by varying the temperature. In fact, moving T, there is a different balance in the free energy between the entropy (that favours the disorder) and the energy (that privileges the order). Hence there may exist a critical value $T = T_c$ in which there is a perfect balance between the two different instances. In order to more precisely distinguish the phases of a system it is necessary to introduce the important concept of *order parameter*.

1.1.3 Order Parameters

To characterize phase transitions we need an order parameter, i.e. a quantity that has a vanishing thermal average in one phase (typically the high temperature phase) and a non-zero average in the other phases. Hence such a quantity characterizes the onset of order at the phase transition. It is worth stressing that there is no general procedure to identify the proper order parameter for each phase transition. Its definition may require, in fact, a certain amount of skill or ingenuity. However, there is a close relation between the order parameter of a system and the symmetry properties of its Hamiltonian. In the example of the magnetic dipoles discussed so far, a physical quantity that has a zero mean value for $T > T_c$ and a finite value for $T < T_c$ is the total magnetization, $\vec{\mathcal{M}} = \sum_i \vec{S}_i$. Hence a local order parameter for such a system is identified by the vector $\vec{S}_i$ since we have

$$\langle \vec{S}_i \rangle = \begin{cases} 0; & T > T_c \\ \vec{S}_0 \neq 0; & T < T_c. \end{cases} \tag{1.1.9}$$

When the system is invariant under translations, the mean value of the spin is the same for all sites.

For what concerns the symmetry properties, it is easy to see that the Hamiltonian (1.1.1) is invariant under an arbitrary global rotation $\mathcal{R}$ of the spins. It is known that the

[2] This is definitely true if the interactions are short range, as we assume hereafter. In the presence of long-range forces the situation is more subtle and the extensivity property of the free energy may be violated.

set of rotations form a group. In the case of vectors with three components,[3] the group is denoted by $SO(3)$ and is isomorphic to the group of the orthogonal matrices 3×3 with determinant equal to 1, with the usual rule of multiplication of the matrices.

In the range $T > T_c$, there is no magnetization and the system does not have any privileged direction: in this phase the symmetry of its Hamiltonian is perfectly respected. On the contrary, when $T < T_c$, the system acquires a special direction identified by the vector $\vec{S}_0 = \langle \vec{S}_i \rangle$, along which the majority of the spins is aligned. In this case, the system is in a phase that has less symmetry of its Hamiltonian and which results in a *spontaneous symmetry breaking*. More precisely, in this phase the symmetry of the system is restricted to the sub-class of rotations along the axis identified by the vector $\vec{S}_0$, i.e. to the group $SO(2)$. One of the tasks of the theory of the phase transitions is to provide an explanation for the phenomenon of the spontaneously symmetry breaking and to study its consequences.

1.1.4 Correlation Functions

The main source of information on the phase transitions comes from scattering experiments. They consists of the study of scattering processes of some probe particles sent on the system (they can be photons, electrons or neutrons). In liquid mixtures, near the critical point, the fluid is sufficiently hot and diluted that the distinction between the liquid and gaseous phases is almost non-existent. The phase transition is signalled by the remarkable phenomenon of critical opalescence, a milky appearance of the liquid due to density fluctuations at all possible wavelengths and to the anomalous diffusion of light.[4] For magnetic systems, neutrons provide the best way to probe these systems: first of all, they can be quite pervasive (enabling us to ignore at the first approximation their multiple scattering processes) and, secondly, they couple directly to the spin of the magnetic dipoles. The general theory of the scattering processes involves in this case the *two-point correlation function* of the dipoles

$$G^{(2)}(\vec{i},\vec{j}) = \langle \vec{S}_i \cdot \vec{S}_j \rangle. \tag{1.1.10}$$

When there is a translation invariance, this function depends on the distance difference $\vec{i} - \vec{j}$. Moreover, if the system is invariant under rotations, the correlator is a function of the absolute value of the distance $r = | \vec{i} - \vec{j} |$ between the two spins, so that $G^{(2)}(\vec{i},\vec{j}) = G^{(2)}(r)$. Strictly speaking, any lattice is never invariant under translations and rotations, but we *can* make use of these symmetries as far as we analyse the system at the distance scales much larger than the lattice space a.

[3] It will become useful to generalize this example to the situation in which the spins are made of n components. In this case the corresponding group of symmetry is denoted by $SO(n)$.

[4] Smoluchowski and Einstein were the first to understand the reason for this phenomenon: the fluctuations in the density of the liquid produces analogous fluctuations in its refraction index. In particular, Einstein showed how these fluctuations can be computed and pointed out their anomalous behaviour near the critical point.

As it is evident by its own definition, $G^{(2)}(r)$ measures the degree of the relative alignment between two spins separated by a distance r. Since for $T < T_c$ the spins are predominantly aligned along the same direction, to study their fluctuations is convenient to subtract their mean value, defining the *connected correlation function*

$$G_c^{(2)}(r) = \langle(\vec{S}_i - \vec{S}_0) \cdot (\vec{S}_j - \vec{S}_0)\rangle = \langle\vec{S}_i \cdot \vec{S}_j\rangle - |\vec{S}_0|^2. \tag{1.1.11}$$

When $T > T_c$, the mean value of the spin vanishes and $G_c^{(2)}(r)$ coincides with the original definition of $G^{(2)}(r)$.

Nearby spins usually tend to be correlated. Away from the critical point, $T \neq T_c$, their correlation extends till a certain distance ξ, called the *correlation length*. This is the typical size of the regions in which the spins assume the same value (Figure 1.3). The correlation length can be defined more precisely in terms of the asymptotic behaviour of the correlation function[5]

$$G_c^{(2)}(r) \simeq e^{-r/\xi}, \qquad r \gg a, \qquad T \neq T_c. \tag{1.1.12}$$

At the critical point $T = T_c$, there is a significant change in the system and the two-point correlation function takes instead a power-law behaviour

$$G_c^{(2)}(r) \simeq \frac{1}{r^{d-2+\eta}}, \qquad r \gg a, \qquad T = T_c. \tag{1.1.13}$$

The parameter η in this formula is the *anomalous dimension* of the order parameter. This is the first example of *critical exponents*, a set of quantities discussed thoroughly in the next section. The power-law behaviour of $G_c^{(2)}(r)$ clearly shows that, at the critical point, fluctuations of the order parameter are significantly correlated on all distance scales. Close to a phase transition, the correlation length diverges:[6] denoting by t the relative

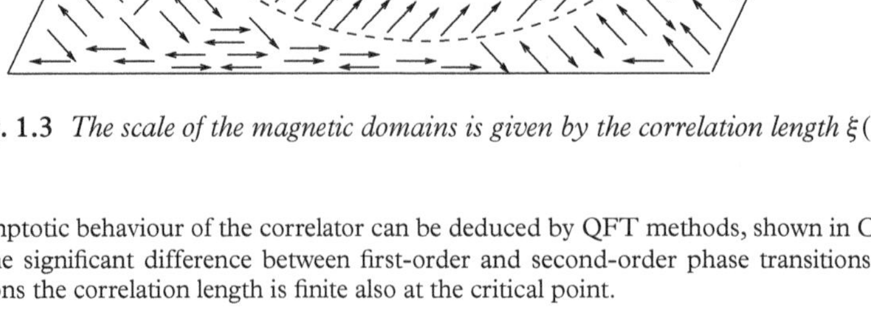

Fig. 1.3 *The scale of the magnetic domains is given by the correlation length $\xi(T)$.*

[5] This asymptotic behaviour of the correlator can be deduced by QFT methods, shown in Chapter 8.

[6] This is the significant difference between first-order and second-order phase transitions. In first-order phase transitions the correlation length is finite also at the critical point.

displacement of the temperature from the critical value, $t = (T - T_c)/T_c$, we observe that near the Curie temperature ξ behaves as

$$\xi(T) = \begin{cases} \xi_+ \, t^{-\nu}, & T > T_c; \\ \xi_- \, (-t)^{-\nu}, & T < T_c, \end{cases} \tag{1.1.14}$$

where ν is another critical exponent.

The two different behaviours of the correlation functions—at the critical point and away from it—can be summed up in a single expression

$$G_c^{(2)}(r) = \frac{1}{r^{d-2+\eta}} f\left(\frac{r}{\xi}\right). \tag{1.1.15}$$

This formula involves the *scaling function* $f(x)$ that depends only on the dimensionless ratio $x = r/\xi$. For large x, this function has the asymptotic behaviour $f(x) \sim e^{-x}$, while its value at $x = 0$ simply fixes its normalization of this quantity, which can always be chosen as $f(0) = 1$. It is worth stressing that the temperature enters the correlation functions only through the correlation length $\xi(T)$.

Aspects of phase transitions. It is now useful to stop and highlight the aspects of phase transitions that have emerged so far. The most important property is that, at $T = T_c$, the fluctuations of the order parameter extend significantly to the entire system, while they are exponentially small away from the critical point. This means that the phase transition taking place at T_c is the result of an extraordinary collective phenomenon that involves at once all the spins of the system.

This observation poses the obvious theoretical problem of understanding how the short range interactions of the spins can give rise to an effective interaction that extends to the entire system when $T = T_c$. There is also another consideration: if we regard

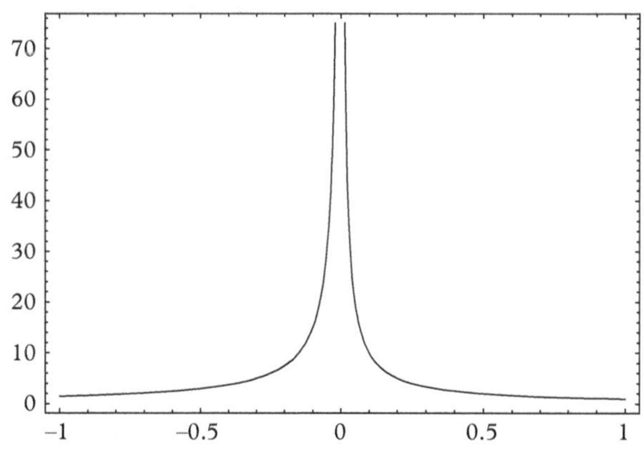

Fig. 1.4 *Behaviour of the correlation length as a function of the temperature near* $T = T_c$.

the correlation length ξ as a measure of the effective degrees of freedom involved in the dynamics, its divergence at the critical point implies that the study of the phase transitions cannot be faced with standard perturbative techniques. Despite these apparent difficulties, however, the study of phase transitions presents some conceptual simplifications that are worth emphasizing. The first simplification concerns the *scale invariance* present at the critical point, namely the symmetry under a dilatation of the length scale

$$a \to \lambda a.$$

Under this transformation, the distance between two points of the system is reduced as

$$r \to r/\lambda.$$

The correlation function (1.1.13), thanks to its power-law behaviour, is invariant under this transformation as far as the order parameter transforms as

$$\vec{S} \to \lambda^{(d-2+\eta)/2}\,\vec{S}. \tag{1.1.16}$$

Expressed differently, at the critical point there is a complete equivalence between a change of the length scale and the normalization of the order parameter. The divergence of the correlation length implies that the system becomes insensitive to its microscopic scales[7] and becomes scale invariant. Moreover, in Chapter 11 we will prove that, under a set of general hypothesis, the global dilatation symmetry expressed by the transformation $a \to \lambda a$ can be further extended to the local transformations $a \to \lambda(\vec{x})\,a$ that change the lengths of the vectors but leave invariant their relative angles. These are the *conformal transformations*. Notice that in the two-dimensional case, the conformal transformations coincide with the mappings provided by the analytic functions of complex variable: studying the irreducible representations of the associated infinite dimensional algebra, we can reach an exact characterization of the bi-dimensional critical phenomena.

The second simplification—strictly linked to the scaling invariance of the critical point—is the *universality* of phase transitions. It is an experimental fact that physical systems of different nature and different composition often show the same critical behaviour: it is sufficient, in fact, that they share the same symmetry group $\mathcal{G}$ of the Hamiltonian and the dimensionality of the lattice space. Hence the critical properties are amply independent from the microscopic details of the various interactions, so that the phenomenology of the critical phenomena falls in different *classes of universality*. Moreover, thanks to the sensitivity of the microscopic details, it is possible always to characterize a given class of universality by studying its simplest representative. We show

[7] Although the system has fluctuations on all possible scales, it is actually impossible to neglect completely the existence of a microscopic scale. In the final formulation of the theory of the phase transitions this scale is related to the renormalization of the theory and, as a matter of fact, is responsible of the anomalous dimension of the order parameter.

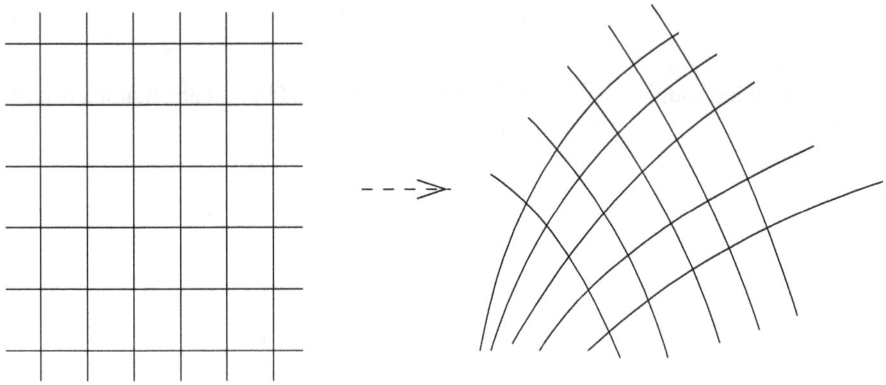

Fig. 1.5 *Conformal transformation. It leaves invariant the angles between the lines.*

later on that all these remarkable universal properties find their elegant justification in the renormalization group formulation. In the meantime, let us complete our discussion on the anomalous behaviour nearby the critical point by introducing other critical exponents.

1.1.5 Critical Exponents

Close to a critical point, the order parameter and the response functions of a statistical system show an anomalous behaviour. Directly supported by a large number of experimental data, these anomalous behaviours are usually expressed in terms of power laws, whose exponents are called *critical exponents*. In addition to the quantities η and ν previously defined, there are other critical exponents directly related to the order parameter. To define them, it is useful to couple the spins to an external magnetic field $\vec{B}$

$$\mathcal{H} = -\frac{\mathcal{J}}{2}\sum_{\langle ij \rangle} \vec{S}_i \cdot \vec{S}_j - \vec{B} \cdot \sum_i \vec{S}_i. \tag{1.1.17}$$

To simplify the notation, let us assume that $\vec{B}$ is along the z axis, with its modulus equal to B. In the presence of B, there is a net magnetization of the system along the z axis with a mean value given by[8]

$$\mathcal{M}(B,T) = \langle S_i^z \rangle \equiv \frac{1}{Z}\sum_{\mathcal{C}} S_i^z \, e^{-\beta\mathcal{H}} = -\frac{\partial F}{\partial B}. \tag{1.1.18}$$

The *spontaneous magnetization* is a function of T alone, defined by

[8] By translation invariance, the mean value is the same for all spins of the system.

$$M(T) = \lim_{B \to 0} \mathcal{M}(B, T), \tag{1.1.19}$$

and its typical behaviour is shown in Figure 1.6. Near T_c, M has an anomalous behaviour, parameterized by the critical exponent β

$$M = M_0(-t)^{\beta}, \tag{1.1.20}$$

where $t = \frac{T - T_c}{T_c}$.

Another critical exponent δ is defined by the anomalous behavior of the magnetization when the temperature is kept fixed at the critical value T_c but the magnetic field is different from zero

$$\mathcal{M}(B, T_c) = \mathcal{M}_0 B^{1/\delta}. \tag{1.1.21}$$

The *magnetic susceptibility* is the response function of the system when we switch on a magnetic field

$$\chi(B, T) = \frac{\partial \mathcal{M}(B, T)}{\partial B}. \tag{1.1.22}$$

This quantity presents a singularity at the critical point, expressed by the critical exponent γ

$$\chi(0, T) = \begin{cases} \chi_+ \, t^{-\gamma}, & T > T_c; \\ \chi_- \, (-t)^{-\gamma}, & T < T_c. \end{cases} \tag{1.1.23}$$

Finally, the last critical exponent that is relevant for our example of a magnetic system is associated to the critical behaviour of the specific heat. This quantity, defined by

$$C(T) = \frac{\partial U}{\partial T}, \tag{1.1.24}$$

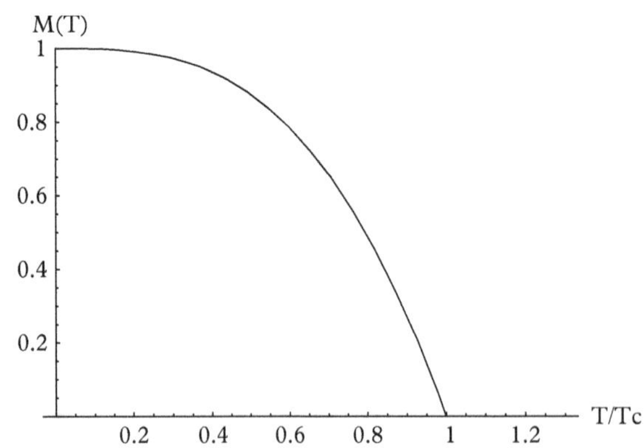

Fig. 1.6 *Spontaneous magnetization versus temperature.*

Exponent	Definition	Condition
α	$C \sim \mid T - T_c \mid^{-\alpha}$	$B = 0$
β	$M \sim (T_c - T)^{-\beta}$	$T < T_c,\ B = 0$
γ	$\chi \sim \mid T - T_c \mid^{-\gamma}$	$B = 0$
δ	$B \sim \mid M \mid^{\delta}$	$T = T_c$
ν	$\xi \sim \mid T - T_c \mid^{-\nu}$	$B = 0$
η	$G_c^{(2)} \sim r^{-(d-2+\eta)}$	$T = T_c$

Table 1.1 *Definition of the critical exponents.*

has a singularity near the Curie temperature parameterized by the exponent α

$$C(T) = \begin{cases} C_+ \, t^{-\alpha}, & T > T_c; \\ C_- \, (-t)^{-\alpha}, & T < T_c. \end{cases} \tag{1.1.25}$$

The summary of the critical exponents of a typical magnetic system is given in Table 1.1.

The critical exponents assume the same value for all statistical systems that belong to the same universality class while, varying the class of universality, they change correspondingly. Hence they are important fingerprints of the various universality classes. Chapter 8 demonstrates that the universality classes can be also identified by the so-called *universal ratios*. These are dimensionless quantities defined in terms of the various response functions: simple examples of universal ratios are given by ξ_+/ξ_i, χ_+/χ_-, or C_+/C_-. Other universal ratios will be defined and analysed in Chapter 8.

Let us end our discussion on critical behaviour with an important remark: a statistical system can present a phase transition (i.e. an anomalous behaviour of its free energy and its response functions) only in its thermodynamical limit $N \to \infty$, where N is the number of particles of the system. Indeed, if N is finite, the partition function is a regular function of the temperature, without singular points at finite value of T, since it is expressed by a sum of a finite number of terms.

1.1.6 Scaling Laws

The exponents α, β, γ, δ, η and ν, previously defined, are not all independent. Already at the earlier stage of the study on the phase transitions, it was observed that they satisfy the algebraic conditions[9]

$$\begin{aligned} &\alpha + 2\beta + \gamma = 2; \\ &\alpha + \beta\delta + \beta = 2; \\ &\nu(2 - \eta) = \gamma; \\ &\alpha + \nu\, d = 2, \end{aligned} \tag{1.1.26}$$

[9] The last of these equations, that which involves the dimensionality d of the system, generally holds for d less of d_c, known as *upper critical dimensions*.

so that it is sufficient to determine only two critical exponents in order to fix all the others.[10] Moreover, the existence of these algebric equations suggests that the thermodynamical quantities of the system are functions of B and T in which these variables enter only *homogeneous combinations*, i.e. they satisfy scaling laws.

An example of scaling law is provided by the expression of the correlator, eqn. (1.1.15). It is easy to see that this expression, together with the divergence of the correlation length (1.1.14), leads directly to the third equation in (1.1.26). To prove it, one needs we need to use a general result of statistical mechanics, known as the *fluctuation-dissipation theorem*, that permits to link the response function to an external field (e.g. the magnetic susceptibility) to the connected correlation function of the order parameter coupled to such a field. For the magnetic susceptibility, the fluctuation-dissipation theorem leads to the identity

$$\begin{aligned}\chi &= \frac{\partial \mathcal{M}(B,T)}{\partial B} = \frac{\partial}{\partial B}\left[\frac{1}{Z}\sum_{C} S_i^z e^{-\beta\mathcal{H}}\right] = \\ &= \beta\sum_j \left(\langle S_j^z S_i^z\rangle - |\langle S_i^z\rangle|^2\right) = \beta\sum_r G_c^{(2)}(r),\end{aligned} \tag{1.1.27}$$

that can be derived by using eqns. (1.1.17) and (1.1.5) for the Hamiltonian and the partition function, together with the definition of the mean value, given by eqn. (1.1.4).

Substituting in (1.1.27) the scaling law (1.1.15) of the correlation function, we have

$$\begin{aligned}\chi &= \beta\sum_r G_c^{(2)}(r) = \beta\sum_r \frac{1}{r^{d-2+\eta}} f\left(\frac{r}{\xi}\right) \simeq \\ &\simeq \int dr\, r^{d-1}\frac{1}{r^{d-2+\eta}} f\left(\frac{r}{\xi}\right) = A\,\xi^{2-\eta},\end{aligned} \tag{1.1.28}$$

where A is a constant given by the value of the integral obtained by the substitution $r \to \xi z$

$$A = \int dz\, z^{1-\eta} f(z).$$

Using the anomalous behaviour of $\xi(t)$ given by (1.1.14), we have

$$\chi \simeq \xi^{2-\eta} \simeq t^{-\nu(2-\eta)}, \tag{1.1.29}$$

[10] As discussed in Chapter 8, the critical exponents are not the most fundamental theoretical quantities. As a matter of fact, they can all be derived by a smaller set of data given by the *scaling dimensions* of the relevant operators.

and, comparing to the anomalous behaviour of χ expressed by (1.1.23), we arrive at the relation

$$\nu(2-\eta)=\gamma.$$

A scaling law can be similarly written for the singular part of the free energy $F_s(B,T)$, expressed by a homogeneous function of the two variables

$$F_s(B,T) = t^{2-\alpha}\mathcal{F}\left(\frac{B}{t^{\beta\delta}}\right). \qquad (1.1.30)$$

It is easy to see that this expression implies the relation

$$\alpha+\beta\delta+\beta = 2, \qquad (1.1.31)$$

i.e. the second equation in (1.1.26). In fact, the magnetization is given by the derivative of the free-energy $F_s(B,T)$ with respect to B

$$M = \left.\frac{\partial F_s}{\partial B}\right|_{B=0} = t^{2-\alpha-\beta\delta}F'(0).$$

When compared to eqn. (1.1.20), we recover eqn. (1.1.31). Scaling relations for other thermodynamical quantities can be obtained in a similar way.

Chapter 8 shows that the homegeneous form assumed by the thermodynamical quantities in the vicinity of critical points has a theoretical justification in the renormalization group equations that control the scaling properties of the system.

1.1.7 Dimensionality of Space and Order Parameters

Although the world in which we live is three dimensional, it is much more convenient to change our perspective and to consider instead the dimensionality d of the space as a variable like another. There are various reasons to adopt this point of view.

The first reason is of phenomenological nature: there are many systems that, for the particular nature of their interactions or their composition, present either a one-dimensional or two-dimensional behaviour. Systems that can be considered one dimensional are those given by long chains of polymers, for instance, in particular if the object of the study is the monomers along the chain. Two-dimensional systems are given by those solids composed of weakly interacting layers, as happens in graphite. Another notable example of a two-dimensional system is provided by the quantum Hall effect, where the electrons of a thin metallic bar are subjected to a strong magnetic field in the vertical direction at very low temperatures. Examples of two-dimensional critical phenomena are also those relative to surface processes of absorption, or phenomena that involve the thermodynamics of liquid films.

It is necessary to emphasize that the effective dimensionality shown by the critical phenomena can depend on the thermodynamic state of the system. Namely there could be a dimensional transmutation induced by the variation of the thermodynamic parameters, e.g. temperature: there are materials that in some thermodynamic regimes appear as if they were bi-dimensional, while in other regimes they have instead three-dimensional dynamics. Consider, for instance, a three-dimensional magnetic system in which the interaction along the vertical axis $\mathcal{J}_z$ is much smaller than the interaction $\mathcal{J}$ among the spins of the same plane, i.e. $\mathcal{J}_z \ll \mathcal{J}$. In high-temperature phase (where the correlation length $\xi(T)$ is small), we can neglect the coupling between the next neighbour planes, so that the system appears to be a two-dimensional one. However, decreasing the temperature, the correlation length $\xi(T)$ increases and, in each plane, there will be large areas in which the spins become parallel and behave as a single spin but of a large value. Hence even though the coupling $\mathcal{J}_z$ between the planes was originally small, their interaction can be quite strong for the large values of the effective dipoles: correspondingly, the system presents at low temperatures a three-dimensional behaviour.

However, there is a more theoretical reason to regard the dimensionality d of a system as an additional parameter. First of all, the existence of a phase transition of a given Hamiltonian depends on the dimensionality of the system. The fluctuations become stronger by decreasing d and, because they disorder the system, the critical temperature decreases correspondingly. Each model with a given symmetry selects a *lower critical dimension* d_i such that, for $d < d_i$ its phase transition is absent. For the Ising model (and, more generally, for all models with a discrete symmetry) $d_i = 1$. For systems with a continuous symmetry, the fluctuations can much more easily disorder the system since the order parameter continuously can change its value without significantly altering the energy. Hence for many of these systems we have $d_i = 2$.

The critical exponents depend on d and, for each system, there is also a *higher critical dimension* d_s: for $d > d_s$, the critical exponents take the values obtained in the mean field approximation that is discussed in Chapter 3. For the Ising model, we have $d_s = 4$. The range

$$d_i < d < d_s$$

of a given system is therefore the most interesting interval of dimensions, for it is the range of d in which we observe the strongly correlated nature of the fluctuations. This is another reason to regard d as a variable of the statistical systems. In fact, the analysis of their critical behaviour usually deals with divergent integrals coming from the large fluctuations of the critical point. To regularize such integrals, a particular elegant method is provided by the so-called *dimensional regularization*, as discussed in a problem at the end of the chapter. This method permits, in particular, to define an expansion parameter $\epsilon = d - d_s$ and to express the critical exponents in power series in ϵ. Further elaboration on these series permits to obtain the critical exponents for finite values of ϵ, i.e. those that correspond to the actual value of d for the system under consideration.

1.2 The Ising Model

After the discussion on the phenomenology of the phase transitions of the previous section, let us now introduce the Ising model. This is the simplest statistical model that has a phase transition. The reason to study this model comes from two different instances: the first is the need to simplify the nature of the spins in order to obtain a system sufficiently simple to be solved exactly, while the second concerns with the definition of a model sufficiently realistic to be compared with the experimental data.

The simplification is obtained by considering the spins σ_i as scalar quantities with values ± 1 rather than the vector quantities $\vec{S}_i$ previously introduced. In this way, the Hamiltonian of the Ising model is given by

$$\mathcal{H} = -\frac{\mathcal{J}}{2}\sum_{\langle i,j\rangle}\sigma_i\sigma_j - B\sum_i \sigma_i, \quad \sigma_i = \pm 1. \tag{1.2.1}$$

When $B = 0$, it has a global discrete symmetry Z_2, implemented by the transformation $\sigma_i \to -\sigma_i$ on all the spins.

Even though the Ising model may appear as a caricature of the actual ferromagnetic substances, it has nevertheless a series of advantages: it is able to provide useful information on the nature of phase transition, on the effects of the cooperative dynamics, and on the role of the dimensionality d of the lattice. For example, the next chapters show that the model has a phase transition at a finite value T_c of the temperature when $d \geq 2$ while it does not have any phase transition when $d = 1$. Moreover, the study of this model helps to clarify the aspects of the phase transitions that occur in lattice gases or, more generally, in all those systems in which the degrees of freedom have a binary nature.

The elucidation of the mathematical properties of the Ising model has involved a large number of scientists since 1920, i.e. when it was originally introduced by Wilhelm Lenz.[11] The first theoretical results are due to Ernst Ising, a PhD student of Lenz at the University of Hamburg, who in 1925 published a short article based on PhD studies in which he showed the absence of a phase transition in the one-dimensional case. Since then, the model is known in the literature as the Ising model.

After this first result, it is necessary to move to 1936 to find an ulterior progress in the understanding of the model where, using an elementary argument, Peierls showed the existence of a critical point in the two-dimensional case, so that the Ising model became a valid and realistic tool for investigating the phase transitions. The exact value of the critical temperature T_c on a two-dimensional square lattice was found by Kramers

[11] We have only to read Ising's original paper to learn that the model was previously proposed by Ising's research supervisor, Wilhelm Lenz. It is rather curious that Lenz's priority has never been recognized by later authors. Lenz himself apparently never made any attempt later on to claim credit for suggesting the model and also never published any papers on it.

and Wannier in 1941, making use of an ingenious technique. They showed that the partition function of the model can be expressed in a systematic way as a series expansion both in the high- and in the low-temperature phases, showing that the two series were related by a duality transformation. In more detail, in the high-temperature phase the variable entering the series expansion is given by $\beta\mathcal{J}$, while in the low-temperature phase the series is in the powers of the variable $e^{-2\beta\mathcal{J}}$. The singularity present in both series, together with the duality relation that links one to the other, allowed them to determine the exact value of the critical temperature of the model on the square lattice, given by the equation $\sinh\left(\frac{\mathcal{J}}{kT_c}\right) = 1$.

The advance of Kramers and Wannier was followed in February 1943 by the fundamental contribution of Lars Onsager, who, at the New York Academy of Science meeting, announced the solution for the partition function of the two-dimensional Ising model at zero magnetic field. The details were published two years later. The contribution of Onsager constitutes a milestone in the field of the phase transitions. The original solution of Onsager, quite complex from a mathematical point of view, has been simplified with the contribution of many authors and, in this respect, it is important to mention Kaufman and Baxter. Since then, there have been many other results concerning several aspects, as the analysis of different two-dimensional lattices, the computation of the spontaneous magnetization, the magnetic susceptibility and, finally, the correlation functions of the spins. In 1976 McCoy, Wu, Tracy, and Barouch, in a remarkable theoretical *tour de force*, showed that the correlation functions of the spins can be determined by the solution of a non-linear differential equation, known in the literature as the Painlevé equation. A similar result was also obtained by Miwa and Jimbo and their collaborators in Kyoto: in particular, they showed that the monodromy properties of a particular class of differential equation can be analysed by using the spin correlators of the Ising model.

In the years immediately after Onsager's proposed solution, the research community felt optimistic about being able to extend his method to the three-dimensional lattice as well as to the bi-dimensional case but in a presence of an external magnetic field. However, despite of the numerous efforts and numerous attempts that finally proved to be premature or wrong, for many years only modest progress has been made on both the arguments.

An exact solution of the three-dimensional case is still unknown, although many of its proprieties are widely known thanks to numerical simulations and series expansions—methods that have been improved during the years with the aid of faster and more efficient computers. The critical exponents or the equations of state, for instance, are nowadays known very accurately and their accuracy increases systematically with new publications on the subject. It is a common opinion among physicists that the exact solution of the three-dimensional Ising model is one of the most interesting open problems of theoretical physics.

Conversely, the analysis of the two-dimensional Ising model in the presence of a magnetic field has received renewed interest since 1990, and has witnessed considerable progress in the understanding of its properties. This development has been possible thanks to methods of QFT and analytic S-matrix, that have been originally proposed in this context by Alexander Zamolodchikov. By means of these methods it was possible to achieve the exact determination of the spectrum of excitations of the Ising model in a magnetic field and the identification of their interactions. Subsequently, Delfino and Mussardo determined the two-point correlation function of the spins of the Ising model in a magnetic field while Delfino and Simonetti calculated the correlation functions that involve the energy operator of the model. In a successive work, Delfino, Mussardo, and Simonetti systematically studied the proprieties of the model by varying the magnetic field and the temperature. This analysis was further refined in a following paper by Fonseca and Zamolodchikov, which led to the thorough study of the analytic structure of the free energy in the presence of a magnetic field and for values of temperature different from the critical value. Besides these authors, many others have largely contributed to the developments of the subject and, in the sequel, there will be ample possibility to give them the proper credit.

The next chapters discuss the important aspects of the Ising model and its generalizations. We emphasize their physical proprieties and provide evidence of their mathematical elegance. As we will see Additionally, this study brings to the fore many important arguments of theoretical physics and mathematics.

1.3 Ernst Ising

The Ising model is one of the best-known models in statistical mechanics, with over 12000 articles published on it or referring to it between 1969 to 2002. Therefore, it may appear quite paradoxical that the extraordinary notoriety of the model is not accompanied by an analogous notoriety of the scientist to whom the model owes its name. The short biographical notes that follow underline the singular history of this humble scientist who, while entangled with the most dramatic events of the twentieth century, became famous by chance and remained unaware of his reputation for many years of its life.

Ernst Ising was born in Cologne on 10 May 1900 into a family of Jewish origin, and who later moved to Bochum in Westfalia where Ernst finished his high school studies. In 1919 he started his university studies at Goettingen in mathematics and physics and later moved to Hamburg. Here, under the supervision of Wilhelm Lenz, he started the study of the ferromagnetic model proposed by Lenz. In 1925 he defended his PhD thesis, devoted to the analysis of the one-dimensional case of the model that bears nowadays his name, and in 1926 he published his results in the journal *Zeitschrift fur Phyisk*. After his PhD, Ising moved to Berlin and between 1925 and 1926 he worked at the Patent Office of the Allgemeine Elektrizitatsgesell Schaft. He then decided to take up teaching and taught for one year at a high school school in Salem, near the Lake of Costanza. In 1928 he decided to return to the university to study philosophy and pedagogy.

In 1930 he married Johanna Ehmer and moved to Crossen as a teacher in the local grammar school. However, when Hitler came to power in 1933, citizens of Jewish origin were removed from the public posts and Ising lost his job in March of that year. He remained unemployed for approximately one year, except for a short period spent in

Paris as a teacher in a school for foreign children. In 1934 he found a new job as a teacher at the school opened in the Jewish community near Caputh, a city close to Potsdam, and in 1937 he became the dean of the same school. On 10 November 1938 he witnessed the devastation of the premises of the school by the boys and the inhabitants of Caputh, who were urged by local politicians to follow the example of the general treatment the Hebrew population throughout Germany.

In 1939 Ernst and Johanna Ising were caught in Luxembourg while they were trying to emigrate to the United States. Their visa applications were rejected due to the limits put on the immigration flows. On Ising's fourtieth birthday, the Germans invaded Luxembourg, and all consular offices were closed: this cut off any possibility of expatriation. Despite everything, Ising and his family survived the horrors of the war, even though Ising was forced to work for the German army on the railway lanes.

It was only two years after the end of the war that Ising and his wife left Europe on a cargo ship bound for United States. He initially taught at the State Teachers' College of Minot, North Dakota, and then at the Bradley University in Peoria, Illinois, where he was Professor of Physics from 1948 till 1976. He became an American citizen in 1953 and in 1971 he was named Teacher of the Year. Ising died 11 May 1998 at home.

The life and the career of Ernst Ising were seriously marked by the events of the Second World War: after his PhD thesis, he never returned to research. He lived quite isolated for many years, almost unaware of the new scientific developments. However, his article published in 1925 had a different fate. It was firstly quoted in an article by Heisenberg in 1928, devoted to the study of exchange forces between magnetic dipoles. Importantly, the true boost to his reputation came from Peierls's famous article published in 1936, titled '*On the Model of Ising for the Ferromagnetic*'. Since then, the scientific literature has seen a large proliferation of articles on this model.

In closing, it is worth adding that it was only in 1949 that Ising became aware of his own importance, and that of his model, within the scientific community.

Appendix 1.A. Ensembles in Classical Statistical Mechanics

Statistical mechanics is the field of physics mainly interested in the thermodynamical properties of systems made of an enormous number of particles, typically of the order of the Avogadro number $N_A \sim 10^{23}$. To study such systems, it is crucial to make use of probabilistic methods as it is generally impossible to determine the trajectory of each particle and is nevertheless meaningless to use them for deriving the thermodynamic properties. On the contrary, the approaches based on probability permits to compute in a easier way the mean values of the physics quantities and their fluctuations.

The statistical mechanics of system at equilibrium can be formulated in three different ways, that are based on *microcanonical ensemble, canonical ensemble, or grand-canonical ensemble*. For macroscopic systems, the three different ensembles give the same final results. The choice of one or another of them is then just a question of what is the most convenient for the problem at hands. In this appendix we recall the formulation of the

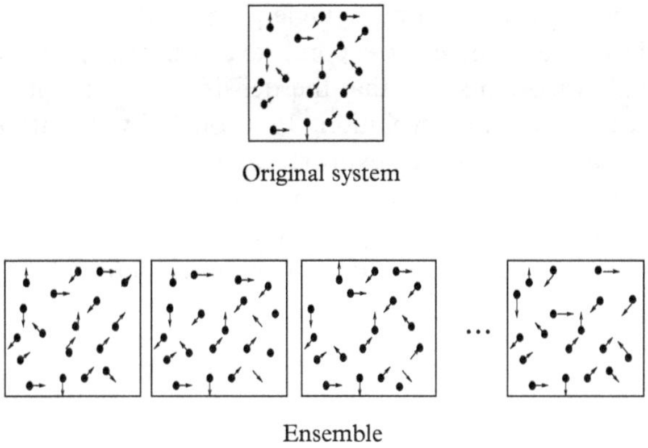

Fig. 1.7 *From the initial system to the ensemble.*

three ensembles of classical statistical mechanics while in the next Appendix we discuss their quantum version.

It is convenient to introduce the phase space Γ of the system. Let us assume that the system is made of N particles, each of them identified by a set of d coordinates q_i and d momenta p_i. The phase space Γ is the vector space of $2d \times N$ dimensions, given by the tensorial product of the coordinates and momenta of all the particles. In the phase space, the system is identified at any given time by a point and its motion is associated to a curve in this space. If the system is isolated, its total energy E is conserved: in this case the motion takes place along a curve of the surface of Γ defined by the equation $H(q_i, p_i) = E$, where $H(q_i, p_i)$ is the Hamiltonian of the system.

For a system with a large number of particles, not only it is impossible to follow its motion, but it is also useless. The only thing that matters is the possibility to predict the average properties of the system that are determined by the macroscopic constraints to which the system is subjected, as its volume V, the total number N of particles, and its total energy E. Since there is generally a huge number of microscopic states compatible with a given set of macroscopic constraints, it is natural to assume that the system will visit all of them during its temporal evolution.[12] Instead of considering the time evolution of the system, it is more convenient to consider an infinite number of copies of the same system, with the same macroscopic constraints. This leads to the idea of the *statistical ensembles*. By using an analogy, this is equivalent to looking at an infinite number of snapshots of a single movie rather than the movie itself. The ensembles provide then a statistical sampling of the system.

[12] The validity of these considerations is based on an additional assumption, namely the ergodicity of the system under consideration. By definition a system is ergodic if its motion passes arbitrarily close to all points of the surfaces of the phase space identified by the macroscopic conditions alone. The motion of the systems that have additional conservation laws is usually not ergodic, since it takes place only on particular regions of these surfaces.

Since each system is represented by a single point in the phase space, the set of the systems associated to the ensemble corresponds to a swarm of points in the phase space. Because the Liouville theorem states that the density of the points at any given point remains constant during the time evolution,[13] a probability density $\tilde{\rho}_i(q,p)$ is naturally defined in Γ. Hence we can determine expectation values of physical quantities in terms of expectation values on the ensemble (a procedure that is relatively easy) rather than as time average of an individual system (a procedure that is instead rather complicated). If the system is ergodic we have in fact the fundamental identity

$$\langle A\rangle = \lim_{t\to\infty} \frac{1}{t}\int_0^t d\tau\, A\,[q(\tau),p(\tau)] = \int dq\,dp A(p,q)\,\tilde{\rho}(q,p).$$

The different ensembles are defined by the different macroscopic conditions posed the system. Let us discuss the three cases that are used more often.

Microcanonical ensemble. The microcanonical ensemble is defined by the following macroscopic conditions: a fixed number N of particles, a given volume V, and a given value of the energy in the range E and $E+\Delta$. In this ensemble the mean values are computed in terms of the probability density $\rho(q,p)$ defined by

$$\rho(q,p) = \begin{cases} 1 & \text{if} E < H(p,q) < E+\Delta, \\ 0 & \text{otherwise} \end{cases} \tag{1.A.1}$$

i.e. for any physical quantity A we have

$$\langle A\rangle = \frac{\int dq\,dp A(q,p)\,\rho(q,p)}{\int dq\,dp\,\rho(q,p)}.$$

The fundamental physical quantity in this formulation is the entropy. Once this quantity is known, we can recover all the rest of the thermodynamics. The entropy is a function of E and V, defined by

$$S(E,V) = k\log\Omega(E,V), \tag{1.A.2}$$

where k is the Boltzmann constant and Ω is the volume in the phase space Γ of the microcanonical ensemble

$$\Omega(E,V) = \int dq\,dp\,\rho(q,p).$$

[13] According to the theorem by Liouville, $\frac{dD}{dt} = 0$, hence the density D satisfies the differential equation $\frac{\partial D}{\partial t} = -\{H,D\}$. At the equilibrium, the density D does not vary with time and then satisfies $\{H,D\} = 0$. This means that it is only a function of the integrals of motion of the system.

The absolute temperature is then given by

$$\frac{1}{T} = \frac{\partial S(E,V)}{\partial E},$$

while the pressure P is defined by

$$P = T\frac{\partial S(E,V)}{\partial V}.$$

For the differential of S we have

$$dS(E,V) = \frac{\partial S}{\partial E}dE + \frac{\partial S}{\partial V}dV = \frac{1}{T}(dE + PdV),$$

i.e. the first law of the thermodynamics.

Canonical ensemble. The canonical ensemble permits us to deal with the statistical properties of a system that is in contact with a thermal bath much larger than the system itself. In this ensemble, the assigned macroscopic conditions are given by the total number N of the particles, the volume V of the system, and its temperature T. In this ensemble we cannot fix a priori the value of the energy, for it can be freely exchanged between the system and the thermal bath. These conditions are considered to be more closely related to the actual physical situations, since the temperature of a system can be easily tuned while it is more difficult to ensure the isolation of a system and the constant value of its energy. The probability density of the canonical ensemble takes the form of the Gibbs distribution

$$\rho(q,p) = e^{-\beta H(q,p)},$$

with $\beta = 1/kT$. The partition function is given by

$$Z_N(V,T) = \int dq\, dp\, e^{-\beta H(q,p)}.$$

The mean values are computed according to the formula

$$\langle A \rangle = \frac{1}{Z_N}\int dq\, dp A(q,p) e^{-\beta H(q,p)}.$$

As discussed in the text, the partition function Z_N permits to recover the thermodynamics of the system. The equivalence between the microcanonical and the canonical ensembles can be proved by analysing the fluctuations of the energy

$$\Delta E^2 = \langle H^2 \rangle - \langle H \rangle^2.$$

A simple calculation gives

$$\langle H^2\rangle - \langle H\rangle^2 = kT^2 \frac{\partial\langle H\rangle}{\partial T} = kT^2 C_V,$$

where C_V is the specific heat. Since in a macroscopic system $\langle H\rangle \propto N$ but also $C_V \propto N$ (for the extensive nature of both quantities), the fluctuations of the energy are of Gaussian type, namely in the limit $N \to \infty$ we have

$$\lim_{N\to\infty} \frac{\Delta E^2}{\langle H\rangle^2} = 0.$$

In other words, even though in the canonical ensemble the energy is a quantity that is not fixed but is subjected to fluctuations, as a matter of fact it assumes the same value in the utmost majority of the systems of the ensemble. This proves the equivalence between the two ensembles.

Grand canonical ensemble. Along the reasoning that we used to introduce the canonical ensemble, i.e. the possibility to control the temperature rather than its conjugate variable given by the energy, to introduce the grand canonical ensemble one argues that it is not realistic to assume that the total number N of the particles of a system is known a priori. In fact, the experiments can usually determine only the mean value of this quantity. Hence in the grand canonical ensemble we posit that the system can have an arbitrary number of particles, with its mean value determined by its macroscopic conditions. By introducing the quantity $z = e^{\beta\mu}$, where μ is the *fugacity*, the probability density of the grand canonical ensemble is given by

$$\rho(q,p,N) = \frac{1}{N!} z^N e^{-\beta H(q,p)}. \qquad (1.A.3)$$

The term $N!$ in this formula takes into account the identity of the configurations obtained by the permutation of N identical particles. By integrating on the coordinates and the momenta present in (1.A.3), we arrive at the probability density relative to N particles. In its normalized form, it is expressed by

$$\rho(N) = \frac{1}{\mathcal{Z}} \frac{z^N}{N!} Z_N(V,T),$$

where $Z_N(V,T)$ is the partition function of the canonical ensemble with N particles, whereas the denominator of this formula defines the grand canonical partition function

$$\mathcal{Z}(z,V,T) = \sum_{N=0}^{\infty} \frac{z^N}{N!} Z_N(V,T).$$

The mean value of the particles of the system can be computed by the formula

$$\langle N \rangle = \sum_{N=0}^{\infty} N \rho(N) = z \frac{\partial}{\partial z} \log \mathcal{Z}(z, V, T). \tag{1.A.4}$$

The fundamental formula of the grand canonical ensemble links the pressure P to the partition function $\mathcal{Z}$

$$P = \frac{1}{\beta V} \log \mathcal{Z}(z, V, T). \tag{1.A.5}$$

The equation of state, i.e. the relationship among P, V and $\langle N \rangle$, is obtained by expressing z by using eqn. (1.A.4) and substituting it in (1.A.5).

The equivalence of this ensemble to the previous ones can be proved by showing that the fluctuations of the number of particles are purely Gaussian. It is easy to prove that, in the infinite volume and away from the critical points of the system, we have in fact

$$\lim_{V \to \infty} \frac{\langle N^2 \rangle - \langle N \rangle^2}{\langle N \rangle^2} = 0.$$

This equation shows that, even though the number of particles of the system is not fixed a priori, it has the same value in almost all copies of the ensemble.

Appendix 1.B. Ensembles in Quantum Statistical Mechanics

This Appendix recalls the main formulae of statistical mechanics in the context of quantum theory. In quantum mechanics any observable A is associated with a Hermitian operator that acts on a Hilbert space. At each time t, the state of an isolate system is identified by a vector $\mid \Psi(t)\rangle$ that evolves according to the Schrödinger equation

$$i\hbar \frac{\partial}{\partial t} \mid \Psi(t)\rangle = H \mid \Psi(t)\rangle, \tag{1.B.1}$$

where H is the Hamiltonian. By using the linear superposition principle, each state of the system can be expressed in terms of a complete set of states $\mid \psi_n\rangle$ provided by the orthonormal eigenvectors of any observable A

$$A \mid \psi_n\rangle = a_n \mid \psi_n\rangle, \quad \langle \psi_n \mid \psi_m\rangle = \delta_{n,m}.$$

This means that $\mid \Psi\rangle$ is given by

$$\mid \Psi\rangle = \sum_n c_n \mid \psi_n\rangle. \tag{1.B.2}$$

For the completeness relation of these states,

$$\sum_n |\, \psi_n\rangle \langle \psi_n \,| = 1.$$

The coefficients c_n of the expansion (1.B.2) are expressed by the scalar product $c_n = \langle \Psi \mid \psi_n \rangle$ and the square of their modulus $|\, c_n \,|^2$ expresses the probability to obtain the eigenvalues a_n as a result of the measurement of the observable A on the state $|\, \Psi\rangle$. Hence

$$\langle \Psi \mid \Psi \rangle = \sum_n |\, c_n \,|^2 = 1.$$

Let us now discuss the statistical properties of the quantum systems. As in the classical case, in the presence of a large number of particles is highly unrealistic to determine the behaviour of a system by solving the Schrödinger equation: first of all, this is an impossible goal to pursue in almost all systems and, secondly, it cannot be used to predict the thermodynamic properties. Hence also in the quantum case, we need to use a statistical formulation: take into account the incomplete information on the state of the system and extract the predictions only on the mean values of the observables. To do so, let us imagine that the system under study can be considered as a sub-system of a larger one (external world) and in thermodynamic equilibrium. A basis of the Hilbert space can be given by all possible tensor product states $|\alpha_p\rangle \otimes |\varphi_n\rangle$, where $|\varphi_n\rangle$ refers to the eigenvalues of the Hamiltonian $\mathcal{H}$ of the sub-system (with spectrum E_n) while $|\alpha_p\rangle$ is a vector basis of the Hamiltonian of the external world system. Within this basis, a generic state $|\Psi\rangle$ at time t is written as

$$|\Psi\rangle(t) = \sum_{p,n} \gamma_{p,n}(t) |\alpha_p\rangle \otimes |\varphi_n\rangle. \tag{1.B.3}$$

Let us consider now the quantum mean value of an observable $\mathcal{O}$ of the sub-system on the state $|\, \Psi\rangle$ at a given time. According to the rules of quantum mechanics, this is given by the expectation value

$$\begin{aligned} \langle \Psi(t) \mid \mathcal{O} \mid \Psi(t) \rangle &= \sum_{p,q,n,m} \gamma^*_{q,m}(t)\, \gamma_{p,n}(t)\, \langle \alpha_q | \alpha_p \rangle \, \langle \varphi_n \mid \mathcal{O} \mid \varphi_m \rangle \qquad (1.B.4) \\ &= \sum_{p,n,m} \gamma^*_{p,m}(t)\, \gamma_{p,n}(t)\, \mathcal{O}_{n,m}, \end{aligned}$$

where $\mathcal{O}_{n,m} = \langle \varphi_n \mid \mathcal{O} \mid \varphi_m \rangle$ and we have use the orthogonality relation $\langle \alpha_q | \alpha_p \rangle = \delta_{p,q}$. Since we are interested only on the sub-system, we can take a statistical average on the external world data. Under the hypothesis of ergodicity,[14] this is equivalent to take the *time average* of (1.B.4). Defining

[14] In quantum mechanics this implies the absence of non-trivial integrals of motion, i.e. a set of observables that commute with the Hamiltonian and that can be simultaneously diagonalized with it.

$$\rho_{m,n} = \sum_p \overline{\gamma^*_{p,m}(t)\, \gamma_{p,n}(t)} \equiv \lim_{t\to\infty} \frac{1}{t} \int_0^t \sum_p \gamma^*_{p,m}(\tau)\, \gamma_{p,n}(\tau)\, d\tau, \tag{1.B.5}$$

the *statistical average* of the observable $\mathcal{O}$ can be expressed by the formula

$$\overline{\langle \mathcal{O} \rangle} = \overline{\langle \Psi \mid \mathcal{O} \mid \Psi \rangle} = \sum_{n,m} \rho_{m,n} \mathcal{O}_{n,m} = \mathrm{Tr}(\rho\, \mathcal{O}), \tag{1.B.6}$$

where the operator ρ, defined by its matrix elements (1.B.5), is the *density matrix*. Since the trace of an operator is independent from the basis, the final result (1.B.6) does not depend on the basis of the eigenvectors that we used to expand the state $|\Psi\rangle$. It should be stressed that the average (1.B.6) that involves the density matrix has two aspects: from one side, it includes the quantum average on the state, but, on the other hand, it performs the statistical average on the wave functions of the environment. Both averages are simultaneously present in the formula (1.B.6).

In quantum statistical mechanics, the density matrix corresponds to the probability distribution of classical statistical mechanics. Hence also in this case, we can introduce three different ensembles.

Microcanonical ensemble. As in the classical case, the microcanonical ensemble is defined by the following macroscopic conditions: a fixed number N of the particles, a fixed volume V, and the energy of the system in the range E and $E + \Delta$. Correspondingly, the density matrix assumes the form

$$\rho_{n,m} = \delta_{n,m}\, w_n, \quad w_n = \begin{cases} 1; & E < E_n < E + \Delta \\ 0; & \text{otherwise} \end{cases}$$

and the thermodynamics is derived starting from the entropy

$$S(E, V) = k \log \Omega(E, V),$$

where

$$\Omega(E, V) = \mathrm{Tr}\, \rho.$$

Canonical ensemble. In this ensemble the macroscopic variables are given by the fixed number N of the particles, the volume V, and the temperature T. The corresponding expression of the density matrix is given by

$$\rho_{n,m} = \delta_{n,m}\, e^{-\beta E_n},$$

with the partition function expressed by

$$Z_N(V,T) = \mathrm{Tr}\,\rho = \sum_n e^{-\beta E_n}.$$

In this ensemble, the thermodynamics is derived starting from the free energy

$$F_N(V,T) = -\beta^{-1} \log Z_N(V,T).$$

Grand canonical ensemble. In the grand canonical ensemble the macroscopic variables are the volume V and the temperature T. In this case, the density matrix acts on a Hilbert space with an indefinite number of particle. Denoting by $E_{n,N}$ the nth energy level with N particles, the density matrix is expressed by

$$\rho_{n,N} = z^N e^{-\beta E_{n,N}},$$

where $z = e^{\beta\mu}$. The equation of state is similar to the classical one

$$P = \frac{1}{\beta V} \log \mathcal{Z}(z,V,T),$$

where $\mathcal{Z}(z,V,T)$ is the grand canonical partition function

$$\mathcal{Z}(z,V,T) = \sum_{N,n} z^N e^{-\beta E_{n,N}}.$$

Indistinguishable particles and statistics. A central idea of quantum theory is the concept of indistinguishable particles: for a system with many identical particles, an operation that exchanges two of them, swapping their positions, leaves the physics invariant. This symmetry is represented by a unitary transformation acting on the many-body wave function. In three spatial dimensions, there are only two possible symmetry operations: the wave function of bosons is symmetric under exchange while that of fermions is anti-symmetric. The limitation to one of the two possible kinds of quantum symmetry comes from a simple topological argument: a process in which two particles are adiabatically interchanged twice is equivalent to a process in which one of the particles is adiabatically taken around the other. Wrapping one particle around another is then topologically equivalent to a loop. In three dimensions, such a loop can be safely reduced to zero. Therefore, the wave function should be left unchanged by two such interchanges of particles. The only two possibilities are that the wave function changes by $\pm$ sign under a single interchange, corresponding to the cases of bosons and fermions, respectively.

For the same topological reason, the concept of identical-particle statistics becomes ambiguous in one spatial dimension. In this case, for swapping the positions of two particles, they need to pass through one another and it becomes impossible to disentangle the statistical properties from the interactions. If the wave function changes sign when two identical particles swap their positions, one could say that the particles are non-interacting fermions or, equivalently, that the particles are interacting bosons, where the change of sign is induced by the interaction as the particles pass through one another. This the main reason at the root of the possibility to adopt the *bosonization procedure* for describing one-dimensional fermions in terms of bosons, and vice versa, as we show in Chapter 12.

In two dimensions, a remarkably rich variety of particle statistics are possible: here there are indistinguishable particles that are neither bosons nor fermions, and they are called *anyons*. In *abelian anyons*, the two-particle wave function can change by an arbitrary phase when one particle is exchanged with the other

$$\psi(r_1,r_2) \rightarrow e^{i\theta}\, \psi(r_1,r_2). \tag{1.B.7}$$

There could be also *non-abelian anyons*. In this case, there is a degenerate set of g states $\psi_a(r_1,\ldots,r_n)$ $(a = 1,2,\ldots,g)$, with anyons at the positions $r_1,r_2,\ldots r_n$. The interchanges of two particles are elements of a group, called *the braid group* (see Problem 1.15). If β_i is the operation that interchanges particles i and $i+1$, it can be represented by a $g\times g$ unitary matrix $\gamma(\beta_i)$ that acts on these states as

$$\psi_a \rightarrow [\gamma(\beta_i)]_{ab}\, \psi_b.$$

The set of the $(n-1)$ matrices $\gamma(\beta_i)$ $(i = 1,2,\ldots,n-1)$ satisfy the Artin relations, discussed in Problem 1.15.

The situation of non-abelian anyons is realized, for instance, by trapping electrons in a thin layer between two semiconductor slabs. At a sufficiently strong magnetic field in the orthogonal direction and at a sufficiently low temperature, the wave function of the two-dimensional electron gas describes a deeply entangled ground state. The excitations above the ground state carry electron charges that are fractions of the original electron charge and have unusual statistical properties under the interchange of two of them. The anyons of this system give rise to the spectacular transport effects of the fractional quantum Hall effect.

Free particles. An important example of quantum statistical mechanics is provided by a system of free particles. This system can be described by the states of a single particle, here denoted by the index ν. Since the particles are indistinguishable at the quantum level, to specify a state of the system is sufficient to state the occupation number n_ν of each of its modes. If ϵ_ν is the energy of the νth mode, the total energy of the system is given by

$$E = \sum_\nu n_\nu\, \epsilon_\nu,$$

while the total number of particles is

$$N = \sum_{\nu} n_{\nu}.$$

For the three-dimensional systems, there are only two cases: the first is relative to the Fermi–Dirac (FD) statistics, the second to the Bose–Einstein (BE) statistics. In the first case, each mode can be occupied at most by one particle, so that the possible values of n_ν are

$$n_{\nu} = 0, 1 \qquad \text{Fermi–Dirac}$$

while, in the second case, each mode can be occupied by an arbitrary number of particles. In this case, the possible values of n_ν coincide with the natural numbers

$$n_{\nu} = 0, 1, 2, \ldots \qquad \text{Bose–Einstein}$$

The most convenient ensemble to describe the thermodynamics of this system is the grand canonical one. The corresponding partition function is

$$\mathcal{Z}(z, V, T) = \sum_{N=0}^{\infty} \sum_{\substack{\{n_\nu\} \\ \sum n_\nu = N}} z^N e^{-\beta \sum n_\nu \epsilon_\nu} = \sum_{N=0} \sum_{\substack{\{n_\nu\} \\ \sum n_\nu = N}} \prod_{\nu} \left(z e^{-\beta \epsilon_\nu}\right)^{n_\nu}.$$

To perform the double sums, it is sufficient to sum *independently* on each index n_ν, for every term in one case appears once and only once in the other, and vice versa. Hence

$$\begin{aligned} \mathcal{Z}(z, V, T) &= \sum_{n_0} \sum_{n_1} \cdots \left[\left(z e^{-\beta \epsilon_0}\right)^{n_0} \left(z e^{-\beta \epsilon_1}\right)^{n_1} \cdots \right] = \\ &= \left[\sum_{n_0} \left(z e^{-\beta \epsilon_0}\right)^{n_0} \right] \left[\sum_{n_1} \left(z e^{-\beta \epsilon_1}\right)^{n_1} \right] \cdots = \\ &= \prod_{\nu} \left[\sum_{n} \left(z e^{-\beta \epsilon_\nu}\right)^{n} \right], \end{aligned} \tag{1.B.8}$$

where the final sum is on the values 0, 1 for the fermionic case and on all the integers for the bosonic case. In the first case we have

$$\mathcal{Z}_F(z, V, T) = \prod_{\nu} \left[1 + z e^{-\beta \epsilon_\nu}\right],$$

while, in the second case, we have a geometrical series

$$\mathcal{Z}_B(z, V, T) = \prod_{\nu} \left[\frac{1}{1 - z e^{-\beta \epsilon_\nu}} \right].$$

The two expressions can be unified by the formula

$$\mathcal{Z}(z, V, T) = \prod_{\nu} \left(1 \pm z e^{-\beta \epsilon_\nu}\right)^{\pm 1},$$

where the + sign referes to the FD statistics whereas the − sign to the BE. The equation of state of both cases is

$$\beta P V = \log \mathcal{Z}(z, V, T) = \pm \sum_{\nu} \log\left(1 \pm z e^{-\beta \epsilon_\nu}\right),$$

where the variable z is related to the average number of particles by the equation

$$N = z \frac{\partial}{\partial z} \log \mathcal{Z}(z, V, T) = \sum_{\nu} \frac{z e^{-\beta \epsilon_\nu}}{1 \pm z e^{-\beta \epsilon_\nu}}. \tag{1.B.9}$$

The last expression shows that the occupation average of each mode is given in both cases by

$$\langle n_\nu \rangle = \frac{z e^{-\beta \epsilon_\nu}}{1 \pm z e^{-\beta \epsilon_\nu}}. \tag{1.B.10}$$

Let us briefly discuss the main features of the FD and BE distributions.

Fermi-Dirac. As is well known, the FD distribution of free particles turns out to be a surprisingly good model for the behaviour of conduction electrons in metal or for understanding, in the relativistic case, the existence of an upper limit of the mass of the dwarf stars (Chandrasekhar limit).

In order to discuss the fermion system in more detail, let us put $z = e^{\beta\mu}$ and let us consider the occupation average $n(\epsilon)$ in the limit $T \to 0$

$$n(\epsilon) = \frac{1}{e^{(\epsilon - \mu)/kT} + 1} \longrightarrow \begin{cases} 1, & \text{if } \epsilon < \mu \\ 0, & \text{if } \epsilon > \mu. \end{cases} \tag{1.B.11}$$

Note that in general the chemical potential depends on temperature. Its zero temperature value is the called the *Fermi energy*, $\epsilon_F = \mu(T = 0)$. The physical origin of the sharp shape of the limit expression (1.B.11) is the Pauli exclusion principle that posits that no two particles can be in the same level of the system. At zero temperature, the particle occupy the lowest possible energy levels up to a finite energy level ϵ_F. In momentum space, the particles fill a sphere of radius p_F, called the *Fermi sphere*. In this regime the gas is said to be *degenerate*. To compute ϵ_F, let us consider the gas inside a cube of side L with periodic boundary conditions, for simplicity. The energy of a single particle is just the kinetic energy $E = \frac{\mathbf{p}^2}{2m}$ and the components p_i of the momentum are quantized as

$$p_i = \frac{2\pi\hbar}{L} q_i, \quad q_i = 0, \pm 1, \pm 2, \ldots$$

For large L it is natural to substitute the sum (1.B.9) with an integral, according to the rule

$$\sum_q \longrightarrow \frac{V}{(2\pi\hbar)^3} \int d\vec{p}, \tag{1.B.12}$$

where $V = L^3$. If the spin of a particle is s, for a given momentum $\vec{p}$ there are $2s+1$ single particle states with the same energy $\epsilon(p)$ and the normalization condition at $T = 0$ becomes

$$N = (2s+1) \frac{V}{(2\pi\hbar)^3} \int_{\epsilon<\epsilon_F} d^3p = (2s+1) \frac{V}{(2\pi\hbar)^3} \frac{4\pi}{3} p_F^3. \tag{1.B.13}$$

Hence,

$$\epsilon_F = \frac{\hbar^2}{2m} \left(\frac{6\pi^2}{2s+1} \frac{N}{V} \right)^{2/3}. \tag{1.B.14}$$

We can define a *Fermi temperature* T_F by $\epsilon_F \equiv kT_F$. The Fermi energy and temperature provide useful energy and temperature scales for understanding the properties of fermion system. For instance, the conduction electron density for metals are typically of order 10^{22} per cubic centimeter, which corresponds to a Fermi temperature of order 10^5 Kelvin. This implies that a room temperature the system can be reasonably approximated by the degenerate distribution (1.B.11). Furthermore, notice that the Fermi energy (1.B.14) increases by increasing the density of the gas and, at sufficiently high density, ϵ_F can be higher than any energy scale ϵ_I associated to the interactions between the particles. This means that, counter-intuitively, in fermion systems the free particle approximation becomes better at higher values of the density!

At finite temperatures but smaller than the Fermi temperature $T < T_F$, $n(\epsilon)$ differs from its zero-temperature form only in a small region about μ of width a few kT, as shown in Figure 1.8. In computing integrals of the form $\mathcal{J} = \int_0^\infty f(\epsilon) n(\epsilon) d\epsilon$, the way they differ from the zero temperature values $\int_0^{\mu=\epsilon_F} f(\epsilon) n(\epsilon) d\epsilon$ depends on the form of $f(\epsilon)$ near μ. Integrating by parts, such integrals can be expressed as

$$\mathcal{J} = -\int_0^\infty g(\epsilon)\, n'(\epsilon)\, d\epsilon, \tag{1.B.15}$$

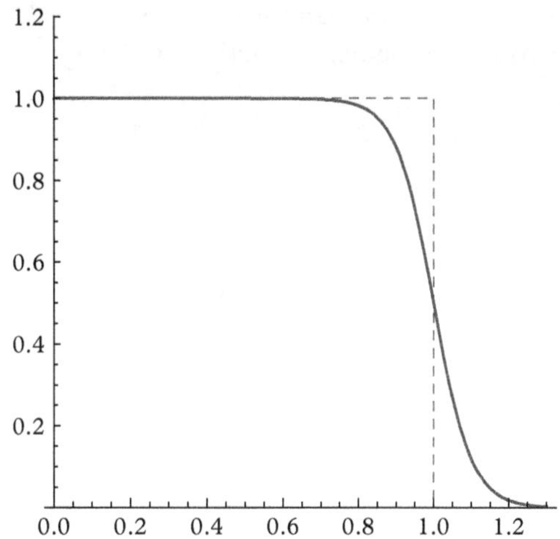

Fig. 1.8 *Fermi-Dirac distribution at $T = 0$ (dashed line) and at $T \neq 0$ (continuous line).*

where $g(\epsilon) = f'(\epsilon)$. Note that $n'(\epsilon)$ is sharply peaked at $\epsilon = \mu$, particularly at low temperature. If $g(\epsilon)$ does not vary rapidly in an interval of order kT near μ, the value of the integral can be thus estimated by replacing $g(\epsilon)$ with the first few terms of its Taylor expansion about $\epsilon = \mu$

$$g(\epsilon) = \sum_{n=0}^{\infty} \frac{1}{n!} \frac{d^n g(\mu)}{d\epsilon^n} (\epsilon - \mu)^n.$$

Substituting this expansion the integral (1.B.15) becomes

$$\mathcal{J} = -\sum_{n=0}^{\infty} \frac{1}{n!} \frac{d^n g(\mu)}{d\epsilon^n} \int_0^{\infty} n'(\epsilon)\,(\epsilon - \mu)^n \, d\epsilon.$$

The various integrals can be evaluated with the substitution $x = (\epsilon - \mu)/kT$ and since n' vanishes away from $\epsilon = \mu$, the lower limit of the integrals can be enlarged to $-\infty$ without a significant error. So

$$\int_0^{\infty} n'(\epsilon)(\epsilon - \mu)^n \, d\epsilon = -(kT)^n I_n$$

where

$$I_n = \int_{-\infty}^{\infty} \frac{x^n e^x}{(e^x + 1)^2} \, dx.$$

Since $e^x/(e^x+1) = 2/\cosh(x/2)$ is an even function, for n odd I_n vanish. The even ones can be expressed in terms of the Riemann function $\zeta(s) = \sum_{n=1}^{\infty} \frac{1}{n^s}$ as

$$I_{2n} = (2n)!(2-2^{2-2n})\zeta(2n)$$

The first representatives are

$$I_0 = 1, \quad I_2 = \frac{\pi^2}{3}, \quad I_4 = \frac{7\pi^4}{15}.$$

In this way we recover the so-called *Sommerfeld expansion* of the integral $\mathcal{J}$ (where we have inserted the original function $f(\epsilon)$)

$$\begin{aligned}&\int_0^{\infty} f(\epsilon)n(\epsilon)d\epsilon \\ &= \int_0^{\mu} f(\epsilon)d\epsilon + \frac{\pi^2}{6}(kT)^2 f'(\mu) + \frac{7\pi^4}{360}(kT)^4 f'''(\mu) + \mathcal{O}\left(\frac{kT}{\mu}\right)\end{aligned}$$

Applying the formula above, it is possible to compute the dependence of the chemical potential from the temperature

$$\mu = \epsilon_F \left[1 - \frac{\pi^2}{12}\left(\frac{kT}{\epsilon_F}\right)^2 - \frac{\pi^4}{80}\left(\frac{kT}{\epsilon_F}\right)^4 + \cdots\right],$$

and the expression of the internal energy

$$U = \frac{3}{5}N\epsilon_F \left[1 + \frac{5}{3}\left(\frac{kT}{\epsilon_F}\right)^2 + \cdots\right],$$

For the pressure we have

$$P = \frac{2}{5}\frac{N}{V}\epsilon_F \left[1 + \frac{5\pi^2}{12}\left(\frac{kT}{\epsilon_F}\right)^2 + \cdots\right].$$

This formula shows that even at zero temperature there is a non-zero value of the pressure, another manifestation of the Pauli principle.

Bose–Einstein. In three dimensions the boson gas presents the interesting phenomenon of BE condensation, i.e. a first-order phase transition. This phenomenon was predicted by Einstein in 1924. The condensation was achieved for the first time in atomic gases in 1995: Cornell, Wieman, were first, with ^{87}Rb atoms, followed by Ketterle and colleagues with ^{23}Na atoms and Hulet and colleagues with ^{7}Li atoms. In these experiments the atomic gas was confined by magnetic and/or optical trap to a relatively small region of space and at temperature of order nano-Kelvin. In order to discuss this remarkable aspect of bosons in more detail, let us consider, as before, the gas inside a cube of side L with periodic boundary conditions. The components p_i of the momentum are quantized as

$$p_i = \frac{2\pi\hbar}{L} q_i, \quad q_i = 0, \pm 1, \pm 2, \ldots$$

and the energy of a single particle is $E = \frac{\mathbf{p}^2}{2m}$. Since the mean value (1.B.10) of the number of particles for each mode ν has to be positive (in particular, the mode relative to the zero energy), for the variable z it holds the condition

$$0 \leq z \leq 1.$$

To compute the mean value of the density of the particle in the limit $L \to \infty$, it seems natural to substitute the sum (1.B.9) with an integral, according to the rule (1.B.12). In this way, we have

$$\frac{N}{V} = \int \frac{d\vec{p}}{\hbar^3} \frac{1}{z^{-1}\, e^{\beta p^2/2m} - 1}, \tag{1.B.16}$$

that, by a change of variable, can be written as

$$N = \frac{V}{\lambda^3} g(z), \tag{1.B.17}$$

where

$$g(z) = \frac{4}{\sqrt{\pi}} \int_0^\infty dx \frac{x^2\, e^{-x^2}}{z^{-1} - e^{-x^2}} = \sum_{n=1}^{\infty} \frac{z^n}{n^{3/2}}.$$

The quantity

$$\lambda = \sqrt{\frac{2\pi\hbar^2}{mkT}}$$

has the dimension of a length and it is called the *thermal wave length*, for it expresses the order of magnitude of the de Broglie wave length associated to a particle of mass m and energy kT. λ can be regarded as the position uncertainty associated with the thermal momentum distribution. The lower the temperature, the longer λ. When atoms are cooled to the point where λ is comparable to the inter-atomic separation, the atomic wave-packets overlap and the indistinguishability of particles becomes an important physical effect. The function $g(z)$ is an increasing function of z, as shown in Figure 1.9.

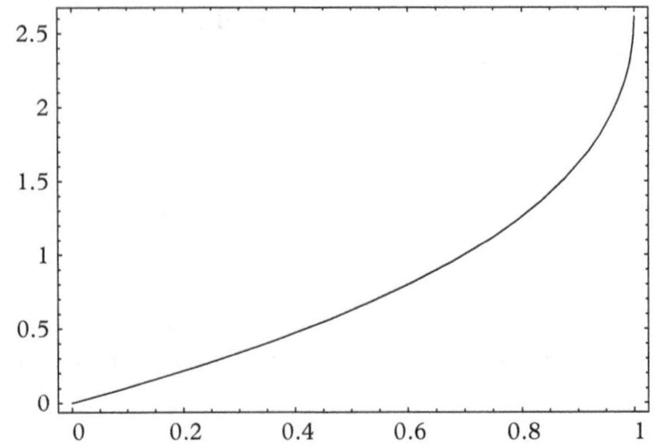

Fig. 1.9 *Plot of the function g(z).*

At $z = 1$ the function reaches its highest value, expressed in terms of the Riemann function $\zeta(x)$ by

$$g(1) = \sum_{n=1}^{\infty} \frac{1}{n^{3/2}} = \zeta\left(\frac{3}{2}\right) \simeq 2.612\ldots$$

and, for all values of z between 0 and 1, the function $g(z)$ satisfies the inequality

$$g(z) \leq g(1) = 2.612\ldots$$

From eqn. (1.B.17) the conclusion seems then to be that there exists a critical density of the system given by

$$N_{max} = g(1)\frac{V}{\lambda^3}.$$

But this is impossible due to the bosonic nature of the gas. In fact, if we had reached this critical density, what prevents us from adding further particles to the system? Hence there should be a mistake in the previous derivation, particularly in the substitution of the sum (1.B.9) with the integral (1.B.16). The cure of this drawback is to isolate the zero-mode before making the substitution of the sum with the integral. This is given by

$$n_0 = \frac{z}{1-z},$$

and for $z \to 1$, it is evident that it can be arbitrarily large, i.e. comparable with the sum of the entire series. Instead of (1.B.17), the correct version of the formula is then

$$N = \frac{V}{\lambda^3} g(z) + \frac{z}{1-z}.$$

Expressing it as

$$\lambda^3 \frac{n_0}{V} = \lambda^3 \frac{N}{V} - g(z),$$

it is easy to see that $n_0/V > 0$ when the temperature and the density of the particles satisfy the condition

$$\lambda^3 \frac{N}{V} \geq g(1) = 2.612\ldots \tag{1.B.18}$$

In this case, a finite fraction of the total number of the particles occupies the lowest energy level and a condensation phenomenon takes place. The system undergoes a phase transition from a normal gas state to a BE condensation, in which there is a macroscopic manifestation of the quantum nature of the system. The phase transition (that is of the first order) is realized when we have

$$\lambda^3 \frac{N}{V} = g(1).$$

This equation defines a curve in the space of the variables P-n-T. In particular, keeping fixed the density $d = N/V$, this equation identifies a critical temperature T_c given by

$$kT_c = \frac{2\pi\hbar^2}{m[d\,g(1)]^{2/3}},$$

Notice that T_c decreases when the mass of the particles increases. As previously mentioned, the BE condensation was realized for the first time in 1995 by using alkaline gases and, since then, it has become a rapidly developing research field.

REFERENCES

Statistical mechanics enjoys a surfeit of excellent texts. We especially recommend the following books as an introduction to many basic ideas and applications:

Feynman, R. P. (1972). Statistical Mechanics, W.A. Benjamin, Inc.
Huang, K (1963). Statistical Mechanics, John Wiley & Sons, Inc.
Kubo, R. (1965). Statistical Mechanics, North Holland Publishing Company.
Landau, L.D., Lifsits, E. M., Pitaevskij, L. P. (1978). Statistical Physics, Pergamon Press.
Mattis, D. C. (2006). The Theory of Magnetism Made Simple, World Scientific.
Reichl, L. E. (1980). Modern Course in Statistical Physics, Arnold Publishers Ltd.

A superb introduction to the modern theory of phase transitions is

Patasinskij, A.Z. and Pokrovskij, V.L. (1979). *Fluctuation Theory of Phase Transitions*, Pergamon Press.

The classical volume by Baxter is an excellent indepth introduction to a large class of exactly solvable models of statistical mechanics and to the methods of solution.

Baxter, R.J. (1982). *Exactly Solved Models in Statistical Mechanics*, Academic Press Limited.

The book by H.E. Slanley is a standard reference for the phenomenology of the critical phenomena and a general overview of the subject.

Stanley, H.E. (1971). *Introduction to Phase Transitions and Critical Phenomena*, Oxford Science Publications, 1971

Some relevant papers and books about the Ising model are

Ising, E. (1925). 'Beitrag zur Theorie des Ferromagnetismus', *Zeitschrift für Physik* 31: 253.

Kaufman, B. (1949). Crystal Statistics II: Partition Function Evaluated by Spinor Analysis, *Physical Review* 76: 1232.

Onsager, L. (1944). 'Crystal Statistics I: A Two-Dimensional Model with an Order-Disorder Transition', *Physical Review* 65: 117.

Peierls, R. (1936). 'On Ising's Model of Ferromagnetism', *Proceedings of the Cambridge Philosophical Society* 32: 477.

Delfino, G. and Mussardo, G (1995). 'The Spin-spin Correlation Function in the Two-Dimensional Ising Model in a Magnetic Field at $T = T_c'$, *Nuclear Physics B* 455: 724.

Delfino, G. Mussardo, G. and Simonetti, P. (1996). 'Non-integrable Quantum Field Theories as Perturbations of Certain Integrable Models', *Nuclear Physics B* 473: 469.

Delfino, G. and Simonetti, P. (1996). 'Correlation Functions in the Two-dimensional Ising Model in a Magnetic Field at $T = T_c'$, *Physics Letters B* 383: 450.

Fonseca P. and Zamolodchikov, A. (2003). 'Ising Field Theory in a Magnetic Field: Analytic Properties of the Free Energy', *Journal of Statistical Physics* 110: 527.

McCoy, B. M. and Wu, T. T. (1973). *The Two-dimensional Ising Model*, Harvard University Press.

Sato, M., Miwa, T., and Jimbo, M. (1980). *Aspects of Holonomic Quantum Fields*, Lecture Notes in Physics vol. 126, Springer Berlin/Heidelberg.

Wu, T. T., McCoy, B. M., Tracy, C., and Barouch, E. (1976). 'Spin-spin Correlation Functions for the Two-dimensional Ising Model: Exact Theory in the Scaling Region', *Physics Review B* 13: 316.

Zamolodchikov, A. B. (1989). 'Integrals of Motion of the (Scaled) $T = T_c$ Ising Model with Magnetic Field', *International Journal of Modern Physics A* 4: 4235.

The computational tractability frontier for the partition functions of several Ising models and their relationship with NP-complete problems is discussed in

Istrail, S., *Statistical Mechanics, Three-dimensionality and NP-completeness: I. Universality of Intractability of the Partition Functions of the Ising Model across Non-planar Lattices* in Proceedings of the 32nd ACM Symposium on the Theory of Computing (STOC00). Portland, Oregon, May 2000, ACM Press, pp. 87–96.

The history of the Ising model is discussed in

Niss, M. (2005). 'History of the Lenz–Ising Model 1920–1950: From Ferromagnetic to Cooperative Phenomena', *Archive for History of Exact Sciences*. 59: 267.
Brush, S. (1967). 'History of the Lenz–Ising Model, *Reviews of Modern Physics*. 39: 883.

Our presentation of the statistical properties of quantum particles is very schematic. A more complete treatment, in particular of the two-dimensional case, can be found in the book

Wilczek, F. (1990). *Fractional Statistics and Anyon Superconductivity*, World Scientific Press (1990) and the references therein.

Bose-Einstein condensation is a rapidly developing field. For an overall view on this subject, see

Pethick, C. and Smith, H. (2008). *Bose–Einstein Condensation in Dilute Gases*, Cambridge University Press.
Pitaevskii, L. and Stringari, S. (2003). *Bose–Einstein Condensation*, Oxford University Press.

PROBLEMS

1.1. Lattice Gas

Consider a lattice gas in which the particles occupy the sites of a d-dimensional lattice, with the constraint that each site cannot be occupied by more than one particle. Let e_i equal a variable that takes values $\{0, 1\}$: 0 when is the site is vacant and 1 when it is occupied. The interaction energy of each configuration is given by

$$\mathcal{H} = \mathcal{J} \sum_{\langle ij \rangle} e_i e_j.$$

Show that the grand canonical partition function of the lattice gas can be put in correspondence with the canonical partition function of the Ising model. Argue that the phase transition of the lattice gas, consists of the condensation of the particles, belongs to the same universality class of the Ising model.

1.2. Potts model

In the Potts model, the spin variable σ_i assume q value, as $\{0, 1, \ldots, q-1\}$. The energy of the configurations is given by

$$\mathcal{H} = -\mathcal{J} \sum_{\langle ij \rangle} \delta_{\sigma_i, \sigma_j},$$

where

$$\delta_{a,b} = \begin{cases} 1 & \text{if } a = b \\ 0 & \text{if } a \neq b. \end{cases}$$

a. Identify the symmetry transformations of the spins that leave the Hamiltonian invariant.

b. Show that for $q = 2$, the Potts model is equivalent to the Ising model.

c. Discuss the configuration of the minimum energy in the anti-ferromagnetic limit $\mathcal{J} \to -\infty$.

1.3. Theorem of equi-partition

Consider a classical one-dimensional harmonic oscillator, with Hamiltonian

$$H = \frac{p^2}{2m} + \frac{m\omega^2 x^2}{2},$$

a. Determine the surface $E = constant$ in the phase space and derive the thermodynamics of the system by using the microcanonical ensemble.

b. Put the system in contact with a thermal bath at temperature T. Compute the partition function in the canonical ensemble and show that the mean value of the energy is independent both from the frequency and the mass of the particle, i.e.

$$\left\langle \frac{p^2}{2m} \right\rangle = \left\langle \frac{m\omega^2 x^2}{2} \right\rangle = \frac{1}{2}\langle H \rangle = \frac{1}{2} kT.$$

c. Show that

$$\langle (E - \langle E \rangle)^2 \rangle = (kT)^2.$$

1.4. Equation of state for homogeneous potentials

Consider a system of classical particles whose interaction potential is given by a homogeneous function of degree η

$$U(\lambda \vec{r}_1, \lambda \vec{r}_2, \ldots, \lambda \vec{r}_N) = \lambda^\eta \, U(\vec{r}_1, \vec{r}_2, \ldots, \vec{r}_N).$$

Show that the equation of state of such a system assumes the form

$$P T^{-1+3/\eta} = f\left(\frac{V}{N} T^{-3/\eta} \right),$$

where, in principle, the function $f(x)$ can be computed once the explicit expression U of the potential is known.

1.5. Zeros of the partition function

Consider a classical system with only two states of magnetization, both proportional to the volume V of the system $M = \pm\alpha V$. In the presence of an external magnetic field B, the Hamiltonian is given by

$$\mathcal{H} = BM.$$

a. Compute the partition function in the canonical ensemble and determine its zeros in the complex plane of the temperature. Show that in the thermodynamical limit $V \to \infty$, there is an accumulation of the zeros at $T = \infty$.
b. Compute $\langle M \rangle$ as a function of B and study the limit of this function when $V \to \infty$.

1.6. Two-state systems

Consider a system of N free classical particles. The energy of each particle can take only two values: 0 and $E (E > 0)$. Let n_0 and n_1 the occupation numbers of the two energy levels and U the total energy of the system.

a. Determine the entropy of the system.
b. Determine $\langle n_0 \rangle$, $\langle n_1 \rangle$ and their fluctuations.
c. Express the temperature T as a function of U and show that it can take negative values.
d. Discuss what happens when a system at negative temperature is put in thermal contact with a system at positive temperature.

1.7. Scaling laws

Given the equation of state of a magnetic system in the form

$$B = M^{\delta}\, \mathcal{Q}\left(\frac{t}{M^{1/\beta}}\right),$$

a. Prove that the parameters β and δ in the expression above are the critical exponents of the system, as defined in the chapter;
b. Show the identity $\gamma = \beta(\delta - 1)$.

1.8. First-order phase transitions

In the second-order phase transitions, the state with the lowest value of the free energy changes continuously when the system crosses its critical point. On the contrary, in the first-order phase transition, the order parameter changes discontinuously.

a. Study the behaviour of the minima of the free energy

$$F(x) = a(T)x^2 + x^4$$

by varying the temperature T as $a(T) = (T - T_c)$ and determine if we are in presence of a first- or second-order phase transition.

b. Analyze the same questions for the free energy given by

$$F(x) = (x^2 - 1)^2\,(x^2 + a(T))$$

with the same expression for $a(T)$.

1.9. Ergodic system

Consider a classical dynamical system with a phase space $(0 < q < 1;\ 0 < p < 1)$ and equation of motion given by

$$q(t) = q_0 + t \ ;\ p(t) = p_0 + \alpha\, t.$$

a. Discuss the trajectories in the phase space when α is a rational and irrational number.
b. Show that the system is ergodic when α is irrational, i.e. the time averages of all functions $f(q,p)$ coincide with their average on the phase space.

Hint: Use the fact that the volume of the phase space is finite to expand any function of the coordinate and momentum in a Fourier series.

1.10. Density of states

Determine the number of quantum states with energy less than E for a free particle in a cubic box of length L. Compare this quantity with the volume of the classical phase space and find the corresponding density of states of the system.

1.11. Quantum harmonic oscillator

The one-dimensional quantum oscillator has the energy spectrum given by $E_n = \hbar\omega\,(n + 1/2)$, $n = 0, 1, 2, \ldots$.

a. Compute the partition function in the canonical ensemble.
b. Compute the specific heat as a function of the temperature and discuss how this quantity differs from the analogous classical expression.

1.12. Riemann function

The Riemann function $\zeta(\beta)$ is defined by

$$\zeta(\beta) = \sum_{n=1}^{\infty} \frac{1}{n^{\beta}}.$$

a. Interpret this expression as the partition function in the canonical ensemble of a quantum system and identify the discrete spectrum of the energies.
b. Compute the density of states and the entropy of the quantum system. Interpret the singularity of $\zeta(\beta)$ at $\beta = 1$ as a phase transition.

1.13. Bose–Einstein Condensation

Appendix 1.B showed that, in three dimensions, an ideal gas with a bosonic statistics presents a Bose–Einstein condensation for sufficiently low temperature. Discuss if the same phenomenon can take place in one and two dimensions. Explain if a Bose–Einstein condensation can happen for a harmonic oscillator in dimension $d = 1, 2, 3$.

1.14. Dimensional regularization

Let d the dimension of the space. Discuss the convergence of the integral

$$I(d) = \int_0^\infty \frac{r^{d-1}}{r^2 + 1} \, dr$$

by varying d.

a. Determine, in its convergent domain, the exact expression of the integral as a function of d and identify the position of its poles.
b. Analytically continue the definition of the integral in any other domain.
c. Compute its value for $d = \frac{1}{3}$ and $d = \pi$.

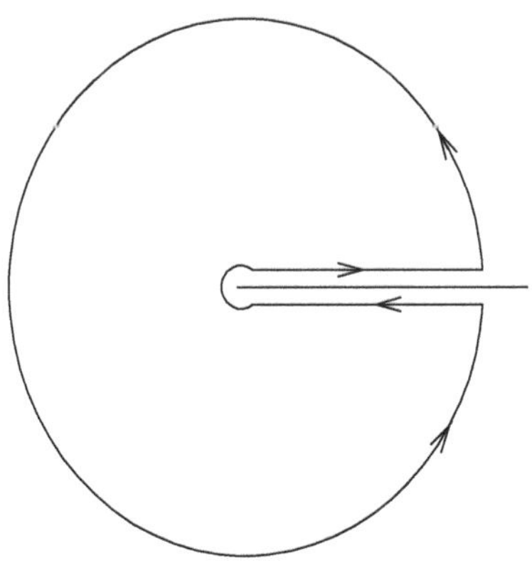

Fig. 1.10 *Integration contour C.*

Hint: Consider the integral in the complex plane

$$\oint_C \frac{z^{d-1}}{z^2+1}\, dz$$

where C is the contour shown in Figure 1.10.

1.15. Braid group

The braid group on n strands, denoted by B_n, is a set of operations has an intuitive geometrical representation, and in a sense generalizes the symmetric group S_n. Here, n is a natural number. Braid groups find applications in knot theory, since any knot may be represented as the closure of certain braids. From the algebraic point of view, the braid group is represented in terms of generators β_i, with $1 \leq i \leq (n-1)$; β_i is a counter-clockwise exchange of the ith and $(i+1)^{th}$ strands. β_i^{-1} is therefore a clockwise exchange of the ith and $(i+1)^{th}$ strands. The generators β_i satisfy the defining relations, called Artin relations (see Figure 1.11).

$$\beta_i \beta_j = \beta_j \beta_i \qquad \text{for } |\, i-j \,| \geq 2$$
$$\beta_i \beta_{i+1} \beta_i = \beta_{i+1} \beta_i \beta_{i+1} \qquad \text{for } 1 \leq i \leq n-1$$

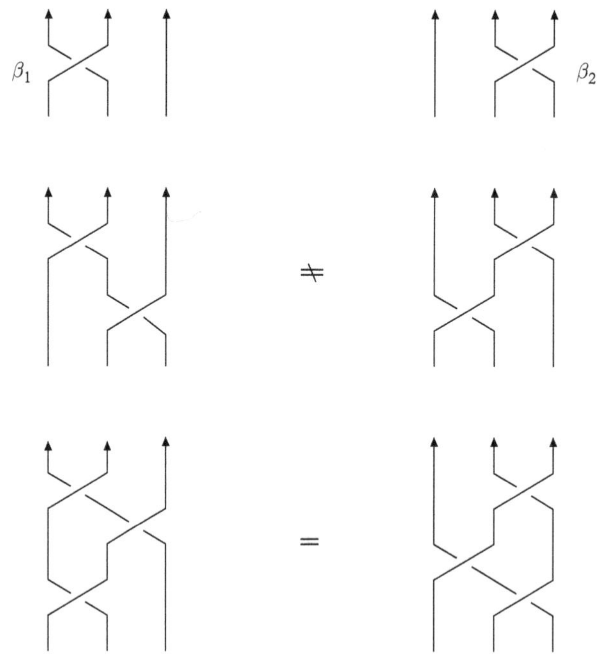

Fig. 1.11 Top: *The two elementary braid operations β_1 and β_2.* **Middle**: *Graphical proof that $\beta_2\beta_1 \neq \beta_1\beta_2$, hence the braid group is not abelian.* **Bottom**: *the Yang–Baxter relation of the braid group.*

The second is also called Yang–Baxter equation. The only difference from the permutation group is that $\beta_i^2 \neq 1$, but this is an enormous difference: while the permutation group is finite (the dimension is $n!$), the braid group is infinite, even for just two strands.

The irreducible representation of the braid group can be given in terms of $g \times g$ dimensional unitary matrices, $\beta_i \to \gamma_i$, where the matrices γ_i satisfy the Artin relations.

a. Consider the group B_3. Prove that

$$\gamma_1 = \begin{pmatrix} e^{-7i\pi/10} & 0 \\ 0 & -e^{-3i\pi/10} \end{pmatrix}, \quad \gamma_2 = \begin{pmatrix} -\tau e^{-i\pi/10} & -i\sqrt{\tau} \\ -i\sqrt{\tau} & -\tau e^{i\pi/10} \end{pmatrix}$$

provide a representation of the Artin relations. Above $\tau = (\sqrt{5}-1)/2$, which satisfies $\tau^2 + \tau = 1$.

b. Both matrices γ_i $(i = 1, 2)$ are matrices of $SU(2)$ and can be written as

$$\gamma_i = \exp\left[i \frac{\theta_i \vec{n}_i}{2} \cdot \vec{\sigma} \right]$$

where σ_j $(j = 1, 2, 3)$ are the Pauli matrices and θ_i is the angle of rotation around the axis $\vec{n}_i$. Identify the angles and the axes of rotation that correspond to γ_1 and γ_2.

c. By multiplying the γ_i (and their inverse) in a sequence of L steps, as in the example below

$$A_L = \underbrace{\gamma_1 \gamma_2 \gamma_1^{-1} \gamma_2 \dots \gamma_1}_{L},$$

one generates another matrix A_L of SU(2), identified by the angle α of rotation around an axis $\vec{n}$, $A_L = \exp\left[i \frac{\alpha \vec{n}}{2} \cdot \vec{\sigma} \right]$. Argue that, by making L sufficiently large, find a string of γ_i and their inverse that approximates with an arbitrary precision any matrix of $SU(2)$.

2

One-dimensional Systems

If our highly pointed Triangles of the Soldier class are formidable, it may be readily inferred that far more formidable are our Women. For, if a Soldier is a wedge, a Woman is a needle.

Edwin A. Abbott, *Flatland*

This chapter presents several approaches to get the exact solution of the one-dimensional Ising model. As already mentioned, the one-dimensional case does not present a phase transition at a finite value of the temperature. However, we show that the origin $T = B = 0$ of the phase diagram nevertheless may be regarded as a critical point: by using appropriate variables, we can define the set of critical exponents and verify that the scaling relations are indeed satisfied.

This chapter also discusses three different generalizations of the Ising model: the first is given by the q-state Potts model—a system that is invariant under the permutation group $\mathcal{S}_q$ of q objects; the second is provided by a system of spins with n-components, invariant under the continuum group of the transformations $O(n)$; the third one is the so-called $Z(n)$ model, i.e. a spin system that is invariant under the set of the discrete rotations associated to the nth roots of unity. We compute the partition function of all these models, pointing out their interesting properties. Finally, we analyse the thermodynamics of the so-called *Feynman's gas*, i.e. a one-dimensional gas of particles with short range potential $V(\mid x_i - x_j \mid)$: the results of this analysis will be useful when we in later chapters we face the study of the correlation functions of the two-dimensional models.

2.1 Recursive Approach

The first method we introduce is based on a recursive approach: it permits obtaining the exact solution of the one-dimensional Ising model in absence of an external magnetic field.

Consider a linear chain of N Ising spins (see Figure 2.1) in absence of an external magnetic field, with free boundary conditions on the first and the last spin of the chain.

Statistical Field Theory: An Introduction to Exactly Solved Models in Statistical Physics. Second Edition. Giuseppe Mussardo, Oxford University Press (2020). © Giuseppe Mussardo. DOI: 10.1093/oso/9780198788102.001.0001

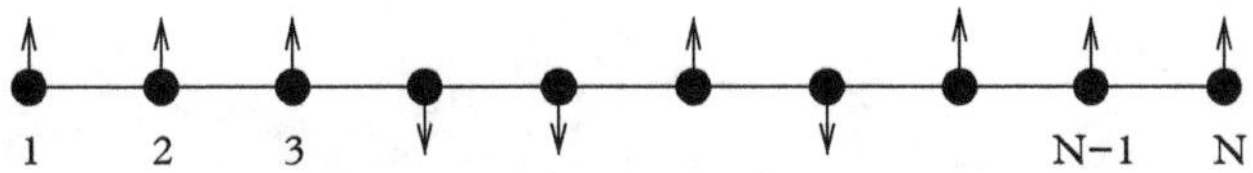

Fig. 2.1 *Linear chain of N Ising spins.*

The more general Hamiltonian of such a system is given by

$$\mathcal{H} = -\sum_{i=1}^{N-1} \mathscr{J}_i \sigma_i \sigma_{i+1},$$

with an interaction $\mathscr{J}_i$ that may change from site to site. The partition function is expressed by

$$Z_N = \sum_{\sigma_1=-1}^{1} \sum_{\sigma_2=-1}^{1} \cdots \sum_{\sigma_N=-1}^{1} \exp\left(\sum_{i=1}^{N-1} \mathcal{J}_i \sigma_i \sigma_{i+1} \right), \tag{2.1.1}$$

where we have introduced the notation $\mathcal{J}_i = \beta \mathscr{J}_i$. The recursive method consists of adding an extra spin to the chain and expressing the resulting partition function Z_{N+1} in terms of the previous Z_N. By adding another spin, we have

$$Z_{N+1} = \sum_{\sigma_1=-1}^{1} \sum_{\sigma_2=-1}^{1} \cdots \sum_{\sigma_N=-1}^{1} \exp\left(\sum_{i=1}^{N-1} \mathcal{J}_i \sigma_i \sigma_{i+1} \right) \sum_{\sigma_{N+1}=-1}^{1} \exp\left(\mathcal{J}_N \sigma_N \sigma_{N+1}\right). \tag{2.1.2}$$

The last sum can be easily computed

$$\sum_{\sigma_{N+1}=-1}^{1} \exp\left(\mathcal{J}_N \sigma_N \sigma_{N+1}\right) = e^{\mathcal{J}_N \sigma_N} + e^{-\mathcal{J}_N \sigma_N} = 2\cosh(\mathcal{J}_N \sigma_N) = 2\cosh \mathcal{J}_N,$$

and the result is *independent* from σ_N, which allows us to rewrite eqn. (2.1.2) as

$$Z_{N+1} = (2\cosh \mathcal{J}_N) Z_N,$$

and the iteration of this relation leads to

$$Z_{N+1} = \left(2^N \prod_{i=1}^{N} \cosh \mathcal{J}_i \right) Z_1.$$

Since the partition function Z_1 of an isolated spin is equal to the number of its states, i.e. $Z_1 = 2$, the exact expression of the partition function of N spins is given by

$$Z_N = 2^N \prod_{i=1}^{N-1} \cosh \mathcal{J}_i. \tag{2.1.3}$$

To see whether there is a critical value T_c of the temperature (below which the system presents a magnetized phase), it is useful to compute the two-spin correlation function

$$G^{(2)}(r) = \langle \sigma_k \sigma_{k+r} \rangle = Z_N^{-1} \sum_{\{\sigma\}} \sigma_k \sigma_{k+r} \exp\left(\sum_{i=1}^{N-1} \mathcal{J}_i \sigma_i \sigma_{i+1} \right), \tag{2.1.4}$$

where the first sum remains a concise way of expressing the sum on the ± 1 values of all the N spins. If $r = 1$, the correlation function is obtained by taking the derivative

$$G^{(2)}(1) = \langle \sigma_k \sigma_{k+1} \rangle = Z_N^{-1} \frac{\partial}{\partial \mathcal{J}_k} \sum_{\{\sigma\}} \exp\left(\sum_{i=1}^{N-1} \mathcal{J}_i \sigma_i \sigma_{i+1} \right).$$

Thanks to the identity $\sigma_i^2 = 1$, valid for the Ising spins, the formula generalized to arbitrary r

$$Z_N G^{(2)}(r) = \frac{\partial}{\partial \mathcal{J}_k} \frac{\partial}{\partial \mathcal{J}_{k+1}} \cdots \frac{\partial}{\partial \mathcal{J}_{k+r-1}} Z_N. \tag{2.1.5}$$

Substituting in this formula eqn. (2.1.3), we have

$$G^{(2)}(r) = \prod_{i=1}^{r} \tanh \mathcal{J}_{k+i-1}. \tag{2.1.6}$$

This expression allows us easily to check the validity of simple physical intuitions. It correctly predicts that, by taking the limit $\mathcal{J}_i \to 0$ that breaks the chain in two separate blocks, if the site i is placed between k and $k+r$, the correlation function vanishes, and vice versa, if the site i is external to the interval $(k, k+r)$, the correlation fucntion is unaffected by the limit $\mathcal{J}_i \to 0$.

If the system is homogeneous, with the same coupling constant $\mathcal{J}$ for all spins, we have the simpler expression

$$G^{(2)}(r) = (\tanh \mathcal{J})^r, \tag{2.1.7}$$

that can be written in a scaling form as

$$G^{(2)}(r) = \exp\left[-r/\xi\right].$$

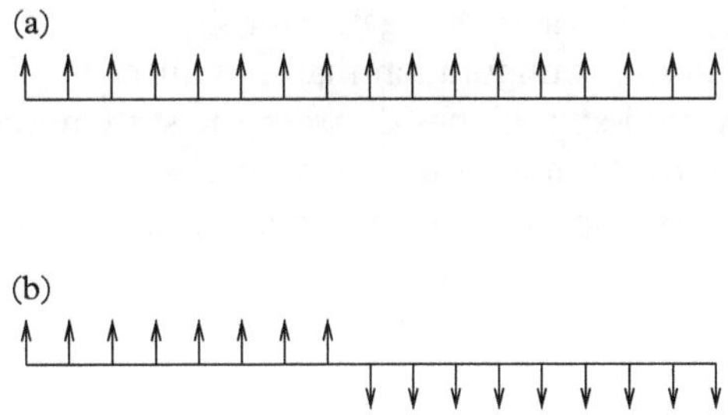

Fig. 2.2 *(a): Ordered low-energy state; (b): Excited state.*

The correlation length ξ, in unit of the lattice space a, is given by

$$\xi(\mathcal{J}) = -\frac{1}{\log \tanh \mathcal{J}}. \tag{2.1.8}$$

We can use this expression of ξ to identify the possible critical points of the system, since ξ diverges at a phase transition. It is easy to see that ξ has only one singular point, given by

$$\mathcal{J} = \beta \mathfrak{J} = \frac{\mathfrak{J}}{kT} \to \infty,$$

i.e. $T = 0$ (if $\mathfrak{J}$ is a finite quantity). One arrives to the same conclusion by analysing the possibility of a non-zero expectation value of the spin, i.e. a non-vanishing limit

$$| \langle \sigma \rangle |^2 = \lim_{r \to \infty} G^{(2)}(r). \tag{2.1.9}$$

Since for finite $\beta \mathfrak{J}$ the hyperbolic tangent entering $G^{(2)}(r)$ is always less than 1, the spontaneous magnetization always vanishes, except for the limiting case $\beta \mathfrak{J} = \infty$, i.e. $T = 0$.

The absence of an ordered phase in a finite interval of the temperature T of the one dimensional Ising model can be readily explained by some simple thermodynamical considerations. In fact, let us assume that at a sufficiently low temperature the system is a complete ordered state, i.e. with all spins aligned up, e.g. $\sigma_i = 1$. The energy of this configuration is $E_0 = -(N-1)\mathfrak{J}$. The configurations of the system with the next higher energy are those in which an entire spin block is inverted at an arbitrary point of the chain (see Figure 2.2). Their number is $N - 1$ (equal to the number of sites where this inversion of the spins can take place) and their energy is $E = E_0 + 2\mathfrak{J}$. At a temperature T, the variation of the free energy induced by these excitations is expressed by

$$\Delta F = \Delta E - T\Delta S = 2\mathfrak{J} - kT \ln(N-1), \tag{2.1.10}$$

and, for N sufficiently large, it is always negative for all value of $T \neq 0$. Hence the ordered state of the system is not the configuration that minimizes the free energy. Because the configurations with inverted spin blocks disorder the system, the ordered phase of the one-dimensional Ising model is always unstable for $T \neq 0$.

However, the absence of a spontaneous magnetization at a finite T does not imply the absence of a singularity at $T = 0$. For example, let us compute the magnetic susceptibility at $B = 0$ by using the fluctuation-dissipation theorem

$$\chi(T, B=0) = \frac{\beta}{N}\sum_{i=1}^{N}\sum_{j=1}^{N}\langle\sigma_i\sigma_j\rangle. \tag{2.1.11}$$

For simplicity, consider the homogeneous case $\langle\sigma_i\sigma_j\rangle = v^{|i-j|}$, with $v = \tanh\beta\mathcal{J}$. In the sum above, there are

- N terms, for which $|i-j| = 0$. Each of them gives rise to a factor $v^0 = 1$.
- $2(N-1)$ terms, for which $|i-j| = 1$. They correspond to the $N-1$ next neighbouring pairs of spins of the open chain and each of them brings a term v^1.
- $2(N-2)$ terms, for which $|i-j| = 2$ and a term v^2, and so on, until we arrive at the last 2 terms for which $|i-j| = N-1$, each of them bringing a factor v^{N-1}.

Hence the double sum (2.1.11) can be expressed as

$$\chi(T, B=0) = \frac{\beta}{N}\left(N + 2\sum_{k=1}^{N-1}(N-k)\,v^k\right).$$

By using

$$\sum_{k=1}^{N-1} v^k = \frac{1-v^N}{1-v},$$

$$\sum_{k=1}^{N-1} k v^k = v\frac{\partial}{\partial v}\sum_{k=1}^{N-1} v^k,$$

we arrive at

$$\chi(T, B=0) = \frac{\beta}{N}\left[N\left(1 + \frac{2v}{1-v}\right) - \frac{2v(1-v^N)}{(1-v)^2}\right].$$

This expression can be simplified by taking the thermodynamical limit $N \to \infty$

$$\chi(T,B=0) = \beta \frac{1+v}{1-v} = \beta\, e^{2\mathcal{J}/kT},$$

and this expression presents an essential singularity for $T \to 0$.

It is also interesting to study the case $\mathcal{J} < 0$, which corresponds to the *anti-ferromagnetic* situation. In such case, the minimum of the energy of the system is realized by those configurations where the spins alternate their values by moving from one site to the next one. The two-point correlation function of the spins is given by eqn. (2.1.7) also in the anti-ferromagnetic case. However, for negative values of $\mathcal{J}$, it changes its sign by changing the lattice sites, as shown in Figure 2.3. The oscillating behaviour of this function is responsible for a a partial cancellation of the terms entering the series (2.1.11) of the magnetic susceptibility that remains indeed finite for all values of temperature.

Using the previous formulae, we can explicitly compute the mean energy U and the specific heat C at $B = 0$. For the mean energy we have

$$\langle U\rangle = -\frac{\partial}{\partial\beta}(\ln Z_N(T,B=0)) = -\sum_{i=1}^{N-1} \mathcal{J}_i \tanh \mathcal{J}_i = -\mathcal{J}(N-1)\tanh\mathcal{J},$$

where the last identity holds in the homogeneous case, while for the specific heat we get

$$C(T,B=0) = \frac{\partial\langle U\rangle}{\partial T} = k(N-1)\left(\frac{\mathcal{J}}{\cosh\mathcal{J}}\right)^2. \tag{2.1.12}$$

The plot of this function is shown in Figure 2.4. Similar functions, with a pronounced maximum, are obtained for the specific heat of all those substances have only one energy gap ΔE and, in the literature, are known as Schottky curves. The reason why the

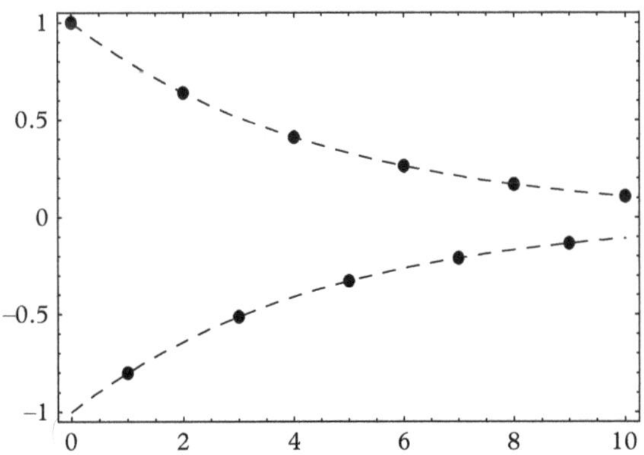

Fig. 2.3 *Two-point correlation function of the spins in the ferromagnetic case (upper curve) and in the anti-ferromagnetic case (lower curve).*

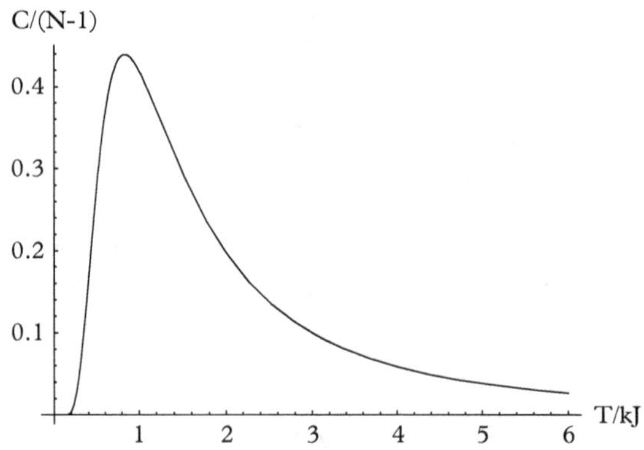

Fig. 2.4 *Specific heat of the one-dimensional Ising model versus temperature.*

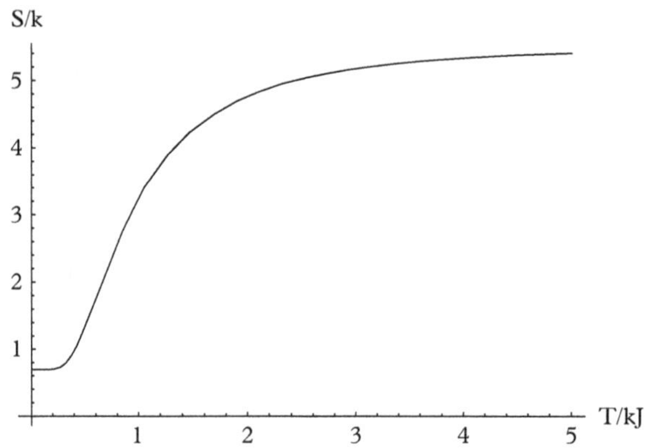

Fig. 2.5 *Entropy versus the temperature.*

one-dimensional Ising model is equivalent to a system with only one energy gap ΔE will become clear after the discussion in the next section on the transfer matrix of the model.

By using eqn. (2.1.3), we can also compute the entropy of the system

$$\begin{aligned} S(T,B=0) &= \frac{\partial}{\partial T}\left[\frac{1}{\beta}\ln Z_N\right] = \\ &= k\,[N\ln 2 + (N-1)\ln\cosh\mathcal{J} - (N-1)\mathcal{J}\tanh\mathcal{J}]. \end{aligned} \tag{2.1.13}$$

The plot of the entropy is in Figure 2.5. For $T \to 0$, the entropy goes correctly to the value $k\ln 2$: in fact, at $T=0$, there are only two effective states of the system—the one in which all spins are up, and the other one in which all spins are down. For $T \to \infty$, we have instead $S \to Nk\ln 2$: in this limit all spins are free to fluctuate in an independent way and, correspondingly, the available number of states of the systems is given by 2^N.

2.2 Transfer Matrix

The exact solution of one-dimensional Ising model can be obtained by using the alternative method of the transfer matrix. This method presents a series of advantages: unlike the recursive method, it can also be applied when there is an external magnetic field. Moreover, it has many points in common with a discrete formulation of quantum mechanics, in particular, the Feynman formulation in terms of a path integral. The transfer matrix method relies on a set of ideas that go beyond the application to the one-dimensional case and allows us to see the remarkable relationship that links classical systems of statistical mechanics in d dimensions with quantum systems in $(d-1)$, which is discussed in more detail Chapter 7. In the two-dimensional case, for instance, it allows us obtain the exact solution of the Ising model in the absence of an external magnetic field (see Chapter 6).

To study the one-dimensional case, let us consider once again a chain of N spins. For simplicity, we consider here the homogeneous case, in which there is only one coupling constant $\mathcal{J}$, with Hamiltonian

$$\mathcal{H} = -\mathcal{J}\sum_{i=1}^{N-1}\sigma_i\sigma_{i+1} - B\sum_{i=1}^{N}\sigma_i. \tag{2.2.1}$$

First we analyse the periodic boundary condition case while more general boundary conditions will be considered later.

2.2.1 Periodic Boundary Conditions

Assuming periodic boundary conditions, the chain has a ring geometry, implemented by the condition

$$\sigma_i \equiv \sigma_{N+i}.$$

The transfer matrix method is based on the observation that the sum on the spin configurations can be equivalently expressed in terms of a product of 2×2 matrices, as follows

$$Z_N = \sum_{\{\sigma\}} V(\sigma_1,\sigma_2)\,V(\sigma_2,\sigma_3)\cdots\, V(\sigma_N,\sigma_1), \tag{2.2.2}$$

where the matrix elements of $V(\sigma,\sigma')$ are defined by

$$V(\sigma,\sigma') = \exp\left[\mathcal{J}\sigma\sigma' + \frac{1}{2}\mathcal{B}(\sigma+\sigma')\right], \tag{2.2.3}$$

with $\mathcal{J} = \beta J$ and $\mathcal{B} = \beta B$. Explicitly

$$\begin{aligned}
\langle +1 \mid V \mid +1 \rangle &= e^{\mathcal{J}+\mathcal{B}}; \\
\langle -1 \mid V \mid +1 \rangle &= e^{-\mathcal{J}}; \\
\langle +1 \mid V \mid -1 \rangle &= e^{-\mathcal{J}}; \\
\langle -1 \mid V \mid -1 \rangle &= e^{\mathcal{J}-\mathcal{B}},
\end{aligned}$$

and therefore V can be written as

$$V = \begin{pmatrix} e^{\mathcal{J}+\mathcal{B}} & e^{-\mathcal{J}} \\ e^{-\mathcal{J}} & e^{\mathcal{J}-\mathcal{B}} \end{pmatrix}. \tag{2.2.4}$$

It is easy to see that the product of the matrix V correctly reproduces the Boltzmann's weights of the Ising model configurations.

In this approach, the configuration space of a single spin may be regarded as the Hilbert space of a *two-state quantum system*: the states will be denoted by $\mid +1\rangle$ e $\mid -1\rangle$, and the completeness relation is expressed by the formula

$$\sum_{\sigma=\pm 1} \mid \sigma\rangle\langle\sigma \mid = 1. \tag{2.2.5}$$

The original one-dimensional lattice can be seen as a temporal axis, along which the quantum dynamics of the two-state system takes place. In more detail, the transfer matrix V plays the role of the quantum time evolution operator for the time interval $\Delta t = a$ (see Figure 2.6)

$$\mid \sigma_{i+1}\rangle = V \mid \sigma_i\rangle \equiv e^{-a\mathbf{H}} \mid \sigma_i\rangle. \tag{2.2.6}$$

In this formula **H** expresses the quantum Hamiltonian, that has not to be confused with the original classical Hamiltonian $\mathcal{H}$ given in eqn. (2.2.1). By adopting this scheme based on a two-state Hilbert space, it becomes evident that the one-dimensional Ising model presents only one energetic gap $\Delta\mathcal{E}$: thus we have a natural explanation of the Schottky form of the specific heat, discussed in the previous section.

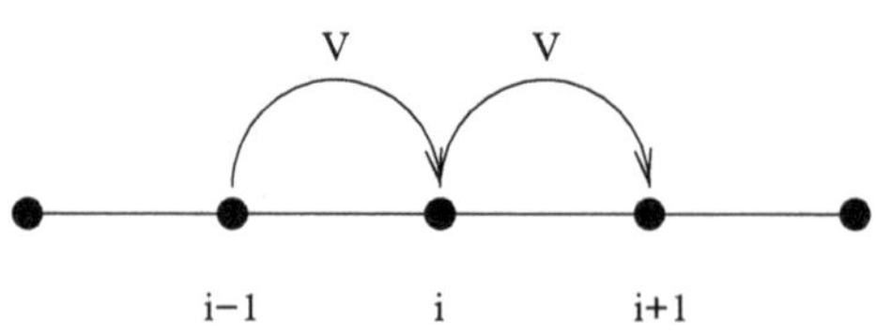

Fig. 2.6 *Transfer matrix as quantum time evolution operator.*

Quantum Hamiltonian. It is an interesting exercise to find the explicit expression of the quantum Hamiltonian $\mathbf{H}$. Let us recall that, in the linear space of the 2×2 matrices, a basis is provided by the identity matrix $\mathbf{1} = \begin{pmatrix} 1 & 0 \\ 0 & 1 \end{pmatrix}$ and by the Pauli matrices $\hat{\sigma}_i$

$$\hat{\sigma}_1 = \begin{pmatrix} 0 & 1 \\ 1 & 0 \end{pmatrix}, \quad \hat{\sigma}_2 = \begin{pmatrix} 0 & -i \\ i & 0 \end{pmatrix}, \quad \hat{\sigma}_3 = \begin{pmatrix} 1 & 0 \\ 0 & -1 \end{pmatrix}. \tag{2.2.7}$$

They satisfy

$$\{\hat{\sigma}_k, \hat{\sigma}_l\} = 2\delta_{kl}, \quad [\hat{\sigma}_k, \hat{\sigma}_l] = 2i\epsilon_{klm}\hat{\sigma}_m \tag{2.2.8}$$

where $\{a, b\} = ab + ba$, $[a, b] = ab - ba$ and ϵ_{klm} is the anti-symmetric tensor in all the three indices, with $\epsilon_{123} = 1$.

In terms of these matrices, V can be written as

$$V = \left(e^{\mathcal{J}} \cosh \mathcal{B}\right) \mathbf{1} + e^{-\mathcal{J}}\, \hat{\sigma}_1 + \left(e^{\mathcal{J}} \sinh \mathcal{B}\right) \hat{\sigma}_3. \tag{2.2.9}$$

Let us determine the constants C, c_1, c_2, c_3 so that V is expressed as

$$V = C \exp\left[c_1\hat{\sigma}_1 + c_2\hat{\sigma}_2 + c_3\hat{\sigma}_3\right]. \tag{2.2.10}$$

By making a series expansion of the exponential

$$\exp\left[c_1\hat{\sigma}_1 + c_2\hat{\sigma}_2 + c_3\hat{\sigma}_3\right] = \sum_{k=0}^{\infty} \frac{\left(c_1\hat{\sigma}_1 + c_2\hat{\sigma}_2 + c_3\hat{\sigma}_3\right)^k}{k!}, \tag{2.2.11}$$

and using the anti-commutation rule (2.2.8), it is easy to see that we arrive at

$$\begin{aligned} \left(c_1\hat{\sigma}_1 + c_2\hat{\sigma}_2 + c_3\hat{\sigma}_3\right)^{2n} &= r^{2n}, \\ \left(c_1\hat{\sigma}_1 + c_2\hat{\sigma}_2 + c_3\hat{\sigma}_3\right)^{2n+1} &= \left(c_1\hat{\sigma}_1 + c_2\hat{\sigma}_2 + c_3\hat{\sigma}_3\right) r^{2n}, \end{aligned}$$

where $r = \sqrt{c_1^2 + c_2^2 + c_3^2}$. By summing the series (2.2.11), eqn. (2.2.10) becomes

$$V = C\left[\cosh r\, \mathbf{1} + \frac{\sinh r}{r}\left(c_1\,\hat{\sigma}_1 + c_2\,\hat{\sigma}_2 + c_3\,\hat{\sigma}_3\right)\right].$$

Comparing this expression with eqn. (2.2.9), we have

$$
\begin{aligned}
C \cosh r &= e^{\mathcal{J}} \cosh \mathcal{B}, \\
C \frac{\sinh r}{r} c_1 &= e^{-\mathcal{J}}, \\
C \frac{\sinh r}{r} c_2 &= 0, \\
C \frac{\sinh r}{r} c_3 &= e^{\mathcal{J}} \cosh \mathcal{B},
\end{aligned}
$$

from which it immediately follows $c_2 = 0$. From the ratio between the fourth and the second equation, we have

$$c_3 = c_1 e^{2\mathcal{J}} \sinh \mathcal{B}.$$

Summing the square of the second and the fourth equations and subtracting the square of the first equation, we get

$$C^2 = 2 \sinh 2\mathcal{J},$$

i.e. $C = \sqrt{2 \sinh 2\mathcal{J}}$. Finally, by taking the ratio of the square of the first and the second equations and using eqn. (2.2.1), c_1 is given by the solution of the transcendental equation

$$\tanh \left[c_1 \sqrt{1 + e^{4\mathcal{J}} \sinh^2 \mathcal{B}} \right] = \frac{\sqrt{1 + e^{4\mathcal{J}} \sinh^2 \mathcal{B}}}{e^{2\mathcal{J}} \cosh \mathcal{B}}. \tag{2.2.12}$$

Hence the quantum Hamiltonian $\mathbf{H}$ is given by

$$\mathbf{H} = -\frac{1}{a} \left[\left(\frac{1}{2} \log(\sinh 2\mathcal{J}) \right) + c_1 \left(\hat{\sigma}_1 + e^{2\mathcal{J}} \sinh \mathcal{B} \hat{\sigma}_3 \right) \right], \tag{2.2.13}$$

where c_1 is the solution of (2.2.12). This expression simplifies when $B = 0$

$$\mathbf{H} = -\frac{1}{a} \left[\left(\frac{1}{2} \log(\sinh 2\mathcal{J}) \right) + c_1 \, \hat{\sigma}_1 \right], \tag{2.2.14}$$

with $\tanh c_1 = e^{-2\mathcal{J}}$. It is interesting to study the limit $a \to 0$ of this expression, the so-called *Hamiltonian limit*. To do that, it is convenient to subtract the first term of the Hamiltonian (2.2.14), which corresponds to an additive constant. We can find a finite expression for $\mathbf{H}$ in the limit $a \to 0$ only by taking the simultaneous limit $\mathcal{J} \to \infty$, with the combination $y \equiv a e^{2\mathcal{J}}$ remaining as fixed. This relationship between the coupling constant $\mathcal{J}$ and the lattice space a is perhaps the simplest equation of the renormalization group: it guarantees that the physical properties of the system remain the same even in the limit $a \to 0$. Consider, for instance, the correlation length ξ

$$\xi = -\frac{a}{\log(\tanh \mathcal{J})},$$

ξ remains finite in the limit $a \to 0$ only by increasing correspondingly the coupling constant among the spin, keeping fixed their combination y.

Let us return to the computation of the partition function. By using eqn. (2.2.2) and the completeness (2.2.5),

$$\begin{aligned} Z_N &= \sum_{\sigma_1=\pm1}\sum_{\sigma_2=\pm1}\cdots\sum_{\sigma_N=\pm1} \langle\sigma_1 \mid V \mid \sigma_2\rangle\langle\sigma_2 \mid V \mid \sigma_3\rangle\cdots\langle\sigma_N \mid V \mid \sigma_1\rangle = \\ &= \sum_{\sigma_1=\pm1} \langle\sigma_1 \mid V^N \mid \sigma_1\rangle = \mathrm{Tr}\, V^N. \end{aligned} \tag{2.2.15}$$

The fact that Z_N is expressed in terms of the trace of the Nth power of the operator V is clearly due to the periodic boundary conditions adopted. The simplest way to compute the trace of V^N is by bringing V in a diagonal form. Being a Hermitian matrix, it can be diagonalized by means of a unitary matrix U

$$U^{-1}\, V\, U = \mathcal{D} = \begin{pmatrix} \lambda_+ & 0 \\ 0 & \lambda_- \end{pmatrix},$$

with $\lambda_+ \geq \lambda_-$. If we define the quantity ϕ by the relation

$$\cot 2\phi = e^{2\mathcal{J}} \sinh \mathcal{B}, \tag{2.2.16}$$

the explicit expression of U is given by

$$U = \begin{pmatrix} \cos\phi & -\sin\phi \\ \sin\phi & \cos\phi \end{pmatrix}. \tag{2.2.17}$$

Since the trace of a product of matrices is cyclic, by inserting into (2.2.15) the identity matrix $\mathbf{1}$ in the form $UU^{-1} = \mathbf{1}$, we have

$$\mathrm{Tr}\, V^N = \mathrm{Tr}\, U\, U^{-1}\, V^N = \mathrm{Tr}\, U^{-1}\, V^N\, U = \mathrm{Tr}\, \mathcal{D}^N = \lambda_+^N + \lambda_-^N. \tag{2.2.18}$$

We need now to determine explicitly the two eigenvalues: by an elementary computation, they are given by

$$\lambda_{\pm} = e^{\mathcal{J}} \cosh \mathcal{B} \pm \sqrt{e^{2\mathcal{J}} \cosh^2 \mathcal{B} - 2\sinh(2\mathcal{J})}. \tag{2.2.19}$$

The free energy per unit of spin is then expressed by

$$F(\beta,B) = -\frac{1}{\beta N} \ln Z_N = -\frac{1}{\beta}\left\{\ln \lambda_+ + \frac{1}{N}\ln\left[1+\left(\frac{\lambda_-}{\lambda_+}\right)^N\right]\right\}. \tag{2.2.20}$$

In the thermodynamical limit $N \to \infty$, taking into account that $\lambda_+ > \lambda_-$ for any value of B, the free energy is determined only by the larger eigenvalue λ_+

$$F(\beta,B) = -\frac{1}{\beta} \ln\left[e^{\mathcal{J}} \cosh \mathcal{B} + \sqrt{e^{2\mathcal{J}} \cosh^2 \mathcal{B} - 2\sinh(2\mathcal{J})}\right]. \tag{2.2.21}$$

Taking the derivative with respect to B of this expression, we obtain the mean value of the magnetization

$$\langle \sigma \rangle = \frac{e^{\mathcal{J}} \sinh \mathcal{B}}{\sqrt{e^{2\mathcal{J}} \cosh^2 \mathcal{B} - 2\sinh 2\mathcal{J}}}. \tag{2.2.22}$$

Figure 2.7 shows the graph of this function, for different values of the temperature.

The free energy (2.2.21) is an analytic function of B and T for all real values of B and for positive values of T. The magnetization is an analytic function of B that vanishes if $B = 0$. Thus, the system does not present any phase transition at finite values of T, as shown previously. However, in the limit $T \to 0$ at B finite, the magnetization presents a discontinuity, expressed by

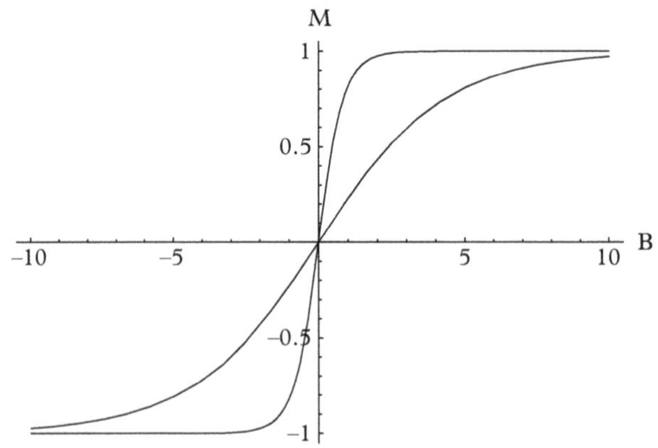

Fig. 2.7 *Magnetization versus the magnetic field B, for different values of the temperature.*

$$\langle \sigma \rangle = \epsilon(B), \tag{2.2.23}$$

where the function $\epsilon(x)$ is defined by

$$\epsilon(x) = \begin{cases} 1 & \text{if } x > 0; \\ 0 & \text{if } x = 0; \\ -1 & \text{if } x < 0. \end{cases}$$

Correlation function. The transfer matrix method can be also applied to compute the correlation functions of the spins. To this aim, it is convenient to write the correlator as

$$\langle \sigma_1 \sigma_{r+1} \rangle = Z_N^{-1} \sum_{\{\sigma\}} \sigma_1 V(\sigma_1, \sigma_2) \cdots \sigma_{r+1} V(\sigma_{r+1}, \sigma_{r+2}) \cdots V(\sigma_N, \sigma_1). \tag{2.2.24}$$

Introducing the diagonal matrix $\mathcal{S}$, with matrix elements

$$\mathcal{S}_{\sigma,\sigma'} = \sigma\, \delta_{\sigma,\sigma'},$$

i.e.

$$\mathcal{S} = \begin{pmatrix} 1 & 0 \\ 0 & -1 \end{pmatrix},$$

eqn. (2.2.24) can be written as

$$\langle \sigma_1 \sigma_{r+1} \rangle = Z_N^{-1} \mathrm{Tr} \left(\mathcal{S}\, V^r\, \mathcal{S}\, V^{N-r} \right). \tag{2.2.25}$$

For the expectation value of σ, we have

$$\langle \sigma \rangle = Z_N^{-1} \mathrm{Tr}\, \mathcal{S}\, V^N. \tag{2.2.26}$$

Using the unitary matrix U that diagonalizes V, we get

$$U^{-1} \mathcal{S}\, U = \begin{pmatrix} \cos 2\phi & -\sin 2\phi \\ -\sin 2\phi & -cos 2\phi \end{pmatrix}.$$

Substituting this expression and the diagonal form of V in eqns. (2.2.25) and (2.2.26), in the limit $N \to \infty$ we have

$$\langle \sigma_i \sigma_{i+r} \rangle = \cos^2 2\phi + \sin^2 2\phi \left(\frac{\lambda_-}{\lambda_+} \right)^r,$$

$$\langle \sigma_i \rangle = \cos 2\phi.$$

Hence the connected two-point correlation function is given by

$$G_c^{(2)}(r) = \langle \sigma_i \sigma_{i+r} \rangle - \langle \sigma_i \rangle \langle \sigma_j \rangle = \sin^2 2\phi \left(\frac{\lambda_-}{\lambda_+} \right)^r . \tag{2.2.27}$$

Besides its elegance, this formula points out an important conceptual aspect of general validity: the correlation length of a statistical system is determined by the ratio of the two largest eigenvalues of the transfer matrix

$$\xi = \frac{1}{\ln \lambda_+ / \lambda_-} . \tag{2.2.28}$$

2.2.2 Other Boundary Conditions: Boundary States

Let us investigate the computation of the partition function of the one-dimensional Ising model with N spins but with boundary conditions of type (a, b) relative to the two spins at the end of the chain. The quantum mechanics interpretation given for the transfer matrix is particularly useful to solve this problem. In fact, the boundary condition of type (a) for the first spin of the chain can be implemented by associating to this spin a special state $| \, a \rangle$ of the Hilbert space. Analogously, the boundary condition of type (b) for the last spin of the chain can be put in relation to another vector $| \, b \rangle$. These two vectors play the role of the initial and final states, respectively of the time evolution of the corresponding quantum system and, for that reason, they are called *boundary states.* Hence in order to compute the partition function $Z^{(a,b)}$, we have simply to evaluate the matrix element of the quantum time evolution operator between the initial $\langle a \, |$ and the final state $| \, b \rangle$

$$\begin{aligned} Z_N^{(a,b)} &= \sum_{\sigma_2 = \pm 1} \cdots \sum_{\sigma_{N-1} = \pm 1} \langle a \mid V \mid \sigma_2 \rangle \langle \sigma_2 \mid V \mid \sigma_3 \rangle \cdots \langle \sigma_{N-1} \mid V \mid b \rangle = \\ &= \langle a \mid V^{N-1} \mid b \rangle . \end{aligned} \tag{2.2.29}$$

This expression can be made explicit by using the unitary matrix U that diagonalizes V. By inserting in (2.2.29) both at right and left sides of the operator V the identity operator as $UU^{-1} = \mathbf{1}$, we have

$$Z^{(a,b)} = \langle a \mid U U^{-1} V^{N-1} U U^{-1} \mid b \rangle = \langle a \mid U D^{N-1} U^{-1} \mid b \rangle . \tag{2.2.30}$$

It is interesting to consider some explicit examples. Consider, for instance, the partition function with boundary conditions $\sigma_1 = \sigma_N = 1$. In this case, we have

$$| a\rangle = | b\rangle = | +\rangle = \begin{pmatrix} 1 \\ 0 \end{pmatrix}.$$

Using the expressions of $U, \mathcal{D}$, and $| +\rangle$ to compute the matrix element (2.2.30), we have

$$\begin{aligned} Z_N^{++} &= \langle + | U\mathcal{D}^{N-1} U^{-1} | +\rangle = \qquad (2.2.31) \\ &= (1,0) \begin{pmatrix} \cos\phi & -\sin\phi \\ \sin\phi & \cos\phi \end{pmatrix} \begin{pmatrix} \lambda_+^{N-1} & 0 \\ 0 & \lambda_-^{N-1} \end{pmatrix} \begin{pmatrix} \cos\phi & \sin\phi \\ -\sin\phi & \cos\phi \end{pmatrix} \begin{pmatrix} 1 \\ 0 \end{pmatrix} = \\ &= \lambda_+^{N-1} \cos^2\phi + \lambda_-^{N-1} \sin^2\phi. \end{aligned}$$

It is easy to obtain the partition functions also in other cases: for instance, with an obvious choice of the notation, we have

$$\begin{aligned} Z_N^{--} &= \lambda_+^{N-1} \sin^2\phi + \lambda_-^{N-1} \cos^2\phi; \\ Z_N^{+-} &= Z_N^{-+} = \sin\phi \cos\phi\, (\lambda_+^{N-1} - \lambda_-^{N-1}), \end{aligned} \qquad (2.2.32)$$

where the boundary condition $\sigma = -1$ is expressed by the vector

$$| -\rangle = \begin{pmatrix} 0 \\ 1 \end{pmatrix}.$$

For free boundary condition, the corresponding vector is given by

$$| \mathrm{f}\rangle = \begin{pmatrix} 1 \\ 1 \end{pmatrix},$$

and the corresponding partition functions is

$$Z_N^{ff} = Z_N^{++} + Z_N^{--} + 2Z_N^{+-} = \lambda_+^{N-1} + \lambda_-^{N-1} + \sin 2\phi \left(\lambda_+^{N-1} - \lambda_-^{N-1}\right). \qquad (2.2.33)$$

When $B = 0$ (that corresponds to $\phi = \pi/4$), this expression coincides with (2.1.3), obtained by the recursive method.

The boundary conditions should not affect the bulk properties of the system when N is very large. Indeed, in the thermodynamical limit $N \to \infty$, they only enter the correction of order $\mathcal{O}\left(\frac{1}{N}\right)$ to the free energy. In the case of fixed boundary conditions, for instance, in the large N limit we have

$$F^{(++)} = -\frac{1}{\beta N}\ln Z_N^{(++)} = -\frac{1}{\beta}\ln \lambda_+ - \frac{1}{\beta N}\ln \cos^2 \phi.$$

The first term is the same for all boundary conditions and coincides with the free energy per unit volume of the system, whereas the second term is associated to the free boundary condition.

2.3 Series Expansions

This section discusses another method to compute the partition function of the one-dimensional Ising model. It is worth mentioning that the nature of this method is quite general: it can be applied to higher-dimensional lattices and it is presently one of the most powerful approach used to analyse three-dimensional cases. The proposal consists of identifying a perturbative parameter in the high-temperature region and expressing the partition function as a series expansion in this small parameter. In the one-dimensional case the application of this method is particularly simple.

Let us consider once again the partition function in the absence of the magnetic field and, initially, with periodic boundary conditions. It can be written as

$$Z_N(T) = \sum_{\{\sigma\}} e^{-\beta \mathcal{H}} = \sum_{\{\sigma\}} \prod_{i=1}^{N} e^{\mathcal{J}\sigma_i \sigma_{i+1}}. \tag{2.3.1}$$

For any pair of Ising spins, there is the identity

$$e^{\mathcal{J}\sigma_i\sigma_j} = \cosh \mathcal{J} + \sigma_i\sigma_j \sinh \mathcal{J} = \cosh \mathcal{J}\left(1 + \sigma_i\sigma_j \tanh \mathcal{J}\right) \tag{2.3.2}$$

that permits expressing eqn. (2.3.1) as

$$Z_N(T) = \cosh^N \mathcal{J} \sum_{\{\sigma\}} \prod_{i=1}^{N} (1 + \sigma_i\sigma_{i+1}\, v), \tag{2.3.3}$$

where $v \equiv \tanh \mathcal{J}$. The parameter v is always less than 1 for all temperatures (except for $T = 0$) and, in particular, it is quite small in the high temperature phase. Once the product in (2.3.3) is developed, a polynomial of order N in the variable v, whose coefficients are expressed in terms of combinations of the spins σ_i. Consider, for example, a lattice made of 3 spins. In this case, we have

$$\prod_{i=1}^{3}(1+\sigma_i\sigma_{i+1}\,v) = (1+v\sigma_1\sigma_2)(1+v\sigma_2\sigma_3)(1+v\sigma_3\sigma_1) =$$
$$1+v(\sigma_1\sigma_2+\sigma_2\sigma_3+\sigma_3\sigma_1)+v^2(\sigma_1\sigma_2\sigma_2\sigma_3+\sigma_1\sigma_2\sigma_3\sigma_1+\sigma_2\sigma_3\sigma_3\sigma_1)+$$
$$+v^3(\sigma_1\sigma_2\sigma_2\sigma_3\sigma_3\sigma_1).$$

We can associate a graph to each of the eight terms of this expression by simply drawing a line for each pair of spin entering the product. The whole set of such graphs is shown in Figure 2.8. As v appears each time that a term $\sigma_i\sigma_{i+1}$ is involved, it follows that all graphs of order v^l contain exactly l lines. However, in order to compute the partition function, we need to sum on all values ± 1. Thanks to the following properties of the spins of the Ising model

$$\sum_{\sigma_j=-1}^{1} \sigma_j^l = \begin{cases} 2 & \text{if } l \text{ is even} \\ 0 & \text{if } l \text{ is odd} \end{cases}$$

the only non-vanishing contributions come from those graphs where all vertices are of even order (i.e. with an even number of lines). These are the *closed graphs*.

The observation made is completely general and applies to lattices of arbitrary dimension. In the one-dimensional case, it leads to a particularly simple result: in fact, among the 2^N initial graphs, the only ones that give rise to a non-vanishing result are the graph of order v^0 (i.e. the one without any line) and the graph of order v^N (i.e. the one in which the lines link all sites and give rise to a ring). Hence, in the one-dimensional case of a lattice with N sites and periodic boundary conditions, we have

$$Z_N(T) = \cosh^N \mathcal{J}\,(2^N + 2^N v^N) = 2^N\,(\cosh^N \mathcal{J} + \sinh^N \mathcal{J}), \tag{2.3.4}$$

which coincides with obtained by the transfer matrix method: eqn. (2.2.15).

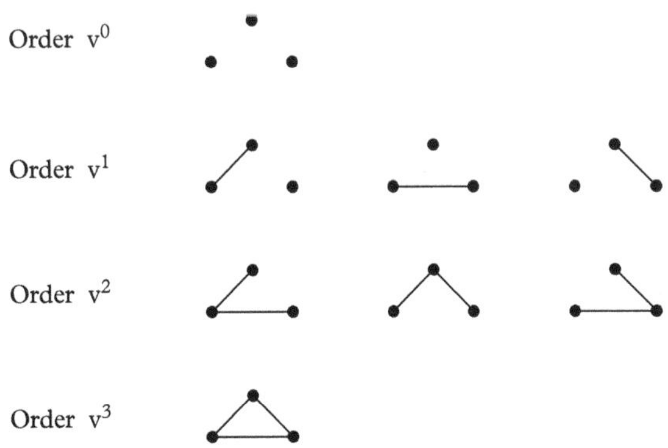

Fig. 2.8 *Graphs relative to a lattice with three spins.*

It is easy to see the difference between the case in which the chain is close (periodic boundary conditions) and the case in which the chain is open (free boundary conditions). In absence of periodic boundary conditions, the only graph that has all even vertices is the one without lines, i.e. the graph of order v^0. Hence, for the free boundary conditions, this method leads directly to the result that was previously obtained by the recursive method

$$Z_N(T) = 2^N \cosh^{N-1} \mathcal{J}. \tag{2.3.5}$$

For an arbitrary lattice in which the interaction is restricted to the next-neighbour spins, the series expansion approach allows us to express the partition function in the following form

$$Z_N(T) = 2^N (\cosh \mathcal{J})^P \sum_{l=0}^{P} h(l)\, v^l, \tag{2.3.6}$$

where P is the total number of segments of the lattice and $h(l)$ is the number of graphs that can be drawn on it by using l lines, with the condition that each vertex is of even order. Hence in the series expansion approach, the solution of the Ising model on an arbitrary lattice results in solving the geometrical problem of counting the closed graphs on the lattice under investigation.

2.4 Critical Exponents and Scaling Laws

The one-dimensional Ising model does not have a phase transition at a finite value of the temperature. However, the point $B = T = 0$ of the phase diagram can be considered a critical point of the system, for the correlation length ξ diverges in correspondence of these values. This leads to a definition of critical exponents that verifies the scaling relations (1.1.26). In Chapter 1, we adopted the variables $t = (T - T_c)/T_c$ and B in order to characterize the displacement from the critical point. In this case, in view of the condition $T_c = 0$, it is more convenient to use the variables $\mathcal{B} = B/kT$ and

$$t = \exp(-2\mathcal{J}) = \exp(-2J/kT). \tag{2.4.1}$$

Looking at the divergence ξ with respect to the new variable t, $\xi \sim (2t)^{-1}$, we have

$$\nu = 1.$$

Analogously, the divergence of the magnetic susceptibility, given by $\chi \sim t^{-1}$, fixes the value of the critical exponent γ

$$\gamma = 1.$$

At the critical point, the correlation function of the spins is constant, hence

$$\eta = 1.$$

Since the spontaneous magnetization always vanishes for $B = 0$, the exponent β is identically null

$$\beta = 0.$$

In an external magnetic field, the magnetization at $T = 0$ is a discontinous function and therefore the critical exponent δ is infinite

$$\delta = \infty.$$

Finally, in the vicinity of the critical point, the singular part of the free energy can be written as

$$F_{sing} \sim t\sqrt{1 + \frac{\mathcal{B}^2}{t^2}}.$$

Comparing with the scaling law (1.1.30) of the free energy, we obtain the two relations

$$\alpha = 1, \quad \beta\delta = 1.$$

We can easily check that the critical exponents derived above satisfy the scaling laws (1.1.26).

2.5 The Potts Model

We can generalize the Ising model in several ways. One possibility is provided by the Potts model, which consists of a statistical model where, at each site of a lattice, there is a variable σ_i that takes q discrete values, $\sigma_i = 1, 2, \ldots, q$. In this model, two adjacent spins have an interaction energy given by $-\mathcal{J}\delta(\sigma_i.\sigma_j)$, where

$$\delta(\sigma, \sigma') = \begin{cases} 1 & \text{if } \sigma = \sigma'; \\ 0 & \text{if } \sigma \neq \sigma', \end{cases}$$

and the Hamiltonian reads

$$\mathcal{H} = -\mathcal{J}\sum_{\langle ij \rangle} \delta(\sigma_i, \sigma_j). \tag{2.5.1}$$

This expression is invariant under the group $\mathcal{S}_q$ of the permutations of q objects. This is a non-abelian group if $q \geq 3$. For the type of interaction, it is clear that the nature of the values taken by the spins is completely inessential: instead of the q values listed, we can consider other q distinct numbers or variables of other nature, for example, that the q values stay for q different colours. When $q = 2$, the two distinct values we can take are ± 1: thanks to the identity $\delta(\sigma,\sigma') = \frac{1}{2}(1+\sigma\sigma')$, making the change $\mathfrak{J} \to 2\mathfrak{J}$, the Potts model is equivalent to the original Ising model.

The partition function of the Potts model defined on a lattice of N sites is expressed by a sum of q^N terms ($\mathcal{J} = \beta\mathfrak{J}$)

$$Z_N = \sum_{\{\sigma\}} \exp\left[\mathcal{J} \sum_{\langle ij \rangle} \delta(\sigma_i, \sigma_j)\right]. \tag{2.5.2}$$

In the one-dimensional case, it can be computed exactly by using either the recursive method or the transfer matrix approach.

Recursive method. Consider a chain of N spins with free boundary conditions at the last spins of the chain. Adding an extra spin, the partition function becomes

$$Z_{N+1} = \left(\sum_{\sigma_{N+1}=1}^{q} e^{\mathcal{J}\,\delta(\sigma_N,\sigma_{N+1})}\right) Z_N. \tag{2.5.3}$$

Making use of the identity

$$e^{x\delta(a,b)} = 1 + (e^x - 1)\,\delta(a,b), \tag{2.5.4}$$

the sum in (2.5.3) can be expressed as

$$\sum_{\sigma_{N+1}=1}^{q} e^{[\mathcal{J}\,\delta(\sigma_N,\sigma_{N+1})]} = \sum_{\sigma_{N+1}=1}^{q} \left[1 + (e^{\mathcal{J}} - 1)\,\delta(\sigma_N,\sigma_{N+1})\right] =$$
$$= q + (e^{\mathcal{J}} - 1).$$

The recursive equation is expressed by

$$Z_{N+1} = \left(q - 1 + e^{\mathcal{J}}\right) Z_N.$$

Since $Z_1 = q$, the iteration of the formula leads to the exact result

$$Z_N = q\left(q - 1 + e^{\mathcal{J}}\right)^{N-1}. \tag{2.5.5}$$

In the thermodynamical limit, the free energy per unit of spin is given by

$$F(T) = -\lim_{N\to\infty} \frac{1}{\beta N} \ln Z_N = -\frac{1}{\beta} \ln\left(e^{\mathcal{J}} + q - 1\right). \tag{2.5.6}$$

Transfer matrix. Equally instructive is the computation of the partition function done using the transfer matrix method. For simplicity, let us assume periodic boundary conditions, i.e. $\sigma_{N+1} \equiv \sigma_1$. In the transfer matrix formalism, the spins are associated to vector of a q-dimensional Hilbert space, with the completeness relation given by

$$\sum_{\sigma=1}^{q} \mid \sigma\rangle \langle \sigma \mid = 1.$$

Analogously to the Ising model, the partition function can be expressed as

$$Z_N = \text{Tr } V^N, \tag{2.5.7}$$

where the transfer matrix V is a $q \times q$ matrix whose elements are

$$\langle \sigma \mid V \mid \sigma' \rangle = \exp\left[\mathcal{J}\delta(\sigma,\sigma')\right]. \tag{2.5.8}$$

Hence V has the diagonal elements equal to $e^{\mathcal{J}}$ whereas all the other off-diagonal elements are equal to 1

$$V = \begin{pmatrix} e^{\mathcal{J}} & 1 & 1 & \cdots & 1 & 1 \\ 1 & e^{\mathcal{J}} & 1 & \cdots & 1 & 1 \\ 1 & 1 & e^{\mathcal{J}} & \cdots & 1 & 1 \\ \cdots & \cdots & \cdots & e^{\mathcal{J}} & \cdots & 1 \\ 1 & 1 & \cdots & \cdots & e^{\mathcal{J}} & 1 \\ 1 & 1 & \cdots & \cdots & 1 & e^{\mathcal{J}} \end{pmatrix}. \tag{2.5.9}$$

To compute the trace of V^N it is useful to determine the eigenvalues of V, which are solutions of the equation

$$\mathcal{D} = \| V - \lambda \mathbf{1} \| = 0. \tag{2.5.10}$$

Denote $x \equiv e^{\mathcal{J}} - \lambda$. The determinant (2.5.10) can be computed by using the well-known property that a determinant does not change by summing or subtracting rows and columns. Subtracting the second column from the first one, the third column from the second one and so on, we have

$$\mathcal{D} = \begin{Vmatrix} x-1 & 0 & 0 & \cdots & 0 & 1 \\ 1-x & x-1 & 0 & \cdots & 0 & 1 \\ 0 & 1-x & x-1 & \cdots & 0 & 1 \\ \cdots & \cdots & \cdots & x-1 & 0 & 1 \\ 0 & 0 & \cdots & \cdots & x-1 & 1 \\ 0 & 0 & \cdots & \cdots & 1-x & x \end{Vmatrix}.$$

Adding the first row to the second one, we get

$$\mathcal{D} = \begin{Vmatrix} x-1 & 0 & 0 & \cdots & 0 & 1 \\ 0 & x-1 & 0 & \cdots & 0 & 2 \\ 0 & 1-x & x-1 & 0 & \cdots & 1 \\ 0 & \cdots & \cdots & x-1 & \cdots & 1 \\ 0 & 0 & \cdots & \cdots & x-1 & 1 \\ 0 & 0 & \cdots & \cdots & 1-x & x \end{Vmatrix}.$$

If we now add the second row to the third, the third row to the fourth, and so on, we have the final expression

$$\mathcal{D} = \begin{Vmatrix} x-1 & 0 & 0 & \cdots & 0 & 1 \\ 0 & x-1 & 0 & \cdots & 0 & 2 \\ 0 & 0 & x-1 & 0 & \cdots & 3 \\ 0 & \cdots & 0 & x-1 & \cdots & 4 \\ 0 & 0 & \cdots & \cdots & x-1 & q-1 \\ 0 & 0 & 0 & \cdots & 0 & x+q-1 \end{Vmatrix}.$$

So the determinant of the secular equation is given by

$$\| V - \lambda \mathbf{1} \| = \left(e^{\mathcal{J}} - 1 - \lambda\right)^{q-1} (e^{\mathcal{J}} + q - 1 - \lambda) = 0, \tag{2.5.11}$$

and the roots are expressed by

$$\lambda_+ = e^{\mathcal{J}} + q - 1 \quad , \quad \lambda_- = e^{\mathcal{J}} - 1. \tag{2.5.12}$$

For $q \geq 0$, we have $\lambda_+ \geq \lambda_-$. The eigenvalue λ_+ is not degenerate, while λ_- is $(q-1)$ degenerate. The physical origin of this degeneration is obvious, since the interaction of the Potts model only distinguishes if two sites are in the same state or not: there is only one way in which they can be equal, but $(q-1)$ ways in which they can be different.

Once the eigenvalues of V are known, the partition function (2.5.7) can be expressed as

$$Z_N = \mathrm{Tr}\, V^N = \lambda_+^N + (q-1)\lambda_-^N. \tag{2.5.13}$$

In the thermodamical limit, the free energy per unit of spin depends only on the largest eigenvalue λ_+

$$F(T) = -\lim_{N\to\infty} \frac{1}{\beta N} \ln Z_N = -\lim_{N\to\infty} \frac{1}{\beta N} \left[N \ln \lambda_+ + \ln \left[1 + (q-1) \left(\frac{\lambda_-}{\lambda_+} \right)^N \right] \right] =$$
$$= -\frac{1}{\beta} \ln \left(e^{\mathcal{J}} + q - 1 \right). \tag{2.5.14}$$

This result coincides with (2.5.6).

Series expansion. Let us consider now the solution of the Potts model obtained in terms of the high-temperature series expansion. Since this method points out some interesting geometrical properties, it is convenient to study the general case of a Potts model defined on an arbitrary lattice $\mathcal{L}$ as, for instance, the one shown in Figure 2.9. Putting

$$v \equiv e^{\mathcal{J}} - 1,$$

and using the identity (2.5.4), the partition function (2.5.2) can be written as

$$Z_N = \sum_{\{\sigma\}} \prod_{\langle ij \rangle} \left[1 + v\,\delta(\sigma_i, \sigma_j) \right]. \tag{2.5.15}$$

Note that v is a small parameter when the temperature T is very high. Let E be the total number of links of the graph $\mathcal{L}$. Inside the sum (2.5.15) there is a product of E factors,

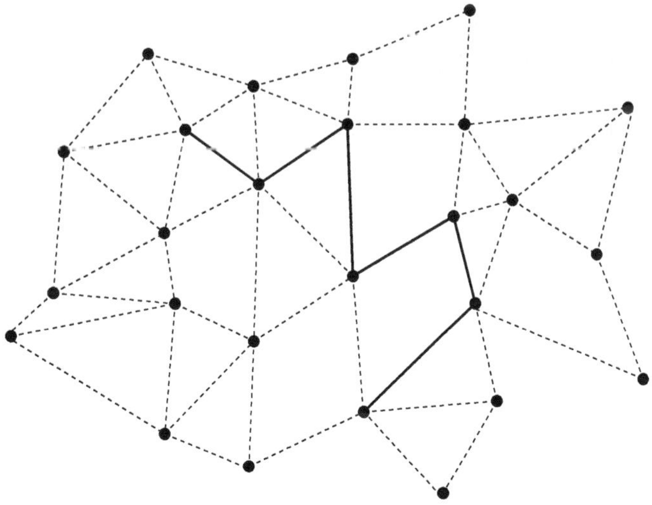

Fig. 2.9 *Lattice $\mathcal{L}$ and graph $\mathcal{G}$.*

each of them being either 1 or $v\delta(\sigma_i,\sigma_j)$. Expanding the product above, there are 2^E terms: their graphical representation is obtained by drawing a line on the link between the sites i and j when the factor $v\delta(\sigma_i,\sigma_j)$ is present. In such a way, there is a one-to-one correspondence between the terms in (2.5.15) and the graphs that can be drawn on the lattice $\mathcal{L}$. Let us now consider one of these graphs $\mathcal{G}$, made of l links and C connected components (an isolated site is considered as a single component). The corresponding term in Z_N contains a factor v^l and, thanks to the factor $\delta(\sigma,\sigma')$ that accompanies v, all spins that belong to the same component have the same value. Summing on all possible values of σ_i, the contribution of this graph to the partition function amounts to $q^C v^l$. Considering all graphs $\mathcal{G}$ of the lattice $\mathcal{L}$, the partition function can be thus expressed in terms of a sum on graphs

$$Z_N = \sum_{\mathcal{G}} q^C v^l. \tag{2.5.16}$$

For the analytic form of this expression, q has not to be necessarily an integer and therefore this formula can be used to define the Potts model for arbitrary values of q. This observation is useful, for instance, in the study of the percolation[1] (associated to the limit $q \to 1$ of the Potts model) or in the analysis of the effective resistance between two nodes of an electric circuit made of linear resistances (expressed in terms of the limit $q \to 0$ of the model).

Chromatic polynomial. It is interesting to study the Potts model in the limit $\mathcal{J} \to -\infty$, i.e. when the temperature T goes to zero and the model is anti-ferromagnetic. In such a limit, neighbour sites should necessarily take different values in order to contribute to the partition function Z_N: hence this quantity provides in this case the number of ways in which it is possible to colour the sites of $\mathcal{L}$ with q colours, with the constrain that two neighbour sites do not have the same colour. The expression obtained by substituting $v = -1$ in Z_N is a polynomial $\mathcal{P}_N(q)$ in the variable in q, called *chromatic polynomial* of the graph $\mathcal{L}$.

In the one-dimensional case, taking the limit $\mathcal{J} \to -\infty$ in the partition function (2.5.5) associated to the free boundary condition of the chain, we get

$$\mathcal{P}_N^a(q) = q(q-1)^{N-1}. \tag{2.5.17}$$

The zeros $q = 0$ and $q = 1$ of this polynomial clearly show that, if we wish to distinguish neighbour sites by means of different colours, it is impossible to colour a one-dimensional lattice by having only one colour or none. The combinatoric origin of (2.5.17) is simple: in fact, the first site can be colour in q different ways but, once a colour is chosen, the next site can be distinguished by employing one of the $(q-1)$ remaining colours, and this argument repeats for the other sites.

For periodic boundary conditions, taking the limit $\mathcal{J} \to -\infty$ in the corresponding expression (2.5.13) of the partition function, we have

[1] For the elaboration of this topic, see the suggested text at the end of the chapter.

$$\mathcal{P}_N^c(q) = (q-1)^N + (-1)^N (q-1) = (q-1)\left[(q-1)^{N-1} + (-1)^N\right]. \qquad (2.5.18)$$

Although this expression differs from (2.5.17), it is easy to see that it has the same real roots $q=0$ and $q=1$. It is an exercise left to the reader to derive it by using a combinatoric argument.

For planar two-dimensional lattices, the limit ($\mathcal{J} \to -\infty$) of the Potts model is deeply related to a famous problem of topology, i.e. the *four-colour problem*. It consists to prove the conjecture that any geographical planar map, in which different neighbour nations are distinguished by different colours, can be drawn using only four colours. If we assume the validity of this result, the conclusion is that the partition function of the Potts model for any planar graph, in the limit $\mathcal{J} \to -\infty$, does not ever have $q=4$ among the set of its zeros. A brief discussion of the four-colour problem appears in Appendix C.

2.6 Models with $O(n)$ Symmetry

Another interesting generalization of the Ising model is provided by the $O(n)$ model, in which each spin $\vec{S}_i$ is a n component vector associated to a point of the n-dimensional sphere

$$|\vec{S}_i|^2 = \sum_{k=1}^{n} (S_i)_k^2 = 1.$$

In the one-dimensional case, the Hamiltonian of the model is given by

$$\mathcal{H} = -\sum_{i=1}^{N-1} \mathcal{J}_i \vec{S}_i \cdot \vec{S}_{i+1}, \qquad (2.6.1)$$

and this expression is clearly invariant under the rotations of the vectors $\vec{S}_i$ associated to the $O(n)$ group. In this formulation, the Ising model is obtained in the limit $n \to 1$. The sum on the configurations of the $O(n)$ model consists of the integrals of the solid angles of the n dimensional spins

$$Z_N(T) = \int d\Omega_1^{(n)} \int d\Omega_2^{(n)} \cdots \int d\Omega_N^{(n)} \exp\left[\sum_{i=1}^{N-1} \mathcal{J}_i \vec{S}_i \cdot \vec{S}_{i+1}\right], \qquad (2.6.2)$$

where

$$\begin{gathered} d\Omega^{(n)} = \sin^{n-2}\theta_{n-1} d\theta_{n-1} \sin^{n-3}\theta_{n-2} d\theta_{n-2} \cdots d\theta_1, \\ 0 \leq \theta_1 \leq 2\pi, \\ 0 \leq \theta_k \leq \pi. \end{gathered}$$

The solid angle is given by

$$\Omega(n) = \int d\Omega^{(n)} = \frac{2\pi^{n/2}}{\Gamma\left(\frac{n}{2}\right)}, \tag{2.6.3}$$

where $\Gamma(x)$ is the function that generalizes the factorial to arbitrary real and complex numbers[2]

To prove (2.6.3), let us consider the well-known identity

$$I = \int_{-\infty}^{+\infty} dx\, e^{-x^2} = \sqrt{\pi}.$$

By taking the product of n of such integrals, we have $(r^2 = x_1^2 + x_2^2 + \cdots + x_n^2)$

$$I^n = \left[\int_{-\infty}^{+\infty} dx\, e^{-x^2}\right]^n = \int d^n x\, e^{-r^2} = \frac{\Omega(n)}{2}\int_0^{\infty} dt\, t^{\frac{n}{2}-1} e^{-t} = \frac{\Omega(n)}{2}\Gamma\left(\frac{n}{2}\right).$$

On the other hand, $I^n = \pi^{n/2}$ and therefore we arrive at (2.6.3). Using

$$\Gamma\left(\frac{1}{2}\right) = \sqrt{\pi},$$

it is easy to check that we obtain the known values of planar and three-dimensional solid angle when $n = 2$ and $n = 3$. For $n = 1$, it correctly reproduces the sum of the states of the Ising model, i.e. $\Omega(1) = 2$, since a *one-dimensional sphere* consists of two points. Other interesting properties of the n-dimensional solid angle are discussed in Appendix B.

To compute (2.6.2), we can use the recursive method. Let us add an extra spin to the system, so that

$$Z_{N+1}(T) = \left(\int d\Omega_{N+1}^{(n)} \exp\left[\mathcal{J}_N \vec{S}_N \cdot \vec{S}_{N+1}\right]\right) Z_N(T).$$

Since the nth axis can always be chosen along the direction of the spin $\vec{S}_n$, we have $\vec{S}_N \cdot \vec{S}_{N+1} = \cos\theta_{n-1}$. Integrating on the remaining angles $\theta_1, \theta_2, \ldots, \theta_{n-2}$, we get

[2] The properties of the function $\Gamma(x)$ and the Bessel functions $I_\nu(x)$ that enter the discussion of this model are discussed in Appendix A.

$$Z_{N+1}(T) = \left(\Omega(n-1)\int_0^{\pi} d\theta_{n-1}\,\sin^{n-2}\theta_{n-1} e^{\mathcal{J}_N\cos\theta_{n-1}}\right) Z_N(T). \tag{2.6.4}$$

Although this is not an elementary integral, it nevertheless can be expressed as a closed formula in terms of the $\Gamma(x)$ function and the Bessel functions $I_\nu(z)$

$$\int_0^{\pi} d\theta_{n-1}\,\sin^{n-2}\theta_{n-1} e^{\mathcal{J}_N\cos\theta_{n-1}} = \frac{\sqrt{\pi}\,\Gamma\left(\frac{n-1}{2}\right)}{\left(\frac{\mathcal{J}_N}{2}\right)^{\frac{n-2}{2}}} I_{\frac{n-2}{2}}(\mathcal{J}_N).$$

Substituting $\Omega(n-1)$ in (2.6.4) and simplifying the resulting expression, we get

$$\int d\Omega^{(n)}_{N+1}\,\exp\left[\mathcal{J}_N\vec{S}_N\cdot\vec{S}_{N+1}\right] = \left[(2\pi)^{n/2}\frac{I_{\frac{n-2}{2}}(\mathcal{J}_N)}{\mathcal{J}_N^{\frac{n-2}{2}}}\right] \equiv \lambda_1(\mathcal{J}_N). \tag{2.6.5}$$

The recursive equation is then given by

$$Z_{N+1} = \lambda_1(\mathcal{J}_N)\,Z_N.$$

Let us consider, for simplicity, the case of equal couplings. By iterating (2.6), we obtain

$$Z_N(T) = [\lambda_1(\mathcal{J})]^{N-1}\,Z_1,$$

where Z_1 is the partition function of a single spin. This is simply expressed by the phase space of the configuration of a single spin, i.e. by the n-dimensional solid angle (2.6.3), so that the final expression is

$$Z_N(T) = \frac{2\pi^{n/2}}{\Gamma\left(\frac{n}{2}\right)}\,[\lambda_1(\mathcal{J})]^{N-1} = \frac{2\pi^{n/2}}{\Gamma\left(\frac{n}{2}\right)}\left[(2\pi)^{n/2}\frac{I_{\frac{n-2}{2}}(\mathcal{J})}{\mathcal{J}^{\frac{n-2}{2}}}\right]^{N-1}, \tag{2.6.6}$$

The free energy, per unit of spin, of the $O(n)$ model is

$$\beta F(\beta) = -\frac{1}{N}\log Z_N = -\frac{N-1}{N}\log\left[\frac{I_{\frac{n-2}{2}}(\mathcal{J})}{\mathcal{J}^{\frac{n-2}{2}}}\right] - \frac{1}{N}\log\Omega(n).$$

and in the thermodynamic limit $N\to\infty$

$$\beta F(\beta) = -\log\left[\frac{I_{\frac{n-2}{2}}(\mathcal{J})}{\mathcal{J}^{\frac{n-2}{2}}}\right]. \tag{2.6.7}$$

As for the Ising model, it is also possible to obtain the exact expression of the two-point correlation function for the $O(n)$ model

$$\langle \vec{S}_i \cdot \vec{S}_{i+r} \rangle = \left[\frac{I_{\frac{n}{2}}(\mathcal{J})}{I_{\frac{n-2}{2}}(\mathcal{J})} \right]^r. \tag{2.6.8}$$

Expressed as

$$\langle \vec{S}_i \cdot \vec{S}_{i+r} \rangle \equiv e^{-r/\xi},$$

we can determine the correlation length of the model that, in unit of the lattice space a, is given by

$$\xi(\mathcal{J}) = -\frac{1}{\log\left[\frac{I_{\frac{n}{2}}(\mathcal{J})}{I_{\frac{n-2}{2}}(\mathcal{J})} \right]}. \tag{2.6.9}$$

The proof of (2.6.8) comes from the following identity of the Bessel functions

$$\frac{d}{dx}\left[x^{-\mu} I_\mu(x)\right] = x^{-\mu} I_{\mu+1}(x).$$

Taking the derivative with respect to $\mathcal{J}$ of eq (2.6.5), this identity to compute the integral

$$\int d\Omega^{(n)} \vec{S} \exp[\mathcal{J}\vec{S} \cdot \vec{S}'] = \left[(2\pi)^{n/2} \frac{I_{\frac{n}{2}}(\mathcal{J})}{\mathcal{J}^{\frac{n-2}{2}}} \right] \vec{S}' \equiv \lambda_2(\mathcal{J})\vec{S}'.$$

We have then

$$\int d\Omega_1^{(n)} \int d\Omega_2^{(n)} \cdots \int d\Omega_N^{(n)} \vec{S}_i \cdot \vec{S}_{i+r} \exp\left[\sum_{i=1}^{N-1} \mathcal{J}\vec{S}_i \cdot \vec{S}_{i+1} \right] =$$

$$[\lambda_1(\mathcal{J})]^{i-1} \, [\lambda_2(\mathcal{J})]^r \, [\lambda_1(\mathcal{J})]^{N-i} \frac{2\pi^{n/2}}{\Gamma\left(\frac{n}{2}\right)}.$$

Dividing this expression by the partition function Z_N, given by (2.6.6), we arrive the final result (2.6.8) of the correlators.

Interestingly, to observe that all expressions considered so far are *analytic functions* of the parameter n and, for that reason, they can be used to study the behaviour of the

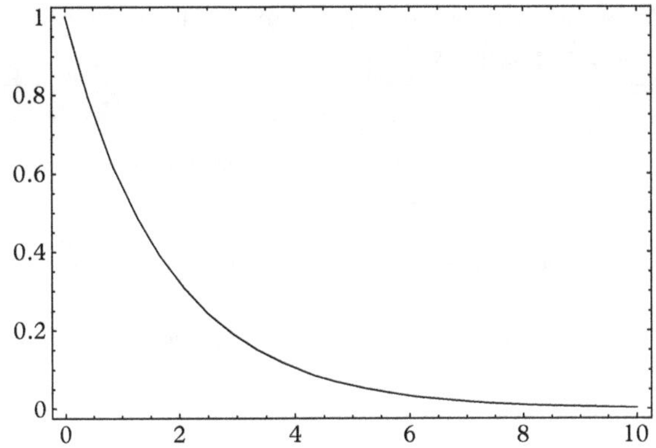

Fig. 2.10 *Typical behaviour of the two-point correlation function of the spins, as a function of the distance r between the spin, for* $n \geq 1$.

$O(n)$ model for arbitrary values of n, which are not necessarily integers. This is a useful observation because it extends to higher dimensions and allows us to study, for instance, the statistical properties of the polymers,[3] whose dilute phase is described by the limit $n \to 0$.

Importantly, $n < 1$ could contain surprising behaviours that, in order to interpret them correctly, may need further consideration. The following analysis aims to study the nature of the model by varying the parameter n. It is convenient to define the quantity

$$\Lambda(\mathcal{J}) \equiv \frac{\lambda_2}{\lambda_1} = \left[\frac{I_{\frac{n}{2}}(\mathcal{J})}{I_{\frac{n-2}{2}}(\mathcal{J})}\right],$$

and to distinguish the cases: (i) $n \geq 1$, (ii) $0 \leq n \leq 1$ and (iii) $n \leq 0$.

- In the first interval, $n \geq 1$, using eqns. (2.A.9) and (2.A.13) from Appendix A, it is easy to check that for all values of $\mathcal{J}$, i.e. of the temperature, we have

$$\lambda_1(\mathcal{J}) > 0, \quad \Lambda(\mathcal{J}) < 1.$$

The first condition, as shown in (2.6.6), implies that the partition function of the model is a positive quantity and, consequently, that the free energy is a real function. The second condition, using eqn. (2.6.8), implies that the correlator has the usual behaviour of a decreasing exponential, as a function of the distance r between the spins.

[3] The relation between the $O(n)$ model and the statistics of the polymers was discovered by De Gennes. Readers interested in further development of this issue are referred to the references given at the end of the chapter.

Both results agree with expectations on the basis of physical considerations. When $n = 1$, using the identity

$$\frac{I_{\frac{1}{2}}(\mathcal{J})}{\mathcal{J}^{\frac{1}{2}}} = \sqrt{\frac{2}{\pi}} \cosh \mathcal{J},$$

we recover the previous expressions of the partition function and correlator of the one-dimensional Ising model.

To study the limit $n \to \infty$, we need to use the asymptotic expressions of the Bessel functions

$$I_\nu(\nu x) \simeq \frac{1}{\sqrt{2\pi\nu}} \frac{e^{\nu(\sqrt{1+x^2} - \xi^{-1}}}{(1+x^2)^{1/4}}, \qquad \nu \to \infty,$$

with

$$\xi^{-1} = \ln\left(\frac{1+\sqrt{1+x^2}}{x}\right).$$

When $n \to \infty$ we can obtain an interesting result by taking simultaneously the limit $\mathcal{J} \to \infty$. We introduce $x \equiv 2\mathcal{J}/(n-2)$ and express all thermodynamic quantities in terms of this variable. Consider, for instance, the ratio of the two eigenvalues λ_2 and λ_1 in this double limit

$$\frac{\lambda_2(x)}{\lambda_1(x)} = e^{-\xi^{-1}}.$$

This allows us to identify the parameter ξ with the correlation length of the model. This quantity diverges for $T \to 0$, whereas it vanishes for $T \to \infty$. The last limit corresponds to the full disordered state of the system, where each spin is independent and completely uncorrelated to the others. The internal energy is given by

$$U = -\frac{\partial}{\partial x} \ln \lambda_1(x),$$

and, using the asymptotic expression of the Bessel functions, it can be expressed as

$$\frac{U(x)}{n} = \frac{1 - \sqrt{1+x^2}}{x}. \tag{2.6.10}$$

This formula shows that the internal energy, relative to each component of the spin, remains finite in the double limit $n \to \infty$, $\mathcal{J} \to \infty$, with x finite.

- In the second interval, $0 \leq n < 1$, using (2.A.9) and (2.A.13), it is easy to see that, for all values of $\mathcal{J}$, we have

$$\lambda_1(\mathcal{J}) > 0.$$

However, the inequality

$$\Lambda(\mathcal{J}) < 1,$$

is not always true: in this interval of values of n, it is always possible to find a value $\mathcal{J}_c$ such that, for $\mathcal{J} > \mathcal{J}_c$, we have $\Lambda(\mathcal{J}) > 1$ (Figure 2.11).

From eqn. (2.6.9), the correlation length $\xi(\mathcal{J})$ is positive for $\mathcal{J} < \mathcal{J}_c$, while for $\mathcal{J} > \mathcal{J}_c$, it becomes negative! Moreover, it diverges at $\mathcal{J}_c$ (Figure 2.12). The critical value $\mathcal{J}_c$ moves toward the origin by decreasing n and, when $n \to 0$, we have $\mathcal{J}_c = 0$. In such a limit, taking into account the factor $\Gamma\left(\frac{n}{2}\right)$ in the denominator of (2.6.6), the partition function vanishes linearly[4] in the variable n, while the correlation function is finite and takes the form

$$\lim_{n\to 0} \langle \vec{S}_i \cdot \vec{S}_{i+r} \rangle = \left[\frac{I_0(\mathcal{J})}{I_{-1}(\mathcal{J})}\right]^r. \tag{2.6.11}$$

Note that, for all the values of temperature, this is an exponentially increasing function of the distance of the spins! Namely, increasing the separation between the spins, their correlation increases exponentially, instead of decreasing—a behaviour that is quite anti-intuitive from a physical point of view.

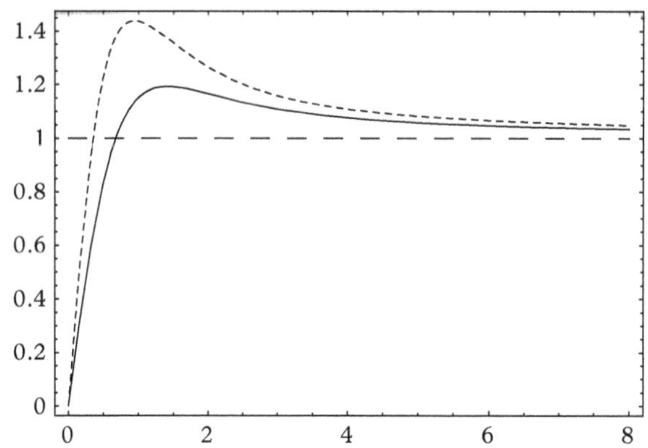

Fig. 2.11 *Λ as a function of $\mathcal{J}$. The dashed line corresponds to $n = 0.3$, the other curve to $n = 0.6$.*

[4] This implies that exists the finite limit $\lim_{n\to 0} \frac{\partial Z}{\partial n}$.

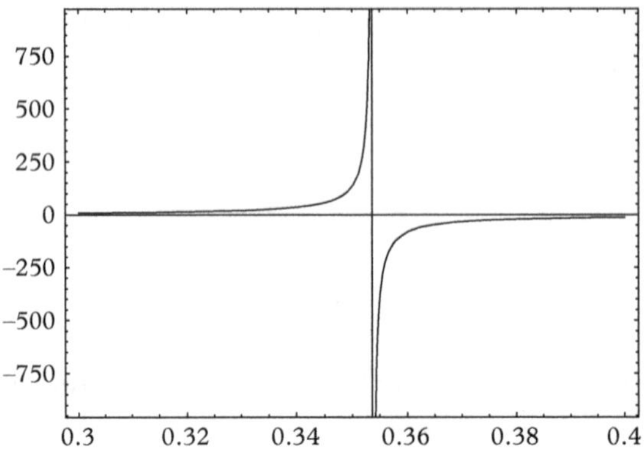

Fig. 2.12 *Plot of the correlation length in the vicinity of $\mathcal{J} = \mathcal{J}_c$.*

- Let us consider the last interval, $n < 0$. Using eqns. (2.A.9) and (2.A.13), the Bessel function $I_{\frac{n-2}{2}}(\mathcal{J})$ is always positive (as function of $\mathcal{J}$), in the following ranges of n

$$-4k < n < -4k+2, \quad k = 1,2,3,\ldots. \tag{2.6.12}$$

In the other intervals

$$-4k-2 < n < -4k, \quad k = 0,1,2,3,\ldots \tag{2.6.13}$$

there are instead values of $\mathcal{J}$ where $I_{\frac{n-2}{2}}(\mathcal{J})$ assumes negative values. Correspondingly, the free energy per unit of spin, given by (2.6.7), is a real function of $\mathcal{J}$ only in

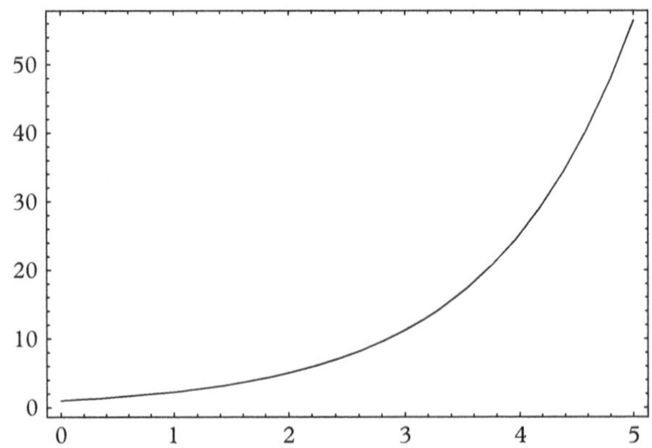

Fig. 2.13 *Correlation function $\langle \vec{S}_i \cdot \vec{S}_{i+r} \rangle$ as a function of the separation r, for n = 0.*

Fig. 2.14 *The continuous intervals and the points identified by the circles are those in which the free energy is real.*

the intervals (2.6.12), whereas in the other intervals it develops an imaginary part that signals the thermodynamic instability of the system. Finally, for $n = -2k$, con $k = 0, 1, 2, \ldots$, $I_{\frac{n-2}{2}}(\mathcal{J})$ is always positive and therefore the free energy is real for those values. The behaviour of the free energy of the model is given in Figure 2.14.

Let us now analyse the ratio $\Lambda(\mathcal{J})$ by starting with the study of the positivity of such a quantity. This is determined by the positivity of the functions $I_{\frac{n-2}{2}}(\mathcal{J})$ and $I_{\frac{n}{2}}(\mathcal{J})$. The investigation of the first function coincides with what has been done previously with the free energy. Concerning the second function, this is positive for all values of $\mathcal{J}$ in the intervals

$$-4k - 2 < n < -4k, \quad k = 1, 2, 3, \ldots. \tag{2.6.14}$$

In the other intervals of n, there are instead values of $\mathcal{J}$ where this function takes negative values. In conclusion, there is no interval of n where the two functions are both positive. This implies that, for any negative n with $n \neq -2k$ ($k = 0, 1, 2, \ldots$), there is always a value $\mathcal{J}_c$ in which the correlation length diverges, assuming complex values in an interval $\mathcal{J} < \mathcal{J}_c$ near the origin. For $n = -2k$, instead, $\Lambda(\mathcal{J})$ is real but larger than 1, so that the correlation length is *negative* for all values of the temperature: the correlation function of the spin thus increases by increasing their separation.

This analysis aims to help interpret the behaviour of the model by varying continuously the number n of the components of the vector $\vec{S}_i$. From this point of view, the one-dimensional $O(n)$ model is a paradigm of an important class of models we meet again in this book and that will allow us to make progress in important fields of theoretical physics.

2.7 Models with Z_n Symmetry

Beside the generalizations of the Ising models given by the Potts and the $O(n)$ models, there is another possible extension provided by the Z_n models. In this case, the spins are planar vectors of unitary length can be identified by their discrete angles θ_i with respect to the horizontal axes

$$\alpha^{(k)} = \frac{2\pi k}{n}, \quad k = 0, 1, 2, \ldots, n-1. \tag{2.7.1}$$

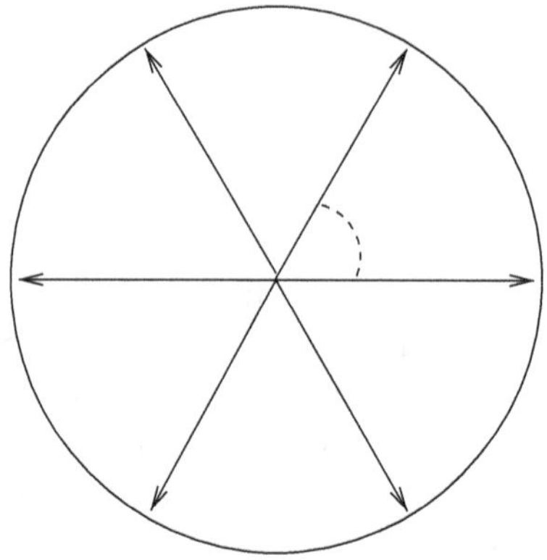

Fig. 2.15 *Possible values of the spins in the Z_6 model.*

They can be associated to the n (complex) roots of unit (Figure 2.15). The Hamiltonian of the Z_n model is defined by

$$\mathcal{H} = -\mathcal{J} \sum_{\langle ij \rangle} \vec{S}_i \cdot \vec{S}_j = -\mathcal{J} \sum_{\langle ij \rangle} \cos(\theta_i - \theta_j), \tag{2.7.2}$$

and is invariant under the abelian group Z_n generated by the discrete rotations of the angles θ_i. In terms of the index k defined in (2.7.1), this symmetry is implemented by the transformations

$$k \to k + m \ (\text{mod}\, n), \quad m = 0, 1, \ldots n. \tag{2.7.3}$$

For some particular values of n, the Z_n models coincide with previously defined models. For instance, when $n = 2$, we observe the familiar Ising model or, equivalently, the 2-state Potts model. When $n = 3$, the Z_3 model is equivalent to the 3-state Potts model: it is sufficient to put $\mathcal{J} = \frac{2}{3}\mathcal{J}_{Potts}$ to have the coincidence of the two Hamiltonians. Finally, when $n \to \infty$ the Z_n model becomes equivalent to the $O(2)$ model, i.e. that model is invariant under an arbitrary rotation of the spins.

In the one-dimensional case, the solution of the Z_n model is achievable by using the recursive method. Let us consider first the partition function of N spin

$$Z_N = \sum_{\theta_1=0}^{n-1} \cdots \sum_{\theta_N=0}^{n-1} \exp\left[\mathcal{J} \sum_{i=0}^{N-1} \cos\left(\frac{2\pi}{n}(\theta_i - \theta_{i+1})\right)\right]. \tag{2.7.4}$$

For $N = 1, Z_1$ is equal to the number of possible states of the system, i.e. $Z_1 = n$. Adding a new spin to the chain, we then have

$$Z_{N+1} = Z_N \sum_{\theta_{N+1}=0}^{n-1} \exp\left[\mathcal{J} \cos\left(\frac{2\pi}{n}(\theta_N - \theta_{N+1})\right)\right],$$

where the last sum is independent from θ_N. Indeed, whatever the value taken by this variable, the sum on the angle θ_{N+1} in the argument $(\theta_N - \theta_{N+1})$ implies that this quantity spans all possible values (2.7.1), i.e. θ_N can be eliminated by a simple change of variable. Hence the partition function satisfies the recursive equation

$$Z_{N+1} = \mu_1(\mathcal{J}, n)\, Z_N, \tag{2.7.5}$$

where we have defined

$$\mu_1(\mathcal{J}, n) \equiv \sum_{k=0}^{n-1} \exp\left[\mathcal{J} \cos\frac{2\pi k}{n}\right]. \tag{2.7.6}$$

By iterating (2.7.5), with the initial condition $Z_1 = n$, we get

$$Z_N = n\, [\mu_1(\mathcal{J}, n)]^{N-1}. \tag{2.7.7}$$

It is easy to compute the correlation function of two spins

$$G(r) = \langle \vec{S}_i \cdot \vec{S}_{i+r} \rangle = \langle \cos(\theta_i - \theta_{i+r}) \rangle.$$

For this, we need the identity

$$\sum_{\{\vec{S}\}} \vec{S}\, e^{\mathcal{J}\vec{S}\cdot\vec{S}'} = \mu_2(\mathcal{J}, n)\, \vec{S}',$$

where the sum is on all discrete values of the vector $\vec{S}$ of the $Z(n)$ model and

$$\mu_2(\mathcal{J}, n) = \frac{\partial}{\partial \mathcal{J}} \mu_1(\mathcal{J}, n).$$

Following the same steps of the Ising and the $O(n)$ models, we have

$$G(r) = \left(\frac{\mu_2}{\mu_1}\right)^r. \tag{2.7.8}$$

When $n = 2$, both the partition function and the correlator coincide with those of the Ising model. When $n \to \infty$, a finite result is obtained by properly rescaling the sum on the states, i.e. multiplying the sum by $2\pi/n$ and then taking the limit. In this way, the previous formula becomes

$$\lim_{n\to\infty} \frac{2\pi}{n} \sum_{k=0}^{\infty} [\ldots] \longrightarrow \int_0^{2\pi} d\alpha\, [\ldots].$$

Hence

$$\lim_{n\to\infty} \mu_1(\mathcal{J},n) = 2\pi I_0(\mathcal{J}), \quad \lim_{n\to\infty} \mu_2(\mathcal{J},n) = 2\pi I_1(\mathcal{J}), \tag{2.7.9}$$

where $I_0(x)$ and $I_1(x)$ are the Bessel functions. It is evident that we recover the results of the $O(2)$ model.

2.8 Feynman Gas

This section discusses a particular one-dimensional gas, known as *Feynman's gas*. Even though it does not belong to the class of systems related to the Ising model, Chapter 20 shows that the thermodynamics of this system provides useful information on the spin–spin correlation function of the bi-dimensional Ising model! For that reason, but also for the peculiarity of this gas, it is useful to present its exact solution.

Let us consider a set of N particles, forced to move along an interval of length L. Let $x_1, x_2, \ldots, x_n$ be their coordinates, while $V(|\, x_i - x_j \,|)$ their interaction potential. We assume that $V(r)$ is a short-range potential, so that we will consider only the interactions among particles which are close to each other, neglecting all the rest. In this case, the partition function of the system can be written as[5]

$$Z_N(L) = \int_{0<x_1<x_2 \cdot x_N<L} e^{-\beta[V(x_1-x_2)+V(x_2-x_3)+\cdots V(x_{N-1}-x_N)]}\, dx_1 dx_2 \cdots dx_N. \tag{2.8.1}$$

In order to solve the model and find out its equation of state, it is natural to find a recursive equation that links Z_N to Z_{N+1}. It is convenient to modify slightly the original problem:[6] this consists of inserting an extra particle in the system but kept fixed at the position y. The partition function of the new version of the problem, denoted by $P_N(y,L)$, is expressed by

[5] The integral on the moments p_i of the particles is Gaussian and can be done straightforwardly. It leads to a normalization constant of the partition function, put equal to 1 for simplicity.

[6] It is easy to prove that, in the thermodynamic limit, this new version does not change the macroscopic properties of the system.

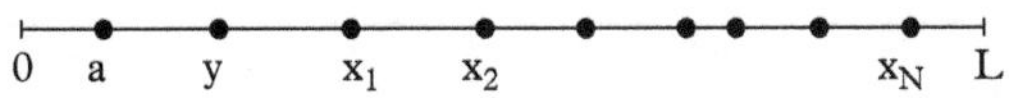

Fig. 2.16 *Feynman gas.*

$$P_N(y, L) = \int_{y<x_1<x_2 \cdot x_N<L} e^{-\beta[V(y-x_1)+V(x_1-x_2)+\cdots V(x_{N-1}-x_N)]} dx_1 \cdots dx_N. \qquad (2.8.2)$$

Let us add a new particle in the position a. It only couples to the particle placed at y, a variable that can now vary on the total interval. Hence we obtain the recursive equation

$$\begin{aligned} P_{N+1}(a, L) &= \int_{a<y} e^{-\beta V(a-y)} dy \int_{y<x_1<x_2 \cdot x_N<L} e^{-\beta[V(y-x_1)+V(x_1-x_2)+\cdots]} dx_1 \cdots dx_N \\ &= \int_{a<y<L} e^{-\beta V(a-y)} P_N(y, L)\, dy. \qquad (2.8.3) \end{aligned}$$

The grand canonical partition function of the gas of particles placed in the interval (y, L) is given by[7] (see Appendix A of Chapter 1)

$$\mathcal{Z}(y, L, z) = \sum_{N=0}^{\infty} z^N P_N(y, L), \qquad (2.8.4)$$

where $z = e^{\beta\mu}$ (μ is the fugacity of the gas). In the grand-canonical ensemble we have the equation of state

$$\mathcal{Z} = e^{\beta P(z) V},$$

where $P(z)$ is the pressure of the gas and V its volume. Since $V = L - y$, we have

$$e^{\beta P(z)(L-y)} = \sum_{N=0}^{\infty} z^N P_N(y, L). \qquad (2.8.5)$$

Multiply both the left and right terms of (2.8.3) by z^{N+1} and sum on N. By using (2.8.4) and its expression given by (2.8.5), we obtain the integral equation

$$e^{\beta P(z)(L-a)} = z \int_x^L e^{\beta P(z)(L-y)} e^{-\beta V(a-y)} dy.$$

[7] In this expression the term $1/N!$ is absent for the ordering of the coordinates in the integrals.

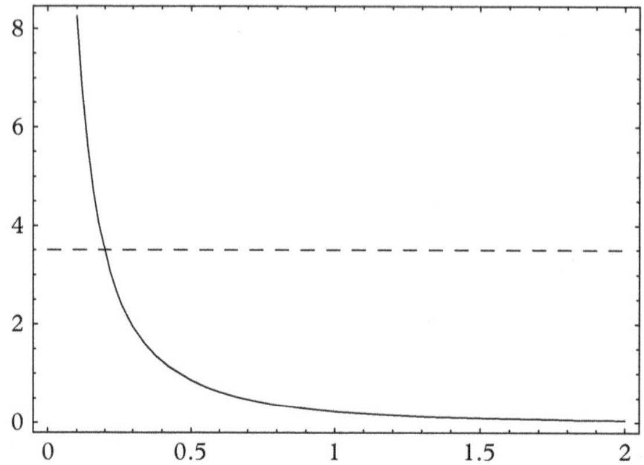

Fig. 2.17 *Graphical solution of (2.8.6). The filled line is the right hand side of this equation, whereas the dashed line corresponds to a given value of the left-hand side of (2.8.6), i.e.* z^{-1}.

Simplifying the common terms present in the left and right sides, in the thermodynamic limit $L \to \infty$ this expression can be written as

$$z^{-1} = \int_0^\infty e^{-\beta P(z) x} e^{-\beta V(x)} \, dx. \tag{2.8.6}$$

This is the central equation of the Feynman's gas: it permits to find the pressure P as a function of z and then to derive all the thermodynamic quantities. For instance, the mean density of particles per unit of length is expressed by

$$\rho(z) = \lim_{L \to \infty} \frac{\langle N \rangle}{L} = \frac{z}{\beta} \frac{dP(z)}{dz}. \tag{2.8.7}$$

For physical considerations, $\rho(z) > 0$. Hence, $P(z)$ is a monotonic increasing function of z and varying z^{-1} on the left-hand side of (2.8.6), there is only one solution for $P(z)$, as shown in (Figure 2.17).

Appendix 2.A. Special Functions

In this appendix we present some properties of the special functions used in the text, namely the $\Gamma(z)$ function and the Bessel functions $I_\nu(x)$.

The $\Gamma(z)$ function

The $\Gamma(z)$ function is an analytic function of the complex variable z. For $\text{Re}\, z > 0$ it is defined by the integral representation

$$\Gamma(z) = \int_0^\infty dt\, t^{z-1} e^{-t}. \tag{2.A.1}$$

To obtain $\Gamma(z)$ for other values of z, we can use its analytical continuation. This can be implemented by using the functional equation

$$\Gamma(z+1) = z\Gamma(z), \tag{2.A.2}$$

that is satisfied in the domain of convergence of the integral (2.A.1). In fact, by integrating by part the expression in (2.A.1), we have

$$\Gamma(z) = \left.\frac{t^z e^{-t}}{z}\right|_0^\infty + \frac{1}{z}\int_0^\infty dt\, t^z e^{-t},$$

and the first term on the right-hand side vanishes when $\mathrm{Re}\, z > 0$, so that we arrive at (2.A.2). By using eqn (2.A.2), we have

$$\Gamma(z) = \frac{\Gamma(z+1)}{z}.$$

Since $\Gamma(z+1)$ is defined when $\mathrm{Re}\, z > -1$, we have obtained the analytic continuation of the $\Gamma(z)$ function in the strip $-1 < \mathrm{Re}\, z < 0$. Repeating the same reasoning, we can further extend its definition in the next strip $-2 < \mathrm{Re}\, z < -1$, and in all other points of the half-plane $\mathrm{Re}\, z < 0$ as well. For instance, if we wish to compute $\Gamma\left(-\frac{1}{2}\right)$, by using (2.A.2), we have

$$\Gamma\left(-\frac{1}{2}\right) = -2\Gamma\left(\frac{1}{2}\right),$$

and the right-hand side can be computed by using (2.A.1). Since $\Gamma\left(\frac{1}{2}\right) = \sqrt{\pi}$, we have $\Gamma\left(-\frac{1}{2}\right) = -2\sqrt{\pi}$.

Note that (2.A.2) implies $\Gamma(n+1) = n!$, when n is a positive integer number: since $\Gamma(1) = \Gamma(2) = 1$, the iterative application of (2.A.2) gives

$$\Gamma(n+1) = n\Gamma(n) = n(n-1)\cdots\Gamma(1) = n!.$$

For that reason, the $\Gamma(z)$ function is considered the generalization, to all real and complex numbers, of the factorial.

The $\Gamma(z)$ function has a pole at $z = 0$. In fact, $\Gamma(1) = 1$, and from (2.A.2), we have

$$\Gamma(z) \simeq \frac{1}{z}, \quad z \to 0$$

We can again use (2.A.2) to prove that $\Gamma(z)$ has other simple poles at $z = -n$, with $n = 1, 2, 3, \ldots$. This result can also be obtained as follows: since the convergence of the integral (2.A.1) only depends on the behaviour nearby the origin, it is convenient to split the interval of the integration as

$$\Gamma(z) = \int_0^1 dt\, t^{z-1} e^{-t} + \int_1^\infty dt\, t^{z-1} e^{-t}.$$

The second integral is always convergent, while the first one converges only when $\mathrm{Re}\, z > 0$. In this interval we can use the series expansion

$$e^{-t} = \sum_{k=0}^{\infty} (-1)^k \frac{t^k}{k!},$$

and, if we integrate term by term, we have

$$\Gamma(z) = \sum_{k=0}^{\infty} (-1)^k \frac{1}{k!(z+n)} + \int_1^\infty dt\, t^{z-1} e^{-t}. \tag{2.A.3}$$

This expression coincides with the original definition of $\Gamma(z)$ in the domain $\mathrm{Re}\, z > 0$, but it is also valid for other valid of z, i.e. it provides the analytic continuation of the $\Gamma(z)$ function for all complex values of z. Eqn (2.A.3) makes explicit the presence of the poles of this function for all negative integers, with residues given by

$$\lim_{z \to -n} (z+n)\, \Gamma(z) = \frac{(-1)^n}{n!}.$$

The plot of $\Gamma(z)$, for real values of z, is given in Figure (2.18).

When z is very large, $\Gamma(z)$ admits the asymptotic expansion

$$\Gamma(z) \simeq z^{z-1/2} e^{-z} \sqrt{2\pi}, \quad z \to \infty \tag{2.A.4}$$

The simplest way to derive this formula is to start from the integral representation (2.A.1) and use the saddle point method, discussed in Appendix 3.A. With the change of variable $t = zx$ in (2.A.1), we have

$$\Gamma(z+1) = z^{z+1} \int_0^\infty dx\, e^{z(\log x - x)}. \tag{2.A.5}$$

The expression in the exponential

$$\varphi(x) \equiv \log x - x,$$

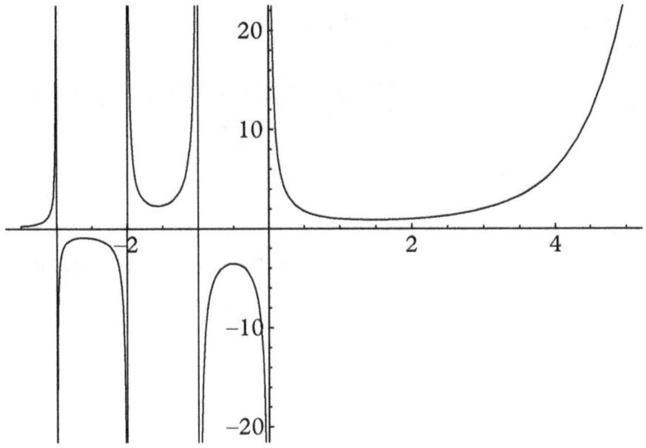

Fig. 2.18 *Plot of the $\Gamma(x)$ function for real values of x.*

goes to minus infinity both at $x \to 0$ and $x \to \infty$ and has a maximum at $x = 1$. When $z \to \infty$, the integral (2.A.5) is then dominated from the values around the maximum of $\varphi(x)$. Expanding around $x = 1$

$$\varphi(x) = -1 - \frac{(x-1)^2}{2} + \cdots$$

and substituting in (2.A.5), we have

$$\Gamma(z+1) \simeq z^{z+1} e^{-z} \int_0^\infty e^{-z\frac{(x-1)^2}{2}}\, dx \simeq \sqrt{2\pi}\, z^{z+1/2} e^{-z}.$$

Using $\Gamma(z+1) = z\Gamma(z)$, we arrive at eqn (2.A.4). When $z = n$, with n integer, we have

$$n! \simeq n^n e^{-n} \sqrt{2\pi n},$$

i.e. the Stirling approximation of the factorial.

The $\Gamma(z)$ function satisfies many mathematical identities, that can be often proved by making use of the functional equation (2.A.2) and its analytic properties. One of them is the so-called *reflection formula*

$$\Gamma(z)\Gamma(1-z) = \frac{\pi}{\sin \pi z}. \tag{2.A.6}$$

To prove it, let us define $\Phi(z) \equiv \Gamma(z)\Gamma(1-z)$. Using (2.A.2), it is easy to see that $\Phi(z)$ is a periodic function, with a period equal to 2, i.e. $\Phi(z+2) = \Phi(z)$. Moreover, $\Phi(z)$ has simple poles at all integer values of z, i.e. $z = 0, \pm 1, \pm 2, \ldots$. Hence, the function obtained by multiplying $\Phi(z)$ with the infinite product

$$z \prod_{k=1}^{\infty} \left(1 - \frac{z^2}{k^2}\right),$$

(that has precisely simple zeros at $z = 0, \pm 1, \pm 2, \ldots$) is a function without any singularities in the complex plane. Therefore, according to Liouville theorem, it is a constant, in this case equal to 1, as can be seen by taking the limit $z \to 0$. Hence, using the identity

$$\frac{\sin \pi z}{\pi} = z \prod_{k=1}^{\infty} \left(1 - \frac{z^2}{k^2}\right), \tag{2.A.7}$$

we arrive at eqn. (2.A.6). It is interesting to note that this formula permits to prove

$$\sum_{k=1}^{\infty} \frac{1}{k^2} = \frac{\pi^2}{6}, \tag{2.A.8}$$

In fact, using the series expansion

$$\sin \pi z \simeq \pi z - \frac{(\pi z)^3}{3!} + \cdots$$

and comparing the z^3 terms of the right- and left-hand sides of (2.A.7), we get (2.A.8). This result was originally obtained by Euler.

The Bessel functions $I_\nu(x)$

The Bessel functions $I_\nu(z)$ are defined by the series expansion

$$I_\nu(z) = \left(\frac{z}{2}\right)^\nu \sum_{k=0} \frac{1}{k!\Gamma(k + \nu + 1)} \left(\frac{z}{2}\right)^{2k}. \tag{2.A.9}$$

For generic ν, they are regular function of z in the complex plane, with a branch cut along the real negative semi-axis. When $\nu = \pm n$, they are instead entire functions of z. The functions $I_\nu(z)$ are solutions of the differential equation

$$z^2 \frac{d^2 w}{dz^2} + z \frac{dw}{dz} - (z^2 + \nu^2) w = 0. \tag{2.A.10}$$

$I_\nu(z)$ and $I_{-\nu}(z)$ are linearly independent except if $\nu = n$, when

$$I_{-n}(z) = I_n(z) \tag{2.A.11}$$

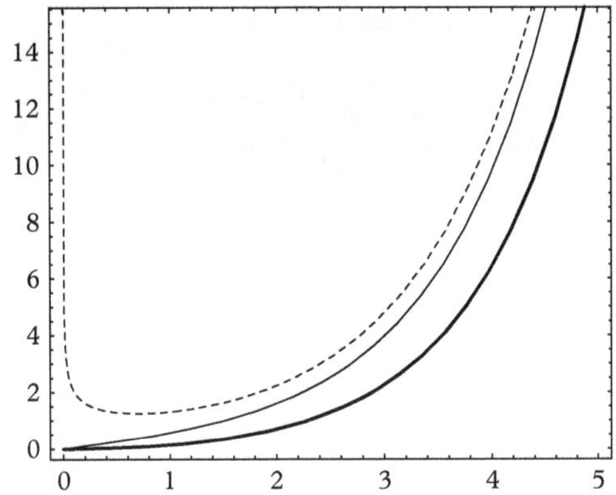

Fig. 2.19 $I_\nu(x)$ *for real values of* x *(*$\nu = -1/3$ *dashed line,* $\nu = 1$ *central line,* $\nu = 2$ *thick line.*

In fact, substituting $\nu = -n$ in (2.A.9), the first n terms of the series expansion vanish, since these are the poles of $\Gamma(x)$ and therefore we have the identity (2.A.11).

For $\mathrm{Re}\,\nu > \frac{1}{2}$, $I_\nu(z)$ admits the integral representation

$$I_\nu(z) = \frac{\left(\frac{z}{2}\right)^\nu}{\sqrt{\pi}\,\Gamma\left(\nu + \frac{1}{2}\right)} \int_0^\pi e^{\pm z\cos\theta} \sin^{2\nu}\theta \, d\theta. \tag{2.A.12}$$

When $z \to 0$ (with fixed ν) $I_\nu(z)$ has the behaviour

$$I_\nu(z) \simeq \frac{1}{\Gamma(\nu+1)} \left(\frac{z}{2}\right)^\nu, \quad z \to 0 \quad (\nu \neq -n) \tag{2.A.13}$$

as can be seen from the series expansion (2.A.9), whereas for $z \to \infty$ grandi valori di z (with fixed ν)

$$I_\nu(z) \simeq \frac{e^z}{\sqrt{2\pi z}} \left[1 - \frac{(4\nu^2 - 1)}{8z} + \frac{(4\nu^2-1)(4\nu^2-9)}{2(8z)^2} + \cdots\right], \quad z \to \infty \tag{2.A.14}$$

This behaviour can be easily checked by the asymptotic expansion around $z = \infty$

$$I_\nu(z) \simeq e^z \sum_{n=0}^{\infty} \frac{a_n}{z^{n+\alpha}}, \quad z \to \infty$$

where α and the coefficients a_n are fixed by substituting this expression into the differential equation (2.A.10). The plot of these functions is shown in Figure (2.19).

For large values of the index ν, we have

$$I_\nu(\nu x) \simeq \frac{1}{\sqrt{2\pi\nu}} \frac{e^{\nu(\sqrt{1+x^2}-\xi^{-1}}}{(1+x^2)^{1/4}}, \quad \nu \to \infty, \tag{2.A.15}$$

with

$$\xi^{-1} = \ln\left(\frac{1+\sqrt{1+x^2}}{x}\right).$$

Using (2.A.9), it is easy to prove that the Bessel functions satisfy the recursive equations

$$\begin{aligned} I_{\nu-1}(z) - I_{\nu+1}(z) &= \frac{2\nu}{z} I_\nu(z) \\ I_{\nu-1}(z) + I_{\nu+1}(z) &= 2\frac{dI_\nu}{dz}(z) \end{aligned} \tag{2.A.16}$$

For $\nu = n$, an integer number, their generating function is given by

$$e^{(z/2)(t+1/t)} = \sum_{n=-\infty}^{+\infty} I_n(z)\, t^n. \tag{2.A.17}$$

The Bessel functions $K_\nu(x)$

In other chapters of the book we will need the modified Bessel functions $K_\nu(x)$. These are also solutions of the differential equation (2.A.10) and can be expressed as linear combination of the $I_\nu(x)$

$$K_\nu(x) = \frac{I_{-\nu}(x) - I_\nu(x)}{\sin \pi\nu}. \tag{2.A.18}$$

When ν is an integer, their definition involves the limit $\nu \to n$ of the above expression. The series expansion of these functions directly comes from the one of the $I_\nu(x)$. The lowest orders are given by

$$\begin{aligned} K_0(x) &= -\log x - \gamma + \log 2 + \cdots \\ K_\nu(x) &= 2^{\nu-1}\Gamma(\nu) x^{-\nu} + \cdots \end{aligned} \tag{2.A.19}$$

For large values of x, their aymptotic behaviour is

$$K_\nu(x) \simeq \sqrt{\frac{\pi}{2x}} e^{-x} \left[1 + \frac{(4\nu^2-1)}{8x} + \frac{(4\nu^2-1)(4\nu^2-9)}{2(8x)^2} + \cdots\right] \tag{2.A.20}$$

They satisfy the recursive equations

$$K_{\nu-1}(x) - K_{\nu+1}(x) = -\frac{2\nu}{x} K_\nu(x)$$
$$K_{\nu-1} + K_{\nu+1}(x) = -2K'_\nu(x) \tag{2.A.21}$$

Their integral representation is given by

$$K_\nu(x) = \int_0^\infty e^{-z\cosh t}\cosh(\nu t)\,dt. \tag{2.A.22}$$

Appendix 2.B. *n*-dimensional Solid Angle

The solid angle in the n-dimensional space is expressed by the formula

$$\Omega(n) = \int d\Omega^{(n)} = \frac{2\pi^{n/2}}{\Gamma\left(\frac{n}{2}\right)}. \tag{2.B.1}$$

This is an analytic function of n and, for this reason, can be computed for arbitrary values of n, not necessarily integers. However, its behaviour is rather peculiar: for the presence of the $\Gamma(x)$ in its denominator (that diverges both at $x = 0$ and $x \to \infty$) the n-dimensional solid angle goes to zero at the two edges of the positive semi-axis, and presents a maximum at $n_m \simeq 7.2$ (see Figure 2.20).

While it is easy to understand that $\Omega(n)$ should vanish when $n \to 0$ (if there is no space, the solid angle between its axes must vanishes as well), it is apparently paradoxical that it also vanishes when the number of the dimensions increases at infinity : our geometrical intuition would rather suggests that the solid angle should increase by increasing the

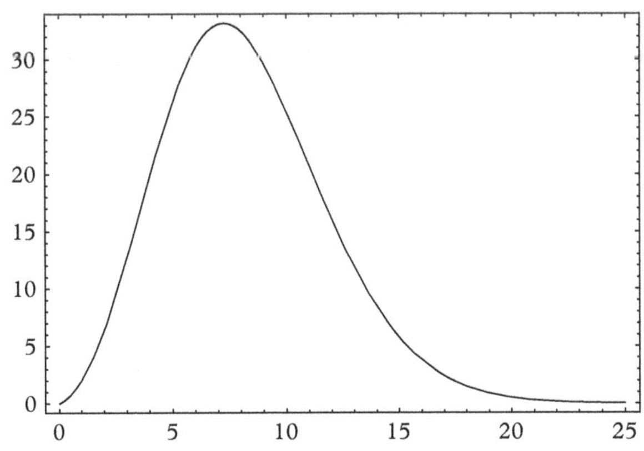

Fig. 2.20 *The solid angle $\Omega(n)$ as a function of n.*

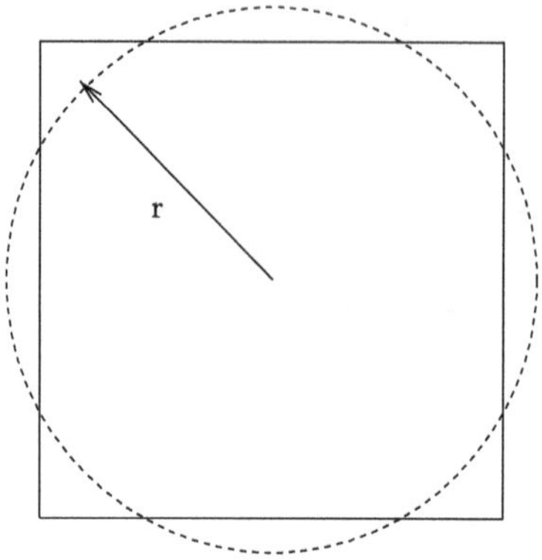

Fig. 2.21 *Cube and sphere of unitary volume.*

dimensions, in sharp contrast however with the formula (2.B.1). There is a way to understand the asymptotic behaviour of $\Omega(n)$ for large n. Using the asymptotic expansion

$$\Gamma(x) \simeq x^{x-1/2} e^{-x} \sqrt{2\pi}, \quad x \to \infty$$

we have

$$\Omega(n) \simeq \frac{2\pi^{n/2}}{\left(\frac{n}{2}\right)^{\frac{n-1}{2}}} \frac{e^{n/2}}{\sqrt{2\pi}}, \quad n \to \infty \tag{2.B.2}$$

Let us consider a n-dimensional cube, with site of length 1: its volume is given by

$$V = 1 \times 1 \times 1 \cdots \times 1 = 1.$$

On the other hand, the estimate of such a volume can be reached by using a sphere of radius r, of the order of the distance from the centre of the cube to one its vertexes,[8] namely

$$r \simeq \sqrt{\frac{1}{2} + \frac{1}{2} + \cdots + \frac{1}{2}} = \sqrt{\frac{n}{2}} \tag{2.B.3}$$

Let us call a_n the proportional constant

$$r(n) = a_n \sqrt{\frac{n}{2}}, \tag{2.B.4}$$

[8] This formula shows the counter-intuitive behaviour of the n-dimensional spaces: in fact, increasing the number of the dimension, the volume of the unitary cube is always equal to 1, whereas its diagonal diverges as $\sqrt{n}$.

Since the volume of a n-dimensional sphere of radius R is given by

$$V(n) = \frac{R^n}{n}\Omega(n),$$

posing $V(n) = 1$ and using the asymptotic behaviour of $\Omega(n)$, it is easy to see that the decreasing behaviour of the solid angle is compensated by the increasing of the term $[r(n)]^n$. This argument provides the explanation of the unusual behaviour of the solid angle $\Omega(n)$ for large n. It is interesting to note that, for $n \to \infty$, we have

$$a_\infty = \frac{1}{\sqrt{\pi e}}.$$

Appendix 2.C. The Four-colour Problem

The four-colour problem has been, for long time, one of the most famous open problems of mathematics and it has played an important role in the development of modern applied mathematics. Since 1852, several ideas pursued for its solution have contributed to the progress of many other fields of mathematics: graph theory, for instance, or topology have both received an enormous stimulus from this problem.

As many other famous problems (the last Fermat theorem, for instance), its statement is rather simple: the *four-colour conjecture* states that four colorus are enough to colour any geographical map (drawn on a plan or a sphere), with the constraint that regions with a common border have a different colour.

Let us make few comments. First of all, the conjecture refers to maps consisting of contiguous regions, i.e. regions that share a border not made of a single point. Otherwise, as a map consisting of pie slices for instance, it would be necessary to use as many colours as the number of slices (see Figure 2.22).

The second comment is that the four-colour conjecture does not refer to maps that consist of disconnected regions. In this case, in fact, it is easy to show examples of maps with five regions which need five colours to distinguish them. One of these examples is in Figure 2.23: the region indicated by the letter E is considered a unique region, even though it is made by two disconnected parts.

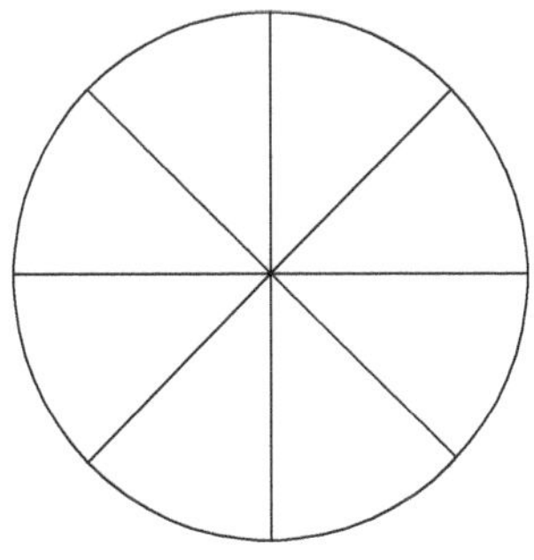

Fig. 2.22 *This map is excluded from the four-colour conjecture.*

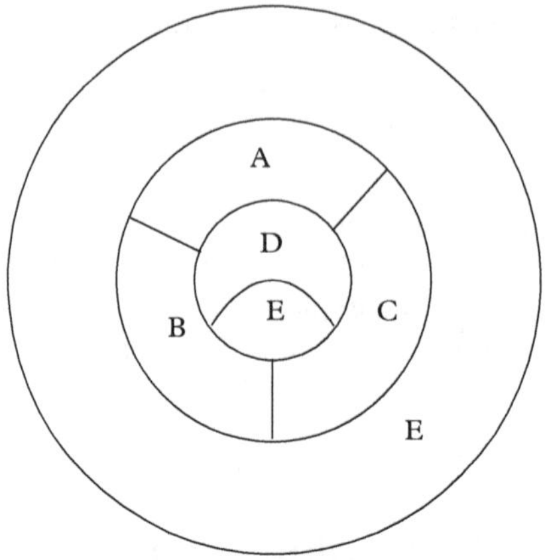

Fig. 2.23 *Map with disconnected regions that does not fall in the four-colour theorem.*

After these remarks, let us discuss in detail the four-colour conjecture, starting from its history. The problem dues its origin to the mathematician Francis Guthrie in 1852: few months after he graduated from University College London, he wrote a letter to his brother Frederick, who was completing his studies at the same college under the supervision of the mathematician Augustus De Morgan. In this letter, he expressed his feeling that four colours seem sufficient to distinguish the regions of a planar geographical map and asked him whether he was aware of any proof of that statement. Frederick's answer was negative and also De Morgan, asked about the problem, was not aware of any proof of the statement. However, De Morgan quickly showed that four colours are indeed necessary to colour any planar map, pointing out an explicit example for which three colours are not sufficient. This map, shown in Figure 2.24, is made of four regions, each of them adjacent to all the others, so that three colours are not sufficient.

As discussed below, using Euler's formula for the graphs, De Morgan was also able to prove that it is impossible to draw a planar map with five regions, each of them adjacent to the remaining four. Following this result, he stated that the case in which it is necessary to use five colours never occurs, namely that the four-colour conjecture is true. However,

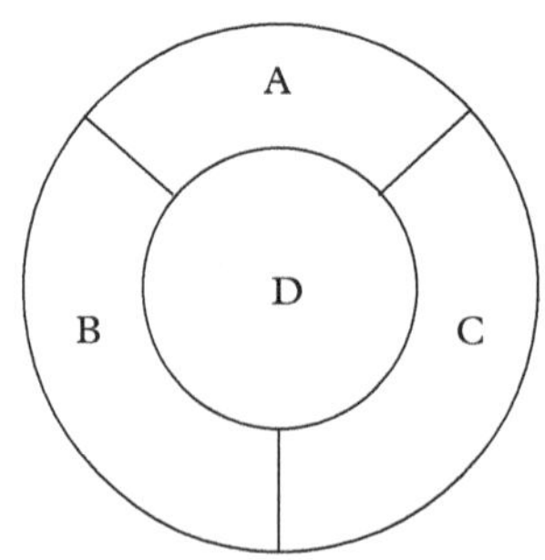

Fig. 2.24 *Map that needs four colours.*

he also realized that his conclusion was wrong: the fact that five mutual adjacent regions in the plane cannot exist is not equivalent to prove the four-colour conjecture.

It is easy to see that it is not correct to assume, as it was done by De Morgan, that the number of colours necessary to draw a map is equal to the number of countries mutually adjacent. A counter-example is shown in Figure 2.25. Although there are no more than three regions mutually adjacent in this map, nevertheless it is necessary to use four colours: three for the external ring and one for the central region.

To summarize: the map of Figure 2.24 points out a *local obstruction* to draw a map using three colours, whereas the map shown in Figure 2.25 provides an example of a *global obstruction* to colour a map with only three colours. Therefore, the result by De Morgan on the impossibility to draw a map in which five regions are mutually adjacent, only addresses the local aspect of the problem and not the global one. Many solutions of the four-colour theorem that have been proposed during the years show the same mistake made by De Morgan.

The official birthday of the four-colour conjecture is the 13 June 1878, when the English mathematician Arthur Cayley discussed the problem at the London Mathematical Society. His contribution was published in the proceedings of the Society and this was the first written version of the problem and its official birthday.

As shown, the four-colour conjecture is easy to state but this simplicity is deceptive. It has been surely deceptive for many mathematicians, even famous, who have attempted to prove it. Hermann Minkowski, for instance, told his students that, in his opinion, the conjecture was not proved yet simply because only mediocre mathematicians had been involved in and he would have been surely able to achieve its solution. But, after a long period spent on the problem, he was forced to admit his failure. *God punishes me for my arrogance, my proof has unfortunately many flaws*, was his last comment.

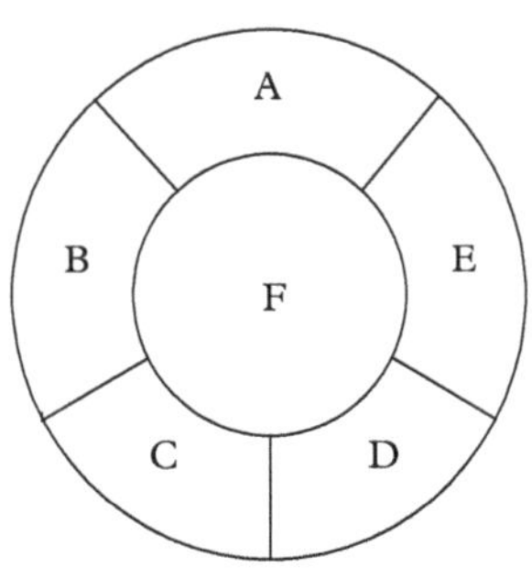

Fig. 2.25 *A counter-example to the presumed proof by De Morgan of the four-colour theorem.*

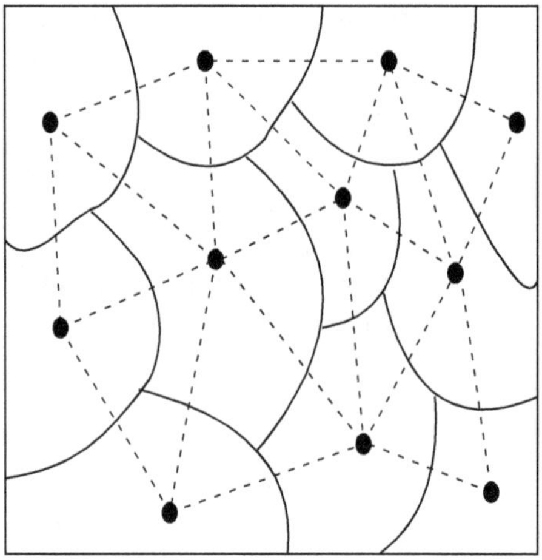

Fig. 2.26 *Dual graph of a map.*

The difficulty of the proof is because the statement refers to *all possible maps*. From this point of view, to establish that a given map can be coloured by four colours is not particularly useful. Let us see how topology and graph theory can help in a clearer formulation of the problem.

As far as the theorem is concerned, it is obvious that the shapes and the actual sizes of the regions are completely irrelevant: they can be deformed with continuity without altering the nature of the problem. The only thing that matters is the relation among the different regions, i.e. the topology of the map. This can be highlighted by means of the *dual graph*. As we will see in other parts of the book, this concept plays an important role in statistical physics and enters the solution of the Ising model and other similar models (see, for instance, Chapter 4). The dual graph of a map is obtained as follows. Firstly, inside each region let us draw a point, denoted as a *vertex*; secondly, let us join the vertices by a line according to the rule that two vertices will be linked each other if and only if their regions share a common border. At the end of this procedure, we have obtained the *dual graph* of the original map. The dual graph shows immediately the topological properties of the initial map and, at the same time, points out the close relationship between the four-colour problem and the Potts model discussed in the text. In fact, the colouring problem of a map can be easily formulated in terms of the dual graph as follows: colour the vertices of the dual graph in such a way that each pair of connected vertices has different colour. Note that an important property of the dual graphs is that all its links can never cross. For that reason, they are only a sub-class of more general graphs.

Graph theory allows us to obtain some useful information on the dual graph. Let F be the number of *faces* (i.e. the different regions of the area of the plane spanned by the graph), V the number of vertices and E the number of edges of a given graph. Euler found a formula that links these quantities together

$$V - E + F = 1. \tag{2.C.1}$$

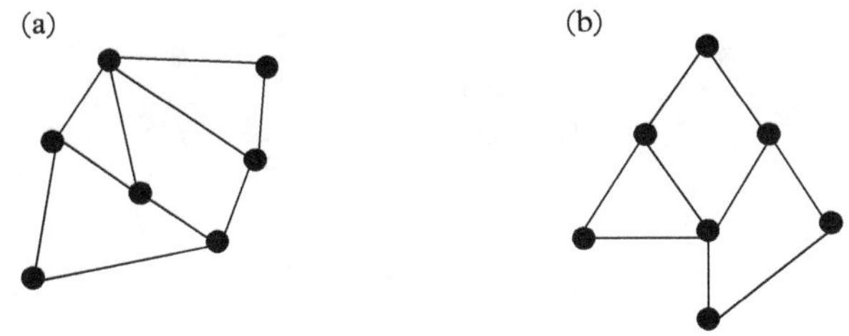

Fig. 2.27 *For every graph, it holds $V - E + F = 1$. In the graph (a), we have $V = 7, E = 10$ e $F = 4$, while in the second $V = 7, E = 9$ e $F = 3$.*

It is easy to check the validity of Euler's formula by analysing few examples, as those shown in Figure 2.27. The proof of this theorem is not difficult and it is interesting to follow its main steps. It consists of simplifying the graph by means of two procedures which never touch the combination $V - E + F$, until we arrive at the most elementary graph made of only one point, for which it trivially holds $V - E + F = 1$.

Starting from an arbitrary graph, the first procedure consists of removing one of its external edges. In this way, E decreases by 1, the same happens for F, while F remains unchanged. Hence, in this procedure the expression $V - E + F$ does not change.

Let us imagine that, by iterating this procedure we end up with a vertex that is linked to the rest of the graph by only one edge. At this point, there is the second procedure to use: it consists of removing both the vertex and the edge. In this operation V decreases by 1, the same happens for E, while F remains instead unchanged. Hence, also in this case, the combination $V - E + F$ is the same before and after the procedure.

Starting from the initial graph and acting by the two procedures, we can systematically remove all the external edges and vertices, until we reach a graph made of only one vertex. The final graph has $V = 1$ ed $E = F = 0$. Since in all these operations the quantity $V - E + F$ is always the same, we thus arrive at Euler's formula (2.C.1).

By using Euler's formula, it is easy to prove the theorem by De Morgan previously mentioned: it does not exist any planar map in which there are five countries, each of them that shares a border with the remaining four. This means that it is impossible to draw a graph with five vertices, each of them linked to the other four, without crossing its edges.

The proof is by a reductio to absurd. Let us assume that there exists a graph made of five vertices, each of them linked to the other four. It is convenient to consider the region outside the graph as an additional face, so that each edge always separates two faces. If we do so, Euler's formula gets modified as follows

$$V - E + F = 2. \tag{2.C.2}$$

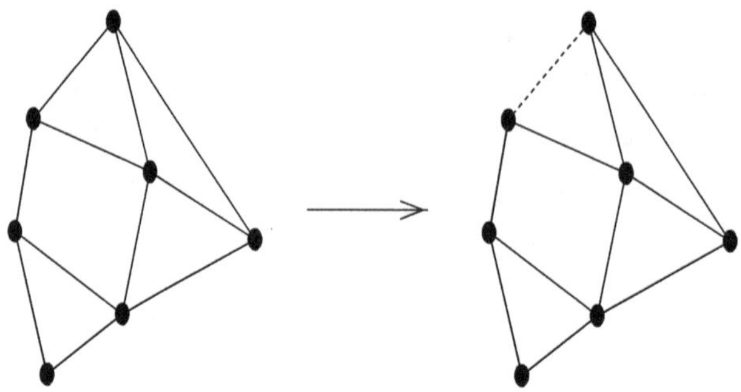

Fig. 2.28 *When an external edge is removed, the quantity $V-E+F$ does not change its value.*

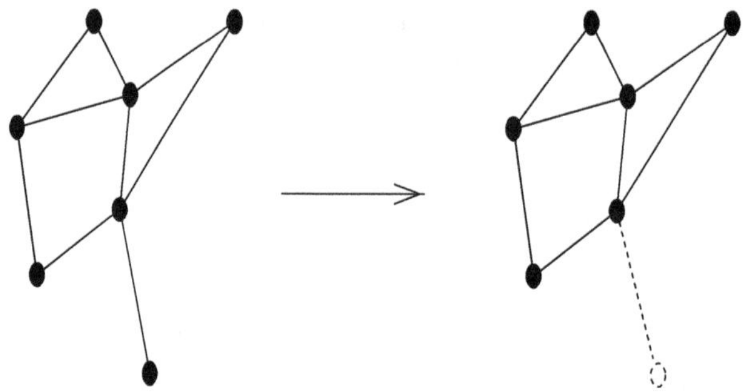

Fig. 2.29 *If an isolated vertex and the edge that links it to the rest of the graph are removed, the quantity $V-E+F$ does not change its value.*

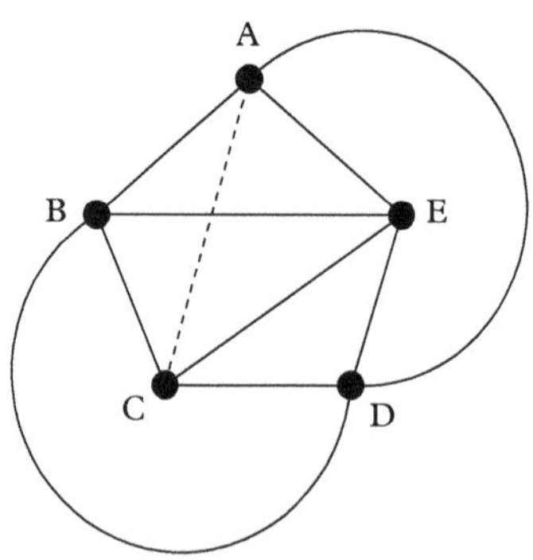

Fig. 2.30 *De Morgan theorem: It is impossible to draw a planar graph of five vertices so that each vertex is linked to the others. Whatever the procedure, there will always be two vertices that cannot be linked by an edge without crossing the previous edges (in the figure these are the A and C vertices).*

In the graph under investigation $V = 5$. Since, by hypothesis, each of them is linked to the other four vertices, we have $E = 5 \cdot 4/2 = 10$. Hence, from Euler's formula one should find $F = 7$. Let us use this last result to do the computation differently: since there are seven faces, each of them with three edges (the external face is included as well in this computation), one should have $3 \cdot 7 = 21$ edges. However, since each of them separates two faces, the previous number should be twice the number of the edges. But 21 is an odd number, so that we arrive at a contradiction. This implies that the initial hypothesis is false, i.e. that it is impossible to draw a planar graph with five vertices, each of them linked to the others.

The result found by De Morgan is not enough to prove the four-colour theorem. The history of all the different attempts to prove this theorem is an interesting chapter of mathematics, too long to be summarized.[9] Here we will only mention two important developments. The first refers to the study by Beraha that, during his PhD thesis, noticed that the chromatic polynomial of a planar graph has often zeros close to the sequence of numbers

$$B_n = 4\cos^2(\pi/n), \quad n = 2,3,4 \tag{2.C.3}$$

known in the literature as *Beraha numbers*. He was able to prove that any Beraha number that is not an integer (with the possible exception of B_{10} for which his proof did not apply) could be a zero of the chromatic polynomial of a finite graph. Vice versa, he also proved that certain graphs with a shape of a strip of length L and finite width have zeros that converge to each Beraha number (not an integer), when $L \to \infty$. In another work made in collaboration with Kahane, he also observed that some families of graphs have the zeros of their chromatic polynomials that converge to the value $q = 4$, that seems to be their accumulation point when $L \to \infty$.

The original idea of Beraha and Kahane was to find a planar graph with a zero of the chromatic polynomial at $q = 4$. The discovery of such a graph would have been a counter-example of the four-colour conjecture. For a twist of fate, between the submission of the article (1976) and its publication, the conjecture turned into a theorem! In fact, in 1976 two mathematicians of the University of Illinois, Kenneth Appel and Wolfgang Haken, announced that they were able to prove the validity of the four-colour conjecture, although along a different path with respect to the traditional mathematical demonstrations. Indeed, one of the main steps of the proof relied on the crucial use of the computer analysis. Beside the remarkable result on the four-colour problem, the important aspect of the proof by Appel and Haken is precisely the unusual way that was adopted for the proof. This has radically changed the concept of *mathematical proof*. According to their own words, 'Obviously it would be possible that, some day, someone will find a shorter proof of the four color theorem. However, it may also be that such a demonstration does not exist at all: in this case, a new kind of theorem has emerged,

[9] We refer the interested reader to Appel, K. and Haken, W. (1977). 'The Solution of the Four-color Problem', *Scientific American*, 237: 108. Additional references are given at the end of the chapter.

a theorem that does not admit a traditional demonstration. Even though the four color theorem will not belong to such a category, nevertheless it provides a good example for a theorem of such a kind and there is no reason to think that this is an isolate case, there may be a large number of problems where such an analysis is indeed needed.'

REFERENCES

In his book, quoted at the beginning of this Chapter, E.A. Abbott provides a vivid testimony of the Victorian era in which he lived. This book is a classic of the scientific literature, filled with social satire as well as *Gulliver's Travels*, and rich in deep theoretical hints on the nature of the space.

Abbott, E.A. (1963). *Flatland. A Romance of Many Dimensions*, Barnes & Noble.

To learn about the percolation phenomenon, we refer to

Stauffer, D. and Aharony, A. (1991). *Introduction to Percolation*, Taylor & Francis.

The physics of polymers is discussed in

de Cloizeaux, J. and Jannick, G. (1989). *Polymers in Solution*, Clarendon Press.
de Gennes, P.G. (1979). *Scaling Concepts in Polymer Physics*, Cornell University Press.

Further references on the four-colour problem are the following

Saary, T.L. and Kainen, P.C. (1977). *The Four-Color Problem*, Dover Publications.
Wilson, R. (2002). *Four Colours Suffice*, Allen Lane/Penguin Press.

For the relation between the Potts model and the four-colour problem see

Saleur, H. (1990). 'Zeros of Chromatic Polynomials: A New Approach to Beraha Conjecture Using Quantum Groups', *Communications in Mathematical Physics*, 132: 657.
Saleur, H. (1991). 'The Anti-ferromagnetic Potts Model in Two Dimensions: Berker–Kadanoff phase, Anti-ferromagnetic Transition, and the Role of Beraha Numbers', *Nuclear Physics B*, 360: 219.

PROBLEMS

2.1. Ising model with vacancies

Consider the one-dimensional Ising model with the familiar Hamiltonian

$$\mathcal{H} = -\mathcal{J}\sum_{i=1}^{N} S_i S_{i+1}$$

but with the values of the spins $S_i = \{0, \pm 1\}$. The null value may be interpreted as a vacancy of the corresponding site. Use the transfer matrix method to compute the partition function of the system.

2.2. Multiple couplings

Consider the one-dimensional Ising model with N spins $\sigma_i = \pm 1$, with the Hamiltonian given by

$$\mathcal{H} = -\mathcal{J}_1 \sum_{i=1}^{N-1} \sigma_i \sigma_{i+1} - \mathcal{J}_2 \sum_{i=1}^{N-2} \sigma_i \sigma_{i+2}$$

Compute the free energy of such a system and analyse its behaviour by varying the parameters $\mathcal{J}_1$ and $\mathcal{J}_2$.

Hint. It is convenient to define $\tau_i = \sigma_i \sigma_{i+1}$ *that assume values* $\tau_i = \pm 1$.

2.3. Grand canonical partition function

Consider a one-dimensional lattice gas on M sites. The variable t_i has values 0 and 1, according whether the relative site is occupied or not. The Hamiltonian is given by

$$\mathcal{H} = -\mathcal{J} \sum_{i=1}^{M} t_i t_{i+1}$$

with $t_{i+M} = t_i$. Moreover, we assume that

$$\sum_{i-1}^{M} t_i = N$$

a. Compute the grand canonical partition function $\mathcal{Z}(z, M, T)$ of such a system.
b. Show that for $\mathcal{J} > 0$ the zeros of $\mathcal{Z}(z, M, T)$ are along the circle $|z| - e^{-\mathcal{J}}$ of the complex plane of the variable z, where $\mathcal{J} = \beta \mathcal{J}$.

2.4. Ising model in a magnetic field

Let us consider the one-dimensional Ising model with nearest-neighbour interactions in the presence of a magnetic field. Show that the partition function $Z_N(T, B)$ is an even function of B and compute its first non-vanishing term of the series expansion with respect to B.

2.5. Ising model in a purely imaginary magnetic field

Consider the one-dimensional Ising model with next-neighbour interaction in the presence of a purely imaginary magnetic field $B = ih$. Study the correlation length of the model, expressed by eqn. (2.2.28) and show that there exist an infinite number of pairs $(\mathcal{J}, h)$ for which it diverges.

2.6. Correlation functions

For the one-dimensional Ising model with next-neighbour interaction:

a. Compute the correlation function $\langle \sigma_i \sigma_j \sigma_k \sigma_l \rangle$, with $i > j > k > l$.
b. Use the expression of the quantity above to compute the derivative of the correlation function $\langle \sigma_i \sigma_j \sigma_k \rangle$ with respect to the magnetic field B, for $B = 0$.

2.7. Potts Model

Consider the one-dimensional q-state Potts model, whose Hamiltonian is given by

$$\mathcal{H} = -\mathcal{J} \sum_{i=1}^{N} \delta(\sigma_i, \sigma_{i+1}).$$

a. Compute the expectation value $\langle \delta(\sigma_k, \sigma_{k+1}) \rangle$.
b. Study this quantity for negative values of q.

2.8. One-dimensional Tonks gas

Consider a gas made of molecules of size a, that can move freely along a line of length L. Their interaction potential is given by

$$U(x_i - x_j) = \begin{cases} +\infty, & \text{if } | x_i - x_j | < a \\ 0, & \text{if } | x_i - x_j | > a \end{cases}.$$

Compute the exact expression of the partition function of the system and its equation of state.

2.9. One-dimensional gas

Considerar a one-dimensional gas of N particles, with potential interaction

$$V(x_1, x_2, \ldots, x_N) = -\sum_{i<j} \log \tanh^2 \left(\frac{x_i - x_j}{2} \right).$$

a. Discuss the validity of the next-neighbour approximation for this potential.
b. Give a numerical estimate of the gas pressure at $z = \frac{1}{2\pi}$.
c. Compute the mean values of the number of particle per unit length and check the validity of the next-neighbour approximation.

2.10. Thermodynamics of one-dimensional oscillators

Consider a set of N particles of mass m placed along a line of length L. Let p_i and q_i be their momentum and coordinate respecively. Their Hamiltonian is

$$H = \sum_{i=1}^{N} \frac{p_i^2}{2m} + \frac{1}{2}k \sum_{i=1}^{N-1} (q_i - q_{i+1})^2.$$

a. Determine the frequencies of the normal modes of the system.

b. Compute the partition function.

3

Approximate Solutions

The approximate computations are a fundamental part of the physical science.

Steven Weinberg

The exact solutions of one-dimensional systems discussed in Chapter 2 are particularly simple and mathematically elegant. Unfortunately, they are more an exception than a rule: for higher-dimensional lattices, the computation of the partition functions of statistical models usually poses a formidable problem from a mathematical point of view. For that reason, it is of the utmost importance to develop some approximate methods that allow us to analyse the most relevant physical aspects and to extract an estimate of critical exponents or other thermodynamical quantities. This chapter discusses several approximate solutions of the Ising model and its generalizations.

3.1 Mean Field Theory of the Ising Model

In the Ising model, each spin interacts both with the external magnetic field and the one created by neighbour spins. The magnetic field created by the spins is obviously a dynamical variable, that cannot be controlled by external knobs, for its value changes with the fluctuations of the configurations. The mean field approximation consists of replacing the magnetic field created by the spins by its thermal average. This substitution gives rise to an interaction among all spins, so that the mean field solution is essentially equivalent to solve the model in the limit $d \to \infty$: for a infinite-dimensional lattice, the mean field solution is then an exact one. Although this limit may appear artificial and distant from the actual physical features of magnets, we shall nevertheless see that the mean field solution is able to capture the main properties of phase transition in the Ising model, in a particularly simple way that is also sufficiently accurate.

Consider the Hamiltonian of a d-dimensional lattice with N spins

$$\mathcal{H} = -\frac{\mathcal{J}}{2} \sum_{\langle i,j \rangle} \sigma_i \sigma_j - B \sum_i \sigma_i. \tag{3.1.1}$$

Statistical Field Theory: An Introduction to Exactly Solved Models in Statistical Physics. Second Edition. Giuseppe Mussardo, Oxford University Press (2020). © Giuseppe Mussardo. DOI: 10.1093/oso/9780198788102.001.0001

Let us introduce the magnetization, defined by

$$m = \frac{1}{N}\left\langle \sum_{i=1}^{N} \sigma_i \right\rangle, \tag{3.1.2}$$

and let us express the product of the spins as

$$\begin{aligned}\sigma_i\sigma_j &= (\sigma_i - m + m)(\sigma_j - m + m)\\ &= m^2 + m(\sigma_i - m) + m(\sigma_j - m) + (\sigma_i - m)(\sigma_j - m).\end{aligned}$$

The last term is quadratic in the spins, of the form $(\sigma_i - \langle\sigma\rangle)(\sigma_j - \langle\sigma\rangle)$. The mean field approximation consists of completly neglecting this term, substituting the previous Hamiltonian with

$$\begin{aligned}\mathcal{H} &= -\frac{\mathcal{J}}{2}\sum_{\langle i,j\rangle}\sigma_i\sigma_j - B\sum_i \sigma_i =\\ &\simeq -\frac{\mathcal{J}}{2}\sum_{\langle i,j\rangle}[-m^2 + m(\sigma_i + \sigma_j)] - B\sum_i \sigma_i.\end{aligned} \tag{3.1.3}$$

Let z be the il *coordination number* of the lattice, i.e. the number of neighbour spins with whom each spin interacts.[1] The first term in (3.1.3) can be expressed as

$$-\frac{\mathcal{J}}{2}\sum_{\langle i,j\rangle}(-m^2) = \frac{1}{2}\mathcal{J}zm^2\sum_i = \frac{1}{2}\mathcal{J}Nzm^2,$$

while

$$-\frac{\mathcal{J}}{2}m\sum_{\langle i,j\rangle}(\sigma_i + \sigma_j) = -\mathcal{J}zm\sum_i \sigma_i.$$

In the mean field approximation, the Hamiltonian thus becomes

$$\mathcal{H}_{cm} = \frac{1}{2}N\mathcal{J}zm^2 - (\mathcal{J}zm + B)\sum_i \sigma_i. \tag{3.1.4}$$

Since all spins are decoupled, it is simple to compute the partition function

[1] For a two-dimensional square lattice, $z = 4$, whereas in a tri-dimensional cubic lattice $z = 6$.

$$Z_N^{cm}(T,B) = \sum_{\{\sigma\}} e^{-\beta\mathcal{H}_{cm}} = e^{-\frac{1}{2}N\mathcal{J}zm^2}\left(\sum_{\sigma=\pm 1} e^{(\mathcal{J}zm+\mathcal{B})\sigma}\right)^N =$$
$$= e^{-\frac{1}{2}N\mathcal{J}zm^2}\,[2\cosh(2\mathcal{J}m+\mathcal{B})]^N. \tag{3.1.5}$$

The free energy per spin is expressed by

$$F^{cm}(T,B) = -\frac{1}{\beta N}\ln Z_N = \frac{1}{2}\mathcal{J}zm^2 - \frac{1}{\beta}\ln[2\cosh(\mathcal{J}m+\mathcal{B})]. \tag{3.1.6}$$

Since the magnetization is given by the derivative of F with respect to B, it must satisfy the self-consistency equation

$$m = -\frac{\partial F}{\partial B} = \tanh(\mathcal{J}zm+\mathcal{B}). \tag{3.1.7}$$

This equation was firstly obtained by Bragg and Williams and, for this reason, the mean fieldsolution is also known as Bragg–Williams approximation. To see whether there is a spontaneous magnetization at $B \to 0$, one needs to look for a non-zero solutions of the trascendental equation

$$m = \tanh(\mathcal{J}zm). \tag{3.1.8}$$

This problem can be solved by a graphical method, i.e. by checking whether the plot of the function on the right-hand side of (3.1.8) has intersections or not with the straight line $y = m$, as shown in Figure 3.1. The value $m = 0$ is obviously always a solution of (3.1.8). However, when the derivative at the origin of the function $\tanh(\mathcal{J}zm)$ on the right-hand side of (3.1.8) is greater than 1, there are other two solutions (of different sign but equal

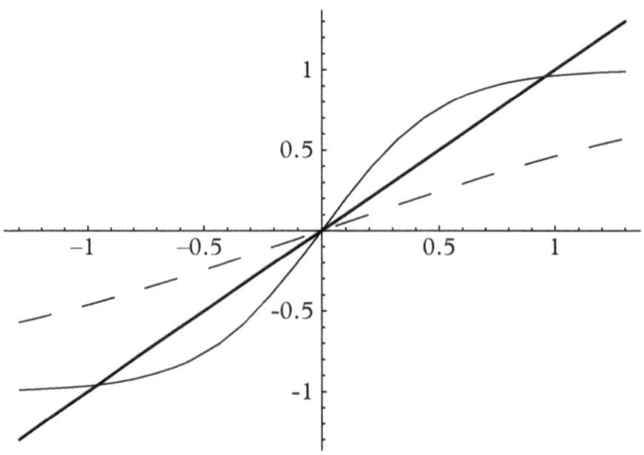

Fig. 3.1 *A graphical solution of eqn. (3.1.8).*

absolute value) $\pm m_0$. Hence, for $\mathcal{J}z = 1$, i.e. when the temperature becomes equal to

$$T_c = \frac{\mathcal{J}z}{k}, \tag{3.1.9}$$

the system undergoes through a phase transition, from a disordered to an ordered phase. For $T < T_c$ the spins are oriented along the direction of the external magnetic field before it will be set to zero: hence, we have $m = m_0$ if $B \to 0^+$ while $m = -m_0$ in the other limit $B \to 0^-$. If the mean value of the magnetization is different from zero, the Z_2 symmetry of the model is spontaneously broken: in the ordered phase, the only effect left from the original Z_2 symmetry of the Hamiltonian is the mapping between the two different solutions, i.e. $m_0 \to -m_0$ and vice versa.

Let us now compute, in the mean field approximation, the critical exponents. Using the variable

$$t = \frac{T - T_c}{T_c},$$

with T_c given in (3.1.9), eqn. (3.1.7) can be written more conveniently as

$$m = -\frac{B}{kT_c} + (1+t)\,\mathrm{arcth}\, m. \tag{3.1.10}$$

$B = 0$. Close to the critical point $t \approx 0$ and the spontaneous magnetization is also small. By expanding the right-hand side of the equation above, we have

$$m_0 = (1+t)\,\mathrm{arcth}\, m_0 = (1+t)\left[m_0 + \frac{1}{3}m_0^3 + \frac{1}{5}m_0^5 + \cdots\right]$$

Solving with respect to m_0, we obtain

$$m_0 = (-3t)^{\frac{1}{2}}\{1 + \mathcal{O}(t)\}. \tag{3.1.11}$$

Hence, the critical exponent β is given by

$$\beta = \frac{1}{2}. \tag{3.1.12}$$

In order to compute the magnetic susceptibility, we derive both terms of (3.1.10) with respect to B. Since $\chi = \frac{\partial m_0}{\partial B}$, we have

$$\chi = -\frac{1}{kT_c} + (1+t)\left(\frac{1}{1+m^2}\right)\chi.$$

For $B = 0$ and $T > T_c$, we have $m_0 = 0$ and therefore χ satisfies the equation

$$\chi = -\frac{1}{kT_c} + (1+t)\chi .$$

Hence

$$\chi = \frac{1}{kT_c} t^{-1}.$$

In the same way, at $B = 0$ but $T < T_c$, we obtain

$$\chi = \frac{1}{2kT_c} (-t)^{-1}.$$

Hence, for the critical exponent γ we get the value

$$\gamma = 1, \tag{3.1.13}$$

With the above computation we can also determine the universal ratio

$$\frac{\chi_+}{\chi_-} = 2.$$

To obtain the exponent δ, consider the equation of state (3.1.10) at $t = 0$. By using the series expansion of the hyperbolic function and simplifying the result, we have

$$\frac{B}{kT_c} \simeq \frac{1}{3} m^3 + \mathcal{O} m^5,$$

i.e. $m \simeq B^{1/3}$ and therefore

$$\delta = 3. \tag{3.1.14}$$

Finally, to obtain the exponent α it is convenient to consider the free energy (3.1.6) in the vicinity of the critical point. Using eqn. (3.1.7) and the identity

$$\cosh x = \frac{1}{(1 - \tanh^2 x)^{1/2}},$$

the free energy can be equivalently expressed as

$$F^{cm}(T, B) = \frac{1}{2} \mathcal{J} z m^2 - \frac{1}{2\beta} \ln \left[\frac{4}{(1 - m_0^2)} \right]. \tag{3.1.15}$$

Let us take $B = 0$. For $T > T_c$, we have $m_0 = 0$ and the free energy is simply equal to

$$F^{cm}(T,0) = -\frac{1}{\beta}\ln 2.$$

For $T < T_c$, $m_0 \neq 0$ and by series expanding (3.1.15) we have

$$F^{cm}(T,0) = -\frac{1}{\beta}\ln 2 - \frac{1}{2\beta}m_0^2(1 - \mathcal{J}z) + \cdots$$

Using (3.1.11), for t sufficiently small and negative, the free energy is given by

$$F^{cm}(T,0) \simeq -\frac{1}{\beta}\ln 2 - \frac{3}{4}t^2 + \cdots$$

Since $F \simeq t^{2-\alpha}$, for the critical exponent α we get

$$\alpha = 0. \tag{3.1.16}$$

Note that in the mean field approximation both the free energy and the mean value of the internal energy do not have a discontinuity at $T = T_c$, while the specific heat has a jump. Since each spin interacts with all the others, the spin–spin correlation function does not depend on their separation, so that $\eta = 0$. The last critical exponent ν can be extracted by the scaling laws and its value is $\nu = 1/2$.

In summary, the mean field approximation is efficient in showing the existence of a phase transition in the Ising model and in predicting its qualitative features. However, there are many aspects that are unsatisfactory from a quantitative point of view. For instance, it predicts the occurrence of a phase transition even for the case $d = 1$, that is excluded by the exact analysis of Chapter 2. Moreover, even when there is a phase transition, as in $d = 2$ or $d = 3$, the mean field theory gives an estimate of the critical temperature that is higher than its actual value and the critical exponents differ from their known values in the both cases, as shown in Table 3.1.

Exponents	Mean field	Ising $d = 1$	Ising $d = 2$	Ising $d = 3$
α	0	1	0	$0.119 \pm .006$
β	1/2	0	1/8	$0.326 \pm .004$
γ	1	1	7/4	$1.239 \pm .003$
δ	3	∞	15	$4.80 \pm .05$
ν	1/2	1	1	$0.627 \pm .002$
η	0	1	1/4	$0.024 \pm .007$

Table 3.1 *Critical exponents of the Ising model for various lattice dimensions.*

The universality of the results obtained in this approximation is due to the absence of spin fluctuations: once we substitute the dynamical magnetization of the spins with its thermal average, the long range correlation among all spins suppresses in fact their fluctuations with respect to their mean value. This long range order favors the energy contribution in the free energy but does not take into the proper account the entropy contribution: for this reason, one obtains a value of the critical temperature T_c higher than the actual one.

3.2 Mean Field Theory of the Potts Model

The mean field approximation for the q-state Potts model shows a novel aspect with respect to the Ising model: a second-order phase transition for $q \leq 2$ but a first-order phase transition for $q > 2$.

In the mean field theory, each of the N spins of the lattice interacts with all the remaining $(N-1)$ ones. In this approximation the Hamiltonian can be written as

$$\mathcal{H}_{mf} = -\frac{1}{N}\mathcal{J}z\sum_{i<j}\delta(\sigma_i,\sigma_j), \tag{3.2.1}$$

where z is the coordination number of the lattice and we have introduced, for convenience, a factor $1/N$ in the coupling constant. To solve the model, let us firstly write the free energy of the system

$$F[\mathcal{C}] = U[\mathcal{C}] - TS[\mathcal{C}] \tag{3.2.2}$$

as a function of the configurations $\mathcal{C}$ of the spins and then let us proceed to determine its minimum. Such a computation can be simplified by noticing that the energy is a highly degenerate function of the system configurations. Consequently, it is more convenient to express the Hamiltonian in terms of a proper set of variables that makes explicit such a degeneracy. Given a configuration $\mathcal{C}$ of the spins, let us denote by $x_i = N_i/N$ the fraction of the spins that are in the ith state, with $i = 1, 2, \ldots, q$. Obviously

$$\sum_{i=1}^{q} x_i = 1. \tag{3.2.3}$$

Since there are $\frac{1}{2}N_i(N_i-1)$ couplings of the type i in the Hamiltonian (3.2.1), the energy $U[\mathcal{C}]$ of this configuration is given by

$$U[\mathcal{C}] = -\frac{1}{2N}\mathcal{J}z\sum_{i=1}^{q} N_i(N_i-1).$$

Dividing by the number of spins and taking the limit $N \to \infty$, we have

$$\frac{U[C]}{N} \simeq -\frac{1}{2}\mathcal{J}z\sum_{i=0}^{q} x_i^2. \tag{3.2.4}$$

Since there are

$$\frac{N!}{N_1!N_2!\ldots N_q!}$$

different ways of organizing the spins without changing the energy, there is an entropy equal to

$$S[C] = k\log\left(\frac{N!}{N_1!N_2!\ldots N_q!}\right).$$

By using the Stirling approximation for each of these terms $\log z! \simeq z\log z$ (with $z \gg 1$) together with the definition of the x_i's, we have

$$\frac{S[C]}{N} \simeq -k\sum_{i=1}^{q} x_i \log x_i. \tag{3.2.5}$$

Hence, the free energy per spin reads

$$F(x_i) = -\sum_{i=1}^{q}\left[\frac{\mathcal{J}z}{2}x_i^2 - kx_i\log x_i\right].$$

This expression has to be minimized but taking into account the normalization condition (3.2.3). The latter constraint can be identically satisfied by using the parameterization

$$\begin{aligned} x_1 &= \frac{1}{q}\left[1+(q-1)s\right], \\ x_i &= \frac{1}{q}(1-s), \; i=2,3,\ldots,q, \end{aligned}$$

with $0 \leq s \leq 1$. In the ferromagnetic case ($\mathcal{J} > 0$) this position takes into account the possible symmetry breaking of the permutation group $\mathcal{S}_q$ in the low temperature phase. Substituting the expressions of x_i in (3.2.4) and (3.2.5), we have

$$\begin{aligned} \frac{\beta}{N}\left[F(s)-F(0)\right] &= \frac{q-1}{2q}\mathcal{J}zs^2 - \frac{1+(q-1)s}{q}\log\left[1+(q-1)s\right] - \frac{q-1}{q}(1-s)\log(1-s) \\ &\simeq -\frac{q-1}{2q}(q-\mathcal{J}z)s^2 + \frac{1}{6}(q-1)(q-2)s^3 + \cdots \end{aligned} \tag{3.2.6}$$

where $\mathcal{J} = \beta J$. Expanding this function in powers of s, we see that for $q = 2$ the cubic term changes its sign: it is negative for $q < 2$ while positive for $q > 2$. This means that there could be a first-order phase transition. Let us consider the two cases separately:

- $q < 2$. The minimum condition for the function in (3.2.6) is expressed by the equation

$$\mathcal{J} z s = \log\left[\frac{1 + (q-1)s}{1-s}\right], \qquad (3.2.7)$$

that has always $s = 0$ as solution. For $\mathcal{J} z > q$ (where q is the derivative of the right-hand side at $s = 0$), there is however another solution $s \neq 0$, as it can be easily seen graphically by plotting both terms of (3.2.7) as done in Figure 3.2. The two solutions coincide when

$$\mathcal{J} = \mathcal{J}_c = \frac{q}{z}.$$

This condition identifies the critical value of the second-order phase transition, that occurs for $q \leq 2$. Note that, for $q = 2$, we recover the critical temperature of the Ising model in the mean field approximation,[2] given by eqn. (3.1.9).

- $q > 2$. In this case we have a different situation: varying $\mathcal{J}$, there is a critical value at which the minimum of the free energy jumps from $s = 0$ to $s = s_c$, as shown in Figure 3.4. This discontinuity is the fingerprint of a first-order phase transition. In this case the critical values $\mathcal{J}_c$ and s_c are obtained by simultaneously solving the equations $F'(s) = 0$ and $F(s) = F(0)$, i.e.

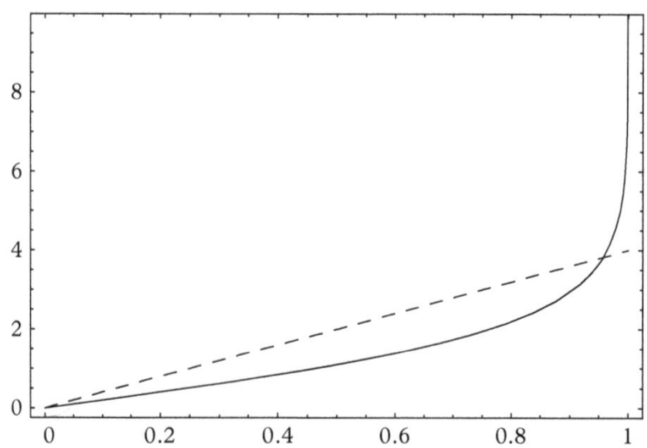

Fig. 3.2 *Graphical analysis of eqn. (3.2.7).*

[2] To obtain the Ising model we have to make the substitution $\mathcal{J} \to 2\mathcal{J}$.

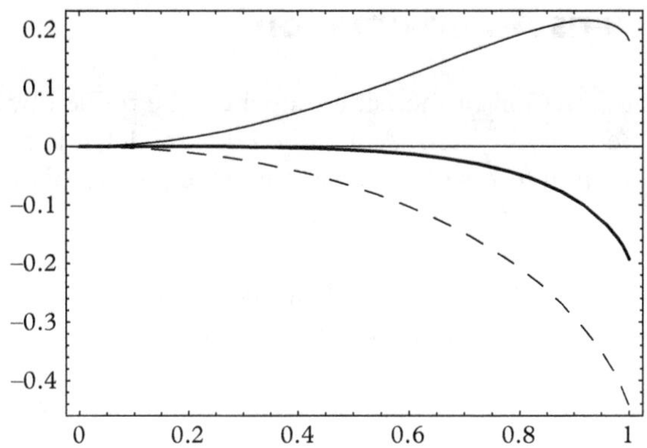

Fig. 3.3 *Plot of the free energy for $q < 2$: $\mathcal{J} > \mathcal{J}_c$ (upper curve), $\mathcal{J} = \mathcal{J}_c$ (black curve) and $\mathcal{J} < \mathcal{J}_c$ (lower curve).*

$$
\begin{aligned}
z\,\mathcal{J}_c &= \frac{2(q-1)}{q-2}\log(q-1), \\
s_c &= \frac{q-2}{q-1}.
\end{aligned}
$$

Computing the internal energy of the system, given by

$$
U = -\mathcal{J}z\frac{q-1}{2q}\,s_{min}^2,
$$

we see that at $\mathcal{J} = \mathcal{J}_c$ this function has a jump that corresponds to a latent heat $\mathcal{L}$ per unit of spin equals to

$$
\mathcal{L} = \mathcal{J}z\,\frac{(q-2)^2}{2q(q-1)}.
$$

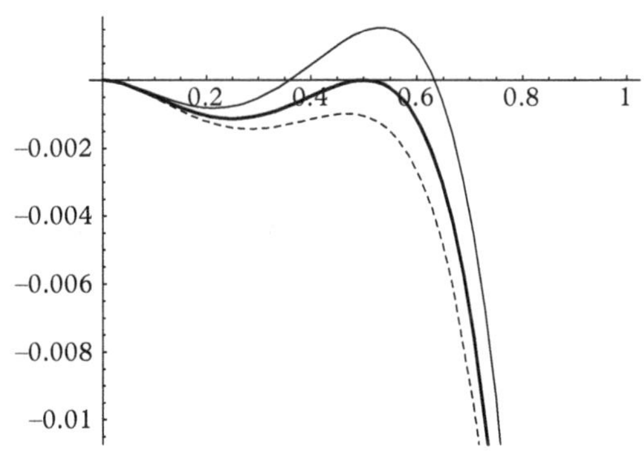

Fig. 3.4 *Plot of the free energy for $q > 2$: $\mathcal{J} > \mathcal{J}_c$ (upper curve), $\mathcal{J} = \mathcal{J}_c$ (black curve) and $\mathcal{J} > \mathcal{J}_c$ (lower curve).*

3.3 Bethe–Peierls Approximation

The mean field approximation of the Ising model can be refined by adopting a formulation proposed by Bethe and Peierls. As for the Potts model, it is convenient to initially express the Hamiltonian in terms of variables that take into account its degeneracy. For a given configuration of the spins, let us define

$$\begin{aligned} N_+ &= \text{total number of spins with value } +1 \\ N_- &= \text{total number of spins with value } -1. \end{aligned}$$

Each couple of nearest-neighbour spins can only be one of the following types: $(++)$, $(--)$, or $(+-)$. Denote by N_{++}, N_{--} and N_{+-} the total number of these pairs. These quantities are not independent: beside the obvious relationship

$$N_+ + N_- = N,$$

we also have

$$\begin{aligned} zN_+ &= 2N_{++} + N_{+-}; \\ zN_- &= 2N_{--} + N_{+-}, \end{aligned} \tag{3.3.1}$$

where z is the coordination number of the lattice. These identities can be proved as follows: once a site where the spin with value $+1$ is selected, draw a line that links this site to all the nearest-neighbour ones, so that there are z lines. Repeating the same procedure for all those sites where the spins have value 1, we have then zN_+ lines. However, the pairs of next-neighbour spins of the type $(++)$ will have *two* lines while those of the type $(+-)$ *only one*, so that we reach the first formula in (3.3.1). Repeating the same argument for the spins with value -1, one obtains the second relationship. Eliminating N_{+-}, N_{--} and N_- from the previous equations we have

$$\begin{aligned} N_{+-} &= zN_+ - 2N_{++}; \\ N_- &= N - N_+; \\ N_{--} &= \frac{z}{2}N + N_{++} - zN_+. \end{aligned}$$

Since

$$\sum_{\langle ij \rangle} \sigma_i \sigma_j = N_{++} + N_{--} - N_{+-} = 4N_{++} - 2zN_+ + \frac{z}{2}N,$$

$$\sum_i \sigma_i = N_+ - N_-,$$

the Hamiltonian of the model can be expressed as

$$\mathcal{H} = -\mathcal{J}\sum_{\langle ij\rangle}\sigma_i\sigma_j - B\sum_i \sigma_i = \qquad (3.3.2)$$

$$= -4\mathcal{J}N_{++} + 2(\mathcal{J}z - B)N_{+} - \left(\frac{1}{2}\mathcal{J}z - B\right)N.$$

The energy of the system depends only on the two quantities N_{++} e N_{+} (the total number of the spins N is considered fixed) and therefore it is a degenerate function of the spin configurations. It is convenient to define an order parameter L (relative to the large distance properties of the system) and an order parameter c (relative to its short distances)

$$\frac{N_{+}}{N} \equiv \frac{1}{2}(L+1) \quad (-1 \leq L \leq +1) \qquad (3.3.3)$$

$$\frac{N_{++}}{\frac{1}{2}zN} \equiv \frac{1}{2}(c+1) \quad (-1 \leq c \leq 1). \qquad (3.3.4)$$

In terms of these order parameters we have

$$\sum_{\langle ij\rangle}\sigma_i\sigma_j = \frac{1}{2}zN(2c - 2L + 1),$$

$$\sum_{i=1}^{N}\sigma_i = NL,$$

and the energy per unit of spin can be written as

$$\frac{1}{N}E(L,c) = -\frac{1}{2}\mathcal{J}z(2c - 2L + 1) - BL. \qquad (3.3.5)$$

After these general considerations, let us discuss the Bethe–Peierls method, focusing on the case $B = 0$. Consider an elementary cell of the lattice, i.e. a site where the spin is in a state s together with its z neighbour sites. Denote by $P(s,n)$ the probability that n of these spins are in the state $+1$. If $s = +1$, then $P(s,n)$ is also equal to the probability to have n pairs $(++)$ and $(z-n)$ pairs $(+-)$. Vice versa, if $s = -1$, $P(s,n)$ is the probability to have n pairs $(+-)$ and $(n-z)$ of the type $(--)$. Given n, there are $\binom{z}{n} = z!/(n!\,(z-n)!)$ ways of selecting n among the z next-neighbour spins. Let us assume that these probabilities can be written as

$$P(+1,n) = \frac{1}{q}\binom{z}{n} e^{\mathcal{J}(2n-z)}\rho^n; \qquad (3.3.6)$$

$$P(-1,n) = \frac{1}{q}\binom{z}{n} e^{\mathcal{J}(z-2n)}\rho^n, \qquad (3.3.7)$$

where q is a normalization factor while ρ is a quantity that takes into account the overall effects of the lattice. While ρ will be determined later, q is obtained by imposing the normalization of the total probability

$$\sum_{n=0}^{z}[P(+1,n)+P(-1,n)] = 1,$$

namely

$$\begin{aligned} q &= \sum_{n=0}^{z}\left[\left(\rho e^{2\mathcal{J}}\right)^n e^{-\mathcal{J}z} + \left(\rho e^{-2\mathcal{J}}\right)^n e^{\mathcal{J}z}\right] = \\ &= \left(e^{\mathcal{J}} + \rho e^{-\mathcal{J}}\right)^z + \left(\rho e^{\mathcal{J}} + e^{-\mathcal{J}}\right)^z. \end{aligned}$$

Using the order parameters L and c defined by eqns. (3.3.3) and (3.3.4), and employing $P(+1,n)$, we have

$$\frac{N_+}{N} = \frac{1}{2}(L+1) = \sum_{n=0}^{z} P(+1,n) = \frac{1}{q}\left(e^{\mathcal{J}} + \rho e^{-\mathcal{J}}\right)^z, \tag{3.3.8}$$

$$\frac{N_{++}}{\frac{1}{2}zN} = \frac{1}{2}(c+1) = \frac{1}{z}\sum_{n=0}^{z} nP(+1,n) = \frac{\rho}{q}e^{\mathcal{J}}\left(e^{-\mathcal{J}} + \rho e^{\mathcal{J}}\right)^{z-1}. \tag{3.3.9}$$

We can now proceed to directly compute the magnetization. Note that

$$\sum_{n=0}^{z} P(+1,n) = \begin{array}{l}\text{probability to find a spin}\\ \textit{with value}{+}1 \text{ in the centre}\end{array}$$

$$\frac{1}{z}\sum_{n=0}^{z} n\,[P(+1,n)+P(-1,n)] = \begin{array}{l}\text{probability to find a spin with value}\\ +1 \text{ among the next-neighbour sites}\end{array}$$

To have a consistent formulation, these two probabilities must be equal. Using (3.3.6) and (3.3.7), we arrive at the equation for the variable ρ

$$\rho = \left(\frac{1+\rho e^{2\mathcal{J}}}{\rho + e^{2\mathcal{J}}}\right)^{z-1}. \tag{3.3.10}$$

Assuming we have solved this equation and found the value of ρ, $\langle L\rangle$ and $\langle c\rangle$ can be obtained through eqns. (3.3.8) and (3.3.9)

$$\langle L\rangle = \frac{\rho^x - 1}{\rho^x + 1}, \tag{3.3.11}$$

$$\langle c \rangle = \frac{2\rho^2}{(1+\rho e^{-2\mathcal{J}})(1+\rho^x)} - 1, \tag{3.3.12}$$

where $x \equiv \frac{z}{z-1}$. The internal energy is given by

$$\frac{1}{N}U(T) = -\frac{1}{2}\mathcal{J}z\,(2\langle c \rangle - 2\langle L \rangle + 1), \tag{3.3.13}$$

whereas the spontaneous magnetization is expressed by

$$\frac{1}{N}\left\langle \sum_{i=1}^{N} \sigma_i \right\rangle = \langle L \rangle. \tag{3.3.14}$$

It is now necessary to solve eqn. (3.3.10). Note that this equation has the following properties:

1. $\rho = 1$ is always a solution.
2. If ρ_0 is a solution, then also $1/\rho_0$ is a solution.
3. Interchanging ρ with $1/\rho$ is equivalent to interchange $\langle L \rangle \to -\langle L \rangle$.
4. $\rho = 1$ corresponds to $\langle L \rangle = 0$, while $\rho = \infty$ corresponds to $\langle L \rangle = 1$.

To find the solution of eqn. (3.3.10), it is useful to use a graphical method, similarly to the mean field solution: we plot the right- and left-hand side functions of eqn. (3.3.10) and determine the points of their intersection, as shown in Figure 3.5.

An important quantity is the value of the derivative of the function on the right-hand side, computed at $\rho = 1$

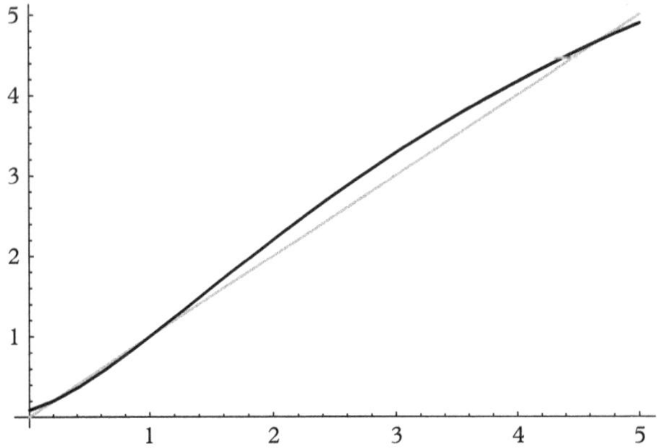

Fig. 3.5 *Graphical solution of eqn. (3.3.10).*

$$g = \frac{(z-1)(e^{4\mathcal{J}}-1)}{(1+e^{2\mathcal{J}})^2}. \tag{3.3.15}$$

In fact, if $g < 1$, the only solution consists of $\rho = 1$. Vice versa, if $g > 1$, there are three solutions, $\rho = 1$, ρ_0 and $1/\rho_0$. Excluding the solution $\rho = 1$ (which corresponds to $\langle L \rangle = 0$) and $1/\rho_0$ (which is equivalent to exchanging the $+1$ spins with -1 spins), the only physical relevant solution is given by ρ_0. In this approach, the critical temperature is given precisely by the condition $g = 1$, namely

$$kT_c = \frac{2\mathcal{J}}{\ln[z/(z-2)]}. \tag{3.3.16}$$

For $T > T_c$ we have

$$\begin{aligned} \rho &= 1; \\ \langle L \rangle &= 0; \\ \langle c \rangle &= \frac{1}{2(1+e^{-2\mathcal{J}})}. \end{aligned} \tag{3.3.17}$$

For $T < T_c$, we have instead $\rho > 1$ and $\langle L \rangle > 0$, i.e. there is a spontaneous magnetization in the system.

The expression (3.3.16) of the critical temperature predicted by the Bethe–Peierls approximation correctly predicts that, in one dimension where $z = 2$, $T_c = 0$. In two dimensions, for a square lattice ($z = 4$), it provides the estimate $kT_c/\mathcal{J} = 2/\ln 2 = 2.885$, which is smaller than obtained in the mean field approximation $kT_c/\mathcal{J} = 4$, but still higher than the exact value $kT_c/\mathcal{J} = 2/\ln(1+\sqrt{2}) = 2.269$, which we will determine in Chapter 4.

3.4 The Gaussian Model

In the Ising model, the computation of the partition function is based on the sums of the discrete variables $\sigma_i = \pm 1$. Notive that such a discrete sum can be written as an integral on the entire real axis by using the Dirac delta function[3]

$$\sum_{\sigma_i = \pm 1} [\cdots] = \int_{-\infty}^{+\infty} d\sigma_i \, \delta(\sigma_i^2 - 1) \, [\cdots].$$

[3] The Dirac delta function $\delta(x)$, with x real, satisfies the properties:

$$\delta(x) = \begin{cases} 0 & x \neq 0 \\ +\infty & x = 0 \end{cases}$$

with $\int_{-\infty}^{+\infty} \delta(x) = 1$. Moreover, $\delta[f(x)] = \sum_i \frac{1}{|f'(x_i)|} \delta(x - x_i)$, where x_i are the roots of the equation $f(x) = 0$.

Using the properties of $\delta(x)$, the Ising model can then be regarded as a statistical model where the spins assume all continuous values of the real axis but with a probability density given by

$$P_I(\sigma_i) = \frac{1}{2}\,[\delta(\sigma_i - 1) + \delta(\sigma_i + 1)]. \tag{3.4.1}$$

With the above notation, the sum on the configurations of a single spin assumes the form

$$\sum_{\sigma_i=\pm 1} [\cdots] = \int_{-\infty}^{+\infty} d\sigma_i\, P_I(\sigma_i)\, [\cdots],$$

and the usual mean values of the Ising model are given by

$$\langle \sigma_i \rangle \equiv \int_{-\infty}^{+\infty} d\sigma_i\, P_I(\sigma_i)\, \sigma_i = 0, \tag{3.4.2}$$

$$\langle \sigma_i^2 \rangle \equiv \int_{-\infty}^{+\infty} d\sigma_i\, P_I(\sigma_i)\, \sigma_i^2 = 1.$$

We can now approximate the Ising model by substituting the probability density $P_I(\sigma_i)$—given by eqn. (3.4.1)—with another probability density $P(\sigma_i)$ that shares the mean values $\langle \sigma_i \rangle$ and $\langle \sigma_i^2 \rangle$ of (3.4.2). A function with such a property is, for instance, the Gaussian curve[4]

$$P_G(\sigma) = \sqrt{\frac{1}{2\pi}}\, \exp\left[-\frac{\sigma^2}{2}\right]. \tag{3.4.3}$$

The spin model defined by this new probability density is known as the *Gaussian model.*

Since thermal averages are computed according to the formula

$$\langle A \rangle = \frac{1}{Z} \int_{-\infty}^{+\infty} \cdots \int_{-\infty}^{+\infty} \prod_{i=1}^{N} P(\sigma_i)\, A\, e^{-\beta\mathcal{H}}\, d\sigma_1 \cdots d\sigma_N,$$

where

$$Z = \int_{-\infty}^{+\infty} \cdots \int_{-\infty}^{+\infty} \prod_{i=1}^{N} P(\sigma_i)\, e^{-\beta\mathcal{H}}\, d\sigma_1 \cdots d\sigma_N,$$

[4] Although $P_I(\sigma)$ and $P_G(\sigma)$ give rise to the same mean values of eqn. (3.4.2), they nevertheless differ from the mean values of the higher powers of the spins. For $P_I(\sigma)$ we have $\langle \sigma^{2n} \rangle = 1$, while for $P_G(\sigma)$, $\langle \sigma^{2n} \rangle = [1 \cdot 3 \cdot 5 \cdots (2n-1)]$.

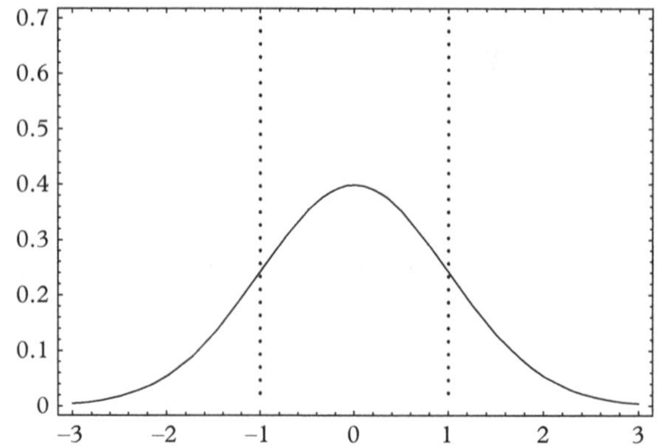

Fig. 3.6 *Probability density $P(\sigma)$: From the Ising model to the Gaussian model.*

it is obvious that the presence of $P(\sigma)$ in the sum over the states can be equivalently interpreted as a new term in the Hamiltonian, so that

$$\mathcal{H} \longrightarrow \mathcal{H}' = \mathcal{H} - \frac{1}{\beta}\sum_{i=1}^{N} \log[P(\sigma_i)].$$

The thermal averages computed with the new Boltzmann's factor

$$\langle A \rangle = \frac{1}{Z}\int_{-\infty}^{+\infty} \cdots \int_{-\infty}^{+\infty} A\, e^{-\beta\mathcal{H}'}\, d\sigma_1 \cdots d\sigma_N,$$

clearly coincide with the previous ones. Hence, we can reformulate the Gaussian model as a system where the spins assume a continuous set of values, with an interaction given, up to a constant, by the Hamiltonian

$$\mathcal{H} = \frac{1}{2\beta}\sum_{i=1}^{N}\sigma_i^2 - \mathfrak{J}\sum_{\langle ij\rangle}\sigma_i\sigma_j - B\sum_{i=1}^{N}\sigma_i. \tag{3.4.4}$$

Let us now proceed to the computation of its partition function

$$Z_N = \int_{-\infty}^{+\infty} \cdots \int_{-\infty}^{+\infty} \exp\left[-\frac{1}{2}\sum_{i=1}^{N}\sigma_i^2 + \mathcal{J}\sum_{\langle kl\rangle}\sigma_k\sigma_l + \mathcal{B}\sum_{l}\sigma_l\right] d\sigma_1 \cdots d\sigma_N,$$

where $\mathcal{J} = \mathfrak{J}/kT$ and $\mathcal{B} = B/kT$. To simplify the formulae, it is convenient to introduce a matrix notation: let σ be a N-component vector $\sigma = (\sigma_1, \sigma_2 \ldots \sigma_N)$, and let $\mathbf{V}$ be a $N \times N$ matrix defined by

$$\sigma^T \mathbf{V} \sigma = \frac{1}{2} \sum_l \sigma_l^2 - \mathcal{J} \sum_{\langle kl \rangle} \sigma_k \sigma_l.$$

Moreover, let $\mathcal{B}$ be a N dimensional vector, with all its components equal to $\mathcal{B}$. In terms of these new notations, the partition function is written as

$$Z_N = \int_{-\infty}^{+\infty} \cdots \int_{-\infty}^{+\infty} \exp\left[-\sigma^T \mathbf{V} \sigma + \mathcal{B}^T \sigma\right] d\sigma_1 \cdots d\sigma_N.$$

The integral on the variables σ_j is Gaussian and can be performed using the formula[5]

$$\int_{-\infty}^{+\infty} \cdots \int_{-\infty}^{+\infty} \exp\left[-x^T \mathbf{V} x + h^T x\right] dx_1 \cdots dx_N = (\pi)^{N/2} \left[\det \mathbf{V}\right]^{-\frac{1}{2}} \exp\left[\frac{1}{4} h^T \mathbf{V}^{-1} h\right] \tag{3.4.5}$$

so that we arrive at

$$Z_n = \pi^{\frac{N}{2}} \left[\det \mathbf{V}\right]^{-\frac{1}{2}} \exp\left[\frac{1}{4} \mathcal{B}^{\mathbf{T}} \mathbf{V}^{-1} \mathcal{B}\right].$$

Cyclic matrix. It is necessary, however, to verify if there are eigenvalues of the matrix V with a real part that is either zero or negative. Their explicit expression clearly depends on the nature of the matrix V, namely from the underlying lattice structure of the Gaussian model. Consider, for simplicity, a d-dimensional cubic lattice of length L in all its direction, with periodic boundary conditions. In this case, $N = L^d$. For the (discrete) translation invariance of the lattice, the matrix elements of V depend only on the difference of its indices

$$V_{\vec{i},\vec{j}} = V(\vec{i} - \vec{j}).$$

For a cubic lattice, the only components that are different from zero are those for which $\vec{i} - \vec{j} = 0$ and

$$\vec{i} - \vec{j} = \begin{cases} (\pm 1, 0, 0, 0, \ldots) \\ (0, \pm 1, 0, 0, \ldots) \\ (0, 0, \pm 1, 0, \ldots) \\ (0, 0, 0, \pm 1, \ldots) \\ \ldots \end{cases}$$

[5] The validity of this formula relies on the condition that all the eigenvalues of V are positive. As we will see, when this condition is not satisfied, the system undergoes a phase transition.

The periodic boundary conditions put the additional constraints

$$V(i+L,j) = V(i,j+L) = V(i,j).$$

A matrix that satisfies all the above properties is called *cyclic matrix* and its eigenvalues can be easily determined by using the Fourier series. The result is

$$\lambda(\omega_1,\ldots,\omega_d) = \frac{1}{2} - \mathcal{J}(\cos\omega_1 + \ldots + \cos\omega_d), \tag{3.4.6}$$

where each frequency ω_j can take one of the L possible values $0, 2\pi/L, 4\pi/L, \ldots, 2\pi(L-1)/L$. From eqn. (3.4.6) it follows that the eigenvalues have a positive real part only if

$$|\mathcal{J}| < \frac{1}{2d}. \tag{3.4.7}$$

This condition determines the range of validity of the Gaussian model and identifies the critical temperature of the model, given by

$$\frac{\mathcal{J}}{kT_c} = \frac{1}{2d}. \tag{3.4.8}$$

It is easy to understand the origin of this critical point by directly analysing the Hamiltonian (3.4.4). Note that the coupling constant of the first term explicitly depends on the temperature through the parameter $1/\beta$. At high temperatures, i.e. $\beta \to 0$, the minimum of the Hamiltonian is reached by the configuration in which all spins have a zero value, $\sigma_i = 0$. In the opposite limit, $\beta \to \infty$, i.e. in the low-temperature phase, the second term may prevail on the first one. In such a case, the Hamiltonian density per unit of spin $\mathcal{H}/N$ can be made arbitrarily negative by allowing the spins to align with each other and grow without any bound on their module. Hence, in the low-temperature phase, the energy of the model is not bounded from below. The two physical pictures, so qualitatively different, obtained in the high- and low-temperature phases clearly signal the existence of a phase transition, with the critical temperature given in (3.4.8). The low-temperature phase of model is however pathological: the model is only defined for $T > T_c$ and therefore it only has a high-temperature phase. The next section explains how to get around this difficulty by posing a bound on the higher values of the spins.

Transfer matrix in 1-D. The analysis done is completely general and applies to arbitrary lattices. However, it is an interesting exercise to solve the one dimensional Gaussian model by using the transfer matrix. To this purpose, consider the Hamiltonian of the one-dimensional Gaussian model

$$\mathcal{H} = \frac{1}{2\beta}\sum_{i=1}^{N}\sigma_i^2 - \mathcal{J}\sum_{i=1}^{N}\sigma_i\sigma_{i+1}.$$

The transfer matrix T of the model has a set of continuous indices: denoting by x and y the values of the spin of two neighbour sites, we have

$$\langle x|T|y\rangle = T(x,y) = \exp\left[-\frac{1}{4}(x^2+y^2)+\mathcal{J}xy\right].$$

To compute the partition function, we have to diagonalize this matrix by solving the integral equation

$$\int_{-\infty}^{+\infty} T(x,y)\,\psi(y)\,dy = \lambda\,\psi(x). \tag{3.4.9}$$

Note that the norm of the integral operator is finite only if

$$|\mathcal{J}| < \frac{1}{2}. \tag{3.4.10}$$

In fact, it is only in this interval that the kernel of the integral operator is a square integrable

$$||T||^2 \equiv \int_{-\infty}^{+\infty}\int_{-\infty}^{+\infty} \langle x|T|y\rangle\,\langle y|T|x\rangle\,dx\,dy = \tag{3.4.11}$$

$$\int_{-\infty}^{+\infty}\int_{-\infty}^{+\infty} \exp\left[-\frac{1}{2}(x^2+y^2)+2\mathcal{J}xy\right]dx\,dy = \frac{2\pi}{\sqrt{1-4\mathcal{J}^2}}.$$

The integral operator is symmetric in its indices $T(x,y) = T(y,x)$ and, for its reality, it is then a Hermitian operator. The general theory of the integral equations[a] tells us that the eigenvalues λ_n of such an operator are real and discrete, while the corresponding eigenfuctions ψ_n form a complete set of orthonormal functions. The operator T admits the spectral decomposition

$$T(x,y) = \sum_{n=0}^{\infty} \lambda_n\,\psi_n(x)\psi_n(y), \tag{3.4.12}$$

and its eigenvalues satisfy the identity

$$||T||^2 = \sum_{n=0}^{\infty} \lambda_n^2. \tag{3.4.13}$$

To determine the spectrum of the integral equation (3.4.9), we can use a very elegant algebraic method similar to that used to find the spectrum of the quantum one-dimensional harmonic oscillator.[b] Let us introduce the differential operator **A**

$$\mathbf{A} = \frac{1}{\sqrt{2}}\left(ux + u^{-1}\frac{d}{dx}\right), \tag{3.4.14}$$

and its Hermitian conjugate[c] $\mathbf{A}^{\dagger}$

$$\mathbf{A}^{\dagger} = \frac{1}{\sqrt{2}}\left(ux - u^{-1}\frac{d}{dx}\right). \tag{3.4.15}$$

These operators satisfy the commutation relation

$$\left[\mathbf{A}, \mathbf{A}^{\dagger}\right] = 1. \tag{3.4.16}$$

Choosing

$$u^2 \equiv u_*^2 = \frac{1}{2}\sqrt{1 - 4\mathcal{J}^2}, \tag{3.4.17}$$

the differential operator $\mathbf{A}$ satisfies the operatorial equation

$$\mathbf{A}\,T = \xi\, T\,\mathbf{A}, \tag{3.4.18}$$

with the constant ξ given by

$$\xi = \frac{1 - 2u_*^2}{2\mathcal{J}} = \frac{1 - \sqrt{1 - 4\mathcal{J}^2}}{2\mathcal{J}}. \tag{3.4.19}$$

To prove (3.4.18), let us initially apply this relation to an arbitrary function $\psi(x)$

$$\int_{-\infty}^{\infty}\left(ux + u^{-1}\frac{d}{dx}\right) T(x,y)\,\psi(y)\,dy = \xi \int_{-\infty}^{\infty} T(x,y)\left(uy + u^{-1}\frac{d}{dy}\right)\psi(y)\,dy.$$

Integrating by parts the last term on the right-hand side, and keeping into mind the arbitrariness of the function $\psi(x)$, we get

$$\left(ux + u^{-1}\frac{d}{dx}\right) T(x,y) = \xi\left(uy - u^{-1}\frac{d}{dy}\right) T(x,y).$$

Equating the terms that are proportional to x and y in both, we obtain

$$\left(u - \frac{1}{2u}\right) = -\xi\,\frac{\mathcal{J}}{u}, \tag{3.4.20}$$

$$\xi\left(u + \frac{1}{2u}\right) = \frac{\mathcal{J}}{u}.$$

Hence, wearrive at eqn. (3.4.19) with u_*, given by (3.4.17), a value that comes from the consistency condition of the systems of equations (3.4.20).

Taking the Hermitian conjugate of (3.4.18), the operator T also satisfies

$$T\mathbf{A}^\dagger = \xi\,\mathbf{A}^\dagger T. \qquad (3.4.21)$$

The two functional equations (3.4.18) and (3.4.21) satisfied by T enable us to reach some important conclusions on its spectrum. Suppose we have identified a real eigenfunction $\psi(x)$ of this operator, with eigenvalue λ. Applying to $\psi(x)$ the operator $\mathbf{A}^\dagger$, we see that $\psi'(x) = \mathbf{A}^\dagger\,\psi(x)$ is also an eigenfunction of T but with the eigenvalue $\lambda' = \xi\,\lambda$. Hence, the iterated application of $\mathbf{A}^\dagger$ to the eigenfunction $\psi(x)$ gives rise to the sequence of eigenfunctions $\left(\mathbf{A}^\dagger\right)^n\,\psi(x)$. Since for $|\mathcal{J}| < 1/2$ we have $\xi < 1$, we obtain the sequence of decreasing eigenvalues

$$\lambda > \lambda\xi > \lambda\xi^2 >, \ldots > \lambda\xi^n. \qquad (3.4.22)$$

Vice versa, making use of (3.4.18), the iterated application of the operator $\mathbf{A}$ to the eigenfunction $\psi(x)$ also generates a sequence of eigenfunctions $\tilde{\psi}_n = \mathbf{A}^n\psi$, but this time with a sequence of *increasing* eigenvalues

$$\lambda < \lambda\xi^{-1} < \lambda\xi^{-2} <, \ldots < \lambda\xi^{-n}. \qquad (3.4.23)$$

Since the T is bounded in the interval (3.4.10), a maximum eigenvalue $\lambda_{\text{max}} \equiv \lambda_0$ must necessarily exist. This implies that the sequence (3.4.23) must stop and the eigenfunction that corresponds to the maximum eigenvalue λ_0 satisfies the equation $\mathbf{A}\,\psi_0(x) = 0$, i.e.

$$\left(u_* x + \frac{1}{u_*}\frac{d}{dx}\right)\psi_0(x) = 0. \qquad (3.4.24)$$

Therefore

$$\psi_0(x) = A_0 \exp\left[-\frac{u_*^2 x^2}{2}\right],$$

where the constant $A_0 = \sqrt{\frac{u_*^2}{\pi}}$ is fixed by the normalization condition

$$\int_{-\infty}^{+\infty} \psi_0^2(x)\,dx = 1.$$

We could directly compute the maximum eigevalue λ_0 by substituting $\psi_0(x)$ in the integral equation. However, in order to control all the results obtained above, it is convenientto proceed in a more general way. Note that the application of the operator T to a generic Gaussian function $g(x) = A\exp[-\lambda^2 x^2/2]$ produces another Gaussian function

$$T g(x) = A \int_{-\infty}^{+\infty} dy \exp\left[-\frac{1}{4}(x^2+y^2)+\mathcal{J}xy\right]\exp\left[-\frac{\lambda^2}{2}y^2\right] =$$
$$= A_0 \int_{-\infty}^{+\infty} dy \exp\left[-\frac{1}{4}x^2 - \frac{1}{4}(1+2\lambda^2)y^2)+\mathcal{J}xy\right] = \tilde{A}\exp\left[-\frac{\tilde{\lambda}^2}{2}x^2\right],$$

with a new exponent $\tilde{\lambda}^2$, given by

$$\tilde{\lambda}^2 = \frac{1}{2} - \frac{\mathcal{J}^2}{\lambda^2+1/2},$$

and a new normalization constant

$$\tilde{A}_0 = A_0\sqrt{\frac{2\pi}{\lambda^2+1/2}}. \tag{3.4.25}$$

If the Gaussian $g(x)$ is an eigenfunction of T, $\tilde{\lambda}^2$ should be obviously equal to the previous one λ^2 and we get the equation

$$\lambda^2 = \frac{1}{2} - \frac{\mathcal{J}^2}{\lambda^2+1/2}. \tag{3.4.26}$$

The solution is given by

$$\lambda^2 = u_*^2 = \frac{1}{2}\sqrt{1-4\mathcal{J}^2},$$

and coincides with the condition (3.4.17), previously obtained for the eigenfunction $\psi_0(x)$.

Substituting in (3.4.25) the value of u_*^2, we obtain the maximum eigenvalue

$$\lambda_0 = \sqrt{\frac{2\pi}{u_*^2+1/2}} = \sqrt{\frac{4\pi}{1+\sqrt{1-4\mathcal{J}^2}}}. \tag{3.4.27}$$

The sequence of eigenvalues is now given by eqn. (3.4.22), with $\lambda = \lambda_0$. In particular, it is easy to check the validity of the identity (3.4.13): in fact, for the right-hand side of this equation we have

$$\sum_{k=0}^{\infty} \lambda_k^2 = \lambda_0^2 \sum_{k=0}^{\infty} \xi^{2k} = \frac{\lambda_0^2}{1-\xi^2},$$

and, substituting the two expressions (3.4.27) and (3.4.19), we precisely obtain the norm of the operator T, expressed by eqn. (3.4.11).

Once the maximum eigenvalue is known, the free energy per unit spin of the model is given by

$$\beta F = -\lim_{N\to\infty} \frac{1}{N} \log Z_N = -\log \lambda_0(\mathcal{J}). \tag{3.4.28}$$

Note that when the temperature tends to its critical value

$$\mathcal{J}_c \to \frac{1}{2},$$

correspondingly $\xi \to 1$. This implies a collapse of all eigenvalues of the transfer matrix. Since the transfer matrix of a classical statistical system can be associated to a Hamiltonian $\mathbf{H}$ of a quantum system by means of the formula

$$T \equiv e^{-a\mathbf{H}},$$

the collapse of all eigenvalues of T corresponds to a very singular point of degeneracy of the quantum Hamiltonian $\mathbf{H}$. At $\mathcal{J} = \mathcal{J}_c$ we have a significant mixing of all eigenstates of $\mathbf{H}$, with a drastic and discontinous change of the fundamental state of the system: the system then undergoes a phase transition.

Using the spectral decomposition[d] of the operator T, eqn. (3.4.12), is easy to see that the quantum Hamiltonian $\mathbf{H}$ assumes the form

$$\mathbf{H} = -\frac{1}{a}\left[\left(\log\frac{1-\sqrt{1-4\mathcal{J}^2}}{2\mathcal{J}}\right)\mathbf{A}^\dagger\mathbf{A} + \frac{1}{2}\log\left(\frac{4\pi}{1+\sqrt{1-\mathcal{J}^2}}\right)\right]. \tag{3.4.29}$$

In the limit $\mathcal{J} \to \mathcal{J}_c$, the coefficient in front of $\mathbf{A}^\dagger\mathbf{A}$ vanished and, as expected, there is an infinite degeneration of the eigenvalues of $\mathbf{H}$.

---

[a] See, for instance, D. Porter, D. Stirling, *Integral Equations*, Cambridge University Press, 1990.

[b] See, for instance, C. Cohen–Tannoudji, B. Diu, F. Laloë, *Quantum Mechanics*, J. Wiley & Sons, 2006.

[c] It is important to recall that the differential operator $\frac{d}{dx}$ is anti-Hermitian.

[d] The normalized eigenfunctions $\psi_n(x)$ are given by $\psi_n(x) = \frac{1}{\sqrt{n!}}\left(\mathbf{A}^\dagger\right)^n \psi_0(x)$, and we have $\langle \psi_m \mid \mathbf{A}^\dagger\mathbf{A} \mid \psi_n\rangle = n\delta_{nm}$.

To cure the pathological features of the low-temperature phase of the Gaussian model, Berlin and Kac proposed a more sophisticated version of the model, the so-called *spherical model.* This model has the additional advantage of being more similar to the Ising model than the Gaussian model itself.

3.5 The spherical model

The spherical model, introduced and solved by Berlin and Kac in 1952, consists of an interesting variant of the Ising model, or better, of the Gaussian model. Like the last one, the N spins of the spherical model interact with their first neighbours and an eventual external field, and assume all real values. However, they are subject to the condition

$$\sum_{j=1}^{N} \sigma_j^2 = N. \tag{3.5.1}$$

When there is homogeneity in the spins, this condition is equivalent to $\langle \sigma_i^2 \rangle \simeq 1$, just like in the original Ising model. However, it is obvious that there is a difference between these two models: in fact, while in the Ising model the sum on the spin configurations corresponds to sum on all the vertices of an $N-$ dimensional hypercube, in the spherical model this sum is replaced by an integral on the N-dimensional spherical surface that passes through them.

Besides its intrinsic interest,[6] we could, however, doubt its physical content, inasmuch as the condition (3.5.1) depends on the dimension N of the system. This is in fact equivalent to having an interaction between all the spins. This objection has found, however, a valid answer in the equivalence (shown by Stanley in 1968) between the spherical model and a spin model with $O(n)$ symmetry and nearest-neighbour interactions, in the limit in which $n \to \infty$. Namely, Stanley proved that the model with Hamiltonian

$$\mathcal{H} = -\mathcal{J} \sum_{\langle ij \rangle} \vec{\sigma}_i \cdot \vec{\sigma}_j,$$

where each spin is a n-dimensional vector satisfying

$$| \vec{\sigma}_i |^2 = n$$

in the limit $n \to \infty$ is equivalent to the spherical model.[7]

[6] As we will see below it is exactly solvable, with a different behaviour with respect to the mean field solution for $d \geq 3$, while for $d = 1$ and $d = 2$ it does not have a phase transition.

[7] Note that the model considered by Stanley differs from the one discussed in Section 2.6 since the module of the spin is $n^{1/2}$ instead of 1.

Let us now compute the partition function of the model and its equation of state. Although not particularly demanding, the following calculations require however a certain mathematical skill. The partition function is given by the multi-dimensional integral

$$Z_N = \int_{-\infty}^{+\infty} \cdots \int_{-\infty}^{+\infty} d\sigma_1 \cdots d\sigma_N \delta\left(N - \sum \sigma_j^2\right) \exp\left[\mathcal{J} \sum_{\langle kl \rangle} \sigma_k \sigma_l + \mathcal{B} \sum_l \sigma_l\right], \quad (3.5.2)$$

with $\mathcal{J} = \mathfrak{J}/kT$ and $\mathcal{B} = B/kT$. The constraint (3.5.1) is enforced by the Dirac delta function. Using

$$\delta(x) = \frac{1}{2\pi} \int_{-\infty}^{+\infty} e^{isx}\, ds,$$

and noting that we can insert in the integral the term

$$e^{\mu(N - \sum_l \sigma_l^2)}$$

(that is equal to 1, thanks to eqn. (3.5.1)), the partition function can be rewritten as

$$Z_N = \frac{1}{2\pi} \int_{-\infty}^{+\infty} \cdots \int_{-\infty}^{+\infty} d\sigma_1 \cdots d\sigma_N \int_{-\infty}^{+\infty} ds \quad (3.5.3)$$
$$\exp\left[\mathcal{J} \sum_{\langle kl \rangle} \sigma_k \sigma_l + \mathcal{B} \sum_l \sigma_l + (\mu + is)(N - \sum_l \sigma_l^2)\right].$$

It is convenient to adopt the compact notation of the previous section. Let us define a $N \times N$ matrix V by means of

$$\sigma^T \mathbf{V} \sigma = (\mu + is) \sum_l \sigma_l^2 - \mathcal{J} \sum_{\langle kl \rangle} \sigma_k \sigma_l.$$

Hence

$$Z_N = \frac{1}{2\pi} \int_{-\infty}^{+\infty} \cdots \int_{-\infty}^{+\infty} d\sigma_1 \cdots d\sigma_N \int_{-\infty}^{+\infty} ds \exp\left[-\sigma^T \mathbf{V} \sigma + \mathcal{B}^T \sigma + (\mu + is) N\right]. \quad (3.5.4)$$

We can choose a sufficiently great value of the arbitrary constant μ in such way that all the eigenvalues of the matrix V have a positive real part (we will specify this condition in more detail ahead, see eqn. (3.5.7)). Under these conditions, we can exchange the integration order on the variables σ_j and s: the integration on the variable σ_j is Gaussian and can be carried out thanks to the formula (3.4.5), so that

$$Z_n = \frac{1}{2}\pi^{\frac{N}{2}-1} \int_{-\infty}^{+\infty} ds\, [\det \mathbf{V}]^{-\frac{1}{2}} \exp\left[(\mu + is)N + \frac{1}{4}\mathcal{B}^T \mathbf{V}^{-1} \mathcal{B}\right]. \tag{3.5.5}$$

To proceed further, it is necessary to specify the nature of the matrix V. For simplicity, also in this case we choose a cubic lattice lattice with $N = L^d$ and with periodic conditions along all directions. V is therefore a cyclic matrix and we can repeat the main steps of the analysis of the previous section. The eigenvalues of V are obtained in terms of the Fourier series, with the final result given by

$$\lambda(\omega_1, \ldots, \omega_d) = \mu + is - \mathcal{J}(\cos\omega_1 + \ldots + \cos\omega_d), \tag{3.5.6}$$

where each frequency ω_j assumes the L values $0, 2\pi/L, 4\pi/L, \ldots, 2\pi(L-1)/L$. From (3.5.6) it is easy to see that the real part is positive if the constant μ satisfies

$$\mu > \mathcal{J} d. \tag{3.5.7}$$

Since the determinant of a matrix is given by the product of its eigenvalues, we have

$$[det\, \mathbf{V}] = \exp\left[\ln \det \mathbf{V}\right] = \exp\left[\sum_{\omega_1} \ldots \sum_{\omega_d} \ln \lambda(\omega_1, \ldots, \omega_d)\right].$$

In the thermodynamic limit $L \to \infty$, the eigenvalues become dense and the sum over them can be converted into an integral

$$\ln \det \mathbf{V} = N\left[\ln \mathcal{J} + g(z)\right],$$

where we have defined

$$z = (\mu + is)/\mathcal{J},$$

and

$$g(z) = \frac{1}{(2\pi)^d} \int_0^{2\pi} \ldots \int_0^{2\pi} d\omega_1 \ldots d\omega_d \ln\left[z - \sum_{j=1}^{d} \cos\omega_j\right]. \tag{3.5.8}$$

The function $g(z)$ is analytic when $\mathrm{Re}\, z > d$ and has a singular point at $z = d$.

We can take further advantage of the cyclic nature of the matrix V to show that the constant vector $\mathcal{B}$ is the eigenvector of V corresponding to its minimum eigenvalue $\mu + is - \mathcal{J} d = \mathcal{J}(z - d)$. Hence

$$\mathcal{B}^T \mathbf{V}^{-1} \mathcal{B} = \mathcal{B}^T \frac{1}{\mathcal{J}(z-d)} \mathcal{B} = \frac{N\mathcal{B}^2}{\mathcal{J}(z-d)}.$$

Putting together the last formulae and making a change of variable from s to z, the partition function can be expressed as

$$Z_N = \left(\frac{\mathcal{J}}{2\pi i}\right)\left(\frac{\pi}{\mathcal{J}}\right)^{\frac{N}{2}} \int_{c-i\infty}^{c+i\infty} dz \exp[N\phi(z)], \tag{3.5.9}$$

where the function $\phi(z)$ is defined by

$$\phi(z) = \mathcal{J}z - \frac{1}{2}g(z) + \frac{\mathcal{B}^2}{4\mathcal{J}(z-d)}. \tag{3.5.10}$$

For the condition on the eigenvalues of V, the integration contour γ is chosen as in Figure 3.7, with $c = (\mu - \mathcal{J}d)/\mathcal{J} > 0$. Since $\phi(z)$ is an analytic function in the semi-plane $\text{Re}\, z > d$, the value of the integral in eqn. (3.5.9) does not depend on the value of the constant c, as far as this constant is positive. In the thermodynamic limit $N \to \infty$, Z_N can be estimated by using the saddle point method, discussed in Appendix A of this chapter. Consider the behaviour of $\phi(z)$ when z is real and positive, in the case in which $\mathcal{J} > 0$ and $\mathcal{B} \neq 0$. It is easy to see that the function diverges both for $z \to d$ and $z \to \infty$, assuming positive values in between. Therefore, the function $\phi(z)$ must have a *minimum* at some positive point z_0 and since $\phi''(z) > 0$, this is the only minimum. Let us choose the constant c to be *exactly* equal to z_0. Since $\phi(z)$ is an analytic function, along the direction of the new path of integration γ it will present a *maximum* at $z = z_0$. Such a maximum rules the behaviour of the integral in the limit $N \to \infty$ and therefore the free energy is given by

$$-F/kT = \lim_{N\to\infty} \frac{1}{N} \ln Z_N = \frac{1}{2} \ln\left(\frac{\pi}{\mathcal{J}}\right) + \phi(z_0). \tag{3.5.11}$$

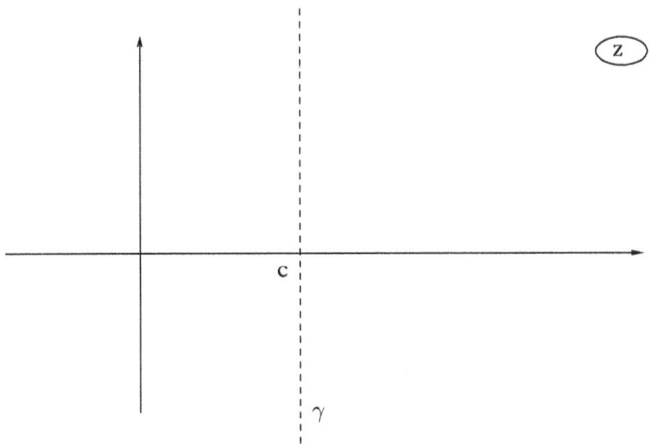

Fig. 3.7 *Contour of integration in the complex plane.*

The value z_0 is determined by the zero of the first derivative of $\phi(z)$ and is solution of the saddle point equation

$$\mathcal{J} - \frac{\mathcal{B}^2}{4\mathcal{J}(z_0 - d)^2} = \frac{1}{2} g'(z_0). \tag{3.5.12}$$

Since there is a unique positive solution of this equation, it allows us to define F as a function of $\mathcal{J}$ and $\mathcal{B}$, for $\mathcal{J} > 0$ and $\mathcal{B} \neq 0$. In Appendix B we show that this equation permits to establish an interesting relation between the spherical model and the brownian motion on a lattice.

Equation of state. The equation of state of the spherical model can be derived as follows. Let us first take a derivative of (3.5.11) with respect to $\mathcal{B}$, keeping $\mathcal{J}$ fixed. Based on (3.5.12) and taking into account that z_0 also depends on $\mathcal{B}$, we have

$$-\frac{d}{d\mathcal{B}}\left(\frac{F}{kT}\right) = \frac{\mathcal{B}}{2\mathcal{J}(z_0 - d)} + \phi'(z_0)\frac{dz_0}{dh}.$$

However, z_0 is exactly the value where the first derivative of $\phi(z)$ vanishes. Using the thermodynamic relation

$$M(B,T) = -\frac{\partial}{\partial H} F(B,T),$$

we have

$$M = \frac{\mathcal{B}}{2\mathcal{J}(z_0 - d)} = \frac{B}{2J(z_0 - d)}.$$

We can now eliminate the variable $(z_0 - d)$ by using the saddle point equation (3.5.12), with the result

$$2J(1 - M^2) = kTg'\left(\frac{B}{2JM}\right).$$

This is the exact equation of state of the spherical model that links the quantities M, B and T.

Let us now discuss in more details the saddle point equation (3.5.12) to see if there is a phase transition in the spherical model. The function $g'(z)$ is expressed by the multi-dimensional integral

$$g'(z) = \frac{1}{(2\pi)^d} \int_0^{2\pi} \cdots \int_0^{2\pi} \frac{1}{z - \sum_{j=1}^{d} \cos \omega_j}\, d\omega_1 \ldots d\omega_d. \tag{3.5.13}$$

Using the identity

$$\frac{1}{a} = \int_0^{\infty} e^{-at}\, dt,$$

and the integral representation (2.A.12) of the Bessel function $I_0(t)$, given in Appendix A of Chapter 2,

$$I_0(t) = \frac{1}{2\pi} \int_0^{2\pi} e^{t\cos\omega}\, d\omega,$$

$g'(z)$ can be expressed in a more convenient form as

$$g'(z) = \int_0^{\infty} e^{-tz}\, [I_0(t)]^d\, dt. \tag{3.5.14}$$

This formula has the advantage of showing the explicit dependence of the dimension d of the lattice, which can be regarded as a continuous variable and not necessarily restricted to integer values.

Let us study the main properties of $g'(z)$. From the asymptotic behaviour of $I_0(t)$,

$$I_0(t) \simeq \frac{e^t}{\sqrt{2\pi t}}, \quad t \to \infty \tag{3.5.15}$$

it follows that the integral (3.5.14) converges when $\mathrm{Re}\, z > d$. Consequently, $g'(z)$ is an analytic function in this semi-plane. For real z, $g'(z)$ is a positive function that monotonically decreases toward its null value when $z \to \infty$. For $z \to d$, using once again (3.5.15), the integral diverges when $d \leq 2$, while it converges when $d > 2$

$$\lim_{z \to d} g'(z) = \begin{cases} \infty, & 0 < d < 2 \\ g'(d) < \infty, & d > 2. \end{cases}$$

This implies that there is a phase transition for $\mathcal{B} = 0$ only when $d > 2$. Consider, in fact, eqn. (3.5.12) when $\mathcal{B} = 0$

$$2\mathcal{J} = g'(z). \tag{3.5.16}$$

If $g'(z)$ diverges for $z \to d$, however we vary the value of $\mathcal{J}$ (i.e. the value of the temperature), there is always a root z_0 of the equation that varies with continuity, as shown by its graphical solution of Figure 3.8.

Vice versa, if $g'(z)$ converges towards the finite value $g'(d)$ when $z \to d$, there is a solution z_0 that varies with continuity as far as $\mathcal{J} < g'(d)$. However, when the function reaches the value $\mathcal{J} = g'(d)$, there is a discontinous change in the nature of the equation. Since the function $g'(z)$ cannot grow more than its limit value $g'(d)$,

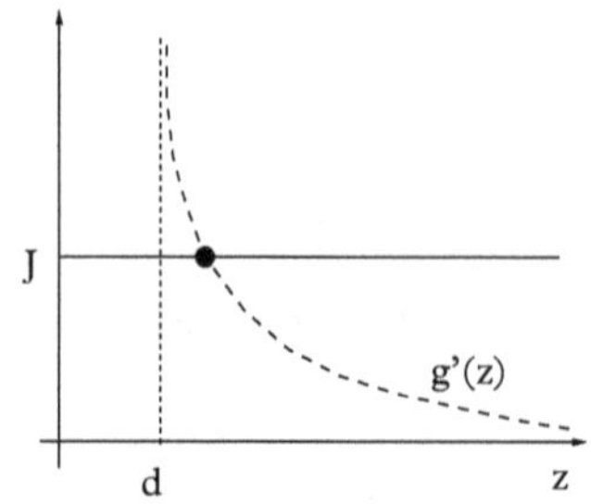

Fig. 3.8 *Graphical solution of the saddle point equation for d < 2.*

further increasing $\mathcal{J}$ the root z_0 of the equation remains fixed at its value $z_0 = d$ (Figure 3.9). The appearance of a spontaneous magnetization below the critical temperature may be regarded qualitatively as a condensation phenomenon akin to Bose–Einstein condensation of integer spins atoms (see Appendix B of Chapter 1). The phase transition point is identified, for $d > 2$, by the condition

$$\mathcal{J}_c = \frac{\mathcal{J}}{kT_c} = \frac{1}{2} g'(d). \tag{3.5.17}$$

From the detailed analysis of the model, as proposed in one of the problems at the end of the chapter, we arrive at the following conclusions: first of all, there is no phase transition for $d \leq 2$, while for $d > 2$ there is a phase transition with the values of the critical exponents as follows: α assumes the value

$$\alpha = \begin{cases} -(4-d)/(d-2), & 2 < d < 4, \\ 0, & d > 4. \end{cases}$$

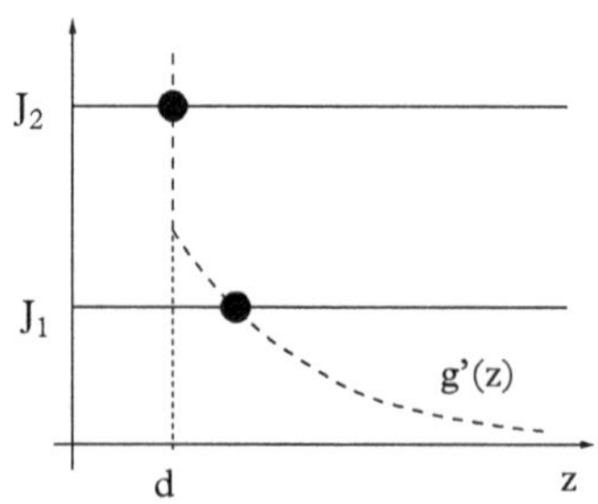

Fig. 3.9 *Graphical solution of the saddle point equation for d > 2. There is a phase transition when* $\mathcal{J} = g'(d)$.

while β is given by

$$\beta = \frac{1}{2},$$

For the critical exponent γ we have

$$\gamma = \begin{cases} 2/(d-2), & 2 < d < 4, \\ 1, & d > 4. \end{cases}$$

Finally, the value of the critical exponent δ is

$$\delta = \begin{cases} (d+2)/(d-2), & 2 < d < 4, \\ 3, & d > 4. \end{cases}$$

Using these results, it is easy to establish the validity of the first two scaling laws (1.1.26). The other two scaling laws allow us to determine the critical exponent ν

$$\nu = \begin{cases} 1/(d-2), & 2 < d < 4, \\ 1/2, & d > 4. \end{cases}$$

and the critical exponent η

$$\eta = 0.$$

In conclusion, the spherical model has the interesting property that its critical exponents vary with the dimensionality d of the lattice in the range $2 < d < 4$, while they assume the values predicted by the mean field theory for $d > 4$. We expect to find the same behaviour in the critical exponents of the Ising model, obviously with a different set of values for the two models.

Appendix 3.A. The Saddle Point Method

In many mathematical situations, we face the problem of estimating the asymptotic behaviour of a function $\mathcal{J}(s)$ when $s \to \infty$. Chapter 2 provided some examples, e.g. the asymptotic behaviour of the $\Gamma(s)$ function or the Bessel functions $I_\nu(s)$, as does this chapter, e.g. the partition function of the spherical model. This appendix explores how to solve this problem when the function $\mathcal{J}(s)$ is expressed as an integral, of a general form

$$\mathcal{J}(s) = \int_C g(z)\, e^{sf(z)}\, dz. \tag{3.A.1}$$

The following considers the case in which s is a real variable. The contour $\mathcal{C}$ is chosen in such a way that the real part of $f(z)$ goes to $-\infty$ at both points of integration (so that the integrand vanishes in these regions) or as a close contour in the complex plane.[8]

If the variable s assumes quite large positive values, the integrand is large when the real part of $f(z)$ is also large and, vice versa, is small when the real part of $f(z)$ is either small or negative. In particular, for $s \to +\infty$, the significant contribution of the integral comes from those regions in which the real part of $f(z)$ assumes its maximum positive value. To see this, expressing $f(z)$ as

$$f(z) = u(x,y) + i\,v(x,y),$$

we have

$$\mathcal{J}(s) = \int_{\mathcal{C}} g(z)\, e^{su(x,y)}\, e^{isv(x,y)}\, dz.$$

If we suggest that the imaginary part of the exponent, $iv(x,y)$, is approximatly constant in the region where the real part has its maximum, i.e. $v(x,y) \simeq v(x_0,y_0) = v_0$, we can approximate the integral as follows

$$\mathcal{J}(s) \simeq e^{isv_0} \int_{\mathcal{C}} g(z)\, e^{su(x,y)}\, dz.$$

Far from the point of the maximum of the real part, the imaginary part can oscillate in an arbitrary way because the integrand is small and the phase factor becomes irrelevant.

Let us now discuss the properties of the maximum point of $sf(z)$. The real part of $sf(z)$ has a maximum, at a given s, in correspondence to the maximum of the real part of $f(z)$, i.e. $u(x,y)$. This point is determined by

$$\frac{\partial u}{\partial x} = \frac{\partial u}{\partial y} = 0.$$

For the Cauchy–Riemann equations satisfied by the analytic functions, these equations can be expressed as

$$\frac{df(z)}{dz} = 0. \tag{3.A.2}$$

It is important to stress that the maximum of $u(x,y)$ is such *only along a particular contour*. In fact, for all points of the complex plane at a finite distance from the origin, neither the real nor the imaginary parts of an analytic function have an absolute maximum or an

8 In the following we assume that the function $g(z)$ is significantly smaller than the term $e^{sf(z)}$ in the regions of interest.

absolute minimum. This is a direct consequence of the Laplace equation satisfied by both functions u and v

$$\frac{\partial^2 u}{\partial x^2} + \frac{\partial^2 u}{\partial y^2} = 0;$$
$$\frac{\partial^2 v}{\partial x^2} + \frac{\partial^2 v}{\partial y^2} = 0.$$

If the second derivative with respect to x of one of the functions u or v is positive, its second derivative with respect to y is necessarily negative. Hence, none of the two functions can have an absolute maximum or minimum. The vanishing of the first derivative of $f(z)$, eqn. (3.A.2), implies that we are in the presence of a saddle point: this is a stationary point that is a maximum of $u(x,y)$ along a given contour, but a minimum along another (Figure 3.10).

The problem is then how to choose a path of integration $\mathcal{C}$ that satisfies the following conditions: a) there exists a maximum of $u(x,y)$ along $\mathcal{C}$; b) the contour passes through the saddle point, so that the imaginary part $v(x,y)$ has the smallest variation. Complex analysis revealed that the curves associated to the equations $u =$ constant and $v =$ constant form a system of orthogonal curves, and the curve $v = c$ (where c is a constant) is always tangent to the gradient ∇u of u. Hence, this is the curve along which we have the maximum decreasing of the function each time we move away from the saddle. Therefore, this is the curve to select as contour of integration $\mathcal{C}$.

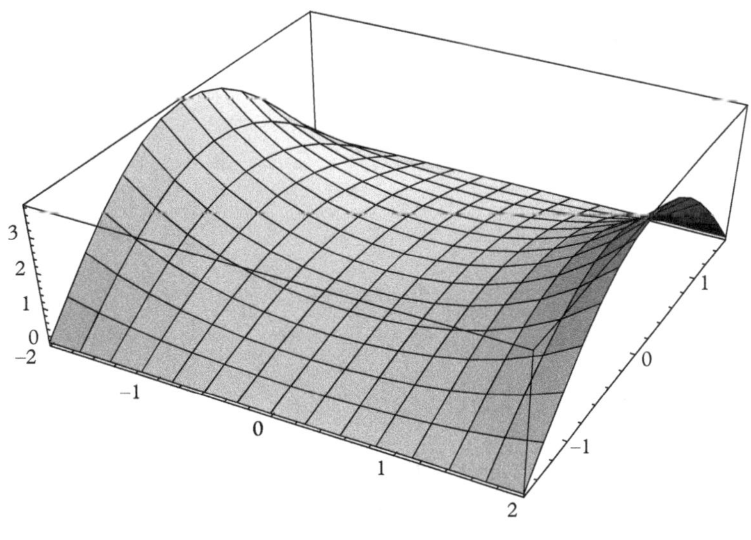

Fig. 3.10 *Saddle point of an analytic function.*

At the saddle point, the function $f(z)$ can be expanded in its Taylor series

$$f(z) \simeq f(z_0) + \frac{1}{2}(z - z_0)^2 f''(z_0) + \cdots$$

Along C, the quadratic correction of the function is both real (the imaginary part is constant along the chosen path) and negative (since we are moving along the path of fastest decrease from the saddle point). Assuming $f(z_0) \neq 0$, we have

$$f(z) - f(z_0) \simeq \frac{1}{2}(z - z_0)^2 f''(z_0) \equiv -\frac{1}{2s} t^2,$$

where we have defined the new variable t. Expressing $(z - z_0)$ in polar coordinates

$$(z - z_0) = \delta \, e^{i\alpha},$$

(with the phase being fixed), we get

$$t^2 = -sf''(z_0)\, \delta^2 \, e^{2i\alpha}.$$

Since t is real, we have

$$t = \pm\delta \mid sf''(z_0) \mid^{1/2},$$

and substituting in (3.A.1), we obtain[9]

$$\mathcal{J}(s) \simeq g(z_0)\, e^{sf(z_0)} \int_{-\infty}^{+\infty} e^{-t^2/2} \frac{dz}{dt}\, dt. \tag{3.A.3}$$

Since

$$\frac{dz}{dt} = \left(\frac{dt}{dz}\right)^{-1} = \left(\frac{dt}{d\delta}\frac{d\delta}{dz}\right)^{-1} = \mid sf''(z_0) \mid^{-1/2} e^{i\alpha},$$

eqn. (3.A.3) becomes

$$\mathcal{J}(s) \simeq \frac{g(z_0)\, e^{sf(z_0)}\, e^{i\alpha}}{\mid sf''(z_0) \mid^{1/2}} \int_{-\infty}^{+\infty} e^{-t^2/2}\, dt. \tag{3.A.4}$$

[9] The integral has been extended to $\pm\infty$ since the integrand is small when t is large.

The integral is now Gaussian (equal to $\sqrt{2\pi}$) so the asymptotic behaviour of $\mathcal{J}(s)$ is given by

$$\mathcal{J}(s) \simeq \frac{\sqrt{2\pi}\, g(z_0)\, e^{sf(z_0)}\, e^{i\alpha}}{|\, sf''(z_0)\,|^{1/2}}, \quad s \to +\infty \tag{3.A.5}$$

Two comments are in order. Sometimes the integration contour passes through two or more saddle points. In such cases, the asymptotic behaviour of $\mathcal{J}(s)$ is obtained by summing all the contributions (3.A.5) relative to the different saddle points. The second comment is about the validity of the method: in our discussion we have assumed that the only significant contribution to the integral comes from the region nearby the saddle point $z = z_0$. This means that one should always check that the condition

$$u(x,y) \ll u(x_0,y_0)$$

holds along the entire contour $\mathcal{C}$ away from$z_0 = x_0 + iy_0$.

Appendix 3.B. Brownian Motion on a Lattice

This appendix recalls the basic notions of the Brownian motion on a d-dimensional lattice (Figure 3.11). It also shows the interesting relation between this problem and the spherical model discussed in the text.

Binomial coefficients. Let us initially consider the one-dimensional case, with lattice sites identified by the variable s, with $s = 0, \pm 1, \pm 2, \ldots$: the problem consists of studying the motion of a particle that, at each discrete time step t_n, has a probability p and $q = 1 - p$ to move respectively to the neighbour site on its right or on its left. Suppose that at $t_0 = 0$ the particle is at the origin $s = 0$: what is the probability $P_n(s)$ that at the time t_n (after n steps) the particle is at the site s? There are several ways to determine such a quantity. One of the most elegant methods consists of assigning a weight $e^{i\phi}$ to the jump toward the right site and a weight $e^{-i\phi}$ to the one toward the left site and to consider the binomial

$$\left(pe^{i\phi} + qe^{-i\phi}\right)^n.$$

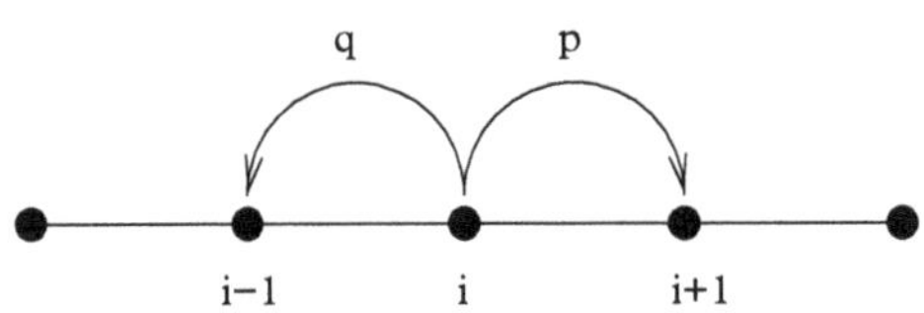

Fig. 3.11 *Brownian motion on a one-dimensional lattice.*

For $n = 1$, we have

$$\left(pe^{i\phi} + qe^{-i\phi}\right),$$

from which we can see that the coefficient p in front of $e^{i\phi}$ represents the probability that, after the first step, the particle is at the site $s = 1$, placed to the right of the original point, whereas the coefficient q in front of the other exponential $e^{-i\phi}$ gives the probability that the particle is at the site $s = -1$ to the left of the origin. Similarly, considering the expression

$$\left(pe^{i\phi} + qe^{-i\phi}\right)^2,$$

and expanding the binomial, the coefficient p^2 in front of the term $e^{2i\phi}$ gives the probability that the particle is at the site $s = 2$ after two steps, the coefficient $2pq$ in front of $e^{0i\phi}$ gives the probability to find the particle at the origin, while the coefficient q^2 in front of $e^{-2i\phi}$ expresses the probability to find the particle at the site $s = -2$. More generally, we have

$$P_n(s) = \text{coefficient in front of} \, e^{is\phi} \text{ in } \left(pe^{i\phi} + qe^{-i\phi}\right)^n.$$

Thanks to the identity

$$\frac{1}{2\pi}\int_{-\pi}^{\pi} e^{-i\phi a}\, d\phi = \delta_{a,0},$$

such a coefficient can be filtered by means of the Fourier transform, so that

$$P_n(s) = \frac{1}{2\pi}\int_{-\pi}^{\pi} \left(pe^{i\phi} + qe^{-i\phi}\right)^n e^{-is\phi}\, d\phi. \tag{3.B.1}$$

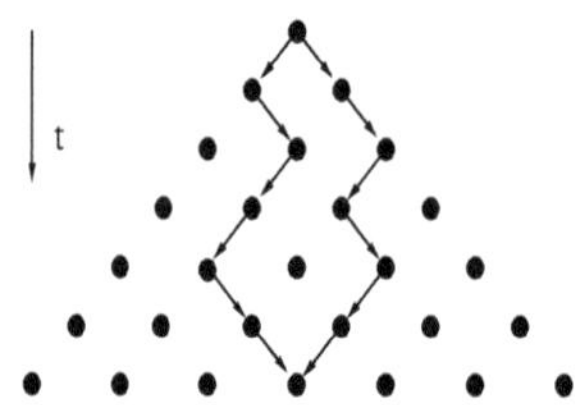

Fig. 3.12 *Two paths that lead to the same point after n steps.*

In the symmetric case, $p = q = \frac{1}{2}$ we have

$$P_n(s) = \frac{1}{2\pi}\int_{-\pi}^{\pi} (\cos\phi)^n\, e^{-is\phi}\, d\phi = \frac{1}{2^n}\frac{n!}{\left[\frac{1}{2}(n+s)\right]!\left[\frac{1}{2}(n-s)\right]!}. \tag{3.B.2}$$

In this case, if n is even, the only possible values of s are also even, with $|\, s \,| \leq n$, while if n is odd, then s is also odd, with $|\, s \,| < n$.

The origin of the binomial coefficient becomes evident by looking at Figure 3.12. In fact, the computation of $P_n(s)$ is equivalent to counting the number of different paths that start from the origin and reach the point s after n steps. In these paths each turn to the right or to the left is weighted by p and q, respectively.

Continuous probability. When $n \to \infty$ the integral (3.B.6) can be estimated by the saddle point method. In this limit, the dominant term of the integral comes from the values of ϕ_i near the origin, so that expanding in series the term $\left[\frac{1}{d}(\cos\phi_1 + \cdots \cos\phi_d)\right]^n$ and keeping only the quadratic terms, we have

$$\left[\frac{1}{d}(\cos\phi_1 + \cdots \cos\phi_d)\right]^n = \exp\left[n \log\frac{1}{d}(\cos\phi_1 + \cdots \cos\phi_d)\right] \simeq$$
$$\simeq \exp\left[-\frac{n}{2d}\left(\phi_1^2 + \phi_2^2 + \cdots + \phi_d^2\right)\right].$$

Changing variables $x_i = \phi_i n^{1/2}$ and performing the integral (3.B.6), we obtain the Gaussian distribution

$$P_n(\vec{s}) \simeq \left(\frac{d}{2\pi n}\right)^{d/2} \exp\left[-\frac{d}{2n}\vec{s}\cdot\vec{s}\right]. \tag{3.B.3}$$

If we now denote by a the lattice space and by τ the time interval between each transition, the variable $\vec{x} = a\vec{s}$ is the distance of the particle from the origin after the time $t = n\tau$. The function $P_n(\vec{s})$ in (3.B.3) is related to the *continuous probability density* $P(\vec{x}, t)$ to find the particle in the volume $d\vec{x}$ nearby the point $\vec{x}$

$$P(\vec{x}, t) = \frac{1}{(4\pi \mathcal{D} t)^{d/2}} \exp\left[-\frac{\vec{x}\cdot\vec{x}}{4\mathcal{D}t}\right], \tag{3.B.4}$$

where $\mathcal{D} = \frac{a^2}{2d\tau}$ is the diffusion constant. In fact, the function $P(\vec{x}, t)$ satisfies the differential equation of the diffusion process

$$\left(\frac{\partial}{\partial t} - \mathcal{D}\nabla^2\right) P(\vec{x}, t) = 0,$$

where ∇^2 isthe Laplacian operator in d dimensions. The dispersion of the probability density $P(\vec{x},t)$ is expressed by the mean value $\langle | \vec{x} |^2 \rangle$ (computed with respect to the probability distribution (3.B.4)): this quantity grows linearly with time

$$\langle | \vec{x} |^2 \rangle = 2\mathcal{D}t. \tag{3.B.5}$$

Generalization. The analysis of the one-dimensional case can be easily generalized in higher dimensional lattices. Consider, for instance, a d-dimensional cubic lattice in which, at each discrete temporal step, there are $2d$ possible transitions to the neighbour sites. For simplicity, let us assume that all these probabilities are the same and equal to $\frac{1}{2d}$. Assigning the weight $e^{i\phi_i}$ for the jump ahead and $e^{-i\phi_i}$ for the jump back along the ith direction, the probability to find the walker at the site s with coordinates $\vec{s} = (s_1, s_2, \ldots, s_d)$ after n steps is expressed by the d-dimensional Fourier transform

$$P_n(\vec{s}) = \frac{1}{(2\pi)^d} \int_{-\pi}^{\pi} \cdots \int_{-\pi}^{\pi} \left[\frac{1}{d} (\cos\phi_1 + \cos\phi_2 + \cdots \cos\phi_d) \right]^n e^{-i\vec{s}\cdot\vec{\phi}}\, d^d\phi. \tag{3.B.6}$$

The problem can be easily generalized to the cases where there are transitions between arbitrary sites, not necessarily next-neighbour. Let $\vec{s}_i$ and $\vec{s}_j$ be two sites of a d-dimensional lattice, with total number of sites equal to L^d. Assuming periodic boundary conditions along all directions, we have the equivalence relationships

$$(s_1, s_2, \ldots, s_d) \equiv (s_1 + L, s_2, \ldots) \equiv (s_1, s_2 + L, s_3, \ldots) \equiv \ldots$$

Let $p(\vec{s}_i - \vec{s}_j)$ be the probability of the transition $\vec{s}_j \longrightarrow \vec{s}_i$. For simplicity we assume that this probability is time independent and a function only on the distance between the two sites. Let us denote, as before, by $P_n(\vec{s})$ the probability that the particle is at the site $\vec{s}$ after n steps. This function satisfies the recursive equation

$$P_{n+1}(\vec{s}) = \sum_{\vec{s}_j} p(\vec{s} - \vec{s}_j)\, P_n(\vec{s}_j), \tag{3.B.7}$$

with the initial condition

$$P_0(\vec{s}) = \delta_{\vec{s},\mathbf{0}}. \tag{3.B.8}$$

Due to their probabilistic nature, $P_n(\vec{s})$ and $p(\vec{s})$ satisfy the normalization conditions

$$\sum_{\vec{s}} P_n(\vec{s}) = 1, \quad \sum_{\vec{s}} p(\vec{s}) = 1. \tag{3.B.9}$$

To solve the recursive equation (3.B.7) let us introduce the generating function[10]

$$G(\vec{s},w) = \sum_{n=0}^{\infty} P_n(\vec{s})\, w^n. \tag{3.B.10}$$

Let us multiply now eqn. (3.B.7) by w^{n+1} and sum on n. Taking into account the definition of $G(\vec{s},w)$ and the initial condition (3.B.8), the generating function $G(\vec{s},w)$ satisfies the equation

$$G(\vec{s},w) - w\sum_{\vec{s}'} p(\vec{s}-\vec{s}')\, G(\vec{s}',w) = \delta_{\vec{s},0}, \tag{3.B.11}$$

where the convolution term comes from the translation invariance of the lattice. This suggests to find its solution by expanding $G(\vec{s},w)$ in Fourier series. Let $g(\vec{k},w)$ and $\lambda(\vec{k})$ the Fourier transform of $G(\vec{s},w)$ and $p(\vec{s})$

$$\begin{aligned} g(\vec{k},w) &= \sum_{\vec{s}} G(\vec{s},w)\exp\left[i\vec{k}\cdot\vec{s}\right]; \\ \lambda(\vec{k}) &= \sum_{\vec{s}} p(\vec{s})\exp\left[i\vec{k}\cdot\vec{s}\right], \end{aligned} \tag{3.B.12}$$

with $\vec{k} = \frac{2\pi}{L}\vec{r}$ and $r_j = 0,1,2,\ldots(L-1)$. In terms of these quantities, eqn. (3.B.11) can be written as

$$g(\vec{k},w) - w\lambda(\vec{k})\, g(\vec{s},w) = 1,$$

from which

$$g(\vec{k},w) = \frac{1}{1-w\lambda(\vec{k})}. \tag{3.B.13}$$

Taking now the inverse Fourier transform, the solution of (3.B.11) is

$$G(\vec{s},w) = \frac{1}{L^d}\sum_{\{r_j=0\}}^{L-1} \frac{\exp(-2\pi i\vec{r}\cdot\vec{s}/L)}{1-w\lambda(2\pi\vec{r}/L)}. \tag{3.B.14}$$

Since $P_n(\vec{s})$ is the coefficient of w^n in $G(\vec{s},w)$, expanding in series the expression above we obtain

[10] A Brownian motion is *transient* if $G(0,1)$ is a finite quantity, while is *recurrent* if $G(0,1)$ is instead divergent. Further on, we clarify the origin of this terminology.

$$P_n(\vec{s}) = \frac{1}{L^d} \sum_{\{r_j=0\}}^{L-1} \left[\lambda \left(\frac{2\pi \vec{r}}{L} \right) \right]^n \exp\left(-2\pi i \vec{r} \cdot \vec{s}/L\right). \tag{3.B.15}$$

When $L \to \infty$, the generating function $G(\vec{s}, w)$ is expressed by the integral

$$G(\vec{s}, w) = \frac{1}{(2\pi)^d} \int_0^{2\pi} \cdots \int_0^{2\pi} \frac{\exp(-i\vec{s} \cdot \vec{k})}{1 - w\lambda(\vec{k})} d\vec{k}. \tag{3.B.16}$$

If the transitions are only those between next-neighbour sites of a cubic lattice, the function $\lambda(\vec{k})$ is given by

$$\lambda(\vec{k}) = \frac{1}{d} \sum_{j=1}^{d} \cos k_j,$$

and $G(\vec{s}, w)$ can be written as

$$G(\vec{s}, w) = \frac{1}{(2\pi)^d} \int_{-\pi}^{\pi} \cdots \int_{-\pi}^{\pi} \frac{\exp(-i\vec{s} \cdot \vec{k})}{1 - w d^{-1} \sum_{j=1}^{d} \cos k_j} d\vec{k}. \tag{3.B.17}$$

Note that $\tilde{G}(\vec{s}, w) \equiv wG(\vec{s}, w)$ satisfies the equation

$$\left[-\nabla_s^2 + (w^{-1} - 1) \right] \tilde{G}(\vec{s}, w) = \delta_{\vec{s},0},$$

where ∇_s^2 is the discrete version of the Laplacian operator on the d-dimensional lattice

$$\nabla_s^2 f(\vec{s}) \equiv \frac{1}{2d} \sum_{\mu=1}^{d} \left[f(\vec{s} + \vec{e}_\mu) + f[\vec{s} - \vec{e}_\mu - 2f(\vec{s}) \right].$$

This function is analogous to the Euclidean propagator of a free bosonic field of mass m: in fact, rescaling the quantities by the lattice space a according to $\vec{s} \to \vec{s}/a$, $\vec{k} \to \vec{k}a$ and posing

$$w^{-1} = 1 + m^2 \frac{a^2}{2d},$$

we have

$$\mathcal{D}(\vec{s}, m^2) = \lim_{a \to 0} \frac{1}{2da^{d-2}} \tilde{G}\left(\frac{\vec{s}}{a}, w \right) = \int_{-\infty}^{+\infty} \frac{d^d \vec{k}}{(2\pi)^d} \frac{e^{i\vec{k} \cdot \vec{s}}}{k^2 + m^2}. \tag{3.B.18}$$

The relationship between the spherical model and the Brownian motion should now be clear. In fact, the function $g'(z)$ defined by (3.5.13) and entering the saddle point equation of the model (3.5.16) is nothing else but the generating function of the Brownian motion on a cubic lattice! More generally, the spherical model with coupling constants $\mathcal{J}_{ij}$ is related to the Brownian motion with a probability transitions $p(\vec{s}_i - \vec{s}_j)$ proportional to $\mathcal{J}_{ij}$. There is, in fact, the following identity

$$g'(z) = \frac{1}{z}\, G(0, dz^{-1}). \tag{3.B.19}$$

Transient and recurrent Brownian motion. As discussed, there is a phase transition in the spherical model only if $g'(d)$ is finite. For the Brownian motion, this condition implies that the corresponding Brownian motion is *transient* and not *recurrent*. For $d = 1$ and $d = 2$ the Brownian motion is always recurrent:[11] this means that a Brownian motion that starts from the origin will always come back to the origin with probability equal to 1. For $d \geq 3$, $G(0, 1)$ is a finite quantity and this implies that the Brownian motion is *transient*: i.e. there is a finite probability that the walker never comes back to the origin. These results are part of Polya's famous problem about the probability of the random walk to return to a given site and its dependence on the dimensionality of the lattice. To derive general these results, it is useful to introduce the following functions:

- $P_n(\vec{s}, \vec{s}_0) =$ probability to be at the site $\vec{s}$ after n steps, where $\vec{s}_0$ is the starting point;
- $F_n(\vec{s}, \vec{s}_0) =$ probability to be at the site $\vec{s}$ *for the first time* after n steps, where $\vec{s}_0$ is the starting point,

together with their corresponding generating functions

$$\begin{aligned} G(\vec{s}, \vec{s}_0; w) &= \delta_{\vec{s},\vec{s}_0} + \sum_{n=1}^{\infty} P_n(\vec{s}, \vec{s}_0)\, w^n, \\ \mathcal{F}(\vec{s}, \vec{s}_0; w) &= \sum_{n=1} F_n(\vec{s}, \vec{s}_0)\, w^n. \end{aligned} \tag{3.B.20}$$

The functions P_n and F_n satisfy

$$\begin{aligned} P_0(\vec{s}, \vec{s}_0) &= \delta_{\vec{s},\vec{s}_0}, \\ P_n(\vec{s}, \vec{s}_0) &= \sum_{k=1}^{n} P_{n-k}(\vec{s}, \vec{s})\, F_k(\vec{s}, \vec{s}_0). \end{aligned} \tag{3.B.21}$$

[11] In the two-dimensional case, this result gives support to the popular saying *All roads lead to Rome.*

In fact, the particle can reach the site $\vec{s}$ for the first time after k steps and can come back later to the same site in the remaining $(n-k)$ steps. So, the sum on k corresponds to all independent ways to implement the transition $\vec{s}_0 \to \vec{s}$ in n steps. Multiplying these equations by w^n, summing on n and using the generating functions, we have

$$\begin{aligned} G(\vec{s},\vec{s}_0;w) &= \sum_{n=0} P_n(\vec{s},\vec{s}_0)\, w^n = \\ &= \delta_{\vec{s},\vec{s}_0} + \sum_{n=1}^{\infty}\sum_{k=1}^{n} w^k\, F_k(\vec{s},\vec{s}_0)\, w^{n-k}\, P_{n-k}(\vec{s},\vec{s}) = \\ &= \delta_{\vec{s},\vec{s}_0} + G(\vec{s},\vec{s},w)\,\mathcal{F}(\vec{s},\vec{s}_0,w). \end{aligned} \tag{3.B.22}$$

Hence

$$\begin{aligned} \mathcal{F}(\vec{s}_0,\vec{s}_0,w) &= 1 - [G(\vec{s}_0,\vec{s}_0,w)]^{-1}, \\ \mathcal{F}(\vec{s},\vec{s}_0,w) &= G(\vec{s},\vec{s}_0,w)/G(\vec{s},\vec{s},w) \quad \text{if } \vec{s} \neq \vec{s}_0. \end{aligned} \tag{3.B.23}$$

These formulae can be now used to study the nature of the Brownian motion on different lattices. If we have translation invariance, $G(\vec{s},\vec{s}_0;w) = G(\vec{s}-\vec{s}_0;w)$ and analogously for $\mathcal{F}$. Note that $\mathcal{F}(0,1)$ is exactly the probability that a particle comes back sooner or later to its starting point. In fact

$$\mathcal{F}(0,1) = F_1(0) + F_2(0) + \cdots \tag{3.B.24}$$

and therefore this quantity corresponds to the sum of the probabilities of all independent ways to come back to the origin, i.e. for the first time after one step, two steps, etc. On the other hand, from (3.B.23) we have

$$\mathcal{F}(0,1) = 1 - [G(0,1)]^{-1}, \tag{3.B.25}$$

so that the particle has probability equal to 1 to come back to the origin if $G(0,1)$ is a divergent quantity, as we saw in $d=1$ and $d=2$. On the contrary, in three dimensions and for a cubic lattice we have

$$\begin{aligned} G(0,1) &= \frac{1}{(2\pi)^3}\int_0^{2\pi}\int_0^{2\pi}\int_0^{2\pi} \frac{d^3\vec{k}}{1-\frac{1}{3}(\cos k_1+\cos k_2+\cos k_3)} = \\ &= \left(\frac{\sqrt{6}}{32\pi^3}\right)\Gamma\left(\frac{1}{24}\right)\Gamma\left(\frac{5}{24}\right)\Gamma\left(\frac{7}{24}\right)\Gamma\left(\frac{11}{24}\right) = \\ &= 1.516386059\ldots \end{aligned} \tag{3.B.26}$$

so that the probability of return to the origin is equal to

$$\mathcal{F}(0,1) = 0.34053733\ldots \tag{3.B.27}$$

Number of distinct points visited in the Brownian motion. Denoting by S_n the mean value of the distinct points visited by the walker after n steps, let us now derive the following asymptotic value when $n \to \infty$ for various lattices

$$S_n \simeq \begin{cases} \left(\frac{8n}{\pi}\right)^{\frac{1}{2}} & d = 1, \\ \frac{\pi n}{\log n} & d = 2, \\ C_d\, n & d \geq 3, \end{cases} \tag{3.B.28}$$

where the constant C_d depends on the structure of the lattice. For their derivation, observe that

$$S_n = 1 + \sideset{}{'}\sum_{\vec{s}} \left[F_1(\vec{s}) + F_2(\vec{s}) + \cdots F_n(\vec{s})\right],$$

where the sum is on all sites of the lattice but the origin. The first term of this expression is related to the origin, i.e. to the initial condition of the particle. With the definition previously given for $F_n(\vec{s})$, each term in the sum represents the probability that a site of the lattice has been visited at least once in the first n steps. Consider now

$$\Delta_k = S_k - S_{k-1}, \quad k = 1, 2, \ldots$$

Since $S_0 = 1$ and $S_1 = 2$, we have $\Delta_1 = 1$. Moreover

$$\Lambda_n = \sideset{}{'}\sum_{\vec{s}} F_n(\vec{s}) = -F_n(0) + \sum_{\vec{s}} F_n(\vec{s}),$$

where the sum is now extended to all lattice sites. The generating function of Δ_n is given by

$$\Delta(w) = \sum_{n=1}^{\infty} w^n\, \Delta_n = -\mathcal{F}(0,w) + \sum_{\vec{s}} \mathcal{F}(\vec{s},w).$$

From (3.B.22) we have

$$\mathcal{F}(\vec{s},w) = \frac{G(\vec{s},w) - \delta_{\vec{s},0}}{G(0,w)}.$$

Since for any n

$$\sum_{\vec{s}} P_n(\vec{s}) = 1,$$

we get

$$\sum_{\vec{s}} G(\vec{s}, w) = 1 + w + w^2 + \cdots = \frac{1}{1-w}.$$

Therefore

$$\Delta(w) = -1 + \frac{1}{(1-w)\, G(0,w)}. \tag{3.B.29}$$

Taking into account that

$$\begin{aligned} S_0 &= 1, \quad S_1 = 2 \\ S_n &= 1 + \Delta_1 + \Delta_2 + \cdots + \Delta_n, \quad n \geq 1 \end{aligned}$$

the generating function of S_n is expressed by

$$\begin{aligned} S(w) &= \sum_{n=0}^{\infty} w^n\, S_n = \\ &= (1-w)^{-1}\left[1 + w\Delta_1 + w^2\Delta_2 + \cdots\right] = \\ &= (1-w)^{-1}\,[1-\Delta(w)]^{-1} = \left[(1-w)^2\, G(0,w)\right]^{-1}. \end{aligned} \tag{3.B.30}$$

Consider $G(0, w)$ for various lattices. For $d = 1$, we have

$$G(0,w) = \frac{1}{2\pi}\int_{-\pi}^{\pi} \frac{dk}{1 - w\cos k} = \frac{1}{\sqrt{1-w^2}}. \tag{3.B.31}$$

For $d = 2$ we have

$$G(0,w) = \frac{1}{(2\pi)^2}\int_{-\pi}^{\pi}\int_{-\pi}^{\pi} \frac{dk_1\, dk_2}{1 - \frac{1}{2}w(\cos k_1 + \cos k_2)} = \frac{2}{\pi}K(w), \tag{3.B.32}$$

where

$$K(w) = \int_0^{\pi/2} \frac{d\alpha}{\sqrt{1 - w^2\sin^2\alpha}} \tag{3.B.33}$$

is the elliptic integral of first kind. For $w \to 1, K(w)$ has a logarithmic singularity, so that

$$G(0,w) \simeq -\frac{1}{\pi}\log(1-w) + \mathcal{O}(1), \quad z \to 1. \tag{3.B.34}$$

For $d \geq 3$, $G(0,w)$ has a finite limit for $w \to 1$, in particular for $d = 3$ it is given by (3.B.26).

To derive the asymptotic behaviour of S_n we need the following theorem.

Theorem 3.1 Let $U(y) = \sum_n a_n e^{-ny}$ be a convergent series for all values $y > 0$, with $a_n > 0$. If for $y \to 0$ $U(y)$ behaves as

$$U(y) \sim \Phi(y^{-1}),$$

where $\Phi(x) = x^\sigma L(x)$ is an increasing positive function of x that goes to infinity when $x \to \infty$, with $\sigma \geq 0$ and $L(cx) \sim L(x)$ for $x \to \infty$, then

$$a_1 + a_2 + \cdots + a_n \sim \frac{\Phi(n)}{\Gamma(\sigma+1)}. \tag{3.B.35}$$

If we now substitute the different expressions of $G(0,1)$ in (3.B.29), put $z = e^{-y}$ and study the limit $y \to 0$, we have

$$\Delta(y) \sim \begin{cases} (2/y)^{1/2} & d = 1, \\ \pi/(y\log 1/y) & d = 2, \\ 1/yG(0,1) & d \geq 3. \end{cases} \tag{3.B.36}$$

Hence

$$\begin{array}{lll} d = 1, & \sigma = \frac{1}{2}, & L(x) = 2^{1/2}, \\ d = 2, & \sigma = 1, & L(x) = \pi/\log x, \\ d \geq 3, & \sigma = 1, & L(x) = 1/G(0,1). \end{array} \tag{3.B.37}$$

Putting now $a_i = \Delta_i$ in (3.B.35), we obtain the asymptotic behaviour (3.B.28) of the mean value of the distinct sites visited after n steps, in the limit $n \to \infty$.

Relation with prime numbers. It is interesting to note that in $d = 2$ the number of distinct sites visited in n steps is proportional to the prime numbers less than the integer n. This quantity has been estimated originally by Gauss: denoting by $\Pi(n)$ the number of primes less than n, Gauss found the asymptotic form of such a function

$$\Pi(n) \simeq \frac{n}{\log n}. \tag{3.B.38}$$

The coincidence between this aspect of number theory and the brownian motion has a elementary explanation, that clarifies some important aspects of the prime numbers.

Gauss's law can be derived in a simple way by employing the sieve of Eratosthenes. Let us denote by $P(n)$ the probability that an integer n is a prime number. Since a generic integer n has probability $1/p_i$ of being divisible by p_i (this comes directly by the sieve of Eratosthenes), the probability that the number n is not divisible by p_i is equal to $(1 - 1/p_i)$. Assuming that there is no correlation between the prime numbers, the probability that the number n is not divisible for all prime p_i less than $n/2$ (i.e. the probability that n is itself a prime) is given by

$$P(n) \simeq \left(1 - \frac{1}{2}\right)\left(1 - \frac{1}{3}\right)\left(1 - \frac{1}{5}\right)\cdots = \prod_{p_i<n}\left(1 - \frac{1}{p_i}\right). \tag{3.B.39}$$

By taking the logarithm of both terms, we have

$$\log P(n) \simeq \sum_{p_i<n} \log\left(1 - \frac{1}{p_i}\right) \simeq -\sum_{p_i<n} \frac{1}{p_i}.$$

The last sum over the primes p_i can be written as a sum over the integers by using the probability $P(k)$ and therefore

$$\log P(n) \simeq -\sum_{k}^{n} \frac{P(k)}{k} \simeq \int_1^n \frac{P(k)}{k}\, dk.$$

Taking the derivative with respect to n in both terms we see that $P(n)$ satisfies the differential equation

$$\frac{dP^1(n)}{dn} = -\frac{P^2(n)}{n},$$

whose solution is the celebrated probability of the prime numbers $P(n) = 1/\log n$.

The key point of this derivation is the iterative nature of the sieve of Eratosthenes. From an algorithmic point of view, this procedure consists of the following steps: (a) once a prime number p_i has been identified, we cancel all its multiples; (b) the next prime p_{i+1} is then the first integer number after p_i that has survived step (a). We then repeat iteratively steps (a) and (b). In the light of this observation, any other sieve on the natural numbers implemented with the same rules will produce the same probability law $P(n) \simeq 1/\log n$. We can adopt, for instance, a sieve purely based on probabilistic laws. For instance, if we select the number 2 as a prime number, we can proceed to cancel all next numbers with probability $1/2$. After this procedure, we take the first number n_i that has survived this sieve and proceed to cancel all numbers $n > n_i$ with probability $1/n_i$, and so on. After all steps of this procedure have been completed, the set $\mathcal{E}$ of the survived numbers are prime numbers only *in average*, although they follows the same Gauss's law (3.B.38). In other words, the probability that a generic integer number n belongs to the set $\mathcal{E}$ is given by Gauss's law (3.B.38).

The arithmetic situation then looks similar to statistical mechanics, where there are general laws that are easy to establish if we consider a large number of similar ensembles but whose origin may appear mysterious if analysed on a particular sample.

Note that, for large n, eqn. (3.B.39) can be cast as the probability that the next two numbers belong to the same set, i.e.

$$P(n+1) \simeq P(n)\left[1 - \frac{1}{n}P(n)\right]. \tag{3.B.40}$$

This equation has the following interpretation. For large n we expect that $P(n+1)$ and $P(n)$ are almost equal, but the difference is that the number n can cancel the next number with probability $1/n$, but the reverse is not true! The probability that n belongs to the set and simultaneously cancels $(n+1)$ is given by the product of the probabilities $\frac{1}{n} \times P(n)$. Therefore, the probability that this does not happen is given by $\left[1 - \frac{1}{n}P(n)\right]$, i.e. the factor present on the right-hand side of (3.B.40).

It should be noted that (3.B.40) is the equation satisfied by the probability to visit a new site in the two-dimensional Brownian motion. In fact, in a Brownian motion the mean square distance from the origin after n steps is $\overline{r^2} \simeq n$. Therefore, in the two-dimensional Brownian motion, after n steps the particle is localized in a circle of radius $\sqrt{n}$. Let $P(n)$ be the probability that the particle visits a new site at the nth step. What is the probability that the particle will visit a new site at the $(n+1)$th step? Going from n to $(n+1)$, the area A of the circle varies as $\Delta A/A = 1/n$ and the probability that these sites have already been visited is given by the product $P(n) \times \Delta A/A = P(n) \times 1/n$. The probability that these sites have not been already visited is given $1 - \frac{1}{n}P(n)$. Since for large n, $P(n+1)$ and $P(n)$ must be proportional, we arrive at eqn. (3.B.40) and therefore $P(n) \simeq 1/\log n$. The average number of distinct sites that have been visited is given by the total area $A \simeq \pi n$ mutiplied by the probability $P(n) = 1/\log n$, i.e. $S_n \simeq \pi n/\log n$.

We leave as an exercise the elementary derivation of the asymptotic behaviour of S_n in $d = 1$ and $d \geq 3$.

REFERENCES

The mean field approximation is discussed in many books of statistical mechanics. We recommend:

Reichl, L.E. (1980). *Modern Course in Statistical Mechanics*, Arnold Publishers LTD.

The Bethe–Peierls approximation is discussed in the following:

Bethe, H.A. (1935). 'Statistical Theory of Superlattices', *Proceedings of the Royal Society A* 150: 552.

Peierls, R. (1936). 'Statistical Theory of Superlattices with Unequal Concentration', *Proceedings of the Royal Society A* 154: 207.

The transfer matrix solution of the one-dimensional Gaussian model is discussed in:

Kadanoff, L.P. (2000). *Statistical Physics. Statics, Dynamics and Renormalization*, World Scientific.

The spherical model has been proposed in:

Berlin, T.H. and Kac, M. (1952). 'The spherical model of a ferromagnet', *Physics Reviews* 86: 821.

The equivalence between the spherical model and a model with $O(n)$ symmetry is discussed in:

Stanley, E. (1968). 'Spherical Model as the Limit of Infinite Spin Dimensionality', Physics Reviews 176: 718

The Brownian motion is one of the most important topics of statistical mechanics. Interested readers are referred to the following:

Itzykson, C. and Drouffe, J.M. (1989). *Statistical Field Theory*, Cambridge University Press.

Montroll, E.W. and West, B.J. (1979). *Studies in Statistical Mechanics VII, Fluctuation Phenomena*, Amsterdam.

Chandrasekhar, S. (1943). 'Stochastic Problems in Physics and Astronomy', *Reviews in Modern Physics* 15: 1.

The magnum opus on the Brownian motion is the two-volume book:

Hughes, B.D. (1995). *Random Walks and Random Environments*, Oxford Science Publications.

PROBLEMS

3.1. Mean field theory for anti-ferromagnetic systems

a. Prove that the mean field theory can be applied to anti-ferromagnetic systems defined on a square lattice.

b. Generalize this result to all lattices that admit two sub-lattices such that the next-neighbour sites of one are sites of the other, and vice versa. This occurs, for instance, in the two dimensional exagonal or square lattices.

c. What happens if we consider a triangular lattice?

3.2. Mean field theory for coupled lattices

Consider a system of Ising spins made of two-dimensional square lattices A and B, coupled together. Let $\mathcal{J}_a$ and $\mathcal{J}_b$ be the coupling constants between next-neighbour spins in the lattices A and B respectively, and $\mathcal{J}$ the coupling constant between the next-neighbour spins of the two lattices (Figure 3.13).

a. Generalize the mean field theory to this system and show that the spontaneous magnetizations of the two lattices satisfy the coupled system of equations

$$M_a = f(\mathcal{J}_a M_a + \mathcal{J} M_b),$$
$$M_b = f(\mathcal{J}_b M_b + \mathcal{J} M_a)$$

Determine the explicit form of the function f.

b. Estimate the critical temperature T_c in the mean field approximation.
c. Show that, for $T > T_c$, the magnetic susceptibility is expressed by the ratio of linear and quadratic polynomials in T.
d. Discuss the case $\mathcal{J}_a = \mathcal{J}_b > 0$ while $\mathcal{J} < 0$ and study the magnetic susceptibility for these values of the couplings.
e. Discuss the limits $\mathcal{J} \to \pm\infty$.

3.3. Spontaneous magnetization in low temperature

Show that, when T is much less than T_c, the mean field theory of a ferromagnet predicts a spontaneous magnetization that differs from its saturation value for terms that are exponentials in $-1/T$.

3.4. Quantum magnets

Consider the Hamiltonian

$$\mathcal{H} = -\frac{1}{2}\sum_{R,R'} \mathcal{J}(R-R')\,\vec{S}(R)\cdot\vec{S}(R')$$

where $\mathcal{J}(R-R') > 0$ and $\vec{S}(R)$ is the quantum operator of spin S.

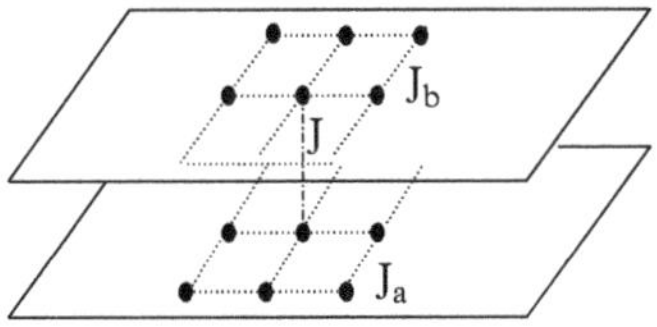

Fig. 3.13 *Layers of Ising spins.*

a. Prove initially the following result: the largest (smallest) diagonal element that a Hermitian operator can have is equal to its largest (smallest) eigenvalue.

b. Use this result to prove that, for $R \neq R'$, $\langle \vec{S}(R) \cdot \vec{S}(R') \rangle \leq S^2$.

c. Let $\mid S\rangle_R$ be the eigenvectors $S_z(R)$ with the maximum eigenvalue

$$S_z(R) \mid S\rangle_R = S \mid S\rangle_R.$$

Prove that the state $\mid 0\rangle = \prod_i \mid S\rangle_R$ is eigenvector of the Hamiltonian with eigenvalue

$$E_0 = -\frac{1}{2} S^2 \sum_{R,R'} \mathcal{J}(R - R').$$

Hint. Express the Hamiltonian in terms of the ladder operators $S_\pm(R) = (S_x \pm iS_y)(R)$ *and use the condition* $S_+(R) \mid S\rangle_R = 0$.

d. Using the result of **a**, show that E_0 is the smallest eigenvalue of the Hamiltonian.

3.5. Critical exponents of the spherical model

Use the equation of state and the other relations discussed in the text to derive the critical exponents of the spherical model.

3.6. Brownian motion with boundary conditions

Let

$$P(s,t) = \frac{1}{\sqrt{4\Pi \mathcal{D}t}} \exp[-s^2/(4\mathcal{D}t)]$$

be the probability distribution in the continuum limit of the one-dimensional Brownian motion.

a. Find the probability distribution $P_a(s,t)$ when the origin is an absorbent point, i.e. when $P_a(0,t) = 0$ for all t.

b. Find the probability distribution $P_r(s,t)$ when the origin is a pure reflecting point, i.e. when it holds the condition $\frac{\partial P_r}{\partial x}(0,t) = 0$ for all t.

Hint. Use $P(s,t)$ *and the linearity of the problem to set up the method of solution.*

3.7. Distinct points visited in the Brownian motion

Give a physical argument for the mean value of the number of distinct sites visited in the Brownian motion and show that this mean value depends on the dimensionality of the lattice as predicted by eqn. (3.B.28).

3.8. Markov processes

Brownian motion is a particular example of a general class of stochastic processes known as *Markov processes*, characterized by a transition probability $w(i \to j) = w_{ij}$ between the discrete states $\{A\} = \{a_1, a_2, a_3, \ldots, a_n\}$ of a stochastic variable A ($w_{ij} \geq 0$ and $\sum_{j=1}^{n} w_{ij} = 1$). These transitions take place at discrete time steps $t_n = n$. Denoting by $P_i(n)$ the probability to be in the ith state at the time n, it satisfies the recursive equation

$$P_i(n+1) = \sum_{j=1}^{n} w_{ij} P_j(n).$$

Using a matrix formalism, it can be expressed as $P(n+1) = WP(n)$.

a. Prove that the eigenvalues of the matrix W satisfy the condition $| \lambda_i | \leq 1$.
b. Show that the system reaches an equilibrium distribution $P_i(\infty)$ for $t \to \infty$ that is independent from the initial condition if and only if the matrix W has only one eigenvalue of modulus 1.
c. Assuming that the conditions of point **b** are satisfied, prove that

$$\lim_{n\to\infty} (W)^n = M = \begin{pmatrix} m_1, & m_2, & \ldots & , m_n \\ m_1, & m_2, & \ldots & , m_n \\ \cdot & & & \\ \cdot & & & \\ \cdot & & & \\ m_1, & m_2, & \ldots & , m_n \end{pmatrix}$$

with $\lim_{n\to\infty} P_i(n) = m_i$.

3.9. Brownian motion on a ring

Consider a Brownian motion on a ring of N sites, with a transition rate to next-neighbour sites equal to $1/2$. Let $P_n(s)$ be the probability to find the walker at the site s at the time n ($s = 1, 2, \ldots, N$). Show that if N is an odd number, there is a unique stationary probability distribution $n \to \infty$. Vice versa, if N is an even number, for $n \to \infty$, the distribution probability can oscillate between two different probability distributions.

3.10. Langevin equation and Brownian motion

Consider a particle of mass m in motion in a fluid (for simplicity we consider a one-dimensional motion), subjected to a friction force proportional to the velocity and a random force $\eta(t)$ due to the randomfluctuations of the fluid density. Denoting by $x(t)$

and $v(t)$ the position and the velocity of the particle at time t, its equations of motion of the particle

$$\frac{dv(t)}{dt} = -\frac{\gamma}{m}v(t) + \frac{1}{m}\eta(t),$$
$$\frac{dx(t)}{dt} = v(t),$$

where $\gamma > 0$ is the friction coefficient. Assume that $\eta(t)$ is random variable, with zero mean and delta-correlated

$$\langle \eta(t) \rangle_\eta = 0, \quad \langle \eta(t_1)\eta(t_2) \rangle_\eta = 2\gamma k_B T \delta(t_1 - t_2)$$

where k_B is the Boltzmann constant, T is the temperature and the average $\langle\rangle_\eta$ is with respect to the probability distribution of the stochastic variable $\eta(t)$.

a. Let x_0 and v_0 the position and velocity of the particle at $t = 0$. Integrating the equations of motion and taking the average with respect to η, show that the correlation function of the velocity is

$$\langle v(t_2)v(t_1) \rangle_\eta = \left(v_0^2 - \frac{k_B T}{m} \right) e^{-(\gamma/m)(t_1+t_2)} + \frac{k_B T}{m} e^{-(\gamma/m)(t_2-t_1)}.$$

with $t_2 > t_1$.

b. Compute the variance of the displacement and show that

$$\langle (x(t) - x_0)^2 \rangle_\eta = \frac{m^2}{\gamma} \left(v_0^2 - \frac{k_B T}{m} \right) \left(1 - e^{-(\gamma/m)t} \right)^2 + \frac{2k_B T}{\gamma} \left[t - \frac{m}{\gamma} \left(1 - e^{-(\gamma/m)t} \right) \right].$$

c. Assuming that the particle is in thermal equilibrium, we can now average over all possible initial velocities v_0. Let us denote this thermal average by $\langle\rangle_T$. By the equipartion theorem we have $\langle v_0^2 \rangle_T = k_B T/m$. Show that, for $t \gg m/\gamma$, the thermal average of the variance of the displacement becomes

$$\langle \langle (x(t) - x_0)^2 \rangle_\eta \rangle_T \simeq (2k_B T/\gamma)t.$$

Part 2

Bi-dimensional Lattice Models

4

Duality of the Two-dimensional Ising Model

Being dual is in the nature of things.
Elias Canetti

This chapter begins our study of the Ising model on the two-dimensional lattice. In two dimensions the model has a phase transition, with critical exponents that have different values from those obtained in the mean field approximation. For this reason, it provides an important example of critical phenomena. As we will see in great detail in this chapter and in the next, among all exactly solved models of statistical mechanics, the two-dimensional Ising model is not only the most studied but it also the model that has evidenced a series of deep mathematical and physical results. Many solutions of the model stand out for the ingenious methods used, as the theory of determinants, the combinatorial approaches, the Grassmann variables or the elliptic functions. Many results have deeply influenced the understanding of the critical phenomena and have strongly stimulated new fields of research. Ideas matured within the study of the two-dimensional Ising model, e.g. the duality between its high- and low-temperature phases, have been ready generalized to other systems of mechanical statistics and have also found important and fundamental applications also in other important areas as, for instance, QFT. Equally fundamental is the discovery that in the vicinity of the critical point, the dynamics of the model can be described from the relativistic Dirac equation for Majorana fermions.

This chapter is devoted to the study of some properties of the model that can be established by means of elementary considerations. We discuss, in particular, the argument by Peierls that shows the existence of a phase transition in the model. We also present the duality relation that links the expressions of the partition functions in the low- and high-temperature phase of a square lattice, and the partition functions of the triangle and exagonal lattices. In the last case, it is necessary to make use of an identity, known as *star–triangle equation*, that will be useful later on to study the commutativity properties of the transfer matrix. At the end of the chapter, we also discuss the general formulation of the duality transformations for lattice statistical models.

Statistical Field Theory: An Introduction to Exactly Solved Models in Statistical Physics. Second Edition. Giuseppe Mussardo, Oxford University Press (2020). © Giuseppe Mussardo. DOI: 10.1093/oso/9780198788102.001.0001

4.1 Peierls Argument

In the 1936 Peierls published an article with the title *On the Model of Ising for the Ferromagnetism* in which he proved that the Ising model in two or higher dimensions has a low temperature region in which the spontaneous magnetization is different from zero. Since in high temperature the system is disordered, it follows that there must exist a critical value of the temperature at which a phase transition takes place.

The Peierls argument starts with the initial observation that to each configuration of spins corresponds a set of closed lines that separate the regions in which the spins assume values $+1$ from those in which they assume values -1 (Figure 4.1). If it is possible to prove that at sufficiently low temperatures the mean value of the regions enclosed by the closed lines is only a small fraction of the total volume of the system, we also prove that the majority of the spins is prevalently in the state in which there is a spontaneous magnetization.

There are several versions of Peierls's original argument. The simplest generalizes the argument already used in the one-dimensional case (see Section 2.1) and concerns the stability of the state with a spontaneous magnetization. Let us consider the two-dimensional Ising model at low temperatures and suppose that is in the state of minimal energy in which all the spins have values $+1$. The thermal fluctuations create domains in which there are spin flips, as the domain in Figure 4.1. The creation of such domains clearly destabilizes the original ordered state. There is an energetic cost to the creation of domain shown in Figure 4.1, given by

$$\Delta E = 2\mathcal{J}L, \tag{4.1.1}$$

where L is the total length of the curve. There are, however, many ways of creating a closed curve of a given total perimeter L. In fact, the domain in which the spins are flipped can be placed everywhere in the lattice and moreover can assume different shapes. To

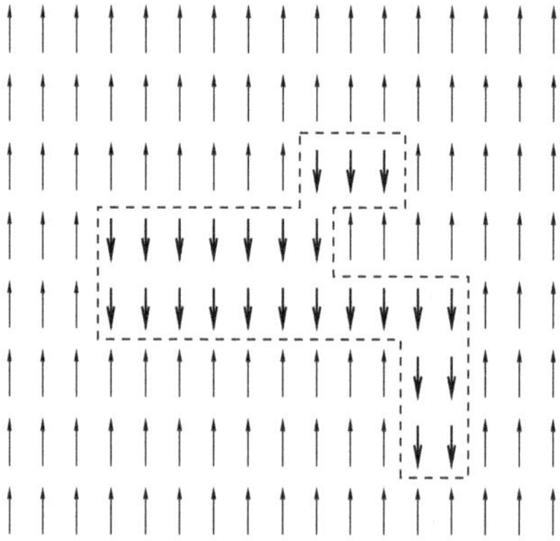

Fig. 4.1 *Closed lines that enclose region with a flipped value of the spin.*

estimate the number of such configurations, imagine that the closed line is created by a random motion on the lattice of total number of steps equal to L. If we assume that at each step of this motion there are only two possibilities,[1] we have 2^L ways of drawing a closed curve of length L. The corresponding variation of the entropy is given by

$$\Delta S = k \ln(2^L) \tag{4.1.2}$$

Hence the total variation of the free energy associated to the creation of such a domain is

$$\begin{aligned} \Delta F &= \Delta E - T\Delta S = 2\mathcal{J}L - kT\ln(2^L) = \\ &= L(2\mathcal{J} - kT\ln 2). \end{aligned} \tag{4.1.3}$$

Therefore the system is stable with respect to the creation of such domains of arbitrary length L (i.e. $\Delta F \geq 0$) if

$$T \leq T_c = \frac{2\mathcal{J}}{k\ln 2} = 2.885\frac{\mathcal{J}}{k}. \tag{4.1.4}$$

Note that such an estimate is surprisingly close to the exact value of the critical temperature $T_c = 2.269...\mathcal{J}/k$, which we determine in the next section.

4.2 Duality Relation in Square Lattices

The Peierls's argument shows that the two-dimensional Ising model has two different phases: the high-temperature phase in which the system is disordered and the low-temperature phase in which the system is ordered, with a non-zero spontaneous magnetization. The exact value of the critical temperature at which the phase transition happens has been firstly determined by Kramers and Wannier by using a duality relation between the high- and the low-temperature partition functions.[2] The self-duality of the two-dimensional Ising model on a square lattice is one of its most important properties, with far-reaching consequences on its dynamics. To prove it, we need to study the series expansions of the high/low-temperature phase of the model. We will see that these expansions have an elegant geometrical interpretation in terms of a counting problem of the polygons that can be drawn on a lattice. The next section considers the square lattice and later sections, the triangle and hexagonal lattices.

[1] On a square lattice, starting from a given site, we can move in four different directions. However, taking four instead of two as possible directions of the motion gives an upper estimate of the entropy, since it does not take into account that the final curve is a close contour.

[2] The self-duality of the model that we discuss only holds in the absence of an external magnetic field.

4.2.1 High Temperature Series Expansion

Consider a square lattice $\mathcal{L}$ with M horizontal links and M vertical links. In the thermodynamical limit $M \to \infty$, M coincides with the total number N of the lattice sites. The following considers a Hamiltonian with different coupling constants along the horizontal and vertical directions. Let $\mathcal{J}$ and $\mathcal{J}'$ be these coupling constants respectively. For the partition function of the model at zero magnetic field we have

$$Z_N = \sum_{\{\sigma\}} \exp\left[K\sum_{(i,j)}\sigma_i\sigma_j + L\sum_{(i,k)}\sigma_i\sigma_k\right], \tag{4.2.1}$$

where the first sum is on the spins along the horizontal links and the second sum along the vertical links, with

$$K = \beta\mathcal{J}; \quad L = \beta\mathcal{J}'.$$

By using the identity

$$\exp\left[x\sigma_i\sigma_l\right] = \cosh x\,(1+\sigma_i\sigma_l\tanh x), \tag{4.2.2}$$

the partion function can be written as

$$Z_N = (\cosh K\,\cosh L)^M \sum_{\{\sigma\}}\prod_{(i,j)}(1+v\sigma_i\sigma_j)\prod_{(i,k)}(1+w\sigma_i\sigma_k), \tag{4.2.3}$$

with

$$v = \tanh K; \quad w = \tanh L.$$

Both parameters v and w are always less than 1 for all values of the temperature, except for $T = 0$ when their value is $v = w = 1$. In particular, they are small parameters in high-temperature phase and it is natural to look for a series expansion of the partition function near $T = \infty$.

If we expand the two products in (4.2.3), we have 2^{2M} terms, since there are $2M$ factors (one for each segment), and each of them has two terms. We can set up a graphical representation for this expansion associating a line drawn on the horizontal link (i,j) to the factor $v\sigma_i\sigma_j$ and a line on the vertical link (i,k) to the factor $w\sigma_i\sigma_k$. No line is drawn if there is instead the factor 1. Repeating this operation for the 2^{2M} terms, we can establish a correspondence between these terms and a graphic configuration on the lattice $\mathcal{L}$. The generic expression of these terms is

$$v^r w^s \sigma_1^{n_1}\sigma_2^{n_2}\sigma_3^{n_3}\dots$$

where r is the total number of the horizontal lines, s the total number of the vertical lines, while n_i is the number of lines where i is the final site. It is now necessary to sum on all spins of the lattice in order to obtain the partition function. Since each spin σ_i assumes values ± 1, we have a null sum unless all $n_1, n_2, \ldots, n_N$ are even numbers and, in this case, the result $2^N v^r w^s$. Based on these considerations, the partition function can be expressed as

$$Z_N = 2^N (\cosh K \cosh L)^M \sum_P v^r w^s, \tag{4.2.4}$$

where the sum is on all the line configurations on $\mathcal{L}$ with an even number of lines at each site, i.e. *all closed polygonal lines* P of the lattice $\mathcal{L}$. Therefore, apart of a prefactor, the partition function is given by the geometrical quantity

$$\Phi(v, w) = \sum_P v^r w^s. \tag{4.2.5}$$

It is easy to compute the first terms of this function. The first term is equal to 1 and corresponds to the case in which there are no polygons on the lattice. The second term corresponds to the smallest closed polygon on the lattice $\mathcal{L}$, i.e. a square with a unitary length (Figure 4.2). The number of such squares is equal to N, since they can be placed on any of the N sites of the lattice. Each of them has a weight $(vw)^2$, hence the second term of the sum (4.2.5) is equal to $N(vw)^2$. The next closed polygonal curve is a rectangle of six sides: there are two kinds (Figure 4.3), each with a degeneracy equal to N, and width $v^4 w^2$ for the first and $v^2 w^4$ for the second.

Using the first terms, the function $\Phi(v, w)$ is given by

$$\Phi(v, w) = 1 + N(vw)^2 + N(v^4 w^2 + v^2 w^4) + \cdots. \tag{4.2.6}$$

The computation of the next terms becomes rapidly involved although it can be clearly performed in a systematic way: presently, the first 40 terms of such a series are known. For our purposes it is not necessary to introduce all these terms, since the duality properties can be established just exploiting the geometrical nature of the sum (4.2.5).

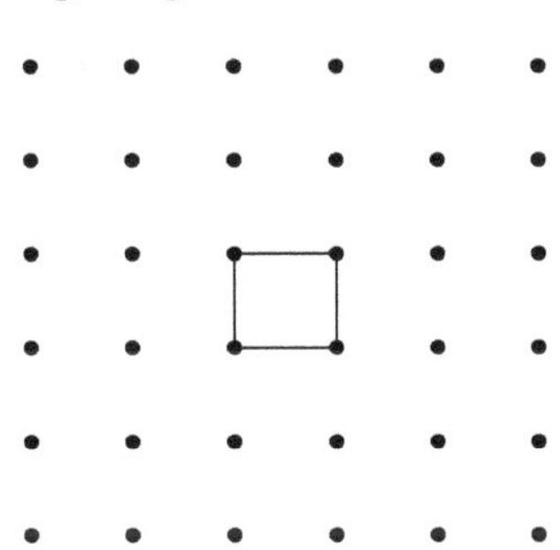

Fig. 4.2 *Second term of the high-temperature expansion.*

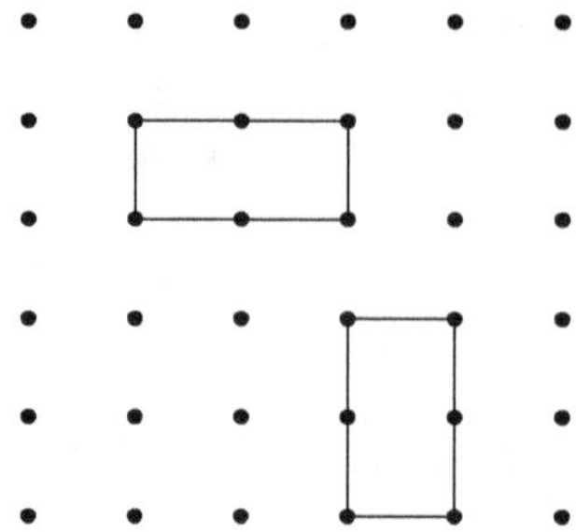

Fig. 4.3 *Third term of the high-temperature expansion.*

4.2.2 Low-temperature Series Expansion

In the low-temperature phase, according to Peierls's argument, the spins tend to align to one another. The series expansion of the partition function in this phase can be obtained as follows. For a given configuration of the spins, let r and s be the numbers of vertical and horizontal links in which the two adjacent spins are anti-parallel. Since M is the total number of vertical links as well as of the horizontal ones, we have $(M - r)$ vertical links and $(M - s)$ horizontal links in which the adjacent spins are parallel. The contribution to the partition function of such a configuration is

$$\exp\left[K(M - 2s) + L(M - 2r)\right].$$

Besides a constant, this expression depends only on the number of links in which the spins are anti-parallel. These segments will be called *anti-parallel links*.

It is now convenient to introduce the concept of *dual lattice*. This notion, familiar in crystallography, has already appeared when discussing the four-colour problem (see

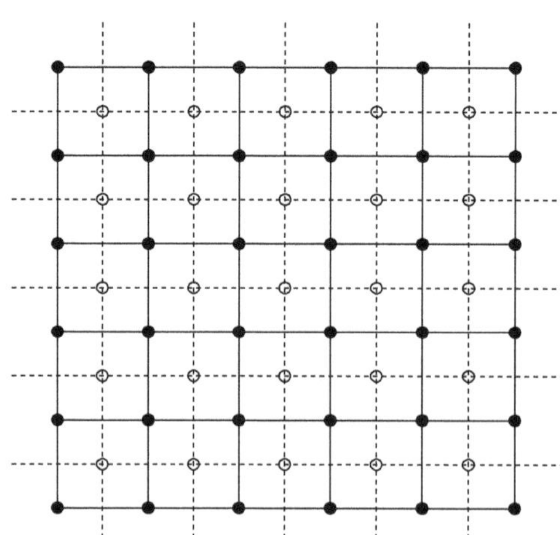

Fig. 4.4 *Dual square lattices.*

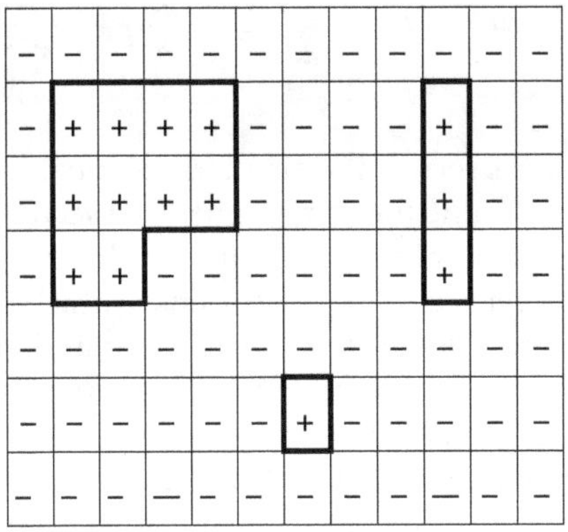

Fig. 4.5 *Polygons that separate the domains with spins* $+1$ *and* -1.

Appendix 2.C). For any planar lattice $\mathcal{L}$, we can define another lattice $\mathcal{L}_D$, that is obtained placing its sites at the centre of the original lattice $\mathcal{L}$ and joining pairwise those relative to adjacent faces, i.e. those sharing a common segment. It is easy to see that the dual lattice of a square lattice is also a square lattice, simply displaced by half-lattice space with respect to the original (see Figure 4.4), while the dual lattice of a triangle lattice is hexagonal, and vice versa.

Given the geometrical relation between the dual and the original lattices, it is easy to see that the spins can be equivalently regarded as defined on the sites of the original lattice $\mathcal{L}$ or at the centre of the faces of the dual lattice $\mathcal{L}_D$. This allows us to introduce a useful graphical formalism. Given a configuration, we can associate to its anti-parallel links a set of lines of the dual lattice by the following rule: if two next-neighbour spins are anti-parallel, then draw a line along the segment of $\mathcal{L}_D$ that crosses them, draw no line if they are parallel. By applying this rule, on the dual lattice $\mathcal{L}_D$ there will be r horizontal lines and s vertical lines. However, it is easy to see that there should always be an even number of lines passing through each site, since there is an even number of next successive changes among the adjacent faces. The drawn lines must therefore form closed polygons on the dual lattice $\mathcal{L}_D$, (Figure 4.5).

It is evident that the closed polygons that have been obtained in this way are nothing else that the perimeters of the different magnetic domains where, inside them, all spins are aligned in the same direction. Since for any given set of polygons there are two corresponding configurations (obtained one from the other by flipping all the spins), the partition function can be written as

$$Z_N = 2\exp[M(K+L)] \sum_{\tilde{P}} \exp[-(2Lr+2Ks)], \tag{4.2.7}$$

where the sum is on all closed polygons $\tilde{P}$ on the dual lattice $\mathcal{L}_D$. This is the low-temperature expansion, because when $T \to 0$, both K and L are quite large and the dominant terms are given by small values of r and s. Therefore, also in this case the partition function is expressed by a geometrical quantity

$$\tilde{\Phi}\left(e^{-2L}, e^{-2L}\right) = \sum_{\tilde{P}} \exp[-(2Lr + 2Ks)]. \tag{4.2.8}$$

Consider the first terms of this series. The first term is equal to 1 and corresponds to the situation in which all spins assume the same value. The second term corresponds to the configuration in which there is only one spin flip: in this case there are two horizontal anti-parallel links and two vertical anti-parallel links, which altogether form a square. The degeneracy of this term is equal to N, since the spin that has been flipped can be placed on any of the N sites of the lattice. The next term is given by the rectangle with six segments, which can be elongated either horizontally or vertically: these rectangles correspond to next-neighbour spins that are anti-parallel to all other spins of the lattice. Kept into account the degeneracy N and the orientation of the rectangle, the contribution of this term to the partition function is $N(e^{-4L-8K} + e^{-8L-4K})$. With these first terms, the function $\tilde{\Phi}(e^{-2L}, e^{-2K})$ is expressed by

$$\tilde{\Phi}\left(e^{-2L}, e^{-2L}\right) = 1 + Ne^{-4L-4K} + N(e^{-4L-8K} + e^{-8L-4K}) + \cdots. \tag{4.2.9}$$

From above, it should be now clear that all terms of the function $\tilde{\Phi}$ have the same origin as those of the function Φ.

4.2.3 Self-duality

The last two sections showed that the partition function of the two-dimensional Ising model on a square lattice can be expressed in two different series expansions, one that holds in high-temperature phase, the other in low-temperature phase, given in eqns. (4.2.4) and (4.2.7), respectively. The final expressions involve a function that has a common geometric nature, i.e. a sum on all the polygonal configurations that can be drawn on the original lattice and its dual. For finite lattices, $\mathcal{L}$ and $\mathcal{L}_D$ differ only at the boundary. In the thermodynamical limit this difference disappears and the two expressions can be obtained one from the other simply by a change of variables. For $N \to \infty$ we have $M/N = 1$: substituting K and L in eqn. (4.2.5) with $\tilde{K}$ and $\tilde{L}$ given by

$$\tanh K = e^{-2\tilde{L}}; \quad \tanh L = e^{-2\tilde{K}}, \tag{4.2.10}$$

and comparing with eqn. (4.2.8), we have in fact

$$\tilde{\Phi}\left(e^{-2\tilde{K}}, e^{-2\tilde{L}}\right) = \Phi(v, w). \tag{4.2.11}$$

This implies the following identity for the partition function

$$\frac{Z_N[K, L]}{2^N (\cosh K \cosh L)^N} = \frac{Z_N[\tilde{K}, \tilde{L}]}{2 \exp[N(\tilde{K} + \tilde{L})]}. \tag{4.2.12}$$

Eqn. (4.2.10) can be expressed in a more symmetrical form

$$\sinh 2\tilde{K} \sinh 2L = 1; \quad \sinh 2\tilde{L} \sinh 2K = 1. \tag{4.2.13}$$

Analogously, eqn. (4.2.12) can be written as

$$\frac{Z_N[K,L]}{(\sinh 2K \sinh 2L)^{N/4}} = \frac{Z_N[\tilde{K},\tilde{L}]}{(\sinh 2\tilde{K} \sinh 2\tilde{L})^{N/4}}. \tag{4.2.14}$$

These equations show the existence of a symmetry of the two-dimensional Ising model and establish the mapping between high- and low-temperature phases of the model. Large values of K and L are equivalent to small values of $\tilde{K}$ and $\tilde{L}$, and vice versa large values of $\tilde{K}$ and $\tilde{L}$ correspond to small values of K and L. It must be stressed that this correspondence between the two phases can be also useful from a computational point of view.

We can now identify the critical point. Let us consider first the isotropic case, i.e. $K = L$ and, correspondingly, $\tilde{K} = \tilde{L}$. At the critical point the partition function presents a divergence: assuming that this happens at the value K_c, the same should happen also at $\tilde{K} = K_c$ thanks to eqn. (4.2.14). These two values can be different but, making the further hypothesis that there is only one critical point—a hypothesis that is fully justified by the physical point of view—these two values must coincide and the critical point is thus identified by the condition

$$\sinh 2K_c = 1; \quad T_c^{square} = 2.26922\ldots \mathcal{J} \tag{4.2.15}$$

Kramers and Wannier originally presented these arguments.

Let us consider now the general case in which there are two coupling constants. Note that combining the eqns. (4.2.13), we have

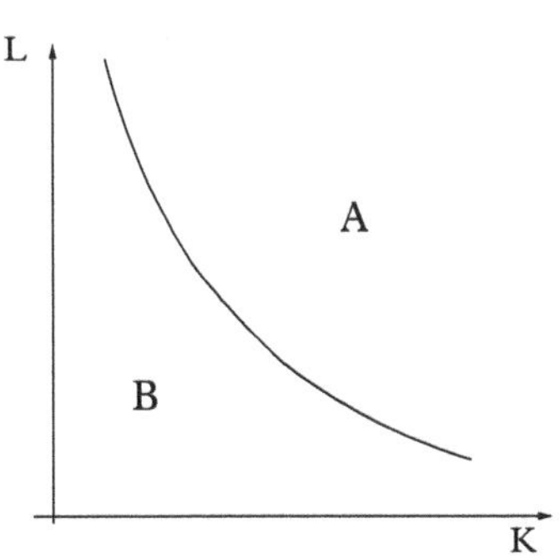

Fig. 4.6 *Critical curve.*

$$\sinh 2K \sinh 2L = \frac{1}{\sinh 2\tilde{K} \sinh 2\tilde{L}}. \tag{4.2.16}$$

This equation implies that, under the mapping $(K,L) \to (\tilde{K},\tilde{L})$, the region A in Figure 4.6 is transformed in the region B, and vice versa, leaving invariant the points along the curve

$$\sinh 2K \sinh 2L = 1. \tag{4.2.17}$$

If there is a line of fixed points in A, there should be another line of fixed points also in B. Assuming that there is only one line of fixed points, this is expressed by eqn. (4.2.17). Therefore, this is the condition that ensures the criticality of the Ising model with different coupling constants along the horizontal and vertical directions. This equation plays an important role both in the solution proposed by Baxter for the Ising model and in the discussion of its Hamiltonian limit.

4.3 Duality Relation: Hexagonal and Triangular Lattices

The duality transformation of the square lattice can be generalized to other lattices. In this section we discuss This section discusses the mapping between low- and high-temperature phases of the Ising model defined on the triangle and hexagonal lattices (Figure 4.7).

Let us introduce the coupling constants K_i and L_i $(i = 1,2,3)$ relative to the triangle and hexagonal lattices, respectively (Figure 4.8). In the absence of a magnetic field, the partition function of the hexagonal lattice is given by

$$Z_N^{\mathrm{H}}(\mathcal{L}) = \sum_{\{\sigma\}} \exp\Big[\mathcal{L}_1 \sum \sigma_l\sigma_i + \mathcal{L}_2 \sum \sigma_l\sigma_j + \mathcal{L}_3 \sum \sigma_l\sigma_k\Big], \tag{4.3.1}$$

with $\mathcal{L}_i = L_i/kT$. In the exponential term, the sums refer to all next-neighbour pairs of spins along the three different directions of the hexagonal lattice. Similarly, in the absence of the magnetic field, we can write the partition function on the triangle lattice as

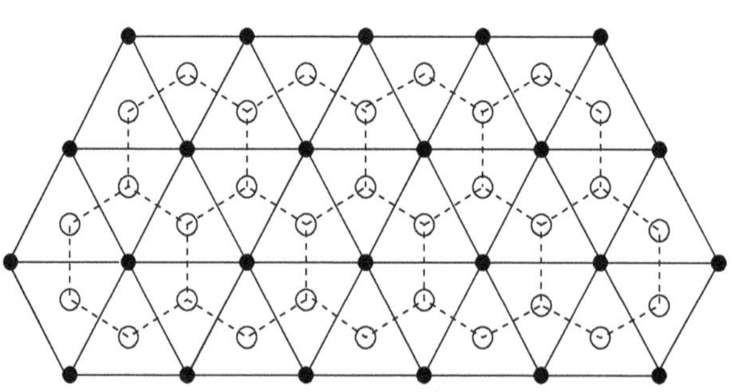

Fig. 4.7 *Dual lattices: hexagonal and triangle lattices.*

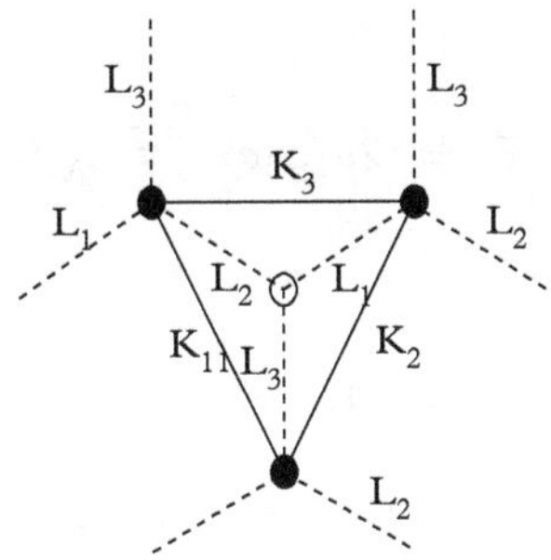

Fig. 4.8 *Coupling constants on the triangle and hexagonal lattices.*

$$Z_N^{\rm T}(\mathcal{K}) = \sum_{\{\sigma\}} \exp\left[\mathcal{K}_1 \sum \sigma_l\sigma_i + \mathcal{K}_2 \sum \sigma_l\sigma_j + \mathcal{K}_3 \sum \sigma_l\sigma_k\right], \tag{4.3.2}$$

with $\mathcal{K}_i = K_i/kT$ and the sums in the exponentials on all next-neighbour pairs of spins in the three different directions of the triangle lattice.

Let us consider the high temperature expansion of the partition function on the triangle lattice. Put $v_i = \tanh \mathcal{K}_i$, we have

$$Z_N^{\rm T}(\mathcal{K}) = (2\cosh \mathcal{K}_1 \cosh \mathcal{K}_2 \cosh \mathcal{K}_3)^M \sum_P v_1^{r_1} v_2^{r_2} v_3^{r_3}, \tag{4.3.3}$$

where the sum in on all closed polygons on the triangle lattice, with a number of sides equal to r_i $(i = 1, 2, 3)$ along the three different directions.

Consider now the low-temperature expansion of the partition function on the hexagonal lattice. This is obtained by drawing the lines corresponding to the anti-parallel links on the dual lattice. Since the triangle lattice of N sites is the dual of the hexagonal lattice with $2N$ sites, in this case we have[3]

$$Z_{2N}^{\rm H}(\mathcal{L}) = e^{[N(\mathcal{L}_1+\mathcal{L}_2+\mathcal{L}_3)]} \sum_P \exp[-2\mathcal{L}_1 r_1 + \mathcal{L}_2 r_2 + \mathcal{L}_3 r_3], \tag{4.3.4}$$

where the sum is on the closed polygons of the triangular lattice with the number of sides r_i $(i = 1, 2, 3)$ along the three directions.

Since in both expressions there is the same geometrical function given by the sum on polygons drawn on the triangle lattice, posing

$$\tanh \mathcal{K}_i^* = \exp[-2\mathcal{L}_i], \quad i = 1, 2, 3 \tag{4.3.5}$$

[3] For large N, the number of links along each of the three directions is equal to N.

the two partition functions are related as

$$Z_{2N}^{\mathrm{H}}(\mathcal{L}) = (2a_1a_2a_3)^{N/2}\, Z_N^{\mathrm{T}}(\mathcal{K}^*), \tag{4.3.6}$$

where

$$a_i = \sinh 2\mathcal{L}_i = 1/\sinh 2\mathcal{K}_i^*, \quad i = 1,2,3.$$

The relation (4.3.5) can be written in a more symmetrical way as

$$\sinh 2\mathcal{L}_i \sinh 2\mathcal{K}_i^* = 1. \tag{4.3.7}$$

As in the square lattice, the duality relation (4.3.7) implies that when one of the coupling constants is small, the other is large, and vice versa. However, the duality relation alone cannot determine in this case the critical temperature of the two lattices, since they are not self-dual. Fortunately, there exists a further important identity between the coupling constants of the two lattices that allows us to identify the singular points of the free energies of both models. This identity is the *star-triangle identity* and it is worth a detailed discussion.

4.4 Star-triangle Identity

The star-triangle identity plays an important role in the two-dimensional Ising model. In addition to the exact determination of the critical temperature for triangular and hexagonal lattices, this identity also enables us to establish the commutativity of the transfer matrix of the model for special values of the coupling constants. This aspect will be crucial for the exact solution of the model discussed in Chapter 6.

To prove such an identity, we first observe that the sites of the hexagonal lattice split into two classes, i.e. the hexagonal lattice is bipartite. The sites of type A interact only with those of type B, and vice versa, while there is no direct interaction between sites of the same type (see Figure 4.9). The generic term that enters the sum in the partition function (4.3.1) can be written as

$$\prod_b W(\sigma_b;\sigma_i,\sigma_j,\sigma_k), \tag{4.4.1}$$

where the product is on all sites of type B and the above quantity is expressed by the Boltzmann factor

$$W(\sigma_b;\sigma_i,\sigma_j,\sigma_k) = \exp\left[\sigma_b(\mathcal{L}_1\sigma_i + \mathcal{L}_2\sigma_j + \mathcal{L}_3\sigma_k\right]. \tag{4.4.2}$$

Since each spin of type B appears only once in (4.4.1), it is simple to sum over them in the expression of the partition function, with the result

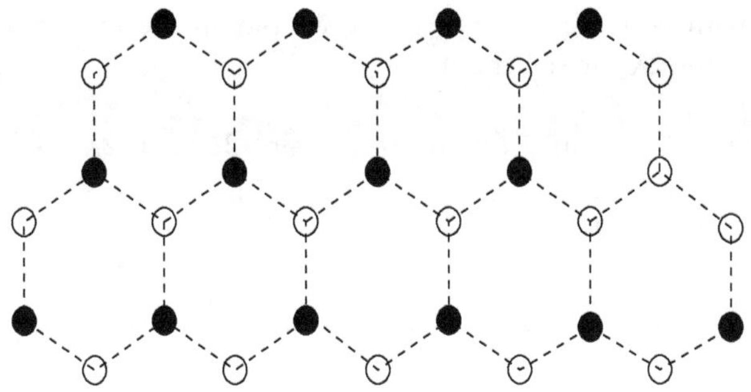

Fig. 4.9 *Bipartition of the hexagonal lattice: site of type A (black sites) and type B (white sites).*

$$Z_N^{\rm E}(\mathcal{L}) = \sum_{\sigma_a} \prod_{i,j,k} w(\sigma_i, \sigma_j, \sigma_k), \tag{4.4.3}$$

where

$$w(\sigma_i, \sigma_j, \sigma_k) = \sum_{\sigma_b = \pm 1} W(\sigma_b; \sigma_i, \sigma_j, \sigma_k) = 2\cosh(\mathcal{L}_1\sigma_i + \mathcal{L}_2\sigma_j + \mathcal{L}_3\sigma_k). \tag{4.4.4}$$

The values of each spin is ± 1 and using the identity

$$\cosh[\mathcal{L}\sigma] = \cosh\mathcal{L}, \quad \sinh[\mathcal{L}\sigma] = \sigma \sinh\mathcal{L}$$

we have

$$\begin{aligned} w(\sigma_i, \sigma_j, \sigma_k) = {} & c_1c_2c_3 + +\sigma_j\sigma_k c_1 s_2 s_3 + \\ & +\sigma_i\sigma_j s_1 s_2 c_3 \sigma_i\sigma_j s_1 c_2 s_3, \end{aligned} \tag{4.4.5}$$

where we have defined

$$c_i \equiv \cosh\mathcal{L}_i, \quad s_i \equiv \sinh\mathcal{L}_i.$$

It is important to note that the quantity $w(\sigma_i, \sigma_j, \sigma_k)$ can be written in such a way to be proportional to the Boltzmann factor of the triangular lattice! This means that there should exist some parameters $\mathcal{K}_i$ and a constant $\mathcal{D}$ such that

$$w(\sigma_i, \sigma_j, \sigma_k) = \mathcal{D}\exp\left[\mathcal{K}_1\sigma_j\sigma_k + \mathcal{K}_2\sigma_i\sigma_k + \mathcal{K}_3\sigma_j\sigma_k\right]. \tag{4.4.6}$$

These parameters can be determined by expanding the exponential as

$$\exp[x\sigma_a\sigma_b] = \cosh x + \sigma_a\sigma_b \sinh x,$$

and comparing with eqn. (4.4.5). Doing so, we obtain the important result that the products $\sinh 2\mathcal{L}_i \sinh 2\mathcal{K}_i$ are all equal

$$\sinh 2\mathcal{L}_1 \sinh 2\mathcal{K}_1 = \sinh 2\mathcal{L}_2 \sinh 2\mathcal{K}_2 = \sinh 2\mathcal{L}_3 \sinh 2\mathcal{K}_3 \equiv h^{-1} \tag{4.4.7}$$

with the constant h equal to

$$h = \frac{(1-v_1^2)(1-v_2^2)(1-v_3^2)}{4\left[(1+v_1v_2v_3)(v_1+v_2v_3)(v_2+v_1v_3)(v_3+v_1v_2)\right]^{1/2}}, \tag{4.4.8}$$

where $v_i = \tanh \mathcal{K}_i$, while the constant $\mathcal{D}$ is expressed by

$$\mathcal{D}^2 = 2h \sinh 2\mathcal{L}_1 \sinh 2\mathcal{L}_2 \sinh 2\mathcal{L}_3.$$

The identity (4.4.6) admits a natural graphical interpretation (Figure 4.8), summing on the spin of the type B at the centre of the hexagonal lattice (the one at the centre of the star), a direct interaction is generated between the spins of type A placed at the vertices of a triangle. In this way we can switch between the Boltzmann factor of the star of the hexagonal lattice and the Boltzmann factor of the triangular lattice.

4.5 Ising Model Critical Temperature: Triangle and Hexagonal Lattices

By using the star-triangle identity, it is now easy to determine the critical temperatures of the Ising model on triangular and hexagonal lattices. In fact, substituting the identity (4.4.6) in (4.4.3), the consequent expression is precisely the partition function of the Ising model on a triangular lattice made of $N/2$. Hence, rescaling $N \to 2N$, we have

$$Z_{2N}^{\mathrm{H}}(\mathcal{L}) = \mathcal{D}^N Z_N^{\mathrm{T}}(\mathcal{K}). \tag{4.5.1}$$

Using this equation, together with the duality relation (4.3.6), we obtain a relation that involves the partition function alone of the triangular lattice

$$Z_N^{\mathrm{T}}(\mathcal{K}) = h^{-N/2} Z_N^{\mathrm{T}}(\mathcal{K}^*), \tag{4.5.2}$$

with

$$\sinh 2\mathcal{K}_i^* = h \sinh 2\mathcal{K}_i, \quad i = 1,2,3, \tag{4.5.3}$$

and h given in (4.4.8). Thanks to (4.5.3), there is a one-to-one correspondance between the point $(\mathcal{K}_1, \mathcal{K}_2, \mathcal{K}_3)$ (relative to the high-temperature phase of the model) and the point $(\mathcal{K}_1^*, \mathcal{K}_2^*, \mathcal{K}_3^*)$ (relative to the low-temperature phase). If in the space of the coupling constants there is a line of fixed point under this mapping, this clearly corresponds to the value $h = 1$. For equal couplings ($\mathcal{K}_1 = \mathcal{K}_2 = \mathcal{K}_3 \equiv \mathcal{K}$), from (4.4.8) we have the equation

$$\frac{(1 - v^2)^3}{4\left[(1 + v^3)v^3(1 + v)^3\right]^{1/2}} = 1, \tag{4.5.4}$$

with $v = \tanh \mathcal{K}$. Taking the square of both terms of this equation and simplifying the expression, we arrive at

$$(1 + v)^4(1 + v^2)^3(v^2 - 4v + 1) = 0.$$

The only solution that also satisfies (4.5.4) and has a physical meaning is given by

$$v_c = 2 - \sqrt{3}.$$

This root determines the critical temperature of the homogeneous triangular lattice

$$\tanh \frac{K}{kT_c} = 2 - \sqrt{3},$$

or, equivalently

$$\sinh \frac{2K}{kT_c} = \frac{1}{\sqrt{3}}. \tag{4.5.5}$$

Numerically

$$T_c^{tr} = 3.64166...K. \tag{4.5.6}$$

Using eqn. (4.3.7) we can obtain the critical temperature of the Ising model on a homogeneous hexagonal lattice

$$\sinh \frac{2L}{kT_c} = \sqrt{3}. \tag{4.5.7}$$

Its numerical value is given by

$$T_c^{hex} = 1.51883...L. \tag{4.5.8}$$

It is interesting to compare the value of the critical temperatures (4.5.6) and (4.5.8) with the critical temperature of the square lattice, $T_c^{square} = 2.26922\mathcal{J}$, given by eqn. (4.2.15). At a given coupling constant, the triangular lattice is the one with the higher critical temperature, followed by the square lattice and then the hexagonal lattice. The reason is simple: the triangular lattice has the higher coordination number, $z = 6$, the hexagonal lattice has the lower coordination number, $z = 3$, while the square lattice is in between the two, with $z = 4$. The higher number of interactions among the spins of the triangular lattice implies that such a system tends to magnetize at higher temperatures than those of the other lattices.

4.6 Duality in Two Dimensions

The previous sections showed that the duality property of the Ising model, both for the square lattice and the hexagonal/triangular lattices, can be established on the basis of a geometrical argument, i.e. counting the closed polygons on the original lattice and its dual. However, the duality properties of a statistical model can be characterized in a purely algebrical way by considering particular transformation of the statistical variables entering the partition function. A particularly instructive example is the following. Consider the expression

$$Z(\beta) = \beta^{1/4} \sum_{n=-\infty}^{\infty} e^{-\pi\beta n^2}. \tag{4.6.1}$$

This can be interpreted as the partition function of a quantum system with energy levels given by $E_n = \pi n^2$. This expression is obviously useful for determining the numerical value of the partition function in the low-temperature phase ($\beta \gg 1$), since in this regime the sum is dominated by the first terms. In the high-temperature phase ($\beta \ll 1$), the situation is rather different and many terms are actually needed to reach a sufficient degree of accuracy. However, using the Poisson resummation formula discussed in Appendix 4.B, it is easy to see that we have

$$\begin{aligned} Z(\beta) &= \beta^{1/4} \sum_{n=-\infty}^{\infty} e^{-\pi\beta n^2} = \\ &= \beta^{1/4} \sum_{m=-\infty}^{\infty} \int_{-\infty}^{\infty} dx\, e^{-\pi\beta x^2} e^{2\pi i m x} = \\ &= \beta^{-1/4} \sum_{m=-\infty}^{\infty} e^{-\pi m^2/\beta}. \end{aligned} \tag{4.6.2}$$

Hence this partition function satisfied the important duality relation

$$Z(\beta) = Z\left(\frac{1}{\beta}\right). \tag{4.6.3}$$

In view of this identity, the partition function in the high-temperature phase can be efficiently computed by employing its dual expression: for $\beta \ll 1$ few terms of (4.6.2) are indeed enough to saturate the entire sum. This example shows that, sometimes, simple algebraic transformations allow us to establish important duality relations of the partition functions. This section focuses on these aspects of the two-dimensional statistical models.

Curl and divergence. In two dimensions, the duality relation is strictly related to the curl and the divergence of a vector field. In fact, a two-dimensional vector field $\vec{v}$ with vanishing line integral along a closed loop C

$$\oint_C d\vec{s} \cdot \vec{v} = 0, \tag{4.6.4}$$

satisfies the equation

$$\nabla \wedge \vec{v} = 0. \tag{4.6.5}$$

In this case, $\vec{v}$ can be expressed as the gradient of a scalar function Φ, i.e. $\vec{v} = \vec{\nabla}\,\Phi$. Going to the components, we have

$$\vec{v} = (v_1, v_2) = \left(\frac{\partial \Phi}{\partial x}, \frac{\partial \Phi}{\partial y}\right), \text{ if } \vec{\nabla} \wedge \vec{v} = 0. \tag{4.6.6}$$

Vice versa, a vector field $\vec{v}$ with vanishing flux across a closed surface S

$$\oint_S d\vec{\Sigma} \cdot \vec{v} = 0, \tag{4.6.7}$$

satisfies the equation

$$\vec{\nabla} \cdot \vec{v} = 0, \tag{4.6.8}$$

and it can always be expressed as $\vec{v} = \vec{\nabla} \wedge \vec{\Psi}$, where in two dimensions $\vec{\Psi} = (\psi, \psi)$ is a vector function of equal components. Explicitly

$$\vec{v} = (v_1, v_2) = \left(\frac{\partial \psi}{\partial y}, -\frac{\partial \psi}{\partial x}\right), \text{ if } \vec{\nabla} \cdot \vec{v} = 0. \tag{4.6.9}$$

The comparision between eqn. (4.6.6) and eqn. (4.6.9) shows that we can swap between them by exchanging $x \longleftrightarrow -y$.

Curl and divergence on a lattice. The above equations have a counterpart for variables that live on a lattice. Consider a square lattice and its dual, where the sites of the first lattice are identified by the coordinates (i, j) while those of the dual by the coordinates $\left(i + \frac{1}{2}, j + \frac{1}{2}\right)$. Suppose that there are some statistical variables defined along the links of the original lattice: denote by $\rho_{i+\frac{1}{2},j}$ the variable defined along the horizontal segment

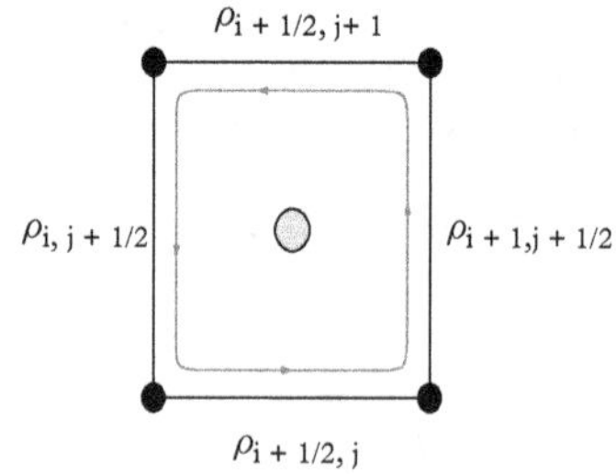

Fig. 4.10 *Circulation along the links of the original lattice. The site at the centre belongs to the dual lattice.*

that links the site (i,j) to the site $(i+1,j)$ and by $\rho_{i,j+\frac{1}{2}}$ the one defined along the vertical segment that links (i,j) to $(i,j+1)$. If the circulation along the perimeter $\mathcal{S}$ of the elementary cell of the lattice is zero (Figure 4.10), we have

$$\rho_{i+\frac{1}{2},j}+\rho_{i+1,j+\frac{1}{2}}-\rho_{i+\frac{1}{2},j+1}-\rho_{i,j+\frac{1}{2}}=0.$$

This is the discrete version of the curl-free equation on the sites of dual lattice. It can be identically satisfied in terms of a variable $\phi_{i,j}$ defined on the sites of the original lattice, by posing

$$\begin{aligned}\rho_{i+\frac{1}{2},j}&=\phi_{i+1,j}-\phi_{i,j},\\ \rho_{i,j+\frac{1}{2}}&=\phi_{i,j+1}-\phi_{i,j}.\end{aligned}$$

Vice versa, the discrete version on a lattice of the divergence-free condition (4.6.8) is given by

$$\rho_{i+\frac{1}{2},j}-\rho_{i-\frac{1}{2},j}+\rho_{i,j+\frac{1}{2}}-\rho_{i,j-\frac{1}{2}}=0.$$

This can be satisfied by expressing the variables ρ in terms of a discrete curl of a variable $\psi_{i+\frac{1}{2},j+\frac{1}{2}}$ defined on the dual lattice

$$\begin{aligned}\rho_{i+\frac{1}{2},j}&=\psi_{i+\frac{1}{2},j+\frac{1}{2}}-\psi_{i+\frac{1}{2},j-\frac{1}{2}},\\ \rho_{i,j+\frac{1}{2}}&=-\psi_{i+\frac{1}{2},j+\frac{1}{2}}+\psi_{i-\frac{1}{2},j+\frac{1}{2}}.\end{aligned}\tag{4.6.10}$$

After these general considerations, let us look at two examples.

4.6.1 Self-duality of the p-state Model

Consider a statistical model with scalar variables $\phi_{i,j}$ defined on the $N\times N$ sites of a square lattice, with periodic boundary conditions. Assume that these variables take discrete values on the interval (p an integer)

$$1 \leq \phi_{ij} \leq p,$$

and their Hamiltonian is a function of the differences of the next-neighbour values

$$\mathcal{H} = -\sum_{i,j}^{N} \left[K_1 \left(\phi_{i+1,j} - \phi_{i,j} \right) + K_2 \left(\phi_{i,j+1} - \phi_{i,j} \right) \right]. \tag{4.6.11}$$

Introducing the notation

$$\begin{aligned} \rho_{i+\frac{1}{2},j} &= \phi_{i+1,j} - \phi_{i,j}; \\ \rho_{i,j+\frac{1}{2}} &= \phi_{i,j+1} - \phi_{i,j}, \end{aligned}$$

together with $K_1 = \beta \mathcal{J}_1$, $K_2 = \beta \mathcal{J}_2$ for the coupling constants along the horizontal and vertical directions respectively, the partition function is given by

$$Z[K] = \mathrm{Tr}_\phi \exp \left[K_1 \rho_{i+\frac{1}{2},j} + K_2 \rho_{i,j+\frac{1}{2}} \right]. \tag{4.6.12}$$

In this expression we adopt the notation[4]

$$\mathrm{Tr}_\phi \equiv \prod_{i-1}^{N} \prod_{j-1}^{N} \frac{1}{\sqrt{p}} \sum_{\phi_{i,j}=1}^{p}$$

and we have taken into account the periodic boundary conditions

$$\begin{aligned} \phi_{i+N,j} &- \phi_{i,j}, \\ \phi_{i,j+N} &= \phi_{i,j}. \end{aligned}$$

There are then N^2 variables $\phi_{i,j}$ on which it is necessary to sum in order to obtain $Z[K]$. However, since the Hamiltonian depends on them only through the variables ρ, it would be more convenient to use directly these quantities. Since their number is equal to $2N^2$, we need to implement the N^2 conditions of vanishing circulation

$$R_{i+\frac{1}{2},j+\frac{1}{2}} \equiv \rho_{i+\frac{1}{2},j} + \rho_{i+1,j+\frac{1}{2}} - \rho_{i+\frac{1}{2},j+1} - \rho_{i,j+\frac{1}{2}} = 0, \quad (\mathrm{mod}\, p). \tag{4.6.13}$$

This can be done by introducing N^2 variables $\psi_{i+\frac{1}{2},j+\frac{1}{2}}$, that take p integer values, conjugated to each of $R_{i+\frac{1}{2},j+\frac{1}{2}}$ and defined on the sites of the dual lattice. We can insert in the partition function the N^2 expressions

[4] We have inserted the factor $1/\sqrt{p}$ in order to make symmetric the final expressions of the partition function.

$$\Delta_{i+\frac{1}{2},j+\frac{1}{2}} = \frac{1}{p} \sum_{\psi_{i+\frac{1}{2},j+\frac{1}{2}}=1}^{p} \exp\left[-\frac{2\pi i}{p} \psi_{i+\frac{1}{2},j+\frac{1}{2}} R_{i+\frac{1}{2},j+\frac{1}{2}}\right].$$

They are equal to 1 if the condition (4.6.13) is satisfied, and 0 otherwise. Hence, the partition function can be equivalently written as

$$Z[K] = \mathrm{Tr}_\rho \, \Delta_{i+\frac{1}{2},j+\frac{1}{2}} \exp\left[K_1 \rho_{i+\frac{1}{2},j} + K_2 \rho_{i,j+\frac{1}{2}}\right],$$

namely

$$Z[K] = \mathrm{Tr}_\rho \, \mathrm{Tr}_\psi \exp\left[K_1 \rho_{i+\frac{1}{2},j} + K_2 \rho_{i,j+\frac{1}{2}} - \frac{2\pi i}{p} \psi_{i+\frac{1}{2},j+\frac{1}{2}} R_{i+\frac{1}{2},j+\frac{1}{2}}\right]. \qquad (4.6.14)$$

where

$$\mathrm{Tr}_\psi \equiv \frac{1}{p} \sum_{\psi_{i+\frac{1}{2},j+\frac{1}{2}}=1}^{p}.$$

Notice that the sum on ρs can be explicitly performed. Each variable ρ appears in three terms: for instance, considering $\rho_{i+\frac{1}{2},j}$, its contribution to the partition function is equal to

$$G = \frac{1}{p} \sum_{\rho_{i+\frac{1}{2},j}=1}^{p} \exp\left[\rho_{i+\frac{1}{2},j}\left(K_1 - \frac{2\pi i}{p}(\psi_{i+\frac{1}{2},j+\frac{1}{2}} - \psi_{i+\frac{1}{2},j-\frac{1}{2}})\right)\right]. \qquad (4.6.15)$$

If we now define the dual coupling constant in terms of the Fourier transform of the original coupling constant

$$e^{\tilde{K}\sigma} = \frac{1}{\sqrt{p}} \sum_{b=1}^{p} e^{Kb} \exp\left(\frac{-2\pi i \sigma b}{p}\right), \qquad (4.6.16)$$

eqn. (4.6.15) can be expressed as

$$G = \frac{1}{\sqrt{p}} \exp\left[\tilde{K}_2 \left(\psi_{i+\frac{1}{2},j+\frac{1}{2}} - \psi_{i+\frac{1}{2},j-\frac{1}{2}}\right)\right]. \qquad (4.6.17)$$

By summing on all variables ρ in (4.6.14), the partition function can be equivalently expressed in terms of the variables ψ of the dual lattice and it fulfills the important self-duality relation

$$Z[K] = Z[\tilde{K}] \tag{4.6.18}$$

with

$$Z[\tilde{K}] = \text{Tr}_{\psi} \prod_{i=1}^{N} \prod_{j=1}^{N} \exp\left[\tilde{K}_2 \left(\psi_{i+\frac{1}{2},j+\frac{1}{2}} - \psi_{i+\frac{1}{2},j-\frac{1}{2}}\right)\right]$$
$$\times \exp\left[\tilde{K}_1 \left(\psi_{i+\frac{1}{2},j+\frac{1}{2}} - \psi_{i-\frac{1}{2},j+\frac{1}{2}}\right)\right]. \tag{4.6.19}$$

In conclusion, the dual coupling constants are defined by the Fourier transform of the original couplings, eqn. (4.6.16). More precisely, the coupling $\tilde{K}_2$ relative to the vertical links is determined by the original coupling K_1 of the horizontal links, while the coupling $\tilde{K}_1$ of the horizontal links depends on the coupling K_2 of the vertical links of the original lattice. This procedure can be clearly implemented also when the couplings are not constant but change along the sites of the lattice.

4.6.2 Duality Relation between XY Model and SOS Model

The application of the duality transformation does not necessarily lead to the same model. Even though in these cases we cannot predict the critical temperature of the model, the duality relation that links two different models can be nevertheless useful for studying the excitations in their high- and low-temperature phases, respectively. For instance, this is the case of the XY model that is related by duality to the SOS (Solid-on Solid) model. The statistical variables of the XY model are the angles θ_i (with values between $-\pi$ and π) defined on each site of the lattice. The Hamiltonian is

$$\mathcal{H} = -\sum_{\langle r,r' \rangle} \hat{f}(\theta_r - \theta_{r'}), \tag{4.6.20}$$

where $\hat{f}(\theta)$ is a periodic function, with period 2π

$$\hat{f}(\theta + 2\pi) = \hat{f}(\theta).$$

The usual choice is[5]

$$\hat{f}(\theta_r - \theta_{r'}) = \mathcal{J}\left[1 - \cos(\theta_r - \theta_{r'})\right]. \tag{4.6.21}$$

The partition function is given by

[5] For simplicity in the sequel we only consider the homogeneous case.

$$Z[K] = \prod_r \int_{-\pi}^{\pi} \frac{d\theta_r}{2\pi} \prod_{\langle r,r' \rangle} e^{f(\theta_r - \theta_{r'})}, \tag{4.6.22}$$

with $f = \beta \hat{f}$. Since every term of this sum is a periodic function of the angles, it can be expanded in the Fourier series

$$e^{f(\theta-\theta')} = \sum_{n=-\infty}^{\infty} e^{\tilde{f}(n)} \exp(2\pi i n(\theta - \theta')). \tag{4.6.23}$$

For the inverse formula we have

$$e^{\tilde{f}(n)} = \int_{-\pi}^{\pi} \frac{d\theta}{2\pi} e^{f(\theta)} \exp(-2\pi i n\theta).$$

Using eqn. (4.6.23), the partition function becomes

$$Z[K] = \prod_r \int_{-\pi}^{\pi} \frac{d\theta_r}{2\pi} \prod_{\langle r,r' \rangle} \sum_{\{n_{r,r'}\}} e^{\tilde{f}(n_{r,r'})} \exp(2\pi i n_{r,r'}(\theta_r - \theta_{r'})), \tag{4.6.24}$$

where $n_{r,r'}$ are variables with integer values defined on the links between the next-neighbour sites r and r'. In two dimensions, every angle θ_r enters the expression of four different terms, i.e. those relative to the segments that link the site r to its four next-neighbour sites. By adopting the previous notation for the coordinates of the sites and for the variables defined along the links, the term in which is present $\theta_{i,j}$ is given (Fig. 4.11) by

$$\exp\left[\theta_{i,j}(n_{i+\frac{1}{2},j} + n_{i,j+\frac{1}{2}} - n_{i-\frac{1}{2},j} - n_{i,j-\frac{1}{2}}\right].$$

Thanks to the identity

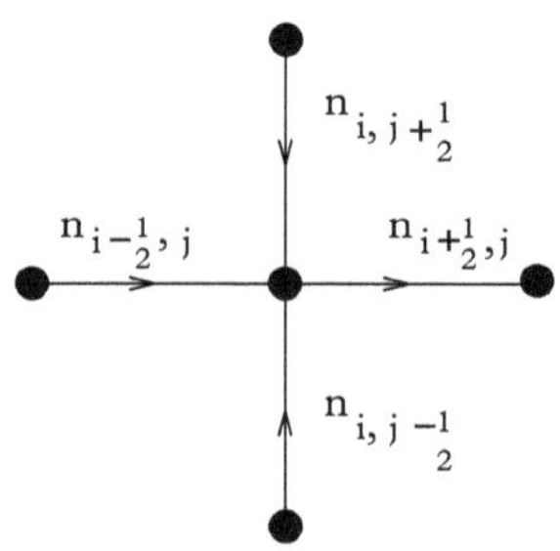

Fig. 4.11 *Condition of the vanishing divergence of the variables $n_{r,r'}$.*

$$\frac{1}{2\pi}\int_{-\pi}^{\pi} d\alpha\, e^{i\alpha x} = \delta_{x,0},$$

by integrating on θ_r in (4.6.24), we have

$$\frac{1}{2\pi}\int_{-\pi}^{\pi} d\theta_r e^{i\theta_r \sum_{r'} n_{r,r'}} = \delta_{\sum_{r'} n_{r,r'},0},$$

i.e. the variable $n_{r,r'}$ defined along the links has zero divergence. Referring to the general discussion of the previous section, the variables $n_{r,r'}$ can be then expressed in terms of the differences of the integer value variables m_s defined on the sites of the dual lattice. In such a way, the original definition (4.6.22) of the partition function becomes a sum on all possible integer values of the variables m_s, defined on the sites s of the dual lattice

$$Z[K] = \sum_{\{m_s\}} \prod_{\langle s,s'\rangle} e^{\tilde{f}(m_s - m_{s'})}. \tag{4.6.25}$$

Hence, the dual model corresponding to the XY model is the SOS Model, names thus because the integer variables m_s can be regarded as the heights (either positive or negative) of a surface of a solid.

Appendix 4.A. Numerical Series

In this Appendix we discuss briefly a numerical method for extracting useful information on the critical behaviour of the thermodynamical quantities by using the first terms of their perturbative series. Let us consider a thermodynamical quantity, the partition function for instance, and suppose that such a quantity is expressed by a series expansion in the parameter x

$$f(x) = \sum_{n=0} a_n x^n. \tag{4.A.1}$$

The problem consists of obtaining the parameters x_c and γ relative to its behaviour close to the critical point x_c

$$f(x) \sim b\,(x_c - x)^{-\gamma} = b x_c^{-\gamma}\left(1 - \frac{x}{x_c}\right)^{-\gamma}, \tag{4.A.2}$$

if the only information available is the first k terms of the series (4.A.1).

The solution of this problem is the following. First of all, the estimate of the critical point x_c can be done by means of the convergence radius of the series (4.A.1) by assuming that there is no other singularity (also complex) closer to the origin. Expanding the right-hand side of (4.A.2) in power series we have

$$f(x) \sim b x_c^{-\gamma}\left[1+\gamma\left(\frac{x}{x_c}\right)+\frac{\gamma(\gamma+1)}{2!}\left(\frac{x}{x_c}\right)^2+\dots\frac{\gamma(\gamma+1)\cdots(\gamma+k-1)}{k!}\left(\frac{x}{x_c}\right)^k+\dots\right] \tag{4.A.3}$$

Considering the ratio of two next coefficients of this series and comparing with the corresponding ratio of the series (4.A.1) we have

$$R_n = \frac{a_n}{a_{n-1}} = \frac{1}{x_c}\left[1+\left(\frac{\gamma-1}{n}\right)\right]. \tag{4.A.4}$$

Hence, with the hypothesis done above on the singularities of the function $f(x)$, a plot of the ratios R_n versus the variable $1/n$ should show a linear behaviour, whose slope provides an estimate of the quantity $x_c^{-1}(\gamma-1)$, whereas its value at the origin gives an estimate of x_c^{-1}.

As an example of this method, let us consider the susceptibility of the Ising model on a two-dimensional triangular lattice. The high-temperature series expansion of this quantity is known up to the 12th term and it is given by ($v = \tanh\beta\mathcal{J}$)

$$\begin{aligned}\chi(T) = {} & 1+6v+30v^2+138v^3+606v^4+258v^5+ \\ & +10818v^6+44574v^7+181542v^8+732678v^9+ \\ & +2.935.218v^{10}+11.687.202v^{11}+46.296.210v^{12}+\cdots\end{aligned} \tag{4.A.5}$$

Employing the ratios R_n obtained by these coefficients, we arrive at the following estimates of the critical temperature and the coefficient γ

$$v_c^{-1} \simeq 3.733 \pm 0.003; \quad \gamma \simeq 1.749 \pm 0.003$$

that are remarkably close to their exact values

$$v_c^{-1} = 2+\sqrt{3} = 3.73205...; \quad \gamma = \frac{7}{4} = 1.75.$$

Appendix 4.B. Poisson Sum Formula

Consider the series

$$f(x) = \sum_{m=-\infty}^{\infty} G(x+mT), \tag{4.B.1}$$

where $G(x)$ is a function that admits Fourier transform. Since $f(x)$ is a periodic function

$$f(x) = f(x+T),$$

it can be expressed in a Fourier series

$$f(x) = \sum_{n=-\infty}^{\infty} c_n \exp\left[\frac{2\pi inx}{T}\right],$$

with the coefficients given by

$$c_n = \frac{1}{T}\int_0^T dy f(y) \exp\left[-\frac{2\pi iny}{T}\right].$$

Substituting the expression of $f(x)$, we have

$$c_n = \frac{1}{T}\sum_{m=-\infty}^{\infty}\int_0^T dy\, G(y+mT) \exp\left[-\frac{2\pi iny}{T}\right].$$

By making the change of variable $y + mT \to z$, we obtain

$$\begin{aligned} c_n &= \frac{1}{T}\sum_{m=-\infty}^{\infty}\int_{mT}^{(m+1)T} dz\, G(z) \exp\left[-\frac{2\pi inz}{T}\right] \\ &= \frac{1}{T}\int_{-\infty}^{\infty} dz\, G(z) \exp\left[-\frac{2\pi inz}{T}\right] = \frac{1}{T}\hat{G}\left(\frac{2\pi n}{T}\right), \end{aligned}$$

where $\hat{G}(p)$ is the Fourier transform of the function $G(x)$

$$\hat{G}(p) = \int_{-\infty}^{\infty} dz\, G(z)\, e^{-ipz}.$$

In such a way, the original series (4.B.1) can be expressed as

$$\sum_{m=-\infty}^{\infty} G(x+mT) = \frac{1}{T}\sum_{m=-\infty}^{\infty} \hat{G}\left(\frac{2\pi m}{T}\right) \exp\left[\frac{2\pi im}{T}\right]. \tag{4.B.2}$$

This equation is known as the *Poisson sum formula.* It is also equivalent to the following identity for the $\delta(x)$ function

$$\sum_{m=-\infty}^{\infty} \delta(x - mT) = \frac{1}{T}\sum_{m=-\infty}^{\infty} \exp\left[\frac{2\pi imx}{T}\right]. \tag{4.B.3}$$

REFERENCES

The famous articles by Peierls and Kramers-Wannier are

Kramers, H.A. and Wannier, G.H. (1941). 'Statistics of the Two-Dimensional Ferromagnet. Part I', *Physics Reviews*, 60: 252.

H.A. Kramers, G.H. Wannier, 'Statistics of the Two-Dimensional Ferromagnet. Part II', *Physics Reviews*, 60: 263.

Peierls, R.E. (1936). 'On Ising's Model of Ferromagnetism', *Proceedings of the Cambridge Philosophical Society*, 32: 477.

Two important references on the duality properties in statistical mechanics and field theory are

Kadanoff, L. (1978). 'Lattice Coulomb Gas Representations of Two-dimensional Problems', *Journal of Physics A*, 11: 1399.

Savit, R. (1980). 'Duality in Field Theory and Statistical Mechanics', *Reviews in Modern Physics*, 52: 453.

It is worth mentioning the studies on the duality properties of the three-dimensional Ising model. See

Balian, R., Drouffe, J.M. and Itzykson, C. (1975). 'Gauge Fields on a Lattice. II Gauge-invariant Ising Model', *Physics Reviews*, 11: 2098.

Wegner, F. (1971). 'Duality in Generalized Ising Models and Phase Transitions without Local Order Parameters', *Journal of Mathematics and Physics*, 12: 2259.

The duality transformation plays an important role also in the theory of the fundamental interactions. Interested readers are referred to

Seiberg, N. and Witten, E. (1994). 'Electric-magnetic Duality, Monopole Condensation and Confinement in N = 2 Supersymmetric Yang–Mills Theory', *Nuclear Physics B*, 426: 19.

PROBLEMS

4.1. Three-dimensional lattices

Generalize Peierls's argument to the Ising model on three-dimensional lattices and prove that the model admits a phase transition.

4.2. Low-temperature series in the presence of a magnetic field

Consider the two-dimensional Ising model on a square lattice with equal coupling along the horizontal and vertical links and in presence of an external magnetic field B. Generalize the discussion on the series expansion of the free energy in the low-temperature phase and show that Z_N can be written as

$$Z_N = \exp[2NK + N\beta B] \sum_{r,s=1}^{\infty} n(r,s) \exp[-2Kr] \exp[-s\beta B]$$

where $K = \beta \mathcal{J}$ and $n(r,s)$ is the number of closed graphs made of r links on the dual lattice, having in their internal region s points of the original lattice.

4.3. Free energy

Consider the high-temperature series expansion of a homegeneous Ising model on a two dimensional square lattice

$$\begin{aligned} Z_N &= (2\cosh\beta\mathcal{J})^{2N} \\ &\times \left[1 + Nv^4 + 2Nv^6 + 2Nv^8 + \frac{1}{2}N(N-9)v^8 + \cdots\right] \end{aligned}$$

($v = \tanh\beta\mathcal{J}$). In the thermodynamic limit, we should have $Z_N \simeq (Z_1)^N = e^{-N\beta f}$, where f is the free-energy per unit site. Using the formula above to find the high-series expansion of Z_1

$$Z_1 = 2(\cosh\beta\mathcal{J})^2 \, (1 + v^4 + 2v^6 - 2v^8 + \cdots).$$

4.4. Poisson sum rule

a. Generalize the Poisson sum rule to the d-dimensional case.
b. Using the Poisson sum rule show

$$\sum_{n=0}^{\infty} \frac{1}{x^2 + n^2} = \frac{1}{2x^2} - \frac{\pi}{2x} + \frac{\pi}{x} \frac{1}{1 - e^{-2Pix}}$$

4.5. Self-duality

Consider the function

$$Z(K) = K^{1/4} \sum_{n=-\infty}^{\infty} e^{-\pi K n^2}$$

that satisfies

$$Z(K) = Z\left(\frac{1}{K}\right).$$

a. Show that in this case the duality relation does not imply a phase transition at $K = 1$.

b. How many terms are necessary in the original expression to compute $Z(K)$ with a precision 10^{-4} for $K = 0.01$? How many terms are needed to reach the same precision by using its dual expression?

4.6. Critical temperature of the 3-state Model

Using the self-duality of the 3-state model to determine its critical temperature.

4.7. Quadratic model

Let

$$\beta\mathcal{H} = -\frac{\mathcal{J}}{2}\sum_{\langle r,r'\rangle}(\Phi_r - \Phi_{r'})^2 + \frac{1}{4}\ln\mathcal{J}$$

be the Hamiltonian of a two-dimensional system where its variables assume all real values. Show that the model is self-dual under the transformation $\mathcal{J} \longleftrightarrow \frac{1}{\mathcal{J}}$.

5

Combinatorial Solutions of the Ising Model

To make a correct conjecture on an event, it seems that it is necessary to calculate the number of all the possible cases exactly and to determine their combinatorics.

Jacob Bernoulli, *Ars Conjectandi*

There are many methods to solve exactly the two-dimensional Ising model at zero magnetic field. Some of these methods have proved to be quite general and they have been employed in the solution of other important models of statistical mechanics. This is the case, for instance, for the method of commuting transfer matrices, based on the solution of the Yang–Baxter equations (Chapter 6). On the contrary, other methods prove to be applicable only to the Ising model, i.e. the two combinatorial approaches we discuss in this chapter. Both methods are quite ingenious and original, which alone justifies their detailed analysis. The first method, which starts from the high-temperature series expansion of the Ising model, finally reduces the free energy computation to a problem of a random walk on a lattice. The second method, which also starts from the high-temperature series, transforms the problem of computing the free energy of the Ising model in a counting problem of dimer configurations on a lattice.

5.1 Combinatorial Approach

5.1.1 Partition Function

The combinatorial solution of the Ising model, originally proposed by Kac and Ward, has its starting point in the high-temperature series expansion of the partition function (Section 4.2.1). The elegant solution presented here is from Vdovichenko. For simplicity, the following considers only the homogeneous case in which there is only one coupling constant, so that only the parameter $v = \tanh \beta \mathcal{J}$ enters the partition function. The partition function on a square lattice is given by

$$Z_N = 2^N (1 - v^2)^{-N} \Phi(v). \tag{5.1.1}$$

Statistical Field Theory: An Introduction to Exactly Solved Models in Statistical Physics. Second Edition. Giuseppe Mussardo, Oxford University Press (2020). © Giuseppe Mussardo. DOI: 10.1093/oso/9780198788102.001.0001

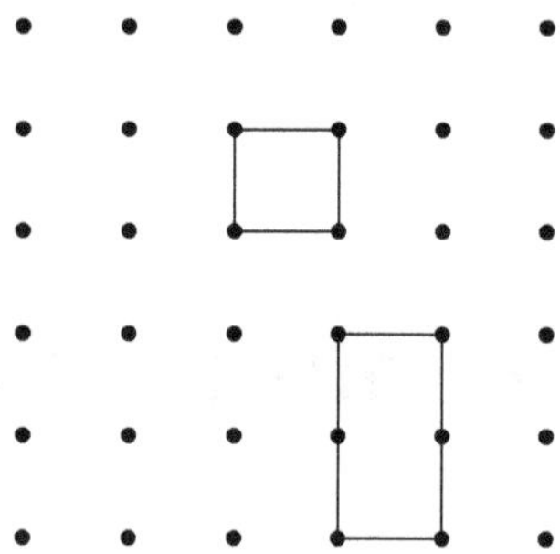

Fig. 5.1 *Graph of order v^{10}.*

with

$$\Phi(v) = \sum_r g_r v^r, \tag{5.1.2}$$

where g_r is the number of the closed graphs, not necessarily connected, given by an even number r of links. For example, the graph in Figure 5.1, is one of the terms of order v^{10} present in the summation (5.1.2).

There are three steps in the Vdovichenko's method of solution: a) expressing the sum on the polygons as a sum on the closed loops without intersections; b) transforming the sum on the closed loops without intersections in a sum on all possible closed loops; c) a random walk on a lattice that can be easily solved.

Let us discuss the implementation of the first step, i.e. how to organize the sum on the polygons in terms of their connected parts. Let us observe that each graph consists of one or more connected parts. For non-self intersecting graphs this statement is obvious: the graph in Figure 5.1, consists of two disconnected parts. But for self-intersecting graphs the statement can be ambiguous and there could be different connected parts according to the different decompositions. In order to clarify this issue, consider the graph in Figure 5.2. This can be decomposed in three different ways, as shown in Figure 5.3: it can be decomposed into one or two connected parts without intersections or into one connected

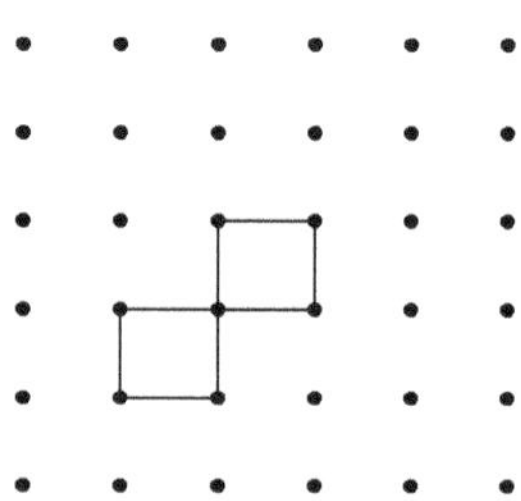

Fig. 5.2 *Self-intersecting graph.*

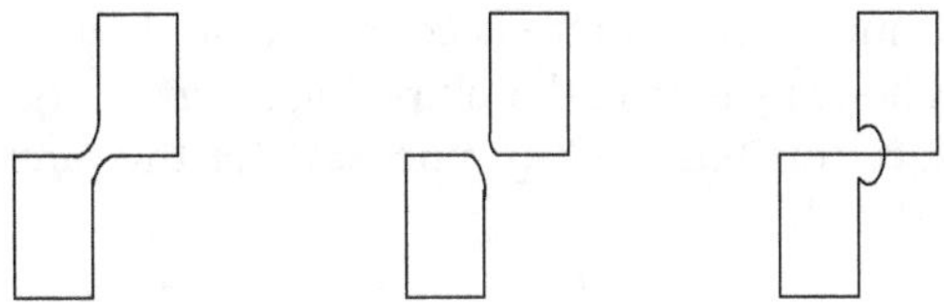

Fig. 5.3 *Three different decompositions in the connected parts for a self-intersecting graph.*

part but with an intersection. It is easy to show that this rule is quite general, because there are always three possible decompositions for all the self-intersections of a graph.

The sum on the polygons given in eqn. (5.1.2) can be organized in a sum on the connected parts of the graphs but we have to be careful to count properly the different terms, in particular, not counting more than the same configuration. This problem can be solved by weighting each graph by a factor $(-1)^n$, where n is the total number of self-intersections of a loop. In this way, all extra terms in the sum disappear. In Figure 5.3, the first two terms are weighted by $+1$, while the last term by -1, so that in the final expression there is correctly only one term.

Notice that, by adopting the prescription given above to perform the sum on the closed loops, we can include in the sum also the graphs with repeated bonds, the simplest of which is shown in Figure 5.4. These graphs are obviously absent in the original formulation of the high-temperature expansion of the model, since in some of their sites there are an odd number of links. However, with the new weight associated to the diagrams, it is easy to see that these terms are cancelled in the sum. In fact, in the connected decomposition part of these graphs, each common link can be passed through in two different ways, one without intersection (Figure 5.4.b), the other with a self-intersection (Figure 5.4c). Hence, the connected parts of this graph have equal and opposite signs and therefore they cancel in the sum.

There is still a disadvantage to the procedure of assigning a weight to the graphs because it depends on a global property of the graph as the number of its intersections. It would be more convenient to express the weight $(-1)^n$ in a local way. This is possible thanks to the familiar geometrical property that the total angle of rotation spanned by the tangent going around a closed plane loop is $2\pi(l+1)$ where l is an integer (positive or negative), with a parity that coincides with the number ν of the self-intersection of the loop. Hence, we can assign a phase factor $e^{i\alpha/2}$ to each point of the loop, where the angle of rotation α takes values $\alpha = 0, \pm\frac{\pi}{2}$ in correspondence with the angle of the change of direction to the next bond, so that the product of all these factors going around the loop gives $(-1)^{\nu+1}$. For a set of s loops we have $(-1)^{n+s}$, with $n = \sum \nu$.

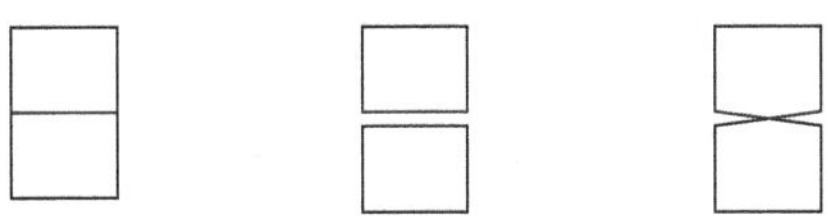

a b c

Fig. 5.4 *Graph with repeated bonds.*

Finally, we can take automatically into account the number of self-intersections of a loop by weighting each node by $e^{i\alpha/2}$ and multiplying the graph (given by a set of s loops) by the factor $(-1)^s$, since this term will compensate for the same factor present in the previous expression $(-1)^{n+s}$.

Let us now denote by f_r the sum over single loops of r links, each loop weighted according to the prescription above. The sum on all pairs of loops with total number r of links is then given by

$$\frac{1}{2!}\sum_{r_1+r_2=r} f_{r_1} f_{r_2},$$

where the factor 2! in the denominator takes into account that the permutation of the two indices gives rise to the same pair of loops. An analogous factor $n!$ is present in the denominator for the sum on n loops. Therefore, the function Φ can be written as

$$\Phi(v) = \sum_{s=0} (-1)^s \frac{1}{s!} \sum_{r_1, r_2, \cdots = 1}^{\infty} v^{r_1+r_2+\cdots+r_s} f_{r_1} \ldots f_{r_s}. \tag{5.1.3}$$

Since in Φ there are terms corresponding to set of loops with any possible total length[1] $r = r_1 + r_2 + \cdots$, in the sum (5.1.3) the indices $r_1, r_2, \ldots$ assume independently all values from 1 to ∞, so that

$$\sum_{r_1, r_2, \cdots = 1}^{\infty} v^{r_1+r_2+\cdots+r_s} f_{r_1} \ldots f_{r_s} = \left(\sum_{r=1}^{\infty} v^r f_r\right)^s.$$

Hence Φ is expressed as

$$\Phi(v) = \exp\left[-\sum_{r=1}^{\infty} v^r f_r\right]. \tag{5.1.4}$$

With this expression we have completed the steps (a) and (b) of the Vdovichenko's method. It remains then to evaluate explicitly the quantity f_r.

Since in a square lattice there are four different directions in which to move, it is convenient to number them by the index $\mu = 1, 2, 3, 4$ (Figure 5.5). Let us introduce a new function $W_r(i, j, \mu)$, defined as the sum over all possible paths of length r that start from a given point of coordinates i_0, j_0 along a direction μ_0 and arrive at a point of coordinates i, j along the direction μ. The paths entering the definition of $W_r(i, j, \mu)$ are weighted with the factors $e^{i\alpha/2}$ previously introduced. If we now choose (i_0, j_0) as initial

[1] The loops with a number of sites larger than the number N of the sites of the lattice do not contribute to the sum, since they necessarily contain repeated bonds.

Fig. 5.5 *Possible directions of movement on a square lattice.*

point, $W_r(i_0, j_0, \mu_0)$ becomes the sum over all loops leaving and returning to same point.[2] We have then the identity

$$f_r = \frac{1}{2r} \sum_{i_0, j_0, \mu} W_r(i_0, j_0, \mu), \tag{5.1.5}$$

where the term $1/(2r)$ takes into account the fact that in the sum on the right-hand side each loop can be crossed in two opposite directions and can have any of its r nodes as a starting point. Thanks to its definition, the function $W_r(i, j, \mu)$ satisfies the recursive equations

$$\begin{aligned}
W_{r+1}(i,j,1) &= W_r(i-1,j,1) + e^{-i\frac{\pi}{4}}\, W_r(i,j-1,2) + 0 + e^{i\frac{\pi}{4}}\, W_r(i,j+1,4),\\
W_{r+1}(i,j,2) &= e^{i\frac{\pi}{4}}\, W_r(i-1,j,1) + W_r(i,j-1,2) + e^{-i\frac{\pi}{4}}\, W_r(i+1,j,3) + 0,\\
W_{r+1}(i,j,3) &= 0 + e^{i\frac{\pi}{4}}\, W_r(i,j-1,2) + W_r(i+1,j,3) + e^{-i\frac{\pi}{4}}\, W_r(i,j+1,4),\\
W_{r+1}(i,j,4) &= e^{-i\frac{\pi}{4}}\, W_r(i-1,j,1) + 0 + e^{i\frac{\pi}{4}}\, W_r(i+1,j,3) + W_r(i,j+1,4).
\end{aligned} \tag{5.1.6}$$

Let us consider, for instance, the first of them. We can reach the point $i, j, 1$ by taking the last $(r+1)$th step from the left, from below or from above, but not from the right. The coefficients present in the equation come from the phase factors relative to the change of the directions. With the same argument we can derive the other equations in (5.1.6). Introducing the matrix Λ of the coefficients, the recursive equations can be written as

$$W_{r+1}(i,j,\mu) = \sum_{i',j',\mu'} \Lambda(ij\mu \mid i'j'\mu')\, W_r(i', j', \mu'), \tag{5.1.7}$$

which admits a suggestive interpretation: this equation can be interpreted as a Markov process associated to a random walk on the lattice, with the transition probability between two next-neighbour sites expressed by the relative matrix element of Λ. Since there are four possible directions for this motion, keeping fixed all other parameters, Λ is a 4×4 matrix in the indices μ' and μ.

In the light of the interpretation given here of the recursive equations, the *transition probability* relative to a path of total length r is expressed by the matrix Λ^r. Notice that the diagonal elements of this matrix express the probability to return to the initial point after traversing a loop of length r, i.e. they coincide with $W_r(i_0, j_0, \mu_0)$. Therefore we have

[2] It is understood that these closed loops cannot pass through the same links in opposite direction. This means that the last step of these walks cannot be along the opposite direction of μ_0.

$\Lambda =$

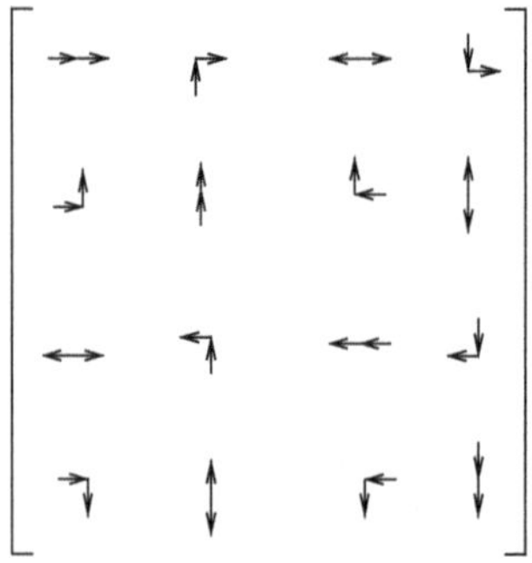

Fig. 5.6 *Matrix elements of* Λ.

$$\mathrm{Tr}\,\Lambda^r = \sum_{i_0,j_0,\mu} W_r(i_0,j_0,\mu),$$

and, comparing with eqn. (5.1.5), we arrive

$$f_r = \frac{1}{2r}\mathrm{Tr}\,\Lambda^r = \frac{1}{2r}\sum_a \lambda_a^r, \tag{5.1.8}$$

where λ_a are the eigenvalues of the matrix Λ. Using this expression in (5.1.4) and interchanging the indices of the sum, we have

$$\begin{aligned}\Phi(v) &= \exp\left[-\frac{1}{2}\sum_i\sum_{r=1}^{\infty}\frac{1}{r}v^r\lambda_i^r\right] = \\ &= \exp\left[\frac{1}{2}\sum_i \log(1-v\lambda_i)\right] = \prod_i \sqrt{1-v\lambda_i}.\end{aligned} \tag{5.1.9}$$

Finally, we need to determine the eigenvalues of Λ. The diagonalization of this matrix with respect the coordinates k and l of the lattice can be easily done by using the Fourier transformation. In fact, defining

$$W_r(p,q,\mu) = \sum_{k,l=0}^{L} e^{-\frac{2\pi i}{L}(pk+ql)}\, W_r(k,l,\mu),$$

with $N = L^2$, and taking the Fourier transform of (5.1.6), we have

$$\begin{aligned}W_{r+1}(p,q,1) &= \epsilon^{-p}\, W_r(p,q,1) + \epsilon^{-q}\alpha^{-1}\, W_r(p,q,2) + \epsilon^{q}\alpha\, W_r(p,q,4),\\ W_{r+1}(p,q,2) &= \epsilon^{-p}\alpha\, W_r(p,q,1) + \epsilon^{-q}\, W_r(p,q,2) + \epsilon^{p}\alpha^{-1}\, W_r(p,q,3),\end{aligned} \tag{5.1.10}$$

$$W_{r+1}(p,q,3) = \epsilon^{-q}\alpha\, W_r(p,q,2) + \epsilon^{p}\, W_r(p,q,3) + \epsilon^{q}\alpha^{-1}\, W_r(p,q,4),$$
$$W_{r+1}(p,q,4) = \epsilon^{-p}\alpha^{-1}\, W_r(p,q,1) + \epsilon^{p}\alpha\, W_r(p,q,3) + \epsilon^{q}\, W_r(p,q,4), \quad (5.1.11)$$

(where $\epsilon = e^{2\pi i/L}$ and $\alpha = e^{i\pi/4}$). Since $W_r(p,q,\mu)$ appears with the same indices p and q both on the left- and right-hand sides of these equations, the Fourier transform of the matrix Λ is diagonal with respect to these indices and we have

$$\Lambda(p,q,\mu \mid p,q,\mu') = \begin{pmatrix} \epsilon^{-p} & \alpha^{-1}\epsilon^{-q} & 0 & \alpha\epsilon^{q} \\ \alpha\epsilon^{-p} & \epsilon^{-q} & \alpha^{-1}\epsilon^{p} & 0 \\ 0 & \alpha\epsilon^{-q} & \epsilon^{p} & \alpha^{-1}\epsilon^{q} \\ \alpha^{-1}\epsilon^{-p} & 0 & \alpha\epsilon^{p} & \epsilon^{q} \end{pmatrix}. \quad (5.1.12)$$

An easy computation shows that

$$\prod_{i=1}^{4}(1 - v\lambda_i) = Det(\mathbf{1} - v\Lambda) = \quad (5.1.13)$$
$$= (1+v^2)^2 - 2v(1-v^2)\left(\cos\frac{2\pi p}{L} + \cos\frac{2\pi q}{L}\right).$$

Coming back to the original expression (5.1.1), we have then

$$Z_N = 2^N(1-v^2)^{\,N}\prod_{p,q}^{L}\left[(1+v^2)^2 - 2v(1-v^2)\left(\cos\frac{2\pi p}{L} + \cos\frac{2\pi q}{L}\right)\right]^{1/2}, \quad (5.1.14)$$

and the free energy of the Ising model is expressed as

$$-\frac{F(T)}{kT} = \log Z_N = N\log 2 - N\log(1-v^2) + \quad (5.1.15)$$
$$+\frac{1}{2}\sum_{p,q=0}^{L}\log\left[(1+v^2)^2 - 2v(1-v^2)\left(\cos\frac{2\pi p}{L} + \cos\frac{2\pi q}{L}\right)\right].$$

When $L \to \infty$, the sum becomes an integral

$$-\frac{F(T)}{kT} = \log Z_N = N\log 2 - N\log(1-v^2) + \quad (5.1.16)$$
$$+\frac{N}{2(2\pi)^2}\int_0^{2\pi}\int_0^{2\pi}\log\left[(1+v^2)^2 - 2v(1-v^2)(\cos\omega_1 + \cos\omega_2)\right]d\omega_1 d\omega_2.$$

This expression shows that $F(T)$ is an extensive quantity, since it is proportional to the total number N of the sites of the lattice. Beside the value $v=1$ (that corresponds to $T=0$), $F(T)$ has a singular point at a finite value of T when the argument of the logarithm inside the integral vanishes. As a function of ω_1 and ω_2, the argument of the logarithm has a minimum when $\cos\omega_1 = \cos\omega_2 = 1$ and the corresponding value is

$$(1+v^2)^2 - 4v(1-v^2) = (v^2+2v-1)^2.$$

It is easy to see that this expression has a minimum, with a null value, only for the positive value

$$v = v_c = \sqrt{2} - 1.$$

The corresponding critical temperature T_c, fixed by

$$\tanh \frac{\mathcal{J}}{kT_c} = v_c, \quad kT_c = 2.26922...\mathcal{J}, \tag{5.1.17}$$

determines the phase transition point. The expansion of the function $F(T)$ in power series in $t = k(T - T_c)/\mathcal{J}$ around this critical point shows that it has both a singular and a regular part. The regular part is simply obtained by substituting $t = 0$ in its expression. In order to determine the singular part, it is sufficient to expand the argument of the logarithm in power series in t, in ω_1 and ω_2. In this way, the integral in (5.1.16) becomes

$$\int_0^{2\pi}\int_0^{2\pi} \log\left[a_1 t^2 + a_2(\omega_1^2 + \omega_2^2)\right] d\omega_1 d\omega_2, \tag{5.1.18}$$

where a_1 and a_2 are two constants expressed by

$$a_1 = 32(3-2\sqrt{2})\left(\frac{\mathcal{J}}{kT_c}\right)^2, \quad a_2 = 2(3-2\sqrt{2}).$$

Computing the integral, the behaviour of the free energy in the vicinity of the phase transition is given by

$$F(T) \simeq A - \frac{B}{2}(T-T_c)^2 \log|T - T_c|, \tag{5.1.19}$$

where A and B are two other constants, with $B > 0$. The specific heat, expressed by the second derivative of $F(T)$ with respect to T, has in this case a logarithmic singularity rather than a power-law behaviour

$$C \sim B \log|T - T_c|.$$

Correspondingly, the critical exponent α of the two-dimensional Ising model is

$$\alpha = 0.$$

5.1.2 Correlation Function and Magnetization

This section briefly discusses the main steps that lead to the computation of the two-point correlation function of the Ising model in terms of the combinatorial method. For the mathematical intricacy of the formulae employed in this method, we will present only the final result. The following chapters show that the computation of the correlation functions can be done in a more efficient way (in the continuum limit) by using the methods of QFT.

In order to simplify the notation, in the following the coordinates of a generic site of the lattice will be denoted by one index alone, i.e. $i \equiv (i_1, i_2)$. Observe that the knowledge of the two-point correlation function

$$G(\mid i-j \mid) = \langle \sigma_i \sigma_j \rangle, \tag{5.1.20}$$

can be used to see whether or not the system possesses a non-zero magnetization

$$M^2 = \lim_{|i-j| \to \infty} \langle \sigma_i \sigma_j \rangle. \tag{5.1.21}$$

Let us focus the attention on the computation of $G(\mid i-j \mid)$, defined by

$$\langle \sigma_i \sigma_j \rangle = \frac{1}{Z} \sum_{\{\sigma\}} \sigma_i \sigma_j \exp\left[K \sum_{(k,l)} \sigma_k \sigma_l \right], \tag{5.1.22}$$

with $K = \beta \mathcal{J}$. Using the familiar identity

$$\exp[x\sigma_k \sigma_l] = \cosh x (1 + \sigma_k \sigma_l \tanh x),$$

the numerator of (5.1.22) can be written as

$$\cosh K^{2N} \sum_{\{\sigma\}} \sigma_i \sigma_j \prod_{(k,l)} (1 + v\sigma_k \sigma_l), \tag{5.1.23}$$

where $v = \tanh K$. Expanding the product 2^{2N} terms. Using the same graphical method discussed in Section 4.2.1, we draw a line along the segment (k, l) if this enters one of the terms of the expansion. This line has a weight equal to v. Once all the lines are drawn, we sum over the values of the spins. The difference with the computation of the partition function in this case consists of the presence of the spins σ_i and σ_j and we have a non-vanishing result only if there is at least one curve that starts from the site i and ends at the site j (Figure 5.7), where all other contributions are made of closed graphs.

Clearly the closed graphs that appear in the expansion of the numerator are the same that enter the expression of the partition function Z_N and therefore they simplify with the term Z_N in the denominator. Hence the correlation function can be expressed by the series

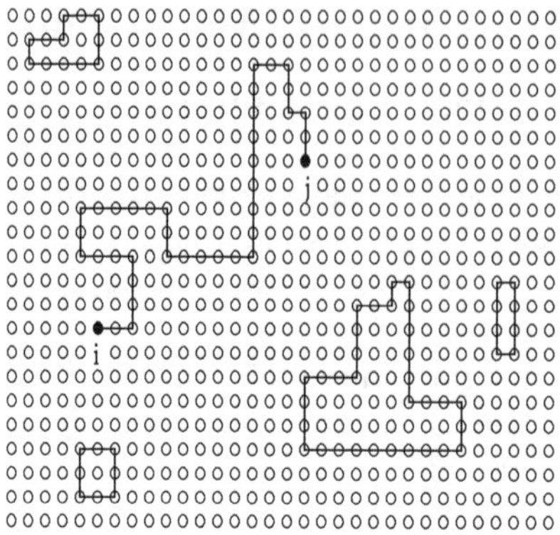

Fig. 5.7 *Graphs that enter the computation of the correlator* $\langle\sigma_i\sigma_j\rangle$.

$$\langle\sigma_i\sigma_j\rangle = \sum_k h_k v^k, \tag{5.1.24}$$

where h_k is the number of graphs of length k (also self-intersecting) that connect the two ending points. A simple example helps in clarifying the content of such a formula.

Consider the correlator of two nearest-neighbour spins. The graphs relative to the lowest orders in v, i.e. $v^1,\ldots,v^5$ are shown in Figure 5.8. Therefore, in this case, the first terms of the series are

$$v + 2v^3 + 6v^5 + \cdots.$$

This example highlights a general and important aspect of the problem. Since the correlation function is nothing else but a conditional probability that the two spins σ_i and σ_j have the same value, for two neighbour spins such a probability is determined by two different effects: 1) the direct interaction between σ_i and σ_j, with a weight v; 2) the sum of all indirect interactions between the two spins, with a weight v^k for those indirect interactions that involving k spins.

Although it is generally difficult to compute the generic coefficient h_k of the series (5.1.24), for their geometrical origin it is however easy to determine the first non-vanishing coefficient. Denoting by $s_1 = \mid j_1 - i_1 \mid$ and $s_2 = \mid j_2 - i_2 \mid$ the horizontal and vertical distances between the spins σ_i and σ_j, the number of paths of total length $s_1 + s_2$ (made of s_1 horizontal steps and s_2 vertical steps) is given by $\frac{(s_1+s_2)!}{s_1!s_2!}$ and therefore

$$\langle\sigma_i\sigma_j\rangle \simeq \frac{(s_1+s_2)!}{s_1!s_2!}\, v^{s_1+s_2} + \cdots. \tag{5.1.25}$$

A further analysis of the series (5.1.24) (that is not discussed here) allows us to reach the following conclusions: for $T \neq T_c$, the two-point correlation function decays exponentially at large distances as

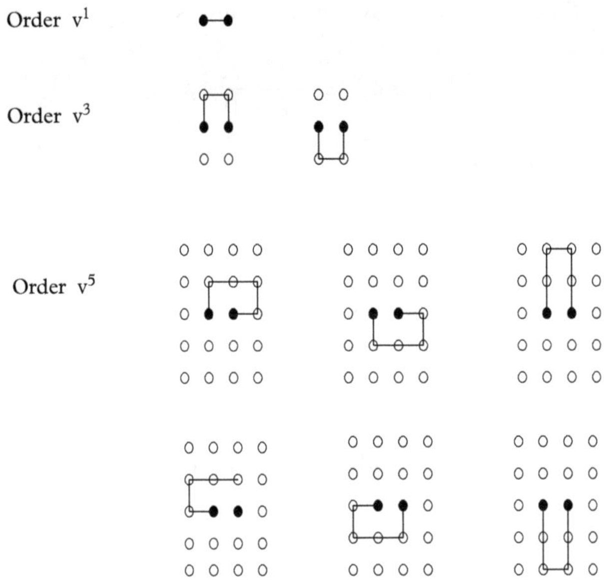

Fig. 5.8 *First-order terms of the correlation function of two nearest-neighbour spins.*

$$\langle \sigma_i \sigma_j \rangle \simeq M^2 + A \exp\left[-\frac{|\, i-j \,|}{\xi} \right], \tag{5.1.26}$$

where $A > 0$ is a constant. Nearby T_c, the correlation length ξ diverges as

$$\xi \simeq |\, T - T_c \,|^{-1}, \tag{5.1.27}$$

and the critical exponent ν of the two-dimensional Ising model is

$$\nu = 1.$$

The spontaneous magnetization M^2 can be extracted by the limit (5.1.21) and its exact expression, originally obtained by Yang, is

$$M^2 = \begin{cases} \left[1 - \left(\frac{1-v^2}{2v} \right)^4 \right]^{1/4}, & T < T_c \\ 0, & T > T_c. \end{cases} \tag{5.1.28}$$

Hence the exact value of the critical exponent β is

$$\beta = \frac{1}{8}.$$

Finally, at $T = T_c$, the correlator decays algebraically as

$$\langle \sigma_i \sigma_j \rangle \simeq \frac{1}{|\, i - j \,|^{1/4}}, \tag{5.1.29}$$

and for the critical exponent η we have

$$\eta = \frac{1}{4}.$$

The remaining critical exponents δ and γ can be obtained by the scaling laws (1.1.26)

$$\delta = 15; \quad \gamma = \frac{7}{4}.$$

These are the exact expressions of all the critical exponents of the two-dimensional Ising model.

5.2 Dimer Method

For the geometrical nature of its high-temperature series expansion, the two-dimensional Ising model can be put in correspondence with the problem of counting the number of dimer configurations on a particular lattice. As we will see, this is a problem of combinatorial nature that can be solved by evaluating the Pfaffian of an anti-symmetrical matrix A.

The Pfaffian of a anti-symmetric $2N \times 2N$ matrix

$$A = \begin{pmatrix} 0, & a_{1,2}, & \cdots, & a_{1,2N} \\ -a_{1,2}, & 0, & \cdots, & a_{2,2N} \\ \cdot & & & \\ \cdot & & & \\ \cdot & & & \\ -a_{1,2N}, & -a_{2,2N}, & \cdots, & 0 \end{pmatrix}$$

is defined as

$$\text{Pf}A = \sum_P{}' \delta_P \, a_{p_1,p_2} \, a_{p_3,p_4} \cdots a_{p_{2N-1},p_{2N}}, \tag{5.2.1}$$

where $p_1,\ldots,p_{2N}$ is a permutation of the set of the numbers $1, 2, \ldots, 2N$, δ_P is the parity of the permutation (± 1 if the permutation P is obtained by an even/odd number of transpositions) and the sum $\sum'_P$ is over all permutations that satisfy the conditions

$$\begin{aligned} p_{2m-1} &< p_{2m}, & 1 < m < N; \\ p_{2m-1} &< p_{2m+1}, & 1 < m < N-1. \end{aligned} \tag{5.2.2}$$

For instance, if $2N = 4$, we have

$$\text{Pf}A = a_{12}a_{34} - a_{13}a_{24} + a_{14}a_{23}.$$

Notice that, for the anti-symmetry of the matrix A, its Pfaffian can be also expressed as

$$\text{Pf}A = \frac{1}{N!2^N} \sum_P \delta_P \, a_{p_1,p_2} \, a_{p_3,p_4} \cdots a_{p_{2N-1},p_{2N}}, \tag{5.2.3}$$

where the sum is on all possible permutations.

The computation of the Pfaffian of a matrix is simplified thanks to this important identity

$$\text{Pf}A = (\det A)^{1/2}. \tag{5.2.4}$$

Unlike the Pfaffians, the determinants are in fact easier to compute, in particular for the property that the determinant of a product of matrices is equal to the product of the determinants.

A dimer is an object that can cover the links between nearest-neighbour sites, with the condition that a given site cannot be occupied by more than one dimer. The combinatorial nature of the dimer problem consists of determining the number of possible dimers covering a lattice, such that all sites are occupied and none of them is occupied more than one. If the lattice is made of N sites, the number of dimers is $N/2$, hence N must be an even number. Before addressing the study of the Ising model in terms of the dimer formulation, it is convenient to study initially the dimer covering of a square lattice.

5.2.1 Dimers on a Square Lattice

The relationship between the dimer covering of a square lattice and the Pfaffian of a matrix can be highlighted by considering a 4×4 lattice. If we number the sites as shown in Figure 5.9, the dimer configuration can be identified by the pairs of numbers

$$(1,2), (3,7), (4,8), (5,6), (9,13), (10,11), (12,16), (14,15),$$

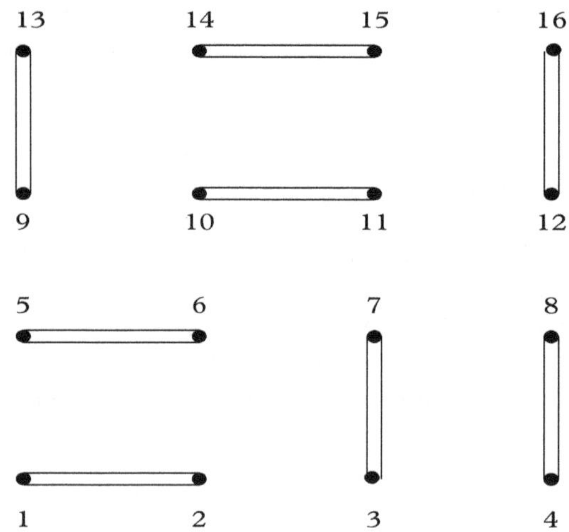

Fig. 5.9 *Dimer configuration of a* 4×4 *square lattice.*

or, more generally, by

$$(p_1,p_2),\,(p_3,p_4),\,(p_5,p_6),\cdots(p_{2N-1},p_{2N}),$$

where $(p_1,p_2,\ldots,p_{2N})$ is a permutation of $(1,2,\ldots,2N)$ that satisfies the constraints (5.2.2) relative to the Pfaffian of a matrix. Assigning the matrix elements according to the rule

$$|\,A_{p,p'}\,| = \begin{cases} z_1, & \text{if } p > p', \text{ where } p \text{ and } p' \text{ are horizontal nearest-neighbour sites} \\ z_2, & \text{if } p > p', \text{ where } p \text{ and } p' \text{ are vertical nearest-neighbour sites} \\ 0, & \text{otherwise} \end{cases} \tag{5.2.5}$$

it is easy to see that there is a one-to-one correspondence between the dimer configurations and the terms present in the definition of the Pfaffian of the matrix A defined above. If we introduce the generating function of the dimers, defined by the formula

$$\Phi(z_1,z_2) = \sum_{n_1,n_2} g(n_1,n_2)\, z_1^{n_1}\, z_2^{n_2}, \tag{5.2.6}$$

where $g(n_1,n_2)$ is the number of dimers that cover completely the lattice, with n_1 placed horizontally and n_2 placed vertically ($n_1+n_2=N/2$), it seems natural to put

$$\Phi(z_1,z_2) = \text{Pf}A. \tag{5.2.7}$$

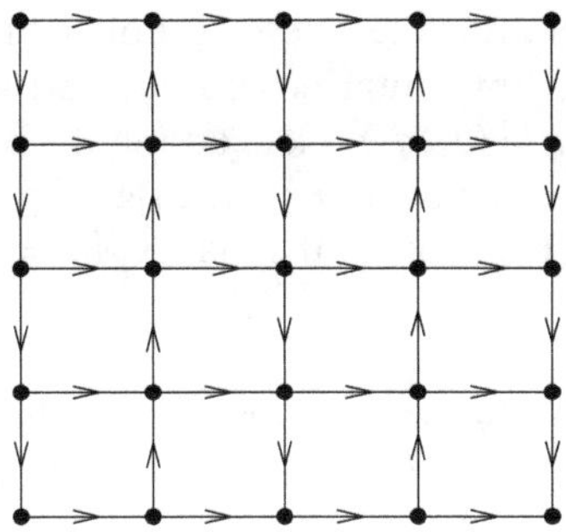

Fig. 5.10 *Assignment of the arrows in the dimer problem on a square lattice. The arrows in the up and the right directions correspond to the positive links, while the others to the negative links.*

There is however an obstacle: in fact, while $g(n_1, n_2)$ in the generating function of the dimers is a positive quantity, the definition of the Pfaffian of A also involves negative terms, i.e. those relative to the odd permutations of the indices. Hence, in order to validate eqn. (5.2.7), in addition to the module (5.2.5) of the matrix elements $A_{p,p'}$, it is also necessary to introduce a phase factor that ensures the positivity of all terms present in Pf affian of A. Thanks to Kasteleyn's theorem, this task can be accomplished for all planar lattices, i.e. for those lattices that do not have crossings of the links. For instance, in the case of a square lattice, an assignment that ensures the validity of eqn. (5.2.7) is given by

$$A_{p,p'} = \begin{cases} z_1, & \text{for the horizontal links that are nearest-neighbour} \\ (-1)^p z_2, & \text{for the vertical links that are nearest-neighbour} \\ 0, & \text{otherwise} \end{cases} \tag{5.2.8}$$

Notice that the definition of $A_{p,p'}$ given in (5.2.8) is equivalent to assign a set of arrows along the links of the lattice, as shown in Figure 5.10. In this way, the original lattice becomes an *oriented lattice*. In the presence of the arrows, the lattice acquires a periodicity along the horizontal axes under a translation of *two* lattice steps. It is therefore convenient

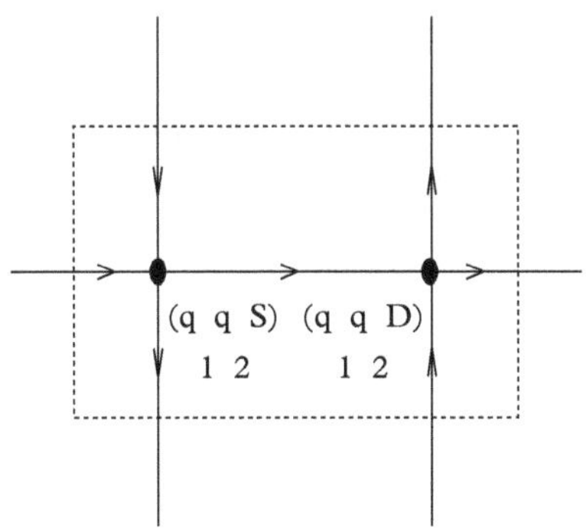

Fig. 5.11 *Elementary cell in the oriented square lattice.*

to assume, as elementary cell, not the one of a unitary length but the one drawn in Figure 5.11, identified by its horizontal position q_1 and its vertical position q_2: these coordinates form the vector $\vec{q} = (q_1, q_2)$. Concerning its internal points, the one on the left is identified by (q_1, q_2, S) while the one on the right by (q_1, q_2, D).

Let us consider the matrix elements of the matrix $A_{\vec{q},\vec{p}} = A(\vec{q}, \vec{p})$. They are themselves 2×2 matrices, given by

$$A_{\vec{q},\vec{p}} = A(q_1, q_2; p_1, p_2) = \begin{pmatrix} a(q_1, q_2, S; p_1, p_2, S) & a(q_1, q_2, S; p_1, p_2, D) \\ a(q_1, q_2, D; p_1, p_2, S) & a(q_1, q_2, D; p_1, p_2, D) \end{pmatrix}. \tag{5.2.9}$$

The only non-vanishing matrix elements of $A_{\vec{q},\vec{p}}$ are given by

$$\begin{aligned} A(q_1, q_2; q_1, q_2) &= \begin{pmatrix} 0 & z_1 \\ -z_1 & 0 \end{pmatrix} = \alpha(0,0), \\ A(q_1, q_2; q_1 + 1, q_2) &= \begin{pmatrix} 0 & 0 \\ z_1 & 0 \end{pmatrix} = \alpha(1,0) \\ A(q_1, q_2; q_1 - 1, q_2) &= \begin{pmatrix} 0 & -z_1 \\ 0 & 0 \end{pmatrix} = \alpha(-1,0), \\ A(q_1, q_2; q_1, q_2 + 1) &= \begin{pmatrix} -z_2 & 0 \\ 0 & z_2 \end{pmatrix} = \alpha(0,1), \\ A(q_1, q_2; q_1, q_2 - 1) &= \begin{pmatrix} z_1 & 0 \\ 0 & -z_2 \end{pmatrix} = \alpha(0,-1). \end{aligned} \tag{5.2.10}$$

It is important to stress that the matrix A only depends on the difference of the indices

$$A(\vec{q}; \vec{p}) = A(\vec{p} - \vec{q}).$$

Imposing periodic boundary conditions along the two directions

$$A(\vec{q} + \vec{N}) = A(\vec{q}),$$

where $\vec{N} = (N_1, N_2)$, A becomes a *cyclic matrix* that can be easily diagonalized with respect to the indices $\vec{q}$ and $\vec{p}$ by Fourier transform.[3] The matrix elements of $A_{\vec{q},\vec{p}}$ are 2×2 matrices and consequently, its diagonal form with respect to $\vec{q}$ and $\vec{p}$ consists of 2×2 matrices placed along its main diagonal. Denoting the latter matrices by $\lambda(\beta_1, \beta_2)$ we have

$$\lambda(\beta_1, \beta_2) = \sum_{\vec{q}} A(\vec{q})\, e^{i\vec{q}\cdot\vec{\beta}},$$

[3] The procedure is similar to the one employed in the Gaussian and spherical models, discussed in Chapter 3.

where each frequency β_i can have the N_i values $0, 2\pi/N_i, 4\pi/N_1, \ldots 2\pi(N_i-1)/N_i$. Hence, the determinant of A is expressed by the product of the determinants of the 2×2 matrices λ

$$\frac{1}{N_1 N_2} \log \mathrm{Det} A = \frac{1}{N_1 N_2} \sum_{k_1=0}^{N_1-1} \sum_{k_2=0}^{N_2-1} \log \mathrm{Det} \lambda \left(\frac{2\pi k_1}{N_1}, \frac{2\pi k_2}{N_2} \right). \tag{5.2.11}$$

In the thermodynamical limit $N_i \to \infty$ the sum can be converted to an integral

$$\lim_{N_i \to \infty} \frac{1}{N_1 N_2} \log \mathrm{Det} A = \frac{1}{(2\pi)^2} \int_0^{2\pi} \int_0^{2\pi} d\beta_1\, d\beta_2 \log \mathrm{Det} \lambda(\beta_1, \beta_2), \tag{5.2.12}$$

where the matrix $\lambda(\beta_1, \beta_2)$ is explicitly given by

$$\begin{aligned} \lambda(\beta_1, \beta_2) &= \sum_{q_1, q_2} \alpha(q_1, q_2)\, e^{iq_1\beta_1 + iq_2\beta_2} = \\ &= \alpha(0,0) + \alpha(1,0)\, e^{i\beta_1} + \alpha(-1,0)\, e^{-i\beta_1} + \\ &\quad \alpha(0,1)\, e^{i\beta_2} + \alpha(0,-1)\, e^{-i\beta_2} = \\ &= \begin{pmatrix} z_2\, e^{-i\beta_2} - z_2\, e^{i\beta_2} & z_1 - z_1\, e^{-i\beta_1} \\ z_1\, e^{i\beta_1} - z_1 & z_2\, e^{i\beta_2} - z_2\, e^{-i\beta_2} \end{pmatrix}. \end{aligned} \tag{5.2.13}$$

Computing the determinant of this matrix and using the important identity (5.2.4), we have

$$\begin{aligned} \lim_{N_i \to \infty} \frac{2}{N_1 N_2} \log \mathrm{Pf} A &= \frac{1}{(2\pi)^2} \int_0^{2\pi} \int_0^{2\pi} d\beta_1\, d\beta_2 \log \left[4 \left(z_1^2 \sin^2 \frac{\beta_1}{2} + z_2^2 \sin^2 \frac{\beta_2}{2} \right) \right] - \\ &= \frac{1}{(2\pi)^2} \int_0^{2\pi} \int_0^{2\pi} d\beta_1\, d\beta_2 \log 2 \left[\left(z_1^2 + z_2^2 \right) - z_1^2 \cos \beta_1 - z_2^2 \cos \beta_2 \right]. \end{aligned} \tag{5.2.14}$$

From the relation $\Phi(z_1, z_2) = \mathrm{Pf} A$ that links the generating function of the dimers to the Pfaffian of the matrix A, by plugging in (5.2.14) the values $z_1 = z_2 = 1$, we obtain the total number of dimer covering of a square lattice. The computation of the integral (proposed as Problem 5.4), gives

$$\lim_{N_i \to \infty} \frac{2}{N_1 N_2} \log \Phi(1,1) = \frac{4G}{\pi}, \tag{5.2.15}$$

where G is the *Catalan constant*, whose numerical value is

$$G = 1 - \frac{1}{3^2} + \frac{1}{5^2} - \frac{1}{7^2} + \cdots = 0.9159655\ldots$$

In conclusion, the number of dimer coverings of a square lattice of N sites, with periodic boundary conditions on both directions, in the limit $N \to \infty$ is given by,[4]

$$D \simeq \exp\left[\frac{NG}{\pi}\right], \quad N \to \infty. \tag{5.2.16}$$

By using the same method, employing the sum instead of the integral, one can obtain the dimer covering of finite lattices. For instance, for a 8×8 lattice, as the one of a chessboard, the number of dimers is 32 and the number of their coverings of the lattice is $D = 2^4 (901)^2 = 12088816$, as it was originally shown by Michael Fisher.

5.2.2 Dimer Formulation of the Ising Model

For the two-dimensional Ising model on a square lattice there is a one to one correspondence between the closed graphs of the high-temperature expansion and the dimer configurations relative to the lattice shown in Figure 5.12, known as Fisher lattice. Both lattices have, as a building block, an elementary cell with four external lines, see Figure 5.13. We can associate to the eight possible configurations of the lines of the Ising model in the elementary cell eight possible dimer configurations on the Fisher lattice, as shown in Figure 5.14 (by rotation, the configuration (c) gives rise to three other configurations whereas the configuration (d) only one). In such a way, to each closed graph of the high temperature expansion of the Ising model on a square lattice corresponds a dimer configuration on the Fisher lattice, and vice versa.

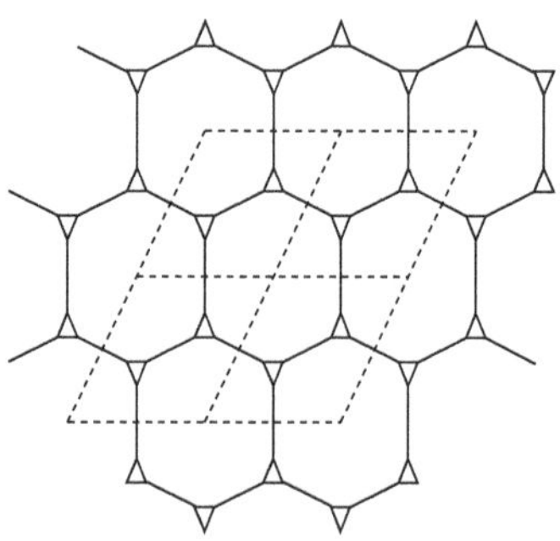

Fig. 5.12 *Fisher lattice.*

[4] Since the elementary cell of the oriented lattice is double the elementary cell of the ordinary lattice, we have $N = 2N_1 N_2$.

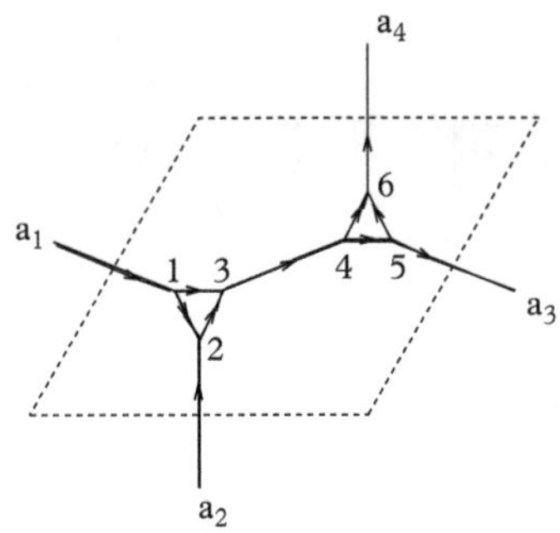

Fig. 5.13 *Elementary cells of the square and Fisher lattices.*

Let us consider the high-temperature expansion of the partition function of the model, given in eqn. (4.2.4), here written as

$$(2\cosh K \cosh L)^{-N} Z_N = \sum_{r,s=0}^{\infty} n(r,s) v^r w^s, \tag{5.2.17}$$

where $n(r,s)$ is the number of closed graphs having r horizontal and s vertical links. Assigning weight v to the dimers along the segments a_1 and a_3, weight w to the dimers placed on the segments a_2 and a_4, and weight 1 to all internal dimers of the cell, it is easy to see that the right-hand side of eqn. (5.2.17) may be interpreted as the generating function of the dimer configurations on the Fisher lattice. In turn, this function can be expressed in terms of the Pfaffian of an opportune anti-symmetric matrix A. Hence we can follow the same steps for the computation of the dimer covering on the square lattice, with the only difference that, instead of the two internal points of the square lattice, this time the elementary cell has the six internal points shown in 5.13, with the corresponding orientation of the links. However, as in the previous case, the only matrices different from zero are $\alpha(0,0)$, $\alpha(\pm 1,0)$ and $\alpha(0,\pm 1)$, so that the matrix of the eigenvalues is given in this case by

$$\lambda(\beta_1,\beta_2) = \begin{pmatrix} 0 & 1 & 1 & 0 & -v e^{-i\beta_1} & 0 \\ -1 & 0 & 1 & 0 & 0 & -w e^{-i\beta_2} \\ -1 & -1 & 0 & 1 & 0 & 0 \\ 0 & 0 & -1 & 0 & 1 & 1 \\ v e^{i\beta_1} & 0 & 0 & -1 & 0 & 1 \\ 0 & w e^{i\beta_2} & 0 & -1 & -1 & 0 \end{pmatrix}, \tag{5.2.18}$$

and therefore

$$\mathrm{Det}\,\lambda(\beta_1,\beta_2) = (1+v^2)(1+w^2) - 2v(1-w^2)\cos\beta_1 - 2w(1-v^2)\cos\beta_2. \tag{5.2.19}$$

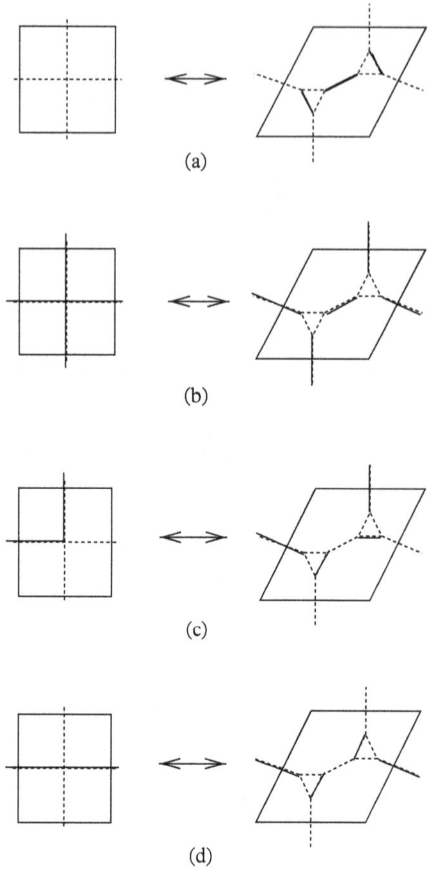

Fig. 5.14 *Correspondence between the lines of the Ising model on a square lattice and the dimers on the Fisher lattice.*

Computing the Pfaffian of the matrix A, we obtain the free energy of the Ising model: in the thermodynamic limit and in the homogeneous case $v = w$, it is given by

$$-\frac{F(T)}{kT} = \lim_{N\to\infty} \log Z_N = -\log 2 + \log(1 - v^2) + \tag{5.2.20}$$
$$-\frac{1}{2(2\pi)^2}\int_0^{2\pi}\int_0^{2\pi} \log\Big[(1+v^2)^2 - 2v(1-v^2)(\cos\phi_1 + \cos\phi_2)\Big]\, d\beta_1\, d\beta_2.$$

This expression coincides with eqn. (5.1.16).

REFERENCES

The combinatorial methods in the solution of the two-dimensional Ising model are discussed in

Kac, M. and Ward, J.C. (1952). 'A Combinatorial Solution of the Two-dimensional Ising Model', *Physics Review*, 88: 1332.

Vdovichenko, N.V. (1965). 'A Calculation of the Partition Function for a Plane Dipole Lattice', *Journal of Experimental and Theoretical Physics*, 20: 477.

The exact expression of the magnetization of the two-dimensional Ising model is found in

Yang, C.N. (1952). 'The Spontaneous Magnetization of a Two-dimensional Ising Model', *Physics Reviews*, 85: 808.

For the dimer solution see

Fisher, M. (1996). 'On the Dimer Solution of the Ising Models', *Journal of Mathematical Physics*, 7: 1776.

Kasteleyn, P.W. (1961). 'The Statistics of Dimers on a Lattice: The Number of Dimer Arrangements on a Quadratic Lattice', *Physica* 27: 1664.

For more recent developments on the dimer formalism, we

Fendley, P. (2002). 'Classical Dimers on the Triangular Lattice', *Physics Reviews B*, 66: 214513

Moessner, R. and Sondhi, S.L. (2003). 'Ising and Dimer Models in Two and Three Dimensions', *Physics Reviews B*, 68: 054405.

PROBLEMS

5.1. High-temperature series

Determine the first three terms of the high-temperature expansion of the correlation function $\langle \sigma_{i+2,j} \sigma_{i,j} \rangle$ of two spins separated by two lattice sites.

5.2. Pfaffian and Determinant

Prove that for a $2N \times 2N$ anti-symmetric matrix A it holds the identity

$$(\text{Pf}A)^2 = \det A.$$

5.3. Number of dimers

a. Give an argument to justify the exponential growth of eqn. (5.2.16) for the dimer coverings in N, where N is the number of sites of a lattice.

b. Use eqn. (5.2.16) to estimate the number of dimer coverings of a 4×4 square lattice.

5.4. Generating function of dimers on a square lattice

Consider the function

$$F(x,y) = \frac{1}{(2\pi)^2} \int_0^{2\pi} \int_0^{2\pi} d\beta_1 \, d\beta_2 \log\left[x + y - x\cos\beta_1 - y\cos\beta_2\right].$$

Its value at $x = y =$, i.e. $F(2,2)$, provides the solution to the problem of the dimer covering of a square lattice, eqn. (5.2.14).

a. Prove that

$$F(x,0) = \log\frac{x}{2}.$$

b. Show that it holds the identity

$$\frac{\partial F}{\partial x} = \frac{2}{\pi x} \arctan\sqrt{\frac{x}{y}}.$$

c. Expanding in power series the term $\arctan\sqrt{\frac{x}{y}}$ and integrating term by term, show that

$$F(2,2) = \frac{4G}{\pi}$$

where G is the Catalan constant.

6

Transfer Matrix of the Two-dimensional Ising Model

I did much of the work in the writing room of the P & O liner Arcadia, in the Atlantic and Indian Oceans. This was good for concentration, but not for communication.

Rodney J. Baxter

In this chapter we study the solution of the two-dimensional Ising model by means of the transfer matrix. Unlike the methods discussed in the previous chapter, the transfer matrix approach has a greater generality and can used to solve exactly other two-dimensional models. While the general ideas behind this approach have been explained in Chapter 2 by means of the one-dimensional case, their application to the two-dimensional cases require more powerful and sophisticate mathematical tools: for instance, the study the eigenvalues of the transfer matrix in the Ising model for $T \neq T_c$ needs to employ the elliptic functions. The same is also true for other models. In order to present in the simplest possible way the main lines of this method, in the following we focus the attention only on the solution of the model at $T = T_c$ because this case can be analysed in terms of simple trigonometric functions.

An important condition is required for implementing efficiently the method: the commutativity of the transfer matrix for different values of the coupling constants. In the Ising model, for instance, this condition can be satisfied by the transfer matrix $T_D(K,L)$ along the diagonal of the square lattice. If the coupling constants K and L fulfill the condition

$$\sinh 2K \sinh 2L = \sinh 2K' \sinh 2L'. \tag{6.0.1}$$

the transfer matrix has the property[1]

$$\left[T_D(K,L), T_D(K',L')\right] = 0, \tag{6.0.2}$$

[1] $[A,B]$ denotes the commutator of the two matrices A and B and it is given by $[A,B] = AB - BA$.

Statistical Field Theory: An Introduction to Exactly Solved Models in Statistical Physics. Second Edition. Giuseppe Mussardo, Oxford University Press (2020). © Giuseppe Mussardo. DOI: 10.1093/oso/9780198788102.001.0001

Equation (6.0.2) implies that the eigenvectors of the transfer matrix do not change if the coupling constants vary along the curve given by eqn. (6.0.1). This is a crucial circumstance for the exact diagonalization of $T_D(K,L)$. Equally important is the possibility to implement the commutativity of the transfer matrices by means of a particular conditions (of local nature) satisfied by the Boltzmann weights. These conditions are known as the Yang–Baxter equations and they play an important role in all exactly solvable models: they enter not only the solution of the statistical models but also the S-matrix theory, the formalism of quantum groups and the classification of the knots.

6.1 Baxter's Approach

There are several ways to define a transfer matrix for the two-dimensional Ising model and each of them shows certain advantages. The transfer matrix that we discuss in this section is associated to the square lattice rotated of 45^0 degrees, as shown in Figure 6.1. The coupling constants K and L, originally placed along the horizontal and vertical directions, are now defined along the diagonals. This lattice is particularly useful to establish the commutativity properties of the transfer matrix defined on it. As it is evident from Figure 6.1, the sites of this lattice can be divided in two classes, A and B, identified by the empty and filled circles: each row of type A is followed by one of type B, and vice-versa. Let m be the total number of rows: assuming periodic boundary conditions along the vertical direction, m is necessarily an even integer. Moreover, imposing periodic boundary conditions also along the horizontal direction, it is easy to see that there is an equal number n of sites both for the rows of type A and type B. For each row, there are 2^n possible spin configurations and in the following they will be simply denoted by μ_r

$$\mu_r = \{\sigma_1, \sigma_2, \ldots, \sigma_n\}\text{row r}$$

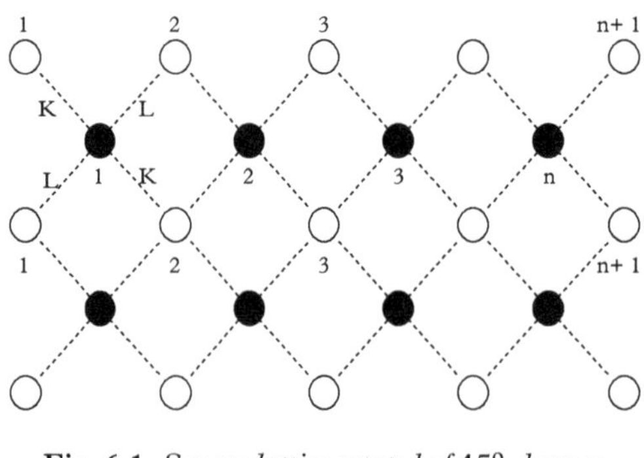

Fig. 6.1 *Square lattice rotated of* 45^0 *degrees.*

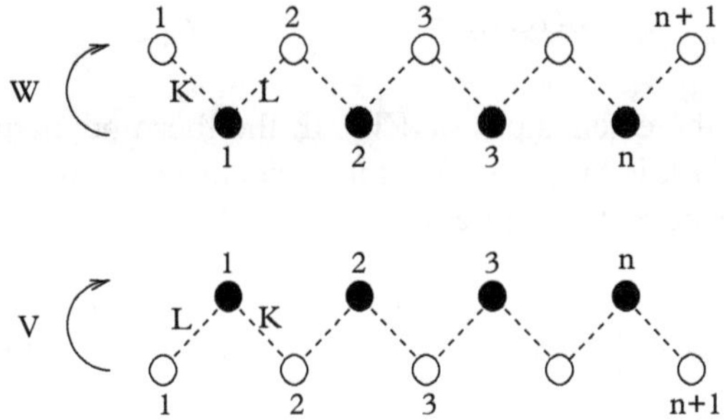

Fig. 6.2 *Transfer matrices V and W.*

Since the spins of type A interact only with those of type B, and vice-versa, it is convenient to introduce two transfer matrices V e W, both of dimension $2^n \times 2^n$ (see Figure 6.2). Denoting collectively by μ the spins of the lower row and by μ' those of the upper row, the operators $V(K,L)$ and $W(K,L)$ are defined by their matrix elements[2]

$$V_{\mu,\mu'}(K,L) = \exp\left[\sum_{i=1}^{n}(K\sigma_{i+1}\sigma_i' + L\sigma_i\sigma_i')\right], \tag{6.1.3}$$

$$W_{\mu,\mu'}(K,L) = \exp\left[\sum_{i=1}^{n}(K\sigma_i\sigma_i' + L\sigma_i\sigma_{i+1}')\right]. \tag{6.1.4}$$

In both formulae we have assumed the periodic boundary conditions $\sigma_{n+1} \equiv \sigma_1$ and $\sigma_{n+1}' \equiv \sigma_1'$.

All statistical weights of the model are generated by the iterated application of the operators V and W to the configuration of the first row. The partition function is thus expressed as

$$Z_N(K,L) = \sum_{\mu_1}\sum_{\mu_2}\cdots\sum_{\mu_m} V_{\mu_1,\mu_2} W_{\mu_2,\mu_3} V_{\mu_3,\mu_4}\cdots W_{\mu_m,\mu_1},$$

namely

$$Z_N(K,L) = \mathrm{Tr}\,(VWVW\ldots VW) = \mathrm{Tr}\,(VW)^{m/2}. \tag{6.1.5}$$

Since the trace of a matrix is independent from its representation, the most convenient way to compute the partition function (6.1.5) consists of diagonalizing the matrix VW, so that

[2] With this choice of the matrix elements, the application of two transfer matrices A and B, one after the other, corresponds to their multiplication in the order AB.

$$Z(K,L) = \lambda_1^m + \lambda_2^m + \cdots \lambda_{2^n}^m, \tag{6.1.6}$$

where λ_1^2, λ_2^2, ... are the eigenvalues of VW. In the thermodynamic limit (where both m and n go to infinity) it is only the maximum eigenvalue that matters because, taking initially the limit $m \to \infty$, with n finite, we have[3]

$$Z(K,L) = (\lambda_{max})^m \left[1 + \left(\frac{\lambda_1}{\lambda_{max}}\right)^m + \left(\frac{\lambda_2}{\lambda_{max}}\right)^m + \cdots\right] \simeq (\lambda_{max})^m. \tag{6.1.7}$$

So we arrive at a formula that is quite analogous to the one-dimensional Ising model. From an algebraic point of view, though, there is a substantial difference between the two cases: while in the one-dimensional case the problem consists of diagonalizing a 2×2 matrix, in the two-dimensional case it is necessary to find the eigenvalues of a $2^n \times 2^n$ matrix, in the limit $n \to \infty$. The mathematical difficulty of such a problem can be faced by taking advantages of some important properties of the transfer matrices.

6.1.1 Commutativity of the Transfer Matrices

The operators V and W explicitly depend on the coupling constants of the lattice, as shown by their definition (6.1.3) and (6.1.4). Consider now the product of V with W but with different coupling constants, as shown in Figure 6.3

$$V(K,L)\, W(K',L'). \tag{6.1.8}$$

Denoting by $\mu = \{\sigma_1, \ldots, \sigma_n\}$ the spins of the lower row, by $\mu' = \{\sigma_1', \ldots, \sigma_n'\}$ the spins of the upper row and by $\mu'' = \{\sigma_1'', \ldots, \sigma_n''\}$ those of the half-way row, the matrix elements of this operator between the states μ and μ' are obtained according to the usual rule of the product of matrices, namely as a sum on the intermediate states μ''

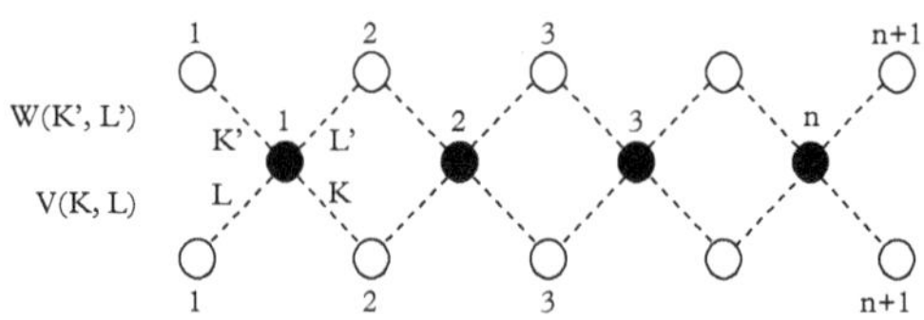

Fig. 6.3 *Product of V and W with different coupling constants.*

[3] With real coupling constants, the matrix VW has all matrix elements positive. The matrices that share such a property are knows as *positive matrices*. The Perron–Frobenius theorem, whose proof is proposed as a problem at the end of the chapter, states that any finite-dimensional positive matrix has a unique maximum eigenvalue, also positive. The corresponding eigenvector has all its components positive as well.

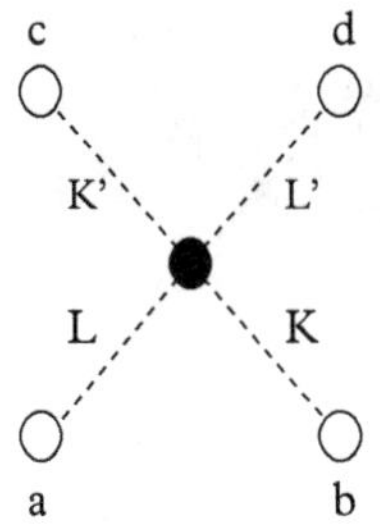

Fig. 6.4 *Elementary statistical weight* $X(a,b;c,d)$.

$$\left(V(K,L)\, W(K',L')\right)_{\mu,\mu'} = \sum_{\{\sigma''\}} \prod_{j=1}^{n} \exp\left[\sigma_j''(K\sigma_{j+1} + L\sigma_j + K'\sigma_j' + L'\sigma_{j+1}')\right].$$

Since each intermediate spin σ_j'' appears only in a single term of the expression,[4] the sum on these spins is particularly simple and the matrix elements of the operator (6.1.8) assumes the factorized form

$$\left(V(K,L)\, W(K',L')\right)_{\mu,\mu'} = \prod_{j=1}^{n} X(\sigma_j, \sigma_{j+1}; \sigma_j', \sigma_{j+1}'), \tag{6.1.9}$$

with the elementary statistical weight $X(a,b,c,d)$ explicitly given by (see Figure 6.4)

$$\begin{aligned} X(a,b,c,d) &= \sum_{\sigma''=\pm 1} \exp\left[\sigma''(La + Kb + K'c + L'd)\right] = \\ &= 2\cosh\left[La + Kb + K'c + L'd\right], \quad a,b,c,d = \pm 1 \end{aligned} \tag{6.1.10}$$

Exchanging the role of the coupling constans (K,L) and (K',L'), one obtains, in general, a different result for the product VW. There is however the identity

$$V(K,L)\, W(K',L') = V(K',L')\, W(K,L), \tag{6.1.11}$$

if the coupling constants satisfy the equation

$$\sinh 2K \sinh 2L = \sinh 2K' \sinh 2L'. \tag{6.1.12}$$

To prove this result, let us observe that for the factorized form (6.1.9) of the product VW, the transformation

$$X(a,b,c,d) \longrightarrow e^{Mac}\, X(a,b,c,d)\, e^{-Mbd}$$

[4] This is one of the mathematical advantages of the transfer matrix defined on the diagonal of the lattice.

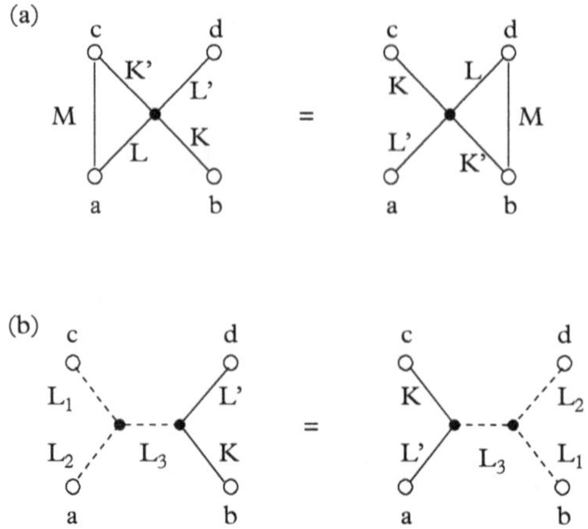

Fig. 6.5 *Star-triangle transformation of eqn. (6.1.13), where the sum on the spins is represented by the black circles.*

does not change the expression (6.1.9). This observation permits to satisfy eqn. (6.1.11) by solving a simpler problem, i.e. the problem to find a number M such that

$$e^{Mac}\, X(a,b;c,d) = X'(a,b;c,d)\, e^{Mbd}, \tag{6.1.13}$$

where X' is the statistical weight obtained by changing $K \to K'$ and $L \to L'$ in the original X. In summary, in order to satisfy the *global* commutativity condition (6.1.11), it is sufficient to find a solution to the *local* condition (6.1.13). This problem can be solved by using the star-triangle identity discussed in Section 4.3.1. Let us consider, in fact, the graphical representation of eqn. (6.1.13) given in Figure 6.5.a: both in the right and left diagrams there is a triangle, given by the interaction of the relative spins. Posing

$$K_1 = L,\;\; K_2 = K',\;\; K_3 = M$$

and changing each triangle into a star, with the relative coupling constants L_i given by eqn. (4.4.7), it is easy to see by looking at Figure 6.5.b that the two expressions are equal if

$$L_1 = K,\;\; L_2 = L'$$

namely, if the coupling constants satisfy the condition

$$\sinh 2K \sinh 2L = \sinh 2K' \sinh 2L'. \tag{6.1.14}$$

Equation (6.1.11) can be further elaborated and entirely expressed in terms of the matrix V. Thanks to the periodic boundary conditions, it is in fact evident that W differs from V simply by a translation of a lattice spacing. With the help of the operator T, with matrix elements

$$T_{\mu,\mu'} = \delta(\sigma_1,\sigma_2')\,\delta(\sigma_2,\sigma_3')\ldots\delta(\sigma_n,\sigma_1'), \tag{6.1.15}$$

and whose effect is to move the lattice of a lattice spacing to the right, one can verify that

$$W(K,L) = V(K,L)\,T. \tag{6.1.16}$$

Moreover

$$V(K,L) = T^{-1}\,V(K,L)\,T, \quad W(K,L) = T^{-1}\,W(K,L)\,T. \tag{6.1.17}$$

Using (6.1.16), eqn. (6.1.11) becomes

$$V(K,L)\,V(K',L') = V(K',L')\,V(K,L), \tag{6.1.18}$$

where the coupling constants satisfy eqn. (6.0.1).

6.1.2 Commutativity of the Transfer Matrices: Graphical Proof

The commutativity relation (6.1.18) can be proved in a graphical way. To this aim we must first consider the square lattice in its usual orientation and then define two sets of operators $P_i(K)$ and $Q_i(L)$ by means of their matrix elements

$$\begin{aligned}(P_i(K))_{\mu,\mu'} &= \exp[K\sigma_i\sigma_{i+1}]\,\delta(\sigma_1,\sigma_1')\ldots\delta(\sigma_n,\sigma_n')\\ (Q_i(L))_{\mu,\mu'} &= \delta(\sigma_1,\sigma_1')\ldots\delta(\sigma_{i-1},\sigma_{i-1}')\exp[L\sigma_i\sigma_i']\\ &\quad\times\delta(\sigma_{i+1},\sigma_{i+1}')\ldots\delta(\sigma_n,\sigma_n').\end{aligned} \tag{6.1.19}$$

$P_i(K)$ creates the statistical weight of the spins σ_i and σ_{i+1} placed on the same horizontal row (without changing their values from the row μ to the next one), while $V_i(L)$ creates the statistical weight of the spins σ_i and σ_i', placed on two rows next neighbour. The result of these operators is visualized in Figure 6.6. It is possible to adopt a uniform notation by defining the operators $U_i(K,L)$

$$U_i(K,L) = \begin{cases} P_j(K), & i = 2j \\ Q_j(L), & i = 2j-1. \end{cases} \tag{6.1.20}$$

These operators satisfy

$$U_i(K,L)\,U_j(K',L') = U_j(K',L')\,U_i(K,L), \quad |\,i-j\,| \geq 2. \tag{6.1.21}$$

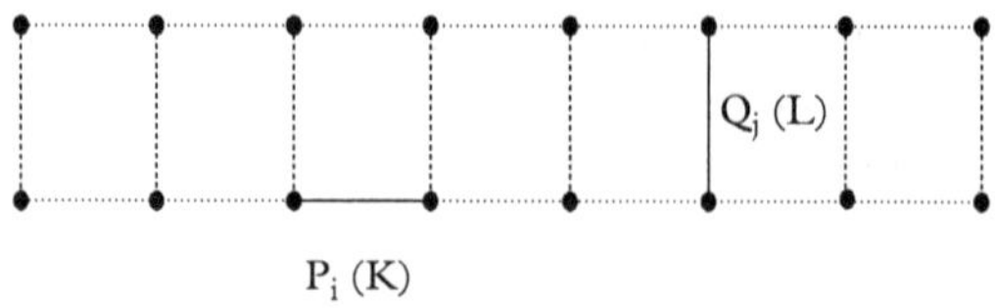

Fig. 6.6 *Action of the operators $P_i(K)$ and $Q_j(L)$ on the square lattice.*

Suppose we are dealing with a set of coupling constants (K_1, K_2, K_3) and (L_1, L_2, L_3) linked one another by the star-triangle relation (4.4.7)

$$\sinh 2K_1 \sinh 2L_1 = \sinh 2K_2 \sinh 2L_2 = \sinh 2K_3 \sinh 2L_3 = h^{-1}. \tag{6.1.22}$$

Using the explicit expression of the matrix elements of P_i and Q_i, it is easy to show that

$$U_{i+1}\, U_i'\, U"_{i+1} = U"_i\, U_{i+1}'\, U_i, \tag{6.1.23}$$

where we have introduced the notation $U_i = U_i(K_1, L_1)$, $U_i' = U_i(K_2, L_2)$ and $U_i'' = U_i(K_3, L_3)$. The graphical interpretation of this equation is given in Figure 6.7. Let us consider now the operator $V(K, L)$ given by the product

$$V(K,L) = U_1(K,L)\, U_2(K,L) \dots U_N(K,L), \tag{6.1.24}$$

with $N = 2n$: its action consists of introducing the statistical weights along the main diagonal of the lattice as shown in Figure 6.8. It is easy to see that $V(K, L)$ coincides with the transfer matrix considered in the previous sections.

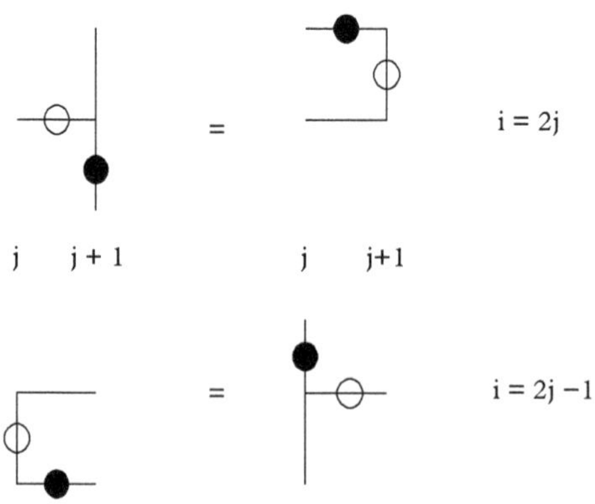

Fig. 6.7 *Graphical form of eqn. (6.1.23). The full circle corresponds to the couplings (K_1, L_1), the empty circle to (K_2, L_2) and the line to (K_3, L_3).*

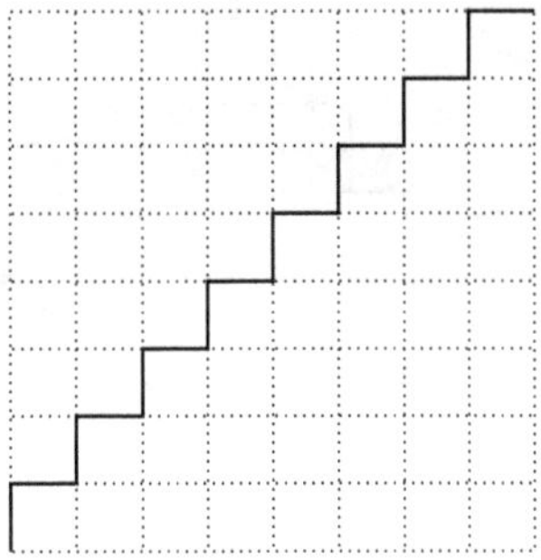

Fig. 6.8 *The operator $V(K,L)$ on the square lattice.*

Let (K_i, L_i) $(i = 1, 2, 3)$ be three different pairs of coupling constants that satisfy the star-triangle equation (6.1.22) and let us define

$$V = U_1\, U_2 \ldots U_N, \quad V' = U_1'\, U_2' \ldots U_N'.$$

Using iteratively eqns. (6.1.21) and (6.1.23), one can show that these operators satisfy the condition

$$V\, V'(U_N'^{-1}\, U_N''\, U_N) - (U_1\, U_1''\, U_1'^{-1})\, V'\, V. \tag{6.1.25}$$

The graphical proof is given in Figure 6.7, where the sequence of the diagrams is generated by the repeated application of the graphical identities of Figure 6.9. The terms within the parenthesis of eqn. (6.1.25) refer to the spins at the boundary and they disappear if we adopt periodic boundary conditions. In this case we have then the commutativity relation (6.1.18)

$$V(K,L)\, V(K',L') = V(K',L')\, V(K,L). \tag{6.1.26}$$

6.1.3 Functional Equations and Symmetries

The factorized form (6.1.9) of VW allows us to write down a functional equation for the matrix elements of this operator. Consider the elementary statistical weight $X(a,b;c,d)$, given by the formula (6.1.10). For the values

$$K' = L + \frac{i\pi}{2}, \quad L' = -K, \tag{6.1.27}$$

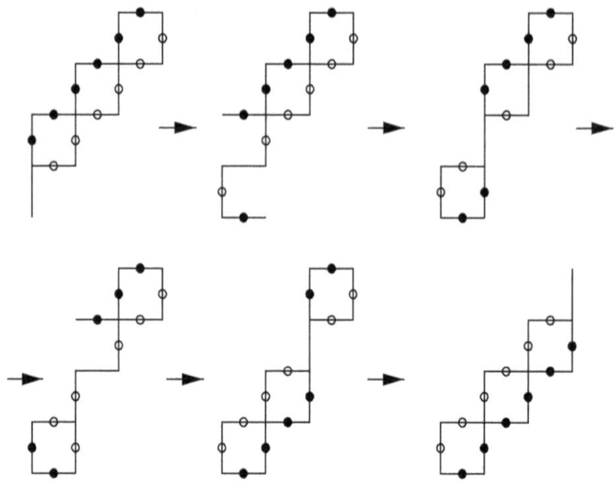

Fig. 6.9 *Graphical proof of the commutativity relation of the transfer matrices along the diagonal of the square lattice.*

we have

$$X(a,b;c,d) = 2\cosh\left[L(a+c)+K(b-d)+\frac{i\pi c}{2}\right] = ic\sinh\left[L(a+c)+K(b-d)\right]. \tag{6.1.28}$$

Hence, it is different from zero only in two cases:

- $a=c$ and $b=d$, where we have

$$X(a,b;a,b) = 2ia\sinh 2La = 2i\sinh 2L.$$

- or $a \neq c$ and $b \neq d$, and in this case

$$X(a,b;-a,-b) = -2ia\sinh 2Kb = -2iab\sinh 2K.$$

In correspondence of the particular values (6.1.27) of the coupling constants, the matrix elements of VW are expressed as

$$\left(V(K,L)\,W(L+\frac{i\pi}{2},-K)\right)_{\mu,\mu'} = (2i\sinh 2L)^n\,\delta(\sigma_1,\sigma_1')\,\delta(\sigma_2,\sigma_2')\ldots\delta(\sigma_n,\sigma_n') + $$
$$+(-2i\sinh 2K)^n\,\delta(\sigma_1,-\sigma_1')\,\delta(\sigma_2,-\sigma_2')\ldots\delta(\sigma_n,-\sigma_n'). \tag{6.1.29}$$

If we introduce the identity operator $\mathbf{I}$, with matrix elements

$$\mathbf{I}_{\mu,\mu'} = \delta(\sigma_1,\sigma_1')\,\delta(\sigma_2,\sigma_2')\ldots\delta(\sigma_n,\sigma_n'), \tag{6.1.30}$$

and the operator **R**, with matrix elements

$$\mathbf{R}_{\mu,\mu'} = \delta(\sigma_1,-\sigma_1')\,\delta(\sigma_2,-\sigma_2')\ldots\delta(\sigma_n,-\sigma_n'), \tag{6.1.31}$$

(both matrices have dimension $2^n \times 2^n$), eqn. (6.1.29) can be written in an operatorial form as

$$V(K,L)\,W(L+\frac{i\pi}{2},-K) = (2i\sinh 2L)^n\,\mathbf{I} + (-2i\sinh 2K)^n\,\mathbf{R}. \tag{6.1.32}$$

The next section shows how this formula is extremely useful to determine the eigenvalues of the matrices V and W, and to find the inverse of the matrix $V(K,L)$ (see Problem 6.2).

Let us discuss the symmetry properties of the matrices V and W. Interchanging K with L and σ_i with σ_i', the matrix W becomes the transpose of V

$$W(K,L) = V^T(L,K), \tag{6.1.33}$$

$$V(K,L)\,W(K,L) = [V(L,K)\,W(L,K)]^T. \tag{6.1.34}$$

Since changing the sign of K and L is equivalent to change the sign of $\sigma_1,\ldots,\sigma_n$ or $\sigma_1',\ldots,\sigma_n'$, we have also

$$V(-K,-L) = R\,V(K,L) = V(K,L)\,R, \tag{6.1.35}$$

with a similar relation for the matrix W.

Let p be the number of spin pairs (σ_{j+1},σ_j') with opposite value and q the number of spin pairs (σ_j,σ_j') with opposite values. Hence, $p+q$ counts the total number of changes of signs that we have in the sequence $\sigma_1,\sigma_1',\sigma_2,\sigma_2',\ldots,\sigma_n'$. So $p+q$ is an even number and, from the definition (6.1.3), it follows

$$V_{\mu,\mu'}(K,L) = \exp\left[(n-2p)\,K + (n-2q)\,L\right]. \tag{6.1.36}$$

In the thermodynamic limit $n \to \infty$ there is no difference whether n is an even or an odd number and, posing for simplicity

$$n = 2s, \tag{6.1.37}$$

where s is an integer, eqn. (6.1.36) can be written in terms of two numbers p' and q' that belong to the interval $(0,s)$

$$V_{\mu,\mu'}(K,L) = \exp\left[\pm 2p' K \pm 2q' L\right]. \tag{6.1.38}$$

The variables p' and q' are either both even or odd, so the matrix $V(K,L)$ satisfies the relation

$$V\left(K \pm i\frac{\pi}{2}, L \pm i\frac{\pi}{2}\right) = V(K,L), \tag{6.1.39}$$

with a similar relation for $W(K,L)$.

6.1.4 Functional Equations for the Eigenvalues

Let us proceed to the determination of the eigenvalues of $V(K,L)$ by using the functional equations satisfied by this operator. Suppose that K and L are two complex numbers subject to the condition

$$h^{-1} = \sinh 2K \sinh 2L, \tag{6.1.40}$$

where h is a given real number. In this case, thanks to eqn. (6.1.40), there is an infinite number of transfer matrices that commute each other, see eqn. (6.1.18). They also commute with T, eqn. (6.1.17), and with R, eqn. (6.1.35). These commutation properties imply that, for all the values of K and L that satisfy eqn. (6.1.40), the transfer matrices have a common basis of eigenvectors. These eigenvectors can depend neither on K nor on L, but they can be functions of h. Denoting by $y(h)$ one of these eigenvectors and by $v(K,L)$, t and r the eigenvalues of the matrices $V(K,L)$, T and R, we have

$$\begin{aligned} V(K,L)\,y(h) &= v(K,L)\,y(h); \\ T\,y(h) &= t\,y(h); \\ R\,y(h) &= r\,y(h). \end{aligned} \tag{6.1.41}$$

The eigenvalues t and c also satisfy

$$t^n = r^2 = 1, \tag{6.1.42}$$

so they are complex number of unite module, independent from K and L. Notice that if K and L satisfy eqn. (6.1.40), the same happens with K' and L' defined in (6.1.27). Hence, applying the functional relation (6.1.32) to the vector $y(h)$, we have

$$v(K,L)\,v(L+\frac{i\pi}{2},-K)\,t = (2i\sinh 2L)^n + (-2i\sinh 2K)^n\,r. \tag{6.1.43}$$

Let $\lambda^2(K,L) \equiv \lambda_i^2$ be one of the eigenvalues[5] of the matrix $V(K,L)\,W(K,L)$. Since $y(h)$ is also an eigenvector of this matrix and $W = V\,T$, we have

$$\lambda^2(K,L) = v^2(K,L)\,t. \tag{6.1.44}$$

With the definition

$$\lambda(K,L) = v(K,L)\,\sqrt{t}, \tag{6.1.45}$$

eqn. (6.1.43) becomes a functional equation that has to be satisfied by the eigenvalues of the transfer matrix

$$\lambda(K,L)\,\lambda(L+\frac{i\pi}{2},-K) = (2i\sinh 2L)^n + (-2i\sinh 2K)^n\,r. \tag{6.1.46}$$

6.2 Eigenvalue Spectrum at the Critical Point

This section shows how it is possible to determine the spectrum of the transfer matrix only using the commutativity property and the analytic structure of the eigenvalues, together with the functional equation (6.1.46). A crucial aspect of the solution is the appropriate parameterization of the coupling constants K and L that satisfy eqn. (6.1.40): a clever parameterization will allow us to take advantage of the powerful theorems of complex analysis and to extract the analytic properties of the eigenvalues.

The actual implementation of this program presents a different level of complexity according to the value of the parameter h. In order to highlight the main steps of such a method, it is convenient to discuss the simplest case:[6] this corresponds to the value $h = 1$ for which the system is at the critical point

$$\sinh 2K\,\sinh 2L = 1, \tag{6.2.47}$$

(see Chapter 4 and, in particular, Section 4.2.3). The eqn. (6.2.47) can be identically satisfied by posing

$$\begin{aligned}\sinh 2K &= \tan u,\\ \sinh 2L &= \cot u.\end{aligned} \tag{6.2.48}$$

The coupling constants K and L are both real and positive for the values u that fall in the range $(0,\frac{\pi}{2})$. The parameterization (6.2.48) allows us to write $\exp(\pm 2K)$ and $\exp(\pm 2L)$ as

[5] In the following we skip, for brevity, the index i. The different eigenvalues are identified by the different solutions of the functional equation (6.1.46).

[6] In the general case we have to use a parameterization in terms of elliptic functions.

$$\begin{aligned}\exp(2K) &= (1+\sin u)/\cos u,\\ \exp(-2K) &= (1-\sin u)/\cos u,\\ \exp(2L) &= (1+\cos u)/\sin u,\\ \exp(-2L) &= (1-\cos u)/\sin u.\end{aligned} \tag{6.2.49}$$

These expressions have the following important properties:

1. they are periodic functions of u, with period period 2π
2. they are meromorphic functions[7] of u, with simple poles.

Since the eigenvalues $\lambda(K,L)$ of the transfer matrix can be regarded as functions of u, it is convenient to adopt the notation $\lambda(u)$ and write the functional equation (6.1.46) as

$$\lambda(u)\,\lambda(u+\frac{\pi}{2}) = (2i\cot u)^n + (-2i\tan u)^n\, r. \tag{6.2.50}$$

Expressing $\exp(\pm 2K)$ and $\exp(\pm 2L)$ in terms of the functions (6.2.49), the matrix elements of $V_{\mu,\mu'}$ assume the form

$$V_{\mu,\mu'} = \frac{A(u)}{(\sin u\,\cos u)^s}, \tag{6.2.51}$$

where $A(u)$ is a polynomial in $\sin u$ and $\cos u$, of total degree $2s$. Hence its general expression is given by

$$A(u) = e^{-2isu}\left(a_0 + a_1\, e^{iu} + \cdots + a_{2n}\, e^{4isu}\right). \tag{6.2.52}$$

Let us now consider the first equation in (6.1.41), which actually consists of 2^n equations. Using known theorems of linear algebra, the eigenvalues $v(K,L)$ are expressed as linear combination of the matrix elements of $V(K,L)$, whose coefficients are given by ratios of the components of the eigenvectors $y(h)$. For the commutativity of all matrices involved in the problem, such ratios are functions only of the variable h *but totally independent on* u. This is a crucial property for the considerations that follow because it implies that each eigenvalue $v(K,L)$ is expressed by a linear combination of terms as (6.2.51) and therefore it has the same form. The same is true for $\lambda(u)$, defined in (6.1.45).

Notice that substituting u with $u+\pi$ is equivalent to change K in $-K\pm i\frac{\pi}{2}$ and L in $-L\pm i\frac{\pi}{2}$, as evident from eqns. (6.2.49). However, these substitutions are equivalent to multiply V by R, as we can see from eqn. (6.1.35). Hence, denoting $v(K,L)$ by $v(u)$, the first of the equations (6.1.41) becomes

$$V(K,L)\,R\,y(h) = v(u+\pi)\,y(h), \tag{6.2.53}$$

[7] A meromorphic function has only poles as singularities in the complex plane.

where we have taken into account once again the independence of $y(h)$ from the variable u. Using the first and the last equation in (6.1.41), we have

$$v(u+\pi) = r\,v(u),$$

namely

$$\lambda(u+\pi) = r\lambda(u). \tag{6.2.54}$$

Since the generic form of $\lambda(u)$ is given by (6.2.51) and $r = \pm 1$, for the periodicity (6.2.54) the corresponding polynomial $A(u)$ in (6.2.52) only has the even coefficients c_{2k} different from zero when $r = 1$, while it only has the odd coefficients c_{2k+1} different from zero when $r = -1$. Then the eigenvalues $\lambda(u)$ can be expressed as

$$\lambda(u) = \rho\,(\sin u \cos u)^{-s} \prod_{j=1}^{l} \sin(u-u_j) \tag{6.2.55}$$

where ρ and $u_1, u_2, \ldots u_l$ are constants to be determined, with

$$l = \begin{cases} 2s & , \quad \text{if } r = +1 \\ 2s-1 & , \quad \text{if } r = -1. \end{cases}$$

Substituting this expression into the functional equation (6.2.50), we have

$$\rho^2 \prod_{j=1}^{l} \sin(u-u_j)\,\cos(u-u_j) = 2^{2s}\left(\cos^{4s} u + r \sin^{4s} u\right). \tag{6.2.56}$$

This identity must be satisfied for all values of u. This expression can be simplified by the substitution

$$x = e^{2iu}, \quad x_j = e^{2iu_j}.$$

We have then

$$\rho^2 \left(\frac{i}{4}\right)^l \prod_{j=1}^{l} \frac{(x^2 - x_j^2)}{x_j} = 2^{-2s} x^{l-2s} \left[(x+1)^{4s} + r(x-1)^{4s}\right]. \tag{6.2.57}$$

Both polynomials on the right- and left-hand sides are of degree l in the variable x^2 and therefore the constants ρ and $x_1, \ldots, x_l$ are determined by the identity of these two polynomials. Since $x_1^2, \ldots, x_l^2$ are the l distinct zeros of the left term, the same should hold for the term on the right-hand side. So, they are fixed by the condition

$$\left[(x+1)^{4s}+r(x-1)^{2s}\right]=0,$$

whose solutions are given by

$$x_j^2=-\tan^2\frac{\theta_j}{2},\ \ j=1,\ldots,l$$

where

$$\theta_j=\begin{cases}\pi\,(j-\frac{1}{2})/2s, & \text{if } r=+1\\ \pi j/2s, & \text{if } r=-1\end{cases}$$

All these values of θ_j fall in the range $(0,\pi)$, so that, defining

$$\varphi_j=\frac{1}{2}\ln\tan\frac{\theta_j}{2},\ \ j=1,\ldots,l$$

we have

$$u_j=\mp\frac{\pi}{4}-i\varphi_j,\ \ j=1,\ldots,l. \tag{6.2.58}$$

Since the sign ∓ 1 of each solution can be chosen independently, there are 2^l possible solutions. There is, however, an extra condition coming from the limits $u\to\pm i\infty$, where $\exp(2K)=\exp(2L)\to\pm i$. Since the matrix elements of the transfer matrix does not change if we alter the sign of $\exp(2K)$ and $\exp(2L)$, we have

$$\lambda(i\infty)=\lambda(-i\infty).$$

From the general expression of the eigenvalues, eqn. (6.2.55), one can check that this condition is automatically satisfied when $r=-1$, while if $r=1$, it leads to the condition

$$(u_1+\cdots+u_{2s})/\pi=N+\frac{1}{2}s,$$

where N is an integer. This implies that only $2s-1$ among the possible signs of the solutions (6.2.58) can be chosen in an independent way. Therefore, as expected, in both cases $r=\pm 1$ there are 2^{2s-1} eigenvalues λ.

To summarize, the eigenvalues $\lambda(u)$ are given by

$$\lambda(u)=\rho\,(\sin u\cos u)^{-s}\prod_{j=1}^{l}\sin(u+i\varphi_j+\frac{1}{4}\eta_j\pi), \tag{6.2.59}$$

where $\eta_1, \ldots, \eta_l$ have values ± 1 and, for $r = 1$ there is the further condition

$$\eta_1 + \cdots + \eta_{2s} = 2s - 4M, \tag{6.2.60}$$

where M is an integer.

6.3 Away from the Critical Point

The analysis done for the eigenvalues at the critical point $T = T_c$ can be also performed for generic value of T. As previously mentioned, this requires a parameterization in terms of the elliptic functions and will not be pursued here. We only mention that this analysis leads to the determination of the maximum eigenvalues of the transfer matrix whose final expression is given by

$$\log \lambda_{max} = \frac{1}{2} \sum_{j=1}^{2s} \mathcal{F}[\pi (j - \frac{1}{2})/2s], \tag{6.3.61}$$

where the function $\mathcal{F}(\theta)$ is

$$\mathcal{F}(\theta) = \log \left\{ 2 \left[\cosh 2K \cosh 2L + h^{-1} (1 + h^2 - 2h \cos 2\theta)^{1/2} \right] \right\}. \tag{6.3.62}$$

In the thermodynamic limit, when $s \to \infty$, the free energy is given by

$$-F/k_B T = \frac{1}{2\pi} \int_0^{\pi} \mathcal{F}(\theta) \, d\theta. \tag{6.3.63}$$

The analysis of the singularity that arises in this expression when $h \to 1$ is proposed as an exercise.

6.4 Yang–Baxter Equation and R-matrix

At the heart of solvability of many lattice statistical models there is the commutativity of the transfer matrix that, as a sufficient condition, needs the Yang–Baxter equation satisfied by the Boltzmann weights R. Let us elaborate on this problem in more abstract terms. Consider Figure 6.10, where each of the lines stands for a vector space spanned by the statistical variables. Let us denote the three vector spaces by $V_{p_1}^{\mu}$, $V_{p_2}^{\nu}$ and $V_{p_3}^{\lambda}$, with μ, ν and λ that label the different multiplets and the p_is that denote the spectral parameters of the Boltzmann weights. Two or more adjacent lines, for example those representing the spaces $V_{p_1}^{\mu}$ and $V_{p_2}^{\nu}$ are tensor products of those spaces, $V_{p_1}^{\mu} \otimes V_{p_2}^{\nu}$. The Boltzmann weight R, associated to the operation of crossing the lines in the diagram, can

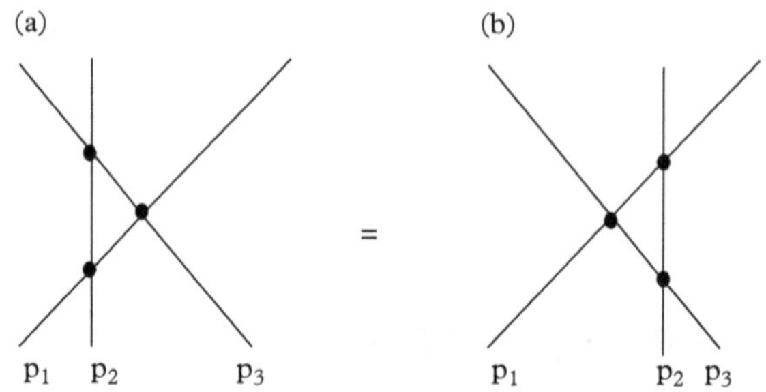

Fig. 6.10 *Yang–Baxter equation satisfied by the Boltzmann weights (here represented by the dots) as functions of the spectral parameter p.*

be abstractly described as a mapping from a vector space of the initial states to the vector space of the final state,

$$R^{\mu\nu}(p_1 - p_2) : V^{\mu}_{p_1} \otimes V^{\nu}_{p_2} \rightarrow V^{\nu}_{p_2} \otimes V^{\mu}_{p_1}. \tag{6.4.64}$$

Here it is assumed that, for the homogeneity of the lattice, the Boltzmann weights depends only on the difference $p_1 - p_2$ of the spectral parameters. This matrix is usually referred to as the R-matrix and satisfy the Yang–Baxter equation of Figure 6.10

$$\begin{aligned}(R^{\mu\nu}(p_1 - p_2) \otimes 1)(1 \otimes R^{\mu\lambda}(p_1 - p_3))(R^{\nu\lambda}(p_2 - p_3) \otimes 1) =\\ (1 \otimes R^{\nu\lambda}(p_2 - p_3))(R^{\mu\lambda}(p_1 - p_3) \otimes 1)(1 \otimes R^{\mu\nu}(p_1 - p_2))\end{aligned} \tag{6.4.65}$$

The Yang–Baxter equation is non-linear and usually it it is difficult to solve directly. Nevertheless its solution has been found for many lattice models, leading to the exact determination of their free energy. An essential property is the invariance of R under a quantum group symmetry, a topic that will be discussed in more detail in Section 18.11, whereas further aspects of R-matrices and Yang–Baxter equation can be found throughout the literature quoted at the end of the chapter. Here we present the main features of this formalism through the study of a significant example.

6.4.1 Six-vertex model

Consider a square lattice $N \times N$ where the fluctuating variables α are attached to each bond connecting the nearest-neighbour lattice site. The vertex Boltzmann weight $R^{\gamma\delta}_{\alpha\beta}$ corresponds to to each configuration around any lattice site

$$R^{\gamma\delta}_{\alpha\beta} = \gamma - \overset{\delta}{\underset{\beta}{\Big|}} - \alpha$$

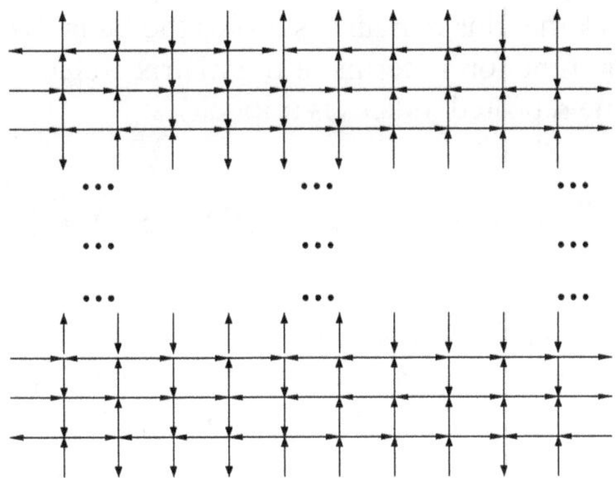

Fig. 6.11 *A configuration of the six-vertex model with periodic boundary consitions.*

Denoting the energy of the vertex by $\epsilon(\alpha,\beta,\gamma,\delta)$, we have $R^{\gamma\delta}_{\alpha\beta} = \exp\left[-\epsilon(\alpha,\beta,\gamma,\delta)\,/\,k_B T\right]$. In the six-vertex model each bond can accept one of the two states characterized by an incoming or outgoing arrow associated to the values $\alpha = \pm$. Furthermore, the only allowed configurations of this model are those in which there are two incoming and two outgoing arrows at each vertex, i.e.

$$
\begin{array}{ll}
R^{++}_{++} = \leftarrow \begin{smallmatrix}\uparrow\\ \uparrow\end{smallmatrix} \leftarrow, & R^{--}_{--} = \rightarrow \begin{smallmatrix}\downarrow\\ \downarrow\end{smallmatrix} \rightarrow \\[2ex]
R^{+-}_{+-} = \leftarrow \begin{smallmatrix}\downarrow\\ \downarrow\end{smallmatrix} \leftarrow, & R^{-+}_{-+} = \rightarrow \begin{smallmatrix}\uparrow\\ \uparrow\end{smallmatrix} \rightarrow \\[2ex]
R^{+-}_{-+} = \leftarrow \begin{smallmatrix}\downarrow\\ \uparrow\end{smallmatrix} \rightarrow, & R^{-+}_{+-} = \rightarrow \begin{smallmatrix}\uparrow\\ \downarrow\end{smallmatrix} \leftarrow
\end{array}
$$

A configuration of the system in shown in Figure 6.11. Assuming invariance under $+ \Leftrightarrow -$, we can parameterize the Boltzmann weights as

$$
\begin{aligned}
R^{++}_{++} &= R^{--}_{--} = a = \sin(\gamma - p)\\
R^{+-}_{+-} &= R^{-+}_{-+} = b = \sin p\\
R^{+-}_{-+} &= R^{-+}_{+-} = c = \sin\gamma
\end{aligned}
$$

where p is the spectral parameter whereas γ is the coupling constant. The weights can be arranged as a 4×4 - matrix

$$
R^{\gamma\delta}_{\alpha\beta} = \begin{pmatrix}
\leftarrow \begin{smallmatrix}\uparrow\\ \uparrow\end{smallmatrix} \leftarrow & & & \\
& \leftarrow \begin{smallmatrix}\downarrow\\ \downarrow\end{smallmatrix} \leftarrow & \leftarrow \begin{smallmatrix}\downarrow\\ \uparrow\end{smallmatrix} \rightarrow & \\
& \rightarrow \begin{smallmatrix}\uparrow\\ \downarrow\end{smallmatrix} \leftarrow & \rightarrow \begin{smallmatrix}\uparrow\\ \uparrow\end{smallmatrix} \rightarrow & \\
& & & \rightarrow \begin{smallmatrix}\downarrow\\ \downarrow\end{smallmatrix} \rightarrow
\end{pmatrix} = \begin{pmatrix}
a & & & \\
& b & c & \\
& c & b & \\
& & & a
\end{pmatrix} \qquad (6.4.66)
$$

It is not difficult to check that this R-matrix satisfies the Yang–Baxter equation (6.4.65). To express the partition function in terms of the matrix R, let us define the *monodromy matrix* (the sum over the repeated indices is implicit)

$$L^{\gamma\{\delta\}}_{\alpha\{\beta\}}(p,\gamma) \equiv R^{\gamma\delta_1}_{\alpha_2\beta_1}(p,\gamma)\, R^{\alpha_2\delta_2}_{\alpha_3\beta_2}(p,\gamma)\cdots R^{\alpha_N\delta_N}_{\alpha\beta_N}(p,\gamma) \equiv \begin{pmatrix} A & B \\ C & D \end{pmatrix} \tag{6.4.67}$$

In this formula we have a matrix product with respect to the *horizontal space* but a tensor product with respect to the N *vertical space*. Therefore the final result is a 2×2 matrix with entries that are operators in $V_N = \otimes_{k=1}^{N} V_v^{(k)}$ ($V_v^{(k)}$ is the vertical space associated to the k-th column, in our case $V_v^{(k)} = \mathbf{C}^2$). The graphical form of the monodromy matrix is

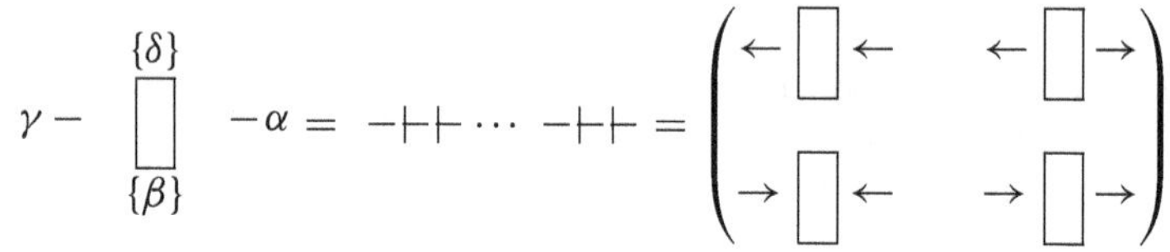

With periodic boundary conditions along the horizontal and vertical axes, the transfer matrix of the model is

$$T(p,\gamma) = \mathrm{Tr}_h L(p,\gamma), \tag{6.4.68}$$

and the partition function is the trace in the tensor product of the vertical space

$$Z(p,\gamma) = \mathrm{Tr}_v[T(p,\gamma)]^N. \tag{6.4.69}$$

Since the R-matrix satisfies the Yang–Baxter equation (6.4.65), the monodromy matrix satisfies

$$R^{\alpha''\beta''}_{\alpha'\beta'}(p-p')\, L^{\alpha'\{\gamma''\}}_{\alpha\{\gamma'\}}(p)\, L^{\beta'\{\gamma'\}}_{\beta\{\gamma\}}(p') = L^{\beta''\{\gamma''\}}_{\beta\{\gamma'\}}(p')\, L^{\alpha''\{\gamma'\}}_{\alpha'\{\gamma\}}(p)\, R^{\alpha'\beta'}_{\alpha\beta}(p-p') \tag{6.4.70}$$

This implies that the operators A, B, C and D of the monodromy matrix satisfy the commutation relations

$$\begin{aligned} A(p)\,B(p') &= \frac{a(p'-p)}{b(p'-p)}\,B(p')\,A(p) - \frac{c(p'-p)}{b(p'-p)}\,B(p)\,A(p') \\ D(p)\,B(p') &= \frac{a(p-p')}{b(p-p')}\,B(p')\,D(p) - \frac{c(p'-p)}{b(p'-p)}\,B(p)\,D(p') \\ \left[C(p),B(p')\right] &= \frac{c(p-p')}{b(p-p')} - A(p)\,D(p') \end{aligned} \tag{6.4.71}$$

Eqn. (6.4.70) also reflects the integrability of the model since yields the commutativity of the transfer matrix for different spectral parameters

$$[T(p), T(p')] = 0, \tag{6.4.72}$$

whose proof of (6.4.72) is similar to the one given in Section 6.1.2. Notice that this equation represents an infinite set of conservation laws for the operators t_n

$$[t_n, t_m] = 0, \quad \log T(p) = -\sum_n t_n p^n. \tag{6.4.73}$$

The lowest conserved operators can be identified with the momentum and the Hamiltonian of the associated quantum system[8]

$$t_0 = iP, \quad t_1 = H$$

Using the commutativity of the transfer matrices, their maximal eigenvalue can be found, in principle, along the lines discussed for the Ising model in previous sections. Equivalently, the solution of the model can be addressed by the Bethe ansatz approach, as sketched in Problem 6.4. Here we simply report the final result for the free energy per unit site

$$\begin{aligned} -F/k_B T &= \log \lambda_{max}(p, \gamma) \\ &= \log \sin(\gamma - p) + \int_{-\infty}^{\infty} \frac{dt}{t} \frac{\sinh[(\pi - \gamma)t] \sinh[2pt]}{2 \cosh \gamma t \sinh \pi t}. \end{aligned} \tag{6.4.74}$$

Let us conclude by outlining the origin of the quantum group symmetry of the model. First, let us write the R-matrix (6.4.66) in terms of the Pauli matrices σ_3 and $\sigma_\pm = \frac{1}{2}(\sigma_1 \pm i\sigma_2)$ as

$$R = \begin{pmatrix} a & & & \\ & b & c & \\ & c & b & \\ & & & a \end{pmatrix} = \begin{pmatrix} \sin\left(\frac{1}{2}\gamma - \sigma_3(p - \frac{1}{2}\gamma\right) & \sigma_- \sin \gamma \\ \sigma_+ \sin \gamma & \sin\left(\frac{1}{2}\gamma + \sigma_3(p - \frac{1}{2}\gamma\right) \end{pmatrix}. \tag{6.4.75}$$

It is easy to see that the Yang–Baxter equation (6.4.65) satisfied by the R-matrix implies the usual $SU(2)$ relations of the Pauli matrix, i.e.

$$[\sigma_3, \sigma_\pm] = \pm 2\sigma_\pm, \quad [\sigma_+, \sigma_-] = \sigma_3.$$

[8] The one-dimensional quantum system associated to the classical two-dimensional six-vertex model is the Heisenberg chain and its continuum limit is described by the Sine–Gordon model.

Taking the limits of the spectral parameter $p \to \pm i\infty$, let us write the monodromy matrix similarly to eqn. (6.4.75)

$$L = \begin{pmatrix} A & B \\ C & D \end{pmatrix} = \begin{pmatrix} \sin\left(\frac{1}{2}\gamma - \mathcal{J}_3(p - \frac{1}{2}\gamma\right) & \mathcal{J}_- \sin\gamma \\ \mathcal{J}_+ \sin\gamma & \sin\left(\frac{1}{2}\gamma + \mathcal{J}_3(p - \frac{1}{2}\gamma\right) \end{pmatrix}. \tag{6.4.76}$$

From the Yang–Baxter equation (6.4.70) satisfied by the monodromy matrix, we can obtain the commutation relations for the $\mathcal{J}$'s

$$[\mathcal{J}_3, \mathcal{J}_\pm] = \pm 2\mathcal{J}_\pm\,, \quad [\mathcal{J}_+, \mathcal{J}_-] = \frac{\sin(\gamma \mathcal{J}_3)}{\sin\gamma}. \tag{6.4.77}$$

These are the commutation relations of the quantum group $SU_q(2)$ that will be discussed in further detail in Section 18.11. Notice that we recover the usual $SU(2)$ commutation relations when $\gamma \to 0$.

REFERENCES

The most important book on the transfer matrix of the two-dimensional system is by Rodney Baxter

Baxter, R. J. (1982). *Exactly Solved Models in Statistical Mechanics*, Academic Press.

Exact solutions of the two-dimensional systems can be also obtained by means of the *Bethe ansatz*. A thorough discussion of this method is given in

Sutherland, B. (2004). *Beautiful Models. 70 Years of Exactly Solved Quantum Many-body Problems*, World Scientific.

Gaudin, M. (1983). *La fonction d'onde de Bethe*, Masson.

For solutions of Yang–Baxter equation and R-matrix see

Faddev, L. (1984). In: Les Houches, Session XXXIX, 1982, *Recent Advances in Field Theory and Statistical Mechanics*, Elsevier.

Andrew, G. E., Baxter, R. J. and Forrester, P. J. (1984). 'Eight-vertex SOS and Generalized Rogers-Ramanujan-type Identities', *Journal of Statistical Physics* 35, 193.

Bazhanov, V. V. and Reshetikin, N. Y. (1989). 'Critical RSOS and Conformal Field Theory', *International Journal of Modern Physics A* 4, 115.

Takahashi, M. and Suzuki, M. (1972). 'One-Dimensional Anisotropic Heisenberg Model at Finite Temperatures', *Progress of Theoretical Physics* 48, 2187.

Tarasov, V. O. (1985). 'Irreducible Monodromy Matrices for the R matrix of the XXZ Model and Local Lattice Quantum Hamiltonians', *Theoretical and Mathematical Physics* 63, 440.

For further studies on the common root of the Yang–Baxter equation in many areas of physics and mathematics (including the knot theory),

Wadati, M., Deguchi, T. and Akutsu, Y. (1989). 'Exactly Solvable Models and Knot Theory, *Physics Reports* 180, 247.

PROBLEMS

6.1. Perron–Frobenius Theorem

Consider a finite dimensional positive matrix M, i.e. with all its matrix elements positive $M_{ij} > 0$. Assume, for simplicity, that M is also a symmetrical matrix. Prove that its maximum eigenvalue is positive and non-degenerate. Moreover, prove that the corresponding eigenvectors have all the components with the same sign (which, therefore, can be chosen to be all positive).

6.2. Inverse of the matrix V

Consider the operator $\mathbf{R}$ defined in eqn. (6.1.31). Using the property $\mathbf{R}^2 = \mathbf{I}$, prove that the inverse of the operator $\mathbf{A} = (2i\sinh 2L)^n\mathbf{I} + (-2i\sinh 2K)^n\mathbf{R}$ is given by

$$A^{-1} = \frac{1}{(2\sinh 2K)^{2n} - (2\sinh 2L)^{2n}}\left[(2i\sinh 2L)^n\mathbf{I} - (-2i\sinh 2L)^n\mathbf{R}\right].$$

Use this expression and the functional equation (6.1.32) to determine the inverse of the operator $V(K,L)$.

6.3. Free energy

Analyse the expression of the free energy of the Ising model, given in eqn. (6.3.63), as a function of the parameter h. Show that, with $t = \frac{T-T_c}{T_c} = h-1$, for $t \to 0$ we have

$$F \simeq t^2 \log|t|.$$

6.4. Bethe ansatz equation

The solution of the six-vertex model consists in finding the eigenvalues of the transfer matrix

$$T(p)\,\psi = (A(p) + D(p))\,\psi = \lambda\,\psi.$$

This problem can be solved by the algebraic Bethe ansatz, whose main steps are as follows: define the *pseudo-vacuum* ϕ, as the state annihilated by the operator $C(p)$

$$C(p)\phi = 0, \qquad \forall p.$$

a. Prove that

$$\phi^{\{\beta\}} = \prod_{k=1}^{N} \delta_{\beta_k,+} = \uparrow \cdots \uparrow.$$

b. Prove that $\phi^{\{\beta\}}$ is an eigenstate of A and D with eigenvalues

$$\begin{aligned} A(p)\phi &= a^N(p)\phi \\ D(p)\phi &= b^N(p)\phi. \end{aligned}$$

However, applying B to ϕ results in neither an eigenvector nor zero, $B(p)\phi \neq \phi, 0$. This suggests an eigenstate of the transfer matrix in the form

$$\psi = B(p_1)\ldots B(p_n)\phi$$

where the parameters p_i are to be determined.

c. Show that, applying $A(p)$ and $D(p)$ to ψ and pushing them through all the Bs by the commutation relations (6.4.71),

$$(A(p) + D(p))\psi = (\lambda_A(p) + \lambda_B(p))\psi + \text{unwanted terms}$$

where

$$\lambda_A(p) = a^N(p) \prod_{k=1}^{n} \frac{a(p_k - p)}{b(p_k - p)}, \qquad \lambda_B(p) = b^N(p) \prod_{k=1}^{n} \frac{a(p_k - p)}{b(p_k - p)}$$

The *unwanted terms*, coming from the second term in eqn. (6.4.71), contain a $B(p)$ and so they can never give a vector proportional to ψ, unless they vanish. Show that this happens if the Bethe ansatz equations hold

$$\left(\frac{b(p_j)}{a(p_j)}\right)^N \prod_{k=1}^{n} \frac{a(p_j - p_k)}{b(p_j - p_k)} \frac{b(p_k - p_j)}{a(p_k - b_j)} = -1, \quad j = 1, 2, \ldots n.$$

Notice that the eigenvalue problem of the transfer matrix has been transformed in a set of transcendental equations above for the spectral parameters $p_1, \ldots p_n$. A further elaboration of the solution of the Bethe ansatz equations leads to the expression (6.4.74) of the free energy of the model.

Part 3

Quantum Field Theory and Conformal Invariance

7

Quantum Field Theory

Surely you are joking Mr. Feynman!

7.1 Motivations

The statistical models we have analysed so far are defined on a lattice and they have a microscopic length scale given by the lattice spacing a. However, in all these models there is another length scale provided by the correlation length ξ: this is a function of the coupling constants and can be varied by varying the external parameters of the systems. When the system is sufficiently close to its critical point, the correlation length is much larger than the microscopic scale, $\xi \gg a$. It is then natural to assume that the configurations of the system are sufficiently smooth on many lattice spacings and to adopt a formalism based on continous quantities like a field $\varphi(x)$.

This book shows the QFT formulation of statistical models has the important advantage to greatly simplify the study of critical phenomena: it helps to select the most important aspects of phase transitions—those related to the symmetries and the dimensionality of the system—and to reach results of great generality. It is worth stressing that the advantage of this method is not only limited to these technical aspects, for the use of QFT in statistical mechanics permits to achieve a theoretical synthesis of wide scope. QFT has been originally developed to describe elementary particles and to reconcile the principles of special relativity with those of quantum mechanics. After the quantization of the electromagnetic field, the subject has witnessed a rapid evolution and has been applied to the analysis of weak interactions, responsible of many radioactive decays, and of strong interactions, responsible of the forces of the quarks inside the hadrons. The degree of refinement reached by this formalism is proved by the incredible precision by which we are able to control nowadays the physical effects on sub-atomic scale. Moreover, its exceptional theoretical richness has lead to extraordinary advances in several fields of physics and mathematics. String theory—a subject developed in the recent years in the attempt to unify all fundamental interactions including the gravitational one—can be considered, for instance, as a natural and elegant development of the QFT.

The reason QFT plays a central role both in the context of elementary particles and critical phenomena is due, in a nutshell, to the principle of universality. This is a primary aspect of all local interactions and it is noteworthy that it naturally emerges from the

Statistical Field Theory: An Introduction to Exactly Solved Models in Statistical Physics. Second Edition. Giuseppe Mussardo, Oxford University Press (2020). © Giuseppe Mussardo. DOI: 10.1093/oso/9780198788102.001.0001

analysis of the renormalization group. Besides, there is a more fundamental reason for it is possible to show that *any* relativistic quantum theory will look at sufficiently low energy like a QFT.[1] In short, this is the most general theoretical framework to describe a set of excitations above a ground state[2] of a system with an infinite degrees of freedom.

Transfer matrix formalism. An obvious question at this point is how it can be possible that a classical statistical system with short range interactions is equivalent to a relativistic quantum theory. The answer is in the transfer matrix formalism. Notice that the partition function of a statistical system with short range interactions can be seen in two equivalent ways: either as a sum on classical variables in a d–dimensional Euclidean space with a classical Hamiltonian $H(\{s_i\})$, or as the trace of a time evolution operator $T = e^{-\tau \mathbf{H}(\{\Phi_i\})}$ associated to a quantum Hamiltonian $\mathbf{H}$ in $(d-1)$ dimensions of certain appropriate variables Φ_i. This equivalence is expressed by the identity[3]

$$Z = \sum_{\{s_i\}} e^{-H(\{s_i\})} = \mathrm{Tr}_{\Phi_i} \prod_{\tau} e^{-\tau \mathbf{H}(\{\Phi_i\})}. \tag{7.1.1}$$

The quantum Hamiltonian $\mathbf{H}(\{\Phi_i\})$ is the first step toward QFT. Translation and rotation invariance of the quantum theory emerge in fact when the lattice spacing goes to zero. Finally, making a change of the time variable $\tau \to -it$, one arrives to a relativistic theory in $(d-1)$ space dimensions and 1 time dimension. Vice-versa, we can start from a QFT that is relativistic invariant in d space-time dimensions and, with the transformation of the time coordinate $t \to i\tau$, define an Euclidean QFT. Once discretized, this theory can be considered for all purposes as a statistical model in d dimensions. In summary, at the root of the equivalence of the formalisms that describe the elementary particles and the

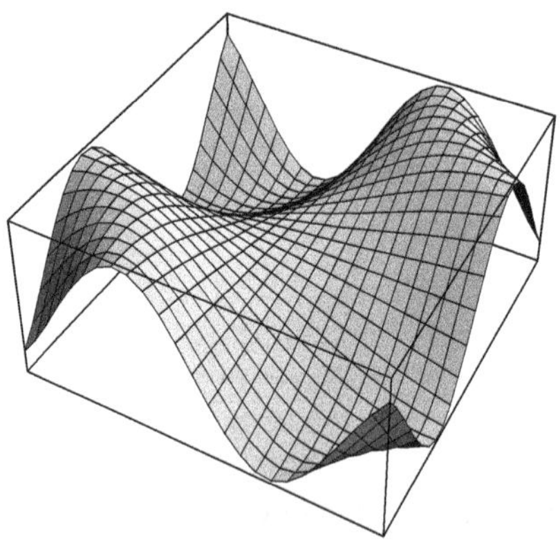

Fig. 7.1 *Continous formulation in term of a field theory.*

[1] See S. Weinberg, *The Quantum Theory of Fields*, Vol. I Foundations, Cambridge Univ. Press, 1995.
[2] The ground state is also called the *vacuum state* of the system.
[3] In the following we will always skip the Planck constant $\hbar$ (considered to be equal to 1) in all formulae.

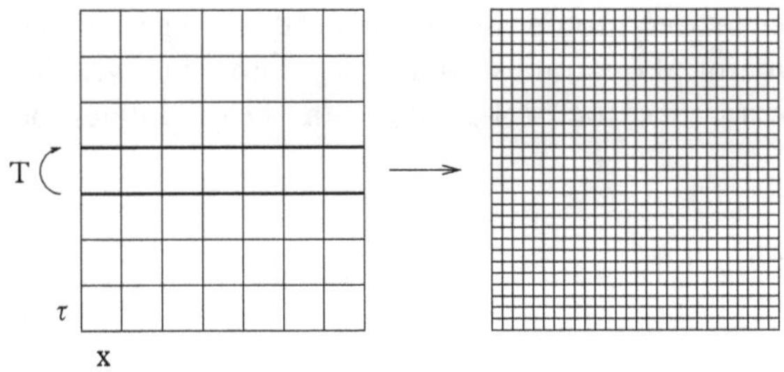

Fig. 7.2 *A classical statistical system in d dimension and the corresponding quantum system in d − 1 dimensions. When the lattice spacing goes to zero the result is a continuous theory both isotropically and translationally invariant.*

critical phenomena there is the possibility to adopt either an operatorial or a functional integral approach to a QFT.

This chapter is an introduction to the main concepts of QFT based on the two approaches mentioned. It is beyond the scope of this book to cover all aspects of such a large subject, so we focus only on those aspects that are useful here. Interested readers are referred to the references at the end of this chapter.

7.2 Order Parameters and Lagrangian

Let us start our discussion by the functional formalism of the Euclidean QFT that is at the root of the continuous formulation of statistical models. This formalism relies on the ability to substitute the sum on the classical *discrete* variables $\{s_i\}$ in terms of a functional integral on the *continuous* variables $\varphi(x)$. This happens near a phase transition point, when the correlation length ξ is much larger than the lattice spacing a

$$Z = \sum_{\{s_i\}} e^{-H(\{s_i\})} \simeq \int \mathcal{D}\varphi(x)\, e^{-S(\{\varphi\})}, \quad \xi \gg a. \tag{7.2.1}$$

Let us comment this expression. The first problem that arises in the functional approach is the identification of the order parameter of the statistical system. As already discussed in Chapter 1, to solve this problem we have to rely on the symmetry of the Hamiltonian and on some physical intuition. For instance, in the presence of a Z_2 symmetry, the role of order parameter can be played by a scalar quantity $\varphi(x)$ that takes values on all the real axis, odd under the Z_2 transformation, $\varphi(x) \to -\varphi(x)$. For a system that is instead invariant under the $O(n)$ symmetry, just to make another example, we can take as order parameter a field with n components $\Phi(x) = [\phi_1(x), \phi_2(x), \ldots, \phi_n(x)]$ that transforms as a vector under the $O(n)$ transformations.

Action and Lagrangian. Once the order parameter is identified, we then introduce the Boltzmann weight associated to its different configurations. Only in this way, in fact, one we further proceed to compute statistical averages, correlation functions and all other thermodynamical quantities. In analogy with what was done for the statistical systems defined on a lattice, the probability of the field configuration is assumed to be proportional to[4]

$$W(\varphi, \{g\}) = \exp[-\mathcal{S}(\varphi, \{g\})] = \exp\left[-\int d\mathbf{x}\, \mathcal{L}(x)\right], \tag{7.2.2}$$

where $\mathcal{S}$ is the *action* of the theory, given by an integral on a Lagrangian density $\mathcal{L}(x)$. The latter is a local quantity, generically expressed in terms of a polynomial of the fields and their derivatives. To simplify the notation, we focus our attention on a QFT of a scalar field $\varphi(x)$, odd under the Z_2 symmetry. In this case, restricting the attention to those terms that are at most of degree 2 in the derivatives,[5] the most general expression of the action is given by

$$\mathcal{S} = \int d\mathbf{x} \left[\frac{1}{2}(\partial_j\varphi)^2 + g_1\varphi + \frac{g_2}{2}\varphi^2(x) + \cdots + \frac{g_n}{n!}\varphi^n(x) + \cdots\right]. \tag{7.2.3}$$

In the d-dimensional Euclidean space, the definition of the derivative term is meant to be a sum on the repeated indices

$$(\partial_j\varphi)^2 \equiv (\partial_j\varphi)(\partial_j\varphi) = \sum_{i=1}^{d}\left(\frac{\partial\varphi}{\partial x_i}\right)^2.$$

The Lagrangian theory (7.2.3) is also known as Landau–Ginzburg theory. To cope with the perturbative analysis of such an action, the custom is to isolate firstly its free part, expressed by the quadratic terms

$$\mathcal{S}_0 = \int d\mathbf{x}\, \mathcal{L}_0 = \int d\mathbf{x}\left[\frac{1}{2}(\partial_j\varphi)^2 + \frac{m^2}{2}\varphi^2(x)\right]. \tag{7.2.4}$$

and consider the remaining terms in (7.2.3) as the interactive part, denoted by $\mathcal{S}_I = \int d\mathbf{x}\, \mathcal{L}_I$. In the expression above, m is the mass parameter.[6] It is also convenient to introduce the concept of the *manifold of the coupling constants*, defined as the space spanned by the set of all couplings $\{g\} = (g_1, g_3, \ldots, g_n, \ldots)$.

Once the Lagrangian is given, the partition function of the system is obtained by summing up all possible configurations of the order parameter

$$Z[\{g\}, a] = \int \mathcal{D}\varphi\, \exp[-\mathcal{S}(\varphi, \{g\})]. \tag{7.2.5}$$

[4] In the following we often use the notation $\mathbf{x}$ to denote a vector. Similarly, we use $d\mathbf{x} = d^d x$.

[5] This can be justified by demanding the causality of the theory.

[6] In the canonical quantization of the theory, m can be indeed identified with the mass of the particle created by the field $\varphi(x)$.

In writing this expression we have emphasized that the partition function depends both on the coupling constants g_i and the microscopic cut-off a provided by the lattice spacing of the original theory. Even if we have adopted a continuous formalism to describe a statistical model, it is in fact necessary to take into account the microscopic scales of the systems. and we will see later several effects of such a dependence. Notice that an obvious reason to introduce the microscopic scale a is related to the definition of the measure $\mathcal{D}\varphi$: with this notation we mean a measure on all possible values of the field $\varphi(x)$. Since φ is a continuous quantity defined on *each* point of the space, $\mathcal{D}\varphi$ is not a-priori well-defined. In order to make sense of it, one can proceed in two equivalent ways.

The measure. The first approach to define a measure consists of considering the field as a collection of discrete quantities φ_i, defined only on N sites of a lattice with spacing length a, so that $\mathcal{D}\varphi$ can be expressed as a product of the differentials of all these variables, whose number can be enormously large but in any case finite

$$\mathcal{D}\varphi = \prod_i^N d\varphi_i. \tag{7.2.6}$$

The second equivalent approach makes use of the translation invariance of the system. This invariance allows us to decompose the field in its Fourier components

$$\varphi(x) = \frac{1}{\sqrt{N}} \sum_k \varphi(k)\, e^{ikx}.$$

When N is finite, the frequencies are discrete. Furthermore, in presence of a microscopic scale a, they satisfy the condition

$$|k| \leq \Lambda \simeq \frac{1}{a}.$$

The lattice space a acts then as an ultraviolet cut-off. This turns out to be a very useful quantity, since it permits also to regularize the divergent terms coming from the perturbative formulation of the theory. In the second approach the measure $\mathcal{D}\varphi$ is also given by a differential of a finite number of variables

$$\mathcal{D}\varphi = \prod_{0\leq|k|\leq 1/a} d\varphi(k). \tag{7.2.7}$$

Notice that in both cases it remains still open the problem to control the behaviour of $\mathcal{D}\varphi$ when $N \to \infty$, or, equivalently, $a \to 0$. This is a problem not only of the measure but of the entire QFT.

Engineering dimensions. As a matter of fact, the ultraviolet cut-off a enters also other key aspects. Consider, for instance, the engineering dimensions of the coupling constants in the action (7.2.3). To determine such quantities, it is necessary to fix initially the dimension of the scalar field φ. Since $\mathcal{A}$ is a dimensionless quantity, each term of the

Lagrangian should have dimension a^{-d}. Consider then the kinetic term $(\partial_j \varphi)^2$: posing the dimension of the field equals to $[\varphi] = a^{x_\varphi}$, we have the condition $a^{-2}\, a^{2x_\varphi} = a^{-d}$ and therefore

$$[\varphi] = a^{1-d/2}. \tag{7.2.8}$$

Once the dimension of $\varphi(x)$ is known, it is easy to obtain the dimensions of the various coupling constants

$$[g_m] = a^{md/2-m-d} \equiv a^{\delta_m}. \tag{7.2.9}$$

It is interesting to observe that each coupling constant has a particular dimension $d_s^{(m)}$ (the so-called upper critical dimension) in which it is dimensionless. For instance, g_3 is dimensionless for $d = 6$, g_4 for $d = 4$ and so on. Notice that the quantity δ_m is positive when

$$d \geq d_s^{(m)} = \frac{2m}{m-2}. \tag{7.2.10}$$

Critical behaviour. On the basis of this information, we can already formulate some educated guesses on the critical behaviour of the theory, although they must be refined by a further analysis. For a Lagrangian with higher coupling constant given by g_n, the corresponding statistical theory is expected to present two different regimes by varying d:

(a) for $d > d_s^{(n)}$, the critical behaviour is expected to be described by the mean field theory, with a classical value for the critical exponents;

(b) $d < d_s^{(n)}$ the system is instead expected to present strong fluctuations with a corresponding significant change of its thermodynamic singularities.

The simplest way to understand these two different critical behaviours is to study the sign of the exponent δ_n: when $\delta_n > 0$ (i.e. $d > d_s^{(n)}$), sending to zero the lattice space a the corresponding coupling constant becomes smaller, while when $\delta_n < 0$ ($d < d_s^{(n)}$) the coupling constant becomes larger. Consequently, as concerns the critical behaviour, in the first case the microscopic fluctuations are expected to be irrelevant, while in the second case to be relevant. Anticipating the results and the terminology of the renormalization group that is discussed in Chapter 8, the coupling constants g_n with $\delta_n > 0$ are called *irrelevant*, those with $\delta_n < 0$ are called *relevant* and, finally, those with $\delta_n = 0$, *marginal*.

The previous analysis was carried out for a theory invariant under a Z_2 symmetry, but the same scenario holds for other theories with different internal symmetry. Namely, each theory has a lower critical dimension d_i, below which there is no longer a phase transition, and a upper critical dimension d_s, beyond which the critical exponents take classical values. The strong fluctuation regime of the order parameters is expected to occur in between, i.e. in the range of dimensions d satisfying

$$d_i \leq d \leq d_s. \tag{7.2.11}$$

For systems with a short range interactions and a discrete symmetry, e.g. the Ising or the Potts models, the lower critical dimension is always $d_i = 1$, whereas for those with a continuous symmetry, e.g. the $O(n)$ model, $d_i = 2$. In the range (7.2.11) the critical exponents assume values that are different from their mean field solution and their determination requires more sophisticate theoretical tools.

7.3 Field Theory of the Ising Model

In order to clarify the formulation of a statistical model in terms of an Euclidean QFT, it is instructive to study in some detail the case of the Ising model. Consider the partition function of this model, generally expressed as

$$Z = \sum_{\{s_i\}} \exp\left[\sum_{i,j} \mathcal{J}_{ij}\, s_i\, s_j + \sum_i h_i\, s_i\right]. \tag{7.3.1}$$

Let us use an identity valid for the Gaussian integral

$$\int_{-\infty}^{+\infty} \prod_i d\phi_i \exp\left[-\frac{1}{4}\sum_{i,j}\phi_i \mathcal{J}_{ij}^{-1}\,\phi_j + \sum_i \phi_i\, s_i\right] = A \exp\left[\sum_{i,j}\mathcal{J}_{ij} s_i\, s_j\right], \tag{7.3.2}$$

(where A is a normalization constant that we will disregard from now on). This identity allows us to express the partition function (7.3.1) in terms of a Lagrangian of a bosonic field ϕ_i, thus swapping from the formulation based on the discrete variables $s_i = (\pm 1)$ to the one based on the continuous variables $\phi_i = (-\infty, +\infty)$. Substituting the identity (7.3.2) in eqn. (7.3.1), we have

$$\begin{aligned} Z &= \sum_{\{s_i\}} \exp\left[\sum_{i,j} \mathcal{J}_{ij}\, s_i\, s_j + \sum_i h_i\, s_i\right] \\ &= \sum_{\{s_i\}} \int_{-\infty}^{+\infty} \prod_i d\phi_i \exp\left[-\frac{1}{4}\sum_{i,j}\phi_i \mathcal{J}_{ij}^{-1}\,\phi_j + \sum_i (\phi_i + h_i)\, s_i\right] \\ &= \int_{-\infty}^{+\infty} \prod_i d\phi_i \exp\left[-\frac{1}{4}\sum_{i,j}(\phi_i - h_i)\mathcal{J}_{ij}^{-1}\,(\phi_j - h_j)\right] \sum_{\{s_i\}} \exp\left[\sum_i \phi_i\, s_i\right]. \end{aligned} \tag{7.3.3}$$

The sum on the spin configurations in the last term can be now explicitly performed because the spins are decoupled

$$\sum_{\{s_i\}} \exp\left[\sum_i \phi_i s_i\right] = \prod_i (2\cosh\phi_i) = A' \exp\left[\sum_i \log[\cosh\phi_i]\right]$$

(where A' is another constant). By means of the linear transformation

$$\phi_i \to \frac{1}{2}\mathcal{J}^{-1}_{ij}\,\phi_j,$$

we arrive (up to multiplicative constants) to the expression

$$Z = e^{-\frac{1}{4}\sum_{i,j} h_i \mathcal{J}^{-1}_{ij} h_j} \times \tag{7.3.4}$$
$$\times \int \mathcal{D}\phi \, \exp\left[-\sum_{i,j} \mathcal{J}_{ij}\phi_i\phi_j + \sum_i \log[\cosh(2\mathcal{J}_{ik}\phi_k)]\right].$$

Quadratic part. Notice that the dependence on the magnetic fields is factorized in the prefactor. To understand the nature of the field theory obtained above, it is useful to study its quadratic part. Using the Fourier transform both for the ϕ_i and the coupling constants

$$\phi_i = \phi(\mathbf{r}_i) = \frac{1}{\sqrt{N}}\sum_{\mathbf{k}} \phi(\mathbf{k})\, e^{i\mathbf{k}\cdot\mathbf{r_i}},$$
$$\mathcal{J}_{ij} = \mathcal{J}(\mathbf{r}_i - \mathbf{r}_j) = \frac{1}{N}\sum_{\mathbf{k}} \mathcal{J}(\mathbf{k})\, e^{i\mathbf{k}\cdot(\mathbf{r_i}-\mathbf{r_j})},$$

we have

$$\sum_{i,j} \mathcal{J}_{ij}\phi_i\phi_j = \sum_{\mathbf{k}} \mathcal{J}(\mathbf{k})\,\phi(\mathbf{k})\,\phi(-\mathbf{k}) = \sum_{\mathbf{k}} \mathcal{J}(\mathbf{k})\,|\phi(\mathbf{k})|^2.$$

One should be careful that a quadratic term is also present in the expansion of

$$\log[\cosh x] = \frac{1}{2}x^2 - \frac{1}{12}x^4 + \cdots$$

explicitly given by

$$2\sum_i (\mathcal{J}_{ij}\phi_j)^2 = 2\sum_{\mathbf{k}} |\mathcal{J}(\mathbf{k})|^2\,|\phi(\mathbf{k})|^2.$$

Putting together the two quadratic terms, the free part of the Lagrangian reads

$$\int d\mathbf{x}\,\mathcal{L}_0 = \sum_{\mathbf{k}} \left[\mathcal{J}(\mathbf{k}) - 2\,|\mathcal{J}(\mathbf{k})|^2\right] |\phi(\mathbf{k})|^2. \tag{7.3.5}$$

Let us now expand this expression in powers of k to the second order[7]

$$\mathcal{J}(\mathbf{k}) \simeq \mathcal{J}_0\,(1 - \rho^2 k^2).$$

If the model has a next-neighbour coupling $\tilde{\mathcal{J}}$ and the lattice has a coordination number z, we have

$$\mathcal{J}_0 = \sum_r \mathcal{J}(r) = (z\beta\tilde{\mathcal{J}})/2, \tag{7.3.6}$$

where $\beta = 1/kT$. The coefficient ρ is of the same order of the lattice spacing a for it is defined by the average

$$\mathcal{J}_0\,\rho^2 k^2 = \frac{1}{2}\sum_{\mathbf{r}} \mathcal{J}(\mathbf{r})\,(\mathbf{k}\cdot\mathbf{r})^2 \simeq \mathcal{J}_0\,k^2 a^2.$$

Coming back to eqn. (7.3.5), we have

$$\int d\mathbf{x}\,\mathcal{L}_0 = \mathcal{J}_0 \sum_{\mathbf{k}} \left[(1 - 2\mathcal{J}_0) + (4\mathcal{J}_0 - 1)\,\rho^2 k^2\right] |\phi(\mathbf{k})|^2. \tag{7.3.7}$$

When the temperature T decreases, $\mathcal{J}_0$ increases and therefore there is a critical value T_c of T for which the term $(1 - 2\mathcal{J}_0)$ vanishes[8]

$$T_c - z\tilde{\mathcal{J}}/k, \tag{7.3.8}$$

that coincides with the critical temperature of the mean field solution of the Ising model. At $T = T_c$ the zero mode of the field becomes unstable, because the corresponding integral on this variable in the functional integral (7.3.4) is no longer damped. Hence,

[7] In the inverse Fourier transform, higher orders give rise to higher derivative terms, whose coupling constants are irrelevant.

[8] Notice that, increasing T, there is another value of the temperature for which the other term $(4\mathcal{J}_0 - 1)$ vanishes and then changes sign. This happens because the original matrix $\mathcal{J}_{ij}$ is ill-defined since it has negative eigenvalues (all its diagonal terms are zero and correspondingly the sum of its eigenvalues vanishes). Since $s_i^2 = 1$, this drawback can be cured as in the spherical model by adding the identity matrix I to $\mathcal{J}_{ij}$ with a proper coefficient in front to ensure the positivity of the eigenvalues. Notice, however, that this operation has the effect to spoil the simple lattice relation (7.3.6) above.

T_c signals a phase transition. Posing

$$\begin{aligned} 1 - 2\mathcal{J}_0 &= \frac{T - T_c}{T_c} \\ 4\mathcal{J}_0 - 1 &= 1 + \mathcal{O}(T - T_c) \\ \mathcal{J}_0 &= \frac{1}{2} + \mathcal{O}(T - T_c) \end{aligned}$$

and substituting in eqn. (7.3.7), we have

$$\int d\mathbf{x}\, \mathcal{L}_0 = \frac{1}{2} \sum_{\mathbf{k}} \left(\frac{T - T_c}{T_c} + \rho^2 k^2 \right) |\phi(\mathbf{k})|^2 .$$

Finally, defining

$$\varphi(x) = \rho\, \phi(x), \quad m^2 = \frac{1}{\rho^2} \frac{T - T_c}{T_c}$$

we arrive at

$$\mathcal{S}_0 = \int d\mathbf{x}\, \mathcal{L}_0 = \int d\mathbf{x} \frac{1}{2} \left[(\partial_j \varphi)^2 + m^2 \varphi^2 \right] = \frac{1}{2} \sum_{\mathbf{k}} (k^2 + m^2) |\varphi(\mathbf{k})|^2 . \tag{7.3.9}$$

Further interaction terms of the action can be recovered by taking into account the higher terms from the expansion of the term $\log[\cosh x]$, which will be discussed later in this chapter.

7.4 Correlation Functions and Propagator

Once the Boltzmann weight of the field configurations is defined, one can proceed to define the correlation functions. They are expressed by the functional integral

$$G^{(n)}(\mathbf{x_1}, \ldots, \mathbf{x_n}) = \langle \varphi(\mathbf{x_1}) \ldots \varphi(\mathbf{x_n}) \rangle = \frac{1}{Z} \int \mathcal{D}\varphi\, \varphi(\mathbf{x_1}) \ldots \varphi(\mathbf{x_n}) \exp\left[-\mathcal{S}(\varphi, \{g\}) \right] . \tag{7.4.1}$$

For a compact expression of these quantities, it is sufficient to couple the field $\varphi(\mathbf{x})$ to an external current $\mathcal{J}(\mathbf{x})$, defining a new partition function

$$Z[\mathcal{J}] = \int \mathcal{D}\varphi \exp\left[-\mathcal{S}(\varphi, \{g\}) + \int d\mathbf{x}\, \mathcal{J}(\mathbf{x}) \varphi(\mathbf{x}) \right] . \tag{7.4.2}$$

In this way

$$G^{(n)}(\mathbf{x_1},\ldots,\mathbf{x_n}) = \frac{1}{Z[\mathcal{J}]}\frac{\delta^n Z[\mathcal{J}]}{\delta\mathcal{J}(\mathbf{x_1})\ldots\delta\mathcal{J}(\mathbf{x_n})}\bigg|_{\mathcal{J}=0}. \tag{7.4.3}$$

One can similarly define the correlation functions in the momentum space, given by

$$\hat{G}^{(n)}(\mathbf{k_1},\ldots,\mathbf{k_n}) = \int d\mathbf{x_1}\ldots d\mathbf{x_n}\, e^{-i(\mathbf{k_1}\cdot\mathbf{x_1}+\cdots\mathbf{k_n}\cdot\mathbf{x_n})}\, G^{(n)}(\mathbf{x_1},\ldots,\mathbf{x_n}). \tag{7.4.4}$$

Since

$$\int d\mathbf{x}\mathcal{J}(\mathbf{x})\,\varphi(\mathbf{x}) = \int \frac{d\mathbf{k}}{(2\pi)^d}\mathcal{J}(-\mathbf{k})\,\varphi(\mathbf{k}),$$

we have

$$\hat{G}(\mathbf{k_1},\ldots,\mathbf{k_n}) = (2\pi)^{nd}\,\frac{1}{Z[\mathcal{J}]}\,\frac{\delta^n Z}{\delta\mathcal{J}(-\mathbf{k_1})\ldots\mathcal{J}(-\mathbf{k_n})}. \tag{7.4.5}$$

It is interesting to determine the scale dimensions of the quantities given above: for the correlation functions in real space we have

$$[G^{(n)}(x_1,\ldots,x_n)] = [\varphi]^n = a^{n(1-d/2)} = \Lambda^{n(d/2-1)}, \tag{7.4.6}$$

while for those in the momentum space

$$[G^{(n)}(k_i)] = \Lambda^{-nd}\,[G^{(n)}(x_i)] = \Lambda^{-n(1/2d+1)}. \tag{7.4.7}$$

For the translation invariance, the Fourier transform (7.4.4) has always a prefactor $\delta^d(\sum_i^n k_i)$. Dividing by this term and denoting by $\bar{G}^{(n)}(k_i)$ the remaining expression, we have

$$[\bar{G}^{(n)}(k_i)] = \Lambda^{d-n(1/2d+1)}. \tag{7.4.8}$$

The propagator. A special role is played by the two-point correlation function of the free theory

$$G_0^{(2)}(\mathbf{x}_1-\mathbf{x}_2) = \Delta(\mathbf{x}_1-\mathbf{x}_2) = \langle\varphi(\mathbf{x}_1)\varphi(\mathbf{x}_2)\rangle_0. \tag{7.4.9}$$

This is the so-called *propagator* of the theory for a reason that will be immediately clear. Its computation is elementary: expressing the free action as

$$S_0 = \int d\mathbf{x} \left[\frac{1}{2}(\partial_j \varphi)^2 + \frac{m^2}{2}\varphi^2 \right] = \frac{1}{2}\int d\mathbf{x}\, \varphi(\mathbf{x}) \left[-\partial^2 + m^2 \right] \varphi(\mathbf{x}), \qquad (7.4.10)$$

and computing the Gaussian integral in (7.4.2), we arrive at

$$Z_0[\mathcal{J}] = \exp\left[\frac{1}{2}\int d\mathbf{x}\, d\mathbf{y}\, \mathcal{J}(\mathbf{x})\, \Delta(\mathbf{x}-\mathbf{y})\, \mathcal{J}(\mathbf{y}) \right], \qquad (7.4.11)$$

where $\Delta(\mathbf{x}-\mathbf{y})$ is, formally, the inverse matrix $(-\partial^2 + m^2)$ in the coordinate space

$$\Delta(\mathbf{x}-\mathbf{y}) \equiv \langle \mathbf{x} | \frac{1}{-\partial^2 + m^2} | \mathbf{y} \rangle.$$

A more transparent form is given by its Fourier transform

$$\Delta(\mathbf{x}-\mathbf{y}) = \int_{0 \leq |k| \leq \Lambda} \frac{d\mathbf{k}}{(2\pi)^d} \frac{\exp\left[i\mathbf{k}\cdot(\mathbf{x}-\mathbf{y})\right]}{k^2 + m^2}, \qquad (7.4.12)$$

where $\Lambda = 1/a$ is the ultraviolet cut-off. The Euclidean propagator can be computed for any dimension d (and for $\Lambda = \infty$) as follows. Going to radial coordinates and denoting by r and k the modulus of the distance and momentum, we have

$$\Delta(r) = \int \frac{d^d k}{(2\pi)^d} \frac{e^{i\mathbf{k}\cdot\mathbf{x}}}{k^2 + m^2} = \frac{\Omega(d-1)}{(2\pi)^d} \int_0^\infty dk \frac{k^{d-1}}{k^2 + m^2} \int_0^\pi d\theta \, \sin^{d-2}\theta \, e^{ikr\cos\theta},$$

where $\Omega(d-1)$ is the solid angle comes from the integration on the $(d-1)$ remaining angles (its explicit expression is given in eqn. (2.B.1)). In order to proceed further, we need some integrals involving the Bessel functions

$$\int d\theta \sin^{2\nu}\theta \, e^{ikr\cos\theta} = \frac{\Gamma\left(\nu + \frac{1}{2}\right)\Gamma\left(\frac{1}{2}\right)}{\left(\frac{kr}{2}\right)^{\nu}} \mathcal{J}_\nu(kr),$$

$$\int_0^\infty dk\, k^{\nu+1} \frac{\mathcal{J}_\nu(ak)}{k^2 + m^2} = m^\nu K_\nu(ma).$$

Using these formulae and simplifying the expressions coming from the Γ functions, the final result is

$$\Delta(r) = (2\pi)^{-d/2} \left(\frac{m}{r}\right)^{d-2} K_{\frac{d-2}{2}}(mr). \qquad (7.4.13)$$

Substituting in this formula the relevant values of d (using for $d = 1$ and $d = 3$ the known expressions for $K_{\pm\frac{1}{2}}(x)$) easily we recover the results shown in Table 7.1.

Let us comment other properties of the propagator. It is easy to see that, for any dimension d, $\Delta(x)$ decreases exponentially for $x \to \infty$ as e^{-mx}. Hence, for distance separations of few units of m^{-1}, the fluctuations of the order parameter are essentially

d	$\Delta(x)$ $(\Lambda = \infty)$	$\Delta(0)$ $(\Lambda \gg m)$
1	$\frac{1}{2m} e^{-mx}$	$\frac{1}{2m}$
2	$\frac{1}{2\pi} K_0(mx)$	$\frac{1}{2\pi} \log\left(\frac{\Lambda}{m}\right)$
3	$\frac{1}{4\pi x} e^{-mx}$	$\frac{\Lambda}{2\pi^2}$
4	$\frac{m}{2\pi^2 x} K_1(mx)$	$\frac{\Lambda^2}{16\pi^2}$

Table 7.1 *Propagator, by varying the dimension d, in the limit $\Lambda \gg m$ and for $x = |\mathbf{x}| \gg \Lambda^{-1}$. In the third column there is the value at the origin when $\Lambda \gg m$. $K_0(r)$ and $K_1(r)$ are the modified Bessel functions.*

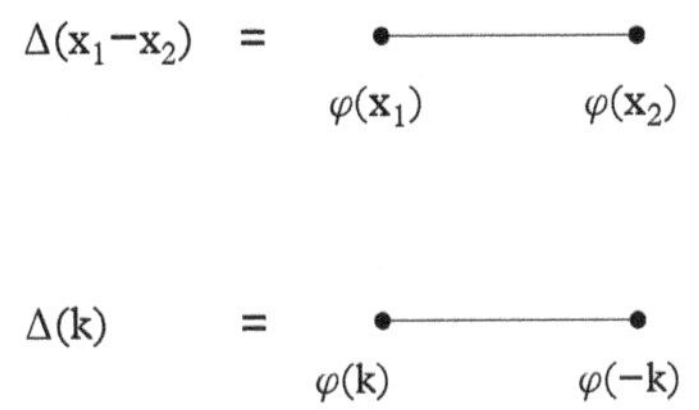

Fig. 7.3 *Propagator of the free theory and its graphical representation.*

uncorrelated. This means that the correlation length of the system can be identified with the inverse of the mass parameter m

$$\xi = \frac{1}{m}. \tag{7.4.14}$$

When m decreases the correlation length ξ increases and for $m \to 0$ its divergence can be interpreted as the onset of a phase transition.

Notice that the value of $\Delta(x)$ at the origin depends on the dimensionality of the system and on the cut-off. If for $d = 1$ the dependence is rather weak, for $d \geq 2$ there is instead a divergence when $\Lambda \to \infty$. This makes once more evident the crucial role played by the ultraviolet cut-off a and by the dimension d of the system.

Finally, since $\Delta(x)$ satisfies the differential equation

$$(-\partial^2_{\mathbf{x}_1} + m^2)\,\Delta(\mathbf{x}_1 - \mathbf{x}_2) = \delta^d(\mathbf{x}_1 - \mathbf{x}_2), \tag{7.4.15}$$

this quantity is also the Green function of the system. From a physical point of view, it describes the propagation of a fluctuation of the field $\varphi(x)$ from the position x_1 to x_2. It is convenient to assign to it a graphical representation in terms of a line that connects the two points x_1 and x_2, as shown in Figure 7.3. An analogous representation is also associated to its Fourier transform

$$\Delta(\mathbf{k}) = \langle \varphi(\mathbf{k})\varphi(-\mathbf{k}) \rangle = \frac{1}{\mathbf{k}^2 + m^2}. \tag{7.4.16}$$

7.5 Perturbation Theory and Feynman diagrams

In presence of interactions, it is often impossible to compute exactly the functional integral (7.4.2). For this reason it is important to develop a perturbative formalism based on a power expansion in the coupling constants. It should be stressed that such an approach has some limitations: the most obvious one is that is restricted to small values of the coupling constants and therefore unable to catch the strong coupling behaviour of the theory. Unfortunately, this is not the only limitation: in most cases, the perturbative series have zero radius of convergence and, at best, they can be asymptotic series (see Problem 7.2). Furthermore, in some QFTs there are non-perturbative aspects associated for instance to topological excitations, as solitons or vortices, that are totally inaccessible to the perturbative approach (see Problem 7.7). Despite all these drawbacks, it is nevertheless important to study the perturbative formulation since it provides useful information on the analytic nature of the various amplitudes and on the corrections to the free theory behaviour.

For the sake of simplicity, we focus our attention on a Lagrangian that has only one interaction term, given by φ^4. Isolating the free part, the action can be written as

$$\mathcal{S} = \mathcal{S}_0 + \frac{g}{4!} \int d\mathbf{x}\, \varphi^4(\mathbf{x}) = \mathcal{S}_0 + \mathcal{S}_I. \tag{7.5.1}$$

As for the propagator, we can also associate a graphical representation to the interaction term $\frac{g}{4!}\varphi^4$: this is given by a vertex with four external lines.

The perturbative definition of the theory is obtained by expanding the Boltzmann weight in powers of g

$$e^{-\mathcal{S}_0 - \mathcal{S}_I} = e^{-\mathcal{S}_0}\left[1 - \mathcal{S}_I + \frac{1}{2}\mathcal{S}_I^2 + \cdots\right].$$

Consider for instance the perturbative definition of the partition function

$$Z[g] = \int \mathcal{D}\varphi\, e^{-\mathcal{S}_0}\left[1 - \mathcal{S}_I + \frac{1}{2}\mathcal{S}_I^2 - \cdots\right]. \tag{7.5.2}$$

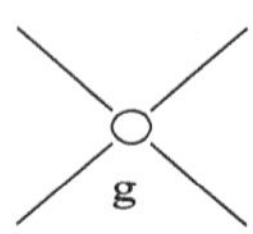

Fig. 7.4 *Vertex of the interaction in correspondence of* $\frac{g}{4!}\varphi^4$.

Wick theorem. Order by order in g, all integrals that enter the expression above are of Gaussian nature and can be explicitly computed by a generalization of the following Gaussian integral in n variables

$$\begin{aligned}\langle x_{k_1} \ldots x_{k_m} \rangle &\equiv \mathcal{N} \int \prod_i dx_i\, x_{k_1} \ldots x_{k_m}\, e^{-\frac{1}{2}\sum_{i,j} x_i A_{ij} x_j} \\ &= \sum_P A^{-1}_{k_{p_1} k_{p_2}} \ldots A^{-1}_{k_{p_{m-1}} k_{p_m}}\end{aligned} \tag{7.5.3}$$

where $\mathcal{N}$ is a constant that ensures the correct normalization of the integral, whereas the last sum is on all possible ways of pairing the indices $k_1, \ldots, k_m$. This expression expresses the content of the Wick theorem in field theory.

Partition function. The partition function (7.5.2) can be written in a compact way as

$$Z[g,\mathcal{J}] = \exp\left[-\frac{g}{4!}\int d\mathbf{x} \frac{\delta^4}{\delta \mathcal{J}^4(\mathbf{x})}\right] \exp\left[\frac{1}{2}\int d\mathbf{x}\, d\mathbf{y}\, \mathcal{J}(\mathbf{x})\, \Delta(\mathbf{x}-\mathbf{y})\mathcal{J}(\mathbf{y})\right]. \tag{7.5.4}$$

an expression that, in the more general case of interaction term $\mathcal{L}_I$, generalizes as

$$Z[\{g\},\mathcal{J}] = \exp\left[-\int d\mathbf{x}\, \mathcal{L}_I\left[\frac{\delta}{\delta \mathcal{J}(\mathbf{x})}\right]\right] Z_0[\mathcal{J}]. \tag{7.5.5}$$

Let us come back to the analysis of eqn. (7.5.4). For the presence of the fourth derivative with respect to the current $\mathcal{J}(x)$, the first correction is obtained by expanding $Z_0[\mathcal{J}]$ up to the second order and then taking the functional derivative with respect to the external currents $\mathcal{J}(x_1), \ldots, \mathcal{J}(x_4)$ by using the functional relation

$$\frac{\delta^4}{\delta \mathcal{J}^4(z)} [\mathcal{J}(z_1)\mathcal{J}(z_2)\mathcal{J}(z_3)\mathcal{J}(z_4)] = 4!\, \delta^d(z-z_1)\, \delta^d(z-z_2)\, \delta^d(z-z_3)\, \delta^d(z-z_4).$$

The result is

$$\begin{aligned}\delta Z/Z_0 = -g\,\Bigg(&\frac{1}{8}\int d\mathbf{z}_1\, \Delta(0) + \frac{1}{4}\int d\mathbf{z}_1\, d\mathbf{z}_2\, d\mathbf{z}_3\, \Delta(0)\, \Delta(z_1-z_2)\mathcal{J}(z_2)\, \Delta(z_1-z_3)\mathcal{J}(z_3) \\ &+ \frac{1}{4!}\int d\mathbf{z}_1 \ldots d\mathbf{z}_5\, \Delta(z_1-z_2)\mathcal{J}(z_2)\, \Delta(z_1-z_3)\mathcal{J}(z_3)\, \Delta(z_1-z_4)\mathcal{J}(z_4)\, \Delta(z_1-z_5)\mathcal{J}(z_5)\Bigg).\end{aligned}$$

This expression, as well as all the others relative to higher perturbative orders, can be easily put in a graphical form, as shown in Figure 7.5: in this figure each empty circle (having four external legs) is associated to an integration variable and to the coupling constant g, each line that connects the points x and y is associated to $\Delta(x-y)$ and each black circle relative to the point z_i corresponds to the insertion of a current $\mathcal{J}(z_i)$. From

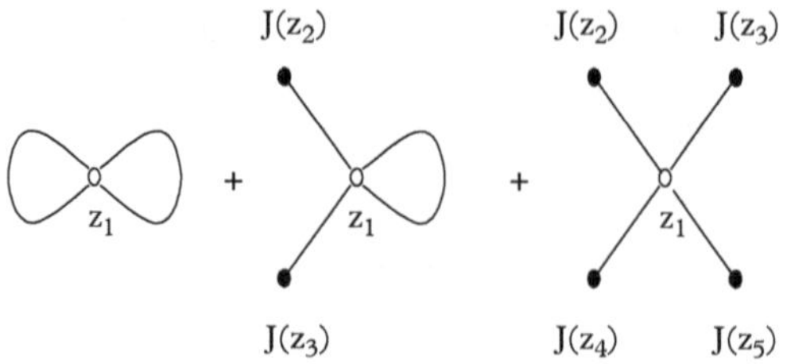

Fig. 7.5 *First perturbative terms of the partition function.*

the Wick theorem, all currents must be contracted among them: in the first diagram of Figure 7.5, for instance, this is realized by contracting pairwise the four currents present at the vertex, in the second diagram by contracting two of the currents of the vertex with two external currents and the remaining ones among themselves and, finally, in the last diagram, by contracting all four currents of the vertex with the external currents. In this procedure, there are certain combinatorial terms that is necessary to take into account on which we shall comment soon.

Sending to zero all the currents, the only term that survives is the first one. In Fourier transform, the first diagram in Figure 7.5 corresponds to

$$\delta Z = -\frac{g}{8} V \left[\int_{0<|k|<1/a} \frac{d^d k}{(2\pi)^d} \frac{1}{k^2 + m^2} \right]^2, \tag{7.5.6}$$

where V is the volume of the system. Note that, in the absence of the cut-off $1/a$, this gives rise to a divergent correction for $d \geq 2$.

The graphical representation that we used, originally proposed by Feynman, is extremely useful for a systematic book-keeping of the perturbative expansion. The perturbative order is given by the number of vertices of the graph: the graphs in Figure 7.6, for instance, correspond to two different corrections of order g^4 to the partition function.

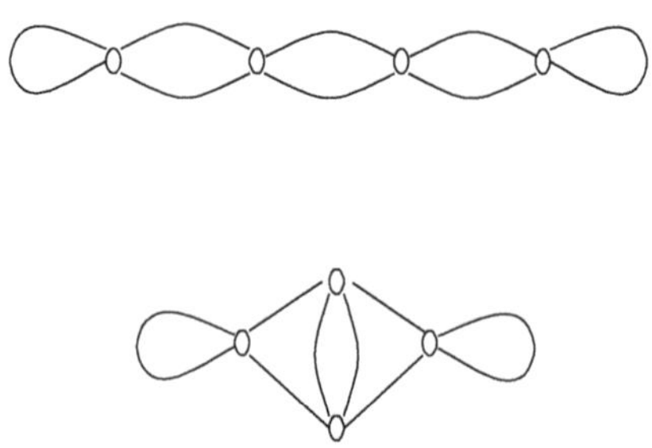

Fig. 7.6 *Two different corrections of order g^4 to the partition function.*

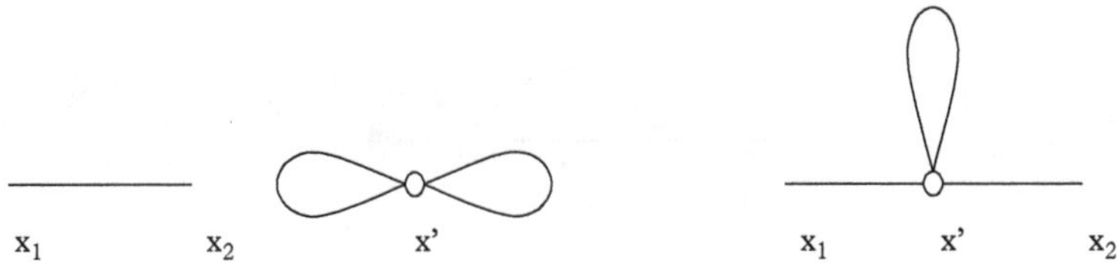

Fig. 7.7 *First-order correction to the two-point correlation function.*

Correlation functions. As for the partition function, a similar expansion also exists for the correlation functions. Since their definition is

$$G(\mathbf{x}_1,\ldots,\mathbf{x}_n) = \frac{1}{Z[\mathcal{J}]}\left.\frac{\delta^n Z}{\delta\mathcal{J}(\mathbf{x}_1)\ldots\delta\mathcal{J}(\mathbf{x}_n)}\right|_{\mathcal{J}=0},$$

in addition to the functional derivatives with respect to the currents of the interaction term, in this case we have also the derivatives with respect to the currents $\mathcal{J}(x_1),\ldots,\mathcal{J}(x_n)$ coupled to the fields. Note that, at each perturbative order there are also the disconnected graphs coming from the expansion of the partition function of the denominator. For instance, at the first order, the two-point correlation function is expressed by the graphs in Figure 7.7.

Connected correlation functions. In order to eliminate the disconnected graphs coming from the partition function and define instead the connected correlation functions $G_c(x_1,\ldots,x_n)$, it is convenient to introduce the functional $F[\mathcal{J}]$ defined as

$$Z[\mathcal{J}] = \exp[F[\mathcal{J}]]. \tag{7.5.7}$$

This functional corresponds, up to a sign, to the free energy of the statistical systems and it is easy to show that

$$G_c(\mathbf{x}_1,\ldots,\mathbf{x}_n) = \left.\frac{\delta^n F[\mathcal{J}]}{\delta\mathcal{J}(\mathbf{x}_1)\ldots\mathcal{J}(\mathbf{x}_n)}\right|_{\mathcal{J}=0}. \tag{7.5.8}$$

Using this formula, for the first representatives of the connected correlation functions we have

$$\begin{aligned}
&G_c^{(1)}(\mathbf{x}) = \langle\varphi(\mathbf{x})\rangle\\
&G_c^{(2)}(\mathbf{x}_1,\mathbf{x}_2) = \langle\varphi(\mathbf{x}_1)\varphi(\mathbf{x}_2)\rangle - \langle\varphi(\mathbf{x}_1)\rangle\langle\varphi(\mathbf{x}_2)\rangle\\
&G_c^{(3)}(\mathbf{x}_1,\mathbf{x}_2,\mathbf{x}_3) = \langle\varphi(\mathbf{x}_1)\varphi(\mathbf{x}_2)\varphi(\mathbf{x}_3)\rangle - \langle\varphi(\mathbf{x}_1)\varphi(\mathbf{x}_2)\rangle\langle\varphi(\mathbf{x}_3)\rangle\\
&\quad\langle\varphi(\mathbf{x}_2)\varphi(\mathbf{x}_3)\rangle\langle\varphi(\mathbf{x}_1)\rangle - \langle\varphi(\mathbf{x}_1)\varphi(\mathbf{x}_3)\rangle\langle\varphi(\mathbf{x}_2)\rangle\\
&\quad+2\langle\varphi(\mathbf{x}_1)\rangle\langle\varphi(\mathbf{x}_2)\rangle\langle\varphi(\mathbf{x}_3)\rangle
\end{aligned} \tag{7.5.9}$$

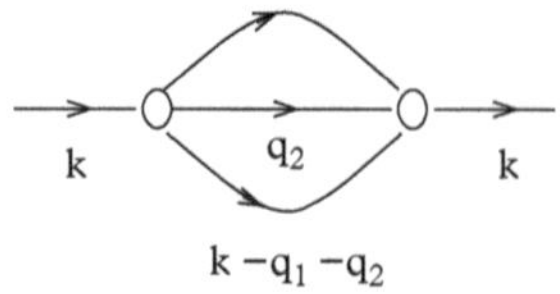

Fig. 7.8 *Correction of order g^2 to the propagator.*

Similar expressions are obtained for the connected correlation function in the momentum space defined by

$$G_c(\mathbf{k}_1,\ldots,\mathbf{k}_n) = \left.\frac{\delta^n F[\mathcal{J}]}{\delta\mathcal{J}(-\mathbf{k}_1)\ldots\mathcal{J}(-\mathbf{k}_n)}\right|_{\mathcal{J}=0}. \tag{7.5.10}$$

There are of course divergent terms also in the perturbative expansion of the connected correlators. For instance, at the first order in g, the correction of the connected two-point function $\langle\varphi(k)\varphi(-k)\rangle$ is given by the diagram on the right side in Figure 7.7, whose explicit form reads

$$-\frac{g}{2}\,[\Delta(k)]^2\int_{0\leq|k|\leq\Lambda}\frac{d^dk}{(2\pi)^d}\frac{1}{k^2+m^2}.$$

This expression is divergent for $d\geq 2$. More complicate terms, which may also present a nested structure of divergences of the internal loops, appear to higher order. Such a graph (of order g^2) is shown in Figure 7.8 and its analytic expression apart of the two external propagators $\Delta(k)$ is given by

$$\frac{g_4^2}{6}\int_{0\leq|q_i|\leq 1/a}\frac{d^dq_1}{(2\pi)^d}\frac{d^dq_2}{(2\pi)^d}\frac{1}{(q_1^2+g_2)(q_2^2+g_2)[(k-q_1-q_2)^2+g_2]}.$$

Neglecting the possible divergences coming from each loop and focusing, instead, on a simple counting of the power of all the momenta, it is easy to see that, in the absence of the cut-off $1/a$, this integral diverges when $d\geq 3$.

Combinatorial factors. Let us now stop a moment and discuss briefly the combinatorial factors of the perturbative terms. They come from the number of equivalent ways of contracting the currents. Let us see, for instance, how to arrive at the factor $\frac{1}{8}$ in the first correction to the partition function, eqn. (7.5.6): expanding $Z_0[\mathcal{J}]$ to the second term

$$Z_0^{(2)} = \frac{1}{2}\left(\frac{1}{2}\right)^2\int d\mathbf{x}_1\ldots d\mathbf{x}_4\mathcal{J}(\mathbf{x}_1)\,\Delta(\mathbf{x}_1-\mathbf{x}_2)\mathcal{J}(\mathbf{x}_2)\mathcal{J}(\mathbf{x}_3)\Delta(\mathbf{x}_3-\mathbf{x}_4)\mathcal{J}(\mathbf{x}_4).$$

To this expression we have to apply the differential operator

$$-\frac{g}{4!}\int d\mathbf{x}\left[\frac{\delta}{\delta\mathcal{J}(\mathbf{x})}\right]^4.$$

The first derivative of this operator can act on any of the four possible currents $\mathcal{J}(\mathbf{x}_1)..\mathcal{J}(\mathbf{x}_4)$ of (7.5): hence, there are four possibilities. The second derivative can act on any of the three remaining terms, the third derivative on any of the two remaining terms and finally, the last one has only one choice. Putting together all these factors, we have: $4 \times 3 \times 2$ equivalent ways of contracting the currents, there is in addition a coefficient $\left(\frac{1}{2}\right)^3$ coming from $Z_0^{(2)}$ and, finally, a factor $1/4!$ coming from the interaction vertex. Hence the final coefficient is given by

$$4 \times 3 \times 2 \times \left(\frac{1}{2}\right)^3 \times \frac{1}{4!} = \frac{1}{8}. \tag{7.5.11}$$

After this brief excursus in the perturbative series expansion, it is worth formulating more precisely the Feynman rules that enable us to computing the correlation functions either in the coordinate or momentum spaces. Below we state the Feynman rules for the φ^4 theory since their generalization to theories with other vertices φ^r is rather straighforward.

Correlation functions in coordinate space

In order to compute the g^n correction of the connected correlation function $G_c(\mathbf{x}_1,\ldots,\mathbf{x}_m)$ one must do:

- Draw all connected and distinct graphs with n vertices (of 4 legs) and m external legs. The latter are identified by the coordinates $\mathbf{x}_1,\ldots,\mathbf{x}_m$ of the external points. Connect all the legs by the lines relative to the propagators. Two diagrams are equivalent if they can be deformed one into the other by moving the vertices of the external points but without cutting any line. For the computation of each diagram, the steps are as follows

  1. Assign to each graph a factor $(-g)^n$.
  2. Associate to each vertex a coordinate $\mathbf{z}_i$.
  3. Assign to each line that connects the points $\mathbf{x}$ and $\mathbf{y}$ the propagator $\Delta(\mathbf{x}-\mathbf{y})$.
  4. Integrate on all coordinates of the internal points.
  5. Multiply by $1/2$ for each line whose final points arrive at the same vertex.
  6. Multiply by $(p!)^{-1}$ for each set of p lines that link two vertices.
  7. Divide by r if the internal points of the graph can be transformed in r ways, leaving the graph invariant.

- Finally, sum on all diagrams.

Correlation functions in the momentum space

To compute the g^n corrections to the connected correlation function $G_c(\mathbf{k}_1,\ldots,\mathbf{k}_m)$ the rules are similar

- Draw all connected and distinct graphs with n vertices (with four legs) and m external legs, the last ones identified by the coordinates $\mathbf{k}_1,\ldots,\mathbf{k}_m$ of the external momenta, and connect all these legs by means of the lines of the propagators. Two diagrams are equivalent if they can be deformed one to the other by moving the vertices or the external points without cutting any lines. To compute explicitly the contribution of the each diagram, the steps are the following:

  1. Assign to each of them a factor $(-g)^n$.
  2. Substitute each line with the expression of the propagator $1/(q_i^2+m^2)$, where $\mathbf{q}_i$ is the momentum of the line, making sure that the conservation of the momenta is enforced at each vertex where several lines meet. Assign the momenta $\mathbf{k}_i$ to the external lines.
  3. Integrate on all the momenta that have not be fixed by the conservation laws, with a factor $(2\pi)^{-d}$ for each integral. The integrals are computed with a cut-off Λ.
  4. Multiply by $1/2$ for each line whose final points arrive at the vertex.
  5. Multiply by $(p!)^{-1}$ for each set of p lines that link two identical vertices.
  6. Finally, divide by r if the internal points of the diagrams can be transformed in r ways by leaving the graph invariant.

- Finally, sum up all the diagrams.

7.6 Legendre Transformation and Vertex Functions

The expectation value of the field $\varphi(x)$, in the presence of an external current, is given by

$$\langle\varphi(\mathbf{x})\rangle \equiv \bar{\varphi}(\mathbf{x}) = \frac{\delta F[\mathcal{J}]}{\delta \mathcal{J}(\mathbf{x})}. \tag{7.6.1}$$

We can define the Legendre transform $\Gamma[\bar{\varphi}]$ of the functional $F[\mathcal{J}]$ by

$$\Gamma[\bar{\varphi}] + F[\mathcal{J}] = \int d\mathbf{x}\,\bar{\varphi}(\mathbf{x})\mathcal{J}(x). \tag{7.6.2}$$

Γ is a function of $\bar{\varphi}$ after the elimination of $\mathcal{J}(\mathbf{x})$ by means of eqn. (7.6.1). It is immediate to prove that

$$\frac{\delta\Gamma[\bar{\varphi}]}{\delta\bar{\varphi}(\mathbf{x})} = \mathcal{J}(\mathbf{x}). \tag{7.6.3}$$

This expression, that can be regarded as the inverse formula of (7.6.1). The formulae above have the following interesting property. If the system is in its ordered phase, we have

$$\bar{\varphi}(\mathbf{x}) = \left.\frac{\delta F[\mathcal{J}]}{\delta \mathcal{J}(\mathbf{x})}\right|_{\mathcal{J}=0} = v \neq 0 \tag{7.6.4}$$

and this implies

$$\left.\frac{\delta \Gamma}{\delta \bar{\varphi}(\mathbf{x})}\right|_{\bar{\varphi}(\mathbf{x})=v} = 0, \tag{7.6.5}$$

i.e. $\Gamma[\bar{\varphi}]$ has a stationary point for a non-vanishing constant value of the field.

Differentiating with respect to $\bar{\varphi}(y)$ the terms of both sides in (7.6.1), one finds

$$\begin{aligned} \delta(\mathbf{x}-\mathbf{y}) &= \frac{\delta^2 F[\mathcal{J}]}{\delta \mathcal{J}(\mathbf{x}) \delta \bar{\varphi}(\mathbf{y})} = \int dz \frac{\delta^2 F[\mathcal{J}]}{\delta \mathcal{J}(\mathbf{x}) \delta \mathcal{J}(\mathbf{z})} \frac{\delta \mathcal{J}(\mathbf{z})}{\delta \bar{\varphi}(\mathbf{y})} \\ &= \int d\mathbf{z} \frac{\delta^2 F[\mathcal{J}]}{\delta \mathcal{J}(\mathbf{x}) \delta \mathcal{J}(\mathbf{z})} \frac{\delta^2 \Gamma}{\delta \bar{\varphi}(\mathbf{z}) \delta \bar{\varphi}(\mathbf{y})}. \end{aligned} \tag{7.6.6}$$

When $\mathcal{J} \to 0$, $\frac{\delta^2 F[\mathcal{J}]}{\delta \mathcal{J}(\mathbf{x}) \delta \mathcal{J}(\mathbf{z})} = G_c^{(2)}(\mathbf{x},\mathbf{z})$, and therefore the expression

$$\frac{\delta^2 \Gamma[\bar{\varphi}]}{\delta \bar{\varphi}(\mathbf{x}) \delta \bar{\varphi}(\mathbf{y})} \equiv \Gamma^2(\mathbf{x},\mathbf{y}) \tag{7.6.7}$$

is the "inverse matrix" of the connected two point function. This statement becomes more evident going to Fourier transform, because eqn. (7.6.6) becomes

$$\Gamma^{(2)}(\mathbf{k})\, G_c^{(2)}(\mathbf{k}) = 1. \tag{7.6.8}$$

The function $\Gamma^{(2)}$ is also called the *vertex function* of the two-point correlation function: it corresponds to the sum of all Feynman graphs of the two-point correlation function that cannot be separated by removing one line. These graphs are called *one-particle irreducible* (see Figure 7.9)

There is an interesting graphical representation of (7.6.8): denoting by $X(k)$ the sum of all one-particle irreducible graphs that remain once the external legs are removed, the perturbative expression of the connected two-point function is expressed by the geometrical series

$$G_c^{(2)}(\mathbf{k}) = \Delta(\mathbf{k}) + \Delta(\mathbf{k})\, X(\mathbf{k})\, \Delta(\mathbf{k}) + \cdots = \frac{1}{\Delta^{-1}(\mathbf{k}) - X(\mathbf{k})} = \frac{1}{\Gamma^{(2)}(\mathbf{k})} \tag{7.6.9}$$

whose graphical representation is shown in Figure 7.10.

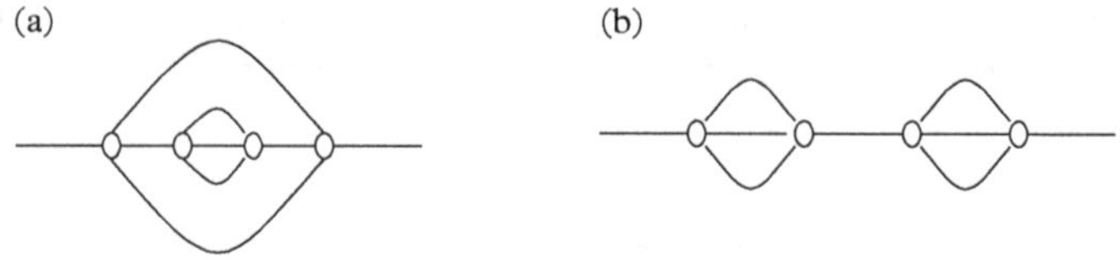

Fig. 7.9 *The graph (a) is one-particle irreducible whereas the graph (b) is not one-particle irreducible: the latter breaks in two by cutting the intermediate line.*

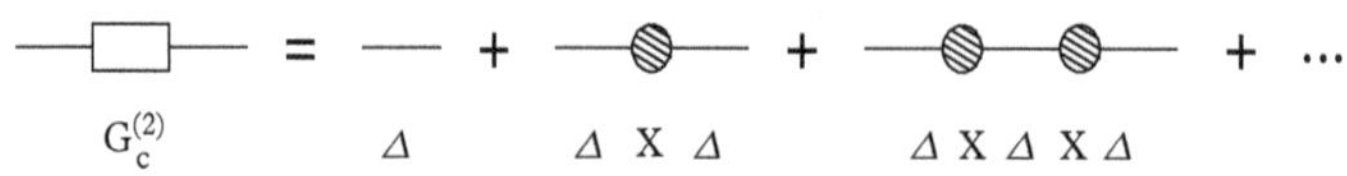

Fig. 7.10 *The connected two-point function is expressed by the infinite series made of the one-particle irreducible diagrams $X(k)$ without the external legs.*

We can proceed in a similar was to define the higher vertex functions, given in the coordinate and momentum space by

$$\Gamma^{(n)}(\mathbf{x}_1,\ldots,\mathbf{x}_n) = \frac{\delta^n \Gamma[\bar{\varphi}]}{\delta\bar{\varphi}(\mathbf{x}_1)\ldots\delta\bar{\varphi}(\mathbf{x}_n)},$$
$$\Gamma^{(n)}(\mathbf{k}_1,\ldots,\mathbf{k}_n) = \frac{\delta^n \Gamma[\bar{\varphi}]}{\delta\bar{\varphi}(-\mathbf{k}_1)\ldots\delta\bar{\varphi}(-\mathbf{k}_n)}.$$

For the translation invariance, $\Gamma^{(n)}(\mathbf{k}_1,\ldots,\mathbf{k}_n)$ has always a prefactor $\delta^d\left(\sum_{i=1}^n \mathbf{k}_i\right)$. The vertex functions where we have factorized this δ function are denoted by $\bar{\Gamma}^{(n)}(\mathbf{k}_i)$. These functions can be seen as the most fundamental objects of the theory since their knowledge enables to obtain all other correlation functions. They are related to the connected correlation functions $\bar{G}_c^{(n)}(\mathbf{k}_1,\ldots,\mathbf{k}_n)$ by the relationship

$$\bar{G}_c^{(n)}(\mathbf{k}_1,\ldots,\mathbf{k}_n) = -G_c^{(2)}(\mathbf{k}_1)\ldots G_c^{(2)}(\mathbf{k}_n)\,\bar{\Gamma}^{(n)}(\mathbf{k}_1,\ldots,\mathbf{k}_n) + Q^{(n)}, \tag{7.6.10}$$

where the $Q^{(n)}$ are the terms coming from the graphs that are reducible by a cut of a line. The graphical representation of the first representatives of the vertex function is shown in Figure 7.11. The dimension of these quantities is

$$[\Gamma^{(n)}(x_i)] = [G^{(n)}(x_i)]\,[V]^{-n}\,[G^{(2)}]^{-n} = \Lambda^{n(d/2+1)}, \tag{7.6.11}$$

where V is the volume of the system. For their Fourier transform we have

$$\begin{aligned}[\Gamma^{(n)}(k_i)] &= \Lambda^{-n(d/2-1)}, \\ [\bar{\Gamma}^{(n)}(k_i)] &= \Lambda^{n+d-nd/2}.\end{aligned} \tag{7.6.12}$$

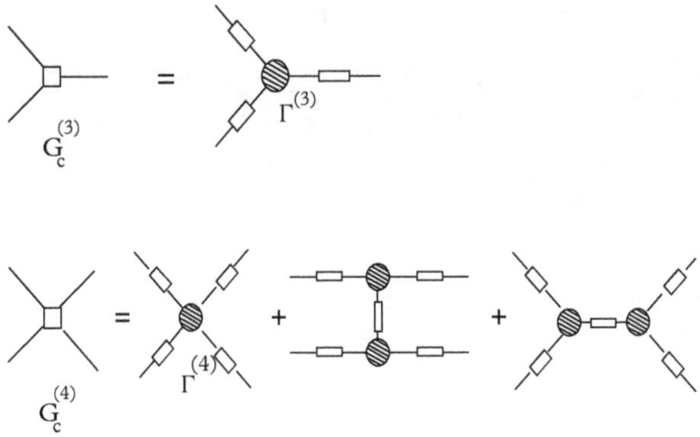

Fig. 7.11 *Vertex functions* $\Gamma^{(3)}$ *and* $\Gamma^{(4)}$*, together with their relation to the corresponding connected correlation functions.*

7.7 Spontaneous Symmetry Breaking and Multi-criticality

The vertex functions enter the expansion of the functional $\Gamma[\bar{\varphi}]$

$$\Gamma[\bar{\varphi}] = \sum_n \frac{1}{n!} \int d\mathbf{x}_1 \ldots d\mathbf{x}_n \, \Gamma^{(n)}(\mathbf{x}_1, \ldots, \mathbf{x}_n) \, \bar{\varphi}(\mathbf{x}_1) \ldots \bar{\varphi}(\mathbf{x}_n). \tag{7.7.1}$$

In order to determine whether or not there is spontaneous symmetry breaking in the system, we need to check if there is a non-vanishing uniform solution v of the field for eqn. (7.6.5). Substituting this value for $\bar{\varphi}$ in 7.7.1, this equation becomes

$$\Gamma[v] = \sum_n \frac{1}{n!} \left[\int d\mathbf{x}_1 \ldots d\mathbf{x}_n \, \Gamma^{(n)}(\mathbf{x}_1, \ldots, \mathbf{x}_n) \right] v^n. \tag{7.7.2}$$

Taking the Fourier transform of $\Gamma^{(n)}(\mathbf{x}_1, \ldots \mathbf{x}_n)$

$$\begin{aligned} \Gamma^{(n)}(\mathbf{k}_1, \ldots \mathbf{k}_n) &= \int \prod_i^n \frac{d\mathbf{k}_i}{(2\pi)^d} e^{i\sum_i \mathbf{k}_i \cdot \mathbf{x}_i} \, \Gamma^{(n)}(\mathbf{x}_1, \ldots, \mathbf{x}_n) \\ &= (2\pi)^d \delta^d\big(\sum \mathbf{k}_i\big) \, \bar{\Gamma}^{(n)}(\mathbf{k}_1, \ldots, \mathbf{k}_n) \end{aligned}$$

(where the $\delta^d(\sum \mathbf{k}_i)$ comes from the translation invariance) and substituting in (7.7.2) we have

$$\Gamma[v] = (2\pi)^d \delta^d(0) \sum_n \frac{1}{n!} \bar{\Gamma}^{(n)}(0, \ldots, 0) v^n \equiv (2\pi)^d \delta^d(0) \, U(v). \tag{7.7.3}$$

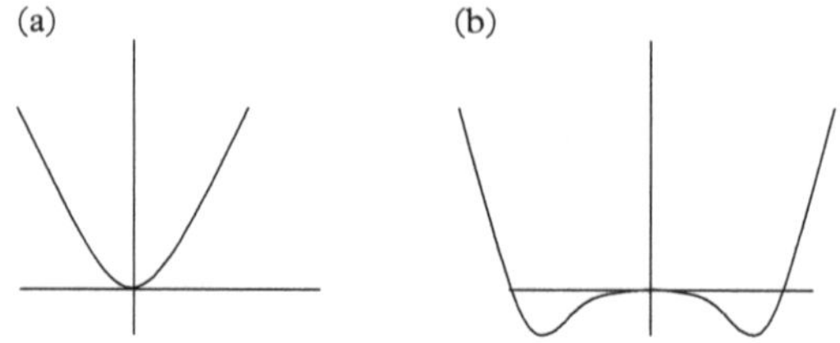

Fig. 7.12 *(a) $m^2 > 0$, there is no spontaneous symmetry breaking; (b) $m^2 < 0$, there is a spontaneous symmetry breaking, with an expectation value of the field different from zero.*

The term $(2\pi)^d \delta^d(0)$ denotes the proportionality of this quantity to the volume of the system. The quantity $U(v)$ is called the *effective potential* of the theory.

At zero order in perturbation theory, the effective potential $U(v)$ exactly coincides with the polynomial terms of the action, with the substitution $\varphi \to v$. Let us analyse the physical scenario that comes from the expression of $U(v)$ at this order, keeping in mind that higher perturbative corrections may alter the values of the coupling constants through the renomalization procedure.

7.7.1 Universality Class of the Ising model

Consider the effective potential at the zero order for the φ^4 theory

$$U_0(v) = hv + \frac{m^2}{2}v^2 + \frac{g}{4!}v^4. \tag{7.7.4}$$

Consider the case $h = 0$. For $m^2 > 0$, this function has a unique minimum at the origin: in this case the Z_2 symmetry of the theory is not broken for the expectation value of the field is zero. However, when $m^2 < 0$, the effective potential has two minima at $v_c = \pm\sqrt{-6m^2/g}$. In this case there is an expectation value for the field different from zero. The choice of any of the two values implies spontaneous symmetry breaking of the Z_2 invariance. This can be seen explicitly by choosing, for instance, $\bar{\varphi} = v_c$ and making the substitution $\varphi(x) \to \varphi'(x) = \varphi(x) - v_c$ in the Lagrangian. Disregarding an additive constant, the Lagrangian of the new field is given by

$$\mathcal{L} = \frac{1}{2}(\partial_j\varphi')^2 - m^2(\varphi')^2(x) + \frac{g\,v_c}{3!}(\varphi')^3(x) + \frac{g}{4!}(\varphi')^4(x). \tag{7.7.5}$$

The mass parameter of the new field is twice the value of the original expression (m^2 is negative in this case) and moreover there is the term $(\varphi')^3$ that is no longer invariant under the transformation $\varphi' \to -\varphi'$. Hence, the fluctuations around any of the two vacua v_c do not respect any longer the original Z_2 symmetry of the model.

One may erroneously think that the Z_2 symmetry of the theory can be re-established by tunneling effect, as it happens in quantum mechanics when a particle is in a double well symmetric potential. However, in field theory, this is impossible because the effective potential given in eqn. (7.7.3) contains a term that is proportional to the volume V of the system. In the thermodynamic limit $V \to \infty$, the potential barrier between the two minima is therefore infinite and consequently the symmetry cannot be restored by tunnelling effect between the vacua.

Notice that the field theory with an interactive term φ^4 has all the essential features of the class of universality of the Ising model. More specifically: (i) a Z_2 symmetry, under which the order parameter is odd and (ii) the possibility to have a non-zero vacuum expectation value of the order parameter when the mass term changes its sign. The identification between the two theories become more evident if we make the assumption that the mass parameter depends on the temperature as

$$m^2(T) \simeq (T - T_c). \tag{7.7.6}$$

The upper critical dimension of the φ^4 theory is $d = 4$ and, indeed, beyond this dimension the Ising model has critical exponents that coincide with their mean field values. For $1 < d < 4$, on the contrary, the Ising model has non-trivial values of critical exponents.

It is important to anticipate that in $d = 2$, in addition to the φ^4 bosonic theory, the Ising model also admits a formulation in term of a fermionic theory. Such a fermionic formulation of the model is discussed in detail in Chapter 9.

7.7.2 Universality Class of the Tricritical Ising Model

A different class of universality from the Ising model is described by the so-called Blume–Capel model. It involves two statistical variables defined on each site of a lattice:

- a spin variable s_k, with values ± 1
- a vacancy variable t_k, with values 0 and 1. This variable specifies whether the site is empty (0) or occupied (1).

The more general lattice Hamiltonian for these variables (with only next-neighbour interactions) is given by

$$\begin{aligned} \mathcal{H} = & -\mathcal{J} \sum_{\langle i,j \rangle}^{N} s_i s_j t_i t_j + \Delta \sum_{i=1}^{N} t_i - H \sum_{i=1}^{N} s_i t_i + \\ & - H_3 \sum_{\langle i,j \rangle}^{N} (s_i t_i t_j + s_j t_j t_i) - K \sum_{\langle i,j \rangle}^{N} t_i t_j . \end{aligned} \tag{7.7.7}$$

In this expression H is an external magnetic field, H_3 is an additional staggered magnetic field, $\mathcal{J}$ is the coupling constant between two next-neighbour spins of occupied sites and, finally, Δ is the chemical potential of the vacancies. When $H = H_3 = 0$ the solution of the

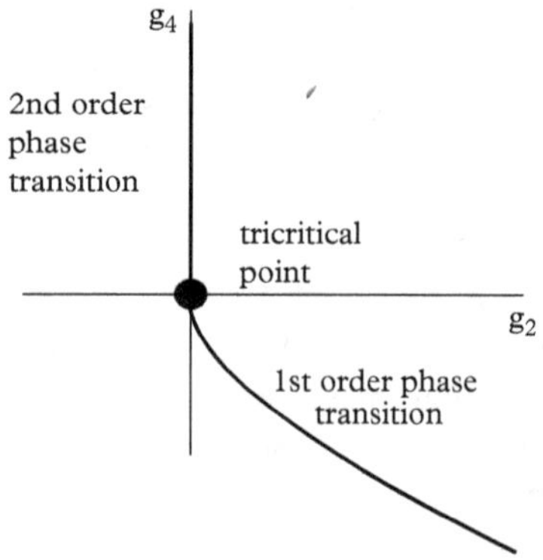

Fig. 7.13 *Phase diagram of the tricritical Ising model in the sub-space of the even coupling constants.*

Blume–Capel model on the lattice shows that there is a tricritical point at $(\mathcal{J}_c, \Delta_c)$. At a tricritical point a line of first-order phase transition meets the line of a second order phase transition. Let us see how these physical aspects are captured by a bosonic Lagrangian theory with the higher power of interaction given by φ^6. The most general action of this theory is

$$\mathcal{S} = \int d^d x \left[\frac{1}{2} (\partial_j \varphi)^2 + g_1 \varphi + g_2 \varphi^2 + g_3 \varphi^3 + g_4 \varphi^4 + \varphi^6 \right], \tag{7.7.8}$$

where the tricritical point is identified by the conditions $g_1 = g_2 = g_3 = g_4 = 0$. Comparing with the Blume–Capel model, the statistical interpretation of the coupling constants is as follows: g_1 plays the role of an external magnetic field h (the equivalent of H), g_2 measures the displacement of the temperature from its critical value $(T - T_c)$ (the equivalent of $\mathcal{J} - \mathcal{J}_c$), g_3 plays the role of the staggered magnetic field (the equivalent of H_3) and, finally, g_4 corresponds to $(\Delta - \Delta_c)$.

From the study of the effective potential, it is easy to see that this theory presents a tricritical point. Putting equal to zero all coupling constants of the odd powers of the field, in the remaining even sector we have

$$U_0(\Phi) = g_2 v^2 + g_4 v^4 + v^6. \tag{7.7.9}$$

The critical line of the second-order phase transition is identified by the condition of zero mass (i.e. infinite correlation length)

$$g_2 = 0, \quad g_4 > 0. \tag{7.7.10}$$

At a line of first-order phase transition there is an abrupt collapse of the vacua. To identify such a line, let us look at the sequence of the potentials (d) and (e) of Figure 7.14. This sequence shows that, moving with continuity the parameters of the model, the two farest

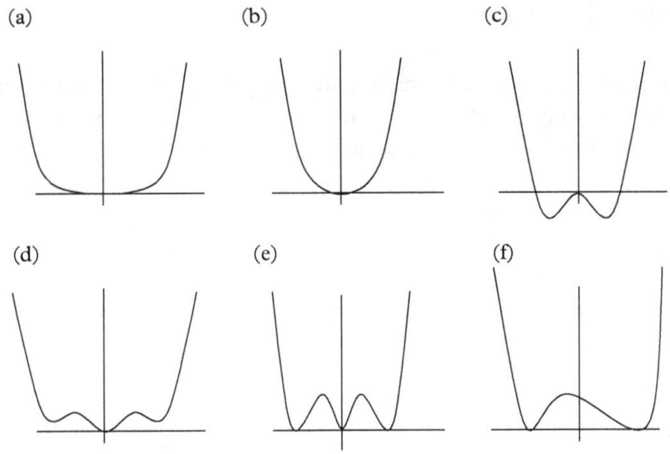

Fig. 7.14 *Some examples of the effective potential of the tricritical Ising model by varying its couplings: (a) critical point; (b) high-temperature phase; (c) low-temperature phase; (d) metastable states; (e) first-order phase transition; (f) asymmetric vacua in the presence of magnetic fields.*

external vacua become suddenly degenerate with the central one. Hence, the line of the first-order phase transition is characterized by the presence of three degenerate vacua and therefore is identified by the condition

$$g_2 > 0, \quad g_4 = -2\sqrt{g_2}. \tag{7.7.11}$$

In conclusion, the point $g_1 = g_2 = g_3 = g_4 = 0$ is indeed a tricritical point.

By varying the parameters in eqn. 7.7.8, the effective potential of this model can take different shapes and consequently its phenomenology can be rather rich. A dimensional analysis shows that the upper critical dimension of the Lagrangian theory (7.7.8) is $d = 3$. At this dimension and beyond, the critical exponents take their classical mean field values, while for $1 < d < 3$ they change significantly their values for the strong fluctuations of the order parameters. The exact solution of this model for $d = 2$ are discussed in detail in Chapter 14.

7.7.3 Multicritical Points

Statistical systems that are invariant under a Z_2 symmetry and with multi-critical behaviour can be described by bosonic field theory with interaction φ^{2n} $(n \geq 3)$. The criticality of these model is reached by fine tuning $2(n-1)$ parameters: in the Lagrangian description this procedure corresponds to put equal to zero all coupling constants of the powers of the field less than φ^{2n} (except the one of φ^{2n-1}, that can be always eliminated by a shift of the field φ, as suggested in Problem 7.5). The detailed description of these classes of universality in $d = 2$ presented in Chapter 11.

7.8 Renormalization

In the previous sections we have seen that the perturbative expansion give rise to expressions that typically diverge when the lattice spacing **a** is sent to zero. This is a well-known problem in QFT. Even though its complete analysis is beyond the scope of this book, we nevertheless draw attention to the main aspects of this topic, using as a guide the Landau–Ginzburg Lagrangians.

The renormalization of a theory consists of the possibility to eliminate the physical effects coming from the lattice spacing **a**—after all, an arbitrary parameter—by an appropriate choice of the coupling constants. For a given dimensionality d of the system, this procedure can be implemented only for certain Lagrangians, and not for others. To present the main results of this analysis, it is sufficient to focus our attention on the vertex functions $\bar{\Gamma}^{(E)}(k_i)$. It is useful to introduce initially the following concept.

Degree of superficial divergence. The Feynman diagrams that enter the vertex functions $\bar{\Gamma}^{(E)}(k_i)$ are generally expressed by multiple integrals. The degree of superficial divergence D of these expressions is defined as the difference between the number of momenta of the numerator, coming from the differentials $d^d k_i$, and the number of momenta of the denominator, the latter coming from the powers k^2 of the propagators. Denoting by L the number of integration variables and by I the number of internal lines of the graph, the superficial divergence D is given by

$$D = Ld - 2I. \tag{7.8.1}$$

If $D = 0$ the diagram is logarithmic divergent, if $D = 1$ it is linear divergent and so on, while if $D < 0$ the diagram is superficial convergent. The reason to distinguish betwen the actual divergent nature of the integral and its superfical divergence comes from the possibility of having nested divergencies: when this happens, the integral can have an actual divergence that is different from the one indicated by its index D. An example is provided by the last diagram in Figure 7.15: in $d = 4$ this diagram has a degree of superficial divergence $D = -2$ but it has actually an internal loop that is logarithmic divergent.

The key point to introduce such a concept is that the superficial divergence D of an amplitude can be fixed only using considerations of graph theory. Let us denote by E the number of external lines and by n_r the number of vertices corresponding to the interaction φ^r. There is an elementary relationship between these two quantities: since a vertex of type r has r lines that start from it and each external line has only one ending point, we have

$$E + 2I = \sum_r r n_r,$$

namely

$$I = \frac{1}{2}(\sum_r r n_r - E). \tag{7.8.2}$$

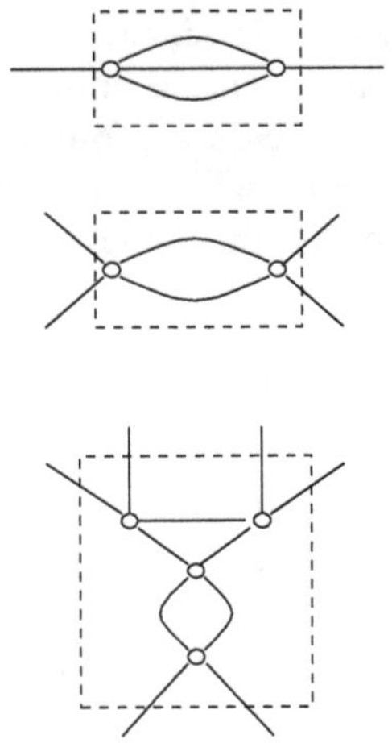

Fig. 7.15 *Degree of superficial divergence of some graphs of the vertex functions (in the dashed box) with $E = 2, E = 4$ and $E = 6$, for the φ^4 theory in $d = 4$.*

The number L of integrals coincides with the number of loops of the graph. In turn, this is equal to the number of internal lines I minus the number of conservation laws of the momenta. Each interaction carries a δ function but we must be careful in considering the one of them that corresponds to the prefactor $\delta\big(\sum_j^E k_j\big)$ associated to the total conservation law of the momenta of the E external lines. Hence

$$L = I - (n_r - 1). \tag{7.8.3}$$

Substituting this expression and eqn. (7.8.2) in (7.8.1) we have

$$\begin{aligned} D &= \left(d + E - \frac{1}{2}Ed\right) + \sum_r n_r \left(\frac{1}{2}rd - d - r\right) \\ &= \left(d + E - \frac{1}{2}Ed\right) + \sum_r n_r \delta_r, \end{aligned} \tag{7.8.4}$$

where the exponent δ_r is the one defined in eqn. (7.2.9). In conclusion, the degree of superficial divergence of an amplitude is given by the sum of two terms: the first is independent of the perturbative order while the second, on the contrary, depends on the type of interaction and on the perturbative order. It is worth noting that the origin of the two terms in (7.8.4) can be traced back by a dimensional analysis: the first term, in fact, simply expresses the dimensionality of the vertex function $\bar{\Gamma}^{(E)}(k_i)$ while the second term keeps into proper account the dimensionality of the coupling constants and the perturbative order in which they are involved.

Renormalizable Lagrangian. Fixed the dimensionality d of the system, if we require that *independently from the perturbative order* only a finite number of vertex functions are divergent, the coupling constant has to be dimensionless, i.e. $\delta_r = 0$. This condition determines which of the Lagrangians is renormalizable in d dimensions: this Lagrangian corresponds to a Landau–Ginzburg one with the highest interaction power φ^r equals to

$$r = \frac{2d}{d-2}. \tag{7.8.5}$$

Vice-versa, if we start by a Lagrangian with φ^r as its highest interaction term, there is a critical dimension, identified by the upper critical dimension d_s given in eqn. (7.2.10), in which this Lagrangian is renormalizable. Obviously the presence of terms with $\delta_r < 0$ can only decrease the superficial divergence of the amplitudes. For this reason we can focus our attention only where $\delta_r = 0$.

Consider, for instance, the Lagrangian theory

$$\mathcal{L} = \frac{1}{2}(\partial_j \varphi)^2 + \frac{m^2}{2}\varphi^2 + \frac{g_4}{4!}\varphi^4. \tag{7.8.6}$$

where $d_s = 4$. If we choose the dimension d of the system exactly equal to d_s, its divergent amplitudes (with $D \geq 0$) correspond to diagrams with external lines $E \leq 4$ (eqn. (7.8.4)). Since the amplitudes with an odd number of external legs vanish[9] for the symmetry $\varphi \to -\varphi$, it remains to consider only those with $E = 2$ and $E = 4$. Note that the divergent vertex functions are those coming from the terms φ^2 and φ^4 already present in the Lagrangian! Such a theory is therefore renormalizable since it is possible to the divergences of the vertex functions $\bar{\Gamma}^{(2)}$ and $\bar{\Gamma}^{(4)}$ by adjusting a set of counterterms that have exactly the same form of the original Lagrangian

$$\mathcal{L} \to \mathcal{L} + \frac{A}{2}(\partial_j \varphi)^2 + \frac{B}{2}\varphi^2 + \frac{C}{4!}\varphi^4. \tag{7.8.7}$$

Bare quantities. The coefficients A, B, C are (divergent) functions of the cut-off a, chosen in such a way to cancel order by order the divergences of the perturbative series. Observe that, defining

$$\begin{aligned} \varphi_0 &= (1+B)^{1/2}\varphi, \\ m_0^2 &= (m^2 + A)(1+B)^{-1}, \\ g_0 &= (g_4 + C)(1+B)^{-2}, \end{aligned} \tag{7.8.8}$$

9 This is certainly true in the symmetric phase of the theory. In the broken symmetry phase of the model the argument has to be modified accordingly but it remains still true that the model is renormalizable.

the modified Lagrangian (7.8.7) can be written as

$$\mathcal{L} = \frac{1}{2}(\partial\varphi_0)^2 + \frac{m_0^2}{2}\varphi_0^2 + \frac{g_0^4}{4!}\varphi_0^4, \tag{7.8.9}$$

that is similar to the initial one. However, this transformation changes radically the meaning of the parameters. All quantities, including the field itself, depend now on the cut-off and are non-universal. For these reasons they are called *bare quantities*. They only serve the scope to remove the infinities. In order to link the bare quantities to the physical parameters of the theory, such as the physical value of the mass or the coupling constant, it is necessary to determine (say, experimentally) the latter quantities at a given value of the momenta of the vertex functions (for instance, at zero momenta) and then use eqn. (7.8.8) for inverting these relations. It is only after the knowledge of the experimental values m_{exp}^2 and λ_{exp} that the theory acquires its predictive power, since it is only then that the formalism is able to determine uniquely all other amplitudes. These quantities become finite functions of m_{sp}^2 and λ_{sp} and, of course, of the external momenta.

From what said above, it should be clear that not all the Lagrangians are renormalizable. For instance, adding an interaction term φ^5 to the φ^4 theory in $d = 4$, with $\delta_5 = 1$, this term produces an infinite sequence of divergent vertex functions. The perturbative cure of these terms relentlessly leads to the addition of counterms with arbitrary powers of φ^n in the Lagrangian, i.e. we arrive at a theory with an infinite number of parameters. In this case we lose any predictive power of the theory defined in the limit $a \to 0$.

Effective theories. On the other hand, it should be say that if there are reasons to consider the lattice spacing as a *finite physical quantity* that plays an important role in the problem under consideration, a priori there is no reason to exclude non-renormalizable Lagrangians. This is, in particular, the modern view about the renormalization problem in QFT and it can be perfectly justified by the renormalization group approach. In conclusion, the final meaning of the QFTs is the one of *effective theories*, i.e. theories that present a dependence on the length scale L or, equivalently, on the energy scale E at which we are analysing the physical systems. From this point of view, the important point is the possibility to control how the physical properties vary by varying the length or the energy scales. As we will see in the next chapter, in the (infinite-dimensional) manifold of the couplings a change of these scales has the effect to induce a motion of the point that represents the system. The properties of this motion will be the object of the renormalization group analysis.

7.9 Field Theory in Minkowski Space

QFTs describe the excitations of a physical systems. These excitations share the same properties of the elementary particles: they can be created at a given point of the system and annihilated at another, or they can propagate for a given time interval causing scattering processes in the meantime. The next two sections highlight these aspects, which are closely related to elementary particles. For doing so, it is necessary firstly to define the QFT in the Minkowski space and, secondly, to adopt an operatorial formalism. We illustrate these features using the Landau–Ginzburg Lagrangians, in particular, the φ^4 theory. Let us begin from the measure with which we have weighted the configurations of the field φ in the d-dimensional Euclidean space

$$W(\{\varphi\}) = \exp[-\mathcal{S}] = \exp\left[-\int d^d x \mathcal{L}(x)\right], \tag{7.9.1}$$

with

$$\mathcal{L} = \frac{1}{2}(\partial_j \varphi)^2 + U(\varphi), \tag{7.9.2}$$
$$U(\varphi) = \frac{m^2}{2}\varphi^2 + \frac{g}{4!}\varphi^4.$$

Let us now select one of the d coordinates, say $x_0 = \tau$, and promote it to the role of Euclidean time variable. Finally, let us make the transformation $\tau \to -it$. As discussed below, this innocent transformation changes completely the meaning of the theory.

Making the same transformation $\tau \to -it$ in the derivative term $(\partial_j \varphi)^2$ of the Lagrangian, we get a new expression of $W(\{\varphi\})$, that we denote by $\tilde{W}(\{\varphi\})$

$$\tilde{W}(\{\varphi\}) = \exp[i\tilde{\mathcal{S}}] \equiv \exp\left[i\int d^{d-1}x\, dt\, \tilde{\mathcal{L}}\right], \tag{7.9.3}$$

where

$$\tilde{\mathcal{L}} = \frac{1}{2}\left[\left(\frac{\partial \varphi}{\partial t}\right)^2 - (\nabla \varphi)^2\right] - U(\varphi). \tag{7.9.4}$$

Comparing $\tilde{\mathcal{L}}$ with the quantity in (7.9.2) we note two differences: the first is that there is a relative sign between the derivatives concerning the spatial coordinates and the one relative to the time variable; the second is that all polynomial terms have changed sign. However, the most important effect is in the quantity $\tilde{W}$, that now is a complex quantity. Hence this quantity has lost the original meaning of probability, acquiring instead the meaning of *amplitude*, in the usual meaning of quantum mechanics. To clarify this point, we will briefly remind the quantization of a particle that moves in a n-dimensional space.

Quantum mechanics of a particle

Let

$$L(q) = \frac{1}{2}\sum_{i=1}^{n}(\dot{q}_i)^2 - V(q), \tag{7.9.5}$$

be the Lagrangian of a particle ($\dot{q}_i = \frac{dq_i}{dt}$),

$$A = \int_0^t dt\, L(q), \tag{7.9.6}$$

its action and H the Hamiltonian, defined by the Legendre transformation

$$H(q,p) = \sum_{i=1}^{n} p_i q_i - L = \sum_{i=1}^{n} \frac{p_i^2}{2} + V(q). \tag{7.9.7}$$

The components of the momentum

$$p_i = \frac{\delta L}{\delta \dot{q}_i} = \dot{q}_i,$$

together with the coordinates q_i, are now operators that satisfy the commutation relations

$$[q_k, p_l] = i\hbar \delta_{k,l}, \quad [q_k, q_l] = 0, \quad [p_k, p_l] = 0. \tag{7.9.8}$$

Denoting by E_n the eigenvalues of the Hamiltonian and $|E_n\rangle$ its eigenvectors, the amplitude that such a particle moves in a time interval t from the point q_0 (where it is localized at the time $t = 0$) to the point q_f, is given by the time evolution of the unitary operator $e^{-itH/\hbar}$

$$\langle q_f, t \,|\, q_0, 0\rangle = \langle q_f \,|\, e^{-itH/\hbar} \,|\, q_0\rangle = \sum_{n=0}^{\infty} \langle q_f \,|\, E_n\rangle\langle E_n \,|\, q_0\rangle\, e^{-itE_n/\hbar}, \tag{7.9.9}$$

where we have used the completeness relation

$$\sum_{i=1}^{\infty} |E_n\rangle\langle E_n| = 1.$$

However this is not the only way to compute such an amplitude: as shown by Feynman (see Appendix A), it can also be obtained by means of a path integral on all the classical trajectories that connect the points $(q_0, 0)$ and (q_f, t). In this approach each path is weighted by $\exp(iA/\hbar)$, namely[10]

[10] Also in this case, to define the measure $\mathcal{D}q$ is necessary to make the variable q discrete on the slides $t_k = k\epsilon$ $(k = 0, 1, \ldots, N)$ of the time interval t, with $\epsilon = t/N$, so that $\mathcal{D}q = \prod_{k=0}^{N} \frac{dq_k}{\sqrt{2\pi\epsilon}}$.

$$\langle q_f, t \,|\, q_0, 0 \rangle = \int_{\substack{q(0)=q_0 \\ q(t)=q_f}} \mathcal{D}q \, \exp(iA/\hbar). \tag{7.9.10}$$

In the semi-classicallimit $\hbar \to 0$, the integral can be estimated by the saddle point: the most important contribution comes from the trajectory for which the action is stationary, $\delta A = 0$, i.e. the trajectory that satisfies the classical equation of motion

$$\frac{d}{dt}\left(\frac{\delta L}{\delta \dot{q}_i}\right) - \frac{\delta L}{\delta q_i} = 0. \tag{7.9.11}$$

As shown in Appendix A, by means of the path integral we can also compute the time-ordered correlation function of the operators

$$\langle q_f t | T[Q(t_1)\ldots Q(t_k)] \, |q_0, 0\rangle = \int_{\substack{q(0)=q_0 \\ q(t)=q_f}} \mathcal{D}q \, q(t_1)\ldots q_k(t) \exp(iA/\hbar), \tag{7.9.12}$$

with $t_1 > t_2 > \ldots t_k$.

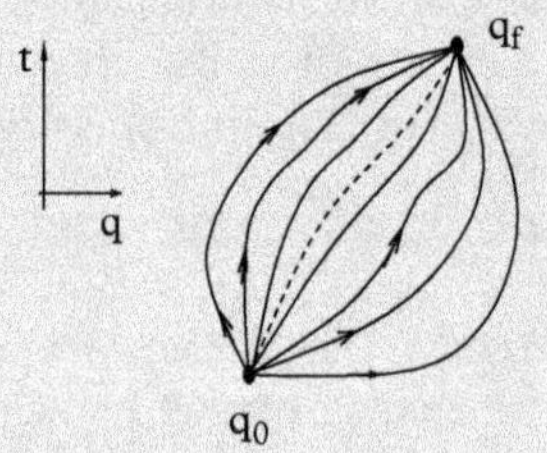

Fig. 7.16 *Feynman integral, namely a sum on the classical trajectories that link the initial and the final points, each trajectory weighted by e^{iA} where A is the action of each trajectory. The dashed line corresponds to the classical trajectory, solution of the classical equation of motion.*

Coming back to the field theory, and in particular to eqn. (7.9.3), we see then that $\tilde{W}(\{\varphi\})$ can be interpreted as the weight of a *classical configuration* of the field $\varphi(x,t)$ in the computation of a *quantum amplitude* (we have posed $\hbar = 1$)

$$\langle \varphi_f(x,t) \,|\, \varphi_0(x,0) \rangle = \int_{\substack{\varphi(x,t)=\varphi_f(x) \\ \varphi(x,0)=\varphi_0(x)}} \mathcal{D}\varphi \, \exp\left[i\tilde{S}\right]. \tag{7.9.13}$$

With this interpretation of $\tilde{W}(\{\varphi\})$, we can now proceed as in the quantum mechanics of a particle but, this time, back-forwardly: instead of using the path integral, we will adopt the operatorial approach to describe the dynamics associated with the Lagrangian (7.9.4). In QFT the role of the operators $q_i(t)$ is played by the field $\varphi(x,t)$, regarded as an operator that acts at each point (x,t) of the space-time. The operator formalism just defined is a relativistic invariant Appendix B. In this appendix one can also find the relevant definitions used in the following.

The field $\varphi(x,t)$ satisfies the operator differential equation coming from the Euler–Lagrange equation of motion

$$\partial_\mu \left(\frac{\partial \tilde{\mathcal{L}}}{\partial(\partial_\mu \varphi)} \right) - \frac{\partial \tilde{\mathcal{L}}}{\partial \varphi} = 0 \tag{7.9.14}$$

that, for the φ^4 theory, reads

$$\left(\Box + m^2\right) \varphi(x,t) = -\frac{g}{3!} \varphi^3(x,t), \tag{7.9.15}$$

where

$$\Box = \frac{\partial^2}{\partial t^2} - \nabla^2.$$

The conjugate momentum is defined by

$$\pi(x,t) = \frac{\delta \tilde{\mathcal{L}}}{\delta \dot{\varphi}(x,t)} = \frac{\partial \varphi}{\partial t}. \tag{7.9.16}$$

As $\varphi(x,t)$, also $\pi(x,t)$ is an operator. In analogy with quantum mechanics, we postulate that these operators satisfy the equal-time commutation relation

$$\begin{aligned} [\varphi(x,t), \pi(y,t)] &= i\delta^d(x-y), \\ [\varphi(x,t), \varphi(y,t)] &= 0, \\ [\pi(x,t), \pi(y,t)] &= 0. \end{aligned} \tag{7.9.17}$$

In terms of $\pi(x)$ we can define the Hamiltonian density by the Legendre transform

$$\mathcal{H}(x,t) = \pi(x,t)\dot{\varphi}(x,t) - \tilde{\mathcal{L}} = \frac{1}{2}\left[\left(\frac{\partial \varphi}{\partial t}\right)^2 + (\nabla\varphi)^2 \right] + U(\varphi). \tag{7.9.18}$$

The Hamiltonian and the momentum are given by

$$H = \int d^{d-1}x \, \mathcal{H}(x,t), \tag{7.9.19}$$
$$\mathbf{P} = -\int d^{d-1}x \, \pi(x,t) \, \nabla \varphi(x,t).$$

As a consequence of the equation of motion (7.9.15), both are conserved quantities

$$\frac{dH}{dt} = \frac{d\mathbf{P}}{dt} = 0 \tag{7.9.20}$$

and can be expressed in terms of the *stress-energy tensor* $T^{\mu\nu}(x)$. This quantity is defined by (see Appendix C)

$$T^{\mu\nu}(x) = \frac{\partial \tilde{\mathcal{L}}}{\partial(\partial_\mu \varphi)} \, \partial^\nu \varphi - g^{\mu\nu} \, \tilde{\mathcal{L}}. \tag{7.9.21}$$

$T^{\mu\nu}$ is a conserved quantity: it satisfies

$$\partial_\mu \, T^{\mu\nu}(x) = 0, \tag{7.9.22}$$

and therefore

$$H = \int T^{00}(x) \, d^{d-1}x, \quad P^i = \int T^{0i}(x) \, d^{d-1}x. \tag{7.9.23}$$

7.10 Particles

To understand the nature of the excitations $\varphi(x,t)$ it is sufficient to consider the free theory ($g = 0$). In this case the operatorial equation satisfied by the field is

$$\left(\Box + m^2\right) \varphi(x,t) = 0. \tag{7.10.1}$$

A plane wave $e^{i\,(\mathbf{kx} - Et)}$ is a solution of this equation if

$$E^2 = \mathbf{k}^2 + m^2. \tag{7.10.2}$$

One easily recognizes that this is the dispersion relation of a relativistic particle. Taking into account the two roots of this equation, the most general solution of (7.10.1) is given by a linear superposition of plane waves

$$\varphi(x,t) = \int d\Omega_k \left[A_{\mathbf{k}} e^{i\mathbf{k}\cdot\mathbf{x}-iE_k t} + A_{\mathbf{k}}^{\dagger} e^{-i\mathbf{k}\cdot\mathbf{x}+iE_k t} \right]. \tag{7.10.3}$$

In this expression and in the next ones that follow $E_k = \sqrt{\mathbf{k}^2 + m^2}$. The coefficients $A_{\mathbf{k}}$ and $A_{\mathbf{k}}^{\dagger}$ are a set of operators, called *annihilation and creation operators* respectively. In writing this solution we have adopted a relativistic invariant differential measure

$$d\Omega_k \equiv \frac{d^{d-1}k}{(2\pi)^{d-1} 2E_k}.$$

For the quadratic nature of the relativistic dispersion relation, there are both positive and negative frequencies in the mode expansion of the field. The negative frequencies can be interpreted as the propagation, back in time, of an anti-particle, a statement that becomes evident if one considers a complex scalar field[11] (see Problem 7.8). Using (7.9.16), we obtain the conjugate momentum

$$\pi(x,t) = -i \int d\Omega_k E_k \left[A_{\mathbf{k}} e^{i\mathbf{k}\cdot\mathbf{x}-iE_k t} - A_{\mathbf{k}}^{\dagger} e^{-i\mathbf{k}\cdot\mathbf{x}+iE_k t} \right]. \tag{7.10.4}$$

The commutation relations of $A_{\mathbf{k}}$ and $A_{\mathbf{k}}^{\dagger}$ can be recovered by imposing the validity of eqn. (7.9.17)

$$\begin{aligned} [A_{\mathbf{k}}, A_{\mathbf{p}}^{\dagger}] &= (2\pi)^{d-1} 2E_k \delta^{d-1}(\mathbf{k}-\mathbf{p}), \\ [A_{\mathbf{k}}, A_{\mathbf{p}}] &= [A_{\mathbf{k}}^{\dagger}, A_{\mathbf{p}}^{\dagger}] = 0 \end{aligned} \tag{7.10.5}$$

Beside the relativistic normalization of these operators, they are the exact analogous of the annihilation and creation operators of the harmonic oscillator.

Substituting the expressions of $\varphi(x,t)$ and $\pi(x,t)$ in H we have

$$H = \frac{1}{2} \int d\Omega_k \left(A_{\mathbf{k}}^{\dagger} A_{\mathbf{k}} + A_{\mathbf{k}} A_{\mathbf{k}}^{\dagger} \right) E_k = \int d\Omega_k \left(A_{\mathbf{k}}^{\dagger} A_{\mathbf{k}} + \frac{1}{2} \right) E_k, \tag{7.10.6}$$

where we have used the commutation relation (7.10.5). The term

$$E_0 = \frac{1}{2} \int d\Omega_k E_k$$

[11] For a real scalar field, the particle coincides with its anti-particle.

is infinite and corresponds to the *vacuum energy*. Since this quantity is a constant, it can be safely subtracted, so that the new definition of the Hamiltonian is

$$H = \int d\Omega_k A^\dagger_{\mathbf{k}} A_{\mathbf{k}} E_k, \tag{7.10.7}$$

This redefinition employs the *normal product* of the operators: a product of operators is normal ordered if all the creation operators are on the left side of the annihilation operators. Denoting the normal order by : :, the new Hamiltonian can be expressed as

$$H = \frac{1}{2}\int d^d x : \left(\pi^2 + (\nabla\varphi)^2 + m^2\varphi^2\right): \tag{7.10.8}$$

Note that the annihilation and creation operators are associated to the plane waves with positive and negative time frequency respectively. Indicating by $\varphi^{(+)}(x)$ and $\varphi^{(-)}(x)$ these two terms in the decomposition of $\varphi(x)$

$$\varphi(x) = \varphi^{(+)}(x) + \varphi^{(-)}(x),$$

we have, for instance

$$:\varphi(x)\varphi(y): = \varphi^{(+)}(x)\varphi^{(+)}(y) + \varphi^{(-)}(x)\varphi^{(+)}(y) + \varphi^{(-)}(x)\varphi^{(-)}(y) + \varphi^{(-)}(y)\varphi^{(+)}(x)$$

Substituting the expressions of $\varphi(x)$ and $\pi(x)$ in the momentum operator, we have

$$P = \int d\Omega_k \left(A^\dagger_{\mathbf{k}} A_{\mathbf{k}} + \frac{1}{2}\right)\mathbf{k} = \int d\Omega_k A^\dagger_{\mathbf{k}} A_{\mathbf{k}} \mathbf{k} \tag{7.10.9}$$

Notice that in this case the zero point of the momentum is absent since, in the integration, it is cancelled by the equal and opposite contributions coming from $\pm\mathbf{k}$.

In the expression of both the energy and the momentum there is the operator

$$N_{\mathbf{k}} = A^\dagger_{\mathbf{k}} A_{\mathbf{k}}. \tag{7.10.10}$$

This is the key observation that supports an interpretation of the QFT in terms of particles. In the following we prove that the operators $N_{\mathbf{k}}$ are simultaneously diagonalizable and their eigenvalues are the integer numbers

$$n_{\mathbf{k}} = 0, 1, 2, \ldots. \tag{7.10.11}$$

In this way, the energy and the momentum associated to the field φ can be written as

$$E = \int d\Omega_k n_{\mathbf{k}} E_k, \quad \mathbf{P} = \int d\Omega_k n_{\mathbf{k}} \mathbf{k}. \tag{7.10.12}$$

From this expressions it is clear that these quantities coincide with the energy and momentum of a set of scalar particles of mass m, with a relativistic dispersion relation. This set contains $n_{\mathbf{k}_1}$ particles of momentum $\mathbf{k}_1$, $n_{\mathbf{k}_2}$ particles of momentum $\mathbf{k}_2$ etc.

The statement that all $N_{\mathbf{k}}$ commute is a simple consequence of the commutation relation (7.9.17)

$$
\begin{aligned}
[N_{\mathbf{k}_1}, N_{\mathbf{k}_2}] &= A^\dagger_{\mathbf{k}_1}\,[A_{\mathbf{k}_1}, A^\dagger_{\mathbf{k}_2}]A_{\mathbf{k}_2} + A^\dagger_{\mathbf{k}_2}\,[A^\dagger_{\mathbf{k}_1}, A_{\mathbf{k}_2}]A_{\mathbf{k}_1} \\
&= \left(A^\dagger_{\mathbf{k}_1} A_{\mathbf{k}_1} - A^\dagger_{\mathbf{k}_1} A_{\mathbf{k}_1}\right) 2E_k\,\delta^{d-1}(\mathbf{k}_1 - \mathbf{k}_2) = 0. \qquad (7.10.13)
\end{aligned}
$$

As in the familiar harmonic oscillator, the spectrum (7.10.11) derives from the commutation relations

$$
[N_{\mathbf{k}}, A^\dagger_{\mathbf{k}}] = A^\dagger_{\mathbf{k}}, \quad [N_{\mathbf{k}}, A_{\mathbf{k}}] = -A_{\mathbf{k}}. \qquad (7.10.14)
$$

These expressions say that $A^\dagger_{\mathbf{k}}$ creates a particle of momentum $\mathbf{k}$ while $A_{\mathbf{k}}$ annihilates such a particle. The state with the minimum energy is the vacuum state, in which there is no particles

$$
N_{\mathbf{k}}\,|\,0\,\rangle = 0. \qquad (7.10.15)
$$

This implies $A_{\mathbf{k}}\,|\,0\,\rangle = 0$ and the multi-particle states with momenta $\mathbf{k}_1 \ldots \mathbf{k}_n$ are given by

$$
|\,n_{\mathbf{k}_1}, n_{\mathbf{k}_2}, \ldots\rangle = \frac{1}{(n_{\mathbf{k}_1}!\,n_{\mathbf{k}_2}!\ldots)^{1/2}}\left(A^\dagger_{\mathbf{k}_1}\right)^{n_{\mathbf{k}_1}}\left(A^\dagger_{\mathbf{k}_2}\right)^{n_{\mathbf{k}_2}}\ldots|\,0\,\rangle, \qquad (7.10.16)
$$

Since the operators $A^\dagger_{\mathbf{k}_j}$ commute one with the other, these states are symmetric under an exchange of the indices and therefore satisfy the Bose-Einstein statistics. The Hilbert space constructed in this way is called the *Fock space* of the theory.

In the light of this discussion, let us see what is the interpretation of the state $\varphi(x)\,|\,0\,\rangle$. From the field expansion and the action of the operators A and $A^\dagger$, we have

$$
\varphi(x, 0)\,|\,0\,\rangle = \int d\Omega_k\, e^{-i\mathbf{k}\cdot\mathbf{x}}\,|\,\mathbf{k}\rangle, \qquad (7.10.17)
$$

where we have indicated by $|\,\mathbf{k}\rangle = A^\dagger_{\mathbf{k}}\,|\,0\,\rangle$ the one-particle state of momentum $\mathbf{k}$. Therefore the state $\varphi(x)0\,\rangle$ is given by a linear superposition of one-particle states of various momentum. In other words, applying the field to the vacuum state *we have created a particle at the point* $\mathbf{x}$. This interpretation is further supported by computing the matrix element

$$
\langle\,0\,|\,\varphi(\mathbf{x})\,|\,\mathbf{k}\rangle = e^{i\mathbf{p}\cdot\mathbf{x}}. \qquad (7.10.18)
$$

This is the coordinate representation of the wave function of one-particle state, as in quantum mechanics $\langle x|p\rangle = e^{ipx}$ is the wave function of the state $|p\rangle$.

7.11 Correlation Functions and Scattering Processes

In defining the correlation functions in the Minkowski space we shall take into account that the fields are operators and therefore they do not generally commute. Quantities of interest are the vacuum expectation values of the T-ordered product of operators.[12] In the free case, the only non-zero correlation function of the field φ is the two-point correlators. It can be computed by using the commutation relations (7.10.5) and the relations $A\,|\,0\rangle = 0$ and $\langle 0\,|\,A^\dagger = 0$

$$\begin{aligned}\Delta_F(x-y) &= \langle 0\,|\,T[\varphi(x)\varphi(y)]\,|\,0\,\rangle \\ &= \langle 0\,|\varphi^{(+)}(x)\varphi^{(-)}(y)]\,|\,0\,\rangle\,\theta(x_0-y_0) + \langle 0\,|\varphi^{(+)}(y)\varphi^{(-)}(x)]\,|\,0\,\rangle\,\theta(y_0-x_0) \\ &= \int d\Omega_k \Big[e^{-ik\cdot(x-y)}\,\theta(x_0-y_0) + e^{ik\cdot(x-y)}\,\theta(y_0-x_0)\Big]. \end{aligned} \tag{7.11.1}$$

This quantity is the so-called *Feynman propagator* that can be written in a relativistic invariant as

$$\Delta_F(x-y) = \int \frac{d^d k}{(2\pi)^d} \frac{i}{k^2 - m^2 + i\epsilon}\, e^{-ik\cdot(x-y)}. \tag{7.11.2}$$

In this formula $k^2 = k_0^2 - \mathbf{k}^2$ and the $i\epsilon$ term in the denominator is equivalent to a prescription in computing the integral on the time component of the momentum: using the residue theorem for the integral on dk_o it is easy to see that one obtains the previous formula. Note that using the analytic continuation $k^0 = ik_E^0$, the so-called Wick rotation, the Feynman propagator becomes (up to a factor i) the propagator of the Euclidean QFT, previously analysed.

The Feynman propagator can be also obtained generalizing the formula (7.9.12) in the limit $T \to \infty$, where T is the time separation between the two vacuum states on the

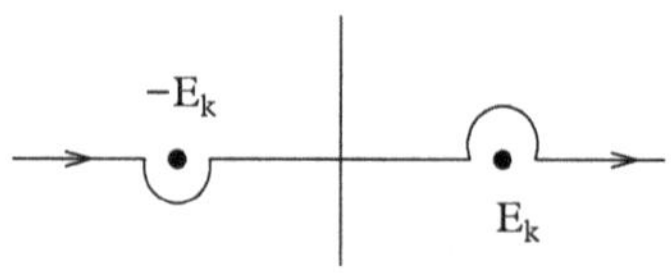

Fig. 7.17 *Integration contour of the variable k_0, that is equivalent to the $i\epsilon$ prescription in the denominator of (7.11.2).*

[12] In the next formulae, all vectors are d-dimensional with the Minkowski metric. Hence x denotes $(x^0, \mathbf{x})$ and $p \cdot x = p^0 x^0 - \mathbf{p}\cdot\mathbf{x}$.

right- and left-hand side

$$\langle 0|T[\varphi(x)\varphi(y)]|0\rangle = \frac{1}{Z_0}\int \mathcal{D}\varphi\, \varphi(x)\,\varphi(y)\, e^{i\int d^{d-1}xdt\,\tilde{\mathcal{L}}_0}. \tag{7.11.3}$$

As usual, Z_0 gives the proper normalization

$$Z_0 = \int \mathcal{D}\varphi\, e^{i\int d^{d-1}xdt\,\tilde{\mathcal{L}}_0}.$$

Analogously to the Euclidean case, we can couple the field to an external current and define ($d^d x \equiv d^{d-1}xdt$)

$$Z_0[\mathcal{J}] = \int \mathcal{D}\varphi \,\exp\left\{ i\int \left[\tilde{\mathcal{L}}_0 + \mathcal{J}(x)\,\varphi(x)\right] d^d x\right\}. \tag{7.11.4}$$

The integral is Gaussian

$$Z_0[\mathcal{J}] = \exp\left\{\frac{i}{2}\int\int \mathcal{J}(x)\,\Delta_F(x-y)\,\mathcal{J}(y)\, d^d x\, d^d y\right\},$$

and then

$$\Delta_F(x-y) - (-i)^2 \frac{\delta^2 Z[\mathcal{J}]}{\delta\mathcal{J}(x)\delta\mathcal{J}(y)}. \tag{7.11.5}$$

In the interactive case, the partition function is given by

$$Z[\mathcal{J}] = \exp\left\{ i\int \tilde{\mathcal{L}}_I\left[-i\frac{\delta}{\delta\mathcal{J}(x)}\right] d^d x\right\} Z_0[\mathcal{J}], \tag{7.11.6}$$

and the correlation functions are defined by

$$G(x_1,\ldots,x_n) = \langle 0|T[\varphi(x_1)\ldots\varphi(x_n)]|0\rangle = (-i)^n \frac{\delta^n Z[\mathcal{J}]}{\delta\mathcal{J}(x_1)\ldots\delta\mathcal{J}(x_n)}. \tag{7.11.7}$$

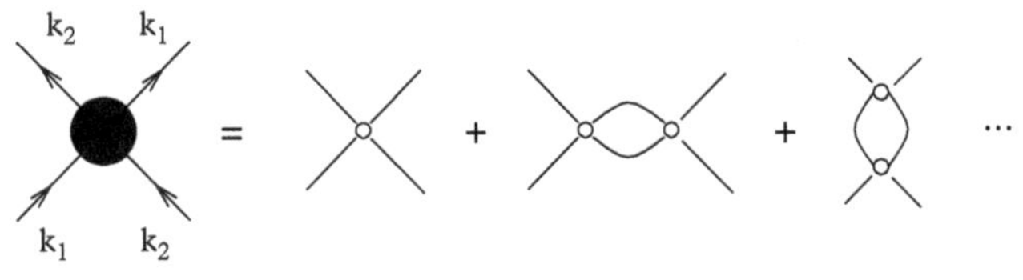

Fig. 7.18 *Elastic scattering amplitude of two particles, given by the infinite sum of all elementary interaction processes ruled by the interaction vertices.*

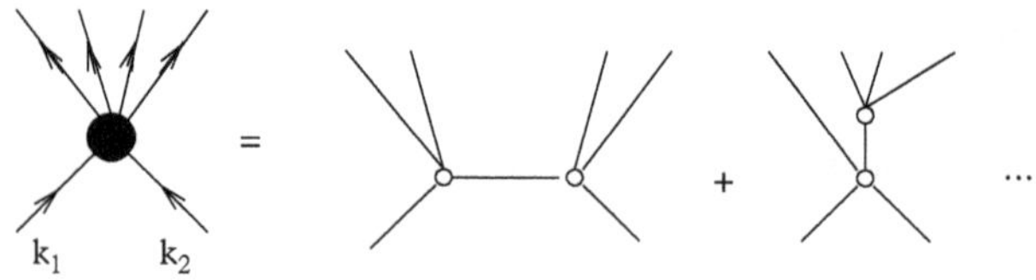

Fig. 7.19 *Production amplitude of a multi-particle state following a collision of two initial particles.*

They admit an expansion in terms of Feynman graphs, analogously to the one previously analized. One should take into account, though, an extra factor i for each vertex and a different expression for the propagator. The perturbative properties of the correlation functions are similar to those previously discussed.

Finally, we comment different interpretation of the Feyman graphs in the Minkowski space. Since the lines are now associated to the propagation of the particles, the various interaction vertices can be considered as the points of the scattering processes. For instance, the connected four-point function (Figure 7.18) can be employed to compute the probability of the elastic scattering of two in-going particle of momenta k_1 and k_2 in out-coming particles with the same momenta.

Analogously, the connected n point correlation functions can be used to compute the production processes of $(n-2)$ particles that originate from the collision of two initial particles, if they have enough energy in their centre of mass (equal or larger than the sum of the mass of the $(n-2)$ particles Figure 7.19). In the absence of conservation laws, all these processes are allowed by the relativistic laws. They are studied in detail in Chapter 17.

Appendix 7.A. Feynman Path Integral Formulation

Let $Q(t)$ be the coordinate operator of a quantum particle in the Heisenberg representation and $|q,t\rangle$ its eigenstates

$$Q(t)\,|q,t\rangle = q\,|q,t\rangle.$$

In the Schrödinger representation Q_S is a time-independent operator, related to $Q(t)$ by the unitary relation $Q(t) = e^{itH/\hbar}\,Q_S\,e^{-itH/\hbar}$. Q_S has time-independent eigenstates, $Q_S\,|q\rangle = q\,|q\rangle$, and their relation to the previous one is given by $|q\rangle = e^{-itH/\hbar}\,|q,t\rangle$. These states satisfy the completeness relation

$$\int dq\,|q\rangle\,\langle q| = 1.$$

It is also useful to introduce the eigenstates of the momentum operator in the Schrödinger representation

$$P\,|p\rangle = p\,|p\rangle.$$

They satisfy the completeness relation

$$\int \frac{dp}{2\pi}\,|p\rangle\langle p| = 1,$$

and their scalar product matrix elements with the states $|q\rangle$, is $\langle q|p\rangle = e^{ipq/\hbar}$. Let us compute the amplitude

$$F(q',t';q,t) = \langle q',t'|q,t\rangle = \langle q'|e^{-i(t'-t)H/\hbar}|q\rangle \tag{7.A.1}$$

dividing the interval $T = (t'-t)$ in $(n+1)$ time slides

$$t = t_0,\, t_1, \ldots, t_{n+1} = t', \quad t_k = t_0 + k\epsilon$$

In the limit $n \to \infty$, we have

$$e^{i(t'-t)H/\hbar} \simeq e^{-i\epsilon H/\hbar} e^{-i\epsilon H/\hbar} \ldots e^{-i\epsilon H/\hbar}.$$

Inserting n times the completeness relation of the eigenstates $|q\rangle$ into eqn. (7.A.1) we get

$$F(q',t';q,t) = \int \prod_{k=1}^{n} dq_k\,\langle q'|e^{-i\epsilon H/\hbar}|q_n\rangle\langle q_n|e^{-i\epsilon H/\hbar}|q_{n-1}\rangle \ldots \langle q_1|e^{-i\epsilon H/\hbar}|q\rangle. \tag{7.A.2}$$

These matrix elements can be computed exactly in the limit $\epsilon \to 0$. With the Hamiltonian given by $H = \frac{p^2}{2m} + V(q)$, inserting the completeness relation of the $|p\rangle$ state we have

$$\begin{aligned} \langle q_k|e^{-i\epsilon H/\hbar}|q_{k-1}\rangle &= \int \frac{dp}{2\pi}\frac{dp'}{2\pi}\langle q_k|p\rangle\langle p|e^{-i\epsilon H/\hbar}|p'\rangle\langle p'|q_{k-1}\rangle \\ = \int \frac{dp}{2\pi}\frac{dp'}{2\pi} e^{i(pq_k - p'q_{k-1})/\hbar}\, e^{-i\epsilon/\hbar\left(\frac{p^2}{2m}+V(\frac{q_k+q_{k-1}}{2})\right)}\,\delta(p-p') \\ = \int \frac{dp}{2\pi} e^{ip(q_k-q_{k-1})}\, e^{-i\epsilon/\hbar\left(\frac{p^2}{2m}+V(\frac{q_k+q_{k-1}}{2})\right)} &= \frac{1}{\sqrt{2\pi\epsilon}}\, e^{i\epsilon/\hbar\left[\frac{(q_k-q_{k-1})^2}{2\epsilon^2} - V(\frac{q_k+q_{k-1}}{2})\right]}. \end{aligned} \tag{7.A.3}$$

Making the hypothesis that in the limit $\epsilon \to 0$, q_{k-1} tends to q_k, we have

$$\frac{(q_k - q_{k-1})^2}{\epsilon^2} \to (\dot{q})^2$$

and therefore the matrix element is expressed by the Lagrangian associated to this part of the trajectory

$$\langle q_k | e^{-i\epsilon H/\hbar} | q_{k_1} \rangle \simeq \frac{1}{\sqrt{2\pi\epsilon}} e^{i\epsilon/\hbar L(\dot{q}_k, q_k)}. \tag{7.A.4}$$

Coming back to (7.A.2), we thus have

$$\begin{aligned} F(q', t'; q, t) &= \lim_{n\to\infty} \int \prod_{k=1}^{n} \frac{dq_k}{\sqrt{2\pi\epsilon}} e^{i\epsilon/\hbar \sum_{k=1}^{n} \left[\frac{1}{2}\dot{q}_k^2 - V(q_k)\right]} \\ &\equiv \int \mathcal{D}q\, e^{i/\hbar \int_t^{t'} dt\, L(q,\dot{q})} = \int \mathcal{D}q\, e^{i/\hbar A}. \end{aligned} \tag{7.A.5}$$

Let us now consider the correlation function of two operators $Q(t_1)Q(t_2)$, with $t_1 > t_2$. Repeating the same argument given above, we arrive at a representation of this quantity in terms of a path integral

$$\langle q', t' | Q(t_1) Q(t_2) | q, t \rangle = \int \mathcal{D}q\, q(t_1)\, q(t_2)\, e^{i/\hbar A}. \tag{7.A.6}$$

However, notice that on the right-hand side the order of the two variables is irrelevant. The path integral expression is then equal to the matrix elements on the left-hand side with the established order, in which the only important thing is that $t_1 > t_2$. If t_1 was less than t_2, the right-hand side would be equal to the matrix element of the two operators, but in a reversed order. This leads to the definition of the time ordering of the operators

$$T[Q(t_1)Q(t_2)] = \begin{cases} Q(t_1)Q(t_2), & t_1 > t_2 \\ Q(t_2)Q(t_1), & t_2 > t_1 \end{cases} \tag{7.A.7}$$

with an obvious generalization for an arbitrary number of them. In such a way, we arrive

$$\langle q', t' | T[Q(t_1) \dots Q(t_k)] | q, t \rangle = \int \mathcal{D}q\, q(t_1) \dots q(t_k)\, e^{i/\hbar A}. \tag{7.A.8}$$

Appendix 7.B. Relativistic Invariance

The Lorentz transformations in $(d+1)$ dimensions leave invariant the front line of the light, defined by

$$s^2 = t^2 - x_1^2 - \dots - x_{d-1}^2.$$

The speed of light c has been posed equal to 1 and we have also used $t = x_0$ to make the notation uniform. More generally, with the definition of the metric tensor

$$g^{\mu\nu} = g_{\mu\nu} = \begin{pmatrix} 1 & 0 & 0 & 0 & \dots & 0 \\ 0 & -1 & 0 & 0 & \dots & 0 \\ 0 & 0 & -1 & 0 & \dots & 0 \\ \dots & \dots & \dots & \dots & \dots & \dots \\ 0 & 0 & 0 & 0 & \dots & -1 \end{pmatrix}$$

the Lorentz transformations Λ^{μ}_{ν} are defined by the condition to leave invariant the metric, i.e.

$$g_{\mu\nu}\,\Lambda^{\mu}_{\rho}\,\Lambda^{\nu}_{\sigma} = g_{\rho\sigma}. \tag{7.B.1}$$

(with a sum on the repeated indices). Thanks to the metric tensor we can raise or lower the indices of a vector or of a tensor. We have

$$x^{\mu} = (t, \mathbf{x}), \quad x_{\mu} = g_{\mu\nu}x^{\nu} = (t, -\mathbf{x})$$

For the derivative we have

$$\partial_{\mu} = \frac{\partial}{\partial x^{\mu}} = \left(\frac{\partial}{\partial x^0}, \nabla\right).$$

The space with such a metrix is called the *Minkowski space*. In order to characterize the infinitesimal form of these transformations, let us pose

$$\Lambda^{\mu}_{\nu} \simeq \delta^{\mu}_{\nu} + \omega^{\mu}_{\nu}.$$

Substituting into (7.B.1), we have

$$\omega_{\mu\nu} + \omega_{\nu\mu} = 0. \tag{7.B.2}$$

In d dimensions the number of free parameters of an anti-symmetric matrix is equal to $d(d-1)/2$. If we add to these transformations also the translations $x^{\mu} \to x^{\mu} + a^{\mu}$, we arrive at the group of Poincaré.

Invariant expressions under the Poincaré group are generically given by scalar products with respect to the metric tensors, such as

$$p \cdot x = g_{\mu\nu}\,p^{\mu}\,x^{\nu} = p^{\mu}\,x_{\mu} = p^0\,x^0 - \mathbf{p}\cdot\mathbf{x}.$$

Another invariant quantity is given by

$$\partial^\mu \partial_\mu = \Box = \frac{\partial^2}{\partial (x^0)^2} - \nabla^2.$$

The momentum of a massive particle satisfies

$$p^2 = p^\mu p_\mu = E^2 - \mathbf{p}^2 = m^2, \tag{7.B.3}$$

where m is the mass.

Since the norm of the vectors (with respect to the metric g) is an invariant, the distance between two points can be classified as follows: (a) if $(x_1 - x_2)^2 > 0$, this is a *time-like* separation; if $(x_1 - x_2)^2 < 0$ this is a *space-like* separation; and if $(x_1 - x_2)^2 = 0$ we have a *light-like* separation (Figure 7.20). Time-like points are related each other by a causality relation while the space-like points are not. In the latter case, in fact, to have a causal relation between them, a signal should travel faster than the speed of light. For light-like events, the temporality of two events is given by the time that is necessary to the light to travel from x_1 to x_2.

It is easy to prove that the volume elements

$$d^d x \equiv dx_0\, dx_1 \ldots dx_{d-1}, \tag{7.B.4}$$

and the momentum volume elements

$$d^d p \equiv dp_0\, dp_1 \ldots dp_{d-1}, \tag{7.B.5}$$

are both invariant: under a Lorentz transformation, to a dilatation of the time component corresponds a contraction of the space component, and the two term compensate each

Fig. 7.20 *With x_2 placed at the origin, the point x_1 can be in one of the three positions shown in the figure. For time-like distances, the event x_2 can be in a causal relation with the event x_1. For space-like distances, the two events cannot be linked by a causal relation, since their time separation is larger than the time that the light would spend to cover their space distance.*

other. Using the invariance of the momentum infinitesimal volume, one can prove the invariance of the measure $d\Omega_k$. In fact, it can be written as

$$d\Omega_k = \frac{d^{d-1}k}{(2\pi)^{d-1}2E_p} = \frac{d^d p}{(2\pi)^d}(2\pi)\,\delta(k^2 - m^2)\Big|_{k^0>0}. \tag{7.B.6}$$

The Lagrangian that appears in (7.9.3) is a scalar density. It gives rise to the equation of motion thanks to the principle of minimum action

$$0 = \delta\tilde{S} = \int d^d x \left\{ \frac{\partial \tilde{\mathcal{L}}}{\partial \varphi}\delta\varphi + \frac{\partial \tilde{\mathcal{L}}}{\partial(\partial_\mu \varphi)}\delta(\partial_\mu \varphi) \right\} \tag{7.B.7}$$

$$= \int d^d x \left\{ \left[\frac{\partial \tilde{\mathcal{L}}}{\partial \varphi} - \partial_\mu \frac{\partial \tilde{\mathcal{L}}}{\partial(\partial_\mu \varphi)} \right] \delta\varphi + \partial_\mu \left(\frac{\partial \tilde{\mathcal{L}}}{\partial(\partial_\mu \varphi)}\delta\varphi \right) \right\}.$$

The last term is a total divergence and it gives rise to a surface integral. This vanishes if we assume that the variation of the field is zero at the boundary. In this way, we arrive at the Euler–Lagrange equation of the field

$$\frac{\partial \tilde{\mathcal{L}}}{\partial \varphi} - \partial_\mu \frac{\partial \tilde{\mathcal{L}}}{\partial(\partial_\mu \varphi)} = 0. \tag{7.B.8}$$

Appendix 7.C. Noether Theorem

There is a deep relation between the symmetries and the conservation laws of a system. This is the content of the Noether theorem. Suppose we change infinitesimally the field

$$\varphi(x) \to \varphi'(x) + \alpha\,\delta\varphi, \tag{7.C.1}$$

where α is an infinitesimal parameter and $\delta\varphi$ is a deformation of the field. Such a transformation is a symmetry of the system if it leaves invariant the equations of motion. To guarantee this condition is sufficient that the action remains invariant under the transformations (7.C.1). More generally, the action is allowed to change up to a surface term, since the latter does not effect the equation of motion. Hence, under (7.C.1), the Lagrangian can change at most by a total divergence

$$\tilde{\mathcal{L}} \to \tilde{\mathcal{L}} + \alpha\,\partial_\mu \mathcal{J}^\mu(x).$$

Comparing this expression with the expression that is explicitly obtained by varying the field in the Lagrangian according to (7.C.1) we have

$$\alpha\,\partial_\mu \mathcal{J}^\mu = \frac{\partial\tilde{\mathcal{L}}}{\partial\varphi}(\alpha\delta\varphi) + \left(\frac{\partial\tilde{\mathcal{L}}}{\partial(\partial_\mu\varphi)}\right)\partial_\mu(\alpha\delta\varphi) \tag{7.C.2}$$

$$= \alpha\,\partial_\mu\left(\frac{\partial\tilde{\mathcal{L}}}{\partial(\partial_\mu\varphi)}\delta\varphi\right) + \alpha\left[\frac{\partial\tilde{\mathcal{L}}}{\partial\varphi} - \partial_\mu\frac{\partial\tilde{\mathcal{L}}}{\partial(\partial_\mu\varphi)}\right]\delta\varphi.$$

The last term vanishes for the equation of motion and therefore we arrive at the conservation law

$$\partial_\mu j^\mu(x) = 0, \quad j^\mu(x) \equiv \frac{\partial\tilde{\mathcal{L}}}{\partial(\partial_\mu\varphi)}\delta\varphi - \mathcal{J}^\mu. \tag{7.C.3}$$

Let us see the consequence of this result if the system is invariant under the translations $x^\mu \to x^\mu - a^\mu$. The field changes as

$$\varphi(x) \to \varphi(x+a) = \varphi(x) + a^\mu\,\partial_\mu\,\varphi(x). \tag{7.C.4}$$

Since the Lagrangian is a scalar quantity, it transforms in the same way

$$\tilde{\mathcal{L}} \to \tilde{\mathcal{L}} + a^\mu\partial_\mu\tilde{\mathcal{L}} = \tilde{\mathcal{L}} + a^\nu\,\partial_\mu(\delta^\mu_\nu\,\tilde{\mathcal{L}}).$$

Using (7.C.3), we obtain the so-called stress-energy tensor

$$T^\mu_\nu \equiv \frac{\partial\tilde{\mathcal{L}}}{\partial(\partial_\mu\varphi)}\partial_\nu\varphi - \tilde{\mathcal{L}}\,\delta^\mu_\nu \tag{7.C.5}$$

that satisfies

$$\partial_\mu\,T^\mu_\nu = 0. \tag{7.C.6}$$

The energy and the momentum of the system is given by

$$E = \int d^dx\,T^{00}(x,t), \quad P^\nu = \int d^dx\,T^{0\nu}(x,t). \tag{7.C.7}$$

....

REFERENCES

The path integral formulation of quantum mechanics is due to Richard Feynman. For a detailed discussion consult

Feynman, R.P. and Hibbs, A.R. (1965). *Quantum Mechanics and Path Integrals*, McGraw-Hill.

There are many superb texts on QFT. We recommend:

Barton, R. (1963). *Introduction to Advanced Field Theory*, J. Wiley & Sons.
Zee, A. (2003). *Quantum Field Theory in a Nutshell*, Princeton University Press.

The functional formalism and its relation with the statistical mechanics can be found in

Amit, D. (1978). *Field Theory, the Renormalization Group and Critical Phenomena*, Mc-Graw Hill.
Binney, J.J., Dowrick, N.J., Fisher A.J. and Newman, M.E.J. (1992). *The Theory of Critical Phenomena*, Oxford University Press.
Parisi, G. (1988). *Statistical Field Theory*, Benjamin/Cumming.
Zinn-Justin, J. (1971). *Quantum Field Theory and Critical Phenomena*, Oxford University Press.

The operatorial approach is discussed in

Bogoliubov, N.N. and Shirkov, D.V. (1976). *Introduction to the Theory of Quantized Fields*, J. Wiley & Sons.
Weinberg, S. (1995). *The Quantum Theory of Fields*, Cambridge University Press.
The application of QFT to elementary particles is discussed in
Peskin, M.E. and Schroeder, D.V. (1995). *An Introduction to Quantum Field Theory*, Addison Wesley.

PROBLEMS

7.1. Lagrangian theory with Z_3 symmetry

Consider a Lagrangian of a complex field $\Phi(x)$ and its conjugate $\Phi^\dagger(x)$ that under a Z_3 transformation transform as

$$\Phi(x) \to e^{2\pi i/3}\,\Phi(x), \quad \Phi^\dagger(x) \to e^{-2\pi i/3}\,\Phi^\dagger(x).$$

Write down the most general Lagrangian that is invariant under these transfomations.

7.2. Perturbative series

Consider the one-dimensional integral

$$I(\lambda) = \int_{-\infty}^{+\infty} dx\, e^{-\alpha x^2 + \lambda x^4}.$$

Write the perturbative series of this expression expanding the term $e^{-\lambda x^4}$ in power series of λ. Compute the perturbative coefficients and show that the series has zero radius of convergence. Give a simple argument of this fact.

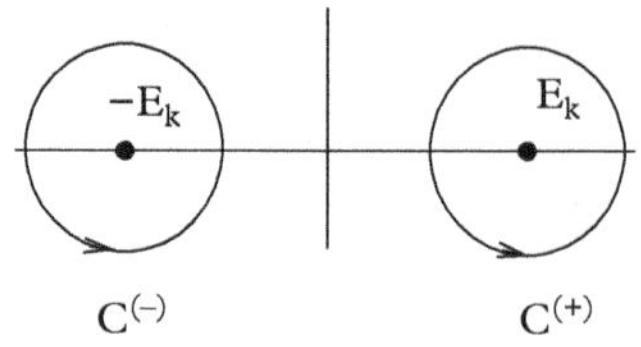

Fig. 7.21 *Contours of integration $C^{(+)}$ and $C^{(-)}$ for $\Delta^{(+)}(x)$ and $\Delta^{(-)}(x)$.*

7.3. Correlation functions and Feynman graphs

Draw the Feynman diagrams relative to the g^2 correction of the 4-point correlation function $G(x_1, \ldots, x_4)$ for the φ^4 theory. Discuss the convergence of the integrals as functions of the dimensionality d of the system.

7.4. φ^3 Lagrangian theory

Calculate the first non-vanishing perturbative order of the partition function for the Lagrangian theory with interaction $\frac{g}{3!}\varphi^3$. Determine the upper critical dimension d_s and discuss the renormalization of this theory.

7.5. Dimensional regularization

An alternative way to regularize the integrals encountered in perturbative series of QFT consists of the dimensional regularization. The main idea behind this approach is to consider the integrals as functions of the dimensionality d of the system, regarded as a continuous variable. Once they are evaluated in the region of the complex plane d where they converge, their values in other domains are obtained by analytic continuation. Prove the validity of the formula

$$\int \frac{d^d p}{(2\pi)^d} \frac{1}{(p^2 + \Delta)^n} = \frac{1}{(4\pi)^{d/2}} \frac{\Gamma\left(n - \frac{d}{2}\right)}{\Gamma(n)} \left(\frac{1}{\Delta}\right)^{n-d/2}.$$

Discuss the analytic structure of this expression as a function of d.

7.6. Invariant functions

Consider the functions

$$\Delta^{(\pm)}(x) = \frac{i}{(2\pi)^d)} \int d^{d-1}k \int_{C^{(\pm)}} dk_0 \frac{e^{ik\cdot x}}{k^2 - m^2},$$

where the contours of integration are shown in Figure 7.21:

a. Show that the correlation function of the commutator of the field is given by

$$\langle 0 | [\varphi(x), \varphi(y)] | 0 \rangle = \Delta(x - y),$$

where

$$\Delta(x-y) = \Delta^{(+)}(x-y) + \Delta^{(-)}(x-y).$$

b. Prove that $\Delta(x)$ vanishes for equal times

$$\Delta(\mathbf{x}-\mathbf{y},0) = 0.$$

Using Lorentz invariance, argue that the relation above implies the vanishing of $\Delta(\mathbf{x}-\mathbf{y})$ for all space-like intervals. From a physical point of view, the commutativity of the field for space-like intervals is the consequence of the causality principle: since space-like points cannot be related by light signals, the measurements done at the two points cannot interfere and therefore the operators commute.

c. Prove that $\Delta(x)$ and $\Delta^{(\pm)}(x)$ satisfy the homogenous equation

$$(\Box + m^2)\,\Delta(x) = (\Box + m^2)\Delta^{(\pm)}(x) = 0,$$

while the Feynman propagator, which corresponds to an infinite contour of integration, satisfies

$$(\Box + m^2)\,\Delta_F(x) = -i\delta^d(x).$$

7.7. Field theories with soliton solutions

Consider the Lagrangian field theory in $1+1$ dimensions

$$\tilde{\mathcal{L}} = \frac{1}{2}(\partial_\mu\varphi)^2 + \frac{m^2}{\beta^2}\,[\cos(\beta\varphi) - 1].$$

a. Expand in powers of β and show that this model corresponds to a Landau–Ginzburg theory with an infinite number of couplings.
b. Write the equation of motion of the field $\varphi(x,t)$.
c. Prove that the configurations

$$\varphi^{(\pm)}(x,0) = \pm 4\arctan\left[\exp(x-x_0)\right],$$

(where x_0 is an arbitrary point), are both classical solutions of the static version of the equation of motion.

d. Show that these configurations interpolate between two next-neighbour vacua. These configurations correspond to topological excitations of the field, called *soliton* and *anti-soliton*.
e. Compute the stress-energy tensor and use the formula $H = \int T^{00}(x)\,dx$ to determine the energy of the solitons. Since they are static, their energy corresponds to their M. Prove that

$$M = \frac{8m}{\beta}.$$

Note that the coupling constant is at the denominator, so that this is a non-perturbative expression.

7.8. Anti-particles

Consider the free theory of a complex field $\phi(x)$. In the Minkowski space the action is

$$\mathcal{S} = \int d^d x \left(\partial_\mu \phi^* \partial^\mu \phi - m^2 \phi^* \phi \right).$$

a. Show that the Hamiltonian is given by

$$H = \int d^{d-1}x \, (\pi^* \pi + \nabla\phi^* \cdot \nabla\phi + m^2 \phi^* \phi).$$

b. Prove that the system is invariant under the continuous symmetry

$$\phi \to e^{i\alpha} \phi, \quad \phi^* \to e^{-i\alpha} \phi.$$

Use the Noether theorem to derive the conserved charge

$$Q = -i \int d^{d-1}x (\pi^* \phi^* - \pi \phi).$$

c. Diagonalize the Hamiltonian by introducing the creation and annihilation operators. Show that the theory contains two sets of operators that can be distinguished by the different eigenvalues of the charge Q: the first set describes the creation and the annihilation of a particle A, while the second set describes the same processes for an anti-particle $\bar{A}$.

d. Show that the propagation of a particle in a space-like interval is the same as the propagation of an anti-particle back in time.

7.9. Conserved currents.

Consider a multiplet of n scalar fields, $\Phi = (\phi_1, \ldots, \phi_n)$.

a. Write the most general Lagrangian that is invariant under a rotation of the vector Φ

$$\Phi_k \to (\mathcal{R})_{kl} \Phi_l.$$

b. Use the Noether theorem to derive the conserved currents associates to this symmetry.

8

Renormalization Group

Everything must change so that nothing changes.
Giuseppe Tomasi di Lampedusa, *Il Gattopardo*

8.1 Introduction

At a critical point, the correlation length ξ diverges: the statistical fluctuations extend on all scales of the system and any attempt to solve the dynamics by keeping into account only a finite degrees of freedom fails. In absence of an exact solution of the model under consideration, the computation of the critical exponents is often obtained only by numerical methods and Montecarlo simulations.

Leaving apart the problem of computing the critical exponents, there is however a general approach to phase transitions that has the advantage to conceptually simplify many of their aspects. This approach goes under the name of *renormalization group* (RG). Beside its practical use, the fundamental ideas of the RG provide a theoretical scheme and a proper language to face critical phenomena and, in particular, to understand their universal properties and scaling laws.

It is worth stressing that the terminology is unlike and for two reasons: (i) the transformations of the RG are irreversible and therefore they do not form a group, as usually meant in mathematics; (ii) moreover, they do not necessarily concern with the renormalization of a theory, i.e. with the cure of the divergencies of the perturbative series. As a matter of fact, the main concepts of the RG has a wider spectrum of validity.

There are many specialized books on the RG and its technical aspects. The interested reader can find a small list of them at the end of the chapter. The aim of this chapter is to present in the simplest possible way the physical scenario provided by the RG, introducing the appropriate terminology and emphasizing the main concepts with the help of some significant examples. Other important aspects of the RG will be discussed in more detail in Chapter 15, in relation with the two dimensional QFTs near to their critical points.

Which is the key idea behind the RG? The answer to this question is: a continuous family of transformations of the coupling constants in correspondence to a change of the length scale of a physical system. In any physical system there are various length scales and the main assumption of the RG is that they are coupled together in a local way. If one is interested to study, for instance, the fluctuations of a magnetic system on a scale

Statistical Field Theory: An Introduction to Exactly Solved Models in Statistical Physics. Second Edition. Giuseppe Mussardo, Oxford University Press (2020). © Giuseppe Mussardo. DOI: 10.1093/oso/9780198788102.001.0001

of order of 1000 Å, it is reasonable to assume that it would be sufficient to consider only the degrees of freedom with comparable wave lengths L, say those in the range 800 Å$< L <$ 1200 Å. The degrees of freedom with very short wave length, of the order of few atomic spacings, should not matter. If this is indeed the case, one is led to the conclusion that the interactions have a shell structure: the fluctuations of the system on scales of 1–2 Å only influence those on scales 2-4 Å, the last ones influence those on scales 4–8 Åand so on. This sequence is ruled by a family of effective Hamiltonians associated to the degrees of freedom that are relevant to each shell of length scale (Figure 8.1).

There are two important aspects that emerge in this cascade scenario. The first aspect concerns the scaling invariant properties. For the absence of a characteristic length to compare with, the fluctuations of the intermediate lengths tend generally to be the same, beside a simple rescaling. There is however no scale invariance for those fluctuations with wave lengths comparable to a length parameter, as the one provided for instance by the lattice spacing. The second aspect is the amplification or de-amplification phenomena that take place in the course of the cascade process. A small change of temperature may have a negligible effect of atomic scale but, if this effect gets amplified to the large scales of the system, it may produce significant macroscopic changes. This is precisely what happens at the critical value T_c of the temperature, when the correlation length diverges inducing all other thermodynamical singularities. Concerning the de-amplification effects, they are at the root of the universality properties of the critical phenomena: it is thanks to them that two magnetic materials, with quite different atomic compositions, may nevertheless share the same critical behaviour.

To implement the RG ideas, the first steps consist of isolating a particular shell of lenght scale and defining a procedure that permits to pass to the next one. In the case of the critical phenomena, this procedure involves a statistical average of all fluctuations within a certain range of lengths. This is equivalent to study the behaviour of the system under a length scale $x \to x' = x/b$ or, tantamountly, under a rescaling of the lattice spacing $a \to a' = ba$. In doing so, one is simply looking at the system under a different magnifying

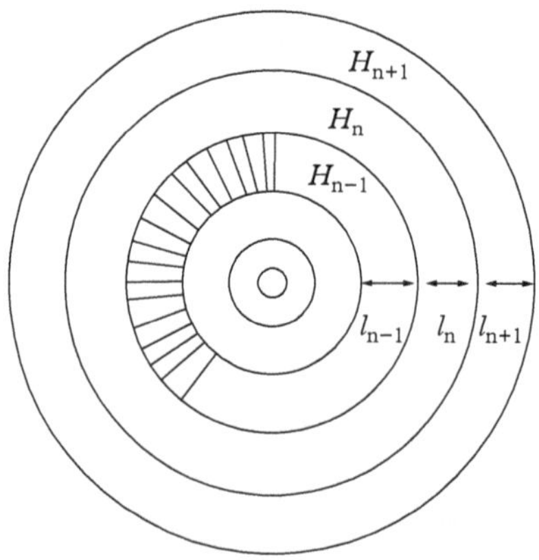

Fig. 8.1 *Length scales and sequence of the effective Hamiltonians for the degrees of freedom of each length shell.*

glass. As a result, an effective Hamiltonian is defined for the degrees of freedom that were not averaged. Implementing iteratively this average procedure, one is able to determine the amplification and de-amplification factors λ_i. These are the eigenvalues of the linearized version of the iterative procedure: under a small change of the initial interactions, the λ_i are the quantities responsible for their amplification/deamplification to the next iteration. If a coupling constant gets amplified, it is called *relevant coupling*. If, on the contrary, it gets deamplified, is is called *irrelevant coupling*. The instability nature of a critical point is determined by the number of its relevant couplings.

Roughly speaking, there are two different ways to implement the RG ideas. The first method is usually employed in contexts of QFT to deal with the divergence of the Feynman diagrams discussed in the previous chapter. Since these computations are usually carried on in momentum space, this implementation of the RG goes under the heading of RG *in the k space*. The second way is known as *real space* RG. This approach is more relevant in a statistical mechanics context, in particular in the discussion of systems defined on a lattice. Moreover, it is more intuitive. For this reason, in the following we will mainly follow this approach.

8.2 Reducing the Degrees of Freedom

Let us consider a statistical system defined on a d-dimensional regular lattice of lattice spacing a, with degrees of freedom s_i placed on its sites. Let $H(\{s_i\}, g_k)$ be the Hamiltonian of the system, where g_k are the coupling constants of the various interactions among the spins s_i. For reasons that will become clear later, it is convenient to include in the Hamiltonian all the possible coupling constants that are compatible with the nature of the degrees of freedom. For instance, if the s_i are Ising variables, the Hamiltonian H can be written as $H = H^{(+)} + H^{(-)}$, where the $\pm$ signs refer to the even and odd sector of the Z_2 symmetry of the model. In the even sector, the most general Hamiltonian is given by

$$H^{(+)}(\{s_i\}, g_k) = \sum_{i,j} g_{ij}^{(2)}\, s_i\, s_j + \sum_{i,j,k,l} g_{ijkl}^{(4)} s_i\, s_j\, s_k\, s_l + \cdots \tag{8.2.1}$$

whereas in the odd sector, the most general Hamiltonian is expressed by

$$H^{(-)}(\{s_i\}, g_k) = \sum_{i} g_i^{(1)}\, s_i + \sum_{i,j,k} g_{ijk}^{(3)}\, s_i\, s_j\, s_k + \ldots \tag{8.2.2}$$

In the formulae above, the indices are not necessarily restricted to next-neighbour sites. The partition function is given by

$$Z(\{g_k\}) = \sum_{\{s_i\}} \exp\left[-H(\{s_i\}, g_k)\right], \tag{8.2.3}$$

where we have included the factor $\beta = 1/KT$ in the definition of the coupling constants of the Hamiltonian. At given values of the g_k, the system has a correlation length $\xi(g_k)$, that is a function of the couplings. This quantity measures the number of degrees of freedom effectively coupled together and one expects that, smaller is $\xi(g_k)$, more effective and accurate is a perturbative study of the model. This observations suggests to look for a scale transformation $a \to b\,a$ that establishes a correspondence between the system with correlation length ξ and the one with correlation length $\xi' = \xi/b < \xi$. The idea is that, if such a transformation exists, its implementation may lead to a solvable or, at least, to a simpler model. Note however that, if the initial system is exactly at the critical point, this transformation will leave it invariant: in this case $\xi = \infty$ and therefore it remains a divergent quantity under any rescaling of the lattice spacing.

Spins within a sphere of radius ξ are correlated each other. Therefore, those within a length shell ba ($b > 1$) satisfying

$$a \ll b\,a \ll \xi,$$

act somehow as a single unit. We can imagine to make a zooming on the system, organizing the variables in spin blocks. Namely, let us divide the original lattice in blocks, denoted by $\mathcal{B}_k$, each of them made of b^d spins. If N is the total number of sites, there are $N b^{-d}$ blocks. Once this partition has been done, let us assign to each block a new variable $\sigma_i^{(1)}$ according to a certain law that involves the spins s_i present in each block

$$\sigma_i^{(1)} = f(\{s_i\}), \;\; \text{con}\; i \in \mathcal{B}_k \tag{8.2.4}$$

Postponing later the discussion on the nature of this law, for the time being let us note that the effect of this transformation is to change the model into a new one, defined on a lattice with a new lattice spacing $a' = b\,a$. After all the dynamical variables have been changed according to the transformation (8.2.4), it is convenient to scale the new lattice by factor b^{-1} (without alter, though, the spins), so that we come back to a lattice equal to the original one. What we have described above is the implementation of the real space RG, that therefore consists of the iteration of the series of transformations

$$\sigma_k^{(n+1)} = f(\{\sigma_i^{(n)}\}), \;\; \text{with}\; i \in \mathcal{B}_k \tag{8.2.5}$$

where $\sigma_i^{(n)}$ denote the spin variables of the n-th step of this procedure.

An important aspect of this transformation is its local nature: the definition of the variables $\sigma_k^{(n+1)}$ only involves the variables $\sigma_i^{(n)}$ and not the original spins s_i. Note that at each step of the procedure we loose information on the fluctuations of the spins that occur on the factor scale b.

8.3 Transformation Laws and Effective Hamiltonians

There are several reasonable choices of the transformation laws $f(\{\sigma_i\})$ for updating the spin variables and each of them gives rise to different RG coarse grainings of the

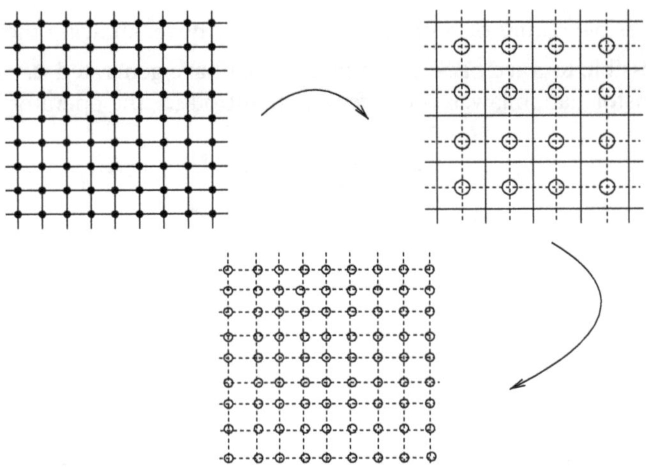

Fig. 8.2 *Sequence of the renormalization group transformation in real space: from the original lattice, with lattice spacing* **a** *and variables* s_i*, to a new lattice with* $\mathbf{a}' = \mathbf{b}\mathbf{a}$ *and block spins* σ_i*. Finally, a scale transformation restores the original lattice spacing* **a**.

system. However, one should realize that what really matters is the asymptotic behaviour of the adopted iterative procedure. In the limit $n \to \infty$ the difference between the transformation laws may be washed out, leading to the same physical scenario. In the real space version of the RG, the two versions mostly used are the following:

- **Decimation**. This law assigns to the spin $\sigma_k^{(n+1)}$ the value of one of the spin $\sigma_i^{(n)}$ of the block $\mathcal{B}_k$, say the central one

$$\sigma_k^{(n+1)} = \sigma_j^{(n)}, \ j \in \mathcal{B}_k$$

- **Majority rule**. This law assigns to the spin $\sigma_k^{(n+1)}$ the value of the majority of the spins $\sigma_i^{(n)}$ of the block $\mathcal{B}_k$, namely

$$\sigma_k^{(n+1)} = A^{(n)} \sum_{i \in \mathcal{B}_k} \sigma_i^{(n)},$$

  where $A^{(n)}$ is a normalization constant. We have to be careful to implement the latter procedure for it depends on the nature of the spins σ_i. For instance, if they are Ising variables with values ± 1, it is convenient to choose blocks with an odd number of spins, in order to avoid the possibility of generating a null value for the next block spins. The normalization constant $A^{(n)}$ is useful to reestablish the correct range of values ± 1 for the new variable.

Note the irreversible nature of both transformations above: knowing the value $\sigma_k^{(n+1)}$ it is indeed impossible to trace back the spins $\sigma_i^{(n)}$ that have generated it.

Given a transformation law, it is convenient to introduce the operator

$$T(\sigma_k^{(n+1)}, \sigma_i^{(n)}) = \begin{cases} 1, & \text{if } \sigma_k^{(n+1)} = f(\{\sigma_i^{(n)}\}) \\ 0, & \text{otherwise} \end{cases} \tag{8.3.1}$$

It satisfies

$$\sum_{\{\sigma_k^{(n+1)}\}} T(\sigma_k^{(n+1)}, \sigma_i^{(n)}) = 1. \tag{8.3.2}$$

By fixing the transformation law of the spin, we determine the *effective Hamiltonian* $H^{(n+1)}(\{\sigma_i^{(n+1)}\}, g_k^{(n+1)})$ for the new block spins. Since the transformation (8.2.5) depends only on the configurations of the spins $\sigma_i^{(n)}$, the new Hamiltonian will be determined by the nth step Hamiltonian $H^{(n)}(\{\sigma_i^{(n)}\}, g_k^{(n)})$ as follows. Let us denote by

$$P(\{\sigma_i^{(n)}\}) = \exp\left[-H^{(n)}\left(\{\sigma_i^{(n)}\}, g_k^{(n)}\right)\right],$$

the probability to realize a configuration $\sigma_i^{(n)}$ and define the new Hamiltonian by means of the conditional probability

$$\begin{aligned} &\exp\left[-H^{(n+1)}\left(\{\sigma_k^{(n+1)}\}, g_k^{(n+1)}\right)\right] \\ &= \sum_{\{\sigma_i^{(n)}\}} \prod_{blocks} T(\sigma_k^{(n+1)}, \sigma_i^{(n)}) \exp\left[-H^{(n)}\left(\{\sigma_i^{(n)}\}, g_i^{(n)}\right)\right]. \end{aligned} \tag{8.3.3}$$

In other words, assigned the new block spins $\sigma_k^{(n+1)}$ according to the transformation law, the spins $\sigma_i^{(n)}$ of the previous step are averaged using as weight their Boltzmann factor. The result is the Boltzmann factor of the new block spins.

To avoid the introduction of some additive constants as we go on in the iteration of the procedure, it may be useful to fix a normalization condition for the sequence of Hamiltonians, as for instance

$$\sum_{\{\sigma_i^{(n)}\}} H^{(n)}\left(\{\sigma_i^{(n)}\}, g_i\right) = 0.$$

Using this normalization, we have

$$\sum_{\{\sigma_k^{(n+1)}\}} \exp\left[-H^{(n+1)}\left(\{\sigma_k^{(n+1)}\}, g_k^{(n+1)}\right)\right] = \sum_{\{\sigma_i^{(n)}\}} \exp\left[-H^{(n)}\left(\{\sigma_i^{(n)}\}, g_i^{(n)}\right)\right] \quad (8.3.4)$$

and the same value of the partition functions

$$Z^{(n+1)}(g_k^{(n+1)}) = Z^{(n)}(g_i^{(n)}). \quad (8.3.5)$$

This equality also holds for the expectation value of any function X of the variables $\sigma_k^{(n+1)}$: this is independent whether we compute it by using $H^{(n+1)}$ or $H^{(n)}$ in view of the identity

$$\begin{aligned} \langle X\rangle &= \frac{1}{Z^{(n+1)}} \sum_{\{\sigma_k^{(n+1)}\}} X(\{\sigma_k^{(n+1)}\}) \exp\left[-H^{(n+1)}\left(\{\sigma_k^{(n+1)}\}, g_k^{(n+1)}\right)\right] \\ &= \frac{1}{Z^{(n)}} \sum_{\{\sigma_i^{(n)}\}} X(\{\sigma_k^{(n+1)}\}) \exp\left[-H^{(n)}\left(\{\sigma_i^{(n)}\}, g_i^{(n)}\right)\right]. \end{aligned} \quad (8.3.6)$$

This allows us to refer to the expectation values without specifying which effective Hamitonian has been used.

Manifold of the coupling constants. To implement successfully the procedure of the RG it is obviously important that the new effective Hamiltonian $H^{(n+1)}$ has the same functional form of $H^{(n)}$, so that the model remains the same at each step of the sequence, beside a change in the value of its coupling constants. As a matter of fact this is impossible if we restrict the attention to the Hamiltonians with a finite number of couplings, since at each step new couplings are generated: for instance, starting from an Hamiltonian with interaction among the next-neighbour spins, the new Hamiltonian has a new interaction among the spins separated by more than a lattice spacing and, furthermore, interactions that involve more than two spins. For this reason, it is convenient to start from the very beginning with the ensemble of all possible coupling constants that are compatible with the symmetry of the model and the nature of the statistical variables. Let us introduce then the *manifold of the coupling constants* and denote by $\{g^{(n)}\} \equiv (g_1^{(n)}, g_2^{(n)}, \ldots)$ the set of all the couplings of the effective Hamiltonian $H^{(n)}$. In such a manifold, the application of eqn. (8.3.3) can be interpreted as a motion of the point $\{g\}$ that identifies the system. This motion is made in discrete time steps and ruled by

$$\{g^{(n+1)}\} = \mathcal{R}(\{g^{(n)}\}), \quad (8.3.7)$$

where $\mathcal{R}$ is, in general, a complicate non-linear transformation. Starting from a point $\{g^{(0)}\}$ and applying (8.3.7), the point of the system evolves in the sequence $\{g^{(1)}\}$, $\{g^{(2)}\}, \ldots$, giving rise in this way to an RG *trajectory* (Figure 8.3). It is important to stress that all points of the trajectory describe the *same physical situation*: they simply correspond to an observation of the system with a different magnifying glass. Note that

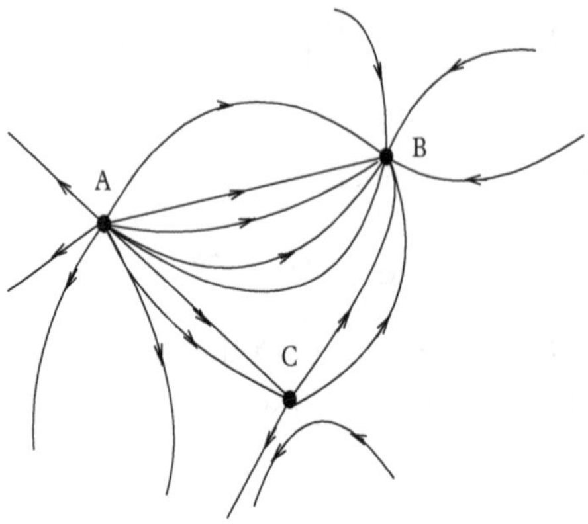

Fig. 8.3 *Trajectories of the renormalization group and fixed points: A is a repulsive fixed point, B is an attractive fixed point, whereas C is a mixed fixed point.*

under the transformation (8.3.7), the correlation length has to be measured with respect to the new lattice spacing and therefore it changes as

$$\xi(g^{(n+1)}) = b^{-1}\,\xi(g^{(n)}). \tag{8.3.8}$$

Hence, it shrinks of a factor b at each step of the procedure.

8.4 Fixed Points

The mathematical nature of the RG transformations is the same of the *dynamical systems*, an important subject of physics and mathematics. An example of a dynamical system is provided by the logistic map discussed in Problem 8.1. A priori, one could expect an arbitrary behaviour for the trajectories that starts from a point P in the space of the coupling constants, with oscillations, discontinuities or a zig-zag behaviour. However, in all cases of physical relevance, one observes a smooth convergence toward some fixed points. A *fixed point* is a point in the manifold of the coupling constants that remains invariant under the mapping (8.3.7)

$$g^* = \mathcal{R}(g^*). \tag{8.4.1}$$

At a fixed point, the correlation length either diverges or vanishes, as it can be easily seen from eqn. (8.3.8) evaluated at $g = g^*$

$$\xi(g^*) = b^{-1}\,\xi(g^*). \tag{8.4.2}$$

Nature of fixed points. The fixed points where $\xi = \infty$ are called *critical points*, whereas those $\xi = 0$ are called *trivial fixed points*. The fixed points can be further classified by their stability nature: they can be *attractive*, *repulsive* or *mixed*. We have an attractive fixed point if, in a neighbour of g^*, the iteration of the transformations $g^{(n)}$ converges to g^*. We have instead a repulsive fixed point if the iteration of the RG transformations that start nearby g^* moves the point away from g^*. A mixed fixed point has both kinds of trajectories in its vicinity.

Linearization. The nature of the fixed points can be determined by studying the linear version of the transformation (8.3.7): putting $g = g^* + \delta g$, we have

$$g^* + \delta g' = \mathcal{R}(g^* + \delta g) \simeq \mathcal{R}(g^*) + \mathcal{K}\,\delta g = g^* + \mathcal{K}\,\delta g,$$

namely

$$\delta g_a' = \mathcal{K}_{ab}\,\delta g_b, \tag{8.4.3}$$

where the matrix $\mathcal{K}_{ab}$ is defined as

$$\mathcal{K}_{ab} = \frac{\partial \mathcal{R}_a}{\partial g_b}. \tag{8.4.4}$$

This matrix is not necessarily symmetric and for this reason it is necessary to distinguish between the right and the left eigenvectors. Denoting by λ^i its eigenvalues and by Δ^i its left eigenvectors $\mathcal{K}$, we have

$$\sum_a \Delta_a^i \mathcal{K}_{ab} = \lambda^i \Delta_a^i. \tag{8.4.5}$$

In terms of Δ_a^i let us now define the linear combination of the displacements δg_a

$$u_i = \sum_a \Delta_a^i \delta g_a. \tag{8.4.6}$$

These linear combinations are called *scaling variables*. They have the important quality to transform in a multiplicative way under the RG transformations

$$\begin{aligned} u_i' &= \sum_a \Delta_a^i \delta g_a' = \sum_{a,b} \Delta_a^i \mathcal{K}_{ab}\,\delta g_b \\ &= \sum_b \lambda^i \Delta_b^i \delta g_b = \lambda^i u_i. \end{aligned} \tag{8.4.7}$$

If b is the rescaling parameter of the block spins, it is common to parameterize λ_i as $\lambda^i = b^{y_i}$ where the quantities y_i are improperly called the eigenvalues of the RG: Section

8.9 shows that they determine the critical exponents of the statistical model. Disregarding the case in which y_i is a complex number,[1] we can have the following cases:

1. $y_i > 0$. In this case the corresponding u_i is a *relevant variable*. A repeated application of the transformations moves its value away from the critical point.
2. $y_i < 0$. In this case u_i is an *irrelevant variable*. Starting sufficiently close to the fixed point, the iteration of the transformation shrinks the initial value to zero.
3. $y_i = 0$. In this case u_i is a *marginal variable*. Iterating the transformation, the value of this variable does not change.

Critical surface. To continue the analysis, let us assume that the dimension of the space of the coupling constants is m and let us consider a fixed point g^* with n relevant variables and $(m-n)$ irrelevant variables. This means that there exists a $(m-n)$-dimensional surface $\mathcal{C}$, called the *critical surface*, that is the attractive basin to the fixed point g^*. On this surface the correlation length is infinite. The coupling constants g_k of the system depend generally on the external parameters of the system, as temperature, pressure or magnetic field. Varying these external parameters, the point $\{g\}$ of the coupling constants varies correspondingly. When there are n relevant variables, in order to intercept the critical surface it is necessary to choose appropriately n external control parameters. In all cases of physical interest, the temperature is one of these parameters and its value has to be tuned to its critical value $T = T_c$ to hit the critical surface. This may be not enough: if there are magnetic fields, they must be switched off and it may be also necessary to tune appropriately the chemical potential. Once such a fine tuning of the n experimental parameters has been done, the point $\{g\}$ in on the critical surface. If we now apply the RG transformations, their iterations of the RG move the point toward the critical point g^*, *independently from its initial position on* $\mathcal{C}$ (Figure 8.4). This is, in a nutshell, the origin

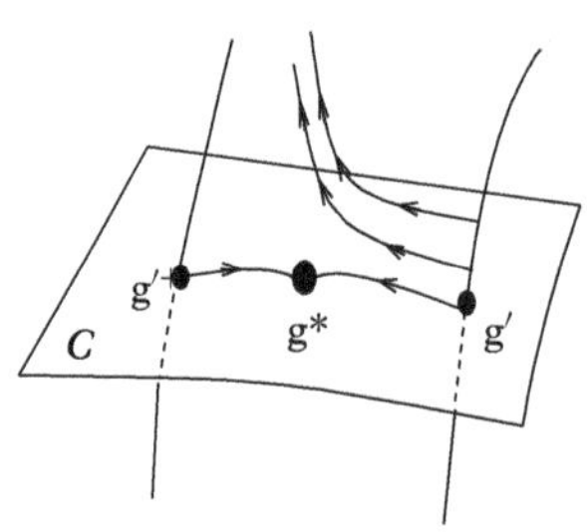

Fig. 8.4 *Once the trajectory reaches the critical surface, the evolution of the coupling constants under the renormalization group converges to the fixed point g^*, independently from the initial position g' or g''. Points outside the critical surface move away from it.*

[1] In this case the trajectories are spirals, that converge to the fixed point g^* if $\mathrm{Re}\,y_i < 0$ or diverge from it if $\mathrm{Re}\,y_i > 0$.

of the universal behaviour of the critical phenomena: Hamiltonians that differ only for their irrelevant operators give rise to the same critical behaviour.

Let us now prove that the correlation length diverges on the critical surface. Suppose that the physical system is represented by the point $\{g\}$ in the space of the coupling constants and, after n iterations, by $\{g^{(n)}\}$. Using eqn. (8.3.8), we have the sequence of identities

$$\xi(g) = b\xi(g^{(1)}) = b^2\xi(g^{(2)}) = \cdots = b^n\xi(g^{(n)})$$

If the initial point $\{g\}$ was on the critical surface, in the limit $n \to \infty$ the sequence of $\{^{(n)}$ converges to $\{g^*\}$, i.e. $\lim_{n\to}\{g^{(n)}\} = \{g^*\}$: since $\xi(g^*) = \infty$ and $b > 1$, we have that $\xi(g) = \infty$ for all points of the critical surface.

Properties of the RG flows. The physical nature of the problem is quite helpful to clarify both the geometrical nature of the trajectories and some of their properties. For instance:

- the RG trajectories cannot intersect, because if this would happen it means that the definition of the successive Hamiltonians would be ambiguous – a circumstance that is impossible from the physical point of view.
- Switching on a relevant variable in an Hamiltonian that is at a fixed point g_i^*, the corresponding flow moves the system away from it. At the end of this motion, the point $\{g\}$ reaches either a trivial fixed point (with a zero correlation length) or another critical point g_f^*. The approach to both final points is obviously along one of their irrelevant directions.
- During the motion along the RG flows, the point may pass close to other fixed points g_a^* $(a = 1, 2, \ldots)$ (Figure 8.5). If the trajectory is sufficiently close to them, there could be a series of interesting cross-over phenomena. According to the scale by which we monitor the system, one can observe the following behaviours: (i) on a short distance, the critical behaviour ruled by the original fixed point g_i^*; (ii) on intermediate scales, the scaling behaviour associate to the nearest fixed points met along the flow; (iii) on large distance, the scaling behaviour ruled by the final fixed point g_f^*.

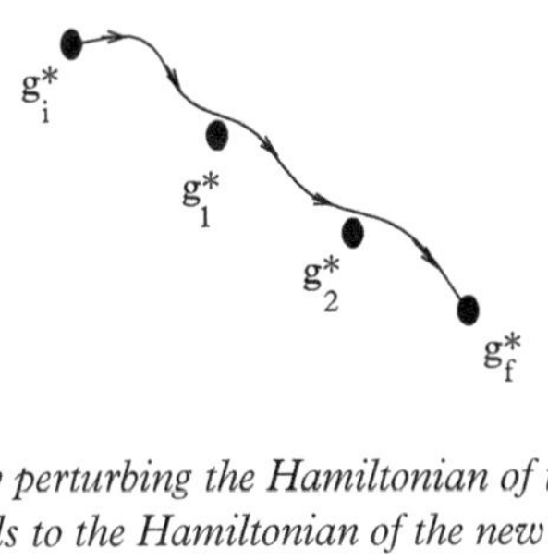

Fig. 8.5 *RG trajectory obtained by perturbing the Hamiltonian of the fixed point g_i^* with a relevant variable. The final point corresponds to the Hamiltonian of the new fixed point g_f^*, while the cross-over phenomena are ruled by the intermediate fixed points met along the trajectory.*

In order to clarify the concepts introduced so far, it is useful to discuss some simple examples.

8.5 The Ising Model

The first example is the one-dimensional Ising model. As the initial Hamiltonian we take the one with the nearest-neighbour interaction

$$H(s_i; \mathcal{J}) = -\mathcal{J} \sum_i s_i s_{i+1}. \tag{8.5.1}$$

Each pair of spins has the Boltzmann weight

$$W(s_i, s_{i+1}; v) = e^{\mathcal{J} s_i s_{i+1}} = \cosh \mathcal{J} (1 + v s_i s_{i+1}), \tag{8.5.2}$$

with $v = \tanh \mathcal{J}$. To apply the RG transformations, we divide the system in blocks, each made of three spins, and then we apply the decimation rule: for each block we choose as a spin of the new system the one that is at the centre (Figure 8.6).

Consider two neighbour blocks. To implement the RG procedure, it is necessary to sum on the spins s_3 and s_4, keeping fixed though the values of the spins at the centre of the two blocks, here denoted as $\sigma_1 \equiv s_2$ and $\sigma_2 = s_5$. In the partition function the terms that involve the degrees of freedom of two neighbour blocks are

$$e^{\mathcal{J}\sigma_1 s_3}\, e^{\mathcal{J} s_3 s_4}\, e^{\mathcal{J} s_4 \sigma_2}.$$

Using the identity $e^{\mathcal{J} x_a x_b} = \cosh \mathcal{J} (1 + v x_a x_b)$ for all the three terms of the previous equation, we have

$$(\cosh \mathcal{J})^3 (1 + v \sigma_1 s_3)(1 + v s_3 s_4)(1 + v s_4 \sigma_2).$$

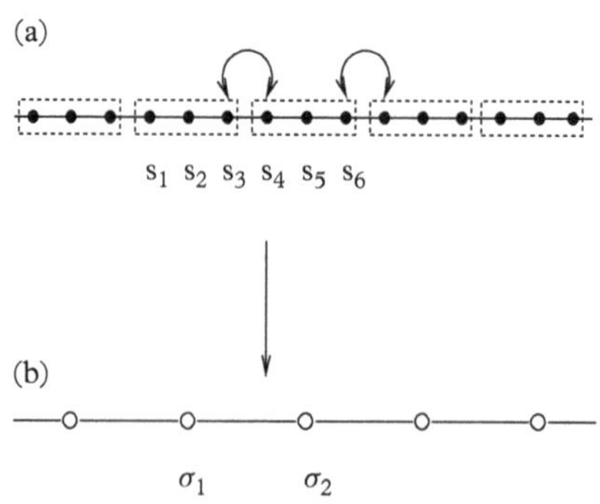

Fig. 8.6 *Spin blocks in the one-dimensional Ising model and the decimation transformation.*

Expanding this product and summing on s_3 and s_4, we get

$$2^2(\cosh \mathcal{J})^3(1+v^3\sigma_1\sigma_2).$$

Beside a multiplicative normalization constant (independent on the spins), this expression is of the same form of (8.5.2) and therefore it defines the new Boltzmann weight $W(\sigma_1, \sigma_2; v')$ of the block spins σ_1 and σ_2 with

$$v' = v^3. \tag{8.5.3}$$

The new Hamiltonian of the system is thus given by

$$H(\sigma_i; \mathcal{J}') = N' p(\mathcal{J}) - \mathcal{J}' \sum_i \sigma_i \sigma_{i+1}, \tag{8.5.4}$$

where $N' = N/3$ is the number of sites of the new lattice, while the value of the new coupling constant is

$$\mathcal{J}' = \tanh^{-1}\left[(\tanh \mathcal{J})^3\right], \tag{8.5.5}$$

$p(\mathcal{J})$ is the contribution to the free energy coming from the degrees of freedom on which we have summed, and it ensures the correct normalization of the partition functions of the two systems

$$p(\mathcal{J}) = -\frac{1}{3}\log\left[\frac{(\cosh \mathcal{J})^3}{\cosh \mathcal{J}'}\right] - \frac{2}{3}\log 2.$$

Let us now use the transformation law of the coupling constants, eqn. 8.5.3), to study the physical content of the model. It is useful to make a plot of this mapping. It is easy to see that the mapping has two fixed points: $v_1^* = 0$ and $v_2^* = 1$. The first is an attractive fixed point, while the second is repulsive: unless v is exactly $v = 1$, each iteration moves the values of v to the origin. Remind that we have absorbed in $\mathcal{J}$ a factor $\beta = 1/kT$. This means that the high-temperature phase around $T \to \infty$ corresponds to the values close to $v \to 0$, while the low-temperature phase around $T \to 0$ corresponds to values $v \to 1$, with $v = 1$ when $T = 0$.

Since the effective coupling constant moves toward smaller values at each iteration, the large scale degrees of freedom are described by an effective Hamiltonian whose temperature increases: this is the region where the system is in its paramagnetic phase and has a finite correlation length. This happens for all values of v (except $v = 1$) and therefore we are led to the conclusion that the one-dimensional Ising model is always in its disorder phase. As discussed in Chapter 2, this conclusion is indeed confirmed by the exact solution of this model. It is also easy to derive how the correlation length depends on the coupling constant: one simply needs to employs the transformation law

$$\xi(v') = \frac{1}{3}\xi(v), \tag{8.5.6}$$

and substitute v' with eqn. (8.5.3). Hence, the correlation length satisfies the functional equation

$$\xi(v^3) = \frac{1}{3}\xi(v), \tag{8.5.7}$$

whose solution is given by

$$\xi(v) = -\frac{\xi_0}{\log v} = -\frac{\xi_0}{\log \tanh \mathcal{J}}. \tag{8.5.8}$$

This expression is in agreement with the behaviour of $\xi(v)$ (discussed in Chapter 2). Note that ξ is always finite, except when $\mathcal{J} \to \infty$ $(T \to 0)$, where it diverges as $\xi \simeq e^{1/T}$. This gives a further evidence that the one-dimensional Ising model is always in a paramagnetic phase, expect when $T = 0$.

Even in the absence of simple analytic expressions, the arguments presented above help to understand the phase diagram of the Ising model on higher dimensional lattices. First, let us consider closely the one-dimensional case: if we refer to the spin variables of two neighbour blocks, in the limit $\mathcal{J} \to \infty$ the equation that fixes the new coupling constant can be written as

$$\mathcal{J}' \simeq \mathcal{J} \langle s_3 \rangle_{\sigma_1 = 1} \langle s_4 \rangle_{\sigma_2 = 1}, \tag{8.5.9}$$

where $\langle s_3 \rangle_{\sigma_1 = 1}$ is the mean value of the spin at the edge of the block, with the condition that the spin in the middle of the block assumes value 1. Since these mean values are

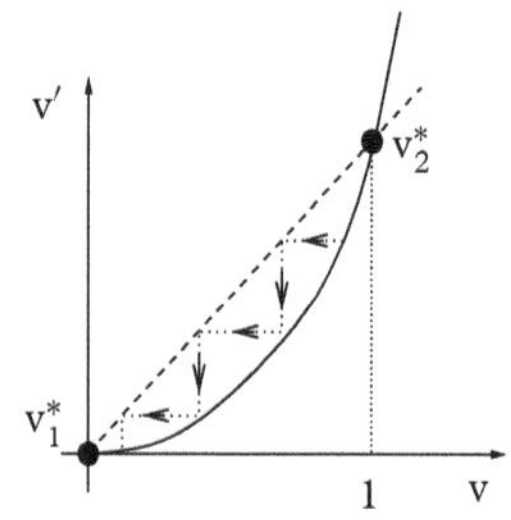

Fig. 8.7 *Renormalization group equation for the one-dimensional Ising model.* $v_1^* = 0$ *and* $v_2^* = 1$ *are the two fixed points, the former an attractive one, the latter a repulsive one. Starting from any value* $v \neq 1$*, the next iterations move the value of* v *toward the origin.*

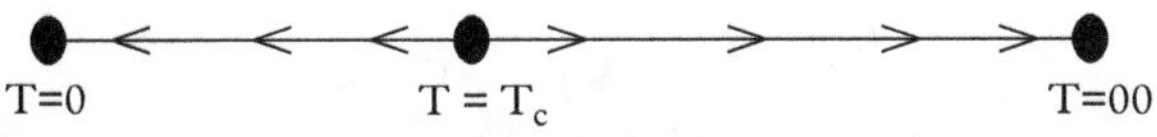

Fig. 8.8 *Phase diagram and renormalization group flows of the d-dimensional Ising model, with $d > 1$. In this case, both low- and high-temperature fixed points are attractive, while the fixed point between them is unstable with respect to the scaling variable associated to the temperature.*

always less than 1 (except at $\mathcal{J} = \infty$), we have $\mathcal{J}' < \mathcal{J}$ and therefore the low-temperature fixed point is always unstable.

However, for d-dimensional lattices (with $d > 1$), the situation is different. Consider once again the transformation law of the couplings in the limit $\mathcal{J} \to \infty$. The value of the new coupling constants is essentially determined by the expectation values of the spins along the boundary of the blocks. Since there are b^{d-1} of them, we have

$$\mathcal{J}' \simeq b^{d-1}\mathcal{J}, \quad \mathcal{J} \to \infty. \tag{8.5.10}$$

For $d > 1$, we have then $\mathcal{J}' > \mathcal{J}$, i.e. the low-temperature fixed point is now attractive! On the other hand, it is easy to convince ourselves that the high-temperature fixed point is also attractive. The attractive nature of both fixed points implies that the Ising model in $d > 1$ should have a critical value at a finite value of the coupling constant, i.e. there should exist a critical temperature T_c at which the model undergoes a phase transition (see Figure 8.8).

8.6 The Gaussian Model

Another simple example of RG transformations is given by the Gaussian model, whose variables $s_i = \varphi_i$ take values on all the real axis. The Hamiltonian of this model, expressed in the k-space, is

$$H = \frac{1}{2}\int_{|k|<1/a} (g_2 + k^2)\,|\varphi(k)|^2\, d^d k. \tag{8.6.1}$$

The microscopic origin of the model is encoded in the cut-off $1/a$ present in the integration on the momenta. The partition function is

$$Z = \int \prod_{|k|<1/a} d\varphi(k)\, e^{-H}. \tag{8.6.2}$$

In order to implement the Rg procedure, let us integrate on the degrees of freedom of the field in the shell of the momenta $1/ba < |k| < 1/a$. This is equivalent to define the new block spins, with scale parameter equals to b. The new Hamiltonian is determined by the equation

$$e^{-H'} = \int \prod_{1/ba<|k|<1/a} d\varphi(k)\, e^{-H}. \tag{8.6.3}$$

Since each variable is decoupled from the others and each integral is Gaussian, disregarding an inessential additive constant, the new Hamiltonian is easily computed

$$H' = \frac{1}{2}\int_{|k|<1/ba} (g_2 + k^2)\, |\varphi(k)|^2\, d^d k. \tag{8.6.4}$$

To restore the initial lattice spacing, let us make the change of variable $k \to k/b$, so that

$$H' = \frac{1}{2}\int_{|k|<1/a} (g_2 + k^2 b^{-2})\, b^{-d}\, |\varphi(k)|^2\, d^d k. \tag{8.6.5}$$

We need to rinormalize the new block variables. This can be done by requiring that the kinetic term $k^2\, |\varphi(k)|^2$ keeps the same form also in the new Hamiltonian. Making the scale transformation on the field

$$\varphi' = b^{(d+2)/2}\, \varphi,$$

we arrive at

$$H' = \frac{1}{2}\int_{|k|<1/a} (g_2' + k^2)\, |\varphi'(k)|^2\, d^d k. \tag{8.6.6}$$

We have thus obtained the transformation law of the coupling constant

$$g_2' = b^2\, g_2, \tag{8.6.7}$$

shows the irrelevant nature of this variable. The fixed points of the transformation law are: $g_2 = 0$, that is a critical point of the model, and $g_2 \to \infty$, that is a trivial fixed point. In fact, the correlation length is given by $\xi(r_0) = 1/\sqrt{g_2}$, and in the former case it diverges, while in the latter it goes to zero.

8.7 Operators and Quantum Field Theory

As discussed in Chapter 7, close to a critical point, and where the correlation length ξ is much larger than the lattice spacing, it is natural to adopt the QFT formalism to describe the dynamics of the statistical systems. For the sake of simplicity, let us focus our attention on theory with a Z_2 internal symmetry. The order parameter is the scalar field $\varphi(x)$ that transforms as $\varphi \to -\varphi$ under the Z_2 symmetry. The more general action of such a model is given by

$$\mathcal{S} = \int d^d x \left[\frac{1}{2}(\partial_\mu \varphi)^2 + g_1 \varphi + \frac{g_2}{2}\varphi^2 + \cdots + \frac{g_n}{n!}\varphi^n + \cdots \right]. \tag{8.7.1}$$

The manifold of the coupling constants is described by the set $\{g\} = (g_1, g_2, \ldots, g_n, \ldots)$. The partition function of the system is expressed by the functional integral

$$Z[\{g\}, a] = \int \mathcal{D}\varphi \exp[-\mathcal{S}[\varphi, \{g\}]]. \tag{8.7.2}$$

Chapter 7 also stressed the fundamental role played by the lattice spacing a even in the continuum theory. In terms of such a parameter, the 'engeneering' dimensions of the coupling constants

$$[g_n] = a^{nd/2-n-d} \equiv a^{\delta_n} \tag{8.7.3}$$

can be regarded as their scaling dimensions with respect to the Gaussian fixed point, identified by the condition $g_1 = g_2 = g_3 = \cdots = 0$. At the Gaussian point, the relevant coupling constants are those with $\delta_n < 0$, the irrelevant ones those with $\delta_n > 0$ while the marginal ones are associated to the condition $\delta_n = 0$.

In addition to the Gaussian fixed point, there may be other fixed points. They can be reached, for instance, perturbing the Gaussian action by some relevant operators, as shown in an explicit example discussed in Section 8.10. Let us then suppose that we are at a new fixed point, characterized by the action $\mathcal{S}^*$ and by a new set of m relevant variables, denoted by λ_i $(i = 1, 2, \ldots, m)$. If $\phi_i(x)$ are the fields conjugated to these variables, in the vicinity of the new fixed point the action can be written as

$$\mathcal{S} = \mathcal{S}^* + \lambda_1 \int d^d x \phi_1(x) + \cdots \lambda_m \int d^d x \phi_m(x). \tag{8.7.4}$$

Under a scaling transformation $x \to x/b$, λ_i scale as $b^{y_i} \lambda_i$ whereas the volume element $d^d x$ scales as $b^{-d} d^d x$. If we require that the action does not depend on the arbitrary parameter b, the fields $\phi_i(x)$ must then scale as[2]

$$\phi_i \to b^{x_i} \phi_i, \tag{8.7.5}$$

with

$$x_i = d - y_i. \tag{8.7.6}$$

[2] The same argument also holds for the conjugate fields of the irrelevant variables, with their scaling dimension also given by (8.7.6). For simplicity we focus the attention only on the relevant terms.

It is easy to prove that, at the fixed point, the two-point correlation functions of the fields $\phi_i(x)$ have the scaling form

$$G_i(r_1 - r_2) = \langle \phi_i(r_1)\phi_i(r_2)\rangle = \frac{1}{|r_1 - r_2|^{2x_i}}. \tag{8.7.7}$$

Let us denote by $G_i(r_1 - r_2, \mathcal{S})$ the correlation function computed by the formula

$$\langle \phi_i(r_1)\phi_i(r_2)\rangle = \frac{1}{Z}\int \mathcal{D}\phi\, \phi_i(r_1)\,\phi_i(r_2)\, e^{-\mathcal{S}}, \tag{8.7.8}$$

using the action $\mathcal{S}$. Making a RG transformation that changes the physical lengths by b, the action $\mathcal{S}$ becomes $\mathcal{S}'$, while the fields ϕ_i scale according to eqn. (8.7.5). We arrive then to the functional equation

$$G_i((r_1 - r_2)/b, \mathcal{S}') = b^{2x_i}\, G_i(r_1 - r_2, \mathcal{S}). \tag{8.7.9}$$

If the initial action $\mathcal{S}$ is the one relative to the fixed point $\mathcal{S}^*$, we have $\mathcal{S}' = \mathcal{S}^*$ and therefore (8.7.46) becomes

$$G_i((r_1 - r_2)/b, \mathcal{S}^*) = b^{2x_i}\, G_i(r_1 - r_2, \mathcal{S}^*). \tag{8.7.10}$$

The solution, except for a multiplicative factor, is provided by eqn. (8.7.7).

Role of the microscopic scale. The expression (8.7.7) for the scaling form of the correlation function gives us the opportunity to make an important comment on the role played by the microscopic scale a. Suppose that the field $\phi_i(x)$ in (8.7.7) coincides with the field $\varphi(x)$ that appears in the action (7.2.3). Since the engineering dimension of this field is $a^{1-d/2}$, we could expect that its correlation function should take the form

$$G_i(r) = \frac{1}{r^{d-2}}. \tag{8.7.11}$$

If $2x_i \neq (d-2)$ obviously this expression does not coincide with the one given in eqn. (8.7.7). This seems to preclude the possibility that the field $\varphi(x)$ could ever have an anomalous dimension. There is however a way out: the scaling law (8.7.7) can be conciliated with the engineering dimension of the field if it exists a length scale a able to absorb the canonical dimension of the field. In other words, the exact expression of the propagator of the scaling field is not that in eqn. (8.7.7) but

$$G_i(r) = \frac{1}{a^{d-2}}\left(\frac{a}{r}\right)^{2x_i}. \tag{8.7.12}$$

If we make a scaling transformation of *all* the dimensional quantities of the problem, the correlator scales according to the canonical dimension of the field

$$G_i \to a^{-(d-2)} G_i.$$

Vice versa, if we want observe the system under a different magnifying glass, we can rescale the length scales but keep fixed the lattice spacing of the system: in this case, we obtain instead the anomalous behaviour expressed by eqn. (8.7.7). In conclusion, the expression (8.7.7) that is usually assumed as the propagator of the scaling fields has to be regarded as a shortening of the general formula (8.7.12), where the lattice spacing a was taken to be 1.

Note that the functional equation (8.7.9) can be employed to find the general form of the correlation functions of the scaling fields in the vicinity of the fixed point. Suppose, for simplicity, that $\mathcal{S}^*$ is perturbed by only one relevant field, with coupling constant λ_k). In this case, eqn. (8.7.9) becomes

$$G_i(r/b, b^{y_k}\lambda_k) = b^{2x_i}\, G_i(r, \lambda_k), \tag{8.7.13}$$

whose general solution can be written as

$$G_i(r, \lambda_k) = \frac{1}{r^{2x_i}} f_i\left(\lambda^{1/y_k} r\right). \tag{8.7.14}$$

In this formula f_i is a homogeneous function of the distance r and of the coupling constant λ_k, whose explicit form can be obtained only studying the details of the model.

8.8 Functional Form of the Free Energy

The linearized form of the RG equations permits to easily derive the scaling form of the free energy in the vicinity of the fixed point and the relationships between the critical exponents. Consider a statistical system with n relevant coupling constants λ_i. In the field theory formulation, in the vicinity of the fixed point the action is given by

$$\mathcal{S} = \mathcal{S}^* + \sum_i^n \lambda_i \int d^d x \phi_i(x). \tag{8.8.1}$$

In the Ising model, for instance, there are two relevant variables, given by the magnetic field $h \equiv \lambda_1$ and by the displacement of the temperature from the critical value $T - T_c \equiv \lambda_2$: the conjugate fields are $\phi_1(x)$, which correspond to the continuum limit of the spin variable s_i, and $\phi_2(x)$, associated to the continuum limit of the energy density, given on the lattice by $\sum_j s_i s_{i+\hat{e}_j}$.

Since the variables λ_j in the action (8.8.1) have dimensions $[\lambda_j] = a^{y_j}$, the theory has a finite correlation length. Selecting one of the couplings, say λ_i, in the thermodynamics limit the correlation length can be expressed as

$$\xi(\{\lambda_j\}) = a\,(K_i\lambda_i)^{-\frac{1}{y_i}}\,L_i\left(\frac{K_1\lambda_j}{(K_i\lambda_i)^{\phi_{1i}}},\ldots,\frac{K_j\lambda_j}{(K_i\lambda_i)^{\phi_{ji}}},\ldots\right), \tag{8.8.2}$$

where $K_i \simeq 1/\lambda_i^{(0)}$ are some non-universal metric terms that depend on the unity by which we measure the coupling constants, L_i are universal homogeneous functions of the $(n-1)$ ratios $\frac{K_j\lambda_j}{(K_i\lambda_i)^{\phi_{ji}}}$, with $j \neq i$ and finally

$$\phi_{ji} = \frac{y_j}{y_i}, \tag{8.8.3}$$

are the so-called *crossover exponents*. There are many (equivalent) ways of expressing this scaling law of the correlation length, according to which coupling constant we choose as prefactor. Each way selects each own scaling function L of the above ratio of the couplings. When $\lambda_k \to 0$ $(k \neq i)$ with $\lambda_i \neq 0$, eqn. (8.8.2) can be written as

$$\xi_i = a\xi_i^0\,\lambda_i^{-\frac{1}{y_i}}, \quad \xi_i^0 \sim K_i^{-\frac{1}{y_i}}. \tag{8.8.4}$$

Consider now the free energy of the system, $f[\lambda_i]$, defined by

$$Z[\{\lambda_i\}] = \int \mathcal{D}\phi_i e^{-\left[S^* + \sum_{i=1}^n \lambda_i \int \phi_i(x)\, d^dx\right]} \equiv e^{-Nf(\lambda_i)}. \tag{8.8.5}$$

In virtue of the identity (8.3.4) of the partition functions, making a RG transformation we have

$$e^{-Nf(\{\lambda\})} = e^{-Np(\{\lambda\}) - N'f(\{\lambda'\})},$$

where $p(\{\lambda\})$ is an additive constant related to the degrees of freedom on which we have integrated. Since the new number of sites is $N' = b^{-d}N$, we have the functional equation

$$f(\{\lambda\}) = p(\{\lambda\}) + b^{-d} f(\{\lambda'\}). \tag{8.8.6}$$

The function $p(\{\lambda\})$ is an analytic function of the coupling constant, since it involves a sum on a finite number of spins. If we are interested in studying the singular behaviour of the free energy, we can safely discard this term and arrive at a functional equation that involves only the singular part of f

$$f_s(\{\lambda\}) = b^{-d} f_s(\{\lambda'\}). \tag{8.8.7}$$

Substituting the expression of the new coupling constants given by the RG transformations, we have

$$f_s(\{\lambda_k\}) = b^{-d} f_s(\{b^{y_k}\lambda_i\}). \tag{8.8.8}$$

Iterating this equation, the irrelevant variables go to zero (this is a manifestation of the universality of the critical behaviour) and the free energy, as function of the relevant variables alone, satisfies

$$f_s(\{\lambda_j\}) = b^{-nd} f_s(\{b^{ny_j}\lambda_i)\}. \tag{8.8.9}$$

As for the correlation length, there are many ways to express the general solution of this equation. Selecting once more one of the couplings, say λ_i, we have

$$f(\{\lambda_i\}) = f_i[\{\lambda_j\}] \equiv (K_i\lambda_i)^{-\frac{d}{y_i}} F_i\left(\frac{K_1\lambda_j}{(K_i\lambda_i)^{\phi_{1i}}}, \dots, \frac{K_j\lambda_j}{(K_i\lambda_i)^{\phi_{ji}}}, \dots\right)^{-\frac{d}{y_i}}. \tag{8.8.10}$$

The functions F_i are universal homogeneous functions of the $(n-1)$ ratios $\frac{K_j g_j}{(K_i g_i)^{\phi_{ji}}}$. We next discuss how there are some obvious advantages in considering different expressions for these scaling functions, obtained by changing the selecting variable λ_i. In fact, in several physical applications, there is only one coupling constant kept different from zero till the end, and the best choice of expressing the free energy depends on this situation. Even in the absence of an explicit expression of the F_is (that can be explicitly found only by solving exactly the model by other methods), the functional dependence of the free energy is sufficient to obtain useful information on the critical behaviour of the model.

8.9 Critical Exponents and Universal Ratios

Let us discuss the definition of several thermodinamical quantities associated to the derivates of the free energy. Here, we adopt the notation $\langle ... \rangle_i$ to denote the expectation values computed with an action that has, at the end, only λ_i as coupling constant different from zero. The first quantities of interest are the expectation values of the fields ϕ_j that can be parameterized as

$$\langle \phi_j \rangle_i = -\frac{\partial f_i}{\partial \lambda_j}\bigg|_{\lambda_k=0} \equiv B_{ji}\,\lambda_i^{\frac{d-y_j}{y_i}}, \tag{8.9.1}$$

with

$$B_{ji} \sim K_j K_i^{\frac{d-y_j}{y_i}}. \tag{8.9.2}$$

Equivalently

$$\lambda_i = D_{ij}\left(\langle \phi_j \rangle_i\right)^{\frac{y_i}{d-y_j}}, \tag{8.9.3}$$

with

$$D_{ij} \sim \frac{1}{K_i K_j^{\frac{y_i}{d-y_j}}}. \tag{8.9.4}$$

The generalized susceptibilities are defined by

$$\hat{\Gamma}^i_{jk} = \frac{\partial}{\partial \lambda_k} \langle \phi_j \rangle_i = -\frac{\partial^2 f_i}{\partial \lambda_k \partial \lambda_j}. \tag{8.9.5}$$

These quantities are obviously symmetric with respect to the lower indices. For the fluctuation-dissipation theorem, they are related to the off-critical correlation functions as

$$\hat{\Gamma}^i_{jk} = \int dx \, \langle \phi_k(x) \phi_j(0) \rangle_i. \tag{8.9.6}$$

Taking out the dependence on the coupling constant λ_i, we have

$$\hat{\Gamma}^i_{jk} = \Gamma^i_{jk} \lambda_i^{\frac{d-y_j-y_k}{y_i}}, \tag{8.9.7}$$

with

$$\Gamma^i_{jk} \sim K_j K_k K_i^{\frac{d-y_j-y_k}{y_i}}. \tag{8.9.8}$$

As shown by these formulae, the various quantities contain the metric factors K_i and their expressions are therefore not universal. However, we can consider some special combinations of these quantities in which the metric factors are cancelled out. Here are a few examples of the so-called *universal ratios*:

$$(R_c)^i_{jk} = \frac{\Gamma^i_{ii} \Gamma^i_{jk}}{B_{ji} B_{ki}}; \tag{8.9.9}$$

$$(R_\chi)^i_j = \Gamma^i_{jj} D_{jj} B_{ji}^{\frac{D-4\Delta_j}{2\Delta_j}}; \tag{8.9.10}$$

$$R^i_\xi = \left(\Gamma^i_{ii} \right)^{1/D} \xi^0_i; \tag{8.9.11}$$

$$(R_A)^i_j = \Gamma^i_{jj} D_{ii}^{\frac{4\Delta_j + 2\Delta_i - 2D}{D-2\Delta_i}} B_{ij}^{\frac{2\Delta_j - D}{\Delta_i}}; \tag{8.9.12}$$

$$(Q_2)^i_{jk} = \frac{\Gamma^i_{jj}}{\Gamma^k_{jj}} \left(\frac{\xi^0_k}{\xi^0_j} \right)^{D-4\Delta_j}. \tag{8.9.13}$$

As the critical exponents, these pure numbers characterize the universality class of a given model. It is worth emphasizing that, from an experimental point of view, it should be simpler to measure universal amplitude ratios rather than critical exponents: in fact, to determine the former quantities we need to perform several measurements at a single, fixed value of the coupling which drives the system away from criticality whereas to determine the latter, we need to make measurements over several decades along the axes of the off-critical couplings. Moreover, although not all of them are independent, the universal ratios are a set of numbers larger than the critical exponents and therefore permit a more precise determination of the class of universality. Finally, being universal quantities, they can be theoretically computed by analysing the simplest representative of the class of universality under scrutiny.

Some of the quantities here have a familiar meaning in the context of the Ising model. In particular, they permit expression of all critical exponents in terms of rational functions of the eigenvalues y_i of the RG. To compare with the formulae of Chapter 1, it is convenient to use the notation $h = \lambda_1$ and $t = \lambda_2$. Consider, for instance, the specific heat

$$C(T) = \frac{\partial^2 f}{\partial t^2}|_{h=0} = \begin{cases} C_+ \, t^{d/y_t - 2}, & T > T_c \\ C_-(-t)^{d/y_t - 2}, & T < T_c. \end{cases} \tag{8.9.14}$$

From the definition of the critical exponent α, $C(T) \simeq |t|^{-\alpha}$, we get

$$\alpha = 2 - d/y_t. \tag{8.9.15}$$

The spontaneous magnetization of the system is obtained by

$$M(T) = \frac{\partial f}{\partial h}|_{h=0} = M_0 \, (-t)^{(d-y_h)/y_t}. \tag{8.9.16}$$

Comparing with the definition of the critical exponent β, $M(T) \simeq (-t)^\beta$, we have

$$\beta = \frac{d - y_h}{y_t}. \tag{8.9.17}$$

The susceptibility at zero magnetic field is given by

$$\chi(T) = \frac{\partial^2 f}{\partial h^2}|_{h=0} = \begin{cases} \chi_+ t^{(d-2y_h)/y_t}, & T > T_c \\ \chi_-(-t)^{(d-2y_h)/y_t}, & T < T_c. \end{cases} \tag{8.9.18}$$

Comparing with the definition of the critical exponent γ, $\chi(T) \simeq |t|^{-\gamma}$,

$$\gamma = \frac{d - 2y_h}{y_t}. \tag{8.9.19}$$

Finally, in order to derive the exponent δ, we must consider the general expression of the magnetization

$$M(T,h) = \frac{\partial f}{\partial h} = \frac{1}{h_0}\left|\frac{t}{t_0}\right|^{(d-y_h)/y_t} \mathcal{F}'\left(\frac{h/h_0}{|t/t_0|^{y_h/y_t}}\right), \tag{8.9.20}$$

with the choice of the scaling form of the free energy given by (t_0 and h_0 are the metric factors)

$$f(t,h) = \left|\frac{t}{t_0}\right|^{d/y_t} \mathcal{F}\left(\frac{h/h_0}{|t/t_0|^{y_h/y_t}}\right). \tag{8.9.21}$$

To have a finite limit of this expression when $|t| \to 0$, it is necessary that $\mathcal{F}'(x)$ behaves as $\mathcal{F}'_0 x^{d/y_h - 1}$ when $x \to \infty$, so that

$$M(T=0,h)\,\mathcal{M}_0\, h^{(d-y_B)/y_t}. \tag{8.9.22}$$

Comparing the definition of the exponent δ, $M(h) \simeq h^{1/\delta}$, we have

$$\delta = \frac{y_h}{d-y_h}. \tag{8.9.23}$$

Hence, all critical exponents are expressed in terms of the eigenvalues y_h and y_t of the RG equations of this model. Moreover, there is a natural explanation of the existence of the scaling laws

$$\begin{aligned} \alpha + 2\beta + \gamma &= 2, \\ \alpha + \beta\delta + \beta &= 2. \end{aligned} \tag{8.9.24}$$

Using the scaling form (8.7.14) of the correlation function, with $\lambda_k = t$, we have

$$\xi = \begin{cases} \xi_+ t^{-1/y_t}, & T > T_c \\ \xi_- (-t)^{-1/y_t}, & T < T_c \end{cases} \tag{8.9.25}$$

and comparing with the definition of $\xi \simeq |t|^{-\nu}$,

$$\nu = 1/y_t. \tag{8.9.26}$$

Comparing, instead, with the critical expression of the correlator $G_i(r) = 1/r^{d-2+\eta}$, we can extract the last critical exponent

$$\eta = d + 2 - 2y_h. \tag{8.9.27}$$

In terms of these expressions we can easily check the validity of the other scaling laws

$$\begin{aligned} \alpha + d\nu &= 2 \\ \gamma &= \nu(2-\eta). \end{aligned} \tag{8.9.28}$$

Concerning the universal ratios of the Ising model, their analytic expressions are available only for the two-dimensional case.[3] They have been computed by using the QFT methods away from the critical point, which we discuss further on in the book chapters. Here, we simply report the exact values of some of these quantities

$$\begin{aligned} C_+/C_- &= 1, \\ \chi_+/\chi_- &= 37.6936..., \\ \xi_+/\xi_- &= 2, \\ C_+\,\chi_+/M_0^2 &= 0.318569..., \\ C_+^{1/2}\,\xi_+ &= 1/\sqrt{2\pi}, \\ \chi_+\,\mathcal{M}_0^{-\delta}\,M_0^{\delta-1} &= 6.7782.... \end{aligned} \tag{8.9.29}$$

8.10 β-Functions

The previous sections demonstrated that a scaling transformation of the lengths corresponds to a change of the coupling constants that leaves the physical content of a theory unchanged. In a continuum formulation of critical phenomena, we can make a scaling transformation $x \to x' = x/b$, where b is an arbitrary real parameter. If b is infinitesimally close to 1, $b \simeq 1 + \delta l$, the coupling constants change infinitesimally as well

$$g_a \to g_a' = g_a + \frac{dg_a}{dl}\,\delta l + \mathcal{O}((\delta l)^2), \tag{8.10.1}$$

and the RG transformation become the differential equation

$$\frac{dg_a}{dl} = \beta_a(\{g\}). \tag{8.10.2}$$

We have thus introduced the $\beta_a(\{g\})$-functions. They are the vector fields that fix the RG flow. These functions are fundamental quantities of the theory for the following reasons. First of all, their zeros identify the fixed point of the theory, since these are the points where the coupling constants do not vary. Secondly, their derivatives, computed at the fixed points, are directly related to the eigenvalues y_i of the RG: in fact, at the fixed

[3] The one-dimensional model has only the paramagnetic disorder phase and therefore there exists only a sub-set of the previously defined universal ratios.

points, the linearized matrix of the Rg is given by $\mathcal{K}_{ab} = \delta_{ab} + \frac{\partial \beta_a}{\partial g_b}\delta l$, with eigenvalues $(1+\delta l)^{y_i} \simeq 1 + y_i \delta l$. Hence, the y_is are nothing else but the derivatives of the β_a functions at the fixed points g^*.

In most cases, the $\beta_a(\{g\})$ functions are only known perturbatively as series expansions in the coupling constants. If g_a corresponds to a scaling variable with eigenvalue y_a with respect to the fixed point $g^* = (0,0,0,\ldots)$ (here chosen as the origin in the manifold of the couplings), the first term of their expansion is simply

$$\beta_a(\{g\}) = y_a g_a + \cdots \tag{8.10.3}$$

Solving at this order the differential equation (8.10.2), it is easy to arrive at the result previously discussed: the coupling constants with $y_a > 0$ grow and move away from the origin, those with $y_a < 0$ shrink to zero.

It is interesting to evaluate the higher orders of the β-functions. It should be said that this computation becomes more and more difficult by increasing the orders of the perturbation theory and it is beyond the scope of this book. However, Chapter 15 reveals a particularly elegant result for the quadratic terms. At this order, in fact, the β_a-functions are expressed as

$$\beta_a(\{g\}) = y_a g_a - \sum_{b,c} C_{abc}\, g_b\, g_c + \cdots \tag{8.10.4}$$

where C_{abc} are the same coefficients that enter the three-point correlation function of the scaling fields $\phi_a i$ (conjugate to the variables g_i) at the critical point. Thanks to the scaling properties of the fields ϕ_i, the general form of these correlators is

$$\langle \phi_a(r_1)\phi_b(r_2)\phi_c(r_3)\rangle = C_{abc} r_{12}^{x_c - x_a - x_b}\, r_{13}^{x_b - x_a - x_c}\, r_{23}^{x_a - x_b - x_c}, \tag{8.10.5}$$

where $r_{ij} = |r_i - r_j|$ and $x_a = d - y_a$.

ϵ-expansion. The case in which the new zero of the β_a functions (8.10.4) is close to the origin is particularly significant. Here, we can deduce the universal properties of the new fixed point in terms of the fixed point at the origin. This happens, for instance, in the Lagrangian theory

$$\mathcal{S} = \int d^d x \left[\frac{1}{2}(\partial_\mu \Phi)^2 + g_2 \Phi^2 + g_4 \Phi^4\right],$$

where $\Phi(x) = [\phi_1(x), \phi_2(x), \ldots, \phi_n(x)]$ is a field with n components. For this model, the lowest orders of the two β-functions are given by

$$\begin{aligned} \beta_2(\{g\}) &= \frac{dg_2}{dl} = 2g_2 - 8(n+2) g_2 g_4 + \cdots \\ \beta_4(\{g\}) &= \frac{dg_4}{dl} = (4-d) g_4 - 8(n+8) g_2^2 + \cdots \end{aligned} \tag{8.10.6}$$

Clearly, in addition to the origin, there is a new fixed point at $g^* = (g_2, g_4) = (0, (4-d)/8(n+8)$. This has a perturbative nature (i.e. compatible with the perturbative approach used to computed the β-functions) if $\epsilon \equiv (4-d) \ll 1$. Computing the derivatives of the β_a at this new fixed point, we can extract the relative critical exponents

$$\begin{aligned}
\alpha &= \frac{4-n}{2(n+8)}\epsilon + \cdots \\
\beta &= \frac{1}{2} - \frac{3}{2}\frac{\epsilon}{n+8} + \cdots \\
\gamma &= 1 + \frac{n+2}{2(n+8)}\epsilon \\
\delta &= 3 + \epsilon \\
\nu &= \frac{1}{2} + \frac{n+2}{4(n+8)}\epsilon.
\end{aligned} \tag{8.10.7}$$

The first significant feature of these expressions is their difference with respect to the classical values obtained by the mean field theory discussed in Chapter 3. The second relevant feature is the analytical nature of the parameter n that allows us to compute the critical exponents for any value of this parameter, not necessarily an integer. Note that for $n = 1$ we obtain a Z_2 invariant theory, while for $n \to \infty$ we obtain the spherical model. A third aspect worth mentioning is the nature of these series expansions: it is generally believed that they are only asymptotic series, i.e. with zero radius of convergence. There are, however, efficient techniques to cure the problem and sum this kind of series (e.g. Borel resummation method). Note that, by substituting in the formulae above $n = 1$ and $\epsilon = 1$, we get an estimate of the critical exponents of the three-dimensional Ising model. We leave as exercises the comparison of these values with those obtained by the numerical simulations discussed in Chapter 3.

REFERENCES

The idea of *block spins* in statistical models has been put forward by Leo Kadanoff in the paper

Kadanoff, L. P. (1966): 'Scaling Laws for Ising Models near T_c', *Physics* 2: 263.

For discussion of the running of the coupling constant, together with the introduction of the β-function, see:

Callan, C. (1970), 'Broken Scale Invariance in Scalar Field Theory', *Physical Review D* 2(8), 1541–7.

Symanzik,K. (1970). 'Small Distance Behaviour in Field Theory and Power Counting', *Communications in Mathematical Physics* 18(3), 227–46.

The version of the RG discussed in this chapter has been proposed by Kenneth Wilson.

Wilson, K. (1975). 'The Renormalization Group: Critical Phenomena and the Kondo Problem', *Reviews of Modern Physics* 47, 773.

Wilson, K. and Kogut, J. (1974). 'The Renormalization Group and the Epsilon Expansion', *Physics Reports* 12, 75.

For the applications of RG ideas in field theory and statistical mechanics, it is useful the reading of

Amit, D. (1978). *Field Theory, the Renormalization Group and Critical Phenomena*, McGraw Hill, Inc.

Cardy, J. (1996). *Scaling and Renormalization in Statistical Physics*, Cambridge University Press.

Fisher, M. (1974), 'The Renormalization Group in the Theory of Critical Behavior', *Reviews of Modern Physics* 46(4): 597.

Goldenfeld, N. (1992). *Lectures on Phase Transitions and the Renormalization Group*. Frontiers in Physics, Vol. 85. Westview Press.

Pfeuty, P, Toulouse, G. (1977). *Introduction to the Renormalization Group and to Critical Phenomena*, John Wiley & Sons.

For a critical history of the subject see:

K. Huang, *A Critical History of Renormalization*, Int. J. Mod. Phys. A28 (2013) 1330050.

It is also useful to mention that a very successful way of applying the RG ideas in a variational way has produced the density matrix Rg method, proposed originally in

White, S. R. (1992). 'Density Matrix Formulation for Quantum Renormalization Groups', *Physical Review Letters* 69, 2863.

The reader can find a discussion on the *exact RG equation* in which no approximation is made and also no expansion is involved with respect to some small parameter of the action in a series of reviews, such as

Bagnuls, C. and Bervillier, C. (2001). 'Exact Renormalization Group Equations. An Introductory Review', *Physics Reports* 348, 91.

Berges, J, Tetradis, N., and Wetterich, C. (2002). 'Nonperturbative Renormalization Flow in Quantum Field Theory and Statistical Physics, *Physics Reports* 363, 223.

Universal ratios of the universality classes of the Ising and tricritical Ising models in two-dimensions have been determined

Delfino, G. (2004). 'Integrable Field Theory and Critical Phenomena: The Ising Model in a Magnetic Field, *Journal of Physics A* 37, R 45.

Fioravanti, D., Mussardo, G. and Simon, P. (2001). 'Universal Amplitude Ratios of the Renormalization Group: Two-dimensional Tricritical Ising Model', *Physics Review* E63, 016103.

PROBLEMS

8.1. Logistic map

Consider the map

$$x_{n+1} = f_r(x_n) = r x_n(1 - x_n).$$

Introduced by Verhulst in 1845, the map models the growth of a population in a region of a finite area: the population at time $n+1$, expressed by x_{n+1}, is proportional to the population x_n at time n, but at the same time is also proportional to the remaining area. This quantity is decreased proportional to x_n, namely $(1 - x_n)$. Despite the innocent aspect, this map has a remarkable mathematical structure.

a. Prove that for $r < 1$ the map has a unique (stable) fixed point at the origin.
b. Prove that for $1 < r < 3$, there exist two fixed points, at $x = 0$ and $x = 1 - 1/r$, where the first is unstable and the second stable.
c. Prove that the second fixed point becomes unstable when $r > 3$. Show that for $3 < r < r_1$ there are two stable fixed points x_1 and x_2 of the map $f_r[f_r[x]] \equiv f_r^{(2)}(x)$, i.e. they are solutions of the equations $x_2 = f_r(x_1)$ and $x_1 = f_r(x_2)$.
d. Show that there is a value $r_2 > r_1$ when the two previous fixed points x_1 and x_2 become unstable. More generally, prove that there exists a sequence of values r_n such that for $r_{n-1} < r < r_n$ there is a set of 2^{n-1} points characterized by the conditions

$$f_r(x_i^*) = x_{i+1}^*, \quad f_r^{(2^{n-1})}(x_i^*) = x_i^*.$$

e. Define the family of functions $g_i(x)$ by the limit

$$g_i(x) = \lim_{n\to\infty} (-\alpha)^n f_{r_{n+i}}^{(2^n)}\left[\frac{x}{(-\alpha)^n}\right].$$

Show that they satisfy the functional equation

$$g_{i-1}(x) = (-\alpha) g_i\left[g_i\left(-\frac{x}{\alpha}\right)\right] \equiv T g_i(x).$$

Study the features of the function $g(x)$, defined as the 'fixed point' of the transformation law T

$$g(x) = T g(x) = -\alpha g\left[g\left(-\frac{x}{\alpha}\right)\right].$$

Prove, in particular, that α is a universal parameter.

8.2. Universal Ratios

Consider the mean field solution of the Ising model and, with the notations used in Chapter 1, compute the universal ratios

$$C_+ \chi_+ / M_0^2, \quad \chi_+ M_0^{\delta-1} \mathcal{M}_0^{-\delta}.$$

8.3. Approximated values of the critical exponents

Use the formulae (8.10.7) to obtain an approximate values of the three-dimensional Ising model and the spherical model. Compare with the numerical values obtained for the three-dimensional Ising model and with the exact expressions of the spherical model.

8.4. β-functions

Consider a statistical system with the space of the coupling constants described by the variables (x, y^2). The fixed point is identified by the origin $(0,0)$. Suppose that the β functions of these coupling constants are given by

$$\frac{dx}{db} = -y^2,$$
$$\frac{dy^2}{db} = -2xy^2.$$

Study the RG flows, with initial conditions (x_0, y_0^2), and show that they are hyperboles in the plane (x, y). Identify the nature of the coupling constants, i.e. if they are relevant, irrelevant or marginal.

9

Fermionic Formulation of the Ising Model

There are different kinds of scientists, as second- or third rank physicists who try their best but do not get too far. There are also first-class scientists, who make discoveries of a great importance. But then there is the genius. Majorana was one of them. He had what nobody else in the world has... unfortunately he was lacking the most natural quality, the simple common sense.

Enrico Fermi

9.1 Introduction

This chapter studies the continuum formulation of the two-dimensional Ising model, starting from the Hamiltonian limit of its transfer matrix. It firstly derives the quantum Hamiltonian of the model and then studies its most important properties, e.g. duality transformation. This symmetry involves the order and disorder operators and we clarify their physical interpretation. Afterwards, it discusses how to diagonalize the quantum Hamiltonian by means of particular fermionic fields. The operator mapping between the order/disorder operators and the fermionic fields is realized by the so-called *Wigner–Jordan* transformation: this brings the original Hamiltonian to a quadratic form in the creation and annihilation operators of the fermions. The determination of the spectrum is then obtained by a Bogoliubov transformation, a technique extremely useful also in other contexts, e.g. superconductivity phenomena. In the limit in which the lattice spacing goes to zero, the Ising model becomes a theory of free Majorana fermions. They satisfy a relativistic dispersion relation and their mass is a direct measurement of the displacement of the temperature from the critical value T_c.

It is important to stress that the fermionic formulation of the two-dimensional Ising model is crucial for the understanding of many of its physical properties and for the computation of its correlation functions. This formulation will be used in other parts of the book to illustrate several other aspects of this model. Given the importance of this subject, in the final section of this chapter we present another approach to show the fermionic content of this model and to derive the Dirac equation satisfied by the Majorana fermion.

Statistical Field Theory: An Introduction to Exactly Solved Models in Statistical Physics. Second Edition. Giuseppe Mussardo, Oxford University Press (2020). DOI: 10.1093/oso/9780198788102.001.0001

9.2 Transfer Matrix and Hamiltonian Limit

This section considers the transfer matrix on a square lattice with the standard orientation of the lattice. Consider a square lattice with $N = n^2$ spins, made of n rows and n columns. The lattice spacing along the vertical and horizontal directions are τ and α respectively. The spins, here denoted by $\sigma_{i,j}$, satisfy periodic boundary conditions

$$\sigma_{i+n,j} = \sigma_{i,j}, \quad \sigma_{i,j+n} = \sigma_{i,j}.$$

Below we will also use the notation σ_i and σ_i' to denote spins of next-neighbour rows, where the index i labels in this case the position of the spins along these rows. Denoting with μ_a ($a = 1,2,\ldots,n$) the set of all spins that belong to the row a

$$\mu_a = \{\sigma_1, \sigma_2, \ldots, \sigma_n\}_{a-row},$$

a configuration of the system is specified by the ensemble $\{\mu_1, \ldots \mu_n\}$. The a-th row interacts only with the next-neighbour rows, namely μ_{a-1} and μ_{a+1}. Let $E(\mu_a, \mu_{a+1})$ be the interaction energy between two next-neighbour rows and $E(\mu_a)$ the energy coming from the interactions of the spins placed on the ath row, eventually also subjected to an external magnetic field B. Assuming the usual Hamiltonian of the model, we have

$$\begin{aligned} E(\mu, \mu') &= -\mathcal{J}' \sum_{k=1}^{n} \sigma_k \sigma_k', \\ E(\mu) &= -\mathcal{J} \sum_{k=1}^{n} \sigma_k \sigma_{k+1} - B \sum_{k=1}^{n} \sigma_k, \end{aligned}$$

where $\mathcal{J}'$ and $\mathcal{J}$ are the couplings along the vertical and horizontal directions respectively. The total energy of a configuration of the system is then

$$E(\mu_1, \ldots, \mu_n) = \sum_{a=1}^{n} \left[E(\mu_a, \mu_{a+1}) + E(\mu_a) \right],$$

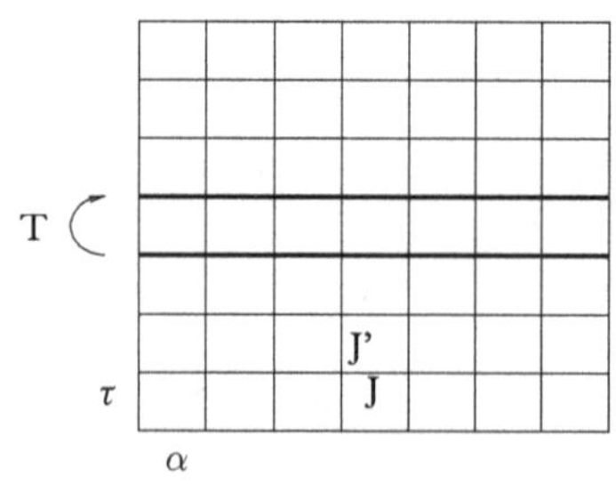

Fig. 9.1 *Lattice parameters and transfer matrix.*

and its partition function is given by

$$Z = \sum_{\mu_1} \sum_{\mu_2} \cdots \sum_{\mu_n} \exp\left[-\beta E(\mu_1, \ldots, \mu_n)\right]. \tag{9.2.1}$$

Let us introduce the transfer matrix T. It is a $2^n \times 2^n$ matrix, with elements given by

$$\langle \mu \mid T \mid \mu' \rangle = \exp\left[-\beta \left(E(\mu, \mu') + E(\mu)\right)\right]. \tag{9.2.2}$$

In terms of T, the partition function is expressed as

$$\begin{aligned} Z &= \sum_{\mu_1} \sum_{\mu_2} \cdots \sum_{\mu_n} \langle \mu_1 \mid T \mid \mu_2 \rangle \langle \mu_2 \mid P \mid \mu_3 \rangle \ldots \langle \mu_n \mid T \mid \mu_1 \rangle \\ &= \sum_{\mu_1} \langle \mu_1 \mid T^n \mid \mu_1 \rangle = \mathrm{Tr}\, T^n. \end{aligned} \tag{9.2.3}$$

The operator T can be further decomposed in terms of three operators

$$T = V_3 V_2 V_1,$$

where the V_i are $2^n \times 2^n$ matrices whose elements are given by

$$\langle \sigma_1 \ldots \sigma_n \mid V_1 \mid \sigma_1' \ldots \sigma_n' \rangle = \prod_{k=1}^{n} e^{L \sigma_k \sigma_k'}, \tag{9.2.4}$$

$$\langle \sigma_1 \ldots \sigma_n \mid V_2 \mid \sigma_1' \ldots \sigma_n' \rangle = \delta_{\sigma_1 \sigma_1'} \ldots \delta_{\sigma_n \sigma_n'} \prod_{k=1}^{n} e^{K \sigma_k \sigma_{k+1}}, \tag{9.2.5}$$

$$\langle \sigma_1 \ldots \sigma_n \mid V_1 \mid \sigma_1' \ldots \sigma_n' \rangle = \delta_{\sigma_1 \sigma_1'} \ldots \delta_{\sigma_n \sigma_n'} \prod_{k=1}^{n} e^{\beta B \sigma_k}. \tag{9.2.6}$$

In these formulae we have introduced the notations $K = \beta \mathcal{J}$ and $L = \beta \mathcal{J}'$. To have a more convenient expressions, let us introduce the operators

$$\tilde{\sigma}_1(a) = \mathbf{1} \times \mathbf{1} \times \ldots \times \overbrace{\sigma_1}^{a} \times \mathbf{1} \ldots \times \mathbf{1} \tag{9.2.7}$$

$$\tilde{\sigma}_2(a) = \mathbf{1} \times \mathbf{1} \times \ldots \times \overbrace{\sigma_2}^{a} \times \mathbf{1} \ldots \times \mathbf{1} \tag{9.2.8}$$

$$\tilde{\sigma}_3(a) = \mathbf{1} \times \mathbf{1} \times \ldots \times \overbrace{\sigma_3}^{a} \times \mathbf{1} \ldots \times \mathbf{1}. \tag{9.2.9}$$

They are defined by the direct product of 2×2 matrices, where the σ_i are the usual Pauli matrices, whereas $\mathbf{1}$ is the unity matrix. For $a \neq b$ it is easy to see that these operators commute each other

$$\left[\tilde{\sigma}_i(a), \tilde{\sigma}_j(b)\right] = 0. \tag{9.2.10}$$

When $a = b$, they satisfy instead the commutation and anti-commutation relations of the Pauli matrices

$$\left[\tilde{\sigma}_i(a), \tilde{\sigma}_j(a)\right] = 2i\epsilon_{ijk}\tilde{\sigma}_k(a), \tag{9.2.11}$$

$$\{\tilde{\sigma}_i(a), \tilde{\sigma}_j(a)\} = 2\delta_{ij}. \tag{9.2.12}$$

In terms of the $\tilde{\sigma}_i(a)$'s, the transfer matrix P can be put in an operatorial form as follows

$$T = \prod_{a=1}^{n} \left[e^{\beta B\tilde{\sigma}_3(a)}\, e^{K\tilde{\sigma}_3(a)\tilde{\sigma}_3(a+1)}\, e^{L\tilde{\sigma}_1(a)}\right]. \tag{9.2.13}$$

As discussed in Chapters 2 and 7, we can associate to a transfer matrix T of a classical statistical system in d dimensions a quantum Hamiltonian H in $(d-1)$ dimensions. In our case we have

$$T \equiv e^{-\tau H}, \tag{9.2.14}$$

where τ is the lattice spacing along the vertical direction of the lattice and H is the one-dimensional quantum Hamiltonian. To obtain the explicit expression of H it is necessary to use the Baker–Hausdorff formula for the exponential of two non-commuting operators

$$e^A e^B = e^{A+B+\frac{1}{2}[A,B]+\frac{1}{12}([A,B],B]+[A,[A,B]])+\cdots}.$$

Its expression, for a finite value of τ, is neither convenient nor particularly illuminating. To have a better insight, it is useful to consider the so-called Hamiltonian limit, i.e. the situation that arises when $\tau \to 0$. For simplicity, we will deal below only with the case when the magnetic field is absent, $B = 0$.

Matrix elements. In taking the Hamiltonian limit, we shall be careful that the physical content of the system does not change: from the RG analysis we know that this can be achieved rescaling appropriately the coupling constants (see Figure 9.2). To determine their dependence on the lattice spacing and, correspondingly, the expression of H that emerges in this limit, we can proceed as follows. For $\tau \to 0$, expanding the expression (9.2.14), we have

$$T \simeq 1 - \tau H. \tag{9.2.15}$$

Let us now consider explicitly some matrix elements of the operator T. If there are nonspin flips going from the a-row to the $(a+1)$-row, we have

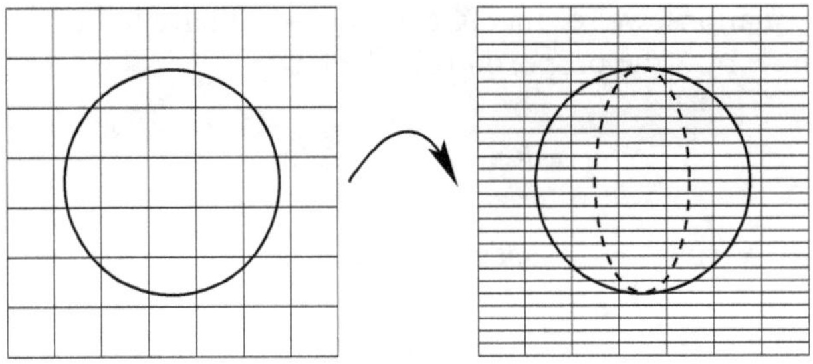

Fig. 9.2 *Hamiltonian limit. The circle in the lattice on the right is the set of points in which the correlation function $\langle \sigma(r)\sigma(0)\rangle$ is constant. If in the new lattice the coupling constants were not rescaled, the circle becomes an ellipse. Only an appropriate rescaling of the couplings leaves invariant the physical content of the original model.*

$$\begin{aligned} T(0\,\text{spin}-\text{flips}) &= \exp\left[K\sum_i \sigma_i\sigma_{i+1}\right] = 1 + K\sum_i \sigma_i\sigma_{i+1} + \cdots \\ &\simeq 1 - \tau H_{0\,\text{spin}-\text{flips}}. \end{aligned} \tag{9.2.16}$$

When there is only one spin flip in going from a row to the next one, we have

$$\begin{aligned} T(1\,\text{spin}-\text{flip}) &= \exp(-2L)\exp\left\{\frac{K}{2}\sum_i\left[\sigma_i\sigma_{i+1} + \sigma_i'\sigma_{i+1}'\right]\right\} \\ &\simeq -\tau H_{1\,\text{spin}-\text{flip}}, \end{aligned} \tag{9.2.17}$$

and, finally, when there are k spin flips the matrix element is

$$\begin{aligned} T(k\,\text{spin}-\text{flips}) &= \exp(-2kL)\exp\left\{\frac{K}{2}\sum_i\left[\sigma_i\sigma_{i+1} + \sigma_i'\sigma_{i+1}'\right]\right\} \\ &\simeq -\tau H_{k\,\text{spin}-\text{flips}}. \end{aligned} \tag{9.2.18}$$

From eqn. (9.2.16), we infer that

$$K \sim \tau, \tag{9.2.19}$$

while from eqns. (9.2.17) and (9.2.18)

$$\exp(-2L) \sim \tau. \tag{9.2.20}$$

From these two equations, we see that K and $\exp(-2L)$ have to be proportional to each other and we denote by λ the proportionality factor

$$K = \lambda \exp(-2L). \tag{9.2.21}$$

We can identity the vertical lattice spacing with

$$\tau = \exp(-2L) \tag{9.2.22}$$

and put the horizontal coupling constant of the spins equals to

$$K = \lambda \tau. \tag{9.2.23}$$

In summary, the physical content of the model does not change in the limit $\tau \to 0$ if we rescale the vertical coupling constant as in eqn. (9.2.22) and the horizontal one as in eqn. (9.2.23). These formulae show that, in the Hamiltonian limit, L grows very large while K becomes extremely small.

Critical value. The value $\lambda = 1$ identifies the critical point of the model. In fact, the scenario that emerges from the rescaling of the couplings can be easily derived from the discussion of Chapter 4, where we have seen that the critical line is given by

$$\sinh 2K \sinh 2L = 1. \tag{9.2.24}$$

Any pair of the coupling constants K and L that satisfies this equation corresponds to a critical situation of the Ising model, with an infinite value of the correlation length. If we consider the $L \to \infty$ asymptotic expression of this equation, when $\sinh 2L \to \frac{1}{2}\exp(2L)$ and $\sinh 2K \to 2K$, we get

$$K = \exp(-2L)$$

Comparing with eqn. (9.2.21), we see that $\lambda = 1$ identifies the critical point of the model. Concerning the other values of λ, as already shown in Chapter 4, $\lambda > 1$ identifies the ordered phase of the model while $\lambda < 1$ corresponds to the disordered phase. The phase diagram is shown in Figure 9.3. The parameter $1/\lambda$ provides a measurement of the displacement of the temperature from its critical value, as we will see in detail in the next section.

Once the lattice spacing τ has been identified with (9.2.22), we can proceed to derive the expression of the quantum Hamiltonian that emerges in the limit $\tau \to 0$

$$H = -\lim_{\tau \to 0} \frac{1}{\tau} \log T.$$

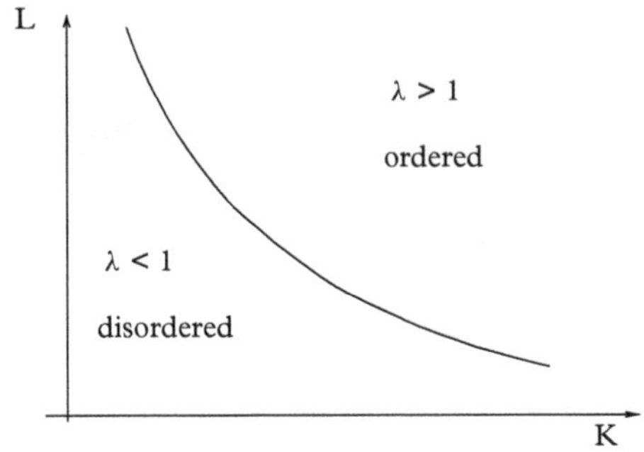

Fig. 9.3 *Phase diagram of the Ising model.*

The only processes that survive in this limit are those without a spin flip or those which induce only one spin flip, and correspondingly H is given by

$$H = -\sum_{a=1}^{n} [\tilde{\sigma}_1(a) + \lambda \tilde{\sigma}_3(a) \tilde{\sigma}_3(a+1)]. \tag{9.2.25}$$

In the Hamiltonian limit, the two dimensional classical Ising model is thus described by a simple one-dimensional quantum Hamiltonian. In the basis in which the operators $\tilde{\sigma}_3(a)$ are diagonal, the term responsable for their spin flips is the operator $\tilde{\sigma}_1(a)$.

9.3 Order and Disorder Operators

In the thermodynamic limit, the sum is extended on all sites between $-\infty$ and $+\infty$ and the Hamiltonian becomes

$$H = -\sum_{a=-\infty}^{\infty} [\tilde{\sigma}_1(a) + \lambda \tilde{\sigma}_3(a) \tilde{\sigma}_3(a+1)]. \tag{9.3.1}$$

To find its spectrum, let us first introduce the so-called *disorder operators*

$$\tilde{\mu}_3\left(r+\frac{1}{2}\right) = \prod_{\rho=-\infty}^{r} \tilde{\sigma}_1(\rho), \tag{9.3.2}$$

$$\tilde{\mu}_1\left(r+\frac{1}{2}\right) = \tilde{\sigma}_3(r)\tilde{\sigma}_3(r+1). \tag{9.3.3}$$

These operators are defined on the sites of the dual lattice, placed between two next-neighbour sites of the original lattice. From their definition, $\tilde{\mu}_1(r+1/2)$ is sensitive to the alignment of two next-neighbour spins. The other operator $\tilde{\mu}_3(r+1/2)$, acting on

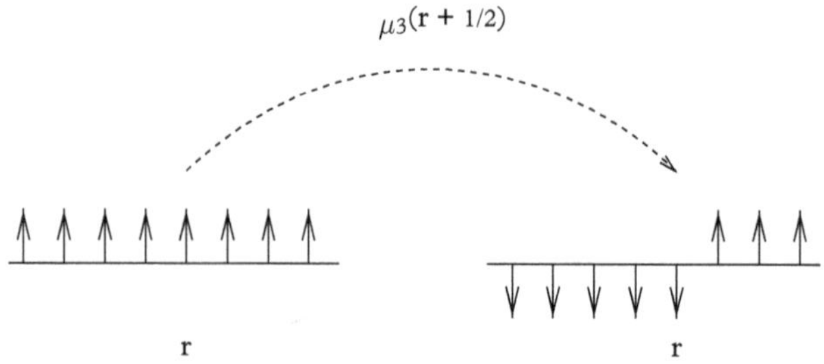

Fig. 9.4 *Action of $\tilde{\mu}_3$ on an ordered configuration of spins and creation of a kink state.*

the original spins of the lattice, makes a spin flip of all those placed on the left-hand side of the point r, as shown in Figure 9.4. Hence, starting from an ordered configuration of the spins, $\tilde{\mu}_3$ creates a *kink* excitation. This is a topological configuration that interpolates between the two states in which all spins are aligned either up or down, i.e. the ground states of the system. Since a kink changes the boundary conditions of the system, inspecting the values of the spins at the edge of the chain, one can easily infer whether there is an even or odd number of kinks in the system. It is also evident that the kink configurations tend to disorder the system and this justifies the terminology adopted for such an operator.

It is to check that the disorder operators $\tilde{\mu}_i$ satisfy the same algebra of the operators $\tilde{\sigma}_i$. Moreover, we have the algebraic relations

$$\begin{aligned}
&\tilde{\mu}_3^2 = \tilde{\mu}_1^2 = 1,\\
&\tilde{\mu}_3\left(r-\frac{1}{2}\right)\tilde{\mu}_3\left(r+\frac{1}{2}\right) = \tilde{\sigma}_1(r),\\
&\prod_{m<n}\tilde{\mu}_1\left(m+\frac{1}{2}\right) = \tilde{\sigma}_3(n+1),\\
&\left[\tilde{\mu}_1\left(r+\frac{1}{2}\right),\tilde{\mu}_3\left(r'+\frac{1}{2}\right)\right] = 2\delta_{r,r'},\\
&\left[\tilde{\mu}_3\left(r+\frac{1}{2}\right),\tilde{\mu}_3\left(r'+\frac{1}{2}\right)\right] = 0,\\
&\left[\tilde{\mu}_3\left(r+\frac{1}{2}\right),\tilde{\sigma}_1(r')\right] = 0.
\end{aligned} \tag{9.3.4}$$

We can now use the disorder operators to express the Hamiltonian (9.3.1) as

$$\begin{aligned}
H &= -\sum_r\left[\tilde{\mu}_3\left(r-\frac{1}{2}\right)\tilde{\mu}_3\left(r+\frac{1}{2}\right)+\lambda\,\mu_1\left(r+\frac{1}{2}\right)\right]\\
&= -\lambda\sum_r\left[\lambda^{-1}\tilde{\mu}_3\left(r-\frac{1}{2}\right)\tilde{\mu}_3\left(r+\frac{1}{2}\right)+\mu_1\left(r+\frac{1}{2}\right)\right].
\end{aligned} \tag{9.3.5}$$

This can be written as

$$H(\tilde{\sigma};\lambda) = \lambda H(\tilde{\mu};\lambda^{-1}). \tag{9.3.6}$$

Since the variables $\tilde{\sigma}_i$ and $\tilde{\mu}_i$ satisfy the same algebra, this expression simply expresses a symmetry of the system. This symmetry is nothing else but the Kramers-Wannier duality of the Ising model, that expresses the invariance of the model under the substitutions

$$\begin{aligned} \tilde{\mu}_1 &\leftrightarrow \tilde{\sigma}_1 \\ \tilde{\mu}_3 &\leftrightarrow \tilde{\sigma}_3 \\ \lambda &\leftrightarrow \lambda^{-1} \end{aligned} \tag{9.3.7}$$

Eqn. (9.3.6) implies that each eigenvalue H satisfies the functional equation

$$E(\lambda) = \lambda E(\lambda^{-1}). \tag{9.3.8}$$

Hence, there is a correspondence of the spectra for $\lambda > 1$ and $\lambda < 1$. Eqn. (9.3.8) leads to some important consequences, as the exact value of λ for which the quantum Hamiltonian is critical. To find this value, it is necessary to look for the vanishing of the mass gap, i.e. the difference between the two lowest eigenvalues. Denoting by $m(\lambda)$ the mass gap of the model, eqn. (9.3.8) implies that, if $m(\lambda_*) = 0$ at a given critical value λ_*, then $m(\lambda)$ must also vanish at λ_*^{-1}. Assuming that there is only one critical point, the two values above must coincide and therefore

$$\lambda_* = \lambda_*^{-1} \quad \rightarrow \quad \lambda_* = 1. \tag{9.3.9}$$

As we previously mentioned, the critical value is indeed $\lambda_* = 1$.

9.4 Perturbation Theory

The function $m(\lambda)$ can be explicitly found by using perturbation theory. By adding a constant, let us firstly write the Hamiltonian as

$$H = \sum_a \left[(1 - \tilde{\sigma}_1(a)) - \lambda \tilde{\sigma}_3(a)\tilde{\sigma}_3(a+1)\right]. \tag{9.4.1}$$

In the high temperature phase, λ is a small parameter and the Hamiltonian can be splitted as

$$H = H_0 + \lambda V,$$

where

$$H_0 = \sum_a [1 - \tilde{\sigma}_1(a)],$$
$$V = -\sum_a \tilde{\sigma}_3(a)\tilde{\sigma}_3(a+1).$$

To determine the first energy level in perturbation theory in λ, initially we have to identify the ground state of H_0 and its energy. It is easy to see that such a state has zero energy and it is characterized by the condition

$$\tilde{\sigma}_1(a)\,|\,0\,\rangle = |\,0\,\rangle, \quad \forall a. \tag{9.4.2}$$

In the basis in which $\tilde{\sigma}_3(a)$ is diagonal, the ground state is expressed by the tensor product of the vectors

$$|v_1(a)\rangle = \frac{1}{\sqrt{2}}\begin{pmatrix}1\\1\end{pmatrix},$$

each of them defined at the corresponding site of the lattice. The other eigenstate of $\tilde{\sigma}_1(a)$ (with eigenvalue -1) is expressed by the vector

$$|v_2(a)\rangle = \frac{1}{\sqrt{2}}\begin{pmatrix}1\\-1\end{pmatrix}.$$

Note that the operator $\tilde{\sigma}_3(a)$, that enters the perturbation V, maps one state to the other, $v_1(a) \leftrightarrow v_2(a)$.

With the ground state given by the tensor product $|\,0\,\rangle = \otimes_a |\,v_1(a)\,\rangle$, one can obtain an excited state by substituting, at an arbitrary point a of the system, consists of having, the vector $v_2(a)$ to $v_1(a)$. Since the localization of this vector is arbitrary, this energy level has a degeneracy equal to the number n of the lattice sites.[1] One can take care of this degeneracy by introducing states with a well defined quantum number of the lattice momentum. The state at zero momentum is obviously the only one invariant under translation and it is given by the linear combination

$$|\,-1\,\rangle = \frac{1}{\sqrt{n}}\sum_a \tilde{\sigma}_3(a)\,|\,0\,\rangle, \tag{9.4.3}$$

with $\langle -1\,|\,-1\,\rangle = 1$. The energy of this excited state can be computed perturbatively

$$E = \mathcal{E}_0 + \lambda\,\mathcal{E}_1 + \lambda^2\mathcal{E}_2 + \cdots, \tag{9.4.4}$$

[1] n provides here an infrared cut-off, to be sent to infinity in the thermodynamic limit.

where

$$\begin{aligned}
\mathcal{E}_1 &= \langle -1\,|V|-1\,\rangle, \\
\mathcal{E}_2 &= \langle -1\,|VgV|-1\,\rangle, \qquad (9.4.5)\\
\mathcal{E}_3 &= \langle\, -1\,|VgVgV|-1\,\rangle - \langle -1\,|V|-1\,\rangle\langle -1\,|Vg^2V|-1\,\rangle, \\
\cdots &= \cdots \qquad (9.4.6)
\end{aligned}$$

with the operator g defined by

$$g = \frac{1}{\mathcal{E}_0 - H}\,[1 - |-1\,\rangle\langle -1\,|].$$

It is easy to see that $\mathcal{E}_0 = 2$. For the next term we have

$$\begin{aligned}
\mathcal{E}_1 &= \langle -1\,|V|-1\,\rangle \qquad (9.4.7)\\
&= -\frac{1}{n}\sum_{a,a'}\langle 0\,|\tilde{\sigma}_3(a)\sum_b[\tilde{\sigma}_3(b)\,\tilde{\sigma}_3(b+1)]\;\tilde{\sigma}_3(a')\,|\,0\,\rangle.
\end{aligned}$$

There are only two terms that contribute to this expression: $a = b$ and $b+1 = a'$, or $a' = b$ and $b+1 = a$. Since $\tilde{\sigma}_3^2 = 1$, both terms give a factor n, that cancels the normalization factor. Hence $\mathcal{E}_1 = -2$.

If one carries on the computation to higher order (a task that we do not report here), there is a remarkable result: all of them are zero! In other words, the perturbative series truncates and coincides with its first two terms. For $\lambda < 1$, the exact mass gap is thus given by

$$m(\lambda) = 2(1-\lambda).$$

This expression explicitly confirms that it vanishes at $\lambda = 1$. We can then use the duality relation, $m(\lambda) = \lambda\, m(\lambda^{-1})$, to obtain the mass gap for $\lambda > 1$. So, for all values of λ, the mass gap is expressed by

$$m(\lambda) = 2\,|1-\lambda|. \qquad (9.4.8)$$

9.5 Expectation Values of Order and Disorder Operators

In the ordered phase of the Ising model, described by $\lambda > 1$, the Hamiltonian $H(\tilde{\sigma}\,;\lambda)$ with periodic boundary conditions has two possible vacuum states. The simplest way to obtain this result is to consider the limit $\lambda \to \infty$, in which the states with the minimum energy of the Hamiltonian (9.3.1) are the ones in which the expectation values of $\tilde{\sigma}_3(a)$

and $\tilde{\sigma}_3(a+1)$ coincide. Denoting by

$$|w_1(a)\rangle = \begin{pmatrix} 1 \\ 0 \end{pmatrix}, \quad |w_2(a)\rangle = \begin{pmatrix} 0 \\ 1 \end{pmatrix}$$

the two eigenvectors of $\tilde{\sigma}_3(a)$ at the site a (with eigenvalues ± 1), there are then *two* degenerate states of minimum energy, given by

$$|0^+\rangle_{\lambda=\infty} = \otimes_a |w_1(a)\rangle, \quad |0^-\rangle_{\lambda=\infty} = \otimes_a |w_2(a)\rangle. \tag{9.5.1}$$

The system will choose one of them, say $|0^+\rangle$, by the mechanics of spontaneously symmetry breaking: this can be induced by switching on a positive magnetic field B on all sites of the system and, once the spins are polarized in the direction of the field, switching B off.

For $\lambda > 1$ but finite, the corresponding vacuum states cannot be expressed by a simple expression as those given above. However, even in the absence of an explicit formula, let us denote by $|0\rangle_\lambda$ the vacuum state of $H(\tilde{\sigma};\lambda)$ after the spontaneously symmetry breaking (this is the corresponding of $|0^+\rangle$). On this state, the operator $\tilde{\sigma}_3(r)$ has a non vanishing expectation value

$${}_\lambda\langle 0|\sum_a \tilde{\sigma}_3(a)|0\rangle_\lambda \neq 0. \tag{9.5.2}$$

The self-duality of the model allows us to interpret this result in an interesting way. Consider, in fact, the following Hamiltonian

$$\lambda^{-1} H(\tilde{\sigma};\lambda) + B\sum_a \tilde{\sigma}_3(a). \tag{9.5.3}$$

Applying a duality transformation, we have

$$\lambda^{-1} H(\tilde{\sigma};\lambda) + B\sum_a \tilde{\sigma}_3(a) = H(\tilde{\mu};\lambda^{-1}) + B\sum_a \prod_{b<a} \tilde{\mu}_1(b+1/2).$$

Expanding at the first order in B, we obtain

$${}_\lambda\langle 0|\sum_a \tilde{\sigma}_3(a)|0\rangle_\lambda = {}_{\lambda^{-1}}\langle 0|\sum_a \prod_{b<a} \tilde{\mu}_1(b+1/2)|0\rangle_{\lambda^{-1}}, \tag{9.5.4}$$

where $|0\rangle_{\lambda^{-1}}$ is the vacuum state of $H(\tilde{\mu};\lambda^{-1})$. Since the operators $\tilde{\sigma}_1$ and $\tilde{\sigma}_3$ are equivalent to $\tilde{\mu}_1$ and $\tilde{\mu}_3$, eqn. (9.5.4) establishes that the operator $\tilde{\mu}_3(a+1/2) = \prod_{b<a} \tilde{\sigma}_1(m)$ has a non-zero expectation value on the high temperature ground state of the model. This operator creates a kink excitation in the system and therefore the high-temperature ground state can be regarded as a *kink condensate*. Notice that the energy of the kink state

is mostly localized around the point a and consequently the kink behaves as a localized particle. The kink state with the lowest energy is the one at zero momentum, which can be written as

$$|1\,\text{kink}\rangle = \frac{1}{\sqrt{n}}\sum_a \prod_{b<a} \tilde{\sigma}_1(b)\,|\,0\,\rangle_{\lambda^{-1}=0}. \tag{9.5.5}$$

From the duality relation, its mass coincides with the expression previously computed, $m(\lambda) = 2|1-\lambda|$.

For $\lambda < 1$, i.e. in the high-temperature phase, there is a similar argument: the operator $\tilde{\sigma}_3$ has a non-zero expectation value on the ground state of the system (that, in this phase, is unique) and consequently we have a vanishing expectation value of the disordered operator $\tilde{\mu}_3$ in the low-temperature phase.

9.6 Diagonalization of the Hamiltonian

The quantum Hamiltonian of the Ising model can be explicitly diagonalized by means of the Wigner-Jordan transformation. To simplify the following expression, it is convenient to start from the Hamiltonian

$$H = -\sum_a \left[\tilde{\sigma}_3(a) + \lambda\,\tilde{\sigma}_1(a)\,\tilde{\sigma}_1(a+1)\right] \tag{9.6.1}$$

that is unitarily equivalent to the previous one (9.3.1). Denoting the site of the lattice by $a = -n, -n+1, \ldots, n$ (we will consider lately the limit $n \to \infty$), let us introduce the operators

$$c(a) = \prod_{b=-n}^{n} e^{i\pi\,\tilde{\sigma}^+(b)\,\tilde{\sigma}^-(b)}\;\tilde{\sigma}^-(a), \tag{9.6.2}$$

$$c^\dagger(a) = \tilde{\sigma}^+(a)\prod_{b=-n}^{n} e^{\;i\pi\,\tilde{\sigma}^+(b)\,\tilde{\sigma}^-(b)}, \tag{9.6.3}$$

where

$$\begin{aligned}\tilde{\sigma}^+(a) &= \frac{1}{2}\left[\tilde{\sigma}_1(a) + i\tilde{\sigma}_2(a)\right],\\ \tilde{\sigma}^-(a) &= \frac{1}{2}\left[\tilde{\sigma}_1(a) - i\tilde{\sigma}_2(a)\right].\end{aligned} \tag{9.6.4}$$

It is easy to show that $c(a)$ and $c^\dagger(a)$ are fermionic operators, that satisfy the anti-commutation relations

$$\left\{c(a), c^\dagger(b)\right\} = \delta_{a,b}, \quad \{c(a), c(b)\} = 0. \tag{9.6.5}$$

We can now write the Hamiltonian (9.6.1) in a convenient form in terms of these new operators. As shown in Problem 9.2, we have in fact

$$\begin{aligned}\tilde{\sigma}_3(a) &= 2c^\dagger(a)c(a) - 1, \\ \tilde{\sigma}_1(a)\tilde{\sigma}_1(a+1) &= \left[c^\dagger(a) - c(a)\right]\left[c^\dagger(a+1) - c(a+1)\right].\end{aligned} \tag{9.6.6}$$

so that

$$H = -2\sum_a c^\dagger(a)c(a) - \lambda\sum_a\left[c^\dagger(a) - c(a)\right]\left[c^\dagger(a+1) - c(a+1)\right] \tag{9.6.7}$$

is a quadratic expression of the fermionic operators. To diagonalize it, the first step is to take the Fourier transform

$$c(a) = \frac{1}{\sqrt{2n+1}}\sum_k e^{-ika}\,c_k, \quad c^\dagger(a) = \frac{1}{\sqrt{2n+1}}\sum_k e^{ika}\,c_k^\dagger, \tag{9.6.8}$$

where c_k and $c_k^\dagger$ are the fermionic annihilation and creation operators in momentum space, and the momenta k take the discrete values

$$k = 0, \pm\frac{2\pi}{2n+1} \pm \frac{4\pi}{2n+1}, \ldots, \pm\frac{2\pi n}{2n+1}.$$

Substituting eqn. (9.6.8) into (9.6.1), we have

$$H = -2\sum_{k>0}(1+\lambda\cos k)\,(c_k^\dagger c_k + c_{-k}^\dagger c_{-k}) + 2i\lambda\sum_{k>0}\sin k\,(c_k^\dagger c_{-k}^\dagger + c_k c_{-k}). \tag{9.6.9}$$

This Hamiltonian is quadratic in c_k e $c_k^\dagger$ but this is still not sufficient to determine its spectrum. In fact, the ground state of H is not the state annihilated by the operators c_k, since in the Hamiltonian above is also present a term $(c_k^\dagger c_{-k}^\dagger + c_k c_{-k})$. We need then to re-write the Hamiltonian in a canonical form in terms of some opportune creation and annihilation fermionic operators $\eta_k^\dagger$ and η_k

$$H = \sum_k [\cdots]\,\eta_k^\dagger\eta_k, \tag{9.6.10}$$

Bogoliubov transformation. To achieve such a goal, it is necessary to make a Bogoliubov transformation: we introduce the operators

$$\begin{aligned}\eta_k &= U_k c_k + iV_k c_{-k}^\dagger, & \eta_{-k} &= U_k c_{-k} - iV_k c_k^\dagger \\ \eta_k^\dagger &= U_k c_k^\dagger - iV_k c_{-k}, & \eta_{-k}^\dagger &= U_k c_{-k}^\dagger + iV_k c_k,\end{aligned} \tag{9.6.11}$$

where, in all these formulae, $k > 0$. The real coefficients U_k and V_k of this transformation are determined by the following requests:

- the operators η_k and $\eta_k^\dagger$ must be fermionic operators.
- the Hamiltonian must be in a diagonal form in these new variables.

Imposing the validity of the relations

$$\{\eta_k, \eta_p^\dagger\} = \delta_{k,p}, \quad \{\eta_k, \eta_p\} = \left\{\eta_k^\dagger, \eta_p^\dagger\right\} = 0, \tag{9.6.12}$$

we arrive at the equation

$$U_k^2 + V_k^2 = 1. \tag{9.6.13}$$

This allows us to parameterize these functions as

$$U_k = \cos\theta_k, \quad V_k = \sin\theta_k. \tag{9.6.14}$$

Inverting now eqn. (9.6.11)

$$\begin{aligned} c_k &= U_k \eta_k - iV_k \eta_{-k}^\dagger, \qquad & c_{-k} &= U_k \eta_{-k} + iV_k \eta_k^\dagger \\ c_k^\dagger &= U_k \eta_k^\dagger + iV_k \eta_{-k}, \qquad & c_{-k}^\dagger &= U_k \eta_{-k}^\dagger - iV_k \eta_k, \end{aligned} \tag{9.6.15}$$

and inserting these formulae into the Hamiltonian (9.6.9), we have

$$\begin{aligned} H = &\sum_{k>0} \left[-2(1+\lambda\cos k)(U_k^2 - V_k^2) + 4\lambda \sin k\, U_k V_k\right](\eta_k^\dagger \eta_k + \eta_{-k}^\dagger \eta_{-k}) \\ &+ \sum_{k>0} \left[4i(1+\lambda\cos k)\, U_k V_k + 2i\lambda \sin k\,(U_k^2 - V_k^2)\right](\eta_k^\dagger \eta_{-k}^\dagger + \eta_k \eta_{-k}). \end{aligned} \tag{9.6.16}$$

In order to bring the Hamiltonian in the form (9.6.10), we need to impose

$$4(1+\lambda\cos k)\, U_k V_k + 2\lambda \sin k\,(U_k^2 - V_k^2) = 0. \tag{9.6.17}$$

Using (9.6.14), we have

$$2U_k V_k = \sin 2\theta_k, \quad U_k^2 - V_k^2 = \cos 2\theta_k,$$

and eqn. (9.6.17) becomes

$$4(1+\lambda\cos k)\sin 2\theta_k + 2\lambda\, \sin k \cos 2\theta_k = 0. \tag{9.6.18}$$

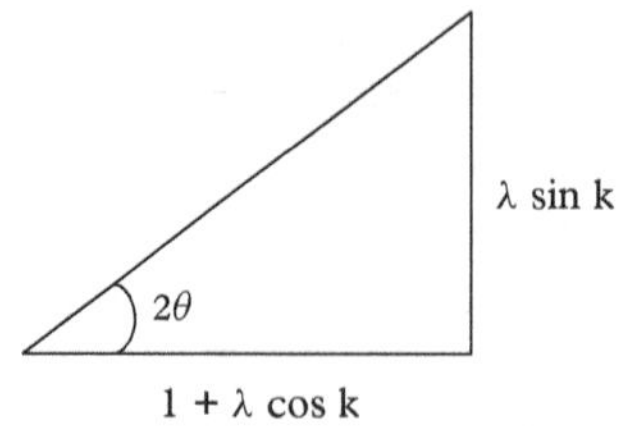

Fig. 9.5 *Relation between the parameters of the Bogoliubov transformation.*

We have then the following equation for the angle θ_k

$$\tan 2\theta_k = -\frac{\lambda \sin k}{1+\lambda \cos k}. \tag{9.6.19}$$

The geometric interpretation of this equation is given in Figure 9.5; taking into account the $(-)$ sign, its solution is expressed by

$$\begin{aligned} \sin 2\theta_k &= \frac{\lambda \sin k}{\sqrt{1+2\lambda \cos k+\lambda^2}}, \\ \cos 2\theta_k &= -\frac{1+\lambda \cos k}{\sqrt{1+2\lambda \cos k+\lambda^2}}. \end{aligned} \tag{9.6.20}$$

Spectrum of the Hamiltonian. With this determination of U_k and V_k, coming back to eqn. (9.6.16), we have

$$H = 2\sum_k \Lambda_k\, \eta_k^\dagger \eta_k + \text{costante} \tag{9.6.21}$$

where

$$\Lambda_k = \sqrt{1+2\lambda \cos k+\lambda^2}. \tag{9.6.22}$$

The plot of this function is given in Figure 9.6. Its minimum is at $k = \pm\pi$, with a value

$$\Lambda_{\pm\pi} = 2|1-\lambda|$$

that is in agreement with the perturbative analysis done in the previous section.

In taking the continuum limit, it is convenient to restore the lattice spacing α and measure the momentum with respect to its minimum value, i.e.

$$k = \pi + k'\alpha.$$

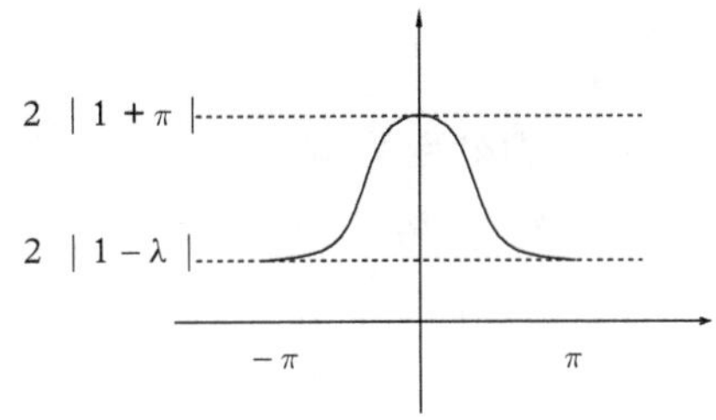

Fig. 9.6 *Dispersion relation of the fermionic particle.*

Let us also define the energy with the correct physical dimension

$$E(k') = \frac{2\Lambda_k}{\alpha}.$$

In the continuum limit $\alpha \to 0$, we shall take $k \to \pi$ in order to have a finite value of the momentum, and therefore

$$E(k') = \sqrt{\frac{4(1-\lambda)^2}{\alpha^2} + 4\lambda k'^2}. \tag{9.6.23}$$

We arrive in this way to a relativistic dispersion relation. If λ is sufficiently close to the critical value, $\lambda \simeq 1$, we have the dispersion relation of a fermionic particle with mass

$$m = \frac{2(1-\lambda)}{\alpha}. \tag{9.6.24}$$

At $\lambda = 1$, it becomes the dispersion relation of a massless particle

$$E(k') \simeq v|k'|, \tag{9.6.25}$$

with $v = 2\sqrt{\lambda}$. In terms of the fields

$$\eta(a) = \frac{1}{\sqrt{2n+1}} \sum_k e^{ika}\, \eta_k, \quad \eta^\dagger(a) = \frac{1}{\sqrt{2n+1}} \sum_k e^{-ika}\, \eta_k^\dagger$$

we can define the new fermionic fields

$$\psi_1(a) = \frac{1}{2}(\eta(a) + \eta^\dagger(a)), \quad \psi_2(a) = \frac{1}{2i}(\eta(a) - \eta^\dagger(a)). \tag{9.6.26}$$

They satisfy the relations

$$\begin{aligned} \psi_i^\dagger(a) &= \psi_i(a), \\ \{\psi_i(a), \psi_j(b)\} &= \delta_{ij}\,\delta_{ab}, \\ \psi_i^2(a) &= \frac{1}{2}, \end{aligned} \tag{9.6.27}$$

Hence they are neutral fermionic fields, also known in the literature as *Majorana fermions.*

9.7 Dirac Equation

In this section we present another way to show the fermionic content of the two-dimensional Ising model. Note that, using eqn. (9.3.1), we can determine the equation of motion of the operator $\tilde{\sigma}_3(r)$

$$\frac{\partial}{\partial \tau}\tilde{\sigma}_3(r) = [H, \tilde{\sigma}_3(r)] = -i\tilde{\sigma}_2(r) = \tilde{\sigma}_1(r)\tilde{\sigma}_3(r). \tag{9.7.1}$$

Using the dual expression (9.3.5) of the Hamiltonian, we can also derive the equation of motion of the operator $\tilde{\mu}_3(r+1/2)$

$$\begin{aligned} \frac{\partial}{\partial \tau}\tilde{\mu}_3\left(r+\frac{1}{2}\right) &= \left[H, \tilde{\mu}_3\left(r+\frac{1}{2}\right)\right] = -i\lambda\,\tilde{\mu}_2\left(r+\frac{1}{2}\right) \\ &= \lambda\,\tilde{\mu}_1\left(r+\frac{1}{2}\right)\tilde{\mu}_3\left(r+\frac{1}{2}\right) = \lambda\,\tilde{\sigma}_3(r)\,\tilde{\sigma}_3(r+1)\,\tilde{\mu}_3\left(r+\frac{1}{2}\right). \end{aligned} \tag{9.7.2}$$

These equations of motion are non-linear and difficult to solve. However, they can be put in a linear form by defining

$$\psi_1(r) = \tilde{\sigma}_3(r)\,\tilde{\mu}_3\left(r+\frac{1}{2}\right), \tag{9.7.3}$$

$$\psi_2(r) = \tilde{\sigma}_3(r)\,\tilde{\mu}_3\left(r-\frac{1}{2}\right). \tag{9.7.4}$$

Using the algebraic properties of the variables $\tilde{\sigma}_i$ and $\tilde{\mu}_i$, the equation of motion for these new variables can be written as

$$\frac{\partial \psi_1(r)}{\partial \tau} = -\psi_2(r) + \lambda\,\psi_2(r+1), \tag{9.7.5}$$

$$\frac{\partial \psi_2(r)}{\partial \tau} = -\psi_1(r) + \lambda\,\psi_1(r-1). \tag{9.7.6}$$

Restoring the lattice spacing α with the substitution $(r \pm 1) \to (r \pm \alpha)$ and going to the continuum limit $\alpha \to 0$, we have

$$\frac{\partial \psi_1(r)}{\partial \tau} = -(1-\lambda)\,\psi_2(r) + \lambda\,\frac{\partial \psi_2(r)}{\partial r}\alpha, \tag{9.7.7}$$

$$\frac{\partial \psi_2(r)}{\partial \tau} = -(1-\lambda)\,\psi_1(r) - \lambda\,\frac{\partial \psi_1(r)}{\partial r}\alpha. \tag{9.7.8}$$

The two fields $\psi_1(r)$ and $\psi_2(r)$ can be organized in a spinorial field

$$\psi(r) = \begin{pmatrix} \psi_1(r) \\ \psi_2(r) \end{pmatrix}, \tag{9.7.9}$$

with anti-commutation relations

$$\{\psi_1(r), \psi_1(r')\} = 2\delta_{r,r'} \tag{9.7.10}$$

$$\{\psi_2(r), \psi_2(r')\} = 2\delta_{r,r'}. \tag{9.7.11}$$

A compact expression of the equation of motion is then given by

$$\left(\gamma^0 \frac{\partial}{\partial t} + \gamma^3 \frac{\partial}{\partial r} + m\right)\psi = 0, \tag{9.7.12}$$

with $t = \alpha\tau$, $m = (1-\lambda)/\alpha$ and with the Euclidean γ matrices given by

$$\gamma^0 = \begin{pmatrix} 0 & 1 \\ 1 & 0 \end{pmatrix}, \quad \gamma^3 = \begin{pmatrix} 1 & 0 \\ 0 & -1 \end{pmatrix}.$$

We arrive in this way to a Dirac equation for a free fermionic neutral field, i.e. a Majorana fermion.

Note that the equation of motion can be derived from the Euclidean action

$$\mathcal{S} = \int d^2x\,\bar{\psi}\,(\gamma^\mu \partial_\mu + m)\,\psi, \tag{9.7.13}$$

where $\bar{\psi} \equiv \psi\,\gamma^0$. For reasons that will become more clear later, it is convenient to introduce the complex coordinates $z = x + it$ e $\bar{z} = x - it$, with $\partial_z = \frac{1}{2}(\partial_x - i\partial_t)$ e $\partial_{\bar{z}} = \frac{1}{2}(\partial_x + i\partial_t)$, and define two new fermionic components by

$$\Psi(z,\bar{z}) = \frac{\psi_1 + i\psi_2}{\sqrt{2}}, \quad \bar{\Psi}(z,\bar{z}) = \frac{\psi_1 - i\psi_2}{\sqrt{2}}. \tag{9.7.14}$$

In these new variables, the action becomes

$$S = \int d^2z \left[\Psi\, \partial_{\bar{z}} \Psi + \bar{\Psi}\, \partial_z \bar{\Psi} + im\, \bar{\Psi}\, \Psi\right]. \tag{9.7.15}$$

with the equation of motion given by

$$\partial_{\bar{z}} \Psi = \frac{im}{2} \bar{\Psi}, \quad \partial_z \bar{\Psi} = -\frac{im}{2} \Psi. \tag{9.7.16}$$

When the mass of the fermion field vanishes, Ψ becomes a purely analytic field while $\bar{\Psi}$ a purely anti-analytic one. The duality of the Ising model is expressed by the invariance of this fermionic theory under the transformations

$$\begin{aligned} m &\to -m \\ \Psi &\to \Psi \\ \bar{\Psi} &\to -\bar{\Psi}. \end{aligned} \tag{9.7.17}$$

As we prove in Chapter 14), in the continuum limit of the model the order and disorder operators satisfy the operator relation

$$\sigma(z,\bar{z})\, \mu(z',\bar{z}') = \frac{1}{\sqrt{2}|z-z'|^{1/4}} \left[\omega\,(z-z')^{1/2}\, \Psi(z',\bar{z}') + \bar{\omega}\,(\bar{z}-\bar{z}')^{1/2}\, \bar{\Psi}(z',\bar{z}')\right], \tag{9.7.18}$$

when $|z-z'| \to 0$, with $\omega = \exp(i\pi/4)$, $\bar{\omega} = \exp(-i\pi/4)$.

The interpretation of the fermionic field in terms of particle excitations is obtained by making an analytic continuation to the Minkowski space. In two dimensions it is convenient to parameterize the relativistic dispersion relations in terms of the rapidity θ as follows: $E = m\cosh\theta$ and $p = m\sinh\theta$. The mode expansion of the fermionic field becomes

$$\Psi(x,t) = \int \frac{d\theta}{2\pi} \left[\omega e^{\frac{\theta}{2}} A(\theta)\, e^{-im(t\cosh\theta - x\sinh\theta)} + \bar{\omega} e^{\frac{\theta}{2}} A^\dagger(\theta)\, e^{im(t\cosh\theta - x\sinh\theta)}\right] \tag{9.7.19}$$

$$\bar{\Psi}(x,t) = \int \frac{d\theta}{2\pi} \left[\bar{\omega} e^{-\frac{\theta}{2}} A(\theta)\, e^{-im(t\cosh\theta - x\sinh\theta)} + \omega e^{-\frac{\theta}{2}} A^\dagger(\theta)\, e^{im(t\cosh\theta - x\sinh\theta)}\right],$$

$A(\beta)$ and $A^\dagger(\beta)$ are, respectively, the annihilitation and creation operators of a neutral fermionic particle. They satisfy the anti-commutation relations

$$\left\{A(\theta), A^\dagger(\theta)\right\} = 2\pi\,\delta(\theta - \theta). \tag{9.7.20}$$

Using this mode expansion and the anti-commutation relations it is easy to compute the correlation functions of the fermionic field (see Problem 9.4).

REFERENCES

The important role of the order and disorder operators is discussed in

Fradkin, E. and Susskind, L. (1978). 'Order and Disorder in Gauge Systems and Magnets', *Physics Review D*, 17: 2637.

Kadanoff, L.P. and Ceva, H. (1971). 'Determination of an Operator Algebra for the Two-Dimensional Ising Model', *Physics Review B*, 3: 3918.

For the fermionic formulation of the Ising model we recommend

Bander, M. and Itzykson, C. (1977). 'Quantum Field Theory Calculation of the Two-dimensional Ising Model Correlation Function', *Physics Review D*, 15: 463.

Lieb, E.H., Schultz, T.D. and Mattis, D.C. (1961). 'Two Soluble Models of Anti-ferromagnetic Chain', *Annals of Physics*, 16: 407.

Pfeuty, P. (1970). 'The One-dimensional Ising Model in a Transverse Field', *Annals of Physics*, 57: 79.

Schultz, T.D., Mattis, D.C. and Lieb, E.H. (1964). 'Two-Dimensional Ising Model as a Soluble Problem of Many Fermions', *Reviews of Modern Physics*, 36: 856.

Zuber, J.B. and Itzykson, C. (1977). 'Quantum Field Theory and the Two-dimensional Ising Model', *Physics Review D*, 15: 2875.

For discussion on extending the fermionic formulation to the three-dimensional Ising model see

Itzykson, C. (1982). 'Ising Fermions (II). Three Dimensions', *Nuclear Physics B*, 210: 477.

PROBLEMS

9.1. Anti-commutation relations

Prove that the operators $c(a)$ and $c^{\dagger}(b)$ satisfy the anti-commutation relations

$$\left\{c(a), c^{\dagger}\right\} = \delta_{a,b}.$$

9.2. Fermion identities

Prove that

$$c(a)\,c^{\dagger}(a+1) = -\tilde{\sigma}^{-}(a)\,\tilde{\sigma}^{+}(a+1)$$
$$c^{\dagger}(a)\,c^{\dagger}(a+1) = \tilde{\sigma}^{+}(a)\,\tilde{\sigma}^{+}(a+1)$$
$$c(a)\,c(a+1) = -\tilde{\sigma}^{-}(a)\,\tilde{\sigma}^{-}(a+1).$$

Moreover, show that the order operators of the Ising model can be expressed in terms of the fermion operators $c(a)$ and $c^{\dagger}(a)$ as

$$\tilde{\sigma}_3(a) = 2c^\dagger(a)c(a) - 1,$$
$$\tilde{\sigma}_1(a)\tilde{\sigma}_1(a+1) = \left[c^\dagger(a) - c(a)\right]\left[c^\dagger(a+1) - c(a+1)\right].$$

9.3. Duality

Show that the Dirac equation (9.7.12) with $m > 0$ is linked by a unitary transformation to the one $m < 0$. The map between the two Hamiltonians establishes the duality relation of the Ising model in its fermionic formulation.

9.4. Correlation functions

Use the mode expansion of the fermion field, given in eqn. (9.7.19), and the anti-commutation relations of the operators $A(\theta)$ and $A^\dagger(\theta)$ to compute the correlation functions

$$G_1(x,t) = \langle 0|\Psi(x,t)\Psi(0,0)|0\rangle$$
$$G_2(x,t) = \langle 0|\bar{\Psi}(x,t)\Psi(0,0)|0\rangle.$$

9.5. XXZ model

Consider the quantum spin chain of the XXZ model with Hamiltonian

$$H = \mathcal{J}_1 \sum_{k=1}^{N} (S_k^x S_{k+1}^x + S_k^y S_{k+1}^y) + \mathcal{J}_2 \sum_k S_k^z S_{k+1}^z.$$

Assume periodic boundary condition. The spin operator $\vec{S}$ is in the $1/2$ representation and given by $\vec{S} = 1/2\vec{\sigma}$, where $\vec{\sigma}$ is the set of Pauli matrices.

a. Show that the sign of $\mathcal{J}_1$ in H can be chosen freely without changing the physical properties of the model. Show that the same is not true for the sign of $\mathcal{J}_2$.
b. Discuss the symmetry of the system when $\mathcal{J}_2/\mathcal{J}_1 \to 0$ and $\mathcal{J}_2/\mathcal{J}_1 \to \infty$.
c. Use the fermionic representation of the operators $S_k^\pm = S_k^x \pm iS_k^x$ and S_k^3 to write the Hamiltonian as

$$H = \frac{\mathcal{J}_1}{2} \sum_{a=1}^{N} \left[c^\dagger(a)c(a+1) + c^\dagger(a+1)c(a)\right]$$
$$+\mathcal{J}_2 \sum_{a=1}^{N} \left(c^\dagger(a)c(a) - \frac{1}{2}\right)\left(c^\dagger(a+1)c(a+1) - \frac{1}{2}\right).$$

d. Consider the case $\mathcal{J}_2 = 0$, the so-called XY model. Use the Bogoliubov transformation to find in this case the spectrum of the fermionic form of the Hamiltonian.

10

Conformal Field Theory

A physical law must possess mathematical beauty.
P.A.M. Dirac

10.1 Introduction

The previous chapters showed that, coming close to a critical point, the correlation length of a statistical system diverges and consequently there are fluctuations on all possible scales. In such a regime, the properties of the statistical systems can be efficiently described by a QFT. Right at the critical point, the correlation length is infinite: the corresponding field theory is therefore massless and becomes invariant under a dilation of the length scales

$$x_a \to \lambda x_a.$$

Under this transformation the fields Φ_i associated to the order parameters transform as

$$\Phi_i \to \lambda^{d_i} \Phi_i,$$

where d_i here denote their anomalous dimensions. Finding the spectrum of the anomalous dimensions is a central problem of the theory: in fact, from Chapter 8 we know that they determine the critical exponents of the various thermodynamical quantities. The singularities of these thermodynamical functions are associated to the fixed points of the RG. In turns, in the vicinity of the fixed points there is a surface of instability that is spanned by the relevant operators present at the fixed points. In order to determine all fixed points and the spectrum of the operators nearby, Polyakov has put forward the hypothesis that the critical fluctuations are invariant under the set of conformal transformations. These transformations have the distinguished property of stretching locally the lengths of the vectors but leaving their relative angles invariant. It is important to stress that in systems with local interactions that are invariant under translations and rotations, the invariance under a global dilation automatically implies an invariance under the conformal transformations. From this point of view, the classification of the fixed points of the RG is equivalent to find all possible QFTs with conformal symmetry.

It is worth to emphasize that the construction of such theories is based on an approach that is completely different from the lagrangian formalism that is usually used in QFT

Statistical Field Theory: An Introduction to Exactly Solved Models in Statistical Physics. Second Edition. Giuseppe Mussardo, Oxford University Press (2020). © Giuseppe Mussardo. DOI: 10.1093/oso/9780198788102.001.0001

discussed in Chapter 7. The approach presented here is based on the algebra of local fields. We assume first of all the existence of a basis of local operators, that include among others the order parameters. Secondly, we suggest that any other quantity, such as products of order parameters, can be expanded in terms of the local operators of the basis. In this way, one is naturally led to introduce the concept of the operator product expansion (OPE) and its corresponding algebra.

This chapter studies the general properties of these conformal invariant theories, pointing out the peculiar aspects that arise in two dimensions. Chapter 11 is devoted to the analysis of an important class of bi-dimensional CFTs, the so called *minimal models*: their remarkable property is to have an operatorial algebra that closes within a finite number of conformal families. Other aspects of the two-dimensional conformal theories will be the subject of subsequent chapters.

10.2 The Algebra of Local Fields

The main goal of a QFT is the determination of the correlation functions

$$G(x_1, x_2, \ldots, x_n) = \langle A_1(x_1)A_2(x_2)\ldots A_n(x_n)\rangle,$$

where $A_i(x_i)$ are regular local functions that, for simplicity, we will assume are constructed in terms of only one fundamental field $\varphi(x)$. Below we consider the Euclidean formulation of the theory and our definition of the vacuum expectation values is provided by the functional integral, with Boltzmann weight given by the action $S[\varphi]$ of the field φ

$$G(x_1, \ldots, x_n) \equiv \frac{1}{Z}\int \mathcal{D}\varphi\, A_1(x_1)\ldots A_n(x_n)\, e^{-S[\varphi]}, \qquad (10.2.1)$$

The correct normalization is ensured by the partition function Z

$$Z = \int \mathcal{D}\varphi\, e^{-S[\varphi]}.$$

In order to clarify the important concept of the algebra of the local field, it is useful to initially consider the free bosonic field $\phi(x)$. In this case, using $\phi(x)$ and the normal ordered product of its powers,[1] we can define the local scalar densities

$$\phi(x), :\phi^2(x):, :\phi^3(x):, \ldots, :\varphi^n(x): \qquad (10.2.2)$$

[1] As normal ordered product we mean here the subtraction of the divergent term coming from the propagator of the product of two operators, i.e. $:\phi^2(x) := \lim_{y\to x}\phi(x)\phi(y) - \langle 0|\phi(x)\phi(y)|0\rangle$. All other normal ordered products can be iteratively constructed starting from this relation.

In a Euclidean D dimensional space-time, the scalar field $\phi(x)$ has dimension equal to $d_\varphi = (D-2)/2$ (in mass unity) and the dimensions of the composite fields (10.2.2) are given by[2]

$$d_\phi,\, 2d_\phi,\, 3d_\phi, \ldots,\, nd_\phi \tag{10.2.3}$$

as easily seen by applying the Wick theorem. For instance

$$\langle :\phi^2(x)::\phi^2(y):\rangle = 2\,[\langle\phi(x)\phi(y)\rangle]^2$$

and the singularity $1/r^{2d_\phi}$ of the correlator $\langle\phi(x)\phi(y)\rangle$ for $r = |x-y| \to 0$, gives rise to

$$\langle :\phi^2(x)::\phi^2(y):\rangle \sim \frac{1}{r^{4d_\phi}}.$$

An analogous result holds for the composite operators $:\phi^n(x):$, so that we arrive at the sequence of dimensions (10.2.3).

The set of fields $\{:\phi^n(x):\}$, to which we have to add the identity operator $\mathbf{I} \equiv \phi^0$, can be used to express any other regular density $A(x)$ of the free bosonic field. This is done by the series expansion of $A(x)$

$$A(x) = \sum_{n=0}^{\infty} a_n :\phi^n(x):. \tag{10.2.4}$$

Basis of local operators. As in the free case, we make the hypothesis that there is a similar set of fields also in the interacting theories. Namely, we assume the existence of a numerable set of fields[3] $\varphi_i(x)$, that are eigenvectors of the dilation operator, whose dimensions are defined by the condition

$$\varphi_i(x) = \lambda^{d_i}\,\varphi_i(\lambda x). \tag{10.2.5}$$

Moreover, we assume that any other operator can be expressed as linear combination of the fields above

$$A(x) = \sum_{n=0}^{\infty} a_n \varphi_n(x). \tag{10.2.6}$$

[2] In the following discussion we assume $D \neq 2$. The $D = 2$ case will be discussed in details later.
[3] For simplicity we consider below only scalar fields. Quantities with spin will be considered later.

At the critical point, the propagator of these operators is

$$\mathcal{D}_{ij}(x_1 - x_2) \equiv \langle \varphi_i(x_1)\varphi_j(x_2)\rangle = \frac{\mathcal{A}_{ij}}{|\, x_1 - x_2\,|^{d_i+d_j}}, \tag{10.2.7}$$

where the numerical constant $\mathcal{A}_{ij}$ expresses their normalization.

In an interactive theory, however, the dimensions d_i are not expressed in a simple way as in the free theory. As a matter of fact, one of the fundamental problem is the determination of their values.

It is worth saying that the validity of the operatorial identity (10.2.6), as well as other identities of similar nature that we will meet later on, has to be understood in a weak sense: this means that it can be used straighforwardly in expressions that involve correlation functions but it can give rise to inconsistencies if interpreted strictly as an operatorial identity (see Problem 10.1). So, for instance, to calculate the correlation function $\langle A(x)B(y)\ldots C(z)\rangle$, we can use the expansion (10.2.6) to express any of these fields – say $A(x)$ – arriving in this way to the expression

$$\langle A(x)B(y)\ldots C(z)\rangle = \sum_{n=0}^{\infty} a_n \,\langle \varphi_n(x)B(y)\ldots C(z)\rangle.$$

In the vicinity of the critical point, the theory has a mass scale m that is a small quantity (it is related to the correlation length ξ by the relation $\xi = m^{-1}$), and the vacuum expectation values of the fields φ_n can be expressed as

$$\langle \varphi_n(x)\rangle = \mu_n\, m^{d_n}, \tag{10.2.8}$$

where μ_n are dimensionless quantities. At the critical point $m \to 0$ and all fields $\varphi_n(x)$ $(n = 1,2,\ldots)$ but the identity operator, have zero expectation value[4]

$$\langle \varphi_n(x)\rangle = 0, \quad n = 1,2,\ldots \tag{10.2.9}$$

Operator algebra. So far we have discussed the operatorial expressions that refer to a given point. Consider now the product of two fluctuating fields $A(x_1)B(x_2)$ placed at two distinct points. If their separation is much less than the correlation length ξ of the system, in particular if we consider the limit $x_1 \to x_2$, it is a natural hypothesis that the properties of this composite operator become those of a local operator, so that it can be expanded in the same basis $\{\varphi_n\}$ given above

[4] This result is obvious if $d_n > 0$. We will see later that, for the conformal invariance of the theories of the fixed points, the same conclusion also holds when there are fields with negative dimensions.

$$A(x_1)B(x_2) = \sum_{k=0}^{\infty} \beta(x_1, x_2)\, \varphi_k(x_2), \tag{10.2.10}$$

where the coefficients $\beta(x_1, x_2)$ contain the dependence on the coordinates x_1 and x_2. If we specialize this relation to the case in which both $A(x_1)$ and $B(x_2)$ are themselves members of the basis, we arrive at the operator algebra

$$\varphi_p(x_1)\varphi_q(x_2) = \sum_{r=0}^{\infty} C_{pq}^{r}(x_1, x_2)\, \varphi_r(x_2). \tag{10.2.11}$$

The function $C_{pq}^{r}(x_1, x_2)$ can be further constrained. If the system is invariant under translations, it can only depends on the separation $| x_1 - x_2 |$ of the two points. At the critical point, the system is also invariant under the scale transformation $x \to \lambda x$ and, for the transformation law of the fields φ_n, it is easy to see that $C_{pq}^{r}(| x_1 - x_2 |)$ is a homogeneous function of degree $d_r - d_p - d_q$. Hence, it can be written as

$$C_{pq}^{r}(x_1, x_2) = c_{pq}^{r} \frac{1}{| x_1 - x_2 |^{d_p + d_q - d_r}}, \tag{10.2.12}$$

where c_{pq}^{r} are pure numbers, known as the *structure constants of the operator algebra.*

To summarize, the hypothesis of an algebra of the local fields consists of: (i) the existence of a basis made of scaling operators $\varphi_i(x)$ with dimensions d_i; (ii) the validity of the operatorial algebra

$$\varphi_p(x_1)\varphi_q(x_2) = \sum_{r=0}^{\infty} c_{pq}^{r} \frac{1}{| x_1 - x_2 |^{d_p + d_q - d_r}}\, \varphi_r(x_2), \tag{10.2.13}$$

that holds at the critical point of the theory. The solution of the field theories that describe the fixed points of the RG consists of finding the spectrum of the dimensions d_i of the scaling fields, together with the set of the structure constants c_{pq}^{r}: once all these

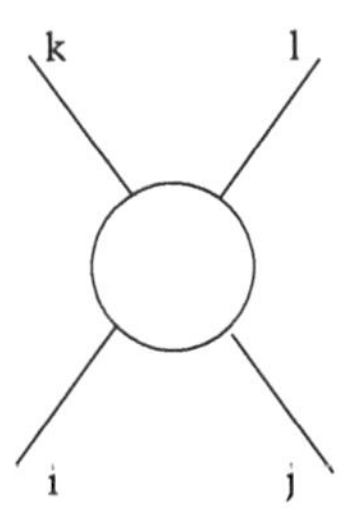

Fig. 10.1 *Four-point correlation functions of the scaling fields.*

quantities are known, one can compute in principle any other correlation functions (see Problem 10.2).

Some consequences. Note that there some immediate consequences of the operator algebra: first of all, to be consistent, the algebra (10.2.13) has to be associative. Consider the 4-point correlation functions of the fields $\varphi_i(x)$ shown in Figure 10.1

$$G_{ijkl}(x_1, x_2, x_3, x_4) = \langle \varphi_i(x_1)\varphi_j(x_2)\varphi_k(x_3)\varphi_l(x_4)\rangle. \qquad (10.2.14)$$

As shown in Figure 10.2, using the operator algebra this correlator can be computed in two equivalent ways:

- expanding the product $\varphi_i(x_1)\varphi_j(x_2)$ and then contracting the resulting field $\varphi_m(x_2)$ with the field $\varphi_n(x_4)$ that comes from an analogous expansion of the product $\varphi_k(x_3)\varphi_l(x_4)$ (with a sum on the intermediate indices m and n);
- alternatively, expanding the two pairs $\varphi_i(x_1)\varphi_k(x_3)$ and $\varphi_j(x_2)\varphi_l(x_4)$, with a final sum on the indices of the propagators of the resulting fields.

The equivalence of these two procedures is known as the *duality symmetry* of the 4-point correlation function. It corresponds to the associativity condition of the operatorial algebra and consists of the infinite number of constraints

$$\sum_{m,n} C_{ij}^{m}(x_1 - x_2)\,\mathcal{D}_{m,n}(x_2 - x_4)\,C_{kl}^{n}(x_3 - x_4) = \sum_{m,n} C_{ik}^{m}(x_1 - x_3)\,\mathcal{D}_{m,n}(x_3 - x_4)\,C_{jl}^{n}(x_2 - x_4). \qquad (10.2.15)$$

Using the expressions (10.2.7) and (10.2.12), this set of equations can be in principle enforced to determine all the scaling dimensions d_i and the structure constants c_{pq}^{r} of the algebra. For this reason they are known as *conformal bootstrap equations* dell'algebra:[5] if we were able to solve them, the dynamical data would be determined self-consistently from

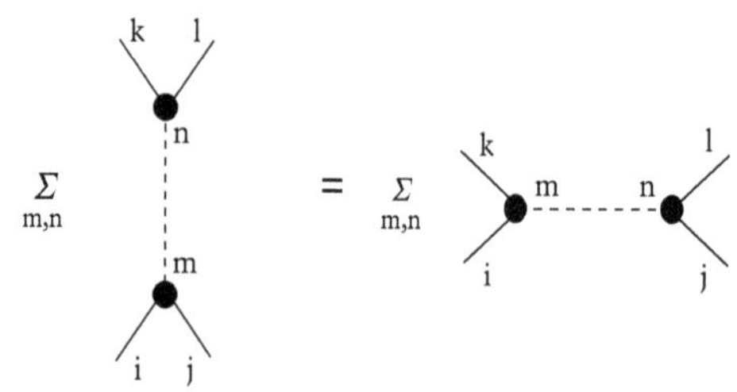

Fig. 10.2 *Duality of the four-point correlation function, that corresponds to the associativity condition of the algebra (10.2.13).*

[5] The term 'bootstrap' denotes the circumstance in which a physical theory owes its validity to its internal consistency. Later we will meet other theories based on bootstrap methods.

the theory itself. They were proposed originally by Polyakov as an alternative approach to solve the QFTs of the critical points. Unfortunately their direct solution proved to be extremely difficult and no much progress has been achieved by their analysis. An important step forward can be instead obtained by studying the important consequences of an additional symmetry of the fixed points, namely the conformal symmetry.

10.3 Conformal Invariance

At the critical point, a statistical system is invariant under a global dilatation of the length scale, $x \rightarrow \lambda x$. Consider now two sub-systems, separated by a distance L considerably larger than their linear dimension l: in this case, it is obvious that the fluctuations will be more uncorrelated when the ratio L/l becomes large as in Figure 10.3. However, nearby the critical point, it does not exist any length scale and the ratio above can be made arbitrarily large. In this way we arrive at the conclusion that the two sub-systems are statistically independent, i.e. that the system is not only invariant under a global dilatation but also under the *local* scale transformations

$$x \rightarrow \lambda(x)\, x \tag{10.3.1}$$

also called *conformal transformations*. The previous considerations can be formulated as a theorem, developed originally by Polyakov:[6] a physical system with local interactions, that is invariant under translations, rotations and a global dilatation, it is also automatically invariant under the larger class of the conformal transformations. Before presenting the

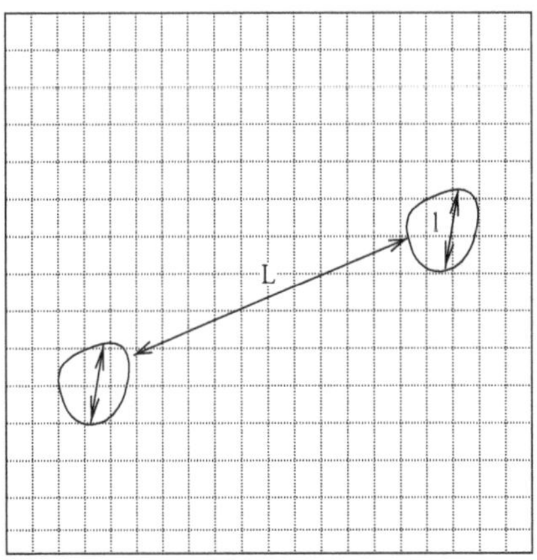

Fig. 10.3 *Sub-systems of linear dimension l separated by a distance $L \gg l$.*

[6] Polyakov, A.M. (1970). 'Conformal Symmetry of Critical Fluctuations', *Journal of Experimental and Theoretical Physics Letters*, 12: 381.

detailed analysis of this important result, it is useful to discuss the main properties of the conformal transformations.

10.3.1 Conformal Transformations in *D* Dimensions

By definition, a conformal transformation of the coordinates $x \to x'$ is an invertible mapping that leaves invariant the metric tensor $g_{\mu\nu}(x)$ up to a scale local factor

$$g'_{\mu\nu}(x') = \Lambda(x)\, g_{\mu\nu}(x). \tag{10.3.2}$$

It is useful to characterize the infinitesimal form of such transformations: using the tensor properties of the metric $g_{\mu,\nu}(x)$, under the transformation

$$x^{\mu} \to x'^{\mu} = x^{\mu} + \epsilon^{\mu}(x) \tag{10.3.3}$$

we have

$$g_{\mu\nu} \to g'_{\mu\nu} = g_{\mu\nu} + (\partial_{\mu}\epsilon_{\nu}(x) + \partial_{\nu}\epsilon_{\mu}(x)).$$

If we now impose the conformal invariance of the metric, eqn. (10.3.2), we have

$$\partial_{\mu}\epsilon_{\nu}(x) + \partial_{\nu}\epsilon_{\mu}(x) = \rho(x)\, g_{\mu\nu}, \tag{10.3.4}$$

where the local function $\rho(x)$ can be easily determined by taking the trace of both terms of this expression

$$\rho(x) = \frac{2}{D}\partial \cdot \epsilon$$

where D is the dimension of the space. If we now take the limit of a flat Euclidean space, with the usual metric $\delta_{\mu\nu} = \text{diag}(1, 1, \ldots, 1)$, we obtain the differential equations that characterize the infinitesimal conformal transformations

$$\partial_{\mu}\epsilon_{\nu}(x) + \partial_{\nu}\epsilon_{\mu}(x) = \frac{2}{D}\partial \cdot \epsilon\, \delta_{\mu\nu}. \tag{10.3.5}$$

Note that from these equations it follows

$$\left[\delta_{\mu\nu}\Box + (D-2)\partial_{\mu}\partial_{\nu}\right]\partial \cdot \epsilon = 0. \tag{10.3.6}$$

The two previous equations imply that the third derivative of ϵ must vanish, so that ϵ can be at most a quadratic function of x. Hence we have the following cases:

- If ϵ is of zero order in x

$$\epsilon^\mu = a^\mu \tag{10.3.7}$$

we recover the usual translation transformations.

- If ϵ is of first order in x, there are two different situations:

$$\epsilon^\mu = \omega^{\mu\nu} x^\nu \tag{10.3.8}$$

with $\omega^{\mu\nu}$ an anti-symmetric tensor $\omega^{\mu\nu} = -\omega^{\nu\mu}$, or

$$\epsilon^\mu = \lambda x^\mu. \tag{10.3.9}$$

The first case corresponds to the rotations while the second one is associated to the global dilatation.

- When ϵ is a quadratic function of x we have

$$\epsilon^\mu = b^\mu x^2 - 2x^\mu b \cdot x. \tag{10.3.10}$$

Its finite form gives rise to the so-called *special conformal transformation*

$$\frac{x'^\mu}{x'^2} = \frac{x^\mu}{x^2} + b^\mu. \tag{10.3.11}$$

This can be interpreted as the result of an inversion plus a translation.

Since in the D-dimensional Euclidean space there are D translation axes and $D(D-1)/2$ possible rotations, it is easy to see that the set of all conformal transformations forms a group, with the number of generators equal to

$$\frac{(D+1)(D+2)}{2}. \tag{10.3.12}$$

All these transformations can be characterized by a very simple geometrical property: they translate, rotate or stretch the vectors placed at a given point but leaving invariant their relative angle.

Constraints on correlation functions. It is instructive to study the constraints on the functional form of the scalar functions $G(x_1, \ldots, x_N)$ that are conformal invariant. To be invariant under translations and rotations, these functions can only depend on the relative distances $|x_i - x_j|$. Furthermore, to be invariant under a dilatation, the dependence on the distances can only be through their ratio, such as

$$\frac{|x_i - x_j|}{|x_k - x_l|}.$$

Since under the special conformal transformation (10.3.11) we have

$$|x_i' - x_j'| = \frac{|x_i - x_j|}{(1 + 2b \cdot x_i + b^2 x_i^2)(1 + 2b \cdot x_j + b^2 x_j^2)},$$

it is evident that, for $N \leq 3$, it is impossible to define conformal invariant transformations of the distances and therefore, in this case, the only conformal invariant functions with $N \leq 3$ variables are only the constants. On the contrary, for $N \geq 4$, by using four different points, we can define the so-called *harmonic ratios*

$$R_{ikmn} \equiv \frac{|x_i - x_k||x_m - x_n|}{|x_i - x_m||x_k - x_n|}. \tag{10.3.13}$$

These quantities are invariant under all conformal transformations. For N points the number of independent harmonic ratios is $N(N-3)/2$ and any arbitrary functions of them is consequently conformal invariant.

10.3.2 Polyakov's Theorem

Let us now present the theorem due to Polyakov on the conformal symmetry of the physical systems with local interactions that are invariant under translations, rotations and a global dilatation. Its proof is simple.

Due to the locality of the theory, there exists a local field $T_{\mu\nu}(x)$, the stress-energy tensor,[7] defined by the variation of the local action $S[\varphi]$ under the transformation (10.3.3)

$$\delta S = \frac{1}{(2\pi)^{D-1}} \int d^D x \, T_{\mu\nu}(x) \, \partial^\mu \epsilon^\nu(x). \tag{10.3.14}$$

Let us derive the equations fulfilled by the stress-energy tensor for the conformal invariance of the theory, expressed by the condition $\delta S = 0$. The translation invariance implies the conservation law

$$\partial_\mu T^{\mu\nu}(x) = 0, \tag{10.3.15}$$

obtained by integrating by parts eqn. (10.3.14) and using the invariance of S under an arbitrary variation of the parameter a^μ in the expression (10.3.7) for ϵ. The rotational invariance implies that the stress-energy tensor is symmetric with respect to its indices

$$T^{\mu\nu}(x) = T^{\nu\mu}(x), \tag{10.3.16}$$

[7] The factor $(2\pi)^{D-1}$, that is unusual with respect to the definition of this field as derived from the Noether theorem, is introduced here for later convenience.

as it can be seen by substituting in eqn. (10.3.14) the transformation (10.3.8). The invariance under dilatation, given by the transformation (10.3.9), finally leads to a zero-trace condition

$$T^{\mu}_{\mu}(x) = 0. \tag{10.3.17}$$

In view of eqns. (10.3.15), (10.3.16) and (10.3.17), the action is automatically invariant under the conformal transformations that satisfy eqn. (10.3.5).

10.4 Quasi-primary Fields

The conformal invariance of a statistical system at a critical point permits to prove a series of important results on the correlation functions of a special class of its operators. These operators are called *quasi-primary* and in this section they will be denoted as $Q_n(x)$. They have the property to transform under a generic conformal mapping as

$$Q_n(x) = \left|\frac{\partial x'}{\partial x}\right|^{d_n/D} Q_n(x'), \tag{10.4.1}$$

where $\left|\frac{\partial x'}{\partial x}\right|$ is the Jacobian of the mapping. For the transformations associated to the translations and rotations we have

$$\left|\frac{\partial x'}{\partial x}\right| = 1,$$

while for the dilatation and the special conformal transformations we have respectively

$$\left|\frac{\partial x'}{\partial x}\right| = \lambda^D, \quad \left|\frac{\partial x'}{\partial x}\right| = \frac{1}{(1+2b\cdot x+b^2x^2)^D}.$$

The transformation law (10.4.1) of the primary fields is obviously more specific than the simple scaling law (10.2.5) and therefore it poses more restrictive conditions on the correlation functions of these fields. They satisfy the equation

$$\langle Q_1(x_1)Q_2(x_2)\dots Q_n(x_n)\rangle = \left|\frac{\partial x'}{\partial x}\right|^{d_1/D}_{x=x_1}\cdots\left|\frac{\partial x'}{\partial x}\right|^{d_n/D}_{x=x_n}\langle Q_1(x_1')Q_2(x_2')\dots Q_n(x_n')\rangle. \tag{10.4.2}$$

Let us consider the two-point correlation functions

$$G^{(2)}_{ab}(x_1,x_2) = \langle Q_a(x_1)Q_b(x_2)\rangle. \tag{10.4.3}$$

Due to the translation and rotational invariance, it depends on the relative distance $| x_1 - x_2 |$. The dilatation invariance implies that $G^{(2)}_{ab}$ behaves as $| x_1 - x_2 |^{-d_a-d_b}$. Finally, using the invariance under the special conformal transformation (10.3.11), we arrive at the condition

$$\delta G^{(2)}_{ab} = -[d_a (b \cdot x_1) + d_b (b \cdot x_2)] \, G^{(2)}_{ab}. \tag{10.4.4}$$

This imposes that $G^{(2)}_{ab}(\mid x_1 - x_2 \mid)$ is different from zero only when $d_a = d_b$. Hence, the two-point correlation functions of the quasi-primary fields satisfy an orthogonality condition. By an appropriate normalization of these fields, their general expression is then

$$\langle Q_a(x_1) Q_b(x_2)\rangle = \frac{\delta_{ab}}{\mid x_1 - x_2 \mid^{2d_a}}. \tag{10.4.5}$$

Consider now the three-point correlation functions of the quasi-primary fields

$$G^{(3)}_{abc}(x_1, x_2, x_3) = \langle Q_a(x_1) Q_b(x_2) Q_c(x_3)\rangle. \tag{10.4.6}$$

As usual, the invariance under translations and rotations implies that this correlator depends on the relative distances $x_{ij} \equiv \mid x_i - x_j \mid$ (con $i, j = 1, 2, 3$). The invariance under the infinitesimal transformations of eqn. (10.3.11) gives rise to the homogeneous equations

$$\frac{1}{2} \sum_{i<j} x_{ij} \frac{\partial G^{(3)}}{\partial x_{ij}} \left[(b \cdot x_i) + (b \cdot x_j)\right] = -\sum_i d_i \, (b \cdot x_i) \, G^{(3)},$$

whose solution is given by

$$\langle Q_a(x_1) Q_b(x_2) Q_c(x_3)\rangle = \frac{C_{abc}}{x_{12}^{d_3 - d_1 - d_2} \, x_{13}^{d_2 - d_1 - d_3} \, x_{23}^{d_1 - d_2 - d_3}}. \tag{10.4.7}$$

The numerical constant C_{abc} cannot be fixed by the conformal invariance. However, using the orthogonality condition (10.4.5), it is easy to see that it coincides with the structure constant c_{abc} of the algebra (10.2.13), relative to these three fields.

In conclusion, thanks to the conformal invariance, one can determine the exact expression of the two and three-point correlation functions of the quasi-primary operators. The same is not true for the higher point correlation functions. In this case, in fact, the N-point correlation functions can have an arbitrary dependence on the $N(N-3)/2$ harmonic ratios. For instance, the general form of the four-point function can be expressed as

$$G^{(4)}(x_1, x_2, x_3, x_4) = F\left(\frac{x_{12} x_{34}}{x_{13} x_{24}}, \frac{x_{12} x_{34}}{x_{23} x_{41}}\right) \prod_{i<j} x_{ij}^{-(d_i + d_j) + d/2}, \tag{10.4.8}$$

with $d = \sum_{i=1}^{4} d_i$ whereas F is an arbitrary function of the two independent harmonic ratios. For $D \neq 2$ and in absence of additional information, the explicit expression of F cannot be determined. As we will see in the next sections, this goal can be however achieved when $D = 2$.

10.5 Two-dimensional Conformal Transformations

Let us consider in more detail the conformal transformations in an Euclidean space when $D = 2$. In this case, something remarkable happens: in fact, eqns. (10.3.5) become

$$\partial_1 \epsilon_1 = \partial_2 \epsilon_2, \quad \partial_1 \epsilon_2 = -\partial_2 \epsilon_1, \tag{10.5.1}$$

i.e. the Cauchy–Riemann conditions relative to the functions of a complex variable! In other words, if we introduce the notation

$$\begin{aligned} \epsilon(z) &= \epsilon_1 + i\epsilon_2 \\ \bar{\epsilon}(\bar{z}) &= \epsilon_1 - i\epsilon_2 \end{aligned} \tag{10.5.2}$$

and consider the new variables

$$\begin{aligned} z &= x_1 + i x_2 \\ \bar{z} &= x_1 - i x_2 \end{aligned}$$

in two dimensions the conformal transformations coincide with the analytic mappings of the complex plane

$$z \to f(z), \quad \bar{z} \to \bar{f}(\bar{z}). \tag{10.5.3}$$

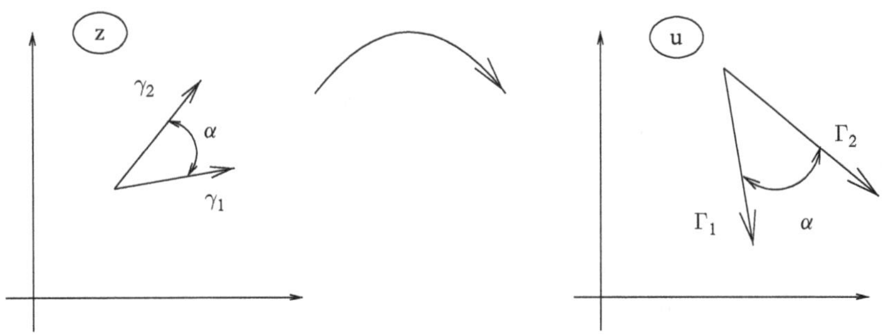

Fig. 10.4 *Conformal transformation of two vectors.*

It is well known that the analytic maps of the complex plane have the property to preserve the angle between two vectors placed at the same point. The proof is simple. Consider an analytic map $u = f(z)$ and suppose that $f'(z_0) \neq 0$, where z_0 is the intersection point of the two vectors γ_1 and γ_2, as shown in Figure 10.4. We have

$$f'(z_0) = \lim_{\Delta z \to 0} \frac{\Delta u}{\Delta z} \equiv k e^{i\beta}.$$

Suppose now to choose the direction in which the variation Δz goes to zero along the curve γ_1, so that Δu moves along the transformed curve Γ_1. In this case we have

$$\beta = \arg f'(z_0) = \lim_{\Delta u \to 0} \arg \Delta u - \lim_{\Delta z \to 0} \arg \Delta z = \Phi_1 - \phi_1,$$

where ϕ_1 and Φ_1 are the angles of the curves γ_1 and Γ_1 with respect to the horizontal axis of the corresponding planes. However, the same quantity is obtained by taking the limit along the direction of the curve γ_2, since the derivative of a function of a complex variable is independent from the the path along which it is computed. We have then

$$\beta = \Phi_2 - \phi_2,$$

where ϕ_2 and Φ_2 are respectively the angles of the curves γ_2 and Γ_2 with respect to the horizontal axis of the corresponding planes. Taking the difference between these two equations we have

$$\phi_2 - \phi_1 = \Phi_2 - \Phi_1 = \alpha.$$

Hence, the angle between two arbitrary curves that pass through the point z_0 at which $f'(z_0) \neq 0$ is equal to the angle between the two transformed curves. The modulus of the two vectors is instead dilated by the factor k.

As we will see later, in the two-dimensional CFTs there is a natural splitting in the variables z and $\bar{z}$. To simplify the notations, in the following we focus the attention only on the transformations that involve the variable z, keeping in mind that the same analysis applies to the variable $\bar{z}$ as well. Moreover, thanks to the splitting of the two variables, it is useful to consider z and $\bar{z}$ as two independent quantities. This is equivalent to regard the original coordinates (x_1, x_2) as variables of C^2 rather than of $\Re^2$ and to consider the new quantities $(z, \bar{z})$ simply as a change of variables. Obviously the original real variables are recovered posing in C^2 the equation $z = \bar{z}$. As we show below, this procedure permits a remarkable efficiency in the development of the formalism. From the definition of z and $\bar{z}$ we have

$$\begin{array}{ll} \partial_z = 1/2(\partial_1 - i\partial_2) & \partial_1 = (\partial_z + \partial_{\bar{z}}) \\ \partial_{\bar{z}} = 1/2(\partial_1 + i\partial_2) & \partial_2 = i(\partial_z - \partial_{\bar{z}}). \end{array} \tag{10.5.4}$$

In the Euclidean two-dimensional space, the element of line is

$$ds^2 = g_{\mu\nu} dx^\mu \, dx^\nu = dx^2 + dy^2 = dz \, d\bar{z},$$

The complex metric is then $g_{zz} = g_{\bar{z}barz} = 0$ and $g_{z\bar{z}} = g_{\bar{z}z} = \frac{1}{2}$ and therefore the components of the stress-energy tensor in this coordinates are given by

$$\begin{array}{l} T_{zz} \equiv T(z,\bar{z}) = \frac{1}{4}(T_{11} - T_{22} + 2iT_{12}), \\ T_{\bar{z}\bar{z}} \equiv \bar{T}(z,\bar{z}) = \frac{1}{4}(T_{11} - T_{22} - 2iT_{12}), \\ T_{z\bar{z}} = T_{\bar{z}z} \equiv \frac{1}{4}\Theta(z,\bar{z}) = \frac{1}{4}(T_{11} + T_{22}) = \frac{1}{4}T^\mu_\mu, \end{array} \tag{10.5.5}$$

In the complex coordinates, the conservation law (10.3.15) becomes

$$\begin{array}{l} \partial_{\bar{z}} T(z, \bar{z}) + \frac{1}{4}\partial_z \Theta(z, \bar{z}); \\ \partial_z \bar{T}(z, \bar{z}) + \frac{1}{4}\partial_{\bar{z}} \Theta(z, \bar{z}). \end{array} \tag{10.5.6}$$

Furthermore, at the critical point the trace $\Theta(z,\bar{z})$ of the stress-energy tensor vanishes

$$\partial_{\bar{z}} T(z, \bar{z}) = \partial_z \bar{T}(z, \bar{z}) = 0. \tag{10.5.7}$$

Hence, at criticality, T depends only on z while $\bar{T}$ depends on $\bar{z}$: the former is then an analytic operator while the latter an anti-analytic one. These properties will be important later.

Let us now use the complex coordinate to investigate some aspects of the two-dimensional conformal invariance. Posing

$$z' = z + \epsilon(z), \quad \epsilon(z) = \sum_{n=-\infty}^{\infty} a_n z^{n+1},$$

every conformal transformation can be characterized by the coefficients a_n of the Taylor–Laurent expansion of $\epsilon(z)$ at the origin. An analogous result holds for the variable $\bar{z}$. As a result of these transformations, at the first order in ϵ a scalar function $S(z,\bar{z})$ changes as

$$\begin{aligned} S(z', \bar{z}') - S(z, \bar{z}) &\equiv \delta S = -(\epsilon(z)\, \partial + \bar{\epsilon}(\bar{z})\, \bar{\partial})\, S(z, \bar{z}) = \\ &= \sum_{n=-\infty}^{\infty} (a_n l_n + \bar{a}_n \bar{l}_n)\, S(z, \bar{z}), \end{aligned}$$

where we have introduced the operators

$$l_n = -z^{n+1}\,\partial, \quad \bar{l}_n = -\bar{z}^{n+1}\,\bar{\partial}, \tag{10.5.8}$$

with $\partial = \partial_z$ and $\bar{\partial} = \partial_{\bar{z}}$. They satisfy the commutation relations

$$\begin{aligned} [l_n, l_m] &= (n-m)\, l_{n+m}; \\ \left[\bar{l}_n, \bar{l}_m\right] &= (n-m)\, \bar{l}_{n+m}; \\ \left[l_n, \bar{l}_m\right] &= 0. \end{aligned} \tag{10.5.9}$$

The conformal symmetry is then expressed in terms of the direct sum of two isomorphic infinite-dimensional algebras.

An apparent paradox. The above result may appear paradoxical: how can we have in two dimensions an infinite-dimensional symmetry when, in the last section, we have established that the number of real parameters of the conformal symmetry is finite and equal, in D dimensions, to $(D+1)(D+2)/2$? In the light of this, for $D=2$ we should have only 6 real parameter transformations and not an infinite number of them.

The explanation of this paradox is as follows: in two dimensions there is only *one* particular class of analytic functions that are well-defined and invertible on all the Riemann sphere (i.e. the complex plane plus the point at infinity). These are the Moebius transformations, given by

$$w(z) = \frac{az+b}{cz+d}, \tag{10.5.10}$$

with $a, b, c, d \in C$ and $ad - bc \neq 0$. Since a uniform rescaling of the parameters does not change the final expression of $w(z)$, we can always assume that they satisfy the condition $ad - bc = 1$. The set of all these transformations is isomorphic to the group $SL(2, C)/Z_2$, where the quotient with respect to Z_2 comes from the invariance of the

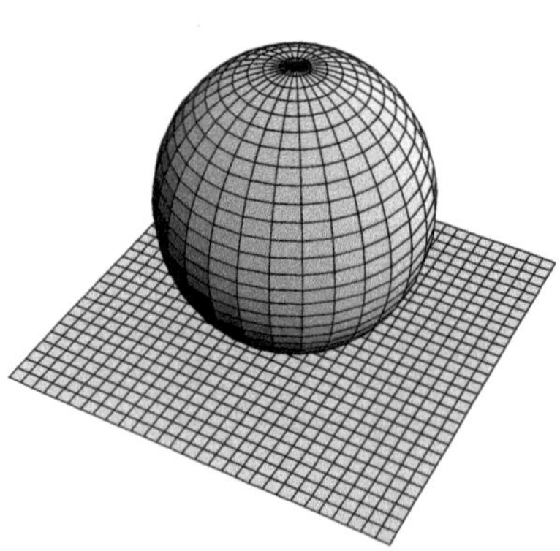

Fig. 10.5 *The Riemann sphere, i.e. the complex plane plus the point at infinity.*

mapping (10.5.10) under the multiplication of all the parameters for -1. Hence there are only three complex parameters, i.e. six real parameters, in agreement with the general result of the previous section.[8] The detailed study of these functions—an argument that is itself interesting—can be found in Appendix 10.A. Here we simply note that a Moebius transformation is uniquely defined by the correspondence between three different points of the z-plane and three different points of the w-plane. If we denote the chosen three points of the z-plane z_1, z_2 and z_3 and the corresponding points of the w-plane w_1, w_2 and w_3, it is easy to prove the identity

$$\frac{(w_1 - w)(w_3 - w_2)}{(w_1 - w_2)(w_3 - w)} = \frac{(z_1 - z)(z_3 - z_2)}{(z_1 - z_2)(z_3 - z)}. \tag{10.5.11}$$

This relation is an implicit expression of the desired Moebius function. By solving eqn. (10.5.11) for w, we get an explicit expression of the coefficients a, b, c, d of the Moebius transformation in terms of the given numbers z_1, z_2, z_3 and w_1, w_2, w_3. This result also shows that $w(z)$ depends at most by three complex parameters. Note that eqn. (10.5.11) states that the harmonic ratio of four points of the complex plane is invariant under a Moebius transformation.

An equivalent way to achieve the same result consists of considering the vector field that generates the conformal transformations in the space of the functions $g(z)$

$$v(z) = -\sum_n a_n l_n = \sum_n a_n z^{n+1} \partial_z.$$

Requiring that the vector field is regular nearby the origin, we have $a_n \neq 0$ only for $n \geq -1$. We can study its behaviour in a neighbour of the point at infinity by using the transformation $z = -1/y$

$$v(z) = \sum_n a_n \left(-\frac{1}{y}\right)^{n+1} \left(\frac{dz}{dy}\right)^{-1} \partial_y = \sum_n a_n \left(-\frac{1}{y}\right)^{n-1} \partial_y.$$

Requiring that the vector field is also regular at infinity, we have $a_n \neq 0$ only for $n \leq 1$. In conclusion, the only conformal transformations that are globally defined on the Riemann sphere are those expressed by a quadratic expression z. They correspond to $a_n l_n$ with $n = 0, \pm 1$, with three arbitrary complex parameters. The same result is obtained for the anti-analytic vector field. However the number of parameters does not double since the generators of the two algebras (10.5.9) that preserve the real section of C^2 are expressed by the linear combinations

$$l_n + \bar{l}_n, \quad i(l_n - \bar{l}_n). \tag{10.5.12}$$

[8] The real projection of C^2 is obtained by taking (10.5.10) together with its complex conjugate $\bar{w}(\bar{z})$. Obviously this procedure does not enlarge the number of independent parameters.

Note that the generators $l_0, l_{\pm 1}$ form a finite sub-algebra of (10.5.9), where $l_{-1} = -\partial_z$ is the generator of the translations, $l_0 = -z\partial_z$ the generator of the dilations and finally $l_1 = -z^2\partial_z$ the generator of the special conformal transformations. In particular, $l_0 + \bar{l}_0$ is the operator that generates the global dilations on the real section of C^2, while $i(l_0 - \bar{l}_0)$ is the operator of the rotations.

To summarize, in two dimensions, it is necessary to distinguish between the *global* and the *local* conformal transformations: the former are given by the Moebius mappings whereas the latter are expressed by all other analytic functions. Note that, except the Moebius transformations, any other analytic function is either not uniquely invertible or not well-defined on all the complex plane.[9] For instance, the function $f(z) = z^3$ is defined on all the complex plane but is not uniquely invertible. Vice-versa, $f(z) = \log z$ has a branch cut along the positive axis and jumps by $2\pi i$ each time we cross it. It is also worth remembering that, for the well-known Liouville theorem, an analytic function cannot be small everywhere unless it is a constant: we have to keep in mind this result for properly interpreting the Polyakov theorem, that is based on the infinitesimal nature of the transformations.

The considerations above should clarify some basic aspects of the conformal invariance in two dimensions but it is still open the question how to interpret the apparent symmetry of the theory under the infinite dimensional class of the local conformal transformations. As we will see in the next sections, the answer to this question can be summarized as follows: the local conformal transformations $z \to f(z)$ present a quantum anomaly. In particular, the two algebras (10.5.9) need a central extension for the fluctuations of a QFT near the phase transition point. For this anomaly, the theory is invariant only under the Moebius transformations and not under all conformal transformations. But, even in the absence of an exact invariance of the physical systems under a generic analytic map, we will show that we can nevertheless control how the system behaves under a local conformal map and this proves to be the key to achieve the exact solution of many statistical models.

10.6 Ward Identity and Primary Fields

The correlation functions of the local fields, for the definition given by eqn. (10.2.1), satisfy a Ward identity. To derive it, consider an arbitrary infinitesimal change of the coordinates

$$x^\mu \to x'^\mu = x^\mu + \epsilon^\mu(x)$$

[9] The analytic functions are also called *holomorphic functions*. The analytic functions with a finite number of poles are called *meromorphic functions* whereas those having only zeros in any finite region of the complex plane are called *entire functions*.

and denote by δA_i the corresponding variation of the fields under this transformation

$$A_i(x) \to A_i(x) + \delta A_i(x). \tag{10.6.1}$$

Under this change of variables, the action changes as

$$\delta S = \frac{1}{(2\pi)^{D-1}} \int d^D x \, T_{\mu\nu}(x) \, \partial^\mu \epsilon^\nu(x).$$

Since a change of coordinates does not alter the description of the physical system, we have $\delta \langle A_1(x_1) \ldots A_n(x_n) \rangle = 0$. This leads to the Ward identity

$$\sum_{i=1}^{n} \langle A_1(x_1) \ldots \delta A_i(x_i) \ldots A_n(x) \rangle + \\ - \frac{1}{(2\pi)^{D-1}} \int d^D x \langle T_{\mu\nu}(x) A_1(x_1) \ldots A_n(x_n) \rangle \, \partial^\mu \epsilon^\nu(x) = 0. \tag{10.6.2}$$

Let us now specialize this equation to the case of two dimensional QFT. First of all, let us define the *primary operators* $\phi_i(z, \bar{z})$: they have the distinguished property to change as follows

$$\phi_i(z, \bar{z}) = \left(\frac{df}{dz} \right)^{\Delta_i} \left(\frac{d\bar{f}}{d\bar{z}} \right)^{\bar{\Delta}_i} \phi_i'(z', \bar{z}'), \tag{10.6.3}$$

under an *arbitrary* conformal transformation $z \to z' = f(z)$, $\bar{z} \to \bar{z}' = \bar{f}(\bar{z})$, where the two real quantities Δ_i and $\bar{\Delta}_i$ are the *conformal weights*. Note that the expression above can be written as

$$\phi_i'(z', \bar{z}') dz'^{\Delta_i} d\bar{z}'^{\bar{\Delta}_i} = \phi_i(z, \bar{z}) dz^{\Delta_i} d\bar{z}^{\bar{\Delta}_i}.$$

This enable us to consider the primary field $\phi_i(z, \bar{z})$ as a 'tensor' of degree $(\Delta_i, \bar{\Delta}_i)$ under the conformal transformations. Obviously a primary field is also a quasi-primary field but it is not true the reverse, since the quasi-primary operators transform as 'tensors' only under the global conformal transformations (expressed, in two dimensions, by the Moebius mappings). To derive the infinitesimal transformation law of the primary fields, pose $z' = z + \epsilon(z)$ and $\bar{z}' = \bar{z} + \bar{\epsilon}(\bar{z})$, and expand up to the first order

$$\delta \phi_i \equiv \phi(z, \bar{z}) - \phi'(z', \bar{z}') = [(\Delta_i \partial \epsilon + \epsilon \partial) + (\bar{\Delta}_i \bar{\partial} \bar{\epsilon} + \bar{\epsilon} \bar{\partial})] \, \phi_i(z, \bar{z}). \tag{10.6.4}$$

Consider the n-point correlation function $\langle \phi_1(x_1) \ldots \phi_n(x_n) \rangle$ of the primary fields in the two-dimensional Euclidean plane. Denote by S a circle of radius sufficiently large to include all the points $x_1, x_2, \ldots x_n$ as shown in Figure 10.6. Suppose now to make an

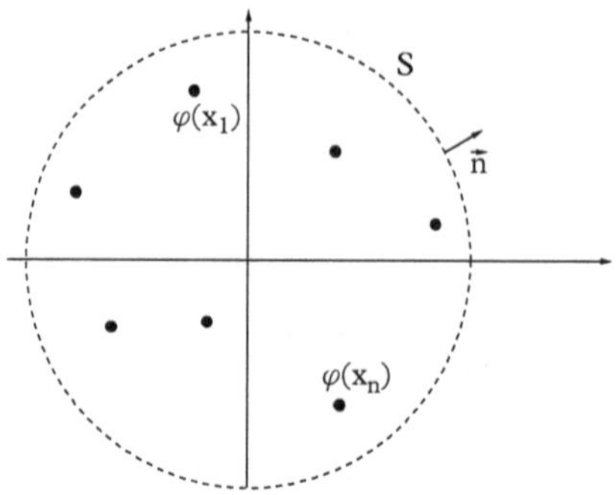

Fig. 10.6 *Correlation functions of the primary fields.*

infinitesimal change of coordinates $x^\mu \to x^\mu + \epsilon^\mu(x)$, where $\epsilon^\mu(x)$ is a regular analytic function for $|z| \leq R$ whereas, outside this region, only a differentiable function but fastly decreasing to zero at infinity. $\epsilon^\mu(x)$ is then an arbitrary small quantity in any region of the plane. Such a transformation is obviously controlled by the Ward identity (10.6.2). At the critical point of the theory $\langle T_{\mu\nu}(x)\rangle = 0$ and therefore we can neglect the last term in (10.6.2). For the second term, the only contribution to the integral comes from the region $|x| > R$ since, inside the circle, the theory is conformal invariant and we can apply the theorem by Polyakov. If we denote generically by X the product of the n primary field, we have

$$\frac{1}{2\pi}\int d^2x\,\langle T_{\mu\nu}(x)\,X\rangle \partial^\mu \epsilon^\nu(x) = \frac{1}{2\pi}\int_{|x|>R} d^2x\,\langle T_{\mu\nu}(x)X\rangle \partial^\mu \epsilon^\nu(x). \tag{10.6.5}$$

Integrating by parts the right-hand side of this expression and neglecting the surface term at infinity (that vanishes for the rapidly decreasing behaviour of ϵ^ν), we have

$$\begin{aligned}\frac{1}{2\pi}\int_{|x|>R} d^2x\,\langle T_{\mu\nu}(x)X\rangle \partial^\mu \epsilon^\nu(x) &= -\frac{1}{2\pi}\int_{|x|>R} d^2x\,\epsilon^\nu\,\langle \partial^\mu T_{\mu\nu}(x)X\rangle \\ &+ \frac{1}{2\pi}\int_C d\Sigma\, n^\mu \epsilon^\nu\,\langle T_{\mu\nu}X\rangle,\end{aligned}$$

where n^μ is the unit vector orthogonal to the surface Σ of the circle C. The first term in the right-hand side vanishes for the conservation law of the stress-energy tensor. By using the complex coordinates $z, \bar{z}$ and the corresponding components of the stress-energy tensor, we can express the second term on the right-hand side as

$$\frac{1}{2\pi}\int_C d\Sigma\, n^\mu \epsilon^\nu\,\langle T_{\mu\nu}X\rangle = \frac{1}{2\pi i}\oint_C dz\,\epsilon(z)\,\langle T(z)X\rangle - \frac{1}{2\pi i}\oint_C d\bar{z}\,\bar{\epsilon}(\bar{z})\,\langle \bar{T}(\bar{z})X\rangle,$$

with $\epsilon(z) = \epsilon^1 + i\epsilon^2$ and $\bar{\epsilon}(\bar{z}) = \epsilon^1 - i\epsilon^2$.

Consider now the first term of the Ward identity (10.6.2) that, for the primary fields, is expressed by

$$\sum_{i=1}^{n}[(\Delta_i\partial_i\epsilon(z)+\epsilon(z)\partial_i)+(\bar{\Delta}_i\bar{\partial}_i\bar{\epsilon}(\bar{z})+\bar{\epsilon}(\bar{z})\bar{\partial}_i)]\langle\phi_1(x_1)\ldots\phi_n(x_n)\rangle.$$

We can use the complex coordinates $(z_i, \bar{z}_i)$ to identify the points in the plane and apply the Cauchy theorem to write the analytic and anti-analytic terms as

$$\sum_{i=1}^{n}[(\Delta_i\partial_i\epsilon(z)+\epsilon(z)\partial_i)]\langle\phi_1(x_1)\ldots\phi_n(x_n)\rangle$$

$$=\frac{1}{2\pi i}\oint_C dz\epsilon(z)\sum_{i=1}^{n}\left[\frac{\Delta_i}{(z-z_i)^2}+\frac{1}{z-z_i}\partial_i\right]\langle\phi_1(z_1,\bar{z}_1)\ldots\phi_n(z_n,\bar{z}_n)\rangle,$$

$$\sum_{i=1}^{n}[(\bar{\Delta}_i\bar{\partial}_i\bar{\epsilon}(z)+\bar{\epsilon}(z)\bar{\partial}_i)]\langle\phi_1(x_1)\ldots\phi_n(x_n)\rangle$$

$$=-\frac{1}{2\pi i}\oint_C d\bar{z}\bar{\epsilon}(z)\sum_{i=1}^{n}\left[\frac{\bar{\Delta}_i}{(\bar{z}-\bar{z}_i)^2}+\frac{1}{\bar{z}-\bar{z}_i}\bar{\partial}_i\right]\langle\phi_1(z_1,\bar{z}_1)\ldots\phi_n(z_n,\bar{z}_n)\rangle.$$

Putting now together all the terms of the Ward identity, we arrive at

$$\frac{1}{2\pi i}\oint_C dz\,\epsilon(z)\left[\sum_{i=1}^{n}\left[\frac{\Delta_i}{(z-z_i)^2}+\frac{1}{z-z_i}\partial_i\right]\langle\phi_1(z_1,\bar{z}_1)\ldots\rangle-\langle T(z)\phi_1(z_1,\bar{z}_1)\ldots\rangle\right]$$

$$-\frac{1}{2\pi i}\oint_C d\bar{z}\bar{\epsilon}(\bar{z})\left[\sum_{i=1}^{n}\left[\frac{\bar{\Delta}_i}{(\bar{z}-\bar{z}_i)^2}+\frac{1}{\bar{z}-\bar{z}_i}\bar{\partial}_i\right]\langle\phi_1(z_1,\bar{z}_1)\ldots\rangle-\langle\bar{T}(\bar{z})\phi_1(z_1,\bar{z}_1)\ldots\rangle\right]=0.$$

Since the two functions $\epsilon(z)$ and $\bar{\epsilon}(\bar{z})$ can be varied independently, the two terms of this equation must vanish separately. So, we have the conformal Ward identity for the analytic sector

$$T(z_1)\phi_i(z_2,\bar{z}_2)=\frac{\Delta_i}{(z-z_i)^2}\phi_i(z_2,\bar{z}_2)+\frac{1}{z-z_i}\partial\phi(z_2,\bar{z}_2)+\text{regular terms},\qquad(10.6.6)$$

as well as for the anti-analytic part

$$\bar{T}(\bar{z}_1)\phi_i(z_2,\bar{z}_2)=\frac{\bar{\Delta}_i}{(\bar{z}-\bar{z}_i)^2}\phi_i(z_2,\bar{z}_2)+\frac{1}{\bar{z}-\bar{z}_i}\partial\phi(z_2,\bar{z}_2)+\text{regular terms}\qquad(10.6.7)$$

Notice that, for what concerns the Ward identity, the primary field $\phi_i(z,\bar{z})$ may be regarded as made of a product of two *chiral primary fields* $\Phi(z)$ and $\bar{\Phi}(\bar{z})$, the first depending only on z while the second only on $\bar{z}$, $\phi_i(z,\bar{z}) = \Phi_i(z)\,\bar{\Phi}_i(\bar{z})$. This factorization is extremely useful to deal with the algebraic properties of the primary fields but is not a faithful representation of the actual nature of the primary fields. As we shall show later, the correlation functions of the primary field $\phi_i(z,\bar{z})$ are not simply given by the product of the correlation functions of the chiral primary fields $\Phi_i(z)$ and $\bar{\Phi}(\bar{z})$.

The primary fields play an important role in CFT. As a matter of fact, their transformation law (10.6.3) is the simplest possible and leads to an OPE with $T(z)$ and $\bar{T}(\bar{z})$ in which there are at most second-order poles. Any other fields of the theory has an OPE with the stress-energy tensor of higher-order poles: to see this, it is sufficient to consider the OPE of $T(z)$ with a derivative of the primary field. In addition to the simplicity of their OPE, the primary fields are also the building blocks of the representation theory of the conformal symmetry. As we show in the next sections, *all* conformal fields of the theory are organized in *conformal families* that are uniquely identified by the primary fields. These families form the irreducible representations of the quantum version of the conformal algebra.

10.7 Central Charge and Virasoro Algebra

This section analyses the quantum version of the conformal algebra that is deeply related to the stress-energy tensor. First it is necessary to note that the role played by the stress-energy tensor in the theory is twofold: on one side, it is the generator of the conformal transformations, on the other side it is a conformal field itself. Since it satisfies a conservation law, its scaling dimension coincides with its canonical dimension, equal to $d_T = d_{\bar{T}} = 2$. The two-point correlation function of its analytic part is generically different from zero and can be expressed as

$$\langle T(z_1)T(z_2)\rangle = \frac{c/2}{(z_1 - z_2)^4}, \tag{10.7.1}$$

where the real coefficient c is the *central charge* of the conformal algebra. The same holds for the anti-analyitic component

$$\langle \bar{T}(\bar{z}_1)\bar{T}(\bar{z}_2)\rangle = \frac{\bar{c}/2}{(\bar{z}_1 - \bar{z}_2)^4}. \tag{10.7.2}$$

For a relativistic and parity invariant theory, it is easy to show that $c = \bar{c}$. From now on we focus the attention only on $T(z)$, keeping in mind that the same results will also hold for $\bar{T}(\bar{z})$. The quantity c is generally different from zero, as it can be seen by the analysis of two simple but significant examples of CFTs.

10.7.1 Example 1. Free Neutral Fermion

Consider the Lagrangian of a neutral bi-dimensional fermion (Majorana fermion)

$$\mathcal{L} = \frac{\lambda}{2\pi}\left[\psi\frac{\partial}{\partial\bar{z}}\psi + \bar{\psi}\frac{\partial}{\partial z}\bar{\psi}\right].$$

The equations of motion are

$$\frac{\partial}{\partial\bar{z}}\psi = \frac{\partial}{\partial z}\bar{\psi} = 0.$$

Hence $\psi(z)$ is a purely analytic field (with conformal weight $\Delta = \frac{1}{2}$, as it can be easily seen directly from the Lagrangian) while $\bar{\psi}(\bar{z})$ is a purely anti-analytic field with conformal weight $\bar{\Delta} = \frac{1}{2}$. Their two-point correlation functions are

$$\begin{aligned}\langle\psi(z_1)\psi(z_2)\rangle &= \frac{1}{\lambda}\frac{1}{z_1 - z_2}; \\ \langle\bar{\psi}(\bar{z}_1)\bar{\psi}(\bar{z}_2)\rangle &= \frac{1}{\lambda}\frac{1}{\bar{z}_1 - \bar{z}_2}.\end{aligned} \tag{10.7.3}$$

The analytic part of the stress-energy tensor is obtained by the Noether theorem

$$T(z) = -\frac{\lambda}{2} : \psi(z)\frac{\partial}{\partial z}\psi(z) : . \tag{10.7.4}$$

The two-point correlator of $T(z)$ can be obtained by the correlator (10.7.3) applying the Wick theorem

$$\begin{aligned}\langle T(z_1)T(z_2)\rangle &= \frac{\lambda^2}{4}\langle : \psi(z_1)\partial_1\psi(z_1) :: \psi(z_2)\partial_2\psi(z_2) :\rangle = \\ &\frac{\lambda^2}{4}\left[\langle\psi(z_1)\partial_2\psi(z_2)\rangle\langle\partial_1\psi(z_1)\psi(z_2)\rangle - \langle\psi(z_1)\psi(z_2)\rangle\langle\partial_1\psi(z_1)\partial_2\psi(z_2)\rangle\right] \\ &= \frac{1}{4}\left[-\frac{1}{(z_1 - z_2)^4} + \frac{2}{(z_1 - z_2)^4}\right] = \frac{1}{4(z_1 - z_2)^4}.\end{aligned} \tag{10.7.5}$$

For this system we have then $c = \frac{1}{2}$.

10.7.2 Example 2. Free Bosonic Field

Consider now the Lagrangian of a neutral free boson

$$\mathcal{L} = \frac{g}{4\pi}(\partial_\mu\Phi)^2.$$

The correlation function of this field is given by

$$G(z,\bar{z}) = \langle \Phi(z_1,\bar{z}_1)\Phi(z_2,\bar{z}_2)\rangle = -\frac{1}{2g}\log z_{12} - \frac{1}{2g}\log \bar{z}_{12}. \tag{10.7.6}$$

Note that the free bosonic field $\Phi(z,\bar{z})$ is not a scaling operator itself. However, we can construct scaling operators as the fields $\partial_z\Phi$ or $e^{i\alpha\Phi}$ using the field Φ. For the analytic part of the stress-energy tensor, derived from the Noether theorem, we have

$$T(z) = -g : (\partial_z\Phi)^2 : . \tag{10.7.7}$$

Its two-point correlation function can be computed by (10.7.6) using the Wick theorem

$$\langle T(z_1)T(z_2)\rangle = g^2 \langle : (\partial_1\Phi(z_1))^2 :: (\partial_2\Phi(z_2) :)^2\rangle =$$
$$= g^2 \left[2\left(\langle\partial_1\Phi(z_1)\partial_2\Phi(z_2)\rangle\right)^2\right] = \frac{1}{2(z_1-z_2)^4}.$$

Therefore the central charge of this system is $c = 1$.

Conformal anomaly. Let us come back to the general discussion on the stress-energy tensor. In the presence of a central charge different from zero, the OPE of $T(z)$ with itself has the singular terms

$$T(z_1)T(z_2) = \frac{c/2}{(z_1-z_2)^4} + \frac{2}{(z_1-z_2)^2}T(z_2) + \frac{1}{z_1-z_2}\partial T(z_2) + \cdots . \tag{10.7.8}$$

Therefore, the infinitesimal conformal transformation of $T(z)$ is given by

$$\delta T(z) = (2\partial\epsilon + \epsilon\partial)T(z) + \frac{c}{12}\partial^3\epsilon(z). \tag{10.7.9}$$

The term proportional to c may be interpreted as a quantum anomaly. Consider, in fact, the Ward identity for the one-point function of this operator

$$\delta\langle T(w)\rangle = \frac{1}{2\pi i}\oint dz\,\epsilon(z)\,\langle T(z)T(w)\rangle = \frac{c}{12}\partial^3\epsilon(z),$$

where, in the last line, we used the expression (10.7.1) of the correlator and then the Cauchy theorem. This term is obviously zero for the *global* conformal transformations[10] but is instead different from zero for all the *local* conformal mappings. This means that, passing from the Euclidean plane in which $\langle T(z)\rangle_{plane} = 0$ to another geometry with the conformal transformation $z \to f(z)$, in the new geometrical domain the energy density is

[10] For these transformations $\epsilon(z)$ is at most quadratic in z.

different from zero! It is for this reason that the central charge is also called 'conformal anomaly'. It is also called 'trace anomaly', because in a CFT defined on a curved manifold it is no longer true that the trace Θ of the stress-energy tensor vanishes as a consequence of the scaling invariance of the theory: in fact the curvature R of the manifold introduces a length scale and in this case it is possible to prove that holds

$$\langle\Theta\rangle = -\frac{c}{12}R. \tag{10.7.10}$$

A non-zero value of c gives rise to a measurable physical effect, known as *Casimir effect*, which is analysed in Section 9, using the main properties of $T(z)$.

Properties of the stress-energy tensor. The first property is its transformation law under a local conformal mapping $z \to \eta$

$$T(z) = T(\eta)\left(\frac{d\eta}{dz}\right)^2 + \frac{c}{12}\{\eta, z\}, \tag{10.7.11}$$

where the last term is the *Schwartz derivative*

$$\{\eta, z\} \equiv \frac{\frac{d^3\eta}{dz^3}}{\frac{d\eta}{dz}} - \frac{3}{2}\left(\frac{\frac{d^2\eta}{dz^2}}{\frac{d\eta}{dz}}\right)^2. \tag{10.7.12}$$

The second important aspect of the stress-energy tensor is its Taylor–Laurent expansion (say, around the origin) in terms of the operators L_n

$$T(z) = \sum_{n=-\infty}^{\infty} \frac{L_n}{z^{n+2}}. \tag{10.7.13}$$

The Schwartz derivative. The Schwartz derivative of a function of the complex variable has the following properties:

1. $\{\eta, z\} = 0$ if and only if $\eta(z)$ is a Moebius transformation $\eta(z) = \frac{az+b}{cz+d}$.
2. it satisfies

$$\left\{\frac{a\eta+b}{c\eta+d}, z\right\} = \{\eta, z\}$$

$$\left\{\eta, \frac{az+b}{cz+d}\right\} = \{\eta, z\}\,(cz+d)^4.$$

3. under the sequence of transformations $z \to \eta \to \gamma$

$$\{\gamma, z\} = \{\gamma, \eta\} \left(\frac{d\eta}{dz}\right)^2 + \{\eta, z\}.$$

The last equation ensures the correct transformation properties of the stress-energy tensor. In fact, for the two individual mappings we have

$$T(z) = T(\eta) \left(\frac{d\eta}{dz}\right)^2 + \frac{c}{12}\{\eta, z\},$$
$$T(\eta) = T(\gamma) \left(\frac{d\gamma}{d\eta}\right)^2 + \frac{c}{12}\{\gamma, \eta\},$$

and therefore, substituting the second of these equations into the first

$$T(z) = \left[T(\gamma) \left(\frac{d\gamma}{d\eta}\right)^2 + \frac{c}{12}\{\gamma, \eta\}\right] \left(\frac{d\eta}{dz}\right)^2 + \frac{c}{12}\{\eta, z\} =$$
$$= T(\gamma) \left(\frac{d\gamma}{dz}\right)^2 + \frac{c}{2}\{\gamma, z\}.$$

Their action on a generic conformal field $A(z, \bar{z})$ is defined as follows

$$L_n A(z_1, \bar{z}_1) = \frac{1}{2\pi i} \oint_{C_1} dz\, (z - z_1)^{n+1}\, T(z) A(z_1, \bar{z}_1), \tag{10.7.14}$$

where C_1 is a closed contour around the point z_1. In other words, the application of L_n to $A(z_1, \bar{z}_1)$ filters the conformal field that appears in front of the power $(z - z_1)^{-n}$ in the OPE of $T(z)$ with $A(z_1, \bar{z}_1)$. The application of two of these operators is given by

$$L_n L_m A(z_1, \bar{z}_1) = \left(\frac{1}{2\pi i}\right)^2 \oint_{C_1'} dz' \oint_{C_1} dz\, (z' - z_1)^{n+1}\, (z - z_1)^{m+1}\, T(z') T(z) A(z_1, \bar{z}_1), \tag{10.7.15}$$

where both the two contours C_1' and C_1 have the point z_1 inside, with C_1'' external to C_1.

It is interesting to note that the operator expansion (10.7.8) can be equivalently expressed in terms of the commutator $[L_n, L_m]$. To compute this quantity, we need to exchange the two integration contourns, paying attention to the singular terms of the

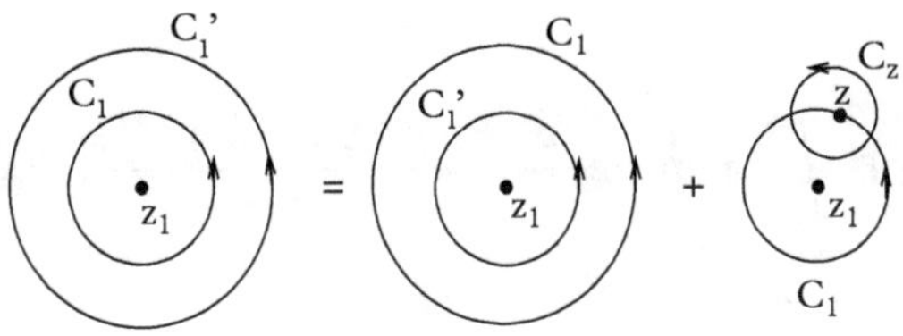

Fig. 10.7 *Exchange of the integration contours for the commutator* $[L_n, L_m]$.

OPE encountered in this exchange. The situation (Figure 10.7) involves[11] an integral on the contour C_z around the singularities and an integral on the contour C_1 of the point z_1

$$[L_n, L_m] = \left(\frac{1}{2\pi i}\right)^2 \oint_{C_z} dz' \oint_{C_1} dz\, (z'-z_1)^{n+1} (z-z_1)^{m+1}\, T(z')T(z) = \left(\frac{1}{2\pi i}\right)^2$$

$$\times \oint_{C_z} dz' \oint_{C_1} dz\, (z'-z_1)^{n+1} (z-z_1)^{m+1} \left(\frac{c/2}{(z'-z)^4} + \frac{2}{(z'-z)^2} T(z) + \frac{1}{z'-z} \partial T(z)\right).$$

Let us consider separately the results of integration of each term. For the first term we have

$$\frac{c}{2}\left(\frac{1}{2\pi i}\right)^2 \oint_{C_z} dz' \oint_{C_1} dz\, \frac{(z'-z_1)^{n+1} (z-z_1)^{m+1}}{(z'-z)^4} =$$

$$= \frac{c}{2\cdot 3!}(n+1)n(n-1)\, \frac{1}{2\pi i} \oint_{C_1} dz (z-z_1)^{n+m-1} =$$

$$= \frac{c}{12} n(n^2-1)\, \delta_{n+m,0}.$$

For the second term

$$2\left(\frac{1}{2\pi i}\right)^2 \oint_{C_z} dz' \oint_{C_1} dz\, \frac{(z'-z_1)^{n+1} (z-z_1)^{m+1}}{(z'-z)^2}\, T(z) =$$

$$= 2\,(n+1)\, \frac{1}{2\pi i} \oint_{C_1} dz (z-z_1)^{n+m+1}\, T(z) =$$

$$= 2(n+1)\, L_{n+m}.$$

[11] In the integrals we have omitted the field $A(z_1, \bar{z}_1)$ since it appears on both sides of the equation.

For the last term

$$\left(\frac{1}{2\pi i}\right)^2 \oint_{C_z} dz' \oint_{C_1} dz \frac{(z'-z_1)^{n+1}(z-z_1)^{m+1}}{z'-z} \partial T(z) =$$
$$= \frac{1}{2\pi i} \oint_{C_1} dz (z-z_1)^{n+m+2} \partial T(z) =$$
$$= -\frac{1}{2\pi i} \oint_{C_1} dz \partial (z-z_1)^{n+m+2} T(z) = -(n+m+2) L_{n+m}.$$

Putting now together all the expressions above and keeping in mind that analogous results hold for the anti-analytic part of the stress-energy tensor, we arrive at the commutation relations

$$\begin{aligned} [L_n, L_m] &= (n-m) L_{n+m} + \frac{c}{12} n(n^2-1) \delta_{n+m,0}, \\ [\bar{L}_n, \bar{L}_m] &= (n-m) \bar{L}_{n+m} + \frac{c}{12} n(n^2-1) \delta_{n+m,0}, \\ [L_n, \bar{L}_m] &= 0. \end{aligned} \tag{10.7.16}$$

These relations define the so-called *Virasoro algebra*: it provides the quantum version of the classical conformal algebra (10.5.9) and it reduces to it when $c = 0$.

An important remark. As a result of this analysis we have made a very important conceptual point that is worth: in two dimensions the problem of classifying all possible universality classes of critical phenomena simply consists of identifying all irreducible representations of the Virasoro algebra. From this point of view, the numerous variety of critical phenomena is on the same footing of the different behaviour of the irreducible representations of the rotation group where, according to the value of the angular momentum, the phenomenology is different but the underlying algebraic structure is the same.

10.8 Representation Theory

Let us discuss the representations of the Virasoro algebra. They can be equivalently analysed in terms of the conformal fields or in terms of vectors in a Hilbert space. There is in fact an isomorphism between the two pictures that can be established as follows.

Radial quantization. Consider the conformal transformation

$$\eta = \frac{L}{2\pi} \log z \tag{10.8.1}$$

that maps the entire complex plane into the infinite cylinder strip of width L, as it can be seen by writing $\eta = \tau + i\sigma$ and expressing z as $z = \rho e^{i\alpha}$

$$\tau = \frac{L}{2\pi}\log\rho, \quad \sigma = \frac{L}{2\pi}\alpha.$$

Circles of the z-plane are mapped in orthogonal sections of the cylinder. In particular, the origin is transformed in the section of the cylinder placed at $-\infty$, whereas the point at infinity of the z-plane is mapped in the section of the cylinder at $+\infty$. For this reason, the map (10.8.1) gives rise to the so-called *radial quantization scheme* in which the longitudinal direction of the cylinder is regarded as *time*, while the transversal direction as (compactified) *space*. Circles in the z-plane correspond to surfaces of equal time on the cylinder. Note that the temporal inversion $\tau \to -\tau$ is implemented by the transformation $z \to 1/\bar{z}$.

In the radial quantization scheme we can introduce the R-ordered product of the fields, analogously to the usual T-ordered product

$$R[\phi_1(z)\phi_2(w)] = \begin{cases} \phi_1(z)\phi_2(w), & \text{if} \quad |z| < |w| \\ \phi_2(w)\phi_1(z), & \text{if} \quad |w| < |w|. \end{cases} \tag{10.8.2}$$

We can also relate their OPE with the commutation relations, as we have already seen for the Virasoro generators. To this aim, let $\beta(z)$ and $\gamma(z)$ be two analytic fields and consider the integral

$$\oint_w \beta(z)\gamma(w)\, dz, \tag{10.8.3}$$

around the point w, taken in anti-clockwise direction. By using the radial quantization of these operators, (10.8.3) can be expressed as a difference of integrals computed along the circular contours of radius $|w| \pm \epsilon$ (10.9). These contours correspond to two time instants slightly different and therefore

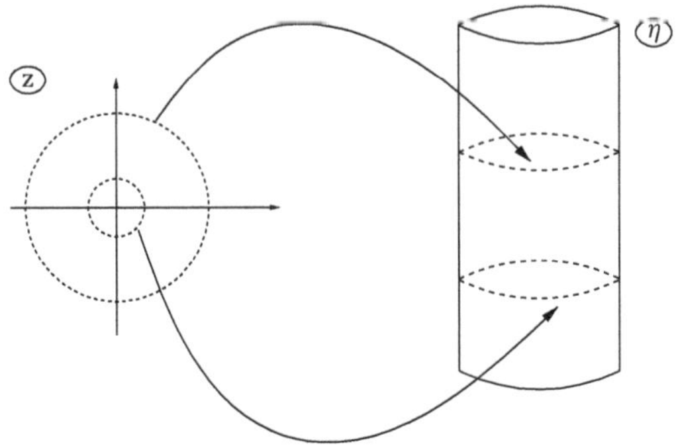

Fig. 10.8 *Conformal transformation of the complex plane onto the cylinder. Circles of different radius are mapped onto different sections of the cylinder.*

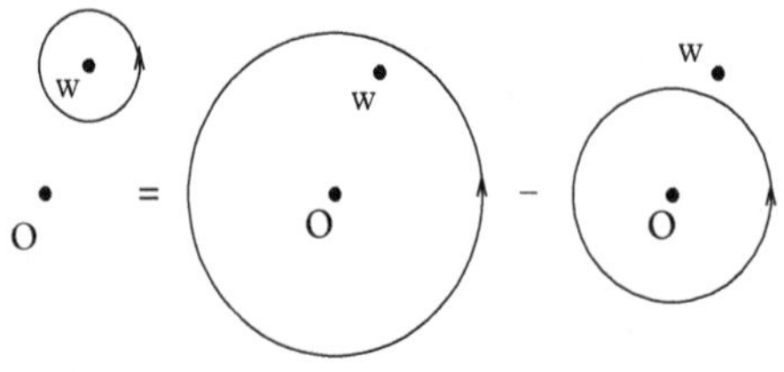

Fig. 10.9 *Commutator as difference of circular contours.*

$$\oint_w \beta(z)\,\gamma(w)\,dz = \oint_{C_1} \beta(z)\gamma(w)\,dz - \oint_{C_2} \gamma(w)\beta(z)dz = [\Gamma,\beta(w)], \tag{10.8.4}$$

where the operator Γ is given by the integral

$$\Gamma = \oint \gamma(z)\,dz$$

taken along a circle around the origin. In the limit $\epsilon \to 0$ the commutator so obtained corresponds to the usual equal time commutator of QFT.

Eqn. (10.8.4) allows us to compute the commutator of the generators of the Virasoro algebra with any primary field of conformal weight Δ. Using eqn. (10.6.6), we have

$$\begin{aligned} [L_n,\phi(w,\bar{w})] &= \frac{1}{2\pi i}\oint_w dz\,z^{n+1}\,T(z)\phi(w,\bar{w}) = \\ &= \frac{1}{2\pi i}\oint_w dz\,z^{n+1}\left[\frac{\Delta\,\phi(w,\bar{w})}{(z-w)^2} + \frac{\partial\phi(w,\bar{w})}{z-w} + \ldots\right] \\ &= \Delta(n+1)w^n\phi(w,\bar{w}) + w^{n+1}\partial\phi(w,\bar{w}), \end{aligned} \tag{10.8.5}$$

A similar expression holds for the anti-analytic generators $\bar{L}_n$.

Hilbert space of conformal states. In cylinder geometry it is possible to introduce a Hilbert space and a Hamiltonian H that will rule the (Euclidean) time evolution of the states. The explicit form of H will be given later. H also determines the time evolution of the fields $A(\sigma,\tau)$

$$A(\sigma,\tau) = e^{H\tau} A(\sigma,0)\,e^{-H\tau}. \tag{10.8.6}$$

To construct the Hilbert space we assume firstly the existence of a vacuum state $|\,0\rangle$. Any other state of this space can be constructed by acting on the vacuum state with certain operators, as it happens for the creation operators of usual QFTs. The initial states are those at $t \to -\infty$ and, thanks to the conformal mapping on the cylinder, they can be defined as

$$|\, A_{in}\rangle = \lim_{z,\bar{z}\to 0} A(z,\, \bar{z})\,|\, 0\rangle. \tag{10.8.7}$$

To introduce the final state, we need to define the adjoint operator of a conformal operator, here given by

$$[A(z,\, \bar{z})]^{\dagger} = A\left(\frac{1}{\bar{z}}, \frac{1}{z}\right) \frac{1}{z^{2\bar{\Delta}}}\, \frac{1}{\bar{z}^{2\Delta}}. \tag{10.8.8}$$

The reason of this definition stays in the relationship that exists between the usual definition of the adjoint operator in the present formulation and in the formulation done in the Minkowski space: the factor i, that is present in the Minkowski formulation and is instead absent in the time evolution (10.8.6), must be compensated by an explicit transformation of the time inversion $\tau \to -\tau$. The other extra factors in (10.8.8) are necessary to preserve the transformation properties of the adjoint operator under the conformal transformations. In fact, parameterize the point at infinity in terms of the map $w = 1/z$ and denote by $\hat{A}(w, \bar{w})$ the operator in these new coordinates. It is natural to pose

$$\langle A_{fin}\,| = \lim_{w,\bar{w}\to 0} \langle 0\,|\, \hat{A}(w, \bar{w}). \tag{10.8.9}$$

For the primary and quasi-primary fields, it is now possible to link $\hat{A}(w, \bar{w})$ to $A(z, \bar{z})$ since

$$\hat{A}(w, \bar{w}) = A(z,\, \bar{z}), \left(\frac{\partial z}{\partial w}\right)^{\Delta} \left(\frac{\bar{\partial}\, \bar{z}}{\bar{\partial}\, \bar{w}}\right)^{\bar{\Delta}} = A\left(\frac{1}{w}, \frac{1}{\bar{w}}\right) (-w^{-2})^{\Delta}\, (-\bar{w}^{-2})^{\bar{\Delta}},$$

and therefore

$$\begin{aligned}\langle A_{fin}\,| &= \lim_{w,\bar{w}\to 0} \langle 0\,|\, \hat{A}(w, \bar{w}) = \lim_{z,\bar{z}\to 0} \langle 0\,|\, A\left(\frac{1}{z}, \frac{1}{\bar{z}}\right) \frac{1}{z^{2\Delta}}\, \frac{1}{\bar{z}^{2\bar{\Delta}}} = \\ &= \lim_{z,\bar{z}\to 0} \langle 0\,|\, [A(z,\, \bar{z})]^{\dagger} = \left[\lim_{z,\, \bar{z}\to 0} A(\bar{z},\, z)\,|\, 0\rangle 0\right] = |\, A_{in}\rangle^{\dagger}.\end{aligned}$$

The final states can be thus defined as

$$\langle A_{fin}\,| \equiv \lim_{z,\, \bar{z}\to \infty} \langle 0\,|\, A(z,\, \bar{z})\, z^{2\Delta}\, \bar{z}^{2\bar{\Delta}}. \tag{10.8.10}$$

For the stress-energy tensor, the definition (10.8.8) of the adjoint operator implies the equality of the expressions

$$T^{\dagger}(z) = \sum_{-\infty}^{\infty} \frac{L_n^{\dagger}}{\bar{z}^{n+2}} \quad \text{and} \quad T\left(\frac{1}{\bar{z}}\right)\frac{1}{\bar{z}^4} = \sum_{-\infty}^{\infty} \frac{L_n}{\bar{z}^{-n+2}},$$

namely

$$L_n^{\dagger} = L_{-n}. \tag{10.8.11}$$

An analogous formula holds for the generators $\bar{L}_n$ of $\bar{T}(\bar{z})$. Applying now $T(z)$ ($\bar{T}(\bar{z})$) to the vacuum state

$$T(z)\,|0\rangle = \sum_{-\infty}^{\infty} \frac{L_n}{z^{n+2}}\,|0\rangle,$$

and demanding their regularity at the origin, we arrive at the conditions that identify this state

$$\begin{aligned} L_n\,|0\rangle &= 0, \quad n \geq -1. \\ \bar{L}_n\,|0\rangle &= 0, \quad n \geq -1. \end{aligned} \tag{10.8.12}$$

In particular, the conditions $L_{0,\pm 1}|0\rangle = 0$ and $\bar{L}_{0,\pm 1}|0\rangle = 0$ establish that the vacuum state is invariant under the $SL(2,C)$ transformations: the vacuum state is the same for all the observers related by the global conformal transformations. Moreover, these relations imply that the vacuum expectation values of T and $\bar{T}$ vanish

$$\langle 0 \mid T(z) \mid 0\rangle = \langle 0 \mid \bar{T}(\bar{z}) \mid 0\rangle = 0. \tag{10.8.13}$$

10.8.1 Representation Theory: the Space of the Conformal States

For the sake of simplicity, let us focus our attention only on the analytic sector of the theory (similar results hold for the anti-analytic one). Consider the state created by the analytic component of the primary field $\phi_{\Delta}(z)$ of conformal weight Δ

$$\mid \Delta\,\rangle \equiv \phi_{\Delta}(0) \mid 0\,\rangle. \tag{10.8.14}$$

Using the OPE (10.6.6) and the definition (10.7.13) of $T(z)$ in terms of the L_ns, it is easy to show that

$$\begin{aligned} L_0 \mid \Delta\rangle &= \Delta \mid \Delta\rangle \\ L_n \mid \Delta\rangle &= 0, \quad n > 0. \end{aligned} \tag{10.8.15}$$

Hence, $| \Delta \rangle$ is an eigenstate of L_0 (with eigenvalue Δ). It can be normalized as

$$\langle \Delta | \Delta \rangle = 1. \tag{10.8.16}$$

Consider now the *descendant states* of $| \Delta \rangle$, i.e. the states that are obtained by acting on $| \Delta \rangle$ by the operators L_n with a negative index. To avoid an over-counting of these states,[12] it is convenient to introduce an ordering, for instance

$$\begin{gathered} | \Delta; n_1, n_2, \ldots n_k \rangle \equiv L_{-n_1} L_{-n_2} \ldots L_{-n_k} \mid \Delta \rangle \\ n_1 \leq n_2 \leq \ldots \leq n_k. \end{gathered} \tag{10.8.17}$$

Using the commutation relations of the Virasoro algebra, we have

$$L_0 \mid \Delta; n_1, n_2, \ldots n_k \rangle = (\Delta + N) \mid \Delta; n_1, n_2, \ldots n_k \rangle, \quad N = \sum_{i=1}^{k} n_i. \tag{10.8.18}$$

This shows that the descendant states are also eigenstates of L_0 with an eigenvalue related to their level N. The negative modes L_{-m} of the Virasoro algebra behave then as the creation operators of the familiar quantum harmonic oscillator, the only difference is that they move by m the eigenvalues of the state they act on. This situation is graphically represented in Figure 10.10.

Structure of the Hilbert space. The Hilbert space of the conformal states has a nested structure. To reach the level N, for instance, we can act directly with the operator L_{-N} on the state $| \Delta \rangle$ or we can act on any descendant of a level M ($M < N$) with L_{M-N} or with

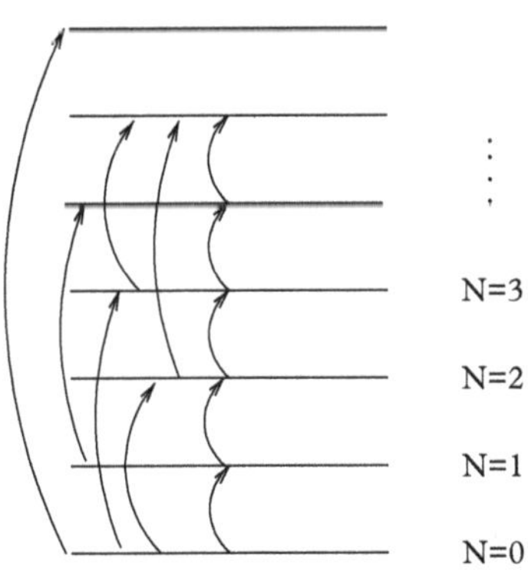

Fig. 10.10 *Levels of different N and action of the operators L_{-m}.*

[12] For the commutation relations of the Virasoro algebra, any other ordering can be expressed as a linear combination of the states given in the text.

any other ordered sequence of operators $L_{-n_j}\dots L_{-n_k}$ that satisfy the condition $\sum_{i=j}^{k} n_i = M-N$. This nested structure gives rise to an exponential growth of the dimensions of the L_0-eigenspaces. These dimensions can be computed noting that the problem consists of to determine in how many ways a positive integer number N can be expressed as sum of all possible integer numbers less that it. This combinatorial problem can be solved in terms of the generating function

$$f(q) = \prod_{n=1}^{\infty} \frac{1}{1-q^n}. \tag{10.8.19}$$

Denoting by $P(N)$ the dimension of the space at level N, we have

$$\sum_{N=0}^{\infty} P(N)\, q^N = \prod_{n=1}^{\infty} \frac{1}{1-q^n}, \tag{10.8.20}$$

To check the validity of this expression is sufficient to expanding each factor on the right-hand side in terms of the geometrical series and then gather together the various terms to form the power of q^N. Expanding in series the function $f(q)$ we have

$$f(q) = 1+q+2q^2+3q^3+5q^4+7q^5+11q^6+15q^7+22q^8+30q^9+42q^{10}+\dots$$

from which we can read the values of $P(N)$. As anticipated, they grow extremely fast and their asymptotic estimate is given by the Hardy–Ramanujan formula

$$P(N) \simeq \frac{\exp\left[\pi\sqrt{\frac{2N}{3}}\right]}{4\sqrt{3}N}. \tag{10.8.21}$$

Let us now investigate in more detail the descendent states. At the level $N=1$ there is only the state $L_1 \mid \Delta\rangle$ and its norm is easily obtained using eqn. (10.8.11), the commutation relations (10.7.17) and the properties (10.8.15) of the state $\mid \Delta\rangle$

$$\begin{aligned}
&\langle \Delta \mid L_{-1}^{\dagger} L_{-1} \mid \Delta\rangle = \langle \Delta \mid L_1 L_{-1} \mid \Delta\rangle = \\
&= \langle \Delta \mid [L_1, L_{-1}] \mid \Delta\rangle = \langle \Delta \mid 2L_0 \mid \Delta\rangle = 2\Delta\langle \Delta|\Delta\rangle = 2\Delta.
\end{aligned}$$

We can also easily compute the norm of the descendant state $L_{-m} \mid \Delta\rangle$

$$\begin{aligned}
&\langle \Delta \mid L_m L_{-m} \mid \Delta\rangle = \langle \Delta \mid [L_m, L_{-m}] \mid \Delta\rangle = \\
&= 2m\,\langle \Delta \mid L_0 \mid \Delta\rangle + \frac{c}{12} m(m^2-1)\,\langle \Delta|\Delta\rangle = \\
&= 2m\Delta + \frac{c}{12} m(m^2-1).
\end{aligned}$$

The computation soon becomes more involved for the other matrix elements of the $P(N) \times P(N)$ matrix, called the Gram matrix, given by the scalar product of the various descendants of the level N

$$M^{(N)} = \begin{pmatrix} \langle\Delta|L_1^N L_{-1}^N|\Delta\rangle & . & . & \langle\Delta|L_1^N L_{-N}|\Delta\rangle \\ . & . & . & . \\ . & . & . & . \\ . & . & . & . \\ \langle\Delta|L_N L_{-1}^N|\Delta\rangle & . & . & \langle\Delta|L_N L_{-N}|\Delta\rangle \end{pmatrix}. \quad (10.8.22)$$

As an explicit example, we present here the computation of the Gram matrix of level $N = 2$, given by

$$M^{(2)} = \begin{pmatrix} 4\Delta(2\Delta+1) & 6\Delta \\ 6\Delta & 4\Delta + c/2 \end{pmatrix}. \quad (10.8.23)$$

When the determinant of all the Gram matrices $M^{(N)}$ is different from zero, all the descendant states are linearly independent and their set provides, by construction, an irreducible representation of the Virasoro algebra. The space of states V_Δ so constructed is called the *conformal family* (or the *Verma module*) of the primary field $\phi_\Delta(z)$ where the seed state $|\,\Delta\rangle$ behaves as *highest state vector* of the Virasoro algebra.

10.8.2 Representation Theory: The Space of Conformal Fields

The representation theory of the Virasoro algebra can be also developed on the space of the conformal fields, similarly to the one in the Hilbert space of the states. This study provides however useful information on the structure of the conformal families.

Given a conformal field $A(z)$, for the definition of the operators L_n, we have

$$T(z)A(w) = \sum_{\infty}^{\infty} \frac{1}{(z-w)^{n+2}} (L_n A)(w). \quad (10.8.24)$$

If we specialize this expression to the case where $A(z)$ is a primary field of conformal weight Δ, with an OPE given by (10.6.6), we can easily extract the action of L_n on this primary field

$$\begin{aligned} (L_0\phi)(z) &= \Delta\phi(z), \\ (L_{-1}\phi)(z) &= \partial\phi(z), \\ (L_n\phi)(z) &= 0 \quad n \geq 1. \end{aligned} \quad (10.8.25)$$

The other L_m with negative index create the descendant fields

$$\phi^{(m)}(z) \equiv (L_{-m}\phi)(z),$$

and we can recover all other fields by recurrence

$$\phi^{(n_1,n_2,\ldots,n_k)}(z) \equiv (L_{-n_1}L_{-n_2}\ldots L_{-n_k}\phi)(z), \tag{10.8.26}$$

adopting the usual ordering $n_1 \leq n_2 \leq \ldots \leq n_k$. These fields are also eigenvectors of L_0 with eigenvalues given by

$$L_0\,\phi^{(n_1,n_2,\ldots,n_k)}(z) = (\Delta + n_1 + n_2 + \ldots n_k)\,\phi^{(n_1,n_2,\ldots,n_k)}(z). \tag{10.8.27}$$

Note that a significant example of a descendant field is provided by the stress-energy tensor! In fact, taking the identity operator I as a primary field, we have

$$(L_{-2}I)(w) = \frac{1}{2\pi i}\oint dz \frac{1}{z-w} T(z) I = T(w). \tag{10.8.28}$$

This explains the more complicated transformation law of $T(z)$, given by eqn. (10.7.11), with respect to the one of a primary field: it is because it is a descendant field.

When the descendant fields (10.8.26) are all linearly independent, they form together with the primary field $\phi(z)$ an irreducible representation of the Virasoro algebra. Since $(L_{-1}\phi)(z) = \partial\phi(z)$, we also deduce that in the conformal family of the operator $\phi(z)$ there are automatically all the derivatives of the primary fields and its descendants.

Let us now prove a result that will be extremely important for the development of the formalism:

All correlation functions of the descendant fields can be obtained by acting with *linear differential operators* $\mathcal{L}_a$ on the correlation functions of the primary fields. The operators $\mathcal{L}_a$ are uniquely fixed by the Virasoro algebra.

We present this result for the simplest case of a correlation function of the primary fields $\phi_i(z_i)$ $(i = 1, \ldots, n-1)$ and the descendant field $(L_{-k}\phi_n)(z)$ of the primary field ϕ_n, where we have

$$\langle \phi_1(z_1)\ldots\phi_{n-1}(z_{n-1})(L_{-k}\phi_n)(z)\rangle = \mathcal{L}_{-k}\,\langle \phi_1(z_1)\ldots\phi_{n-1}(z_{n-1})\phi(z)\rangle. \tag{10.8.29}$$

The linear differential operator $\mathcal{L}_{-k}$ is expressed as

$$\mathcal{L}_{-k} = -\sum_{i=1}^{n-1}\left[\frac{(1-k)\Delta_i}{(z_i - z)^k} + \frac{1}{(z_i - z)^{k-1}}\frac{\partial}{\partial z_i}\right]. \tag{10.8.30}$$

To prove eqn. (10.8.29) is convenient to start from the Ward identity

$$\langle T(z)\phi_1(z_1)\dots\phi_n(z_n)\rangle = \sum_{i=1}^{n}\left[\frac{\Delta_i}{(z-z_i)^2} + \frac{1}{z-z_i}\frac{\partial}{\partial z_i}\right]\langle\phi_1(z_1)\dots\phi_n(z_n)\rangle,$$

and consider the limit $z \to z_n$. Using eqn. (10.8.24), the left-hand side of the Ward identity becomes

$$\sum_{k\geq 0}^{\infty}(z-z_n)^{k-2}\langle\phi_1(z_1)\dots(L_{-k}\phi_n)(z_n),$$

and, for the Cauchy formula

$$\langle\phi_1(z_1)\dots\phi_{n-1}(z_{n-1})(L_{-k}\phi_n)(z)\rangle = \frac{1}{2\pi i}\oint_{z_n} dz(z-z_n)^{1-k} \times$$
$$\left[\sum_{i=1}^{n}\left[\frac{\Delta_i}{(z-z_i)^2} + \frac{1}{z-z_i}\frac{\partial}{\partial z_i}\right]\langle\phi_1(z_1)\dots\phi_n(z_n)\rangle\right].$$

Since the residue at infinity of this expression vanishes, we can use the residue theorem of complex analysis to express the contour integral around the point z_n in terms of the contour integrals (taken clockwise) around the points z_i $(i = 1,\dots,n-1)$. However, the last quantities are simply the opposite of the contour integrals taken in the usual anti-clockwise direction (Figure 10.11). Hence

$$\langle\phi_1(z_1)\dots\phi_{n-1}(z_{n-1})(L_{-k}\phi_n)(z)\rangle = -\frac{1}{2\pi i}\sum_{j=1}^{n-1}\oint_{z_j} dz(z-z_n)^{1-k} \times$$
$$\times\left[\sum_{i=1}^{n}\left[\frac{\Delta_i}{(z-z_i)^2} + \frac{1}{z-z_i}\frac{\partial}{\partial z_i}\right]\langle\phi_1(z_1)\dots\phi_n(z_n)\rangle\right] =$$
$$= -\sum_{j-1}^{n-1}\left[\frac{(1-k)\Delta_j}{z_j-z_n)^k} + \frac{1}{z_j-z_n)^{k-1}}\frac{\partial}{\partial z_j}\right]\langle\phi_1(z_1)\dots\phi_n(z_n)\rangle.$$

Thus we obtain eqn. (10.8.29). Similar formulae can be easily derived for all other descendant fields.

There are several important consequences of eqn. (10.8.29) and alike.

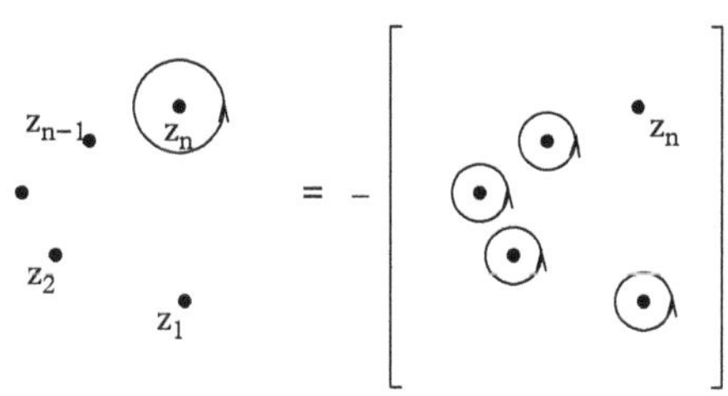

Fig. 10.11 *Theorem of the residues applied to eqn. (10.8.31).*

Orthogonality of conformal families. The first consequence is on the orthogonality condition of the two point correlation functions of the descendant fields. In fact, the orthogonality condition (10.4.5) of the primary fields automatically implies that also the two-point correlation functions of the descendant fields of two different families vanish. Hence, there is a complete orthogonality between two different conformal families.

Structure constants of descendant fields. The second important consequence concerns the structure constants of the descendant fields in the operator algebra (10.2.13). As an outcome of the existence of the linear differential operators $\mathcal{L}_a$, these quantities are proportional to the structure constants c_{ijp} of the primary fields, with a proportionality coefficient that is *uniquely* determined by the Virasoro algebra. In more detail, denoting by $C_{ijp}^{(k,\bar{k})}$ the structure constant of two primary fields ϕ_i, ϕ_j with a descendant $\phi_p^{(k,\bar{k})}$ at the levels k and $\bar{k}$ of the primary field ϕ_p, we have

$$C_{ijp}^{(k,\bar{k})} = C_{ijp}\,\beta_{ijp}^{(k)}\,\bar{\beta}_{ijp}^{(\bar{k})}, \tag{10.8.31}$$

where $\beta_{ijp}^{(k)}$ is a rational expression of the conformal weights Δ_i (alias) and the central charge c. The same for $\bar{\beta}_{ijp}^{(k)}$, a function of $\bar{\Delta}_i$ and c. Both quantities can be computed in a purely algebraic way by applying the relative operators $\mathcal{L}_a$ of the descendant field to its the three-point correlation functions with the two primary fields.

Notice that eqn. (10.8.31) implies that, if $C_{ijp} = 0$ then all other structure constants of the descendant fields vanish as well. In another words, if two primary fields ϕ_i and ϕ_j do not couple to the primary field ϕ_p, they do not couple either to any of its descendants. Hence, in two-dimensional CFTs, the determination of the structure constance of the operatorial algebra (10.2.13) simply reduces to determine only the structure constants of the primary fields.

10.9 Hamiltonian on a Cylinder Geometry and Casimir Effect

Consider a conformal theory defined on a cylinder of width L with periodic boundary conditions. The coordinates along the cylinder are given by $-\infty < \tau < +\infty$ and $0 \leq \sigma \leq L$. This theory can be analysed in terms of the conformal transformation

$$w \equiv \tau + i\sigma = \frac{L}{2\pi}\log z, \tag{10.9.1}$$

that maps the plane into the cylinder. Using the transformation law (10.7.11) of the stress-energy tensor, we have

$$T_{cyl}(w) = \left(\frac{2\pi}{L}\right)^2 \left[T_{pl}(z)\,z^2 - \frac{c}{24}\right], \tag{10.9.2}$$

with an analogous expression for $\bar{T}$. We can now define the Hamiltonian of this conformal theory in terms of the space integral of $\hat{T}_{\tau\tau}$

$$\begin{aligned} H &= \frac{1}{2\pi}\int_0^L \hat{T}_{\tau\tau}(\sigma)\,d\sigma = \frac{1}{2\pi}\int_0^L (T(\sigma)+\bar{T}(\sigma))\,d\sigma = \\ &= \frac{2\pi}{L}(L_0+\bar{L}_0) - \frac{\pi c}{6L}, \end{aligned} \tag{10.9.3}$$

where we have used eqn. (10.9.2) and the definition of the Virasoro generators in the complex plane

$$L_0 = \frac{1}{2\pi i}\oint zT(z)\,dz, \quad \bar{L}_0 = -\frac{1}{2\pi i}\oint \bar{z}\bar{T}(\bar{z})\,d\bar{z}.$$

The theory on the cylinder has also a translation invariance along the σ axis and therefore we can also define the momentum operator P

$$P = \frac{1}{2\pi}\int_0^L \hat{T}_{\tau\sigma}\,d\sigma = \frac{2\pi}{L}(L_0-\bar{L}_0). \tag{10.9.4}$$

This operator commutes with H. From the explicit expressions of H and P it can be seen that their eigenvectors are in one-to-one correspondance with the eigenvectors of $L_0+\bar{L}_0$ and $L_0-\bar{L}_0$. The minimum value E_0 of the energy is

$$E_0 = -\frac{\pi c_{eff}}{6L}, \tag{10.9.5}$$

where

$$c_{eff} = c - 24\Delta_{min}, \tag{10.9.6}$$

is the *effective central charge*, given by the central charge c and the minimum eigenvalue Δ_{min} of L_0. For unitary theories $\Delta_{min} = 0$ and therefore $c_{eff} = c$. Futhermore, for unitary theories $c > 0$. For non-unitary theories, Δ_{min} is generically negative as well as the central charge c. However, as we shall see in the next chapter, an interesting observation is that the effective central charge of all minimal conformal models (either unitary or not) is always positive

$$c_{eff} = c - 24\Delta_{min} > 0. \tag{10.9.7}$$

The finite expression (10.9.5) of the ground state energy on a cylinder is known as *Casimir effect*: it depends on its width L and vanishes in the limit $L \to \infty$ when the cylinder reduces to a plane. In addition to its conceptual relevance, this formula is useful to identify which conformal theory is behind the critical behaviour of a statistical model defined on a lattice: in fact, it is sufficient to study its ground state energy on a cylinder geometry as a function of L and extract accordingly its effective central charge.

The previous expressions of H and P are also useful to determine the transfer matrix of the conformal models. For simplicity, let us focus the attention on a unitary conformal models, with $c > 0$ and $\Delta_i > 0$. In the plane, the two-point correlation function of a primary field is

$$\langle \phi(z,\bar{z})\phi(z',\bar{z}')\rangle = (z-z')^{-2\Delta}(\bar{z}-\bar{z}')^{-2\bar{\Delta}}. \tag{10.9.8}$$

Using the transformation law (10.6.3) of the primary fields under a conformal transformation and the map (10.9.1), we can immediatly write down the correlation function on the cylinder

$$\langle \phi(w,\bar{w})\phi(w',\bar{w}')\rangle = \left(\frac{\pi}{L}\right)^{2(\Delta+\bar{\Delta})} \frac{1}{(\sinh\pi(w-w)/L)^{2\Delta}(\sinh\pi(\bar{w}-\bar{w}')/L)^{2\bar{\Delta}}}.$$

Posing $w = \tau + i\sigma$ and $w' = \tau' + i\sigma'$, for $\tau > \tau'$ this expression can be expanded as

$$\left(\frac{\pi}{L}\right)^{2x} \sum_{N,\bar{N}=0}^{\infty} a_N a_{\bar{N}} \exp[-2\pi(x+N+\bar{N})(\tau-\tau')/L]$$
$$\times \exp[2\pi i(s+N+\bar{N})(\sigma-\sigma')/L], \tag{10.9.9}$$

where $x = \Delta + \bar{\Delta}$ is the scaling dimension of the operator, $s = \Delta - \bar{\Delta}$ is its spin and the coefficients a_N are given by

$$a_N = \frac{\Gamma(x+N)}{\Gamma(x)N!}.$$

The expression above can be compared with the one computed using the transfer matrix. In the transfer matrix approach, the conformal fields $\phi(u,v)$ are regarded as operators that act on states of the Hilbert space on the cylinder, with the time evolution provided by eqn. (10.8.6). Hence

$$\langle \phi(w,\bar{w})\phi(w',\bar{w}')\rangle \equiv \langle 0 \mid e^{H\tau}\phi(0,\sigma)e^{-H\tau}e^{H\tau'}\phi(0,\sigma')e^{-H\tau'} \mid 0\rangle.$$

On the other hand, we can use the momentum operator P to express $\phi(0,\sigma)$ as

$$\phi(0,\sigma) = e^{-iP\sigma}\phi(0,0)e^{iP\sigma}.$$

Inserting now in the expression of the correlation function the completeness relation of the eigenstates $| n, k\rangle$ of the energy and the momentum, we have

$$\langle \phi(w,\bar{w})\phi(w',\bar{w}')\rangle = \sum_{n,k} | \langle 0 | \phi(0,0) | n,k\rangle |^2 \; e^{-(E_n-E_0)(\tau-\tau')+ik(\sigma-\sigma')}. \tag{10.9.10}$$

Comparing with (10.9.9), we see that the energy and the momentum of these states are given, as expected, by

$$E_n = E_0 + 2\pi(x+N+\bar{N})/L, \quad p_n = 2\pi(s+N+\bar{N})/L \tag{10.9.11}$$

with $E_0 = -\pi c/(6L)$. The matrix element of the operator ϕ on the ground state, here denoted by $| \phi\rangle$, is

$$\langle 0 | \phi(0,0) | \phi\rangle = \left(\frac{2\pi}{L}\right)^x, \tag{10.9.12}$$

while the matrix elements on the descendant states are given by

$$| \langle 0 | \phi(0,0) | \phi, N, \bar{N}\rangle |^2 = \left(\frac{2\pi}{L}\right)^{2x} a_N a_{\bar{N}}. \tag{10.9.13}$$

The same can be done for the three-point functions of the primary fields. Trasforming their expression from the plane to the cylinder and expanding it for $\tau_1 \gg \tau_2 \gg \tau_3$, we have

$$\begin{aligned}\langle \phi_i(\tau_1,\sigma_1)\phi_j(\tau_2,\sigma_2)\phi_k(\tau_3,\sigma_3)\rangle = C_{ijk} \left(\frac{2\pi}{L}\right)^{x_i+x_j+x_k} & e^{-2\pi x_i(\tau_1-\tau_2)/L} e^{-2\pi x_k(\tau_2-\tau_3)/L} \\ & \times e^{2\pi i s_i(\sigma_1-\sigma_2)/L} e^{2\pi i s_k(\sigma_2-\sigma_3)/L}. \end{aligned} \tag{10.9.14}$$

Comparing this expression with the one obtained by the operatorial formalism, one can conclude that the structure constants C_{ijk} of the primary fields is given by the matrix element of the lowest energy states of the conformal families

$$\langle \phi_i | \phi(0,0) | \phi_k\rangle = c_{ijk} \left(\frac{2\pi}{L}\right)^{x_j}. \tag{10.9.15}$$

Its derivation is left as an exercise.

10.10 Entanglement Entropy

Interestingly enough, in addition to the Casimir energy, the central charge also determines the entanglement entropy of the underlying CFT. Here we present a brief discussion of the problem together with some relevant formulae: interested readers are referred to the chapter references.

Quantum entanglement is one of the most striking fingerprints of quantum mechanics and its measure is an important theoretical problem. For an extended quantum system one of these measures involves the so-called *entanglement entropy* defined as follows. Let us assume that the whole system $\mathcal{S}$ is in a pure quantum state $|\Psi\rangle$, therefore with density matrix $\rho = |\Psi\rangle\langle\Psi|$, and an observer A performs measurements of a complete set of observables only on a sub-set A, while another observer B does the same for the remainder part of the system $B = \mathcal{S} - A$. The reduced density matrix for the observer A is $\rho_A = \mathrm{Tr}_B\, \rho$. The entanglement entropy is then given by the von Neumann entropy $S_A = -\mathrm{Tr}_A\, \rho_A \log \rho_A$ associated to this reduced density matrix. One can also show that $S_A = S_B$, so that the two observers are on the same footing. For an untangled product state, we have $S_A = 0$ while S_A assumes its maximum value for a maximally entangled state.

Path integral formulation. Consider for simplicity a one-dimensional quantum system in some domain of length L described by a bosonic field $\phi(x)$ and Hamiltonian H. For a thermal configuration at temperature β the density matrix element between two field configurations $\phi''(x'')$ and $\phi'(x')$ is given by

$$\rho(\phi'(x')|\phi''(x'')) = Z^{-1}(\beta)\,\langle\phi''(x'')|e^{-\beta H}|\phi'(x')]\rangle \tag{10.10.1}$$

and can be expressed through a Euclidean path integral

$$\rho = Z^{-1}\int \mathcal{D}\phi \prod_x \delta(\phi(x,0) - \phi'(x')) \prod_x \delta(\phi(x,\beta) - \phi''(x''))\, e^{-S_E} \tag{10.10.2}$$

where S_E is the Euclidean action. The normalization factor ensures that $\mathrm{Tr}\rho = 1$ and is found by posing $\phi'(x) = \phi''(x)$ and integrating over these configurations: this sews together the edges along the Euclidean times $\tau = 0$ and $\tau = \beta$ with the corresponding geometry given by a cylinder of circumference β.

Imagine now that the sub-system A is made of several disjoint intervals (open cuts) with extrema $(u_1, v_1), \ldots, (u_M, v_M)$. To compute the entanglement entropy for this sub-system we can use the replica trick, which consists of the following steps: first consider n copies of the original system, labelled by an integer k ($k = 1, 2, \ldots, n$) and sew them together cyclically along the cuts so that $\phi'_k(x) = \phi''_{k+1}(x)$, with periodic boundary conditions along the Euclidean time axis expressed by the condition $\phi'_n(x) = \phi''_1(x)$. This procedure defines an n-sheeted Riemann surface $\mathcal{R}_n$ as the one in Figure 10.12. Denoting the path integral on this n-sheeted geometry as $Z_n(A)$ we have

$$\mathrm{Tr}\,\rho_A^n = \frac{Z_n(A)}{Z^n}. \tag{10.10.3}$$

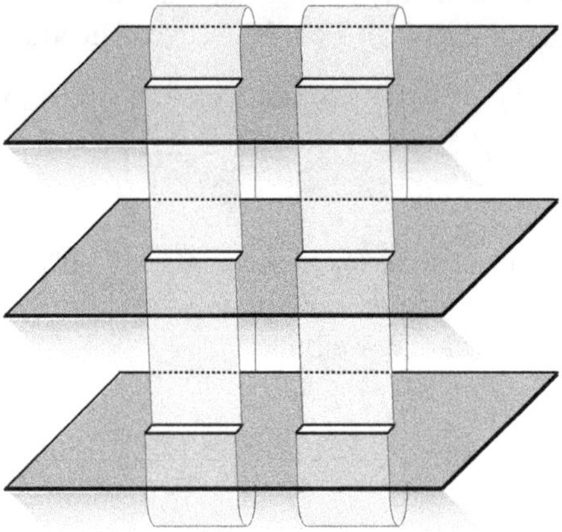

Fig. 10.12 *A representation of the n-replicated quantum system where the sub-system A consists of two disjoint intervals. Figure courtesy of Erik Tonni.*

If λ are the eigenvalues of ρ_A (with $0 \leq \lambda \leq 1$) then $\mathrm{Tr}\rho_A = \sum_\lambda \lambda = 1$) and $\mathrm{Tr}\,\rho_A^n = \sum_\lambda \lambda^n$. Therefore the left-hand side in eqn. (10.10.3) is absolutely convergent and analytic for all $\mathrm{Re}\,n > 1$. This implies that it exists the derivative of this quantity with respect to n and if the entropy $\rho_A = -\sum_\lambda \log\lambda$ is finite, the limit of $n \to 1^+$ of the first derivative of $\mathrm{Tr}\,\rho_A^n$ converges to this value. Therefore for the entanglement entropy there is the suggestive formula

$$S_A = -\lim_{n\to 1} \frac{\partial}{\partial n} \frac{Z_n(A)}{Z^n}. \tag{10.10.4}$$

Single interval and twist operators. Let us consider a QFT nearby its critical point, with its universal properties described by a CFT. Notice that

$$\mathrm{Tr}\,\rho_A^n == \langle 0|0\rangle_{\mathcal{R}_n}, \tag{10.10.5}$$

where on the right-hand side there is the vacuum expectation value in the n-sheeted surface. In a CFT to determine this quantity it is enough to control how it transforms under a general conformal transformation, which is equivalent to insert the holomorphic part of the stress-energy tensor. Hence we have

$$\langle 0|0\rangle_{\mathcal{R}_n} \simeq \langle T(w)\rangle_{\mathcal{R}_n}. \tag{10.10.6}$$

We present the computation of this quantity in the simplest case consisting of a single interval of length l (of extrema u and v) in an infinitely long one-dimensional quantum system at zero temperature. Using the conformal mapping $w \to \zeta = (w-u)/(w-v)$, the two branch points are mapped to 0 and ∞. Moreover, the analytic transformation

$$\zeta \to z = \zeta^{1/n} = \left(\frac{w-u}{w-v}\right)^{1/n}, \tag{10.10.7}$$

maps the n-sheeted Riemann surface $\mathcal{R}_n$ to the z-plane $\mathbb{C}$. Using now the transformation properties (10.7.11) of the stress-energy tensor and the fact that $\langle T(z)\rangle_{\mathbb{C}} = 0$ by rotational and translation invariance, we have

$$\langle T(w)\rangle_{\mathcal{R}_n} = \frac{c}{12}\{z,w\} = \frac{c(1-(1/n)^2)}{24}\frac{(v-u)^2}{(w-u)^2(w-v)^2}. \tag{10.10.8}$$

This expression can be compared with a Ward identity coming from the OPE of the stress-energy tensor in the presence of two *twist fields* $\Phi_n(u)$ and $\Phi_{-n}(v)$ placed at the extrema u and v of the cut with scaling dimensions $\Delta_n = \overline{\Delta}_n = (c/24)(1-(1/n)^2)$

$$\begin{aligned}\langle T(w)\Phi_n(u)\Phi_{-n}(v)\rangle_{\mathbb{C}} &= \frac{\Delta_n}{(w-u)^2(w-v)^2(v-u)^{2\Delta_n-2}(\overline{v}-\overline{u})^{2\overline{\Delta}_n}}\\ &= \left(\frac{\Delta_n}{(w-u)^2}+\frac{\Delta_n}{(w-v)^2}+\frac{1}{w-u}\frac{\partial}{\partial u}+\frac{1}{w-v}\frac{\partial}{\partial v}\right)\langle\Phi_n(u)\Phi_{-n}(v)\rangle_{\mathbb{C}}\end{aligned} \tag{10.10.9}$$

with normalization chosen as

$$\langle\Phi_n(u)\Phi_{-n}(v)\rangle_{\mathbb{C}} = |u-v|^{-2\Delta_n-2\overline{\Delta}_n}. \tag{10.10.10}$$

On each sheet the insertion of $T(w)$ is given by (10.10.9) and since this has to be inserted on all sheets, we need an extra factor n in front of this formula. The formulae above imply that $\mathrm{Tr}\,\rho_A^n = Z_n(A)/Z^n$ behaves under a generic conformal transformation as the nth power of the two-point function of the twist operator

$$\mathrm{Tr}\,\rho_A^n = d_n\left(\frac{u-v}{a}\right)^{-c/6(n-1/n)}, \tag{10.10.11}$$

where the overall constant d_n cannot be determined by CFT and the microscopic lattice spacing a has been inserted to make the result dimensionless. Employing now the formula (10.10.4) the entanglement entropy for a single interval is given by

$$S_A = \frac{c}{3}\log\frac{l}{a} + d, \tag{10.10.12}$$

where c is the central charge of the system, a its microscopic lattice spacing and d a non-universal constant.

Finite temperature. A CFT at a finite temperature corresponds to a theory defined on an infinitely long cylinder of circumference β. The conformal transformation that realizes a map from an original w plane into the cylinder is $w \to w' = (\beta/2\pi)\log w$. This leads to the following expression of the entanglement entropy

$$S_A(\beta) = \frac{c}{3}\log\left(\frac{\beta}{\pi a}\sinh\frac{\pi l}{\beta}\right) + d. \qquad (10.10.13)$$

Notice that for $l \ll \beta$, as before $S_A \simeq (c/3)\log(l/a)$ while in the opposite limit $l \gg \beta$, we have $S_A \simeq (c/3)(l/\beta)$. In the latter limit, the von Neumann entropy is extensive and its density is in agreement with the standard thermodynamic entropy of an isolated system of length l. Hence, eqn. (10.10.13) describes the crossover between the quantum entanglement entropy (realized for $\beta \to \infty$) and the thermodynamic entropy of a system (for finite values of β).

Finite systems. Making the replacement $\beta \to L$ and orienting the branch cut perpendicular to the axis of the cylinder, it is possible to use the previous mapping to compute the entanglement entropy of a sub-system of length l in a finite system of total length L subjected to periodic boundary conditions. The formula reads in this case

$$S_A = \frac{c}{3}\log\left(\frac{L}{\pi a}\sin\frac{\pi l}{L}\right) + d. \qquad (10.10.14)$$

Notice that this formula is correctly symmetric under the transformation $l \to (L - l)$ and reaches its maximum when $l = L/2$.

Non-critical systems. When the system is not exactly at the critical point but possesses instead a finite correlation length ξ, smaller than the length of each interval, the various disjoint parts of the systems do not essentially interact and this results in an entanglement entropy given by

$$S_A \simeq \frac{c}{6}\log\frac{\xi}{a}. \qquad (10.10.15)$$

Remarks. We close this section with several remarks. The first concerns the ubiquity of the central charge as a fundamental characteristic of the CFT: notice that, using eqn. (10.10.12), its value can be extract measuring the entanglement entropy of a system as a function of the lenght l of its sub-interval. The second remark is about the non-extensive behaviour of S_A as a function of the length l of the sub-system: eqn. (10.10.12) is indeed the one-dimensional analog of the area-law behaviour that seems to hold for all local quantum systems. Namely, in a d-dimensional system, the leading term of the entanglement entropy grows at most proportionally with the area of the boundary separating the two partitions A and B

$$S_A \simeq \left(\frac{l}{a}\right)^{d-1}. \tag{10.10.16}$$

This scaling law of the entanglement entropy has the important consequence to greatly reduces the complexity of quantum many-body systems. Finally, the computation of entanglement entropy for systems with generic multi-partitions still constitutes an open and very difficult mathematical problem.

Appendix 10.A. Moebius Transformations

This appendix discusses some important aspects of the Moebius trasformations. They are closely related to the group of isometries of the hyperbolic plane and the tri-dimensional hyperbolic surfaces. An important sub-group of these transformations is given by the *modular group* that plays an important role in the classification of the partition functions of the conformal theories.

As discussed in the text, the Moebius transformations are given by

$$w(z) = \frac{az+b}{cz+d}, \tag{10.A.1}$$

with a, b, c, d complex numbers that satisfy $ad - bc \neq 0$. Since multiplying all these numbers by a common factor does not alter the mapping (10.A.1), we can always assume that they satisfy the condition

$$ad - bc = 1. \tag{10.A.2}$$

Any Moebius trasformation that is not simply a linear function can be obtained as the composition of two linear transformations and one inversion. In fact, if $c = 0$, the map is linear. If, on the contrary, $c \neq 0$, it can be written

$$w(z) = \frac{a}{c} + \frac{bc - ad}{c(cz+d)}. \tag{10.A.3}$$

This expression shows that the original mapping can be decomposed in the sequence of the three transformations

$$z_1 = cz + d, \quad z_2 = \frac{1}{z_1}, \quad w = \frac{a}{c} + \frac{bc - ad}{c} z_2. \tag{10.A.4}$$

Group structure. The Moebius transformations form a group. This means that the class of these functions contains the identity and the inverse transformations and, furthermore, that the product of two Moebius transformations belongs to the same set. It is easy to prove this statement. With the choice $b = c = 0$, $a = d = 1$, we obtain the identity transformation $w(z) = z$. To determine the inverse, we need to solve the equation $w(z) = f(z)$ for the variable z in terms of in w, with the final result (expressed in the variable z) given by

$$\frac{dz - b}{-cz + a}. \tag{10.A.5}$$

This corresponds to the substitutions $a \to d$, $b \to -b$, $c \to -c$ and $d \to a$. As a by-product of this computation, we find that the combination $ad - bc$ is an invariant quantity. Consider now the product of two transformations: let $z_2 = f_2(z)$ and $w = f_1(z_2)$ the two transformations with parameters a_i, b_i, c_i, d_i $(i = 1, 2)$. The final result is given by

$$f_3(z) = \frac{a_3 z + b_3}{c_3 z + d_3}, \tag{10.A.6}$$

with

$$\begin{aligned} a_3 &= a_1 a_2 + b_1 c_2, \qquad & b_3 &= a_1 b_2 + b_1 d_2 \\ c_3 &= c_1 a_2 + d_1 c_2, \qquad & d_3 &- c_1 b_2 + d_1 d_2. \end{aligned} \tag{10.A.7}$$

These composition laws can be elegantly expressed in terms of a matrix algebra, associating to the transformation (10.A.1) the matrix

$$W = \begin{pmatrix} a & b \\ c & d \end{pmatrix}. \tag{10.A.8}$$

The condition (10.A.2) becomes $W = 1$. Hence the inverse matrix exists and is given by

$$W^{-1} = \begin{pmatrix} d & -b \\ -c & d \end{pmatrix}, \tag{10.A.9}$$

that corresponds to (10.A.5). It is also simple to see that the composition law (10.A.7) corresponds to the usual matrix multiplication law. The decomposition (10.A.4) implies that any Moebius transformation is either linear or it can be decomposed as $W_1 W_2 W_3$, where W_i are expressed by the matrices

$$W_1 = \begin{pmatrix} a_1 & b_1 \\ 0 & 1 \end{pmatrix}, \quad W_2 = \begin{pmatrix} 0 & 1 \\ 1 & 0 \end{pmatrix}, \quad W_3 = \begin{pmatrix} a_3 & b_3 \\ 0 & 1 \end{pmatrix}. \tag{10.A.10}$$

Harmonic ratio. It is immediate to show that the harmonic ratio of four distinct points $z_1, \ldots, z_4$ is invariant under a Moebius map, namely

$$\frac{(w_1 - w_4)(w_3 - w_2)}{(w_1 - w_2)(w_3 - w_4)} = \frac{(z_1 - z_4)(z_3 - z_2)}{(z_1 - z_2)(z_3 - z_4)}, \tag{10.A.11}$$

where the w_i are the images of the points z_i under the mapping (10.A.1). Note that this equation has an important consequence. Namely, posing $w_4 = w$ e $z_4 = z$, we have

$$\frac{(w_1 - w)(w_3 - w_2)}{(w_1 - w_2)(w_3 - w)} = \frac{(z_1 - z)(z_3 - z_2)}{(z_1 - z_2)(z_3 - z)}, \tag{10.A.12}$$

that can be written in the form (10.A.1), where the coefficients a, b, c, d are uniquely fixed in terms of the points z_i and w_i. This means that a Moebius trasformation is uniquely determined once we fix the mapping of *three* different points in the complex plane. A close look at eqn. (10.A.1) shows that the point $z = -b/a$ is mapped onto the point $w = 0$, the point $z = -d/c$ onto the point at infinity $w = \infty$ and, finally, the point at infinity of the z-plane onto the point $w = a/c$.

Circles onto circles. An important geometrical property of the Moebius transformations is that they map circles onto circles, including in this terminology also the straight lines, regarded as circles of infinite radius.[13] To prove it, it is sufficient to show that each of the three elementary transformations (10.A.4) in which any Moebius transformation can be decomposed, has this property. Let us write initially the general expression of a line and a circle in complex coordinates: for a straight line we have

$$ax + by + c = 0, \quad a, b, c \in \Re \tag{10.A.13}$$

and using $z = x + iy$, $\bar{z} = x - iy$, it reads

$$Az + \bar{A}\bar{z} + c = 0, \quad A = \frac{a - ib}{2}. \tag{10.A.14}$$

For a circle of radius r, whose centre is in z_0, we have $(z - z_0)(\bar{z} - \bar{z}_0) = r^2$, namely

$$z\bar{z} + B\bar{z} + \bar{B}z + C = 0, \quad B = -z_0, \quad C = |B|^2 - r^2. \tag{10.A.15}$$

Under a translation and a rotation, expressed generally by the transformation $z \to az + b$, both (10.A.14) and (10.A.15) keep their form. Under the inversion $z = 1/w$, $\bar{z} = 1/\bar{w}$, eqn. (10.A.14) becomes

$$cw\bar{w} + A\bar{w} + \bar{A}w = 0. \tag{10.A.16}$$

[13] This is a natural assumption on the Riemann sphere associated to the complex plane.

If $c = 0$ (the original line passes through the origin), this equation defines a new straight line that passes through the origin. Vice versa, if $c \neq 0$, the equation above defines a circle of radius $|A|/|c|$, centred at $-A/c$. Acting now with an inversion transformation on (10.A.15), it becomes

$$Cw\bar{w} + Bw + \bar{B}\bar{w} + 1 = 0. \tag{10.A.17}$$

If $C = 0$ (this corresponds to the original circle that passes through the origin) we have a straigh line. Otherwise, it defines a new circle, with centre at $z_0 = -\bar{B}/C$ and radius $r^2 = |B|^2/|C|^2 - 1/C$.

Closely related to the property discussed above, there is the transformation law that involve the internal and external points of the circles. Let $\mathcal{D}_i$ be the set of internal points of the circle C in the z-plane and $\mathcal{D}_e$ the set of its external points, with an analogous definition of $\mathcal{D}'_i$ and $\mathcal{D}'_e$ for the points relative to the circle C' in the w-plane, in which the circle C is mapped. There can be only two cases: (i) the first, in which $\mathcal{D}_i$ is mapped onto $\mathcal{D}'_i$ and correspondingly $\mathcal{D}_e$ onto $\mathcal{D}'_e$; (ii) the second, in which $\mathcal{D}_i$ is mapped onto $\mathcal{D}'_e$ while $\mathcal{D}_e$ onto $\mathcal{D}'_i$. The proof is left as an exercise.

Symmetric points. We also mention, without the proof, another characteristic property of the Moebius map: it transforms symmetric points with respect a circle onto symmetric points of the image circle. Two points p and q are symmetric with respect a circle of centre z_0 and radius r if z_0, p and q are aligned in the given order, with the distances $|z_0 - p|$ and $|z_0 - q|$ that satisfy the condition

$$|z_0 - p|\,|z_0 - q| = r^2. \tag{10.A.18}$$

Denoting by w_0 the centre of the image circle, R its radius and p' and q' the image points of p and q, we find that

$$|w_0 - p'|\,|w_0 - q'| = R^2. \tag{10.A.19}$$

Fixed points. It is interesting to observe that the Moebius transformations can be also characterized by the properties of their fixed points. These are the points left invariant by the map (10.A.1)

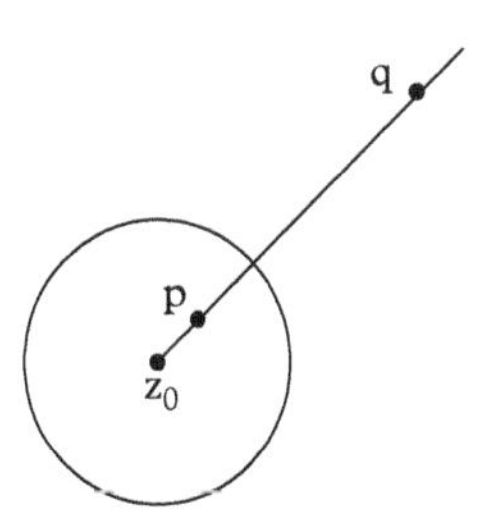

Fig. 10.13 *Symmetric points p and q with respect to a circle of radius r.*

$$z = w(z). \tag{10.A.20}$$

They can be of four different types: parabolic, elliptic, hyperbolic and lossodromic. This classification has both a geometrical and algebraic meaning, as shown by the figures given below. The different classes can be distinguished by the trace $\text{Tr}M = a + d$ of the matrix M. In more detail, the Moebius transformations are

- *parabolic*, if $a + d = \pm 2$.
- *elliptic*, if $a + d$ is a real number, with $|\, a + d \,| \leq 2$.
- *hyperbolic*, if $a + d$ is a real number, with $|\, a + d \,| \geq 2$.
- *lossodromic*, if $a + d$ is a complex number.

Since the trace of a matrix is invariant under a conjugation transformation

$$M \to U^{-1} M U, \tag{10.A.21}$$

where U is also a Moebius transformation, all members of a conjugate class are of the same type.

Solving the algebraic equation of the second order (10.A.20) and denoting the two roots as $\gamma_{1,2}$, we have

$$\gamma_{1,2} = \frac{(a-d) \pm \sqrt{(a-d)^2 + 4bc}}{2} = \frac{(a-d) \pm \sqrt{(a+d)^2 - 4}}{2}, \tag{10.A.22}$$

where we have used the relation $ad - bc = 1$. Except the trivial cases $c = 0$, and $a = d$, or $b = c = 0$, in which there is an infinite number of fixed points (since the transformation is the identity), there are in general two distinct fixed points. However, they coalesce when

$$(\text{Tr}M)^2 = (a+d)^2 = 4. \tag{10.A.23}$$

Let us consider the two cases separately. When the two fixed points coincide, we are in the presence of a *parabolic transformations*. All these transformations are conjugated to the matrix

$$M_p = \begin{pmatrix} 1 & 1 \\ 0 & 1 \end{pmatrix}. \tag{10.A.24}$$

If γ denotes the only fixed point, their general form is

$$\frac{1}{w - \gamma} = \frac{1}{z - \gamma} + \beta, \tag{10.A.25}$$

where β is a free parameter related to the translations. In the parabolic case we have that: (i) any circle that passes through the fixed point is transformed onto a tangent circle that passes through the fixed point; (ii) any family of tangent circles is then transformed into itself; (iii) the internal region of each circle is transformed onto itself. Under this class of transformations the way in which the plane changes is shown in Figure 10.14.

When there are two distinct fixed points, eqn. (10.A.12) implies

$$\frac{w-\gamma_1}{w-\gamma_2} = \kappa\,\frac{z-\gamma_1}{z-\gamma_2}, \tag{10.A.26}$$

where κ is a constant that depends on γ_1, γ_2, z_2 and w_2. Hence, the general expression of a Moebius trasformation with two distinct fixed points depends on an additional constant κ. Using the conjugation transformation, the two points $\gamma_{1,2}$ can be mapped one at 0 and the other to ∞, Consequently, all these transformations are conjugated to the matrix

$$M = \begin{pmatrix} \lambda & 0 \\ 0 & \lambda^{-1} \end{pmatrix} \tag{10.A.27}$$

with $\lambda^2 = \kappa$. This matrix corresponds to the mapping $w = \kappa z$. For this reason, the constant κ is called the *multiplier* of the transformation. We have an *elliptic transformation* when

$$0 < (a+d) < 4. \tag{10.A.28}$$

This condition is equivalent to $|\kappa| = 1$, namely $\kappa = e^{i\alpha}$, with α a real parameter.[14] In this case we have the following properties: (i) an arc of a circle passing through the fixed

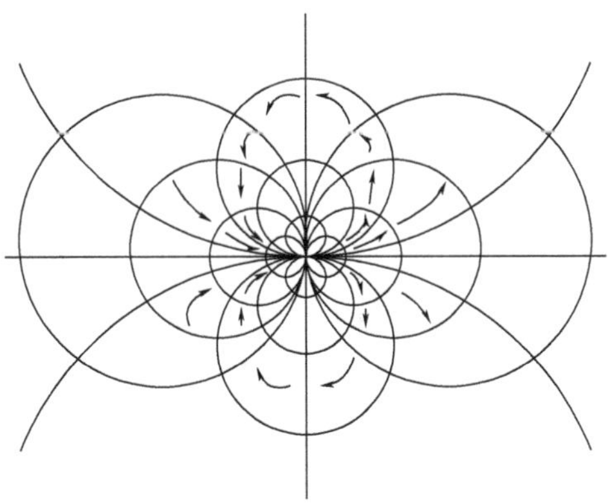

Fig. 10.14 *Transformation of the complex plane under a Moebius map of parabolic type.*

[14] Since the multiplier of M^n is κ^n, the only Moebius transformations of a finite order are those elliptic and they correspond to rational values of α.

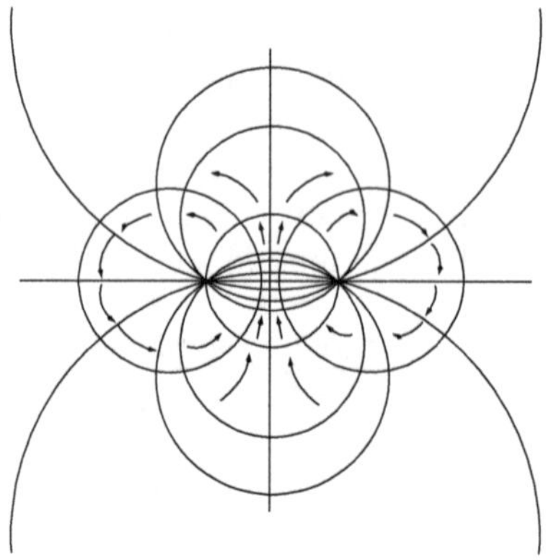

Fig. 10.15 *Transformation of the complex plane under a Moebius map of elliptic type.*

points is transformed in another arc of a circle passing through them but rotated of an angle α with respect to the original one; (ii) each circle orthogonal to the circles passing through the fixed points is transformed onto itself and the same holds for its internal region (Figure 10.15).

We have a *hyperbolic transformation* when

$$(a+d)^2 \geq 4, \tag{10.A.29}$$

namely when κ is a real number. Note that $w'(\gamma_1) = \kappa$ whereas $w'(\gamma_2) = \kappa^{-1}$ so that, if $\kappa > 1$, γ_1 is a repulsive point, whereas γ_2 is an attractive point. Their role is swapped if $\kappa < 1$. For the hyperbolic transformations we have: (i) each circle that passes through the fixed points is transformed onto itself, namely each of the two arcs of which the circle is composed is mapped on itself; (ii) the internal region of a circle passing through the fixed points is mapped onto itself; (iii) each circle that is orthogonal to a circle passing through the fixed points is transformed in an analogous circle. The way the hyperbolic transformations act is shown in Figure 10.16.

Finally, we have a *lossodromic transformation* in the remaining cases, namely when $(\text{Tr}M)^2$ does not belong to the interval $[0,4]$. In this case the multiplier is given by $\kappa = Ae^{i\alpha}$. Hence its action is a combination of the motions shown in Figure 10.15 and Figure 10.16. Each arc passing through the fixed points is transformed in a similar arc but rotated by α, while a circle orthogonal to the circles passing through the fixed points is transformed onto another orthogonal circle. The lossodromic transformations do not have, in general, fixed circles expect in the case in which $\alpha = \pi$.

Let us now discuss two particular examples of Moebius transformations that may be useful later on. The first is the transformation that maps the upper half-plane $\text{Im}\, z > 0$ in the internal region of the circle $|w| < 1$. Its general expression is

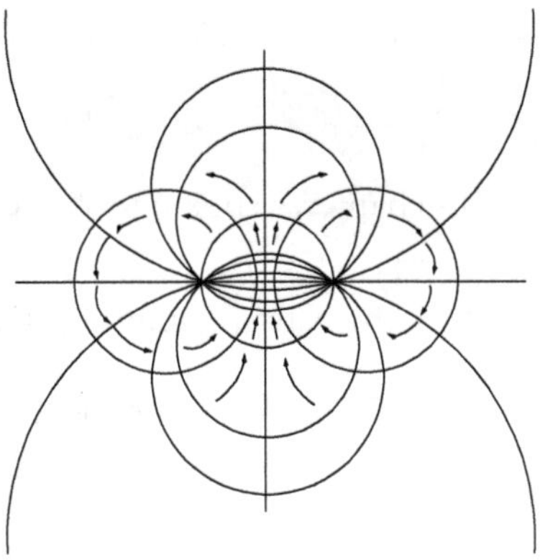

Fig. 10.16 *Transformation of the complex plane under a Moebius map of the hyperbolic type.*

$$w(z) = \lambda \frac{z-\alpha}{z-\bar{\alpha}}, \quad |\lambda| = 1, \quad \text{Im}\,\alpha > 0. \tag{10.A.30}$$

To prove that the upper half-plane is mapped onto the internal points of the circle, consider the points along the real axis. For those points we have $|z-\alpha| = |z-\bar{\alpha}|$, and therefore they are mapped in the points of the circle $|w| = 1$. On the other hand, the point $z = \alpha$ is transformed onto the origin $w = 0$. For the properties of the Moebius map discussed above, this is sufficient to conclude that any other point of the domain $\text{Im}\,z > 0$ is mapped inside the circle. Note that the point $z = \bar{\alpha}$ is mapped onto $w = \infty$ and this is enough to conclude that the lower half-plane is transformed onto the external region of the circle $|w| = 1$.

The second map we consider maps the disc $|z| < 1$ onto the disc $|w| < 1$. Its general expression is

$$w(z) = \lambda \frac{z-\alpha}{\bar{\alpha}z-1}, \quad |\lambda| = 1, \quad |\alpha| < 0. \tag{10.A.31}$$

Note, in fact, that the points of the circle in the z-plane are expressed by $z = e^{i\phi}$ and for those points we have

$$|w| = |\lambda| \left| \frac{e^{i\phi}-\alpha}{\bar{\alpha}e^{i\phi}-1} \right| = \frac{|\alpha - e^{i\phi}|}{|\bar{\alpha} - e^{-i\phi}|} = 1. \tag{10.A.32}$$

Since $z = 0$ is mapped onto the point $\lambda\alpha$, with $|\lambda\alpha| < 1$, this is sufficient to conclude that all internal points of the circle in the z-plane are mapped onto the internal point of the circle in the w-plane.

REFERENCES

For the mathematical part of this chapter, a superb text on complex analysis is

Ablowitz, M. and Fokas, A. (1977). *Complex Variables. Introduction and Applications*, Cambridge Texts in Applied Mathematics, Cambridge University Press.

There are several review articles on CFTs. We recommend

Cardy, J. (1988). 'Conformal Invariance and Statistical Mechanics', Les Houches Session XLIX, *Field, Strings and Critical Phenomena.*

Ginsparg, P. (1988). 'Applied Conformal Field Theory', Les Houches Session XLIX, *Field, Strings and Critical Phenomena.*

Zamolodchikov, A.B. and Zamolodchikov, Al.B. (1989). 'Conformal Field Theory and Critical Phenomena in Two-Dimensional Systems', *Soviet Science Review A. Physics*, 10: 269.

For an exhaustive and pedagogical analysis of the CFT, particularly for their mathematical aspects, we

Di Francesco, P., Mathieu, P. and Senechal, D. (1977). *Conformal Field Theory*, Springer-Verlag.

Everyone should read the original article in which the two-dimensional CFTs were firstly introduced

Belavin, A., Polyakov, A. and Zamolodchikov, A.B. (1984). 'Infinite Conformal Symmetry in Two-dimensional Quantum Field Theory', *Nuclear Physics B*, 241: 333.

For the entanglement entropy in CFT we recommend

Calabrese, P. and Cardy, J. (2004). 'Entanglement Entropy and Quantum Field Theory', *Journal of Statistical Mechanics*, P060002.

Calabrese, P., Cardy, J. and Tonni, E. (2009). 'Entanglement Entropy of Two Disjoint Intervals in Conformal Field Theory', *Journal of Statistical Mechanics*, P11001.

Holzhey, C., Larsen, F. and Wilczek, F. (1994). 'Geometric and Renormalized Entropy in Conformal Field Theory', *Nuclear Physics B*, 424: 44.

For a conformal bootstrap in higher dimensions, see

Dolan, F.A. and Osborn, H. (2001). 'Conformal Four-point Functions and the Operator Product Expansion', *Nuclear Physics B*, 599: 459.

El-Showk, S., Paulos, M.F., Poland, D., Rychkov, V.S., Simmons-Duffin, D. and Vichi, A. (2012). 'Solving the 3D Ising Model with the Conformal Bootstrap', *Physics Review D*, 86: 025022.

Osborn, H. and Petkou, A.C. (1994). 'Implications of Conformal Invariance in Field Theories for General Dimensions', *Annals of Physics*, 231: 311.
Rattazzi, R., Rychkov, V.S., Tonni, E. and Vichi, A. (2008). 'Bounding Scalar Operator Dimensions in 4D CFT', *Journal of High Energy Physics*, 0812: 031.

PROBLEMS

10.1. Operatorial identities

There is a simple example that shows the necessity of considering the operatorial identities only in a weak sense, i.e. true only for the matrix elements. Consider an interacting scalar field $\phi(x)$ and suppose that for $x_0 \to -\infty$ its interactions vanish. In this case it seems natural to pose the operatorial identity

$$\lim_{x_0 \to -\infty} \phi(x) = \phi_{in}(x)$$

where $\phi_{in}(x)$ is a free bosonic field. However, this leads to a contradiction. In fact, it the relation above were true, we would have

$$\lim_{x_0 \to -\infty} \lim_{y_0 \to -\infty} \langle 0 \mid \phi(x)\phi(y) \mid 0 \rangle = \langle 0 \mid \phi_{in}(x)\phi_{in}(y) \mid 0 \rangle .$$

Since $\phi_{in}(x)$ is a free field, the right-hand side is the usual progagator $G_{free}(x-y)$ of a scalar free field

$$G_{free}(x-y) = \int \frac{d^d p}{(2\pi)^d} \frac{1}{p^2 - m^2} e^{ip(x-y)} .$$

a. Use the Lorentz invariance to fix the dependence on the coordinates of the propagator $G(x-y)$ of the interacting field $\phi(x)$.

b. Argue that the propagator does not coincide in the limit $x_0 \to -\infty$ with $G_{free}(x-y)$.

10.2. Correlation functions

Assuming the validity of the OPE, show that all the correlation functions of a massless field theory can be expressed in terms of the propagators and the structure constants.

10.3. Laplace equation and conjugate harmonic functions

a. Show that the real and imaginary parts of an analytic function of a complex variable z

$$f(z) = \Omega(x, y) + i\Psi(x, y)$$

are both harmonic functions, i.e. they satisfy the Laplace equation

$$\nabla^2\Omega = \nabla^2\Psi = 0.$$

b. Vice-versa, use the Cauchy–Riemann equations to show that if $\Omega(x, y)$ is a function that satisfies the Laplace equation, then there exists another harmonic function $\Psi(x, y)$ (called the conjugate function of Ω) such that $f(z) = \Omega + i\Psi$ is an analytic function of complex variable.

10.4. Hydrodynamics of an ideal fluid in two dimensions

Consider the stationary motion of an incompressible and irrotational fluid in two dimensions. Denoting by $\vec{v}(x, y) = (v_1, v_2)$ the vector field of its velocity at the point (x, y) of the plane, it satisfies

$$\vec{\nabla}.\vec{v} = 0, \quad \vec{\nabla} \wedge \vec{v} = 0$$

a. Show that these conditions implies the existence of a potential Ω that satisfies the Laplace equation. Moreover, show that, by introducing the conjugate function Ψ and defining $f(z) = \Omega + i\Psi$ (the so-called *complex potential*), we have

$$\frac{df}{dz} = \frac{\partial\Omega}{\partial x} + i\frac{\partial\Psi}{\partial x} = \frac{\partial\Omega}{\partial x} - i\frac{\partial\Omega}{\partial y} = v_1 - iv_2 = \bar{v}.$$

The complex vector field of the velocity is then given by

$$v = \overline{\left(\frac{df}{dz}\right)}.$$

b. Study the flux lines of the velocity associated to the analytic function

$$f(z) = \frac{i\gamma}{2\pi} \ln z$$

and show that the vector field of the velocity corresponds to a vortex, localized at the origin. Give an interpretation of the parameter γ.

c. Study the flux lines of the velocity relative to the potential

$$f(z) = v_0\left(z + \frac{a^2}{z}\right) + \frac{i\gamma}{2\pi} \ln z.$$

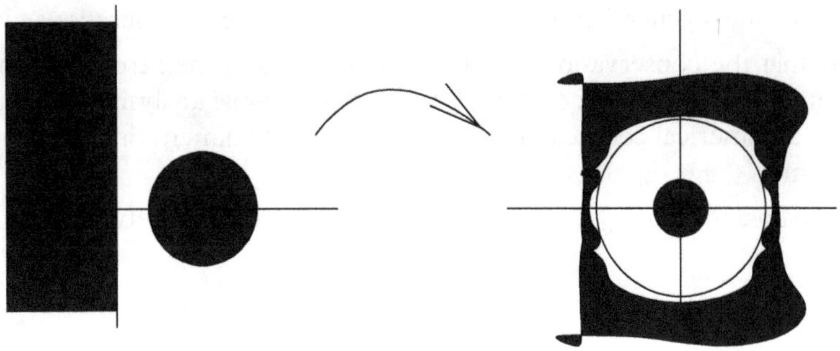

Fig. 10.17 *Conformal map of two domains.*

Determine the points where the velocity vanish and study their location by varying the parameter γ.

10.5. Moebius transformations

Show that the transformation $w = (z-a)/(z+a)$, $a = \sqrt{c^2-\rho^2}$ with c and ρ real and $0 < \rho < c$, maps the domain delimitated by the circle $|z-c| = \rho$ and the imaginary axis, onto the annulus domain delimitated by $|w| = 1$ and a concentric circle of radius δ (Figure 10.17). Find, in particular, the value of δ.

10.6. Operator product expansion in the channel of the identity operator

Let $\phi_i(z)$ a primary field of a CFT with central charge c. Let Δ_i its conformal weight. Prove that the Ward identity uniquely fixes the first terms of the operator expansion in the channel of the identity operator, namely

$$\phi_i(z)\phi_i(w) = \frac{1}{(z-w)^{2\Delta_i}}\left[\mathbf{I} + \frac{2\Delta_i}{c}T(w) + \cdots\right].$$

10.7. Casimir effect

Consider two parallel horizontal planes, separated by a distance **a** along the axis z. Suppose that a massless field theory is defined between the two planes, with boundary conditions that ensure a non-zero value of the expectation value of the stress-energy tensor $T_{\mu\nu}$, $t_{\mu\nu}(t, \vec{x}) \equiv \langle 0 \mid T_{\mu\nu}(t, \vec{x}) \mid 0\rangle$. The system is assumed to be time invariant. Thanks to the symmetry of the problem, $t_{\mu\nu}$ can be written in terms of the metric tensor $g_{\mu\nu}$ and the tensors made of the unit vector $\hat{z}_\mu = (0,0,0,1)$.

a. Write the most general expression of $t_{\mu\nu}$ based on the considerations given above.
b. Show that the conservation law $\partial^\mu T_{\mu\nu}(t, \vec{x}) = 0$ and the zero-trace condition of $T_{\mu\nu}$ uniquely $t_{\mu\nu}$ up to a constant. Use the dimensional analysis to fix this constant (up to a numerical coefficient) in terms of the only dimensional parameter of the problem.
c. Use the final form of $t_{\mu\nu}$ to compute the force per unit area between the two planes.

10.8. Central extension

Consider commutation relations given by

$$[L_n, L_m] = (n-m)L_{n+m} + f(n,m)C,$$

where $f(n,m) = -f(m, n)$ and $[C, L_n] = 0$.

a. Show that these commutation relations satisfy the Jacobi identities if and only if

$$f(n,m) = \eta\, n(n^2-1)\,\delta_{n+m,0} + (n-m)\lambda(n+m)$$

where η is a constant and $\lambda(n)$ a function.
b. Show that the second term in $f(n,m)$ can be absorbed by a redefinition of the generators L_n, i.e. $L_n \to L_n - Cg(n)$.

10.9. Entanglement entropy

Consider a system of two spin–$\frac{1}{2}$ degrees of freedom and a pure state given by

$$|\Psi\rangle = \cos\theta\,|\uparrow\downarrow\rangle + \sin\theta\,|\downarrow\uparrow\rangle.$$

The observer A measures the first spin while B measures the second one.

a. Compute the entanglement entropy S_A.
b. Check that is takes its larger value $S_A = \log 2$ when $\cos^2\theta = \frac{1}{2}$.

11

Minimal Conformal Models

Small is beautiful (Anonymous)

11.1 Introduction

This chapter discusses a particular class of conformal theories, the so-called *minimal models*. The peculiarity of these models consists in the *finite* number of their conformal families that close an OPE. The anomalous dimensions of the conformal fields and the central charge of these theories can be computed exactly and, in particular, assume rational values. Furthermore, of the minimal models we can explicitly compute both the correlation functions of the order parameters and the partition function on a torus, i.e on a cylinder with periodic boundary conditions on both directions. Their mathematical elegance is accompanied by an important physical interpretation, discussed in more detail in Chapter 14, the conformal minimal models describes the scaling limit of an infinite number of statistical models with a discrete symmetry, among which we find the Ising model, the tricritical Ising model, the Potts model, the Yang–Lee edge singularity, etc. In addition, the unitary minimal models can be put in correspondence with the critical Landau–Ginzburg theories with power interaction $\phi^{2(p-1)}$ ($p = 3, 4, \ldots$): as a matter of fact, they provide the exact solution of these theories at their multi-critical point. For all these reasons, the minimal conformal models play a crucial role in the modern understanding of the critical phenomena.

This chapter focuses on the general discussion of the minimal models of the Virasoro algebra. We initially highlight the presence of null vectors in the representations of the Virasoro algebra in correspondence of discrete values of the conformal dimensions and the central charge, encoded in the Kac determinant of the so-called *degenerate fields*. Later we discuss the fusion rules that derive from the particular structure of the Verma module of the degenerate fields and the Coulomb gas formalism that allows us to compute the exact expressions of the correlation functions. Finally, we study the modular invariance of these model and the partition functions compatible with this symmetry. Further aspects of these models will be addressed in other chapters.

Statistical Field Theory: An Introduction to Exactly Solved Models in Statistical Physics. Second Edition. Giuseppe Mussardo, Oxford University Press (2020). © Giuseppe Mussardo. DOI: 10.1093/oso/9780198788102.001.0001

11.2 Null Vectors and Kac Determinant

The starting point in the study of the minimal models is the presence of particular null vectors inside the conformal families. This circumstance is of the utmost importance not only for the study of the conformal theories at the critical point but also for their off-critical deformations. For this reason, it deserves to be investigated in detail.

From Section 10.8, we know that a conformal family[1] is identified by the vector $\mid \Delta\rangle$ associated to the primary field ϕ_Δ. This vector satisfies the conditions

$$\begin{aligned} L_0 \mid \Delta\rangle &= \Delta \mid \Delta\rangle, \\ L_n \mid \Delta\rangle &= 0 \quad n = 1,2,\ldots \\ \langle \Delta \mid \Delta\rangle &= 1, \end{aligned} \tag{11.2.1}$$

A conformal family is build on such a vector and on all its descendants obtained by applying to it an ordered string of creation operators L_{-n}. The vector $\mid \Delta\rangle$, as already noticed in the previous chapter, plays the role of *highest-weight vector* of the Virasoro algebra. For arbitrary values of Δ and c, all the descendant vectors are linearly independent and the set of all these vectors form then an irreducible representation of the Virasoro algebra. However, for particular values of Δ and c, there are some null vectors: in such a case, to have an irreducible representation we have to factorize with respect to these states. Before we expose the general case, it is convenient to familiarize with some simple examples of null vectors at the lowest levels of the conformal families.

Let us start from the level $N = 1$. Given the primary state $\mid \Delta\rangle$, at this level there is only one descendant state given $\mid X_1\rangle = L_{-1} \mid \Delta\rangle$. If we request that this is a null vector, its norm must vanish

$$\begin{aligned} \langle X_1 \mid X_1\rangle &= \langle \Delta \mid L_1 L_{-1} \mid \Delta\rangle = \\ &= 2\langle \Delta \mid L_0 \mid \Delta\rangle = 2\Delta\langle \Delta \mid \Delta\rangle = 0. \end{aligned}$$

This equation has the only solution $\Delta = 0$. In other words, the only conformal family that has a null vector at level $N = 1$ is the family of the identity operator **I**.

A more interesting situation occurs at the level $N = 2$. In this case there are two possible descendant states, the first given by $L_{-1}^2 \mid \Delta\rangle$ and the second by $L_{-2} \mid \Delta\rangle$. Let us determine the conditions for which a linear combination of these states

$$\mid X_2\rangle = (L_{-2} + aL_{-1}^2) \mid \Delta\rangle \tag{11.2.2}$$

gives rise to a null vector. If $\mid X_2\rangle = 0$, the same is true for the vectors obtained by applying to it either L_1 or L_2. In the first case, using the commutation relations of the Virasoro modes and the properties of the primary state $\mid \Delta\rangle$, we have

[1] In this section we focus our attention only on the analytic sector of the theory. As usual, analogous considerations can be done for the anti-analytic sector.

$$L_1 \mid X_2\rangle = (L_{-2}L_1 + 3L_{-1} + 2a(L_{-1}L_0 + L_0L_{-1})) \mid \Delta\rangle =$$
$$= (3 + 2a(2\Delta + 1))L_{-1} \mid \Delta\rangle = 0.$$

This condition fixes then the coefficient a of the linear combination (11.2.2)

$$a = -\frac{3}{2}\frac{1}{2\Delta + 1}. \tag{11.2.3}$$

Applying now L_2 to $\mid X_2\rangle$ and making again use of the commutation relations of the Virasoro modes and the conditions of $\mid \Delta\rangle$, we get

$$[L_2, L_{-2}] \mid \Delta\rangle + a[L_2, L_{-1}^2] \mid \Delta\rangle = \left(4L_0 + \frac{c}{2}\right) \mid \Delta\rangle + 3aL_1L_{-1} \mid \Delta\rangle =$$
$$\left(4L_0 + \frac{c}{2} + 6aL_0\right) \mid \Delta\rangle = \left(4\Delta + \frac{c}{2} + 6a\Delta\right) \mid \Delta\rangle = 0$$

namely

$$c = -4\Delta(2 + 3a) = \frac{2\Delta(5 - 8\Delta)}{2\Delta + 1}. \tag{11.2.4}$$

Summarizing the result of this computation, if the central charge c of the conformal model and the conformal dimension Δ of the field under scrutiny are related by the condition (11.2.4), then there exists a linear combination of the descendants at the level $N = 2$ of this primary field ϕ_Δ that leads to a null vector.

It is worth pointing out that there is a general way to determine whether or not a null vector at the level N of a conformal family exists. It consists of computing the zeros of the determinant of the Gram matrix at the level N (see Section 10.8.1). For $N = 2$, the Gram matrix is given by

$$M^{(2)} = \begin{pmatrix} 4\Delta(2\Delta + 1) & 6\Delta \\ 6\Delta & 4\Delta + c/2 \end{pmatrix}$$

and its determinant can be written as

$$||M^{(2)}|| = 2(16\Delta^3 - 10\Delta^2 + 2c\Delta^2 + c\Delta) = 32(\Delta - \Delta_{1,1})(\Delta - \Delta_{1,2})(\Delta - \Delta_{2,1})$$

where

$$\Delta_{1,1} = 0, \quad \Delta_{(1,2),(2,1)} = \frac{1}{16}(5 - c) \pm \sqrt{(1 - c)(25 - c)}. \tag{11.2.5}$$

Note the appearance of the solution $\Delta_{1,1} = 0$, whose presence was expected. It corresponds to the possibility of having a null vector at level $N = 1$ that, obviously, will also give rise to a null vector at the level $N=2$, if we act on it by the arising operator L_{-1}. The other two zeros $\Delta_{1,2}$ and $\Delta_{2,1}$ correspond to the condition (11.2.4) previously derived.

Kac determinant. Remarkably, the zeros of the Gram matrix of level N can be computed exactly. Kac's important 1978 mathematical result is a crucial step in the development of the two-dimensional conformal theories. The corresponding formula, the so-called *Kac determinant*, is given by

$$\det M^{(N)} = A_N \prod_{r,s\geq 1; rs\leq N} \left[\Delta - \Delta_{r,s}\right]^{P(N-rs)}, \tag{11.2.6}$$

where $P(N-rs)$ is the number of partitions of the integer number $(N-rs)$ and A_N is a positive constant that is not important for the discussion that follows. The matrix $M^{(N)}$ was defined in eqn. (10.8.22). The zeros $\Delta_{r,s}$ can be parameterized in different ways. One of them, particularly useful for the Coulomb gas formulation that we will discuss later, is expressed in terms of two parameters, called *charges* $\alpha_\pm$

$$\begin{aligned} \Delta_{r,s}(c) &= \Delta_0 + \frac{1}{4}(r\alpha_+ + s\alpha_-)^2, \\ \Delta_0 &= \frac{1}{24}(c-1), \\ \alpha_\pm &= \frac{\sqrt{1-c}\pm\sqrt{25-c}}{\sqrt{24}}. \end{aligned} \tag{11.2.7}$$

Equivalently, posing

$$\alpha_+ = \sqrt{t}, \quad \alpha_- = -\frac{1}{\sqrt{t}}$$

the previous conformal quantities can be expressed as

$$\begin{aligned} c &= 13 - 6\left(t + \frac{1}{t}\right) \\ \Delta_{r,s} &= \frac{1}{4}(r^2-1)t + \frac{1}{4t}(s^2-1) - \frac{1}{2}(rs-1). \end{aligned} \tag{11.2.8}$$

Note that, fixed the value of the central charge, there are two possible values of t

$$t = 1 + \frac{1}{12}\left[1 - c \pm \sqrt{(1-c)(25-c)}\right].$$

but we can choice indifferently any of the two, since this does not change the Kac determinant. The parameter t is real only in the cases $c<1$ or $c>25$, while it is generally complex in the interval $1<c<25$.

A third way to write the conformal data consists of the parameterization of the central charge and the zeros of the Kac determinant given by

$$
\begin{aligned}
c &= 1 - \frac{6}{q(q+1)} \\
\Delta_{r,s} &= \frac{[(q+1)r - qs]^2 - 1}{4q(q+1)}
\end{aligned}
\tag{11.2.9}
$$

where the real parameter q is related to the central charge c by

$$
q = -\frac{1}{2} \pm \frac{1}{2}\sqrt{\frac{25-c}{1-c}}. \tag{11.2.10}
$$

Note that the Kac formula does not predict the eigenvalues of the matrix $M^{(N)}$ but only their product. In fact, at each level N, the number of roots $\Delta_{r,s}$ is larger that the number $P(N)$ of its eigenvalues. Another important observation is that the first null vector of the conformal family $V(c, \Delta_{r,s})$ occurs at the level $N = rs$, since the combinatoric function $P(N - rs)$ vanishes, by definition, for $N < rs$. The multiplicity of the zeros, given by $P(N - rs)$, has the same origin as the one previously pointed out in the explicit computation of the null vectors at the level $N = 2$: namely, among the zeros of the polynomial at the level N, there are also those corresponding to the null vectors of lower levels. At the level N, there are in fact the null vectors generated by the $P(N - rs)$ combinations of $L_{-n_1} \ldots L_{-n_k}$, with $\sum n_i = N - rs$, applied to the null vectors of level rs.

11.3 Unitary Representations

With the explicit formula of the Kac determinant, we can identify the values of c and Δ that give rise to the unitary irreducible representantions of the Virasoro algebra, in which there are no states with negative norm. Before proceeding with the mathematical analysis of this problem, strictly speaking, the unitary condition is not necessary in statistical mechanics: many non-unitary models find their applications in the discussion of interesting statistical mechanics, also providing a useful generalization of ordinary QFTs. In the sections to come, we show that there are certain statistical models that require the presence of negative anomalous dimensions.

Coming back to the problem of determining the unitary representations, from the mathematical point of view we determine those regions of c and Δ in which the Kac determinant is negative: in these regions there are definitely states whose norm is negative and the corresponding representations are not unitary. Vice-versa, in the regions where the determinant is positive, a further analysis is needed to exclude the presence of such negative norm states, since an even number of them ends up in a positive value of the determinant.

It is easy to see that for $c < 0$ the corresponding conformal theories are non-unitary: in fact it is sufficient to consider the stress-energy tensor of these theory, associated to the descendant $L_{-2} \mid 0\rangle$ of the identity family, to see that the norm of this state is given by

$$\langle 0 \mid L_2 L_{-2} \mid 0\rangle = \frac{c}{2} \tag{11.3.1}$$

and, for $c < 0$, this is a negative quantity.

For $c > 1$, all representation with $\Delta > 0$ are unitary. It is necessary to distinguish two cases: (i) $1 < c < 25$ and (ii) $c > 25$. In the first case, expressing $\Delta_{r,s}(c)$ as

$$\Delta_{r,s} = \frac{1-c}{96}\left[\left((r+s)+(r-s)\sqrt{\frac{25-c}{1-c}}\right)^2 - 4\right],$$

we can see that $\Delta_{r,s}$ either has an imaginary part or, for $r = s$, is a negative quantity. For $c > 25$, they are instead all negative. The non-zero value of the Kac determinant in the region $\{c > 1; \Delta > 0\}$ implies that all eigenvalues of $M^{(N)}$ are positive. In fact, for large values of Δ, the Gram matrix is dominated by its diagonal elements, i.e. those with the higher powers of Δ. Since these elements are all positive in this region, this shows that the eigenvalues of $M^{(N)}$ are all positive for large values of Δ. Moreover, the determinant never vanishes in the region $c > 1$ and $\Delta > 0$, implying that all its eigenvalues remain positive in the entire region.

For $c = 1$, the Kac determinant vanishes at the $\Delta_n = \frac{n^2}{4}$, with n an integer number, but otherwise it is never negative; even in this case, there is no problem in having unitary representations for $\Delta > 0$.

Hence, the only subtle case is posed by the analysis of the region $0 < c < 1$ and $\Delta > 0$. This problem has been studied by Friedan, Qiu and Shenker,[2] and their results can be summarized as follows: all point of the region $R : \{(c, \Delta) \mid 0 < c < 1; \Delta > 0\}$ correspond to non-unitary representations, except the *discrete* series associated to these values of the central charge and the conformal dimensions

$$\begin{aligned} c = c(q) &= 1 - \frac{6}{m(m+1)}, \quad q = 2, 3, 4, \ldots \\ \Delta = \Delta_{r,s}(q) &= \frac{[(q+1)r - qs]^2 - 1}{4q(q+1)}, \quad (1 \le r \le q\,,\ 1 \le s \le q+1), \end{aligned} \tag{11.3.2}$$

where m is an integer number. These discrete values of the central charges and conformal dimensions define the so-called *conformal minimal unitary models*, in the following denoted by $\mathcal{M}_m$

[2] Friedan, D. Qiu, Z. and Shenker, S. (1984). 'Conformal Invariance, Unitarity And Two-Dimensional Critical Exponents', *Physics Review Letters*, 52: 1575.

11.4 Minimal Models

In the interval $0 < c < 1$, the unitary condition selects the discrete set of values (11.3.2). In this section we show that is possible to introduce a more general class of minimal models, from now on denoted by $\mathcal{M}_{p,q}$, whose central charge and conformal dimensions are expressed by the rational values

$$c = 1 - 6\frac{(p-q)^2}{pq}, \quad (p,q) = 1 \tag{11.4.1}$$
$$\Delta_{r,s} = \frac{[(pr-qs]^2 - (p-q)^2]}{4pq}, \quad (1 \leq r \leq q-1\,,\, 1 \leq s \leq p-1)$$

where p and q are two co-prime integers, i.e. without common divisors. Note that in all these models we have $\Delta_{1,1} = 0$ and this conformal dimension corresponds to the identity operator $\mathbf{I}$. The unitary minimal models are recovered by the choice $p = q + 1$ in eqn. (11.4.1). In all other cases, the minimal conformal theories are non-unitary, characterized by a negative value of the central charge and some of its conformal dimensions. The lowest negative conformal dimension is given by

$$\Delta_{min} = \Delta_{1,n} = \Delta_{q-1,p-n} = \frac{1-(p-q)^2}{4pq}, \tag{11.4.2}$$

that, even though the central charge of these minimal models is negative, their *effective central charge*

$$c_{eff} = c - 24\Delta_{min} = 1 - \frac{6}{pq} \tag{11.4.3}$$

is instead always a positive quantity.

As anticipated in the introduction of this chapter, the conformal minimal models satisfy a series of important properties and they are nowadays the most studied and understood conformal theories. In particular, they play an essential role both in the qualitative and quantitative analysis of the phase transitions that take place in two dimensional systems. To orientate the reader in the discussion to come, it is convenient to briefly summarize their main features.

1. In the minimal models, the number of conformal families is finite and the conformal dimensions are expressed by the rational numbers given in eqn. (11.4.1).
2. The OPE of any pair of conformal fields of these theories involves only a finite number of the operators of the same minimal model.
3. The correlation function of all the conformal fields satisfy a set of linear differential equations that can be exactly solved.

4. The structure constants of the conformal algebra can be exactly computed.
5. Their partition functions on a torus geometry can be exactly determined.

Finally, these minimal conformal models provide the exact solution, at criticality, of a significant series of statistical models, such as the Ising model, the tricritical Ising model, the Potts model, etc., and among the non-unitary models, the Yang–Lee edge singularity, the self-avoinding random walks, the percolation, etc. Thanks to them, there has been a great advance in the comprehension of the classes of universality. Let us now go on with the detailed discussion of the aspects summerized above.

11.4.1 Kac Table

The zeros of the Kac determinant, expressed for instance by eqn. (11.2.7), can be graphically associated to a set of points with integer coordinates (r, s) of the first quadrant of a cartesian plane. For the nature of these points, it is naturally to define a lattice on this plane (Figure 11.1). This graphical representation is extremely useful to illustrate some remarkable properties of the Kac formula of the minimal models.

The dashed line in Fig. 11.1 has a slope $\tan\theta = -\alpha_+/\alpha_-$. If $\delta_{r,s}$ stays for the distance of a point (r, s) of the lattice from this straight line, the zeros of the Kac determinant can be written as

$$\Delta_{r,s} = \Delta_0 + \frac{1}{4}(\alpha_+ + \alpha_-)^2\,\delta^2_{r,s}. \qquad (11.4.4)$$

When the slope is irrational, the line obviously never meets a point of the lattice. On the contrary, if it is rational, there exists two co-prime integers p and q, with $p > q$, such that

$$p\alpha_- + q\alpha_+ = 0. \qquad (11.4.5)$$

In this case, the line passes through the point (q, p). In the rational case, it is easy to see that the zeros of the Kac determinant satisfy the properties

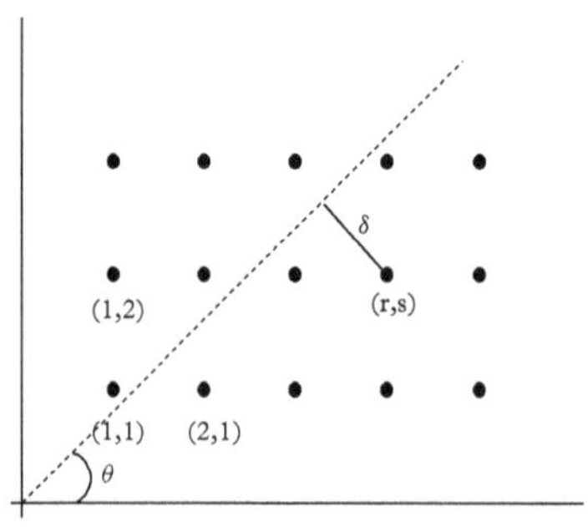

Fig. 11.1 *Kac table.*

$$\begin{aligned}\Delta_{r,s} &= \Delta_{r+q,s+p},\\ \Delta_{r,s} &= \Delta_{q-r,p-s}.\end{aligned} \tag{11.4.6}$$

These relations can be easily interpreted from a geometrical point of view: the point (r, s) of the lattice has the same distance from the line of slope p/q of the infinite series of points $(r+nq,\ s+np)$ obtained by reflection with respect to the same line.

We can express the central charge and conformal dimensions according to the formula (11.4.1) that identifies the most general minimal models. Note that, in addition to eqns. (11.4.6), there are also the relations

$$\begin{aligned}\Delta_{r,s} + rs &= \Delta_{q+r,p-s} = \Delta_{q-r,p+s}\\ \Delta_{r,s} + (q-r)(p-s) &= \Delta_{r,2p-s} = \Delta_{2q-r,s}.\end{aligned} \tag{11.4.7}$$

These expressions implies that the null vector at the level $N = rs$ of the conformal family $V_{r,s}$ is itself a *highest-weight vector* of the Virasoro algebra, because its conformal dimension is also expressed in terms of the Kac table! Moreover, beside the null vector at the level rs, the conformal family $V_{r,s}$ also contains another null vector at the level $(q-r)(p-s)$. In turns, these two null vectors generate additional null vectors, and so on. Therefore, inside the conformal family of the primary field $\phi_{r,s}$, there is an infinite nested structure of null vectors. This null vector hierarchy deeply influences both the correlation functions and the characters of such primary operators.

11.4.2 Differential Equations

For the minimal models, either unitary or non-unitary, the conformal dimensions coincide with the zeros of the Kac determinant. Let us study how this circumstance leads to a result of a great relevance for the correlation functions of their primary fields.

The primary field associated to $\Delta_{r,s}$ has its first null vector at the level $N = rs$: this vector is expressed by a particular linear combination of the $P(rs)$ descendant states $\phi_{r,s}^{(n_1,n_2,\ldots)}$ of $\phi_{r,s}$ present at that level. Denoting the null vector by $\phi_{r,s}^{null}$, its general expression is

$$\phi_{r,s}^{null}(z) = [a_1 L_{-1}^{rs} + a_2 L_{-1}^{rs-2} L_{-2} + \cdots a_{rs} L_{-rs}]\phi_{r,s}(z) = \sum a_i \phi_{r,s}^{(n_1,n_2,\ldots)}, \tag{11.4.8}$$

where all the coefficients a_i can be fixed by imposing the linear dependence of the vectors involved in the expression above. Any correlation functions in which such a null vector enters obviously vanishes

$$\langle \phi_{r,s}^{null}(z)\phi_1(z_1)\ldots\phi_n(z_n)\rangle = 0. \tag{11.4.9}$$

On the other hand, Section 10.8.2 showed how the correlation functions of the descendant fields $\phi_{\Delta}^{(n_1,n_2,\ldots)}$ at the level N of a primary field are obtained applying a linear differential operator of order N to the correlation functions of the primary fields alone. Since the null vector is also expressed by a linear combination of descendant field, once we identify the linear differential operator associated to each of them and collect all the terms, we arrive at the following important conclusion: in virtue of the null vector at the level rs, the correlation functions of the primary field $\phi_{r,s}(z)$ are solutions of a *linear differential equation* of order rs

$$\mathcal{D}_{rs}\langle\phi_{r,s}(z)\phi_1(z_1)\ldots\phi_n(z_n)\rangle = 0. \tag{11.4.10}$$

Previously, we observed that the null vectors of the conformal family $V_{r,s}$ are in infinite number and organized in a nested structure. There is, for instance, another null vector at the level $(q-r)(p-s)$ and this implies that the correlation functions of the field $\phi_{r,s}$ are also solutions of another linear differential equation of order $(q-r)(p-s)$. All other null vectors lead to an infinite hierarchy of linear differential equations satisfied by these correlators

$$\mathcal{D}_a\langle\phi_{r,s}(z)\phi_1(z_1)\ldots\phi_n(z_n)\rangle = 0, \tag{11.4.11}$$

whose order a is equal to the level a of the various null vectors: the explicit form can be determined making use of the linear combination of the null vector in terms of the descendant fields at the level a.

For this underlying structure of linear differential operators, not surprisingly the OPE of the primary fields of the minimal models are severely constrained.

11.4.3 Operator Product Expansion and Fusion Rules

Let us initially focus our attention on the conformal field $\phi_{1,2}$ of the minimal models. Its first null vector occurs at the level $N = 2$ and its explicit expression is

$$\phi_{1,2}^{null}(z) = \left[L_{-2} - \frac{3}{2(2\Delta_{1,2}+1)}L_{-1}^2\right]\phi_{1,2}(z). \tag{11.4.12}$$

Hence the correlation functions of this field satisfy the linear differential equation

$$\left\{\frac{3}{2(2\Delta_{1,2}+1)}\frac{\partial^2}{\partial z^2} - \sum_{i=1}^{n}\left[\frac{\Delta_i}{(z-z_i)^2} + \frac{1}{z-z_i}\frac{\partial}{\partial z_i}\right]\right\}\langle\phi_{1,2}(z)\phi_1(z_1)\ldots\phi_n(z_n)\rangle = 0. \tag{11.4.13}$$

Consider now the OPE of the field $\phi_{1,2}(z)$ with any other primary field $\phi_\Delta(z_1)$

$$\phi_{1,2}(z)\phi_\Delta(z_1) = \sum_{\Delta'} C_{(1,2),\Delta}^{\Delta'}(z-z_1)^{\Delta'-\Delta-\Delta_{1,2}}\left[\phi_{\Delta'}(z_1)+\cdots\right]. \tag{11.4.14}$$

This expansion has to be compatible with the differential equation satisfied by the field $\phi_{1,2}(z)$. Inserting this operator expansion in eqn. (11.4.13) and posing equal to zero the most singular term in this expression, we arrive at the characteristic equation associated to the differential equation

$$\frac{3x(x-1)}{2(2\Delta_{1,2}+1)} - \Delta + x = 0, \tag{11.4.15}$$

where $x = \Delta' - \Delta - \Delta_{1,2}$. This is a second-order algebraic equation that shows that the OPE of $\phi_{1,2}$ with any other conformal field cannot have more than two conformal families. Furthermore, it allows us to determine the conformal dimension of the primary field $\phi_{\Delta'}$ that is generated by the operator expansion with ϕ_Δ. Remarkably enough, if Δ is expressed by one value of the Kac formula, i.e. $\Delta = \Delta_{r,s}$, then the solutions of the characteristic equation also belong to the set of values of the Kac table! Namely, if $\Delta = \Delta_{r,s}$, the two solutions of the quadratic equations are given by

$$\Delta' = \left\{\Delta_{r,s-1}, \Delta_{r,s+1}\right\}. \tag{11.4.16}$$

Simplifying the notation of the OPE to its skeleton form, we can write

$$\phi_{1,2} \times \phi_{r,s} = [\phi_{r,s-1}] + [\phi_{r,s+1}], \tag{11.4.17}$$

and, in particular

$$\phi_{1,2} \times \phi_{1,2} = [\mathbf{I}] + [\phi_{1,3}]. \tag{11.4.18}$$

In other words, only degenerate fields enter the OPE of $\phi_{1,2}$ with any of the degenerate field $\phi_{r,s}$. It must be stressed, though, that the result above does not take into account the actual value of the structure constant: as it is, it only states which conformal families may possibly enter the OPE. As we show later, the vanishing of one or more of the structure constants further reduces the number of the conformal families. In this case we say that a *truncation* of the OPE has occurred.

Repeating the same analysis for the degenerate field $\phi_{2,1}$ the same conclusions are reached, the only difference being the swapping of the relative indices. With the same notation introduced above, we have in fact

$$\begin{aligned} \phi_{2,1} \times \phi_{r,s} &= [\phi_{r-1,s}] + [\phi_{r+1,s}] \\ \phi_{2,1} \times \phi_{2,1} &= [\mathbf{I}] + [\phi_{3,1}]. \end{aligned} \tag{11.4.19}$$

The graphical interpretation of these results is immediate: by using iteratively the OPE of the operators $\phi_{1,2}$ and $\phi_{2,1}$ it is possible to generate all the other degenerate fields of the minimal models, i.e. we can move horizontally and vertically along the Kac lattice, visiting all its points, as shown in Figure 11.2.

An explicit example of the phenomenon of truncation is provided by the OPE of the fields $\phi_{1,2}$ and $\phi_{2,1}$. Using the formulae above, either with respect to the first field and the second one, we have

$$\begin{aligned} \phi_{1,2} \times \phi_{2,1} &= [\phi_{0,2}] + [\phi_{2,2}] \\ \phi_{1,2} \times \phi_{2,1} &= [\phi_{2,0}] + [\phi_{2,2}] \end{aligned} \tag{11.4.20}$$

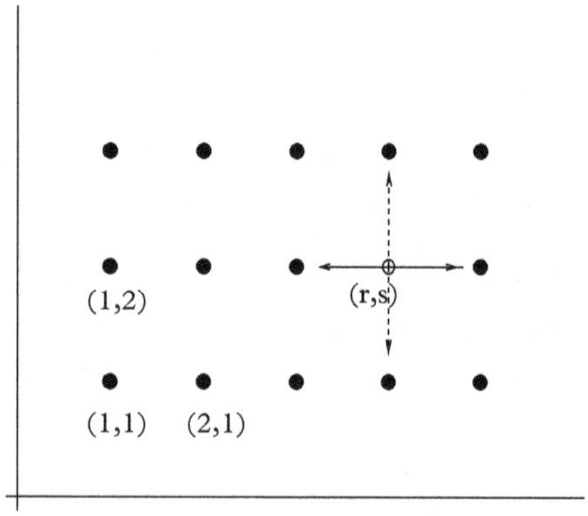

Fig. 11.2 *Action of the operators $\phi_{1,2}$ (dashed line) and $\phi_{2,1}$ (continous line).*

Since the two different ways should lead to the same result, the structure constants that involve both the fields $\phi_{0,2}$ and $\phi_{2,0}$ must vanish. So, we are in the presence of a truncation of the OPE of $\phi_{1,2} \times \phi_{2,1}$ that reduces then to the expression

$$\phi_{1,2} \times \phi_{2,1} = [\phi_{2,2}]. \tag{11.4.21}$$

We can now iteratively insert the operators $\phi_{1,2}$ and $\phi_{2,1}$, using at the same time the associativity of the operator algebra, to compute the fusion rules of the other degenerate fields. Consider, for instance, the product of three fields $\phi_{2,1} \times \phi_{2,1} \times \phi_{r,s}$. Applying twice the fusione rules (11.4.19), we get

$$\phi_{3,1} \times \phi_{r,s} = [\phi_{r+2,s}] + [\phi_{r,s}] + [\phi_{r-2,s}]. \tag{11.4.22}$$

Analogously, consider $\phi_{1,2} \times \phi_{1,2}\phi_{r,s}$. Using eqn. (11.4.17) twice, we arrive at

$$\phi_{1,3} \times \phi_{r,s} = [\phi_{r,s+2}] + [\phi_{r,s}] + [\phi_{r,s-2}]. \tag{11.4.23}$$

It is easy to check that these are precisely the fusion rules that are compatible with the linear differential equations of the third order satisfied by the fields $\phi_{3,1}$ and $\phi_{1,3}$, as proposed in Exercise 11.1.

Fusion rule. Carrying on a similar analysis for the other fields, one can reach the general formula of the *fusion rules* relative to two arbitrary degenerate fields of the Kac table

$$\phi_{r_1,s_1} \times \phi_{r_2,s_2} = \sum_{r_3=|r_1-r_2|+1}^{\min(r_1+r_2-1,2q-1-r_1-r_2)} \quad \sum_{s_3=|s_1-s_2|+1}^{\min(s_1+s_2-1,2p-1-s_1-s_2)} [\phi_{r_3,s_3}] \tag{11.4.24}$$

where both indices are summed in steps of two. These fusion rules can be written in a more transparent form noting that they are similar to the fusion rules of two irreducible representations of spins j and j' of $SU(2)$. To this aim, it is useful to use as indices $r_i = 2j_1 + 1$ and $r_i = 2j'_i + 1$. This similarity explains the null values of the structure constants for all odd values of r (corresponding to the vector representations of $SU(2)$) as well as their vanishing when there are two even indices and one odd (corresponding to two spinor representations and one vector representation). However, there is an important difference between the fusion rules of CFT and those of $SU(2)$, as clearly shown by the upper restriction of the two sums that involve the parameters q and p. In fact, the fusion rules of the minimal models are not those of $SU(2)$ but those of the *quantum group* $SU_q(2)$. In the minimal models there are two quantum groups:[3] the first $SU_{q_1}(2)$ with $q_1 = \exp(i\pi q/p)$ acting on the rows, the second $SU_{q_2}(2)$ with $q_2 = \exp(i\pi p/q)$ acting on the column. Since q_1 and q_2 are both roots of unity, the representations of the corresponding quantum groups get restricted and their composition laws are expressed by the fusion rules given above.

11.4.4 Verlinde Algebra

It is important to formulate in a more abstract way the fusion rules for better analysing their properties. Denoting by ϕ_i a generic primary field, the algebraic structure of the fusion rules can be expressed putting simply to 1 all the non-zero structure constants by

$$\phi_i \times \phi_j = \sum_k N_{ij}^k \phi_k, \tag{11.4.25}$$

where N_{ij}^k are a set of integers that express the number of independent fusions that related ϕ_i and ϕ_j to the field ϕ_k. For the minimal models, these numbers can only be 0 and 1, but for conformal theories with an extended algebra they can be generically integers.

For their same definition, the quantities N_{ij}^k are symmetric with respect to the indices i and j. The associativity condition of the algebra (11.4.25) leads to a quadratic condition for the quantities N_{ij}^k: this can be derived by the two possible ways of applying eqn. (11.4.25) to the product of three fields

$$\phi_i \times \phi_j \times \phi_l = \begin{cases} \sum_k N_{ij}^k \phi_k \times \phi_l = \sum_{k,p} N_{ij}^k N_{kl}^p \phi_p \\ \phi_i \times \sum_k N_{jl}^k \phi_k = \sum_{k,p} N_{jl}^k N_{ik}^p \phi_p. \end{cases} \tag{11.4.26}$$

Using the matrix notation $(N_i)_j^k = N_{ij}^k$ and the symmetry with respect to the indices i,j, the identity of the two expressions above reads

$$N_i N_l = N_l N_i. \tag{11.4.27}$$

[3] For the notations and the theory of quantum group, see Section 18.11.

This condition can be also expressed as

$$N_j N_l = \sum_k N_{jl}^k N_k. \tag{11.4.28}$$

The commutativity of the matrices N_i, shown in eqn. (11.4.27), implies that all these matrices can be diagonalized simultaneously and their eigenvalues $\lambda_i^{(n)}$ form a one-dimensional representation of the fusion rules. Note that the algebra (11.4.25), known as *Verlinde algebra*, is very similar to the formula that appears in the theory of the finite groups and that rules the composition law of their irreducible representations. Further properties of the Verlinde algebra can be found in Problem 11.6.

11.5 Coulomb Gas

We have seen that the correlation functions of the primary fields of the minimal models, for their null vectors, satisfy a series of linear differential equations. Hence, to determine explicitly the correlators, we can adopt the following strategy:

1. find the explicit expression of the null vectors;
2. translate this expression in the corresponding linear differential equation;
3. find its solutions.

All these steps can be explicitly implemented for the minimal models. However, there exists a more efficient way to find the final expressions of the correlators. The method proposed by Dotsenko and Fateev enables us to write down directly the final expression of the correlation functions without passing through the three steps given above. It is based on a modified version of the Coulomb gas in two dimensions. However, first we explain the CFT associated to a free massless bosonic field. Further details on this theory will be given in Section 12.4.

11.5.1 Free Theory of a Bosonic Field

Consider the action of a free massless scalar field in two dimensions

$$A = \frac{g}{2} \int d^2x \, \partial_\mu \varphi \, \partial^\mu \varphi. \tag{11.5.1}$$

The propagator of this theory needs both an ultraviolet and an infrared cut-off, given respectively by a and R, and it can be written as

$$G(z, \bar{z}) = \langle \varphi(z, \bar{z}) \varphi(0, 0) \rangle = \begin{cases} -\frac{1}{4\pi g} \log \frac{z}{a} - \frac{1}{4\pi g} \log \frac{\bar{z}}{a}, & z, \bar{z} \neq 0 \\ -\frac{1}{4\pi g} \log \frac{R}{a}, & z = \bar{z} = 0. \end{cases} \tag{11.5.2}$$

Note that this correlator is also the Green function of a two-dimensional electrostatic problem, and for this reason, the formalism we are going to present is also known as Coulomb gas approach. To simplify the formula to come, here we choose as the coupling constant $g = 8\pi$.

As evident from the form of its propagator, $\varphi(x)$ is not a conformal field. However, conformal fields can be constructed in terms of some of its composite operators, as for instance all derivative fields $\partial_z^n \partial_{\bar{z}}^m \varphi$ or the exponential operators $\tilde{V}_\alpha = e^{i\alpha\varphi}$, also known as *vertex operators*. The quantity α entering the exponential is also called the *charge parameter*. Let us focus the attention of the analytic part of the theory. It is easy to see that the n-point correlation functions of the vertex operators can be computed by means of the Wick theorem or directly by the functional integral for the action is quadratic

$$\prod_{i=1}^{n} \langle \tilde{V}_{\alpha_i}(z_i) \rangle = \left(\frac{a}{R}\right)^{(\sum_i^n \alpha_i)^2} \prod_{i<j} \left(\frac{z_{ij}}{a}\right)^{2\alpha_i \alpha_j}. \tag{11.5.3}$$

We need to get rid of the two cut-offs. To eliminate the dependence from the ultraviolet cut-off, it is sufficient to subtract all the tadpole contributions coming from the contractions of the the field $\varphi(z)$ with itself. This can be implemented defining the renormalized vertex operator

$$\tilde{V}_\alpha(z) \to V_\alpha = \lim_{a \to 0} a^{-\alpha^2/2} e^{i\alpha\varphi} \equiv : e^{i\alpha\varphi} : \tag{11.5.4}$$

To eliminate the dependence on the infrared cut-off, so that to have a non-zero limit of the correlation functions when $R \to \infty$, it is necessary to impose the neutrality conditions of all the charges

$$\sum_{i=1}^{n} \alpha_i = 0. \tag{11.5.5}$$

This is in agreement with the well-known result of statistical mechanics that a system of electric charges is unstable unless it has a zero total charge.

Looking at the two-point correlation functions of the renormalized vertex operators satisfying the neutrality condition

$$\langle V_\alpha(z) V_{-\alpha}(w) \rangle = \frac{1}{(z-w)^{2\alpha^2}}, \tag{11.5.6}$$

we can extract the conformal dimension of the two vertex operators $V_{\pm\alpha}(z)$, given by

$$\Delta_\alpha = \Delta_{-\alpha} = \alpha^2. \tag{11.5.7}$$

The analytic component of the stress-energy tensor associated to the action (11.5.1) is

$$T(z) = -\frac{1}{4} : (\partial_z \varphi)^2 :, \tag{11.5.8}$$

and, as we have previously seen, the central charge of this system is $C = 1$.

One may wonder if it would be possible to modify the Coulomb gas in such a way to have values of the central charge different from $C = 1$ and conformal dimensions equal to those of the Kac table of the minimal models. This is indeed possible, as discussed in the next section.

11.5.2 Modified Coulomb Gas

Consider a vertex operator with charge $-2\alpha_0$ and suppose we insert it in a correlation function, moving its position at infinity (i.e. on the north pole of the Riemann sphere associate to the complex plane) using the prescription (10.8.10). With this procedure we can define a new set of correlators given by

$$<< V_{\alpha_1}(z_1) \ldots V_{\alpha_n}(z_n) >> \equiv \lim_{R \to \infty} R^{8\alpha_0^2} \langle V_{-2\alpha_0}(R)\, V_{\alpha_1}(z_1) \ldots V_{\alpha_n}(z_n) \rangle. \tag{11.5.9}$$

Note that to recover the translation invariant of these quantity (that is expressed by the dependence on the coordinates only through the differences $z_i - z_j$), it is necessary to place the vertex operator $V_{-2\alpha_0}$ just at infinity. These new set of correlation functions for the vertex operators $V_{\alpha_1} \ldots V_{\alpha_n}$ satisfy a different neutrality condition

$$\sum_{i=1}^{n} \alpha_i = 2\alpha_0. \tag{11.5.10}$$

In agreement with that, the two-point correlation function of the vertex operators is now given by[4]

$$\langle V_\alpha(z) V_{2\alpha_0 - \alpha}(w) \rangle = \frac{1}{(z - w)^{2\alpha(\alpha - 2\alpha_0)}}, \tag{11.5.11}$$

The new conformal dimension are then

$$\Delta_\alpha = \Delta_{2\alpha_0 - \alpha} = \alpha(\alpha - 2\alpha_0). \tag{11.5.12}$$

This result can be directly confirmed by the OPEs of the vertex operators with the new expression of the stress-energy tensor. To derive the new stress-energy tensor, we

[4] In the following instead of using the notation $<< \ldots >>$ we switch back to the simplest notation $\langle \ldots \rangle$ to denote the correlation functions also in the modified Coulomb gas system.

should observe that placing a charge at infinity is equivalent to modify the original action (11.5.1) in such a way to make anomalous the original $U(1)$ symmetry implemented by $\varphi \to \varphi + \eta$. In a generalized system of coordinates, this can be realized by coupling the field φ to the scalar curvature R of the space manifold

$$A = \frac{1}{8\pi} \int d^2x \sqrt{g} (\partial_\mu \varphi \partial^\mu \varphi + 2i\alpha_0 R\varphi). \tag{11.5.13}$$

This new action is no longer invariant under a shift of the φ and its variation becomes

$$\delta A = i \frac{\alpha_0}{4\pi} \int d^2x \sqrt{g} R. \tag{11.5.14}$$

In two dimensions this is just a topological term, that can be computed by the Gauss-Bonnet theorem

$$\int d^2 \sqrt{g} R = 8\pi (1 - h), \tag{11.5.15}$$

where h is the number of handles of the Riemann surface on which is defined the field theory (11.5.13) and for a sphere we have $h = 0$. In correspondence to the new action (11.5.13), there is a new version of the stress-energy tensor given by the Noether theorem and its analytic component reads

$$T(z) = -\frac{1}{4} (\partial \varphi)^2 + i\alpha_0 \, \partial^2 \varphi, \tag{11.5.16}$$

Its two-point function is given by

$$\langle T(z) T(w) \rangle = \frac{1 - 24\alpha_0^2}{2(z - w)^4}. \tag{11.5.17}$$

In conclusion, with a charge at infinity, the central charge of the theory assumes a different value from the original $C = 1$ and is now

$$C = 1 - 24\alpha_0^2. \tag{11.5.18}$$

The OPE of the new stress-energy tensor with the vertex operator $V_\alpha(w)$ becomes

$$\begin{aligned} T(z) : e^{i\alpha\phi(w)} : &= \frac{\alpha^2 - 2\alpha_0 \alpha}{(z - w)^2} : e^{i\alpha\phi} : + \frac{i\alpha}{(z - w)} : \partial \varphi \, e^{i\alpha\varphi} : + : T(z) e^{i\alpha\varphi} := \\ &= \frac{\alpha(\alpha - 2\alpha_0)}{(z - w)^2} V_\alpha(w) + \frac{1}{z - w} \partial V_\alpha(w) + \dots \end{aligned} \tag{11.5.19}$$

This formula clearly shows that, in the presence of the charge at infinity, the conformal dimension of the vertex operator V_α is effectively given by eqn. (11.5.12).

It is also important to discuss the conformal trasformation of the scalar field. In the absence of the charge at infinity, under the transformation $z \to f(z)$, $\phi(z)$ trasforms as $\phi(z) \to \phi(f(z))$. But with the charge at infinity, there is a change of the boundary conditions and the field transforms instead as

$$\phi(z) \to \phi(f(z)) + 2i\alpha_0 \ln f'(z), \tag{11.5.20}$$

whose infinitesimal form is

$$\delta\phi(z) = \epsilon(z)\partial_z\phi(z) + 2i\alpha_0\epsilon'(z). \tag{11.5.21}$$

11.5.3 Screening Operators

The modified Coulomb gas formalism allows us to describe the conformal models with central charge less than 1 and, in particular, the minimal models. In this approach the primary fields are associated to the vertex operators. Notice that, assigned the conformal dimension Δ of the primary field, there are however *two* different vertex operators V_α and $V_{2\alpha_0-\alpha}$ that can be put in correspondence with it, because the charges satisfy the quadratic condition $\Delta = \alpha(\alpha - 2\alpha_0)$. The two-point correlation function of the primary field ϕ_Δ is different from zero but, in the formalism of the Coulomb gas, it can be computed in four different way, namely

$$\langle\phi_\Delta(z)\phi_\Delta(0)\rangle \to \begin{cases} \langle V_\alpha(z)V_{2\alpha_0-\alpha}(w)\rangle \\ \langle V_{2\alpha_0-\alpha}(z)V_\alpha(w)\rangle \\ \langle V_\alpha(z)V_\alpha(w)\rangle \\ \langle V_{2\alpha_0-\alpha}(z)V_{2\alpha_0-\alpha}(w)\rangle. \end{cases} \tag{11.5.22}$$

The first two expressions automatically satisfy the neutrality condition (11.5.10): hence they are different from zero and give rise to the usual expression $\langle\phi_\Delta(z)\phi_\Delta(w)\rangle$. On the contrary, the last two expressions do not fulfill the neutrality condition (11.5.10) and are therefore zero. But we must then correct this drawback in such a way that one can equivalently use either $V_\alpha(z)$ or $V_{2\alpha_0-\alpha}(z)$ to represent the primary field $\phi_\Delta(z)$.

The solution to this problem consists of the introduction of the so-called *screening operators*. Such operators, once inserted in the correlators, should be able to absorb the extra charge that spoils the neutrality condition without altering, though, the conformal properties of the correlators. To satisfy such conditions, the screening operators must have a zero conformal dimension but a non-zero charge. It is impossible to fulfill these conditions in terms of local operators but there is no problem in finding them in terms of a closed contour integral of an operator of conformal dimension equal to 1. Hence, denoting by $V_{\alpha_\pm}$ the vertex operators of these fields with conformal dimension $\Delta = 1$, we can pose

$$Q_{\pm} = \oint dz\, V_{\alpha_{\pm}}(z), \tag{11.5.23}$$

where the charges $\alpha_{\pm}$ satisfy the equation

$$\alpha_{\pm}(\alpha_{\pm} - 2\alpha_0) = 1. \tag{11.5.24}$$

The solutions of this equation are

$$\alpha_{\pm} = \alpha_0 \pm \sqrt{\alpha_0^2 + 1}. \tag{11.5.25}$$

Note that

$$\begin{aligned} \alpha_+ + \alpha_- &= 2\alpha_0 \\ \alpha_+ \alpha_- &= -1. \end{aligned} \tag{11.5.26}$$

Inserting an integer number of these operators $Q_{\pm}$ in the correlation functions, we can therefore screen the extra charge present in their vertex operator representation. This cancellation mechanism takes place only if the extra charge is expressible in terms of integer multiples of $Q_{\pm}$. Consider, for instance, the third expression in (11.5.22): inserting now an integer number of the screening operators, it becomes

$$\langle V_{\alpha}(z) V_{\alpha}(w) Q_+^r\, Q_-^s \rangle \tag{11.5.27}$$

and the neutrality conditions translates in the condition

$$2\alpha + r\alpha_+ + s\alpha_- = 2\alpha_0 = \alpha_+ + \alpha_-. \tag{11.5.28}$$

Hence we can screen the extra charge of the original expression only if the charges α present in the system satisfy the quantization condition

$$\alpha = \alpha_{r,s} = \frac{1}{2}(1-r)\alpha_+ + \frac{1}{2}(1-s)\alpha_-. \tag{11.5.29}$$

In this case, there is a complete equivalence of the operators $V_{\alpha_{r,s}}$ and $V_{2\alpha_0 - \alpha_{r,s}}$, after all, the logic consistency of the modified Coulomb gas. In correspondence of the values (11.5.29) the conformal dimensions of the fields are given by

$$\Delta_{r,s} = \frac{1}{4}(r\alpha_+ + s\alpha_-)^2 - \alpha_0^2. \tag{11.5.30}$$

They assume the form (11.2.7) given by the Kac table and the symmetry $\alpha \to 2\alpha_0 - \alpha$ translates in the transformation $(r,s) \to (-r,-s)$. To recover the minimal models, we need however to impose an additional quantization condition on the charges $\alpha_{\pm}$ ($p > q$)

$$q\alpha_+ + p\alpha_- = 0. \tag{11.5.31}$$

where p and q are two co-prime integers. With this last condition, it is easy to see that eqns. (11.5.18) and (11.5.30) reproduce the central charge and the conformal dimensions of the minimal models, eqn. (11.4.1), and we have moreover the periodicity relation

$$\alpha_{r,s} = \alpha_{r+q,s+p}. \tag{11.5.32}$$

In the next section we discuss how to compute the correlation functions of the minimal models using the Coulomb gas formalism.

11.5.4 Correlation Functions

The correlation functions of the primary fields $\phi_{r,s}$ satisfy an infinite number of linear differential equations in coincidence of their null vector hierarchy. The main advantage of the Coulomb gas formalism is to provide the solutions of the differential equations directly in terms of their integral representation. In this section we initially discuss the implementation of this formalism for the simplest cases of the correlators of the fields $\phi_{1,2}$ and $\phi_{2,1}$.

Consider the holomorphic part of the four-point correlation function of the primary field

$$G(z_1, z_2, z_3, z_4) = \langle \phi_{n,m}(z_1)\phi_{1,2}(z_2)\phi_{1,2}(z_3)\phi_{n,m}(z_4)\rangle. \tag{11.5.33}$$

This quantity is surely different from zero since there exists a common conformal channel—given by family of the identity operator—in the OPE of $\phi_{1,2} \times \phi_{1,2}$ and di $\phi_{n,m} \times \phi_{n,m}$. Since the primary fields $\phi_{r,s}$ can be associated either to $V_{\alpha_{r,s}}$ or $V_{2\alpha_0-\alpha_{r,s}}$, the correlation function (11.5.33) admits 16 different expressions in terms of the vertex operators of the Coulomb gas. Out of these expressions, the one that needs fewer screening operators is[5]

$$\langle V_{\alpha_{n,m}}(z_1)V_{\alpha_{1,2}}(z_2)V_{\alpha_{1,2}}(z_3)V_{2\alpha_0-\alpha_{n,m}}(z_4)\rangle.$$

The extra charge present in this representation is $2\alpha_{1,2}$, thus its screening requires only one operator Q_-. This leads to the integral representation

$$\begin{aligned}&\langle \phi_{n,m}(z_1)\phi_{1,2}(z_2)\phi_{1,2}(z_3)\phi_{n,m}(z_4)\rangle = \\ &\oint_C dv\, \langle V_{\alpha_{1,2}}(z_1)V_{\alpha_{1,2}}(z_2)V_{\alpha_{n,m}}(z_3)V_{2\alpha_0-\alpha_{n,m}}(z_4)V_{\alpha_-}(v)\rangle.\end{aligned} \tag{11.5.34}$$

[5] The other expressions lead to the integral representation of the solutions of the higher-order differential equations satisfied by the same correlator.

For the analytic nature of the integrand as a function of v, the integral does not depend on the precise shape of the contour, although it must be chosen to enclose the points $z_1,\ldots,z_4$ otherwise it could be shrunk to a point, with a vanishing result. Performing the expectation values of the vertex operators in the integrand by using the Wick theorem

$$\prod_{i=1}^{n}\langle V_{\alpha_i}(z_i)\rangle = \prod_{i<j}^{n}(z_{ij})^{2\alpha_i\alpha_j}$$

we have

$$\oint_{C_1} du_1\langle V_{\alpha_{n,m}}(z_1)V_{\alpha_{1,2}}(z_2)V_{\alpha_{1,2}}(z_3)V_{2\alpha_0-\alpha_{n,m}}(z_4)V_{\alpha_+}(u_1)\rangle =$$
$$(z_{12}z_{13})^{2\alpha_{1,2}\alpha_{n,m}}(z_{23})^{2\alpha_{1,2}^2}(z_{24})^{2\alpha_{n,m}(2\alpha_0-\alpha_{n,m})}(z_{24}z_{34})^{2\alpha_{1,2}(2\alpha_0-\alpha_{n,m})}\times$$
$$\oint dv(v-z_1)^{2\alpha_-\alpha_{n,m}}[(v-z_2)(v-z_3)]^{2\alpha_-\alpha_{1,2}}(v-z_4)^{2\alpha_-(2\alpha_0-\alpha_{n,m})}.$$

To simplify the expressions, let us use the Moebius invariance to fix three out of the four points of the correlator: we pose then $z_1 = 1$, $z_3 = 0$ and $z_4 = \infty$, leaving as a free position $z_2 = z$. In this way we arrive at

$$\langle\phi_{n,m}(1)\phi_{1,2}(z)\phi_{1,2}(0)\phi_{n,m}(\infty)\rangle = z^{2\alpha_{1,2}^2}(1-z)^{2\alpha_{1,2}\alpha_{n,m}}$$
$$\times\oint_C dv\, v^{2\alpha_-\alpha_{1,2}}(v-1)^{2\alpha_-\alpha_{n,m}}(v-z)^{2\alpha_-\alpha_{1,2}}.$$

Consider now the integral

$$F(z,a,b,c) = \oint dv\, v^a(v-1)^b(v-z)^c, \tag{11.5.35}$$

with

$$a = 2\alpha_-\alpha_{1,2} \quad b = 2\alpha_-\alpha_{n,m} \quad c = 2\alpha_-\alpha_{1,2}. \tag{11.5.36}$$

The integrand, as a function of the complex variable v has branch cuts at the points $v = 0, 1, z$. To be closed, the integration contour must cross each cut twice. There are several ways to choose such a contour, although only two of them are independent: the most convenient choice consists of the paths shown in Figure 11.3. If the integral along these paths converges, the first contour can be restricted to the interval $C_1 = [1,\infty]$ while the second to the interval $C_2 = [0,z]$. With this choice of the paths of integration, we have defined two different functions

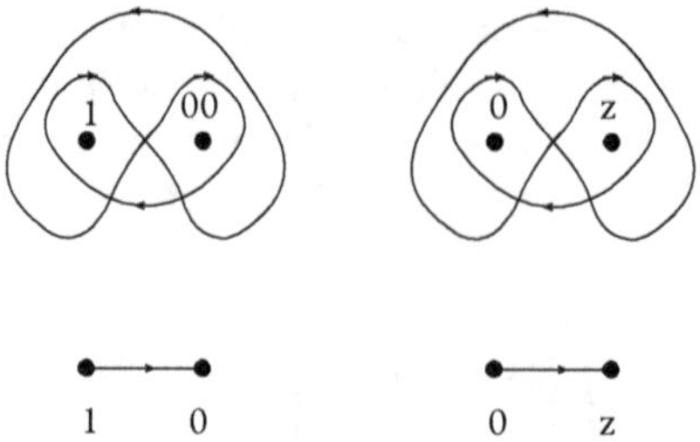

Fig. 11.3 *Independant contours.*

$$
\begin{aligned}
I_1(z, a, b, c) &= \int_1^{\infty} dv\, v^a (v-1)^b (v-z)^c \\
&= \frac{\Gamma(-a-b-c-1)\Gamma(b+1)}{\Gamma(-a-c)} F(z; -c, -a-b-c-1, -a-c), \\
I_2(z, a, b, c) &= \int_0^{z} dv\, v^a (v-1)^b (z-v)^c \\
&= \frac{\Gamma(a+1)\Gamma(c+1)}{\Gamma(a+c+2)} z^{1+a+c} F(z; -b,\, a+1,\, a+c+2),
\end{aligned}
\tag{11.5.37}
$$

where $F(z, \alpha, \beta, \gamma)$ is a hypergeometric function (see Appendix 11.A for its properties). The two functions above set up the vector space of the solutions of the second-order differential equation satisfied by the correlation functions of the primary field $\phi_{1,2}$. An analogous result is obtained if we consider the anti-analytic part of the correlator, so that the general form of the correlation function in the physical plane is expressed as a linear combination of the analytic and anti-analytic solutions of the differential equation

$$
\begin{aligned}
G(z, \bar{z}) &= \langle \phi_{n,m}(1,1)\phi_{1,2}(z,\bar{z})\phi_{1,2}(0,0)\phi_{n,m}(\infty)\rangle \\
&= |z|^{4\alpha_{1,2}} |1-z|^{4\alpha_{1,2}\alpha_{n,m}}\, Y(z,\bar{z}),
\end{aligned}
\tag{11.5.38}
$$

where

$$
Y(z,\bar{z}) = \sum_{i,j=1}^{2} \mathcal{X}_{ij} I_i(z) I_j(\bar{z}).
$$

Monodromy invariance. The explicit expression of the coefficients $\mathcal{X}_{ij}$ can be obtained by the condition that the correlation functions is an unambiguous function of the points in the plane, i.e. independent on the paths by which we reach the points. To implement such a condition, we have to analyse the monodromy group associated to the functions $I_i(z)$. They have the singular points $z = 0, 1, \infty$. If we make an analytic continuation along a closed contour that encloses one of these points (Figure 11.4), they do not return

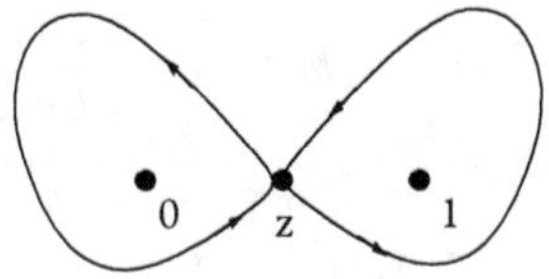

Fig. 11.4 *Analytic continuation of the functions $I_i(z)$ around the singular points $z = 0, 1$.*

to their original value, instead the new function—which is still a solution of the linear differential equation—is expressed as a linear combination of the I_i's

$$\begin{aligned} I_i(z) &\to (g_0)_{ij} I_j(z), \\ I_i(z) &\to (g_1)_{ij} I_j(z). \end{aligned} \tag{11.5.39}$$

Note that it is sufficient to consider the monodromy properties only around the points $z = 0, 1$, since those around the point $z = \infty$ follow from them. The monodromy matrix $(g_0)_{ij}$, in our case, is diagonal

$$g_0 = \begin{pmatrix} 1 & 0 \\ 0 & e^{2\pi i(a+c+1)} \end{pmatrix}. \tag{11.5.40}$$

Imposing the invariance of the correlation function (11.5.38) under this transformation immediately leads to the conditions $\mathcal{X}_{12} = \mathcal{X}_{21} = 0$ and, after that, the function $Y(z, \bar{z})$ reduces to a diagonal form

$$Y(z, \bar{z}) = \sum_{i=1}^{2} \mathcal{X}_{ii} I_i(z) I_i(\bar{z}). \tag{11.5.41}$$

To determine the remaining coefficients $\mathcal{X}_{11}$ e $\mathcal{X}_{22}$ we must impose the invariance under the monodromy transformation g_1. The simplest way to do that is to express initially the functions $I_i(z)$ in terms of another basis that has the series expansion in the variable $(1 - z)$

$$I_i(z) = \sum_{j=1}^{2} a_{ij} \hat{I}_j(1 - z). \tag{11.5.42}$$

For the hypergeometric functions this can be done using the Gauss formulae: with $s(x) \equiv \sin(\pi x)$ we have[6]

[6] Note that exchange of the indices a and b in the functions on the right-hand side of these expressions.

$$I_1(z;a,b,c) = \frac{s(a)}{s(b+c)}\hat{I}_1(1-z;b,a,c) - \frac{s(c)}{s(b+c)}\hat{I}_2(1-z;b,a,c)$$

$$I_2(z;a,b,c) = -\frac{s(a+b+c)}{s(b+c)}\hat{I}_1(1-z;b,a,c) - \frac{s(b)}{s(b+c)}\hat{I}_2(1-z;b,a,c). \tag{11.5.43}$$

Substituting eqn. (11.5.42) in (11.5.41) we get

$$Y(z,\bar{z}) = \sum_{i,k,l=1,2} \mathcal{X}_{ii} a_{ij} a_{ik} \hat{I}_j(1-z)\hat{I}_k(1-\bar{z}). \tag{11.5.44}$$

Since the monodromy matrix of the functions $\hat{I}_i(1-z)$ around the point $z=1$ is diagonal, the monodromy invariance implies that the quadratic form in $\hat{I}_i$ must be diagonal as well. Hence

$$\frac{\mathcal{X}_{11}}{\mathcal{X}_{22}} = -\frac{a_{21}a_{22}}{a_{12}a_{11}} = \frac{s(a+b+c)s(b)}{s(a)s(c)}. \tag{11.5.45}$$

It is clear that, besides an overall constant λ (simply related to the normalization of the operators and that cannot be fixed by the monodromy invariance), the correlation function is given by

$$G(z,\bar{z}) = \langle \phi_{n,m}(1,1)\phi_{1,2}(z,\bar{z})\phi_{1,2}(0,0)\phi_{n,m}(\infty)\rangle = \lambda\,|z|^{4\alpha_{1,2}}|1-z|^{4\alpha_{1,2}\alpha_{n,m}} \times$$
$$\times \left[\frac{s(b)s(a+b+c)}{s(a+c)}|I_1(z;a,b,c)|^2 + \frac{s(a)s(c)}{s(a+c)}|I_2(z;a,b,c)|^2\right] \tag{11.5.46}$$

λ can be fixed once specifying the normalization of the conformal fields that we choose to be

$$\langle \phi_\Delta(z,\bar{z})\phi_\Delta(w,\bar{w})\rangle = 1/|z-w|^{4\Delta}. \tag{11.5.47}$$

With the values of the parameters a, b, c of the hypergeometric functions given in eqn. (11.5.36), it is now easy to see that the function $I_2(z;a,b,c)$ corresponds to the channel of the conformal family of the identity operator present in this correlator. This means that in the limit $z \to 0$ the singularity coming from this term is precisely $1/|z|^{4\Delta_{1,2}}$. Hence, using the numerical factor in the function $I_2(z;a,b,c)$, the value of λ that implements the normalization condition (11.5.47) is given by

$$\lambda = \left(\frac{\Gamma(a+c+2)}{\Gamma(a+1)\Gamma(c+1)}\right)^2 \frac{s(a+c)}{s(a)s(c)}. \tag{11.5.48}$$

Structure constants. We can now extract the exact value of some of the structure constants of the conformal operator algebra. The OPE of the field $\phi_{1,2}$ with itself is

$$\phi_{1,2}(z,\bar{z})\phi_{1,2}(0,0) = \frac{1}{|z|^{4\Delta_{1,2}}}\{\mathbf{I}+\ldots\} + C^{(13)}_{(12,12)}\frac{1}{|z|^{2(2\Delta_{1,2}-\Delta_{1,3})}}\{\phi_{1,3}+\ldots\}. \quad (11.5.49)$$

Substituting this expression on the right-hand side of (11.5.46), in the limit $z \to 0$ we have

$$\frac{1}{|z|^{4\Delta_{1,2}}}\langle\phi_{n,m}(1)\phi_{n,m}(\infty)\rangle + C^{(13)}_{(12,12)}\frac{1}{|z|^{2(2\Delta_{1,2}-\Delta_{1,3})}}\langle\phi_{n,m}(1)\phi_{1,3}(0)\phi_{n,m}(\infty)\rangle. \quad (11.5.50)$$

The three-point correlator $\langle\phi_{n,m}(1)\phi_{1,3}(0)\phi_{n,m}(\infty)\rangle$ in this case is precisely equal to the structure constant $C^{(13)}_{(nm),(nm)}$. Comparing now (11.5.50) with the right-hand side of eqn. (11.5.46) in the limit $z \to 0$, the singularity with power law $1/|z|^{(2(2\Delta_{1,2}-\Delta_{1,3})}$ is reproduced by the function $|I_1(z;a,b,c)|^2$. Keeping into account the value of λ and the numerical factor in the definition of this function we thus arrive at the formula

$$C^{(13)}_{(12,12)}C^{(13)}_{(nm,nm)} = \frac{s(a+b+c)s(b)}{s(a)s(c)}\left(\frac{\Gamma(a+c+2)\Gamma(-a-b-c-1)\Gamma(b+1)}{\Gamma(a+1)\Gamma(c+1)\Gamma(-a-c)}\right)^2. \quad (11.5.51)$$

It is worth stressing that the analysis of the singularities of the four-point correlation functions allows us to determine only the product of the structure constants, in agreement with the associative nature of the OPE discussed in the previous chapter (see eqn. (10.2.15) and Figure 10.2). Plugging in the formula above $n=1$ and $m=2$, we can determine (up to a sign, that we choose to be positive) the structure constant $C^{(13)}_{(12,12)}$

$$C^{(13)}_{(12,12)} = \frac{\Gamma(2-2\rho)}{\Gamma(2\rho)}\left[-\frac{\gamma^3(\rho)}{\gamma(3\rho-1)}\right]^{1/2}\frac{\gamma(1-\rho)}{\gamma(2-3\rho)}, \quad (11.5.52)$$

where we introduce the notation

$$\rho \equiv \alpha_-^2, \quad \gamma[x] \equiv \Gamma(x)/\Gamma(1-x).$$

Once $C^{(13)}_{(12,12)}$ is known, we can now use eqn. (11.5.51) to determine the other structure constant

$$C^{(13)}_{(nm,nm)} = \frac{\Gamma(2-2\rho)}{\Gamma(2\rho)}\left[-\frac{\gamma^3(\rho)}{\gamma(3\rho-1)}\right]^{1/2}\frac{\gamma(n+(1-m)\rho)}{\gamma(1+n-(1+m)\rho)}. \quad (11.5.53)$$

The exact expression of the correlator $\langle\phi_{nm}(1,1)\phi_{12}(z,\bar{z})\phi_{12}(0,0)\phi_{nm}(\infty)\rangle$ permits to derive easily other structure constants. In fact, going in the dual channel and studying the limit $z \to 1$, one can extract the structure constants $C^{(n,m\pm1)}_{(12,nm)}$. The simplest way to do such a computation is to express the functions $I_i(z)$ in terms of the functions $\hat{I}_i(1-z)$ and write the correlator as

$$G(z,\bar{z}) = \langle \phi_{n,m}(1,1)\phi_{1,2}(z,\bar{z})\phi_{1,2}(0,0)\phi_{n,m}(\infty)\rangle = \lambda\,\frac{s(c)s(a+b+c)+s(a)s(b)}{s(a+c)s^2(b+c)} \times \tag{11.5.54}$$
$$|z|^{4\alpha_{1,2}}|1-z|^{4\alpha_{1,2}\alpha_{n,m}}\left[s(a)s(a+b+c)\,|\hat{I}_1(1-z;b,a,c)|^2 + s(b)s(c)\,|\hat{I}_2(1-z;b,a,c)|^2\right].$$

It is now necessary to use the OPE

$$\phi_{1,2}(z,\bar{z})\phi_{n,m}(1,1) = C^{(n,m+1)}_{(12,nm)}\frac{1}{|z-1|^{2\gamma_+}}\{\phi_{n,m+1}+\ldots\} + C^{(n,m-1)}_{(12,nm)}\frac{1}{|z-1|^{2\gamma_-}}\{\phi_{n,m-1}+\ldots\}$$

with

$$\gamma_\pm = \Delta_{1,2} + \Delta_{n,m} - \Delta_{n,m\pm 1}.$$

Substituting this formula on the left-hand side of (11.5.54), the first function $|\hat{I}_1(1-z,b,a,c)|^2$ is easily identified with the intermediate states coming from the conformal family $\phi_{n,m+1}$, whereas the second function $|\hat{I}_2(1-z;a,b,c)|^2$ is associated to the intermediate states of the family of $\phi_{n,m-1}$. Using the value of λ given above and the normalization of the functions $\hat{I}_i$, one obtains the structure constants

$$C^{(n,m+1)}_{(12,nm)} = \left[\frac{\gamma(2-2\rho)\gamma(n-m\rho)}{\gamma(1-\rho)\gamma(1+n-(1+m)\rho)}\right]^{1/2} \tag{11.5.55}$$

$$C^{(n,m-1)}_{(12,nm)} = C^{(n,m+1)}_{(12,nm)}\sqrt{-\frac{\gamma(n+(1-m)\rho)}{\gamma(-n-(m+1)\rho)}\frac{\Gamma^2(m\rho+1-n)}{\Gamma^2(n+1-m\rho)}}. \tag{11.5.56}$$

Other correlators. Let us briefly comment on the computation of the other correlators, referring the reader to the original articles for their derivation. Suppose, for instance, we wish to compute the correlator of the primary field $\phi_{n,m}(z,\bar{z})$

$$G = \langle \phi_{n,m}(z_1,\bar{z}_1)\phi_{n,m}(z_2,\bar{z}_2)\phi_{n,m}(z_3,\bar{z}_3)\phi_{n,m}(z_4,\bar{z}_4)\rangle.$$

The first step consists of expressing the analytic and anti-analytic parts in terms of the vertex operator representation that needs the smallest number possible of screening operators. A possible choice is

$$\oint_{C_1} du_1 \ldots \oint_{C_{n-1}} du_{n-1} \oint_{S_1} dv_1 \ldots \oint_{S_{m-1}} dv_{m-1}$$
$$\langle V_{\alpha_{n,m}}(z_1)V_{\alpha_{n,m}}(z_2)V_{\alpha_{n,m}}(z_3)V_{2\alpha_0-\alpha_{n,m}}(z_4)V_{\alpha_+}(u_1)\ldots V_{\alpha_+}(u_{n-1})V_{\alpha_-}(v_1)\ldots V_{\alpha_-}(v_{m-1})\rangle.$$

In this expression there are $n \times m$ independent contours that, correspondingly, define a similar number of independent functions $I_i(z)$ $(i = 1, 2, \ldots, nm)$. These functions span the vector space of the solutions of the linear differential equation of order $n \times m$ satisfied by the correlation function. Together with the anti-analytic part, we arrive at a linear combination of these functions that provides the most general solution

$$G = \sum_{i,j=1}^{nm} \mathcal{X}_{ij} I_i(z) I_j(\bar{z}). \tag{11.5.57}$$

The coefficients $\mathcal{X}_{ij}$ can be determined by imposing the monodromy invariance under the monodromy group identified by the functions fi$I_i(z)$ and the normalization of the two-point functions. After all these steps, we can obtain the complete determination of all structure constants of the conformal theory. As a further example of their expressions, we report here the value

$$C^{(n,m+2)}_{(13,nm)} = \frac{2\rho - 1}{(m+1)\rho - n} \left[\frac{\gamma(2 - 3\rho)\gamma(n - m\rho)}{\gamma(1 - \rho)\gamma(1 + n - (m+2)\rho)} \right]^{1/2}, \tag{11.5.58}$$

with the notations previously introduced.

11.6 Landau–Ginzburg Formulation

This section aims to show that the minimal unitary models $\mathcal{M}_q$ describe the dynamics of the multi-critical points of a QFT of a massless scalar field φ with polynomial interaction of highest power $\varphi^{2(q-1)}$. This is the so-called Landau–Ginzburg theory, where the Euclidean action is given by

$$\mathcal{A} = \int d^2x \left[\frac{1}{2} (\partial_\mu \varphi)^2 + V(\varphi) \right]$$

and the general form of the potential is expressed by the normal order powers of the field φ

$$V(\varphi) = g_1 \varphi + g_2 :\varphi^2: + \cdots + g_{2(q-2)} :\varphi^{2(q-2)}: + g :\varphi^{2(q-1)}: . \tag{11.6.1}$$

Note the absence of the term $\varphi^{2(q-1)-1}$: this term can be always removed by a shift of the field $\varphi \to \varphi + const$ and absorbed in the linear term $g_1 \varphi$. Below we assume that the higher coupling constant g is fixed to a positive value. By varying the different parameters g_i, the shape of $V(\varphi)$ can greatly vary. Nevertheless, there always exists a set of values of the coupling constants for which the potential presents $(p-1)$ degenerate ground states: for instance in the φ^4 theory, this situation is realized by a family of curves associated to the parameter a, and the potential, shown in Figure 11.5, given by

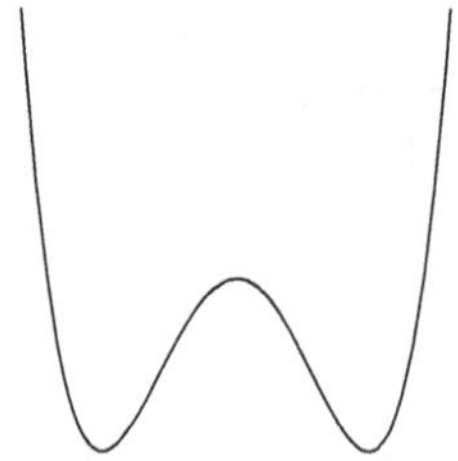

Fig. 11.5 *Potential with two degenerate vacua for the φ^4 theory.*

$$V(\varphi) = g(\varphi^2 - a^2)^2. \tag{11.6.2}$$

Analogously, for the ϕ^6 theory, there is a family of potentials that presents three degenerate vacua by varying the parameter b, as shown in Figure 11.6

$$V(\varphi) = g\varphi^2 (\varphi^2 - b^2)^2. \tag{11.6.3}$$

In the general case, these vacua correspond to the $(q-1)$ different phases of the model. In the space of the coupling constants, the point at which all the coupling constants but g vanish is then a multi-critical point: at this point, characterized by the vanishing of the first $(2q-3)$ derivatives, there is a coalescence of $(q-1)$ different phases.

The operator content of the Landau–Ginzburg theory with potential (11.6.1) consists of $2(q-2)$ scalar relevant fields, associated to the various powers $:\varphi^k:$ $(k = 1,2,\ldots,2(q-2))$, and the irrelevant operators given by all other their derivative fields, as $:\varphi^k \partial_\mu\varphi\partial_\mu\varphi:$. At the multi-critical point, the equation of motion in complex coordinates is given by

$$\partial_z \partial_{\bar{z}}\varphi \sim :\varphi^{2q-3}:. \tag{11.6.4}$$

Such an equation has to be understood as an operator identity once inserted in the correlation functions. Namely, each time that in a correlation function the field $:\varphi^{2q-3}:$ appears, it can be substituted by $\partial_z \partial_{\bar{z}}\varphi$

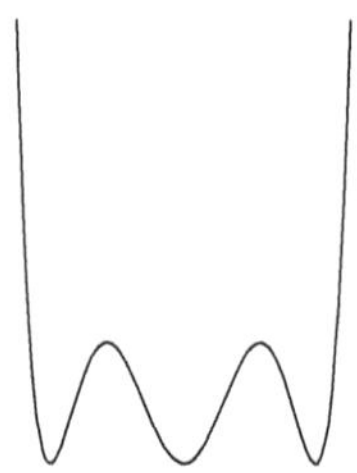

Fig. 11.6 *Potential in the φ^6 theory with three degenerate vacua.*

$$\langle \cdots : \varphi^{2q-3} : \cdots \rangle = \langle \cdots \partial_z \partial_{\bar{z}} \varphi \cdots \rangle.$$

We now show how these features of the Landau–Ginzburg theory are implemented by the minimal unitary models $\mathcal{M}_q$.

Counting the operators. To start with, notice that these models have a number of relevant fields precisely equal to $2(q-2)$: they correspond to the scalar conformal fields $\phi_{\Delta,\Delta}(z,\bar{z})$ with conformal dimension $\Delta < 1$. As already noticed, in these models the formula of the conformal dimensions

$$\Delta_{r,s} = \frac{((q+1)r - qs)^2 - 1}{4q(q+1)} \tag{11.6.5}$$

is proportional to the distance of the points of the Kac lattice to the straight line of slope $q/(q+1)$ and therefore the counting problem simply reduces to determine how many points of the lattice fall inside the strip identified by the condition $\Delta_{r,s} < 1$, as shown in Figure 11.7.

Identification of the operators. The first thing to do is identifying the most relevant field of the conformal model $\mathcal{M}_p$ with the scalar field φ that enters the Landau–Ginzburg Lagrangian. It is easy to check that such a conformal field is the one placed at the position $(2,2)$ of the Kac table, with conformal dimension

$$\Delta = \Delta_{2,2} = \frac{3}{4q(q+1)}. \tag{11.6.6}$$

With the position $\varphi \equiv \phi_{2,2}$, let us now proceed toward the recursive definition of the normal product of the higher powers of φ by means of the OPE provided by the conformal theory. Remember that, in CFT, to define the composite operator $: A^2 : (x)$ of an operator $A(x)$ (with anomalous dimension η_1), we have to consider the OPE of $A(x)$ with itself

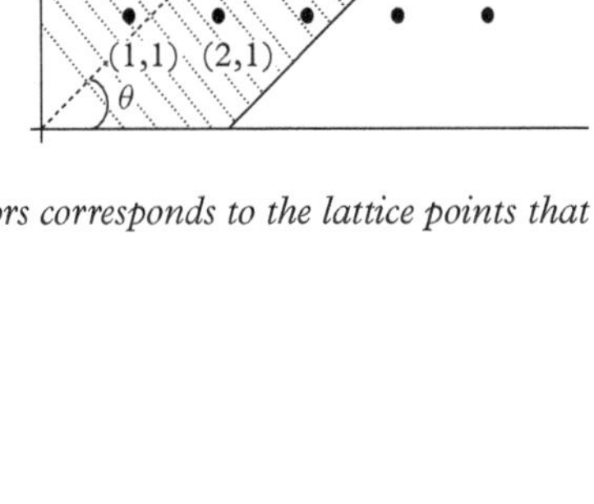

Fig. 11.7 *The relevant operators corresponds to the lattice points that satisfy $\Delta_{r,s} < 1$, i.e. those inside the strip shown in the figure.*

$$A(x)A(0) - \langle A(x)A(0)\rangle = |x|^{\eta_2 - 2\eta_1} A_2(0) + \cdots \tag{11.6.7}$$

Once the most singular terms of this expression have been subtracted, the identification of the operator $: A^2 :$ is made through the limit

$$: A^2(0) : \equiv A_2 = \lim_{x\to 0} |x|^{2\eta_1 - \eta_2} \left(A(x)A(0) - \langle A(x)A(0)\rangle\right). \tag{11.6.8}$$

The higher-power composite operators $: A^{k+1} :$ are defined in an analogous way, and they coincide with those conformal fields selected by the limit

$$: A^{k+1} : (0) = \lim_{x\to 0} |x|^{\eta_1 + \eta_k - \eta_{k+1}} \left[A(x) : A^k(0) : - \sum_{l=1}^{k/2} C_l \, |x|^{\eta_{k-2l} - \eta_1 - \eta_k} : A^{k-2l} : (0) \right]. \tag{11.6.9}$$

Note that the most singular terms of this expansion come from the previous operators $: A^{k-2l} :$, with $l = 1, 2, \ldots, < k/2$, and therefore they must be subtracted. In this expression η_p are the anomalous dimensions of the composite fields $: A^p :$ whereas the coefficients C_p are the relative structure constants that ensure the existence of the limit.

Given the above definition of the normal order and the identification of φ with $\phi_{2,2}$, it is easy to see that the composite operator $: \varphi^2 :$ ends up to be the conformal field $\phi_{3,3}$. In fact, the fusion rules of $\phi_{2,2}$ are

$$\phi_{2,2} \times \phi_{2,2} = [\mathbf{1}] + [\phi_{3,3}] + [\phi_{1,3}] + [\phi_{3,1}] \tag{11.6.10}$$

and, once we have subtracted the contibution coming from the identity family, the most singular term in the expansion and the one that survives in the limit (11.6.8) is given by the conformal field $\phi_{3,3}$. We can then proceed to identitfy the higher powers of φ, with the final result given by (Figure 11.8)

$$: \varphi^k = \begin{cases} \phi_{k+1,k+1} & k = 0, 1, \ldots q-2 \\ \phi_{k-q+3,k-q-2} & k = q-1, q, q+1, 2p-4. \end{cases} \tag{11.6.11}$$

The anomalous dimension $\eta_k = 2\Delta_k$ of these composite operators is obtained using eqn. (11.6.5)

$$\eta_k = \begin{cases} \frac{(k+1)^2 - 1}{2q(q+1)} & k = 0, 1, 2, \ldots, q-2 \\ \frac{(k+3)^2 - 1}{2q(q+1)} & k = q-1, q, \ldots, 2q-4. \end{cases} \tag{11.6.12}$$

The key point in the route to establish the identification of the conformal model $\mathcal{M}_q$ with the $\varphi^{2(q-2)}$ Landau–Ginzburg theory is met when we consider the operator expansion

6					
5					φ^4
4				φ^3	φ^8
3			φ^2	φ^7	
2		φ	φ^6		
1	1	φ^5			
	1	2	3	4	5

Fig. 11.8 *The correspondence between the conformal fields and the composite operators* φ^k *for the minimal model* $\mathcal{M}_6$.

$\varphi : \varphi^{2(q-2)} := \phi_{2,2}\,\phi_{q-2,q-1}$. The most singular contribution in this product comes from the conformal fields $\phi_{q-3,q-2} \equiv: \varphi^{2q-5}$ and $\phi_{q-1,q2} = \phi_{2,2}$, and both must be subtracted in the proper definition of the composite operator φ^{2q-3}. After these subtractions, the first term that remains is the first descendant of the conformal field $\phi_{2,2}$, that is nothing else but $\partial_z\,\partial_{\bar{z}}\varphi$. In this way we arrive at the operator identity

$$: \varphi^{2q-3} := \partial_z\,\partial_{\bar{z}}\,\varphi \tag{11.6.13}$$

that coincides with the equation of motion (11.6.4) of the Landau–Ginzburg theory. Finally, the other conformal fields that enter the Kac table of the minimal model $\mathcal{M}_p$ can be identified with the irrelevant composite fields $: \varphi^k\,\partial_\mu\varphi\partial_\mu\varphi$.

11.7 Modular Invariance

In the two-dimensional conformal theories there is a natural splitting of the analytic and anti-analytic parts of the fields. This algebraic separation is stated by the Ward identity and is quite useful in many contexts (like, finding the irreducible representations, the linear differential equations, etc.). However, to recover the real physical situation we have to combine together the analytic and the anti-analytic parts.

In the previous section we have seen that, in the infinite plane, a way to establish the correct combination of the two sectors is given by the condition of the monodromy invariance of the correlation functions. There is, however, another approach to find the physical content of the conformal theories. Proposed originally by Cardy, this approach consists of studying the properties of a CFT defined on a torus, i.e. a cylinder with periodic boundary conditions along both directions. For the geometrical symmetry of this problem, there are quite severe constraints on the operator content of the theories.

Before exposing in detail the mathematical formalism of this approach, it is useful to present its main idea by means of a simple example. Consider a rectangle of sides L along the vertical direction and R along the horizontal one. There are two ways to compute the partition function of a CFT defined on such a geometry:

- First, we consider the vertical axis as time direction, with the time propagation ruled by a Hamiltonian H_R. This quantization scheme defines the so-called *L-channel* of the theory. In this case the partition function can be expressed as

$$Z_1(L,R) = \mathrm{Tr}\, e^{-LH_R}. \tag{11.7.1}$$

- Second, instead we consider the horizontal axis as time direction, with a time evolution implemented by a Hamiltonian H_L. This quantization scheme defines the so-called *R-channel* of the theory and, correspondingly, the partition function is given by

$$Z_2(R,L) = \mathrm{Tr}\, e^{-RH_L}. \tag{11.7.2}$$

The two ways of computing the partition function are obviously equivalent and this leads to the identity

$$Z_1(L,R) = Z_2(R,L), \tag{11.7.3}$$

that expresses the *modular invariance* of the theory. As we show in the next section, by enforcing the validity of eqn. (11.7.3) we can characterize the operator content of the conformal theories, given by an appropriate combination of their analytic and anti-analytic sectors.

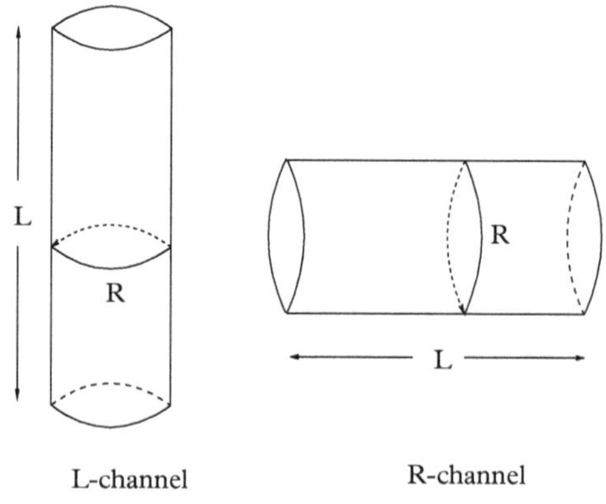

Fig. 11.9 *A torus geometry, i.e. a cylinder with periodic boundary conditions along both directions, and the relative quantization channels.*

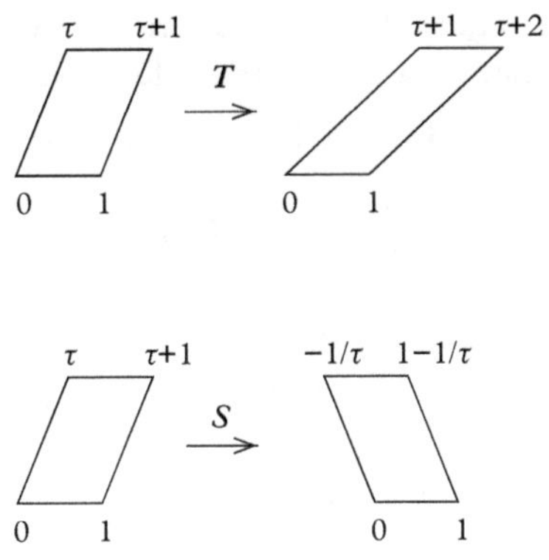

Fig. 11.10 *Transformation of the lattice under the action of the generators $\mathcal{T}$ and $\mathcal{S}$.*

11.7.1 Torus Geometry

Let us now refine the previous considerations by studying the mathematical properties that are relevant for the modular invariance. A torus is defined by specifying two independent vectors in the plane and identifying the points that differ by a linear combination of them with integer coefficients. In this way, the plane takes the periodic structure of a lattice. In the complex plane, such vectors can be specified by two complex numbers ω_1 and ω_2, the so-called *periods* of the lattice, and a torus is therefore defined by the equivalence relation

$$z \equiv z + n\omega_1 + m\omega_2 \qquad n, m \in N. \tag{11.7.4}$$

Such a tiling of the plane is not unique. Let $\omega'_{1,2}$ be in fact two new periods. Since they give to the same lattice structure, they can be expressed as a linear combination with integer coefficients of the original periods ω_1 and ω_2

$$\begin{pmatrix} \omega'_1 \\ \omega'_2 \end{pmatrix} = \begin{pmatrix} a & b \\ c & d \end{pmatrix} \begin{pmatrix} \omega_1 \\ \omega_2 \end{pmatrix} \qquad \begin{array}{l} a, b, c, d \in Z \\ ad - bc = 1. \end{array} \tag{11.7.5}$$

The determinant of such a transformation should not vanish since, by the symmetrical role of the two sets of periods, the linear combination should be invertible. Moreover, its normalization is fixed by the condition that the area of the elementary cell of the lattice is the same either if expressed in the basis given by $\omega'_{1,2}$ or $\omega_{1,2}$.

In the complex plane, two lattices correspond to two conformally equivalent torus geometries if they differ just by a rotation and a dilatation. We can use this freedom to reduce the pair of periods (ω_1, ω_2) to the values $(1, \tau)$, where the complex number $\tau = \omega_2/\omega_1$, with $\mathrm{Im}\,\tau > 0$ is the *modular parameter*. Without losing generality, we can choose as vertices of the torus the points $\{0, 1, \tau, (1+\tau)\}$. The physical request is that the conformal theories defined on such a geometry should not depend either from the scale or from the orientation of the lattice, i.e. the condition that theory presents a *modular invariance.* Since under the change (11.7.5) the modular parameter transforms according to the Moebius map

$$\tau \rightarrow \frac{a\tau + b}{c\tau + d} \qquad ad - bc = 1 \tag{11.7.6}$$

the corresponding symmetry group coincides with the 2×2 linear transformations with integer coefficients and determinant equal to 1. Furthermore, since all parameters a, b, c, d can be changed by sign without affecting the final transformation, the modular group Γ is given by $SL(2,Z)/Z_2$ and consists of the group of the discontinuous diffeomorphisms of the torus, i.e. the set of all those transformations of the torus that cannot be obtained adiabatically starting from the identity transformation. Such a discrete group can be generated by the repeated action of the operators (see Problem 11.5)

$$\begin{aligned} \mathcal{T} &: \tau \rightarrow \tau + 1 \qquad \text{i.e.} \qquad \mathcal{T} = \begin{pmatrix} 1 & 0 \\ 1 & 1 \end{pmatrix} \\ \mathcal{S} &: \tau \rightarrow -1/\tau \qquad \text{i.e.} \qquad \mathcal{S} = \begin{pmatrix} 0 & 1 \\ -1 & 0 \end{pmatrix} \end{aligned} \tag{11.7.7}$$

whose graphical representation is shown in Figure 11.10. These transformations satisfy

$$\mathcal{S}^2 = (\mathcal{S}\mathcal{T})^3 = 1. \tag{11.7.8}$$

The *fundamental domain* of the modular group is defined as that region of the upper half complex plane for which any pair of its points cannot be related by a modular transformation, whereas any other external point can be reached from one of its interior point by a modular transformation. The usual choice of the fundamental domain is the following: $-\frac{1}{2} < \text{Re}\,\tau < \frac{1}{2}, \mid \tau \mid \geq 1$ (Figure 11.11).

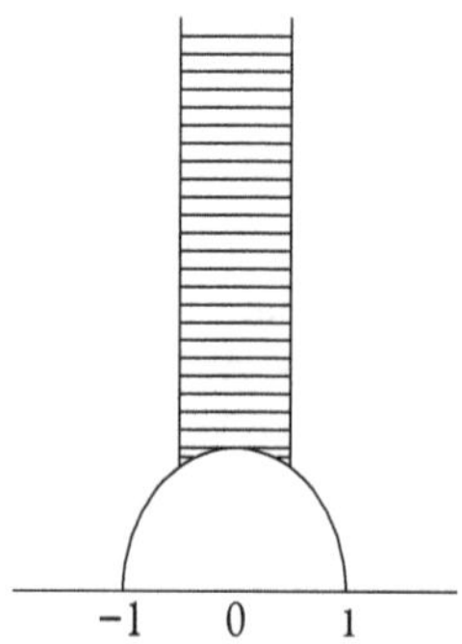

Fig. 11.11 *Fundamental domain of the modular group* Γ *(dashed area).*

11.7.2 Partition Function and Characters

In order to define the partition function on a torus, it is necessary to specify the time and space directions of the lattice. Let us initially choose as space direction the one along the real axis of the complex plane, while as time direction the imaginary axis. This choice introduces the *L*-channel, according to the terminology above. The translations along these axes are implemented by the momentum operator P and by the Hamiltonian H respectively. The partition function is then

$$Z(\tau,\bar{\tau}) = \text{Tr}\exp\{-H\text{Im}\,\tau - iP\text{Re}\,\tau\}. \tag{11.7.9}$$

For H and P we can use the expression previously derived for a cylinder of width R (in the present case $R = 1$), namely

$$H = \frac{2\pi}{R}\left(L_0 + \bar{L}_0 - \frac{c}{12}\right), \quad P = \frac{2\pi}{R}(L_0 - \bar{L}_0) \tag{11.7.10}$$

where L_0 and $\bar{L}_0$ are the generators of the Virasoro algebra. Substituting these expressions in (11.7.9) and collecting the various terms we get

$$Z(\tau,\bar{\tau}) = \text{Tr}\exp\left\{2\pi i\left[\tau(L_0 - \frac{c}{24}) - \bar{\tau}(\bar{L}_0 - \frac{c}{24})\right]\right\}. \tag{11.7.11}$$

Defining the parameters

$$q \equiv e^{2\pi i\tau}, \quad \bar{q} \equiv e^{-2\pi i\bar{\tau}} \tag{11.7.12}$$

the partition functions can be expressed as

$$Z(q,\bar{q}) = \text{Tr}\left(q^{L_0 - c/24}\,\bar{q}^{\bar{L}_0 - c/24}\right). \tag{11.7.13}$$

The eigenstates of $(L_0, \bar{L}_0)$ are organized in terms of the irreducible representations given by the Verma modules of the direct sum of the two Virasoro algebras. Hence we can decompose the trace on these states in the sum of these representation and write then

$$Z(q,\bar{q}) = \sum_{\Delta,\bar{\Delta}} N_{\Delta,\bar{\Delta}}\,\chi_{\Delta}(q)\,\chi_{\bar{\Delta}}(\bar{q}), \tag{11.7.14}$$

where the non-negative integers $N_{\Delta,\bar{\Delta}}$ represent the number of times the representations associated to the conformal weights $(\Delta,\bar{\Delta})$ enter the trace, whereas $\chi_{\Delta}(q)$ are the *characters* of the Virasoro algebra, defined by

$$\chi_\Delta(q) \equiv q^{-c/24}\mathrm{Tr}\, q^{L_0}|_\Delta = q^{-(c/24)+\Delta} \sum_{n=0}^{\infty} d_\Delta(n) q^n. \tag{11.7.15}$$

The coefficients $d_\Delta(n)$ are the dimensions of the vector spaces at the level n in the representation identified by the conformal weight Δ.

The problem is now to determine the set of the integers $N_{\Delta,\bar{\Delta}}$ that ensures the modular invariance of the partition function. This means that Z must be a function invariant both under $\mathcal{T}: \tau \to \tau + 1$ and di $\mathcal{S}: \tau \to -1/\tau$

$$\begin{aligned} Z(\tau) &= Z(\tau+1), \\ Z(\tau) &= Z(-1/\tau). \end{aligned} \tag{11.7.16}$$

For the minimal conformal models, the explicit expression of the characters of the degenerate fields $\varphi_{r,s}$ is provided by the Rocha–Caridi formula

$$\chi_{r,s}(q,c) = \eta^{-1}(q)\, q^{-\frac{(c-1)}{24}+\Delta_{r,s}} \sum_{k=-\infty}^{\infty} q^{pp'k^2}\left(q^{k(rp'-sp)} - q^{k(rp'+sp)}\right), \tag{11.7.17}$$

where $\eta(q)$ is the Dedekind function

$$\eta(q) = q^{1/24} \prod_{k=1}^{\infty}\left(1 - q^k\right). \tag{11.7.18}$$

To find which integers $N_{\Delta,\bar{\Delta}}$ ensure the validity of eqns. (11.7.16) it is necessary to analyse how the characters transform under the action of the two generators of the modular group. $\mathcal{T}$ acts on the characters in a particularly simple way

$$\mathcal{T}:\; \chi_\Delta(q) \to e^{2\pi i(\Delta - c/24)}\, \chi_\Delta(q). \tag{11.7.19}$$

The invariance of the partition function under this transformation implies that $N_{\Delta,\bar{\Delta}} = 0$, unless $\Delta - \bar{\Delta} = k$, where k is an integer.

Consider now the action of $\mathcal{S}$. Note that this transformation implements an exchange of the space and time axes of the theory. This implies that if we had computed the partition function swapping the role of the two directions, we would have obtained an expression similar to the previous (11.7.14)

$$Z(\tilde{q},\bar{\tilde{q}}) = \sum_{\Delta,\bar{\Delta}} N_{\Delta,\bar{\Delta}}\, \chi_\Delta(\tilde{q})\, \chi_{\bar{\Delta}}(\bar{\tilde{q}}), \tag{11.7.20}$$

but with the fundamental difference given by the presence of the quantity $\tilde{q} = e^{-2\pi i/\tau}$ instead of the original variable q. The equality of this expression with the one in eqn. (11.7.14) has two important consequences:

1. There should exist a linear transformation that links the characters expressed in terms of the variables q and $\tilde{q}$;
2. There should exist a stringent condition on the coefficients $N_{\Delta,\bar{\Delta}}$ that ensures the identity of the two expressions of the partition function.

Let us address the first point. Note that the expression of the characters of the degenerate fields in the minimal models is very similar to the infinite series that define the $\theta_i(z)$ functions, given in general by an infinite sum of exponentials, with a quadratic expression of k in the exponent. As for the $\theta_i(z)$ functions, the characters present remarkable properties under the transformation $\tau \to -1/\tau$, whose derivation requires the Poisson resummation formula. Here we only state the final result relative to the minimal models identified by the pair of co-prime integer numbers p, p':

$$\chi_{r,s}(\tilde{q}) = \sum_{r',s'} S_{rs}^{r's'} \chi_{r's'}(q), \tag{11.7.21}$$

where

$$S_{rs}^{r's'} = \left(\frac{8}{pp'}\right)^{1/2} (-1)^{(r+s)(r'+s')} \sin\frac{\pi r r'}{p} \sin\frac{\pi s s'}{p'}. \tag{11.7.22}$$

This formula shows that the characters $\chi_{r,s}$ change according to a finite-dimensional representation of the modular group Γ. Note that the matrix elements[7] $S_{rs}^{r's'}$ are symmetric and real. Moreover, since the transformation $\mathcal{S}$ is unitary, it holds $\mathcal{S}^2 = 1$. Denoting by $\mathcal{R}$ this finite-dimensional representation, the combination of the characters in the partition function transforms as $M \equiv R \otimes R^*$. Therefore to find the modular invariant expressions of the partition functions we have to determine the integer coefficients $N_{\Lambda,\bar{\Lambda}}$ that satisfy the condition (expressed in matrix notation)

$$MN = N. \tag{11.7.23}$$

In other words, we shall find the eigenvectors, with non-negative integer components and with eigenvalue equal to 1, of the matrix M. An additional condition is $N_{0,0} = 1$: this enforces the presence of the identity operator in the partition function with a multiplicity equal to 1.

The general solution of this mathematical problem has been found by Cappelli, Itzykson and Zuber. It has a remarkable structure: in fact, the modular invariant partition functions can be put in correspondence with the series ADE that classify the simply laced Lie algebra.[8] At first sight it may seem surprising to see CFTs being classified by

[7] The pair of indices (rs), as well as $(r's')$, has to be considered as a single index that identifies the corresponding conformal field.

[8] The discussion of the Lie algebra can be found in the Appendix of Chapter 13.

the ADE Lie algebra but, on the other hand, Lie algebras arise whenever integrability and local symmetries are involved. They also enter the classification of regular convex polyadra.

The explicit expressions of the partition functions are reported in Table 11.1. Without claiming a full justification of these formulae (see the references for omre details), it is however possible to understand the origin of some of them. For instance, a natural solution of eqn. (11.7.23) is provided by the diagonal combination, i.e. by the integers $N_{\Delta,\bar{\Delta}} = \delta_{\Delta,\bar{\Delta}}$. The partition functions associated to this solution involve all the scalar primary fields of the Kac table of a given model. To present another class of solutions, consider the case of unitary minimal models, where $p' = p+1$. We assume that the indices r, s run on all possible values of the Kac table ($1 \leq r \leq p; 1 \leq s \leq p+1$) and therefore each primary field appears twice. It is easy to see that if p is an odd number, we have

$$S_{rs}^{r's'} = (-1)^{s-1} S_{rs}^{r',p'-s'} = (-1)^{s'-1} S_{r,p'-s}^{r's'}. \tag{11.7.24}$$

p, p''	$\frac{1}{2}\sum_{r=1}^{p'-1}\sum_{s=1}^{p-1} \mid \chi_{rs} \mid^2$	$(A_{p'-1}, A_{p-1})$
$p' = 4\rho+2$ $p \geq 1$	$\frac{1}{2}\sum_{s=1}^{p-1}\left\{\sum_{\substack{r\,\mathrm{odd}=1\\ r\neq 2\rho+1}}^{4\rho+1} \mid \chi_{rs} \mid^2 + 2 \mid \chi_{2\rho+1,s} \mid^2 + \sum_{r\,\mathrm{odd}=1}^{2\rho-1} (\chi_{rs}\chi^*_{r,p-s} + c.c.)\right\}$	$(D_{2\rho+2}, A_{p-1})$
$p' = 4\rho$ $p \geq 2$	$\frac{1}{2}\sum_{s=1}^{p-1}\left\{\sum_{r\,\mathrm{odd}=1}^{4\rho-1} \mid \chi_{rs} \mid^2 + \mid \chi_{2\rho,s} \mid^2 + \sum_{r\,\mathrm{even}=2}^{2\rho-2} (\chi_{rs}\chi^*_{r,p'-s} + c.c.)\right\}$	$(D_{2\rho+1}, A_{p-1})$
$p' = 12$	$\frac{1}{2}\sum_{s=1}^{p-1}\left\{\mid \chi_{1s} + \chi_{7s} \mid^2 + \mid \chi_{4s} + \chi_{8s} \mid^2 + \mid \chi_{5s} + \chi_{11s} \mid^2\right\}$	(E_6, A_{p-1})
$p' = 18$	$\frac{1}{2}\sum_{s=1}^{p-1}\left\{\mid \chi_{1s} + \chi_{17s} \mid^2 + \mid \chi_{5s} + \chi_{13s} \mid^2 + \mid \chi_{7s} + \chi_{11s} \mid^2 + \right.$ $\left. + \mid \chi_{9s} \mid^2 + [(\chi_{3s} + \chi_{15s})\chi^*_{9s} + cc]\right\}$	(E_7, A_{p-1})
$p' = 30$	$\frac{1}{2}\sum_{s=1}^{p-1}\left\{\mid \chi_{1s} + \chi_{11s} + \chi_{19s} + \chi_{29s} \mid^2 + \mid \chi_{7s} + \chi_{13s} + \chi_{17s} + \chi_{23s} \mid^2\right\}$	(E_8, A_{p-1})

Table 11.1 *Modular invariant partition functions of the conformal minimal models.*

This identity implies that the combination made by the characters

$$\chi_{rs} + \chi_{r,p'-s} \quad (s\,\text{odd}) \tag{11.7.25}$$

defines an invariant sub-space. Therefore, the partition function given by

$$Z = \frac{1}{2} \sum_{r} \sum_{s\,\text{odd}} |\, \chi_{rs} + \chi_{r,p'-s} \,|^2 \tag{11.7.26}$$

is invariant under the $\mathcal{S}$ transformation. It is also easy to check that $\Delta_{rs} - \Delta_{r,p+1-s}$ is always an integer if $p = 1\,(mod\,4)$, and in these case, the above partition function is also invariant under $\mathcal{T}$. Similar invariant expressions can be found for all values of $p \geq 5$.

In addition to these two infinite series of solutions, there are others that are relative to particular values of p', given by $p' = 12, 18, 30$. As we mentioned above, all the modular invariant solutions can be put in correspondence with the ADE algebras that appear in so many branches of mathematics, as in the classification of finite subgroups of the group of rotations or in the classification of the critical points in the theory of the catastrophies. Pasquier has also shown that the modular invariant partition functions can be obtained as a continuum limit of certain discrete lattice statistical models defined in terms of the Dynkin diagrams. The modular invariant partition functions of the conformal minimal models are reported in Table 11.1.

Appendix 11.A. Hypergeometric functions

Let a, b and c be complex numbers. The hypergeometric differential equation

$$z(z-1)\frac{d^2w}{dz^2} + [(a+b+1)z - c]\frac{dw}{dz} + abw = 0, \tag{11.A.1}$$

has three singular regular points at $z = 0, 1, \infty$. When c is different from zero or it is a negative integer, an analytic solution of this equation in the vicinity of $z = 0$ is expressed by the series

$$F(z; a, b, c) = \sum_{n=0}^{\infty} \frac{(a)_n (b)_n}{(c)_n} \frac{z^n}{n!}, \quad c \neq 0, -1, -2, \ldots \tag{11.A.2}$$

where

$$(a)_n \equiv a(a+1)\ldots(a+n-1) = \frac{\Gamma(a+n)}{\Gamma(a)}.$$

If a or b are equal to zero or are negative integers, the series truncates and the hypergeometric function becomes a simple polynomial.

Since the hypergeometric differential equation is of second order, it admits a second solution, usually written in the form

$$w_2(z) = z^{1-c} F(z;\ a-c+1,\ b-c+2,\ 2-c). \tag{11.A.3}$$

It is easy to see that if c is an integer, either the two solutions coincide or one of them diverges. In the second case, the second solution presents a logarithmic contribution.

REFERENCES

The formula for the Kac determinant was proposed in

Kac, V.G. (1980). 'Highest-weight Representations of Infinite Dimensional Lie Algebras', In: O. Lehto (Ed), *Proceedings of the International Congress of Mathematicians*, vol. 1, Helsinki, 1978. American Mathematical Society, pp. 299–304.

The unitary properties of the minimal models are discussed in

Friedan, D., Qiu, Z. and Shenker, S. (1984). 'Conformal Invariance, Unitarity And Two-Dimensional Critical Exponents', *Physics Review Letters*, 52: 1575.

The modified Coulomb gas method is discussed in in

Dotsenko, V.S. and Fateev, V. (1984). 'Conformal Algebra and Multipoint Correlation Functions in 2d Statistical Models', *Nuclear Physics B*, 240: 312.

Dotsenko, V.S. and Fateev, V. (1985). 'Four-point Correlation Functions and the Operator Algebra in the Two-Dimensional Convormal Invariant Theories with the Central Charge $c < 1$', *Nuclear Physics B*, 251: 691.

For the quantum group formulation of the minimal models see

Alvarez-Gaume, L. Gomez, C. and Sierra, G. (1989). 'Hidden Quantum Symmetries in Rational Conformal Field Theories', *Nuclear Physics B*, 319: 155.

Alvarez-Gaume, L. Gomez, C. and Sierra, G. (1989). 'Quantum Group Interpretation of Some Conformal Field Theories', *Physics Letters B*, 220: 142.

The Landau–Ginzburg picture of the unitary minimal models has been proposed in

Zamolodchikov, A.B. (1986). 'Conformal Symmetry and Multicritical Points in Two-dimensional Quantum Field Theory', *Soviet Journal of Nuclear Physics*, 44: 529.

The principle of modular invariance has been discussed in

J.L. Cardy, *Operator Content Of Two-Dimensional Conformally Invariant Theories*, Nucl. Phys. B 270 (1986), 186.

The full classification of modular invariant partition functions is covered in

Cappelli, A., Itzykson, C. and Zuber, J.B. (1987). 'Modular Invariant Partition Functions in Two Dimensions', *Nuclear Physics B*, 280: 445.

Interested readers are referred to discussion about the question of the spectrum of two-dimensional CFT in

Gepner, D. (1987). 'On the Spectrum of 2D Conformal Field Theories', *Nuclear Physics B*, 287: 111.

The identification of the modular invariant partition functions with the statistical modes based on the Dynkin diagrams is covered in

Pasquier, V. (1987). 'Two-dimensional Critical Systems Labelled by Dynkin Diagrams', *Nuclear Physics B*, 285: 162.

A detailed analysis of the hypergeometric functions can be found in

P.T. Bateman, *Compex Variables. Introduction and Applications*, Cambridge Texts in Applied Mathematics, Cambridge University Press, 1977.

For the classification of regular polyhedra and the correspondence with ADE Lie algebras see

Coxeter, H.S.M. (1973). *Regular Polytopes*, Dover Publications.

Slodowy, P. (1980). 'Simple Singularities and Simple Algebraic Groups', Vol. 815, Lecture Notes in Mathematics, Springer.

The Verlinde algebra is deeply related to the modular group. It has found application in the topological quantum computation. The interested reader may consult

Cardy, J.L. (1989). 'Boundary Conditions, Fusion Rules and the Verlinde Algebra', *Nuclear Physics B*, 324: 324.

Das Sarma, S., Freedman, M., Nayak, C., Simon, S.H., and Stern, A. (2008). 'Non-Abelian Anyons and Topological Quantum Computation', *Reviews of Modern Physics*, 80: 1083.

Verlinde, E. (1988). 'Fusion Rules and Modular Transformations in 2d Conformal Field Theory', *Nuclear Physics B*, 300: 360.

PROBLEMS

11.1. Null vector at the third level

a. Show that the linear combination that gives rise to null vectors at the level $N = 3$ is given by

$$\left[L_{-3} - \frac{2}{\Delta+1} L_{-1} L_{-2} + \frac{1}{(\Delta+1)(\Delta+2)} L_{-1}^3\right] \phi_\Delta = 0$$

with $\Delta = \Delta_{1,3}$ or $\Delta = \Delta_{3,1}$.

b. Determine the differential equation satisfied by the correlators of the primary fields $\phi_{1,3}$ and $\phi_{3,1}$.

c. Show that the fusion rules of the fields $\phi_{1,3}$ and $\phi_{3,1}$ given in eqns. (11.4.22) and (11.4.23) are compatible with the differential equation satisfied by their correlation functions.

11.2. Structure constant

For the minimal unitary models, identified by the integer p, compute the limiting value of $C_{(13)(13)}^{(13)}$ for $p \to \infty$.

11.3. Fusion rules

Consider the minimal models $\mathcal{M}_{2,2n+1}$ $(n = 1, 2, \ldots)$. Compute the central charge and the effective central charge by identifying the operator with the lowest conformal dimension. Determine the fusion rules of these models.

11.4. Non-unitarity model $\mathcal{M}_{3,5}$

The non-unitarity model $\mathcal{M}_{3,5}$ has the following Kac table.

$$\begin{array}{|cc|} \hline \frac{3}{4} & 0 \\ \frac{1}{5} & -\frac{1}{20} \\ -\frac{1}{20} & \frac{1}{5} \\ 0 & \frac{3}{4} \\ \hline \end{array}$$

With the identification of the fields

$$1 = \Phi_{0,0}, \qquad \sigma = \Phi_{-\frac{1}{20},-\frac{1}{20}};$$
$$\varphi = \Phi_{\frac{1}{5},\frac{1}{5}}, \qquad \psi = \Phi_{\frac{3}{4},\frac{3}{4}}.$$

prove that the fusion rules of this model are given by

$$\begin{aligned} \psi \times \psi &= 1, & \psi \times \sigma &= \varphi; \\ \sigma \times \sigma &= 1 + \varphi, & \psi \times \varphi &= \sigma; \\ \varphi \times \varphi &= 1 + \varphi, & \varphi \times \sigma &= \sigma + \psi. \end{aligned}$$

Compute the four-point correlation functions involving ψ and σ and determine the exact expressions of the structure constants of the conformal algebra.

11.5. Modular group

Prove that any matrix

$$M = \begin{pmatrix} a & b \\ c & d \end{pmatrix}$$

with integer coefficients and satisfying $ad - bc = 1$, can be obtained by the multiplication of suitable powers of the elementary matrices

$$\mathcal{T} = \begin{pmatrix} 1 & 0 \\ 1 & 1 \end{pmatrix} \quad \mathcal{S} = \begin{pmatrix} 0 & 1 \\ -1 & 0 \end{pmatrix}.$$

11.6. Quantum dimensions

Denote the number of linearly-independent states having $\mathcal{N}$ fields of type a as $\mathcal{H}_a(\mathcal{N})$. The quantum dimension d_a of the excitations of type a is given by studying the behaviour $\mathcal{H}_a(\mathcal{N})$ for large $\mathcal{N}$, which behaves as $\mathcal{H}_a(\mathcal{N}) \simeq d_a^{\mathcal{N}}$. To compute d_a, let us fuse M ϕ_a fields recursively using the Verlinde algebra

$$\phi_a \cdot \phi_a \cdot \ldots \cdot \phi_a = \sum_{\{c_k\}} N_{aa}^{c_1} N_{ac_1}^{c_2} \cdots N_{ac_{M-2}}^{c_{M-1}} \phi_{c_{M-1}}.$$

Observe that this is the product of $(M-1)$ copies of the matrix $(N_a)_b^c = N_{ab}^c$, so that in the limit $M \to \infty$, the product will be dominated by the largest eigenvalue of N_a.

a. If S_a^b is the unitary matrix that simultaneously diagonalizes all the matrices N_a of the Verlinde algebra, show that the fusion coefficients are expressed by

$$N_{ab}^c = \sum_j \frac{S_a^j S_b^j S_j^c}{S_0^j},$$

where 0 denotes the identity field and the sum is over all fields entering the algebra.

b. Show that the largest eigenvalue (and therefore the quantum dimension d_a) is given by

$$d_a = \frac{S_a^0}{S_0^0}.$$

c. Consider the algebra

$$\mathbf{1}\cdot\mathbf{1}, \quad \mathbf{1}\cdot\phi=\phi, \quad \phi\cdot\phi=\mathbf{1}+\phi.$$

Compute the quantum dimension d_ϕ and show that is equal to the golden ratio

$$d_\phi = \frac{\sqrt{5}+1}{2}.$$

12

Conformal Field Theory of Free Bosonic and Fermionic Fields

Science is spectral analysis. Art is light synthesis.
Karl Kraus

12.1 Introduction

This chapter discusses two explicit examples of CFTs. We start our analysis with the free massless bosonic theory discussed in the previous chapter. Then we discuss the CFT of a complex fermion operator (a Dirac fermion) using its decomposition in the real Majorana components. The central charge of both bosonic and fermion theories is $c = 1$ and this suggests the existence of an equivalence between them. The transformation that maps a bosonic into a fermion theory, and vice-versa, is known as *bosonization*: it provides a useful tool both for the comprehension of the conformal theories and for a wide range of applications, in particular in low-dimensional condensed matter systems.

12.2 Conformal Field Theory of Free Bosonic Fields

This section is devoted to the detailed analysis of the CFT of a massless bosonic field employed in Chapter 11 for the Coulomb gas approach. Despite the simple form of the Lagrangian of this model, it presents a rich operator content and a remarkable duality property of its partition function on a torus. Later we also established the equivalence of this theory with the theory of massless Dirac fermion.

12.2.1 Quantization of the Bosonic Field

Let $\varphi(x, t)$ be a free bosonic and real field, with action

$$\mathcal{A} = \frac{g}{2} \int d^2x \, \partial_\mu \varphi \partial^\mu \varphi. \tag{12.2.1}$$

Statistical Field Theory: An Introduction to Exactly Solved Models in Statistical Physics. Second Edition. Giuseppe Mussardo, Oxford University Press (2020). © Giuseppe Mussardo. DOI: 10.1093/oso/9780198788102.001.0001

Let us assume that the model is defined on a cylinder of width L with periodic boundary conditions $\varphi(x+L,t) = \varphi(x, t)$. Expanding the field in its Fourier modes

$$\varphi(x, t) = \sum_n e^{2\pi inx/L}\varphi_n(t)$$
$$\varphi_n(t) = \frac{1}{L}\int_0^L dx\, e^{-2\pi inx/L}\varphi(x, t)$$

and substituting this expression into the action, we have

$$S = \frac{1}{2}gL\int dt\sum_n\left[\dot{\varphi}_n\dot{\varphi}_{-n} - \left(\frac{2\pi n}{L}\right)^2\varphi_n\varphi_{-n}\right]. \tag{12.2.2}$$

Let $\Pi(x, t) = g\partial_t\varphi(x, t)$ be the conjugate momentum of the field, with commutation relations (at a given time t)

$$[\varphi(x, t), \Pi(y,t)] = i\delta(x-y) \quad [\varphi(x, t), \varphi(y,t)] = 0 \quad [\Pi(x, t), \Pi(y,t)] = 0. \tag{12.2.3}$$

Expanding also $\Pi(x, t)$ in the Fourier series and denoting by π_n the conjugate momenta of φ_n, we have

$$\pi_n = gL\dot{\varphi}_{-n}, \quad [\varphi_n, \pi_m] = i\delta_{nm} \tag{12.2.4}$$

with $\varphi_n^\dagger = \varphi_{-n}$ and $\pi_n^\dagger = \pi_{-n}$. We can define the Hamiltonian of the system by using the Legendre transformation

$$H = \frac{1}{2gL}\sum_n[\pi_n\pi_{-n} + (2\pi ng)^2\varphi_n\varphi_{-n}]. \tag{12.2.5}$$

This is the Hamiltonian of a set of decoupled harmonic oscillators of frequencies $\omega_n = 2\pi|n|/L$. Note that the oscillator with $n = 0$ has zero frequency. To keep into account this feature, it is convenient to explicitly separate the zero mode φ_0 of the field and introduce, for the other modes, the operators a_n and $\bar{a}_n$ through the formulae

$$\varphi_n = \frac{i}{n\sqrt{4\pi g}}(a_n - \bar{a}_{-n}), \tag{12.2.6}$$
$$\pi_n = \sqrt{\pi g}\,n(a_{-n} + \bar{a}_n).$$

These operators satisfy the commutation relations

$$[a_n, a_m] = n\delta_{n+m,0} \quad [a_n, \bar{a}_m] = 0 \quad [\bar{a}_n, \bar{a}_m] = n\delta_{n+m,0}. \tag{12.2.7}$$

For the zero mode we have

$$[\varphi_0, \pi_0] = i. \tag{12.2.8}$$

Substituting these new operators into eqn. (12.2.5) we obtain

$$H = \frac{1}{2\pi gL}\pi_0^2 + \frac{2\pi}{L}\sum_{n>0}(a_{-n}a_n + \bar{a}_{-n}\bar{a}_n), \tag{12.2.9}$$

and therefore

$$[H, a_{-n}] = \frac{2\pi}{L}na_{-n}, \quad [H, \bar{a}_{-n}] = \frac{2\pi}{L}n\bar{a}_{-n}, \quad [H, \varphi_0] = 0. \tag{12.2.10}$$

These expressions show that a_{-n} and $\bar{a}_{-n}$ (with $n > 0$) act as raising operators of the theory: applying one of these operators to an energy eigenstate with eigenvalue E, we obtain another energy eigenstate with eigenvalue $E + 2\pi n/L$. The operators a_n and $\bar{a}_n$ (with $n > 0$) act instead as annihilation operators of the theory: their application to an eigenstate with eigenvalue E give rise to another energy eigenstate with eigenvalue $E - 2\pi n/L$. Eqn. (12.2.10) helps us to easily obtain the time evolution of the operators in the Heisenberg representation

$$\begin{aligned} a_n(t) &= a_n(0)\, e^{-2\pi int/L} \\ \bar{a}_n(t) &= \bar{a}_n(0)\, e^{-2\pi int/L} \end{aligned} \quad \varphi_0(t) = \varphi_0 + \frac{1}{gL}\pi_0 t. \tag{12.2.11}$$

Hence, the solution of the equation of motion of the field $\varphi(x, t)$ reads

$$\varphi(x, t) = \varphi_0 + \frac{1}{gL}\pi_0 t + \frac{i}{\sqrt{4\pi g}}\sum_{n\neq 0}\frac{1}{n}\left(a_n e^{2\pi in(x-t)/L} - \bar{a}_{-n} e^{2\pi in(x+t)/L}\right), \tag{12.2.12}$$

where all the operators that appear in this formula are those relative to the time $t = 0$. Equivalently, adopting a Euclidean formulation, with $t = -i\tau$ and introducing the coordinates

$$z = e^{2\pi(\tau - ix)/L}, \quad \bar{z} = e^{2\pi(\tau + ix)/L},$$

we have

$$\varphi(z, \bar{z}) = \varphi_0 - \frac{i}{4\pi g}\pi_0 \ln(z\bar{z}) + \frac{i}{\sqrt{4\pi g}}\sum_{n\neq 0}\frac{1}{n}\left(a_n z^{-n} + \bar{a}_{-n}\bar{z}^{-n}\right). \tag{12.2.13}$$

This expression explicitly shows the decoupling of the analytic and anti-analytic components of the field, both due to the the equation of motion $\bar{\partial}\,\partial\varphi = 0$ and the periodic boundary conditions chosen for the field φ

$$\varphi(z, \bar{z}) = \phi(z) + \bar{\phi}(\bar{z}), \tag{12.2.14}$$

where

$$\begin{aligned}\phi(z) &= \frac{1}{2}\varphi_0 - \frac{i}{4\pi g}\pi_0 \ln z + \frac{i}{\sqrt{4\pi g}}\sum_{n\neq 0}\frac{1}{n}a_n z^{-n} \\ \bar{\phi}(\bar{z}) &= \frac{1}{2}\varphi_0 - \frac{i}{4\pi g}\pi_0 \ln \bar{z} + \frac{i}{\sqrt{4\pi g}}\sum_{n\neq 0}\frac{1}{n}\bar{a}_n \bar{z}^{-n}.\end{aligned} \tag{12.2.15}$$

According to the negative or positive value of the index n, the operators a_n create or annihilate the analytic excitation of the field φ, with a similar situation for $\bar{a}_n$ with respect to the anti-analytic excitations.[1] It is also convenient to define

$$\theta(z, \bar{z}) = \phi(z) - \bar{\phi}(\bar{z}), \tag{12.2.16}$$

the so-called *dual field* of φ. It satisfies $\partial_\mu \varphi = -i\epsilon_{\mu\nu}\partial_\nu \theta$, and in complex coordinates

$$\begin{aligned}\partial_z \varphi &= \partial_z \theta, \\ \partial_{\bar{z}} \varphi &= -\partial_{\bar{z}} \theta.\end{aligned} \tag{12.2.17}$$

It is easy to verify that in the original Minkowski space, it satisfied the commutation relation (at fixed time t)

$$[\varphi(x, t), \theta(y, t)] = -i\epsilon(x - y), \tag{12.2.18}$$

where $\epsilon(v)$ is the step function

$$\epsilon(v) = \begin{cases} 1, & v > 0 \\ 0, & v < 0 \end{cases}.$$

Eqn. (12.2.18) clearly shows the non-local relationship between φ and θ.

We have already noticed that φ is not a scaling field, whereas scaling fields are the two currents $\mathcal{J}(z) = i\partial\varphi$ and $\bar{\mathcal{J}}(\bar{z}) = -i\bar{\partial}\varphi$ that both generate a $U(1)$ symmetry. The expansion of these fields is given by

$$i\partial\varphi = i\partial\phi = \frac{1}{\sqrt{4\pi g}}\sum_{n\neq 0} a_n z^{-n}, \tag{12.2.19}$$

[1] It is interesting to observe that the formulae given in the text appear also in string theory, in particular they enter the quantization of the closed string, with the zero mode φ_0 associated to the centre of mass of the string and π_0 to its total momentum.

(with a similar formula for $i\bar{\partial}\varphi$), where we have introduced the notation

$$a_0 = \bar{a}_0 \equiv \frac{\pi_0}{\sqrt{4\pi g}}.$$

We can now use the previous expression to define the analytic part of the stress-energy tensor

$$T(z) = -2\pi g : \partial\phi(z)\,\partial\phi(z) := \frac{1}{2}\sum_{n,m} z^{-n-m-2} : a_n\, a_m : \tag{12.2.20}$$

and extract the modes L_n of the Virasoro algebra of this theory

$$\begin{aligned} L_n &= \frac{1}{2}\sum_{m=-\infty}^{\infty} a_{n-m}\, a_m \quad (n \neq 0) \\ L_0 &= \frac{1}{2}\alpha_0^2 + \sum_{m=1}^{\infty} a_{-m}\, a_m. \end{aligned} \tag{12.2.21}$$

The Hamiltonian (12.2.9) can be written as

$$H = \frac{2\pi}{L}(L_0 + \bar{L}_0) - \frac{\pi}{6L}. \tag{12.2.22}$$

The term $-\pi/6L$ is obtained by the normal order of a_n by using the regularization given by the Riemann function $\zeta(s) = \sum_{n=1}^{\infty} n^{-s}$ for the divergent series

$$\sum_{n>0} n = \zeta(-1) = -1/12.$$

Comparing this formula with the general expression previously derived for H on a cylinder geometry, eqn. (10.9.3), we see that the central charge of the free bosonic theory is $c = 1$.

Starting from the scalar field φ, in addition to $\partial\varphi$, we can construct an infinite series of scaling operators, given by the vertex operators

$$V_{\alpha,\bar{\alpha}}(z, \bar{z}) =: e^{i\alpha\phi(z) + i\bar{\alpha}\bar{\phi}(\bar{z})} : \tag{12.2.23}$$

whose conformal weights are

$$\Delta_\alpha = \frac{\alpha^2}{8\pi g}, \quad \bar{\Delta}_\alpha = \frac{\bar{\alpha}^2}{8\pi g}. \tag{12.2.24}$$

They satisfy the OPE

$$V_{\alpha,\bar{\alpha}}(z,\bar{z})\,V_{\beta,\bar{\beta}}(w,\bar{w}) = (z-w)^{\alpha\beta/4\pi g}\,(\bar{z}-\bar{w})^{\bar{\alpha}\bar{\beta}/4\pi g}\,V_{\alpha+\beta,\bar{\alpha}+\bar{\beta}}(w,\bar{w}) + \cdots. \quad (12.2.25)$$

An interesting interpretation of the vertex operators is given in Section 12.2.2.

12.2.2 Vertex Operators

Since the Hamiltonian (12.2.5) does not depend on φ_0, it commutes with its conjugate momentum π_0 and this quantity can be used as a quantum number to identify the various eigenstates of H. For the decoupling of the analytic and anti-analytic sectors, we focus the attention on one of them, e.g. the analytic one. Let us consider then the analytic part of the vertex operator (12.2.23), denoting by p_0 the value of the conjugate momentum to the zero mode φ_0 in this sector. Let us introduce the 'ground states' $|\,\alpha\,\rangle$, with $\alpha = p_0/\sqrt{4\pi g}$. They are characterized by the algebraic conditions

$$\begin{aligned} a_n\,|\,\alpha\,\rangle &= 0 \quad (n>0), \\ a_0\,|\,\alpha\,\rangle &= \alpha\,|\,\alpha\,\rangle. \end{aligned} \quad (12.2.26)$$

From eqn. (12.2.21), we see that $|\,\alpha\,\rangle$ has conformal weight $\frac{\alpha^2}{8\pi g}$ and any other states of the Fock space of this theory is obtained by acting on $|\,\alpha\rangle$ with the creation operators a_{-n} $(n>0)$. These ground states are in one-to-one correspondence with the vertex operators. In fact, as shown in Problem 12.1, the ground state $|\,\alpha\rangle$ comes from the application of the vertex operator $V_\alpha(z) =: e^{i\alpha\phi(z)}$ to the conformal vacuum state $|\,0\,\rangle$

$$|\,\alpha\,\rangle = V_\alpha(0)\,|\,0\,\rangle. \quad (12.2.27)$$

Up to now, the real parameter α is a free quantity. However, we can constrain the set of its values by noting that the Lagrangian of the massless scalar field is invariant under the transformation $\varphi \to \varphi + \delta$, where δ is a constant. This is the $U(1)$ symmetry generated by the current $i\partial\varphi$ and allows us to identify the field φ with $\varphi + 2\pi R$: this compacting is equivalent to regard φ as an angular variable along a circle of radius R. In this new interpretation, the most general boundary conditions are given by

$$\varphi(x+L,\,t) \equiv \varphi(x,\,t) + 2\pi m R, \quad (12.2.28)$$

where $m \in \mathbf{Z}$ is the number of times that φ winds in its internal space when the space coordinate reaches the edge of the cylinder.

The compactification of φ induces a quantization in integer multiples of $1/R$ of its conjugate momentum π_0: the operator associated to the zero mode becomes also an angular variable and, with the identification $\varphi_0 \equiv \varphi_0 + 2\pi R$, only the exponentials $e^{ie\varphi_0/R}$, with $e \in \mathbf{Z}$, are well-defined. Since $[\varphi_0, \pi_0] = i$, we have then

$$e^{-ie\varphi_0/R}\,\pi_0\,e^{ie\varphi_0/R} = \pi_0 + \frac{e}{R}. \tag{12.2.29}$$

In complex coordinates and in terms of the integers e and m introduced above, the new expansion of the field is given by

$$\begin{aligned}\varphi(z, \bar{z}) = \varphi_0 - i\left(\frac{e}{4\pi gR} + \frac{mR}{2}\right)\ln z + \frac{i}{\sqrt{4\pi g}}\sum_{p\neq 0}\frac{1}{p}a_p\,z^{-p}\\ -i\left(\frac{e}{4\pi gR} - \frac{mR}{2}\right)\ln\bar{z} + \frac{i}{\sqrt{4\pi g}}\sum_{p\neq 0}\frac{1}{p}\bar{a}_p\,\bar{z}^{-p}.\end{aligned} \tag{12.2.30}$$

For the modes L_0 and $\bar{L}_0$ we have

$$\begin{aligned}L_0 &= \sum_{p>0} a_{-p}a_p + 2\pi g\left(\frac{mR}{2} + \frac{e}{4\pi gR}\right)^2,\\ \bar{L}_0 &= \sum_{p>0} \bar{a}_{-p}\bar{a}_p + 2\pi g\left(\frac{mR}{2} - \frac{e}{4\pi gR}\right)^2.\end{aligned} \tag{12.2.31}$$

In terms of the integers e and m we can now define the most general expression of the vertex operator

$$\begin{aligned}V_{e,m}(z, \bar{z}) &= :\exp\left[i\left(\frac{e}{4\pi gR} + \frac{mR}{2}\right)\phi(z) + i\left(\frac{e}{4\pi gR} - \frac{mR}{2}\right)\bar{\phi}(\bar{z})\right]:\\ &= :\exp\left[i\frac{e}{4\pi gR}\varphi(z, \bar{z}) + i\frac{mR}{2}\theta(z, \bar{z})\right]:\end{aligned} \tag{12.2.32}$$

whose anomalous dimension and spin are given by

$$\begin{aligned}\eta_{e,m} &= \Delta + \bar{\Delta} = \frac{1}{4\pi g}\left(\frac{e^2}{(4\pi gR)^2} + \frac{m^2R^2}{4}\right),\\ S_{e,m} &= \Delta - \bar{\Delta} = \frac{em}{(4\pi g)^2}.\end{aligned} \tag{12.2.33}$$

To simplify the formula below, it is convenient to assume $g = 1/4\pi$. With such a choice, the previous expressions become

$$\begin{aligned}\eta_{e,m} &= \left(\frac{e^2}{R^2} + \frac{m^2R^2}{4}\right),\\ S_{e,m} &= em.\end{aligned} \tag{12.2.34}$$

Note that the simultaneous substitutions $R \leftrightarrow 2/R$ and $m \leftrightarrow e$ leave invariant the spectrum of both the anomalous dimensions and spins. This observation will be useful in the discussion of the partition functions of the next section. In the language of the Coulomb gas, the integers e and m can be identified with the *electric* and *magnetic* charges of the system respectively. The reason of this interpretation becomes evident if we consider the operator expansion

$$\varphi(z_1,\bar{z}_1)\, V_{e,m}(z_2,\bar{z}_2) = -\left[\frac{e}{R}\ln|z_{12}|^2 + \frac{mR}{2}\ln\frac{z_{12}}{\bar{z}_{12}}\right] V_{e,m}(z_2,\bar{z}_2) + \cdots. \qquad (12.2.35)$$

If we consider only the purely electric vertex operator

$$V_{e,0}(z_2,\bar{z}_2) =: e^{i\frac{e}{R}\varphi(z_2,\bar{z}_2)}:$$

and we wind $\varphi(z_1,\bar{z}_1)$ around the point $(z_2,\bar{z}_2)$ at which the vertex operator acts (by making the analytic continuation $z_{12} \to e^{2\pi i} z_{12}$, $\bar{z}_{12} \to e^{-2\pi i}\bar{z}_{12}$), this operation does not induce any discontinuity in the field φ. However, repeating the same operation in the presence of the purely magnetic vertex operator

$$V_{0,m}(z_2,\bar{z}_2) =: e^{i\frac{mR}{2}\theta(z_2,\bar{z}_2)}:$$

the field $\varphi(z_1,\bar{z}_1)$ has a jump equal to $2\pi mR$. The most general vertex operator $V_{e,m}$ is a combination of electric and magnetic vertex operators and its two-point correlation function is given by

$$\langle V_{e,m}(z_1,\bar{z}_1)V_{-e,-m}(z_2,\bar{z}_2)\rangle = \frac{1}{|\,z_{12}\,|^{\eta}}\left(\frac{z_{12}}{\bar{z}_{12}}\right)^{S}. \qquad (12.2.36)$$

Let us now discuss the partition function of the bosonic field on a torus, highlighting its remarkable duality properties.

12.2.3 Free Bosonic Field on a Torus

Let us initially consider the partition function of a Gaussian free bosonic theory. In this case the variable φ takes values on all the real axis. In terms of the path integral, its expression would be

$$Z(\tau) = \int \mathcal{D}\varphi\, e^{-\mathcal{A}}, \qquad (12.2.37)$$

with

$$\mathcal{A} = \frac{g}{2}\int d^2x\,(\partial\varphi)^2 = -\frac{g}{2}\int d^2x\,\varphi\,\Box\,\varphi, \qquad (12.2.38)$$

where we have assumed periodic boundary conditions on both directions of the torus

$$\begin{aligned} \varphi(z+\tau, \bar{z}+\bar{\tau}) &= \varphi(z, \bar{z}), \\ \varphi(z+1, \bar{z}+1) &= \varphi(z, \bar{z}). \end{aligned} \tag{12.2.39}$$

The definition (12.2.37) presents certain drawbacks. For the quadratic expression of the action, the functional integral reduces to the product of the eigenvalues $\lambda_{n,m}$ of the Laplacian δ on the torus

$$Z \sim \prod_{n,m} \left(\frac{1}{\lambda_{n,m}} \right)^{1/2}. \tag{12.2.40}$$

Among the eigenvalues, there is $\lambda_{0,0} = 0$, which corresponds to the zero mode of the field associated to its constant configuration. The original definition of the partition function (12.2.37) is therefore divergent. For the correct definition of this quantity it is necessary to restrict the functional integration only to the non-zero modes of the field φ. To this aim, let us define

$$Z_B(\tau) = 2\pi \int D\varphi \sqrt{A}\,\delta\left(\int d^2x \varphi(x)\, \varphi_0 \right) e^{-S}, \tag{12.2.41}$$

where A is the area of the torus $A = \text{Im}\,\tau$, $\varphi_0 = A^{-1/2}$ is the normalized eigenfunction of the zero mode on this geometry and the prefactor 2π has been inserted for future convenience. The integral $\int d^2x\varphi(x)\,\varphi_0$ filters the zero mode of the field, on which it is no longer necessary to integrate for the delta-function inserted into the functional integral.

To compute (12.2.41), expand φ on the basis of the normalized eigenfunctions $\varphi_{n,m}$ of the operator δ

$$\varphi = \sum_{n,m} c_{n,m}\, \varphi_{n,m}.$$

The eigenvalues corresponding to the boundary conditions (12.2.39) are given by

$$\lambda_{n,m} = (2\pi)^2\, |nk_2 + mk_1|^2,$$

where $k_{1,2}$ are the vectors of the basis of the lattice that is dual to the original lattice defined by the periods $\omega_{1,2}$. With the choice $(\omega_1, \omega_2) = (1, \tau)$, we have

$$k_1 = -i\omega_2/A = -i\tau/A, \qquad k_2 = i\omega_1/A = i/A$$

namely

$$\lambda_{n,m} = \left(\frac{2\pi}{A}\right)^2 |n - m\tau|^2. \tag{12.2.42}$$

For the partition function (12.2.41) we have then

$$Z_B(\tau) = 2\pi\sqrt{A}\int \prod_{n,m}{}' dc_{n,m}\, e^{-\frac{g}{2}\sum_{n,m}\lambda_{n,m}c_{n,m}^2} = \sqrt{\frac{A}{\det'\frac{g}{2\pi}\square}} = \sqrt{A}\prod_{n,m}{}' \left(\frac{1}{\frac{g}{2\pi}\lambda_{n,m}}\right)^{1/2} \tag{12.2.43}$$

where the index in the product means the omission of the term $n = m = 0$. To evaluate this infinite product we use the regularization given by the Riemann zeta function. We recall that, with the usual definition of this function, $\zeta(s) = \sum_{n=1}^{\infty} n^{-s}$, we have $\zeta(-1) = -\frac{1}{12}$, $\zeta(0) = -\frac{1}{2}$ and $\zeta'(0) = \frac{d\zeta(0)}{ds} = -\frac{1}{2}\ln 2$. So, with this regularization, we have for instance

$$\prod_{n=1}^{\infty} a = a^{\zeta(0)} = a^{-1/2}, \qquad \prod_{n=-\infty}^{\infty} a = a^{2\zeta(0)+1} = 1.$$

Other useful formulae are

$$\prod_{n=1}^{\infty} n^{\alpha} = e^{-\alpha\zeta'(0)} = (2\pi)^{\alpha/2},$$

$$\prod_{n=-\infty}^{\infty} (n + a) = a\prod_{n=1}^{\infty}(-n^2)\left(1 - \frac{a^2}{n^2}\right) = 2i\sin\pi a.$$

Applying these expressions, we have

$$\begin{aligned}
\det'\frac{g}{2\pi}\square &= \prod_{(n,m)\neq(0,0)} \left(\frac{\sqrt{g}}{A}\right)^2 (n - m\tau)(n - m\bar{\tau}) \\
&= \left(\frac{A}{\sqrt{g}}\right)^2 \left(\prod_{n\neq 0} n^2\right) \prod_{m\neq 0, n\in\mathbf{Z}} (n - m\tau)(n - m\bar{\tau}) \\
&= \left(\frac{A}{\sqrt{g}}\right)^2 (2\pi)^2 \prod_{m>0, n\in\mathbf{Z}} (n - m\tau)(n + m\tau)(n - m\bar{\tau})(n + m\bar{\tau}) \\
&= \frac{(2\pi A)^2}{g} \prod_{m>0} \left(e^{-i\pi m\tau} - e^{i\pi m\tau}\right)^2 \left(e^{-i\pi m\bar{\tau}} - e^{i\pi m\bar{\tau}}\right)^2
\end{aligned}$$

$$= \frac{(2\pi A)^2}{g} \prod_{m>0} (q\bar{q})^{-m} (1-q^m)^2 (1-\bar{q}^m)^2$$

$$= \frac{(2\pi A)^2}{g} (q\bar{q})^{\frac{1}{12}} \prod_{m>0} (1-q^m)^2 (1-\bar{q}^m)^2 = \frac{(2\pi A)^2}{g} \eta^2 \bar{\eta}^2,$$

where $\eta(q)$ is the Dedekind function

$$\eta(q) = q^{\frac{1}{24}} \prod_{m=1}^{\infty} (1-q^m). \tag{12.2.44}$$

Substituting in (12.2.43), we arrive at the final expression of the partition function of a Gaussian bosonic field on a torus

$$Z_B(\tau) = \frac{g^{1/2}}{(\mathrm{Im}\,\tau)^{1/2} \eta(q)\, \eta(\bar{q})}. \tag{12.2.45}$$

To check that this function is invariant under the modular group, we need the transformations of the Dedekink function under $\mathcal{T}$ and $\mathcal{S}$

$$\begin{aligned} \eta(\tau+1) &= e^{i\pi/12}\, \eta(\tau), \\ \eta(-1/\tau) &= \sqrt{-i\tau}\, \eta(\tau). \end{aligned} \tag{12.2.46}$$

These formulae are derived by using the identity

$$\eta(\tau) = \frac{1}{2} \theta_2(\tau)\theta_3(\tau)\theta_4(\tau), \tag{12.2.47}$$

and the modular tranformations of the Jacobi $\theta_i(\tau)$ functions, discussed in Problem 12.2.

Let us now generalize the previous result (12.2.45) when the scalar field φ is compactified on a circle of radius R. The equations of motion are obviously the same as for the Gaussian case but the boundary conditions are different. Instead of those expressed by (12.2.39), we have in fact

$$\begin{aligned} \varphi(z+\omega_1, \bar{z}+\bar{\omega}_1) &= \varphi(z, \bar{z}) + 2\pi R m, \\ \varphi(z+\omega_2, \bar{z}+\bar{\omega}_2) &= \varphi(z, \bar{z}) + 2\pi R n. \end{aligned} \tag{12.2.48}$$

The integers (m, n) identify a specific topological class of the field configurations of φ and, integrating on these configurations, we define the corresponding partition function $Z_{m,n}$. To compute such a quantity, let us decompose the field in terms of its classical solution $\varphi^{cl}_{m,n}$ (that satisfies the boundary conditions (12.2.48)) and of its fluctuation $\tilde{\varphi}$, that is a fully periodic function

$$\begin{aligned} \varphi &= \varphi^{cl}_{m,n} + \tilde{\varphi}, \\ \varphi^{cl}_{m,n} &= 2\pi R \left[\frac{z}{\omega_1} \frac{m\bar{\tau}-n}{\bar{\tau}-\tau} - \frac{\bar{z}}{\omega_1 *} \frac{m\tau-n}{\bar{\tau}-\tau} \right]. \end{aligned} \tag{12.2.49}$$

Substituting this expression in the action (12.2.38), we can decompose this quantity into the action $\mathcal{A}[\tilde{\varphi}]$ of the periodic field and into the action $\mathcal{A}[\varphi^{cl}_{m,n}]$ relative to the classical configuration of the field. The latter quantity is expressed by

$$\begin{aligned}
\mathcal{A}[\varphi^{cl}_{m,n}] &= \frac{g}{2}\int d^2x (\nabla \varphi^{cl}_{m,n})^2, \\
&= 2g \int dz\, d\bar{z}\, \partial\varphi^{cl}_{m,n}\, \bar{\partial}\varphi^{cl}_{m,n} \\
&= 8\pi^2 g R^2 A \frac{1}{|\omega|^2}\left|\frac{m\tau - n}{\tau - \bar{\tau}}\right| \\
&= 2\pi^2 g R^2 \frac{|m\tau - n|^2}{\mathrm{Im}\,\tau}.
\end{aligned} \tag{12.2.50}$$

The functional integral on the periodic term $\tilde{\varphi}$ of the field gives rise to the prefactor $Z_B(\tau)$ previously computed and therefore

$$Z_{m,n}(\tau) = Z_B(\tau)\exp\left[-2g\pi^2 R^2 \frac{|m\tau - n|^2}{\mathrm{Im}\,\tau}\right]. \tag{12.2.51}$$

Let us determine the transformation properties of this expression under the modular group. For a generic modular transformation, the parameter τ changes as

$$\tau \to (a\tau + b)/(c\tau + d)$$

and therefore

$$\begin{aligned}
\frac{|m\tau - n|^2}{\mathrm{Im}\,\tau} &\to \frac{|(ma\tau + bm)/(c\tau + d) - n|^2\, |c\tau + d|^2}{\mathrm{Im}\,[(a\tau + b)(c\tau + d)]} \\
&= \frac{|(ma - nc)\tau + bm - dn|^2}{\mathrm{Im}\,\tau},
\end{aligned}$$

where we use the formula

$$\mathrm{Im}\,[(a\tau + b)(c\tau + d)] = \mathrm{Im}[(ad - bc)\tau] = \mathrm{Im}\,\tau \quad (ad - bc = 1).$$

Hence, under a modular transformation, the indices (m, n) transform with the matrix

$$\begin{pmatrix} m \\ n \end{pmatrix} \to \begin{pmatrix} a & -c \\ -b & d \end{pmatrix}\begin{pmatrix} m \\ n \end{pmatrix}, \tag{12.2.52}$$

so that

$$\begin{aligned}
Z_{m,n}(\tau + 1) &= Z_{m,n-m} \\
Z_{m,n}(-1/\tau) &= Z_{-n,m}.
\end{aligned} \tag{12.2.53}$$

To have a modular invariant partition function one simply needs to sum on all sectors relative to the different boundary conditions. We arrive then to the final expression

$$Z(R) = \sqrt{2\pi g}\,R \frac{1}{\mathrm{Im}\,\tau|\eta(\tau)|^2} \sum_{m,n} \exp\left[-2g\pi^2 R^2 \frac{|m\tau - n|^2}{\mathrm{Im}\,\tau}\right], \tag{12.2.54}$$

where the prefactor $\sqrt{2\pi g}R$ comes from the integration on the zero mode of the field. This term can be also justified in a different way, i.e. transforming the previous expression with the Poisson resummation formula

$$\sum_{n=-\infty}^{\infty} \exp[-\pi a n^2 + bn] = \frac{1}{\sqrt{a}} \sum_{k=-\infty}^{\infty} \exp\left[-\frac{\pi}{a}\left(k + \frac{b}{2\pi i}\right)^2\right]. \tag{12.2.55}$$

Posing for simplicity $g = 1/4\pi$ and

$$a = R^2/2\tau_2 \qquad b = \pi m R^2 \tau_1/\tau_2 \qquad \tau = \tau_1 + i\tau_2$$

we arrive at

$$Z(R) = \frac{1}{|\eta(\tau)|^2} \sum_{e,m\in \mathbf{Z}} q^{(e/R+mR/2)^2/2}\, \bar{q}^{(e/R-mR/2)^2/2}. \tag{12.2.56}$$

Comparing with eqn. (11.7.13), it is easy to see that the expressions of L_0 and $\bar{L}_0$ coincide with those given in (12.2.31). The spectrum of the anomalous dimensions and spins, given in eqn. (12.2.34), shows that the partition function is symmetric under the simultaneous change of $e \leftrightarrow m$ and $R \leftrightarrow 2/R$. This leads to the duality relation of the partition function (12.2.56)

$$Z(R) = Z(2/R). \tag{12.2.57}$$

The computation of the partition function for the self-dual value $R = \sqrt{2}$ is proposed in Problem 12.4.

12.3 Conformal Field Theory of a Free Fermionic Field

This section discusses the conformal theory of the complex fermion field (Dirac field)

$$\Psi(z, \bar{z}) = \begin{pmatrix} \chi(z, \bar{z}) \\ \bar{\chi}(z, \bar{z}) \end{pmatrix}, \tag{12.3.1}$$

in the Euclidean space. The action is

$$\mathcal{A} = \frac{\lambda}{2\pi} \int d^2x \, \bar{\Psi} \, \gamma^\mu \, \partial_\mu \, \Psi, \tag{12.3.2}$$

where $\bar{\Psi} = \Psi^\dagger \gamma^0$, while the Euclidean Dirac matrices γ^μ satisfy the algebra

$$\{\gamma^\mu, \gamma^\nu\} = 2\delta^{\mu,\nu}. \tag{12.3.3}$$

We choose as representation of the γ^μ matrices

$$\gamma^0 = \sigma_1 = \begin{pmatrix} 0 & 1 \\ 1 & 0 \end{pmatrix}, \quad \gamma^1 = \sigma_2 = \begin{pmatrix} 0 & -i \\ i & 0 \end{pmatrix}, \tag{12.3.4}$$

where σ_i are the usual Pauli matrices. The two-dimensional analog of the γ^5 matrix is here given by σ_3. In the complex coordinates, the Euclidean Dirac operator associated to this fermion is

$$\mathcal{D} = \gamma^0 \partial_\tau + \gamma^1 \partial_x = \begin{pmatrix} 0 & 2\partial_z \\ 2\partial_{\bar{z}} & 0 \end{pmatrix}, \tag{12.3.5}$$

and the equations of motion are

$$\begin{aligned} \partial_{\bar{z}} \chi(z, \bar{z}) &= 0, \\ \partial_z \bar{\chi}(z, \bar{z}) &= 0. \end{aligned} \tag{12.3.6}$$

They show that $\chi(z, \bar{z}) = \chi(z)$ is a purely analytic field whereas $\bar{\chi}(z, \bar{z}) = \bar{\chi}(\bar{z})$ is purely anti-analytic. The two-point correlation functions are

$$\begin{aligned} \langle \chi^\dagger(z_1) \, \chi(z_2) \rangle &= \frac{1}{\lambda} \frac{1}{z_1 - z_2}, \\ \langle \bar{\chi}^\dagger(\bar{z}_1) \, \bar{\chi}(\bar{z}_2) \rangle &= \frac{1}{\lambda} \frac{1}{\bar{z}_1 - \bar{z}_2}, \\ \langle \chi^\dagger(z_1) \, \bar{\chi}(\bar{z}_2) \rangle &= \langle \bar{\chi}^\dagger(\bar{z}_1) \, \chi(z_2) \rangle = 0. \end{aligned} \tag{12.3.7}$$

It should be noticed that the complex fermion field Ψ can be written in terms of the two real Majorana fermions ψ_1 and ψ_2, with $\psi_i = \psi_i^\dagger$

$$\Psi(z, \bar{z}) = \begin{pmatrix} \chi(z) \\ \bar{\chi}(\bar{z}) \end{pmatrix} = \frac{1}{\sqrt{2}} \begin{pmatrix} \psi_1 + i\psi_2 \\ \bar{\psi}_1 + i\bar{\psi}_2 \end{pmatrix}. \tag{12.3.8}$$

Since $\chi^\dagger = (\psi_1 - i\psi_2)/\sqrt{2}$, the analytic component of the stress-energy tensor of this theory is

$$T(z) = \frac{\lambda}{2} : (\partial\Psi^\dagger\,\Psi - \Psi^\dagger\,\partial\Psi) := -\frac{\lambda}{2} : (\psi_1\partial\psi_1 + \psi_2\partial\psi_2) : \qquad (12.3.9)$$

and is given by the sum of the stress-energy tensors relative to the two real fermions ψ_1 and ψ_2. From the correlator $\langle T(z_1)T(z_2)\rangle$ (that can be computed using the results of Chapter 11 for the Majorana fermion), one obtains the value of the central charge, $c = 1$. Analogous formulae hold for the anti-analytic component of the fermion field.

To study the quantization of Ψ it is sufficient to consider the quantization of its Majorana components. We deal with this problem in Section 12.3.1.

12.3.1 Quantization of the Free Majorana Fermion

Here and in the next sections we denote by $\psi(z)$ and $\bar{\psi}(\bar{z})$ the analytic and anti-analytic components of the Majorana fermion, with action[2]

$$\mathcal{A} = \frac{1}{2\pi}\int d^2x\left[\psi\,\partial_{\bar{z}}\,\psi + \bar{\psi}\,\partial_z\,\bar{\psi}\right]. \qquad (12.3.10)$$

The equations of motion

$$\begin{aligned} \partial_z\,\bar{\psi} &= 0, \\ \partial_{\bar{z}}\psi &= 0, \end{aligned} \qquad (12.3.11)$$

show that the two components are decoupled. Moreover, ψ depends only on z, whereas $\bar{\psi}$ only on $\bar{z}$. The conformal weights of the two fields are

$$\psi \to \left(\frac{1}{2}, 0\right) \qquad \bar{\psi} \to \left(0, \frac{1}{2}\right).$$

As seen previously, the analytic and anti-analytic components of the stress-energy tensor associated to the action (12.3.10) are

$$T = -\frac{1}{2} : \psi\partial_z\psi :, \quad \bar{T} = -\frac{1}{2} : \bar{\psi}\partial_{\bar{z}}\bar{\psi} : \qquad (12.3.12)$$

Let us now focus the attention on the analytic sector, since analogous considerations can be applied to the anti-analytic one.

[2] To simplify the notation from now on we take $\lambda = 1$.

Given the conformal weight of $\psi(z)$, the OPE with itself is

$$\psi(z_1)\psi(z_2) = \frac{1}{z_1 - z_2} + \cdots. \tag{12.3.13}$$

In the complex plane, the mode expansion of the Taylor–Laurent series reads

$$\psi(z) = \sum_{n=-\infty}^{\infty} \frac{\psi_n}{z^{n+1/2}}, \tag{12.3.14}$$

where

$$\psi_n = \oint_C \frac{dz}{2\pi i} z^{n-1/2}\, \psi(z), \tag{12.3.15}$$

with C a closed contour around the origin. Using eqn. (12.3.13), we can derive the anti-commutation relations of the modes: we simply need to use the OPE and exchange, as usual, the order of the contours around the origin

$$\begin{aligned}\{\psi_n, \psi_m\} &= \left[\oint \frac{dz}{2\pi i}, \oint \frac{dw}{2\pi i}\right] z^{n-1/2}\, w^{m-1/2}\, \psi(z)\psi(w) \\ &= \oint \frac{dw}{2\pi i} w^{m-1/2} \oint \frac{dz}{2\pi i} z^{n-1/2} \frac{1}{z-w} \\ &= \oint \frac{dw}{2\pi i} w^{m+n-1} = \delta_{n+m,0}.\end{aligned} \tag{12.3.16}$$

Neveu–Schwarz and Ramond sectors. It is worth stressing that we can choose two different monodromy properties of the field $\psi(z)$. In fact, the fermion field is naturally defined on the double covering of the complex plane: with a branch cut that starts from the origin, when the coordinate z goes around the origin

$$\psi(e^{2\pi i}\, z) = \pm\, \psi(z) \tag{12.3.17}$$

we can adopt either periodic (P) or anti-periodic (A) boundary conditions. The first case defines the so-called Neveu–Schwarz (NS) sector, while the second defines the so-called Ramond (R) sector. In the Neveu–Schwarz sector, the mode expansion of the field is given in terms of half-integer indices, while in the R sector the indices n of the (12.3.14) are instead integers

$$\begin{aligned}&\psi(e^{2\pi i}\, z) = \psi(z), \quad n \in \mathbf{Z} + \tfrac{1}{2}, \quad (NS) \\ &\psi(e^{2\pi i}\, z) = -\psi(z), \quad n \in \mathbf{Z}, \quad (R).\end{aligned} \tag{12.3.18}$$

It is also convenient to introduce the operator $(-1)^F$, where F is the fermionic number, defined in terms of its anti-commutation with the field ψ

$$(-1)^F \psi(z) = -\psi(z)(-1)^F.$$

This operator satisfies $\left((-1)^F\right)^2 = 1$ and

$$\left\{(-1)^F, \psi_n\right\} = 0, \ \forall n. \tag{12.3.19}$$

There are some interesting consequences of the integer or half-integer mode expansion of the field both for its correlation functions and for the operator content. Let us analyse first the periodic case: to compute its two-point function of the vacuum state, we can use the anti-commutations of its modes, keeping in mind that

$$\begin{aligned} \psi_n \mid 0\rangle &= 0, \quad n > 0, \\ \langle 0 \mid \psi_n &= 0, \quad n < 0. \end{aligned} \tag{12.3.20}$$

Hence, we have

$$\begin{aligned} \langle 0 \mid \psi(z)\psi(w) \mid 0\rangle &= \langle 0 \mid \sum_{n=1/2}^{\infty} \psi_n z^{-n-1/2} \sum_{m=-1/2}^{-\infty} \psi_m w^{-m-1/2} \mid 0\rangle \\ &= \sum_{n=1/2}^{\infty} z^{-n-1/2} w^{n-1/2} = \frac{1}{z}\sum_{n=0}^{\infty}\left(\frac{w}{z}\right)^n = \frac{1}{z-w}. \end{aligned} \tag{12.3.21}$$

Let us consider now the two-point correlation function when the field satisfies the anti-periodic boundary conditions. In such a case, we have to take into account the presence of the zero mode of the field that satisfy

$$\{\psi_0, \psi_0\} = 1, \quad \{(-1)^F, \psi_0\} = 0. \tag{12.3.22}$$

Applying ψ_0 to an eigenstate of L_0 does not change its eigenvalue. This means that the ground state of the R sector must realize a representation of the two-dimensional algebra given by ψ_0 and $(-1)^F$. The smallest irreducible representation consists of a doublet of operators σ and μ, the so-called *order and disorder operators*, with the same conformal weight. In this space, a 2×2 matrix representation of ψ_0 and $(-1)^F$ is given by

$$\psi_0 = \begin{pmatrix} 0 & 1 \\ 1 & 0 \end{pmatrix}, \quad (-1)^F = \begin{pmatrix} 1 & 0 \\ 0 & -1 \end{pmatrix} \tag{12.3.23}$$

In this representation the fields σ and μ are eigenvectors of $(-1)^F$ with eigenvalue $+1$ and -1 respectively.

In the presence of the order/disorder fields, the OPE of the fermionic field is

$$\psi(z)\sigma(w) \sim (z-w)^{-1/2}\mu(w)+\cdots; \quad \psi(z)\mu(w) \sim (z-w)^{-1/2}\sigma(w)+\cdots. \tag{12.3.24}$$

Therefore we can interpret the two-point correlation function of the field $\psi(z)$ with anti-periodic boundary conditions as their correlation in the presence of these two fields, placed at the origin and at infinity respectively

$$\langle\psi(z)\psi(w)\rangle_A \equiv \langle 0 \mid \sigma(\infty)\psi(z)\psi(w)\sigma(0) \mid 0\rangle = \langle 0 \mid \mu(\infty)\psi(z)\psi(w)\mu(0) \mid 0\rangle. \tag{12.3.25}$$

To compute these corrrelators, we can use the expansion in the integer modes of ψ: separating the zero mode, and using in this computation simply its vacuum expectation value $\psi_0^2 = \frac{1}{2}$, we obtain

$$\begin{aligned}\langle\psi(z)\psi(w)\rangle_A &= \langle\sum_{n=0}\psi_n z^{-n-1/2}\sum_{m=0}^{-\infty}\psi_m w^{-m-1/2}\rangle_A \\ &= \sum_{n=1}^{\infty} z^{-n-1/2}w^{n-1/2} + \frac{1}{2}\frac{1}{\sqrt{zw}} \\ &= \frac{1}{\sqrt{zw}}\left(\frac{w}{z-w}+\frac{1}{2}\right) = \frac{1}{2}\frac{\sqrt{\frac{z}{w}}+\sqrt{wz}}{z-w}.\end{aligned} \tag{12.3.26}$$

Note that, in the limit $z \to w$, this result correctly reproduces the OPE (12.3.13), as expected, because this relation expresses a *local* property of the field and is insensitive to the boundary conditions chosen for it.

It is now easy to compute the conformal dimensions of the fields σ and μ. Let us firstly use the Ward identity

$$T(z)\sigma(0) \mid 0\rangle = \frac{\Delta_\sigma}{z^2}\sigma(0) \mid 0\rangle + \cdots$$

that leads to

$$\langle T(z)\rangle_A \equiv \langle 0 \mid \sigma(\infty)T(z)\sigma(0) \mid 0\rangle = \frac{\Delta_\sigma}{z^2}. \tag{12.3.27}$$

The left-hand side of this equation can be evaluated using both the definition of the normal order

$$T(z) = \lim_{\eta\to 0}\frac{1}{2}\left(-\psi(z+\eta)\partial_z\psi(z)+\frac{1}{\eta^2}\right) \tag{12.3.28}$$

and the correlation function (12.3.26). Hence,

$$\langle T(z)\rangle_A = \lim_{w\to z}\left[-\frac{1}{4}\,\partial_w\left(\frac{\sqrt{z/w}+\sqrt{w/z}}{z-w}\right)+\frac{1}{2(z-w)^2}\right] = \frac{1}{16z^2} \tag{12.3.29}$$

and so

$$\Delta_\sigma = \Delta_\mu = \frac{1}{16}. \tag{12.3.30}$$

Bosonic order/disorder operators. It is interesting to remark that an analogous result for the periodic and anti-periodic boundary conditions also holds for the bosonic field. In fact, in view of the symmetry of the action under $\varphi \to -\varphi$, also in this theory we can adopt anti-periodic boundary conditions. Consider, for instance, $\mathcal{J}(z)$, the analytic component of the current, with expansion

$$\mathcal{J}(z) = i\,\partial_z\Phi(z) = \sum_n a_n\, z^{-n-1}. \tag{12.3.31}$$

When $\mathcal{J}(e^{2\pi i}z) = \mathcal{J}(z)$, $n \in \mathbf{Z}$, while when $\mathcal{J}(e^{2\pi i}z) = -\mathcal{J}(z)$, we have $n \in \mathbf{Z}+1/2$. As for the fermions, in the anti-periodic case we can introduce the order/disorder operators ς and τ, with operator expansion

$$\partial\Phi(z)\,\varsigma(w) - (z-w)^{-1/2}\,\tau(w) + \cdots. \tag{12.3.32}$$

Contrary to the fermionic case, in this case the two fields have different conformal weights, related by $\Delta_\tau = \Delta_\varsigma + \frac{1}{2}$. In the presence of anti-periodic boundary conditions, the two-point correlation function of the current is given by

$$\langle\,\partial\Phi(z)\,\partial\Phi(w)\,\rangle_A \equiv \langle\,0\,|\,\sigma(\infty)\partial\Phi(z)\,\partial\Phi(w)\,\sigma(0)\,|\,0\,\rangle. \tag{12.3.33}$$

Repeating the same computation as in the fermionic field, we have

$$-\,\langle\,\partial\Phi(z)\,\partial\Phi(w)\rangle_A = \frac{\left(\sqrt{\frac{z}{w}}+\sqrt{\frac{w}{z}}\right)}{2(z-w)^2}. \tag{12.3.34}$$

The conformal weight of ς can be derived by the vacuum expectation value of the stress-energy tensor

$$\langle T(z)\rangle_A = -\frac{1}{2}\lim_{z\to w}\left\langle\partial\Phi(z)\partial\Phi(w) + \frac{1}{(z-w)^2}\right\rangle_A = \frac{1}{16z^2} \tag{12.3.35}$$

namely $\Delta_\varsigma = \frac{1}{16}$.

Using eqn. (12.3.14), the stress-energy tensor becomes

$$T(z) = \frac{1}{2}\sum_{n,k}\left(k+\frac{1}{2}\right) z^{-n-2} : \psi_{n-k}\psi_k : \tag{12.3.36}$$

where with the normal order :: we mean an ordering of the operators, with the lowest index place on the left. Since $T(z) = \sum_n L_n z^{-n-2}$, the Virasoro generators are

$$L_n = \frac{1}{2}\sum_{k}\left(k+\frac{1}{2}\right) : \psi_{n-k}\psi_k : . \tag{12.3.37}$$

For the generators L_n (with $n \neq 0$) there is no problem to implement the normal order, since the operators involved in their definition anti-commute. However, we have to pay attention to the definition of L_0, in which there may be an additive constant coming from the anti-commutation of the operators ψ_{-k} and ψ_k. This constant can be determined by the vacuum expectation of $T(z)$. However, we have to distinguish the NS or the R sectors

$$\begin{aligned} L_0 &= \sum_{k>0} k\psi_{-k}\psi_k \qquad \left(\text{NS}, k \in \mathbf{Z}+\frac{1}{2}\right) \\ L_0 &= \sum_{k>0} k\psi_{-k}\psi_k + \frac{1}{16} \quad (\text{R}, k \in \mathbf{Z}). \end{aligned} \tag{12.3.38}$$

12.3.2 Fermions on a Torus

To discuss the partition function of the fermionic theory, let us initially consider the transformation that maps the plan in a cylinder geometry of width L. This is given by

$$w = \frac{L}{2\pi}\log z. \tag{12.3.39}$$

Since the fermionic field has conformal weight $1/2$, the field ψ on the cylinder is related to the field ψ_{pl} on the plane by the transformation

$$\psi(w) = \left(\frac{dz}{dw}\right)^{1/2} \psi_{pl}(z) = \sqrt{\frac{2\pi z}{L}}\,\psi_{pl}(z). \tag{12.3.40}$$

Let x be the space coordinate along the cylinder and τ its Euclidean time variable, such that $w = \tau - ix$. At a fixed τ, using eqn. (12.3.40) and the mode expansion of the field in the plane, we can easily derive the expansion of the field on the cylinder

$$\psi(x) = \sqrt{\frac{2\pi}{L}} \sum_k \psi_k e^{2\pi i k x/L}. \tag{12.3.41}$$

Since it is a free theory, the Euclidean time evolution of the modes is expressed by

$$\psi_k(t) = \psi_k(0)\, e^{-2\pi k \tau/L}, \tag{12.3.42}$$

and therefore, for any x and τ, we have the expansion

$$\psi(w) = \sqrt{\frac{2\pi}{L}} \sum_k \psi_k\, e^{-2\pi k w/L}, \tag{12.3.43}$$

where $\psi_k = \psi_k(0)$. Notice that, for the transformation law (12.3.40), on the cylinder there is a *swapping* of the boundary conditions with respect to those of the plane: the R field, that has an integer mode expansion, now corresponds to periodic boundary conditions while the NS field, with half-integer modes, satisfies anti-periodic boundary conditions

$$\begin{array}{ll} \psi(x+2\pi L) = \psi & \text{(R)} \\ \psi(x+2\pi L) = -\psi & \text{(NS)} \end{array} \tag{12.3.44}$$

It is interesting to compute L_0 for the two boundary conditions

$$(L_0)_{cyl} = \frac{1}{2}\sum_n n : \psi_{-n}\psi_n := \sum_{n>0} n\psi_{-n}\psi_n - \frac{1}{2}\sum_{n>0} n. \tag{12.3.45}$$

The last term is obviously divergent but it can be regularized in terms of the Riemann zeta function. In the R case, the sum is on all the integers

$$\sum_{n=1}^{\infty} n = \zeta(-1) = -\frac{1}{12}.$$

In the NS case, the sum runs on the half-integers $n = \frac{2k+1}{2}$. Such a series can be written as a sum on all the integers, minus the sum on the even numbers

$$\frac{1}{2}\sum_{k=0}^{\infty}(2k+1) = \frac{1}{2}\left[\sum_{m=1}^{\infty} m - \sum_{m=1}^{\infty} 2m\right] = -\frac{1}{2}\zeta(-1) = \frac{1}{24}.$$

Hence

$$(L_0)_{cyl} = \sum_{n>0} \psi_{-n}\psi_n + \begin{cases} \frac{1}{24} & \text{Ramond} \\ -\frac{1}{48} & \text{Neveu–Schwarz} \end{cases} \tag{12.3.46}$$

where, for R, the sum is on the integers and for NS on the half-integers. To interpret the presence of the additional constant, to remember the transformation law of T in passing from the plane to the cylinder

$$T_{cyl}(w) = \left(\frac{dz}{dw}\right)^2 T_{pl}(z) + \frac{c}{12}\{z,w\} = \left(\frac{2\pi}{L}\right)^2 \left[z^2\, T_{pl}(z) - \frac{c}{24}\right]. \tag{12.3.47}$$

Substituting $T(z) = \sum_n L_n z^{-n-2}$, we get

$$T_{cyl}(w) = \left(\frac{2\pi}{L}\right)^2 \left[\sum_n L_n z^{-n} - \frac{c}{24}\right] = \left(\frac{2\pi}{L}\right)^2 \sum_n \left(L_n - \frac{c}{24}\,\delta_{n,0}\right) e^{-2\pi n w/L}$$

namely

$$(L_0)_{cyl} = L_0 - \frac{c}{24}. \tag{12.3.48}$$

Since the Majorana fermion has a central charge $c = \frac{1}{2}$, in the NS sector we correctly recover the ground state energy given by $-\frac{1}{48}$. In the R sector, we have to take into account the conformal weight of the ground states of this sector, equal to $\frac{1}{16}$: the difference $\frac{1}{24} - (-\frac{1}{48}) = \frac{1}{16}$ is precisely the conformal weight of this state.

Calculus for anti-commuting quantities. To proceed, it is necessary to briefly remind the mathematical properties of the anti-commuting variables. Let α_i $(i = 1, \ldots n)$ a set of anti-commuting variables $\{\alpha_i, \alpha_j\} = 0$. Since $\alpha_i^2 = 0$, any function $f(\alpha_1, \ldots, \alpha_n)$ of these variables, once expanded in series, is at most a polynomial of first order α_i. Moreover, for anti-commuting variables the integration rules are

$$\int d\alpha_i = 0 \qquad \int d\alpha_i\, \alpha_j = \delta_{ij} \tag{12.3.49}$$

namely, the integration corresponds to take the derivative. Consider the quantity

$$I = \int d\alpha_1 \ldots d\alpha_n \exp\left[-\alpha_i A_{ij} \alpha_j\right], \tag{12.3.50}$$

where A_{ij} is an anti-symmetric matrix of dimension n, where n is an even number. Expanding in power series the exponential, for the nature of the variables α_i, there is only a finite number of terms

$$I = \int d\alpha_1 \ldots d\alpha_n \prod_{i<j} (1 - \alpha_i A_{ij} \alpha_j). \tag{12.3.51}$$

Expanding the product, the only terms that survive the integration are those in which each variable appears only once, and therefore there are $n/2$ matrix elements of A_{ij}, with the result

$$I = \sum_{p\in S_n} \epsilon(p) A_{p(1)p(2)} A_{p(3)p(4)} \cdots A_{p(n-1)p(n)}, \tag{12.3.52}$$

where $\epsilon(p)$ is the sign of the permutation p. The expression above is the 'Pfaffian' of the matrix A, $Pf(A)$, a quantity that was introduced in Chapter 5. It satisfies the identity

$$(Pf(A))^2 = \det A. \tag{12.3.53}$$

Partition function. Based on the previous formulae, let us now compute the partition function on a torus of the fermion system with action $\mathcal{A}$ given by (12.3.10)

$$Z = \int \mathcal{D}\psi\, \mathcal{D}\bar{\psi}\, e^{-\mathcal{A}} = (\det\partial)^{1/2}(\det\bar{\partial})^{1/2} = (\det\Box)^{1/2}. \tag{12.3.54}$$

It is necessary to specify the boundary conditions on the torus, both along the horizontal and vertical directions

$$\begin{aligned} \psi(z+\omega_1) &= e^{2\pi i\nu}\,\psi(z), \\ \psi(z+\omega_2) &= e^{2\pi i\mu}\,\psi(z). \end{aligned} \tag{12.3.55}$$

We assume that the same conditions are imposed for the anti-analytic component. Requiring that the action is periodic on a torus under a shift of any of its periods, there are only four possibilities

$$\begin{aligned} (\mu,\nu) &= (0,0) && (P,P), \\ (\mu,\nu) &= \left(0,\tfrac{1}{2}\right) && (P,A), \\ (\mu,\nu) &= \left(\tfrac{1}{2},0\right) && (A,P), \\ (\mu,\nu) &= \left(\tfrac{1}{2},\tfrac{1}{2}\right) && (A,A), \end{aligned} \tag{12.3.56}$$

where P and A denote the *periodic* and the *anti-periodic* boundary conditions respectively. If $z_{\mu,\nu}$ denotes the functional integral on the chiral component with boundary condition of type μ along the vertical axis (the time direction) and type ν along the horizontal axis (the space direction), for the partition function we have

$$Z_{\mu,\nu} = |\, z_{\mu,\nu}\,|^2. \tag{12.3.57}$$

The quantities $z_{\mu,\nu}$ can be easily computed in terms of the characters of the fermionic theory. Let us consider, firstly, the simplest case (A,A), in which we have

$$z_{A,A} = q^{-c/24}\,\mathrm{tr}_A\, q^{L_0} = q^{-1/48}\,\mathrm{tr}_A\, q^{L_0}, \tag{12.3.58}$$

where $q = e^{2\pi i\tau}$, $c = 1/2$ and the trace is taken in the anti-periodic sector of the theory, i.e. the Neveu–Schwarz. Similarly, we have

$$z_{A,P} = \frac{1}{\sqrt{2}}\, q^{-1/48}\,\mathrm{tr}_P\, q^{L_0}, \tag{12.3.59}$$

where this time the trace is taken in the periodic sector, i.e. the Ramond sector (the factor $\frac{1}{\sqrt{2}}$ that enters the definition of this quantity is introduced to simplify later its modular properties).

It is necessary to discuss separately the case of periodic boundary conditions along the vertical axis. Since the natural boundary conditions for a fermionic field are the anti-periodic ones, to change them and make them periodic it is necessary to introduce in the trace an operator that anti-commutes with the fermion. Such an operator is just $(-1)^F$, that was previously introduced, and for the remaining partition functions we have then

$$z_{P,A} = q^{-1/48}\,\mathrm{tr}_A\,(-1)^F\, q^{L_0} \tag{12.3.60}$$

$$z_{P,P} = q^{-1/48}\,\mathrm{tr}_P\,(-1)^F\, q^{L_0}. \tag{12.3.61}$$

The computation of these two traces is elementary. For each fermionic mode there are only two states and therefore, for each term of type $q^{n\psi_{-n}\psi_n}$ we have

$$\begin{aligned} \mathrm{tr}\, q^{n\psi_{-n}\psi_n} &= 1 + q^n \\ \mathrm{tr}\,(-1)^F\, q^{n\psi_{-n}\psi_n} &= 1 - q^n \end{aligned} \tag{12.3.62}$$

and therefore

$$\begin{aligned} \mathrm{tr}\, q^{\sum_{n>0} n\psi_{-n}\psi_n} &= \mathrm{tr} \prod_{n>0} q^{n\psi_{-n}\psi_n} = \prod_{n>0}(1+q^n) \\ \mathrm{tr}\,(-1)^F\, q^{\sum_{n>0} n\psi_{-n}\psi_n} &= \mathrm{tr} \prod_{n>0} (-1)^F\, q^{n\psi_{-n}\psi_n} = \prod_{n>0}(1-q^n). \end{aligned} \tag{12.3.63}$$

Using the expression of L_0 given in (12.3.46), in the two anti-periodic (NS) cases (with half-integer mode expansion) and in the periodic (R) case (with integer mode expansion), we get

$$z_{AA}(\tau) = q^{-1/48}\mathrm{tr}_A q^{L_0} = q^{-1/48}\prod_{n=0}^{\infty}(1+q^{n+1/2}) = \sqrt{\frac{\theta_3(\tau)}{\eta}}$$

$$z_{PA}(\tau) = q^{-1/48}\mathrm{tr}_A(-1)^F q^{L_0} = q^{-1/48}\prod_{n=0}^{\infty}(1-q^{n+1/2}) = \sqrt{\frac{\theta_4(\tau)}{\eta}}$$

$$z_{AP}(\tau) = \frac{1}{\sqrt{2}}q^{-1/48}\mathrm{tr}_P q^{L_0} = q^{1/24}\prod_{n=0}^{\infty}(1+q^{n}) = \sqrt{\frac{\theta_2(\tau)}{\eta}}$$

$$z_{PP}(\tau) = \frac{1}{\sqrt{2}}q^{-1/48}\mathrm{tr}_P(-1)^F q^{L_0} = q^{1/24}\prod_{n=0}^{\infty}(1-q^{n}) = 0,$$

where $\theta_i(\tau)$ are the Jacobi functions defined in Problem 12.2. Note that the partition function z_{PP} vanishes, for the zero mode present with these boundary conditions and the integration rules (12.3.49). Under the modular transformation $\tau \to \tau+1$, the partition functions change as

$$\begin{aligned} z_{AA}(\tau+1) &= e^{-i\pi/24}\, z_{PA}(\tau), \\ z_{PA}(\tau+1) &= e^{-i\pi/24}\, z_{AA}(\tau), \\ z_{AP}(\tau+1) &= e^{i\pi/12}\, z_{AP}(\tau), \end{aligned} \tag{12.3.64}$$

while under $\tau \to -1/\tau$

$$\begin{aligned} z_{AA}(-1/\tau) &= z_{PA}(\tau), \\ z_{PA}(-1/\tau) &= z_{AP}(\tau), \\ z_{AP}(-1\tau) &= z_{PA}(\tau). \end{aligned} \tag{12.3.65}$$

In light of these transformations, the modular invariant partition function is obtained by including all the three boundary conditions, namely

$$Z = |\, z_{AA}\,|^2 + |\, z_{AP}\,|^2 + |\, z_{PA}\,|^2 = \left|\frac{\theta_2}{\eta}\right| + \left|\frac{\theta_3}{\eta}\right| + \left|\frac{\theta_4}{\eta}\right|. \tag{12.3.66}$$

In Chapter 14 we show that this partition function corresponds to the square of the partition function of the Ising model.

12.4 Bosonization

As we have seen in Appendix 1.B, in system with one-dimensional space there is no distinction between the statistical and the interaction properties of the particles. The term ‘bosonization’ refers to the possibility of describing a relativistic theory of Dirac fermions in $(1+1)$ dimensions in terms of a bosonic theory. Such a possibility permits in many cases a drastic simplification of the original fermionic theory.

The original idea of this transformation is from Mattis and Lieb, who were able to exactly solve the Thirring model. An important step forward in condensed matter is also thanks to Luther and Peschel. In QFT, the most famous work is Coleman's, who proved the equivalence of the Sine–Gordon and the massive Thirring model.

This section presents the main formulae of the theory that links the fermionic and bosonic fields. Note that the equivalence of these two theories is also suggested by the common value of their central charge, $c = 1$.

12.4.1 Bosonization Rules

The two-point correlation functions of the two components of the complex fermion field $\Psi(z, \bar{z})$ defined in (12.3.8) are given by eqn. (12.3.7). Given the free nature of the theory, the multi-point correlators are computed in terms of the Wick theorem. Focusing the attention on the analytic part of Ψ we have

$$\langle \chi^\dagger(z_1) \dots \chi^\dagger(z_n)\, \chi(w_1) \dots \chi(w_n) \rangle = \det\left(\frac{1}{z_i - w_j}\right). \tag{12.4.1}$$

With the choice $g = 1/4\pi$, the propagators of the bosonic field are $\langle \phi(z_1)\phi(z_2)\rangle = -\ln z_{12}$ and $\langle \bar{\phi}(\bar{z}_1)\bar{\phi}(\bar{z}_2)\rangle = -\ln \bar{z}_{12}$. Let us consider the purely analytic vertex operators $V_{+1} =: e^{i\phi(z)} :$ and $V_{-1} =: e^{-i\phi(z)}$, together with those purely anti-analytic $\bar{V}_{+1} =: e^{i\bar{\phi}(\bar{z})} :$ and $\bar{V}_{-1}(\bar{z}) =: e^{-i\bar{\phi}(\bar{z})} :$. Using these expressions and the Wick theorem, it is easy to prove that

$$\langle e^{i\phi(z_1)} \dots e^{i\phi(z_n)}\, e^{-i\phi(w_1)} \dots e^{-i\phi(w_n)} \rangle = \frac{\prod_{i<j}(z_i - z_j)(w_i - w_j)}{\prod_{i,j}(z_i - w_j)}. \tag{12.4.2}$$

At first sight, this expression seems different from the correlation functions of the fermion fields. However, Cauchy's mathematical identity states the identity of the two expressions! Namely, it holds

$$\frac{\prod_{i<j}(z_i - z_j)(w_i - w_j)}{\prod_{i,j}(z_i - w_j)} = \det\left(\frac{1}{z_i - w_j}\right). \tag{12.4.3}$$

To prove this identity, it is sufficient to check that both expressions have the same poles and zeros. Based on this equality between the correlation functions of the fermionic and bosonic fields, it is natural to establish the following correspondence between the two operators[3]

[3] For $g \neq 1/4\pi$ the formulae change as follows: $\chi \to e^{i\sqrt{4\pi g}\phi}$ and similarly for the others. Strictly speaking, when there are several fermions, it is necessary to update the formulae above by introducing the so-called *Klein factors* that implement the correct anti-commutation relations of the fermion fields. To simplify the discussion here, we ignore this aspect of the problem.

$$\begin{aligned}
\chi(z) &=: e^{i\phi(z)} : \\
\chi^{\dagger}(z) &=: e^{-i\phi(z)} : \\
\bar{\chi}(\bar{z}) &=: e^{-i\bar{\phi}(\bar{z})} : \\
\bar{\chi}^{\dagger}(z) &=: e^{i\bar{\phi}(\bar{z})} : .
\end{aligned} \tag{12.4.4}$$

With these expressions, we can establish the operator identities

$$\begin{aligned}
\bar{\chi}^{\dagger}(\bar{z})\,\chi(z) &= \bar{\Psi}\frac{(1+\sigma_3)}{2}\Psi =: e^{i\varphi(z,\bar{z})} : \\
\chi^{\dagger}(z)\,\bar{\chi}(\bar{z}) &= \bar{\Psi}\frac{(1-\sigma_3)}{2}\Psi =: e^{-i\varphi(z,\bar{z})} :
\end{aligned} \tag{12.4.5}$$

(σ_3 is the Pauli matrix) or, equivalently

$$\bar{\Psi}\,\Psi =: \cos\varphi : \quad ; \quad \bar{\Psi}\,\sigma_3\,\Psi = i : \sin\varphi : . \tag{12.4.6}$$

It is necessary however to establish the bosonic espression of the fermionic current. In fact, a naive application of the bosonization rules would lead to

$$\bar{\Psi}\gamma^0\Psi = \bar{\Psi}\sigma_1\Psi = \chi^{\dagger}\chi + \bar{\chi}^{\dagger}\bar{\chi}$$

and therefore

$$\chi^{\dagger}(z)\chi(z) \to e^{-i\phi(z)+i\phi(z)} = 1,$$

which is clearly false. To understand the origin of the discrepancy, we must remember that the normal order exponential operators do not obey the usual additive rule of the exponentials. In fact,

$$: e^{i\alpha\phi(z_1)} :: e^{i\beta\phi(z_2)} := (z_1 - z_2)^{\alpha\beta} : e^{i\alpha\phi(z_1)+i\beta\phi(z_2)} : . \tag{12.4.7}$$

Taking this into account, it is convenient to compute the fermionic current in correspondence of two points slightly separated, so that

$$\begin{aligned}
\chi^{\dagger}(z_1)\chi(z_2) &= (z_1 - z_2)^{-1} : e^{-i(\phi(z_1)-\phi(z_2))} : \\
&= \frac{1}{z_1 - z_2} - i\partial_z\phi + \cdots .
\end{aligned} \tag{12.4.8}$$

If we omit the first term of this equation (that corresponds to the identity operator present in the OPE of the two fermionic fields), this is equivalent to define the normal order of the operators as

$$: \chi^\dagger(z)\,\chi(z) := \lim_{\eta\to 0}\Big[\chi^\dagger(z+\eta)\,\chi(z) - \langle\,\chi^\dagger(z+\eta)\,\chi(z)\,\rangle\Big].$$

We arrive then at the operator identity

$$: \chi^\dagger(z)\,\chi(z) := -i\,\partial_z\phi(z). \tag{12.4.9}$$

Repeating the same considerations for the anti-analytic part, we get an equivalent formula (with a change of sign)

$$: \bar\chi^\dagger(\bar z)\,\bar\chi(\bar z) := i\,\partial_{\bar z}\bar\phi(\bar z). \tag{12.4.10}$$

Returning to the Euclidean coordinates (x_0, x_1), we have

$$\begin{aligned} j^0 &= :\ \bar\Psi\sigma_1\Psi := -i\,(\partial_z\phi(z) - \partial_{\bar z}\bar\phi(\bar z)) \\ &= -i\,[(\partial_z - \partial_{\bar z})\phi + (\partial_z - \partial_{\bar z})\bar\phi] \\ &= -\partial_{x_1}\varphi(x_0, x_1), \end{aligned} \tag{12.4.11}$$

and similarly

$$j^1 =:\ \bar\Psi\sigma_2\Psi := \partial_{x_0}\varphi \tag{12.4.12}$$

Bosonic theory	Fermionic theory
$\mathcal{A}_B = \frac{g}{2}\int d^2(\partial\varphi)^2$	$\mathcal{A}_F = \int d^2x[\bar\chi^\dagger\partial_z\bar\chi + \chi^\dagger\partial_{\bar z}\chi]$
$\varphi(z, \bar z) = \phi(z) + \bar\phi(\bar z)$ $\theta(z, \bar z) = \phi(z) - \bar\phi(\bar z)$	$\Psi(z, \bar z) = \begin{pmatrix}\chi(z)\\ \bar\chi(\bar z)\end{pmatrix}$
$: e^{i\sqrt{4\pi g}\phi} :$	χ
$: e^{-i\sqrt{4\pi g}\phi} :$	$\chi^\dagger$
$: e^{-i\sqrt{4\pi g}\bar\phi} :$	$\bar\chi$
$: e^{i\sqrt{4\pi g}\bar\phi} :$	$\bar\chi^\dagger$
$-i\sqrt{4\pi g}\,\partial_z\phi$	$: \chi^\dagger\chi(z) :$
$i\sqrt{4\pi g}\,\partial_{\bar z}\bar\phi$	$: \bar\chi^\dagger\bar\chi(z) :$
$: \cos\sqrt{4\pi g}\varphi :$	$: \bar\Psi\Psi :$
$i : \sin\sqrt{4\pi g}\varphi :$	$: \bar\Psi\sigma_3\Psi :$

Table 12.1 *Bosonization formulae.*

where σ_2 is the Pauli matrix. It is easy to prove that this current is conserved $\partial_\mu j^\mu = 0$, and its bosonization expression is

$$: \bar{\Psi}\gamma^\mu\Psi := -\epsilon^{\mu\nu}\,\partial_\nu\varphi. \tag{12.4.13}$$

Other useful bosonization formulae are discussed in Problems 12.6 and 12.7. A summary of the bosonization rules is given in Table 12.1.

REFERENCES

The partition functions of free conformal theories on a torus are studied in

Itzykson C. and Zuber, J.B. (1986). 'Two-dimensional conformal invariant theories on a torus', *Nuclear Physics B*, 275: 580.

Ginsparg, P. (1988). 'Curiosity at c =1', *Nuclear Physics B*, 295: 153.

The bosonization approach has been developed in

Coleman, S. (1975). 'The Quantum Sine–Gordon Equation as the Massive Thirring Model', *Physics Reviews D*, 11: 2088.

Luther, A. and Peschel, I. (1975). 'Calculation of Critical Exponents in Two-dimensions from Quantum Field Theory in One Dimension', *Physics Reviews B*, 12: 3908.

Mattis, D. and Lieb, E.H. (1965). 'Exact Solution of a Many Fermion System and its Associated Boson Field', *Journal of Mathematical Physics*, 6: 304.

The bosonization approach and its applications are discussed in detail in

Giamarchi, T. (2004). *Quantum Physics in One Dimension*, Oxford University Press.

Gogolin, A., Nersesyan, A. and Tsvelik, A. (1998). *Bosonization and Strongly Correlated Systems*, Cambridge University Press.

PROBLEMS

12.1. Vertex operators

In order to prove that the ground states $|\,\alpha\rangle$ of the free bosonic theory are obtained by applying the vertex operators $V_\alpha(z,\bar{0})$ to the conformal vacuum $|\,0\rangle$ it is necessary to show that $V_\alpha(0,0)$ is an eigenstate of π_0 with eigenvalue α and, furthermore, that $a_n\,V_\alpha(0,0)\,|\,0\rangle = 0$, con $n > 0$. To this aim

a. Prove the validity of the formula

$$[B, e^A] = e^A \, [B, A]$$

assuming that the commutator $[B,A]$ is a constant.

b. Pose $B = \pi_0, A = i\alpha\varphi(z, \bar{z})$ and show that

$$[\pi_0, V_\alpha] = \alpha \, V_\alpha$$

Consequently

$$\pi_0 \, V_\alpha(0,0) \mid 0\rangle = \alpha \, V_\alpha(0,0) \mid 0\rangle$$

c. Show that

$$[a_n, V_\alpha(z, \bar{z})] = \alpha \, z^n \, V_\alpha(z, \bar{z})$$

and conclude that a_n annihilates $V_\alpha(0,0) \mid 0\rangle$.

12.2. Modular transformations of the $\theta_i(\tau)$ functions.

Let us denote by $\theta_i(z,\tau)$ the theta functions of Jacobi, with $\theta_1(\tau) = 0$, while the others have the infinite series and product representations

$$\begin{aligned}
\theta_2(\tau) &= \sum_{n\in\mathbf{Z}} q^{(n+1/2)^2/2} = 2\,q^{1/8} \prod_{n=1}^{\infty} (1-q^n)\,(1+q^n)^2 \\
\theta_3(\tau) &= \sum_{n\in\mathbf{Z}} q^{n^2/2} = \prod_{n=1}^{\infty} (1-q^n)\,(1+q^{n+1/2})^2 \\
\theta_4(\tau) &= \sum_{n\in\mathbf{Z}} (-1)^n \, q^{n^2/2} = \prod_{n=1}^{\infty} (1-q^n)\,(1-q^{n-1/2})^2.
\end{aligned}$$

Use these expressions and the Poisson resummation formula (12.2.55) to prove

$$\begin{aligned}
\theta_2(\tau+1) &= e^{i\pi/4}\,\theta_2(\tau), & \theta_2(-1/\tau) &= \sqrt{-i\tau}\,\theta_4(\tau) \\
\theta_3(\tau+1) &= \theta_4(\tau), & \theta_3(-1/\tau) &= \sqrt{-i\tau}\,\theta_3(\tau) \\
\theta_4(\tau+1) &= \theta_3(\tau), & \theta_4(-1/\tau) &= \sqrt{-i\tau}\,\theta_2(\tau).
\end{aligned}$$

12.3. Dedekind function

The Dedekind function is defined by the infinite product ($q = e^{2i\pi\tau}$)

$$\eta(\tau) = q^{1/24} \prod_{n=1}^{\infty} (1 - q^n).$$

a. Use the definition of the $\theta_i(\tau)$ functions in terms of the infinite product to prove the identity

$$\eta^3(\tau) = \frac{1}{2}\theta_2(\tau)\theta_3(\tau)\theta_4(\tau).$$

b. Use the modular transformations of $\theta_i(\tau)$ to prove

$$\begin{aligned} \eta(\tau + 1) &= e^{i\pi/12}\, \eta(\tau) \\ \eta(-1/\tau) &= \sqrt{-i\tau}\, \eta(\tau). \end{aligned}$$

12.4. Bosonic partition function at the self-dual point

Consider the expression (12.2.56) of the partition function of a bosonic field with a compactification radius R equal to the self-dual value $R = \sqrt{2}$

$$Z(\sqrt{2}) = \frac{1}{|\eta(\tau)|^2} \sum_{n,m \in \mathbf{Z}} q^{\frac{1}{4}(n+m)^2}\, \bar{q}^{\frac{1}{4}(n-m)^2}.$$

Prove that this expression can be written as

$$Z(\sqrt{2}) = |C_0|^2 + |C_1|^2$$

where

$$\begin{aligned} C_0(\tau) &= \frac{1}{\eta} \sum_{k \in \mathbf{Z}} q^{m^2} = \frac{\theta_3(2\tau)}{\eta(\tau)} \\ C_1(\tau) &= \frac{1}{\eta} \sum_{k \in \mathbf{Z}} q^{(m+1/2)^2} = \frac{\theta_2(2\tau)}{\eta(\tau)}. \end{aligned}$$

12.5. Jacobi identity

The aim of this exercise is to prove, by physical arguments, the Jacobi identity

$$\prod_{n=1}^{\infty}(1-q^n)(1+q^{n-1/2}w)(1+q^{n-1/2}w^{-1}) = \sum_{n=-\infty}^{\infty} q^{n^2/2}\, w^n$$

that holds for $|q| < 1$ and $w \neq 0$. Consider then the partition function of a free system of f fermions and $\bar{f}$ anti-fermions, with energy levels $E = E_0(n - \frac{1}{2})$, $n \in \mathbf{Z}$, and total fermion number $N = N_f - N_{\bar{f}}$. Let $q = e^{-E_0/T}$ and $w = e^{\mu/T}$.

a. Show that the grand canonical partition function is given by

$$Z(w,q) = \sum_{f,\bar{f}} e^{-E/T+\mu N/T} = \sum_{N=-\infty}^{\infty} w^N Z_N(q) \qquad (12.4.14)$$
$$= \prod_{n=1}^{\infty}(1+q^{n-1/2}w)(1+q^{n-1/2}w^{-1})$$

where $Z_N(q)$ is the partition function at a given number N of the fermions.

b. Consider Z_0. The lowest energy states that contribute to this quantity have all negative energy levels occupied (they form the Dirac sea, with a total energy normalized to the value $E = 0$) whereas the excited states are described by the integers $k_1 \geq k_2 \geq k_3 \ldots k_l > 0$ with $\sum_i^l k_i = M$. The energy of these states is $E = ME_0$. Prove that Z_0 is given by

$$Z_0 = \sum_{M=0}^{\infty} P(M)\, q^M = \prod_{n=1}^{\infty} \frac{1}{1-q^n}$$

where $P(M)$ is the combinatoric function that expresses in how many ways an integer M is expressed as a sum of numbers minor than it.

c. Consider now the sector with fermionic number N, where the first positive levels are occupied. Argue that this sector contributes with the factor

$$q^{1/2}\cdots q^{N-3/2}\, q^{N-1/2} = q^{\sum_{n=1}^{N}(j-1/2)} = q^{N^2/2}$$

in their partition function, while the remaining excitation gives rise to the same partition function Z_0, so that

$$Z_N = q^{N^2/2}\, Z_0.$$

d. Use now eqn. (12.4.14) to prove the Jacobi identity.

12.6. Quantum Pythagorean theorem

A regularization of the normal order $:A(x)B(x):$ of two operators can be obtained by the limit $\lim_{\eta\to 0} :A(x-\eta/2)B(x+\eta/2):$ and an average on all directions of η, so that the final expression is invariant under the rotations. The average is equivalent to the substitution $\eta^{\mu}\eta^{\nu}/|\eta|^2 \to \frac{1}{2}\delta^{\mu\nu}$.

Use this regularization and the bosonization formulae of the text to prove the quantum version of the Pythagorean theorem

$$(:\cos\varphi:)^2 + (:\sin\varphi:)^2 = -\frac{1}{4}(\partial\varphi)^2.$$

(Observe that $(:\cos\varphi:)^2 \neq :\cos^2\varphi:$).

12.7. Equivalence of the Sine–Gordon and Thirring models

Consider the Sine–Gordon model of a scalar bosonic field φ, whose Lagrangian is

$$\mathcal{L} = \frac{1}{2}(\partial\varphi)^2 + \frac{m^2}{\beta^2}(\cos\beta\varphi - 1).$$

Use the bosonization formulae, prove that this Lagrangian can be transformed in the Lagrangian of the Thirring model

$$\mathcal{L} = i\bar{\Psi}\gamma^{\mu}\partial_{\mu}\Psi - M\bar{\Psi}\Psi - \frac{1}{2}g(\bar{\Psi}\gamma^{\mu}\Psi)(\bar{\Psi}\gamma_{\mu}\Psi)$$

where Ψ is a complex fermionic field, with the coupling constants related as

$$\frac{\beta^2}{4\pi} = \frac{1}{1+\frac{g}{\pi}}.$$

Note that $\beta^2 = 4\pi$ is equivalent to $g = 0$, i.e. a free fermionic model!

13

Conformal Field Theories with Extended Symmetries

Ideas are incredibly similar when you have a chance to know them.
Samuel Beckett

13.1 Introduction

This chapter deals with those field theories that present, in addition to conformal invariance, a symmetry under a larger group of transformations. These models can have interesting applications in a wide range of topics, as the study of fundamental interactions, statistical mechanics and condensed matter.

Our first example is the superconformal models that have, in addition to the Virasoro generators, also their fermionic partners. The minimal models of these theories have a finite number of conformal families and rational values of the central charge and conformal dimensions. As in the pure bosonic case, the fusion rules of the unitary superconformal minimal models admits a remarkable interpretation in terms of Landau–Ginzburg theories. We also study the conformal models that are invariant under the discrete $\mathbf{Z}_N$ symmetry, the so-called parafermion models. Finally, our study focuses on the conformal theories invariant under a current algebra based on a Lie group G and their Lagrangian realization provided by the Wess–Zumino–Witten model.

A conformal theory is usually formulated in terms of an associative algebra that involves fields mutually local. However it is also useful to consider theories that have non-local fields. This is the case of both the superconformal and parafermion models. It is therefore convenient to define here the concept of non-local fields and refer to it later on: a field $\mathcal{O}_1(x)$ is γ-local with respect to another field $\mathcal{O}_2(x_2)$ if their product $\mathcal{O}(x_1)\mathcal{O}(x_2)$ acquires a phase $\exp(2\pi i\gamma)$ when the variable x_1 is analytically continued clockwise along a closed contour that encloses the point x_2 (Figure 13.1).

13.2 Superconformal Models

This section presents the main properties of conformal theories in which there is also a supersymmetry, i.e. a symmetry that links the bosonic and fermionic fields. They are a generalization of the conformal theories previously encountered. Since any

Statistical Field Theory: An Introduction to Exactly Solved Models in Statistical Physics. Second Edition. Giuseppe Mussardo, Oxford University Press (2020). © Giuseppe Mussardo. DOI: 10.1093/oso/9780198788102.001.0001

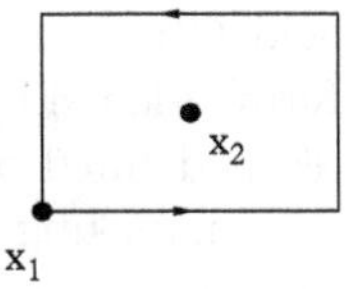

Fig. 13.1 *A close loop of the variable x_1 around the point x_2.*

super symmetric theory is also super conformal on short scales, the classification of the super conformal fixed points gives us useful information on the realization of all possible super symmetric theories. Here we focus our attention only on the two-dimensional super symmetric theories, and interested readers are referred to the reference list for a broader discussion of the super symmetric theories and their application in various fields of physics.

In two dimensions, the super conformal invariance is associated to two super-currents, $G(z)$ and $\bar{G}(\bar{z})$, the former a purely analytic field while the latter a purely anti-analytic one. They are both fermionic fields, with conformal weights $(\frac{3}{2},0)$ and $(0,\frac{3}{2})$, respectively. The algebra of these generators is defined by the singular terms of their OPE: for $G(z)$ we have

$$G(z_1)\, G(z_2) = \frac{2c}{3(z_1 - z_2)^3} + \frac{2}{z_1 - z_2} T(z_2) + \cdots, \tag{13.2.1}$$

with an analogous expression for $\bar{G}$. The parameter c is the central charge, the same quantity that enters the operator expansion of $T(z)$

$$T(z_1)T(z_2) = \frac{c}{2(z_1 - z_2)^2} + \frac{2}{(z_1 - z_2)^2} T(z_2) + \frac{1}{z_1 - z_2} \partial T(z_2) + \cdots. \tag{13.2.2}$$

The field $G(z)$ (and $\bar{G}$) is itself a primary field, with OPE

$$T(z_1)G(z_2) = \frac{3}{2(z_1 - z_2)^2} G(z_2) + \frac{1}{z_1 - z_2} \partial G(z_2) + \cdots. \tag{13.2.3}$$

Let us define the generators L_n and G_n through the expansions

$$T(z) = \sum_{n=-\infty}^{\infty} \frac{L_n}{z^{2+n}}; \quad G(z) = \sum_{m=-\infty}^{\infty} \frac{G_m}{z^{3/2+m}} \tag{13.2.4}$$

namely

$$L_n = \oint_C \frac{dz}{2\pi i} z^{n+1} T(z); \quad G_m(z) = \oint_C \frac{dz}{2\pi i} z^{m+1/2} G(z).$$

Note that, in the expansion of the field $G(z)$, the indices can assume either integer or half-integer value. In fact, $G(z)$ is a fermionic field and, as we have seen in the previous chapter for the free fermionic field ψ, is defined on the double covering of the plane, with a branch cut starting from the origin: making the analytic continuation $z \to e^{2\pi i} z$, we can have two possible boundary conditions

$$G(e^{2\pi i} z) = \pm G(z). \tag{13.2.5}$$

In the periodic case (relative to +), called Neveu–Schwarz (NS) sector, the indices m are half-integers, $m \in \mathbf{Z} + \frac{1}{2}$. In the anti-periodic case (relative to −), the R sector, the indices m are instead integer numbers, $m \in \mathbf{Z}$.

The OPE that involve $T(z)$ and $G(z)$ can be equivalently expressed as algebraic relations of their modes. Exchanging the order of the integration contours and taking into account the singular terms of their expansion (see Section 10.7), we arrive the infinite-dimensional algebra

$$\begin{aligned}
[L_n, L_m] &= (n-m)L_{n+m} + \frac{c}{12} n(n^2-1)\delta_{n+m,0} \\
[L_n, G_m] &= \frac{1}{2}(n-2m)\, G_{n+m} \\
\{G_n, G_m\} &= 2L_{n+m} + \frac{c}{3}\left(n^2 - \frac{1}{4}\right)\delta_{n+m,0}.
\end{aligned} \tag{13.2.6}$$

The peculiar aspect of this algebra is the simultaneous presence of commutation and anti-commutation relations.

As in the pure conformal case, the classification of superconformal theories reduces to finding all irreducible representations of the algebra (13.2.6) with the central charge c as a free parameter. The space $\mathcal{A}$ of these representations is given by the direct sum of the NS and R sub-spaces:

$$\mathcal{A} = \mathcal{A}_{NS} \oplus \mathcal{A}_R.$$

Furthermore, each of the sub-spaces is decomposed into the direct sum of the *superconformal families*

$$\mathcal{A}_{NS} = \oplus_l [\Phi_l]_{NS}; \quad \mathcal{A}_R = \oplus_\lambda [\Phi_\lambda]_R, \tag{13.2.7}$$

where the *primary fields* Φ_l and Φ_λ of this algebra satisfy

$$\begin{aligned}
L_n \Phi_a &= 0 \qquad n > 0 \\
L_0 \Phi_a &= \Delta_a \Phi_a \\
G_m \Phi_a &= 0 \qquad m > 0.
\end{aligned} \tag{13.2.8}$$

As for the Virasoro algebra, the representations are built starting from the primary fields and applying to them the *creation operators* L_n and G_m, with $n, m < 0$. So, the representations are uniquely identified by the conformal weights Δ_a of the primary fields. The same considerations hold for the anti-analytic sector of the theory.

Super space. It is interesting to note that the operators

$$\delta_\epsilon = \oint_C \frac{dz}{2\pi i}\epsilon(z)\, T(z); \quad \delta_\omega = \oint_C \frac{dz}{2\pi i}\omega(z)\, G(z) \tag{13.2.9}$$

can be interpreted as the (holomorphic) generators of the infinitesimal change of the coordinates $(Z,\bar{Z}) = (z,\theta;\bar{z},\bar{\theta})$ of a $2+2$ dimensional *super space*, where z and $\bar{z}$ are the usual complex coordinates, whereas θ and $\bar{\theta}$ are fermionic coordinates. For the analytic part of this super space, we have the following super conformal transformation

$$z \to z + \epsilon(z) - \omega(z)\theta; \quad \theta \to \theta + \frac{1}{2}\epsilon'(z) + \omega(z). \tag{13.2.10}$$

Hence, $\epsilon(z)$ and $\omega(z)$ are the bosonic and fermionic infinitesimal transformation respectively. The peculiar nature of (13.2.10) consists of being the conformal transformation of the 1-form $dz + \theta\, d\theta$. It is therefore convenient to consider $G(z)$ and $T(z)$ as the components of a *super*- stress-energy tensor

$$W(z,\theta) = G(z) + \theta\, T(z) \tag{13.2.11}$$

Neveu–Schwarz sector. In the NS sector the representations are given in terms of the superfields

$$\Phi_l(Z,\bar{Z}) = \Phi_l(z,\bar{z}) + \theta\, \psi_l(z,\bar{z}) + \bar{\theta}\, \bar{\psi}_l(z,\bar{z}) + i\theta\, \bar{\theta}\, \tilde{\Phi}_l(z,\bar{z}) \tag{13.2.12}$$

where the primary fields Φ_l is the first component while $\psi_l = G_{-1/2}\Phi_l$, $\bar{\psi}_l = \bar{G}_{-1/2}\Phi_l$ and $\tilde{\Phi}_l = -iG_{-1/2}\bar{G}_{-1/2}\Phi_l$.

R sector. In the R sector the field $G_m A_r$ cannot be local with respect to the fields of $\mathcal{A}_R$ and consequently the space $\mathcal{A}_R$ naturally decomposes in two locality classes: $\mathcal{A}_R = \mathcal{A}_R^{(+)} \oplus \mathcal{A}_R^{(-)}$, where all fields are mutually local in each class while any field $\mathcal{A}_R^{(+)}$ is semi-local (with a semi-local index equal to $1/2$) with respect to $\mathcal{A}_R^{(-)}$. The operators G_m act in $\mathcal{A}_R$ as $G_m : \mathcal{A}_R^{(\epsilon)} \to \mathcal{A}_R^{(-\epsilon)}$, with $\epsilon = \pm$. This implies, in particular, that the primary fields in the R sector are organized in a doublet of fields $\Phi_\lambda \in \mathcal{A}_R^{(\epsilon)}$, with the operators G_0 and $\bar{G}_0$ that act on them as 2×2 matrices. From the algebraic relations of the modes, we also have

$$G_0^2 = L_0 - \frac{c}{24}. \tag{13.2.13}$$

Hence, for a scalar field Φ_λ with conformal weights $(\Delta_\lambda, \bar{\Delta}_\lambda)$ we get

$$G_0\,\Phi_\lambda^{(\epsilon)} = 2^{-3/2}\,(1+i\epsilon)\,\beta_\lambda\,\Phi_\lambda^{(-\epsilon)};\quad \bar{G}_0\,\Phi_\lambda^{(\epsilon)} = 2^{-3/2}\,(1-i\epsilon)\,\beta_\lambda\,\Phi_\lambda^{(-\epsilon)}$$

where $\tilde{c} = 2/3c$ and the parameter β subjected to the condition

$$\Delta_\lambda - \frac{\tilde{c}}{16} = \frac{1}{4}\beta_\lambda^2.$$

The only exception to these transformation laws is given by the Ramond field $\Phi_{(0)}$ of conformal weight $\Delta_{(0)} = \tilde{c}/16 = c/24$, if such a field actually exists in the theory: in this case, in fact, $G_0\Phi_{(0)} = \bar{G}_0\Phi_{(0)} = 0$ and therefore the second component is not necessarily present.

Irreducible representations and minimal models. The irreducible representations of the superconformal algebra are determined in the same way as those of the Virasoro algebra previously discussed. In this case, the conformal weights can be expressed similarly to (11.2.7), namely

$$\Delta_{r,s} = \Delta_0 + \frac{1}{4}(r\beta_+ + s\beta_-)^2 + \frac{1}{32}[1-(-1)^{r+s}], \tag{13.2.14}$$

where

$$\begin{aligned} \Delta_0 &= (\tilde{c}-1)/16 \\ \beta_\pm &= \frac{1}{4}\left(\sqrt{1-\tilde{c}} \pm \sqrt{9-\tilde{c}}\right); \quad \beta_+\beta_- = -\frac{1}{2}. \end{aligned} \tag{13.2.15}$$

In this formula r and s are two natural numbers: for the NS fields, $r+s \in 2\mathbf{Z}$, whereas for the Ramond fields $r+s \in 2\mathbf{Z}+1$. These degenerate fields has similar properties of the usual degenerate conformal fields, namely in their OPE enters only degenerate fields. Similarly, their correlation functions satisfy linear diffential equations. When the parameter $\rho = -\beta_-/\beta_+$ becomes a rational number, the operator algebra closes within a *finite number* of superconformal families. Particularly interesting are the unitary superconformal series, here denoted by $\mathcal{SM}_p$ $(p = 3,4,5,\ldots)$, with

$$\rho = \frac{p}{p+2}.$$

In this case there are $[p^2/2]$ primary fields $\Phi_{r,s}$, where the indices r and s assume the values $r = 1,2,\ldots,(p-1); s = 1,2,\ldots(p+1)$ (where $[x]$ is the integer part of x). The central charge and the conformal weights take the discrete values

$$c = \frac{3}{2}\left[1 - \frac{8}{p(p+2)}\right], \quad p = 3, 4, \ldots \tag{13.2.16}$$
$$\Delta_{r,s} = \frac{[(p+2)r - ps]^2 - 4}{8p(p+2)} + \frac{1}{32}[1 - (-1)^{r+s}].$$

The Coulomb gas method can be generalized to the superconformal model, both in the Neveu–Schwarz and R sectors, and to determine the exact values of the structure constants of the operator algebra. It is interesting to note that, in the R sector, the representation of the conformal fields can be implemented in terms of the magnetization operator σ of the Ising model, as is discussed in detail in Chapter 14.

Additional symmetry. The operator algebra of the minimal models $\mathcal{SM}_p$ may present additional symmetry, according to the value of p. In fact, if $p \in 2\mathbf{Z}+1$, the spaces $\mathcal{A}_R^{(+)}$ and $\mathcal{A}_R^{(-)}$ are isomorphic and therefore, for these values of p, the models are invariant under the duality transformation $\mathcal{A}_R^{(+)} \to \mathcal{A}_R^{(-)}$, similarly to the Kramers–Wannier duality of the Ising model. If p is instead an even number, the model $\mathcal{SM}_p$ contains the vacuum field $\Phi_{\frac{p}{2},\frac{p}{2}+1}$ of the Ramond sector and therefore it is not invariant under duality. However, it has a symmetry $Z_2 \times Z_2$, espressed in the form

$$\Phi_{r,s} \to (\epsilon_1)^{r+1}(\epsilon_2)^{s+1}\,\Phi_{r,s}$$

where the parameters $\epsilon_{1,2}$ can be either ± 1.

Landau–Ginzburg theory. Using arguments that are similar to those presented in the previous chapter, it can be shown that the unitary superconformal models $\mathcal{S}M_p$ are associated to a supersymmetric Landau–Ginzburg theory. The super potential relative to the minimal models is given by $W(\Phi) = g\Phi^p$ and the action reads

$$\mathcal{A}_p = \int d^2x\, d^2\theta \left[-\frac{1}{2} D\Phi\,\bar{D}\Phi + g\Phi^p\right] \tag{13.2.17}$$

where

$$D = \partial_\theta - \theta\,\partial_z \quad \bar{D} = \partial_{\bar{z}} - \bar{\theta}\,\partial_{\bar{z}}$$

are the covariant derivatives, θ and $\bar{\theta}$ are fermionic variables, while $\Phi(z,\bar{z},\theta,\bar{\theta})$ is a superfield

$$\Phi(z,\bar{z},\theta,\bar{\theta}) = \varphi + \theta\,\psi + \bar{\theta}\,\bar{\psi} + i\theta\bar{\theta}\,\chi.$$

The integration on the fermionic variables θ and $\bar{\theta}$ is done according to the rules of fermionic calculus presented in Section 12.3.

Identifying also in this case the NS superconformal primary field that sits in the position $(2,2)$ of the Kac table with Φ, i.e. $\Phi_{2,2} \equiv \Phi$, and using the fusion rules of the superconformal minimal model, we can recursively define the composite operators $:\Phi^k:$ and show that their fusion rules lead to the operator identity

$$D\bar{D}\Phi \simeq \Phi^{p-1}. \tag{13.2.18}$$

This formula coincides with the equation of motion that can be derived by the supersymmetric action (13.2.17).

As for the minimal models of the Virasoro algebra, also for the superconformal minimal models we can determine the exact expression of the modular invariant partition functions on a torus. On this topic, we refer readers to the original work by Cappelli.

The series of the superconformal minimal models has intersection with the Virasoro minimal models: that the model $\mathcal{SM}_3$ describes the tricritical Ising model, that coincides with the second minimal model of the Virasoro unitary series. The supersymmetry of this model provides a different interpretation of the primary fields and gives reason of the particular relationships that exist among the structure constants of the conformal model. Furthermore, notice that the second minimal superconformal model has central charge $c = 1$ and can be regarded as a particular realization of the Gaussian free bosonic theory analysed in Chapter 12. It is worth to stress that the supersymmetry, so long searched in the particle accelerators, has found its first physical realization in statistical mechanics!

13.3 Parafermion Models

Non-local operators naturally appear in field theories associated the continuum limit of lattice statistical models with a $\mathbf{Z}_N$ symmetry. These theories have been investigated in detail by Fateev and Zamolodchikov. $\mathbf{Z}_N$ is an abelian group, generated by the powers of the generator Ω, and its elements are given by $\Omega, \Omega^2, \ldots, \Omega^{N-1}$, with $\Omega^N = 1$. In these models, the order parameter has $(N-1)$ components, here denoted by σ_k, $k = 1, 2, \ldots (N-1)$: they are scalar fields, with $\sigma_k^\dagger = \sigma_{N-k}$, and conformal weights $d_k = d_{N-k}$. These fields form a representation of $\mathbf{Z}_N$ and satisfy

$$\Omega\,\sigma_k = \omega^k \sigma_k, \quad \omega = \exp(2\pi i/N). \tag{13.3.1}$$

Statistical models that are invariant under a $\mathbf{Z}_N$ symmetry can also be invariant under duality. For the self-dual theories, in addition to the $(N-1)$ order parameters, there are other $(N-1)$ operators μ_l $(l = 1, 2, \ldots N-1)$, with $\mu_l^\dagger = \mu_{N-l}$. These are the disorder operators, with the same conformal weights of the order parameters, $d_l = d_{N-l}$. The fields μ_l and σ_k are mutually local among themselves, but are non-local with respect to each other: the semi-local parameter of the fields σ_k and μ_l is equal to $\gamma_{kl} = kl/N$. The disorder fields form a representation of the dual group $\tilde{\mathbf{Z}}_N$, generated by $\tilde{\Omega}$ and satisfy

$$\tilde{\Omega}\,\mu_l = \omega^l \mu_l. \tag{13.3.2}$$

In light of this operator content, the self-dual models possess an enlarged symmetry $\mathbf{Z}_N \times \tilde{\mathbf{Z}}_N$. This allows us to introduce the concept of *charge*: we say that a field $\mathcal{O}_{(k,l)}$ has a charge (k,l) with respect to the group $\mathbf{Z}_N \times \tilde{\mathbf{Z}}_N$ if

$$\Omega \mathcal{O}_{(k,l)} = \omega^k \, \mathcal{O}_{(k,l)}, \quad \tilde{\Omega} \mathcal{O}_{(k,l)} = \omega^l \, \mathcal{O}_{(k,l)}$$

with the integers k and l that are defined modulo N. Under an OPE, there is an abelian composition law for these fields, given (up to the actual value of the structure constants) by

$$\mathcal{O}^{(i)}_{(k,l)} \, \mathcal{O}^{(j)}_{(k',l')} = \sum_n \mathcal{O}^{(k)}_{(k+k',l+l')}, \tag{13.3.3}$$

where the sums on the indices are modulo N. Using this definition, the fields σ_k have charge $(k,0)$ while μ_l have charge $(0,l)$. In general, the semi-local index of two fields $\mathcal{O}_{(k,l)}$ and $\mathcal{O}_{(k',l')}$ is equal to $\gamma = (kl' + k'l)/N$. In addition to the symmetry $\mathbf{Z}_N \times \tilde{\mathbf{Z}}_N$, we also assume that these theories are invariant under the charge conjugation $\mathcal{C}$ and parity $\mathcal{P}$ transformations, with

$$\begin{aligned} \mathcal{C} &: \sigma_k \to \sigma_k^\dagger; \quad \mu_l \to \mu_l^\dagger; \\ \mathcal{P} &: \sigma_k \to \sigma_k; \quad \mu_l \to \mu_l. \end{aligned} \tag{13.3.4}$$

In the next chapter we will see that the simplest representative of these theories is provided by the Ising model, invariant under the group $\mathbf{Z}_2 \times \tilde{\mathbf{Z}}_2$. In the operator content of this theory there is a Majorana fermion, whose analytic and anti-analytic components are $\psi(z)$ and $\bar{\psi}(\bar{z})$ respectively, which appear in the short-distance expansion of the order and disorder parameters

$$\sigma(z,\bar{z}) \, \mu(0,0) = \frac{1}{\sqrt{2}} (z\bar{z})^{-1/2} \left[z^{1/2} \, \psi(0) + \bar{z}^{1/2} \bar{\psi} + \cdots \right] \tag{13.3.5}$$

These fields satisfy the analiticity and anti-analiticity conditions $\partial_{\bar{z}} \psi = \partial_z \bar{\psi} = 0$. We can now generalize these formulae to the case $\mathbf{Z}_N$: for the OPE of $\sigma_k(x) \, \mu_k(0)$ and $\sigma_k(x) \, \mu_k^\dagger(0)$ we pose

$$\begin{aligned} \sigma_k(z,\bar{z}) \, \mu_k(0,0) &= z^{\Delta_k - 2d_k} \, \bar{z}^{\bar{\Delta}_k - 2d_k} \, \psi_k(0,0) + \cdots \\ \sigma_k(z,\bar{z}) \, \mu_k^\dagger(0,0) &= z^{\bar{\Delta}_k - 2d_k} \, \bar{z}^{\Delta_k - 2d_k} \, \bar{\psi}_k(0,0) + \cdots \end{aligned} \tag{13.3.6}$$

where we have also used the symmetry (13.3.4). The fields ψ_k and $\bar{\psi}_k$ are operators with conformal weights Δ_k and $\bar{\Delta}_k$. From the semi-locality of the operators σ_k and μ_k we can easily derive the condition

$$\Delta_k - \bar{\Delta}_k = -\frac{k^2}{N} \;\; (\text{mod}\, \mathbf{Z}). \tag{13.3.7}$$

Let us assume that in the self-dual critical theory it holds the condition $\bar{\Delta}_k = 0$. The fields ψ_k and $\bar{\psi}_k$ satisfy

$$\partial_{\bar{z}} \psi_k = 0; \quad \partial_z \bar{\psi}_k = 0, \tag{13.3.8}$$

so that $\psi_k = \psi_k(z)$ and $\bar{\psi}_k = \bar{\psi}_k(\bar{z})$. In this case the conformal weights Δ_k coincide with the spins of the fields and their general expression is then

$$\Delta_k = m_k - \frac{k^2}{N}, \tag{13.3.9}$$

where m_k are integer numbers. The operators ψ_k and $\bar{\psi}_k$ have charge equal to (k,k) and $(k,-k)$ respectively, and they are semi-local each other. In contrast with the scalar order and disorder fields previously introduced, these fields have spins and therefore it is natural to call them *parafermions*. The simplest expression of (13.3.9) that also satisfies the condition $\Delta_k = \Delta_{N-k}$ is provided by

$$\Delta_k = \frac{k(N-k)}{N}. \tag{13.3.10}$$

In the following we assume that these are the conformal weights of the parafermions. The fields ψ_k generate a close operatorial algebra

$$\begin{aligned} \psi_k(z_1)\,\psi_l(z_2) &= \mathbf{C}_{k,l}\,(z_{12})^{-2kl/N}\,\psi_{k+l}(z_2) + \cdots \\ \psi_k(z_1)\,\psi_k^{\dagger}(z_2) &= (z_{12})^{-2\Delta_k}\left[1 + \frac{2\Delta_k}{c} z_{12}^2\,T(z_2) + \cdots\right] \end{aligned} \tag{13.3.11}$$

where $T(z)$ is the analytic component of the stress-energy tensor, and $\mathbf{C}_{k,l}$ are the structure constants of this algebra, whereas c is the central charge. These parameters can be fixed by imposing the associativity of this algebra. This condition leads to the values of the structure constants

$$\mathbf{C}_{k,l} = \frac{\Gamma(k+l+1)\,\Gamma(N-k+1)\,\Gamma(N-l+1)}{\Gamma(k+1)\,\Gamma(l+1)\,\Gamma(N-k-l+1)\,\Gamma(N+1)}, \tag{13.3.12}$$

and the central charge

$$c = \frac{2(N-1)}{N+2}. \tag{13.3.13}$$

As for the Virasoro and the superconformal agebras, the fields of the self-dual $\mathbf{Z}_N \times \tilde{\mathbf{Z}}_N$ can be classified by the irreducible representation of the parafermionic algebra (13.3.11). Their Hilbert space is decomposed in *parafermionic conformal families*

$$\mathcal{A} = \oplus_{k=0}^{N-1} \, [\sigma_k]_\psi \,, \tag{13.3.14}$$

whose primary operators are the order parameters σ_k. Their conformal weights can be obtained by expressing $T(z)$ in terms of the normal order of the fields ψ_k and $\psi_k^\dagger$, and then using the OPE (13.3.6). As a result we have

$$d_k = \frac{k(N-k)}{N(N+2)}. \tag{13.3.15}$$

Section 13.3.1 shows that the parafermionic theories also naturally appear in the Kac–Moody algebra $SU(2)_N$. In particular, using the results relative to this theory we can easily derive all the other conformal data of the parafermionic models. For instance, for the structure constants that enter the OPE

$$\sigma_{k_1}(z,\bar{z})\sigma_{k_2}(0,0) = \mathbf{C}_{k_1,k_2}(z\bar{z})^{2d_{k_1+k_2}-d_{k_1}-d_{k_2}}\,\sigma_{k_1+k_2} + \ldots$$

with the operators normalized as

$$\langle \sigma_k(z,\bar{z})\sigma_{k'}^\dagger \rangle = \delta_{k,k'}\,(z\bar{z})^{-2d_k},$$

we have

$$\mathbf{C}_{k_1,k_2} = \frac{\Gamma\left(\frac{1}{N+2}\right)\Gamma\left(\frac{1+k_1+k_2}{N+2}\right)\Gamma\left(\frac{N-k_1+1}{N+2}\right)\Gamma\left(\frac{N-k_2+1}{N+2}\right)}{\Gamma\left(\frac{N+1}{N+2}\right)\Gamma\left(\frac{N-k_1-k_2+1}{N+2}\right)\Gamma\left(\frac{k_1+1}{N+2}\right)\Gamma\left(\frac{k_2+1}{N+2}\right)}. \tag{13.3.16}$$

These quantities can be extracted by the four-point correlation functions of the σ_k operators. The simplest of them is given by

$$\langle \sigma_1(z_1,\bar{z}_1)\sigma_1^\dagger(z_2,\bar{z}_2)\sigma_k(z_3,\bar{z}_3)\sigma_k^\dagger(z_4,\bar{z}_4)\rangle = (z_{12}\bar{z}_{12})^{-2d_1}\,(z_{34}\bar{z}_{34})^{-2d_k}\,\mathcal{G}_{1,k}(x,\bar{x}),$$

where x and $\bar{x}$ are the harmonic ratios

$$x = \frac{z_{12}z_{24}}{z_{14}z_{23}}, \quad \bar{x} = \frac{\bar{z}_{12}\bar{z}_{24}}{\bar{z}_{14}\bar{z}_{23}},$$

and the function $\mathcal{G}_(1,k)(x,\bar{x})$ is expressed by

$$\mathcal{G}_{1,k}(x,\bar{x}) = (x\bar{x})^{-k/N(N+2)}\,\frac{\Gamma\left(\frac{1}{N+2}\right)\Gamma\left(\frac{N}{N+2}\right)}{\Gamma\left(\frac{N+1}{N+2}\right)\Gamma\left(\frac{2}{N+2}\right)} \times$$

$$\times \left[\frac{\Gamma\left(\frac{k+2}{N+2}\right)\Gamma\left(\frac{N-k+1}{N+2}\right)}{\Gamma\left(\frac{N-k}{N+2}\right)\Gamma\left(\frac{k+1}{N+2}\right)} F^{(1)}(k,x)F^{(1)}(k,\bar{x}) + \right. \tag{13.3.17}$$
$$\left. + \frac{\Gamma\left(1-\frac{k}{N+2}\right)\Gamma\left(\frac{k+1}{N+2}\right)}{\Gamma\left(\frac{k}{N+2}\right)\Gamma\left(1-\frac{k+1}{N+2}\right)} \frac{(x\bar{x})^{(N+1-k)/(N+2)}}{(N+1-k)^2} F^{(2)}(k,x)F^{(2)}(k,\bar{x}) \right].$$

In this formula $F^{(i)}$ are the hypergeometric functions

$$F^{(1)}(k,x) = F\left(\frac{k}{N+2}, -\frac{1}{N+2}, \frac{k+1}{N+2}; x\right);$$
$$F^{(2)}(k,x) = F\left(\frac{N+1}{N+2}, \frac{N-k}{N+2}, \frac{2N-k+3}{N+2}; x\right).$$

Similarly we can also obtain the exact expression of the correlators that involve the order and disorder operators, the simplest example being

$$\langle \mu_1(z_1,\bar{z}_1)\mu_1^\dagger(z_2,\bar{z}_2)\sigma_k(z_3,\bar{z}_3)\sigma_k^\dagger(z_4,\bar{z}_4)\rangle = (z_{12}\bar{z}_{12})^{-2d_1}(z_{34}\bar{z}_{34})^{-2d_k}\mathcal{H}_{1,k}(x,\bar{x}),$$

where $\mathcal{H}_(1,k)(x,\bar{x})$ is given by

$$\mathcal{H}_{1,k}(x,\bar{x}) = \bar{x}^{k/N}(x\bar{x})^{-k/N(N+2)} \frac{\Gamma\left(1+\frac{1}{N+2}\right)\Gamma\left(\frac{N}{N+2}\right)}{\Gamma\left(\frac{N+1}{N+2}\right)\Gamma\left(\frac{2}{N+2}\right)} \times$$
$$\times \left[\frac{\Gamma\left(\frac{k+2}{N+2}\right)\Gamma\left(\frac{N-k+1}{N+2}\right)}{\Gamma\left(\frac{N-k}{N+2}\right)\Gamma\left(1+\frac{k+1}{N+2}\right)} F^{(1)}(k,x)F^{(2)}(N-k,\bar{x}) + \right. \tag{13.3.18}$$
$$\left. + \frac{\Gamma\left(1-\frac{k}{N+2}\right)\Gamma\left(\frac{k+1}{N+2}\right)}{\Gamma\left(\frac{k}{N+2}\right)\Gamma\left(1+\frac{N+k+1}{N+2}\right)} x\,(x\bar{x})^{-(k+1)/(N+2)} F^{(1)}(k,x)F^{(2)}(N-k,\bar{x}) \right].$$

This expression clearly shows that moving the point $(z_2,\bar{z}_2)$ along a closed contour that encloses the point $(z_3,\bar{z}_3)$, the correlation function acquires a phase factor, related to the non-locality of the two operators.

13.3.1 Relation to Lattice Models

The formulae of the previous section provide the exact solution of the QFTs of the critical points with a $\mathbf{Z}_N$ symmetry. It is useful to investigate their relation with the exactly solvable theories defined on a lattice that share the same symmetry. These theories are defined in terms of the variables σ_r, defined on any site r of the lattice, that take values ω^q, $q = 0, 1, \ldots (N-1)$. Assuming that their interaction is restricted to the nearest neighbours, the partition function can be written as

$$Z = \sum_{\{\sigma_r\}} e^{-\sum_{r,a=1,2} H(\sigma_r, \sigma_{r+e_a})} = \sum_{\{\sigma_r\}} \prod_{r,a} W(\sigma_r, \sigma_{r+e_a}), \tag{13.3.19}$$

where e_a are the basis vectors of the lattice. The Hamiltonian must be invariant under the $\mathbf{Z}_N$ transformations and the charge conjugation $\mathcal{C}$

$$H(\omega\sigma, \omega\sigma') = H(\sigma, \sigma') = H(\sigma^\dagger, \sigma^{\dagger,\prime}). \tag{13.3.20}$$

Consequently, the Boltzmann weights $W(\sigma, \sigma')$ can be written as

$$W(\sigma, \sigma') = e^{-H(\sigma, \sigma')} = \sum_{k=0}^{N-1} w_k \, (\sigma^\dagger \sigma')^k, \tag{13.3.21}$$

where the real and positive parameters w_k satisfy the condition $w_k = w_{N-k}$. As normalization we will choose $w_0 = 1$. Hence, such models are parameterized by the parameters w_k, with $k = 1, 2, \ldots \leq [N/2]$, where $[x]$ is the integer part of the number x.

The duality transformation of these lattice models can be performed as discussed in Chapter 4: the spins σ_r are replaced by the *dual spins* μ_l, associated to the sites of the dual lattice, with their interaction described by the same type of formulae shown in (13.3.19) and (13.3.21), where the dual parameters $\tilde{w}_k$ are expressed in terms of the original parameters w_i as

$$\tilde{w}_k = \left(1 + \sum_{q=1}^{N-1} w_q \, \omega^{kq}\right) \left(1 + \sum_{q=1}^{N-1} w_q\right)^{-1}. \tag{13.3.22}$$

The system is then self-dual if satisfies the conditions

$$\tilde{w}_k = w_k, \quad k = 1, 2, \ldots (N-1). \tag{13.3.23}$$

For $N = 2, 3$, these lattice models coincide with the Ising and the 3-state Potts models respectively. The equation (13.3.23) identifies in both case their critical temperature. For $N = 4$, the corresponding model is a special case of the Ashkin–Teller model. The self-dual line is described by

$$w_2 + 2w_1 = 1, \tag{13.3.24}$$

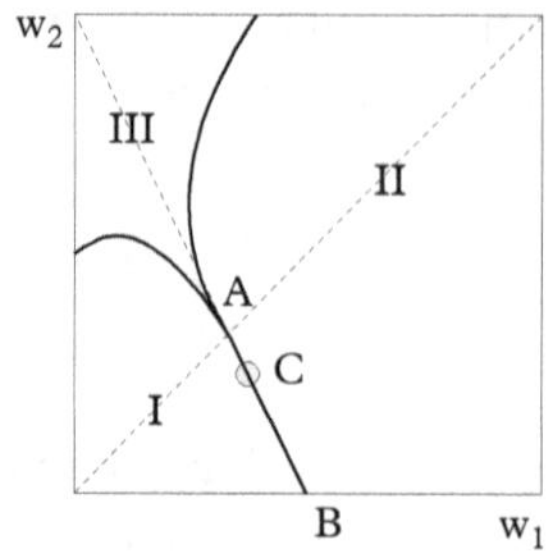

Fig. 13.2 *Phase diagram of the* $\mathbf{Z}_4$ *lattice model.*

and the exact solution of this model can be found in the book by Baxter.[1] Its phase diagram is shown in Figure 13.2. There are three phases, according to the values of the parameters: phase I, where $\langle\sigma\rangle \neq 0$ and $\langle\mu\rangle = 0$; phase II, where $\langle\sigma\rangle = 0$ and $\langle\mu\rangle \neq 0$; phase III, where $\langle\sigma\rangle = \langle\mu\rangle = 0$. The points of the segment AB of the diagram that belong to the line (13.3.24) are all critical points of the system and therefore the critical exponents vary with continuity along AB. There is however a peculiar point C, identified by the equations

$$w_1 = \frac{\sin(\pi/16)}{\sin(3\pi/16)}, \quad w_2 = 1 - 2w_1, \tag{13.3.25}$$

where it is possible to show that the corresponding critical theory is precisely given by the parafermionic conformal theory $\mathbf{Z}_4 \times \tilde{\mathbf{Z}}_4$ previously analysed.

Similarly, for a lattice model with the $\mathbf{Z}_5$ symmetry, we have the phase diagram shown in Figure 13.3. Also in this case there are three distinct phases, with the same characterization used for the previous $\mathbf{Z}_4$ model. The critical line is given by

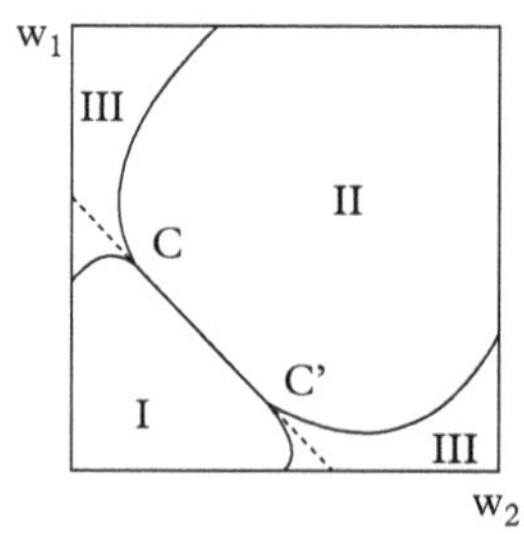

Fig. 13.3 *Phase diagram of the lattice* $\mathbf{Z}_5$ *model.*

[1] Baxter, R.J. (1982). *Exactly Solved Models in Statistical Mechanics*, Academic Press.

$$w_1 + w_2 = \frac{1}{2}(\sqrt{5} - 1). \tag{13.3.26}$$

This line contains, in particular, two symmetric bifurcation points, C and C', whose corresponding theory in the continuum can be shown to coincide with the parafermionic conformal theory $\mathbf{Z}_5 \times \tilde{\mathbf{Z}}_5$.

In general, the points of the critical lines of the self-dual models described by the parafermionic theory have been identified by Fateev and Zamolodchikov. They correspond to the values

$$w_k = \prod_{l=0}^{k-1} \frac{\sin \pi l/N + \pi/4N)}{\sin(\pi(l+1)/N - \pi/4N)}. \tag{13.3.27}$$

13.4 Kac–Moody Algebra

In this section we consider the CFTs characterized by a set of analytic currents $\mathcal{J}^a(z)$ of conformal weights $(\Delta, \bar{\Delta}) = (1, 0)$ and an analogous set of anti-analytic currents $\bar{\mathcal{J}}^a(\bar{z})$ of conformal weights $(0, 1)$. As usual, we focus our attention on the analytic sector, with similar results for the anti-analytic one. Conformal theories based on a current algebra prove to be an important tool in the development of both string theory and condensed matter physics. Moreover, they give rise to one of the most general realization of CFT: the minimal models previously discussed are in fact particular cases of them.

Let us start our discussion with the OPE of the currents. For dimensional reasons, this can be written as

$$\mathcal{J}^a(z_1)\mathcal{J}^b(z_2) = \frac{\tilde{k}^{ab}}{(z_1 - z_2)^2} + \frac{if^{abc}}{z_1 - z_2}\mathcal{J}^c(z_2) + \cdots \tag{13.4.1}$$

where, in the last term, it is meant to be a sum on the index c. The structure constants f^{abc} are obviously anti-symmetric in the indices a and b. For the associativity of this operator expansion, they satisfy the Jacobi identity

$$\sum_d \left(f^{ade} f^{bcd} + f^{cde} f^{abd} + f^{bde} f^{cad} \right) = 0. \tag{13.4.2}$$

Therefore these quantities also play the role of the structure constants of a Lie algebra[2] $\mathcal{G}$. In the following we assume that this algebra is associated to a compact Lie group, characterized by a positive definite Cartan matrix. In this case the indices a, b etc, run on the values $1, \ldots |G| = \dim G$. In the algebra $\mathcal{G}$ it is always possibile to choose a basis such that

[2] The basic properties of the Lie algebras are summarized in the appendix at the end of the chapter.

$$\tilde{k}^{ab} = \tilde{k}\delta^{ab}. \tag{13.4.3}$$

The algebra (13.4.1), defined by the operator expansion of the currents, is called the affine algebra or Kac–Moody algebra. Expanding the currents in modes, for instance at the origin

$$\mathcal{J}^a(z) = \sum_{n=-\infty}^{\infty} \frac{\mathcal{J}_n^a}{z^{n+1}}, \tag{13.4.4}$$

we can translate the operator expansion (13.4.1) into the commutation relations of the modes

$$\left[\mathcal{J}_m^a, \mathcal{J}_n^b\right] = i f^{abc} \mathcal{J}_{m+n}^c + \tilde{k} m \delta^{ab} \delta_{m+n,0}. \tag{13.4.5}$$

Note that the zero modes of the currents, $\mathcal{J}_0^a$, give rise to the usual commutation relations of the generators of the Lie algebras.

The representation theory of the affine algebras can be developed along the lines of the Virasoro algebra. Also in this case, it is possible to define a vacuum state $|0>$, annihilated by all positive modes of the currents

$$\mathcal{J}_n^a |0\rangle = 0 \quad n \geq 0. \tag{13.4.6}$$

There is also the notion of *primary field* $\varphi_{(r)}^l$, in this case made of a field multiplet, that satisfies the operator expansion

$$\mathcal{J}^a(z_1)\varphi_{(r)}^l(z_2) = \frac{(R_{(r)}^a)^{lk}}{z_1 - z_2}\varphi_{(r)}^k(z_2) + \cdots \tag{13.4.7}$$

where $(R_{(r)}^a)^{lk}$ are the matrix of the generators $\mathcal{J}^a$, in the representation labelled by (r). The highest weight vectors of the Kac–Moody algebra are obtained acting with the primary fields on the vacuum state

$$|(r)\rangle = \varphi_{(r)}(0)|0\rangle. \tag{13.4.8}$$

In particular, this multiplet of states gives rise to a representation of the zero modes of the algebra, i.e. of the group G

$$\mathcal{J}_0^a |(r)\rangle = R_{(r)}^a |(r)\rangle, \quad \mathcal{J}_n^a |(r)\rangle = 0 \quad n > 0. \tag{13.4.9}$$

As for the stress-energy tensor and the primary fields of the Virasoro algebra, also in this case it is possible to prove that holds a Ward identity satisfied by the currents

$$\langle \mathcal{J}^a(z)\varphi_{(r_1)}(z_1)\dots\varphi_{(r_n)}(z_n)\rangle = \sum_{j=1}^{n} \frac{R^a_{(r_j)}}{z-z_j}\langle\varphi_{(r_1)}(z_1)\dots\varphi_{(r_n)}(z_n)\rangle. \tag{13.4.10}$$

13.4.1 Virasoro Operators and Sugawara Formula

For the CFTs ruled by a set of currents it is natural to assume that the stress-energy tensor can be expressed as their composite operator. Since the conformal weight of $T(z)$ is equal to 2, while the one of the currents $\mathcal{J}^a$ is 1, it should be possible to express $T(z)$ as a quadratic expression of $\mathcal{J}^a$, invariant under the group. This reasoning leads to the *ansatz*

$$T(z) = \frac{1}{\gamma}\sum_{a=1}^{|G|} :\mathcal{J}^a(z)\mathcal{J}^a(z): = \frac{1}{\gamma}\left(\lim_{w\to z}\sum_{a=1}^{|G|}\mathcal{J}^a(w)\mathcal{J}^a(z) - \frac{\tilde{k}|G|}{(w-z)^2}\right). \tag{13.4.11}$$

Expressing $T(z) = \sum L_n/z^{n+2}$, for the generators of the Virasoro algebra we have

$$L_n = \frac{1}{\gamma}\sum_{m=-\infty}^{\infty} :\mathcal{J}^a_{m+n}\mathcal{J}^a_{-m}:. \tag{13.4.12}$$

The constant γ in these formulae can be fixed demanding that the currents $\mathcal{J}^a(z)$ are themselves primary fields with conformal weights $(1,0)$, fulfilling the OPE

$$T(z_1)\mathcal{J}^a(z_2) = \frac{\mathcal{J}^a(z_2)}{(z_1-z_2)^2} + \frac{\partial\mathcal{J}^a(z_2)}{z_1-z_2} + \cdots. \tag{13.4.13}$$

Note that this relation is equivalent to the commutation relations

$$\left[L_m, \mathcal{J}^a_n\right] = -n\mathcal{J}^a_{m+n}. \tag{13.4.14}$$

To determine γ, consider the expression of L_{-1} and apply this operator to a highest weight state $|(r)\rangle$. Using eqn. (13.4.9), it is easy to check that in this procedure only the first term is different from zero, with the result

$$L_{-1}|(r)\rangle = \frac{2}{\gamma}\mathcal{J}^a_{-1}R^a_{(r)}|(r)\rangle. \tag{13.4.15}$$

Applying to both terms of this expression $\mathcal{J}_1^b$ and using eqn. (13.4.14), we obtain

$$\begin{aligned} R^b_{(r)}\,|\,(r)\,\rangle &= \frac{2}{\gamma}\,(if^{abc}\mathcal{J}_0^c + \tilde{k}\delta^{ab})\,R^a_{(r)}\,|\,(r)\,\rangle \\ &= \frac{2}{\gamma}\left(if^{abc}\frac{1}{2}if^{dca}R^d_{(r)} + \tilde{k}R^b_{(r)}\right)|\,(r)\,\rangle \\ &= \frac{2}{\gamma}\left(\frac{1}{2}C_A + \tilde{k}\right)R^b_{(r)}\,|\,(r)\,\rangle, \end{aligned} \tag{13.4.16}$$

where we have defined the Casimir invariant C_A in the adjoint representation of the algebra through the formula

$$C_A\,\delta^{ab} = \sum_{c,d} f^{acd} f^{bcd}. \tag{13.4.17}$$

From (13.4.16) we arrive at the value of the constant γ

$$\gamma = 2\tilde{k} + C_A. \tag{13.4.18}$$

Once we have determined γ, we can compute the central charge of these theories by the two-point correlator of $T(z)$

$$\langle T(z_1)T(z_2)\rangle = \frac{c_G}{2}\,\frac{1}{(z_1 - z_2)^4}. \tag{13.4.19}$$

Since

$$T(z) = \frac{1/2}{\tilde{k} + C_A/2} \sum_{a=1}^{|G|} :\mathcal{J}^a(z)\mathcal{J}^a(z): \tag{13.4.20}$$

and

$$\langle \mathcal{J}^a(z_1)\mathcal{J}^b(z_2)\rangle = \frac{\tilde{k}\,\delta^{ab}}{(z_1 - z_2)^2}, \tag{13.4.21}$$

this yields

$$c_G = \frac{\tilde{k}\,|G|}{\tilde{k} + C_A/2}. \tag{13.4.22}$$

In the literature the relation that links $T(z)$ to the currents $\mathcal{J}_a$ is known as the *Sugawara formula*.

13.4.2 Maximal Weights

This section discusses the representations of the Kac–Moody algebra associated to the irreducible and unitary maximal weights. These are also the representations that are irreducible for the ordinary Lie algebras and since they have the lowest eigenvalue of L_0, are called *vacuum representations*. The unitary conditions are expressed by $\mathcal{J}^{a\dagger}(z) = \mathcal{J}^a(z)$ and this implies $\mathcal{J}_n^{a\dagger} = \mathcal{J}_{-n}^a$. In the Cartan basis, the generators are given by $H^i(z)$ and $E^{\pm\alpha}(z)$, where $i = 1, \ldots r_G$ are the indices that identify the generators that commute each other, while the positive roots α denote the creation and annihilation operators. On this basis the highest weight states that form a vacuum representation satisfy

$$H_n^i\,|\,\lambda\,\rangle = E_n^{\pm\alpha}\,|\,\lambda\,\rangle = 0, \quad n > 0. \tag{13.4.23}$$
$$H_0^i\,|\,\lambda\,\rangle = \lambda^i\,|\,\lambda\,\rangle, \;\; E_0^{\alpha}\,|\,\lambda\,\rangle = 0, \;\; \alpha > 0.$$

The remaining states are obtained acting on the state $|\,\lambda\,\rangle$ either by $E_0^{-\alpha}$ or by any mode $\mathcal{J}_{-n}^a$, with $n > 0$.

The constant $\tilde{k}$ of the Kac–Moody algebra depends on the chosen normalization of the structure constants. Hence it is convenient to consider the following quantity $k = 2\tilde{k}/\psi^2$, called the level of the affine algebra, that is independent from the normalization of the structure constants. In this expression ψ denotes the highest root of the algebra and ψ^2 its norm. For the unitary conformal theories, the constant k is quantized and takes only integer value. To show this, it is convenient to consider firstly the case $G = SU(2)$. With the normalization $f^{abc} = \sqrt{2}\epsilon^{abc}$ and $\psi^2 = 2$, the generators are given by

$$I^{\pm} = \frac{1}{\sqrt{2}}(\mathcal{J}_0^1 \pm i\mathcal{J}_0^2) \;\; I^3 = \frac{1}{\sqrt{2}}\mathcal{J}_0^3. \tag{13.4.24}$$

They satisfy

$$\left[I^+, I^-\right] - 2I^3, \;\; \left[I^3, I^{\pm}\right] = \pm I^{\pm}. \tag{13.4.25}$$

With the chosen normalization, the operator $2I^3$ has integer eigenvalues on any finite-dimensional representation of the group. There is however another set of operators that fulfill the same algebra $SU(2)$ given by

$$\begin{aligned} \tilde{I}^+ &= \frac{1}{\sqrt{2}}(\mathcal{J}_{+1}^1 - i\mathcal{J}_{+1}^2) \\ \tilde{I}^- &= \frac{1}{\sqrt{2}}(\mathcal{J}_{-1}^1 + i\mathcal{J}_{-1}^2) \\ \tilde{I}^3 &= \frac{1}{2}k - \frac{1}{\sqrt{2}}\mathcal{J}_0^3. \end{aligned} \tag{13.4.26}$$

It is easy to show that they satisfy the relations $[\tilde{I}^+,\tilde{I}^-]=2\tilde{I}^3$, $[\tilde{I}^3,\tilde{I}^\pm]=\pm\tilde{I}^\pm$. These commutation relations imply that also the operator $2\tilde{I}^3 = k-2I^3$ possesses integer eigenvalues and, consequently, k is an integer number, $k\in\mathbf{Z}$.

The argument presented above is not only valid for $SU(2)$ but also for any other algebra $\mathcal{G}$. In fact, the highest root ψ always results in an $SU(2)$ sub-algebra, generated by

$$I^\pm = E_0^{\pm\psi}, \quad I^3 = \psi\cdot H_0/\psi^2. \tag{13.4.27}$$

This sub-algebra is accompanied by another $SU(2)$ sub-algebra given by

$$\tilde{I}^\pm = E_{\pm 1}^{\mp\psi} \quad \tilde{I}^3 = (\tilde{k}-\psi\cdot H_0)/\psi^2 \tag{13.4.28}$$

so that, repeating the steps of the previous argument, we arrive at the conclusion that also in this case the level $k=2\tilde{}/\psi^2=2\tilde{I}^3+2I^3$ can take only integer values.

Conformal weights and constraint thereof. Let us now compute the conformal weights of the vacuum representations. Eqn. (13.4.12) yields

$$\begin{aligned}L_0\,|\,(r)\,\rangle &= \frac{1/2}{\tilde{k}+C_A/2}\sum_{a,m}:\mathcal{J}^a_m\mathcal{J}^a_{-m}:\,|\,(r)\,\rangle \\ &= \frac{1/2}{\tilde{k}+C_A/2}\sum_a R^a_{(r)}R^a_{(r)}\,|\,(r)\,\rangle = \frac{C_r/2}{\tilde{k}+C_A/2}\,|\,(r)\,\rangle,\end{aligned} \tag{13.4.29}$$

where C_r is the Casimir in the (r) representation. Thus, the conformal weight of the multiplet made of the primary fields $\varphi_{(r)}(z)$ is

$$\Delta_r = \frac{C_r/2}{\tilde{k}+C_A/2} = \frac{C_r/\psi^2}{k+\tilde{h}_G}, \tag{13.4.30}$$

where $\tilde{h}_G = C_A/\psi^2$ is the dual Coxeter number. However not all the representations can be accepted. To understand the constraint to which they are subjected, let us consider once again the $SU(2)$ case. For this algebra, the vacuum states transform in the spin j representation and therefore

$$L_0\,|\,(j)\,\rangle = \frac{j(j+1)}{k+2}\,|\,(j)\,\rangle. \tag{13.4.31}$$

Fixed k, the only values of the spin j that can appear in this formula are those that satisfy the condition

$$2j\le k. \tag{13.4.32}$$

To this aim, let us analyse in more detail the (j) representation. The $(2j+1)$ states of this representation are identified by their eigenvalue with respect to I^3, namely $I^3 \,|\,(j),m\rangle = m\,|\,(j),m\rangle$. Consider then the state with the maximum value of m, i.e. $m = j$, and the matrix element

$$\langle j\,|\,\tilde{I}^+\tilde{I}^-\,|j\rangle = \langle j\,|\left[\tilde{I}^+,\tilde{I}^-\right]|j\rangle = \langle j\,|\,(k-2I^3)\,|j\rangle = k-2j \geq 0. \tag{13.4.33}$$

Hence, eqn. (13.4.32) implies that for a fixed value of k, there are only $k+1$ possible values of j, given by $j = 0, \frac{1}{2}, \ldots, \frac{k}{2}$.

We see that the condition can be generalized (13.4.32) to all other groups. Instead of $|j\rangle$, the state $|\,\lambda\,\rangle$, where λ is the highest weight of the vacuum representation. Using the previous argument, we arrive at the constraint

$$2\psi\cdot\lambda/\psi^2 \leq k. \tag{13.4.34}$$

This is the condition that determines the representations that can appear in the algebra at a fixed value of k.

With the identification of the primary fields of the Kac–Moody algebras, the remaining states that form the Verma modules of these theories are obtained by acting on the primary fields by the operators $\mathcal{J}^a_{-n}$. As for the conformal theories with $c < 1$, these representations contain certain null vectors, that is necessary to mode out in order to define the irreducible representations. For the affine algebras, we can show that all the null vectors are descendent of only one primitive null vector. For a generic affine algebra, this state is constructed using the generators (13.4.28) of the sub-algebra $SU(2)$. Note, in fact, that the eigenvalues of $2\tilde{I}^3$ on the state with highest weight $|\,(r),\lambda\,\rangle$ are given by $M = k - 2\psi\cdot\lambda/\psi^2$. The set of states generated acting by subsequent powers of $\tilde{I}^-$ on $|\,(r),\lambda\,\rangle$ form then an irreducible and finite dimensional representation of the algebra (13.4.28). Hence M is an integer number and we have

$$(\tilde{I}^-)^{M+1}\,|\,(r)\,\lambda,\rangle = 0. \tag{13.4.35}$$

This is precisely the primitive null vector of the Verma module. For the group $SU(2)$ and its j representation, this condition translates into the equation

$$(\mathcal{J}^+_{-1})^{k-2j+1}\,|\,(j),j\rangle = 0. \tag{13.4.36}$$

Correlation functions. Let us address now the correlation functions of the primary fields. As shown below, they satisfy a linear differential equation of the first order. Consider, in fact, the Sugawara formula for the generator L_{-1} of the Virasoro algebra[3]

[3] Each term in the normal order product appears twice and this cancels the factor 1/2 in the formula (13.4.12).

$$L_{-1} = \frac{1}{\tilde{k}+C_A/2}(\mathcal{J}^a_{-1}\mathcal{J}^a_0 + \mathcal{J}^a_{-2}\mathcal{J}^a_1 + \cdots). \tag{13.4.37}$$

Acting on a primary field, it yields

$$\left(L_{-1} - \frac{\sum_a \mathcal{J}^a_{-1} R^a_{(r)}}{\tilde{k}+C_A/2}\right)\varphi_{(r)} = 0. \tag{13.4.38}$$

Consider now the Ward identity (13.4.10) and multiply both terms of this expression for $R^a_{(r_k)}$. Taking the limit $z \to z_k$ and using the OPE of the currents, we arrive at the linear differential equation, Kniznik–Zamolodchikov equation

$$\left[\left(\tilde{k}+C_A/2\right)\frac{\partial}{\partial z_k} + \sum_{a,j\neq k}\frac{R^a_{(r_j)}R^a_{(r_k)}}{z_j - z_k}\right]\langle\varphi_{r_1}(z_1)\cdots\varphi_{r_n}(z_n)\rangle = 0. \tag{13.4.39}$$

To obtain the final expression of the correlator we solve this equation using the correct asymptotic expansion, together with the one relative to the anti-analytic part, and impose the monodromy invariant condition on the solutions.

13.4.3 Wess–Zumino–Witten Models

The conformal models that satisfy a Kac–Moody algebra differ from the other conformal models for an important property: they can be consistently defined by a lagrangian formalism based on a non-linear sigma model with a topological term. This section presents the main steps of this derivation. Consider initially the action

$$\mathcal{S}_0 = \frac{1}{4\lambda^2}\int d^2x \,\mathrm{Tr}\,(\partial^\mu g^{-1}\partial_\mu g), \tag{13.4.40}$$

where λ^2 is a dimensionless positive constant. The bosonic field $g(x)$ is a matrix with values in a semi-simple Lie group g. To have a real action, $g(x)$ must belong to a unitary representation of such a group. For the trace, we adopt the normalization

$$\mathrm{Tr}\,(t^a t^b) = 2\delta^{ab}, \tag{13.4.41}$$

where t_a are the generators of the Lie algebra in the representation under consideration. Note that if g is a unitary matrix, $g^{-1}\partial_\mu g$ is an anti-Hermitian matrix since

$$(g^{-1}\partial_\mu g)^\dagger = \partial_\mu g^{-1} g = -g^{-1}\partial_\mu g, \tag{13.4.42}$$

and $\partial_\mu g^{-1} = -g^{-1}\partial_\mu g g^{-1}$, where the last relation comes from the identity $\partial_\mu(gg^{-1}) = 0$.

Although the theory above is conformal invariant at the classical level, it is well known that this invariance is broken at the quantum level by the renormalization procedure. For the ultraviolet divergences we are forced to introduce a length scale and therefore the $\beta(\lambda)$ function is different from zero. At the quantum level, the theory becomes asymptotically free and its spectrum is purely massive.

The breaking of conformal invariance of the action (13.4.40) at the quantum level can be directly checked by the absence of conserved currents that are purely analytic and anti-analytic. Under the variation $g \to g + \delta g$, we have

$$\delta S_0 = \frac{1}{2\lambda^2} \int d^2x \,\mathrm{Tr}\left[g^{-1}\delta g\, \partial^\mu (g^{-1}\partial_\mu g)\right], \tag{13.4.43}$$

and therefore we get the equations of motion

$$\partial^\mu (g^{-1}\partial_\mu g) = 0. \tag{13.4.44}$$

They can be interpreted as the conservation law of the currents

$$\mathcal{J}_\mu = g^{-1}\partial_\mu g. \tag{13.4.45}$$

Switching to complex coordinates and introducing the notation

$$\bar{\mathcal{J}}_z = g^{-1}\partial_z g, \quad \bar{\mathcal{J}}_{\bar{z}} = g^{-1}\partial_{\bar{z}} g, \tag{13.4.46}$$

we have

$$\partial_z \bar{\mathcal{J}}_{\bar{z}} + \partial_{\bar{z}} \bar{\mathcal{J}}_z = 0. \tag{13.4.47}$$

In order to have a separate conservation law of the two compoents of the currents, it is necessary that each of the two terms of this equation vanishes separately. However, this is impossible, because this would lead to some inconsistencies. In fact, assuming that $\partial_z(g^{-1}\partial_{\bar{z}} g) = 0$, we would also have

$$\partial_z \partial_{\bar{z}} g = \partial_{\bar{z}} g\, g^{-1}\, \partial_z g. \tag{13.4.48}$$

The left-hand side is clearly symmetric under the exchange $z \leftrightarrow \bar{z}$ and this would imply the identity

$$\partial_{\bar{z}} g\, g^{-1}\, \partial_z g = \partial_z g\, g^{-1}\, \partial_{\bar{z}} g. \tag{13.4.49}$$

However this identity is generically false for the elements of a non-commutative group, since it would correspond to the equality $ABC = CBA$, with $A = \partial_{\bar{z}} g$, $B = g^{-1}$ and $C = \partial_z g$.

In order to have a separate conservation of the analytic and anti-analytic components, the correct choice is

$$\mathcal{J}_z = \partial_z g g^{-1}, \quad \mathcal{J}_{\bar{z}} = g^{-1} \partial_{\bar{z}} g. \tag{13.4.50}$$

In this case, the conservation of one quantity implies the conservation of the other

$$\partial_z (g^{-1} \partial \bar{z} g) = g^{-1} \partial_{\bar{z}} (\partial_z g g^{-1}) g. \tag{13.4.51}$$

Hence, the question is whether it is possible to modify the action (13.4.40) in such a way that the conserved currents become those defined by eqn. (13.4.50) instead of those given in eqn. (13.4.46). There is indeed a positive answer and the way to implement it is to use the Wess–Zumino term

$$\Gamma = -\frac{i}{24\pi} \int_B d^3 y \epsilon^{ijk} \operatorname{Tr}\left(\tilde{g}^{-1} \frac{\partial \tilde{g}}{\partial y_i} \tilde{g}^{-1} \frac{\partial \tilde{g}}{\partial y_j} \tilde{g}^{-1} \frac{\partial \tilde{g}}{\partial y_k} \right). \tag{13.4.52}$$

This expression needs an explanation. Imagine the original complex plane, with the point at infinity, compactified into the Riemann sphere S. The matrix g is then a map of the surface S onto the group G. However, this map can be extended to a new map $\tilde{g}(y)$, from all the *internal* points of the three-dimensional sphere B, with boundary given by the surface S, onto the group G, as shown in Figure 13.4. The new matrix $\tilde{g}$ is the one that appears in eqn. (13.4.52), where the coordinates of the three-dimensional sphere are denoted by y_1, y_2 and y_3.

The Wess-Zumino term (13.4.52) has the important property to be defined up to an additive quantity that is an integer multiple of 2π. This ambiguity comes from the existence of topological distinct ways of extending the original map g to the map $\tilde{g}$ that involve the internal points of the three-dimensional sphere. Although the expression (13.4.52) is a three-dimensional integral, its integrand is a total derivative and therefore its final value depends only on the values of $\tilde{g}$ at the boundary, i.e. on the original function g. To understand the origin of the ambiguity of Γ, let us consider the case $G = SU(2)$. The parameter space of this group is a three-dimensional sphere, whose parameterization

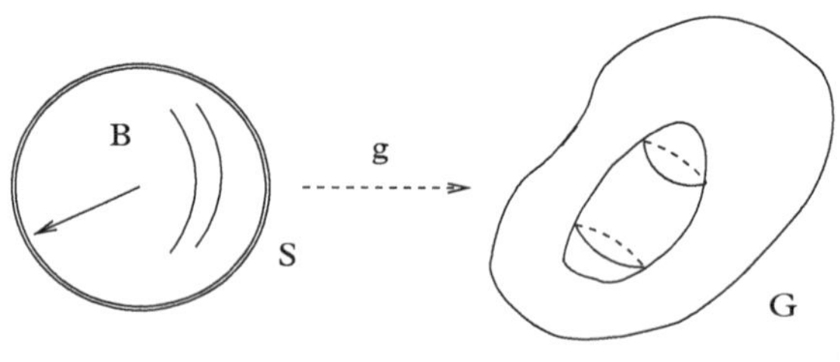

Fig. 13.4 *The map* $\tilde{g}$ *of the three-dimensional sphere B (with boundary given by the two-dimensional surface S) onto the group G.*

is given by the angles ψ, θ and φ, with a line element

$$ds^2 = d\psi^2 + \sin^2\theta\,(d\theta^2 + \sin^2\theta\, d\varphi^2). \tag{13.4.53}$$

Using the parameterization of the matrix g in terms of ψ, θ, φ and the Pauli matrices

$$\begin{aligned} g &= \exp\left(\frac{i}{2}\varphi\,\sigma_3\right) \exp\left(\frac{i}{2}\theta\,\sigma_1\right) \exp\left(\frac{i}{2}\psi\,\sigma_3\right) \\ &= \begin{pmatrix} \cos(\theta/2)\exp[i(\varphi+\psi)/2] & i\sin(\theta/2)\exp[i(\varphi-\psi)/2] \\ i\sin(\theta/2)\exp[i(\psi-\varphi)/2] & \cos(\theta/2)\exp[-i(\varphi+\psi)/2] \end{pmatrix}, \end{aligned} \tag{13.4.54}$$

it is easy to see that the integrand in eqn. (13.4.52) corresponds to the Jacobian of the transformation from the coordinates (ψ, θ, φ) to (y_1, y_2, y_3)

$$\Gamma_{SU(2)} = \frac{i}{4\pi}\int d^3y \frac{\partial(\psi,\theta,\varphi)}{\partial(y_1,y_2,y_3)} = \frac{i}{4\pi}\int d^2x \epsilon_{\mu\nu}\varphi\,\sin\theta\,\partial_\mu\theta\,\partial_\nu\psi \tag{13.4.55}$$

and this explicitly shows that Γ depends only on the boundary values of $\tilde{g}$. However, the result of the integration cannot be expressed in a local form in terms of g. This matrix is in fact periodic in φ, whereas Γ is not: when φ changes of $2\pi n$, Γ changes in

$$\Delta\Gamma = i\frac{n}{2}\int d^2x \epsilon_{\mu\nu}\sin\theta\,\partial_\mu\theta\,\partial_\nu\psi. \tag{13.4.56}$$

The last integral is however an integer, since it expresses the number of times the vector field $\vec{n} = (\cos\theta, \sin\theta\,\cos\psi, \sin\theta\,\sin\psi)$ wraps the three-dimensional sphere.

It is important to stress that the explicit result shown for $SU(2)$ also applies to all other semi-simple Lie groups, for a topological theorem due to Bott. For this ambiguity of the Wess–Zumino term, the coupling constant that multiplies Γ must be necessarily an integer, here denoted by k. Hence, let us consider the new action

$$\mathcal{S} = \mathcal{S}_0 + k\Gamma, \tag{13.4.57}$$

and its variation under $g \to g + \delta g$. For $\delta\mathcal{S}_0$ we have the previous result (13.4.43), whereas for $\delta\Gamma$ we have

$$\delta\Gamma = \frac{i}{8\pi}\int d^2x \epsilon_{\mu\nu}\,\mathrm{Tr}\,(g^{-1}\delta g\,\partial^\mu(g^{-1}\partial^\nu g)). \tag{13.4.58}$$

Put together the two terms, the equation of motion becomes

$$\partial^\mu(g^{-1}\partial_\mu g) + i\frac{\lambda^2 k}{4\pi}\epsilon_{\mu\nu}\partial^\mu(g^{-1}\partial^\nu g) = 0 \tag{13.4.59}$$

that, in complex coordinates, can be written as

$$\left(1+\frac{\lambda^2 k}{4\pi}\right)\partial_z(g^{-1}\partial_{\bar{z}}g)+\left(1-\frac{\lambda^2 k}{4\pi}\right)\partial_{\bar{z}}(g^{-1}\partial_z g)=0. \tag{13.4.60}$$

This equation shows that, choosing

$$\lambda^2=\frac{4\pi}{k}, \tag{13.4.61}$$

we have the desired conservation law

$$\partial_z(g^{-1}\partial_{\bar{z}}g)=0. \tag{13.4.62}$$

Since λ^2 is a positive quantity, the integer k is positive as well. Choosing the other solution, $\lambda^2=-4\pi/k$ with $k<0$, we obtain instead the conservation of the dual current, $\partial_{\bar{z}}(g^{-1}\partial_z g)=0$. With this choice of the coupling constant, the solution of the equation of motion assumes the factorized form

$$g(z,\bar{z})=h(z)\bar{h}(\bar{z}), \tag{13.4.63}$$

where $h(z)$ and $\bar{h}(\bar{z})$ are two arbitrary functions. The separated conservation law of the analytic and anti-analytic components of the currents implies furthermore the invariance of the action under the transformation

$$g(z,\bar{z})\to\mathcal{G}(z)g(z,\bar{z})\bar{\mathcal{G}}^{-1}(\bar{z}), \tag{13.4.64}$$

where $\mathcal{G}$ and $\bar{\mathcal{G}}$ are two arbitrary matrices of the group G, in the same representation of g. For infinitesimal values, we have

$$\mathcal{G}(z)\simeq 1+\omega(z),\quad \bar{\mathcal{G}}(\bar{z})\simeq 1+\bar{\omega}(\bar{z}),$$

and

$$\delta_\omega g=\omega g,\quad \delta_{\bar{\omega}}g=-\bar{\omega}g.$$

With the choice (13.4.61), the variation of the action under $g\to g+\delta_\omega g+\delta_{\bar{\omega}}g$ is given by

$$\begin{aligned}\delta S&=\frac{k}{4\pi}\int d^2x\,\mathrm{Tr}\left(g^{-1}\delta g\left[\partial_z(g^{-1}\partial_{\bar{z}}g)\right]\right)\\&=\frac{k}{2\pi}\int d^2x\,\mathrm{Tr}\left[\omega(z)\partial_{\bar{z}}(\partial_z g g^{-1})-\bar{\omega}(\bar{z})\partial_z(g^{-1}\partial_{\bar{z}}g)\right],\end{aligned} \tag{13.4.65}$$

that clearly vanishes after an integration by parts. Therefore, the original global symmetry $G \times G$ of the sigma model, in the presence of the Wess–Zumino term, is enhanced with the choice (13.4.61) to a local symmetry $G(z) \times G(\bar{z})$. The analytic currents

$$\begin{aligned} \mathcal{J}(z) &\equiv -k\mathcal{J}_z(z) = -k\,\partial_z g g^{-1}, \\ \bar{\mathcal{J}}(\bar{z}) &\equiv k\mathcal{J}_{\bar{z}}(\bar{z}) = kg^{-1}\,\partial_{\bar{z}} g, \end{aligned} \tag{13.4.66}$$

give rise to the Kac–Moody current algebra of the previous section, where k is the same integer that enters the OPE (13.4.1).

This scenario can be explicitly confermed by a perturbative computation of the β function. For instance, for the group $SO(N)$ we get

$$\beta(\lambda) = -\frac{\lambda^2(N-2)}{4\pi}\left[1 - \left(\frac{\lambda^2 k}{4\pi}\right)^2\right], \tag{13.4.67}$$

and this function has a fixed point at $\lambda^2 = \left|\frac{4\pi}{k}\right|$, as shown in Figure 13.5. At these values of the coupling constant the correlation length of the model diverges and the theory acquires a conformal symmetry described by the Kac-Moody algebra.

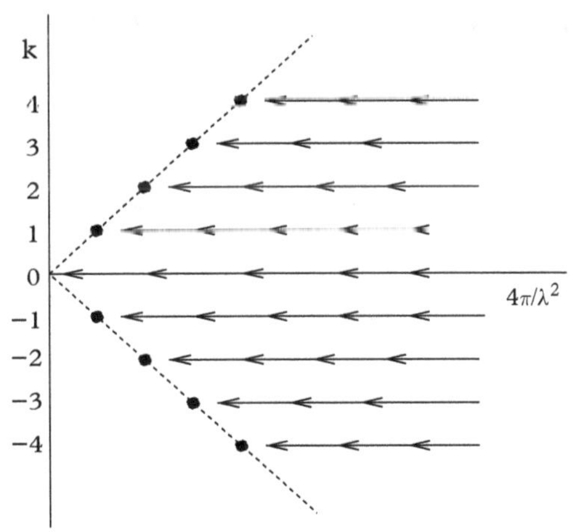

Fig. 13.5 *Renormalization group flows of the coupling constant λ^2. The strong coupling region is on the right of the graph. The coupling constant stops its growing at the fixed points of the β function, i.e. $\lambda^2 = \left|\frac{4\pi}{k}\right|$.*

13.5 Conformal Models as Cosets

The conformal theories associated to the Kac–Moody algebra are useful to construct a vast class of models. The method that we are going to present here, known as *coset approach*, is based on a simple observation. Consider a group G and one of its subgroup H. The currents associated to the original group will be generically denoted by $\mathcal{J}_G^a$, while those of H by $\mathcal{J}_H^i$, where the index i assumes values on the adjoint representation of H, namely $i = 1, \dots, |H|$, where $|H| = \dim H$. Using the Sugawara formula, we can construct the two stress-energy tensors associated to these groups[4]

$$T_G = \frac{1/2}{k_G + \tilde{h}_G} \sum_{a=1}^{|G|} : \mathcal{J}_G^a(z) \mathcal{J}_G^a(z) : \tag{13.5.1}$$

and

$$T_H = \frac{1/2}{k_H + \tilde{h}_H} \sum_{i=1}^{|H|} : \mathcal{J}_H^i(z) \mathcal{J}_H^i(z) : . \tag{13.5.2}$$

For the OPE of the currents $\mathcal{J}_H^i$ with both stress-energy tensors we have

$$\begin{aligned} T_G(z_1)\mathcal{J}_H^i(z_2) &= \frac{\mathcal{J}_H^i(z_2)}{(z_1 - z_2)^2} + \frac{\partial \mathcal{J}_H^i(z_2)}{z_1 - z_2} + \cdots \\ T_H(z_1)\mathcal{J}_H^i(z_2) &= \frac{\mathcal{J}_H^i(z_2)}{(z_1 - z_2)^2} + \frac{\partial \mathcal{J}_H^i(z_2)}{z_1 - z_2} + \cdots . \end{aligned} \tag{13.5.3}$$

As a consequence, the OPE of $(T_G - T_H)$ with $\mathcal{J}_H^i$ does not have singular terms. Since T_H is entirely constructed in terms of the currents $\mathcal{J}_H^i$, $T_{G/H} \equiv T_G - T_H$ has also an operator expansion without singular terms with T_H. Posing

$$T_G = (T_G - T_H) + T_H \equiv T_{G/H} + T_H, \tag{13.5.4}$$

we have an orthogonal decomposition of the original Virasoro algebra—associated to T_G—in two Virasoro algebras that commute each other—associated to $T_{G/H}$ and T_H respectively. The central charge of the Virasoro algebra associated to $T_{G/H}$ is thus given by

$$c_{G/H} = c_G - c_H = \frac{k_G |G|}{k_G + \tilde{h}_G} - \frac{k_H |H|}{k_K + \tilde{h}_H}. \tag{13.5.5}$$

[4] In the following we assume it holds the normalization $\psi^2 = 1$.

A significant class of CFTs is obtained by the coset $(G \times G)/G$, where the group G in the denominator corresponds to the diagonal subgroup of the two groups in the numerators. Denoting by $\mathcal{J}^a_{(1)}$ and $\mathcal{J}^a_{(2)}$ the currents in the two groups of the numerators, for those of the denominator we have $\mathcal{J}^a = \mathcal{J}^a_{(1)} + \mathcal{J}^a_{(2)}$. The most singular part of their operator expansion is given by

$$\mathcal{J}^a(z_1)\mathcal{J}^b(z_2) \simeq \mathcal{J}^a_{(1)}(z_1)\mathcal{J}^a_{(1)}(z_2) + \mathcal{J}^a_{(2)}(z_1)\mathcal{J}^a_{(2)}(z_2) \simeq \frac{(k_1+k_2)\,\delta^{ab}}{(z_1-z_2)^2} + \cdots \tag{13.5.6}$$

and therefore the level of G at the denominator is $k = k_1 + k_2$. An important example of this construction is

$$G/H = SU(2)_{k-1} \times SU(2)_1 / SU(2)_k. \tag{13.5.7}$$

The central charge of these theories is

$$c_{G/H} = \frac{3(k-1)}{k+1} + 1 - \frac{3k}{k+2} = 1 - \frac{6}{(k+1)(k+2)}. \tag{13.5.8}$$

Note that, with the position $q = k+1 = 3, 4, \ldots$, these values coincide with those of eqn. (11.3.2), i.e. the same central charge of the unitary minimal models of the Virasoro algebra!

Another significant example is obtained by considering

$$G/H = SU(2)_{k-1} \times SU(2)_2 / SU(2)_{k+1}$$

whose central charge is

$$c_{G/H} = \frac{3(k-1)}{k+1} + \frac{3}{2} - \frac{3(k+1)}{k+3} = \frac{3}{2}\left(1 - \frac{8}{(k+1)(k+3)}\right) \tag{13.5.9}$$

These are the values of the central charge of the minimal unitary superconformal models, given in eqn. (13.2.16).

Finally, let us analyse how to obtain the states of the model associated to the coset G/H. To this aim, it is necessary to study the decomposition of the representations of G in the splitting (13.5.4) of the stress-energy tensors. Let $|c_G, \lambda_G\rangle$ be the representations of the affine algebra associated to G, where c_G is the central charge relative to the level k_G and λ_G is the highest weight of the vacuum representation. Since $T_G = T_{G/H} + T_H$, these representations decompose into a direct sum of the irreducible representations

$$|c_G, \lambda_G\rangle = \oplus_j \left[|c_{G/H}, \Delta^j_{G/H}\rangle \otimes |c_H, \lambda^j_H\rangle \right], \tag{13.5.10}$$

where $|c_{G/H}, \Delta^{\jmath}_{G/H}\rangle$ denotes the irreducible representation of $T_{G/H}$ with the lowest eigenvalue of L_0 given by $\Delta^{j}_{G/H}$. Some significant examples of this formula will be discussed in the next chapter.

13.5.1 Relation with parafermions

There is an important relationship between the Kac–Moody theories based on the group $SU(2)$ and the parafermionic models. This relationship can be established as follows. Let us initially introduce a free massless boson satisfying the equation $\partial\,\bar{\partial}\varphi = 0$

$$\varphi(z, \bar{z}) = \phi(z) + \bar{\phi}(\bar{z}),$$

with correlators

$$\begin{aligned}
\langle\phi(z)\phi(0)\rangle &= 2\log z, \\
\langle\bar{\phi}(z)\bar{\phi}(0)\rangle &= 2\log\bar{z}, \\
\langle\phi(z)\phi(0)\rangle &= 0
\end{aligned}$$

Its stress-energy tensor $T_b(z) = (\partial\phi)^2$ generates a Virasoro algebra with central charge $c = 1$. Suppose that, in addition to this bosonic field, there are also the parafermionic fields associated to $\mathbf{Z}_N \times \tilde{\mathbf{Z}}_N$, which are decoupled by φ. In terms of the operators of both theories let us construct the currents

$$\begin{aligned}
\mathcal{J}^3(z) &= N\,\partial\phi(z), \\
\mathcal{J}^+(z) &= N\,\psi_1(z) : e^{i/N^{1/2}\,\phi(z)}, \\
\mathcal{J}^-(z) &= N\,\psi^\dagger : e^{-i/N^{1/2}\,\phi(z)}.
\end{aligned} \tag{13.5.11}$$

It is easy to check that the conformal weights of these currents are $(1,0)$: this is obvious for $\mathcal{J}^3$, for the other two currents their conformal weight is given by the sum of the conformal weights of the two fields

$$\Delta_\pm = \frac{N-1}{N} + \frac{1}{N} = 1.$$

Using the operator expansion of the fields ψ_1, $\psi_1^\dagger$ and the vertex operator of the bosonic field ϕ, one can check that these currents satisfy

$$\mathcal{J}^a(z_1)\mathcal{J}^b(z_2) = \frac{N\,q^{ab}}{(z_1 - z_2)^2} + \frac{f^{ab}_c}{z_1 - z_2}\mathcal{J}^c(z_2) + \cdots \tag{13.5.12}$$

where $q^{00} = 1/2q^{+-} = 1/2q^{-+} = 1$ whereas f_c^{ab} are the structure constants of $SU(2)$

$$f_+^{0+} = -f_+^{+0} = -f_-^{0-} = f_-^{-0} = 1,$$
$$f_0^{+-} = -f_-^{-+} = 2.$$

Hence these currents give rise to a Kac-Moody algebra $SU(2)$ of level $k = N$. The stress-energy tensor of such a theory is the sum of the stress-energy of the free bosonic theory and the one of the parafermionic model

$$T_t(z) = T_b(z) + T_{pf}(z). \tag{13.5.13}$$

and the central charge is the sum of the central charges of the two theories

$$c_t = 1 + \frac{2(N-1)}{N+2} = \frac{3N}{N+2}. \tag{13.5.14}$$

This indeed coincides with the central charge of the Kac—Moody algebra $SU(2)$ of level $k = N$. In the light of this result, the parafermionic models $\mathbf{Z}_N$ can be considered as the coset theory $SU(2)_N/U(1)$. This permits to identify the fields of the parafermionic theory in terms of the decomposition of the representations of $SU(2)_N$ with respect to the sub-group $U(1)$, an observation that greatly simplifies the computation of the correlation functions of the parafermionic models.

Appendix 13.A. Lie Algebra

In this appendix we remind the main results of the Lie algebra, and interested readers are referred to the chapter references for more information on the subject. First of all, for any compact Lie group with n parameters there is a Lie algebra of dimension n, and vice versa. For the compact groups there are the following properties: (a) there is always a unitary representation; (b) any irreducible representation is finite-dimensional; (c) in order to find a representation of the group it is sufficient to find a representation of the algebra.

A Lie algebra $\mathcal{G}$ of dimension n is a vector space with an internal composition law given

$$(\lambda_i, \lambda_j) \to [\lambda_i, \lambda_j] = \sum_k c_{ij}^k \lambda_k, \tag{13.A.1}$$

where c_{ij}^k are the structure constants of the algebra and $[,]$ is the commutator. This composition law satisfies the Jacobi identity

$$[x, [y, z]] + [z, [x, y]] + [y, [z, x]] = 0. \tag{13.A.2}$$

A representation of the Lie algebra is obtained by associating each of its elements x to a matrix $M(x)$, with the condition $M([x, y]) = [M(x), M(y)]$. Particularly important is the *adjoint representation* given by $x \to ad(x)$, where $ad(x)$ is a linear application of $\mathcal{G}$ in itself, defined by

$$ad(x)y = [x,y]. \tag{13.A.3}$$

For the Jacobi identity, it holds $[ad(x), ad(y)] = ad([x, y])$. In terms of this representation we can define a bi-linear form, i.e. a scalar product among the elements of the algebra, by the formula

$$\langle x|y\rangle = \mathrm{Tr}\,(ad(x)\,ad(y)). \tag{13.A.4}$$

An invariant sub-space under the adjoint representation is called an *ideal* $\mathcal{I}$ of $\mathcal{G}$, namely $y \in \mathcal{I}$ if $ad(x)y = [x,y] \in \mathcal{I}$, for every $x \in \mathcal{G}$. The ideals are crucial for the further analysis of the Lie algebras. In fact, there are three classes of algebras:

1. The simple Lie algebras, that have no ideals at all.
2. The semi-simple Lie algebras, that do not have abelian ideals.
3. All other algebras

Presently there is a complete mathematical theory only for the first two classes. Let us now introduce another useful concept: a sub-algebra $\mathcal{C}$ is a Cartan sub-algebra if it has the properties: (a) $\mathcal{C}$ is a maximal abelian sub-algebra, i.e. there is no other sub-algebra that contains $\mathcal{C}$; (b) if $h \in \mathcal{C}$, then in any representation of $\mathcal{C}$ on a complex vector space $A(h)$ is a diagonalizable operator. The dimension r di $\mathcal{C}$ is the *rank* of $\mathcal{G}$. Let us now recall, without giving proofs, the theory of semi-simple Lie algebras.

Let $\mathcal{G}$ be a n-dimensional Lie algebra (with complex coefficients) and $\mathcal{C}$ its Cartan sub-algebra of dimension r.

- Any operator $ad(h_i)$ with $h_i \in \mathcal{C}$ is diagonalizable in $\mathcal{G}$. Since $[h_i, h_j] = 0$, there exists a set of common eigenvectors $e_{\alpha_1,\ldots,\alpha_r}$, with

$$ad(h_i)e_{\alpha_1,\ldots,\alpha_r} = \alpha_i\, e_{\alpha_1,\ldots,\alpha_r}.$$

- The h_i can be always chosen (by an appropriate choice of the basis) in such a way that the eigenvalues α_i are all real. The r-dimensional vector $\alpha = (\alpha_1, \ldots, \alpha_r)$ is called *root*. The algebra $\mathcal{G}$ can be written as a direct sum $\mathcal{G} = \mathcal{C} \oplus_a \mathcal{G}_a$, where $\mathcal{C}$ corresponds to the null root $(0, \ldots, 0)$ while $\mathcal{G}_a$ corresponds to the vector sub-space associated to the non vanishing root a. It is possible to prove that this is a one-dimensional space. Hence, there are $n - r$ non-vanishing roots.

- Consider the restriction of the scalar product (A.4) in $\mathcal{C}$, namely

$$g_{ij} = \mathrm{Tr}\,(ad(h_i)\,ad(h_j)).$$

In the basis $\{h_i, e_\alpha\}$, the operators $ad(h_i)$ are diagoonal and therefore $g_{ij} = \sum_\alpha \alpha_i \alpha_j$. Since $g_{ij} = g_{ji}$ and g_{ij} is a real matrix, it can be diagonalized. Moreover, we can show that g_{ij} is a non-singular matrix positive defined. Hence, introducing its inverse by the definition $g_{ij}g^{jk} = \delta_i^k$, we can define a scalar product among the roots

$$\langle \alpha | \beta \rangle = \sum_i \alpha^i \beta_i = \sum_{i,j} g^{ij} \alpha_i \beta_j. \tag{13.A.5}$$

we can always choose a basis in which $g_{ij} = \delta_{ij}$, so that $\langle \alpha|\beta\rangle = \sum_i \alpha_i \beta_i$. As we shall see soon, in the basis $\{h_i, e_\alpha\}$ all the commutation relations of the Lie algebra are fixed by the roots.

Roots. The roots are the building blocks of the Lie algebras. They satisfy a series of properties enumerate below

1. If α is a root, then $k\alpha$ is a root only $k = 0, \pm 1$. Hence the $n - r$ roots come in pair and we have $n - r = 2m$.
2. If α and β are two roots, they uniquely identify two non-negative integers p and q such that $\beta - p\alpha, \beta - (p-1)\alpha, \ldots, \beta + q\alpha$ are the only roots of the form $\beta + k\alpha$. This series of roots is called the *string α containing β*. Exchanging α with β, we can identify two other non-negative integers p' and q' that characterize the string β containing α. These numbers satisfy

$$p - q = 2\frac{\langle \alpha|\beta\rangle}{\langle \alpha|\alpha\rangle}, \quad p' - q' = 2\frac{\langle \alpha|\beta\rangle}{\langle \beta|\beta\rangle}. \tag{13.A.6}$$

Since

$$-p \leq (q - p) \leq q, \quad -p' \leq (q' - p') \leq q'$$

if α and β are two non vanishing roots we have that

$$\beta - 2\frac{\langle \alpha|\beta\rangle}{\langle \alpha|\alpha\rangle}\alpha, \quad \alpha - 2\frac{\langle \alpha|\beta\rangle}{\langle \beta|\beta\rangle}\beta,$$

are also non vanishing roots. Note that the first is obtained by reflecting β with respect to the orthogonal plane to α, while the second reflecting α with respect to the orthogonal plane to β.

3. Since

$$ad(h_i)[e_\alpha, e_\beta] = [h_i, [e_\alpha, e_\beta]] = (\alpha_i + \beta_i)[e_\alpha, e_\beta]$$

there are the following cases

a. $\alpha + \beta \neq 0$, with $\alpha + \beta$ not a root. In this case $[e_\alpha, e_\beta] = 0$, otherwise $[e_\alpha, e_\beta]$ would be an eigenvector of $ad(h_i)$ and $\alpha + \beta$ a root.

b. $\alpha + \beta = 0$, in this case $[e_\alpha, e_{-\alpha}] \in C$ and then it can be written as

$$[e_\alpha, e_{-\alpha}] = \sum_i \lambda^i h_i. \tag{13.A.7}$$

Choosing the normalization $\langle e_\alpha | e_{-\alpha} \rangle = 1$ (that determines the roots up to a factor d_α such that $d_\alpha d_{-\alpha} = 1$), we have $\lambda^i = \alpha^i$.

c. $\alpha + \beta \neq 0$, but with $\alpha + \beta$ a root. Since the space of the eigenvectors is one-dimensional, $[e_\alpha, e_\beta] = N_{\alpha,\beta} e_{\alpha+\beta}$ and the coefficient $N_{\alpha,\beta}$ satisfies the conditions

$$N_{\alpha,\beta} = N_{\beta,-\alpha-\beta} = N_{-\alpha-\beta,\alpha} = -N_{\beta,\alpha}. \tag{13.A.8}$$

From the normalization condition $\langle e_\alpha | e_{-\alpha} \rangle = 1$, we can always choose d_α in such a way that $N_{\alpha,\beta} = -N_{-\alpha,-\beta}$ and, in this case, we arrive at the condition

$$N^2_{\alpha,\beta} = \frac{q(p+1)}{2} \langle \beta | \beta \rangle. \tag{13.A.9}$$

This relation determines $N_{\alpha,\beta}$ up to a sign, which can be chosen to satisfy the relations (13.A.8).

In summary, all the commutation relations of the Lie algebra are encoded in the following formulae

$$\begin{aligned}
[h_i, h_j] &= 0, \\
[h_i, e_{\pm\alpha}] &= \pm\alpha_i e_{\pm\alpha}, \\
[e_\alpha, e_{-\alpha}] &= \sum_i \alpha^i h_i, \\
[e_\alpha, e_\beta] &= \begin{cases} 0 & \text{if } \alpha+\beta \neq 0 \text{ and } \alpha+\beta \text{ is not a root} \\ N_{\alpha,\beta} e_{\alpha+\beta} & \text{if } \alpha+\beta \neq 0 \text{ and } \alpha+\beta \text{ is a root} \end{cases}.
\end{aligned} \tag{13.A.10}$$

As anticipated, the roots of a Lie algebra uniquely fix its structure. Hence the classification of the Lie algebras reduces to study the vector space of dimension r that satisfy the properties discussed.

Simple roots. A root is called *positive* if its first non vanishing component is positive. A root is simple if: (a) it is a positive root; (b) it cannot be written as sum of positive roots. The simple roots have two important properties that are easy to prove: (i) if α and β are simple roots, then $\alpha - \beta$ is not a root; (ii) $\langle\alpha|\beta\rangle \leq 0$ and moreover

$$2\frac{\langle\alpha|\beta\rangle}{\langle\alpha|\alpha\rangle} = p - q = -q, \tag{13.A.11}$$

since, for the point (a), $p = 0$.

The utility of the simple roots is stated by the following theorem: there are exactly r simple roots, all linearly independent, and any other positive root can be written as their linear combination. In addiiton, if α is a positive root but not simple, there always exists a simple root $\alpha^{(k)}$ so that $\alpha - \alpha^{(k)}$ is a positive root. These two properties ensures that all the roots of the algebra can be determined in terms of the simple roots. From eqn. (13.A.6) we infer that there are severe constraints on the angle between two roots and the ratio of their lengths. In fact, since

$$2\frac{\langle\alpha|\beta\rangle}{\langle\alpha|\alpha\rangle} = m, \quad 2\frac{\langle\alpha|\beta\rangle}{\langle\beta|\beta\rangle} = n \tag{13.A.12}$$

we have

$$\frac{(\langle\alpha|\beta\rangle)^2}{\langle\alpha|\alpha\rangle\,\langle\beta|\beta\rangle} = \frac{mn}{4} = \cos^2\varphi_{\alpha,\beta} \leq 1 \tag{13.A.13}$$

and, if $m, n \neq 0$,

$$\frac{\langle\alpha|\alpha\rangle}{\langle\beta|\beta\rangle} = \frac{n}{m}. \tag{13.A.14}$$

If we now specialize these equations to the case in which α and β are simple roots, we have both $m, n < 0$ and there are only the following cases

m	n	φ	$\langle\alpha\vert\alpha\rangle/\langle\beta\vert\beta\rangle$
–1	–1	120°	1
–1	–2	135°	2
–1	–3	150°	3
0	0	90°	arbitrary

The scalar product of the simple roots defines the *Cartan matrix*

$$A_{ij} = \frac{2\langle\alpha_i|\alpha_j\rangle}{\langle\alpha_j|\alpha_j\rangle}. \tag{13.A.15}$$

The matrix elements of A_{ij} are necessarily integers and its diagonal elements are equal to 2. If the roots do not have the same length, A_{ij} is not a symmetric matrix. It is convenient to introduce a special notation for the quantity $2\alpha_i/|\alpha_i|^2$, with $|\alpha_i|^2 = \langle\alpha_i|\alpha_i\rangle$

$$\alpha_i^\vee = \frac{2\alpha_i}{|\alpha_i|^2}. \tag{13.A.16}$$

Hence the Cartan matrix can be elegantly written as $A_{ij} = \langle\alpha_i|\alpha_j^\vee\rangle$. Let us also define the *dual Coxeter number*, given

$$\tilde{h}_G = \sum_{i=1}^{r} \alpha_i^\vee + 1. \tag{13.A.17}$$

Since any semi-simple Lie algebra is the direct sum of simple algebras, it is sufficient to discuss the classification of the latter ones.

Classification of the simple Lie algebras. This problem consists of finding all sets of r simple roots that satisfy the condition discussed, with none of them orthogonal to the others. The fundamental result of the theory can be expressed in a graphical way in terms of the Dynkin diagrams. In fact, since the length of the simple roots can at most take two values, let us associate a circle to each root. Two circles are linked by one, two or three lines according whether their angle is equal to 120°, 135° or 150° respectively. If the two roots are orthogonal the relative circles are not connected. The black circles are associated to the shorter roots. The final classification of the simple Lie algebras is given in Figure 13.6.

These algebras are all distinct when $r \geq 4$. Note that:

1. When $r = 1$ there is only one Lie algebra, A_1.
2. When $r = 2$ the Dynkin diagrams of B_2 and C_2 are identical, therefore the two algebras coincide.
3. When $r = 3$ A_3 and D_3 have the same Dynkin diagram, so $A_3 = D_3$.
4. There are four families of algebras with an arbitrary large number of simple roots: the series A_r, that corresponds to the group of the unitary matrices $SU(r+1)$, the series B_r, relative to the group of the orthogonal matrices $O(2r+1)$, the series C_r relative to the sympletic matrices $Sp(2r)$ (These are the linear transformations U that leave invariant an anti-symmetric non-singular matrix I, namely $U^t I U = I$), the series D_r that corresponds to the group of the orthogonal matrices $O(2r)$. In

Group	Algebra	Dynkin diagram	Dimension
SU(r+1)	A_r		r (r+2) r ≥ 1
O(2 r +1)	B_r	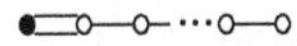	r (2 r +1) r ≥ 2
Sp(2 r)	C_r		r (2 r +1) r ≥ 2
O(2 r)	D_r	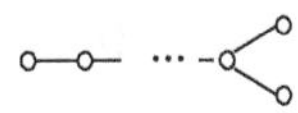	r (2 r −1) r ≥ 3
	G_2		14
	F_4	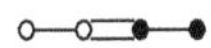	52
	E_6	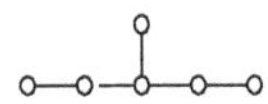	78
	E_7	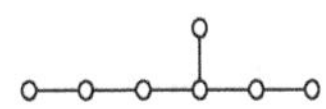	133
	E_8	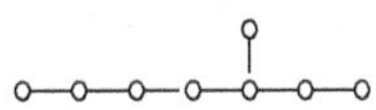	248

Fig. 13.6 *Simple Lie algebras and Dynkin diagrams.*

additional to these families, there are five exceptional algebras, called G_2, F_4, E_6, E_7 and E_8.

5. Among the Lie algebras, only A_n, D_n and the three exceptional algebras E_6, E_7, E_8 have roots all of the same length. These algebras are known as *simply laced algebras*.

Let us discuss now the representation theory.

Representation theory. Let us recall that a representation of a Lie algebra on a complex vector field L is defined by a linear map $x \to T(x)$, where $x \in \mathcal{G}$ and T is an operator that acts in L, such that $T([x,y]) = [T(x), T(y)]$. In the following we only deal with the finite-dimensional representations, for which applies the Weyl theorem: any finite-dimensional representation of a semi-simple Lie algebra is completely reducible. Hence we can restrict our attention only to the irreducile representations.

Choosing a basis $\{h_i, e_\alpha, e_{-\alpha}\}$ in $\mathcal{G}$, let $\{H_i, E_\alpha, E_{-\alpha}\}$ be the corresponding operators in a given representation. It is always possible to implement the conditions $H_i = H_i^\dagger$ and $E_\alpha^\dagger = E_{-\alpha}$. The H_is are a set of Hermitian operators that commute each other. Hence they can be simultaneously diagonalized and their eigevalues are real. Let $M = (M_1, \ldots, M_r)$ the set of eigenvalues on a common eigenvectors of the H_i

$$H_i \,|M\rangle = M_i \,|M\rangle. \tag{13.A.18}$$

M can be regarded as a r-dimensional real vector and it is called the *weight vector*. Denoting by L_M the space of the eigenvectors associated to the weight M, the vector space L decomposes as

$$L = \oplus L_M. \tag{13.A.19}$$

In general, the spaces L_M are not one-dimensional and therefore the operators H_i do not form a complete set of commuting operators. Therefore some of the weights M can be degenerate. There is no a general procedure to remove this degeneracy. However, it is possible to show that the number of operators, that commute with all H_i that permits to remove such a degeneracy is at most equal to $(n-3r)/2$.

Properties of the weight vectors. If $|M\rangle$ is a vector of L_M, from the commutation relations of H_i with E_α we get

$$H_i E_\alpha |M\rangle = (\alpha_i + M) E_\alpha |M\rangle. \tag{13.A.20}$$

Supposing that $E_\alpha|M\rangle \neq 0$, we have that also $M+\alpha = (M_1+\alpha_1,\ldots,M_r+\alpha_r)$ is a weight and $E_\alpha|M\rangle$ belongs to $L_{M+\alpha}$. If $E_\alpha E_\alpha|M\rangle \neq 0$, we can repeat the same reasoning to conclude that also $M+2\alpha$ is a weight, with $E_\alpha^2|M\rangle$ belonging to $L_{M+2\alpha}$. For recurrence, if $E_\alpha^k|M\rangle \neq 0$, then $M+k\alpha$ is a weight and $E_\alpha^k|M\rangle \in L_{M+k\alpha}$. However, since L is a finite dimensional space, such a procedure must stop, i.e. it should exist an integer q such that $E_\alpha^q|M\rangle \neq 0$ (therefore $M+q\alpha$ is a weight) but $E_\alpha^{q+1}|M\rangle = 0$. Repeating the same steps with $E_{-\alpha}$ we can determine an integer p such that $E_{-\alpha}^p|M\rangle \neq 0$ but $E_{-\alpha}^{p+1}|M\rangle = 0$. From this we derive that these vectors

$$M - p\alpha, \ldots, M + q\alpha \tag{13.A.21}$$

are all and only the weight vectors of the form $M+k\alpha$. The operators E_α and $E_{-\alpha}$ are the raising and decreasing operators of the spectrum. In terms of the tensor g^{ij} we can introduce a scalar product among the weights and the roots

$$\langle M|\alpha\rangle = \sum_{ij} g^{ij}\alpha_i M_j,$$
$$\langle M|M'\rangle = \sum_{ij} g^{ij} M_i M'_j.$$

If p and q are the integers previously introduced, we have

$$\frac{2\langle M|\alpha\rangle}{|\alpha|^2} = p - q,$$

and therefore

$$M - \frac{2\langle M|\alpha\rangle}{|\alpha|^2}$$

is a weight. It is worth that the considerations done above are very similar to those used for the roots—a circumstance not surprising since the roots are nothing else but the weight vectors of a particular representation, the adjoint.

Since the r simple roots $\alpha^{(i)}$ form a basis in the r-dimensional space of the real vectors, any weight can be expressed in terms of them as

$$M = \sum_{i=1}^{r} M_i \alpha^{(i)}. \tag{13.A.22}$$

We can introduce an order in this space. We say that $M > M'$ if the first component of the vector $M - M'$ is positive. For the finite number of distinct weights, there exists then a *highest weight vector*, i.e. a weight that is greater than the other. As a consequence of this definition, if α is a positive root and $|\Lambda\rangle$ is an eigenvector belonging to the space of the highest weight, then $E_\alpha|\Lambda\rangle = 0$.

Let R be the representation of $\mathcal{G}$ in the linear space L, and $|\Lambda,1\rangle$, $|\Lambda,2\rangle \ldots |\Lambda,k\rangle$ a set of independent vectors belonging to the space of the highest weight Λ. Consider the sub-space $L^{(1)}$ defined by the vectors

$$E_{-\alpha}E_{-\beta}\ldots|\Lambda,1\rangle, \tag{13.A.23}$$

obtained by applying a finite product of $E_{-\alpha}$ (including the repetition of the same operator) where $\alpha,\beta,\ldots$ are positive roots. It is easy to see that this is an invariant and irreducible space. It is obviously invariant under the action of the operators H_i and $E_{-\alpha}$, while applying one of the operator E_α (with $\alpha > 0$), this can be moved, using its commutation relations, to the end of the product, where we get $E_\alpha|\Lambda,1\rangle = 0$. Doing so, we generate a sequence of vectors having the form (13.A.23). If the representation R is irreducible we have then $R = L^{(1)}$. In $L^{(1)}$ there is only one independent vector with highest weight Λ, all other weights have the form

$$\Lambda - \sum_{\alpha>0} k_\alpha \alpha, \tag{13.A.24}$$

where k_α are integer numbers, equal to the number of times in which the operator $E_{-\alpha}$ appears in (13.A.23).

The importance of the concept of the highest weight is stressed by the following theorems due to Cartan. The first theorem states that two irreducible representation that have the same highest weight are equivalent. The second theorem states that a r dimensional vector Λ is the highest weight vector of an irreducible representation if and only if

$$\Lambda_{\alpha_i} = \frac{2\langle\Lambda|\alpha^{(i)}\rangle}{|\alpha^{(i)}|^2}, \tag{13.A.25}$$

is a non-negative integer for any simple root $\alpha^{(i)}$. Hence, once we choose a set of simple roots $\alpha^{(i)}$, any set of non-negative integers $(\Lambda_{\alpha_1}, \Lambda_{\alpha_2}, \ldots, \Lambda_{\alpha_r})$ uniquely defines an irreducible representation of $\mathcal{G}$ and all representations are obtained in this way. The other weights have the form (13.A.24) and are obtained by applying the decreasing operators $E_{-\alpha}$.

Other useful formulae. Finally, we discuss some formulae entering the formalism of the Kac–Moody algebras. The constant $C_A/2$ that appears in the expression of the stress-energy tensor and the central charge generally depends on the chosen normalization of the structure constants f^{abc}. Let $R^a_{(r)}$ the matrices of a representation (r) of G, with dimension d_r and normalization

$$\mathrm{Tr}\, R^a_{(r)} R^b_{(r)} = l_r \delta^{ab}. \tag{13.A.26}$$

Summing on the indices a and b, in the range $1, \ldots |G|$, we get

$$C_r d_r = l_r |G| \tag{13.A.27}$$

where C_r is the quadratic Casimir in the representation r. Summing instead only on the indices of the Cartan sub-algebra of G $(a, b = 1, \ldots, r_G)$, we get

$$\sum_{j=1}^{d_r} \mu^2_{(j)} = l_r r_G \tag{13.A.28}$$

where r_G is the rank of G and μ are the weights of the representation (r).

For the adjoint representation, we have $d_A = |G|$ and

$$C_A = l_{(A)} = r_G^{-1} \sum_{a=1}^{|G|} \alpha^2_{(a)} \tag{13.A.29}$$

where α are the roots. Denoting by ψ the highest root, the quantity $\tilde{h}_G \equiv C_A/\psi^2$ is independent from the normalization and it is expressed by

$$\tilde{h}_G = \frac{C_A}{\psi^2} = \frac{1}{r_G}\left(n_L + \left(\frac{S}{L}\right)^2 n_S\right) \tag{13.A.30}$$

In this formula $n_{S,L}$ is the number of the short (long) roots of the algebra (the highest root ψ is always a long root) whereas S/L is the ratio of their lengths. As shown, for the Lie algebras the roots can have at most two different lengths. The quantity $\tilde{h}_G$ is the *dual Coxeter number*, previously defined by the formula (13.A.17).

SU(n) $(n \geq 2)$	$\tilde{h}_{SU(n)} = n$	$l_{(n)} = \frac{1}{2}\psi^2$
SO(n) $(n \geq 4)$	$\tilde{h}_{SO(n)} = n-2$	$l_{(n)} = \psi^2$
E_6	$\tilde{h}_{E_6} = 12$	$l_{(27)} = 3\psi^2$
E_7	$\tilde{h}_{E_7} = 18$	$l_{(56)} = 6\psi^2$
E_8	$\tilde{h}_{E_8} = 30$	$l_{(248)} = 30\psi^2$
Sp(2n) $(n \geq 1)$	$\tilde{h}_{Sp(2n)} = n+1$	$l_{2n} = \frac{1}{2}\psi^2$
G_2	$\tilde{h}_{G_2} = 4$	$l_{(7)} = \psi^2$
F_4	$\tilde{h}_{F_4} = 9$	$l_{26} = 3\psi^2$

Table 13.1 *Dual Coxeter numbers of the Lie algebras.*

The simply laces algebras (A, D, E) have simple roots of the same length. The remaining algebras have roots of two different lengths and their ratio L/S is $\sqrt{2}$ for $SO(2n+1), Sp(2n)$ and F_4, while is $\sqrt{3}$ for G_2.

We can now easily compute all dual Coxeter numbers for the compact Lie algebras (Table 13.1).

REFERENCES

For a general introduction to supersymmetry and its application we suggest

Cooper, F., Khare, A. and Sukhatme, U. (2001). *Supersymmetry in Quantum Mechanics*, World Scientific.

Efetov, K. (1999). *Supersymmetry in Disorder and Chaos*, Cambridge University Press.

Sonhius, M.F. (1985). 'Introducing Supersymmetry', *Physics Reports*, 128: 39.

The two-dimesional superconformal models are discussed in

Friedan, D. Qiu, Z. Shenker, S. (1985). 'Superconformal Invariance in 2 Dimensions and the Tricritical Ising Model', *Physics Letters B*, 151: 37.

Cappelli, A. (1987). 'Modular Invariant Partition Functions of Superconformal Theories', *Physics Letters B*, 185: 82.

Mussardo, G., Sotkov, G. and Stanishkov, M. (1989). 'N=2 Superconformal Minimal Models', *International Journal of Modern Physics A*, 4: 1135.

Qiu, Z. (1986). 'Supersymmetry, Two-dimensional Critical Phenomena and the Tricritical Ising Model', *Nuclear Physics B*, 270: 205.

The parafermionic models and the lattice statistical systems with $\mathbf{Z}_N$ are discussed in

Fateev, V. and Zamolodchikov, A.B. Parafermionic currents in the two-dimensional conformal quantum eld theory and self-dual critical points in Z(n) statistical systems, Soviet Physics JETP 62 (1985), 215; Physics Letters 92 A (1982), 37.

QFTs with a Wess–Zumino term have been studied by many authors. We recommend the fundamental articles by

Polyakov, A. and Wiegman, P. (1983). 'Goldstone Fields in Two Dimensions with Multivalued Actions', *Physics Letters B*, 141: 223.

Witten, E. (1984). 'Non-abelian Bosonization in Two Dimensions, *Communications in Mathematical Physics*, 92: 455.

The conformal models with a Kac–Moody symmetry were originally proposed in

Knizhnik, V.G. and Zamolodchikov, A.B. Current algebra and Wess-Zumino model in two dimensions, Nucl. Phys. B 247 (1984), 63.

The construction of the conformal models using the coset approach is found in

Goddard, P., Kent, A. and Olive, D. (1985). 'Virasoro Algebra and Coset Space Models', *Physics Letters B*, 152: 88.

Goddard, P., Kent, A. and Olive, D. (1986). Unitary Representations of the Virasoro and Super-Virasoro Algebras, *Communications in Mathematical Physics*, 103: 105.

For readers interested ni group theory and the Lie algebra we recommend

Cornwell, J.F. (1989). *Group Theory in Physics*, Academic Press Limited.

Wybourne, B. (1973). *Classical Groups for Physicists*, John Wiley & Sons.

PROBLEMS

13.1. Spontaneous supersymmetry breaking

Let Q be the generator of a $N=1$ supersymmetric theory and $Q^\dagger$ its adjoint operator. With a proper normalization we have

$$\{Q, Q^\dagger\} = H,$$

where H is the Hamiltonian of the system.

a. Show that the Hamiltonian of a supersymmetric theory contains no negative eigenvalues.

b. Show that any state whose energy is not zero cannot be invariant under a supersymmetry transformation.

c. Show that supersymmetry is spontaneously broken if and only if the energy of the lowest lying state (the vacuum) is not exactly zero.

d. Consider the two-dimensional superconformal models on a cylinder, for which $Q = G_0$ and $H = Q^2 = L_0 - c/24$. Show that in the first model of the minimal unitary series, given by the tricritical Ising model, supersymmetry is broken while in the second minimal model, given by the Gaussian model, is exact.

13.2. Central charge of the parafermions

On a physical basis argue why in the limit $N \to \infty$ the central charge of the parafermionic systems is equal to $c = 2$.

13.3. Polyakov–Wiegman identity

Consider the action of the sigma model with a Wess-Zumino topological term

$$S(g) = \frac{k}{16\pi} \int d^2x \, \mathrm{Tr}\,(\partial^\mu g^{-1} \partial_\mu g) + k\Gamma.$$

Prove the identity

$$S(gh^{-1}) = S(g) + S(h) + \frac{k}{2\pi} \int d^2x \, \mathrm{Tr}\,(g^{-1} \partial_{\bar{z}} g \, h^{-1} \partial_z h).$$

Show that this identity gives rise to the invariance of the action under the transformation

$$g(z, \bar{z}) \to \mathcal{G}(z) g(z, \bar{z}) \mathcal{G}^{-1}(\bar{z}).$$

13.4. Correlation functions of the currents

For the conformal models with a Kac–Moody algebra, compute the-point correlation functions of the analytic currents

$$\langle \mathcal{J}^a(z_1) \mathcal{J}^b(z_2) \mathcal{J}^c(z_3) \mathcal{J}^d(z_4) \rangle.$$

13.5. Bosonization of the $SU(2)_1$ theory

Verify that the central charge of the theory $SU(2)_1$ is $c = 1$. Compute the spectrum of the conformal weights of this theory and determine a representation of the corresponding conformal fields in terms of the vertex operators of a bosonic field φ.

14

The Arena of Conformal Models

Madamina il catalogo è questo.
(Leporello, in *Don Juan*)

14.1 Introduction

This chapter looks at some significant minimal conformal models. As shown in Chapter 11, these models provide explicit examples of exactly solved QFTs: of these theories we know the operator content, the fusion rules of their fields, the corresponding structure constants, the correlation functions of the order parameters and, finally, their modular invariant partition function on a torus. Despite this large knowledge, there is still an important open problem, namely the identification of the classes of universality they are describing. Is there a way to associate these exactly solved critical theories to the continuum limit of lattice statistical models? Unfortunately there is no a direct method to answer this question: the identification of the various classes of universality can be achieved only by the comparison of the critical exponents predicted by CFT with the values obtained by the exact solution of the models defined on a lattice, further supporting this identification on the basis of the symmetry of the order parameters. This has been the approach followed, for instance, by Huse who identified a particular critical regime of the lattice RSOS models solved by Andrew, Baxter and Forrester with the unitarity minimal models of CFT. Rather than going into a technical analysis of this identification, instead this chapter analyses in detail the first minimal models (in the following denoted, in general cases, by $\mathcal{M}_{p,q}$ and $\mathcal{M}_q$ for the unitary cases), in particular those corresponding to the Ising model, the tricritical Ising model and the Yang–Lee model. We also discuss the 3-state Potts model as an example of a statistical model associated to a partition function of the type (A, D), according to the notation introduced in Chapter 11. Finally, we study the statistical models of geometric type (as, for instance, those that describe the self-avoiding walks) and their formulation in terms of conformal minimal models.

14.2 The Ising Model

Consider the first minimal unitary conformal model, obtained by substituting $q = 3$ in eqn. (11.3.2). Such a model has the central charge $c = \frac{1}{2}$ and the Kac table is reported in the table 14.1 below.

Statistical Field Theory: An Introduction to Exactly Solved Models in Statistical Physics. Second Edition. Giuseppe Mussardo, Oxford University Press (2020). © Giuseppe Mussardo. DOI: 10.1093/oso/9780198788102.001.0001

$\frac{1}{2}$	$\frac{1}{16}$	0
0	$\frac{1}{16}$	$\frac{1}{2}$

Table 14.1 *Kac table of the minimal unitary model $\mathcal{M}_3$.*

To denote the operator content of this theory let us introduce the notation[1]

$$\begin{aligned} \mathbf{1} &= (0,0) \\ \psi &= \left(\tfrac{1}{2},0\right) \\ \bar{\psi} &= \left(0,\tfrac{1}{2}\right) \\ \epsilon &= \left(\tfrac{1}{2},\tfrac{1}{2}\right) \\ \sigma &= \left(\tfrac{1}{16},\tfrac{1}{16}\right) \\ \mu &= \left(\tfrac{1}{16},\tfrac{1}{16}\right). \end{aligned} \tag{14.2.1}$$

Below we present a series of arguments to show that the CFT described by this minimal model corresponds to the exact solution of the two-dimensional Ising model at its critical point.

The first indication comes from the numerical values of the Kac table. Assuming that the scalar field σ can be associated to the continuum limit of the magnetization field of the two-dimensional Ising model, for the corresponding critical index η the value is

$$\eta = \frac{1}{4}, \tag{14.2.2}$$

and coincides with the exact value known for this critical index from the exact lattice solution. Analogously, assuming that the scalar field ϵ describes the continuum limit of the energy operator of the two-dimensional Ising model (i.e. the conjugate operator to the temperature displacement $|T - T_c|$), we can derive the critical exponents ν and α

$$\nu = 2 - 2\Delta_\epsilon = 1, \quad \alpha = 2 - 1/(1 - \Delta_\epsilon) = 0 \tag{14.2.3}$$

Also in this case, these quantities coincide with their known exact values obtained by the lattice solution.

A further support to the hypothesis that the class of universality is the one of the Ising model comes from the skeleton form of the fusion rules. Using the results of Chapter 11, the operator algebra that involves the fields σ, μ and the chiral field ψ (with analogous relations for the anti-chiral field $\bar{\psi}$) is given by

[1] $(\Delta, \bar{\Delta})$ are the conformal weights provided by the Kac table.

$$\begin{aligned} \psi\,\psi &= \mathbf{1} \\ \psi\,\sigma &= \mu \\ \psi\,\mu &= \sigma. \end{aligned} \tag{14.2.4}$$

These relations show that ψ is a fermionic field (here subject to anti-periodic boundary conditions) and that the operators σ and μ play the role of order and disorder fields. The fermionic structure present in the conformal model $\mathcal{M}_3$ perfectly matches the fermionic structure identified in the lattice version of the Ising model, discussed in Chapter 9, where we showed that the continuum limit of the Ising model corresponds to a free fermionic theory for a Majorana field, with central charge $c = \frac{1}{2}$.

For the algebra of the scalar fields we have

$$\begin{aligned} \sigma\,\sigma &= \mathbf{1} + \epsilon \\ \mu\,\mu &= \mathbf{1} + \epsilon \\ \epsilon\,\sigma &= \sigma \\ \epsilon\,\mu &= \mu \\ \epsilon\,\epsilon &= \mathbf{1}. \end{aligned} \tag{14.2.5}$$

This algebra highlights the Z_2 spin symmetry of the Ising model, under which both σ and μ are odd fields ($\sigma \to -\sigma, \mu \to -\mu$) while ϵ is even, $\epsilon \to \epsilon$. Moreover, at its critical point the Ising model is also invariant under the Kramers–Wannier duality transformation, under which $\epsilon \leftarrow -\epsilon$ and $\sigma \leftrightarrow \mu$. The odd parity of ϵ under the duality transformation naturally explains the absence of ϵ in the OPE of this field with itself.

Finally, note that the algebra (14.2.5) of the scalar fields can be also interpreted as the algebra of the composite operators of a φ^4 Landau–Ginzburg theory—a theory notoriously associated to the class of universality of the Ising model. In fact, following the general discussion presented in Section 11.6, let us pose $\sigma \equiv \varphi$. Using the operator expansion, we have $:\varphi^2: = \epsilon$ and $:\varphi^3: = \partial_z\,\partial_{\bar{z}}\varphi$. This shows that this conformal model provides the exact solution of the field theory associated to the Lagrangian

$$\mathcal{L} = \frac{1}{2}(\partial_\mu \varphi)^2 + g\varphi^4$$

Let us now discuss the correlation fucntions and the structure constants of this model.

14.2.1 Operator Product Expansion and Correlation Functions

If we identify the chiral field ψ with the analytic component of the Majorana fermion of the Ising model and $\bar{\psi}$ with its anti-analytic component, the continuum limit of the energy operator ϵ is given by[2] $\epsilon(z,\bar{z}) = i\,\bar{\psi}(\bar{z})\psi(z)$. The fermionic representation of this operator permits to easily compute all its correlators using the Wick theorem. Since the

[2] The i in this definition is necessary for the anti-commutation rule of the fermionic field and the positivity of the correlation function $\langle \epsilon(z,\bar{z})\epsilon(w,\bar{w})\rangle = \frac{1}{|z-w|^2}$.

Wick theorem always involves the contractions pairwise of different fields, it is easy to see that the only non-zero correlators are those with an even number of fields ϵ. The same conclusion can be reached based on the duality property of the model, since under this transformation $\epsilon \to -\epsilon$ and therefore only the correlation functions with an even number of ϵ can be different from zero. Using the factorization in the analytic and anti-analytic components, we have

$$\begin{aligned} G_{2n} &= \langle \epsilon(z_1,\bar{z}_1)\ldots\epsilon(z_n,\bar{z}_n)\rangle \\ &= (-1)^n \langle \psi(z_1)\bar{\psi}(\bar{z}_1)\ldots\psi(z_n)\bar{\psi}(\bar{z}_n)\rangle \\ &= \langle \psi(z_1)\ldots\psi(z_n)\rangle\,\langle\bar{\psi}(\bar{z}_1)\ldots\bar{\psi}(\bar{z}_n)\rangle. \end{aligned} \tag{14.2.6}$$

For each of the two terms, the Wick theorem leads to the sum of all possible two-point correlation functions multiplied by the sign of the corresponding permutation. The final result can be expressed in terms of a Pfaffian of the $(2n)\times(2n)$ anti-symmetric matrix A, with matrix elements $A_{ij} = -A_{ji} = \langle\psi(z_i)\psi(z_j)\rangle = 1/(z_i - z_j)$. We have then

$$\begin{aligned} \langle \epsilon(z_1,\bar{z}_1)\ldots\epsilon(z_n,\bar{z}_n)\rangle &= \left|\mathrm{Pf}\left[\frac{1}{z_i - z_j}\right]_{1\le i,j\le 2n}\right|^2 \\ &= \det\left[\frac{1}{z_i - z_j}\right] \end{aligned} \tag{14.2.7}$$

since the square of the Pfaffian of an anti-symmetric matrix A is equal to its determinant.

For the computation of the correlation functions that involve the fields σ and μ, we can proceed in two different ways.

- The first method consists of applying the general strategy explained in Chapter 12: the operators σ and μ occupy the position $(1,2)$ in the Kac table and therefore their correlators satisfy a second-order linear differential equation, whose explicit solution can be obtained by using the modified Coulomb gas approach. If we consider, for instance the four-point correlation function[3]

$$F(\eta,\bar{\eta}) = \langle\sigma(\infty)\sigma(1,1)\sigma(\eta,\bar{\eta})\sigma(0,0)\rangle,$$

we get

$$F(\eta,\bar{\eta}) = \left(\frac{1}{\eta\bar{\eta}(1-\eta)(1-\bar{\eta})}\right)^{1/8}\left[\,|\,Y_+(\eta)\,|^2 + |\,Y_-(\eta)\,|^2\right], \tag{14.2.8}$$

[3] We use the Moebius invariance to fix three of the four points of this correlator at the positions $z_1 = \infty$, $z_2 = 1$ and $z_4 = 0$.

where

$$Y_{\pm}(\eta) = \sqrt{1 \pm \sqrt{1-\eta}}.$$

From the analysis of the singularity of this expression for $\eta \to 0$ and the operator expansion

$$\sigma(z_1, \bar{z}_1)\sigma(z_2, \bar{z}_2) = \frac{1}{|z_1 - z_2|^{1/4}} [1 + \cdots] + C^{\epsilon}_{\sigma\sigma} |z_1 - z_2|^{3/4} [\epsilon(z_2, \bar{z}_2) + \cdots]$$

one infers that the function $| Y_+(\eta) |^2$ corresponds to the channel of the identity operator **1** while $| Y_-(\eta) |^2$ to the channel of the operator ϵ. Using the decomposition of the correlators in the conformal blocks showed in Figure 14.1, we arrive at the quadratic equation for the structure constant $C^{\epsilon}_{\sigma\sigma}$

$$\left(C^{\epsilon}_{\sigma\sigma}\right)^2 = \frac{1}{4}.$$

Note that this equation cannot fix the sign of the structure constant: hence our choice to take the positive sign

$$C^{\epsilon}_{\sigma\sigma} = \frac{1}{2} \tag{14.2.9}$$

is purely arbitrary. There is a point, though, with the choice of the sign of the structure constants. In order to appreciate this aspect, it is sufficient to observe that the four-point correlation function of the disorder operator

$$F(\eta, \bar{\eta}) = \langle \mu(\infty)\mu(1,1)\mu(\eta, \bar{\eta})\mu(0,0) \rangle,$$

is expressed in terms of the same function (14.2.8) and, from the singular term of its expression, we arrive at the same quadratic equation for the structure constant $C^{\epsilon}_{\mu\mu}$

$$\left(C^{\epsilon}_{\mu\mu}\right)^2 = \frac{1}{4}.$$

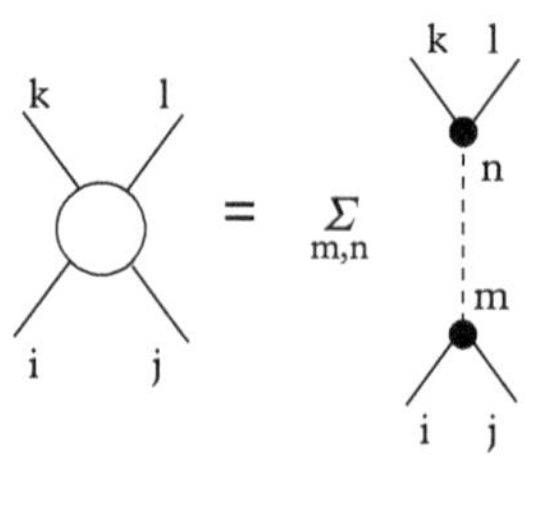

Fig. 14.1 *Expansion of the correlation functions in the conformal blocks.*

However, in this case, we have to choose the negative solution

$$C^{\epsilon}_{\mu\mu} = -\frac{1}{2}. \tag{14.2.10}$$

To prove that this is the right choice, consider the four-point correlation function that involves both field

$$G(\eta, \bar{\eta}) = \langle \mu(\infty)\sigma(1,1)\sigma(\eta,\bar{\eta})\mu(0,0)\rangle.$$

It satisfies the same second-order differential equation fulfilled by the previous correlators. However, its solution must take into account the semi-local property of these fields, i.e. the correlator should acquire a (-1) sign when the variable η is analytically continued along the closed contours that enclose either the origin or the point at infinity. Hence, in this case, the solution is given by

$$G(\eta, \bar{\eta}) = \frac{1}{2}\left(\frac{1}{\eta\bar{\eta}(1-\eta)(1-\bar{\eta})}\right)^{1/8} \left[Y_{+}(\eta)\, Y_{-}(\bar{\eta}) + Y_{-}(\eta)\, Y_{+}(\bar{\eta})\right]. \tag{14.2.11}$$

Studying the singularities that are present in this expression when $\eta \to 1$ we get the equation

$$C^{\epsilon}_{\sigma\sigma}\, C^{\epsilon}_{\mu\mu} = -\frac{1}{4}, \tag{14.2.12}$$

that clearly shows the equal and opposite value of the two structure constants.

Other correlation functions can be computed as well using straightforwardly the Coulomb gas. Instead of presenting these results, let us illustrate another efficient method to compute the correlation functions of the Ising model.

- The second method for computing the correlators of the Ising model is based on the bosonization rules, exploiting the circumstance that the Ising model is a free fermionic theory. As a theory of real Majorana fermions, it cannot be directly bosonized but, if we consider two copies of the same theory, we can define a Dirac fermion theory that can be instead bosonized. Let $i = 1, 2$ the index of each copy of the Ising model. In terms of the two Majorana fermions ψ_1 and ψ_2 (together with their anti-analytic components), we can define the Dirac field as

$$\Psi(z, \bar{z}) = \begin{pmatrix} \chi(z) \\ \bar{\chi}(\bar{z}) \end{pmatrix} = \frac{1}{\sqrt{2}} \begin{pmatrix} \psi_1 + i\psi_2 \\ \bar{\psi}_1 + i\bar{\psi}_2 \end{pmatrix} \tag{14.2.13}$$

and apply the bosonization rule

$$\chi(z) = e^{i\phi(z)}, \quad \bar{\chi}(\bar{z}) = e^{-i\bar{\phi}(\bar{z})}. \tag{14.2.14}$$

It is now essential to provide the bosonization representation of the various fields of the two copies of the Ising model. Let us start from the energy operator of the two-copy model, given by $\tilde{\epsilon} = \epsilon_1 \times \epsilon_2$. Using eqn. (12.4.6) we have

$$\begin{aligned}(\bar{\Psi}\Psi)(z,\bar{z}) &= \psi_1\bar{\psi}_1 + \psi_2\bar{\psi}_2 = \\ &= i(\epsilon_1 + \epsilon_2) = \cos\varphi(z,\bar{z}).\end{aligned} \tag{14.2.15}$$

Since $\psi_1\,\psi_2 = i\partial_z\varphi$ we have also

$$\begin{aligned}\epsilon_1\epsilon_2 &= (i\psi_1\bar{\psi}_1)\,(i\psi_2\bar{\psi}_2) = \psi_1\psi_2\,\bar{\psi}_1\bar{\psi}_2 \\ &= -\partial_z\varphi\,\partial_{\bar{z}}\varphi.\end{aligned} \tag{14.2.16}$$

Using these expressions and the correlators of the bosonic field φ, we easily recover the previous expressions (14.2.7) of the correlators of the ϵ_i operators.

Let us consider now the correlators of the spin fields. For the two-copy model, the spin operator is expressed by the product of the spin operators of each copy, $\tilde{\sigma} = \sigma_1 \times \sigma_2$. Since the two copies do not interact with each other, the correlation functions of $\tilde{\sigma}$ provide the *square* of the correlation functions of the original Ising model. Keeping in mind the conformal weight of the spin field, we can pose

$$\tilde{\sigma} \to \sqrt{2}\cos\frac{\varphi}{2} \tag{14.2.17}$$

and, using the two-point correlation function of this vertex operator, we find

$$\langle\tilde{\sigma}(z,\,\bar{z})\tilde{\sigma}(w,\bar{w})\rangle = \langle\sigma(z,\,\bar{z})\sigma(w,\bar{w})\rangle^2 = \frac{1}{|z-w|^{1/2}}. \tag{14.2.18}$$

Eqn. (14.2.17) enables us to compute all the (square) of the correlators of the field σ of the Ising model. In fact,

$$\begin{aligned}\langle\sigma(z_1,\,\bar{z}_1)\cdots\sigma(z_n,\,\bar{z}_n)\rangle^2 &= 2^{n/2}\left\langle\prod_{i=1}^{n}\cos\frac{\varphi}{2}(z_i,\,\bar{z}_i)\right\rangle \\ &= 2^{-n/2}\sum_{\{\alpha_i=\pm 1\}}\prod_{i<j}|z_i - z_j|^{\alpha_i\alpha_j/2}.\end{aligned} \tag{14.2.19}$$

To characterize the disorder operator $\tilde{\mu} = \mu_1 \times \mu_2$ of the two-copy system, it is necessary to use the duality. Under this transformation $\epsilon \to -\epsilon$, while $\sigma \leftrightarrow \mu$, and therefore in the bosonization formalism this symmetry is implemented by the substitution $\varphi \to \pi - \varphi$. In this way, we arrive at the identification

$$\tilde{\mu}(z, \bar{z}) \to \sqrt{2} \sin \frac{\varphi}{2}(z, \bar{z}). \tag{14.2.20}$$

We can now easily compute the mixed correlator

$$\begin{aligned} &\langle \sigma(z_1, \bar{z}_1)\mu(z_2, \bar{z}_2)\sigma(z_3, \bar{z}_3)\mu(z_4, \bar{z}_4)\rangle^2 = \\ &= \frac{1}{2}\frac{|z_{13}z_{24}|^{1/2}}{|z_{14}z_{23}z_{12}z_{34}|^{1/2}}\left[-1 + \frac{|z_{12}z_{34}|}{|z_{13}z_{24}|} + \frac{|z_{14}z_{23}|}{|z_{13}z_{24}|}\right]. \end{aligned} \tag{14.2.21}$$

Thanks to this expression we can fix the OPE of the order and disorder operators

$$\sigma(z_1, \bar{z}_1)\mu(z_2, \bar{z}_2) = \frac{C^{\psi}_{\sigma\mu}(z_1 - z_2)^{1/2}\,[\psi(z_2) + \cdots] + C^{\bar{\psi}}_{\sigma\mu}(\bar{z}_1 - \bar{z}_2)^{1/2}\,[\bar{\psi}(\bar{z}_2) + \cdots]}{|z_1 - z_2|^{1/4}}.$$

Analysing the limit $z_1 \to z_2$ and $z_3 \to z_4$, we have

$$C^{\psi}_{\sigma\mu} = \frac{e^{i\pi/4}}{\sqrt{2}} \qquad C^{\bar{\psi}}_{\sigma\mu} = \frac{e^{-i\pi/4}}{\sqrt{2}}$$

In conclusion, from the bosonization procedure we get the following OPE

$$\begin{aligned} \psi(z_1)\,\sigma(z_2, \bar{z}_2) &= \frac{e^{i\pi/4}}{\sqrt{2}(z_1 - z_2)^{\frac{1}{2}}}\,\mu(z_2, \bar{z}_2) \\ \psi(z_1)\,\mu(z_2, \bar{z}_2) &= \frac{e^{-i\pi/4}}{\sqrt{2}(z_1 - z_2)^{\frac{1}{2}}}\,\sigma(z_2, \bar{z}_2) \\ \bar{\psi}(\bar{z}_1)\,\sigma(z_2, \bar{z}_2) &= \frac{e^{-i\pi/4}}{\sqrt{2}(\bar{z}_1 - \bar{z}_2)^{\frac{1}{2}}}\,\mu(z_2, \bar{z}_2) \\ \bar{\psi}(\bar{z}_1)\,\mu(z_2, \bar{z}_2) &= \frac{e^{i\pi/4}}{\sqrt{2}(\bar{z}_1 - \bar{z}_2)^{\frac{1}{2}}}\,\sigma(z_2, \bar{z}_2). \end{aligned} \tag{14.2.22}$$

Note that, the basis of σ and μ chosen above, the 2×2 matrix representations of the zero-modes ψ_0 and $\bar{\psi}_0$ of the fermionic field are non-diagonal and given by

$$\psi_0 = \frac{1}{\sqrt{2}}\begin{pmatrix} 0 & e^{i\pi/4} \\ e^{-i\pi/4} & 0 \end{pmatrix}, \qquad \bar{\psi}_0 = \begin{pmatrix} 0 & e^{-i\pi/4} \\ e^{i\pi/4} & 0 \end{pmatrix}.$$

14.2.2 Coset Constructions and E_8 Algebra

At the critical point, remarkably enough, the Ising model can be described by two differents coset conformal models. The first coset is based on the affine algebra of the group $SU(2)$, namely

$$\mathcal{M}_3 = \frac{SU(2)_1 \otimes SU(2)_1}{SU(2)_2}. \qquad (14.2.23)$$

The representations of the affine algebra $SU(2)$ at the level $k=1$ are given by the multiplets of spin $(0)_1$ and $(\frac{1}{2})_1$, with conformal weights equal to 0 and $\frac{1}{4}$, whereas the representations at the level $k=2$ are given by the multiplets of spin $(0)_2$, $(\frac{1}{2})_2$ and $(1)_2$, with conformal weights 0, $\frac{3}{16}$ and $\frac{1}{2}$. The products of the two representations of $SU(2)_1$ decompose as

$$\begin{aligned}
(0)_1 \times (0)_1 &= \left[(0)_{Ising} \otimes (0)_2\right] \oplus \left[\left(\frac{1}{2}\right)_{Ising} \otimes (1)_2\right] \\
(0)_1 \times \left(\frac{1}{2}\right)_1 &= \left(\frac{1}{16}\right)_{Ising} \otimes \left(\frac{1}{2}\right)_2 \\
\left(\frac{1}{2}\right)_1 \times \left(\frac{1}{2}\right)_1 &= \left[(0)_{Ising} \otimes (1)_2\right] \oplus \left[\left(\frac{1}{2}\right)_{Ising} \otimes (0)_2\right],
\end{aligned} \qquad (14.2.24)$$

and therefore one recovers the Kac table of the model.

Quite surprisingly, the second coset construction uses the exceptional group E_8. Consider, in fact, the coset

$$\frac{(E_8)_1 \otimes (E_8)_1}{(E_8)_2}. \qquad (14.2.25)$$

The dual Coxeter number of E_8 is $\tilde{h} = 30$. Using formula (13.5.5), we have $c = \frac{1}{2}$. At level $k=1$ there is only the representation of the identity field, with conformal weight equal to 0. At level 2, there are instead three different representations, here denoted by the symbol Π_i, of the conformal weights

$$(E_8)_2 \quad \rightarrow \quad \{1, \Pi_1, \Pi_7\} = \{0, \tfrac{15}{16}, \tfrac{3}{2}\}. \qquad (14.2.26)$$

In particular, Π_1 is the adjoint representation of the group E_8. Their components, with respect to the basis of the simple roots of E_8 $(n_1, n_2, \cdots, n_8)$ (with n_i integers) are

$$\begin{aligned}
1 &\rightarrow (0,0,0,0,0,0,0,0) \\
\Pi_1 &\rightarrow (1,0,0,0,0,0,0,0) \\
\Pi_7 &\rightarrow (0,0,0,0,0,0,1,0).
\end{aligned} \qquad (14.2.27)$$

The Ising model is obtained by the decomposition

$$(0)_1 \times (0)_1 \quad = \quad \left[(0)_{Ising} \otimes (0)_2\right] \oplus \left[(\tfrac{1}{16})_{Ising} \otimes (\tfrac{15}{16})_2\right] \oplus \left[(\tfrac{1}{2})_{Ising} \otimes (\tfrac{3}{2})_2\right]. \tag{14.2.28}$$

The underlying E_8 structure of the Ising model will be vital to understanding its off-critical behaviour when an external magnetic field is present.

14.2.3 Characters and Partition Function

The Kac table of the minimal model $\mathcal{M}_3$ has three fields and there are correspondingly three different characters of the Virasoro algebra, χ_0, $\chi_{\frac{1}{2}}$ and $\chi_{\frac{1}{16}}$. Their explicit expression can be computed by the Rocha–Caridi formula given in eqn. (11.7.17) but, as we show below, they can be also computed by taking advantage of the fermionic formulation of the model and using the results of Section 12.3.2.

On a cylinder, the fermion has an expansion in half-integer or integer modes according to if it satisfies anti-periodic or periodic boundary conditions along the space direction. Consider initially the anti-periodic case. If $|0>$ is the lowest energy state in this sector, the excited states are given by $\psi_{-n_1}\ldots\psi_{-n_k}|0>$, where $n_i \in \mathbf{Z}+\frac{1}{2}$. We can use the expression of $L_0 = \sum_{n>0} n\,\psi_{-n}\psi_n$ to order their sequence as the growing of their eigenvalues

L_0 eigenvalue	state		
0	$\|0\rangle$		
$\frac{1}{2}$	$\psi_{-1/2}\|0\rangle$		
$\frac{3}{2}$	$\psi_{-3/2}\|0\rangle$		
2	$\psi_{-3/2}\psi_{-1/2}\|0\rangle$		(14.2.29)
$\frac{5}{2}$	$\psi_{-5/2}\|0\rangle$		
3	$\psi_{-5/2}\psi_{-1/2}\|0\rangle$		
$\frac{7}{2}$	$\psi_{-7/2}\|0\rangle$		
4	$\psi_{-7/2}\psi_{-1/2}\|0\rangle$	$\psi_{-5/2}\psi_{-3/2}\|0\rangle$	
$\ldots$	$\ldots$		

We have then

$$\mathrm{Tr}_A\, q^{L_0} = 1+q^{1/2}+q^{3/2}+q^2+q^{5/2}+q^3+q^{7/2}+2q^4+\cdots. \tag{14.2.30}$$

The states (14.2.29) form a representation of the Virasoro algebra with $c=\frac{1}{2}$ but such a representation is reducible for it can be decomposed into the direct sum of the two representations $[0]\oplus\left[\frac{1}{2}\right]$ of the minimal model $\mathcal{M}_3$. First of all, note that the states with conformal weights $\Delta=0$ and $\Delta=\frac{1}{2}$ appear only once in the tower of these states. This means that these conformal families have a multiplicity equal to 1. Furthermore, note that the states that belong to the family [0] are obtained by applying an even number of fermionic fields, while those of the family $\left[\frac{1}{2}\right]$ are obtained by acting on $|0\rangle$ by an odd number of operators ψ_{-n}. These two sets are therefore distinguished by their opposite eigenvalue with respect to the operator $(-1)^F$, and the irreducible representations are recovered by using the projectors $\frac{1}{2}(1 \pm (-1)^F)$

$$\begin{aligned}\chi_0(q) &\equiv q^{-1/48}\,\mathrm{Tr}_{\Delta=0}\,q^{L_0} = q^{-1/48}\,\mathrm{Tr}_A\,\frac{1}{2}(1+(-1)^F)\,q^{L_0}\\ \chi_{\frac{1}{2}}(q) &\equiv q^{-1/48}\,\mathrm{Tr}_{\Delta=\frac{1}{2}}\,q^{L_0} = q^{-1/48}\,\mathrm{Tr}_A\,\frac{1}{2}(1-(-1)^F)\,q^{L_0}.\end{aligned} \tag{14.2.31}$$

Let us now consider the periodic sector of the fermionic field, whose expression of L_0 on the cylinder is given by

$$L_0 = \sum_{n>0} n\psi_{-n}\psi_n + \frac{1}{16} \qquad n \in \mathbf{Z}.$$

The zero mode of the fermionic field has a two-dimensional representation space, spanned by $|\sigma\rangle = |\frac{1}{16}\rangle_+$ and $|\mu\rangle = |\frac{1}{16}\rangle_-$, that have eigenvalues ± 1 with respect to the operator $(-1)^F$. The tower of states in the periodic sector is expressed by

$$\begin{array}{ll} L_0 \text{ eigenvalue} & \text{state} \\ \frac{1}{16}+0 & |\frac{1}{16}\rangle_\pm \\ \frac{1}{16}+1 & \psi_{-1}|\frac{1}{16}\rangle_\pm \\ \frac{1}{16}+2 & \psi_{-2}|\frac{1}{16}\rangle_\pm \\ \frac{1}{16}+3 & \psi_{-3}|\frac{1}{16}\rangle_\pm \quad \psi_{-2}\psi_{-1}|\frac{1}{16}\rangle_\pm. \\ \cdots & \cdots \end{array} \tag{14.2.32}$$

Hence, there are two irreducible representations associated to the two states $|\frac{1}{16}\rangle_\pm$. One may think to separate them using once more the projectors $\frac{1}{2}(1 \pm (-1)^F)$. However in this sector it holds identically $\mathrm{Tr}_R\,(-1)^F\,q^{L_0} = 0$ because at each level there is always the same number of states with equal and opposite fermion number. In conclusion, there is the same expression for the character of the two families (another manifestation of the self-duality of the model) and this is given by

$$\chi_{\frac{1}{16}}(q) \equiv q^{-1/48}\,\mathrm{Tr}_P\,\frac{1}{2}(1 \pm (-1)^F)q^{L_0} = q^{1/24}\,(1+q+q^2+2q^3+\cdots). \tag{14.2.33}$$

Partition functions. We can now use the characters χ_0, $\chi_{\frac{1}{2}}$ and $\chi_{\frac{1}{16}}$ to compute different partition functions on a torus and extract the relative operator content of the model. Adopting the order of the characters given above, the modular matrix $\mathcal{S}$ that implements their transformation under $\tau \to -1/\tau$ is

$$\mathcal{S} = \frac{1}{2}\begin{pmatrix} 1 & 1 & \sqrt{2} \\ 1 & 1 & -\sqrt{2} \\ \sqrt{2} & -\sqrt{2} & 0 \end{pmatrix}. \tag{14.2.34}$$

If we consider the partition function with periodic boundary conditions along both horizontal and vertical axes of the torus, this quantity is given by the diagonal solution of the modular equation

$$Z_{PP}(q) = \mid \chi_0(q) \mid^2 + \mid \chi_{\frac{1}{2}} \mid^2 + \mid \chi_{\frac{1}{16}} \mid^2. \tag{14.2.35}$$

In the presence of these boundary conditions, the operator content of the theory is expressed by the scalar conformal families $\{\mathbf{1}\}$, $\{\epsilon\}$ and $\{\sigma\}$.

We can also use the Z_2 symmetry of the model to implement other boundary conditions. Suppose we want to compute the partition function with periodic boundary conditions along the space axis but with anti-periodic ones along the time axis for the spin field. This corresponds to compute the trace of an operator that implements a change of sign to the conformal family of the spin field $\sigma \to -\sigma$ but that leaves invariant both the identity and energy fields. The final expression is then

$$Z_{AP} = \mid \chi_0 \mid^2 + \mid \chi_{\frac{1}{2}} \mid^2 - \mid \chi_{\frac{1}{16}} \mid^2. \tag{14.2.36}$$

Also in this case the operator content of the theory is expressed by the scalar conformal families $\{\mathbf{1}\}$, $\{\epsilon\}$ and $\{\sigma\}$, with a negative multiplicity of the last family for the given boundary conditions.

We can now use the modular transformation $\tau \to -1/\tau$ that induces a change of the horizontal and vertical axes to compute the partition function with anti-periodic boundary conditions along the horizontal axis and periodic along the vertical axis. Using eqn. (14.2.34) to transform the characters, we have

$$Z_{PA} = \chi^* \chi_{\frac{1}{2}} + \chi_{\frac{1}{2}}^* \chi_0 + \mid \chi_{\frac{1}{16}} \mid^2. \tag{14.2.37}$$

The operator content of the theory with these boundary conditions is expressed by the conformal scalar family $\{\sigma\}$ but, in this case, there are also the chiral and anti-chiral families $\{\psi, \mathbf{1}\}$ and $\{\mathbf{1}, \bar{\psi}\}$.

It is interesting to observe that the combination $Z_{AP} + Z_{PA}$ is invariant, by construction, under the modular transformation $\mathcal{S}$, and it is also invariant under $\mathcal{T}^2$, where $\mathcal{T}$ implements the transformation $\tau \to \tau + 1$. The partition function expressed by this combination

$$Z = Z_{AP} + Z_{PA} = \mid \chi_0 + \chi_{\frac{1}{2}} \mid^2 \tag{14.2.38}$$

corresponds to the operator content of the Ising model given by the fields $\mathbf{1}$, ψ, $\bar{\psi}$ and ϵ that are all mutually local. The spin field is not local with respect to both ψ and $\bar{\psi}$ and is therefore absent in this situation.

14.3 The Universality Class of the Tricritical Ising Model

Let us now discuss the universality class of the tricritical Ising model (TIM), associated to the second unitary minimal model $\mathcal{M}_4$. One of its microscopic realizations is provided by the Blume–Capel model discussed in Section 7.7.2. Equivalently, this class of universality can be associated to a Landau–Ginzburg Lagrangian based on a scalar field φ, a formulation that has the advantage of an easy bookkeeping of the Z_2 symmetry property of each order parameter. The Euclidean action

$$\mathcal{S} = \int d^D x \left[\frac{1}{2} (\partial_\mu \varphi)^2 + g_1 \varphi + g_2 \varphi^2 + g_3 \varphi^3 + g_4 \varphi^4 + \varphi^6 \right], \tag{14.3.1}$$

with the tricritical point identified by the condition $g_1 = g_2 = g_3 = g_4 = 0$. We recall that the statistical interpretation of the coupling constants reads as follows: g_1 plays the role of an external magnetic field h, g_2 measures the displacement of the temperature from its critical value, i.e. $g_2 \sim (T - T_c)$, g_3 may be regarded as a sub-leading magnetic field h' and, finally, g_4 may be interpreted as a chemical potential for the vacancies.

In two dimensions—our current focus—there are strong fluctuations of the order parameters and this implies that the critical exponents and the universal ratios are quite different from their estimates provided by a mean field theory. We can use the conformal theory to obtain an exact solution of this model at its critical point. In fact, it is described by the second unitary minimal model $\mathcal{M}_4$: its central charge is $c = \frac{7}{10}$ and the exact values of its conformal weight are

$$\Delta_{l,k} = \frac{(5l - 4k)^2 - 1}{80}, \qquad \begin{array}{l} 1 \leq l \leq 3 \\ 1 \leq k \leq 4 \end{array} \tag{14.3.2}$$

They are organized in Table 14.2.

$\frac{3}{2}$	$\frac{6}{10}$	$\frac{1}{10}$	0
$\frac{7}{16}$	$\frac{3}{80}$	$\frac{3}{80}$	$\frac{7}{16}$
0	$\frac{1}{10}$	$\frac{6}{10}$	$\frac{3}{2}$

Table 14.2 *Kac table of the unitary minimal model* $\mathcal{M}_4$.

even ∗ *even*	
$\epsilon * \epsilon = [1] + c_1\ [t]$	
$t * t = [1] + c_2\ [t]$	
$\epsilon * t = c_1\ [\epsilon] + c_3\ [\varepsilon'']$	$c_1 = \frac{2}{3}\sqrt{\frac{\Gamma(\frac{4}{5})\Gamma^3(\frac{2}{5})}{\Gamma(\frac{1}{5})\Gamma^3(\frac{3}{5})}}$
	$c_2 = c_1$
even ∗ *odd*	$c_3 = \frac{3}{7}$
$\epsilon * \sigma' = c_4\ [\sigma]$	$c_4 = \frac{1}{2}$
$\epsilon * \sigma = c_4\ [\sigma'] + c_5\ [\sigma]$	$c_5 = \frac{3}{2}c_1$
$t * \sigma' = c_6\ [\sigma]$	$c_6 = \frac{3}{4}$
$t * \sigma = c_6\ [\sigma'] + c_7\ [\sigma]$	$c_7 = \frac{1}{4}c_1$
	$c_8 = \frac{7}{8}$
odd ∗ *odd*	$c_9 = \frac{1}{56}$
$\sigma' * \sigma' = [1] + c_8\ [\varepsilon'']$	
$\sigma' * \sigma = c_4\ [\epsilon] + c_6\ [t]$	
$\sigma * \sigma = [1] + c_5\ [\epsilon] + c_7\ [t] + c_9\ [\varepsilon'']$	

Table 14.3 *The operator product expansion algebra and the relative structure constants.*

There are six scalar primary fields and, out of them, four are relevant operators: the OPE algebra and the relative structure constants are reported in Table 14.3. The correlation functions of these fields can be computed straightforwardly using the Coulomb gas, as proposed in Problem 14.2, and will not be presented here.

Landau–Ginzburg. The six primary fields perfectly match the identification provided by the composite fields of the Landau-Ginzburg theory and by the symmetries of the model. There are two different Z_2 symmetries, one associated to the spin transformation, the other to the duality.

With respect to the Z_2 spin symmetry $\varphi \to -\varphi$ we have

1. two odd fields: the magnetization operator $\sigma = \phi_{\frac{3}{80},\frac{3}{80}} \equiv \varphi$ and the sub-leading magnetic operator $\sigma' = \phi_{\frac{7}{16},\frac{7}{16}} \equiv: \varphi^3 :$
2. four even fields: the identity operator $1 = \phi_{0,0}$, the energy operator $\varepsilon = \phi_{\frac{1}{10},\frac{1}{10}} \equiv: \varphi^2 :$, and the density operator $t = \phi_{\frac{6}{10},\frac{6}{10}} \equiv: \varphi^4 :$, associated to the vacancies. Finally, there is also the irrelevant field $\varepsilon'' = \phi_{\frac{3}{2},\frac{3}{2}}$. The OPE of these fields gives rise to a sub-algebra of the fusion rules.

As for the Ising model, also for the TIM there is another Z_2 associated to the duality transformation, under which the fields change as

- the magnetization order parameters change into the disorder operators

$$\mu = D^{-1}\sigma D = \tilde{\phi}_{\frac{3}{80},\frac{3}{80}}, \quad \mu' = D^{-1}\sigma' D = \tilde{\phi}_{\frac{7}{16},\frac{7}{16}}. \tag{14.3.3}$$

- the even fields transform instead in themselves

$$D^{-1}\varepsilon D = -\varepsilon, \quad D^{-1}tD = t, \quad D^{-1}\varepsilon'' D = -\varepsilon'', \tag{14.3.4}$$

  ε and ε'' are odd fields while t is an even field under this transformation.

Supersymmetry. It is interesting to note that this critical model provides an explicit realization of a supersymmetric field theory. In fact, the TIM is also the first model of the minimal unitary superconformal series: the Z_2 even fields enter the definition of a superfield of the Neveu-Schwarz sector

$$\mathcal{N}(z, \bar{z}, \theta, \bar{\theta}) = \varepsilon(z, \bar{z}) + \bar{\theta}\,\psi(z, \bar{z}) + \theta\,\bar{\psi}(z, \bar{z}) + i\theta\bar{\theta}\,t(z, \bar{z}), \tag{14.3.5}$$

while the Z_2 odd magnetization operators form two irreducible representations of the Ramond sector. The supersymmetric Landau–Ginzburg of the model can be written as

$$\mathcal{S} = \int d^2x d^2\theta \left[\frac{1}{2}\mathcal{D}\mathcal{N}\bar{\mathcal{D}}\mathcal{N} + \mathcal{N}^3\right], \tag{14.3.6}$$

where $\mathcal{D}$ and $\bar{\mathcal{D}}$ are the covariant derivatives

$$\mathcal{D} = \frac{\partial}{\partial\theta} - \theta\frac{\partial}{\partial z}, \quad \overline{\mathcal{D}} = \frac{\partial}{\partial\bar{\theta}} - \bar{\theta}\frac{\partial}{\partial\bar{z}}. \tag{14.3.7}$$

Note that the supersymmetry and the organization of its Z_2 even primary fields in a superfield are at the root of the relationships that link the various structure constants (see, for instance, the identity $c_2 = c_1$).

Exceptional algebra E_7. In addition to the conformal and superconformal invariance, the TIM holds another surprise. In fact, it can be also realized in terms of a coset on the exceptional algebra E_7

$$\mathcal{M}_4 = \frac{(E_7)_1 \otimes (E_7)_1}{(E_7)_2}. \tag{14.3.8}$$

For E_7, the dual Coxeter number is $\tilde{h} = 18$ and therefore the central charge of this coset theory is $c = \frac{7}{10}$. At the level $k = 1$, the possible representations are given by the identity 1 and the representation Π_6, with conformal weights equal to 0 and $\frac{3}{4}$ respectively

$$(E_7)_1 \quad \rightarrow \quad \{1, \Pi_6\} = \{0, \tfrac{3}{4}\}. \tag{14.3.9}$$

Their components with respect to the simple roots of E_7 $(n_1, n_2, \cdots, n_7)$ (with n_i integers) are

$$\begin{array}{lcl} 1 & \rightarrow & (0,0,0,0,0,0,0) \\ \Pi_6 & \rightarrow & (0,0,0,0,0,1,0). \end{array} \tag{14.3.10}$$

At the level $k = 2$, there are instead the representations

$$(E_7)_2 \quad \rightarrow \quad \{1, \Pi_1, \Pi_2, \Pi_5, \Pi_6\} = \{0, \tfrac{9}{10}, \tfrac{21}{16}, \tfrac{7}{5}, \tfrac{57}{80}\}. \tag{14.3.11}$$

with the corresponding fundamental weights given by

$$\begin{array}{lcl} \Pi_1 & \rightarrow & (1,0,0,0,0,0,0) \\ \Pi_2 & \rightarrow & (0,1,0,0,0,0,0) \\ \Pi_5 & \rightarrow & (0,0,0,0,1,0,0). \end{array} \tag{14.3.12}$$

Π_1 is the adjoint representation E_7. We can recover the conformal weights of the TIM by the decomposition of the various representations

$$\begin{aligned} (0)_1 \times (0)_1 &= [(0)_{TIM} \otimes (0)_2] \quad + [(\frac{1}{10})_{TIM} \otimes (\Pi_1)_2] + [(\frac{6}{10})_{TIM} \otimes (\Pi_5)_2] \\ (0)_1 \times (\frac{3}{4})_1 &= [(\frac{7}{16})_{TIM} \otimes (\Pi_2)_2] + [(\frac{3}{80})_{TIM} \otimes (\Pi_6)_2] \\ (\frac{3}{4})_1 \times (\frac{3}{4})_1 &= (\frac{3}{2})_{TIM} \otimes (0)_2. \end{aligned} \tag{14.3.13}$$

Note that the energy operator $\Phi_{\frac{1}{10},\frac{1}{10}}$ is associated to the adjoint representation of E_7, an observation that will be crucial in the analysis of the off-critical model, when the temperature is moved away from its critical value $T \neq T_c$.

14.4 3-state Potts Model

On a square lattice, the Hamiltonian of the 3-state Potts model is given by

$$H = -\frac{\mathcal{J}}{2} \sum_{x,\alpha} (\sigma_x \bar{\sigma}_{x+\alpha} + \bar{\sigma}_x \sigma_{x+\alpha}) = -\mathcal{J} \sum_{x,\alpha} \cos(\eta_x - \eta_{x+\alpha}), \tag{14.4.1}$$

where the discrete spin variables are represented by $\sigma = \exp(i\eta)$, $\bar{\sigma} = \exp(-i\eta)$, with $\eta = 0, \pm\frac{2\pi}{3}$. It is known that this model has a duality symmetry and, at its self-dual point $\mathcal{J}_c = \frac{2}{3}\log(\sqrt{3}+1)$, presents a second-order phase transition. The lattice theory is exactly solvable and consequently all critical exponents are known. In this section we plan to show that the conformal theory that emerges at the critical point coincides with the unitary minimal model $\mathcal{M}_5$. More precisely, the operator content of the 3-state Potts model is given only by a sub-set of the Kac table of the conformal model $\mathcal{M}_5$. The subset of fields are those entering the modular invariant partition function of the type (A, D).

To find the field theory description of the microscopic statistical model, let us assume that there exists the continuum limit of its spin and energy operators, here denoted by $\sigma(x)$, $\bar{\sigma}(x)$ and $\epsilon(x)$. Moreover, let us assume that it holds the OPE

$$\begin{aligned}\sigma(x_1)\bar{\sigma}(x_2)+\bar{\sigma}(x_1)\sigma(x_2) &= \frac{1}{|x_1-x_2|^{2\Delta_\sigma}}+\mathbf{C}^{\epsilon}_{\sigma\bar{\sigma}}\frac{1}{|x_1-x_2|^{2\Delta_\sigma-\Delta_\epsilon}}\epsilon(x_2)+\cdots\\ \epsilon(x_1)\sigma(x_2) &= \mathbf{C}^{\sigma}_{\epsilon\sigma}\frac{1}{|x_1-x_2|^{\Delta_\epsilon}}\sigma(x_2)+\cdots\\ \epsilon(x_1)\epsilon(x_2) &= \frac{1}{|x_1-x_2|^{2\Delta_\epsilon}}.\end{aligned} \tag{14.4.2}$$

From the known expression of the critical exponents $\alpha=\frac{1}{3}$ and $\beta=\frac{1}{9}$ coming from the exact solution of the lattice model, we can infer the conformal weights of the scaling operators

$$\Delta_\sigma=\Delta_{\bar{\sigma}}=\frac{1}{15}, \qquad \Delta_\epsilon=\frac{2}{5}. \tag{14.4.3}$$

Let us now consider the Kac table of the minimal model $\mathcal{M}_5$, reported in Table 14.4

In this table there is the field $\Phi_{3,3}=\Phi_{2,3}$, with conformal weight $\Delta_\sigma=\frac{1}{15}$ and the field $\Phi_{2,1}=\Phi_{3,5}$ with $\Delta_\epsilon=\frac{2}{5}$. It is therefore natural to identify these conformal field with the scaling operators associated to the spin and energy operators of the lattice model. However the exact solution of the lattice model does not have operators with conformal weights $\frac{1}{8}$, $\frac{1}{40}$, $\frac{21}{40}$ and $\frac{13}{8}$. What is then the correct identification of the Z_3 Potts model?

3	$\frac{13}{8}$	$\frac{2}{3}$	$\frac{1}{8}$	0
$\frac{7}{5}$	$\frac{21}{40}$	$\frac{1}{15}$	$\frac{1}{40}$	$\frac{2}{5}$
$\frac{2}{5}$	$\frac{1}{40}$	$\frac{1}{15}$	$\frac{21}{40}$	$\frac{7}{5}$
0	$\frac{1}{8}$	$\frac{2}{3}$	$\frac{13}{8}$	3

Table 14.4 *Kac table of the unitary minimal model $\mathcal{M}_5$.*

To answer this question, we should remember that, for $p \geq 5$, the conformal minimal model $\mathcal{M}_p$ admits *two* different partition functions. The first of them is the purely diagonal partition function, i.e. the one in which appear all fields of the Kac table, each with multiplicity equal to 1. This leads to the expression

$$Z_{diag} = \frac{1}{2} \sum_{r=1}^{4} \sum_{s=1}^{5} |\chi_{r,s}|^2. \tag{14.4.4}$$

The field theory associated to the operator content of this partition function does not correspond to the 3-state Potts model but it rather defines the critical theory of a Landau–Ginzburg scalar field φ, that presents only a Z_2 invariance $\varphi \to -\varphi$. Its action is

$$\mathcal{S} = \int d^2x \left[\frac{1}{2} (\partial_\mu \varphi)^2 + \varphi^8 \right]. \tag{14.4.5}$$

There is however another modular invariant partition function associated to the minimal model $\mathcal{M}_5$ expressed by

$$Z_{Potts} = \sum_{r=1,2} \left\{ |\chi_{r,1} + \chi_{r,5}|^2 + 2|\chi_{r,3}|^2 \right\}. \tag{14.4.6}$$

The operator content identified by this partition function is different from the previous one: it involves only a sub-set of the fields of the Kac table of the minimal model $\mathcal{M}_5$. There are, in fact, only the fields $\Phi_{r,s}$ with $s = 1,5$ and $r = 1,2$. Combining the analytic and the anti-analytic parts, the critical theory described by this partition function have the scalar fields given in Table 14.5. These fields close an operator algebra, also in the absence of the other fields of the Kac table. Their skeleton fusion rules are reported in Table 14.6.

(r,s)	Δ	Field	Interpretation
$(1,1)$ or $(4,5)$	0	$\mathbf{1}$	Identity
$(2,1)$ or $(3,5)$	$\frac{2}{5}$	ϵ	energy
$(3,3)$ or $(2,3)$	$\frac{1}{15}$	σ	spin
$(3,1)$ or $(2,5)$	$\frac{7}{5}$	$\mathbf{X}$	
$(4,1)$ or $(1,5)$	3	$\mathbf{Y}$	
$(4,3)$ or $(1,3)$	$\frac{2}{3}$	$\mathbf{Z}$	

Table 14.5 *Scalar operators of the non-diagonal partition function of the model* $\mathcal{M}_5$.

$\epsilon \times \epsilon = \mathbf{1} + X$	$\epsilon \times \sigma = \sigma + Z$
$\epsilon \times X = \epsilon + Y$	$\epsilon \times Y = X$
$\sigma \times \sigma = \mathbf{1} + \epsilon + \sigma + X + Y + Z$	$\sigma \times X = \sigma + Z$
$\sigma \times Y = \sigma$	$\sigma \times Z = \epsilon + \sigma + X$
$X \times X = \mathbf{1} + X$	$X \times Y = \epsilon$
$X \times Z = \sigma$	$Y \times Y = \mathbf{1}$
$Y \times Z = Z$	$Z \times Z = \mathbf{1} + Y + Z$

Table 14.6 *Fusion rules of the scalar fields of the 3-state Potts model.*

In addition to these scalar fields there are certain fields with spin, here denoted by their conformal weights $\Phi_{(\Delta,\bar{\Delta})}$. They are constructed by combining in a non-diagonal way the analytic and anti-analytic fields: $W = \Phi_{(3,0)}$, $\bar{W} = \Phi_{(0,3)}$, $\mathcal{J} = \Phi_{(\frac{7}{5},\frac{2}{5})}$ and $\bar{\mathcal{J}} = \Phi_{(\frac{2}{5},\frac{7}{5})}$.

It is interesting to observe that the 3-state Potts models at criticality can be also obtained by the parafermionic theory $\mathbf{Z}_N$ with $N = 3$. It is easy to check the equality of the central charges of both theories, as well as the conformal weights of the spin operator $\Delta_\sigma = \frac{1}{15}$. The role of the parafermionic current is here played by the chiral operator $\Phi_{1,3}$, with conformal weight $\Delta_\psi = \frac{2}{3}$.

Generalization. Remarkably, the analysis presented for the 3-state Potts model can be generalized to the Q-state Potts model, where Q is regarded as a continuous variable (see Chapter 2). The range of values of Q for which the Potts model is critical is given by the interval $Q \in (1, 4)$: for $Q = 1$, the Potts model describes the critical phenomenon of percolation, for $Q = 2$ we have the usual Ising model, while for $Q > 4$, the Potts model presents a first-order phase transition that cannot be described by a CFT. The relation that identifies the minimal models $\mathcal{M}_p$ with the Q-state Potts model is

$$Q = 4\cos^2\frac{\pi}{p+1}, \tag{14.4.7}$$

and it is easy to check that it correctly reproduces, for $p = 3$ the Ising model ($with Q = 2$), for $p = 5$ the 3-state Potts model and for $p \to \infty$ the 4-state Potts model. The exact solution of the lattice models is known for generic values of Q and therefore all values of the thermal and magnetic critical exponents are known as well. This permits the identification of the anomalous dimension of several order parameters, as

$$\begin{aligned} X_{T_n} &= 2\Delta_{n+1,1} = \frac{n^2 + ny}{2-y}, \\ X_{H_n} &= 2\Delta_{N-1-n,n} = \frac{(2n_1)^2 - y^2}{4(2-y)}, \end{aligned} \tag{14.4.8}$$

where we have introduced the notation

$$N \equiv \frac{p+1}{2}, \quad y \equiv \frac{1}{N}. \tag{14.4.9}$$

14.5 The Yang–Lee Model

Among the minimal non-unitary models, a simple but particularly significant example is given by the model $\mathcal{M}_{2,5}$. Its central charge is $c = -22/5$ and the Kac table consists of only one row (Table 14.7). In addition to the identity operator, there is only a field φ of conformal weight $\Delta = -1/5$. Hence the effective central charge is $c_{eff} = c - 24\Delta_{min} = 2/5$. As shown originally by Cardy, this model admits a statistical interpretation in terms of a field theory associated to the Yang–Lee zero singularities of the Ising model. Let us discuss the main steps that lead to this conclusion using the Yang–Lee theorem.

The partition function of a statistical model defined on a lattice is an analytic function of its parameters as far as the number N of the fluctuating variables is finite. Its singularities only emerge in the thermodynamical limit $N \to \infty$. Consider the Ising model at a given value T of the temperature and in presence of an external magnetic field B. As a function of B, at finite N, the zeros of the partition function cannot be on the real axis of B, since Z is expressed by a sum of positive terms. Hence they are placed in complex conjugate points of the complex plane B and they tend to accumulate along certain curves in the limit $N \to \infty$. In particular, as shown by Yang and Lee, in the Ising model these zeros accumulate along the imaginary axis $B = ih$. Correspondingly the free energy of the system can be expressed in terms of the density of these zeros along the imaginary axis

$$F(b) = \log Z = \int_{-\infty}^{+\infty} dx\, \rho(x, T) \log(h - ix), \tag{14.5.1}$$

with the magnetization given by

$$M = \frac{\partial F}{\partial B} = \int_{-\infty}^{+\infty} dx\, \frac{\rho(x, T)}{h - ix}. \tag{14.5.2}$$

Below the critical temperature, i.e. for $T < T_c$, the distribution of the zeros extends till the real axis, so that $\rho(0, T) \neq 0$. Consequently the magnetization is a discontinued function of the variable B when it crosses the real axis, and the system presents a first-order

0	$-\frac{1}{5}$	$-\frac{1}{5}$	0

Table 14.7 *Kac table of the minimal non-unitary model $\mathcal{M}_{2,5}$.*

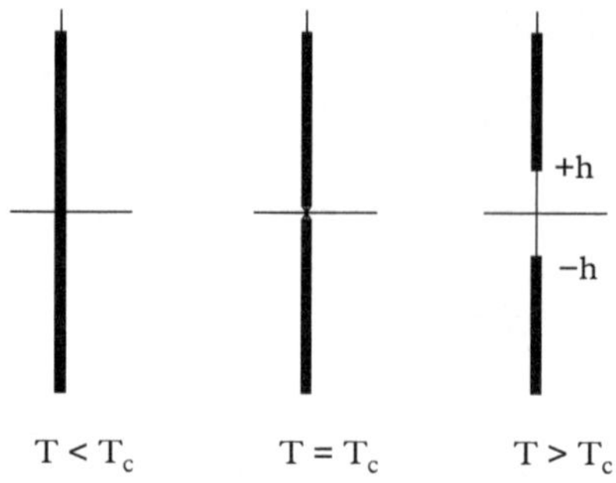

Fig. 14.2 *Density of the zeros of the partition function of the Ising model in the complex plane of the variable B by varying the temperature.*

phase transition. Precisely at $T = T_c$ we have $\rho(0, T_c) = 0$ and there is a second-order phase transition. In the high-temperature phase $T > T_c$, the system is paramagnetic and the distribution of the zeros starts from two symmetric critical values $\pm h_c(T)$ and then extends along the magnetic axis.

In the vicinity of h_c, the density of the zeros has an anomalous behaviour

$$\rho(h, T) = (h - h_c)^{\sigma}. \tag{14.5.3}$$

An analogous anomalous behaviour is present in the magnetization, as a function of the (complex) magnetic field

$$M(ih) \simeq (h - h_c)^{\sigma}. \tag{14.5.4}$$

Thanks to the thermodynamical relations discussed in Chapter 1, we can link the critical exponent σ to the critical exponent η of the operator corresponding to the fluctuations of the model in the presence of an imaginary magnetic field

$$\sigma = \frac{1}{\delta} = \frac{d - 2 + \eta}{d + 2 - \eta}. \tag{14.5.5}$$

Fisher has identified the effective action of the order paramter, given by the Landau–Ginzburg theory

$$\mathcal{S} = \int d^d x \left[\frac{1}{2} (\partial \varphi)^2 + i(h - h_c)\, \varphi + i g \varphi^3 \right]. \tag{14.5.6}$$

Note that the non-unitarity of the model manifests itself in the imaginary value of the coupling constant. In two dimensions, a model that could reproduce the dynamics of such a theory should satisfy two main requests: (i) it must be a non-unitary model; (ii) it must have only one relevant field φ satisfying the fusion rule

$$\varphi \times \varphi = \mathbf{1} + C_{\varphi,\varphi}^{\varphi}\, \varphi. \tag{14.5.7}$$

with a purely imaginary structure constant $C_{\varphi,\varphi}^{\varphi}$.

These are precisely the features of the minimal non-unitary model $\mathcal{M}_{2,5}$ whose structure constant is given by

$$C_{\varphi,\varphi}^{\varphi} = i \left[\frac{\Gamma^2\left(\frac{6}{5}\right)\Gamma\left(\frac{1}{5}\right)\Gamma\left(\frac{2}{5}\right)}{\Gamma\left(\frac{3}{5}\right)\Gamma^3\left(\frac{4}{5}\right)} \right]^{1/2}. \tag{14.5.8}$$

This quantity can be computed by using the exact expression of the four-point correlation function of the field φ. Since this field occupies the position $(1,2)$ of the Kac table, its correlators are given either solving the corresponding second-order differential equation or applying the Coulomb gas method. The result is

$$\begin{aligned} &\langle \varphi(z_1, \bar{z}_1)\varphi(z_2, \bar{z}_2)\varphi(z_3, \bar{z}_3)\varphi(z_4, \bar{z}_4)\rangle = \\ &\left| \frac{z_{13}\, z_{24}}{z_{12}\, z_{23}\, z_{34}\, z_{14}} \right|^{-4/5} \left\{ |F_1(\eta)|^2 + C^2\, |F_2(\eta)|^2 \right\} \end{aligned} \tag{14.5.9}$$

where η is the harmonic ratio $\eta = z_{12}z_{34}/z_{13}z_{24}$ and $F_i(\eta)$ are the hypergeometric functions

$$\begin{aligned} F_1(\eta) &= F\left(\frac{3}{5}, \frac{4}{5}, \frac{6}{5}, \eta\right) \\ F_2(\eta) &= \eta^{-1/5}\, F\left(\frac{3}{5}, \frac{2}{5}, \frac{4}{5}, \eta\right). \end{aligned} \tag{14.5.10}$$

The value of the critical exponent σ predicted by this conformal model is $\sigma = -1/6$, in reasonable agreement with its numerical determination $\sigma = -0.163$.

14.6 Conformal Models with *O*(*n*) Symmetry

Chapter 2 showed that spin models with a continous symmetry $O(n)$ provide a generalization of the Ising model and, in particular, their limit $n \to 0$ describes the class of universality of the self-avoiding random walks. In these theories, the spins are vectors $\vec{S}$ with n components and length $|\vec{S}|^2 = n$. Taking advantage of the universality of critical phenomena, we can choose any microscopic lattice to study their behaviour. The most convenient one turns out to be a lattice with coordination number equal to 3, as for instance the exagonal lattice shown in Figure 14.3.

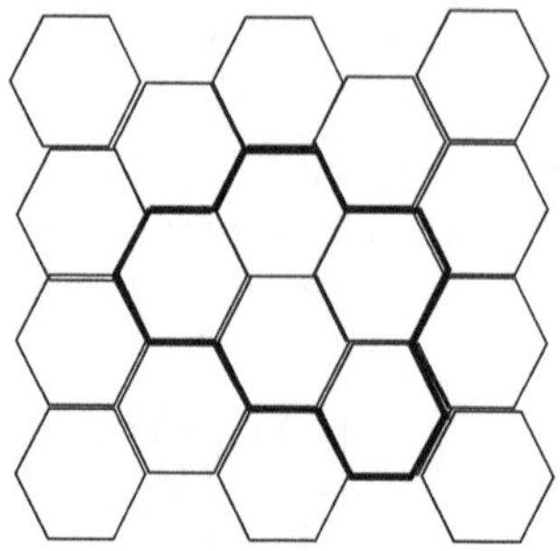

Fig. 14.3 *Hexagonal lattice and one of its closed-spin circuits.*

We assume that the partition function of the system is expressed by

$$Z = \int \prod_k d\vec{S}_k \prod_{\langle i,j \rangle} (1 + x\vec{S}_i \cdot \vec{S}_j), \tag{14.6.1}$$

where the product on i and j is on the nearest-neighbour sites. The integration rules on the spins are

$$\int dS^a\, S^a = 0$$
$$\int dS^a\, (S^a)^2 = 1$$
$$\int d\vec{S}\, \mathbf{S}^2 = n.$$

Expand now the product $\prod_{\langle i,j \rangle}(1 + x\vec{S}_i \cdot \vec{S}_j)$ and integrate on the values of the spins: due to the coordination number of the lattice and the integration rules stated, the only terms that are different from zero are those relative to the self-avoiding closed circuits. Since each of these circuits carries a factor n coming from the integration on the spins and a factor x for each of its segments, the partition function becomes

$$Z = \sum_{closecircuit} n^{N_C} x^{N_S}, \tag{14.6.2}$$

where N_C is the number of the closed circuits, while N_S is the number of the segments. We can use this expression to analytically continue the definition of the model to arbitrary values of n, not necessarily integers. The partition function presents a critical point x_c given by

$$x_c = (2 + \sqrt{2-n})^{-1/2}, \tag{14.6.3}$$

at which there is a second-order phase transition. This is described by a CFT with central charge $c(n) = 1 - 6/k(k+1)$, where the relation that links n and k is expressed by

$$n = 2\cos(\pi/k), \quad k \geq 1. \tag{14.6.4}$$

Note, in particular, that $c = 0$ when $n = 0$ but its derivative $\frac{\partial c}{\partial n}$ at $n = 0$ is different from zero and equal $5/3\pi$. For $n = 1, c = 1/2$ and we recover the Ising model. The anomalous dimension of the energy operator of these theories is

$$\eta_e = 2(k-1)/(k+1), \tag{14.6.5}$$

i.e. $\eta_e = 2/3$ for $n = 0$. This exponent is related to the exponent ν that characterizes the divergence of the correlation length by the relation $\nu = 1/(2 - x_e) = 3/4$. This value is perfectly in agreement with the critical exponent of the exact lattice solution of the self-avoiding random walk found by Nienhuis.

REFERENCES

For exactly solved lattice models and their identification with minimal models of CFT see

Andrew, G.E., Baxter R.J. and Forrester, P.J. (1984). 'Eight-Vertex SOS Model and Generalized Rogers–Ramanujan-Type Identities', *Journal of Statistical Physics*, 35: 193.

Huse, D. (1984). 'Exact Exponents For Infinitely Many New Multicritical Points, *Physics Review B*, 30: 2908.

An excellent review article on the universality class of the tricritical point is

Lawrie, I. and Serbach, S. (1984). 'Theory of Tricritical Points', in C. Domb and J.L. Lebowitz (Eds) *Phase Transitions and Critical Phenomena*, vol. 9, Academic Press, pp. 2–161.

The superconformal invariance of the two-dimensional TIM was pointed out by Friedan, Qiu and Shenker and studied in the articles

Friedan, D. Qiu, Z.A. Shenker, S. 'Superconformal Invariance In Two-Dimensions And The Tricritical Ising Model', Phys. Lett. B 151 (1985), 37

Qiu, Z.A. Supersymmetry, 'Two-Dimensional Critical Phenomena And The Tricritical Ising Model', Nucl. Phys. B 270 (1986), 205

Mussardo, G. Sotkov, G. Stanishkov, M. 'Ramond Sector Of The Supersymmetric Minimal Models', Phys. Lett. B 195 (1987), 397.

The identification of the class of universality of the 3-state Potts model with the unitary minimal model $\mathcal{M}_5$ is discussed in

Dotsenko, V.S. (1984). 'Critical Behavior And Associated Conformal Algebra of the Z(3) Potts Model', *Nuclear Physics B*, 235: 54.

The identification of the minimal non-unitary model $\mathcal{M}_{2,5}$ with the Yang–Lee edge singularity of the zeros of the partition function is considered in

Cardy, J.L. (1985). 'Conformal Invariance and the Yang–Lee Edge Singularity in Two Dimensions', *Physics Review Letters*, 54: 1354.

The Coulomb gas formalism and the relation with the $O(n)$ statistical models are discussed in

Nienhuis, B. (1984). 'Critical Behavior of Two-dimensional Spin Models and Charge Asymmetry in the Coulomb Gas', *Journal of Statistical Physics*, 34: 731.

PROBLEMS

14.1. Correlator of the Ising model

Consider the following correlator of the two-dimensional Ising model

$$H(\eta,\bar{\eta}) = \langle \sigma(\infty)\epsilon(1,1)\epsilon(\eta,\bar{\eta})\sigma(0,0)\rangle.$$

Use the modified Coulomb gas to show that it is given by

$$H(\eta,\bar{\eta}) = \frac{1}{4}\left|\frac{\eta+1}{\eta^{1/2}(1-\eta)}\right|.$$

14.2. Structure constants

Use the modified Coulomb gas to compute the correlation functions of the TIM. Determine the values of the structure constants given in the text.

14.3. Vacua of the multi-critical Ising model

Consider the potential of the multi-critical Ising model

$$V(\varphi) = g_1\varphi + g_2\varphi^2 + g_3\varphi^3 + g_4\varphi^4 + g_5\varphi^5 + g_6\varphi^6 + \varphi^8.$$

a. Show that, opportunely fine tuning the parameters, the model has a phase with four degenerate vacua.
b. Argue that this is an enough information to conclude that the class of universality of this model does not coincide with the one of the 3-state Potts model, although the two models share the same value of the central charge.

Part 4

Away from Criticality

15

In the Vicinity of the Critical Points

Lume v'è dato a bene e a malizia.
Dante Alighieri

15.1 Introduction

The previous chapters have dedicated ample space to the study of two-dimensional statistical systems at criticality, providing their exact solutions in terms of CFTs. This chapter starts investigating the deformations of CFTs that move the statistical systems away from criticality. As pointed out in Chapter 8, in the RG approach the characterization of the classes of universality must include, in addition to the conformal theory of the fixed points, also the description of the *scaling region* nearby.

The scaling region is a multi-dimensional space, parameterized by the coupling constants of the scalar relevant fields $\Phi_{\Delta_i,\Delta_i}(x)$ that are present in the CFT of the fixed point under scrutiny. These operators are identified by the condition $x_i = 2\Delta_i < 2$. The fixed point action is unstable for the insertion of these operators, and any RG flow that starts from a given fixed point can be obtained by a combination of the couplings of these relevant fields. If $\mathcal{S}^*$ is the conformal action of the fixed point and n is the number of its relevant fields, the most general deformation is given by

$$\mathcal{S} = \mathcal{S}^* + \sum_{i=1}^{n} \lambda_i \int \varphi_i(x)\, d^2x. \tag{15.1.1}$$

As discussed in Section 15.2, for what concerns the ultraviolet divergences encountered in the perturbative series of the theory (15.1.1), the QFTs defined by the relevant deformations of a conformal action are of the superrenormalizable type. In other words, the relevant operators do not influence the short-distance behaviour of the system but, on the contrary, they drastically change the large-distance scales. The first effect of their presence is the breaking of the conformal invariance and the generation of a mass scale, function of the coupling constants. The latter are, in fact, dimensional quantities, expressed in terms of a mass scale M by

Statistical Field Theory: An Introduction to Exactly Solved Models in Statistical Physics. Second Edition. Giuseppe Mussardo, Oxford University Press (2020). © Giuseppe Mussardo. DOI: 10.1093/oso/9780198788102.001.0001

$$M = \mathcal{D}_i \lambda_i^{\frac{1}{2-2\Delta_i}}, \tag{15.1.2}$$

where the coefficients $\mathcal{D}_i$ are pure numbers that can be fixed once we choose a normalization of the operators. In the following we adopt the conformal normalization, identified by the short distance behaviour of their two-point correlation function

$$\langle \varphi_i(r)\varphi_j(0)\rangle \simeq \frac{\delta_{ij}}{r^{2x_j}}, \quad r \to 0. \tag{15.1.3}$$

Excluding the possibility of pathological cases, as for instance the presence of limit cycles of the RG, there are in general two different physical scenarios associated to the action (15.1.1):

1. In the first scenario, the final point of the RG flow is also a fixed point associated to another CFT. In this case, the QFT associated to this RG flow has an ultraviolet behaviour ruled by CFT_1 of the starting point, while its infrared behaviour is controlled by CFT_2 of the final point. The occurrence of this scenario can be detected by studying the behaviour of the two-point correlation functions $G_i(r) = \langle \varphi_i(r)\varphi_i(0)\rangle$: in this case they present a power-law behaviour in both regimes $r \to 0$ and $r \to \infty$

$$G_i(r) = \begin{cases} r^{-2x_i^{(1)}}, & r \to 0 \\ r^{-2x_i^{(2)}}, & r \to \infty \end{cases}. \tag{15.1.4}$$

   with $x_i^{(1)} \neq x_i^{(2)}$. These two quantities are the anomalous dimensions of the field φ_i with respect to the initial and final conformal field theories respectively.

   QFTs of this type have massless excitations, i.e. the physical correlation length of the problem is *infinite* all along the RG flow. However, the conformal invariance is broken for the non-vanishing values of the $\beta_i(\{\lambda_j\})$ functions of the coupling constants.

2. In the second scenario, by far the most common one, the system presents a *finite* correlation length ξ. In this case, the infrared behaviour of the theory is ruled by a massive QFT. Once again, the identification of this circumstance can be done by looking at the two-point correlation functions: in this case, for $r \to \infty$ they present an exponential decay while for $r \to 0$ they have a power-law behaviour, determined by the initial conformal theory CFT_1

$$G_i(r) = \begin{cases} r^{-2x_i}, & r \to 0 \\ e^{-m_i r}, & r \to \infty \end{cases}. \tag{15.1.5}$$

   In this expression $m_i = \xi^{-1}$ is the mass of the lightest particle that couples to the field φ_i.

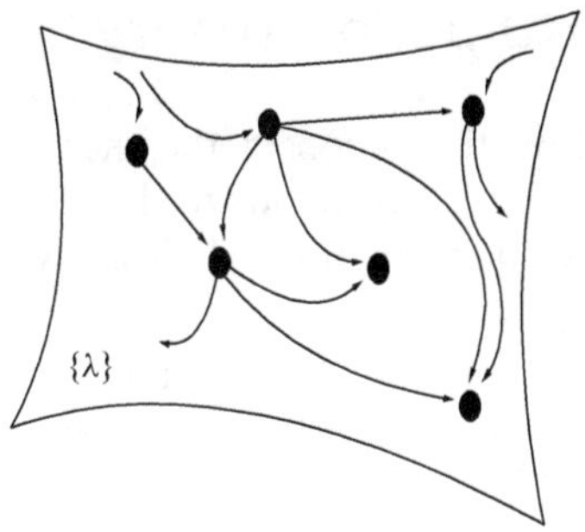

Fig. 15.1 *Renormalization group flows in the coupling constant space.*

From a geometrical point of view, the nature of the RG flows in the multi-dimensional coupling constant space is shown in Figure 15.1.

The analysis of the off-critical theories poses a series of interesting questions:

- Is there a way to predict whether a deformation of a conformal action gives rise to a massless or a massive theory?
- If the off-critical theory is massive, is it possible to determine its mass spectrum?
- Is it possible to characterize the operator content of the off-critical theory and its correlation functions?
- Do the off-critical correlation functions satisfy differential equations? Of what kind?
- Is it possible to determine the thermodynamics of these models?
- What is the relationships between the conformal data, e.g. central charge, anomalous dimensions, structure constant and the off-critical data, such as the mass spectrum?

Presently there is not a general answer to all these questions. However, there is a series of important results that allow us to reach a satisfactory control of the off-critical theories, at least from a perturbative point of view. It should be pointed out that the situation can be undoubtedly better for particular deformations: as we show in the next chapters, certain off-critical theories are in fact severely constrained by the presence of infinite conserved charges. These theories corresponds to integrable models that can be exactly solved by a formalism based on the S-matrix. Their study turns out to be decisive to solve some of the above mentioned questions.

This chapter initially studies the nature of the perturbative series associated to the perturbed action, reformulating the equations to which they give rise. Later we discuss two general results of the RG flows, known as *c-theorem* and Δ-theorem, that allow us to obtain extremely useful information on the theories of the initial and final fixed points.

15.2 Conformal Perturbation Theory

In vicinity of the critical point, the action of the theory can always be expressed as (15.1.1). The unperturbed action corresponds to the CFT of the fixed point, of which we know in principle all correlation functions. This allows us to define the perturbative series for any physical quantity away from criticality. For simplicity, hereafter we consider only cases in which the deformation is made only by one relevant scalar field $\varphi(x)$ of conformal weights (Δ, Δ). The expectation value of any operator A of the perturbed theory is expressed by the series

$$\begin{aligned} \langle A \rangle_\lambda &= \frac{1}{Z_\lambda} \left\langle A \exp\left[-\lambda \int d^2x \varphi(x) \right] \right\rangle_0 \\ &= \frac{1}{Z_\lambda} \sum_{n=0}^{\infty} \frac{(-\lambda)^n}{n!} \int d^2x_1 \ldots d^2x_n \langle A \varphi(x_1) \ldots \varphi(x_n) \rangle_0 \end{aligned} \tag{15.2.1}$$

where

$$Z_\lambda = \left\langle \exp\left[-\lambda \int d^2x \ \varphi(x) \right] \right\rangle_0 . \tag{15.2.2}$$

In this expression, $\langle \cdots \rangle_0$ are the correlation functions of the unperturbed CFT. In the computation of the integrals of the perturbative series, there are however both ultraviolet and infrared divergences. They can be regularized by introducing an ultraviolet cut-off ϵ and an infrared cut-off R—quantities that finally have to be sent to the limits $\epsilon \to 0$ and $R \to \infty$.

For what concerns the ultraviolet properties of the perturbative series, the QFTs defined by deformations of the relevant fields are of the super-normalizable type. Therefore, the ultraviolet divergences can be dealt with the standard renormalization methods: they lead to a redefinition of the local fields and, when the deformation operator is marginal (i.e. with conformal weight $\Delta = 1$) also to a renormalization of the coupling constant.

The infrared divergences are of different type. They cannot be absorbed in the redefinition of the local quantities and, for this reason, they give rise to non-analytic expressions in the coupling constants. The physical origin of this phenomenon is easy to understand. In fact, the vacuum state of the deformed theory (as well as all other excited states) is not adiabatically related to the vacuum states of the conformal theory:[1] if, for instance, the perturbed system corresponds to a massive theory, the new Hilbert space is set by the Fock space of the multi-particle states, whereas the original Hilbert space is

[1] can draw an analogy with a quantum mechanics problem. Consider a free-particle system on a line, in which we switch on a potential like $g|x|$ that cuts off the free asymptotic states. The perturbed system has all and only bound states, which are not adiabatically related to the energy eigenstates of the unperturbed system. In particular, their energies scale as a function of the coupling constant according to the non-analytic law $g^{2/3}$.

spanned by the Verma modules of the conformal states. In particular, the vacuum of the perturbed theory is the state without any particle excitations, whereas the vacuum of the unperturbed theory is characterized in a completely different way, since it is the state annihilated by all L_n with $n \geq -1$.

The different nature of the ultraviolet and infrared divergences permits their separate treatment, providing the key to control the theory in a perturbative way. Let us firstly discuss the ultraviolet properties and later the infrared ones.

Ultraviolet divergences. To understand the ultraviolet structure of the theory, Let us consider initially what we can learn from the first-order calculation. Let $\Phi(0)$ be a field of the perturbed theory (to become later a renormalized field), obtained as a deformation of the field $\tilde{\Phi}(0)$ of the original conformal theory. Denote by X a generic product of other fields and consider the correlator $\langle X\Phi(0)\rangle$. Its perturbative definition is given by

$$\langle X\,\Phi(0)\rangle_\lambda \simeq \langle X\,\tilde{\Phi}(0)\rangle_0 - \lambda\int_{\epsilon<|x|<R} d^2x\,\langle X\,\tilde{\Phi}(0)\varphi(x)\rangle_0 + \cdots. \tag{15.2.3}$$

This integral is ultraviolet divergent only if the OPE

$$\varphi(x)\,\tilde{\Phi}(0) = \sum_k C^k_{\varphi\Phi}\,|x|^{2(\Delta_k-\Delta_\Phi-\Delta)}\,\tilde{A}_k(0) \tag{15.2.4}$$

contains the fields A_k with conformal dimensions Δ_k that fulfil the equation

$$\gamma_k \equiv \Delta_k - \Delta_\Phi - \Delta + 1 \leq 0. \tag{15.2.5}$$

In this case, to obtain a finite expression at the first order of the correlation functions it is sufficient to define the renormalized operator by

$$\Phi = \tilde{\Phi} + \lambda\sum_k b_k\epsilon^{2\gamma_k}\tilde{A}_k + \mathcal{O}(\lambda^2), \tag{15.2.6}$$

where

$$b_k = \pi\,\frac{C^k_{\varphi\Phi}}{\gamma_k}.$$

This formula shows that, in general, the renormalization procedure induces a mixing of the original operators with those of lower conformal dimensions.

Off-critical operator product expansion. Let us now analyse in more detail the renormalization procedure of the ultraviolet and infrared divergences by studying the two-point correlation function of the renormalized field $\Phi(x)$, whose perturbative expression is given by

$$\langle \Phi(x)\Phi(0)\rangle_\lambda = \sum_{k=0}^{\infty} \frac{(-\lambda)^n}{n!} \langle \tilde{\Phi}(x)\tilde{\Phi}(0)\varphi(y_1)\ldots\varphi(y_n)\rangle_0 \, d^2y_1\ldots d^2y_n. \tag{15.2.7}$$

To evaluate the behaviour of this correlator (at least in the limit $|x| \to 0$, even though the final expression also holds for finite values of x), it is convenient to start from the OPE

$$\Phi(x)\Phi(0) = \sum_k C^k_{\Phi\Phi}(x) A_k(0), \tag{15.2.8}$$

where A_k ($k = 0, 1, \ldots$) is a complete set of fields and $C^k_{\Phi\Phi}(x)$ are their relative structure constants with the field Φ. Since the structure constants are local quantities, they are analytic functions of the coupling constant λ and therefore can be expanded in power series theoreof. The fields A_k are the renormalized expressions of the operators $\tilde{A}_k$ present at the critical point: denoting by $(\Delta_k, \bar{\Delta}_k)$ their conformal dimensions, on a basis of a simple dimensional analysis argument, we have

$$C^k_{\Phi\Phi}(x) = x^{\Delta_k - 2\Delta_\Phi} \, \bar{x}^{\bar{\Delta}_k - 2\bar{\Delta}_\Phi} \sum_{n=0}^{\infty} C^{k\,(n)}_{\Phi\Phi} \, (\lambda \, r^{2-2\Delta})^n, \tag{15.2.9}$$

where $r = (x\bar{x})^{1/2}$. However, the price to pay for the analiticity of the structure constants is the presence of the vacuum expectation values of some of the operators A_k. These are non-local quantities and therefore non-analytic with respect to the coupling constant. For dimensional reason, their expression is given by

$$\langle A_k(0)\rangle_\lambda = \lambda^{\frac{\Delta_k}{1-\Delta}} \, Q_k, \tag{15.2.10}$$

where Q_k are pure numbers. The set of these vacuum expectation values encodes important information on the infrared properties of the theory and cannot be determined by perturbation theory. Obviously many of them vanish for symmetry reasons as, for instance, those of fields with spin or those of the derivative fields. Selecting only those with vacuum expectation value different from zero, we have

$$\langle \Phi(x)\Phi(0)\rangle_\lambda = \sum_\nu C^\nu_{\Phi\Phi}(x) \, \langle A_\nu(0)\rangle_\lambda. \tag{15.2.11}$$

It is worth mentioning that the exact vacuum expectation values of the primary fields can be computed for integrable deformations of conformal theories and the relevant formulae are presented in Chapter 20. Together with the perturbative expressions of the structure constants, discussed below, the vacuum expectation values permits the determination of the correlation functions.

Renormalization. Consider now the renormalization of these fields by analysing the matrix elements

$$\tilde{I}_l^k(\lambda, R, \epsilon) = \langle \tilde{A}^k(\infty)\tilde{A}_l(0)\rangle \tag{15.2.12}$$
$$= \sum_{n=0}^{\infty} \frac{(-\lambda)^n}{n!} \int_{\epsilon<|y_i|<R} \langle \tilde{A}^k(\infty)\tilde{A}_l(0)\,\varphi(y_1)\ldots\varphi(y_n)\rangle_0\, d^2y_1\ldots d^2y_n.$$

By means of the ultraviolet and infrared cut-offs, all these integrals are finite and the quantities $\tilde{I}_l^k(\lambda, R, \epsilon)$ are regular functions of λ. Adopting the conformal normalization of the fields, we have

$$\tilde{I}_l^k(\lambda, R, \epsilon) = \delta_l^k + \mathcal{O}(\lambda). \tag{15.2.13}$$

For the invariance under rotations, the matrix $\tilde{I}_l^k$ is diagonal in the spin of the fields and it is therefore possible to consider any spin sector separately. For simplicity we discuss only the spin-zero sector, the one of the scalar fields.

In the limit $\epsilon \to 0$, the matrix elements $\tilde{I}_l^k(\lambda, R, \epsilon)$ become singular and it is natural to assume that they have a factorized form

$$\tilde{I}_l^k(\lambda, R, \epsilon) = \sum_{k'} U_{k'}^k(\lambda, \epsilon)\, I_l^{k'}(\lambda, R) \tag{15.2.14}$$

where $I_l^{k'}(\lambda, R)$ are the elements of the renormalized matrix whereas $U_{k'}^k(\lambda, \epsilon)$ are the elements of the renormalization matrix. Both are regular functions of λ. For dimensional reasons, we have

$$U_l^k(\lambda, \epsilon) = \sum_{n=0}^{\infty} \frac{U_l^{k\,(n)}(\lambda\,\epsilon^{2-2\Delta})^n}{\epsilon^{2(\Delta_l-\Delta_k)}}. \tag{15.2.15}$$

When $\epsilon \to 0$, it is necessary to keep only the terms with negative powers ϵ. For $\Delta < 1$, there is only a finite number of them. Organizing the fields A_i in the increasing order of the conformal dimensions $\Delta_0 \leq \Delta_1 \leq \Delta_2 \ldots$, the matrix U_l^k assumes a triangular form, i.e.

$$U_l^k(\lambda, \epsilon) = 0, \quad \Delta_k > \Delta_l \tag{15.2.16}$$

The inverse matrix of U has obviously the same properties (15.2.15) and (15.2.16) of U and we can then define the renormalized fields as

$$A_k = \sum_l (U^{-1})_k^l \tilde{A}_l. \tag{15.2.17}$$

With the normalization $U_l^k(\lambda,\epsilon) = \delta_l^k + \mathcal{O}(\lambda)$, for the renormalized fields we have

$$A_k = \tilde{A}_k + \cdots \tag{15.2.18}$$

with a finite number of other terms, corresponding to the operators with lower conformal dimensions. In this way, we recover the result previously obtained at the first order.

It is necessary to stress that the analysis done above can become more involved in the presences of 'resonances' of the conformal dimensions, that $\Delta_k - \Delta_l = n(1-\Delta)$. In this case, there are also logarithmic divergences and the factorized form (15.2.14) becomes ambiguous since it depends an an arbitary renormalization point. Although this is an interesting question, it is however outside the scope of the present analysis.

The elements of the renormalized matrix

$$I_l^k(\lambda,R) = \langle \tilde{A}^k(\infty) A_l(0)\rangle_\lambda, \tag{15.2.19}$$

are independent from the ultraviolet cut-off ϵ and they have the same structure of the matrix U

$$I_l^k(\lambda,R) = \sum_{n=0}^{\infty} \frac{I_l^{k\,(n)}\,(\lambda\, R^{2-2\Delta})^n}{R^{2(\Delta_l-\Delta_k)}}, \tag{15.2.20}$$

where, now, it is necessary to keep the terms with positive powers of R. The sum of this series produces a non-trivial dependence on R. Although it is not easy to find their exact expression, it is natural to assume that these series behave in a homogeneous way: this means that exists the limit

$$\lim_{R\to\infty} \frac{I_l^k(\lambda,R)}{I_0^k(\lambda,R)} = \lambda^{\frac{\Delta_k}{1-\Delta}}\, Q_l^{(k)} \tag{15.2.21}$$

and that this limit corresponds to the vacuum expectation values of the operators.

It is also convenient to define the quantities

$$\begin{aligned} G^k_{\Phi\Phi}(\lambda,x,R) &= \langle \tilde{A}^k(\infty)\,\Phi(x)\,\Phi(0)\rangle_\lambda \qquad (15.2.22)\\ &= \sum_{n=0}^{\infty} \frac{(-\lambda)^n}{n!} \int_{|y_i|<R} \langle \tilde{A}^k(\infty)\,\Phi(x)\,\Phi(0)\,\varphi(y_1)\ldots\varphi(y_n)\rangle_0\, d^2y_1\ldots d^2y_n. \end{aligned}$$

Since Φ are the renormalized fields, all the integrals are ultraviolet convergent and it is not necessary to introduce the ultraviolet cut-off ϵ. Substituting this expression in the operator expansion (15.2.8), it yields

$$C^k_{\Phi\Phi}(x) = \sum_l G^l_{\Phi\Phi}(\lambda,x,R)\,(I^{-1})^k_l(\lambda,R). \tag{15.2.23}$$

Structure constants. The quantity above is now finite in the limit $R \to \infty$ and allows us to compute the structrure constants of the renormalized fields. As an explicit example, we present here their first-order term. Let $\mathbf{C}_{\Phi\Phi}^k$, $\mathbf{C}_{\varphi\Phi}^k$ and $\mathbf{C}_{\varphi l}^k$ the structure constants of the conformal theory, namely

$$\begin{aligned} \langle \tilde{A}^k(\infty)\Phi(x)\Phi(0)\rangle_0 &= \mathbf{C}_{\Phi\Phi}^k\,(x\bar{x})^{\Delta_k-2\Delta_\Phi} \\ \langle \tilde{A}^k(\infty)\varphi(x)\Phi(0)\rangle_0 &= \mathbf{C}_{\varphi\Phi}^k\,(x\bar{x})^{\Delta_k-\Delta-\Delta_\Phi} \\ \langle \tilde{A}^k(\infty)\varphi(x)\tilde{A}_l(0)\rangle_0 &= \mathbf{C}_{\varphi l}^k\,(x\bar{x})^{\Delta_k-\Delta_l-\Delta}. \end{aligned} \tag{15.2.24}$$

At the first order, we have

$$\tilde{I}_l^k(\lambda,R,\epsilon) = \delta_l^k - \lambda\pi\,\mathbf{C}_{\varphi l}^k \frac{R^{2(\Delta_k-\Delta_l-\Delta+1)} - \epsilon^{2(\Delta_k-\Delta_l-\Delta+1)}}{\Delta_k-\Delta_l-\Delta+1}, \tag{15.2.25}$$

and, therefore, at the same perturbative order in λ

$$\begin{aligned} I_l^k(\lambda,R) &= \delta_l^k - \lambda\pi\,\mathbf{C}_{\varphi l}^k \frac{R^{2(\Delta_k-\Delta_l-\Delta+1)}}{\Delta_k-\Delta_l-\Delta+1} \\ U_l^k(\lambda,\epsilon) &= \delta_l^k + \lambda\pi\,\mathbf{C}_{\varphi l}^k \frac{\epsilon^{2(\Delta_k-\Delta_l-\Delta+1)}}{\Delta_k-\Delta_l-\Delta+1}. \end{aligned} \tag{15.2.26}$$

Hence, at the first order in λ the structure constants are given by

$$\begin{aligned} C_{\Phi\Phi}^k(x) &= \mathbf{C}_{\Phi\Phi}^k\,(x\bar{x})^{\Delta_k-2\Delta_\Phi} \\ &- \lambda\int_{|y|<R} \langle \tilde{A}^k(\infty)\,\varphi(y)\Phi(x)\Phi(0)\rangle_0\,d^2y + \lambda\pi\sum_l \frac{\mathbf{C}_{\Phi\Phi}^l\,\mathbf{C}_{\varphi l}^k\,R^{2(\Delta_k-\Delta_l-\Delta+1)}}{\Delta_k-\Delta_l-\Delta+1}. \end{aligned} \tag{15.2.27}$$

Substituting the operator expansion (15.2.8), it is easy to see that the last term of the previous expression is the one that cancels the infrared divergences, so that the final expression that is finite at the first perturbative order is

$$\begin{aligned} C_{\Phi\Phi}^k(x) &= r^{2(\Delta_k-2\Delta_\Phi)}\times \\ &\left(\mathbf{C}_{\Phi\Phi}^k + \lambda\pi\,r^{2-2\Delta}\sum_l\left[\frac{\mathbf{C}_{\Phi\Phi}^l\,\mathbf{C}_{\varphi l}^k}{\Delta_k-\Delta_l-\Delta+1} - \frac{\mathbf{C}_{\varphi\Phi}^l\,\mathbf{C}_{\Phi l}^k}{\Delta_l-\Delta_\Phi-\Delta+1}\right] + \mathcal{O}(\lambda^2)\right), \end{aligned} \tag{15.2.28}$$

where $r = |x|$. It should be said that this series is not particularly convenient from a practical point of view. The most efficient way to compute the first-order correction of the structure constants is in fact through the integral

$$C_{\Phi\Phi}^{k\,(1)}(r) = -\lambda \int' \langle \tilde{A}^k(\infty)\,\varphi(y)\Phi(x)\Phi(0)\rangle_0 \, d^2y, \tag{15.2.29}$$

where the index in the integral means that we have to neglect all divergent terms that appear in the limit $R \to \infty$. This is equivalent to regard the integral as an analytic expression of the conformal dimensions of the fields. For instance, the integral of the three-point correlation function of the fields Φ_i, Φ_j and Φ_k of conformal dimensions Δ_i, Δ_j and Δ_k produces the analytic function

$$\begin{aligned}\mathcal{J}_{ijk}(x) = \int' d^2y\,\langle \Phi_i(x)\Phi_j(0)\Phi_k(y)\rangle_0 = \mathbf{C}_{ijk}\, r^{2(1-\Delta_i-\Delta_j-\Delta_k)} \times \\ \times 2\pi\, \frac{\Gamma(\Delta_i-\Delta_j-\Delta_k+1)\,\Gamma(\Delta_j-\Delta_i-\Delta_k+1)\,\Gamma(2\Delta_k-1)}{\Gamma(2-2\Delta_k)\,\Gamma(\Delta_i+\Delta_k-\Delta_j)\,\Gamma(\Delta_j+\Delta_k-\Delta_i)}.\end{aligned} \tag{15.2.30}$$

15.3 Example: The Two-point Function of the Yang–Lee model

A simple application of the formalism developed in the previous section is the computation of the off-critical correlation function of the Yang-Lee model (see Section 14.5). Suppose we perturb the CFT action $\mathcal{S}_0$ of this model by adding the perturbation of the field φ with $\Delta = -1/5$

$$\mathcal{S} = \mathcal{S}_0 + ig \int d^2x\, \varphi(x).$$

For this model a sensible QFT is obtained if the coupling constant is purely imaginary as shown above, with $g > 0$. The off-critical two-point function of the field $\varphi(x)$ can be written as

$$G(r) = \langle \varphi(x)\varphi(0)\rangle = C_{\varphi\varphi}^{I}(r)\,\langle I\rangle + C_{\varphi\varphi}^{\varphi}(r)\,\langle \varphi(0)\rangle + C_{\varphi\varphi}^{\bar{T}T}(r)\,\langle :\bar{T}T:(0)\rangle + \cdots. \tag{15.3.1}$$

The structure constants have a regular perturbative expansion that, taking into account the conformal dimension of the field φ, is given by

$$\begin{aligned} C_{\varphi\varphi}^{I}(r) &= r^{4/5}\left(1 + Q_1^I t + Q_2^I t^2 + \cdots\right) \\ C_{\varphi\varphi}^{\varphi}(r) &= \mathbf{C}_{\varphi\varphi}^{\varphi}\left(1 + Q_1^{\varphi} t + Q_2^{\varphi} t^2 + \cdots\right) \\ C_{\varphi\varphi}^{\bar{T}T}(r) &= \frac{1}{121} r^{24/5}\left(1 + Q_1^{\bar{T}T} t + Q_2^{\bar{T}T} t^2 + \cdots\right) \end{aligned}$$

where $t = gr^{12/5}$ is the dimensionless coupling constant and $\mathbf{C}_{\varphi\varphi}^{\varphi}$ is the (imaginary) structure constant of the CFT, given in eqn. (14.5.8). All coefficients $Q_i^{\mathcal{O}}$ can in principle be computed using the perturbative scheme of the previous section: the numerical values of the first ones are

$$\begin{aligned} Q_1^I &= 0.319800\ldots \\ Q_1^{\varphi} &= 0.02122\ldots. \end{aligned}$$

We need also the vacuum expectation value of the various fields, $\langle I \rangle$, $\langle \varphi(0) \rangle$, etc. entering eqn. (15.3.1). Here we take into account only $\langle I \rangle$ and the vacuum expectation value of the field φ, that can be computed exactly (see Chapter 20) with value

$$\begin{aligned} \langle \varphi \rangle &= i g^{-1/6} \frac{5}{24\sqrt{3}} \left(\frac{12}{25}\right)^{5/6} \left(\frac{12}{\pi}\right)^{1/6} \left(\frac{\Gamma\left(\frac{1}{3}\right)}{\Gamma\left(\frac{5}{6}\right)}\right)^2 \left(\left|\frac{\Gamma\left(-\frac{1}{5}\right)\Gamma\left(\frac{3}{5}\right)}{\Gamma\left(\frac{2}{5}\right)\Gamma\left(\frac{6}{5}\right)}\right|\right)^{5/12} \\ &= 0.840184\ldots i g^{-1/6} \end{aligned}$$

Substituting the expressions given above for the structure constants and the vacuum expectation values, can obtain an estimate of the function $G(r)$ up to order $\mathcal{O}(r^{24/5})$. The plot of this function, with a minus sign in front, is given in Figure 15.2.

This example is particularly significant the role played by the vacuum expectation value of the various fields for the off-critical correlators. Notice that, for the presence of a non-zero value of $\langle \varphi \rangle$, the two-point correlation function $G(r)$ behaves for small r not as $r^{4/5}$, as CFT reasons suggest, but rather as $r^{2/5}$. Furthermore, the competition between the terms coming from the families I and φ, that have opposite sign, produces the curve drawn above, that starts to bend for values of $r \simeq \xi$, where ξ is the finite correlation function of the off-critical model, whose exact value can be determined by the thermodynamic Bethe ansatz analysis of Chapter 19,

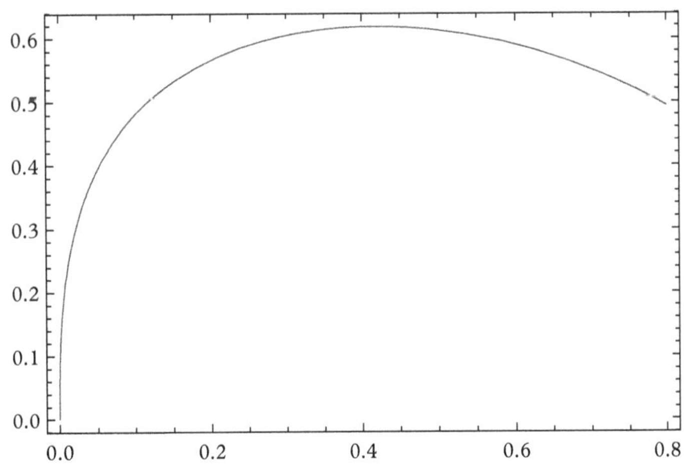

Fig. 15.2 *Plot of $-G(r)$ vs the distance r at $g = 1$.*

$$\xi = g^{-5/12}\left(\frac{25}{12}\right)^{5/12}\left(\frac{\pi}{12}\right)^{1/12}\left(\frac{\Gamma\left(\frac{5}{6}\right)}{\Gamma\left(\frac{1}{3}\right)}\right)\left(\left|\frac{\Gamma\left(\frac{2}{5}\right)\Gamma\left(\frac{6}{5}\right)}{\Gamma\left(-\frac{1}{5}\right)\Gamma\left(\frac{3}{5}\right)}\right|\right)^{5/24} = 0.37836...g^{-5/12}.$$

15.4 Renormalization Group and β-functions

This section reconsiders the RG theory in the perspective pursued in this chapter. We also present the computation at the lowest orders of the β functions.

The key ideas of the RG can be expressed as follows. Each theory is described by an action $\mathcal{S}(g,a)$ that is a function of a certain number of dimensionless coupling constants[2] $g = \{g_1, g_2, \ldots\}$ and a ultraviolet cut-off that cures the divergences coming from the short distance operator expansions. The main hypothesis consists of the existence of a one-parameter families of flows in the manifold G of the coupling constants, $\mathcal{R}_t\, G \to G$, such that the QFT described by the action $\mathcal{S}(\mathcal{R}_t g, e^t a)$ is equivalent to the theory described by the action $\mathcal{S}(g,a)$. In more detail, both theories give rise to the same result on a range of scales $|x| \gg e^t a$. Our main interests here are the two-dimensional theories but the results given below can be easily generalized to QFT in higher dimensions. As a starting point, let us consider the correlation functions of the local fields $A_i(x)$, defined as usual by the functional integral

$$\langle A_1(x_1)\ldots A_n(x_n)\rangle = \int \mathcal{D}\varphi\, A_1(x_1)\ldots A_n(x_n)\, e^{-\mathcal{S}[\varphi]}, \qquad (15.4.1)$$

where we have included in the definition of the action $\mathcal{S}$ an additive constant that ensures the correct normalization of the expression above (this permits to avoid the introduction of the normalization factor Z^{-1}). Consider now the Ward identity coming from the substitution

$$x^\mu \to x'^\mu = x^\mu + \epsilon^\mu(x),$$

Since[3]

$$\delta\mathcal{S} = \int d^2x\, T_{\mu\nu}(x)\, \partial^\mu \epsilon^\nu(x),$$

and

$$A_i(x) \to A_i(x) + \delta A_i(x), \qquad (15.4.2)$$

[2] Dimensionful couplings can be rescaled by appropriate powers of the cut-off a and expressed by their dimensionless version.

[3] With respect to the formula in eqn. (10.3.29), we have absorbed a factor $1/(2\pi)$ in the definition of $T_{\mu\nu}$.

we have

$$\sum_{i=1}^{n}\langle A_1(x_1)\ldots\delta A_i(x_i)\ldots A_n(x)\rangle - \int d^2x \langle T_{\mu\nu}(x) A_1(x_1)\ldots A_n(x_n)\rangle\, \partial^{\mu}\epsilon^{\nu}(x) = 0. \tag{15.4.3}$$

For a global dilation, the variation of the fields is given by

$$\delta A(x) = \epsilon\left(\frac{1}{2}x^{\mu}\,\partial_{\mu} + \hat{D}\right)A(x), \tag{15.4.4}$$

where $\hat{D}$ is the operator that implements the internal transformation of the fields under this transformation. Defining the trace of the stress-energy tensor

$$\Theta(x) \equiv T_{\mu}^{\mu}(x) \tag{15.4.5}$$

and substituting the last two equations in (15.4.3) we get

$$\sum_{i=1}^{n}\left\langle\left(\frac{1}{2}x^{\mu}\partial_{\mu} + \hat{D}_i\right)A_1\ldots A_n\right\rangle - \int d^2x\langle\Theta(x)A_1(x_1)\ldots A_n(x_n)\rangle = 0. \tag{15.4.6}$$

Let us discuss now the nature of the action. We assume that it is given by an integral on a local expression of the fields

$$\mathcal{S}_g = \int d^2x\,\mathcal{L}_g, \tag{15.4.7}$$

which depends on a certain number of dimensionless couplings $\{g_a\} = \{g_1, g_2, \ldots\}$. For instance, the perturbed CFTs (15.1.1) have, as coupling constants, those of the relevant fields, rescaled by the ultraviolet cut-off, i.e. $g_i = a^{2(1-\Delta_i)}\,\lambda_i$. For the property of $\mathcal{L}$, the derivative

$$\varphi_i(x) = \frac{\partial\mathcal{L}}{\partial g_i}, \tag{15.4.8}$$

is a local field of the theory. If we consider only homogeneous and isotropic interactions, these are scalar fields and the space of all of them, $\mathcal{O}^{(0)}$, may be regarded as the tangent space to G at the point $\{g_1, g_2, \ldots\}$. The fields $A_i(x)$ can also depend on the coupling constants and their variation, varying the couplings, is expressed in terms of an operator $\hat{B}_i$

$$\hat{B}_k A_i(x) = \frac{\partial}{\partial g_k}A_i(x). \tag{15.4.9}$$

The necessity to introduce a coupling dependence of the fields is obvious in view of the ultraviolet divergences and the implementation of their renormalization, as discussed in the previous section. For the correlation function (15.4.1) this implies

$$\frac{\partial}{\partial g_a}\langle A_1(x_1)\dots A_n(x_n)\rangle = \sum_{i=1}^{n}\langle A_1(x_1)\dots \hat{B}_a A_i(x_i)\dots A_n(x_n)\rangle - \int d^2x\,\langle \varphi_a(x) A_1(x_1)\dots A_n(x_n)\rangle. \tag{15.4.10}$$

The trace of the stress-energy tensor belongs to the space $\mathcal{O}^{(0)}$ and can be expressed in terms of the fields of this space as

$$\Theta(x) = \sum_a \beta^a(g)\,\varphi_a(x). \tag{15.4.11}$$

The coefficients $\beta_a(g)$ are the β functions of the theory. To see this, note that if dt is the infinitesimal parameter of the dilation, $x_\mu \to (1+dt)x_\mu$, the trace $\Theta(x)$ can be also defined as

$$\Theta(x) = \frac{d\mathcal{L}}{dt}, \tag{15.4.12}$$

an expression that can be written as

$$\Theta(x) = \sum_a \frac{\partial \mathcal{L}}{\partial g_a}\frac{\partial g_a}{\partial t} = \sum_a \frac{\partial g_a}{\partial t}\varphi_a(x). \tag{15.4.13}$$

Hence the β_a functions express the variation of the coupling constant under a change of the length scale

$$\beta^a(\{g\}) = \frac{\partial g_a}{\partial t}. \tag{15.4.14}$$

Combining now eqn. (15.4.11) with eqn. (15.4.11) we arrive at the celebrated Callan–Symanzik equation

$$\sum_{i=1}^{n}\left\langle\left(\frac{1}{2}x_i^\mu\frac{\partial}{\partial x_i^\mu} + \hat{\gamma}^{(i)}(g)\right)A_1(x_1)\dots A_n(x_n)\right\rangle + - \sum_a \beta^a(g)\frac{\partial}{\partial g_a}\langle A_1(x_1)\dots A_n(x_n)\rangle = 0, \tag{15.4.15}$$

where the linear operators $\hat{\gamma}^{(i)}(g)$, defined by

$$\hat{\gamma}(g) = D + \beta^a(g) B_a, \tag{15.4.16}$$

act on the fields $A_i(x)$. The operator $\hat{\gamma}$ is the matrix of the anomalous dimensions. Since the stress-energy tensor is a conserved quantity, it does not renormalize and therefore its anomalous dimension coincides with the canonical one

$$\hat{\gamma}(g)\Theta = 2\Theta. \tag{15.4.17}$$

In this way we easily obtain the action of $\hat{\gamma}(g)$ on the fields of the basis in the space $\mathcal{O}^{(0)}$

$$\hat{\gamma}(g)\,\varphi_a \equiv \gamma_a^b(g)\,\varphi_b = \left(2\delta_a^b - \frac{\partial \beta^b}{\partial g^a}\right)\varphi_b. \tag{15.4.18}$$

In the renormalized theory, every β_a function and any matrix element of the operator $\hat{\gamma}$ do not depend on the initial point of the RG $\mathcal{R}_0$. In particular, two field theories corresponding to coupling constants $g(t_1)$ and $g(t_2)$ that belong to the same integrated curve of their evolution equation

$$dg^a = \beta^a(g)\,dt, \tag{15.4.19}$$

differ only by a scale transformation of the length scale $x_\mu \rightarrow e^{(t_1-t_2)}\,x_\mu$. Therefore the scaling properties of the theory only depend on the vector fields $\beta^a(g)$. The simplest and, at the same time, the most important characteristic of these functions is associated to the fixed points g^*, that satisfy the equation $\mathcal{R}_t g^* = g^*$. These are the points where the β functions vanish, $\beta^a(g^*) = 0$. These conditions identify the critical points of the system, where the correlation length diverges. If, in addition to the location of the fixed points, we also know the derivatives of the β at these points, eqn. (15.4.18) allows us to compute the anomalous dimensions of the various operators. The next section provides an interesting example of this formalism.

Let us now discuss how to determine the first terms of the $\beta_a(g)$ functions by using the perturbative expansion of the partition function of the perturbed conformal theories, with an action given in eqn. (15.1.1). It is convenient to take care firstly of the dimensionality of the coupling constants: introducing a as a length scale and simultaneously as ultraviolet cut-off, they can be expressed as $\lambda_i = g_i a^{-2(1-\Delta_i)}$, where g_i are now the dimensionless couplings. In terms of these quantities, the perturbative expansion of Z is given by

$$Z = \int \mathcal{D}\varphi \,\exp\left[-\mathcal{S}^* - \sum_i g_i \int \frac{d^2x}{a^{2(1-\Delta_i)}}\varphi_i(x)\right] = Z^*\left[1 - \sum_i g_i \int \frac{d^2x}{a^{2(1-\Delta_i)}}\langle \varphi_i(x)\rangle\right.$$
$$\left. + \frac{1}{2}\sum_{i,j} g_i g_j \int_{|x_1-x_2|>a} \langle \varphi_i(x_1)\varphi_j(x_2)\rangle \frac{d^2x_1}{a^{2(1-\Delta_i)}}\frac{d^2x_2}{a^{2(1-\Delta_j)}} + \cdots \right].$$

To find the β functions, we seek how to change the coupling constants g_i under the infinitesimal scale transformation $a \to (1+\delta t)a$ in such a way that the partition function remains invariant.

Observe that, in the perturbative expansion of Z, the length scale a appears both explicitly (in the factors $a^{2-2\Delta_i}$ of the denominators) and implicitly, as cut-off of the integrals. If the rescaling of a is done with the infinitesimal parameter δt, there is an additive effect of the different dependences in the computation of the β functions. Let us consider then the two different terms separately.

The first dependence of the coupling constants from a is simple to take into account. In fact, a change of a is compensated by the substitution

$$g_i \to (1+\delta t)^{2(1-\Delta_i)} g_i \simeq g_i + 2(1-\Delta_i) g_i \, \delta t. \tag{15.4.20}$$

The effect of a change of the cut-off in the integrals can be instead estimated by the OPE. Consider, for instance, the second perturbative order. The integral, after a rescaling of a, can be written as

$$\int_{|x_1-x_2|>a(1+\delta t)} [\cdots] = \int_{|x_1-x_2|>a} [\cdots] - \int_{a<|x_1-x_2|<a(1+\delta t)} [\cdots]. \tag{15.4.21}$$

The first terms produces the original contribution in Z, the second term can be computed through the operator expansion of the conformal theory

$$\sum_k \mathbf{C}_{ijk} a^{2(\Delta_k-\Delta_i-\Delta_j)} \int_{a<|x_1-x_2|<a(1+\delta t)} \langle \varphi_k(x_2) \rangle \frac{d^2x_1}{a^{2(1-\Delta_i)}} \frac{d^2x_2}{a^{2(1-\Delta_j)}}. \tag{15.4.22}$$

The integral on x_1 gives the area of the infinitesimal annulus, i.e. $2\pi a^2 \delta t$. Taking into account the powers of a and the original negative sign of this contribution, see eqn. (15.4.21), in the partition function the term above gives

$$-\pi \delta t \sum_{i,j,k} \mathbf{C}_{ijk} g_i g_j \int \langle \varphi_k(x) \rangle \frac{d^2x}{a^{2(1-\Delta_k)}}. \tag{15.4.23}$$

The presence of this term in Z can be then compensated by a redefinition of the coupling constant g_k

$$g_k \to g_k - \pi \sum_{i,j} \mathbf{C}_{ijk} g_i g_j \, \delta t. \tag{15.4.24}$$

Gathering together the two contributions given in eqns. (15.4.20) and (15.4.24), we arrive at the first terms of the β functions

$$\frac{dg_k}{dt} \equiv \beta_k(g) = 2(1-\Delta_k)\, g_k - \pi \sum_{i,j} \mathbf{C}_{ijk}\, g_i\, g_j + \cdots . \tag{15.4.25}$$

Few comments are in order:

- The β functions rule the ultraviolet behaviour of the field theories and therefore it is not surprising that their first coefficients are expressed by the conformal data, such as the conformal dimensions and structure constants of the primary fields. Note that when the coupling constant g_k corresponds to a relevant operator with $\Delta_k < 1$, the first term is responsible of the repulsive nature of the origin fixed point, $g_i = 0$. Viceversa, if the coupling constant corresponds to an irrelevant operator with $\Delta_k > 1$, the origin becomes an attractive fixed point.
- The formula (15.4.25) can be generalized to higher-dimensional field theories, as long as we know the structure constants and the anomalous dimensions of the operators.
- The higher-order terms of the β functions can be computed, in principle, iterating the argument given above. It is easy, if fact, to see the iterative nature of the renormalization procedure: the terms of order $g_i^{n_i}\, g_j^{n_j}\, g_k^{n_k} \cdots$ influence the renormalization of the terms $g_i^{n_i-1}\, g_j^{n_j-1}\, g_k^{n_k+1} \cdots$. However, their explicit computation becomes soon involved and will not be discussed here.
- Focusing the attention only on the first perturbative terms of the β functions given in eqn. (15.4.25), it is important to establish the range of validity of these expressions. We expect that we can trust them when the coupling constants are those of the quasi-marginal operators, for which $|\epsilon_i| \sim \epsilon \ll 1$, where $\epsilon_i \equiv (1-\Delta_i)$. In these cases, in fact, the non-linear terms become comparable with the linear term when $g_i \sim \epsilon$. This process can give rise to new fixed points in the region $g_i \sim \epsilon$, that is a range compatible with the perturbative nature of the approach itself. The next section presents an interesting realization of this situation.

  When the coupling constant corresponds to a strongly relevant operator, the eventual fixed points are localized at a finite distance from the origin, i.e. outside the perturbative regime of the formalism. In this case, the evolution equations of the coupling constants give, in general, only a qualitative indication of the RG flows and caution must be used in extract quantitative predictions.
- It is interesting to observe that, at the lowest perturbative orders, the evolution equations of the coupling constants are irrotational flows, generated by the gradients of a scalar function. In fact, they can be written as

$$\dot{g}_k = \frac{\partial}{\partial g_k} \tilde{C}(g), \tag{15.4.26}$$

where the function $\tilde{C}(g)$ is defined as

$$\tilde{C}(g) = \sum_i (1 - \Delta_i) g_i^2 - \frac{\pi}{3} \sum_{ijl} \mathbf{C}_{ijl} g_i g_j g_l. \tag{15.4.27}$$

This observation will be useful in the discussion in Section 15.6.

15.5 c-theorem

For the deformations of the unitary conformal models there is an important theorem associated to the flows. The theorem, due to A.B. Zamolodchikov, states the following: for a two-dimensional field theory that is unitary, invariant under rotations and for which it holds the conservation of the stress-energy tensor, there exists a function of the coupling constants $C(\{\lambda_i\})$ that decreases along the flows, being stationary only at the fixed points. Its value at the fixed points coincide with the central charge c of the corresponding CFTs.

There is a simple proof of this theorem. Let T, Θ and $\overline{T}$ be the components of spin 2, 0 and -2 respectively of the stress-energy tensor. For their dimensions and spins, the off-critical correlators of these quantities can be parameterized as

$$\begin{aligned} < T(z,\bar{z}) T(0,0) > &= \frac{F(m\, z\bar{z})}{z^4}, \\ < T(z,\bar{z}) \Theta(0,0) > &= \frac{G(m\, z\bar{z})}{z^3 \bar{z}}, \\ < \Theta(z,\bar{z}) \Theta(0,0) > &= \frac{H(m\, z\bar{z})}{z^2 \bar{z}^2}, \end{aligned} \tag{15.5.1}$$

where m is a mass scale. Using the conservation law of the stress-energy tensor in complex coordinates (given in eqn. (10.5.6))

$$\begin{aligned} \partial_{\bar{z}} T + \tfrac{1}{4}\, \partial_z \Theta &= 0, \\ \partial_z \overline{T} + \tfrac{1}{4}\, \partial_{\bar{z}} \Theta &= 0, \end{aligned} \tag{15.5.2}$$

we obtain the differential equations for the scalar functions F, G and H

$$\begin{aligned} \dot{F} + \tfrac{1}{4}\left(\dot{G} - 3G\right) &= 0; \\ \dot{G} - G + \tfrac{1}{4}\left(\dot{H} - 2H\right) &= 0, \end{aligned} \tag{15.5.3}$$

where $\tau = m^2 z\bar{z} = (mR)^2$ and

$$\dot{F} \equiv \frac{dF(x)}{d\log\tau}.$$

Defining

$$C \equiv 2F - G - \frac{3}{8}H, \tag{15.5.4}$$

we have

$$\dot{C} = -\frac{3}{4}H. \tag{15.5.5}$$

The hypothesis of unitarity implies that H is a positive quantity and therefore C is a non-decreasing function of the distance τ, at fixed values of the coupling constants. At the critical points, the trace vanishes, $\Theta = 0$, and consequently $G = H = 0$ with $F = \frac{1}{2}c$. Hence the function C assumes the value of the central charge of the corresponding CFT.

The same theorem can be reformulated in terms of the coupling constants. In fact, fixed the parameter τ (for instance, $\tau = 1$), the quantities F, G and H become functions of the coupling constants g. For the dimensionless nature of the function $C(R,g)$ and its independence from the cut-off, it satisfies the RG equation

$$\left(\frac{1}{2}R\frac{\partial}{\partial R} - \sum_a \beta^a \frac{\partial}{\partial g_a}\right) C(R,g) = 0. \tag{15.5.6}$$

Using now eqn. (15.4.11), we get

$$\beta^a \frac{\partial}{\partial g^a} C(g) = -\frac{3}{4}\, G_{ab}(g)\, \beta^a(g)\, \beta^b(g), \tag{15.5.7}$$

where

$$G_{ab}(g) = G_{ab}(1,g), \quad G_{ab}(z\bar{z},g) = (m\, z\, \bar{z})^2 \, \langle \varphi_a(z,\bar{z})\varphi(0,0)\rangle$$

is a symmetric matrix, that is positive definite for the unitarity of the theory. As a by-product of this result, we see that $G_{ab}(g)$ may be regarded as a metric tensor in the space G of the coupling constants, with line element $ds^2 = G_{ab}(g) dg^a\, dg^b$.

The c-theorem admits also an integral formulation, that ends up in a sum rule. Integrating eqn. (15.5.5) from the ultraviolet fixed point at $r = 0$ and the infrared fixed point at $r = \infty$, and denoting by $\Delta c = c_1 - c_2$ the difference of central charges of the two conformal theories emerging in these limits, we have the equivalent expressions

$$\begin{aligned}\Delta c &= \frac{3}{4}\int_0^\infty d(r^2)\, r^2 \langle \Theta(r)\Theta(0)\rangle \\ &= \frac{3}{4\pi}\int d^2 r\, r^2 \langle \Theta(r)\Theta(0)\rangle \\ &= \frac{3}{2}\int dr\, r^3 \langle \Theta(r)\Theta(0)\rangle .\end{aligned} \tag{15.5.8}$$

This formula remarkably links the second moment of the off-critical correlation function of Θ to the variation of the central charges along the RG flow (an infrared massive theory corresponds to $c_2 = 0$).

The expression above can be also written in terms of the correlation function of the perturbing field. Consider, in fact, a CFT perturbed, for simplicity, by only one relevant scalar field φ, with conformal dimension Δ

$$\mathcal{S} = \mathcal{S}^* + \lambda \int d^2 x\, \varphi(x). \tag{15.5.9}$$

Let us examine the renormalization of the analytic component $T(z)$ of the stress-energy tensor. At the first order we have

$$\langle T(z)\cdots\rangle_\lambda = \langle T(z)\cdots\rangle_0 - \lambda \int d^2 z_1 \langle T(z)\varphi(z_1,\bar{z}_1)\cdots\rangle + \cdots .$$

From the operator expansion

$$\begin{aligned}T(z)\varphi(z_1,\bar{z}_1) &= \frac{\Delta}{(z-z_1)^2}\varphi(z_1,\bar{z}_1) + \frac{1}{z-z_1}\partial_z\varphi(z_1,\bar{z}_1) + \cdots \\ &= \frac{\Delta}{(z-z_1)^2}\varphi(z,\bar{z}) + \frac{1-\Delta}{z-z_1}\partial_z\varphi(z,\bar{z}) + \cdots\end{aligned}$$

can see that the integral is ultraviolet divergent and needs to be regularized. This can be done by inserting in the integral the step function[4] $H((z-z_1)(\bar{z}-\bar{z}_1)-a^2))$, where a is the ultraviolet cut-off. The most singular term vanishes after the angular integration but, for the presence of the function H and the cut-off a, the quantity $\partial_{\bar{z}} T$ is no longer zero. In fact,

$$\begin{aligned}\partial_{\bar{z}} T &= -\lambda \int \frac{1-\Delta}{(z-z_1)}(z-z_1)\,\partial_z\varphi(z,\bar{z})\,\delta(|z-z_1|^2 - a^2)\, d^2 z_1 \\ &= -\pi\,\lambda\,(1-\Delta)\,\partial_z\varphi .\end{aligned} \tag{15.5.10}$$

[4] $H(x) = 0$ if $x < 0$, while $H(x) = 1$ if $x > 0$.

Since the stress-energy tensor satisfies the conservation law

$$\partial_{\bar{z}} T + \frac{1}{4} \partial_z \Theta = 0,$$

comparing with the equation above we get

$$\Theta = 4\pi \lambda (1 - \Delta) \varphi(z, \bar{z}) + \cdots . \tag{15.5.11}$$

Note that this expression can be directly recovered from eqn. (15.4.12), for Θ is the conjugate field to the scale transformation $x \to tx$: in the action (15.5.9), $d^2x \to t^2 d^2x$ while $\varphi \to t^{-2\Delta} \varphi$, therefore taking the derivative with respect to t, we obtain eqn. (15.5.11), with the additional factor 2π that takes into account the conformal normalization of the operators. Let us look at some interesting applications of the c-theorem.

15.6 Applications of the c theorem

This section studies the RG flow associated to the deformation $\Phi_{1,3}$ of the minimal conformal models. Later, we present two applications of the sum rule of the c-theorem, relative to the Ising model and the lagrangian theory of the Sine–Gordon model.

15.6.1 Minimal Conformal Models $\mathcal{M}_p$ perturbed by the $\Phi_{1,3}$ Operator

The first significant application of the c-theorem is in the study of the unitary minimal models $\mathcal{M}_p$ perturbed by the relevant operator $\Phi_{1,3}$

$$\mathcal{S} = \mathcal{S}_p + \lambda \int d^2x \, \Phi_{1,3}(x). \tag{15.6.1}$$

$\Phi_{1,3}$ is an operator characterized by two specific properties. The first property is related to the OPE with itself, which has the skeleton form

$$\Phi_{1,3} \times \Phi_{1,3} = \mathbf{1} + \mathbf{C}_1 \, \Phi_{1,3} + \mathbf{C}_2 \, \Phi_{1,5}. \tag{15.6.2}$$

Since $\Phi_{1,5}$ is an irrelevant operator, the operator expansion above implies the renormalization of the field $\Phi_{1,3}$, that does not mix with any other fields. The second property is related to its conformal dimension

$$\Delta_{1,3} = 1 - \frac{2}{p+1} \equiv 1 - \epsilon. \tag{15.6.3}$$

For p sufficiently large, $\Phi_{1,3}$ is a quasi-marginal operator and therefore we are in the condition of the validity of eqn. (15.4.25). The structure constant $\mathbf{C}_1$ tends to a finite limit for $p \to \infty$: using its exact expression (obtained substituting in eqn. (11.5.53) the indices (n, m) with $(1, 3)$), we have

$$\mathbf{C}_1(\epsilon) = \frac{4}{\sqrt{3}} \frac{(1-2\epsilon)^2}{(1-\epsilon)(1-3\epsilon/2)} \left[\frac{\Gamma(1-\epsilon/2)\Gamma(1+3\epsilon/2)}{\Gamma(1+\epsilon/2)\Gamma(1-3\epsilon/2)}\right]^{1/2} \times$$
$$\times \frac{\Gamma(1-2\epsilon)\Gamma^2(1+\epsilon)}{\Gamma(1+2\epsilon)\Gamma^2(1-\epsilon)} = \frac{4}{\sqrt{3}}\left(1 - \frac{3\epsilon}{2} + \cdots\right). \tag{15.6.4}$$

The β function of the dimensionless coupling constant g associated to λ is then

$$\dot{g} = \beta(g) = 2\epsilon g - \pi \mathbf{C}_1 g^2 + \cdots. \tag{15.6.5}$$

The plot of this function, given in Figure 15.3, shows the existence of a new fixed point. As shown in Section 15.4, the β function can be written as the gradient of a scalar function

$$\dot{g} = \frac{\partial}{\partial g}\tilde{C}(g),$$

where $\tilde{C}(g)$, in this case, is given by

$$\tilde{C}(g) = \epsilon g^2 - \frac{\pi}{3}\mathbf{C}_1 g^3. \tag{15.6.6}$$

This expression allows us to easily compute the function $C(g)$ entering the c-theorem. In fact, both function have the same stationary points and, at this perturbative order, they must be proportional each other, so

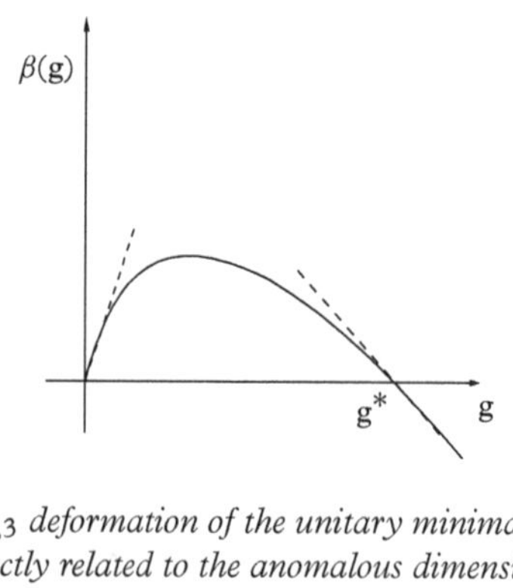

Fig. 15.3 *β function relative $\Phi_{1,3}$ deformation of the unitary minimal models. The slope of the tangents at the fixed points is directly related to the anomalous dimension of the operator.*

$$C(g) = c + \alpha \tilde{C}(g) + \mathcal{O}(g^4).$$

The proportionality constant can be fixed by a perturbative computation of C, using eqns. (15.5.5) and (15.5.11)

$$\begin{aligned} C &= c - \frac{3}{4}(4\pi)^2 (1-\Delta)^2 g^2 \int_0^1 \frac{r^4}{r^{4\Delta}} \frac{d(r^2)}{r^2} + \cdots \\ &= c - 6\pi^2 \epsilon g^2 + \cdots \end{aligned} \tag{15.6.7}$$

i.e. $\alpha = -6\pi^2$.

The fixed point of the β function is at $g^* = 2\epsilon/\pi \mathbf{C}_1$, and this value is compatible with the perturbation expansion. Substituing it in $C(g)$ we get an estimate of the central charge of the new fixed point

$$C(g^*) = c - \frac{8\epsilon^3}{\mathbf{C}_1^2} + \mathcal{O}(\epsilon^4), \tag{15.6.8}$$

Substituting in this expression $\epsilon \simeq 2/p$ and the value of the structure constant given above, we have

$$C(g^*) = c(p) - \frac{12}{p^3}. \tag{15.6.9}$$

At this perturbative order, the new value of the central charge coincides with the one of the unitary minimal model $\mathcal{M}_{p-1}$. Hence, the deformation $\Phi_{1,3}$ associated to a positive sign of the coupling constant gives rise, for large values of p, to a massless RG flow between two nearest CFTs

$$\mathcal{M}_p \to \mathcal{M}_{p-1}. \tag{15.6.10}$$

This scenario is compatible with the Landau–Ginzburg formualtion of these unitary minimal model. In fact, in the Landau–Ginzburg, the field $\Phi_{1,3}$ of the model $\mathcal{M}_p$ corresponds to the operator $:\varphi^{2(p-1)-2}:$. Once we switch on this deformation, the highest power $:\varphi^{2(p-1)}:$ that defines the original conformal theory $\mathcal{M}_p$ becomes irrelevant. Consequently, the infrared dynamics of the perturbed system is described by a Landau–Ginzburg with interaction $:\varphi^{2(p-2)}$, i.e. the one that corresponds to the conformal theory $\mathcal{M}_{p-1}$.

In this RG flow there is also an evolution of the other fields, so that they occupy different positions in the Kac table of the starting and ending CFT. Hence, also the anomalous dimensions of the operator change accordingly, and their variation can be computed using eqn. (15.4.18), evaluated at $g = g^*$. Let us compute, for instance, the anomalous dimension at the new fixed point g^* of the original field $\Phi_{1,3}$: the derivative of β at this fixed point is

$$\frac{\partial \beta}{\partial g}(g^*) = -2\epsilon \tag{15.6.11}$$

and therefore the conformal dimension of the field at this new fixed point is $\Delta' = 1 + 2/(p+1)$. At this perturbative order, it coincides with the conformal dimension of the irrelevant field $\Phi_{3,1}$ of the conformal model $\mathcal{M}_{p-1}$. Hence, in the RG flow, the operator $\Phi_{1,3}^{(p)}$ of $\mathcal{M}_p$ transforms in the operators $\Phi_{3,1}^{(p-1)}$ of $\mathcal{M}_{p-1}$.

In addition to the field $\Phi_{1,3}$, one can also study the evolution of the other fields of the Kac table. Consider, for instance, those along the main diagonal of the Kac table of the starting conformal theory, $\Phi_{n,n}^{(p)}$. To follow their evolution, some preliminary data are needed. Their OPE with $\Phi_{1,3}^{(p)}$ reads

$$\Phi_{n,n}^{(p)}\,\Phi_{1,3}^{(p)} = \mathbf{C}_{(n,m),(1,3)}^{(n,m)}\left[\Phi_{n,n}^{(p)}\right] + \mathbf{C}_{(n,m),(1,3)}^{(n,n+2)}\left[\Phi_{n,n+2}^{(p)}\right] + \mathbf{C}_{(n,m),(1,3)}^{(n,n-2)}\left[\Phi_{n,n-2}^{(p)}\right].$$

For large p, the structure constants tend to

$$\begin{aligned}
\mathbf{C}_{(n,n),(1,3)}^{(n,n)} &= \frac{(n-p)^2(p+1)}{2\sqrt{3}(p-1)} + \mathcal{O}(\epsilon) \;\; (n \leq p) \\
\mathbf{C}_{(n,n),(1,3)}^{(n,n)} &= \frac{(n^2-1)}{8\sqrt{3}}\epsilon^2 + \mathcal{O}(\epsilon^3) \;\; (n \ll p) \\
\mathbf{C}_{(n,n),(1,3)}^{(n,n+2)} &= \left(\frac{p+2}{3p}\right)^{1/2} + \mathcal{O}(\epsilon) \\
\mathbf{C}_{(n,n),(1,3)}^{(n,n-2)} &= \frac{(n^2-1)^{1/2}}{\sqrt{3}n} + \mathcal{O}(\epsilon).
\end{aligned}$$

The conformal dimensions of the operators $\Phi_{n,n}^{(p)}$ are

$$\Delta_{(n,n)}^{(p)} = \frac{n^2-1}{4p(p+1)} = \frac{n^2-1}{16}\epsilon^2\left(1 + \frac{\epsilon}{2} + \cdots\right), \tag{15.6.12}$$

and, for $n \ll p$, they are strongly relevant. Their operator expansion with $\Phi_{1,3}$ shows that they do not mix with any other operators. To determine in which fields they transform along the RG flow, one must compute their anomalous dimensions at $g = g^*$ in terms of the derivative of $\beta^{(n)}$ functions of these fields with respect the coupling constant g. Using the general formula (15.4.25), we get

$$\frac{\partial \beta^{(n)}}{\partial g} = -2\pi\,\mathbf{C}_{(n,n),(1,3)}^{(n,n)}\,g + \cdots \tag{15.6.13}$$

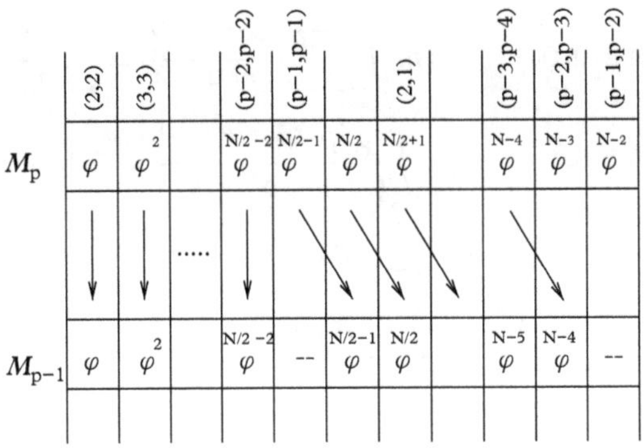

Fig. 15.4 *Mappings of the conformal fields in the massless RG flow $\mathcal{M}_p \to \mathcal{M}_{p-1}$ induced by $\Phi^{(p)}_{1,3}$. Here $N = 2(p-1)$.*

so that, at the new fixed points, the anomalous dimensions are

$$\begin{aligned}\hat{\gamma}(g^*)\,\Phi_{n,n} &= \gamma_{(n,n)}(g^*)\,\Phi_{n,n} = \left[2\Delta_{(n,n)} + 2\pi\,\mathbf{C}^{(n,n)}_{(n,n),(1,3)}\,g^*\right]\Phi_{n,n} \\ &= 2\epsilon^2\,\frac{n^2-1}{16}\left(1+\frac{3\epsilon}{2}+\cdots\right) = \frac{n^2-1}{4p(p-1)} + \mathcal{O}(\epsilon^4).\end{aligned}$$

This implies the mappings

$$\Phi^{(p)}_{n,n} \to \Phi^{(p-1)}_{n,n}. \tag{15.6.14}$$

The analysis can be extended to all other fields and the final result is summarized in Figure 15.4.

15.6.2 Ising model at temperature $T \neq T_c$

Consider the two-dimensional Ising model in its fermionic formulation at $T \neq T_c$. At the critical point the action is

$$\mathcal{S}^* = \int d^2x \left[\psi\,\partial_{\bar{z}}\,\psi + \bar{\psi}\,\partial_z\,\bar{\psi}\right], \tag{15.6.15}$$

and the perturbation that moves the system away from the critical point is given by the mass term $im\int \bar{\psi}\,\psi\, d^2x$. To compute the correlator of $\Theta(x) = im\bar{\psi}\,\psi$ we need the propagator of the massive fermionic field

$$
\begin{aligned}
\langle\, \bar{\psi}(z,\bar{z})\psi(0,0)\,\rangle &= -im\int \frac{d^2p}{(2\pi)^2}\frac{e^{\frac{i}{2}(p\bar{z}+\bar{p}z)}}{p^2+m^2} = -i\frac{m}{2\pi}K_0(mr) \\
\langle\, \psi(z,\bar{z})\psi(0,0)\,\rangle &= -i\int \frac{d^2p}{(2\pi)^2}\frac{\bar{p}e^{\frac{i}{2}(p\bar{z}+\bar{p}z)}}{p^2+m^2} = 2\,\partial_z\frac{1}{2\pi}K_0(mr) \\
&= -\frac{m}{2\pi}\frac{\bar{z}}{z}K_1(mr) \\
\langle\, \bar{\psi}(z,\bar{z})\bar{\psi}(0,0)\,\rangle &= -\frac{m}{2\pi}\frac{z}{\bar{z}}K_1(mr)
\end{aligned}
\tag{15.6.16}
$$

with $r=\sqrt{z\bar{z}}$, where $K_i(x)$ are the modified Bessel functions. Applying the Wick theorem, we have

$$
\begin{aligned}
\langle\Theta(r)\Theta(0,0)\rangle &= -m^2\,\langle\bar{\psi}(r)\psi(r)\bar{\psi}(0)\psi(0)\rangle \\
&= -|\langle\bar{\psi}(r)\psi(0)\rangle|^2 + \langle\psi(r)\psi(0)\rangle\,\langle\bar{\psi}(r)\bar{\psi}(0)\rangle \\
&= \left(\frac{m^2}{2\pi}\right)^2\left[K_1^2(mr)-K_0^2(mr)\right].
\end{aligned}
\tag{15.6.17}
$$

Substituting this expression into the sum rule (15.5.8) and computing the integral, we get

$$
\Delta c = \frac{1}{2}. \tag{15.6.18}
$$

Since $c=\frac{1}{2}$ is the central charge of the critical Ising model, we have an explicit check that the perturbed theory has central charge $c=0$, i.e. a purely massive field theory.

15.6.3 A Lagrangian theory: the Sine–Gordon model

Another interesting example of the c-theorem sum rule comes from a lagrangian theory with a varying coupling constant. Consider the Sine–Gordon model, with a Lagrangian

$$
\mathcal{L} = :\frac{1}{2}(\partial_\mu\varphi)^2 + \frac{m^2}{\beta^2}(\cos\beta\varphi - 1):. \tag{15.6.19}
$$

Let us restrict the attention to the range $\beta^2 < 8\pi$, where $\cos\beta\varphi$ is a relevant operator. This massive model (with $c=0$) can be regarded as a deformation of the free massless bosonic theory with $c=1$. Adopting this interpretation, we have

$$
\lambda\varphi \equiv \epsilon(x) = \frac{m^2}{\beta^2} : (\cos\beta\varphi - 1):
$$

with anomalous dimension

$$2\Delta = \frac{\beta^2}{4\pi}.$$

Eqn. (15.5.8) becomes

$$\Delta c = 3\pi \left(2 - \frac{\beta^2}{4\pi}\right)^2 \int d^2x \mid x \mid^2 < \epsilon(x)\epsilon(0) >. \tag{15.6.20}$$

Since the left-hand side of (15.6.20) does not depend on β (in particular, it is equal identically to 1), the same must hold for the right-hand side. Note that the sum rule is already saturated at the zero order in β by the term that corresponds to the free massive theory. At this order we have in fact

$$\epsilon(r) = \frac{m^2}{2}\varphi^2(r), \qquad \Delta = 0;$$

$$< \epsilon(r)\epsilon(0) > = \frac{m^4}{2} < \varphi(r)\varphi(0) >^2 = \frac{m^4}{8\pi^2} K_0^2(mr),$$

and therefore

$$\Delta c_0 = 3\pi \frac{m^4}{2\pi^2} \int d^2x \mid x \mid^2 K_0^2(m \mid x \mid) = 3 \int_0^\infty dR\, R^3\, K_0^2(R) = 1. \tag{15.6.21}$$

This implies that, expanding in power series with respect to β^2, in the right-hand side of (15.6.20) all coefficients but the constant must vanish. Let us check the validity of this conclusion on the first non-trivial term: taking the first derivative with respect to β^2 and posing $\beta^2 = 0$ we have

$$\frac{1}{3\pi}\frac{d(\Delta c)}{d\beta^2}\bigg|_{\beta^2=0} = -\frac{1}{\pi}\int d^2x \mid x \mid^2 < \epsilon(x)\epsilon(0) > +4 \int d^2x \mid x \mid^2 \frac{d}{d\beta^2} < \epsilon(x)\epsilon(0) >.$$

The first term has been already computed in the free massive theory. To compute the second term, let us expand $\cos(\beta\varphi)$ up to the fourth order and then use the Wick theorem, with the result

$$\frac{1}{3\pi}\frac{d(\Delta c)}{d\beta^2}\mid_{\beta^2=0} = -\frac{1}{12\pi^2} + \frac{m^6}{(2\pi)^4}\int d^2x\, d^2z \mid x \mid^2 K_0^2(m \mid x - z \mid)\, K_0^2(m \mid z \mid). \tag{15.6.22}$$

The last integral can be easily computed: changing variable

$$\mid x - z \mid \rightarrow \mid t \mid,$$

it is expressed by the product of the integrals

$$2\pi \int_0^\infty R^3\, K_0^2(R)\, dR = \tfrac{2\pi}{3};$$
$$2\pi \int_0^\infty R\, K_0^2(R)\, dR = \pi;$$

(the term (15.6.22) with the scalar product $\vec{x}\cdot\vec{z}$ vanishes after the angular intergration). Inserting this expression in (15.6.22), we can see that the variation of the central charge by varying β^2 is effectively zero, as it should be. The perturbative check can be easily generalized to the next order and it is natural to conjecture its validity to all perturbative orders. Hence, using the c-theorem we can generate in this case an infinite number of identities that involve the integrals of the correlation functions of the Sine–Gordon model.

15.7 Δ theorem

The c-theorem provides useful information on the RG flow induced by the relevant fields of a CFT. The Delfino, Simonetti and Cardy theorem that allows us to follow directly the change of the anomalous dimensions of the various fields.

Consider the off-critical correlators of the components $T(z,\bar{z})$ and $\Theta(z,\bar{z})$ of the stress-energy tensor with a field Φ, where the latter is the deformed primary field of the perturbed conformal theory. These correlators can be parameterized as

$$\begin{aligned}\langle T(z,\bar{z})\Phi(0,0)\rangle &= \frac{U(mz\bar{z})}{z^2},\\ \langle \Theta(z,\bar{z})\Phi(0,0)\rangle &= \frac{V(mz\bar{z})}{z\bar{z}}.\end{aligned}$$

Using the conservation of the stress-energy tensor, eqn. (15.5.2), we arrive at the differential equation

$$\dot{D} = \frac{1}{4}V, \tag{15.7.1}$$

where $D = U + \frac{1}{4}V$ and the dot denotes the logarithmic derivative $z\bar{z}\frac{d}{dz\bar{z}}$. Since the trace Θ is related to the perturbing field by the relation (15.5.11), the short-distance expansion of the function V is determined by the OPE with the perturbing field Φ, i.e.

$$V(x) \simeq 2\pi\,\lambda\,(2-2\Delta)\,C^0_{\varphi\Phi}\,|x|^{2(\Delta_0-\Delta_\Phi-\Delta+1)}\,\langle A_0\rangle, \tag{15.7.2}$$

where A_0 is the most relevant operator that appears in this expansion. It is necessary to distinguish two cases.

1. if

$$\Delta_0 - \Delta_\Phi - \Delta + 1 > 0, \tag{15.7.3}$$

then $V(x)$ vanishes in the conformal limit $x \to 0$. In this case the function D is stationary at the fixed point and coincides with U. If the operator Φ does not mix with other fields under renormalization, we can use its operator expansion with T and for the function U we have

$$U(x) \simeq \Delta_\Phi \langle \Phi \rangle. \tag{15.7.4}$$

If the perturbed theory is associated to massless flow to another CFT, the same analysis can be repeated nearby the other fixed point and, integrating on all distance scales, we establish the sum rule

$$\Delta_\Phi^{\mathrm{uv}} - \Delta_\Phi^{\mathrm{ir}} = -\frac{1}{4\pi \langle \Phi \rangle} \int d^2x \, \langle \Theta(x) \Phi(0) \rangle. \tag{15.7.5}$$

If the RG flow lead to a massive theory, we have instead $\Delta_\Phi^{ir} = 0$.

2. If we have instead

$$\Delta_0 - \Delta_\Phi - \Delta + 1 \leq 0, \tag{15.7.6}$$

the function $V(x)$ does not vanish at the origin and the attempt of using eqn. (15.7.5) fails for the divergence of the integral. This is a simple consequence of the mixing of the operator Φ under renormalization. In this case, the function $U(x)$ does not present the behaviour of eqn. (15.7.4) at short distances. Its correct behaviour is obtained by directly integrating eqn. (15.7.1), namely

$$U(x) \simeq \pi\lambda(1-\Delta)\, C^0_{\varphi\Phi} \frac{1-\gamma_0}{\gamma_0} |x|^{2\gamma_0} \quad \gamma_0 < 0$$
$$U(x) \simeq 2\pi\lambda(1-\Delta)\, C^0_{\varphi\Phi} \langle A_0 \rangle \log|x| \quad \gamma_0 = 0$$

where we introduce the notation $\gamma_0 = \Delta_0 - \Delta_\Phi - \Delta + 1$.

It is worth stressing that the Δ-theorem, as expressed by eqn. (15.7.5), can be easily generalized to QFTs of any dimension when the integral converges both in the ultraviolet and the infrared regions. Indeed it simply expresses the Ward identity relative to the trace of the stress-energy tensor of any field theory, i.e. the field responsible of the global scale transformations.

REFERENCES

The splitting of the ultraviolet and infrared properties of a QFT has a long history. It has been applied, for instance, to study the vacuum of quantum chromodynamics. For this topic see

M. A. Shifman, A.I. Vainshtein, V. I. Zakharov, *QCD and Resonance Physics. Sum Rules*, Nucl. Phys. B 147 (1979), 385.

In two dimensions the perturbative formalism has been developed by in

Al.B. Zamolodchikov, *Two point correlation function in scaling Lee-Yang model*, Nucl. Phys. B 348 (1991), 619.

For a discussion of the RG in two-dimensional systems see

A.B. Zamolodchikov, *Renormalization group and perturbation theory about fixed points in two-dimensional field theory*, Sov. J. Nucl. Phys. 45 (1987), 1090.

A.W.W. Ludwig, J.L. Cardy, *Perturbative evaluation of the conformal anomaly at new critical points with applications to random systems*, Nucl. Phys. B 285 (1987), 687.

The computation of the β-functions with the formalism of the OPE is discussed in

J.L. Cardy, *Conformal invariance and statistical mechanics*, in Fields, Strings and Critical Phenomena (Les Houches 1988), ed. E. Brezin and J. Zinn-Justin

The c-theorem is proved in

Zamolodchikov, A.B. (1986). 'Irreversibility of the Flux of the Renormalization Group in a 2-D Field Theory', *Journal of Experimental and Theoretical Physics Letters*, 43: 730.

Its integral version is found in

Cardy, J. (1998). 'Central Charge and Universal Combinations of Amplitudes in Two-Dimensional Theories Away from Criticality', *Physics Review Letters*, 60: 2709.

The Δ-theorem has been proved in the article

Delfino, G., Simonetti, P. and Cardy, J.L. (1996). 'Asymptotic Factorization of Form-factors in Two-dimensional Quantum Field Theory', *Physics Letters B*, 387: 327.

16

Integrable Quantum Field Theories

This belief is handed down in Beersheba: that, suspended in the heavens, there exists another Beersheba, where the city's most elevated virtues and sentiments are poised, and that if the terrestrial Beersheba will take the celestial one as its model the two cities will become one.

Italo Calvino, Invisible Cities

16.1 Introduction

An integrable QFT is characterized by an infinite number of conserved charges. In classical mechanics, the existence of a sufficient number of integrals of motion allows us to pass from the initial coordinates and momenta to the angle-action variables, thus finding the exact solution of the equation of motion by quadrature. Similarly, if in a QFT there are an infinite number of conservation laws, we can derive the exact mass spectrum of its excitations, the S-matrix of the scattering processes, the correlations functions, the thermodynamics, and so on, in short its exact solution. For reasons that will become clearer later, non-trivial integrable QFTs can only occur in $(1+1)$ dimensions.[1] In higher dimensions, in fact, they are either free theories or models with non-local interactions. Hence, we focus our attention only on two-dimensional models.

In $(1+1)$ dimensions, using the complex notation to denote the analytic and anti-analytic indices of tensor quantities, the conservation law of a current with components (T_{s+1}, Θ_{s-1}) is written as

$$\partial_{\bar{z}} T_{s+1} = \partial_z \Theta_{s-1}, \tag{16.1.1}$$

and this leads to the conservation of the charges

$$Q_s = \oint [T_{s+1}\, dz + \Theta_{s-1} d\bar{z}]. \tag{16.1.2}$$

[1] A remark on the notation: in the following we use the terminology '$(1+1)$ dimensions' if we want to stress the Minkowski version of a two-dimensional QFT, while we use the terminology 'two dimensions' to denote both a generic two-dimensional QFT or its Euclidean version.

Statistical Field Theory: An Introduction to Exactly Solved Models in Statistical Physics. Second Edition. Giuseppe Mussardo, Oxford University Press (2020). © Giuseppe Mussardo. DOI: 10.1093/oso/9780198788102.001.0001

The integer index s that identifies the integrals of motion is called the *spin* of the operator $\mathcal{Q}_s$. The value $s=1$ always corresponds to the stress-energy tensor, with $\mathcal{Q}_1=\mathcal{P}$, where $\mathcal{P}=E+P$ is the analytic part of the total momentum of the system. The set of values of s is specific of each integrable model, as we in the examples presented in this chapter.

In the previous chapters we stress that an important aspect of the two-dimensional conformal theories is the splitting of the analytic and anti-analytic sectors. If we take, in this case, as T_{s+1} any independent descendent field at the level $(s+1)$ of the identity conformal family, it is easy to see that the conservation laws (16.1.1) are identically satisfied for, in conformal theories, T_{s+1} is a purely analytic field satisfying $\partial_{\bar{z}} T_{s+1}=0$. Hence, all two-dimensional conformal theories have an infinite number of conservation laws and therefore can be also considered as integrable models.

Perturbing the conformal theories by means of the insertion of one or more relevant fields, there is a breaking of the factorization of the analytic/anti-analytic sectors. Consequently, there is in general the destruction of all the hierarchy of the conserved currents of the conformal point. As Chapter 15 shows, the analysis of the models away from criticality can be always carried on by perturbative techniques, but this approach rarely leads to an exact solution of the model. It is therefore a circumstance of the utmost importance that some particular deformations of the critical action lead to the definition of integrable models also away from criticality. This possibility opens in fact more interesting scenarios than the perturbative approach, since it allows us to solve exactly the statistical models also away from criticality.

In the first part of this chapter we discuss the integrable QFTs that are associated to a Lagrangian density. In the second part, we examine the conditions on the deformations of a conformal theory that lead to the existence of conserved currents and an integrable theory away from the critical point. The physical consequences of this remarkable circumstance will be the objects of our study in the next chapters.

16.2 The Sinh–Gordon Model

Consider the two-dimensional Euclidean space and the Lagrangian theory of the so-called Sinh–Gordon model. Its action is given by[2]

$$\mathcal{S}=\frac{1}{8\pi}\int d^2x\left[(\partial_\mu\phi)^2+\frac{\mu^2}{g^2}\cosh(\sqrt{2}g\phi(x))\right]. \tag{16.2.1}$$

With this normalization, in the limit $m\to 0$ the two-point correlation function is

$$\langle\phi(x)\phi(y)\rangle = -2\log(|x-y|^2). \tag{16.2.2}$$

[2] The numerical factors of this definition are chosen in such a way to match the notation of the general Toda field theories, discussed later on.

The Sinh–Gordon model provides the simplest example of a general class of integrable models, the so-called Toda field theories discussed in more detail in Section 16.6. The action (16.2.1) is invariant under the Z_2 transformation $\phi \to -\phi$. The potential of the Lagrangian

$$V(\phi) = \frac{\mu^2}{8\pi g^2} \cosh(\sqrt{2}g\phi),$$

has a unique minimum at the origin, with a quadratic curvature equal to μ^2. Hence, in the Minkowski space, this theory describes the interactions of a relativistic particle of (bare) mass μ. In the Euclidean space, though, there are several equivalent ways to consider this field theory according to the different splittings of the action

$$\mathcal{S} = \mathcal{S}_0 + \mathcal{S}_I. \tag{16.2.3}$$

In this expression $\mathcal{S}_0$ plays the role of the unperturbed action while $\mathcal{S}_I$ of its deformation. This splitting leads to different expressions of the central charge of the conformal theory that emerges in the ultraviolet regime of the Sinh–Gordon model. Which central charge will be selected is in fact related to the choice of $\mathcal{S}_0$, while the central charge in the infrared regime is always $C_{ir} = 0$ since, independently of the various splittings, the theory is always massive in its infrared regime.

We should not be surprised that the action (16.2.1) can describe different ultraviolet fixed points. As a matter of fact, the value of the central charge is not linked to the Lagrangian but it is instead related to the definition that we assume for the associate stress-energy $T^{\mu\nu}(\mathrm{x})$. This is an operator intrinsically defined up to a total divergence: if we denote by $\tilde{T}^{\mu\nu}$ the stress-energy tensor coming from the Noether theorem, there is in fact a one-parameter family (labelled by the parameter α) of stress-energy tensors associated to the *same* Lagrangian given by

$$\begin{aligned} T_{\mu\nu}(x) &= \tilde{T}_{\mu\nu}(x) + \alpha\left(\partial_\mu\partial_\nu - g_{\mu\nu}\Box\right)\phi(x), \\ \tilde{T}_{\mu\nu}(x) &= \left[\partial_\mu\phi\,\partial_\nu\phi - \eta_{\mu\nu}\left(\frac{1}{2}(\partial\phi)^2 - V(\phi)\right)\right]. \end{aligned} \tag{16.2.4}$$

We recall that, in the Coulomb gas formalism, the presence of α is equivalent to introduce a charge at infinity and, as shown in Chapter 11, this changes the conformal properties of the fields and leads to different values of the central charge. Let us see the four different ways of interpreting the theory (16.2.1), each of them corresponding to a particular choice of the parameter α.

Feynman perturbation approach. The first approach is based on the standard perturbation theory defined by the Feynman graphs. In this case, firstly we expand in power series in g the hyperbolic term present in the Lagrangian, identifying as $\mathcal{S}_0$ the expression

$$\mathcal{S}_0 = \frac{1}{16\pi} \int d^2x \Big[(\partial_\mu \phi)^2 + \mu^2 \phi^2) \Big], \tag{16.2.5}$$

and as $\mathcal{S}_I$ all the other terms of the series expansion. In this approach, the Sinh–Gordon model appears as a Landau–Ginzburg theory with an infinite number of interaction terms, all even in the field ϕ. Since in two dimensions the field ϕ is dimensionless, the theory is renormalizable. The only ultraviolet divergences come from the *tadpole* graphs (Figure 16.1). Called Λ the ultraviolet cut-off, these divergences can be removed at once by redefining the exponentials in terms of the normal order product[3] with respect to an arbitrary mass scale μ

$$e^{\pm\sqrt{2}g\phi(x)} \to :e^{\pm\sqrt{2}g\phi(x)}: = \left(\frac{\Lambda}{\mu}\right)^{g^2} e^{\pm\sqrt{2}g\phi(x)}. \tag{16.2.6}$$

The removal of the divergences is equivalent to the renormalization of the mass term μ^2. This choice of $\mathcal{S}_0$ corresponds to $\alpha = 0$ in eqn. (16.2.4). The corresponding ultraviolet central charge is then $C_{uv} = 1$, derived by applying the c-theorem discussed in Chapter 15.

Deformation of Gaussian action. An alternative way to consider the Sinh–Gordon model consists of taking as $\mathcal{S}_0$ only the kinetic term and as $\mathcal{S}_I$ the Z_2 invariant combination of the vertex operators

$$\mathcal{S}_O = \frac{1}{16\pi} \int d^2x (\partial_\mu \phi)^2, \quad \mathcal{S}_I = \frac{1}{16\pi g^2} \int d^2x \left(e^{\sqrt{2}g\phi} + e^{-\sqrt{2}g\phi} \right). \tag{16.2.7}$$

In this case, $\mathcal{S}_0$ is explicitly associated to the conformal theory of a free Gaussian bosonic field (discussed in Chapter 12), whose ultraviolet central charge is $C_{uv} = 1$. Also in

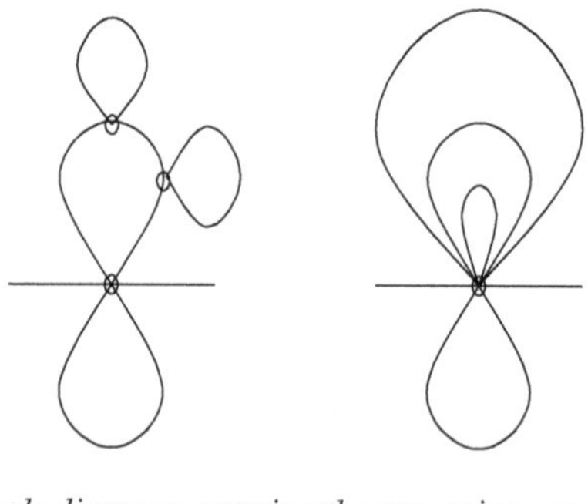

Fig. 16.1 *Tadpole diagrams entering the two-point correlation function.*

[3] This definition of the normal order prohibits that a propagator starts and ends at the same point. It is similar to the definition adopted in Section 11.5.1 for the vertex operators, even though in this case the conformal dimensions of the exponential operators are negative, $\Delta = -g^2$.

this case $\alpha = 0$, but with $\tilde{T}_{\mu\nu}$ only expressed by the kinetic part $\partial_\mu\phi\partial_\nu\phi$, while all the remaining terms considered as part of its trace.

Deformation of Liouville action I. A third way to look at the Sinh–Gordon model is to take as $\mathcal{S}_0$ the Liouville action

$$\mathcal{S}_0 = \int d^2x \left[\frac{1}{16\pi} (\partial_\mu \phi)^2 + \lambda e^{\sqrt{2} g\phi} \right], \tag{16.2.8}$$

and as $\mathcal{S}_I$ the deformation given by the relevant operator $e^{-\sqrt{2}g\phi}$. Although the Liouville action is formally invariant under the conformal transformations, its correct quantization requires the introduction of a charge at infinity

$$Q_+ = \frac{1}{\sqrt{2}} \left(g + \frac{1}{g} \right). \tag{16.2.9}$$

This quantization method is done along the Coulomb gas approach discussed in Section 11.5.1. The charge at infinity shifts the value of the ultraviolet central charge, which is no longer $C_{uv} = 1$ but

$$C_{uv} = 1 + 12\, Q_+^2. \tag{16.2.10}$$

There is also a shift of the conformal dimensions of the vertex operators $e^{\alpha\phi(x)}$, now given by

$$\Delta_+^{(\alpha)} = -\frac{1}{2}\alpha^2 + \alpha\, Q_+. \tag{16.2.11}$$

It should be now clear the reason to quantize the Liouville theory with the charge at infinity: in fact, it is only in this way that the vertex operator $e^{\sqrt{2}g\phi(x)}$, present in the action (16.2.8), acquires a conformal dimension equal to 1, so that the action (16.2.8) becomes conformal invariant at the quantum level. For the perturbing operator $e^{-g\phi(x)}$, its new conformal dimension is instead

$$\Delta_- = -1 - 2g^2. \tag{16.2.12}$$

Deformation of Liouville action II. In this Liouville approach, we can exchange the role played by the two exponentials, namely we can take as $\mathcal{S}_0$ the Liouville theory defined by the other exponential

$$\mathcal{S}_0 = \int d^2x \left[\frac{1}{16\pi} (\partial_\mu \phi)^2 + \lambda e^{-\sqrt{2} g\phi} \right], \tag{16.2.13}$$

and as perturbation S_I the one induced by the vertex operator $e^{\sqrt{2}g\phi(x)}$. As in the previous case, this Liouville action needs a charge at infinity for its correct quantization, this time given by

$$\mathcal{Q}_{-} = -\frac{1}{\sqrt{2}}\left(g+\frac{1}{g}\right). \tag{16.2.14}$$

This charge at infinity modifies the values of the central charge and the conformal dimensions of the vertex operators $e^{\alpha\phi(x)}$, which can be obtained by substituting $\mathcal{Q}_{+} \to \mathcal{Q}_{-}$ in the previous formulae (16.2.10) and (16.2.11). In this scheme the operator $e^{-\sqrt{2}g\phi(x)}$ has conformal dimension equal to 1, while that of $e^{\sqrt{2}g\phi(x)}$ is

$$\Delta_{+} = -1-2g^2. \tag{16.2.15}$$

Integrability. It is worth stressing that, independently of how we interpret the theory in the ultraviolet regime, the Sinh–Gordon model enjoys a fundamental property: it is an integrable model. To obtain the classical expression of the conserved charges $\mathcal{Q}_s$, it is convenient to rescale for simplicity the field as $\sqrt{2}g\phi(x) \to \phi$ and pose $\mu = 1$. Let us also introduce the light-cone coordinates σ and τ

$$\sigma = \frac{1}{2}(x-t)\,; \quad \tau = \frac{1}{2}(x+t).$$

In these coordinates the equation of motion becomes

$$\partial_\sigma\, \partial_\tau\, \phi(\sigma,\tau) = \sinh(\phi). \tag{16.2.16}$$

There is a conserved charge $\mathcal{Q}_s$ if exists a current with components $(\mathcal{J}_s^0, \mathcal{J}_s^1)$ satisfying the equation $\partial_\mu \mathcal{J}_s^\mu = 0$. This can be written in light-cone coordinates defining $\mathcal{J}_s^0 = T_{s+1} + \Theta_{s-1}$ and $\mathcal{J}_s^1 = T_{s+1} - \Theta_{s-1}$. For the densities $T_{s+1}[\phi]$ and $\Theta_{s-1}[\phi]$ we have

$$\frac{\partial}{\partial\sigma} T_{s+1}[\phi] = \frac{\partial}{\partial\tau}\Theta_{s-1}[\phi]. \tag{16.2.17}$$

The index s refers to the spin of this current, related to the difference of the partial derivatives ∂_τ^n and ∂_σ^k present in the expression of the densites, $s = n-k-1$. The charge $\mathcal{Q}_s$

$$\mathcal{Q}_s = \int_{-\infty}^{\infty} dx\, \mathcal{J}_s^0 = \int [T_{s+1}\, d\tau + \Theta_{s-1}\, d\sigma], \tag{16.2.18}$$

is a conserved quantity since, for eqn. (16.2.17), satisfies

$$\frac{dQ_s}{dt} = 0. \tag{16.2.19}$$

To explicitly find the densities $T_{s+1}[\phi]$ and $\Theta_{s-1}[\phi]$, let us define the field $\hat{\phi}(\sigma,\tau)$, solution of the so-called *Bäcklund transformations*

$$\begin{aligned} \partial_\sigma(\hat{\phi}-\phi) &= 2\epsilon \sinh\left(\frac{1}{2}(\hat{\phi}+\phi)\right), \\ \partial_\tau(\hat{\phi}+\phi) &= \frac{2}{\epsilon}\sinh\left(\frac{1}{2}(\hat{\phi}-\phi)\right). \end{aligned} \tag{16.2.20}$$

Assuming that $\phi(\sigma,\tau)$ is a solution of the equation of motion, eqns. (16.2.20) provide another solution. In fact, acting with ∂_τ of the first of them and using the second equation, we have

$$\partial_\tau\,\partial_\sigma(\hat{\phi}-\phi) = 2\,\sinh\frac{1}{2}(\hat{\phi}-\phi)\,\cosh\frac{1}{2}(\hat{\phi}+\phi) = \left[\sinh(\hat{\phi})-\sinh(\phi)\right].$$

The field $\hat{\phi}(z,\bar{z},\epsilon)$ can be expressed in power series of the parameter ϵ

$$\hat{\phi}(\sigma,\tau,\epsilon) = \sum_{n=0}^{\infty}\phi^{(n)}(\sigma,\tau)\,\epsilon^n \tag{16.2.21}$$

where $\phi^{(n)}(\sigma,\tau)$ can be computed by plugging it into (16.2.20) and comparing term to term in ϵ. For the first terms we have[4]

$$\begin{aligned} \hat{\phi}^{(0)} &= \phi, & \phi^{(1)} &= 2\phi_\tau, \\ \hat{\phi}^{(2)} &= 2\phi_{\tau\tau}, & \phi^{(3)} &= 2\phi_{\tau\tau\tau}-\phi_\tau^3/3, \\ \hat{\phi}^{(4)} &= 2\phi_{\tau\tau\tau\tau}-2\phi_\tau^2\,\phi_{\tau\tau}, & \cdots \end{aligned} \tag{16.2.22}$$

The existence of this series expression gives us the possibility to obtain an infinite number of conservation laws starting from a finite number of them. To this aim we can use, for instance

$$\left(\frac{1}{2}\psi_\sigma^2\right)_\tau + (1-\cosh\psi)_\sigma = 0, \tag{16.2.23}$$

[4] $\phi_\tau \equiv \partial_\tau\phi$ e $\phi_\sigma \equiv \partial_\sigma\phi$.

or a similar equation

$$\left(\frac{1}{2}\psi_\tau^2\right)_\sigma + (1-\cosh\psi)_\tau = 0, \tag{16.2.24}$$

whose validity can be easily checked employing the equation of motion (16.2.16) satisfied by ϕ. Using, for instance, eqn. (16.2.24) and substituting eqn. (16.2.22), we obtain an infinite number of conserved densities. The first non-trivial expressions (i.e. those that cannot be expressed as total derivative) are

$$\begin{aligned} T_2 &= \frac{1}{2}\phi_\tau^2 \\ T_4 &= 2\phi_{\tau\tau}^2 + 2\phi_\tau\,\phi_{\tau\tau\tau} \\ T_6 &= 2\phi_{\tau\tau\tau}^2 + 4\phi_{\tau\tau}\,\phi_{\tau\tau\tau\tau} - 6\phi_{\tau\tau}^2\phi_\tau^2 - 2\phi_\tau^3\phi_{\tau\tau\tau} + 2\phi_\tau\phi_{\tau\tau\tau\tau\tau}. \end{aligned} \tag{16.2.25}$$

In general, it can be proved that non-trivial conservation laws are obtained for the all odd values of s

$$s = 1, 3, 5, \ldots. \tag{16.2.26}$$

The set of these values of s constitutes the spectrum of the conserved charges. It is also possible to show that the classical expressions of the conserved currents, opportunely modified, keep their meaning also at the quantum level and that the corresponding charges are in involution, i.e. they commute each other

$$[\mathcal{Q}_s, \mathcal{Q}_{s'}] = 0. \tag{16.2.27}$$

The next chapter shows how the involution nature of the conserved charges and the spectrum of s strongly influence the structure of the massive excitations and their dynamics.

16.3 The Sine–Gordon Model

Under the analytic continuation $g \to ig$, the Sinh–Gordon model becomes the Sine–Gordon model, with the Euclidean action given by

$$\mathcal{S} = \frac{1}{8\pi}\int d^2x\left[(\partial_\mu\phi)^2 - \frac{\mu^2}{g^2}\cos(\sqrt{2}g\phi(x))\right]. \tag{16.3.1}$$

As seen in Chapter 12, this theory enters the bosonization procedure. However, there is a wider range of phenomena where the Sine–Gordon plays a crucial role: it describes, for instance, the dislocation of a crystal subjected to an external force; in quantum optics, it describes the propagation of a light wave in a material made of two quantum levels and, in superconductivity, it plays an important role in the theory of magnetic flux of a Josephson junction.

For the Sine–Gordon model we can also adopt the four different approaches discussed for the Sinh–Gordon to study its ultraviolet limit. For their similarity, we will not repeat their analysis here. However, it is important to note the following circumstance: adopting the quantization scheme of the Liouville theory, since now the exponentials are complex, we get an ultraviolet central charge less than 1

$$C_{uv} = c = 1 - 12Q^2, \quad Q = \pm\frac{1}{\sqrt{2}}\left(g - \frac{1}{g}\right) \tag{16.3.2}$$

(the sign of Q depends on the particular choice of the exponential that enters the Liouville action). This implies that, with an appropriate choice of the coupling constant g, the central charge c can take the values of the minimal models $\mathcal{M}_{p,q}$. When this happens, it is easy to see that the perturbing operator of the Liouville action corresponds to the operator $\Phi_{1,3}$ of the minimal models.

The Sine–Gordon has an infinite sequence of conserved charges in involution that ensures both its classical and quantum integrability. It is useful to provide the explicit sequence of the first quantum densities T_{s+1}, in the quantization scheme of the Liouville theory, adopting the complex coordinates $z = t + ix$ and $\bar{z} = t - ix$. If $T(z)$ is the analytic component of the stress-energy tensor associated to the Liouville theory

$$T(z) = -\frac{1}{2}(\partial_z\phi)^2 + iQ\partial_z^2\phi, \tag{16.3.3}$$

and $(AB)(z)$ denotes the normal order of two operators as defined by their OPE

$$(AB)(z) = \oint_z \frac{dw}{w - z} A(w)B(z),$$

we have

$$\begin{aligned}
T_2 &= T, \\
T_4 &= (T^2) \\
T_6 &= (T(T^2)) + \frac{c+2}{12}(T\partial_z^2 T), \\
T_8 &= (T(T(T^2))) + \frac{c+8}{6}(T(T\partial_z^2 T)) + \frac{1}{180}(c^2 + 4c - 101)(T\partial_z^4 T). \\
\cdots &= \cdots
\end{aligned} \tag{16.3.4}$$

The dynamics of the model is better understood if we consider its formulation in the Minkowski space. To simplify the notations, let us rescale the field and the coupling constant as $\phi \to \phi/\sqrt{4\pi}$ and $g \to g\sqrt{8\pi}$, so that the action in the Minkowski space becomes

$$\mathcal{S} = \int d^2x \left[\frac{1}{2} (\partial_\mu \phi)^2 + \frac{\mu^2}{g^2} \cos g\phi(x) \right].$$

The potential of the theory

$$V(\phi) = \frac{\mu^2}{g^2} [1 - \cos(g\phi)], \tag{16.3.5}$$

(to which we have added, for comodity, a constant) presents an infinite series of degenerate minima placed at $\phi = 2\pi n/g$ $(n = 0, \pm 1, \ldots)$ (Figure 16.2). In the quantum version, they correspond to an infinity family of equivalent vacua, denoted by $|0\rangle_n$. Around each minimum, the potential has a quadratic concavity μ^2 that can be associated to the mass of the scalar particle created out of the vacuum by the field ϕ. However, this scalar particle does not exhaust the spectrum of the excitations of the the model. In the Sine–Gordon there are in fact topological excitations of finite energy, associated to those field configurations that interpolate between two degenerate vacua.

Topological excitations. The topological excitations can be specified by two integers (n_1, n_2) that label the vacua $2\pi n_1/g$ and $2\pi n_2/g$ reached by the field $\phi(x)$ at $x \to \pm\infty$. Let us define then the topological charge

$$\mathcal{T} = n_1 - n_2 = \frac{1}{2\pi g} \int_{-\infty}^{\infty} dx \, \frac{\partial \phi}{\partial x}. \tag{16.3.6}$$

For the periodicity of $V(\phi)$, the field ϕ is defined modulus $2\pi/g$, i.e. at any given point x the value of the field can be changed as $\phi(x) \to \phi(x) + 2\pi k$, maintaining though the continuity of its configurations. The topological charge is insensitive to these transformations as far as we keep fix the final values assumed by the fields.

Let us write down the energy of a generic configuration of $\phi(x,t)$

$$E[\phi] = \int_{-\infty}^{\infty} dx \left[\frac{1}{2} \left(\frac{\partial \phi}{\partial t^2} \right)^2 + \left(\frac{\partial \phi}{\partial x^2} \right)^2 + V(\phi) \right], \tag{16.3.7}$$

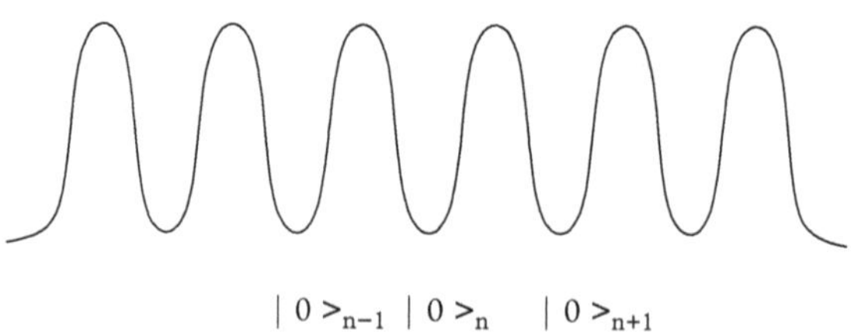

Fig. 16.2 *Potential of the Sine–Gordon model and sequence of the infinite equivalent vacua.*

and the equation of motion of the model

$$\frac{\partial^2\phi}{\partial t^2} - \frac{\partial^2\phi}{\partial x^2} = \frac{\partial V}{\partial \phi}. \tag{16.3.8}$$

The classical expression of the elementary topological configurations, i.e. those associated to $\mathcal{T} = \pm 1$, can be obtained by looking at the static solutions of the equation of motion. In this case the first term on the right-hand side vanishes and the equation of motion reduces to

$$\frac{\partial^2\phi}{\partial x^2} = -\frac{\partial V}{\partial \phi}.$$

This expression coincides, formally, with the equation of motion of classical mechanics of a fictitious particle described by the coordinate $\phi(x)$ and subjected to the potential $-V(\phi)$ (note the change of sign in the potential). In this interpretation, the original variable x in the field $\phi(x)$ plays the role of time coordinate of the classical particle. As any classical system subjected to a conservative force, it has an integral of motion given by its mechanical energy (that must not to be confused with $E[\phi]$), given by

$$W = \frac{1}{2}\left(\frac{d\phi}{dx}\right)^2 - V(\phi). \tag{16.3.9}$$

The value of the constant of motion W can be immediately determined. In fact, if requiring that the static solutions $\phi(x)$ have a finite energy $E[\phi]$, we must have at $x \to \pm\infty$ both $V(\phi) \to 0$ and $(\partial\phi/\partial x) \to 0$. In analogy with the Newtonian motion of the particle, this means that the particle at $x \pm \infty$ has to be in one of the maxima of the potential $-V(\phi)$ and, furthermore, that its velocity has to vanish both at the starting and ending points. For the constant W we have the $W = 0$. Instead of solving the second-order equation of motion (16.3.8), for the static solutions we can take advantage of the mechanical analogy and find the solution by quadrature from eqn. (16.3.9)

$$\frac{d\phi}{dx} = \pm\sqrt{2V(\phi)} \quad \to \quad (x - x_0) = \pm \int_{\phi(x_0)}^{\phi(x)} \frac{d\bar{\phi}}{\sqrt{2V(\bar{\phi})}}, \tag{16.3.10}$$

where x_0 is an arbitrary constant of integration. Performing the integral, with the explicit expression of $V(\phi)$ given in (16.3.5), we get

$$\bar{\phi}(x) = \pm 4/g \arctan\left[\exp m(x - x_0)\right]. \tag{16.3.11}$$

The first solution, the one with the positive sign, has topological charge $\mathcal{T} = 1$ and corresponds to a *soliton* that interpolates between the vacuum $\bar{\phi} = 0$ and the next one at $2\pi/g$ or, equivalently, between a generic pair of vacua $2\pi n/g$ and $2\pi(n+1)/g$. The second solution, the one with the negative sign, has instead $\mathcal{T} = -1$ and corresponds to an *anti-solitons* that interpolates between a generic pair of vacua $2\pi n/g$ e $2\pi(n-1)/g$, as shown in Figure 16.3). The origin of the terminology is in the peculiar form assumed by the energy density $\epsilon(x)$ of these solutions entering the formula

$$E[\bar{\phi}] = \int_{-\infty}^{\infty} dx\, \epsilon(x), \quad \epsilon(x) = \frac{4\mu^2}{g^2} \frac{1}{\cosh^2 m(x - x_0)}. \tag{16.3.12}$$

As shown in Figure 16.3, $\epsilon(x)$ has a shape strongly localized at x_0 that rapidly decreases to zero outside an interval large approximatively $1/m$. For this localization property, the solitonic solutions of the Sine–Gordon can be interpreted as particle excitations of the system and the energy (16.3.12) of the static solution corresponds to the mass M_s of the soliton/anti-soliton

$$M_s = \frac{8\mu^2}{g^2}. \tag{16.3.13}$$

The non-perturbative nature of the solitonic solutions is revealed by the dependence on the coupling constant, placed in the denominator of the expression above. The particle nature of these excitations is further confirmed by using the Lorentz invariance of the equation of motion: given a static solution $\bar{\phi}(x)$, we can use a Lorentz transformation[5] resulting in a solution that moves with velocity v

$$\bar{\phi}(x) \to \bar{\phi}\left[\frac{m(x - x_0) - vt}{\sqrt{1 - v^2}}\right].$$

It is easy to check that this expression indeed satisfies the equation of motion (16.3.8) and substituting it in (16.3.7), we get

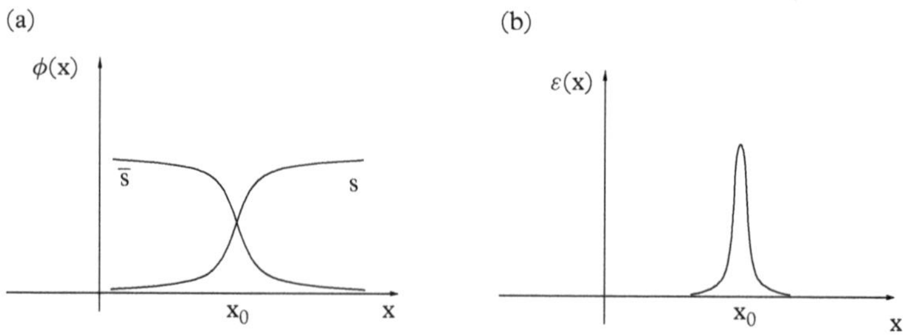

Fig. 16.3 *Solitonic solutions and their energy density* $\epsilon(x)$.

[5] The velocity is measured in unit of the light velocity, so that its limiting value is $v = 1$.

$$E[\bar{\phi}(x,t)] = \frac{M_s}{\sqrt{1-v^2}}.$$

Hence we recover the Einstein relationship that links the mass and the energy of a particle. The solitons are then particle excitations of the system that, in the classical description, appear as waves that propagate in the medium without dispersion or dissipation, always keeping intact their shape.

Time-dependent solutions. The Sine–Gordon admits exact solutions also in other topological sectors, although they are time-dependent expressions. For instance, a solution with $\mathcal{T} = 0$ is

$$\bar{\phi}_{s\bar{s}}(x,t) = \frac{4}{g} \arctan\left(\frac{\sinh(mvt/\sqrt{1-v^2})}{v\cosh(mx/\sqrt{1-v^2})} \right). \tag{16.3.14}$$

It has the peculiar property to tend for $t \to \pm\infty$ to a configuration made of a soliton and an anti-soliton

$$\bar{\phi}_{s\bar{s}}(x,t) \to \bar{\phi}_s\left(\frac{x+v(t\pm\Delta_{s\bar{s}}/2)}{\sqrt{1-v^2}}\right) + \bar{\phi}_{\bar{s}}\left(\frac{x-v(t\pm\Delta_{s\bar{s}}/2)}{\sqrt{1-v^2}}\right), \qquad t \to \pm\infty.$$

When the time varies, this solution describes an elastic scattering process, whose only effect is a negative time shift $\Delta_{s\bar{s}} \equiv (1-v^2)v\log v$ of the propagation of the soliton and the anti-soliton with respect to their free propagation. The elasticity of the scattering processes is a common characteristic in all other topological sectors. For instance, in the sector with topological charge $\mathcal{T} = 2$, a solution of the equation of motion is given by

$$\bar{\phi}_{ss}(x,t) = \frac{4}{g} \arctan\left(\frac{v\sinh(x/\sqrt{1-v^2})}{\cosh(vt/\sqrt{1-v^2})} \right). \tag{16.3.15}$$

At any given time, it interpolates between the vacua $-2\pi/g$ and $2\pi/g$. It can be then interpreted as a configuration made of two solitons. Following the time evolution of this solution, we realize that it corresponds to the elastic scattering of the two solitons, since at $t \to \pm\infty$ it becomes

$$\bar{\phi}_{ss}(x,t) \to \bar{\phi}_s\left(\frac{x+v(t\pm\Delta_{ss}/2)}{\sqrt{1-v^2}}\right) + \bar{\phi}_s\left(\frac{x-v(t\pm\Delta_{ss}/2)}{\sqrt{1-v^2}}\right), \qquad t \to \pm\infty$$

Also in this case, the only effect of the interaction manifests in a time shift Δ_{ss}, although positive this time (see Problem 16.4).

In conclusion, the Sine–Gordon theory is an integrable theory that have multi-solitonic solutions that describe purely elastic scattering processes. The elasticity of these scattering processes is a consequence of the infinite number of conserved charges of the model. To compute the complete spectrum of the excitations of its quantum version it is necessary to study the S-matrix of the scattering processes, a subject addressed in Chapter 18.

16.4 The Bullogh–Dodd Model

The Bullogh–Dodd model is another Lagrangian system that is integrable both at the classical and at the quantum level. Its Euclidean theory is defined by

$$\mathcal{S} = \frac{1}{8\pi} \int d^2x \left\{ (\partial_\mu \phi)^2 + \frac{\mu^2}{12g^2} \left[2\, e^{\sqrt{2}g\phi} + e^{-2\sqrt{2}g\phi} \right] \right\}. \tag{16.4.1}$$

It may be considered as a deformation of the Liouville theory

$$\mathcal{S}_0 = \frac{1}{8\pi} \int d^2x \left\{ (\partial_\mu \phi)^2 + \frac{\mu^2}{6g^2}\, e^{\sqrt{2}g\phi} \right\}, \tag{16.4.2}$$

by means of the exponential $e^{-2\sqrt{2}g\phi}$. As in the Sinh–Gordon model, the quantization of this theory requires the introduction of a charge at infinity, in this case espressed by

$$Q_+ = \frac{1}{\sqrt{2}}(g + 1/g).$$

This leads to an ultraviolet central charge equal to

$$C_{uv} = 1 + 12 Q_+^2. \tag{16.4.3}$$

The conformal dimension $\Delta(\alpha)$ of the exponentials $e^{\alpha\phi}$ is given by

$$\Delta(\alpha) = -\frac{\alpha^2}{2} + \alpha\, Q_+. \tag{16.4.4}$$

In such a way, the exponential in (16.4.2) has a conformal dimension equal to 1, while the other exponential has a conformal dimension

$$\Delta[e^{-2\sqrt{2},g\phi}] = -\frac{1}{2} - \frac{3}{4}g^2.$$

By the analytic continuation $g \to ig$, the central charge (16.4.3) becomes less than 1 and, by an opportune choice of the coupling constant g, can match the central charges of the minimal models. In these case, it is easy to show that the operator $e^{-2\sqrt{2}g\phi}$ corresponds to the operator $\Phi_{1,2}$ of the conformal minimal models. However, contrary to what happens in the Sinh–Gordon model, the substitution $g \to ig$ makes this time the action (16.4.1) a complex quantity and therefore it is not obvious how it can give rise, as it does indeed, to a consistent physical theory.

Vice versa, as a starting Liouville theory we can assume

$$\mathcal{S}_0 = \frac{1}{8\pi}\int d^2x \left\{(\partial_\mu\phi)^2 + \frac{\mu^2}{12g^2}\,e^{-2\sqrt{2}g\phi}\right\}. \tag{16.4.5}$$

In this case the charge at infinity is

$$Q_- = -\frac{1}{\sqrt{2}}\left(2g + \frac{1}{2g}\right).$$

The central charge and the conformal dimensions have the same expressions as above, once we make the substitution $Q_+ \to Q_-$. In this way the exponential present in the action has a conformal dimension equal to 1, while

$$\Delta[e^{\sqrt{2}g\phi}] = -\frac{1}{2} - 3g^2.$$

With the analytic continuation $g \to ig$ and with the choice of g that matches the central charge with the one of conformal minimal models, the perturbing operator $e^{\sqrt{2}g\phi}$ can be identified with the field $\Phi_{2,1}$ of the minimal models.

To discuss the perturbative quantization of the theory, based on the Minkowski space and the Feynman graphs, it is convenient to scale the field and the coupling costant in such a way that the action reads

$$\mathcal{L} = \frac{1}{2}(\partial_\mu\phi)^2 - \frac{\mu^2}{6g^2}\left(2e^{g\phi} + e^{-2g\phi}\right), \tag{16.4.6}$$

where μ is a mass parameter and g the coupling constant. Also this model belongs to the Toda field theories,[6] discussed in Section 16.6. The series expansion of the exponential terms gives rise to the n-leg interaction vertices

$$V(\phi) = \frac{\mu^2}{6g^2}\left(2e^{g\phi} + e^{-2g\phi}\right) = \frac{\mu^2}{2g^2} + \sum_{k=2}^{\infty}\frac{g_k}{k!}\phi^k,$$

where

$$g_k = \frac{\mu^2}{3}g^{k-2}\left[1 + (-1)^k 2^{k-1}\right].$$

[6] This model can be obtained by the folding with respect to a Z_2 symmetry of the affine simply-laced algebra A_2. The resulting algebra is denoted by BC_1 and its Coxeter number is $h = 3$.

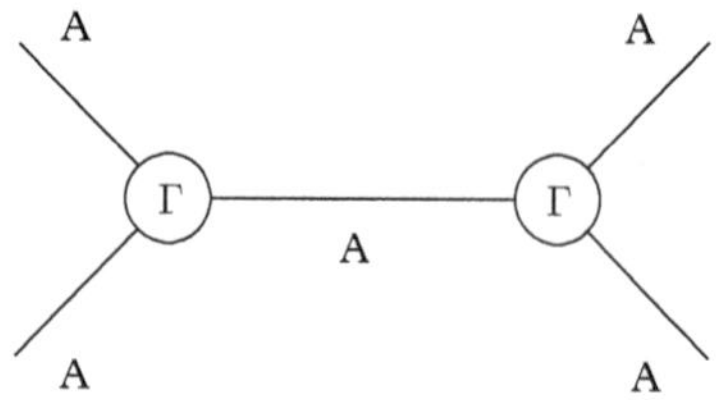

Fig. 16.4 *Scattering amplitudes of the Bullogh-Dodd theory.*

Also in this case the renormalization of the divergences met in the perturbative series reduces to eliminate the tadpoles. This is equivalent to renormalize the mass term μ as

$$\mu^2 \to \mu^2 \left(\frac{\Lambda}{\mu}\right)^{g^2/4\pi}.$$

This theory is supported by an infinite number of conserved charges, whose spectrum of the spin s is given by

$$s = 1, 5, 7, 11, 13, \ldots \tag{16.4.7}$$

i.e. all odd integer numbers but the multiples of 3.

The perturbative particle content of the theory consists of a particle, denoted by A, that takes part in scattering processes in which it appears as bound state of itself. This is a simple consequence of the ϕ^3 vertex in $V(\phi)$, that gives rise to scattering processes as those shown in Figure 16.4. Here we anticipate that this perturbative scenario will be confirmed by the exact S-matrix of this model discussed in Chapter 18.

16.5 Integrability versus Non-integrability

One may wonder what is so special in the Sinh–Gordon, Sine–Gordon or Bullogh–Dodd models with respect to other two-dimensional field theories. The answer to this question can be given both at the classical and the quantum level. Let us first discuss the classical aspects.

The Sine–Gordon model is not the only theory to possess static topological configurations. If we examine further the argument used to find the solitonic solutions, we realize that it is sufficient that the theory simply possesses two degenerate next-neighbour vacua. From this point of view, even the ϕ^4 theory, in the phase in which the Z_2 is spontaneously broken, should have solitonic excitations. The potential $\mathcal{U}(\phi)$ of this theory

$$\mathcal{L} = \frac{1}{2}(\partial_\mu \phi)^2 - \mathcal{U}(\phi), \quad \mathcal{U} = \frac{\lambda}{4}\left(\phi^2 - \frac{\mu^2}{\lambda}\right),$$

has in fact, two degenerate minima at $\phi_\pm = \pm m/\sqrt{\lambda}$. This is indeed the case, and the explicit expression of the solitons of this theory is obtained by inserting $\mathcal{U}(\phi)$ in (16.3.10)

$$\bar{\phi}(x) = \pm\frac{m}{\sqrt{\lambda}}\tanh\left[\frac{m}{\sqrt{2}}(x-x_0)\right]. \tag{16.5.1}$$

The soliton mass is obtained by substituting their energy density

$$\epsilon(x) = \frac{m^4}{2\lambda}\frac{1}{\cosh^4(m(x-x_0)/\sqrt{2})},$$

in $E[\phi]$ given in (16.3.7)

$$M = \frac{2\sqrt{2}\,m^3}{3\;\;\lambda}.$$

Hence, even in the ϕ^4 theory there are solitonic phenomena of non-perturbative nature.

If the configurations of the static solitons of the Sine–Gordon and the ϕ^4 theory may appear very similar,[7] their differences show up when we consider the multi-solitonic configurations. In fact, in the Sine–Gordon these configurations have the properties to preserve the shape that they have at $t \to -\infty$ also at $t \to +\infty$. This is not the case of the ϕ^4 theory. In other words, the scattering processes that take place in the ϕ^4 theory are inelastic: the initial particles, identified as the multi-solitons present at $t \to -\infty$, lose their identity during the time evolution.

This classical situation has a quantum analog: this permits to easily appreciate the drastic difference that exists between the Sine–Gordon and ϕ^4 theory or, more generally, its difference with respect to any other theory invariant under a Z_2 symmetry. Chapter 17 discusses how, quantum integrable theories have the same peculiar features already noticed at the classical level, i.e. the elasticity of the scattering processes. This is indeed a peculiar aspect for relativistic QFTs since, for purely kinematic reasons, the number of particles is not a conserved quantity: for instance, one can always create $4, 6, 8, \ldots$ particles simply increasing the energy in the centre of mass of two colliding particles. In the light of this remark, following Dorey's argument, let us start from the more general two-dimensional Z_2 invariant Lagrangian theory

$$\mathcal{L} = \frac{1}{2}\left[(\partial_\mu\phi)^2 - \mu^2\phi^2\right] - \frac{g_4}{4!}\phi^4 - \frac{g_6}{6!}\phi^6 - \cdots \tag{16.5.2}$$

[7] Note that in the ϕ^4 theory there are no topological sectors with topological charge $|\mathcal{T}| > 1$, since there are only two vacua. The multi-soliton sequences of this theory consist of only alternate configurations of soliton and anti-solitons.

and let us find out the conditions on the coefficients $g_4, g_6, \ldots$ that prevent the production processes. Let us analyse the simplest case, i.e. a process in which two initial particles gives rise to four final particles. The Feynman rules are

——— $= i/(p^2 - \mu^2 + i\epsilon)$

 $= -ig_4$

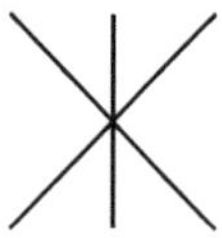 $= -ig_6$

etc. Applying these rules, let us compute the tree-level processes in which we have $2 \to 4$ particles. Suppose, for simplicity, that the initial particles have just the energy to create the four out-coming particles. Going in the centre of mass reference frame, the momenta (p^0, p^1) of the initial particles, satisfying the on-shell relation $(p^0)^2 - (p^1)^2 = \mu^2$, are then $(2\mu, \pm\sqrt{3}\mu)$. The total energy is $E_t = 4\mu$, a value sufficient to create the four final particles. Since these four particles will be all at rest, their common value of the momenta is $(\mu, 0)$. Using the conservation of the total momentum at each vertex, for the graphs of Figure 16.5 we have

$$
\begin{aligned}
(a) &\rightarrow \; i\frac{g_4^2}{32\mu^2} \\
(b) &\rightarrow -i\frac{g_4^2}{96\mu^2} \\
(c) &\rightarrow -i\frac{g_6}{48\mu^2}.
\end{aligned}
$$

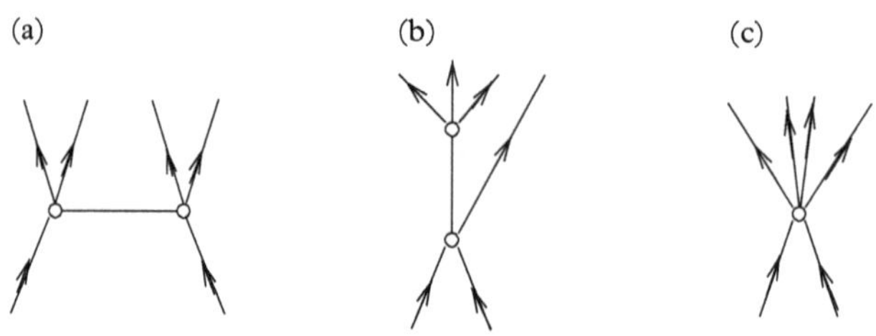

Fig. 16.5 *Feynman graphs at the tree level for the production process* $2 \to 4$.

The total amplitude is

$$(a)+(b)+(c)=\frac{i}{48\mu^2}(g_4^2-g_6), \tag{16.5.3}$$

so that, choosing $g_6=g_4^2$, we can *dynamically* suppress this production process.

Generalizing the analysis to the production process $2\to 6$ and requiring its dynamical absence, one derives the further condition $g_6=g_4^3$. Carrying on the same analysis for the higher particle production processes, one arrives to the following result: the only lagrangian theories with a Z_2 symmetry that *dynamically* suppress all production processes are represented by the potentials

$$U(\phi)=\mu^2\left[\frac{1}{2}\phi^2\pm\frac{g^2}{4!}\phi^4+\frac{g^4}{6!}\phi^6\pm\frac{g^6}{8!}\phi^8+\cdots\right]. \tag{16.5.4}$$

It is immediate to recognize that these potentials either correspond to the one of the Sinh–Gordon model (when all the sign are chosen positive) or of the Sine–Gordon model (with the choice of the alternate signs).

Repeating the same analysis with the most general Landau–Ginzburg potential, which also presents odd powers of the field ϕ, the one that is selected by the absence of production processes is given by

$$\begin{aligned} U(\phi) &= \mu^2\left[\frac{1}{2}\phi^2-\frac{g}{6}\phi^3+\frac{g^2}{8}\phi^4-\frac{g^3}{24}\phi^5+\cdots\right] \\ &= \frac{\mu^2}{6g^2}\left[2e^{g\phi}+2\,e^{-2g\phi}-3\right], \end{aligned} \tag{16.5.5}$$

namely, the Bullogh–Dodd model!

In conclusion, the only relativistic two-dimensional Lagrangian theories which involve only one scalar field and that are integrable both at the classical and at the quantum level are the Sinh–Gordon and the Sine–Gordon theories (if there is a Z_2 symmetry) or the Bullogh–Dodd theory.

16.6 The Toda Field Theories

A generalization of the models encountered so far is provided by the Toda field theories. These theories can be constructed using the Lie algebras discussed in the Appendix of Chapter 13. In the following we focus on the simply-laced algebras A_n, D_n and E_n, i.e. those with simple roots of the same length. The Toda field theories based on the non-simply laced algebras can be defined, at least at the classical level, by an identification of the roots using the symmetry properties of the Dynkin diagram. This is the so-called *folding* procedure, as shown in Table 6.1 (a) and Table 6.1 (b) below.

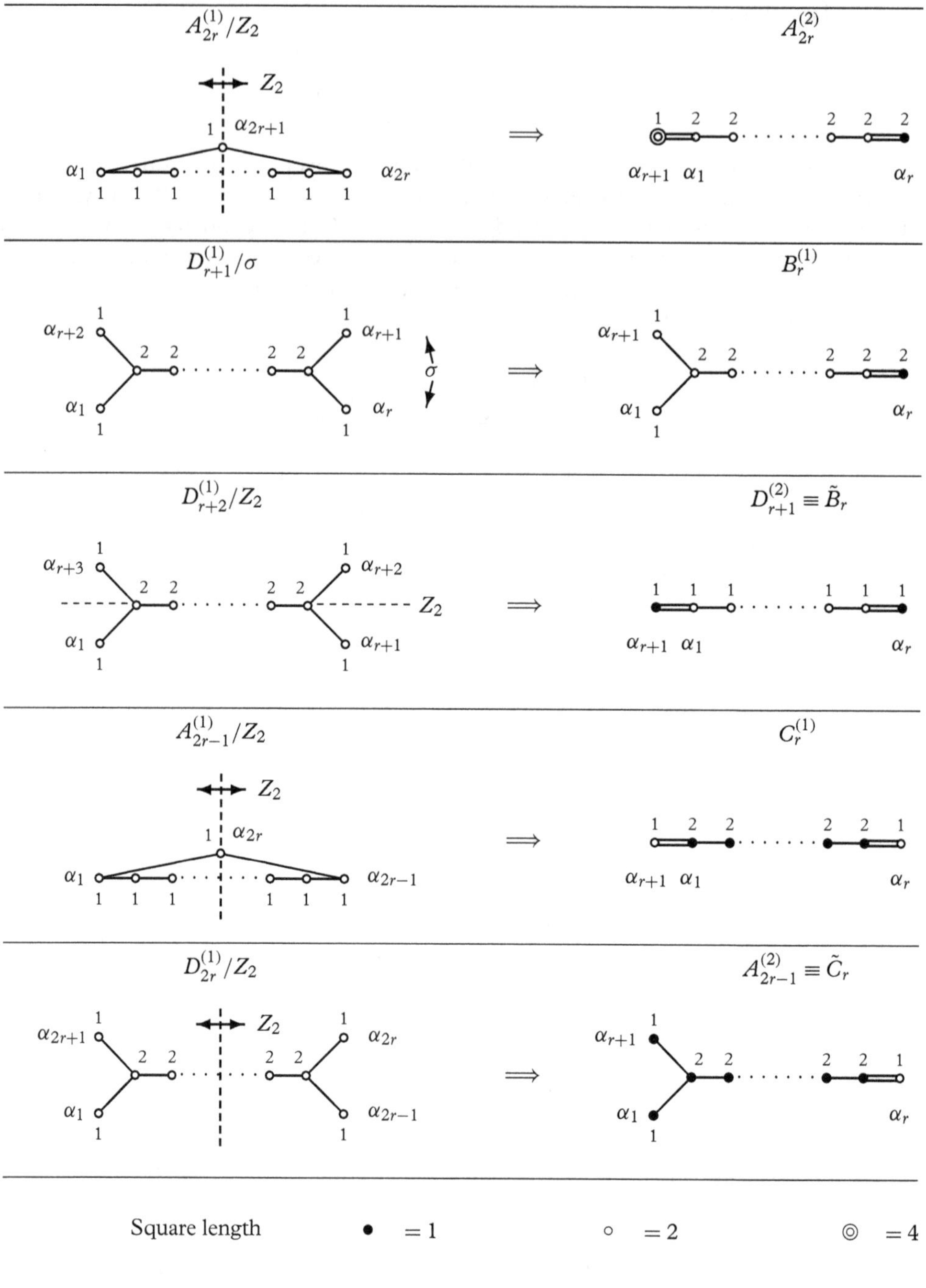

Table 16.1 *(a) Foldings of the Dynkin diagrams of the simply-laced algebras: the principal series. Near the roots there are the numbers q_i.*

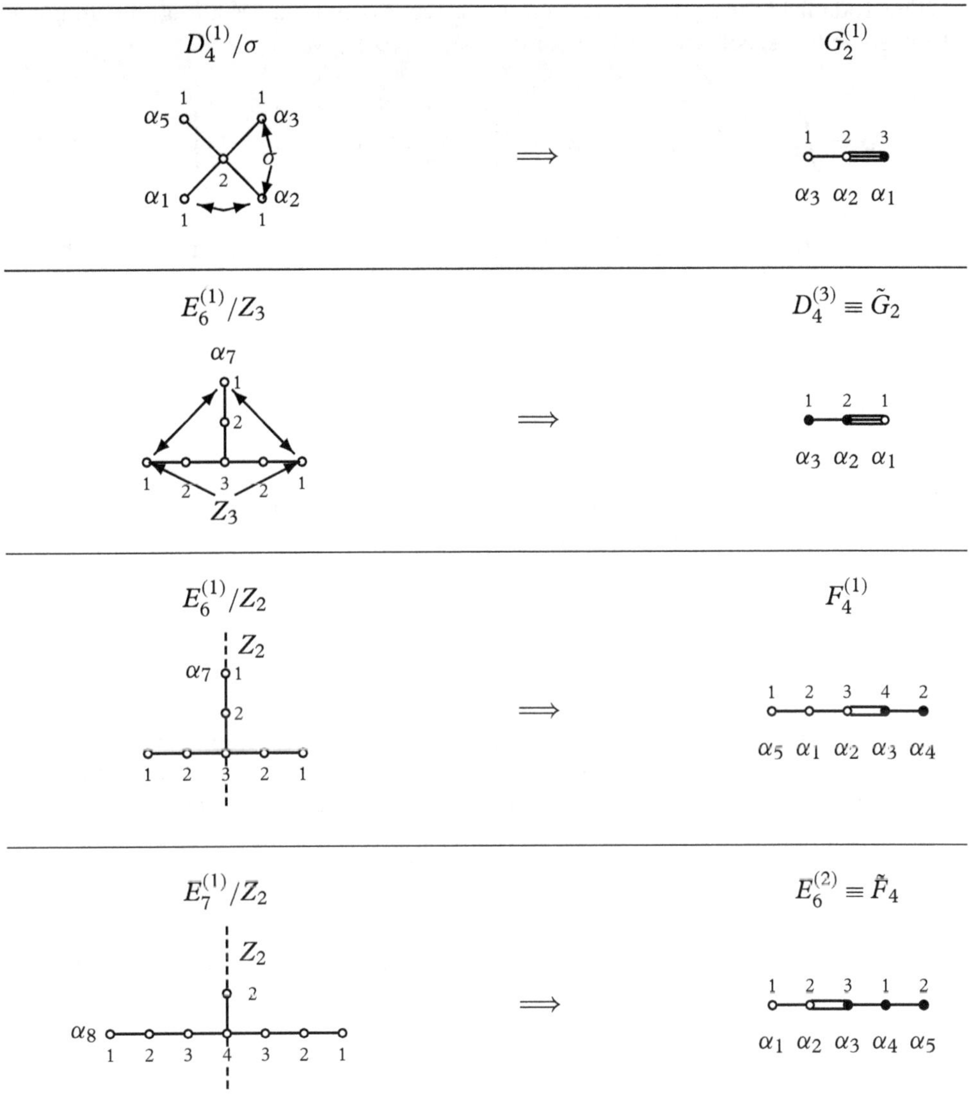

Square length $\bullet$ $=1$ or $2/3$ $\circ$ $=2$

Table 16.1 *(b) Foldings of the Dynkin diagrams of the simply-laced algebras: the exceptional series. Near the roots there are the numbers q_i.*

The Toda field theory associated to a Lie algebra G of rank r is a Lagrangian model of r bosonic fields, collected in a vector $\phi = (\phi_1, \dots, \phi_r)$, given by

$$\mathcal{S} = \int d^2x \left\{ \frac{1}{8\pi} (\partial_\mu \phi) \cdot (\partial^\mu \phi) + \frac{\mu^2}{16\beta^2} \sum_{i=1}^{r+1} q_i [\exp(\beta \alpha_i \cdot \phi) - 1] \right\}. \tag{16.6.1}$$

μ^2 and β are real parameters. The set $\{\alpha_i\}_{i=1}^r$ is given the simple roots of G, with their norm equal to 2. The set of the integer numbers $\{q_i\}$ is different for each algebra and it is related to the definition of the *maximal root* of the algebra, given by[8]

$$\alpha_{r+1} = -\sum_{i=1}^{r} q_i \alpha_i. \tag{16.6.2}$$

The extended set of roots, obtained by adding the maximal root, form the Dynkin diagram of the *affine Lie algebras*. For these systems, posing $q_{r+1} = 1$, we have

$$\sum_{i=1}^{r+1} q_i \alpha_i = 0, \qquad \sum_{i=1}^{r+1} q_i = h \tag{16.6.3}$$

where, for the simply-laced algebras, h coincides with ψ, the Coxeter number of G.

The exponential with the maximal root is responsible for the massive nature of these field theories. Also for the Toda field theories we can adopt two different way of looking at the action (16.6.1), depending on the choice of $\mathcal{S}_0$. So, taking for $\mathcal{S}_0$ the action that excludes the maximal root, we have a generalized Liouville theory

$$\mathcal{S}_0 = \int d^2x \left[\frac{1}{8\pi} (\partial_\mu \phi) \cdot (\partial^\mu \phi) + \frac{\mu^2}{16\beta^2} \sum_{i=1}^{r} q_i [\exp(\beta \alpha_i \cdot \phi) - 1] \right]. \tag{16.6.4}$$

Analogously to the cases previously analysed, these actions describe conformal models. Their quantization requires a set of charges at infinity, encoded in the vector

$$\vec{Q} = (\beta + 1/\beta)\, \vec{\rho}, \quad \vec{\rho} = \frac{1}{2} \sum_{\alpha > 0} \alpha. \tag{16.6.5}$$

[8] For the non-simply laced algebra there is also the possibility to extend the Dynkin diagram of the original theory by adding the shorter maximal root. These are the so-called *twisted* algebras and denoted by $\tilde{G}$. In all the non-twisted models h is equal to the Coxeter number ψ, while for the twisted ones h is equal either to the dual Coxeter number $\tilde{\psi}$ of the same algebra or of another non-simply laced algebra.

The analytic component of the stress-energy tensor, given by

$$T(z) = -\frac{1}{2}(\partial_z\phi)^2 + Q\cdot\partial_z^2\phi,$$

gives rise to the central charge

$$C = r\left[1 + h(h+1)(\beta + 1/\beta)^2\right]. \tag{16.6.6}$$

The second way to approach the Toda field theories consists of using the Feynman perturbation theory. As in previous cases, all the perturbative divergences of these theories come from the tadpole diagrams, that can be eliminated by defining the normal order of the exponential operators. This induces a renormalization of the mass parameter μ^2.

$$\mu^2 \to \mu^2\left(\frac{\Lambda^2}{\mu^2}\right)^{\frac{\beta^2}{4\pi}\frac{\tilde{h}}{h}} \tag{16.6.7}$$

where

$$\tilde{h} = \frac{1}{2}\sum_{a=1}^{r}\sum_{i=1}^{r+1} q_i\alpha_i^a\alpha_i^a. \tag{16.6.8}$$

In the simply laced algebras $\tilde{h} = h$ and these two numbers get simplified in (16.6.7).

In the Feynman perturbative approach it is necessary to determine the classical values of the masses of the various particles A_a, as coming from the quadratic terms of the Lagrangian

$$M_{ab}^2 = \mu^2\sum_{i=1}^{r+1} q_i\alpha_i^a\alpha_i^b. \tag{16.6.9}$$

Mass spectrum

The classical mass spectrum is determined by the zeros of the characteristic equation

$$\|M^2 - x\cdot\mathbf{1}\| = 0. \tag{16.6.10}$$

The left-hand side in (16.6.10) is a polynomial of order r, whose general form is

$$\mathcal{P}(x) = x^r - p_1x^{r-1} - p_2x^{r-2} - \ldots - p_r.$$

The first coefficient p_1 is simply the trace of M^2 and, for the simply laced algebra, this is simply twice their Coxeter number. The other coefficients p_i can be expressed in terms of the trace of higher powers of M^2. To simplify the notation, let us pose $\mathcal{M} = M^2$. Their expression is then

$$k p_k = a_k - p_1 a_{k-1} - \cdots p_{k-1} a_1, \tag{16.6.11}$$

where

$$\begin{aligned} a_1 &= \operatorname{Tr}\mathcal{M} = \sum_i m_i^2 \\ a_2 &= \operatorname{Tr}\mathcal{M}^2 = \sum_i m_i^4 \\ \cdot\cdot & \quad \cdots \\ a_n &= \operatorname{Tr}\mathcal{M}^n = \sum_i m_i^{2n}. \end{aligned} \tag{16.6.12}$$

It is convenient to introduce a matrix $\mathcal{N}$ directly linked to the Dynkin diagrams. Its matrix elements are given by

$$\mathcal{N}_{ij} = (q_i \alpha_i, \alpha_j) = \sum_{k=1}^{r} q_i \alpha_i^k \alpha_j^k.$$

It is easy to prove that

$$\operatorname{Tr}\mathcal{M}^s = \operatorname{Tr}\mathcal{N}^s, \qquad s = 1, 2, \cdots n.$$

Hence, the characteristic equation of $\mathcal{M}$ coincides with the one of $\mathcal{N}$. However, $\mathcal{N}$ is a $(n+1)\times(n+1)$ matrix while $\mathcal{M}$ is a $n \times n$ matrix. Since α_0 is expressed by a linear combination of the other roots, $\mathcal{N}$ is then a singular matrix: one of its eigenvalues vanishes, whereas the others coincide with the eigenvalues of $\mathcal{M}$.

In the basis of its eigenvectors, $M_{ij}^2 = \mu_i^2 \delta^{ij}$. The mass spectrum is degenerate if the group of the automorphisms of the Dynkin diagram is non-trivial. In these case it may be convenient to organize the particles pairwise, associate to complex conjugated fields. In the simply laced algebra, a remarkable result is that the masses can be organized in a vector

$$\mathbf{m} = (m_1, m_2, \cdots m_r),$$

that is, the eigenvectors of the incidence matrix I of the algebra $\mathcal{G}$. It is defined by $I = 2 - C$, where C is the Cartan matrix. In fact, $\mathbf{m}$ is the Perron–Frobenius eigenvector of I and its components can be thus associated directly to the dots of the Dynkin diagram itself. On the other hand, since the dots of the Dynkin diagram are also associated to the fundamental representation of the simply laced algebra, we arrive at the interesting conclusion that there is a correspondence between the particles of mass m_i and the relative representations of $\mathcal{G}$. This will be a useful observation in the future discussion of the scattering processes of these theories.

Let us now discuss in detail the mass spectrum of the various Toda field theories, with the final result of this analysis summarized in the Tables 16.2 and 16.3.

16.6.1 $A_n^{(1)}$ Series

For this series, the matrix $\mathcal{N}$ reduces to the Cartan matrix of the affine Lie algebras. The characteristic equation associated to $\mathcal{N}$ is given by

$$\mathcal{Q}_{n+1}(x) = \begin{vmatrix} 2-x & -1 & 0 & \cdots & 0 & 0 & -1 \\ -1 & 2-x & -1 & \cdots & 0 & 0 & 0 \\ 0 & -1 & 2-x & -1 & \cdots & 0 & 0 \\ \cdots & \cdots & \cdots & \cdots & \cdots & \cdots & \cdots \\ \cdots & \cdots & \cdots & \cdots & \cdots & \cdots & \cdots \\ 0 & 0 & 0 & \cdots & -1 & 2-x & -1 \\ -1 & 0 & 0 & \cdots & 0 & -1 & 2-x \end{vmatrix}$$

Posing $2y = 2 - x$, it is possible to show that

$$\mathcal{Q}_{n+1} = 2\,(\mathcal{T}_{n+1}(y) - 1), \tag{16.6.13}$$

where $\mathcal{T}_{n+1}(y)$ is the Chebyshev polynomial of the first type

$$\mathcal{T}_{n+1}(\cos\theta) = \cos(n+1)\theta.$$

The mass spectrum of the series $A_n^{(1)}$ is given by the n non-vanishing roots of the equation $\mathcal{T}_{n+1}(y) = 1$, namely

$$m_k^2 = 4\sin^2\frac{k\pi}{n+1} \qquad k = 1, 2, \cdots n. \tag{16.6.14}$$

16.6.2 $D_n^{(1)}$ Series

For this series, we have

$$\mathcal{N} = \begin{vmatrix} 4 & -2 & 0 & \cdots & 0 & 0 & -2 & -2 \\ -2 & 4 & 2 & \cdots & 0 & 0 & 0 & 0 \\ \cdots & \cdots & \cdots & \cdots & \cdots & \cdots & \cdots & \cdots \\ \cdots & \cdots & \cdots & \cdots & \cdots & \cdots & \cdots & \cdots \\ \cdots & \cdots & \cdots & \cdots & \cdots & \cdots & \cdots & \cdots \\ 0 & -1 & 0 & \cdots & 0 & 0 & 2 & 0 \\ -1 & 0 & 0 & \cdots & 0 & 0 & 0 & 2 \end{vmatrix}$$

The characteristic equation has the form

$$\|\mathcal{N} - x \cdot \mathbf{1}\| = 2^{n+2}(y-1)(2y-1)^2\mathcal{U}_{n-2}(y) = 0, \tag{16.6.15}$$

where $x = 4(1-y)$ and $\mathcal{U}_m$ is the Chebyshev polynomial of the second kind. The roots of (16.6.15) are given by

$$\begin{array}{rcl} y_{n+1} & = & 1 \quad \rightarrow \quad x_{n+1} = 0 \\ y_n & = & \frac{1}{2} \quad \rightarrow \quad x_n = 2 \\ y_{n-1} & = & \frac{1}{2} \quad \rightarrow \quad x_{n-1} = 2 \end{array} \tag{16.6.16}$$

and by $\mathcal{U}_{n-2}(y) = 0$, i.e.

$$y_k = \cos\frac{k\pi}{n-1} \quad \rightarrow \quad x_k = 8\sin^2\frac{k\pi}{2(n-1)}, \quad k = 1, 2, \cdots n-2. \tag{16.6.17}$$

The first root in (16.6.16) is irrelevant for the spectrum. The spectrum is reported in Table 16.2.

16.6.3 E_n Series

The analysis of these exceptional algebras has to be done separately for each of them.

1. The characteristic equation for the E_6 algebra is

$$\begin{aligned} \|\mathcal{M} - x \cdot \mathbf{1}\| &= x^6 - 24x^5 + 216x^4 - 936x^3 + 2052x^2 - 2160x + 864 = \\ &= [x^2 - 12x + 24]\,[x^2 - 6x + 6]^2. \end{aligned} \tag{16.6.18}$$

   There are two doublets of degenerate particles, plus other two particles of different masse. The spectrum is given in Table 16.2.

2. The characteristic equation of the Toda field theory based on E_7 is

$$\begin{aligned} \|\mathcal{M} - x \cdot \mathbf{1}\| = x^7 - 36x^6 + 504x^5 - 3552x^4 + \\ +13536x^3 - 27648x^2 + 27648x - 10368 = \\ [x-6]\,[x^3 - 18x^2 + 72x - 72] \times \\ [x^3 - 12x^2 + 36x - 24]. \end{aligned} \tag{16.6.19}$$

   The mass spectrum can be found in Table 16.2. Thanks to the Z_2 automorphism of the Dynkin diagram of the affine E_7 algebra, the particles can be classified in even and odd particles with respect this Z_2 symmetry.

3. For the Toda field theory on E_8 we have

$$
\begin{aligned}
\|\mathcal{M} - x\mathbf{1}\| = {} & x^8 - 60x^7 + 1440x^6 - 18000x^5 + 1257440x^4 + \\
& -518400x^3 + 1166400x^2 - 1296000x + 518400 = \\
& [x^4 - 30x^3 + 240x^2 - 720x + 720] \times \\
& [x^4 - 30x^3 + 300x^2 - 1080x + 720].
\end{aligned}
\tag{16.6.20}
$$

The masses of this theory are reported in Table 16.3.

These cases cover all the simply-laced Toda field theories. A similar analysis can be also done for those defined by the non-simply laced algebra by using the foldings and the final results are collected in Table 16.2.

Coupling constants

After the mass term, the next perturbative consists of the three-particle coupling constants

$$
f^{abc} = \mu^2 \beta \sum_i q_i \alpha_i^a \alpha_i^b \alpha_i^c. \tag{16.6.21}
$$

These expressions enjoy a series of interesting geometrical properties. First of all, it is possible to prove that they vanish if it is impossible to draw a triangle with sides made of m_a, m_b, and m_c whose internal angles are a rational fraction of π. This can be seen as a natural consequence of the algebraic nature of the values of the masses. Moreover, the quantities f^{abc} vanish if they do not respect a discrete symmetry of the affine Dynkin diagram. Consider, for instance, the symmetry Z_2 of $E_7^{(1)}$: if two of the indices of f^{abc} refer to two even particles and the third one to an odd particle, this coupling constant clearly vanishes. Finally, when they are different from zero, the quantities f^{abc} are proportional to the area $\mathcal{A}^{abc}$ of the aforementioned mass triangle. For the simply laced algebra we have

$$
\left| f^{abc} \right| = \frac{4\mu^2 \beta}{\sqrt{h}} \mathcal{A}^{abc}. \tag{16.6.22}
$$

Obviously the non-vanishing values of f^{abc} indicate the possible scattering processes into which enter the particles and their bound states. In fact, with $f^{abc} \neq 0$ we can have the process in Figure 16.6: in the collision, the initial particles A_a and A_b form a bound state A_c that decays in the same particles of the final state. From the symmetry of the indices of f^{abc} we immediately infer that the same scenario occurs for the processes of the crossed channels, namely the particle A_a can be regarded as bound state of the particle A_b and A_c, as well as the particle A_b may be regarded as bound state of the particles A_a and A_c.

There are other n-particle vertices of the perturbation theory coming from the series expansions of the exponential terms. Interestingly enough, they admit a geometrical

$A^{(1)}_{2r}$ $2M\sin(\frac{\pi i}{2r+1}), \quad 1 \leq i \leq 2r$	$A^{(2)}_{2r}$ $4M\sin(\frac{\pi i}{2r+1}), \quad 1 \leq i \leq r$
$D^{(1)}_{r+1}$ $M, M, 2M\sin(\frac{\pi i}{2r}), \quad 1 \leq i \leq r-1$	$B^{(1)}_{r}$ $M, 2M\sin(\frac{\pi i}{2r}), \quad 1 \leq i \leq r-1$
$D^{(1)}_{r+2}$ $M, M, 2M\sin(\frac{\pi i}{2r+2}), \quad 1 \leq i \leq r$	$D^{(2)}_{r+1} \equiv \tilde{B}_r$ $\sqrt{2}M\sin(\frac{\pi i}{2r+2}), \quad 1 \leq i \leq r$
$A^{(1)}_{2r-1}$ $2M\sin(\frac{\pi i}{2r}), \quad 1 \leq i \leq 2r-1$	$C^{(1)}_{r}$ $2M\sin\left(\frac{\pi i}{2r}\right), \quad 1 \leq i \leq r$
$D^{(1)}_{2r}$ $M, M, 2M\sin(\frac{\pi i}{2(2r-1)}), 1 \leq i \leq 2r-2$	$A^{(2)}_{2r-1} \equiv \tilde{C}_r$ $\frac{M}{\sqrt{2}}, \sqrt{2}M\sin(\frac{\pi i}{(2r-1)}), 1 \leq i \leq r-1$
$D^{(1)}_{4}$ $M, M, M, \sqrt{3}M$	$G^{(1)}_{2}$ $M, \sqrt{3}M$
$E^{(1)}_{6}$ $m_1 = m_{\bar{1}} = M$ $m_2 = m_{\bar{2}} = 2M\cos(\frac{\pi}{12})$ $m_3 = 2M\cos(\frac{\pi}{4})$ $m_4 = 4M\cos(\frac{\pi}{12})\cos(\frac{\pi}{4})$	$D^{(3)}_{4} \equiv \tilde{G}_2$ m_3, m_4 $F^{(1)}_{4}$ m_1, m_2, m_3, m_4
$E^{(1)}_{7}$ $m_1 = M$ $m_2 = 2M\cos(\frac{5\pi}{18})$ $m_3 = 2M\cos(\frac{\pi}{9})$ $m_4 = 2M\cos(\frac{\pi}{18})$ $m_5 = 4M\cos(\frac{5\pi}{18})\cos(\frac{\pi}{18})$ $m_6 = 4M\cos(\frac{\pi}{9})\cos(\frac{2\pi}{9})$ $m_7 = 4M\cos(\frac{\pi}{18})\cos(\frac{\pi}{9})$	$E^{(2)}_{6} \equiv \tilde{F}_4$ m_2, m_4, m_5, m_6

Table 16.2 *Masses of the Toda field theories related by the folding procedure.*

interpretations in terms of the moments of a distribution of a set of positive electric charges $\{q_i\}$, placed in the points indicated by the vectors α_i. Adopting this interpretation, the total charge of the system is h. The condition (16.6.3) is then nothing else but the definition of the cent reference frame of the charges while the diagonalization of (16.6.9) is equivalent to the choice of the coordinates along the principal axes of the ellipsoid defined by the quadrupole moments.

$E_8^{(1)}$
$m_1 = M$
$m_2 = 2M\cos(\frac{\pi}{5})$
$m_3 = 2M\cos(\frac{\pi}{30})$
$m_4 = 2m_2\cos(\frac{7\pi}{30})$
$m_5 = 2m_2\cos(\frac{2\pi}{15})$
$m_6 = 2m_2\cos(\frac{\pi}{30})$
$m_7 = 4m_2\cos(\frac{2\pi}{5})\cos(\frac{7\pi}{30})$
$m_8 = 4m_2\cos(\frac{\pi}{5})\cos(\frac{2\pi}{15})$

Table 16.3 *Mass spectrum of the Toda field theory* $E_8^{(1)}$.

The Toda field theories possess an infinite set of conserved currents, both at the classical and quantum level. For our scope, rather than their explicit expressions, it is sufficient to know the spectrum of the their spins s. This is given by the Coxeter exponents of the Lie algebra under investigation, modulo the Coxeter number. The sets of these values is given in Table 16.4. The quantum integrability of the Toda field theories has an extremely important consequence for the scattering processes in which are involved the particles A_i of these theory, namely their elasticity. This property is already manifest at the tree level of the scattering processes of two initial particles $A_a A_b$ going into two final particles $A_c A_d$: indeed, the only non-vanishing amplitudes are those in which the final particles coincide with the initial ones. At the lowest order in the coupling constant β, this amplitude is ruled by the sum of the Feynman graphs shown in Figure 16.7 and this sum vanishes unless the final particles $A_c A_d$ are equal to the initial ones (in the propagators of the last three diagrams it enters all particles that are compatible with $f^{ijk} \neq 0$).

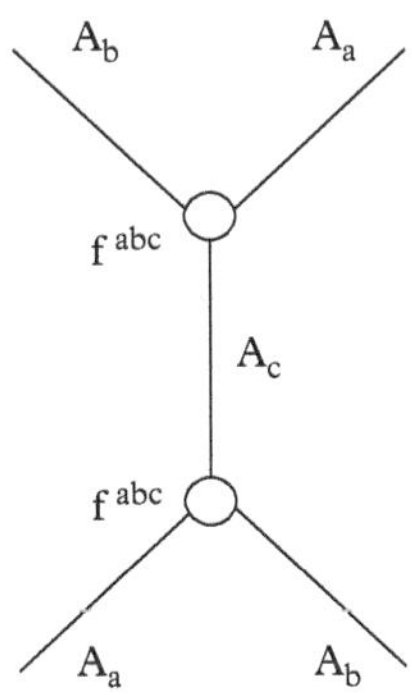

Fig. 16.6 *Scattering process of the particles A_a and A_b that gives rise to the bound state given by the particle A_c.*

Algebra	ψ	Exponents
$A_r^{(1)}$	$r+1$	$1,2,\cdots,r$
$A_{2r}^{(2)} \equiv A_{2r}/Z_2$	$4r+2$	$1,3,5,\cdots,2r-1,2r+3,\cdots,4r+1$
$B_r^{(1)}$	$2r$	$1,3,5,\cdots,2r-1$
$\tilde{B}_r \equiv D_{r+1}^{(2)}$	$2r+2$	$1,3,5,\cdots,2r+1$
$C_r^{(1)}$	$2r$	$1,3,5,\cdots,2r-1$
$\tilde{C}_r \equiv A_{2r-1}^{(2)}$	$4r-2$	$1,3,5,\cdots,4r-3$
$D_r^{(1)}$	$2r-2$	$1,3,5,\cdots,2r-3,r-1$
$E_6^{(1)}$	12	$1,4,5,7,8,11$
$E_7^{(1)}$	18	$1,5,7,9,11,13,17$
$E_8^{(1)}$	30	$1,7,11,13,17,19,23,29$
$G_2^{(1)}$	6	$1,5$
$\tilde{G}_2 \equiv D_4^{(3)}$	12	$1,5,7,11$
$F_4^{(1)}$	12	$1,5,7,11$
$\tilde{F}_4 \equiv E_6^{(2)}$	18	$1,5,7,11,13,17$

Table 16.4 *Coxeter numbers and Coxeter exponents of the affine Dynkin diagrams.*

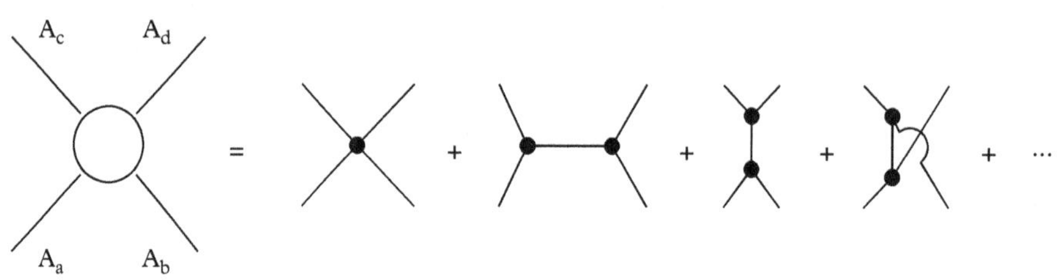

Fig. 16.7 *Scattering process of the particles A_a and A_b with final state given by the particles A_c and A_d. If $A_cA_d \neq A_aA_b$, the four-particle vertex f^{abcd} of the first graph cancels the sum of the other Feynman graphs whose internal propagators are made of all the particles allowed by the three-particle vertices $f^{ijk} \neq 0$.*

16.7 Toda Field Theories with Imaginary Coupling Constant

If we make the analytic continuation $\beta \to i\beta$ in the previous action of the Toda field theories, we arrive, in general, to a complex action (the only real case is for the algebra $SU(2)$, that gives rise to the Sine–Gordon model). Even though the interpretation of these theories having a complex action is problematic from the point of view of a standard QFT quantization, it can be nevertheless shown that, for particular values of β, an opportune restriction of their Hilbert space leads to the definition of consistent models. Note that, with this transformation, the Liouville part of these theories is associated to a CFT with a value of the central charge less than the rank r of the algebra. Choosing the discrete values

$$\beta^2 = \frac{p}{p+1} \quad p = k+h, k+h+1, \ldots$$

for the central charge we have

$$c = r\left[1 - \frac{h(h+1)}{p(p+1)}\right]. \qquad (16.7.1)$$

This value corresponds to the conformal theory constructed on the coset[9]

$$\frac{G_k \times G_1}{G_{k+1}}.$$

In these theories, the vertex operator associated to the maximal root

$$V_{\alpha_{max}} = e^{i\beta\alpha_{r+1}\cdot\phi},$$

i.e. the perturbing operator of the conformal theory has conformal dimension

$$\Delta_{\alpha_{max}} = 1 - \frac{h}{k+h+1}. \qquad (16.7.2)$$

Let us discuss some significant examples.

- Taking the algebra E_8 and $k = 1$, we have

$$c = \frac{1}{2}, \quad \Delta_{max} = \frac{1}{16}. \qquad (16.7.3)$$

Hence the Toda field theory associated to this value of the imaginary coupling constant corresponds to the magnetic deformation of the Ising model.

[9] To compare with the formulae of Chapter 13 that the dimension $|G$ of the algebra is related to its rank r and the Coxeter numberψ by the relation $|G| = r(\psi + 1)$.

- Taking the algebra E_7 and $k=1$, we have

$$c=\frac{7}{10},\quad \Delta_{max}=\frac{1}{10}. \tag{16.7.4}$$

Hence, in this case, the relativeToda field theory with imaginary coupling constant corresponds to the thermal deformation of the TIM.

- With the algebra E_6 and $k=1$, we have

$$c=\frac{6}{7},\quad \Delta_{max}=\frac{1}{7}. \tag{16.7.5}$$

This theory corresponds to the thermal deformation of the tricritical 3-state Potts model.

16.8 Deformation of Conformal Conservation Laws

In this section we set a criterion to establish whether or not a deformation of a conformal theory gives rise to an integrable model away from criticality. At the first order in the coupling constant, this criterion is based on the OPE and on the formula of the conformal characters. When the integrals of motion belong to the conformal family of the identity operator, the corresponding analysis can be carried on in a purely algebraic way.

16.8.1 Operator Product Expansion

Consider a conformal minimal model $\mathcal{M}_{p,q}$ that is deformed by a relevant primary scalar field $\Phi_{lk}(z,\bar{z})=\phi_{lk}(z)\bar{\phi}_{lk}(\bar{z})$, with anomalous dimension $x=2\Delta<2$. The perturbed action is

$$\mathcal{S}=\mathcal{S}_0+\lambda\int \Phi_{lk}(z,\bar{z})\,d^2z.$$

Let $C_{s+1}(z)$ be a conserved current of the conformal model $\mathcal{M}_{p,q}$ ($\partial_{\bar{z}}C_s(z)=0$) of spin $s+1$ (that we assume to be either an integer or fractional number), local with respect to Φ_{lk}:

$$C_{s+1}(z)\Phi_{lk}(w,\bar{w}) = \sum_{n=2}^{m}\frac{d_{lk}^{(n)}}{(z-w)^n}\Phi_{lk}^{(n)}(w,\bar{w})+\frac{1}{z-w}B_{lk}(w,\bar{w})+\cdots \tag{16.8.1}$$

where n is an integer, $\Phi_{lk}^{(n)}$ and B_{lk} are the descendent fields of Φ_{lk}, while $d_{lk}^{(n)}$ denote here the structure constants of this OPE. The Ward identity for the current $C_s(z,\bar{z})$ can be expressed in terms of the conformal Ward identity

$$\begin{aligned}\langle C_{s+1}(z,\bar{z})\cdots\rangle &= \langle C_{s+1}(z)\cdots\rangle_0 \\ &+ \lambda\int dw\, d\bar{w}\, \langle C_{s+1}(z)\Phi_{lk}(w,\bar{w})\cdots\rangle_0 + \mathcal{O}(\lambda^2).\end{aligned} \tag{16.8.2}$$

At the first order in λ, eqns. (16.8.1) and (16.8.2), together with the identity

$$\partial_{\bar{z}}\,\frac{1}{z-w+i\epsilon} = \delta(z-w)\delta(\bar{z}-\bar{w}),$$

give rise to

$$\partial_{\bar{z}}\, C_{s+1}(z,\bar{z}) = \lambda\Big(\, B_{lk}(z,\bar{z}) - d^{(2)}_{lk}\,\partial_z\Phi^{(2)}_{lk}\Big). \tag{16.8.3}$$

The existence of a conservation law away from the critical point only depends on if B_{lk} is a total derivative with respect to z. The simplest example is provided by the stress-energy tensor: if $C_2 = T$, then

$$B_{lk} - d^{(2)}_{lk}\,\partial_z\Phi^{(2)}_{lk} = (1-\Delta)\,\partial_z\Phi_{lk}(z,\bar{z})$$

and, in this case, we have

$$\partial_{\bar{z}}\, T(z,\bar{z}) = -\frac{1}{4}\partial_z\,\Theta, \qquad \Theta = -4\lambda\,(1-\Delta)\Phi_{lk}(z,\bar{z}).$$

The corresponding conserved charge is expressed by

$$\mathcal{Q}_1 = \int (T\,dz + \frac{1}{4}\Theta\, d\bar{z}).$$

Let us look at some other significant examples.

1. The minimal model $\mathcal{M}_{4,5}$ corresponds to the class of universality of the TIM. On the other hand, this model is also the first of the superconformal series. Let us choose then as C_s the supercurrent $G_{3/2}$ of spin $s=\frac{3}{2}$ and as deformation the vacancy density, i.e. the operator $\Phi_{13} = \Phi_{\frac{3}{5},\frac{3}{5}}$. In the following we will use the notation $\Phi_{\Delta,\bar{\Delta}}$ for the conformal fields. The supersymmetric OPE

$$G(z_1)\Phi_{\frac{3}{5},\frac{3}{5}}(z_2,\bar{z}_2) \quad = \left(\frac{1}{5z_{12}^2} + \frac{1}{z_{12}}\partial_2\right)\Phi_{\frac{1}{10},\frac{3}{5}}(z_2,\bar{z}_2) + \cdots$$

($z_{12} \equiv z_1 - z_2$) leads to the conservation law

$$\partial_{\bar{z}} G(z,\bar{z}) = \partial_z \bar{\Psi}(z,\bar{z}) \qquad \bar{\Psi}(z,\bar{z}) = \frac{4}{5}\,\lambda\,\Phi_{\frac{1}{10},\frac{3}{5}}(z,\bar{z}).$$

The corresponding charge has spin $s = \frac{1}{2}$

$$Q_{\frac{1}{2}} \equiv Q = \int (G\,dz \,+\, \bar{\Psi}\,d\bar{z}).$$

Using the OPE

$$G(z_1)G(z_2) = \frac{2}{z_{12}}\,T(z_2) + \cdots$$

$$G(z_1)\,\Phi_{\frac{1}{10},\frac{1}{10}}(z_2,\bar{z}_2) = \frac{1}{z_{12}}\,\Phi_{\frac{3}{5},\frac{1}{10}}(z_2,\bar{z}_2) + \cdots$$

it is easy to show that

$$\begin{aligned} Q^2 &= \int dz_1 dz_2\; G(z_1)G(z_2) + \frac{4}{5}\lambda \int d\bar{z}_1 d\bar{z}_2 \left\{G(z_1\bar{z}_1), \Phi_{\frac{1}{10},\frac{3}{5}}(z_2,\bar{z}_2)\right\} \\ &= \int (2\,T\,dz + \frac{4}{5}\lambda\,\Phi_{\frac{3}{5},\frac{3}{5}}\,d\bar{z}) \;=\; 2\mathcal{P}. \end{aligned}$$

In addition to Q, we can similarly prove the conservation of $\bar{Q}$, that is constructed starting by the anti-analytic component $\bar{G}_{3/2}$ of the supercurrent. In this case we have $\bar{Q}^2 = 2\bar{\mathcal{P}}$, where $\bar{\mathcal{P}} = E - P$. Finally

$$Q\,\bar{Q} + \bar{Q}\,Q = \frac{4}{5}\,\lambda \int \left[\left(\partial_z\,\Phi_{\frac{1}{10},\frac{1}{10}}\right) dz + \left(\partial_{\bar{z}}\,\Phi_{\frac{1}{10},\frac{1}{10}}\right) d\bar{z}\right] = \mathcal{T}. \qquad (16.8.4)$$

The right-hand side of this equation is the topological charge $\mathcal{T}$. In fact, the TIM perturbed by the vacancy density operator is driven, for $\lambda < 0$, in a phase where there are three degenerate vacua, as shown in Figure 16.8. The system has therefore solitonic excitations that interpolate between two nearest vacua and that are characterized by their topological charge. The integrability of this theory implies the elasticity of the scattering processes in which are involved the solitons. The charges $\{Q, \bar{Q}, P, \bar{P}\}$ generate a global supersymmetry of this model away from criticality.

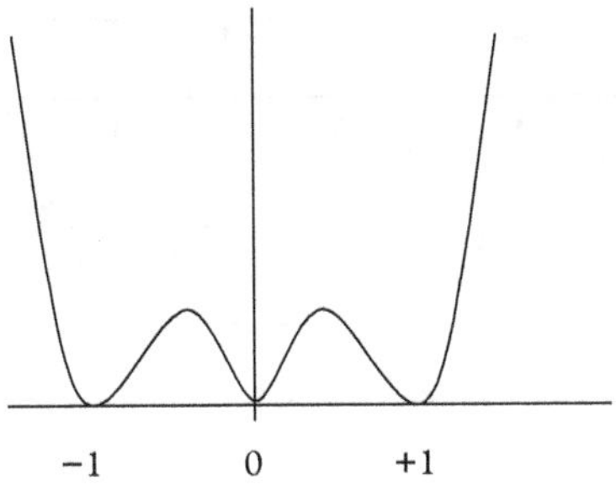

Fig. 16.8 *Effective potential of the TIM perturbed by* $\Phi_{\frac{3}{5},\frac{3}{5}}$ *con* $\lambda < 0$. *The off-critical model has solitonic excitations that interpolate between two nearest vacua.*

2. The universality class of the tricritical 3-state Potts model corresponds to a sub-algebra of the minimal mode $\mathcal{M}_{6,7}$, as the universality class of the 3-state Potts model corresponds to a sub-algebra of $\mathcal{M}_{5,6}$. Let us choose as C_s the chiral field W of spin $s = 5$ and as deformation $\Phi_{12}(z,\bar{z}) = \Phi_{\frac{1}{7},\frac{1}{7}}$. The operator expansion

$$\mathcal{W}(z_1)\ \Phi_{\frac{1}{7},\frac{1}{7}}(z_2) = \left(\frac{w_0}{z_{12}^2} + \frac{1}{z_{12}}\partial_2\right)\Phi_{\frac{22}{7},\frac{1}{7}}(z_2,\bar{z}_2) + \cdots$$

(where w_0 is a constant) gives rise to a conserved charge of spin 4

$$\mathcal{Q}_4 = \int (\mathcal{W}\ dz + \Lambda\ d\bar{z}), \qquad \Lambda = (\,w_0 - \frac{2}{7})\ \Phi_{\frac{22}{7},\frac{1}{7}}.$$

In this case, Φ_{12} is the scaling operator corresponding to the energy density of the lattice model. Hence its insertion into the action moves the temperature of the system away from its critical value. This perturbation preserves the permutation symmetry $S_3 = Z_2 \otimes Z_3$ of the model, generated by C (the charge conjugate operator) and ϑ, with

$$C^2 = \vartheta^3 = 1.$$

$\mathcal{Q}_4$ is an odd operator under C, i.e. $C\,\mathcal{Q}_4\,C = -\mathcal{Q}_4$, while the first conserved charge given by the total momentum $\mathcal{P}$ is an even operator, $C\mathcal{P}C = \mathcal{P}$.

16.8.2 Integrals of Motion of the Identity Family

It is possible to set up an efficient algebraic method to identify the integrals of motion coming from the conformal family of the identity operator. First we recall that in the conformal space the Virasoro operator L_{-1} acts as a derivative with respect to the analytic coordinate, i.e. $L_{-1} \to \partial_z$. Let us define $\hat{\Lambda}_{s+1} = \Lambda_{s+1}/L_{-1}\Lambda_s$ as the space

s	0	1	2	3	4	5	6
dim $\hat{\Lambda}_n$	1	0	1	0	1	0	2
Vectors	$\mathbf{I}$	-	$L_{-2}\mathbf{I}$	-	$T_4 = L_{-2}^2\mathbf{I}$	-	$T_6^{(1)} = L_{-2}^3\mathbf{I}$ $T_6^{(2)} = L_{-3}^2\mathbf{I}$

Table 16.5 *Dimensionality and vectors of the basis of Λ_n for $n \leq 6$.*

of the quasi-primary operators at the level $s+1$ of the conformal family $[\mathbf{I}]$ of the identity. Let $T_{s+1}^{(k)}$ be the vector basis of this space: their expressions consists of opportune polynomials in L_{-n}

$$T_{s+1}^{(k)} = L_{n_1} L_{-n_2} \cdots L_{-n_k}\mathbf{I}, \quad \sum_i n_i = s+1 \tag{16.8.5}$$

with the first representatives given in Table 16.5.

The eventual conserved current will be constructed in terms of linear combinations of these vectors of the basis. Note that the operator $\partial_{\bar{z}}$ can be interpreted as a mapping from the space $\hat{\Lambda}_{s+1}$ to the space of the operators at the level s of the perturbing field[10]

$$\partial_{\bar{z}} T_{s+1}^{(k)}(z,\bar{z}) = \lambda\, R_s^{(k)}(z,\bar{z}), \quad \partial_{\bar{z}} : \hat{\Lambda}_{s+1} \to \Phi_s, \tag{16.8.6}$$

with the operator $R_s^{(k)}$ explicitly expressed by

$$R_s^{(k)}(z,\bar{z}) = \oint_z \frac{d\xi}{2\pi i}\, T_{s+1}^{(k)}(z)\, \Phi_{lk}(\xi,\bar{z}).$$

Since the contour integral of two operators corresponds to compute their commutator (see Chapter 10), we also have

$$R_s^{(k)}(z,\bar{z}) = \left[T_{s+1}^{(k)}(z), \int d\xi\, \Phi_{lk}(\xi,\bar{z}) \right].$$

In addition to $\partial_{\bar{z}}$ we can also introduce an infinite family of operators D_n that map the family Λ of the identity operator into the space of the perturbing field

$$D_n\, \Lambda(z,\bar{z}) \equiv \oint_z \frac{d\xi}{2\pi i} \Lambda(z)\, (\xi - z)^n\, \Phi_{lk}(\xi,\bar{z}), \tag{16.8.7}$$

[10] We remind that the spin s measures the difference between the analytic and anti-analytic indices of the densities.

with $D_0 = \partial_{\bar{z}}$. Since the primary field Φ_{lk} satisfies

$$[L_n, \Phi_{lk}(\xi,\bar{\xi})] = \left[(\xi - z)^{n+1}\,\partial_\xi + \Delta\,(n+1)\,(\xi - z)^n\right]\Phi_{lk}(\xi,\bar{\xi}),$$

it holds the relations

$$\begin{aligned} [L_n, D_m] &= -\,(m+(1-\Delta)(n+1))\,D_{n+m}, \\ D_{-m}\mathbf{I} &= \frac{1}{(m+1)!}\,L_{-1}^{m+1}\,\Phi_{lk}(z,\bar{z}). \end{aligned} \tag{16.8.8}$$

These equations allow us to easily compute $R_s^{(k)}$. For instance, choosing $T_2 = T = L_{-2}\mathbf{I}$, we have

$$\partial_{\bar{z}} T = \lambda D_0\,L_{-2}\mathbf{I} = \lambda(\Delta-1)\,D_{-2}\mathbf{I} = \lambda\,(\Delta-1)\,L_{-1}\,\Phi_{lk}(z,\bar{z}),$$

and, since $L_{-1}\,[\ldots] = \partial_z\,[\ldots]$, we recover the conservation law of the stress-energy tensor.

Consider now the quasi-primary field of spin 4 of the identity family $T_4 = (T^2) = L_{-2}^2\mathbf{I}$. Let us compute $\partial_{\bar{z}} T_4$ with the rules given above

$$\begin{aligned} \partial_{\bar{z}} T_4 &= \lambda\,D_0\,L_{-2}\,L_{-2}\mathbf{I} = \lambda(\Delta-1)\,(D_{-2}L_{-2} + L_{-2}D_{-2})\mathbf{I} = \lambda(\Delta-1) \\ &\times \left(2L_{-2}L_{-1} + \frac{\Delta-3}{6}\,L_{-1}^3\right)\Phi_{lk} = \lambda(\Delta-1)\left(-2L_{-3} + 2L_{-1}L_{-2} + \frac{\Delta-3}{6}\,L_{-1}^3\right)\Phi_{lk}. \end{aligned}$$

For a generic operator Φ_{lk}, the right-hand side is not a total derivative for the presence of the operator L_{-3} and, consequently, there is no a conservation law. However, if the perturbing field coincides with the operator $\Phi_{1,3}$, the null vector equation of this field at the level 3

$$\left(L_{-3} - \frac{2}{\Delta+2}L_{-1}\,L_{-2} + \frac{1}{(\Delta+1)(\Delta+2)}\,L_{-1}^3\right)\Phi_{1,3} = 0, \tag{16.8.9}$$

allows us to re-express L_{-3}, arriving then to the conservation law

$$\partial_{\bar{z}} T_4 = \partial_z\,\Theta_2,$$

with

$$\Theta_2 = \lambda\frac{\Delta-1}{\Delta+2}\left\{2\Delta L_{-2} + \frac{(\Delta-2)(\Delta-1)(\Delta+3)}{6(\Delta+1)}\,L_{-1}^2\right\}\Phi_{1,3}.$$

The conserved charge $\mathcal{Q}_3$ commutes with $\mathcal{Q}_1$, as it can be shown using the commutation relations of the L_ns. Using eqn. (16.8.9) and the other null vector equations satisfied by $\Phi_{1,3}$, it is possible to prove that there are infinite conserved currents for all odd integer values of the spin s. Their expressions coincide with the analogous expressions of the Sine–Gordon model, eqn. (16.3.4), a fact that should not be surprising in the light of the relationship between the Sine–Gordon model and the $\Phi_{1,3}$ deformation of the minimal models.

If the perturbing field is either $\Phi_{1,2}$ or $\Phi_{2,1}$, the first non-trivial conservation law is obtained by the following linear combination of the quasi-primary fields of spin 6

$$T_6 = T_6^{(1)} + a\,T_6^{(2)}, \qquad a = \frac{18}{2\Delta+1} + \Delta - 2. \tag{16.8.10}$$

For the null vector equation of these fields

$$\left(L_{-2} - \frac{3}{2(2\Delta+1)}\right)\Phi = 0$$

we have in fact

$$\partial_{\bar{z}} T_6 = \partial\,\Theta_4,$$

The explicit expression Θ_4 is proposed as exercise in Problem 16.5. For these operators, it can be shown that other conserved currents are obtained for the values of the spin

$$s = 1, 5 \pmod 6). \tag{16.8.11}$$

16.8.3 Counting Method

All the examples discussed above have illustrated the importance of the OPE for defining the conserved currents with lower values of the spin. An extremely powerful method to establish a sufficient condition of their existence, without bothering to explicitly compute them, has been introduced by A.B. Zamolodchikov. It goes under the name of *counting argument*. The following discussion focuses on the conserved currents coming from the identity operator although analogous results can be easily established by considering other conserved currents coming from conformal families of other generators that are local with respect to the perturbing field Φ.

Let $\hat{\Lambda}_{s+1}$ be the space of the quasi-primary descendant fields of the identity operator and $\hat{\Phi}_s$ the quotient space at level s of the perturbing field

$$\hat{\Phi}_s = \Phi_s / L_{-1}\,\Phi_{s-1}.$$

The linear map

$$\partial_{\bar{z}} \;:\; \hat{T}_{s+1} \to \lambda\,\hat{\Phi}_s$$

has clearly a non-zero kernel when

$$\dim \hat{T}_{s+1} > \dim \hat{\Phi}_s. \tag{16.8.12}$$

If this condition is fulfilled, then there are necessarily some fields $T_{s+1}(z,\bar{z}) \in \hat{T}_{s+1}$ and $\Phi_{s-1}(z,\bar{z}) \in \hat{\Phi}_{s-1}$ such that

$$\partial_{\bar{z}}\, T_{s+1}(z,\bar{z}) = \lambda\, \partial_z\, \Phi_{s-1}(z,\bar{z}),$$

i.e., there is a conserved current of spin s. It is easy to check the condition (16.8.12) by computing the dimension of the involved spaces by means of the conformal characters

$$\sum_{s=0}^{\infty} q^s \dim \hat{T}_n = (1-q)\, \tilde{\chi}_{1,1}(q) + q,$$

$$\sum_{s=0}^{\infty} q^{s+\Delta_{kl}} \dim(\hat{\Phi}_{k,l})_s = (1-q)\, \tilde{\chi}_{k,l}(q),$$

where

$$\tilde{\chi}_{r,s}(q) = q^{\frac{(c-1)}{24}-\Delta_{r,s}}\, \chi_{r,s}(q),$$

with $\chi_{r,s}(q)$ the character of the field $\Phi_{r,s}$, whose explicit expression was presented in Chapter 11.

The counting argument provides useful information on the structure of the conserved currents only for values of the spin low enough.[11] Using the counting argument it is easy to prove the existence of non-trivial integrals of motion for the deformations of the minimal models induced by the operators $\Phi_{1,3}$, $\Phi_{1,2}$ and $\Phi_{2,1}$. Hence, these deformations always define integrable models away from criticality.

16.8.4 Examples

This section provides some examples of the application of the counting argument.

- The first example comes from a model analysed in the previous section, i.e. the thermal deformation of the tricritical 3-state Potts model. In this case, there are two classes of conserved charges. The first class has origin from the family of the identity operator, while the second class comes from the descendants of $\mathcal{W}$. These two classes are also distinguished by their quantum number under the charge conjugate operator C. The result is

[11] The reason is that the dimension of the higher-level spaces of $\Phi_{r,s}$ asymptotically grows faster than the dimension of the same spaces coming from the identity operator.

$$\dim \hat{T}_{s+1} > \dim (\hat{\Phi}_{\frac{1}{7},\frac{1}{7}})_s \quad \text{for} \quad s = 1,5,7,11 \qquad (C-\text{even}).$$
$$\dim \hat{W}_{s+1} > \dim (\hat{\Phi}_{\frac{22}{7},\frac{1}{7}})_s \quad \text{for} \quad s = 4,8 \qquad (C-\text{odd}).$$

In the light of these results, it is natural to conjecture that the spectrum of the spin of the conserved charges is given by

$$s = 1,4,5,7,8,11 \quad (\text{mod } 12). \tag{16.8.13}$$

These values of the spin coincide with the Coxeter exponent of E_6, modulo the Coxeter number of this algebra. The presence of this algebra should not be surprising for the additional symmetry of this model, that can be also defined in term of the coset $(E_6)_1 \otimes (E_6)_1/(E_6)_2$.

- An analogous computation for the TIM ($\mathcal{M}_{4,5}$) perturbed by the energy operator $\Phi_{1,2} = \Phi_{\frac{1}{10},\frac{1}{10}}$ gives

$$\dim \hat{T}_{s+1} > \dim (\hat{\Phi}_{\frac{1}{10},\frac{1}{10}})_s \quad \text{for} \quad s = 1,5,7,9,11,13.$$

These values of s coincide with the first Coxeter exponents of E_7. It is natural to conjecture that the full spectrum of the spins of the conserved charges is given in this case by

$$s = 1,5,7,9,11,13,17 \quad (\text{mod } 18) \tag{16.8.14}$$

where 18 is the Coxeter number of E_7. This structure of the spins is obviously related to the coset realization $\frac{(E_7)_1 \otimes (E_7)_1}{(E_7)_2}$ of the model.

- For the Ising model ($\mathcal{M}_{3,4}$) perturbed by the magnetization operator $\Phi_{1,2} = \Phi_{\frac{1}{16},\frac{1}{16}}$, we have

$$\dim \hat{T}_{s+1} > \dim (\hat{\Phi}_{\frac{1}{16}\frac{1}{16}})_s \quad \text{for} \quad s = 1,7,11,13,17,19,$$

namely, the first representatives of the infinite series of the Coxeter exponents of E_8, modulo the Coxeter number $h = 30$ of this algebra.

$$s = 1,7,11,13,17,19,23,29 \quad (\text{mod } 30). \tag{16.8.15}$$

This is not a coincidence, since the Ising model can be also defined in terms of the coset $\frac{(E_8)_1 \otimes (E_8)_1}{(E_8)_2}$.

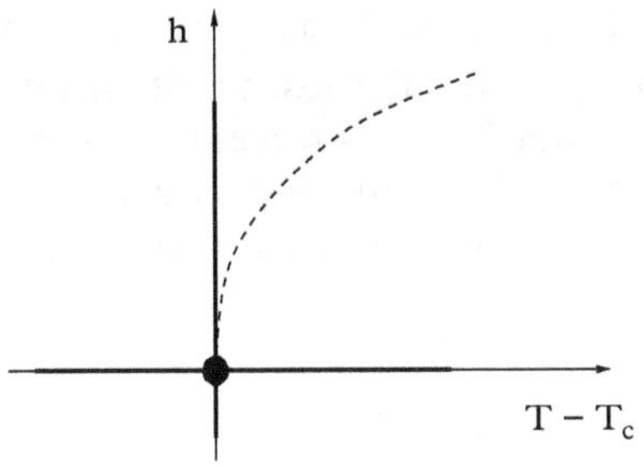

Fig. 16.9 *Space of the coupling constants of the Ising model nearby the critical point, here placed at the origin. The thermal and magnetic axes define two separate integrable models. Another potential integrable point in the plane will belong to a renormalization group flow (dashed line), so that the model would be integrable all along this curve.*

16.9 Multiple Deformations of Conformal Field Theories

Thus far we have analysed the conformal models deformed by only one relevant operators. But can the analysis above be generalized to deformations made of several fields? For instance, in the Ising model, there are two deformations—the thermal and the magnetization deformations—that separately give rise to two different integrable models. Are there in this model other lines[12] that are integrable in the plane $(h, T - T_c)$? The same question can be formulated for other models too, as for instance, for the TIM where the two deformations $\Phi_{1,3}$ and $\Phi_{1,2}$ are individually integrable deformations.

Although presently there is not a final answer to this question, an explicit computation to identify possible conserved currents with low value of the spin s gives a negative answer. The essential reason stays in the different null vector structures that support the single deformations. This negative result leads to be pessimistic on the possibility that there exists conserved currents of higher spin. To present this computation, let us first recall the derivation of a conservation law $C_s \in \hat{T}_s$ when there is a single deformation, restricting the attention on the unitary theory. Considering higher-order perturbation terms, we have in general

$$\partial_{\bar{z}} C_s(z,\bar{z}) = \lambda \, B_{lk}^{(1)}(z,\bar{z}) + \cdots \lambda^n \, B_{lk}^{(n)}(z,\bar{z}) + \ldots \tag{16.9.1}$$

Keeping in mind the dimensionality of the coupling constant, a dimensional analysis fixes the scaling dimensions of the operators $B_{lk}^{(n)}(z,\bar{z})$, given by

$$[s - n(1-\Delta), 1 - n(1-\Delta)].$$

[12] If there exists an integrable point, this necessarily belongs to an RG flow and therefore the Ising model would be integrable also along this line (Figure 16.9).

Since $\Delta < 1$, there exists an integer n_c such that, for all $n > n_c$ the conformal dimension of $B_{lk}^{(n)}(z,\bar{z})$ becomes negative. However, the absence of operators with negative conformal dimensions in the unitary minimal models forces the series (16.9.1) necessarily to stop (as a matter of fact, in most cases only the first term survives). If we now consider the deformations made by two operators with conformal dimensions Δ_1 and Δ_2 (and with corresponding couplings λ_1 and λ_2), the generalization of eqn. (16.9.1) is

$$\partial_{\bar{z}}\, C_s(z,\bar{z}) = \sum_{n,m=1} \lambda_1^n \lambda_2^m \, B_{lk}^{(n,m)}(z,\bar{z}). \tag{16.9.2}$$

The conformal dimensions of the quantities $B_{lk}^{(n,m)}$ are

$$[s - n(1-\Delta_1) - m(1-\Delta_2), 1 - n(1-\Delta_1) - m(1-\Delta_2)].$$

This series must truncate, for the same reason given above. Moreover, at least in the two explicit examples considered here, the Ising and the TIM, the series splits in two independent expressions, one that is only a function of λ_1 while the other of λ_2. The reason is simple: in fact, the analytic conformal dimension must coincide with one of the conformal dimensions present in the Kac tables of the model. For the Ising model perturbed both by the energy and magnetization fields, we must have

$$1 - n\,\frac{1}{2} - m\,\frac{15}{16} = \Delta_r, \tag{16.9.3}$$

for some Δ_r of this model. However possible values of Δ_r are only $\Delta_r = \{0, \frac{1}{2}, \frac{1}{16}\}$ and it is therefore impossible to have both n and m different from zero at the same time. The same situation occurs for the TIM perturbed by the energy and vacancy densities, $\Phi_{\frac{1}{10}\frac{1}{10}}$ and $\Phi_{\frac{3}{5}\frac{3}{5}}$.

Therefore for these models, eqn. (16.9.2) is expressed by the direct sum of the contribution of both terms. If there exists a conserved current, this should appear at the common level of the conserved currents of both deformations. Concerning the field $\Phi_{\frac{1}{2}\frac{1}{2}}$ of the Ising model and the field $\Phi_{\frac{3}{5}\frac{3}{5}}$ of the TIM, both are $\Phi_{1,3}$ operators and therefore their associate conserved currents exist for

$$s = 1, 3, 5, 7, \ldots.$$

For the magnetic deformation of the Ising model the spectrum of the conserved currents is given by the Coxeter exponents of E_8

$$s = 1, 7, 11, 13, 19, 23, 29 \pmod{30}.$$

For the TIM, the spectrum of the conserved currents associated to the second operator $\Phi_{\frac{1}{10},\frac{1}{10}}$ coincides with the Coxeter exponents of E_7

$$s = 1,5,7,9,11,13,17 \pmod{18}.$$

Hence, in both models, the common values of the spins of their double deformation coincide with the Coxeter exponents of the corresponding E_n algebra.

In the following we explicitly show that there are no conserved currents of a double deformation of these models for the lowest values of s. As mentioned, this result underlines the absence of the integrability of these statistical models under their multiple deformations.

16.9.1 The Tricritical Ising Model

We start the analysis from this model because there may exist a conserved current at $s = 5$ level, whereas for the Ising model we shall consider at least $s = 7$.

The explicit expression of the conserved current $C_6^{(1)}$ of the Φ_{13} deformation of the minimal model $\mathcal{M}_{p,p+1}$ coincides with the corresponding expression of the Sine–Gordon

$$T_6^{(1)} = (T(T^2)) + \frac{9}{40}\,(T\partial^2\,T), \tag{16.9.4}$$

where we have substituted $c = 7/10$. Applying $\partial_{\bar{z}}$ to (16.9.4) and using the algebraic formalism of the operators D_n we have

$$\partial_{\bar{z}} C_6 = \lambda_1\,(1+\Delta_{13})[5\,L_{-5} - 4\,L_{-2}L_{-3}]\Phi_{13} + \\ +L_{-1}[\cdots].$$

The first term on the right-hand side is indeed zero for the Φ_{13} deformation, as a consequence of the null vector equation of this operator at level 3. Hence $C_6^{(1)}$ is the conserved quantity under the $\Phi_{1,3}$ deformation. We need to check then if $T_6^{(1)}$ is still a conserved quantity if we perturb the model by means of the second operator $\Phi_{1,2} = \Phi_{\frac{1}{10}\frac{1}{10}}$. Repeating the previous steps, we get

$$\partial_{\bar{z}} C_6 = \lambda_2\,(1+\Delta_{12})\,[9\,L_{-5} - 6\,L_{-2}L_{-3}]\,\Phi_{12} \\ +L_{-1}[\cdots]. \tag{16.9.5}$$

In this case, however, the null vector equation satisfied by the operator Φ_{12}

$$\left[L_{-2} - \frac{5}{42}L_{-1}^2\right]\Phi_{12} = 0$$

does not lead to the vanishing of the right-hand side of eqn. (16.9.5)! As a matter of fact, the explicit expression of the conserved current under the Φ_{12} deformation is given by eqn. (16.8.10)

$$T_6^{(2)} = (T(T^2)) + \frac{131}{10}(T\partial^2\, T), \tag{16.9.6}$$

that does not coincide with (16.9.4). Hence, the final conclusion of this computation is the absence of a conserved current of spin $s = 5$ for a multiple deformation of the TIM. A similar analysis can be done also for the level $s = 7$, with a negative result as well.

16.9.2 The Ising Model

For the Ising model in an external magnetic field and at $T \neq T_c$ the first common value of the spin for both deformation is $s = 7$. The explicit expression of the conserved current under the $\Phi_{1,3}$ deformation is

$$C_8 = (T(T(T^2))) + \frac{c+8}{6}(T(T\partial_z^2 T)) + \frac{1}{180}(c^2 + 4c - 101)(T\partial_z^4 T) \tag{16.9.7}$$

with $c = \frac{1}{2}$. Repeating the same steps of the computation shown for the example above, one can explicitly show that this current is not conserved under the $\Phi_{1,2}$ deformation associated to the magnetization operator.

Both examples clearly show the reason of the absence of common conserved currents, related to the different structure of the null vectors of the different deformations. It would be a major discovery in statistical mechanics if in the future we could show the possibility of a conservation law for the multiple deformations of the Ising model.

REFERENCES

For integrability aspects of the classical systems see

De Luca A. and Mussardo, G. (2016). 'Equilibration Properties of Classical Integrable Field Theories', *Journal of Statistical Mechanics*, 1606(6): 064011.

Dodd, R.K., Eilbeck, J.C., Gibbons, J.D. and Morris, H.C. (1982). *Solitons and Nonlinear Wave Equations*, Academic Press.

Drazin, P.G. and Johnson, R.S. (1989). *Solitons: An Introduction*, Cambridge University Press.

Faddev L. and Takhtajan, L. (2007). *Hamiltonian Methods in the Theory of Solitons*, Springer-Verlag.

Novikov, S., Manakov, S., Pitaevski, L. and Zakharov, V. (1984). *Theory of Solitons*, Consultants Bureau.

Scott, A. Chu, F. McLaughlin, D. (1973). 'The Soliton: A New Concept in Applied Science', *Proceedings of the IEEE*, 61: 1443.

An incisive text on the relationship between the classical and quantum systems is

R. Rajaraman, *Solitons and Instantons*, North Holland, 1982.

A fundamental paper on the integrable deformations of the conformal theories is

Zamolodchikov, A.B. (1989). 'Integrable Field Theory from Conformal Field Theory', *Advanced Studies in Pure Mathematics*, 19: 641.

The relation between the integrable $\Phi_{1,3}$ deformation of the conformal minimal models and the Sine–Gordon theory has been discussed in

Eguchi, T. and Yang, S.K. (1990). 'Deformations of Conformal Field Theories and Solitons Equations', *Physics Letters B*, 235: 202.

A review paper on the Toda Field theories and the integrable theories away from criticality is

Mussardo, G. (1992). 'Off-critical Statistical Models: Factorized Scattering Theories and Bootstrap Program', *Physics Reports* 218: 215.

The quantum integrability of the Sine/Sinh–Gordon and Bullogh–Dodd models using the Feynman diagrams has been discussed in

P. Dorey, (1996). 'Exact S matrices', In: Z. Horvath and L. Palla (Eds), *Proceedings of the Eötvös Summer School in Physics: Conformal Field Theories and Integrable Models*, Springer, pp. 85–125.

PROBLEMS

16.1. Backlund transformations

a. Write down the Backlund transformations for the Sine–Gordon model.
b. Taking as initial solution of the equation of motion $\phi = 0$, determine the new solution $\hat{\phi}(z, \bar{z})$ and show that it coincides with the solitonic solution of the model.
c. Iterate the procedure to determine the other classical solutions of the Sine–Gordon model.

16.2. Scattering processes of the solitons

Analyse the solution (16.3.15) with topological charge $\mathcal{T} = 2$ of the Sine–Gordon model in the limits $t \to \pm\infty$. Determine the time delay Δ_{ss}. Based on the positive sign of this quantity and the negative sign of the analogous quantity $\Delta_{s\bar{s}}$ for the scattering of the soliton and anti-soliton, argue on the nature of the interactions between soliton–soliton and soliton–anti-soliton.

16.3. Lax pair

Consider the pair of first-order differential operators (called Lax pair):

$$L(x,t\mid\theta) = \frac{d}{dx} + i\left(\frac{\beta}{4}\partial_t\phi\,\sigma_3 + m\sinh\theta, \cos\frac{\beta\phi}{2}\sigma_1 + m\cosh\theta\,\sin\frac{\beta\phi}{2}\sigma_2\right)$$

$$M(x,t\mid\theta) = \frac{d}{dt} + i\left(\frac{\beta}{4}\partial_x\phi\,\sigma_3 + m\cosh\theta\,\cos\frac{\beta\phi}{2}\sigma_1 + m\sinh\theta\,\sin\frac{\beta\phi}{2}\sigma_2\right)$$

where σ_i are the Pauli matrices and θ the rapidity variable. If $[L,M]=0$

a. show that the field ϕ satisfies the equation of motion of the Sine–Gordon model

$$\Box\phi + \frac{m^2}{\beta}\sin\beta\phi = 0.$$

b. Take a rectangle domain $-L \leq x \leq L; 0 \leq t \leq T$ and assume periodic boundary condition $\phi(-L) = \phi(L)$. With the notation

$$\mathcal{U}_L(\theta,t) = \overrightarrow{\exp}\left(\int_{-L}^{L} U(x,t\mid\theta)\,dx\right), \quad \mathcal{V}_L(\theta) = \overrightarrow{\exp}\left(\int_0^T V(x,t\mid\theta)\,dt\right)$$

for the ordered integrals, show that

$$\mathcal{U}_L(\theta,T) = \mathcal{V}_L^{-1}(\theta)\mathcal{U}_L(\theta,0)\,\mathcal{V}_L(\theta)$$

so that $\mathrm{Tr}\,\mathcal{U}_L(\theta,t)$ is independent of t; conclude that, θ being arbitrary, there are an infinite number of conserved quantities.

16.4. Derrick theorem

The aim of this exercise is to show that the static solitonic solution of finite energy can only exist for $1+1$ dimensional theories. Consider, in the $(d+1)$ dimensional Minkowski space, the Lagrangian

$$\mathcal{L} = \frac{1}{2}\partial_\mu\phi\partial^\mu\phi - U(\phi),$$

where $U(\phi)$ is a non-negative function, that vanishes at the vacua of the theory. The static energy E can be written as $E = W_1 + W_2$, where

$$W_1 = \frac{1}{2}\int d^dx(\nabla\phi)^2, \quad W_2 = \int d^dx\,U(\phi).$$

Let $\phi(x)$ be a static solution of the equation of motion of the theory.

a. Determine the variation of W_1 and W_2 under the transformation $\phi(x) \to \phi(\lambda x)$.

b. Using the condition that $\phi(x)$ is a solution of the equation of motion, show that the energy $E[\lambda]$ is stationary for $\lambda = 1$

c. Since $W_1 \geq 0$ and $W_2 \geq 0$, show that only non-vanishing solutions exist for $d \leq 2$.

16.5. Liouville theory and minimal models

a. In the quantization scheme of the Sine–Gordon model in terms of the Liouville theory, determine the quantized values of the coupling constant g that reproduce the central charges of the minimal models. Prove that the conformal dimension of the vertex operator that perturbes the Liouville action is equal to $\Delta_{1,3}$.

b. Repeat the same exercise for the two Liouville theories, with complex exponentials, associated to the Bullogh–Dodd model. Show that the perturbations correspond to the operators $\Phi_{1,2}$ and $\Phi_{2,1}$ of the minimal models respectively.

16.6. Conserved currents

Using the algebra of the operators D_n and the null vector equation at the level 2 satisfied by $\Phi_{1,2}$ and $\Phi_{2,1}$, find the linear combination T_6 of the basis vectors $T_6^{(1)}$ and $T_6^{(2)}$ that satisfies

$$\partial_{\bar{z}} T_6 = \partial_z \Theta_4.$$

Determine the density Θ_4.

17

S-matrix Theory

All animals are equal, but some animals are more equal than others.
George Orwell, *Animal Farm*

This chapter presents the S-matrix theory of two-dimensional integrable models. This leads, in particular, to the exact spectrum of the massive excitations away from the critical point. From a mathematical point of view, the two-dimensional nature of the systems and their integrability are the crucial features that lead to important simplifications of the formalism and its successful application. It is worth mentioning that, initially developed to overcome the obstacles encountered by the QFT in dealing with the strong interactions of the hadronic particles[1] (such as protons, neutrons or pions), the S-matrix has achieved its most beautiful intellectual triumph in the analysis of the two-dimensional statistical models away from criticality, particularly when they are described by integrable theories. These significant developments have been pioneered by A.B. Zamolodochikov.

The key point of this formalism is the self-consistent dynamical method for computing the exact expressions of all scattering amplitudes and the mass of the particles. This is the so-called *boostrap approach*,[2] where all particles are democratically on the same footing: there is no distinction between the particles of the asymptotic states and the bound states, and any massive excitation of the theory can equivalently be regarded as asymptotic state or bound state of a pair of particles of the same theory. From this point of view, all particles are composite states and no one is more elementary than another. The only difference between them (apart some internal quantum number) consists of the value of their masses, that may provide a hint on the number of interactions they are involved with. Quoting Orwell, we can then say that the lightest particle of the theory is the one more equal than the others.

In this chapter we firstly address the general principle of the S-matrix theory and secondly we discuss their application to the two dimensional cases. Chapter 18 presents some significant examples of this remarkable formalism, in particular the exact solution of the Ising model in an external magnetic field at $T = T_c$.

[1] We refer the reader interested to these developments to the Appendix of this chapter.

[2] In addition to the conformal bootstrap, this is another example of theory whose solution is based on its own mathematical and physical self-consistency.

Statistical Field Theory: An Introduction to Exactly Solved Models in Statistical Physics. Second Edition. Giuseppe Mussardo, Oxford University Press (2020). © Giuseppe Mussardo. DOI: 10.1093/oso/9780198788102.001.0001

17.1 Analytic Scattering Theory

In a relativistic context, the S-matrix theory is a generalization of the scattering process theory of quantum mechanics, briefly discussed in Appendix 17.B. Its aim is to derive general conditions on the transition amplitudes of the scattering processes involving the multi-particle asymptotic states, in order to arrive at their computation without relying on an underlying lagrangian formalism.

17.1.1 General Properties

The main properties at the root of the S-matrix theory are the following:

1. The short-range of the interactions
2. The superposition principle of quantum mechanics
3. The conservation of probability
4. The invariance under the Lorentz transformations of special relativity
5. The causality principle
6. The analiticity principle.

Let us discuss in more detail each point and work out their consequences for a generic scattering theory in $d \geq 2$. The two-dimensional case will be analysed separately later on.

To adopt the S-matrix formalism to describe the scattering processes is necessary to assume that the interactions are short range, so that the initial and final states, in which the particles are well separated one from another, consist of free particle states. These multi-particle states can be specified assigning the momenta[3] and other possible quantum numbers. For simplicity we focus our attention only on the scattering processes of the scalar particles. Since the scattering processes involve the physical particle states instead of virtual ones, the components of their momenta satisfy the d-dimensional *on-shell* condition

$$p_\mu p^\mu = m^2,$$

where m is the mass of the particle. This equation gives rise to the dispersion relation

$$E^2 - |\vec{p}|^2 = m^2, \tag{17.1.1}$$

that links together the energy $E = p_0$ and the space component $\vec{p}$ of the momentum. The spectrum of the eigenvalues of the spatial momentum is obviously continuum but, to

[3] In the following as momentum we mean the d-dimensional relativistic momentum of the particles, i.e. the set of all its components $(p^0, \vec{p})$. However, using the on-shell condition (17.1.1), it is obvious that the multi-particle states are identified just by the space components of their momenta.

simplify the discussion below, it is useful to use momentarily the compact notation $| n\rangle$ to denote the states of the system. They are made of free particles, and they form a basis of the Hilbert space that satisfy the orthogonal and completeness relations

$$\langle m \mid n\rangle = \delta_{nm}, \quad \sum_n \mid n\rangle\langle n \mid = 1.$$

Later we will specialize the Lorentz invariant normalization condition of the states.

At $t = -\infty$, let $| i\rangle$ be the initial state of the system, given by a certain number of free particles. At $t = +\infty$, i.e. after they have interacted, the final state $|\tilde{f}\rangle$ of the system also consists of free particles, although not in the same number or with the same momenta of the initial state, as shown in Figure 17.1. For the superposition principle of quantum mechanics, the final state can be written as $|\tilde{f}\rangle = S \mid i\rangle$, where S is a linear operator.[4] Hence, the probability that a measure on the final state produces as a result the state $|f\rangle$ is expressed by the modulus square of the matrix element

$$S_{fi} = \langle f \mid S \mid i\rangle. \tag{17.1.2}$$

Consider now an initial normalizable state $|\psi\rangle$, given by a linear superposition of the basis vectors

$$|\psi\rangle = \sum_n a_n \mid n\rangle, \quad \sum_n |a_n|^2 = 1.$$

The total probability that this state evolves as a final state in any basis vectors is obviously equal to 1 and we have therefore

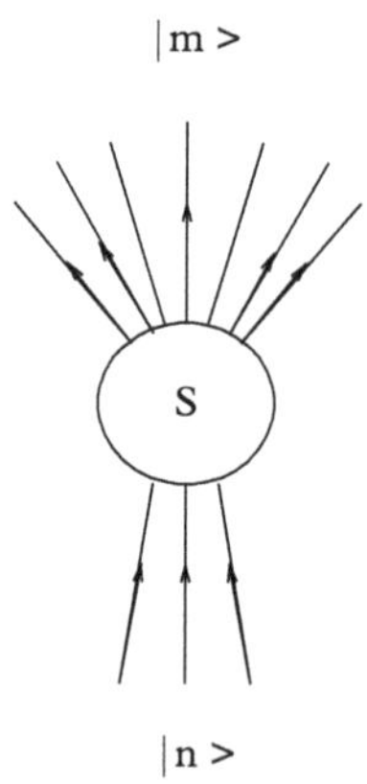

Fig. 17.1 *Quantum transition from an initial n-particle state to a final m-particle state.*

[4] S is the time evolution operator from $t = -\infty$ to $t = +\infty$. If the system admits a QFT formulation, it is expressed as $S = \mathcal{T} \exp[-i \int_{-\infty}^{+\infty} d^d x \mathcal{H}_I(x)]$, where $\mathcal{H}_I$ is the Hamiltonian density relative to the interaction and $\mathcal{T}$ denotes the time-ordering of the expressions obtained by the series expansion of the exponential term.

$$1 = \sum_m |\langle m \mid S \mid \psi\rangle|^2 = \sum_m \langle \psi \mid S^\dagger \mid m\rangle\langle m \mid S \mid \psi\rangle$$
$$= \langle \psi \mid S^\dagger S \mid \psi\rangle = \sum_{n,m} a_n^\star a_m \langle n \mid S^\dagger S \mid m\rangle.$$

Since this identity should hold for arbitrary values of the coefficients a_n, necessarily

$$\langle n \mid S^\dagger S \mid m\rangle = \delta_{nm},$$

or, in operator form,

$$S^\dagger S = 1. \tag{17.1.3}$$

Similarly, imposing equal to 1 the total probability that an arbitrary final state comes from some initial state is, one obtains the condition

$$S S^\dagger = 1. \tag{17.1.4}$$

In conclusion, the probability conservation requires S to be an unitary operator.

Let us now analyse the Lorentz invariance of the scattering theory. Let L be an arbitrary proper Lorentz transformation and $L \mid m\rangle = \mid m'\rangle$. The relativistic invariance of the theory, that ensures the independence of the physical observables from the reference frames, is expressed by the identity

$$|\langle m' \mid S \mid n'\rangle|^2 = |\langle m \mid S \mid n\rangle|^2.$$

This relation cannot fix the relative phase between the two matrix elements but, given the intrinsic arbitrariness of the overall phase of the S-matrix, we can pose the more stringent condition

$$\langle m' \mid S \mid n'\rangle = \langle m \mid S \mid n\rangle. \tag{17.1.5}$$

As we will see later, this equation implies that the S-matrix, once we factorize a delta-function for the conservation of the total momenta, depends on the momenta of the particles only through their Lorentz invariant combinations of their scalar products.

Without interactions, the state of a system does not change and in this case the S-matrix is simply the identity operator. It is a common procedure to separate the free time evolution, given by the identity operator, and write the S-matrix as

$$S_{fi} = \delta_{fi} + i(2\pi)^d \delta^d(P_f - P_i)\, T_{fi}. \tag{17.1.6}$$

The matrix elements T_{fi} define the *scattering amplitudes*. In the second term of this expression we have also explicitly written the factor $\delta^d(P_f - P_i)$ that expresses the conservation law of the total momentum, where P_i and P_f are the sum of the momenta of the initial and final particles respectively. For the non-diagonal matrix elements $i \to f$ the matrix elements of the identity operator vanish and we have

$$S_{fi} = i(2\pi)^d \delta^d(P_f - P_i)\, T_{fi}. \tag{17.1.7}$$

The relative probability is obtained by the modulus square of this amplitude. In computing such a modulus square there is however a problem, whose origin is in the interpretation to assign to the square of the delta-function. This problem can be solved by using initially the following representation of $\delta(x)$

$$\delta^d(P_f - P_i) = \frac{1}{(2\pi)^d}\int e^{i(P_f-P_i)x}\, d^d x.$$

Computing now another integral of this kind at $P_f = P_i$ (just for the presence of the delta-function) and taking the integral on a finite time interval t and on a $(d-1)$-dimensional volume V, sufficiently large but finite, the result is $Vt/(2\pi)^d$. For the modulus square of the matrix element we have then

$$\mid S_{fi} \mid^2 = (2\pi)^d\, \delta^d(P_f - P_i) \mid T_{fi} \mid^2\, Vt.$$

Dividing now for the factor Vt, we get the transition probability per unit volume and unit time

$$P_{i\to f} = (2\pi)^d\, \delta^d(P_f - P_i) \mid T_{fi} \mid^2.$$

The most important cases, both from a theoretical and experimental point of view, are those in which the initial state is made either of a one particle or two particles. The first case concerns the *decay processes*, i.e. when a heavy particle decays into other lighter particles, whereas the second case is relative to the *scattering* of two particles, that can result in an elastic diffusion or in a production process.

It is now useful to specify more precisely the normalization of the states. The more convenient choice is related to the covariant normalization of the one-particle state

$$\langle p' \mid p\rangle = 2E\,(2\pi)^{d-1}\,\delta^{d-1}(\vec{p}\,' - \vec{p}). \tag{17.1.8}$$

This is a Lorentz invariant normalization and it is equivalent to integrate on the mass-shell state of a particle as

$$\int \frac{d^{d-1}p}{(2\pi)^{d-1}2E} \mid p\rangle\langle p \mid p'\rangle = \int \frac{d^d p}{(2\pi)^{d-1}}\, \delta(p^2 - m^2) \mid p\rangle\langle p \mid p'\rangle = \mid p'\rangle,$$

with $E > 0$. Hence, the density of states associated to a *on shell* particle with momentum in the interval $(p, p + dp)$ is given by

$$\frac{d^{d-1}p}{(2\pi)^{d-1}2E}.$$

Decay process. Taking into account the proper normalization of the states, the probability of a decay of a particle of energy E into a n-particle state is expressed by

$$d\Gamma = (2\pi)^d \delta^d(P - p_1 - \cdots - p_n) \mid T_{fi} \mid^2 \frac{1}{2E} \prod_{i=1}^{n} \frac{d^{d-1}p_i}{(2\pi)^{d-1}2E_i}, \tag{17.1.9}$$

Scattering process 2 → *n*. The probability that a collision of two particles of momenta $p_1 = (E_1, \vec{p}_1)$ and $p_2 = (E_2, \vec{p}_2)$ produces their transformation in an arbitary number of other particles with momenta p_i is given by

$$dP = (2\pi)^d \delta^d(P - p_1 - \cdots - p_n) \mid T_{fi} \mid^2 \frac{1}{4E_1E_2} \prod_{i=1}^{n} \frac{d^{d-1}p_i}{(2\pi)^{d-1}2E_i}. \tag{17.1.10}$$

In the last case, rather than the probability, it is often more interesting to compute the Lorentz invariant *cross section* $d\sigma$ of the cross. This is obtained dividing the probability dP by

$$j = \frac{I}{E_1E_2},$$

where I is the scalar quantity

$$I = \sqrt{(p_1 \cdot p_2)^2 - (m_1m_2)^2}.$$

It is easy to see that j is the flux density of the colliding particles. In fact, in the reference frame of the center of mass of the system ($\vec{p}_1 = -\vec{p}_2 = \vec{p}$), we have $I = \mid \vec{p} \mid (E_1 + E_2)$ and then

$$j = \mid \vec{p} \mid \left(\frac{1}{E_1} + \frac{1}{E_2} \right) = v_1 + v_2,$$

where v_1 and v_2 are the velocities of the two colliding particles. Hence the cross-section is the probability transition per unit of the flux of the scattering particles.

Note that in the probability of both decay or scattering processes there is the quantity

$$d\Phi_n = \frac{d^{d-1}p_1}{(2\pi)^{d-1}2E_1} \cdots \frac{d^{d-1}p_1}{(2\pi)^{d-1}2E_1} (2\pi)^d \, \delta^d(P - p_1 - p_2 - \cdots - p_n). \qquad (17.1.11)$$

This is the differential n-particle *phase space*. It expresses the density of states for a n-particle system with total momentum P. This quantity also enters the spectral density of the correlation functions, which are discussed in Chapter 19. Given its relevance in many aspects of the theory, its detailed study is carried on in Appendix 17.C.

Let us now investigate the consequences of the unitarity condition of the S-matrix. Substituting in (17.1.4) eqn. (17.1.6), we get

$$T_{fi} - T^{\star}_{if} = i\,(2\pi)^d \sum_n \delta^d(P_f - P_i)\, T_{fn}\, T^{\star}_{in}, \qquad (17.1.12)$$

where the sum on the index n here denotes, in a compact notation, both a sum and an integral on all intermediate states allowed by the conservation of the total momentum of the process. Note that the left-hand side of this equation is linear with respect to the matrix elements of T, whereas the right-hand side is quadratic. If the theory under investigation has a coupling constant g that can be regarded as a perturbative parameter, the first consequence of eqn. (17.1.12) is the Hermiticity of the matrix T at the first perturbative order

$$T_{fi} \simeq T^{\star}_{if}. \qquad (17.1.13)$$

In fact, the left-hand side of (17.1.12) is of the first order in g, whereas the right-hand side is of the second order.

Optical theorem. Another important consequence of eqn. (17.1.12) is the *optical theorem* relative to the scattering process of two particles. To prove it, let us initially sandwich eqn. (17.1.12) with the states $| \, p_1, p_2\rangle$ and $| \, p_3, p_4\rangle$

$$2\,\mathrm{Im}\,\langle p_3, p_4 \,|\, T \,|\, p_1, p_2\rangle = (2\pi)^d \sum_n \delta^d(P_f - P_i)\, \langle p_3, p_4 \,|\, T \,|\, n\rangle \langle p_1, p_2 \,|\, T^{\star} \,|\, n\rangle. \qquad (17.1.14)$$

If the scattering process is purely elastic, the final state coincides with the initial state and in this case we have

$$2\,\mathrm{Im}\, T_{ii} = (2\pi)^d \sum_n \delta^d(P_f - P_i) \,|\, T_{in} \,|^2.$$

Note that the right-hand side of this expression differs only for a multiplicative factor from the total cross section σ_t of all possible scattering processes obtained by a given initial state i

$$\sigma_t = \left(\frac{(2\pi)^d}{j}\right) \sum_n |\, T_{in} \,|^2 \, \delta^d(P_i - P_n).$$

Therefore we have the *optical theorem*, stated by the relation

$$\sigma_t = \frac{2}{j} \mathrm{Im}\, T_{ii}.$$

This theorem allows us to compute the total cross-section of the theory (that also include all the inelastic processes) in terms of the imaginary part of the purely elastic scattering amplitude of two particles.

Finally, let us discuss the final principles upon which the S-matrix theory is based; namely, the causality and the analyticity principles. We expect that these two aspects should be deeply related each other, on the basis of the well-known example of the dispersion relations satisfied by the Green functions of an ordinary quantum system (see Problem 17.1). However, in the context of relativistic quantum mechanics, it is generally difficult to pin down the precise analytic structure of the S-matrix in terms of the causality principle. Quite often, in fact, the analytic properties of the S-matrix elements are conjectured on the basis of those derived in the non-relativistic scattering processes or encountered in the perturbative diagrams of the associate QFT, when this is known. In short, the basic assumption on which we rely on is encoded in the following statement: *the transition amplitudes coincide with the real boundary values of analytic functions of several complex variables having the minimum number of singularities dictated by specific physical processes.* The study of the two-particle scattering process will help us in clarifying some important aspects of this topic.

17.1.2 Two-body Scattering Process

Let us consider in more detail the diffusive scattering process of two initial scalar particles (with momenta p_1 and p_2) going in two scalar particles (with momenta p_3 and p_4; see Figure 17.2),

$$A_1 + A_2 \to A_3 + A_4. \tag{17.1.15}$$

Once we factorize the delta-function of the conservation of the total momentum

$$\langle p_3, p_4 \mid T \mid p_1, p_2 \rangle = i(2\pi)^d \, \delta^d(p_1 + p_2 - p_3 - p_4)\, \mathcal{T}, \tag{17.1.16}$$

the remaining quantity $\mathcal{T}$ is an analytic function of the relativistic invariants of this process. They can be expressed in terms of the Mandelstam variables s, t and u, given by

$$s = (p_1 + p_2)^2, \quad t = (p_1 - p_3)^2, \quad u = (p_1 - p_4)^2. \tag{17.1.17}$$

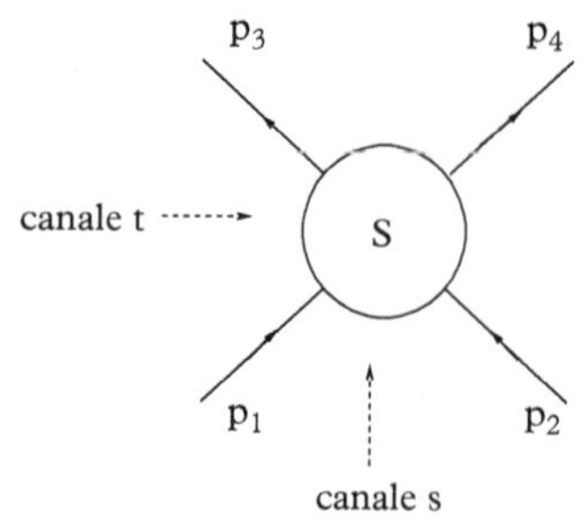

Fig. 17.2 *Two-particle scattering process.*

These quantities are not all independent, since from the conservation law

$$p_1 + p_2 = p_3 + p_4, \quad p_i^2 = m_i^2 \ (i = 1, 2, 3, 4)$$

we have

$$s + t + u = \sum_{i=1}^{4} m_i^2. \tag{17.1.18}$$

It is easy to understand the meaning of s, going in the reference frame of the centre of mass of the process (17.1.15), defined by $\vec{p}_1 + \vec{p}_2 = 0$. In this frame $s = E^2$, where $E = E_1 + E_2$ is the total energy in the frame of the centre of mass. The variable t is instead the square of the energy in the centre of mass of the crossed channel

$$A_1 + \overline{A}_3 \to \overline{A}_2 + A_4, \tag{17.1.19}$$

and the same is true for the variable u, with respect to the crossed channel

$$A_1 + \overline{A}_4 \to \overline{A}_2 + A_3. \tag{17.1.20}$$

In the equations above, $\overline{A}_i$ denotes the anti-particle: going in a cross channel, we must reverse the arrow of the outgoing particle that becomes then the anti-particle.

Production thresholds and branch cuts. In view of eqn. (17.1.18), the amplitude $\mathcal{T}$ is a function of only two of the Mandelstam variables, say s and t. Let us study its analytic structure as a function of s at fixed t, assuming for simplicity that each of the four particles involved in this scattering process has the same mass m. The physical values of s are given by $s \geq s_2 = (2m)^2$: this is the set of values for which exists the physical state of the two asymptotic particles. In the interval

$$(2m)^2 \leq s \leq (3m)^2,$$

corresponding to values of the total energy in the centre of mass less than the threshold of inelastic production, the two-particle states are the only intermediate states that can appear in the right-hand side of eqn. (17.1.14). Therefore

$$2\,\mathrm{Im}\,\langle p_3,p_4\mid T\mid p_1,p_2\rangle = (2\pi)^d \int \frac{d^{d-1}k_1}{(2\pi)^{d-1}2E_1}\frac{d^{d-1}k_2}{(2\pi)^{d-1}2E_2}\delta^d(p_1+p_2-k_1-k_2) \times \langle p_3,p_4\mid T\mid k_1,k_2\rangle\langle p_1,p_2\mid T^\star\mid k_1,k_2\rangle. \quad (17.1.21)$$

But once the threshold value is overcome, in the next interval

$$(3m)^2 < s < (4m)^2,$$

it is necessary to add other terms in the right-hand side of (17.1.21): these terms are those relative to the intermediate states made of three particles, compatible with the conservation law of the energy. In the same way, there are other additional terms due to the N-particle intermediate states each time that s overcomes their threshold of production, $s_N = (Nm)^2$.

The discontinuity in the imaginary part of the amplitude of the elastic scattering by varying s implies that it has certain singularities in correspondence with the threshold values of the inelastic processes. They are branch points of the amplitude $\mathcal{T}$, as it is easy to show using the Feynman diagrams of the perturbative series. In the complex plane of the variable S it is thus convenient to draw a series of cuts starting from the various thresholds to infinity, all along the real axis (Figure 17.3). In this way the scattering amplitude becomes a one-value function on the corresponding Riemann surface. The *physical sheet* is obtained without crossing any cuts of Figure 17.3, whereas the other sheets, called *non-physical sheets* are defined specifying the crossing of one or more cuts of the amplitude $\mathcal{T}(s,t)$.

Bound states and poles. The lowest threshold, at $s_2 = 4m^2$, is associated with the physical state of two particles. As analytic function of s, $\mathcal{T}(s,t)$ can be also evaluated

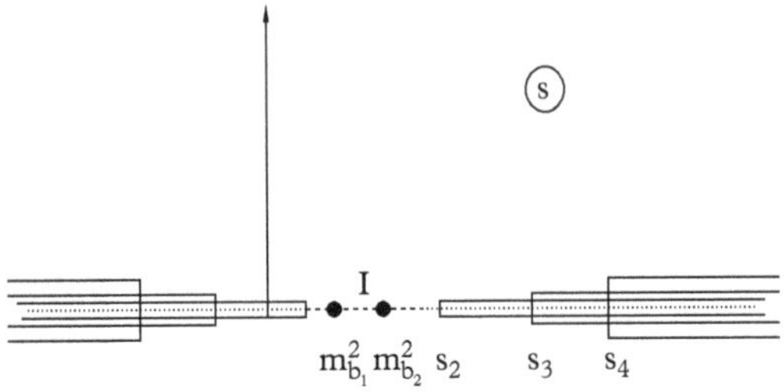

Fig. 17.3 *Analytic structure of the elastic scattering amplitude of two particles. On the right-hand side there are the branch cuts relative to the threshold values in the s-channel, while on the left-hand side there are the branch cuts relative to the threshold values of the t-channel. The circle represent the poles of the amplitude, relative to the bound states.*

for non-physical values of s, as those less than s_2. The possibility to create an arbitrary number of particles starting from the two-particle state can be also considered for $s < s_2$. However, in this case, these are only the one-particle states, with mass $m_{b_i} < 2m$. These are obviously virtual processes since they are precluded by the conservation of the energy that holds for the physical process. However, they determine the *bound states* of the asymptotic particles and, as for the non-relativistic scattering amplitudes (see Appendix 17.B), correspond to simple poles in the amplitude $\mathcal{T}(s,t)$. This analytic structure is confirmed by the perturbative theory based on the Feynman graphs. Consider, for instance, a theory in which there is a ϕ^3 interaction: in the scattering process of two particles there are the graphs shown in Figure 17.4. The first diagram, apart some constants, is given by

$$(a) \longrightarrow \frac{1}{(p_1+p_2)^2 - m^2 + i\epsilon}, \tag{17.1.22}$$

and gives rise to a pole in the s-channel, while the second diagram

$$(b) \longrightarrow \frac{1}{(p_1-p_2)^2 - m^2 + i\epsilon}, \tag{17.1.23}$$

gives rise to a pole in the t-channel.

Physical regions and crossing invariance. The region in which $\mathcal{T}(s,t)$ coincides with the amplitude relative to the physical scattering process (17.1.15) is the one in which there are positive values of the energies of all particles and real values of their momenta. For particles of equal mass, this region is identified by the conditions[5]

$$s \geq 4m^2, \quad t \leq 0, \quad u \leq 0, \tag{17.1.24}$$

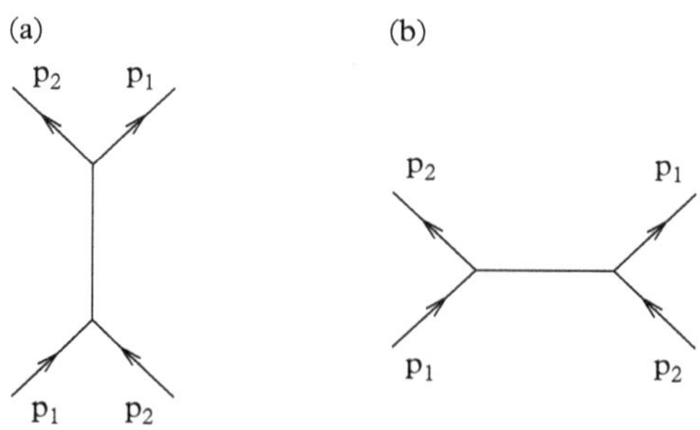

Fig. 17.4 *Feynman diagram relative to (a) the s-channel amplitude and (b) the t-channel amplitude, for the scattering process of two particles in a ϕ^3 theory.*

[5] If the masses are different, the conditions are slightly more complicated.

as it can be seen expressing s, t and u in terms of the momentum $\vec{q}$ and the scattering angle θ in the frame of the centre of mass

$$\begin{aligned} s &= 4(m^2 + q^2), \\ t &= -2q^2(1 - \cos\theta), \\ t &= -2q^2(1 + \cos\theta). \end{aligned}$$

Since $\mathcal{T}(s,t)$ is an analytic function of both variables, it can be analytically continued from the original domain (17.1.24) to the regions

$$t \geq 4m^2, \quad s \leq 0, \quad u \leq 0, \tag{17.1.25}$$

and

$$u \geq 4m^2, \quad s \leq 0, \quad t \leq 0. \tag{17.1.26}$$

The first region corresponds to the physical domain relative to the channel (17.1.19) while the second region to the physical domain of the channel (17.1.20). This implies that the *same analytic function* can be used to describe the three different physical processes given in (17.1.15) (the *s-channel*), in (17.1.19) (the *t-channel*) and in (17.1.20) (the *u-channel*). This fundamental property of the amplitude $\mathcal{T}(s,t)$ expresses the *crossing invariance* of the scattering processes.

t-channel. As we have identified the threshold singularities of $\mathcal{T}(s,t)$ by varying s at fixed t, we can similarly identify the singularities of this amplitude by varying t and u, using the crossing invariance. In the t-channel the threshold singularities are placed at

$$t = 4m^2, \quad 9m^2, \quad 16m^2, \cdots \tag{17.1.27}$$

and analogously in the u-channel

$$u = 4m^2, \quad 9m^2, \quad 16m^2, \cdots \tag{17.1.28}$$

For the relation (17.1.18), fixed the value u_0 of the variable u, the branch points (17.1.27) then appear in the complex plane of the variable s at the points

$$s = -u_0, \quad -u_0 - 5m^2, \quad -u_0 - 12m^2, \cdots \tag{17.1.29}$$

whereas the pole at $t = m_{b_i}^2$ appears in the position

$$s = -u_0 + 3m_{b_i}^2. \tag{17.1.30}$$

The analytic structure (at $u = u_0$, fixed) is shown in Figure (17.3).

Physical amplitude. Since we are in the presence of branch cuts along the real axis of the s-plane, it is necessary to establish which is the limit of the function $\mathcal{T}(s,t)$ associated to the physical amplitude in the s-channel. The physical region of this process is identified by $u_0 < 0$ and by the real values of s, with $s > 4m^2 - u_0$. The perturbative theory shows that the physical amplitude is recovered by taking the limit from the upper half complex plane on the first cut of the function $\mathcal{T}(s,t)$, namely

$$\mathcal{T}_{phys} = \lim_{\epsilon \to 0^+} \mathcal{T}(s + i\epsilon, u_0). \tag{17.1.31}$$

Note that this result is equivalent to the Feynman prescription $i\epsilon$ in the propagators of the particles

$$\frac{1}{p^2 - m^2 + i\epsilon}.$$

In fact, adopting this prescription, any integral on the momenta of the intermediate particles can be computed with real external momenta, i.e. in correspondence to a real value of the variable s. Moreover, eqn. (17.1.31), together with the Hermitian analyticity, implies that the amplitude $\mathcal{T}$ is a real function in the real interval I between the two branch cuts (Figure 17.3), as it can also be proved directly from the Schwartz reflection principle in complex analysis.

17.2 General Properties of Purely Elastic Scattering Matrices

Let us now specialize the general conditions discussed in the previous section to the case of $(1+1)$ scattering theories when there is an infinite number of conserved charges $\mathcal{Q}_s$ in involution. These two circumstances give rise to a drastic simplification of the analytic structure of the S-matrix and will lead to an exact expression of the scattering amplitudes in many interesting cases.

17.2.1 Rapidity variable and asymptotic states

The momenta of the particles involved in scattering processes are *on-shell*. In $(1+1)$ dimensions there exists an efficient parameterization of the dispersion relation $E^2 - p^2 = m^2$ in terms of the rapidity variable θ. For a particle of mass m_i we have in fact

$$p_i^{(0)} = m_i \cosh\theta_i, \quad p_i^{(1)} = m_i \sinh\theta_i. \tag{17.2.32}$$

Note that the Lorentz transformations can be regarded as rotation of a hyperbolic angle α and therefore implemented as $\theta \to \theta + \alpha$. Furthermore, both components of the momentum can be changed by sign with the transformation $\theta_i \to i\pi - \theta_i$. In this way, the momentum of the original particle becomes the one of its own anti-particle. Later it will be also useful to consider the light-cone components, defined by

$$\begin{aligned} p &= p^{(0)} + p^{(1)} = m_i e^{\theta}, \\ \bar{p} &= p^{(0)} - p^{(1)} = m_i e^{-\theta}. \end{aligned} \tag{17.2.33}$$

They satisfy $p\bar{p} = m_i^2$.

The rapidity variable has an interesting geometric intepretation, due to the Italian mathematician Riccati. In a plane with axes given by E and p, the dispersion relation $E^2 = p^2 + m^2$ represents an hyperbole, as shown in Figure 17.5.a. The rapidity is proportional to the area $\mathcal{A}$ that is encompassed between the hyperbole and the straight line that joins the origin to the point of the hyperbole identified by the variable θ. The relation is $\mathcal{A} = \frac{m^2\theta}{2}$. An analogous result is obtained for the angle α that parameterizes the points of a circle $x^2 + y^2 = m^2$, shown in Figure 17.5(b). Posing $x = m\cos\alpha$ and $y = m\sin\alpha$, the area $\mathcal{A}$ between the horizontal axis and the segment the identifies the point on the circles is in fact $\mathcal{A} = \frac{m^2\alpha}{2}$. The two geometrical situations are related by the analytic continuation $\alpha \to i\theta$.

The n-particle states of this theory can be written as

$$| A_{a_1}(\theta_1) A_{a_2}(\theta_2) \ldots A_{a_n}(\theta_n) \rangle, \tag{17.2.34}$$

where by the symbol $A_{a_i}(\theta_i)$ we denote the particle of type a_i that is moving with rapidity θ_i. By means of a linear superposition of these states, we can construct wave packets that are localized both in momentum and coordinate spaces. In this way, we can imagine to assign a well-defined position to the particles above. In the massive theories, the interactions are short range and consequently a state like (17.2.34) represents a collection of free particles except in the time instants in which the wave packets overlap. Here was discuss in more detail how to represent the initial and final states.

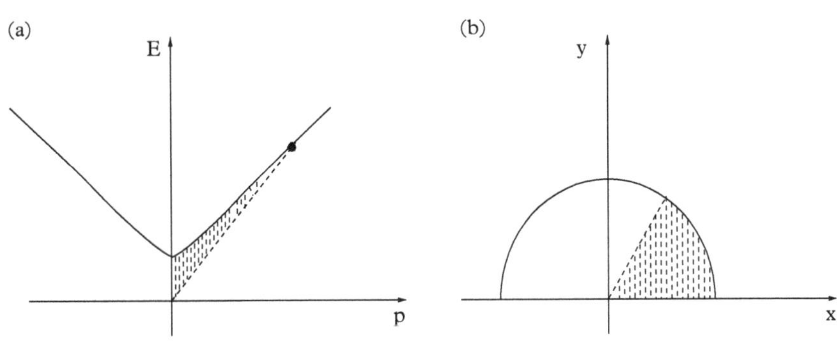

Fig. 17.5 *Geometrical interpretation of the rapidity variable θ.*

An initial asymptotic state is given by a set of free particles at $t \to -\infty$. Since in the $(1+1)$ dimensional theories the actual motion takes place along a line, this means that the fastest particle must be on the farthest left-hand side of all the others, while the slowest must be on the right-hand side of all the others, with the remaining particles ordered according to the value of their rapidities between those two. To express such a situation in a more formal way, it is appropriate to consider the symbols $A_{a_i}(\theta_i)$ as non-commuting variables, whose order is associated to the space ordering of the particles that they represent. In this way, an initial asymptotic state can be written as

$$|\,A_{a_1}(\theta_1)A_{a_2}(\theta_2)\ldots A_{a_n}(\theta_n)\rangle, \tag{17.2.35}$$

where the rapidities are ordered in a *decreasing* way

$$\theta_1 \geq \theta_2 \geq \theta_3 \cdots \geq \theta_n. \tag{17.2.36}$$

Similarly, a final asymptotic state is made of free particles at $t \to +\infty$. Hence each particle must be on the left-hand side of all the others that move faster than it. The final asymptotic states can be then represented by

$$|\,A_{a_1}(\theta_1)A_{a_2}(\theta_2)\ldots A_{a_n}(\theta_n)\rangle, \tag{17.2.37}$$

but this time with an *increasing* order of the rapidities, i.e.

$$\theta_1 \leq \theta_2 \leq \theta_3 \cdots \leq \theta_n. \tag{17.2.38}$$

Obviously each product (17.2.34) can always be ordered in the way we like by means of certain number of commutation of the symbols $A_i(\theta_i)$ between neighbouring particles. As we will see below, each commutation can be interpreted as a scattering process of two particles. It is custom to normalize the states as

$$\langle A_i(\theta_1)\mid A_j(\theta_2)\rangle = 2\pi\,\delta_{ij}\,\delta(\theta_1-\theta_2). \tag{17.2.39}$$

Consequently, the density of states with rapidities $(\theta, \theta + d\theta)$ is given by $\frac{d\theta}{2\pi}$.

17.2.2 Conserved Charges

The existence of an infinite number of conserved charges $\mathcal{Q}_{\pm s}$ in involution has a series of drastic consequences on the scattering processes. As discussed in Chapter 16, the charges can be identified by their *spin* index s and the local ones[6] can be expressed by the integral of their current densities

[6] In some integrable theories, such as the Sine–Gordon model or the non-linear $O(3)$ sigma model, there are also non-local conserved charges, often associated to operators with fractional spin.

$$\mathcal{Q}_s = \int [T_{s+1}(z,\bar{z})\, dz + \Theta_{s-1}(z,\bar{z})\, d\bar{z}], \quad s \geq 1,$$

where $T_{s+1}(z,\bar{z})$ and $\Theta(z,\bar{z})$ are local fields that satisfy the conservation law

$$\partial_{\bar{z}} T_{s+1} = \partial_z \Theta_{s-1}.$$

Analogously, for the charges with negative spins, hereafter denoted also by $\bar{\mathcal{Q}}_s$, we have

$$\bar{\mathcal{Q}}_s = \int \left[\bar{T}_{s+1}(z,\bar{z})\, dz + \bar{\Theta}_{s-1}(z,\bar{z})\, d\bar{z}\right],$$

with

$$\partial_z \bar{T}_{s+1} = \partial_{\bar{z}} \bar{\Theta}_{s-1}.$$

Note that $\mathcal{Q}_{\pm 1}$ coincide with the light-cone components of the momentum

$$\begin{aligned} \mathcal{Q}_1 &= P = P^{(0)} + P^{(1)}, \\ \mathcal{Q}_{-1} &= \bar{P} = P^{(0)} - P^{(1)}. \end{aligned} \qquad (17.2.40)$$

Since, by hypothesis, these charges commute among themselves

$$[\mathcal{Q}_s, \mathcal{Q}_{s'}] = 0,$$

they can be diagonalized simultaneously. The spectrum of the values s of the conserved charges varies by varying the theory and, as we shall see, it is deeply related to the structure of the bound states. Their action of the one-particle states lead to

$$\mathcal{Q}_s \mid A_a(\theta)\rangle = \omega_s^{(a)}(\theta) \mid A_a(\theta)\rangle, \qquad (17.2.41)$$

where the functional dependence of the functions $\omega_s^{(a)}(\theta)$ is determined by the tensor properties of $\mathcal{Q}_s$: under the Lorentz group, $\mathcal{Q}_s$ transforms as s copies of P while $\mathcal{Q}_{-s}$ as s copies of $\bar{P}$, therefore it is natural to regard $\mathcal{Q}_{\pm s}$ as tensors of rank s and pose

$$\omega_s^{(a)}(\theta) = \chi_s^{(a)}\, e^{s\theta}, \qquad (17.2.42)$$

where $\chi_s^{(a)}$ is called the *eigenvalue* of the charge $\mathcal{Q}_s$ on the particle a. The spectrum of these eigenvalues is an interesting problem itself and will be discussed through some examples in Section 17.5.1.

Further restrictions may come from the discrete symmetries of the model. For instance, if the theory in exam is invariant under the charge conjugation C, the conserved charges can be classified as even or odd operators $\mathcal{Q}_s^{(\pm)}$ with respect to C. Furthermore, assuming that the parity P is also a symmetry of the system, we can show the validity of the commutation relations

$$
\begin{aligned}
C\,\mathcal{Q}_s^{(+)}\,C &= \mathcal{Q}_s^{(+)} = (-1)^{s+1}\,\mathcal{Q}_s^{(+)} \\
C\,\mathcal{Q}_s^{(-)}\,C &= -\mathcal{Q}_s^{(-)} = (-1)^{s+1}\,\mathcal{Q}_s^{(-)}.
\end{aligned}
\tag{17.2.43}
$$

They imply that the values of s for the C-even charges are only odd numbers, while those of the C-odd charges are even integers.

Let us now analyse how the infinite conserved charges constrain the scattering processes. Coleman and Mandula showed that in $(3+1)$ dimensional theories the existence of only one conserved charge of tensor rank larger than 2 implies a trivial S-matrix, i.e. $S = 1$. This result does not apply to the $(1+1)$-dimensional theories but, in this case, there is a series of severe constraints listed hereafter.

1. The number of particles with mass m_a remains the same before and after the collision has taken place.
2. The set of the final momenta of the particles is the same of the initial momenta, namely the scattering processes are purely elastic.
3. The scattering amplitude for the process in which are involved n particles can be completely factorized in terms of the $n(n-1)/2$ elastic scattering two-particle amplitudes.

Let us now prove these properties.

17.2.3 Elasticity in the Scattering Processes

In order to prove the elasticity of the scattering processes note that the conserved charges act on the multi-particle states as

$$
\mathcal{Q}_s \mid A_{a_1}(\theta_1)\ldots A_{a_n}(\theta_n)\rangle = \sum_{i=1}^{n} \chi_s^{(a_i)}\, e^{s\theta_i} \mid A_{a_1}(\theta_1)\ldots A_{a_n}(\theta_n)\rangle.
$$

Since

$$
\frac{d\mathcal{Q}_s}{dt} = 0,
$$

there is an infinite sequence of constraints that involve the sum of the higher powers of the momenta of the initial and final particles

$$
\sum_{i\in in} \chi_s^{(a_i)}\, e^{s\theta_i} = \sum_{j\in fin} \chi_s^{(a_j)}\, e^{s\theta_j}. \tag{17.2.44}
$$

The only solution to these infinite number of equations (apart from the permutations of particles with the same quantum numbers) corresponds to the case in which the final and the initial sets of rapidity are equal. Hence, in theories having an infinite number of conserved charges, the annihilation and production processes are absent: all scattering processes are therefore elastic.

17.2.4 Factorization of the Scattering Processes

In addition to elastic, the scattering processes in these theories are also factorized. For a heuristic explanation of this feature, it is necessary to understand the action of the conserved charges $\mathcal{Q}_s$ on a localized wave packet. If $\mathcal{Q}_s$ is the space component of the two charges $\mathcal{Q}_{\pm s}$, assuming for simplicity that $\chi_s^{(a)} = 1$ we have

$$e^{ic\mathcal{Q}_s} \mid A_a(p)\rangle = e^{icp^s} \mid A_a(p)\rangle .$$

Now, let

$$\psi(x) = \int_{-\infty}^{+\infty} dp\, e^{-a(p-p_0)^2}\, e^{ip(x-x_0)},$$

be the wave function of a state that is well localized both in the momentum space (around $p = p_0$) and in the coordinate space (around $x = x_0$). Acting by $e^{ic\mathcal{Q}_s}$ on this state we have

$$\tilde{\psi}(x) = \int_{-\infty}^{+\infty} dp\, e^{-a(p-p_0)^2}\, e^{ip(x-x_0)}\, e^{icp^s}.$$

This new function is now localized at $x = x_0 - scp_0^{s-1}$, as it can be seen by a saddle point computation. Hence, for $s > 1$, the centre of the wave-packet is translated by a quantity that depends on the $(s-1)$ power of its momentum (for $s = 1$, $\mathcal{Q}_s$ coincides with the ordinary momentum operator that shifts equally all wave packets by the same amount). The above result shows that wave-packets with different momenta can be shifted differently acting on them with the conserved charges $e^{ic\mathcal{Q}_s}$ of higher spin.[7]

Consider now the collision of three particles of momenta $p_1 < p_2 < p_3$, associated to wave-packets well-localized both in momentum and coordinate spaces. Depending on the initial positions of the three packets, we can have three types of collisions, as given in Figures 17.6 and 17.7 respectively. The first type consists of the simultaneous collision of the three particles. The other two types are drawn in Figure 17.7, in which the scattering process is made of three distinct *two-particle* collisions, well separated in space and time. Obviously the chronological sequence of these collisions is different in the two graphs of Figure 17.7.

[7] This result clarifies the Coleman–Mandula theorem. In fact, in $(d+1)$ dimensional theories, with $d > 1$, the possibility to translate differently particles of different momenta means that their trajectories can never cross, i.e. the particles move freely without experiencing any collision and therefore $S = 1$.

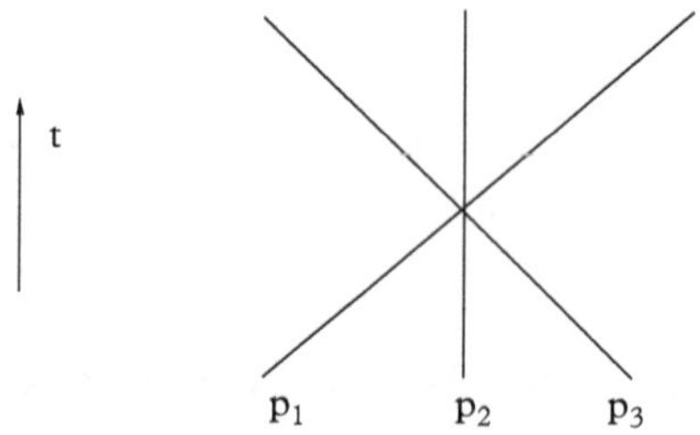

Fig. 17.6 *A simultaneous collision of three particles.*

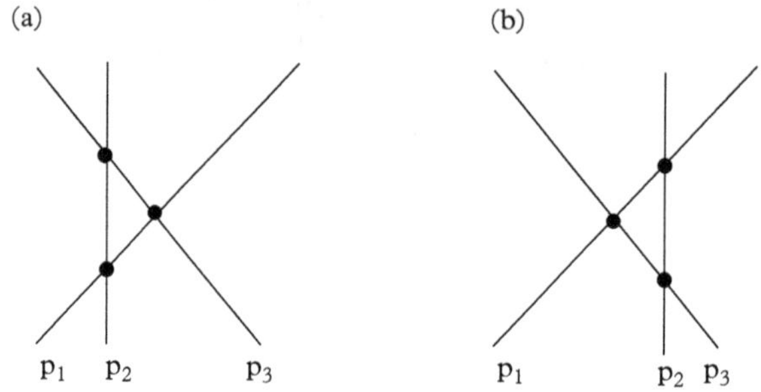

Fig. 17.7 *A three-particle collision realized by a sequence of two-particle collisions. These two cases and the one drawn in the Figure 17.6 are related by a symmetry transformation and therefore have the same amplitude.*

In a generic scattering theory, the processes relative to Figures 17.6 and 17.7 have different amplitudes. However, for integrable theories, the three different situations can be obtained one from the other by an opportune action of the operators e^{icQ_s}. Since these operators commute with the Hamiltonian of the system (associated to $Q_{\pm 1}$), their action must lead to equivalent physical situations. Therefore, in integrable theories, there must be an equality among the three scattering amplitudes of the two processes in Figure 17.7! We have thus achieved two extremely important results:

- Since in an integrable theory the S-matrix of a three-particle process can be factorized in two different but equivalent ways (corresponding to the different sequences of two-particle collisions shown in Figure 17.7), the two-particle scattering amplitudes $S^2(p_a, p_b)$ must satisfy the so-called Yang–Baxter equation[8]

$$S^2(p_2,p_3)\,S^2(p_3,p_1)\,S^2(p_1,p_2) = S^2(p_1,p_2)\,S^2(p_1,p_3)\,S^2(p_2,p_3). \qquad (17.2.45)$$

- The previous result can be easily generalized to the n-particle processes. In fact, it is easy to show that the fulfilment of the Yang–Baxter equations (17.2.45) are the

[8] The detailed matrix structure of this equation will be specified later.

sufficient and necessary conditions for the factorization of this amplitude in terms of the $n(n-1)/2$ two-particle elastic amplitudes. As before, in these collisions a possible exchange of the momenta can occur only between particles with the same mass and the same quantum numbers.

For the properties of elasticity and factorization, the S-matrix theory of two-dimensional system drastically simplifies and the scattering amplitudes can be explicitly found for many important physical models. It is in fact sufficient to find the two-particle scattering amplitudes to have the full control of any other scattering processes. In turn, the two-particle scattering amplitudes can be found as solutions of the Yang–Baxter equation, together with the general requirements of unitarity and crossing symmetry.

17.3 Unitarity and Crossing Invariance Equations

In this section we discuss the unitary and crossing symmetry equations that hold for the two-particle elastic scattering amplitudes of a $(1+1)$-dimensional integrable theory.

Let p_1 and p_2 be the initial and final momenta of the incoming particles A_i and A_j and the outgoing ones A_l and A_k, as shown in Figure 17.8). In addition to the delta function $\delta^{(2)}(p_1+p_2-p_3-p_4)$ of the conservation of the total energy and momentum, the Lorentz invariance imposes that the scattering amplitude depends on the particle momenta only by their invariant combinations, given by the Mandelstam variables s, t and u defined in eqn. (17.1.17). Note that for the $(1+1)$-dimensional systems and for the elasticity of the scattering process u vanishes identically, $u=0$, while s and t can be both expressed in terms of the difference of the rapidites of the particles.[9] In fact, using the parameterization (17.2.32), the Mandelstam variable s of the process

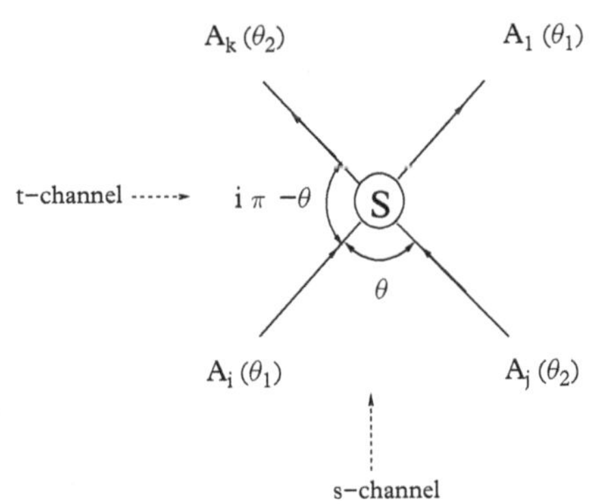

Fig. 17.8 *Elastic scattering process of two particles.*

[9] For the elastic processes in the $(1+1)$-dimensional system there is only one independent Mandelstam variable for the equalities $p_3=p_2$ and $p_1=p_4$ of the momenta.

$$A_i A_j \to A_k A_l,$$

is given by

$$\begin{aligned} s(\theta_{ij}) = (p_1 + p_2)^2 &= m_i^2 + m_j^2 + 2 m_i m_j \cosh\theta_{ij}, \\ \theta_{ij} &= \theta_i - \theta_j. \end{aligned} \tag{17.3.46}$$

For the physical processes θ_{ij} assumes real values and consequently also s is real and takes values $s \geq (m_i + m_j)^2$. The Mandelstam variable t is instead given by

$$t(\theta_{ij}) = (p_1 - p_2)^2 = m_i^2 + m_j^2 - 2 m_i m_j \cosh\theta_{ij}. \tag{17.3.47}$$

Consequently, we can switch between the s to the t-channels by the analytic continuation

$$t(\theta) = s(i\pi - \theta), \tag{17.3.48}$$

that admits the natural geometrical interpretation shown in Figure 17.8, if we regard θ as the (imaginary) angle between the lines of the incoming particles.

In $(1+1)$-dimensional systems, the two-particle S-matrix elements are defined by[10]

$$| A_i(\theta_1) A_j(\theta_2) \rangle = S_{ij}^{kl}(\theta) \mid A_k(\theta_2) A_l(\theta_1) \rangle, \tag{17.3.49}$$

with $\theta = \theta_{12}$ and $\theta_1 > \theta_2$, consistently with the definition of the initial and final asymptotic states previously discussed. In this equation it is implicit a sum on the indices k and l that occurs if the particles with $k \neq i$ and $l \neq j$ are not distinguished by any eigenvalues of the conserved charges. Note that the dependence of the S-matrix on the difference of the rapidities is a consequence of the relativistic invariance of the theory, since a Lorentz transformation changes the value of the rapidity of each particle by a constant. There is a relation between the S-matrix given above and the one written in terms of the original Mandelstam variable s, here denoted by $\mathcal{S}$: this relation is given by the jacobian of the transformation $s(\theta)$

$$\mathcal{S}_{ij}^{kl}(s) = 4 m_i m_j \sinh\theta \, S_{ij}^{kl}(\theta). \tag{17.3.50}$$

Constraints from discrete symmetries. In an elastic scattering theory with r types of particles, the set of the r^4 functions $S_{ij}^{kl}(\theta)$ completely determined the full S-matrix of the problem. However, these functions are not all independent. First of all, the matrix elements $S_{ij}^{kl}(\theta)$ are non-zero only when the particles A_i and A_k (as well as A_j and A_l) have the same quantum numbers with respect to the conserved charges. This implies,

[10] In these theories it is usual to define S as the unitary operator that maps the *initial states* onto the *final states*, i.e. $| \text{in} \rangle = S \mid \text{fin} \rangle$. This is the definition that we will hereafter. Strictly speaking, this definition corresponds to the operator S^{-1} previously introduced.

in particular, the equality of their masses $m_i = m_k$ and $m_j = m_l$. Moreover, assuming the invariance of the theory under the charge conjugation C, the parity P and the time reversal T, there are the further relations

$$\begin{aligned} S_{ij}^{kl}(\theta) &= S_{ji}^{lk}(\theta), \qquad P \\ S_{ij}^{kl}(\theta) &= S_{\bar{i}\bar{j}}^{\bar{k}\bar{l}}(\theta), \qquad C \\ S_{ij}^{kl}(\theta) &= S_{lk}^{ji}(\theta), \qquad T \end{aligned} \tag{17.3.51}$$

where $\bar{a} = Ca$ denotes the anti-particle state.

Yang–Baxter equations. The Yang–Baxter equations impose additional constraints on these amplitudes and their explicit form is (there is a sum on all the repeated indices)

$$S_{ij}^{ab}(\theta_{12})\, S_{bk}^{cl}(\theta_{13})\, S_{ac}^{nm}(\theta_{23}) = S_{jk}^{ab}(\theta_{23})\, S_{ia}^{nc}(\theta_{13})\, S_{cb}^{ml}(\theta_{12}). \tag{17.3.52}$$

These correspond to r^6 equations, in correspondence with the values of the six external indices i, j, k, l, m, n. This is an over-determined set of equations because their number is larger that the r^4 amplitudes to be determined. Hence, solutions of these equations can only be found for special functional forms of the functions $S_{ij}^{kl}(\theta)$. Note that, for their homogeneity, the Yang–Baxter equations (17.3.52) can only fix the ratios of the scattering amplitudes. Some explicit examples of solutions will be considered in later sections.

Let us now focus our attention on the analytic properties of the scattering amplitudes. They can be derived specializing the general considerations presented in the first section of this chapter. Let us initially consider the analytic properties with respect to the Mandelstam variable s, to translate them later in terms of the rapidity θ. We have the following properties

- $S(s)$ is a one-value analytic function in the complex plane of s with two elastic branch cuts, the first for $s \leq (m_i - m_j)^2$ and the second for $s \geq (m_i + m_j)^2$. The physical domains of this function are for values just above the branch cut on the right, i.e. $s^+ = s + i0$ and $s > (m_i + m_j)^2$. The first sheet of the Riemann surface of this function is called the *physical sheet*.
- S is a real analytic function, namely it assumes complex conjugate values at complex conjugate points

$$S_{ij}^{kl}(s^*) = \left[S_{ij}^{kl}(s)\right]^*.$$

In particular this implies that $S(s)$ assumes real values when s is itself real, with $(m_i - m_j)^2 \leq s \leq (m_i + m_j)^2$.

The unitarity equation is expressed by $S(s^+)S^\dagger(s^+) = 1$. This is a matrix relation, with a sum on all intermediate states between S and $S^\dagger$. When s^+ increases, it is energetically possible that states with a higher number of particles enter this sum, giving rise to production processes and consequently to additional branch cuts of $S(s)$. However, this circumstance does not occur in integrable theories and, in this case, the unitarity conditions involve only the two-particle states

$$S_{ij}^{kl}(s^+)\left[S_{kl}^{nm}(s^+)\right]^* = \delta_i^n\,\delta_j^m.$$

Using the real analiticity of these function, this equation can be written as

$$S_{ij}^{kl}(s^+)\,S_{kl}^{nm}(s^-) = \delta_i^n\,\delta_j^m,$$

with $s^- = s - i0$. This equation shows the necessity a branch cut at $s = (m_i + m_j)^2$ and, furthermore, that this branch cut is of the square root type. To prove it, let $S_\gamma(s)$ be the function obtained by the analytic continuation of $S(s)$ after an anti-clock wise path around that point. The unitarity condition imposes the validity of $S(s^+)S_\gamma(s^+) = 1$ for all physical values of s^+. This relation can be analytically continued for all values of s, with the result

$$S_\gamma(s) = S^{-1}(s).$$

In particular, if s^- is a point below the cut, we have

$$S_\gamma(s^-) = S^{-1}(s^-) = S(s^+),$$

where the second equality follows from applying twice the unitarity equation. Since $S_\gamma(s^-)$ is just the analytic continuation of $S(s^+)$ obtained with a double twist around the point $s = (m_i + m_j)^2$, it follows that at this point there is a square root singularity.

Concerning the cut that goes from $s = (m_i - m_j)^2$ to $s = -\infty$, we use the fundamental invariance of the relativistic scattering theories under the crossing transformations. In fact, if one of the incoming particles, say the one with index j, inverts its motion so that it becomes an outgoing particle and the same operation is done with the outgoing particle of index l to transform it in an incoming particle, the original amplitude becomes the amplitude of another scattering channel. For this new amplitude we have then i and $\bar{l}$ as incoming particles, and k and $\bar{j}$ as outgoing particles, where the symbols $\bar{a}$ denotes the anti-particles. This ends up to look at Figure 17.8 from left to right, instead from bottom to top, so that now the direct channel is described by the Mandelstam variable t instead than the original variable s. Since in this new process $p_2 = p_3$, the relation between s and t is simply

$$t = (p_1 - p_2)^2 = 2p_1^2 + 2p_2^2 - (p_1 + p_2)^2 = 2m_i^2 + 2m_j^2 - s.$$

The crossing invariance permits to recover the amplitude relative to this scattering process by means of the analytic continuation of the original amplitude in the region of the s plane where the variable t assumes physical values, i.e. $t \in \Re$ e $t \geq (m_i + m_j)^2$. The physical amplitudes are then related by

$$S_{ij}^{kl}(s^+) = S_{i\bar{l}}^{k\bar{j}}(2m_i^2 + 2m_j^2 - s^+). \tag{17.3.53}$$

Here it is also easy to prove that the point $s = (m_i - m_j)^2$ has a square root branch singularity. This does not imply though that the Riemann surface associated to the function $S(s)$ is only made of two sheets. In fact, by an analytic continuation of this function along a path that crosses the branch cut on the left we may reach a different sheet than the one obtained by an analytic continuation through the cut on the right. Hence, moving up or down these sheets and crossing the left and right cuts, we can span the Riemann surface of the S-matrix, made in general of several sheets, possibly infinite.

We now translate the considerations above in terms of the rapidity variable. Note that the inverse transformation of (17.3.46) given by

$$\theta_{ij} = \log \left[\frac{s - m_i^2 - m_j^2 + \sqrt{[(s - (m_i + m_j)^2)(s - (m_i - m_j)^2]}}{2m_i m_j} \right],$$

maps the physical sheet of the s-plane in the strip $0 \leq \text{Im}\theta_{ij} \leq \pi$. The second sheet is instead mapped in the strip $-\pi \leq \text{Im}\theta_{ij} \leq 0$. This structure repeats with period $2\pi i$, as shown in Figure 17.9. Moreover, as shown in eqn. (17.3.47), the Mandelstam variable t is obtained by substituting in eqn. (17.3.46), $\theta_{ij} \to i\pi - \theta_{ij}$. Hence, the map (17.3.46) realizes an uniformization of the original analytic structure, since in the plane of the variable θ there are no longer branch cuts. This implies that the S-matrix, considered as a function of θ, is an analytic function at the image points of the original cuts, i.e. at 0 and $i\pi$, as well as at all other points $in\pi$ of the other sheets. Since the integrability of the theory guarantees that these are the only branch points of the original amplitude, we

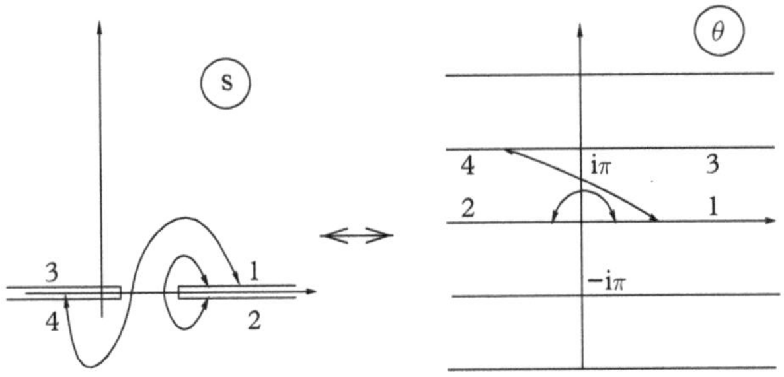

Fig. 17.9 *Map between the s-plane and the θ-plane, together with the unitarity and crossing symmetry conditions.*

arrive at the important result that $S(\theta)$ is a meromorphic function of θ. Since $S(s)$ is a real analytic function, $S(\theta)$ assumes real values on the imaginary axis of θ.

Expressed in terms of θ, the unitarity condition becomes

$$\sum_{n,m} S_{ij}^{nm}(\theta)\, S_{nm}^{kl}(-\theta) = \delta_i^k\, \delta_j^l, \tag{17.3.54}$$

while the crossing invariance condition

$$S_{ij}^{kl}(\theta) = S_{i\bar{l}}^{k\bar{j}}(i\pi - \theta). \tag{17.3.55}$$

There are some important aspects of the formulation of the S-matrix theory in terms of the rapidity variable worth highlighting. The first aspect is that the unitarity and crossing symmetry equations can be analytically continued for arbitrary values of θ and therefore they hold in all the complex plane of this variable. The second aspect concerns the definition itself of the S-matrix that, as a function of θ, can be written in an operator form as

$$A_i(\theta_1)A_j(\theta_2) = S_{ij}^{kl}(\theta)A_k(\theta_2)A_l(\theta_1). \tag{17.3.56}$$

This equation defines an algebra for the symbols $A_a(\theta)$, the so-called *Faddev–Zamolodchikov algebra.* Therefore the scattering processes can be equivalently interpreted as commutation relations among the operators that create the particles. In this respect, the unitarity equation (17.3.54) can be seen as a simple consequence of this algebra. Analogously, the Yang–Baxter equations simply derive by the associativity condition of the Faddev–Zamolodchikov algebra, as shown in Problem 17.4.

17.4 Analytic Structure and Bootstrap Equations

The elastic S-matrices are meromorphic analytic functions in the complex plane of θ. The bound states, originally associated to the simple poles of these amplitudes in the interval of s between $(m_i - m_j)^2$ and $(m_i + m_j)^2$, correspond now to simple poles with positive residue[11] along the imaginary segment $(0, i\pi)$ of the θ variable. Consider a S-matrix with incoming particles A_i and A_j that has a simple pole in the s-channel at $\theta = i u_{ij}^n$. In correspondence of this pole, the amplitude can be expressed as

$$S_{ij}^{kl} \simeq i \frac{R^{(n)}}{\theta - i u_{ij}^n}, \tag{17.4.57}$$

[11] Chapter 18 demonstrates how this concept can be generalized both to the cases of poles with negative residues and higher-order poles.

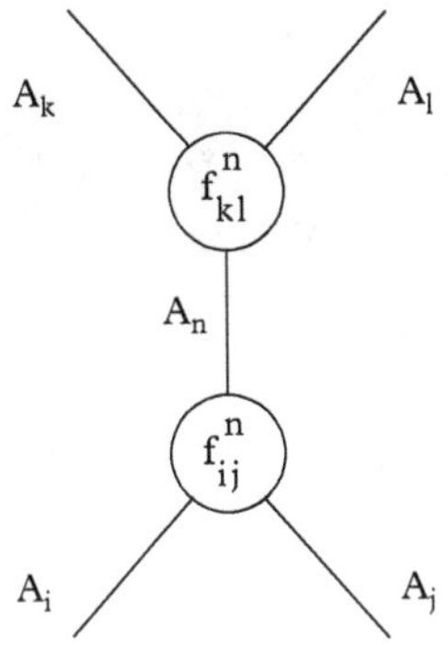

Fig. 17.10 *Residue of the pole and its expression in terms of the* on-shell *coupling constants.*

with the residue $R^{(n)}$ related to the *on-shell* vertex functions of the incoming particles and the bound state A_n, as shown in Figure 17.10

$$R^{(n)} = f_{ij}^n f_{kl}^n. \tag{17.4.58}$$

A non-zero value of f_{ij}^n obviously implies a pole singularity in the other two amplitudes S_{in} and S_{jn} as well, where the poles are now due to the bound states A_j and A_i. Since in the bootstrap approach the bound states are on the same footing than the asymptotic states, there is an important relation among the masses of the system: if $\theta = iu_{ij}^n$ is the position of the pole in the scattering of the particles A_i and A_j, the mass of the bound state is given by

$$m_n^2 = m_i^2 + m_j^2 + 2m_i m_j \cos u_{ij}^n. \tag{17.4.59}$$

This relation is simply obtained by substituting in the Mandelstam variable s given in eqn. (17.3.46) the resonance condition $\theta = iu_{ij}^n$. Notice that this formula expressed a

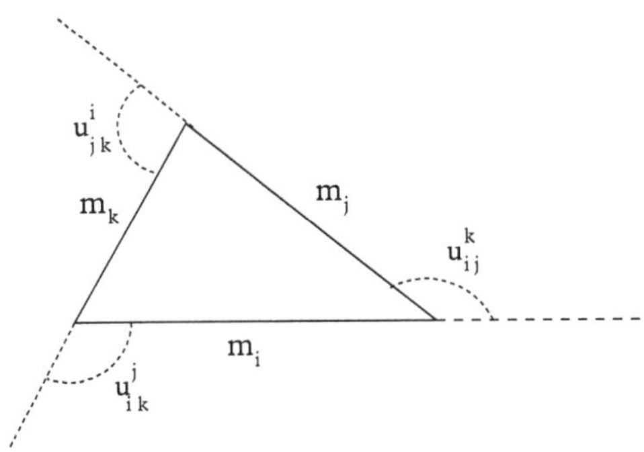

Fig. 17.11 *Mass triangle.*

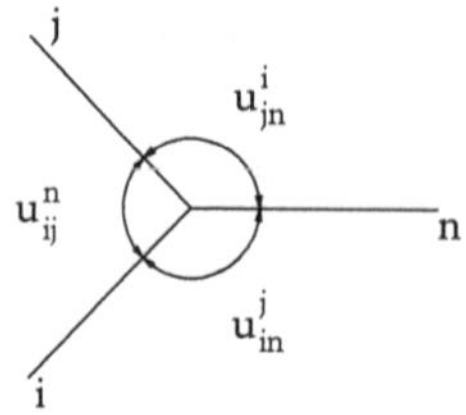

Fig. 17.12 *Relation among the position of the poles.*

well-known geometrical relation, known as Carnot theorem, among the sides of a triangle (here equal to the values of the masses), where u_{ij}^n is one of the external angle as shown in Figure 17.11). This figure clearly highlights the symmetric role played by the three particles.

For the deep geometrical nature of the quantities involved in this formulation and as a consequence of (17.4.59), it is easy to show that the positions of the poles in the three channels satisfy

$$u_{ij}^n + u_{in}^j + u_{jn}^i = 2\pi. \tag{17.4.60}$$

This relation, shown in Figure 17.12, expresses a well-known properties of the external angles of a triangle.

As we are going to see later, the elastic S-matrix of $(1+1)$-dimensional systems may also have higher-order poles, whose interpretation stays in the singularities coming from multiple scattering processes. Instead of an abstract discussion of this issue, we prefer to illustrate later its features by means of some explicit examples.

Diagonal S-matrices. To proceed further in the discussion of the analytic structure of the elastic S-matrices, it is convenient to make an additional simplification in the theory so far presented. This simplification occurs in two cases: (i) when the system has a non-degenerate mass spectrum and (ii) when the system has a degenerate spectrum but with all particles nevertheless uniquely identified in terms of their different eigenvalues with respects to the conserved charges. In both cases, the elasticity of the scattering processes enforces the vanishing of the reflection amplitude (see Problem 17.5): the corresponding S-matrix is then completely diagonal and the Yang–Baxter equations are then identically satisfied. The unitarity and crossing symmetry conditions simplify as follows

$$S_{ab}(\theta)\, S_{ab}(-\theta) = 1, \quad S_{ab}(\theta) = S_{a\bar{b}}(i\pi - \theta), \tag{17.4.61}$$

where $\bar{b}$ is the anti-particle of b. These two equations imply that the amplitudes $S_{ab}(\theta)$ are periodic functions of θ with period $2\pi i$: in this case the Riemann surface of the S-matrix consists of a double covering of the complex plane s.

Remarkably enough, there is a general solution of eqns. (17.4.61) that can be expressed in terms of products of the meromorphic functions

$$s_x(\theta) = \frac{\sinh \frac{1}{2}(\theta + i\pi x)}{\sinh \frac{1}{2}(\theta - i\pi x)}. \tag{17.4.62}$$

For their periodicity, the parameter x can always be chosen as $-1 \leq x \leq 1$. In the double covering of the original variable s, i.e. in the strip $-\pi \leq \mathrm{Im}\theta < \pi$, these functions have a simple pole at $\theta = i\pi x$ and a simple zero at $\theta = -i\pi x$. Moreover, they have the properties

$$\begin{aligned} &s_x(\theta)\, s_x(-\theta) = s_x(\theta) s_{-x}(\theta) = 1, \\ &s_x(\theta) = s_{x+2}(\theta) = s_{-x}(-\theta), \\ &s_0(\theta) = -s_1(\theta) = 1, \\ &s_x(i\pi - \theta) = -s_{1-x}(\theta). \end{aligned} \tag{17.4.63}$$

A suggestive interpretation of these functions is proposed in Problem 17.8.

When the particles involved in the scattering are instead neutral, i.e. when the particles coincide with their anti-particles, the solution of equations (17.4.61) can be expressed in terms of the functions

$$f_x(\theta) = s_x(\theta)\, s_x(i\pi - \theta) = \frac{\tanh \frac{1}{2}(\theta + i\pi x)}{\tanh \frac{1}{2}(\theta - i\pi x)}. \tag{17.4.64}$$

The simple poles of these functions are at $\theta = i\pi x$ and $\theta = i\pi(1-x)$ and they are related by the crossing transformation. They have also simple zeros at $-i\pi x$ and $-i\pi(1-x)$. Important properties of these functions are

$$f_x(\theta) = f_x(i\pi - \theta) = f_{1-x}(\theta), \quad f_x(-\theta) = f_{-x}(\theta) = 1/f_x(\theta). \tag{17.4.65}$$

In summary, as consequences of the unitarity and crossing symmetry equations, any amplitude $S_{ab}(\theta)$ of a diagonal S-matrix can be expressed as

$$S_{ab}(\theta) = \prod_{x \in \mathcal{A}_{ab}} s_x(\theta), \tag{17.4.66}$$

if there are charged particles, or by

$$S_{ab}(\theta) = \prod_{x \in \mathcal{A}_{ab}} f_x(\theta), \tag{17.4.67}$$

if the particles are neutral. Functions such as $s_x(\theta)$ or $f_x(\theta)$, which identically solve the unitarity and the crossing equations, are known as *CDD factors*, from Castillejo, Dalitz and Dyson, the authors who originally analysed the general solution of the scattering equations.

Bootstrap principle. The unitarity and crossing symmetry equations alone, however, are not able to fix the position of the poles of these amplitudes, i.e. to determine the sets $\mathcal{A}_{ab}$. To achieve this aim it is necessary to make use of a dynamical condition. This is provided by the bootstrap principle that posits that the bound states are on the same footing of the asymptotic states. As a consequence, the amplitudes that involve the bound states can be obtained in terms of the amplitudes of the external particles, and vice-versa. This translates in an additional equation satisfied by the scattering amplitudes

$$S_{i\bar{l}}(\theta) = S_{ij}(\theta + i\bar{u}_{jl}^{k})\, S_{ik}(\theta - i\bar{u}_{lk}^{j}), \tag{17.4.68}$$

where

$$\bar{u}_{ab}^{c} \equiv \pi - u_{ab}^{c}. \tag{17.4.69}$$

This equation comes from the commutativity of the two processes shown in Figure 17.13, obtained one from the other by the translation of the world-line of the asymptotic particle A_i (see Problem 17.6).

Rules of the game. To summarize the discussion done so far, in order to determine the S-matrix by the bootstrap approach we have to find a set of poles relative to all amplitudes S_{ab}, that are compatible with the bootstrap equation (17.4.68) and that can be interpreted in terms of bound states or multi-particle scattering processes of the asymptotic particles themselves. The masses of the particles are determined by the relation (17.4.59). In practice this means starting from the amplitude that involves the lightest particle, therefore with the simplest pole structure, and then iteratively applying the bootstrap equations (17.4.68) in order to get the scattering amplitudes involving the bound states of higher mass.

It should be stressed, though, that not all the choices of the initial amplitude gives rise to consistent bootstrap systems. Presently, it is still an open problem to classify in their full generality the integrable models in bootstrap interaction. Valuable information is gained by the spectrum of the conserved charges, as discussed in the next section.

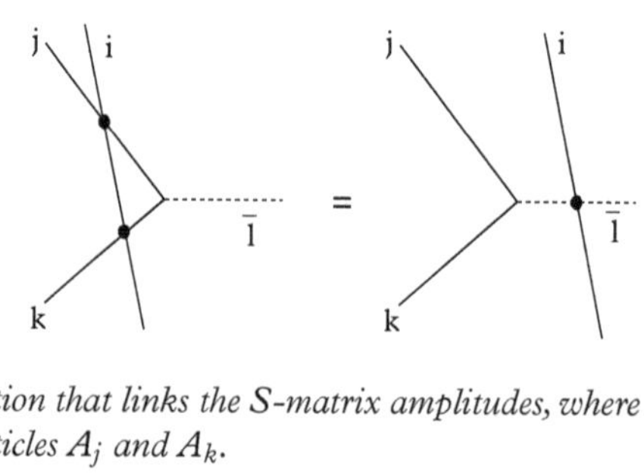

Fig. 17.13 *Bootstrap equation that links the S-matrix amplitudes, where $A_{\bar{l}}$ is the bound state in the scattering process of the particles A_j and A_k.*

Important examples of consistent S-matrices will be given in the next chapter and they are extremely helpful to clarify several aspects of the iterative bootstrap procedure.

To simplify the repeated applications of the bootstrap equations (17.4.68), it is useful to define the operator $\mathcal{R}^y$, whose application to a function $G(\theta)$ is given by

$$\mathcal{R}^y(G(\theta)) = G(\theta + i\pi y)\, G(\theta - i\pi y).$$

Applying $\mathcal{R}^y$ to the functions $s_x(\theta)$ and $f_x(\theta)$ and using their properties, we have

$$\mathcal{R}^y(s_x(\theta)) = s_{x+y}(\theta)\, s_{x-y}(\theta),$$
$$\mathcal{R}^y(f_x(\theta)) = f_{x+y}(\theta) f_{x-y}(\theta).$$

They also have the commutative and distributive properties

$$\mathcal{R}^y(\mathcal{R}^z(G)) = \mathcal{R}^z(\mathcal{R}^y(G)), \quad \mathcal{R}^y(G_1)\, \mathcal{R}^y(G_2) = \mathcal{R}^y(G_1\, G_2).$$

Finally, if a function $G(\theta)$ satisfies the equation

$$G(\theta) = G(i\pi - \theta) = 1/G(-\theta), \tag{17.4.70}$$

the same holds for the function transformed by $\mathcal{R}^y$.

17.5 Conserved Charges and Consistency Equations

This section studies the relation between the spins of the conserved charges and the bound states of a scattering theory. The integrals of motion $\mathcal{Q}_s$ are a set of dynamical data relative to each scattering theory. Had the Lagrangian of the model been known, it would be possible in principle to determine them explicitly. The knowledge of the S-matrix alone leads instead only to some constraints on the values of the spins s. It also leads to the determination of the ratios of the eigenvalues of $\mathcal{Q}_s$. As shown below, these results derive by the bootstrap principle and the locality properties of the conserved charges.

Let $\mathcal{Q}_s$ be the set of all conserved charges. Since they commute each other, they can be simultaneously diagonalized together with the Hamiltonian, and the asymptotic states $A_a(\theta)$ are also eigenvectors of $\mathcal{Q}_s$

$$\mathcal{Q}_s \mid A_a(\theta)\rangle = \chi_s^{(a)}\, e^{s\theta} \mid A_a(\theta)\rangle. \tag{17.5.71}$$

For a conserved charge of spin s, there exists at least an eigenvalue $\chi_s^{(a)}$ different from zero. Note that $\chi_1^{(a)}$ is simply the mass of the particle a, $\chi_1^{(a)} = m_a$. The locality of the conserved charges implies that their action on the multi-particle states is given by

$$\mathcal{Q}_s \mid A_{a_1}(\theta_1) \cdots A_{a_n}(\theta_n)\rangle = (\omega_s^{(a_1)}(\theta_1) + \cdots \omega_s^{(a_n)}(\theta_n)) \mid A_{a_1}(\theta_1) \cdots A_{a_n}(\theta_n)\rangle. \tag{17.5.72}$$

Suppose that the amplitude S_{ab} presents a pole at $\theta = iu_{ab}^c$ corresponding to the bound state $A_{\bar{c}}$. Then, this can be defined as

$$\lim_{\epsilon \to 0} \epsilon \mid A_a(\theta + i\bar{u}_{ac}^b + \epsilon) A_b(\theta - i\bar{u}_{bc}^a)\rangle = \mid A_{\bar{c}}(\theta)\rangle. \tag{17.5.73}$$

Applying Q_s to both terms of this equation and using eqns. (17.5.71) and (17.5.72), one obtains an infinite-dimensional homogeneous system of linear equations for the eigenvalues $\chi_s^{(a)}$

$$\chi_s^{(a)} e^{is\bar{u}_{ac}^b} + \chi_s^{(b)} e^{-is\bar{u}_{bc}^a} = \chi_s^{(\bar{c})}. \tag{17.5.74}$$

A solution of this system is obviously $\chi_s^{(i)} = 0$ $(\forall s, i)$. However, this is not an interesting solution because it implies the absence of all conserved charges. Non-trivial solutions can be found only for particular values of the resonance angles u_{ab}^c of the S-matrix, corresponding to the vanishing of the determinant of the homogeneous linear system (17.5.74).

Consider, for instance, the case in which $a = b$, with $\chi_s^{(a)} \neq 0$. Eqn. (17.5.74) can be written in this case as

$$2\cos(s\bar{u}_{ac}^a) = \frac{\chi_s^{(c)}}{\chi_s^{(a)}}. \tag{17.5.75}$$

If the bound state c corresponds to the same initial particle a, this equation admits the solutions

$$\bar{u}_{aa}^a = \frac{\pi}{3}, \quad s = 1, 5 \ (\mathrm{mod}\, 6). \tag{17.5.76}$$

Note that the exact value of the resonance angle $\bar{u}_{aa}^a = \frac{\pi}{3}$ directly comes from the geometry of the mass triangle, in this case an equilateral triangle. The S-matrix of this example presents the so-called Φ^3 property, since the particle A_a is simultaneously asymptotic state and bound state of itself. Read in reverse, this result hints that each time that the spectrum of the conserved spins consists of integer numbers that are not divisible by both 2 and 3, the particle mass spectrum may present the Φ^3 property.

To proceed in our analysis, it is useful to introduce the notion of *bootstrap fusion*. Let A_a be the operator that creates a particle a in bootstrap interaction. The bound state structure can be encoded in this relation

$$A_i \times A_j = \sum_k n_{ij}^k A_k \tag{17.5.77}$$

n_{ij}^k are boolean variables, with values 0 and 1, different from zero only when A_k is the bound state of the scattering process of the particles A_i and A_j. Even though the strong analogy of this relation with the Verlinde algebra of the CFT, it should be stressed that the bound state relation is not an associative algebra. The reason is that the symbols A_is present in eqn. (17.5.77) actually depend on the rapidities θ_i and the definition of the bound states given in eqn. (17.5.73) involves special values of these rapidities: hence, for instance, starting from an asymptotic state made of three particles, $| A_i(\theta_1)A_j(\theta_2)A_k(\theta_3)\rangle$, we can arrive without contradiction to two *different* results whether we 'fuse' the first two particles on their bound states and these once again with the remaining $A_k(\theta_3)$ or we 'fuse' the last two particles on their bound states and these once again with the remaining $A_i(\theta_1)$, since the values of the rapidities involved in these fusions are different in the two cases.

As mentioned, the full classification of all bootstrap systems is still an open problem, even though there are strong indications that the only consistent systems with diagonal S-matrices are those related to Toda field theories or reduction thereof. Below we present only some simple but instructive examples of consistent bootstrap systems.

17.5.1 Non-degenerate Bootstrap Systems

Let us assume the existence of a non-trivial solutions of the set of equations (17.5.74). For their homogeneous form, we can always choose to normalize to 1 all the non-zero eigenvalues of the lightest particle. For a neutral particle, it is easy to show by induction that all remaining eigenvalues are real. Eqns. (17.5.74) split then in two different sets

$$(\chi_s^{(c)})^2 = (\chi_s^{(a)})^2 + (\chi_s^{(b)})^2 + 2\chi_s^{(a)}\,\chi_s^{(b)}\cos(s u_{ab}^c),$$

$$\chi_s^{(a)}\sin(s\bar{u}_{ac}^b) = \chi_s^{(b)}\sin(s\bar{u}_{bc}^a).$$

The first provides a generalization of the mass triangle equation (17.4.59), while the second generalizes a simple geometrical property of this triangle. It should be stressed that the second equation is particularly useful from a computational point of view: to have a non-zero values of $\chi_s^{(a)}$ and $\chi_s^{(b)}$, the ratio of the two trigonometric functions $\sin(s\bar{u}_{ac}^b)/\sin(s\bar{u}_{bc}^a)$ must be in fact *independent* from any bound state A_c in the channel $| A_a A_b\rangle$. Hence, knowing the resonance angle of any of the bound states in this channel, we can use this equation either to correctly identify the value of the others or, alternatively, to prove that it is impossible to have conserved charges of higher spins compatible with the structure of the bootstrap algebra.

Let us consider some significant examples of bootstrap systems that involve N particles, starting from the simplest case $N = 1$.

- **N=1**. In this case, assuming the existence of only one bound state, the only fusion process is that which sees the particle as a bound state of itself

$$A \times A \to A. \qquad (17.5.78)$$

The resonance angle is $u_{aa}^a = \frac{\pi}{3}$ and the only possible values of the spins of the conserved charges are

$$s = 1,5 \pmod 6). \tag{17.5.79}$$

A physical realization of this system is provided by the off-critical Yang–Lee model or the Bullogh–Dodd lagrangian, as we will see in the next chapter.

- N=2. In addition to the reducible fusion rules $A_a \times A_a \to A_a$, $A_b \times A_b \to A_b$, consider the examples

$$\begin{array}{lll} (i) & A_a \times A_a \to A_b, & A_b \times A_b \to A_a \\ (ii) & A_a \times A_a \to A_a + A_b, & A_b \times A_b \to A_a. \end{array}$$

The consistency equations of the processes (i) are

$$2\,\chi_s^{(a)} \cos(s\bar{u}_{ab}^a) = \chi_s^{(b)}, \quad 2\,\chi_s^{(b)} \cos(s\bar{u}_{ab}^b) = \chi_s^{(a)}.$$

For $\chi_s^{(a,b)} \neq 0$ they become

$$4\cos(s\bar{u}_{ab}^a)\cos(s\bar{u}_{ab}^b) = 1.$$

This equation admits two types of solutions

$$\bar{u}_{ab}^a = \frac{\pi}{12}, \quad \bar{u}_{ab}^b = \frac{5\pi}{12}, \quad s = 1,4,5,7,8,11 \pmod{12}, \tag{17.5.80}$$

$$\bar{u}_{ab}^a = \frac{\pi}{5}, \quad \bar{u}_{ab}^b = \frac{2\pi}{5}, \quad s = 1,3,7,9, \pmod{10}. \tag{17.5.81}$$

Note that the spectrum of s of the first solution coincides with the Coxeter exponents of the Toda field theory on $E_6^{(1)}$.

If we restrict our attention to neutral particles, there are no conserved spins with $s = 2k$. In this case the spectrum of the conserved spins of the first solution becomes

$$s = 1,5,7,11 \pmod{12}. \tag{17.5.82}$$

It coincides with the Coxeter exponents of the Toda field theory based on $\tilde{F}_4 = E_6^{(2)}$, obtained by folding of the original E_6 Dynkin diagram with respect to its Z_2 automorphism.

For the process (ii), as possible values of the spins it is necessary to take those compatible with the one-particle sub-process. For instance, for the solution (17.5.81), we have

$$s = 1,7,11,13,17,19,23,29 \pmod{30}. \tag{17.5.83}$$

This spectrum coincides with the Coxeter exponents of the Toda field theory based on $E_8^{(1)}$.

- **Bootstrap chains.** For a generic bootstrap system of N neutral particles it is easy to analyse the case in which there is a bootstrap chain of bound states

$$A_k \times A_k \to A_{k+1}, \; k = 1, 2, \ldots N, \quad A_{N+1} = A_1.$$

The consistency equation is

$$\prod_{k=1}^{N} 2\cos(s\,\bar{u}_{k,k+1}^{k}) = 1,$$

whose solution is given by

$$\begin{aligned} \bar{u}_{k,k+1}^{k} &= \frac{k\pi}{2N+1} \\ s &= 1, 3, \ldots, 2N-1, 2N+3, \ldots, 4N+1 \; (\mathrm{mod}\; 4N+2). \end{aligned} \tag{17.5.84}$$

In this case the spectrum of conserved spins coincides with the Coxeter exponents of the Toda field theories based on $A_{2N}^{(2)}$.

Appendix 17.A. Historical Developments of the *S*-matrix Theory

The *S*-matrix theory is an interesting elementary particle physics and it is worth mentioning its basic developments. The interested reader is referred to the references at the end of the chapter for a broader perspective on the subject.

Proposed originally by Heisenberg to overcome the difficulties of QFT in dealing with the divergences of the perturbative series, the *S*-matrix theory has received considerable attention during the 1950s and 1960s, in particular for the study of strong interactions of hadronic particles, such as protons, neutrons and pions. The enormous number of particles and hadronic resonances experimentally discovered during those decades made clear the difficulty to call all of them *elementary particles*. A further discovery was that the hadronic resonances present high values of their spin $\mathcal{J}$, related to the square of their mass by a linear relation, $\mathcal{J} = \alpha' m^2$, where the constant $\alpha' \sim 1\;(\mathrm{Gev})^2$ is the *Regge slope*. The first attempts to use the QFT to describe the hadronic phenomena very disastrous. In fact, the difficulty was in incorporating both the unstable particles (the resonances) and the particles with spin higher than 1: the only known examples of consistent QFTs, i.e. renormalizable in perturbation theory, are those limited to stable particles with spin 0, 1/2 and 1. The large values of the effective coupling constants coming from the experiments led also to doubt on the efficiency and validity of the possible perturbative approaches for describing such processes.

In light of all these drawbacks, it was necessary to look for an alternative theory of the hadronic processes that could eventually be extented to other interactions. The new approach, based on a set of principles and the analytic properties of the quantum amplitudes, arrived under the name of *The Analytic Theory of the S-matrix*. Proposed and studied in great detail by the group of physicists in Berkeley, in particular Chew and Mandelstam, the theory developed further with important contributions by Weisskopf, Frautschi, Regge and many others. Since the analysis of the scattering processes is the common and closest point between the theory and the experiments, the expectations were that the results derived by the S-matrix theory should not depend on the existence (or the absence) of an underlying QFT of the interactions. A fundamental theory based on the S-matrix should then provides answers to the following questions:

1. What is the difference between stable and unstable particles? Does a theoretical framework exist for both? As it is well known, the Lagrangian formulation of QFT makes only use of the stable asymptotic particles and therefore does not allow an equally supported discussion of both cases.
2. Is it possible to determine the mass spectrum and the coupling constants of the theory? Remember that, in a Lagrangian theory to the contrary, both masses and coupling constants are free parameters of the model.

The initial studies of the S-matrix as a function of the energy, the momentum, the angular momentum, etc., showed the suggestive circumstance that the analytic structure of the S-matrix appeared to be the simplest possible. This was assumed then as a principle and formalized under the heading of *the principle of maximum analiticity of the S-matrix*. If this hypothesis was correct, the physics of strong interactions should not have arbitrary constants, except for the fundamental constants of nature, such as the speed of light c, the Planck constant h, and one parameter scale. Consequently, all the strong interaction particles would be composite particles and could be considered on the same basis. These were the basis of the *bootstrap principle*.

All these theoretical developments were deeply influenced by the Regge's proposed formalism to analyse the scattering amplitudes as functions in the complex plane of the angular momentum. In particular, using Regge's theory, it was possible to study the asymptotic behaviour of the amplitudes for large values of s and to give an estimate of the high-energy limit of the cross-sections. Among the results obtained as a result of Regge's theory it is worth mentioning that:

1. The prediction on the high-energy asymptotic behaviour of the scattering processes dominated by the exchange of particles (with the relative associated poles) in the t-channel (Figure 17.14)

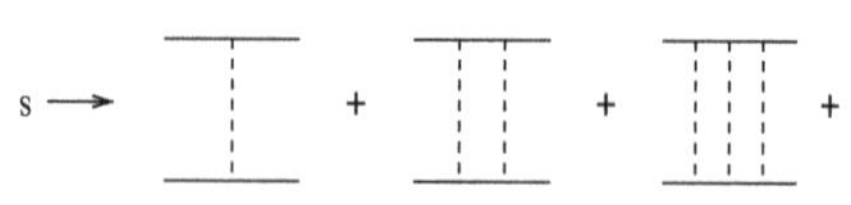

Fig. 17.14 *Amplitudes that determine the high-energy behavior of the scattering process.*

$$\sigma_{tot} \simeq s^{\alpha_0 - 1}.$$

2. The prediction of the relation between the total cross-section of a process with incoming particles $A+B$ and the cross-sections relative to the incoming particles $A+A$ and $B+B$

$$\sigma_{tot}^{(A+B)} = \left[\sigma_{tot}^{(A+A)} \, \sigma_{tot}^{(B+B)}\right]^{1/2}.$$

   This prediction was based on the close relation between the Regge poles and the resonances, with the factorized expression of the amplitude near a Regge pole

$$f_{nm}(l,s) \simeq \frac{\gamma_n \gamma_m}{l - \alpha(s)}.$$

However, the most important result obtained by the analytic S-matrix theory was the scattering amplitude discovered by Veneziano that exactly implements the duality between the s and the t channels. Let us discuss it in more detail. In the presence of particles exchanged in the t-channel, having an increasing values of mass and spin, the amplitude in this channel assumes the form

$$A(s,t) = -\sum_{J} \frac{g_J^2 \, (-s)^J}{t - m_j^2}. \tag{17.A.1}$$

Had this sum only a *finite* number of terms, it would define an amplitude that does not have poles in the s-channel, since at any fixed value of t, $A(s,t)$ is manifestly an integer function[12] of s. However, we arrive at a different conclusion if the series is made of infinite terms, for it could diverge at different values of s, giving rise to poles also in the s-channel. In this case, it would not be obvious that, in implementing the crossing symmetry we should also include the corresponding terms of the s-channel, for they could be already present in the series (17.A.1).

The same conclusion could be also reached by starting by the s-channel to arrive at an analogous formula

$$\tilde{A}(s,t) = -\sum_{J} \frac{g_J^2 \, (-t)^J}{s - m_J^2}. \tag{17.A.2}$$

It is now possible to imagine that, with an appropriate choice of the coupling constants g_J and the masses m_J, the two amplitudes $A(s,t)$ and $\tilde{A}(s,t)$ define the *same function*: if this is the case, the scattering amplitude could be equivalently written as a series on the

[12] Remember that in the Feynman perturbation theory, in order to implement the crossing symmetry we must include both the diagrams of the s and t channels.

infinite poles of the t-channel or the s-channel, with an explicit duality between the two pictures. Veneziano demonstrated it thus, showing the amplitude

$$A(s,t) = \frac{\Gamma[-\alpha(s)]\,\Gamma[-\alpha(t)]}{\Gamma[-\alpha(s)-\alpha(t)]}, \quad \alpha(x) = \alpha_0 + \alpha' x. \tag{17.A.3}$$

For the linear behaviour of $\alpha(x)$, it is easy to show that the singularities of the amplitude (17.A.3) are simple poles to the exchange of particles of mass $m^2 = (n-\alpha_0)/\alpha'$, $n = 0, 1, 2, \ldots$ both in the s and t channels. Moreover, the residue at the pole $\alpha(t) = n$ is a polynomial of order n in s that corresponds to a particle of spin n. The same happens for the poles of the s-channel. Using the asymptotic behaviour of the function $\Gamma(z)$, it is easy to see that the Veneziano amplitude presents a Regge behaviour in both variables

$$\begin{aligned} A(s,t) &\simeq s^{\alpha(t)}, \quad s \to \infty, \quad t \text{ fixed} \\ A(s,t) &\simeq t^{\alpha(s)}, \quad t \to \infty, \quad s \text{ fixed} \end{aligned}$$

The discovery of the Veneziano amplitude has had an enormous influence on the developments in the strong interaction studies. Moreover, it has been the starting point for the development of String Theory.

The Regge theory and the analytic S-matrix theory have dominated the theoretical studies for long time, becoming an extremely sophisticated field, with many subtleties and adjustments introduced to incorporate in the formalism new phenomena discovered along the years in the strong interaction domain. It was also in fierce competition and often in open opposition with the formulation given of the fundamental interactions by QFT. There were violent polemics among the supporters of the two different formulations, just as it was long ago with those who supported the wave or the corpuscolar theory of light. The scientific atmosphere of those years is condensed in this humorous story.

A student was curious to know whether the Mandelstam dispersion relation of the scattering amplitude could be derived by the quantum field theory. He addressed the question to Weisskopf who answered: 'Field theory? What is a field theory?'. He went then to ask the same question to Wigner, who said: 'Mandelstam? Who is Mandelstam?'. Finally, quite discouraged, the student thought to address the question directly to Chew 'who, when he, heard the question, pronounced: 'Proof? What is a proof?'

However, despite the initial triumphs, the S-matrix theory sank into oblivion, not because it was proved wrong but simply because was too complicated to handle, and many years of study have produced only modest advances. Finally, it was supplanted by the QFT, which came back into vogue via the suggestive hints of the *deep inelastic scattering processes*. The new quantum cromodynamics theory, a QFT based on a non-abelian gauge group, has the important feature of the asymptotic freedom that, in addition to its compatibility with all the experimental data, also permits to make new quantitative predictions.

In the light of these historical developments, it is fair to say that the vindication of the basic principles of the S-matrix theory comes from the study of the two-dimensional statistical models, with the solution of important systems, like the Ising model in an external magnetic field, which resisted the theoretical attempts for many years.

Appendix 17.B. Scattering Processes in Quantum Mechanics

This appendix reviews the main formulae of the scattering theory in quantum mechanics. We examine, in particular, the one-dimensional systems, i.e. those closer to the S-matrix theory of the $(1+1)$-dimensional systems studied in the text. In the following we pose $\hbar = 1$. Consider initially a particle of mass m and momentum p that moves freely along the real axis, with Hamiltonian

$$H_0 = \frac{p^2}{2m},$$

Since p commutes with H_0, we can simultaneously diagonalize both operators. The common eigenfunctions are the plane waves

$$\begin{aligned} \psi_k(x) &= e^{ikx} \\ p\,\psi_k(x) &= k\psi_k(x) \\ H_0\,\psi_k(x) &= \tfrac{k^2}{2m}\,\psi_k(x). \end{aligned}$$

The time evolution of these eigenfunctions is

$$\psi_k(x,t) = e^{-iE_k t}\,\psi_k(x) = e^{-itk^2/2m}\,\psi_k(x). \tag{17.B.1}$$

The energy spectrum is continuous and doubly degenerate, since it depends on the square of the momentum. Hence any linear combination of ψ_k and ψ_{-k} is also an eigenfunction of H_0. H_0 also commutes with the parity operator P and therefore we can choose a basis with functions of a given parity

$$\begin{aligned} \psi_{k0}(x) &= \cos kx, \quad P\,\psi_{k0}(x) = \psi_{k0}(x) \\ \psi_{k1}(x) &= \sin kx, \quad P\,\psi_{k1}(x) = -\psi_{k1}(x). \end{aligned} \tag{17.B.2}$$

Let us imagine now to add to the Hamiltonian a potential $V(x)$, finite and different from zero, only inside a region $|\,x\,| < x_0$, as in Figure 17.15. For simplicity, we assume that V is an even function, $V(x) = V(-x)$.

$$\begin{aligned} H &= \tfrac{p^2}{2m} + V(x) \\ V(x) &= 0 \quad \text{for } |\,x\,| > x_0. \end{aligned} \tag{17.B.3}$$

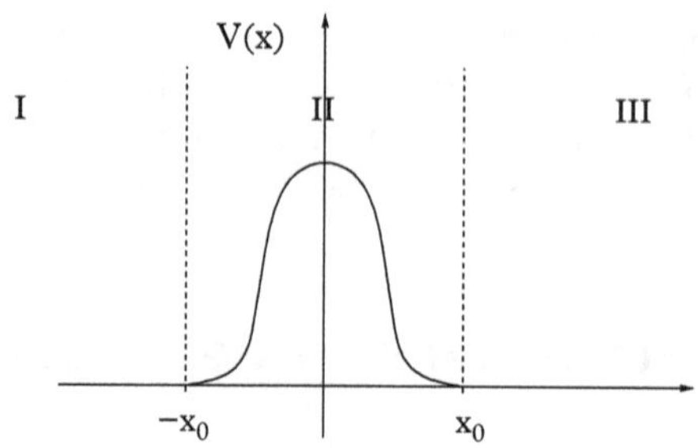

Fig. 17.15 *Potential of the scattering process. In the regions I and III the particle moves freely.*

The spectrum of the eigenvalues with $E \geq 0$ remains invariant, as well as the eigenfunctions in the external regions *I* and *III*

$$\psi(x) = \begin{cases} Ae^{ikx} + Be^{-ikx}, & x < -x_0 \\ Ce^{ikx} + De^{-ikx}, & x > x_0 \end{cases} \tag{17.B.4}$$

The linear relation that links A and B to the coefficients C and D depends on the shape of the potential $V(x)$.

Consider now the *scattering solutions* $\psi_+(x)$ of the Schrödinger problem, i.e. those with $D = 0$

$$\psi_+(x) = \begin{cases} Ae^{ikx} + Be^{-ikx}, & x < -x_0 \\ Ce^{ikx}, & x > x_0 \end{cases} \tag{17.B.5}$$

In this case, A is the coefficient of the incoming wave, B is the amplitude of the reflected wave, while C is the amplitude of the transmitted wave. The reflection and transmission coefficients are given by

$$\begin{aligned} \mathcal{R} &= \frac{B}{A}, \\ \mathcal{T} &= \frac{C}{A}. \end{aligned} \tag{17.B.6}$$

Since the sum of the densities of the reflected and transmitted waves must be equal to the density of the incoming wave, we have

$$|\mathcal{R}|^2 + |\mathcal{T}|^2 = 1. \tag{17.B.7}$$

The reflection and transmission coefficients can be expressed in terms of the *phase shifts* δ_0 and δ_1, defined by the stationary eigenfunctions of the Hamiltonian

$$\begin{aligned} \psi_0 &= \cos(kx+\delta_0),\ (x>x_0); \quad \psi_0 = \cos(kx-\delta_0),\ (x<-x_0) \\ \psi_1 &= \sin(kx+\delta_1),\ (x>x_0); \quad \psi_1 = \sin(kx-\delta_1),\ (x<-x_0). \end{aligned} \tag{17.B.8}$$

The S-matrix in the channels of a given parity is given by

$$S_a = e^{2i\delta_a}, \quad a=0,1. \tag{17.B.9}$$

The linear combination of eigenstates of given parity (17.B.8) that gives rise to the scattering eigenfunction ψ_+ is

$$\psi_+ = \begin{cases} e^{ikx} + \frac{1}{2}(e^{2i\delta_0} - e^{2i\delta_1})\,e^{-ikx}, & (x<-x_0) \\ e^{i\delta_0}\,\psi_0 + i\,e^{i\delta_1}\,\psi_1 = \frac{1}{2}(e^{2i\delta_0} + e^{2i\delta_1})\,e^{ikx}, & (x>x_0) \end{cases} \tag{17.B.10}$$

Hence

$$\begin{aligned} \mathcal{R} &= \tfrac{1}{2}(e^{2i\delta_0} - e^{2i\delta_1}) = \tfrac{1}{2}[(e^{2i\delta_0}-1) - (e^{2i\delta_1}) - 1)] \\ &= \textstyle\sum_{l=0}^{1} i(-1)^l e^{i\delta_l}\sin\delta_l \\ \mathcal{T} &= \tfrac{1}{2}(e^{2i\delta_0} + e^{2i\delta_1}) = \tfrac{1}{2}[(e^{2i\delta_0}-1) + (e^{2i\delta_1}) - 1)] + 1 \\ &= 1 + \textstyle\sum_{l=0}^{1} ie^{i\delta_l}\sin\delta_l, \end{aligned} \tag{17.B.11}$$

and the reflection and transmission coefficients are completely determined by the phase shifts of the even and odd eigenfunctions.

Consider now the case in which the potential is given by

$$V(x) = -2g\,\delta(x). \tag{17.B.12}$$

We start from the even eigenfunctions. Imposing the continuity of the wave function at the origin and the discontinuity of its derivative, ruled by the $\delta(x)$ function

$$\begin{aligned} &\psi_0(0^+) = \psi(0^-) \\ &\frac{d\psi_0(0^+)}{dx} - \frac{d\psi_0(0^-)}{dx} = -2k\sin\delta_0 = -2g\,\psi_0(0) = -g\cos\delta_0 \end{aligned}$$

we can determine the even phase shift

$$\tan\delta_0 = \frac{g}{k}. \tag{17.B.13}$$

The S-matrix in this channel is then

$$e^{2i\delta_0} = \frac{1+i\tan\delta_0}{1-i\tan\delta_0} = \frac{k+ig}{k-ig}. \tag{17.B.14}$$

The variation of the phase is then

$$\delta_0(+\infty) - \delta_0(-\infty) = -2\pi g/|g|,$$

and depends on the sign of g.

The odd solution vanishes at the origin, hence the odd phase shift is identically zero. The corresponding S-matrix is then equal to 1

$$\begin{aligned} \delta_1 &= 0 \\ e^{2i\delta_1} &= 1. \end{aligned} \tag{17.B.15}$$

The expressions δ_0 and δ_1 permits to obtain the ratios (17.B.7) and to define a solution of the Schrödinger equation for all the values of k.

It is interesting to analyse the nature of this solution for complex values of the momentum $k = k_1 + ik_2$. The real part can be always considered positive or zero since it corresponds to the physical momentum of the incoming particle. Substituting k in (17.B.5), we see that the imaginary part k_2 enters the real part of the exponentials. Choosing now k as the value of the pole of the S-matrix, i.e. $k = ig$, one can have a normalizable eigenfunction by imposing $A = 0$. This solution corresponds to a *bound state* of the system, whose energy is $E_b = -g^2/(2m)$. Obviously in this case we should have $g > 0$.

More generally, we can show the following properties of the non-relativistic S-matrix

1. The poles of the S-matrix with positive imaginary values of the momentum, $k_n = ia_n$ $(a_n > 0)$, correspond to the energies $E_n = -a_n^2/(2m)$ of the bound states of the system.
2. There are no poles in the complex plane of the variable $k = k_1 + ik_2$ with a non-vanishing real part k_1 in the half-plane $k_2 > 0$.
3. The poles in the complex plane with negative imaginary part, $k_2 < 0$, correspond instead to the resonances.

The proof of the property (1) follows the one given for the potential $\delta(x)$. For the point (2), let us suppose that the S-matrix has a pole at $k = k_1 + ik_2$, with $k_2 > 0$. Substituting in (17.B.5) and putting to zero the coefficient A, we have a normalizable eigenfunction also in this case. The problem though is in the time evolution of this eigenfunction: using eqn. (17.B.1) we get

$$\psi_+(x,t) = e^{-it(k_1^2-k_2^2)/2m}\, e^{tk_1 k_2/m}\, \psi_+(x) \tag{17.B.16}$$

and, if $k_1 > 0$, the eigenfunction grows exponentially when $t \to +\infty$, leading to a violation of the conservation of the probability.

A pole in the complex plane but with negative imaginary part is however perfectly plausible. It corresponds to a solution whose probability decreases in a given channel.

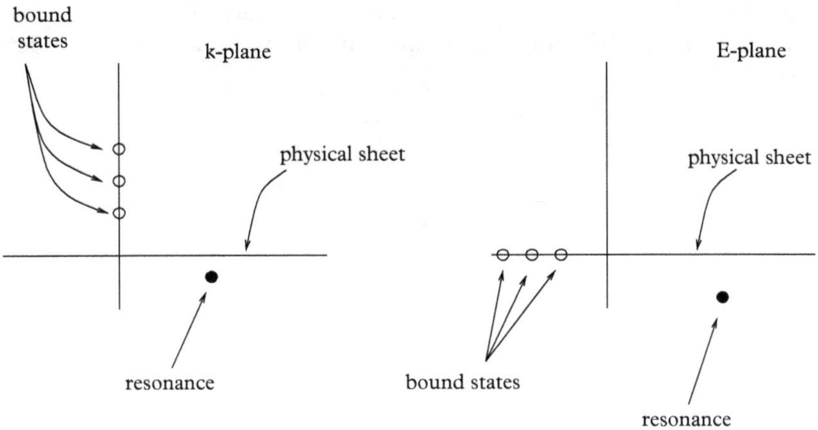

Fig. 17.16 *Analytic structure of the S-matrix in the planes of the complex variables k and E.*

This means that it will grow in other channels so that there is a global conservation of the probability. Poles with negative imaginary parts correspond to the *resonances*. The situation in the plane of the complex variables k and E is shown in Figure 17.16.

Since the S-matrix in any channel of a given parity is a unitary operator, in the vicinity of a pole $\bar{k}$ it can be parameterized as

$$S = e^{2i\delta} = \frac{k - \bar{k}^*}{k - \bar{k}}, \tag{17.B.17}$$

where $\bar{k}^*$ is the complex conjugate of $\bar{k}$. Changing to the energy, $E = E_r - i\Gamma/2$ (with $\Gamma > 0$, since there could be no poles in the upper half-plane), we have

$$S = \frac{E - E_r - i\Gamma/2}{E - E_r + i\Gamma/2}. \tag{17.B.18}$$

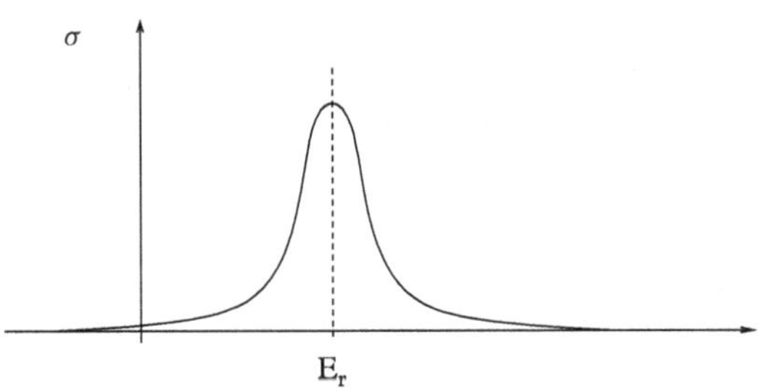

Fig. 17.17 *Cross-section relative to an S-matrix with a resonance pole.*

Note that, close to the energy of the resonance, the phase $\delta(E)$ of the S-matrix has an abrupt jump of 2π. We can now compute the diffusion amplitude T, defined by $S = 1 + iT$

$$T = -\frac{\Gamma}{E - E_r + i\Gamma/2}. \tag{17.B.19}$$

and the cross-section

$$\sigma \sim |T|^2 = \frac{\Gamma^2}{(E - E_r)^2 + \Gamma^2/4}. \tag{17.B.20}$$

As shown in Figure 17.17, the cross-section has the typical bell shape of a resonance phenomenon, with the width determined by the parameter Γ. It is easy to see that this is related to the life-time τ of the resonance state by $\tau = 1/\Gamma$.

Appendix 17.C. *n*-particle Phase Space

An important quantity that enters the probability computation of the scattering and decay processes is the differential n-particle *phase space*

$$d\Phi_n = \frac{d^{d-1}p_1}{(2\pi)^{d-1}2E_1} \cdots \frac{d^{d-1}p_1}{(2\pi)^{d-1}2E_1} (2\pi)^d \delta^d(P - p_1 - p_2 - \cdots - p_n). \tag{17.C.1}$$

The integral of this expression is a relativistic invariant quantity that depends only on the modulus of the total momentum, i.e. P^2

$$\Phi_n(P^2) = \int \frac{d^{d-1}p_1}{(2\pi)^{d-1}2E_1} \cdots \int \frac{d^{d-1}p_n}{(2\pi)^{d-1}2E_n} (2\pi)^d \delta^d(P - p_1 - p_2 - \cdots - p_n). \tag{17.C.2}$$

This quantity has an analogous in statistical mechanics. In fact, its definition recalls the partition function of a statistical model in the micro-canonical ensemble, the role of the total energy being here played by P^2. For the delta function that involves all momenta, its exact computation can be done only in few cases or for particular values of P^2.

Two-particle phase space. Let us study in more detail the properties of $\Phi_n(P^2)$, starting by the computation of the two-particle phase space when the momentum P is time-like (that is the more relavant case). This is the only case in which the phase case can be computed exactly. Since Φ_2 is a relativistic invariant quantity, we can choose a reference frame where $P = (E, 0)$ and

$$\Phi_2(E) = \int \frac{d^{d-1}p_1}{(2\pi)^{d-1}2E_1} \int \frac{d^{d-1}p_2}{(2\pi)^{d-1}2E_2} (2\pi)^d \delta^{d-1}(\vec{p}_1 - \vec{p}_2)\delta(E - E_1 - E_2)$$
$$= \frac{\Omega(d-1)}{4(2\pi)^{d-2}} \int_0^\infty dp \frac{p^{d-2}}{\sqrt{(p^2+m_1^2)(p^2+m_2^2)}} \delta(E - \sqrt{p^2+m_1^2} - \sqrt{p^2+m_2^2})$$
$$= \frac{1}{2^{d-1}\pi^{\frac{d-3}{2}}\Gamma(\frac{d-1}{2})} \frac{|\, p_{cm}\,|^{d-3}}{E_{cm}} \Theta(E - (m_1+m_2)), \tag{17.C.3}$$

where

$$\Theta(x) = \begin{cases} 1, & \text{if } x > 0 \\ 0, & \text{if } x < 0 \end{cases}$$

and $|\, p_{cm}\,|$ is the modulus of the space component of the momentum in the reference frame of the centre of mass, corresponding to the energy E

$$|\, p_{cm}\,| = \frac{1}{2E_{cm}} \sqrt{[E^2 - (m_1+m_2)^2][E^2 - (m_1 - m_2)^2]}. \tag{17.C.4}$$

To arrive to (17.C.3), we used the expression (2.6.3) of the solid angle in $(d-1)$–dimensions.

Recursive equation. The explicit computation of the phase space with a higher number of particle cannot be done exactly. However, its numerical determination can be reached by means of the recursive equation

$$\Phi_n(P^2) = \int \frac{d^{d-1}p_n}{(2\pi)^{d-1}2E_n} \Phi_{n-1}(P - p_n). \tag{17.C.5}$$

By iteration, this formula leads to integrals that involve the two-particle phase space, as shown in Figure 17.18. The proof of (17.C.5) is immediate. From its definition we have

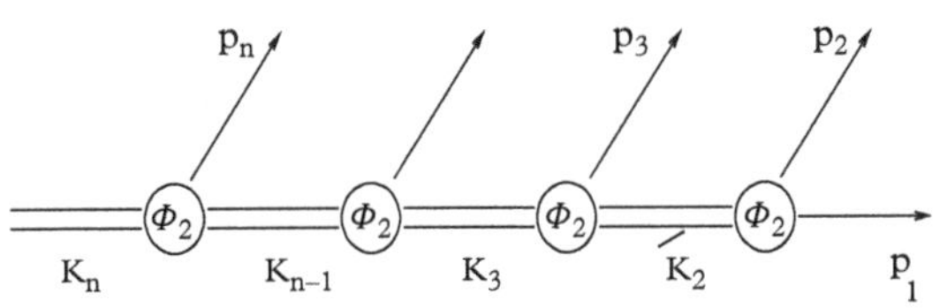

Fig. 17.18 *Iteration of the recursive equation for the n-particle phase space.*

$$\Phi_n(P^2) = (2\pi)^d \int \prod_{i=1}^{n} \frac{d^{d-1}p_i}{(2\pi)^{d-1}2E_i} \delta(P - p_1 - \cdots - p_n)$$
$$= \int \frac{d^{d-1}p_n}{(2\pi)^{d-1}2E_n} \left[(2\pi)^d \int \prod_{i=1}^{n-1} \frac{d^{d-1}p_i}{(2\pi)^{d-1}2E_i} \delta((P - p_n) - p_1 \cdots - p_{n-1}) \right]$$
$$= \int \frac{d^{d-1}p_n}{(2\pi)^{d-1}2E_n} \Phi_{n-1}(P - p_n). \tag{17.C.6}$$

$\Phi_n(P)$ is a function of $P^2 \equiv M_n^2$. Analogously $\Phi_{n-1}(P - p_n)$ is function of

$$(P - p_n)^2 = (p_1 + \cdots p_{n-1})^2 \equiv K_{n-1}^2 \equiv M_{n-1}^2,$$

where M_{n-1}^2 is the square of the invariant mass of the system of particles $1, 2, \ldots, (n-1)$. Since Φ_{n-1} is function only of this last variable, it is convenient to write eqn. (17.C.5) using the identity

$$1 = \int dM_{n-1}^2 \delta(M_{n-1}^2 - K_{n-1}^2),$$
$$1 = \int d^d K_{n-1} \delta^d(P - p_n - K_{n-1}).$$

Hence

$$\Phi_n(M_n^2) = \int dM_{n-1}^2 \delta(K_{n-1}^2 - M_{n-1}^2) \int d^d K_{n-1} \delta^d(P - p_n - K_{n-1})$$
$$\times \int \frac{d^d p_n}{(2\pi)^{d-1}} \delta(p_n^2 - m_n^2)\, \Phi_{n-1}(M_{n-1}^2) \tag{17.C.7}$$
$$= \frac{1}{2\pi} \int_{\mu_{n-1}^2}^{(M_n - m_n)^2} dM_{n-1}^2\, \Phi_2(M_n^2; K_{n-1}, p_n)\, \Phi_{n-1}(M_{n-1}^2),$$

where

$$\mu_i \equiv m_1 + m_2 + \cdots m_i.$$

$\Phi_2(M_n^2; K_{n-1}, p_n)$ is the two-particle phase space of total momentum $P^2 = M_n^2$, relative to the masses of the momenta K_{n-1} and p_n given by (17.C.3).

Laplace transform. It is useful to make use of the Laplace transform to solve the constraint on the momenta given by the delta function. Define

$$\Phi_n(\alpha) = \int d^d P e^{-\alpha \cdot P}\, \Phi_n(P^2), \tag{17.C.8}$$

where α is a Lorentz time-like vector $\alpha = (\alpha_0, \vec{\alpha})$, with $\alpha_\mu \alpha^\mu > 0$. Thanks to this transformation we have

$$\Phi_n(\alpha) = (2\pi)^d \int d^d P \prod_{i=1}^{n} \frac{d^{d-1} p_i}{(2\pi)^{d-1} 2E_i} \delta^d (P - p_1 - \cdots - p_n) e^{-\alpha \cdot P}$$
$$= (2\pi)^d \prod_{i=1}^{n} \int \frac{d^{d-1} p_i}{(2\pi)^{d-1} 2E_i} e^{-\alpha \cdot p_i} = (2\pi)^d \prod_{i=1}^{n} \phi_i(\alpha). \quad (17.C.9)$$

The functions $\phi_i(\alpha)$ can be easily computed choosing the reference frame where $\alpha = (\beta, 0)$ and computing the integral using the spherical coordinates. In fact we have

$$\phi_i(\alpha) = \int \frac{d^{d-1} p_i}{(2\pi)^{d-1} 2E_i} e^{-\alpha \cdot p_i} = \frac{\Omega(d-1)}{(2\pi)^{d-1}} \int_0^\infty dp \frac{p^{d-2}}{2E} e^{-\beta E}$$
$$= \frac{\Omega(d-1)}{(2\pi)^{d-1}} \int_{m_i^2}^\infty dE \, (E^2 - m_i^2)^{\frac{d-3}{2}} e^{-\beta E},$$

where $\Omega(d-1)$ is the solid angle in $(d-1)$ dimensions, given by eqn. (2.6.3). The last integral can be expressed in terms of the Bessel function $K_\nu(z)$, whose integral representation is

$$K_\nu(z) = \frac{\left(\frac{z}{2}\right)^\nu \Gamma\left(\frac{1}{2}\right)}{\Gamma\left(\nu + \frac{1}{2}\right)} \int_1^\infty e^{-zt} (t^2 - 1)^{\nu - \frac{1}{2}} dt. \quad (17.C.10)$$

So, we have

$$\phi_1(\beta) = \frac{2}{(2\pi)^{\frac{d}{2}}} \left(\frac{m}{\beta}\right)^{\frac{d-2}{2}} K_{\frac{d-2}{2}}(\beta m). \quad (17.C.11)$$

On the other side, in the reference frame where $\alpha = (\beta, 0)$, eqn. (17.C.8) can also be expressed as

$$\Phi_n(\beta) = \int d^d p \, e^{-\beta E} \, \Phi_n(p^2)$$
$$= \int ds \int d^d p \, \delta(p^2 - s) \, e^{-\beta E} \, \Phi_n(s)$$
$$= \int ds \int \frac{d^{d-1} p}{2E} e^{-\beta E} \, \Phi_n(s) \quad (17.C.12)$$
$$= \Omega(d-1) \int ds \int_{\sqrt{s}}^\infty dE \, (E^2 - s)^{\frac{d-3}{2}} e^{-\beta E} \, \Phi_n(s).$$

Using also in this case the integral representation (17.C.10) and the $(d-1)$-dimensional solid angle, the last expression can be written as

$$\Phi_n(\beta) = \frac{(2\pi)^{\frac{d}{2}}}{\pi} \frac{1}{\beta^{\frac{d-2}{2}}} \int_0^\infty ds\, s^{\frac{d-2}{4}} K_{\frac{d-2}{2}}(\beta\sqrt{s})\, \Phi_n(s). \tag{17.C.13}$$

Hence, it holds the identity

$$\frac{(2\pi)^{\frac{d}{2}}}{\pi} \frac{1}{\beta^{\frac{d-2}{2}}} \int_0^\infty ds\, s^{\frac{d-2}{4}} K_{\frac{d-2}{2}}(\beta\sqrt{s})\, \Phi_n(s) = (2\pi)^d \prod_{i=1}^n \frac{2}{(2\pi)^{\frac{d}{2}}} \left(\frac{m_i}{\beta}\right)^{\frac{d-2}{2}} K_{\frac{d-2}{2}}(\beta m_i). \tag{17.C.14}$$

Phase space at the threshold. Let us use eqn. (17.C.14) in the limit $\beta \to \infty$ to estimate the behaviour of $\Phi_n(s)$ near the threshold energy $\sqrt{s} = \sum_i^n m_i$. Using the asymptotic behaviour of the Bessel function

$$K_\nu(z) \simeq \left(\frac{\pi}{2z}\right)^{\frac{1}{2}} e^{-z} \qquad z \to \infty$$

substituting this expression in (17.C.14) and simplifying, we have

$$\frac{1}{(2\pi)^{\frac{d+1}{2}}} \frac{1}{\beta^{\frac{d-1}{2}}} \int_0^\infty ds\, s^{\frac{d-3}{4}} e^{-\beta\sqrt{s}}\, \Phi_n(s) = \prod_{i=1}^n \frac{1}{(2\pi)^{\frac{d-1}{2}}} \frac{m_i^{\frac{d-3}{2}}}{\beta^{\frac{d-2}{2}}} e^{-\beta m_i}. \tag{17.C.15}$$

With the change of variable $E = \sqrt{s}$, eqn. (17.C.15) becomes

$$\int_0^\infty dE\, E^{\frac{d-1}{2}} e^{-\beta E}\, \Phi_n(E) = A_n \beta^{\frac{1-n}{2}(d-1)} e^{-\beta \sum_i m_i},$$

with

$$A_n = \frac{1}{2} (2\pi)^{\frac{1}{2}[d+1-n(d-1)]} \left(\prod_{i=1}^n m_i\right)^{\frac{d-3}{2}}.$$

Using the general properties of the Laplace transform $\mathcal{L}$

$$\mathcal{L}[F(s-a)] = e^{-a\beta} \mathcal{F}(\beta), \quad \mathcal{L}[x^\nu] = \frac{\Gamma(\nu+1)}{\beta^{\nu+1}},$$

where $\mathcal{F}$ denotes the Laplace transform $\mathcal{L}$ of the function $F(s)$, it is easy to see that for $E \to \sum_i m_i$, the n-particle phase space goes to zero as

$$\Phi_n(E) \simeq \frac{A_n}{(\sum_i m_i)^{\frac{d+1}{2}}} \frac{1}{\Gamma\left(\frac{1}{2}(d-1)(n-1)\right)} (E - \sum_i m_i)^{\frac{(n-1)(d-1)-2}{2}}. \tag{17.C.16}$$

Phase space at high-energy. Let us use now the formula (17.C.14) to study the behaviour of the n-particle phase space for $m_i \to 0$, i.e. in the massless limit or equivalently at high energy. In this case it is necessary to distinguish two cases: (a) $d \neq 2$ and (b) $d = 2$. Let us consider the first case. Using the series expansion of the Bessel function $K_\nu(x)$ for $\nu \neq 0$

$$K_\nu(x) \simeq \frac{1}{x^\nu}, \quad x \to 0$$

the mass terms in the right-hand side of eqn. (17.C.14) simplify and we have

$$\frac{(2\pi)^{\frac{d}{2}}}{\pi} \frac{1}{\beta^{\frac{d-2}{2}}} \int_0^\infty ds\, s^{\frac{d-2}{4}} K_{\frac{d-2}{2}}(\beta\sqrt{s})\, \Phi_n(s) = (2\pi)^d \frac{2^n}{(2\pi)^{\frac{nd}{2}}} \frac{1}{\beta^{n(d-2)}}. \tag{17.C.17}$$

With the change of variable $E = \sqrt{s}$ in the integral on the left and collecting the terms we have

$$\int_0^\infty dE\, E^{\frac{d}{2}} K_{\frac{d-2}{2}}(\beta E)\, \Phi_n(E) = \pi \frac{2^{n-1}}{(2\pi)^{\frac{(n-1)d}{2}}} \left(\frac{1}{\beta}\right)^{(2n-1)(\frac{d-2}{2})}. \tag{17.C.18}$$

Since

$$\int_0^\infty x^\mu K_\nu(ax)\, dx = 2^{\mu-1} a^{-\mu-1} \Gamma\left(\frac{1+\mu+\nu}{2}\right) \Gamma\left(\frac{1+\mu-\nu}{2}\right),$$

the n-particle phase space behaves for $E \to \infty$ and $d \neq 2$ as

$$\Phi_n(E) \simeq B_n E^{n(d-2)-d}, \tag{17.C.19}$$

where

$$B_n = \pi \frac{2^{n(3-d)+\frac{d}{2}-1}}{(2\pi)^{(n-1)\frac{d}{2}}} \frac{1}{\Gamma\left(\frac{n(d-2)}{2}\right) \Gamma\left(\frac{(n-1)(d-2)}{2}\right)}.$$

Let us consider now the behaviour of the n-particle phase space for large values of the energy when $d = 2$. For dimensional reasons we expect that it scales as

$$\Phi_n(s) \simeq \frac{1}{s}, \quad s \to \infty$$

but there could be logarithmic corrections. On the basis of the cases $n = 2$ and $n = 3$, we pose the ansatz

$$\Phi_n(s) \simeq \alpha_n \frac{1}{s} \left(\ln \frac{s}{m^2} \right)^{n-2}, \tag{17.C.20}$$

where m is a mass scale whereas α_n is a constant to determine. The presence of the logarithms does not allow us to follow the previous computation, where we posed to zero all the masses. Consider now the recursive equation (17.C.7) in the limit $M_n^2 \to \infty$

$$\Phi_n(M_n^2) \simeq \frac{1}{2\pi} \int_\epsilon^{M_n^2} \Phi_2(M_n^2; K_{n-1}, p_n)\, \Phi_{n-1}(M_{n-1}^2),$$

where ϵ is a small but non-zero quantity. Substituting in this formula the expression of the two-particle phase space and the ansatz (17.C.20), we have

$$\begin{aligned} \Phi_n(M_n^2) &\simeq \frac{1}{2\pi} \alpha_{n-1} \int_\epsilon^{M_n^2} \frac{1}{(M_n^2 - M_{n-1}^2)} \frac{1}{M_{n-1}^2} \left(\ln \frac{M_{n-1}^2}{m^2} \right)^{n-3} \\ &= \frac{1}{2\pi M_n^2} \alpha_{n-1} \int_\epsilon^{M_n^2} dM_{n-1}^2 \left[\frac{1}{M_{n-1}^2} - \frac{1}{M_n^2 - M_{n-1}^2} \right] \left(\ln \frac{M_{n-1}^2}{m^2} \right)^{n-3}. \end{aligned}$$

The first term of this equation is responsible for the most singular part and keeping only this, we have

$$\Phi_n(M_n^2) \simeq \frac{\alpha_{n-1}}{2\pi} \frac{1}{n-2} \frac{1}{M_n^2} \left(\ln \frac{M_n^2}{m^2} \right)^{n-2}. \tag{17.C.21}$$

Comparing this expression with the ansatz (17.C.20), we obtain the recursive equation for the constants α_n

$$\alpha_n = \frac{1}{2\pi(n-2)} \alpha_{n-1},$$

whose solution is

$$\alpha_n = \frac{1}{(2\pi)^{n-2}(n-2)!} \lambda,$$

where λ is a constant that has to be fixed by comparing with the actual expression of Φ_3, for instance. It turns out that $\lambda = 3$. Hence, in $d = 2$, the asymptotic expression of the $\ln$–particle phase space for $s \to \infty$ is

$$\Phi_n(s) \simeq \frac{3}{(2\pi)^{n-2}(n-2)!}\,\frac{1}{s}\left(\ln\frac{s}{m^2}\right)^{n-2}. \tag{17.C.22}$$

REFERENCES

For a discussion of the general principles of the S-matrix theory, we recommend

Eden, R.J., Landshoff, P.V., Olive, D.I. and Polkinghorne, J.C. (1966). *The Analytic S-Matrix*, Cambridge University Press.

Landau, L.D. Lifsits, E.M. (1974). *Relativistic Quantum Mechanics*, Pergamon Press.

Landau, L.D. Lifsits, E.M. (1977). *Quantum Mechanics*, 3rd edn, Pergamon Press.

Lipkin, H. (1978). *Quantum Mechanics. New Approaches to Selected Topics*, North Holland.

Taylor, J.R. (1980). *Scattering Theory*, J. Wiley & Sons.

The famous article quoted in the text about the triviality of the S-matrix in $d > 2$ in the presence of conserved charges is

Coleman, S. and Mandula, J. (1967). 'All Possible Symmetries of the S-matrix', *Physics Review*, 159: 1251.

Fundamental papers on the development of the S-matrix theory in $(1+1)$ dimensional systems are

Zamolodchikov, A. (1989). 'Integrable Field Theory from Conformal Field Theory', *Advanced Studies in Pure Mathematics*, 19: 641.

Zamolodchikov, A.B. and Zamolodchikov, Al.B. (1979). 'Factorized S-matrices in Two Dimensions as the Exact Solutions of Certain Relativistic Quantum Field Theories', *Annals of Physics*, 120: 253.

For non-local conserved currents in two-dimensional QFTs see

D. Bernard and A. Leclair, Quantum group symmetries and non-local currents in 2-D QFT, Comm. Math. Phys. 142 (1991), 99.

For further detail on the historical developments of the S-matrix theory and the crossing paths with QFT, we suggest

Weinberg, S. (1986). 'Particle Physics: Past and Future', *International Journal of Modern Physics A*, 1: 135.

White, A.R. (2001). 'The Past and Future of S-Matrix Theory'. In: E.R. Pike and P. Sabatier (Eds), *Scattering: Scattering and Inverse Scattering in Pure and Applied Science*, Academic Press.

For an interpretation of the CDD factors in the scattering amplitude see

Mussardo G. and Simon, P. (2000). 'Bosonic-type S-matrix, Vacuum Instability and CDD Ambiguities', *Nuclear Physics B*, 578: 527.

Smirnov F.A. and Zamolodchikov, A.B. (2017). 'On Space of Integrable Theories', *Nuclear Physics B*, 915: 363.

A classic and popular book on the bootstrap interactions is

Capra, F. *The Tau of Physics*, Bantam Books, Toronto, 1984.

An excellent book on string theory is

Green, B. *The Elegant Universe*, Norton, New York.

Professional books about string theory for interested readers include

Green, M., Schwarz, J.H. and Witten, E. (2000). *String Theory*, Cambridge University Press.

Polchinski, J. (2002). *String Theory*, Cambridge University Press.

PROBLEMS

17.1. Causality and analiticity

Consider a linear system in which the output $b(t)$ depends on the input $a(t)$ as

$$b(t) = \int_{-\infty}^{t} G(t-t')\, a(t')\, dt'.$$

If the system is causal, the Green function $G(t-t')$ vanishes when $t < t'$. Let

$$\hat{G}(\omega) = \int_{-\infty}^{\infty} e^{i\omega\tau}\, G(\tau)\, d\tau = \int_{0}^{\infty} e^{i\omega\tau}\, G(\tau)\, d\tau$$

its Fourier transform. If $a(t)$ and $b(t)$ are both real, also $G(\tau)$ is a real function and

$$\hat{G}^*(\omega) = \hat{G}(-\omega^*).$$

a. Show that, if $G(\tau)$ is a square integrable function, then $\hat{G}(\omega)$ is an analytic function in the upper half-plane $\text{Im}\,\omega > 0$. This implies that $\hat{G}(\omega)$, for real ω, is a function obtained as boundary value of an analytic function.
b. Posing $\hat{G}(\omega) = \hat{G}_1(\omega) + i\hat{G}(\omega)$, use the Cauchy theorem to prove that these functions are related one to the other by the dispersion relations

$$\hat{G}_1(\omega) = \frac{1}{\pi} \mathcal{P} \int_{-\infty}^{+\infty} \frac{1}{\nu - \omega}\, \hat{G}_2(\nu)\, d\nu$$

$$\hat{G}_2(\omega) = -\frac{1}{\pi}\mathcal{P}\int_{-\infty}^{+\infty} \frac{1}{\nu - \omega}\,\hat{G}_1(\nu)\,d\nu$$

where $\mathcal{P}$ denotes the principal part of the integral.

17.2. Decay process

A particle of mass M and three-dimensional momentum P decays in two particles of masses m_1 and m_2.

a. Use the conservation of the energy and the momentum to prove that the total energy of the first particle in the reference frame of the centre of mass is

$$E_1 = \frac{M^2 + m_1^2 - m_2^2}{2M}$$

and that E_2 is obtained from the previous expression exchanging m_1 with m_2.

b. Show that the kinetic energy T_i of the particle i, in the same reference frame, is given by

$$T_i = \Delta M\left(1 - \frac{m_i}{M} - \frac{\Delta M}{2M}\right)$$

where $\Delta M = M - m_1 - m_2$.

17.3. Physical region of the amplitudes

Determine the physical region of the s-channel process when the mass of the particles are different.

17.4. Yang–Baxter equations

Prove that the Yang–Baxter equations given in eqn. (17.3.52) of the text can be obtained as a consequence of the associativity condition of the Faddev–Zamolodchikov algebra.

17.5. Reflection amplitude

Consider the following scattering amplitudes of a particle A and its anti-particle $\overline{A}$

$$\begin{aligned}
|A(\theta_1)A(\theta_2)\rangle &= S(\theta)\;|A(\theta_2)A(\theta_1)\rangle, \\
|A(\theta_1)\overline{A}(\theta_2)\rangle &= t(\theta)\;|\overline{A}(\theta_2)A(\theta_1)\rangle + r(\theta)\;|A(\theta_2)\overline{A}(\theta_1)\rangle.
\end{aligned}$$

a. Prove that it holds

$$\begin{aligned}
&S(\theta)\,S(-\theta) = t(\theta)\,t(-\theta) + r(\theta)\,r(-\theta) = 1 \\
&t(\theta)\,r(-\theta) + r(\theta)\,t(-\theta) = 0 \\
&t(\theta) = S(i\pi - \theta), \quad r(\theta) = r(i\pi - \theta).
\end{aligned}$$

b. Prove that if the particles A and $\overline{A}$ are uniquely distinguishable by their different eigenvalues of the conserved charges, then the reflection amplitude vanishes, i.e. $r(\theta) = 0$.

17.6. Bootstrap equations

Derive the bootstrap equations (17.4.68) imposing the commutativity of the processes shown in Figure (17.13).
Hint. Note that the line of the particle A_i in the second graph is parallel to the same line of the first graph. Identify the angles in the two figures and use the resonance condition.

17.7. Scattering in a potential with two delta functions

Consider a one-dimensional system of quantum mechanics with Hamiltonian given by

$$\mathcal{H} = \frac{p^2}{2m} + V(x)$$

with

$$V(x) = -g_1\,\delta(x+a) - g_2\,\delta(x+a)$$

(g_1 and g_2 positive).

a. Compute the phase shifts δ_0 and δ_1 and the corresponding S-matrix elements.
b. Analyse the analytic structure of the S-matrix by varying the momentum k.
c. Determine the wave function of the bound states.

17.8. Interpretation of the two-dimensional S-matrix.

The non-relativistic S-matrix of a particle of mass $m = 1$ relative to the potential $V(x) = -2a\pi\,\delta(x)$ is given by

$$\tilde{S}(k) = \frac{k + i\pi\,a}{k - i\pi\,a}.$$

If we would like to generalize this result to the relativistic case, we must use the rapidity variable θ. Note that for small values of the momentum, $\theta \simeq k$. Substituting in the expression of S, we have

$$\tilde{S}(\theta) = \frac{\theta + i\pi\,a}{\theta - i\pi\,a}.$$

This expression, however, does not fulfill the important property $S(\theta) = S(\theta \pm 2\pi i)$ of the relativistic S-matrix.

a. Discuss how it can be iteratively implemented the periodicity of the relativistic S-matrix starting from $\tilde{S}(\theta)$.

b. Use the infinite product representation of the hyperbolic function $\sinh x$

$$\sinh x = x \prod_{k=1}^{\infty} \left[1 + \left(\frac{x}{k\pi} \right)^2 \right],$$

to show that the final result can be expressed as

$$S(\theta) = \frac{\sinh \frac{1}{2}(\theta + i\pi\, a)}{\sinh \frac{1}{2}(\theta - i\pi\, a)} = s_a(\theta).$$

17.9. S-matrix with resonances.

Consider an S-matrix for a neutral scalar particle.

a. Show that the unitarity and crossing invariance equations

$$S(\theta)S(-\theta) = 1, \quad S(\theta) = S(i\pi - \theta),$$

imply that that $S(\theta)$ is a periodic function along the imaginary axis of the rapidity variable, i.e.

$$S(\theta) = S(\theta + 2\pi i).$$

b. Assume that $S(\theta)$ also presents a periodic behaviour along the real axis of the rapidity variable θ with a period T

$$S(\theta) = S(\theta + T).$$

Show that in the domain $0 \leq \mathrm{Re}\theta \leq T$ and $-i\pi \leq \mathrm{Im}\theta \leq i\pi$ the S-matrix must have necessarily at least two zeros and two poles unless it is a constant.

c. Argue that the presence of poles with a real part in the physical strip $0 \leq \mathrm{Im}\theta \leq i\pi$ would spoil the causality properties of the scattering theory.

d. Let

$$\theta_{n,m} = i\pi\, a + 2m\pi\, i + nT$$

an infinite sequence of zeros in the physical strip, where $0 < a < 1$ is a parameter linked to the coupling constant of the model. Argue that these zeros correspond to resonances and compute their mass and width.

18

Exact S-Matrices

The particles are nothing else than lumps of energy, they come and go, their own identity is all in this dance of creation and annihilation processes.

Kenneth Ford

This chapter presents the exact S-matrix associated to several two-dimensional statistical models away from their critical point. Closing the bootstrap procedure, we are able to find at the same time the set of all amplitudes and the mass spectrum of the theory. A crucial step in the determination of the scattering amplitudes is the knowledge of the spectrum of the spins relative to the conserved currents. In the first sections we address the minimal S-matrix of several off-critical statistical systems. Many of these examples are related to the Toda field theories previously discussed. Later we use the minimal S-matrices of the statistical models to determine the exact S-matrices of the Lagrangian Toda field theories. The scattering amplitudes of the Toda field theories shows an important symmetry of these models under the weak-strong duality transformation $g \to 8\pi/g$ of their coupling constant. At the end of the chapter we discuss the exact S-matrix of the Sine–Gordon model, with a series of comments on its interesting features, and the quantum group reductions that lead to the general S-matrices of integrable deformations of CFT.

18.1 Yang–Lee and Bullogh–Dodd Models

The CFT associated to the Yang–Lee edge singularity is non-unitary, with central charge $c = -22/5$ and only one relevant field ϕ with conformal dimension $\Delta = -1/5$. As discussed in Section 14.5, this theory describes the critical behaviour of a Ising model in a purely imaginary magnetic field ih. The Landau–Ginzburg Lagrangian is given by

$$\mathcal{L} = \int [\frac{1}{2}(\partial\phi)^2 - i(h - h_c)\phi - ig\phi^3]d^2x \tag{18.1.1}$$

and the scaling region nearby the critical point is a one-dimensional space, spanned by the (purely imaginary) coupling constant of the relevant field ϕ. We can use the characters of the identity family and of the field ϕ to count the dimensions of the quasi-primary

Statistical Field Theory: An Introduction to Exactly Solved Models in Statistical Physics. Second Edition. Giuseppe Mussardo, Oxford University Press (2020). © Giuseppe Mussardo. DOI: 10.1093/oso/9780198788102.001.0001

n	1	2	3	4	**5**	6	7	8	9	10	**11**	12	**13**	14	15	16	**17**
$\hat{\Lambda}_{n+1}$	1	0	0	0	1	0	1	0	1	0	2	0	2	1	2	1	3
$\hat{\phi}_n$	0	0	0	1	0	1	0	1	1	1	1	2	1	2	2	3	2

Table 18.1 *Dimensions of the spaces $\hat{\Lambda}_{n+1}$ and $\hat{\phi}_n$. For each value of n for which the former is larger than the latter there must exist a conserved current.*

fields at level n, as shown for the first representatives in Table 18.1. We can then apply the counting method (see Section 16.8.3) to establish that the off-critical system has conserved charges with spin

$$s = 1, 5, 7, 11, 13, 17, 19, 23. \tag{18.1.2}$$

The sequence of these spins is made of odd numbers not divisible by 3 and is therefore compatible with the existence of a massive excitation associated to a particle A that is the bound state of itself. Hence, its exact S-matrix must have a pole at $\theta = 2i\pi/3$. The crossing symmetry helps in fixing the position of the pole in the t-channel at $\theta = i\pi/3$. Assuming that there are no additional poles, the only solution of the bootstrap equation

$$S_{AA}(\theta) = S_{AA}\left(\theta - \frac{i\pi}{3}\right) S_{AA}\left(\theta + \frac{i\pi}{3}\right) \tag{18.1.3}$$

is given by

$$S_{AA} = \frac{\tanh\frac{1}{2}(\theta + i\frac{2\pi}{3})}{\tanh\frac{1}{2}(\theta - i\frac{2\pi}{3})} = f_{\frac{2}{3}}. \tag{18.1.4}$$

One can extract the value of the *on-shell* renormalized coupling constant[1] comparing with the Feynman diagrams coming from the Lagrangian (18.1.1), as shown in Figure 18.1

$$-ig^2 = 3m^4 \sinh(2i\pi/3) = i\frac{3\sqrt{3}}{2}m^4 \tag{18.1.5}$$

Unitarity paradox and its solution. Notice that the residue has an *opposite sign* with respect to what expected in a unitary theory. On the other hand, the S-matrix S (18.1.4) satisfies by construction the unitarity equation $S(\theta)S(-\theta) = 1$. Hence, it seems we are in presence of an apparent contradiction whose solution and, consequently, the compatibility of the two definitions which express the unitarity condition is the following.

[1] This is defined as i times the residue at the pole of the S-matrix. The factor m^4 is introduced for dimensional reasons.

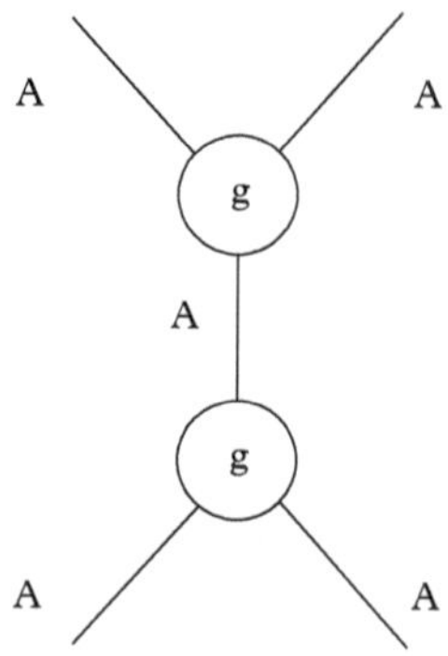

Fig. 18.1 *Residue at the pole expressed in terms of the* on-shell *coupling constant.*

For this model, it is possible to define a charge conjugate operator C ($C^2 = 1$) through the position

$$C\phi C = -\phi.$$

The Hamiltonian associated to the Lagrangian (18.1.1) is not Hermitian but satisfies $H^\dagger = CHC$. The multi-particle states of the Fock space are created by the iterate action of the field ϕ on the vacuum state. They are eigenvectors of C with eigenvalues $(-1)^N$, where N is the number of particles. Since H is not a Hermitian operator, its left eigenvectors $< n_l \mid$ do not coincide with the adjoint right eigenvectors, but are related to them by the relation $< n_l \mid = < n_r \mid C$. The completeness relation of this theory is then

$$\sum_n \mid n_r\rangle\langle n_l \mid = \sum_n \mid n_r\rangle\langle n_r \mid C.$$

The unitarity condition of the S-matrix

$$SS^\dagger = 1, \tag{18.1.6}$$

simply expresses that the initial and final states form a basis of the Hilbert space and it is not sensitive whether the Hamiltonian is Hermitian or not. However, if we insert the completeness relation in (18.1.6), each of the intermediate states is weighted by $(-1)^N$. This is the reason of the negative sign of the residue, for it comes from the one-particle intermediate state. In conclusion, the S-matrix is unitary since it conserves the probability but it has a negative sign of the residue for the negative eigenvalue of C on the one-particle state. For its simplicity, the Yang–Lee model has proved to be the ideal theoretical playground for the analysis of the integrable deformations of the conformal models. A successful check of this S-matrix can be performed by the thermodynamical Bethe ansatz, as discussed in the next chapter.

Bullogh–Dodd model. The S-matrix of the Yang–Lee model is the so-called *minimal part* of the S-matrix of the Bullogh–Dodd model, defined by the Lagrangian

$$\mathcal{L} = \frac{1}{2}(\partial_\mu \phi)^2 - \frac{\mu^2}{6\lambda^2}\left(2e^{\lambda\phi} + e^{-2\lambda\phi}\right). \tag{18.1.7}$$

To determine the S of this theory, notice that both models share the same spectrum of the spins of the conserved charges and have only one particle exitation. The S-matrix of the Lagrangian model can be then obtained by multiplying the minimal S-matrix of the Yang–Lee model with some additional terms, called the *Z factors*, satisfying the following requirements: (i) they must be solutions of the bootstrap equation; (ii) they must not introduce additional poles and, finally, (iii) they must depend on the coupling constant.

As discussed in Problem 18.1, another solution of the bootstrap equation (18.1.3) is given by

$$S(\theta) = f_{\frac{2}{3}}(\theta) f_{-\frac{B}{3}}(\theta) f_{\frac{B-2}{3}}(\theta), \tag{18.1.8}$$

and the quantity B can be determined by comparing the perturbative expansion of the S-matrix with the Feynman diagrams coming from the Bullogh–Dodd Lagrangian. Notice that the two additional Z-factors introduce a set of zeros in the physical sheet of the scattering amplitude and no additional poles. From the perturbative comparison at the lowest orders, one can conjecture that the exact result is expressed by the relation

$$B(\lambda) = \frac{\lambda^2}{2\pi} \frac{1}{1 + \frac{\lambda^2}{4\pi}}. \tag{18.1.9}$$

Note that, assuming the validity of this formula, the exact S-matrix of the Bullogh–Dodd model is invariant under the transformation $B(\lambda) \to 2 - B(\lambda)$, namely, under the weak-strong duality transformation of the coupling constant

$$\lambda \to \frac{4\pi}{\lambda}. \tag{18.1.10}$$

For all values of λ, except $\lambda = 0, \infty$ and the self-dual point $\lambda = \sqrt{4\pi}$, the S-matrix presents a simple pole at $\theta = 2\pi i/3$, that corresponds to the bound state given by the particle itself. The residue allows us to find the on-shell three-particle vertex of this theory

$$\Gamma^2(B) = 2\sqrt{3} \frac{\tan\left(\frac{\pi B}{6}\right)}{\tan\left(\frac{\pi B}{6} - \frac{2\pi}{3}\right)} \frac{\tan\left(\frac{\pi}{3} - \frac{\pi B}{6}\right)}{\tan\left(\frac{\pi B}{6} + \frac{\pi}{3}\right)}. \tag{18.1.11}$$

Notice that $\Gamma(B)$ vanishes at $B=0$ and $B=2$ (both points correspond to the free lagrangian model) but it also vanishes at the self-dual point $B=1$, where the S-matrix becomes

$$S(\theta,1) = f_{-\frac{2}{3}}(\theta). \tag{18.1.12}$$

The vanishing of Γ at the self-dual point is essentially due to the non-simply laced nature of this Toda field theory.[2]

18.2 $\Phi_{1,3}$ Integrable Deformation of the Conformal Minimal Models $\mathcal{M}_{2,2n+3}$

The Yang–Lee model belongs to the series of non-unitary minimal models $\mathcal{M}_{2,2n+3}$, whose Kac table consists of only one column. In these theories, in addition to the identity operator, there are n conformal field with negative conformal dimensions

$$\Delta_{1,r} = \Delta_{1,2n+3-r} = -\frac{(r-1)(2n+2-r)}{2(2n+3)}, \qquad r = 0,1,\ldots n. \tag{18.2.1}$$

The central charge c and the effective central charge c_{eff} are given by

$$c = -\frac{2n(6n+5)}{2n+3}, \qquad c_{eff} = \frac{2n}{2n+3}.$$

The scattering theory defined by the $\Phi_{1,3}$ integrable deformation is supported by the spectrum of the spins of the conserved charges given by

$$s = 1,3,\ldots,2n-1,2n+3,\ldots 4n+1 \quad (\text{mod } 4n+2).$$

This spectrum is compatible with a set of n massive particles with the bootstrap fusions

$$\begin{aligned} A_1 \times A_1 &\to A_2 \\ A_2 \times A_2 &\to A_3 \\ &\cdots \\ A_n \times A_n &\to A_1. \end{aligned} \tag{18.2.2}$$

Using the results of Chapter 17, a solution of the consistency equations for the resonance angles is

$$\overline{u}_{k,k+1}^{k} = \frac{k\pi}{2n+1}, \qquad k = 1,2,\ldots n.$$

[2] This does not happen for all the other Toda field theories based on simply laced algebras.

and consequently, the exact mass spectrum is

$$m_a = \sin\left(\frac{a\pi}{2n+1}\right), \qquad a = 1, 2, \ldots n. \tag{18.2.3}$$

The minimal scattering amplitude of the lowest mass particle A_1 is

$$S_{11}(\theta) = f_{\frac{2}{2n+1}}(\theta), \tag{18.2.4}$$

whereas all other amplitudes can be obtained by applying recursively the bootstrap equations

$$S_{ab}(\theta) = f_{\frac{|a-b|}{2n+1}} f_{\frac{a+b}{2n+1}} \prod_{k=1}^{min(a,b)-1} \left(f_{\frac{|a-b|+2k}{2n+1}}\right)^2. \tag{18.2.5}$$

The simple pole of the first term (for $a \neq b$)

$$\theta = i\, u_{ab}^{|a-b|} = i\left(1 - \frac{|\, a-b\,|}{2n+1}\right)\pi \tag{18.2.6}$$

corresponds to the particle $A_{|a-b|}$ that appears as bound state in this scattering process. The simple pole of the second factor

$$\theta = i\, u_{ab}^{n(a,b)} = i\,(a+b)\frac{\pi}{2n+1} \tag{18.2.7}$$

is due to the particle of type $n(a,b) = \min\,(a+b, 2n+1-a-b)$. The double poles of the remaining functions derive from the bootstrap procedure and are associated to the multiple intermediate scattering processes, as those shown in Figure 18.2. In these processes, two initial particles A and B 'break' in the intermediate particles a, b and c, with the relative angles dictated by the scattering theory

$$\begin{aligned} \varphi &= u_{Ac}^{a} + u_{Bc}^{b} - \pi \\ \eta &= \pi - u_{ac}^{a} - u_{bc}^{B}. \end{aligned} \tag{18.2.8}$$

In order to actually draw these graphs and to have correspondingly a double pole, it is necessary that the resonance angles satisfy the geometrical condition

$$u_{ac}^{A} + u_{bc}^{B} \leq \pi. \tag{18.2.9}$$

This condition puts a dynamical constraint on the set of the resonance angles for having a double pole in the scattering amplitudes.

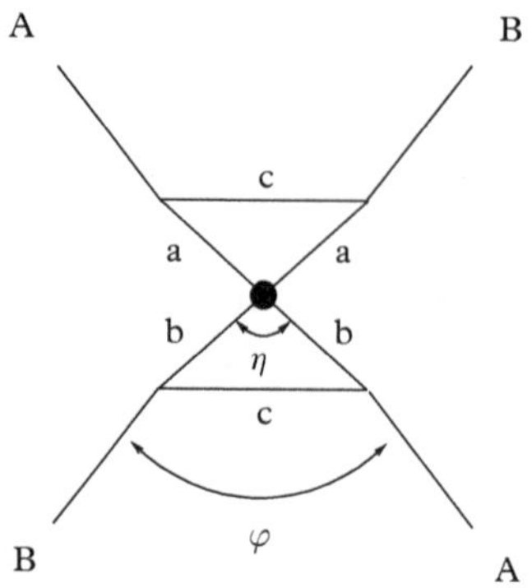

Fig. 18.2 *Multi-scattering process that gives rise to a double pole in the S-matrix* $S_{AB}(\theta)$ *at* $\theta = i(u_{Ac}^a + u_{Bc}^b - \pi)$.

As we show later, these S-matrices can be obtained as RSOS reduction of the Sine–Gordon model, when the solitons disappear from the spectrum and only breathers remain. The amplitudes above are also the minimal S-matrices of the Toda field theories based on A_{2n}^2. In order to obtain the full S-matrix of these Lagrangian models it is necessary to multiply the minimal S-matrix for the Z-factors that do not contain additional poles, solution of the bootstrap equations and functions of the coupling constant. For these theories this scheme can be implemented starting from the scattering amplitude involving the particle with the lowest mass

$$S_{11}(\theta) = f_{\frac{2}{2n+1}}(\theta) f_{-B}(\theta) f_{-\frac{2}{2n+1}+B}(\theta), \tag{18.2.10}$$

where the function $B(\lambda)$ is given by

$$B(\lambda) = \frac{1}{4\pi(2n+1)} \frac{\lambda^2}{1+\frac{\lambda^2}{4\pi}}. \tag{18.2.11}$$

All other scattering amplitudes of the Toda field theories are obtained applying the bootstrap equations.

18.3 Multiple Poles

The S-matrix discussed in Section 18.2 shows the presence of double poles. In the S-matrices that we meet in the next sections there are also higher order poles. This analytic structure is a necessary consequences of the bootstrap equations. However, a consistent interpretation of the scattering theory requires an explanation of these higher order poles in terms of the elementary processes that take place in the system. Let us then briefly discuss the origin of these singularities in order to better understand the scattering theory of the two-dimensional systems.

The simple poles of a S-matrix are associated at the bound states. This identification holds in any dimension and it is one of the key point of the analytic theory of the S-matrix. The higher-order poles, on the other hand, only occur in the two-dimensional S-matrices. In the four-dimensional theories, for instance, the same diagrams that produce the multiple poles of the two-dimensional theories give rise instead to branch cut singularities in the Mandelstam variable s. It is only for the dimensionality of the space-time that these singularities become double-, triple- and higher-order poles instead of branch cuts. In this respect, it is important to notice that two-dimensional scattering processes have the peculiar feature to be in one-to-one correspondence with the geometrical figures that we can draw on a page, i.e. the angles between the world-lines of the particles A_i, A_j and A_k are precisely those associated to the resonance angles u_{ij}^k.

Assuming that the scattering theory is in correspondence with a set of Feynman rules (that for simplicity we assume to be of the $g\phi^3$ Landau–Ginzburg theory), it is possible to prove that there is a very simple rule to determine the order of the pole: a S-matrix has a higher pole of order p

$$S_{ab}(\theta) \simeq \frac{g^{2p} R_p}{(\theta - \theta_0)^p}, \tag{18.3.1}$$

if we can actually draw the Feynman diagrams associated to this scattering process, starting from the resonance angles u_{ij}^k, in which there are P internal propagators and L loops, with the condition

$$p = P - 2L. \tag{18.3.2}$$

Applying this rule for instance to the diagram in Figure 18.2, we see that there are six internal propagators and two loops, and therefore this diagram corresponds to a double order pole. However, as discussed, it should be actually possible to draw such a diagram, i.e. the resonance angles u_{ij}^k should be those that permit the existence of such a geometrical figure. Analogously, if in the scattering theory there are resonance angles that allows us to draw a diagram (Figure18.3 (a)), then there is a third order of poles in the amplitude, whereas the possibility to draw a diagram as in Figure 18.3(b) provides the explanation of a fourth-order pole in the scattering process of the asymptotic particles A and B.

Another general rule concerning the higher-order poles is that those of odd orders can be generally associated to bound states (Figure 18.3(a)) in which there is in the middle the propagator of a one-particle state, whereas those with even order generally describe multi-scattering processes which do not lead to bound state creation, as is the case in Figure 18.3 (b).

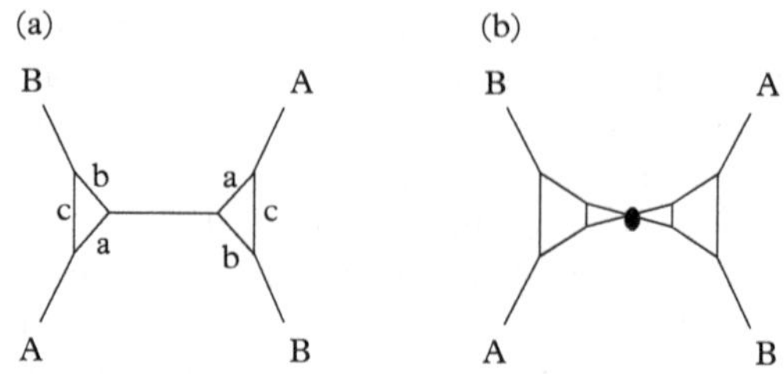

Fig. 18.3 *a: Multiple scattering process responsible for a third-order pole in $S_{AB}(\theta)$. b: multiple scattering process that gives rise to a fourth-order pole in S_{AB}.*

18.4 S-Matrices of the Ising Model

The Ising model has two integrable deformation. The first is the thermal deformation, that moves the system away from its critical temperature at zero magnetic field. The second is the magnetic deformation, obtained by coupling the system to an external magnetic field but keeping the temperature of the system at its critical value. The S-matrices of the two deformations have a completely different structure: the first is the simplest possible S-matrix, while the second is the richest one! Moreover, the first is the minimal S-matrix of the Lagrangian Sinh–Gordon model, whereas the second is the minimal S-matrix of the Toda field theory based on the exceptional algebra E_8. Here we discuss each of them in more detail.

18.4.1 Thermal deformation of the Ising Model

At zero magnetic field, the Ising model away from the critical temperature is a theory of free Majorana fermions, with a Lagrangian given by

$$\mathcal{L} = \psi \frac{\partial}{\partial \bar{z}} \psi + \bar{\psi} \frac{\partial}{\partial z} \bar{\psi} + i m \bar{\psi}\, \psi. \tag{18.4.1}$$

The mass parameter measures the displacement of the temperature from the critical value

$$m = T - T_c.$$

The low-temperature phase is related to the high-temperature phase by duality. In low temperature there are two degenerate vacua: the Z_2 symmetry of the model is spontaneously broken and the massive excitations consists of the kink and the anti-kink that interpolate between the two vacua. These are neutral particles, here denoted by $A(\theta)$, associated to the fermionic field. Since the S-matrix can be regarded as the operator that implements the commutation relation between the operators that create the particles, for the fermionic nature of the problem we have

$$A(\theta_1)A(\theta_2) = -A(\theta_2)A(\theta_1),$$

namely

$$S = -1. \tag{18.4.2}$$

In the high-temperature phase the system has a unique vacuum. The massive excitation $A(\theta)$ of this phase is odd under the Z_2 spin symmetry and can be regarded as a bosonic particle created by the operator $\sigma(x)$, since we have

$$\langle 0|\sigma(0)|A(\theta)\rangle \neq 0.$$

For the self-duality of the model, the S-matrix is, as before, $S = -1$. Notice, however, that in this phase the particle A appears as an interacting particle, otherwise its S-matrix should be the one of a free theory given, for a bosonic theiry, by $S = 1$. In both phases the model does not have additional bound states.

As in the previous section, the S-matrix of the thermal deformation of the Ising model can be regarded as the minimal S-matrix of a Lagrangian system, in this case the Sinh–Godon model with Lagrangian

$$\mathcal{L} = \frac{1}{2}(\partial\phi)^2 - \frac{m^2}{g^2}(\cosh g\phi - 1). \tag{18.4.3}$$

To determine the exact S-matrix of this integrable model, we have to identify the appropriate Z-factor as a function of the coupling constant. The simplest choice leads to the following expression for the exact S of the Sinh–Gordon model

$$S(\beta) = f_{-B}(\theta), \tag{18.4.4}$$

where $B(g)$ is a function of the coupling constant that can be determined comparing the perturbative expansion of (18.4.4) with the Feynman diagrams coming from the Lagrangian (18.4.3). The final result is

$$B(g) = \frac{g^2}{8\pi}\frac{1}{1+\frac{g^2}{8\pi}}. \tag{18.4.5}$$

Section 18.10 shows that this expression can be obtained as the analytic continuation of an analogous formula established for the Sine–Gordon model. Notice the invariance of the S-matrix of the Sinh–Gordon model under the weak-strong duality

$$g \to \frac{8\pi}{g}. \tag{18.4.6}$$

This symmetry is not evident in the Lagrangian of the model and it is only shown up in its exact S-matrix. Presently it is still an open problem to find the proper Lagrangian formulation (if any) that explicitly shows this dynamical invariance of the Sinh–Gordon model.

Coming back to the Ising model, it is interesting to observe that its S-matrix can be obtained as a limiting case of the exact S-matrix of a generic Z_n model, when $T > T_c$, as obtained by Koberle and Swieca. In this theory there $n-1$ particles, with mass spectrum given by

$$m_a = \sin\left(\frac{\pi a}{n}\right), \quad a = 1, 2, \ldots n-1. \tag{18.4.7}$$

The S-matrix of the fundamental particle is

$$S_{11} = \frac{\tanh\frac{1}{2}(\theta + i\frac{2\pi}{n})}{\tanh\frac{1}{2}(\theta - i\frac{2\pi}{3})} = f_{\frac{2}{n}}, \tag{18.4.8}$$

and, substituting $n = 2$ in this formula, we get the S-matrix (18.4.2).

18.4.2 Magnetic Deformation of the Ising Model

The counting argument shows that the magnetic deformation of the Ising model has a spectrum of the first spins of the conserved charges given by

$$s = 1, 7, 11, 13, 17, 19. \tag{18.4.9}$$

Notice the absence of spins that have 3 or 5 as divisors. The lack of multiples of 3 can be easily explained by postulating the existence of a fundamental particle A_1 (with mass m_1) that possesses the 'Φ^3' property, i.e. to be a bound state of itself. In the S-matrix of this particle there should be then a pole at $\theta = u_{11}^1 = \frac{2\pi i}{3}$. This feature is compatible with the explicit breaking of the Z_2 of the model. Concerning the absence of spin s divisible by 5, it can be explained by conjecturing the existence of a second particle A_2 (with mass m_2) that, together with A_1, gives rise to a sub-system of bootstrap fusions

$$\begin{aligned} A_1 \times A_1 &\to A_1 + A_2 \\ A_2 \times A_2 &\to A_1. \end{aligned} \tag{18.4.10}$$

Let u_{11}^2 be the resonance angle corresponding to the bound state A_2 that appears in the amplitude $S_{11}(\theta)$, while let u_{22}^1 be the resonance angle associate to A_1 in the amplitude $S_{22}(\theta)$. Using the variables

$$y_1 = \exp\left(\frac{i}{2}u_{11}^2\right), \quad y_2 = \exp\left(\frac{i}{2}u_{22}^1\right),$$

the consistency equations involving the spins of conserved charges and the resonance angles become

$$\begin{aligned} y_1^s + y_1^{-s} &= \left(\frac{m_2}{m_1}\right)^s \frac{\chi_s^2}{\chi_s^1}, \\ y_2^s + y_2^{-s} &= \left(\frac{m_1}{m_2}\right)^s \frac{\chi_s^1}{\chi_s^2}. \end{aligned} \tag{18.4.11}$$

For the values of s given in (18.4.9), a non-trival solution is given by

$$y_1 = \exp\left(\frac{i\pi}{5}\right), \quad y_2 = \exp\left(\frac{2i\pi}{5}\right).$$

with the mass ratio

$$\frac{m_2}{m_1} = 2\cos\frac{\pi}{5}.$$

In light of these results, we can conclude that in the amplitude $S_{11}(\theta)$ of the fundamental particle there are poles with positive residue at the resonance angles

$$\theta = iu_{11}^1 = \frac{2\pi i}{3}, \qquad \theta = iu_{11}^2 = \frac{2\pi i}{5}, \tag{18.4.12}$$

and poles with negative residue in the crossing channel at

$$\theta = i\overline{u}_{11}^1 = \frac{i\pi}{3}, \qquad \theta = i\overline{u}_{11}^2 = \frac{3\pi i}{5}. \tag{18.4.13}$$

However, as shown in Section 18.1, it is impossible to solve the bootstrap equations

$$S_{11}(\theta) = S_{11}\left(\theta - \frac{i\pi}{3}\right) S_{11}\left(\theta + \frac{i\pi}{3}\right) \tag{18.4.14}$$

in terms of a function that has only the sets of poles (18.4.12) and (18.4.13). In fact, it is necessary to include at least another set of poles, without breaking the conserved currents with spins given in (18.4.9). The minimal way to do so is to introduce an additional pole at $\theta = i\frac{\pi}{15}$ (with positive residue) and its companion of the crossed channel at $\theta = i\frac{14}{15}$ (with negative residue). In such a way, the exact S-matrix of the fundamental particle is expressed by

$$S_{11}(\theta) = f_{\frac{2}{3}}(\theta) f_{\frac{2}{5}}(\theta) f_{\frac{1}{15}}(\theta). \tag{18.4.15}$$

Using this expression as initial seed of the bootstrap equation, we can complete the bootstrap procedure (see Table 18.3 and 18.4). The final theory has eight particles, whose mass spectrum coincides with the one of the Toda field theory based on the exceptional algebra E_8

$$
\begin{aligned}
m_1 &= m \\
m_2 &= 2m_1 \cos\frac{\pi}{5} = (1.6180339887..)\, m_1 \\
m_3 &= 2m_1 \cos\frac{\pi}{30} = (1.9890437907..)\, m_1 \\
m_4 &= 2m_2 \cos\frac{7\pi}{30} = (2.4048671724..)\, m_1 \\
m_5 &= 2m_2 \cos\frac{2\pi}{15} = (2.9562952015..)\, m_1 \\
m_6 &= 2m_2 \cos\frac{\pi}{30} = (3.2183404585..)\, m_1 \\
m_7 &= 4m_2 \cos\frac{\pi}{5} \cos\frac{7\pi}{30} = (3.8911568233..)\, m_1 \\
m_8 &= 4m_2 \cos\frac{\pi}{5} \cos\frac{2\pi}{15} = (4.7833861168..)\, m_1
\end{aligned}
$$

As observed in Chapter 16, the masses can be put in correspondence with the Perron–Frobenius vector of the incidence matrix of the corresponding Dynkin diagram (Table 18.2). Notice that in this bootstrap system only the first three particles have a mass less than the lowest threshold $2m_1$. The stability of the particles with mass higher than the lowest decay threshold $2m_1$ is entirely due to the integrability of the theory.

The complete set of the scattering amplitudes is given below, where we use the notation

$$
(\gamma) \equiv f_{\frac{\gamma}{30}}(\theta).
$$

Several amplitudes have higher-order poles that can be explained in terms of the multi-scattering processes constructed in terms of the resonance angles of the theory.

E_8 Toda theory. The underlying E_8 structure of this scattering theory can be traced back to the coset realization $(E_8)_1 \otimes (E_8)_1/(E_8)_2$ of the critical Ising model and its Liouville quantization based on the set same of simple roots (Section 16.7). This suggests that to

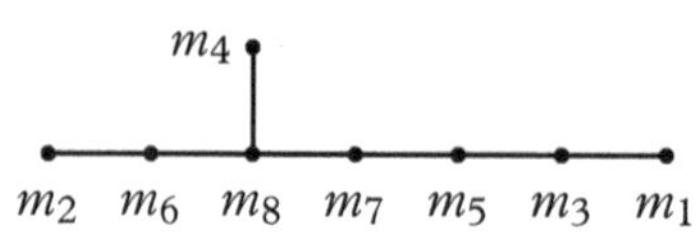

Table 18.2 *Dynkin diagram of E_8, with the association of the masses to the dots of the diagram.*

a	b	S_{ab}
1	1	$\overset{1}{(20)}\,\overset{2}{(12)}\,\overset{3}{(2)}$
1	2	$\overset{1}{(24)}\,\overset{2}{(18)}\,\overset{3}{(14)}\,\overset{4}{(8)}$
1	3	$\overset{1}{(29)}\,\overset{2}{(21)}\,\overset{4}{(13)}\,\overset{5}{(3)}\,(11)^2$
1	4	$\overset{2}{(25)}\,\overset{3}{(21)}\,\overset{4}{(17)}\,\overset{5}{(11)}\,\overset{6}{(7)}\,(15)$
1	5	$\overset{3}{(28)}\,\overset{4}{(22)}\,\overset{6}{(14)}\,\overset{7}{(4)}\,(10)^2\,(12)^2$
1	6	$\overset{4}{(25)}\,\overset{5}{(19)}\,\overset{7}{(9)}\,(7)^2\,(13)^2\,(15)$
1	7	$\overset{5}{(27)}\,\overset{6}{(23)}\,\overset{8}{(5)}\,(9)^2\,(11)^2\,(13)^2\,(15)$
1	8	$\overset{7}{(26)}\,\overset{8}{(16)^3}\,(6)^2\,(8)^2\,(10)^2\,(12)^2$
2	2	$\overset{1}{(24)}\,\overset{2}{(20)}\,\overset{4}{(14)}\,\overset{5}{(8)}\,\overset{6}{(2)}\,(12)^2$
2	3	$\overset{1}{(25)}\,\overset{3}{(19)}\,\overset{6}{(9)}\,(7)^2\,(13)^2\,(15)$
2	4	$\overset{1}{(27)}\,\overset{2}{(23)}\,\overset{7}{(5)}\,(9)^2\,(11)^2\,(13)^2\,(15)$
2	5	$\overset{2}{(26)}\,\overset{6}{(16)^3}\,(6)^2(8)^2(10)^2(12)^2$
2	6	$\overset{2}{(29)}\,\overset{3}{(25)}\,\overset{5}{(19)^3}\,\overset{7}{(13)^3}\,\overset{8}{(3)}\,(7)^2(9)^2(15)$
2	7	$\overset{4}{(27)}\,\overset{6}{(21)^3}\,\overset{7}{(17)^3}\,\overset{8}{(11)^3}\,(5)^2(7)^2(15)^2$
2	8	$\overset{6}{(28)}\,\overset{7}{(22)^3}\,(4)^2(6)^2(10)^4(12)^4(16)^4$

Table 18.3 *S-matrix of the Ising model in a magnetic field at $T = T_c$. The factors $\left(f_{\gamma/30}(\theta)\right)^{p_\gamma}$ in $S_{ab}(\theta)$ correspond to $(\gamma)^{p_\gamma}$ ($p_\gamma = 1$ is omitted). The upper index **c** in (γ) denotes the particle A_c that appears as a bound state of A_aA_b at $\theta = i\pi\gamma/30$ in the amplitudes $S_{ab}(\theta)$.*

obtain the exact S-matrix of the Lagrangian Toda field theory based on E_8 it is sufficient to multiply the minimal S-matrix elements provided by the Ising model in a magnetic field by the appropriate Z-factors that carry the coupling constant dependence on λ, without introducing additional poles in the physical sheet. With the normalization of the Toda field theories given in eqn. (16.6.1), for the amplitude of the fundamental particle, the Z-factor is given by

$$Z_{11}(\theta) = f_{-B/h}(\theta) f_{-\frac{1}{15}+B/h}(\theta) f_{-\frac{2}{3}-B/h}(\theta) f_{-\frac{2}{5}+B/h}(\theta), \tag{18.4.16}$$

where

$$B(\beta) = \frac{\beta^2}{1+\beta^2}, \tag{18.4.17}$$

3	3	$\overset{2}{(22)}\,\overset{3}{(20)^3}\,\overset{5}{(14)}\,\overset{6}{(12)^3}\,\overset{7}{(4)}\;(2)^2$
3	4	$\overset{1}{(26)}\,\overset{5}{(16)^3}\;(6)^2(8)^2(10)^2(12)^2$
3	5	$\overset{1}{(29)}\,\overset{3}{(23)}\,\overset{4}{(21)^3}\,\overset{7}{(13)^3}\,\overset{8}{(5)}\;(3)^2(11)^4(15)$
3	6	$\overset{2}{(26)}\,\overset{3}{(24)^3}\,\overset{6}{(18)^3}\,\overset{8}{(8)^3}\;(10)^2(16)^4$
3	7	$\overset{3}{(28)}\,\overset{5}{(22)^3}\;(4)^2(6)^2(10)^4(12)^4(16)^4$
3	8	$\overset{5}{(27)}\,\overset{6}{(25)^3}\,\overset{8}{(17)^5}\;(7)^4(9)^4(11)^2(15)^3$
4	4	$\overset{1}{(26)}\,\overset{4}{(20)^3}\,\overset{6}{(16)^3}\,\overset{7}{(12)^3}\,\overset{8}{(2)}\;(6)^2(8)^2$
4	5	$\overset{1}{(27)}\,\overset{3}{(23)^3}\,\overset{5}{(19)^3}\,\overset{8}{(9)^3}\;(5)^2(13)^4(15)^2$
4	6	$\overset{1}{(28)}\,\overset{4}{(22)^3}\;(4)^2(6)^2(10)^4(12)^4(16)^4$
4	7	$\overset{2}{(28)}\,\overset{4}{(24)^3}\,\overset{7}{(18)^5}\,\overset{8}{(14)^5}\;(4)^2(8)^4(10)^4$
4	8	$\overset{4}{(29)}\,\overset{5}{(25)^3}\,\overset{7}{(21)^5}\;(3)^2(7)^4(11)^6(13)^6(15)^3$
5	5	$\overset{4}{(22)^3}\,\overset{5}{(20)^5}\,\overset{8}{(12)^5}\;(2)^2(4)^2(6)^2(16)^4$
5	6	$\overset{1}{(27)}\,\overset{2}{(25)^3}\,\overset{7}{(17)^5}\;(7)^4(9)^4(11)^4(15)^3$
5	7	$\overset{1}{(29)}\,\overset{3}{(25)^3}\,\overset{6}{(21)^5}\;(3)^2(7)^4(11)^6(13)^6(15)^3$
5	8	$\overset{3}{(28)}\,\overset{4}{(26)^3}\,\overset{5}{(24)^5}\,\overset{8}{(18)^7}\;(8)^6(10)^6(16)^8$
6	6	$\overset{3}{(24)^3}\,\overset{6}{(20)^5}\,\overset{8}{(14)^5}\;(2)^2(4)^2(8)^4(12)^6$
6	7	$\overset{1}{(28)}\,\overset{2}{(26)^3}\,\overset{5}{(22)^5}\,\overset{8}{(16)^7}\;(6)^4(10)^6(12)^6$
6	8	$\overset{2}{(29)}\,\overset{3}{(27)^3}\,\overset{6}{(23)^5}\,\overset{7}{(21)^7}\;(5)^4(11)^8(13)^8(15)^4$
7	7	$\overset{2}{(26)^3}\,\overset{4}{(24)^5}\,\overset{7}{(20)^7}\;(2)^2(8)^6(12)^8(16)^8$
7	8	$\overset{1}{(29)}\,\overset{2}{(27)^3}\,\overset{4}{(25)^5}\,\overset{6}{(23)^7}\,\overset{8}{(19)^9}\;(9)^8(13)^{10}(15)^5$
8	8	$\overset{1}{(28)^3}\,\overset{3}{(26)^5}\,\overset{5}{(24)^7}\,\overset{7}{(22)^9}\,\overset{8}{(20)^{11}}\;(12)^{12}(16)^{12}$

Table 18.4 *Continuation of S-matrix of the Ising model in a magnetic field at $T = T_c$.*

with $h = 30$, the Coxeter number of this algebra, and β its coupling constant. Also in this case, the S-matrix of the Lagrangian model presents the remarkable symmetry under the weak-strong duality

$$\beta \to \frac{1}{\beta}.$$

Bootstrap fusion rules. The bootstrap fusion rules of both models (Ising and Toda) can be written in a general form once we introduce proper notation. Notice that the squares of the masses $\{m_1, m_6, m_5, m_7\}$ are the roots of the fourth order-polynomial

$$P_1 = x^4 - 30x^3 + 300x^2 - 1080x + 720$$

and for these quantities, let us introduce the notation

$$(m_1, m_6, m_5, m_7) \to \quad (C_1, C_2, C_3, C_4).$$

The squares of the masses $\{m_2, m_3, m_8, m_4\}$ are instead the roots of the fourth-order polynomial

$$P_1 = x^4 - 30x^3 + 240x^2 - 720x + 720$$

and to denote them, let us introduce the notation

$$(m_2, m_3, m_8, m_4) \to \quad (B_1, B_2, B_3, B_4).$$

In this way, the rules of the bootstrap fusions of the bound states related to the E_8 algebra can be written as (with a cyclic notation, i.e. $B_{i+4} \equiv B_i$ and $C_{i+4} \equiv C_i$)

$$\begin{aligned}
C_i \times C_i &= C_i + B_i + B_{i+1} \\
C_i \times C_{i+1} &= C_{i+2} + C_{i+3} + B_{i+3} \\
C_i \times C_{i+2} &= C_{i+1} + C_{i+3} + B_{i+1} + B_{i+3} \\
C_i \times C_{i+3} &= C_{i+1} + C_{i+2} + B_{i+2} \\
B_i \times B_i &= C_i + A_{i+1} + C_{i+2} + B_i + B_{i+3} \\
B_i \times B_{i+1} &= C_i + C_{i+1} + B_{i+1} \\
B_i \times B_{i+2} &= C_{i+1} + C_{i+3} \\
B_i \times B_{i+3} &= C_i + B_i + C_{i+3} \\
C_i \times B_i &= C_i + B_i + B_{i+1} + B_{i+3} \\
C_i \times B_{i+1} &= C_i + C_{i+2} + B_i + B_{i+3} \\
C_i \times B_{i+2} &= B_{i+2} + C_{i+3} \\
C_i \times B_{i+3} &= C_{i+1} + C_{i+2} + B_i + B_{i+1} + B_{i+3}
\end{aligned} \qquad (18.4.18)$$

An explicit check of the S-matrix of the Ising model in a magnetic field is provided by the thermodynamics Bethe ansatz, discussed in more detail in Chapter 20.

18.5 The Tricritical Ising Model at $T \neq T_c$

The TIM away from its critical temperature is described by the integrable deformation $\epsilon = \Phi_{1,2}$ with conformal dimension $\Delta = \frac{1}{10}$. This corresponds to the massive deformation of the Liouville action based on the E_7 algebra. Therefore, we expect that the corresponding scattering theory involves seven particles. Let us explore how this theory can be derived. The perturbed action is

$$\mathcal{S} = \mathcal{S}_{CFT} + \lambda \int d^2z\, \epsilon(z,\bar{z}). \tag{18.5.1}$$

For $\lambda > 0$ the system is in its Z_2 symmetric phase. Its low-temperature phase, $\lambda < 0$, is related by duality to the high-temperature one. Therefore, in the following we focus our attention only on the massive theory (18.5.1) with $\lambda > 0$.

The spins of the conserved charges coincide with the Coxeter exponents of the Toda field theory based on the E_7 algebra, whose Coxeter number is $h = 18$

$$s = 1, 5, 7, 9, 11, 13, 17 \pmod{18}.$$

In computing the mass spectrum and the scattering amplitudes, it is important to notice that the *fundamental particle* cannot be in a bound state of itself for the Z_2 symmetry of the model that can be used to label the particles. We expect that the fundamental particle is odd under this symmetry and therefore cannot fulfill the Φ^3 property. However, the existence of a Z_2 even particle with the Φ^3 property is not in contradiction with the spins of the conserved charges, as the charge Q_9 annihilates this state. In light of this observation, let us assume that the lightest Z_2 even particle, here denoted by A_2, appears as bound state in the scattering amplitude of the fundamental Z_2 odd particle A_1. Since for the eigenvalues of the conserved charges we have $\chi_9^{(1)} \neq 0$ but $\chi_9^{(2)} = 0$, using the consistency equation we obtain the resonance angle u_{11}^2 by the condition

$$\cos\left(\frac{9u_{11}^2}{2}\right) = 0. \tag{18.5.2}$$

The solution that gives rise to a consistent system is identified as $u_{11}^2 = \frac{5\pi}{9}$. This fixes the mass ratio of these two particles to be

$$m_2 = 2\cos\left(\frac{5\pi}{18}\right) m_1.$$

The pole in S_{11} at $\theta = i5\pi/9$ with positive residue implies a pole in S_{12} at $\theta = i5\pi/9$ with negative residue, corresponding to the particle A_1 in the crossed channel. With these data, the bootstrap equations that involve S_{11} and S_{12} become

$$S_{12}(\theta) = S_{11}\left(\theta + i\frac{5\pi}{18}\right) S_{11}\left(\theta - i\frac{5\pi}{18}\right), \tag{18.5.3}$$

$$S_{11}(\theta) = S_{11}\left(\theta + i\frac{4\pi}{9}\right) S_{12}\left(\theta - i\frac{5\pi}{18}\right). \tag{18.5.4}$$

We cannot satisfy these equations only with a pole in S_{11} and S_{12}. The minimal way to satisfy them is to introduce an additional pole at $\theta = i\pi/9$ (with positive residue) in S_{11} and a pole at $\theta = i\,7\pi/18$ (with positive residue) in S_{12}. The new pole at $\theta = i\,\pi/9$ in S_{11} corresponds to a new bound state A_4, a particle that is even under the Z_2 symmetry, with mass

$$m_4 = 2\cos\left(\frac{\pi}{18}\right) m_1.$$

The pole at $\theta = i7\pi/18$ in S_{12} represents another bound state A_3, odd under the Z_2 symmetry, with mass

$$m_3 = 2\cos\left(\frac{\pi}{9}\right) m_1.$$

So, the initial ansatz for these amplitudes is

$$S_{11}(\theta) = -f_{\frac{1}{9}}(\theta) f_{\frac{5}{9}}(\theta), \qquad S_{12}(\theta) = f_{\frac{7}{18}}(\theta) f_{\frac{13}{18}}(\theta) \tag{18.5.5}$$

All other amplitudes can be iteratively computed by employing the bootstrap equations (17.4.68). The bootstrap process closes with seven particles, whose masses and Z_2 quantum numbers are given in Table 18.5. Additionally, as in the Ising model in a magnetic field, the masses can be associated to the dots of the Dynkin diagram of the E_7 algebra. They enter the component of the Perron–Frobenius eigenvector of the

$m_1 = M$	1	odd
$m_2 = 2M\cos(\frac{5\pi}{18})$	1.28557	even
$m_3 = 2M\cos(\frac{\pi}{9})$	1.87938	odd
$m_4 = 2M\cos(\frac{\pi}{18})$	1.96961	even
$m_5 = 4M\cos(\frac{5\pi}{18})\cos(\frac{\pi}{18})$	2.53208	even
$m_6 = 4M\cos(\frac{\pi}{9})\cos(\frac{2\pi}{9})$	2.87938	odd
$m_7 = 4M\cos(\frac{\pi}{18})\cos(\frac{\pi}{9})$	3.70166	even

Table 18.5 *Mass spectrum of the thermal tricritical Ising model, together with their numerical values and the Z_2 quantum numbers.*

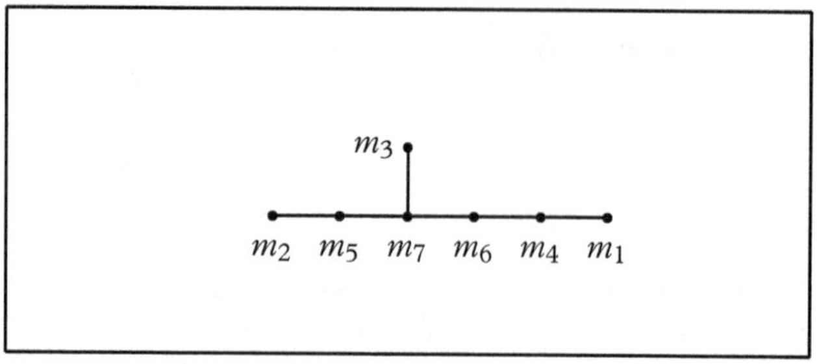

Table 18.6 *Dynkin diagram of the E_7 algebra and correspondence between the masses of the particles and the dots of the diagram.*

incidence matrix of this Dynkin diagram (Table 18.6). The complete set of the scattering amplitudes are in Table 18.7.

Bootstrap fusion rules of the TIM. According to the roots of the algebraic equations that determine the masses of the corresponding Toda field theory (Section 16.6), the seven particles can be organized in two triplets and one singlet

$$\begin{aligned}(Q_1, Q_2, Q_3) &\equiv (m_6, m_3, m_1)\\ (K_1, K_2, K_3) &\equiv (m_2, m_4, m_7)\\ (N) &\equiv (m_5).\end{aligned}$$

The first triplet consists of the odd particles under the Z_2 symmetry. The second triplet and the singlet are made of Z_2 even particles. The bootstrap fusions that involve $[N]$ and $[N, K_i]$ form a closed sub-system of these fusions

$$\begin{array}{ll} N \cdot N = N, & N \cdot K_A = K_1 + K_2 + K_3 \\ K_A \cdot K_{A+1} = K_A + N, & K_A \cdot K_A = K_A + K_{A+1} + N. \end{array} \tag{18.5.6}$$

The other particles couple only to the previous ones

$$\begin{array}{ll} K_A \cdot Q_A = Q_{A+1}, & K_A \cdot Q_{A+1} = Q_1 + Q_2 + Q_3 \\ K_A \cdot Q_{A-1} = Q_{A-1} + Q_{A+1}, & Q_A \cdot Q_A = K_{A-1} + K_{A+1} \\ Q_A \cdot Q_{A+1} = K_A + K_{A-1} + N, & N \cdot Q_A = Q_{A-1} + Q_{A+1}. \end{array} \tag{18.5.7}$$

A check that confirms the validity of this S-matrix description of the thermal deformation of the TIM will be given by the thermodynamical Bethe ansatz.

***E_7* Toda theory.** As for the other S-matrices previously discussed, the S-matrix of the thermal deformation of the TIM can be used as minimal S-matrix of the corresponding Lagrangian model, given by the Toda field theory based on the exceptional E_7 algebra, see eqn. (16.6.1). In this case the Z-factor that enters the amplitude of the fundamental particle is

$$Z_{11}(\theta) = -f_{-B/h}(\theta) f_{-\frac{1}{9}+B/h}(\theta) f_{-\frac{4}{9}-B/h}(\theta), \tag{18.5.8}$$

a	b	S_{ab}
1	1	$-\overset{2}{(10)}\overset{4}{(1)}$
1	2	$\overset{1}{(13)}\overset{3}{(7)}$
1	3	$-\overset{2}{(14)}\overset{4}{(10)}\overset{5}{(6)}$
1	4	$\overset{1}{(17)}\overset{3}{(11)}\overset{6}{(3)}\,(9)$
1	5	$\overset{3}{(14)}\overset{6}{(8)}\,(6)^3$
1	6	$-\overset{4}{(16)}\overset{5}{(12)}\overset{7}{(4)}\,(10)^2$
1	7	$\overset{6}{(15)}\,(9)(5)^2(7)^2$
2	2	$\overset{2}{(12)}\overset{4}{(8)}\overset{5}{(2)}$
2	3	$\overset{1}{(15)}\overset{3}{(11)}\overset{6}{(5)}\,(9)$
2	4	$\overset{2}{(14)}\overset{5}{(8)}\,(6)^2$
2	5	$\overset{2}{(17)}\overset{4}{(13)^3}\overset{7}{(3)}\,(9)(7)^2$
2	6	$\overset{3}{(15)}\,(9)(5)^2(7)^2$
2	7	$\overset{5}{(16)}\overset{7}{(10)^3}\,(4)^2(6)^2$
3	3	$-\overset{2}{(14)}\overset{7}{(2)}\,(12)^2(8)^2$
3	4	$\overset{1}{(15)}\,(9)(5)^2(7)^2$
3	5	$\overset{1}{(16)}\overset{6}{(10)^3}\,(4)^2(6)^2$
3	6	$\overset{2}{(16)}\overset{7}{(8)^3}\overset{5}{(12)^3}\,(4)^2$
3	7	$\overset{3}{(17)}\overset{6}{(13)^3}\,(9)^2(3)^2(7)^4$
4	4	$\overset{4}{(12)}\overset{7}{(4)}\overset{5}{(10)^3}\,(2)^2$
4	5	$\overset{2}{(15)}\overset{4}{(13)^3}\overset{7}{(7)^3}\,(9)$
4	6	$\overset{1}{(17)}\overset{6}{(11)^3}\,(9)^2(3)^2(5)^2$

Table 18.7 *S-matrix of the thermal deformation of the tricritical Ising model. The factors* $(t_{\gamma/18}(\theta))^{p_\gamma}$ *in* $S_{ab}(\theta)$ *correspond to* $(\gamma)^{p_\gamma}$ ($p_\gamma = 1$ *is omitted). The upper index* **c** *in* (γ) *denotes the bound state* A_c *in the amplitude* $S_{ab}(\theta)$, *whose pole is at* $\theta = i\pi\gamma/18$.

a b	S_{ab}
4 7	$\overset{4}{(16)}\,\overset{5}{(14)^3}\,(8)^4(12)^4$
5 5	$\overset{5}{(12)^3}\,(4)^2(2)^2(8)^4$
5 6	$\overset{1}{(16)}\,\overset{3}{(14)^3}\,(6)^4(8)^4$
5 7	$\overset{2}{(17)}\,\overset{4}{(15)^3}\,\overset{7}{(11)^5}\,(9)^3(5)^4$
6 6	$-\,\overset{4}{(14)^3}\,\overset{7}{(10)^5}\,(16)^2(12)^4$
6 7	$\overset{1}{(17)}\,\overset{3}{(15)^3}\,\overset{6}{(13)^5}\,(9)^3(5)^6$
7 7	$\overset{2}{(16)^3}\,\overset{5}{(14)^5}\,\overset{7}{(12)^7}\,(8)^8$

Table 18.7 *Continued*

with

$$B(\beta) = \frac{\beta^2}{1+\beta^2}, \tag{18.5.9}$$

where $h = 18$ is the Coxeter number of the E_7 algebra and β its coupling constant. All the other amplitudes can be obtained by applying the bootstrap equations.

18.6 Thermal Deformation of the 3-state Potts Model

The class of universality of this model is described by a sub-set of operators of the minimal model $\mathcal{M}_{5,6}$ with central charge $c = 4/5$. The Landau–Ginzburg of the critical model is

$$\mathcal{L} = (\partial_\mu \Phi)(\partial_\mu \Phi^*) + \left[(\Phi)^3 + (\Phi^*)^3\right], \tag{18.6.1}$$

where Φ is a complex scalar field. The two most relevant magnetization operators can be identified with Φ and Φ^*, whereas the other two sub-leading magnetic operators correspond to $(\Phi^*)^2\Phi$ and $\Phi^*\Phi^2$. The energy operator is associated to $\Phi^*\Phi$.

Away from the critical temperature, the action of the model can be written as

$$\mathcal{A} = \mathcal{A}_{CFT} + \lambda \int \epsilon(x)\, d^2x. \tag{18.6.2}$$

and it corresponds to an integrable theory. In the Landau–Ginzburg formalism, the thermal deformation is equivalent to add a mass term $m^2 \Phi^* \Phi$ in the Lagrangian (18.6.1). The perturbed theory is still invariant under the permutation group S_3 present at the critical point and therefore the particles can be labelled by the corresponding quantum numbers. An irreducible representation of this discrete symmetry group is given by a doublet of particle–anti-particle $(A,\bar{A})$ of mass m. Under the action of the generators of the group, these states transform as

$$\vartheta A = \omega A; \quad \vartheta \bar{A} = \overline{\omega} \bar{A}; \quad CA = \bar{A},$$

where $\omega = \exp(2\pi i/3)$. In this case, the most general S-matrix is given by

$$\begin{aligned} |\,A(\theta_1)A(\theta_2)\rangle_{in} &= u(\theta_{12})\,|\,A(\theta_1)A(\theta_2)\rangle_{out}; \\ |\,A(\theta_1)\bar{A}(\theta_2)\rangle_{in} &= t(\theta_{12})\,|\,A(\theta_1)\bar{A}(\theta_2)\rangle_{out} + r(\theta_{12})\,|\,\bar{A}(\theta_1)A(\theta_2)\rangle_{out}. \end{aligned}$$

However, as a direct consequence of the infinite conserved charges of this theory, it is easy to show that the reflection amplitude vanishes. Therefore, the S-matrix is completely diagonal. Furthermore, the crossing invariance implies

$$t(\theta) = u(i\pi - \theta),$$

while the unitarity condition leads to

$$t(\theta)\,t(-\theta) = 1; \qquad u(\theta)\,u(-\theta) = 1.$$

The minimal solution of these equations is

$$u(\theta) = \frac{\sinh(\theta/2 + i\pi/3)}{\sinh(\theta/2 - i\pi/3)}, \qquad t(\theta) = \frac{\sinh(\theta/2 + i\pi/6)}{\sinh(\theta/2 - i\pi/6)}. \tag{18.6.3}$$

Notice that the anti-particle $\bar{A}$ appears as bound state of the particle A, and vice versa.

18.6.1 Thermal Deformation of the 3-state Tricritical Potts Model

The tricritical version of the 3-state Potts model can be identified with a sub-set of the fields of the minimal conformal model $\mathcal{M}_{6,7}$. As the ordinary Potts model, its tricritical version is invariant under the permutation group S_3. Its thermal deformation is implemented by adding to the conformal action the energy operator $\Phi_{1,2}$ with conformal dimensions $(\Delta, \bar{\Delta}) = \left(\frac{1}{7}, \frac{1}{7}\right)$. This is the most relevant field of the Kac table that is invariant under the S_3 symmetry.

The off-critical model is integrable. To compute the S-matrix, let us assume the existence of two douplets $(A_a, A_{\bar{a}})$ and $(A_b, A_{\bar{b}})$ with the bootstrap fusions

$$A_a \times A_a \to A_{\bar{a}} + A_{\bar{b}}, \qquad A_b \times A_b \to A_{\bar{a}} + A_{\bar{b}},$$

and masses m_a, m_b ($m_a < m_b$). From the analysis of the consistency equations done in the previous chapter, we arrive at the resonance angles

$$\overline{U}^a_{a\bar{b}} = \frac{\pi}{12}, \qquad \overline{U}^{\bar{b}}_{a\bar{b}} = \frac{5\pi}{12}, \qquad \overline{U}^a_{a\bar{a}} = \frac{\pi}{3}. \tag{18.6.4}$$

Also in this case all reflection amplitudes vanish. The scattering amplitudes relative to the douplet with lower mass $(A_a, A_{\bar{a}})$ are given by

$$\begin{aligned} |\, A_a(\theta_1)A_a(\theta_2)\rangle &= S_{aa}(\theta_{12})\,|\, A_a(\theta_2)A_a(\theta_1)\rangle; \\ |\, A_a(\theta_1)A_{\bar{a}}(\theta_2)\rangle &= S^T_{a,\bar{a}}(\theta_{12})\,|\, A_a(\theta_2)A_{\bar{a}}(\theta_1)\rangle. \end{aligned}$$

The bootstrap fusion $a \times a \to \bar{a}$ implies

$$\begin{aligned} S^T_{a\bar{a}}(\theta) &= S_{aa}(\theta - i\tfrac{\pi}{3})S_{aa}(\theta + i\tfrac{\pi}{3}) \\ S_{aa}(\theta) &= S^T_{a\bar{a}}(\theta - i\tfrac{\pi}{3})S^T_{a\bar{a}}(\theta + i\tfrac{\pi}{3}). \end{aligned}$$

Equivalently

$$S_{aa}(\theta)S_{aa}\left(\theta - i\frac{2\pi}{3}\right)S_{aa}\left(\theta + i\frac{2\pi}{3}\right) = 1.$$

The minimal solution of these equations, that satisfies the unitarity condition is

$$S_{aa}(\theta) = \frac{\sinh(\frac{\theta}{2} + i\frac{\pi}{3})\sinh(\frac{\theta}{2} + i\frac{\pi}{12})\sinh(\frac{\theta}{2} + i\frac{\pi}{4})}{\sinh(\frac{\theta}{2} - i\frac{\pi}{3})\sinh(\frac{\theta}{2} - i\frac{\pi}{12})\sinh(\frac{\theta}{2} - i\frac{\pi}{3})} \equiv s_{\frac{2}{3}}(\theta)s_{\frac{1}{6}}(\theta)s_{\frac{1}{2}}(\theta). \tag{18.6.5}$$

S_{aa} has two simple poles with positive residue: the first, $\theta = i\frac{2\pi}{3}$, corresponds to the particle $A_{\bar{a}}$ while the other, at $\theta = i\frac{\pi}{6}$, corresponds to the particle $A_{\bar{b}}$. Their mass ratio is

$$m_{\bar{b}} = m_b = 2m_a \cos\left(\frac{\pi}{12}\right).$$

The additional pole at $\theta = i\frac{\pi}{2}$ has negative residue and corresponds to a bound state in the crossed channel. In fact,

$$S^T_{a\bar{a}}(\theta) = S_{aa}(i\pi - \theta) = -s_{\frac{1}{3}}(\theta)s_{\frac{1}{2}}(\theta)s_{\frac{5}{6}}(\theta)$$

that has a simple pole with positive residue at $\theta = i\frac{\pi}{2}$. This pole is associated to a new neutral particle A_c, with mass

$$m_c = 2m_a \cos\left(\frac{\pi}{4}\right).$$

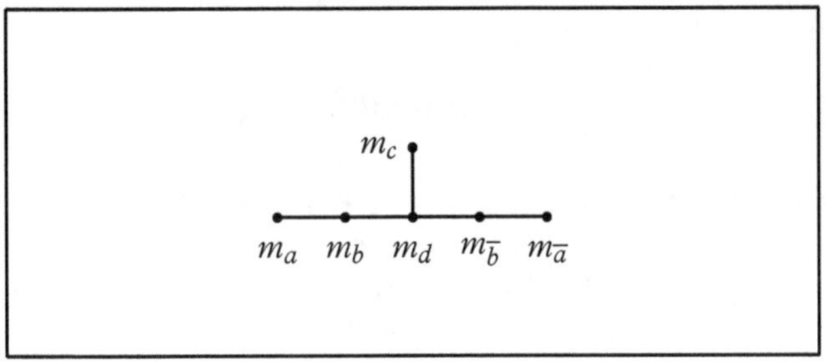

Table 18.8 *Dynkin diagram of E_6, with the relative association between the masses and the dots of the diagram.*

The scattering amplitute $S_{\bar{a}b}$ is recovered by the equation

$$S_{\bar{a}b}(\theta) = S_{\bar{a}\bar{a}}\left(\theta - i\frac{\pi}{12}\right) S_{\bar{a}b}\left(\theta + i\frac{\pi}{12}\right)$$

with the result

$$S_{\bar{a}b}(\theta) = s_{\frac{3}{4}}(\theta)\, s_{\frac{1}{4}}(\theta)\, s_{\frac{1}{12}}(\theta)\, s_{\frac{5}{12}}(\theta)\, s^2_{\frac{7}{12}}(\theta). \tag{18.6.6}$$

The pole analysis of $S_{\bar{a}b}$ shows an additional neutral particle A_d, that enters the bootstrap fusion

$$A_{\bar{a}} \times A_b \to A_c + A_d$$

with mass

$$m_d = 4m_a \cos\left(\frac{\pi}{12}\right) \cos\left(\frac{\pi}{4}\right).$$

It is easy to show that the set of these six particles $\{A_a, A_{\bar{a}}, A_b, A_{\bar{b}}, A_c, A_d\}$ closes the bootstrap procedure. The masses can be put in correspondence with the dots of the Dynkin diagram of E_6 (Table 18.8). The complete set of the scattering amplitudes is

$$
\begin{array}{ll}
S_{aa} = (\tfrac{1}{6})(\tfrac{2}{3})(\tfrac{1}{2}), & S_{\bar{a}\bar{a}} = S_{aa} \\
S^T_{a\bar{a}} = -(\tfrac{1}{3})(\tfrac{5}{6})(\tfrac{1}{2}) & S_{ab} = S_{\bar{a}\bar{b}} = (\tfrac{1}{4})(\tfrac{3}{4})(\tfrac{7}{12})(\tfrac{11}{12})(\tfrac{5}{12})^2 \\
S_{a\bar{b}} = S_{\bar{a}b} = (\tfrac{1}{12})(\tfrac{1}{4})(\tfrac{3}{4})(\tfrac{5}{12})(\tfrac{7}{12})^2, & S_{ac} = S_{\bar{a}c} = (\tfrac{1}{4})(\tfrac{3}{4})(\tfrac{5}{12})(\tfrac{7}{12}) \\
S_{ad} = S_{\bar{a}d} = (\tfrac{1}{6})(\tfrac{5}{6})(\tfrac{1}{3})^2(\tfrac{2}{3})^2(\tfrac{1}{2})^2, & S_{bb} = (\tfrac{5}{6})(\tfrac{1}{6})^2(\tfrac{1}{3})^2(\tfrac{1}{6})^2(\tfrac{2}{3})^3(\tfrac{1}{2})^3 \\
S^T_{b\bar{b}} = -(\tfrac{1}{6})(\tfrac{5}{6})^2(\tfrac{2}{3})^2(\tfrac{1}{3})^3(\tfrac{1}{2})^3, & S_{bc} = S_{\bar{b}c} = (\tfrac{1}{6})(\tfrac{5}{6})(\tfrac{1}{2})^2(\tfrac{2}{3})^2(\tfrac{1}{3})^2 \\
S_{bd} = S_{\bar{b}d} = (\tfrac{1}{12})(\tfrac{11}{12})(\tfrac{1}{4})^3(\tfrac{1}{4})^3(\tfrac{5}{12})^4(\tfrac{7}{12})^4, & S_{cc} = -(\tfrac{1}{6})(\tfrac{5}{6})(\tfrac{1}{3})(\tfrac{2}{3})(\tfrac{1}{2})^2 \\
S_{cd} = (\tfrac{1}{12})(\tfrac{11}{12})(\tfrac{1}{4})^2(\tfrac{3}{4})^2(\tfrac{5}{12})^3(\tfrac{7}{12})^3, & S_{dd} = -(\tfrac{1}{6})^3(\tfrac{5}{6})^3(\tfrac{1}{3})^5(\tfrac{2}{3})^5(\tfrac{1}{2})^6,
\end{array}
$$

where we use the notation

$$(x) \equiv s_x(\theta).$$

E_6 **Toda theory**. The amplitudes above are the minimal S-matrices of the Toda field theory based on the exceptional algebra E_6. The relation between this statistical model and the Toda field theory discussed in Chapter 16. To obtain the exact S-matrix of the Toda field theory with real coupling constant β it is sufficient to multiply the minimal amplitude $S_{aa}(\theta)$ for the Z-factor

$$Z(\theta) = \left(-\frac{B}{h}\right)\left(-\frac{1}{6}+\frac{B}{h}\right)\left(-\frac{1}{2}-\frac{B}{h}\right)\left(-\frac{2}{3}+\frac{B}{h}\right), \tag{18.6.7}$$

where

$$B(\beta) = \frac{\beta^2}{1+\beta^2} \tag{18.6.8}$$

and $h = 12$ is the Coxeter number of E_6.

18.7 General Expression Toda Field Theories

The previous sections provided several examples of S-matrices related to the simply laced affine Toda field theories, particularly those relative to series E_n. It is rather remarkable that it can be given an universal formula for all the amplitudes of the S-matrix of the Toda field theories. For these theories we adopt the normalization given in eqn. (16.6.1). This universal formula involves two quantities: the incidence matrix of these algebra

$$\mathbf{I}_{ij} = (2-C)_{ij} = 2\delta_{ij} - \alpha_i \cdot \alpha_j, \tag{18.7.1}$$

where C is the Cartan matrix of the algebra, given in terms of the scalar products of the simple roots α_i; and the universal function $B(g)$ of the coupling constant

$$B(g) = \frac{g^2}{1+g^2}. \tag{18.7.2}$$

Posing

$$S_{ab}(\theta) = e^{i\delta_{ab}(\theta)}, \tag{18.7.3}$$

the phase-shift $\delta_{ab}(\theta)$ for the scattering of the particles a and b is given by

$$\delta_{ab}(\theta) = \int_0^\infty \frac{dt}{t}\left[8\sinh\left(\frac{\pi Bct}{h}\right)\sinh\left(\frac{\pi(1-B)ct}{h}\right)\left(2\cosh\frac{\pi ct}{h}-\mathbf{I}\right)^{-1}_{ab} - 2\delta_{ab}\right] \times \sin\theta t. \tag{18.7.4}$$

The particles a and b belong to the spectrum of r (the rank of the algebra) excitations whose exact masses were reported in Section 16.6. In the expression of the phase-shifts we have explicitly put the speed of light c for an interesting limit, which we discuss in Section 18.8.

18.8 Non-relativistic Limit of Toda Field Theories

Imagine we restore the speed of light c into the expression of the Lagrangian of Toda field theory previously given in eqn. (16.6.1) so that the exact phase-shifts (18.7.4) of these theories correspond to the Lagrangian

$$\mathcal{L} = \frac{1}{8\pi}(\partial_\mu\phi)\cdot(\partial^\mu\phi) - \frac{c^2m^2}{g^2}\sum_{i=1}^{r+1} q_i[\exp(g\alpha_i\cdot\phi)-1], \tag{18.8.1}$$

where $\partial_0 = c^{-1}\partial_t$. It is interesting to try to identify the non-relativistic integrable field theories that manifest by taking the non-relativistic (NR) limit $c\to\infty$. Notice that for $c\to\infty$ the relative rapidity is approximatively $\theta\sim\frac{v}{c}$ with v the relative velocity of the two particles: to make manifest the NR limit, in the integral of eqn. (18.7.4) we can make the change of variable $\tau = c^{-1}t$, so that

$$S_{ab}(\theta) = \exp\left\{-i\int_0^\infty \frac{d\tau}{\tau}\left[8\sinh\left(\frac{Bc\tau}{2}\right)\sinh\left(\frac{\pi c\tau}{h}-\frac{Bc\tau}{2}\right) \times\left(2\cosh\frac{\pi c\tau}{h}-\mathcal{I}\right)^{-1}_{ab} - 2\delta_{ab}\right]\sin(v\tau)\right\}. \tag{18.8.2}$$

It is pretty clear that in order to have a finite expression in the limit $c\to\infty$ we have also to take simultaneously the limit $g\to 0$, so that $Bc\to\frac{g^2}{4h}$. On the other hand, in the limit $c\to\infty$ the matrix $T_{ab}\equiv\left(2\cosh\frac{\pi c\tau}{h}-\mathcal{I}\right)^{-1}_{ab}$ becomes purely diagonal

$$T_{ab}\simeq\delta_{ab}\,e^{-\frac{\pi c\tau}{h}},\quad c\to\infty,$$

and thus the NR limit of the scattering amplitudes is also expressed by purely diagonal terms

$$S_{ab}(\theta) \to \exp\left[i2\delta_{ab}\int_0^\infty \frac{d\tau}{\tau}e^{-\frac{\beta^2}{4h}\tau}\sin(v\tau)\right] = \exp\left[\delta_{ab}\ln\left(\frac{\frac{\beta^2}{4h}+iv}{\frac{\beta^2}{4h}-iv}\right)\right]. \tag{18.8.3}$$

Thus different particles are decoupled in the NR limit (since their scattering matrix is simply 1) and only the scattering of identical particles survives

$$S_{aa} \to \frac{k - i2m_a\frac{\beta^2}{8h}}{k + i2m_a\frac{\beta^2}{8h}}. \tag{18.8.4}$$

The above scattering amplitudes describe a collision of two particles of mass m_a and coupling $g^2/8h$ associated to a diagonal multi-species Lieb–Liniger model. Let us briefly describe this model.

In the second quantization formalism, it corresponds to a non-linear Schrödinger theory: this consists of a local non-relativistic field theory for r complex Bose fields $\psi_i(t,x)$ (with masses m_i, $i = 1,\ldots,r$) that satisfy the canonical commutation relations

$$[\psi_i(t,x),\psi_j^\dagger(t,x')] = \delta_{ij}\,\delta(x-x'), \quad [\psi_i(t,x),\psi_j(t,x')] = 0. \tag{18.8.5}$$

The most general Hamiltonian of this model is given by

$$H = \int \mathrm{d}x \left(\sum_{i=1}^{r}\frac{\partial_x\psi_i^\dagger\partial_x\psi_i}{2m_i} + \sum_{i,j}\lambda_i\,\psi_i^\dagger\psi_j^\dagger\psi_i\psi_j\right), \tag{18.8.6}$$

with the corresponding Lagrangian density expressed by

$$\mathcal{L} = \sum_{i=1}^{r}\left[i\frac{1}{2}\left(\psi_i^\dagger\partial_t\psi_i - \partial_t\psi_i^\dagger\psi_i\right) - \frac{1}{2m_i}\partial_x\psi_i^\dagger\partial_x\psi_i\right] - \sum_{i,j}^{r}\lambda_{i,j}\,\psi_i^\dagger\psi_j^\dagger\psi_i\psi_j. \tag{18.8.7}$$

In this model the number of particles of each species is individually conserved. This allows us to go to the first quantization formalism, relative to a given number of particles N, and express the quantum Hamiltonian of the model as a coupled set of Lieb–Liniger models

$$H_N = -\sum_{k=1}^{r}\frac{1}{2m_k}\sum_{i_k=1}^{N_k}\frac{\partial^2}{\partial x_{i_k}^2} + 2\sum_{k,l}^{r}\lambda_{k,l}\sum_{i_k<j_l}\delta(x_{i_k}-x_{j_l}), \tag{18.8.8}$$

the amplitudes (18.8.4) are those relative to the *diagonal* multi-species Lieb–Liniger model, i.e. the model in which we have $\lambda_{i,j} = \lambda\,\delta_{i,j}$. In order to recover directly the expression (18.8.7) of the Lieb–Liniger model in the NR limit, it is useful to expand each field of the Toda field theories as

$$\phi_a(t,x) = \frac{1}{\sqrt{2m_a}}\left(e^{im_ac^2t}\psi_a^\dagger(t,x) + e^{-im_ac^2t}\psi_a(t,x)\right). \tag{18.8.9}$$

By means of canonical commutation rules of the fields ϕ_a and their conjugate momenta, it is easy to see that in the NR limit the fields ψ_a are independent NR bosonic species

$$[\psi_a(x),\psi_b(y)] = 0, \quad [\psi_a(x),\psi_b^\dagger(y)] = \delta_{a,b}\delta(x-y). \tag{18.8.10}$$

Once the expressions (18.8.9) of the fields are substituted into the action of the Toda field theories, to obtain the NR limit it is sufficient to neglect unpaired rapidly oscillating terms (as those coming from powers of $e^{\pm im_ac^2t}$) as well as terms in the action multiplied by inverse powers $1/c^k$ of the speed of light: all these terms, once integrated in time, will vanish in the limit $c\to\infty$. The NR equation of motion coming from the action in the double limit $c\to\infty$ and $g\to 0$ can be written as

$$i\partial_t\psi_a^\dagger = \frac{1}{2m_a}\partial_x^2\psi_a^\dagger - \frac{\beta^2}{8}\sum_{a,a'}\frac{\Lambda_{a,a'}}{\mu_a\mu_{a'}}\psi_a^\dagger\psi_{a'}^\dagger\psi_{a'} \tag{18.8.11}$$

where

$$\Lambda_{a,a'} = C^{(4)}_{a,a,a',a'} + \sum_b\left[\frac{[C^{(3)}_{a,a',b}]^2}{(m_{a'}-m_a)^2-m_b^2} + \frac{[C^{(3)}_{a,a',b}]^2}{(m_a+m_{a'})^2-m_b^2} - \frac{C^{(3)}_{a,a,b}C^{(3)}_{b,a',a'}}{m_b^2}\right].$$

In this expression $C^{(3)}_{a,b,c}$ and $C^{(4)}_{a,b,c,d}$ are the three and four couplings of the Toda field theories, respectively. $\Lambda_{a,a'}$ coincides with the sum of the tree level diagrams of Toda field theories (Figure 18.4), where these graphs are computed at zero rapidity and with incoming and outgoing particles of type a and a' which scatter elastically. The NR Hamiltonian associated with these equations of motion, from which (18.8.11) can be derived, is readily written as

$$H = \int dx\left[\sum_a\frac{\partial_x\psi_a^\dagger\partial_x\psi_a}{2m_a} + \frac{\beta^2}{16}\sum_{a,a'}\frac{\Lambda_{a,a'}}{m_am_{a'}}\psi_a^\dagger\psi_{a'}^\dagger\psi_a\psi_{a'}\right]. \tag{18.8.12}$$

Hence, the NR limit of the Toda field theories apparently consists of a set of bosons of different species, coupled together through a density–density interaction. However some of the couplings $\Lambda_{a,a'}$ may vanish. Since in the Toda field theories the scattering is purely transmissive, such a feature must be also true in the NR limit. However, as shown in one of the exercises at the end of this chapter, an inter-species density–density interaction such as in the Hamiltonian (18.8.12) is never purely transmissive and therefore, to be

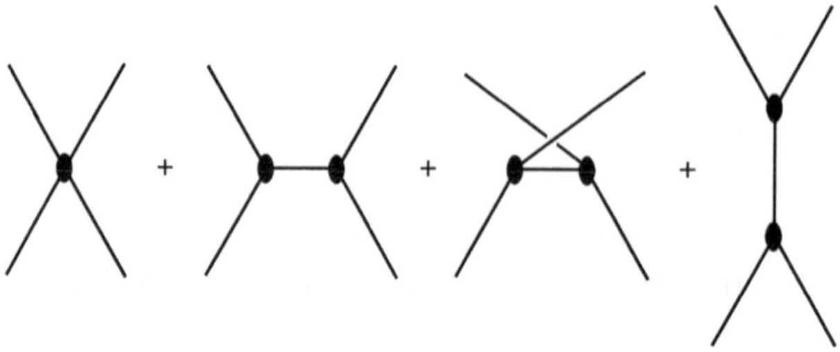

Fig. 18.4 *Representation of the coupling $\Lambda_{a,a'}$ of (18.8.12) through Feynman graphs.*

consistent with the scattering of the Toda theories, different species must be necessarily decoupled. Indeed, for the Toda field theories it holds the identity

$$\Lambda_{a,a'} = \frac{2m_a^2}{h}\delta_{a,a'}. \tag{18.8.13}$$

With this extra piece of information, we finally arrive at the conclusion that the Hamiltonian coming from the NR limit of the Toda field theories consists of a set of r decoupled Lieb–Liniger models, all with the same interaction but different masses

$$H = \int dx \sum_a \left[\frac{\partial_x \psi_a^\dagger \partial_x \psi_a}{2m_a} + \frac{g^2}{8h} \psi_a^\dagger \psi_a^\dagger \psi_a \psi_a \right]. \tag{18.8.14}$$

18.9 Models with Internal *O(n)* Invariance

So far we have been dealing with S-matrices that were completely diagonal in the indices of the particles. The next few sections present several interesting and important examples of non-diagonal S-matrices. We start our analysis with the S-matrices of the theories are invariant under the $O(n)$ transformations. Statistical models that have such a symmetry are characterized by an isotropic ferromagnetic interaction among the n components of the spin variables $\vec{S}_i$. For the elastic S-matrix of these theories it is necessary to distinguish three cases: (i) $n > 2$; (ii) $n < 2$ and (iii) $n = 2$. This section discusses the first two cases, whereas the discussion of the $n = 2$ case can be found in Section 18.10.

18.9.1 $n > 2$

For the symmetry of the system, let us assume that the spectrum of the theory consists of a multiplet of n particles of equal mass, denoted by the symbols A_i $(i = 1, 2, \ldots, n)$. Enforcing the $O(n)$ invariance of the scattering theory, we can decompose the S-matrix elements as

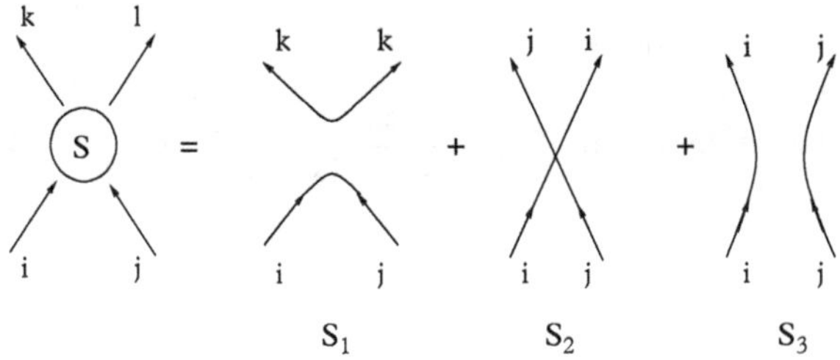

Fig. 18.5 *Decomposition of the S-matrix into the invariant amplitudes under the O(n) group.*

$$
\begin{aligned}
A_i(\theta_1)A_j(\theta_2) = \delta_{ij}\, S_1(\theta) \sum_{k=1}^{n} A_k(\theta_2)A_k(\theta_1) \\
+ S_2(\theta)A_j(\theta_2)A_i(\theta_1) + S_3(\theta)A_i(\theta_2)A_i(\theta_1).
\end{aligned}
\tag{18.9.1}
$$

The functions $S_2(\theta)$ and $S_3(\theta)$ are the transmission and reflection amplitudes respectively, while $S_1(\theta)$ describes the annihilation-creation process $A_i + A_i \to A_j + A_j$, with $i \neq j$. This decomposition is represented in Figure 18.5. These functions satisfy the unitarity equation

$$
\begin{aligned}
&S_2(\theta)S_2(-\theta) + S_3(\theta)S_3(-\theta) = 1 \\
&S_2(\theta)S_3(-\theta) + S_3(\theta)S_2(-\theta) = 0 \\
&n S_1(\theta)S_1(-\theta) + S_1(\theta)S_2(-\theta) + S_1(\theta)S_3(-\theta) \\
&+S_2(\theta)S_1(-\theta) + S_3(\theta)S_1(-\theta) = 0.
\end{aligned}
\tag{18.9.2}
$$

Moreover, they are related by the crossing symmetry relationships

$$
S_2(\theta) = S_2(i\pi - \theta) \tag{18.9.3}
$$

$$
S_1(\theta) = S_3(i\pi - \theta) \tag{18.9.4}
$$

as shown in Figure 18.5 from left to right, rather than from bottom to top.

In addition to these basic conditions, the amplitudes satisfy a non-trivial set of Yang–Baxter equations

$$
\begin{aligned}
&S_2\, S_1\, S_3 + S_2\, S_3\, S_3 + S_3\, S_3\, S_2 = S_3\, S_2\, S_3 + S_1\, S_2\, S_2 + S_1\, S_1\, S_2 \\
&S_3\, S_1\, S_3 + S_3\, S_2\, S_3 = S_3\, S_3\, S_1 + S_3\, S_3\, S_2 + S_3\, S_3\, S_1 \\
&+S_2\, S_2\, S_3 + 2S_1\, S_3\, S_1 + S_1\, S_3\, S_2 + S_1\, S_3\, S_3 + S_1\, S_2\, S_2 + S_1\, S_1\, S_1,
\end{aligned}
$$

where the arguments in each generic term $S_a S_b S_c$ of these equations are θ for the first factor S_a, $\theta + \theta'$ for the second factor S_b and θ' for the third one S_c. The general solution of these equations has the functional form

$$S_3(\theta) = -\frac{i\lambda}{\theta} S_2(\theta), \tag{18.9.5}$$

$$S_1(\theta) = -\frac{i\lambda}{i[(n-2)/2]\lambda - \theta} S_2(\theta).$$

Substituting these expressions in the crossing equations (18.9.3) and (18.9.4), we can determine the parameter λ

$$\lambda = \frac{2\pi}{n-2}. \tag{18.9.6}$$

With this new piece of information, substituting the amplitudes above in the unitarity equations, we arrive at the condition

$$S_2(\theta) S_2(-\theta) = \frac{\theta^2}{\theta^2 + \lambda^2}. \tag{18.9.7}$$

In order to solve this equation, together with (18.9.3) coming from the crossing symmetry, we can follow an iterative strategy. Notice that a solution of (18.9.7) is given

$$Q(\theta) = \frac{\theta}{\theta + i\lambda}.$$

However, this spoils the crossing symmetry equation (18.9.3), which can be re-established by writing

$$Q(\theta) = \frac{\theta}{\theta + i\lambda} \frac{i\pi - \theta}{i\pi - \theta + i\lambda}.$$

In turns, this new expression spoils the unitarity condition (18.9.7), which can be saved by rewriting $Q(\theta)$ as

$$Q(\theta) = \frac{\theta}{\theta + i\lambda} \frac{i\pi - \theta}{i\pi - \theta + i\lambda} \frac{i\pi + \theta + i\lambda}{i\pi + \theta}.$$

Iterating these two operations that, step by step, satisfy the unitarity and crossing symmetry equations, we end up in an infinite product. Using the identity

$$\frac{\Gamma(\alpha)\Gamma(\beta)}{\Gamma(\alpha+\gamma)\Gamma(\beta-\gamma)} = \prod_{k=0}^{\infty} \left[\left(1 + \frac{\gamma}{\alpha+k}\right) \left(1 - \frac{\gamma}{\beta+k}\right) \right],$$

the final result can be concisely expressed in terms of the Γ functions as

$$S_2(\theta) = \mathcal{U}^{(+)}(\theta)\mathcal{U}^{(+)}(i\pi - \theta); \tag{18.9.8}$$

$$\mathcal{U}^{(+)}(\theta) = \frac{\Gamma\left(\frac{\lambda}{2\pi} - i\frac{\theta}{2\pi}\right)\Gamma\left(\frac{1}{2} - i\frac{\theta}{2\pi}\right)}{\Gamma\left(\frac{1}{2} + \frac{\lambda}{2\pi} - i\frac{\theta}{2\pi}\right)\Gamma\left(-i\frac{\theta}{2\pi}\right)}.$$

Now we can use eqn. (18.9.5) to determine the remaining two amplitudes. The S-matrix so obtained does not have poles in the physical sheet. Hence the theory does not present additional bound states and the only excitations are given by the original particle A_i. It is possible to show that this scattering theory is in agreement with the perturbative computations done using the bosonic Lagrangian

$$\mathcal{L} = \frac{1}{2}\partial_\mu \vec{S} \cdot \partial^\mu \vec{S}, \quad |\vec{S}| = 1. \tag{18.9.9}$$

This is a non-linear σ-model: although the Lagrangian looks like the one of a free massless theory, the constraint on the components of the field induces, ipso facto, a mass term and a series of interactions. The non-linear σ-model is renormalizable, asymptotically free and explicitly $O(n)$ symmetric. The simplest way to show that a mass gap is present in this theory is to study its large n limit: introducing a coupling constant g and enforcing the constraint using the Fourier representation of the δ function, we can write the Lagrangian of the model as

$$\mathcal{L} = \frac{n}{2g}\left[(\partial_\mu \vec{S})^2 + i\lambda(x)\,(\vec{S}^2 - 1)\right].$$

$\lambda(x)$ is the Lagrangian multiplier field associated to the constraint. In writing such an expression we have parameterized the coupling constant in such a way to have a factor n in front of the Lagrangian. In the path integral of this theory we can now integrate out $\vec{S}$, since this vector is no longer constrained, obtaining an effective action for the field $\lambda(x)$

$$\mathcal{S}_{\mathit{eff}}(\lambda) = \frac{n}{2}\left[-\int d^2x\left(i\frac{\lambda(x)}{g} + \mathrm{tr}\log(-\partial^2 + i\lambda)\right)\right]. \tag{18.9.10}$$

For the factor n in front of $\mathcal{S}_{\mathit{eff}}$, in the large n-limit we can ignore the fluctuations of $\lambda(x)$ and evaluate it at the action saddle point. This can be done by deforming the functional integration contour of λ into the complex plane and a saddle point is found at a constant, imaginary value of $\lambda = i\lambda_0$. Posing $\lambda_0 = m^2$, the saddle point equation is expressed by

$$\frac{1}{g} = \int \frac{d^2k}{(2\pi)^2}\frac{1}{k^2 + m^2} = \frac{1}{2\pi}\log\frac{\Lambda}{m}$$

where Λ is an ultraviolet cut-off. This equation determines the mass parameter m of the theory, in terms of the cut-off and the bare coupling g:

$$m = \Lambda e^{-2\pi/g}.$$

At lowest order in $1/n$, the theory consists of just n free boson particles of mass m. This results is consistent with the S-matrix formulation given above.

18.9.2 $n < 2$

Section 14.6 showed that, for $-2 < n < 2$ the $O(n)$ model has a critical point. The conformal theory has central charge

$$c = 1 - \frac{6}{p(p+1)},$$

where p is a function of the index n

$$n = 2\cos\left(\frac{\pi}{p}\right). \tag{18.9.11}$$

In the continuum limit, the energy operator corresponds to the primary field $\Phi_{1,3}$. Correspondingly, the off-critical theory

$$\mathcal{S} = \mathcal{S}_{CFT} + \tau \int \Phi_{1,3}(x)\, d^2x.$$

defines an integrable theory. Let us compute the S-matrix based on the following assumptions:

1. As in the previous case, the particles A_i are associated to a vector representation of $O(n)$, even though now n can take continuous values;
2. Their S-matrix can be still decomposed into the invariant amplitudes, as done in eqn. (18.9.1);
3. In this range of values of n, there is also $n = 0$ that corresponds to the self-avoiding random walks. Associating the world-lines of the particles A_i to the lines that enter the high-temperature expansion of the lattice model, it seems natural to conjecture that for all n in the interval $-2 < n < 2$ it holds the condition

$$S_2(\theta) = 0. \tag{18.9.12}$$

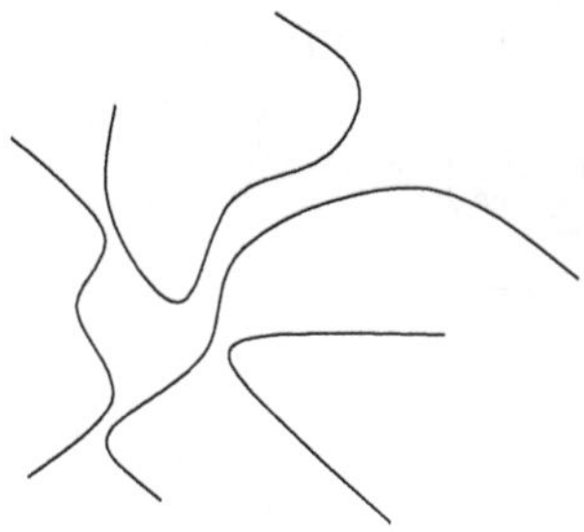

Fig. 18.6 *Configurations of self-avoiding polymers associated to the world-lines of the particles of the $O(n)$ scattering theory for $-2 < n < 2$.*

Notice that this equation is consistent with the crossing symmetry of the problem, while for the other amplitudes we have

$$S_1(\theta) = S_3(i\pi - \theta). \tag{18.9.13}$$

The general solution of the Yang–Baxter equation is given in this case by

$$\begin{aligned} S_1(\theta) &= i \sinh\left(\frac{\theta}{p}\right) R(\theta); \\ S_3(\theta) &= i \sinh\left(\frac{i\pi-\theta}{p}\right) R(\theta), \end{aligned} \tag{18.9.14}$$

where $R(\theta)$ is a function that satisfies

$$R(\theta) = R(i\pi - \theta). \tag{18.9.15}$$

It can be determined by imposing the unitarity equations

$$\begin{aligned} &S_3(\theta)\, S_3(-\theta) = 1 \\ &S_3(\theta)\, S_1(-\theta) + S_1(\theta)\, S_3(-\theta) + n\, S_1(\theta)\, S_1(-\theta) = 0. \end{aligned} \tag{18.9.16}$$

Notice that the second equation of this system is automatically satisfied by virtue of (18.9.11). The first equation implies instead

$$R(\theta)\, R(-\theta) = -\left[\sinh\left(\frac{i\pi - \theta}{p}\right) \sinh\left(\frac{i\pi + \theta}{p}\right)\right]^{-1}. \tag{18.9.17}$$

The minimal solution of both eqns. (18.9.15) and (18.9.17) is given by

$$R(\theta) = \frac{1}{\sinh\left(\frac{i\pi-\theta}{p}\right)} \frac{\Gamma\left(1-\frac{\theta}{i\pi p}\right)}{\Gamma\left(1+\frac{\theta}{i\pi p}\right)} \times \prod_{k=1}^{\infty} \frac{\Gamma\left(\frac{2k}{p}-\frac{\theta}{i\pi p}\right)\Gamma\left(1+\frac{2k}{p}-\frac{\theta}{i\pi p}\right)\Gamma\left(\frac{2k-1}{p}+\frac{\theta}{i\pi p}\right)\Gamma\left(1+\frac{2k-1}{p}+\frac{\theta}{i\pi p}\right)}{\Gamma\left(\frac{2k}{p}+\frac{\theta}{i\pi p}\right)\Gamma\left(1+\frac{2k}{p}+\frac{\theta}{i\pi p}\right)\Gamma\left(\frac{2k-1}{p}-\frac{\theta}{i\pi p}\right)\Gamma\left(1+\frac{2k-1}{p}-\frac{\theta}{i\pi p}\right)}. \tag{18.9.18}$$

As shown in Problem 18.6, the infinite product admits an integral representation, so that $R(\theta)$ can be also expressed as

$$R(\theta) = \frac{1}{\sinh\left(\frac{i\pi-\theta}{p}\right)} \exp\left[i\int_0^{\infty}\frac{dt}{t}\frac{\sinh\frac{\pi(p-1)t}{2}}{\sinh\frac{\pi pt}{2}\cosh\frac{\pi t}{2}}\sin\theta t\right]. \tag{18.9.19}$$

Notice that for $n = 1$

$$S_1(\theta) + S_3(\theta) = -1, \tag{18.9.20}$$

which coincides with the S-matrix of the thermal Ising model, as it should. In the limit $n \to 0$, the two amplitudes can be interpreted as the two possible interactions among the chains of the polymers (Figure 18.6).

18.10 S-matrix of the Sine-Gordon Model

For $n = 2$ the scattering theory of the $O(n)$ model coincides with the one associated to the Sine–Gordon model, whose Lagrangian is

$$\mathcal{L}_{SG} = \frac{1}{2}(\partial_\mu\phi)^2 + \frac{m^2}{\beta^2}(\cos\beta\phi - 1). \tag{18.10.1}$$

In view of the key role that it plays in the whole field of integrable models, we emphasize this model particularly. It is useful to define

$$\xi \equiv \frac{\beta^2}{8}\frac{1}{1-\frac{\beta^2}{8\pi}}. \tag{18.10.2}$$

This holds for the renormalized coupling constant. It is worth mentioning that Coleman discovered that the quantum Sine–Gordon model is equivalent to the massive Thirring model for a Dirac field

$$\mathcal{L}_{MTM} = i\bar{\psi}\gamma^\mu\partial_\mu\psi - m_0\bar{\psi}\psi - \frac{g}{2}(\bar{\psi}\gamma^\mu\psi)^2, \tag{18.10.3}$$

with the position

$$\frac{g}{\pi} = \frac{4\pi}{\beta^2} - 1, \quad \text{for} \quad 0 \leq \frac{\beta^2}{8\pi} < 1. \tag{18.10.4}$$

Under this mapping, the Sine–Gordon soliton is identified with the Thirring fermion. Note that the Thirring interaction becomes

$$\begin{array}{llll} \text{attractive } (g > 0) & \text{for} \quad \beta^2 < 4\pi & \text{i.e.} & \xi < \pi, \\ \text{repulsive } (g < 0) & \text{for} \quad \beta^2 > 4\pi & \text{i.e.} & \xi > \pi. \end{array} \tag{18.10.5}$$

It is important to keep this in mind for understanding the structure of the bound states of the Sine–Gordon model discussed below.

The best way to reveal the structure of the scattering theory of this model is to define its basic excitations by the complex linear combinations

$$A(\theta) = A_1(\theta) + iA_2(\theta), \quad \bar{A}(\theta) = A_1(\theta) - iA_2(\theta),$$

where A_1 and A_2 are the degenerate particles of the original $O(2)$ model. In terms of these new excitations, the scattering amplitudes can be written as

$$\begin{aligned} A(\theta_1)\bar{A}(\theta_2) &= S_T(\theta)\bar{A}(\theta_2)A(\theta_1) + S_R(\theta)A(\theta_2)\bar{A}(\theta_1), \\ A(\theta_1)A(\theta_2) &= S(\theta)A(\theta_2)A(\theta_1), \\ \bar{A}(\theta_1)\bar{A}(\theta_2) &= S(\theta)\bar{A}(\theta_2)\bar{A}(\theta_1). \end{aligned} \tag{18.10.6}$$

They can be collected into a 4×4 matrix with non-zero entries given by

$$S^{SG}(\theta) = \begin{pmatrix} S & & & \\ & S_T & S_R & \\ & S_R & S_T & \\ & & & S \end{pmatrix}, \tag{18.10.7}$$

The quantities above can be interpreted as the scattering amplitudes of the soliton A and the anti-soliton $\bar{A}$. S_T and S_R are the transmission and reflection amplitudes respectively in the soliton–anti-soliton scattering process, while S, for the charge conjugation symmetry, is the common transmission amplitude in the soliton/soliton and anti-soliton/anti-soliton scatterings. Notice the close similarity between the S-matrix (18.10.7) and the R-matrix of the 6-vertex model given in eqn. (6.4.66).

The amplitudes satisfy the crossing symmetry equations

$$S(\theta) = S_T(i\pi - \theta), \quad S_R(\theta) = S_R(i\pi - \theta) \tag{18.10.8}$$

and the unitarity conditions

$$\begin{aligned} S(\theta)S(-\theta) &= 1, \\ S_T(\theta)S_T(-\theta) + S_R(\theta)S_R(-\theta) &= 1, \\ S_T(\theta)S_R(-\theta) + S_R(\theta)S_T(-\theta) &= 0. \end{aligned} \tag{18.10.9}$$

Using the Yang–Baxter equations satisfied by the amplitudes, they can be expressed as

$$\begin{aligned} S_T(\theta) &= \frac{\sinh\frac{\pi\theta}{\xi}}{\sinh\frac{\pi(i\pi-\theta)}{\xi}} S(\theta), \\ S_R(\theta) &= i\frac{\sin\frac{\pi^2}{\xi}}{\sinh\frac{\pi(i\pi-\theta)}{\xi}} S(\theta). \end{aligned} \tag{18.10.10}$$

Substituting them in the unitarity and crossing symmetry equations, we get the equations satisfied by $S(\theta)$

$$\begin{aligned} S(\theta)S(-\theta) &= 1, \\ S(i\pi-\theta) &= \frac{\sinh\frac{\pi\theta}{\xi}}{\sinh\frac{\pi(i\pi-\theta)}{\xi}} S(\theta). \end{aligned}$$

Its solution can be written in terms of an infinite product

$$\begin{aligned} S(\theta) = \prod_{k=0}^{\infty} & \frac{\Gamma\left(1+(2k+1)\frac{\pi}{\xi}-i\frac{\theta}{\xi}\right)\Gamma\left(1+2k\frac{\pi}{\xi}+i\frac{\theta}{\xi}\right)}{\Gamma\left(1+(2k+1)\frac{\pi}{\xi}+i\frac{\theta}{\xi}\right)\Gamma\left(1+2k\frac{\pi}{\xi}-i\frac{\theta}{\xi}\right)} \\ \times & \frac{\Gamma\left((2k+1)\frac{\pi}{\xi}-i\frac{\theta}{\xi}\right)\Gamma\left((2k+2)\frac{\pi}{\xi}+i\frac{\theta}{\xi}\right)}{\Gamma\left((2k+1)\frac{\pi}{\xi}+i\frac{\theta}{\xi}\right)\Gamma\left((2k+2)\frac{\pi}{\xi}-i\frac{\theta}{\xi}\right)}. \end{aligned} \tag{18.10.11}$$

This expression admits the integral representation

$$S(\theta) = -\exp\left[-i\int_0^{\infty}\frac{dt}{t}\,\frac{\sinh\frac{t(\pi-\xi)}{2}}{\sinh\frac{\xi t}{2}\cosh\frac{\pi t}{2}}\sin\theta t\right]. \tag{18.10.12}$$

Another useful representation of this amplitude is

$$S(\theta) = -(-1)^n \prod_{k=1}^{n} \left[\frac{\theta + ik\xi}{\theta - ik\xi} \right] \tag{18.10.13}$$
$$\times \exp\left\{ -i \int_0^\infty \frac{dt}{t} \frac{\left[2\sinh \frac{t(\pi-\xi)}{2} e^{-n\xi t} + \left(e^{-n\xi t} - 1\right)\left(e^{(\xi-\pi)t/2} + e^{-(\pi+\xi)t/2}\right)\right]}{2\sinh\frac{\xi t}{2}\cosh\frac{\pi t}{2}} \sin\theta t \right\}.$$

This mixed representation is particularly helpful for determining the numerical values of $S(\theta)$: notice that the convergence of the integral increases by increasing the integer n, where the only consequence is more power factors in the first product term. Note that the integer n can be varied arbitrarily without changing the value of $S(\theta)$ and, in particular, for $n = 0$ we recover the previous expression (18.10.12). The proof of the integral and the mixed representation of $S(\theta)$ is suggested in Problem 18.6.

The pole structure of the S-matrix is determined by the various terms that enter its expression. It is important to focus attention on the poles that may belong to the physical sheet $0 < \theta < i\pi$: using the results of Appendix 2.A for the Γ functions (or simply looking at the mixed representation (18.10.13)), it is easy to see that $S(\theta)$ has the set of poles at

$$\theta = in\xi, \quad n = 0, 1, \ldots \tag{18.10.14}$$

Other poles of the S-matrix come from the factor $\sinh \frac{\pi(i\pi-\theta)}{\xi}$ in the denominator of the right-hand side of eqn. (18.10.10), placed at

$$\theta = i(\pi - n\xi), \quad n = 0, 1, \ldots. \tag{18.10.15}$$

A necessary condition such that both sets of poles fall in the physical sheet is

$$\xi < \pi. \tag{18.10.16}$$

Thus, if the condition (18.10.16) is satisfied, the poles (18.10.15) lead to the bound states in the s-channel of the SG model whereas the poles (18.10.14) lead to the bound states of the crossed t-channel. The number of these bound states is

$$\bar{N} = \left[\frac{\pi}{\xi} \right], \tag{18.10.17}$$

where $[x]$ is the integer part of x. To support the interpretation of the poles given above, it is convenient to define the amplitudes

$$S_{\pm}(\theta) = (S_R \pm S_T)(\theta). \tag{18.10.18}$$

These quantities correspond to scattering processes with well-defined quantum numbers under the charge conjugation operator: S_- has charge conjugation $C = -1$ while S_+ has $C = +1$, as shown by writing the scattering processes as

$$\begin{aligned}\left[A(\theta_1)\bar{A}(\theta_2) + \bar{A}(\theta_1)A(\theta_2)\right] &= S_+(\theta)\left[A(\theta_2)\bar{A}(\theta_1) + \bar{A}(\theta_2)A(\theta_1)\right],\\ \left[A(\theta_1)\bar{A}(\theta_2) - \bar{A}(\theta_1)A(\theta_2)\right] &= S_+(\theta)\left[A(\theta_2)\bar{A}(\theta_1) - \bar{A}(\theta_2)A(\theta_1)\right].\end{aligned}$$

The explicit expressions of these amplitudes are

$$S_\pm(\theta) = -\frac{1}{\sinh\frac{\pi(\theta - i\pi)}{\xi}}\left[i\sin\frac{\pi^2}{\xi} \pm \sinh\frac{\pi\theta}{\xi}\right] S(\theta), \tag{18.10.19}$$

and their residue at the poles (18.10.15) is

$$S_\pm(\theta) \simeq -\frac{i}{\theta - i\pi + in\xi}(-1)^n \xi \sin\frac{\pi^2}{\xi}\left[1 \pm (-1)^n\right] S(i\pi - in\xi). \tag{18.10.20}$$

Hence, $S_+(\theta)$ only has poles when n is an even number, while $S_-(\theta)$ only when n is an odd number. Both sets of poles have a positive residue[3] and therefore, as anticipated, they correspond to the poles of the s-channel associated to the bound states B_n. These are ordinary scalar particles, called *breathers*, with eigenvalues $C = (-1)^n$ under the charge conjugation. If M is the mass of the solitons, the mass spectrum of the bound states is given by

$$m_n = 2M\sin\frac{k\xi}{2}, \quad k = 1, 2, \ldots, \bar{N}. \tag{18.10.21}$$

The S-matrix elements that involve the bound states can be computed by the bootstrap equations. For their scattering processes with the solitons

$$\begin{aligned}A(\theta_1)B_n(\theta_2) &= S^{(n)}(\theta)\,B_n(\theta_2)A(\theta_1),\\ \bar{A}(\theta_1)B_n(\theta_2) &= S^{(n)}(\theta)\,B_n(\theta_2)\bar{A}(\theta_1),\end{aligned} \tag{18.10.22}$$

we have

$$S^{(n)}(\theta) = \frac{\sinh\theta + i\cos\frac{n\xi}{2}}{\sinh\theta - i\cos\frac{n\xi}{2}}\prod_{k=1}^{n-1}\frac{\sin^2\left(\frac{(n-2k)\xi}{4} - \frac{\pi}{4} + i\frac{\theta}{2}\right)}{\sin^2\left(\frac{(n-2k)\xi}{4} - \frac{\pi}{4} - i\frac{\theta}{2}\right)}. \tag{18.10.23}$$

[3] The mixed representation (18.10.13) is particularly useful in the evaluation of these residues. Attention has to be paid when $\xi = \pi/m$, with m an integer, for the simultaneous presence of a pole in $S(i\pi - in\xi)$ and a zero in $\sin\pi^2/\xi$.

For the scattering processes that involves only the particles B_n

$$B_n(\theta_1)B_m(\theta_2) = S^{(n,m)}(\theta)\, B_m(\theta_2)B_n(\theta_1), \tag{18.10.24}$$

we have the amplitudes

$$\begin{aligned} S^{(n,m)}(\theta) = &\frac{\sinh\theta + i\sin\left(\frac{(n+m)\xi}{2}\right)}{\sinh\theta - i\sin\left(\frac{(n+m)\xi}{2}\right)}\,\frac{\sinh\theta + i\sin\left(\frac{(n-m)\xi}{2}\right)}{\sinh\theta - i\sin\left(\frac{(n-m)\xi}{2}\right)} \\ &\times \prod_{k=1}^{n-1} \frac{\sin^2\left(\frac{(m-n-2k)\xi}{4} + i\frac{\theta}{2}\right)}{\sin^2\left(\frac{(m-n-2k)\xi}{4} - i\frac{\theta}{2}\right)}\,\frac{\cos^2\left(\frac{(m+n-2k)\xi}{4} + i\frac{\theta}{2}\right)}{\cos^2\left(\frac{(m+n-2k)\xi}{4} - i\frac{\theta}{2}\right)} \end{aligned} \tag{18.10.25}$$

with $n \geq m$. From the analysis of the poles of these expressions it is easy to see that the particle B_n can be regarded as a bound state of $B_k + B_l$, with $k + l = n$. Consequently, iterating this relation, the particle B_n is the bound state of n elementary particles B_1. The lowest particle B_1 can be associated to the excitation created by the field ϕ that enters the Lagrangian (18.10.1). When $n = m = 1$, the amplitude of the fundamental particle is given by

$$S^{(1,1)}(\theta) = \frac{\sinh\theta + i\sin\xi}{\sinh\theta - i\sin\xi}. \tag{18.10.26}$$

With the expression of ξ given in (18.10.2), this amplitude can be expanded in power series of β^2 and successfully compared with the perturbative series coming from the Lagrangian (18.10.1). Notice that making the analytic continuation $\beta \to ig$ the Sine–Gordon model becomes the Sinh–Gordon model: the formula (18.10.2) given for ξ becomes the expression (18.4.5) previously obtained for the function $B(g)$ of the latter model, while the amplitude (18.10.26) reduces to the S-matrix (18.4.4) of the Sinh–Gordon model.

Finally, notice that when $\xi > \pi$, the pole in the soliton–anti-soliton amplitude falls outside the physical sheet. Correspondingly, there is no longer the bound state B_1 associated to the field ϕ, despite the fact the Lagrangian is expressed in terms of this field! This observation shows that the spectrum of a QFT is a question of dynamical nature and less intuitive than it could appear.

18.11 S-Matrices for $\Phi_{1,3}$, $\Phi_{1,2}$, $\Phi_{2,1}$ Deformation of Minimal Models

Section 16.8, the $\Phi_{1,3}$, $\Phi_{1,2}$ and $\Phi_{2,1}$ deformations of the minimal models $\mathcal{M}_m$ of CFT generally lead to integrable massive field theories with kink behaviour. This means that such deformations give rise to an effective potential of the theory with a finite number of degenerate vacua and the basic massive excitations are the kinks that interpolate between them. There may also be kink-bound states.

There is a general approach for dealing with such massive theories, deeply related to the Sine–Gordon model (for the $\Phi_{1,3}$ deformation) and to the Bullogh–Dodd model with imaginary coupling (for the $\Phi_{1,2}$ and $\Phi_{2,1}$ deformations) (see Chapter 16). The main idea behind this approach consists of the well-known relation between the S-matrices and the R-matrices entering the transfer matrix of lattice integrable models (see Sections 6.4 and 17.2). An important features of both these quantities is their invariance under the quantum group $SL_q(2)$. The q-parameter is a function of the coupling constant and, when q is a root of unity, it is possible to restrict the Hilbert space of the original models, preserving both the integrability and the locality of an invariant set of operators. Let us explore how this procedure is implemented for the Sine–Gordon model and for the Bullough–Dodd model.

18.11.1 Quantum Group Symmetry of the Sine–Gordon

The quantum group $SL_q(2)$ is the deformation of the algebra of functions over $SL(2)$. It is defined by the universal enveloping algebra $\mathcal{U}_q[sl(2)]$ with the commutation relations

$$[\mathcal{J}_+, \mathcal{J}_-] = \frac{q^H - q^{-H}}{q - q^{-1}}, \qquad [H, \mathcal{J}_\pm] = \pm 2\mathcal{J}_\pm. \tag{18.11.1}$$

If the deformation parameter q goes to 1, eqn. (18.11.1) reduces to the ordinary $SL(2)$ commutation relations and, correspondingly, the quantum group $SL_q(2)$ to the ordinary $SL(2)$ group. $\mathcal{U}_q[sl(2)]$ forms a Hopf algebra with the *co-multiplication* Δ_q defined by

$$\begin{aligned} \Delta_q(H) &= 1 \otimes H + H \otimes 1 \\ \Delta_q(\mathcal{J}_\pm) &= q^{H/2} \otimes \mathcal{J}_\pm + \mathcal{J}_\pm \otimes q^{-H/2}. \end{aligned} \tag{18.11.2}$$

The co-multiplication Δ_q is an algebra homomorphism and is the analogous of addition of angular momentum in $SU(2)$, which reduces when $q \to 1$. The irreducible representations of $SL_q(2)$ are generated by the co-multiplication Δ_q which defines tensor product representations.

Because of the resemblance between the algebra structure of $SL(2)$ and $\mathcal{U}_q[sl(2)]$, the representation theory of the quantum group is quite similar to the classical theory. The irreducible representations of $\mathcal{U}_q[su(2)]$ are labelled by $j = 0, \frac{1}{2}, 1, \ldots$ acting on a Hilbert space V_j with basis vectors $|j, m\rangle$ $(-j \leq m \leq j)$ as follows

$$\mathcal{J}_3 \,|j,m\rangle = m \,|j,m\rangle, \qquad \mathcal{J}_\pm \,|j,m\rangle = \sqrt{[j \mp m]_q \, [j \pm m + 1]_q} \,|j,m\rangle,$$

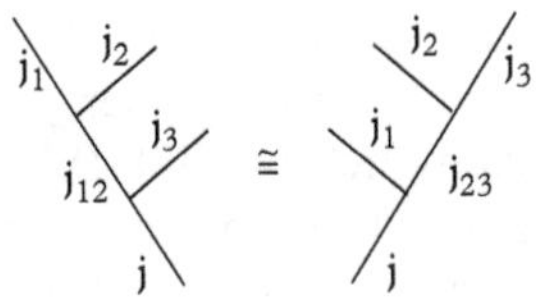

Fig. 18.7 *Equivalence between the two Hilbert spaces $(V_{j_1} \otimes V_{j_2}) \otimes V_{j_3}$ (left-hand side) and $V_{j_1} \otimes (V_{j_2} \otimes V_{j_3})$ (right-hand side) rules by the $6-j$ symbols.*

where all the usual numbers have turned into q-numbers, so defined

$$[n]_q \equiv \frac{q^n - q^{-n}}{q - q^{-1}} \quad \text{and} \quad [n]_q \to n \quad \text{as} \quad q \to 1. \tag{18.11.3}$$

All these representations can be obtained starting with the fundamental representation $j = \frac{1}{2}$ and using eqn. (18.11.2) with the relation

$$|\mathcal{J}, M; j_1, j_2\rangle \sum_{m_1, m_2} \begin{bmatrix} j_1 & j_2 & \mathcal{J} \\ m_1 & m_2 & j \end{bmatrix}_q |j_1, m_1\rangle \otimes |j_2, m_2\rangle, \tag{18.11.4}$$

where the quantum analogue of the Clebsh–Gordon (CG) coefficients appears.[4] To cluster together three representations there are two possibilities: one is related to the configuration $(V_{j_1} \otimes V_{j_2}) \otimes V_{j_3}$, the other one to $V_{j_1} \otimes (V_{j_2} \otimes V_{j_3})$. Both are physically equivalent and related each to the other by the quantum analogue of the 6-j symbols: denoting by $e_m^{j_{12},j}(j_1, j_2 \,|\, j_3)$ an orthonormal basis in $V_{j_1} \otimes V_{j_2} \otimes V_{j_3}$ associated to the left-hand side of Figure 18.7 and by $e_m^{j_{23},j}(j_1 \,|\, j_2, j_3)$ an orthonormal basis associated to the right-hand side of the same figure, we have

$$e_m^{j_{12},j}(j_1, j_2 \,|\, j_3) = \sum_{j_{23}} \begin{Bmatrix} j_1 & j_2 & j_{12} \\ j_3 & j & j_{23} \end{Bmatrix}_q e_m^{j_{23},j}(j_1 \,|\, j_2, j_3).$$

As far as q is not a root of unity, the irreducible representations has dimension $(2j+1)$. However, when q is a root of unity, we can see from eqn. (18.11.3) that some of the q-CG coefficients (and the q-6j symbols) become singular. For this case, the sensible representation theory of the $SL_q(2)$ is obtained by restricting the allowed spins to $\{0, 1/2, \ldots, j_{max}\}$, where j_{max} is determined by the condition

$$[2j_{max} + 1]_q = 0 \quad \to \quad j_{max} = \frac{N}{2} - 1 \quad \text{for} \quad q^N = \pm 1. \tag{18.11.5}$$

[4] For the classical values of the Clesh–Gordon and 6-j coefficients see Landau and Lifshitz (1992). *Quantum Mechanics. Non-relativistic Theory*, Pergamon.

This restriction on the allowed representations of $SU_q(2)$ with q a root of unity leads to the truncation of the Hilbert space. It is this property, in particular, that is responsible for the fusion rules of the minimal models of CFT (Section 11.4).

For the Sine–Gordon, the restriction on the spins may lead to the truncation of the multi-kink Hilbert space. Let us see in more detail how this happens. Notice that the quantum group $SL_q(2)$ can be realized by a constant R-matrix defined by

$$R_{12}(q)\,(g\otimes 1)\,(1\otimes g) = (1\otimes g)\,(g\otimes 1)\,R_{12}(q) \quad \text{with} \quad g\in SL_2(q). \qquad (18.11.6)$$

Using eqn. (18.11.2), this implies $\left[R(q), \Delta_q(g)\right] = 0$ for any $g \in SL_q(2)$. Fundamentally, g is a 2×2 matrix with q-determinant equal to 1 (i.e. $g_{11}g_{22} - q g_{12}g_{21} = 1$) and the non-zero entries of the R_{12}-matrix are

$$R_{12}(q) = \begin{pmatrix} q & & & \\ & 1 & q-q^{-1} & \\ & 0 & 1 & \\ & & & q \end{pmatrix}. \qquad (18.11.7)$$

The spectral parameter λ can be introduced in the R-matrix as follows

$$\hat{R}_{12}(\lambda, q) = \lambda \hat{R}_{12}(q) - \lambda^{-1}\hat{R}_{12}^{-1}(q) \quad \text{with} \quad \hat{R}_{12} = \mathcal{P} R_{12}, \qquad (18.11.8)$$

where the permutation $\mathcal{P}$ is defined as $\mathcal{P}(V_1\otimes V_2) = V_2 \otimes V_1$. The matrix $\hat{R}_{12}(p,q)$ acts on the vector space $\mathbf{C}^2\otimes\mathbf{C}^2$ and is solution to the Yang–Baxter equation.

The $SL_q(2)$ quantum group symmetry of the Sine–Gordon is obtained by noticing that the soliton S-matrix of this model, given in eqn. (18.10.7), can be expressed in terms of $\hat{R}_{12}(p,q)$ as

$$S^{SG}(\theta) = \frac{S(\theta)}{2\sinh\frac{\pi(i\pi-\theta)}{\xi}}\,\hat{R}_{12}(\lambda = e^{\theta}, q), \qquad q = -e^{-i\pi^2/\xi}, \qquad (18.11.9)$$

where $S(\theta)$ is given in eqn. (18.10.12). Notice that the deformation parameter q is related to the coupling constant. From $\left[S^{SG}, \Delta_q\right] = 0$, the soliton and anti-soliton pair forms the fundamental spin–1/2 representation, whereas the multi-soliton states may be regarded as the irreducible representations with higher spins which are created by tensor products like those in eqn. (18.11.4).

18.11.2 Restricted Sine–Gordon model

For arbitrary values of the coupling constant of the Sine–Gordon model, starting from the spin-1/2 representation of the soliton/anti-soliton, we can obtain multi-soliton states with $j = 1, 3/2, \ldots$. Ordinarily there is no limit to the number of solitons. But, if ξ assume rational values, q given in eqn. (18.11.9) becomes root of unit and j is bounded by j_{max}. This means that the Sine–Gordon model at special rational values of the coupling

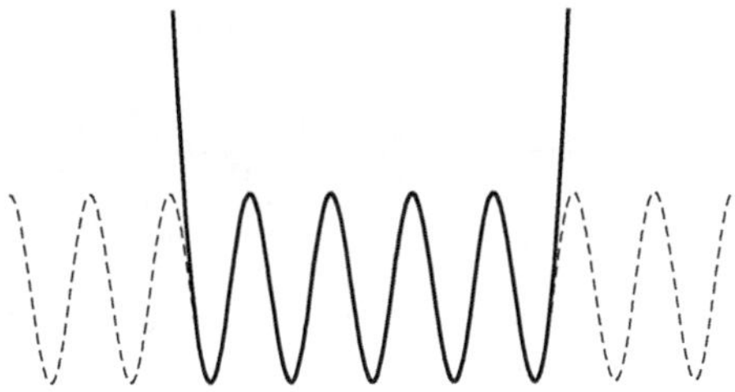

Fig. 18.8 *Effective potential of the Sine–Gordon theory at $\xi = \pi/m$ (for $m = 6$). The dashed line is the original untruncated potential.*

constant cannot sustain solitons exceeding a certain number. Let us consider the various cases of interest.

Φ_{13} deformation of minimal unitary models $\mathcal{M}_m$. The scattering theory of these models is obtained with the condition

$$\frac{\beta^2}{8\pi} = \frac{m}{m+1} \qquad \longrightarrow \qquad \xi = m\pi, \tag{18.11.10}$$

In this case $j_{max} = \frac{m}{2} - 1$ and there are at most $(m-2)$ solitons (or anti-solitons). This peculiar aspect of the model can be understood as follows. In the original Hilbert space of the Sine–Gordon theory, there are many sectors, each containing a certain number of solitons. Then there is the sector containing up to $(m-2)$ solitons. That sector decouples from the rest of the Hilbert space if eqn. (18.11.10) holds and it can be isolated out. Since solitons connect neighbouring vacua, for a system having only up to a certain number of solitons its effective potential is going to be reduced: while there is an infinite degeneracy of vacua in the original Sine–Gordon theory, at the special values (18.11.10) there is a truncation. With $(m-2)$ solitons, the truncated potential has only $(m-1)$ vacua (Figure 18.8), and we can imagine that the effective potential turnes over at the edges. This agrees with the description of the massive $\Phi_{1,3}$ perturbed unitary minimal models $\mathcal{M}_m$: after perturbation, the original multiple vacua split into $(m-1)$ degenerate ones only.

To get the S-matrix of the truncated theory it is necessary to change the basis in the Hilbert space: since a soliton–antisoliton pair forms a spin–1/2 representation $|\frac{1}{2}, \pm\frac{1}{2}\rangle$, we can decompose the multi-soliton Hilbert space $\mathcal{H}$ into the irreducible spaces characterized by the higher spin, as shown in Figure 18.9. In lattice models this is equivalent to change the Boltzmann weights from the fluctuating variables expressed in terms of the vertices to the so-called RSOS (Restricted Solid On Solid) variables

$$\mathcal{H} = \sum_{m_i = \pm 1/2} |\frac{1}{2}, m_1\rangle \otimes |\frac{1}{2}, m_2\rangle \otimes \cdots |\frac{1}{2}, m_N\rangle = \sum_{\substack{0 \le j_1 < \infty \\ |j_i - j_{i+1}| = 1/2}} |j_1, \cdots, j_N\rangle$$

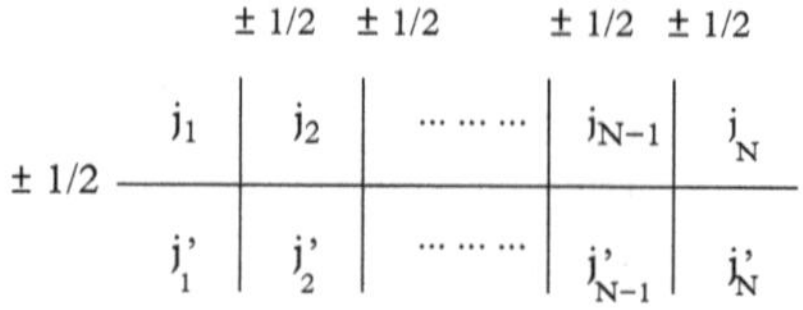

Fig. 18.9 *Change of basis from the vertex to IRF form.*

with appropriate q-CG coefficients. In this new basis, the multi-soliton Hilbert space is spanned by the kink $K_{ab}(\theta)$, where a, b act as the RSOS vacua satisfying the condition $|a-b| = 1/2$. The kink is therefore a domain wall between two vacua, and a multi-kink state $| K_{ab}(\theta_1)K_{bc}(\theta_2)K_{cd}(\theta_3)\dots\rangle$ should also have the next-neighbouring indices equal in order to have no jumps in the field configuration (Figure 18.10). The S-matrix of the two-kink scattering can be derived from a restriction of the original S-matrix. For the scattering process

$$| K_{da}(\theta_1)K_{ab}(\theta_2)\rangle \to | K_{dc}(\theta_2)K_{cb}(\theta_1)\rangle$$

whose graphical representation is

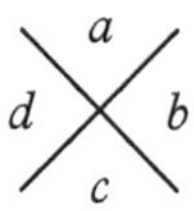

$$= S^{ab}_{dc}(\theta)$$

the explicit form of the RSOS S-matrix is

$$S^{ab}_{dc}(\theta) = \frac{S(\theta)}{2\sinh\frac{\pi(i\pi-\theta)}{\xi}} \left(\frac{[2a+1]]_q\,[2c+1]_q}{[2d+1]_q\,[2b+1]_q}\right)^{-\theta/2\pi i} \tag{18.11.11}$$
$$\times\left[\delta_{db}\sinh\frac{\theta}{\xi}\left(\frac{[2a+1]]_q\,[2c+1]_q}{[2d+1]_q\,[2b+1]_q}\right)^{1/2} + \delta_{ac}\sinh\left(\frac{i\pi-\theta}{\xi}\right)\right].$$

Note that for the values of ξ given in eqn. (18.11.10) the original Sine–Gordon S-matrix and also the S-matrix S^{ab}_{dc} of the kinks do not have bound states. It satisfies the crossing

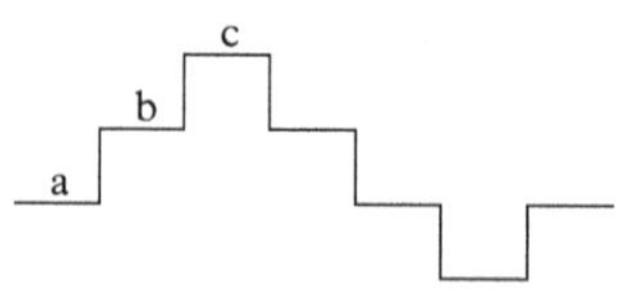

Fig. 18.10 *Multi-kink configuration.*

condition $S_{dc}^{ab}(\theta) = S_{ad}^{bc}(i\pi - \theta)$. Its non-vanishing basic entries are (up to a common prefactor)

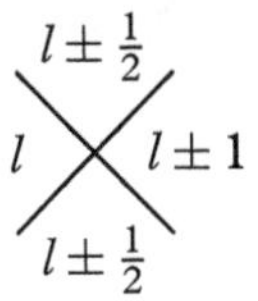

$$= \sinh\left(\frac{\pi(i\pi - \theta)}{\xi}\right)$$

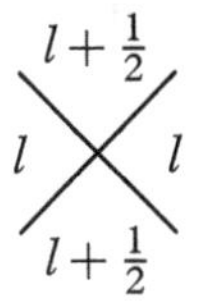

$$= \frac{\sin \pi^2/\xi}{\sin[(2l+1)\pi^2/\xi]} \sinh\left(\frac{\pi\,[i\pi(2l+1)+\theta]}{\xi}\right)$$

(18.11.12)

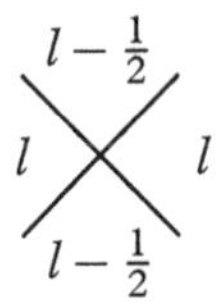

$$= \frac{\sin \pi^2/\xi}{\sin[(2l+1)\pi^2/\xi]} \sinh\left(\frac{\pi\,[i\pi(2l+1)-\theta]}{\xi}\right)$$

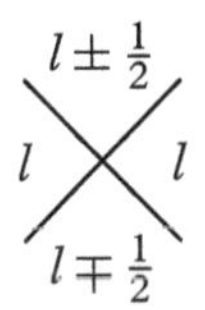

$$= \frac{\sqrt{\sin[2l\pi^2/\xi]\,\sin[\pi^2(2l+2)/\xi]}}{\sin[(2l+1)\pi^2/\xi]} \sinh\left(\frac{\pi\theta}{\xi}\right).$$

In the formulae above, l labels the different vacua states and takes the value

$$l = 0, \frac{1}{2}, \ldots, \frac{(m-2)}{2}. \tag{18.11.13}$$

Φ_{13} deformation of minimal non-unitary models $\mathcal{M}_{p_1 p_2}$. The scattering theory of these models are obtained for these rational values of the coupling constant

$$\frac{\beta^2}{8\pi} = \frac{p_1}{p_2} \quad \longrightarrow \quad \xi = \pi\,\frac{p_1}{p_2 - p_1} \tag{18.11.14}$$

(with $p_2 > p_1$). Notice that for these values of the coupling constant, the S-matrix of the Sine–Gordon has poles corresponding to the bound states. Since the breathers are singlets of $SU_q(2)$, the restriction does not change the breather sector. The S-matrices of the breathers of the restricted Sine–Gordon theory are given exactly by eqns. (18.10.25). Furthermore, notice that for these values $q^{p_1} = -1$, and the maximum value $j_{max} = \frac{(p_1-2)}{2}$ is determined only by p_1. The labels of the vacuum states are

$$l = 0, \frac{1}{2}, \ldots, \frac{(p_1-2)}{2}. \tag{18.11.15}$$

However, in this case there is an additional constraint coming from the unitarity condition of the RSOS S-matrix: the RSOS S-matrix (18.11.11) satisfies the equation

$$\sum_k S_{dk}^{ab}(\theta)\, S_{dc}^{kb}(-\theta) = \delta_{ac} \tag{18.11.16}$$

and, as far as the condition $S^\dagger(\theta) = S(-\theta)$ is satisfied, the scattering theory is unitary. The problem is with the last term in eqn. (18.11.12) that contains square roots. The reality of this term selects the series of values

$$\xi = \pi\,\frac{r}{rk+1}, \qquad \begin{array}{l} r = 2,3,\ldots \\ k = 0,1,\ldots \end{array} \tag{18.11.17}$$

and

$$\xi = \pi\,\frac{3}{3k+2}, \qquad k = 0,1,\ldots. \tag{18.11.18}$$

These are the only values of ξ for which the RSOS S-matrix description of the perturbed conformal models $\mathcal{M}_{p_1p_2}$ admit a self-consistent physical interpretation. For other rational values of ξ one can still use the RSOS S-matrix as monodromy algebra of the asymptotic particles but should be ready to sacrifice some of the usual properties of the S-matrix. Notice that in the series $\xi = \frac{2\pi}{2n+1}$ the solitons disappaer completely and only breathers remain in the spectrum. Their S-matrix was determined in Section 18.2. For the series $\xi = \frac{3\pi}{3n+1}$, there are instead solitons but they they behave as ordinary particles because there are only two vacua.[5]

18.11.3 Quantum Group Symmetry of the Bullough–Dodd Model

As discussed in Section 16.4, the $\Phi_{1,2}$ and $\Phi_{2,1}$ integrable deformations of the minimal models of CFT

$$\mathcal{S}_{\pm}^{(12)} = \mathcal{S}_{CFT} \pm g \int \Phi_{12}(x)\, d^2x \tag{18.11.19}$$

$$\mathcal{S}_{\pm}^{(21)} = \mathcal{S}_{CFT} \pm g \int \Phi_{21}(x)\, d^2x \tag{18.11.20}$$

[5] When there are only two vacua, the kink degrees of freedom are frozen because in the scattering $|\,K_{da}(\theta_1)K_{ab}(\theta_2)\rangle \rightarrow |\,K_{dc}(\theta_2)K_{cb}(\theta_1)\rangle$ the intermediate indices are forced to be equal, $b = c$.

can be associated to the Bullogh–Dodd model with an imaginary coupling and a charge at infinity. Its Lagrangian can be formally written as

$$\mathcal{L} = \frac{1}{2}(\partial_\mu \phi)^2 + \frac{1}{2} e^{i\beta\phi} + e^{-i\beta\phi/2}. \tag{18.11.21}$$

An important difference with respect to the Sine–Gordon model (i.e. the Sinh–Gordon model with imaginary coupling) is that the Lagrangian (18.11.21) is not a Hermitian operator, so the definition itself of a Lagrangian as (18.11.21) seems to be problematic. The solution of this problem and the resulting S-matrices for the massive $\Phi_{1,2}$ and $\Phi_{2,1}$ integrable deformations is one of the most beautiful results of the quantum group approach. Smirnov showed that in this case only the restricted RSOS theories have a physical meaning. Here we review the main steps of this analysis, considering first the Φ_{12} deformation.

$\Phi_{1,2}$ deformation. Since the Bullogh–Dodd model is related to the (non-simply laced) Lie algebra $A_2^{(2)}$, the first step is to consider the R-matrix of this algebra. Similarly to the Sine–Gordon model, it contains a spectral parameter λ but is constructed using the spin–1 representation of $SL_q(2)$. Its expression is

$$\hat{R}_{12}(\lambda, q) = (\lambda^{-1} - 1) q^{3/2} R_{12}(q) + (1 - \lambda) q^{-3/2} R_{21}^{-1}(q) + q^{-5/2}(q^2 - 1)(q^3 + 1)\mathcal{P},$$

where $\mathcal{P}$ is the permutation operator. $\hat{R}_{12}(\lambda, q)$ is an operator acting on the vector space $\mathbf{C}^3 \otimes \mathbf{C}^3$. The matrix $R_{12}(q)$ is the constant solution of the Yang–Baxter equation for the spin–1 representation of the quantum group $SL_q(2)$, given by

$$R_{12}(q) = \exp\left(\frac{H \otimes H}{4}\right)\left[I + (q^2 - 1) E \otimes F + (q - 1)^2 (q + 1) E^2 \otimes F^2\right]$$

where

$$H = \begin{pmatrix} 2 & 0 & 0 \\ 0 & 0 & 0 \\ 0 & 0 & -2 \end{pmatrix}, \quad E = \begin{pmatrix} 0 & 1 & 0 \\ 0 & 0 & q^{-1/2} \\ 0 & 0 & 0 \end{pmatrix}, \quad F = \begin{pmatrix} 0 & 0 & 0 \\ 1 & 0 & 0 \\ 0 & q^{1/2} & 0 \end{pmatrix}.$$

The second step is to identify the hypothetical S-matrix of the three-component kink of the Bullogh–Dodd with imaginary coupling. To interpret the matrix $R_{12}(\lambda, q)$ as an S-matrix, we must relate the spectral parameter λ to the rapidity variable θ and q to the coupling constant β of the model. With the identification

$$q = e^{i\frac{16\pi}{\beta^2}}, \qquad \lambda = e^{\frac{2\pi}{\xi}\theta}, \qquad \xi = \frac{2}{3}\left(\frac{\pi\beta^2}{16\pi - \beta^2}\right), \tag{18.11.22}$$

the hypothetical S-matrix of the three-component kink is given by

$$\hat{S}_{12}(\theta) = S_0(\theta)\,\hat{R}_{12}\left(e^{\frac{2\pi}{\xi}\theta}, e^{i\frac{16\pi}{\beta^2}}\right). \tag{18.11.23}$$

The prefactor $S_0(\theta)$ ensures the validity of the ‘unitarity’ equation $\hat{S}_{12}(\theta)\hat{S}_{21}(-\theta)$ and reads

$$\begin{aligned} S_0(\theta) = &\left[\sinh\frac{\pi}{\xi}(\theta - i\pi)\sinh\frac{\pi}{\xi}\left(\theta - \frac{2\pi i}{3}\right)\right]^{-1} \\ &\times \exp\left[-2i\int_0^\infty \frac{dt}{t}\,\frac{\sinh\left(\frac{\pi t}{3}\right)\cosh\left[\left(\frac{\pi}{6}-\frac{\xi}{2}\right)t\right]}{\cosh\left(\frac{\pi t}{2}\right)\sinh\left(\frac{\xi t}{2}\right)}\sin(\theta t)\right]. \end{aligned} \tag{18.11.24}$$

This prefactor satisfies the crossing relation $S_0(\theta) = S_0(i\pi - \theta)$. It can also be expressed in terms of an infinite product of Γ-functions, using the same procedure of the Sine–Gordon model. For a generic value of ξ, the simple poles that lie on the physical sheet $0 \leq \theta \leq i\pi$ are at the crossing symmetric places

$$\theta = \begin{cases} i\pi - i\xi m, & i\xi m, \qquad m > 0 \\ \frac{2\pi i}{3} - i\xi m, & \frac{i\pi}{3} + i\xi m, \qquad m \leq 0 \end{cases}. \tag{18.11.25}$$

In both sets, the first poles are the singularities of the s-channel, whereas the second poles are those of the crossing t-channel.

For the first set in (18.11.25), the R-matrix degenerates into a one-dimensional projector and the corresponding poles correspond then to the breather bound states. Using the bootstrap equation, the S-matrix of the fundamental breather (corresponding to the pole at $\theta = i\pi - i\xi$) is given by

$$S_{b_1 b_1}(\theta) = f_{\frac{2}{3}}(\theta) f_{\frac{\xi}{\pi}}(\theta) f_{\frac{\xi}{\pi}-\frac{1}{3}}(\theta), \tag{18.11.26}$$

where the function $f_x(\theta)$ are those defined in (17.4.64).

For the second set of poles (18.11.25) the R-matrix degenerates instead into a three-dimensional projector and these poles are associated to the higher kinks. But, from a physical point of view, $\hat{S}_{12}(\theta)$ has some drawbacks that prevent its use as the correct scattering amplitude of the perturbed CFT. For instance, when q is a root of unity, the $\hat{R}_{12}$-matrix does not satisfy the relation $\hat{R}^*_{12}(\lambda) = \hat{R}_{21}(\lambda^{-1})$, that is crucial to correctly implement the unitarity condition of the scattering amplitudes. Therefore, as it is, the S-matrix (18.11.23) cannot be assumed as the scattering amplitude of the Bullogh–Dodd at imaginary coupling. It is only its RSOS restriction that has a physical interpretation and this happens when $q^r = 1$. The RSOS kink states that enter the reduced model

$$|\,\{\theta_1, j_1, a_1\}, \{\theta_2, j_2, a_2\}, \ldots \{\theta_n, j_n, a_n\}\rangle$$

are characterized by their rapidities θ_i, their $SL_q(2)$ spin and by a string of numbers a_i (that identify the vacua) constrained by

$$a_i \leq \frac{1}{2}(r-2), \qquad |\, a_k - 1 \,| \leq a_{k+1} \leq \min(a_k+1, r-3-a_k). \tag{18.11.27}$$

These constraints formally correspond to the decomposition of tensor products of irreducible representations of $SL_q(2)$ for $q^r = 1$:

$$V_{j_1} \otimes V_{j_2} = \sum_{j=|j_1-j_2|}^{\min(j_1+j_2, r-j_1-j_2-2)} V_j, \quad j_1, j_2 \leq \frac{r-2}{2}$$

The S-matrix of the RSOS kinks is

$$S^{a_k a'_k}_{a_{k-1} a_{k+1}}(\theta) = \frac{i}{4} S_0(\theta) \left[\begin{Bmatrix} 1 & a_{k-1} & a_k \\ 1 & a_{k+1} & a'_k \end{Bmatrix}_q \left(e^{\frac{2\pi}{\xi}\theta} - 1 \right) q^{(c_{a_{k+1}} + c_{a_{k-1}} - c_{a_k} - c_{a'_k} + 3)/2} \right.$$
$$\left. - \left(e^{-\frac{2\pi}{\xi}\theta} - 1 \right) q^{-(c_{a_{k+1}} + c_{a_{k-1}} - c_{a_k} - c_{a'_k} + 3)/2} + q^{-5/2}(q^3+1)(q^2-1)\,\delta_{a_k a'_k} \right]. \tag{18.11.28}$$

Herein, $c_a \equiv a(a+1)$ and the $\{...\}_q$ are the quantum 6-j symbols. As for the RSOS restriction of the Sine–Gordon, the above S-matrix is unitary if and only if the 6-j symbols are real. This happens for the following cases:

(i)

$$\frac{\beta^2}{8\pi} = \frac{m}{m+1}, \tag{18.11.29}$$

that corresponds to the $\Phi_{1,2}$ deformation of the minimal unitary models $\mathcal{M}_m$.

(ii)

$$\frac{\beta^2}{8\pi} = \frac{2}{2n+1}, \qquad \frac{\beta^2}{8\pi} = \frac{3\pi}{3n \pm 1}, \tag{18.11.30}$$

related to the Φ_{12} deformation of the minimal models $\mathcal{M}_{2,2n+1}$ and $\mathcal{M}_{3,3n\pm1}$. For these values of $\beta^2/8\pi$ the maximal allowed spins are 0 and $\frac{1}{2}$. Hence, the kinks disappear and the spectrum is only given by the breathers obtained closing the bootstrap with the initial S-matrix (18.11.26).

(iii)

$$\frac{\beta^2}{8\pi} = \frac{4\pi}{4n \pm 1}, \tag{18.11.31}$$

that correspond to the $\Phi_{1,2}$ deformation of the minimal models $\mathcal{M}_{4,4n\pm 1}$. For this series the maximal allowed spin is equal to 1 and, according to the RSOS restriction, the kinks behave as a scalar particle.

For other rational values of $\beta^2/8\pi = r/s$ we can still use the RSOS S-matrix as monodromy algebra of the asymptotic particles, but we should be ready to sacrifice some of the usual properties of the S-matrix. Properly interpreted, they can be assumed as the S-matrix of the Φ_{12} perturbation of the minimal models $\mathcal{M}_{r,s}$.

Let us discuss in more detail the vacuum structure for the values (18.11.29) that correspond to the Φ_{12} deformation of the unitary minimal models. Since the R-matrix is based on spin–1 representation, there are two closed sub-spaces V_m^+ and V_m^- containing respectively half-integer or integer spins out of the set $a = 0, 1/2, 1, \ldots$ $m/2 - 1$. Each of these sub-spaces is associated to the set of vacua and the RSOS restriction gives rise to two quantum field theories. In the Landau–Ginzburg picture, the field $\Phi_{1,2}$ is associated to the composite operator $:\varphi^{m-2}:$. Hence, when m is odd, the field Φ_{12} is odd under the Z_2 spin symmetry and therefore, changing the sign of g in (18.11.19) leads to the same theory. On the contrary, when m is even the two theories $\mathcal{S}_{\pm}^{(12)}$ are expected to be different.[6] It becomes natural to identify V^+ with the vacuum states of the theory $\mathcal{S}_{+}^{(12)}$ and V^- with those of $\mathcal{S}_{-}^{(12)}$. Thus, we have the following situations

- m **odd**. There are $(m-1)/2$ degenerate vacua in both theories $\mathcal{S}_{\pm}^{(12)}$ labelled as

$$
\begin{aligned}
a &= \frac{1}{2}, \frac{3}{2}, \ldots, \frac{m-2}{2}, \quad g > 0 \\
a &= 0, 1, \ldots, \frac{m-3}{2}, \quad g < 0
\end{aligned}
$$

- m **even.** The number of vacua of $\mathcal{S}_{+}^{(12)}$ is equal to $(m-2)/2$, while the number of vacua of $\mathcal{S}_{-}^{(12)}$ is equal to $m/2$. Their label is

$$
\begin{aligned}
\mathbf{a} &= \frac{1}{2}, \frac{3}{2}, \ldots, \frac{m-3}{2}, \quad g > 0, \\
\mathbf{a} &= 0, 1, \ldots, \frac{m-2}{2}, \quad g < 0.
\end{aligned}
$$

Finally, let us discuss some checks of the formalism above. For $m = 3$, there are only breathers and the S-matrix of the first breather, eqn. (18.11.26), correctly coincides with the amplitude (18.4.15) of the Ising model perturbed by a magnetic field. The bootstrap procedure closes with eight breathers. For $m = 4$, the two theories $\mathcal{S}_{\pm}^{(12)}$ are

[6] It can be proved, however, that the two theories are related by duality.

related by duality: for $g < 0$ there are only two vacua and the kinks behave as particles. The RSOS S-matrix of the lowest kink and of the lowest breather correctly coincides with the amplitudes (18.5.5) of the TIM away from its critical temperature and the bootstrap closes with seven particles.

Φ_{21} deformation. The discussion of the RSOS S-matrix of the Φ_{21} deformation is similar to the one above, the only difference being in the definition of the q-parameter and the spectral parameter λ. In this case, the corresponding values are as follows

$$q = e^{i\frac{\pi^4 \beta^2}{4}}, \qquad \lambda = e^{\frac{2\pi}{\tilde{\xi}}\theta}, \qquad \tilde{\xi} = \frac{8}{3}\left(\frac{\pi^2}{\beta^2 - 4\pi}\right) \tag{18.11.32}$$

and in all previous formulae ξ has to be changed to $\tilde{\xi}$. In this case we shall also ensure that $\beta^2/4\pi > 1$, in such a way that the field Φ_{21}, with conformal dimension $\Delta_{21} = \frac{1}{2} + \frac{6\pi}{\beta^2}$, is a relevant operator. In the Landau–Ginzburg picture, the field Φ_{21} is associated to the composite operator $:\varphi^{m-1}:$. Therefore, under the Z_2 spin symmetry, Φ_{21} is odd if m is even and even if m is odd. In this case the structure of the vacua of the theories $\mathcal{S}_{\pm}^{(21)}$ is as follows

- when m is even, for both $\mathcal{S}_{\pm}^{(21)}$ the number of the vacua is $m/2$ and

$$a = \frac{1}{2}, \frac{3}{2}, \ldots, \frac{m-1}{2}, \quad g > 0$$
$$a = 0, 1, \ldots, \frac{m-2}{2}, \quad g < 0.$$

- when m is odd, the number of the vacua is $(m-1)/2$ for $g > 0$, and $(m+1)/2$ for $g < 0$, with

$$a = \frac{1}{2}, \frac{3}{2}, \ldots, \frac{m-2}{2}, \quad g > 0$$
$$a = 0, 1, \ldots, \frac{m-1}{2}, \quad g < 0.$$

Sub-leading magnetization deformation of the Tricritical Ising model. An interesting example of the formalism above is provided by the TIM, i.e. the unitary minimal model $\mathcal{M}_4$, perturbed by the sub-leading magnetization operator Φ_{21}. This is a field odd under the Z_2 spin-reversal transformation: since this deformation explicitly breaks the Z_2 symmetry of the tricritical point, the corresponding massive theory can exhibit the 'Φ^3-property'. The counting argument supports this picture, giving for the spin of the conserved currents the values s = $(1, 5, 7, 11, 13)$. The RSOS picture predicts for such a theory two vacuum states (hereafter denoted by $|\,0\rangle$ and $|\,1\rangle$), which can be associated to the minima of the asymmetric double-well Landau–Ginzburg potential in Figure 18.11.

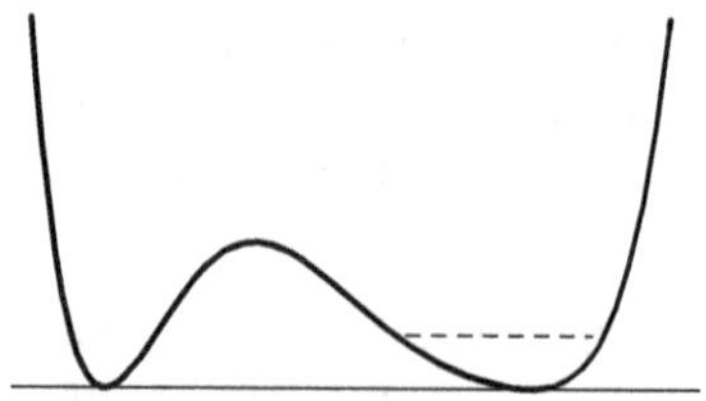

Fig. 18.11 *Landau–Ginzburg potential for the sub-leading magnetic deformation of the Tricritical Ising model.*

The twofold degeneracy of the vacua permits two fundamental kink configurations $|K_+\rangle$ and $|K_-\rangle$ and, possibly, bound states $|B\rangle$ thereof. If the two vacua were related by a symmetry transformation, *i.e.* if we were in the situation of a Z_2 spontaneously broken symmetry, there would be a double degeneracy of the breather-like bound state $|B\rangle$. But the absence of a Z_2 symmetry makes it possible that in this case only one of the two asymptotic states $|K_+K_-\rangle$ or $|K_-K_+\rangle$ couples to the bound state $|B\rangle$! This is confirmed by the explicit solution of the model, given by the RSOS S-matrix. In this case, the only possible values of a_i which label the vacuum states in the RSOS S-matrix are 0 and 1. The one-particle states are thus the vectors: $|K_{01}\rangle$, $|K_{10}\rangle$ and $|K_{11}\rangle$. They correspond to the states that we previously denoted as $|K_+\rangle$, $|K_-\rangle$ and $|B\rangle$, respectively. All of them have the same mass m. Notice that the state $|K_{00}\rangle$ is projected out because of the reduction. The scattering processes are given by

$$\begin{aligned}
|K_{01}(\theta_1)K_{10}(\theta_2)\rangle &= S_{00}^{11}(\theta_1-\theta_2)\,|K_{01}(\theta_2)K_{10}(\theta_1)\rangle,\\
|K_{01}(\theta_1)K_{11}(\theta_2)\rangle &= S_{01}^{11}(\theta_1-\theta_2)\,|K_{01}(\theta_2)K_{11}(\theta_1)\rangle,\\
|K_{11}(\theta_1)K_{10}(\theta_2)\rangle &= S_{10}^{11}(\theta_1-\theta_2)\,|K_{11}(\theta_2)K_{10}(\theta_1)\rangle,\\
|K_{11}(\theta_1)K_{11}(\theta_2)\rangle &= S_{11}^{11}(\theta_1-\theta_2)\,|K_{11}(\theta_2)K_{11}(\theta_1)\rangle+S_{11}^{10}(\theta_1-\theta_2)\,|K_{10}(\theta_2)K_{01}(\theta_1)\rangle,\\
|K_{10}(\theta_1)K_{01}(\theta_2)\rangle &= S_{11}^{00}(\theta_1-\theta_2)\,|K_{10}(\theta_2)K_{01}(\theta_1)\rangle+S_{11}^{10}(\theta_1-\theta_2)\,|K_{11}(\theta_2)K_{11}(\theta_1)\rangle.
\end{aligned} \tag{18.11.33}$$

Explicitly, the above amplitudes are given by

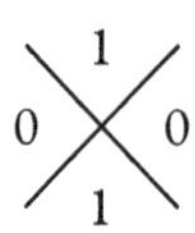

$$= S_{00}^{11}(\theta) = \frac{i}{2}\,S_0(\theta)\,\sinh\left(\frac{9}{5}\theta - i\frac{\pi}{5}\right),$$

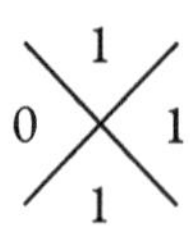

$$= S_{01}^{11}(\theta) = -\frac{i}{2}\,S_0(\theta)\,\sinh\left(\frac{9}{5}\theta + i\frac{\pi}{5}\right),$$

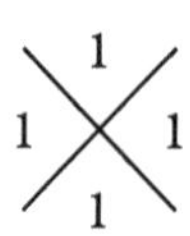

$$= S_{11}^{11}(\theta) = \frac{i}{2}\,S_0(\theta)\,\frac{\sin\left(\frac{\pi}{5}\right)}{\sin\left(\frac{2\pi}{5}\right)}\,\sinh\left(\frac{9}{5}\theta - i\frac{2\pi}{5}\right),$$

$$= S_{11}^{01}(\theta) = -\frac{i}{2}\, S_0(\theta) \left(\frac{\sin\left(\frac{\pi}{5}\right)}{\sin\left(\frac{2\pi}{5}\right)}\right)^{\frac{1}{2}} \sinh\left(\frac{9}{5}\theta\right),$$

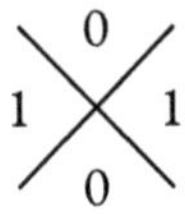

$$= S_{11}^{00}(\theta) = -\frac{i}{2}\, S_0(\theta) \frac{\sin\left(\frac{\pi}{5}\right)}{\sin\left(\frac{2\pi}{5}\right)} \sinh\left(\frac{9}{5}\theta + i\frac{2\pi}{5}\right).$$

The function $S_0(\theta)$ which implements the unitarity condition reads

$$S_0(\theta) = -\left(\sinh\frac{9}{10}(\theta - i\pi)\sinh\frac{9}{10}\left(\theta - \frac{2\pi i}{3}\right)\right)^{-1}$$
$$\times\, w_{-\frac{1}{5}}(\theta))\, w_{\frac{1}{10}}(\theta)\, w_{\frac{3}{10}}(\theta)\, s_{\frac{2}{9}}(\theta)\, s_{-\frac{8}{9}}(\theta)\, s_{\frac{7}{9}}(\theta)\, s_{-\frac{1}{9}}(\theta),$$

where

$$w_x(\theta) = \frac{\sinh\left(\frac{9}{10}\theta + i\pi x\right)}{\sinh\left(\frac{9}{10}\theta - i\pi x\right)}, \qquad s_x(\theta) = \frac{\sinh\frac{1}{2}(\theta + i\pi x)}{\sinh\frac{1}{2}(\theta - i\pi x)}.$$

The amplitudes are periodic along the imaginary axis of θ with period $10\,\pi i$. The whole structure of poles and zeros is quite rich. On the physical sheet, $0 \leq \text{Im}\,\theta \leq i\pi$, the poles of the S-matrix are located at $\theta = \frac{2\pi i}{3}$ and $\theta = \frac{i\pi}{3}$. The first pole corresponds to a bound state in the direct channel whereas the second one is the singularity due to the particle exchanged in the crossed process. The residues at $\theta = \frac{2\pi i}{3}$ are given by

$$\begin{aligned}
r_1 &= \text{Res}_{\theta=\frac{2\pi i}{3}}\; S_{00}^{11}(\theta) = 0\,; \\
r_2 &= \text{Res}_{\theta=\frac{2\pi i}{3}}\; S_{01}^{11}(\theta) = i\left(\frac{s\left(\frac{2}{5}\right)}{s\left(\frac{1}{5}\right)}\right)^2 \omega; \\
r_3 &= \text{Res}_{\theta=\frac{2\pi i}{3}}\; S_{11}^{11}(\theta) = i\,\omega\,; \\
r_4 &= \text{Res}_{\theta=\frac{2\pi i}{3}}\; S_{11}^{01}(\theta) = i\left(\frac{s\left(\frac{2}{5}\right)}{s\left(\frac{1}{5}\right)}\right)^{\frac{1}{2}} \omega; \\
r_5 &= \text{Res}_{\theta=\frac{2\pi i}{3}}\; S_{11}^{00}(\theta) = i\,\frac{s\left(\frac{2}{5}\right)}{s\left(\frac{1}{5}\right)}\,\omega;
\end{aligned} \tag{18.11.34}$$

where $s(x) \equiv \sin(\pi x)$ and

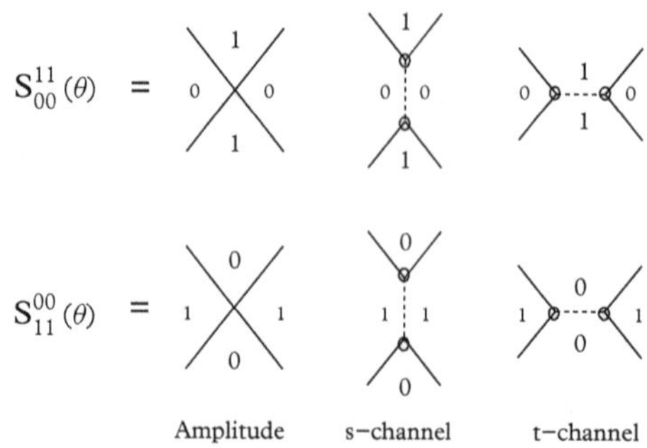

Fig. 18.12 *Elastic scattering amplitudes of the kinks in an asymmetric wells potential and their intermediate states in the s-channel and in the t-channel.*

$$\omega = \frac{5}{9}\,\frac{s\left(\frac{1}{5}\right)s\left(\frac{1}{10}\right)s\left(\frac{4}{9}\right)s\left(\frac{1}{9}\right)s^2\left(\frac{5}{18}\right)}{s\left(\frac{3}{10}\right)s\left(\frac{1}{18}\right)s\left(\frac{7}{18}\right)s^2\left(\frac{2}{9}\right)}.$$

Hence, in the s-channel of the amplitude S_{00}^{11}, there is no bound state related to $\mid K_{00}\rangle$ (a state which does not exist in the model): its only singularity comes from the bound state $\mid K_{11}\rangle$, exchanged however in the t-channel. In the amplitude S_{11}^{00} the situation is reverted (the two amplitudes are related by crossing): there is the s-channel singularity due to the bound state $\mid K_{11}\rangle$ while that of the t-channel is absent. Figure 18.12 shows where the original amplitudes are streched along the vertical direction (s-channel) and along the horizontal one (t-channel).

18.12 Elastic SUSY *S*-matrix

In this last Section we present a simple example of elastic and factorizable S-matrix that involves only one boson and one fermion, both of mass m, and which is invariant under a $N = 1$ supersymmetry. In theoretical physics, such a symmetry is interesting for a series of reasons, mostly related to particle physics and the unification of all the interactions. It is also remarkable in its appearance and role in statistical physics as it provides explanation for the peculiar dynamics of several interesting models. Interested readers are referred to Chapter 24 for a detailed discussion of Lagrangian QFT that are invariant under a $N = 1$ supersymmetry.

The supersymmetry generally leads to the organization of equal-mass particles in multiplets of bosonic and fermionic particles. Here, a one-particle state is denoted as $|b(\theta)\rangle$ or $|f(\theta)\rangle$, depending whether it is a boson or fermion, or generically as $|A(\theta)\rangle$. The SUSY charges act on these states as

$$\begin{aligned} Q_+ \mid b(\beta)\rangle &= \omega\,\sqrt{m}e^{\beta/2}\mid f(\beta)\rangle; \quad Q_- \mid b(\beta)\rangle = \rho\,\sqrt{m}e^{-\beta/2}\mid f(\beta)\rangle; \\ Q_+ \mid f(\beta)\rangle &= \omega^{-1}\sqrt{m}e^{\beta/2}\mid b(\beta)\rangle; \quad Q_- \mid f(\beta)\rangle = \rho^{-1}\sqrt{m}e^{-\beta/2}\mid b(\beta)\rangle, \end{aligned} \tag{18.12.1}$$

i.e. in terms of two matrices

$$Q_+ = \begin{pmatrix} 0 & \omega \\ \omega^{-1} & 0 \end{pmatrix}, \quad Q_- = \begin{pmatrix} 0 & \rho \\ \rho^{-1} & 0 \end{pmatrix}, \tag{18.12.2}$$

satisfying

$$Q_+^2 = Q_-^2 = 1, \tag{18.12.3}$$

$$\{Q_+, Q_-\} = 0. \tag{18.12.4}$$

While the first two conditions are generally true in any $N = 1$ SUSY theory, the commutator of the two charges may be also different from zero (discussed later) and its value related to a topological charge present in the theory. Here, we discuss $\{Q_+, Q_-\} = 0$ for the sake of simplicity. The parameter ω is given by

$$\omega = -i\rho = e^{i\pi/4}. \tag{18.12.5}$$

The action of Q_+ and Q_- on multi-particle states must take into account the fermionic nature of these operators, which therefore involves braiding relations

$$Q_+ \mid A_1(\beta_1)A_2(\beta_2)\ldots A_n(\beta_n)\rangle = \sqrt{m}\sum_{k=1}^{n} e^{\beta_k/2} \tag{18.12.6}$$

$$\mid (Q_F A_1(\beta_1))(Q_F A_2(\beta_2))\ldots(Q_F A_{k-1}(\beta_{k-1})(Q_+ A_k(\beta_k))A_{k+1}(\beta_{k+1})\ldots A_n(\beta_n)\rangle$$

and

$$Q_- \mid A_1(\beta_1)A_2(\beta_2)\ldots A_n(\beta_n)\rangle = \sqrt{m}\sum_{k=1}^{n} e^{-\beta_k/2} \tag{18.12.7}$$

$$\mid (Q_F A_1(\beta_1))(Q_F A_2(\beta_2))\ldots(Q_F A_{k-1}(\beta_{k-1})(Q_- A_k(\beta_k))A_{k+1}(\beta_{k+1})\ldots A_n(\beta_n)\rangle$$

where Q_F is the fermion parity operator, which on the basis $\mid b\rangle$ and $\mid f\rangle$ is represented by the diagonal matrix

$$Q_F = \begin{pmatrix} 1 & 0 \\ 0 & -1 \end{pmatrix}. \tag{18.12.8}$$

Particularly important here is the representation of the two supercharges on the two-particle states $\mid b(\beta_1)b(\beta_2)\rangle$, $\mid f(\beta_1)f(\beta_2)\rangle$, $\mid f(\beta_1)b(\beta_2)\rangle$, $\mid b(\beta_1)f(\beta_2)\rangle$. The first two states belong to the $F = 1$ sector (even number of fermionic particles) whereas the remaining two states belong to the $F = -1$ sector (odd number of fermionic particles). By choosing for convenience $\beta_1 = \beta/2$ and $\beta_2 = -\beta/2$, the operator Q_+ will be represented by the matrix

$$\mathcal{Q}_+(\beta) = \begin{pmatrix} 0 & 0 & \omega x & \omega x^{-1} \\ 0 & 0 & -\omega^{-1}x^{-1} & \omega^{-1}x \\ \omega^{-1}x & -\omega x^{-1} & 0 & 0 \\ \omega^{-1}x^{-1} & \omega x & 0 & 0 \end{pmatrix}, \tag{18.12.9}$$

where $x \equiv e^{\beta/4}$. For $\mathcal{Q}_-$ we have analogously

$$\mathcal{Q}_-(\beta) = \begin{pmatrix} 0 & 0 & \rho x^{-1} & \rho x \\ 0 & 0 & -\rho^{-1}x & \rho^{-1}x^{-1} \\ \rho^{-1}x^{-1} & -\rho x & 0 & 0 \\ \rho^{-1}x & \rho x^{-1} & 0 & 0 \end{pmatrix}. \tag{18.12.10}$$

In the following we also need the representation matrix of the operator $\mathcal{Q}_+\mathcal{Q}_-$ on the above two particle states, given by

$$(\mathcal{Q}_+\mathcal{Q}_-)(\beta) = 2\begin{pmatrix} \frac{\omega}{\rho} & -\omega\rho\sinh\frac{\beta}{2} & 0 & 0 \\ \frac{1}{\omega\rho}\sinh\frac{\beta}{2} & -\frac{\omega}{\rho} & 0 & 0 \\ 0 & 0 & 0 & -\frac{\omega}{\rho}\cosh\frac{\beta}{2} \\ 0 & 0 & -\frac{\omega}{\rho}\cosh\frac{\beta}{2} & 0 \end{pmatrix}. \tag{18.12.11}$$

The amplitudes of the two-particle S-matrix

$$|A_i(\theta_1), A_j(\theta_2)\rangle_{\text{in}} = S_{ij}^{lk}(\theta)|A_l(\theta_2), A_k(\theta_1)\rangle_{\text{out}}, \tag{18.12.12}$$

in addition to satisfy the conditions from unitarity, crossing symmetry, analyticity and Yang–Baxter equations must be also invariant under the action of the supercharges: this leads to the equations

$$\mathcal{Q}_\pm(\theta)\, S_{ij}^{lk}(\theta) = S_{ij}^{lk}(\theta)\, \mathcal{Q}_\pm(-\theta), \tag{18.12.13}$$

$$(\mathcal{Q}_+\mathcal{Q}_-)(\theta)\, S_{ij}^{lk}(\theta) = S_{ij}^{lk}(\theta)\, (\mathcal{Q}_+\mathcal{Q}_-)(-\theta). \tag{18.12.14}$$

Interestingly, examples of S-matrices that satisfy these conditions are those relative to a particular value of the coupling constant of the Sine–Gordon model and also an S-matrix relative to a particular integral deformation of the TIM. The former and the latter models give rise to systems of zero and non-zero SUSY topological charge $\mathcal{Z}$, respectively.

Sine–Gordon model. The simplest possible SUSY S-matrix, solution of the eqns. (18.12.13) and (18.12.14) and satisfying unitarity and crossing symmetry equations is a 4×4 matrix that, written in the $|bb\rangle$, $|ff\rangle$, $|fb\rangle$, $|bf\rangle$ basis, reads

$$S(\theta) = R(\theta) \begin{pmatrix} -1/\cosh(\theta/2) & i\tanh(\theta/2) & 0 & 0 \\ i\tanh(\theta/2) & -1/\cosh(\theta/2) & 0 & 0 \\ 0 & 0 & -1 & 0 \\ 0 & 0 & 0 & -1 \end{pmatrix} \qquad (18.12.15)$$

with

$$R(\theta) = \exp\left[\frac{i}{2}\int_0^\infty \frac{dt}{t}\frac{\sin\frac{\theta t}{\pi}}{\cosh^2\frac{t}{2}}\right].$$

It is notable that this S-matrix is identical to the ordinary Sine–Gordon S-matrix for the soliton/anti-solitons at the special value of the coupling $\beta^2 = 16\pi/3$, with the Lagrangian of the model given in (18.10.1). This Lagrangian does not explicitly exhibit any sign of SUSY, as it is written in terms of a single bosonic field. However, it is interesting to see that a SUSY emerges in this model in terms of operators that transform the field-creating solitons and anti-solitons into each other. These soliton-creating operators are defined as

$$\mathcal{O}_a^n(x) = \lim_{\epsilon\to+0} \exp\left\{-\frac{n}{4\beta}\int_{-\infty}^{x} \partial_y\phi(x,y)dx\right\} \exp\{ia\phi(x+\epsilon,y)\}, \qquad (18.12.16)$$

where n is an integer denoting the topological charge of the operator. Where $n = 0$, these reduce to the well-known vertex operators

$$\mathcal{O}_a^0(x) = \exp\{ia\phi(x,y)\}. \qquad (18.12.17)$$

These soliton-creating fields provide an explanation for the emergence of SUSY at the particular value of the coupling. First of all, the Sine–Gordon has semi-local conserved charges that generate the affine quantum group $U_q(\widehat{sl(2)})$, with

$$q = e^{i8\pi^2/\beta^2}. \qquad (18.12.18)$$

Secondly, these conserved charges can be written in terms of the soliton-creating operators as

$$G_\pm = \frac{1}{N_Q}\int_{-\infty}^{\infty}\left(\mathcal{O}^{\pm2}_{\pm(2\beta/\sqrt{8\pi})^{-1}}(x) + \pi\xi m\mathcal{O}^{\pm2}_{\pm\nu}(x)\right)dx, \qquad (18.12.19)$$

$$\bar{G}_\pm = \frac{1}{N_Q}\int_{-\infty}^{\infty}\left(\mathcal{O}^{\pm2}_{\mp(2\beta/\sqrt{8\pi})^{-1}}(x) + \pi\xi m\mathcal{O}^{\pm2}_{\mp\nu}(x)\right)dx, \qquad (18.12.20)$$

where ξ is the usual renormalized coupling constant of the Sine–Gordon model, see eqn. (18.10.2), while $\nu = (2\beta/\sqrt{8\pi})^{-1} - \beta/\sqrt{8\pi}$ and N_Q is a normalization constant. Thirdly, these semi-local conserved charges satisfy the generalized commutation conditions

$$G_-\bar{G}_+ - q^2\bar{G}_+G_- = \frac{1-q^{2H}}{1-q^{-2}}, \quad G_+\bar{G}_- - q^2\bar{G}_-G_+ = \frac{1-q^{-2H}}{1-q^{-2}}, \tag{18.12.21}$$

where H is the usual topological charge of the Sine–Gordon model

$$H = \frac{\beta}{16\pi^2}\int_{-\infty}^{\infty} \partial_x\phi(x,y)dx. \tag{18.12.22}$$

The SUSY value of the coupling, $\beta^2 = 16\pi/3$ is special because the conditions (18.12.21) simply become the anti-commutation conditions

$$G_-\bar{G}_+ + \bar{G}_+G_- = 0, \quad G_+\bar{G}_- + \bar{G}_-G_+ = 0. \tag{18.12.23}$$

At this value of the coupling these conserved charges have spin-1/2, and their action on one-soliton (anti-soliton) states is given by

$$\begin{aligned} &G_\pm|A_\pm(\theta)\rangle = 0, &&\bar{G}_\pm|A_\pm(\theta)\rangle = 0, \\ &G_\mp|A_\pm(\theta)\rangle = e^{\frac{\theta}{2}}|A_\mp(\theta)\rangle, &&\bar{G}_\mp|A_\pm(\theta)\rangle = e^{-\frac{\theta}{2}}|A_\mp(\theta)\rangle, \end{aligned} \tag{18.12.24}$$

where $|A_+(\theta)\rangle$ and $|A_-(\theta)\rangle$ are one soliton and one anti-soliton state, respectively. It can be even shown that the fermionic charges $G_\pm$, $\bar{G}_\pm$ generate even a larger extended $N = 2$ SUSY. The $N = 1$ SUSY sub-algebra is constructed in terms of the semi-local charges as

$$W_+ = G_+ + G_- \quad W_- = -i\bar{G}_+ + i\bar{G}_-, \tag{18.12.25}$$

Coming back now to the expression of the S-matrix (18.12.15), in the Sine–Gordon model the basis involving solitons and anti-solitons is given by $|S\bar{S}\rangle, |\bar{S}S\rangle, |SS\rangle, |\bar{S}\bar{S}\rangle$. The two basis, one of bosons and fermions and the other of solitons and anti-solitons, can be identified with each other in four possible ways: we can identify, for instance, $|bb\rangle$ with $|S\bar{S}\rangle$ but also with $|\bar{S}S\rangle$; this choice forces a unique identification of $|ff\rangle$ with either $|\bar{S}S\rangle$ or $|S\bar{S}\rangle$. In the same way, and independently, there are two ways of identifying $|bf\rangle$ and $|fb\rangle$ with $|SS\rangle$ and $|bf\rangle$. At any rate, this ambiguity leads to the same physical situation.

Notice that even though the solitons and anti-solitons are topological excitations of the Sine–Gordon model, from the SUSY point of view they are considered instead to have $\mathcal{Z} = 0$, i.e. zero SUSY topological charge. An example of non-zero SUSY topological charge is given instead by the TIM.

Tricritical Ising Model. Section 16.8 demonstrated an interesting statistical field theory which shows a SUSY invariance in the scattering amplitudes of its excitation is the TIM once perturbed by its vacancy density operator. The supersymmetry allows us to organize the operator content of this model (usually classified only in terms of the irreducible representations of the Virasoro algebra) in terms of a superfield

$$\Phi(x,\theta) = \epsilon(x) + \bar{\theta}\psi(x) + \frac{1}{2}\bar{\theta}\theta t(x), \tag{18.12.26}$$

where, in addition to the fermion $\psi(x)$, there are the operator $\epsilon(x)$—the energy density operator—and the field $t(x)$ that describes the vacancy density. All these operators belong to the spin Z_2 even sector of the model. However, there are two additional Z_2 odd fields, $\sigma(x)$ and $\sigma'(x)$ that play the role of magnetization and sub-leading magnetization operators, respectively: they also give rise to other irreducible representations of the SUSY but in the so-called Ramond sector. Perturbed by the vacancy operator $t(x)$, the effective SUSY off-critical action is given by the Landau–Ginzburg action

$$\begin{aligned} \mathcal{A} &= \int d^2x d^2\theta \left[\frac{1}{4} (\bar{D}_\alpha \Phi) D_\alpha \Phi + \frac{1}{3!} \Phi^3 + \lambda \Phi \right] \\ &= \int d^2x \left(\frac{1}{2} \left[(\partial_\mu \epsilon)^2 + i\bar{\psi}\gamma^\mu \partial_\mu \psi \right] - (\epsilon + \lambda)^2 - \epsilon \bar{\psi}\psi \right). \end{aligned} \tag{18.12.27}$$

When $\lambda > 0$ there is a massless flow from TIM to the Ising model. Since $\lambda > 0$ gives rise to a spontaneous breaking of SUSY, the corresponding Goldstino in this case is just the familiar Majorana fermion of the Ising model. When instead $\lambda < 0$, SUSY is exact but there is a degeneracy of the vacuum state, which appears to be doubly degenerate if described in terms of the ϵ variable while, actually, it is triply degenerate, as becomes clear expressing the potential of the theory in terms of the order parameter $\sigma(x)$. The solution of this apparent mismatch in the counting of the vacuum states is easily found once we see the operator identity $\epsilon =: \sigma^2 :$ that connects the two order parameters, so that the vacuum $\epsilon = 1$ corresponds to the *two* vacua $\sigma = \pm 1$. The potential for the field σ is plotted in Figure 18.13: the farthest vacua are denoted by ± 1 while the central one is denoted by 0; also drawn are the kink excitations $|K_{ab}(\theta)$ which connect the various neighbouring vacua. In this case we have four one-particle states: $|K_{-,0}\rangle$, $|K_{+,0}\rangle$, $|K_{0,-}\rangle$ and $|K_{0,+}\rangle$, and the three matrices $\mathcal{Q}_+, \mathcal{Q}_-$ and $\mathcal{Q}_F$ on these states take the form

$$\mathcal{Q}_+ = \begin{pmatrix} 0 & i & 0 & 0 \\ -i & 0 & 0 & 0 \\ 0 & 0 & 1 & 0 \\ 0 & 0 & 0 & -1 \end{pmatrix}, \quad \mathcal{Q}_- = \begin{pmatrix} 0 & i & 0 & 0 \\ -i & 0 & 0 & 0 \\ 0 & 0 & -1 & 0 \\ 0 & 0 & 0 & 1 \end{pmatrix}, \quad \mathcal{Q}_F = \begin{pmatrix} 0 & 1 & 0 & 0 \\ 1 & 0 & 0 & 0 \\ 0 & 0 & 0 & 1 \\ 0 & 0 & 1 & 0 \end{pmatrix}. \tag{18.12.28}$$

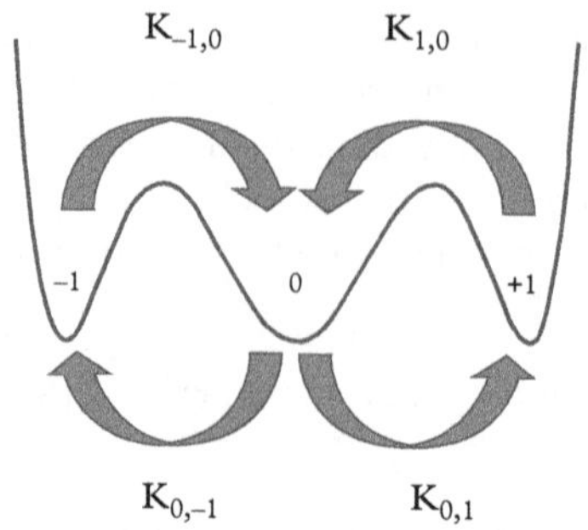

Fig. 18.13 *Effective potential for the order parameter $\sigma(x)$ of the tricritical Ising model along the first-order phase transition line. There are three degenerate vacua connected by the kink excitations $|K_{ab}(\theta)\rangle$.*

Notice that the topological charge $\mathcal{Z}$ on the one-particle states this time is different from zero

$$\mathcal{Z} = \{\mathcal{Q}_+, \mathcal{Q}_-\} = \begin{pmatrix} 2 & 0 & 0 & 0 \\ 0 & 2 & 0 & 0 \\ 0 & 0 & -2 & 0 \\ 0 & 0 & 0 & -2 \end{pmatrix} \tag{18.12.29}$$

and, indeed, on the kink $|K_{ab}\rangle$ connecting the vacua a and b the topological charge $\mathcal{Z}$ takes the value $2(a^2 - b^2)$. The two-body elastic S-matrix for the kink excitations is defined by $(\theta_{12} \equiv \theta_1 - \theta_2)$

$$|K_{ac}(\theta_1)K_{cb}(\theta_2)\rangle = S_{ab}^{cd}(\theta_{12})\ |K_{ad}(\theta_2)K_{db}(\theta_2)\rangle, \tag{18.12.30}$$

and the non-zero amplitudes—fixed in terms of SUSY invariance, continuity of the vacuum indices in the kink states, crossing and unitarity equations – are given by

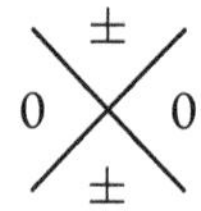

$$= S_{00}^{\pm\pm}(\theta) = \gamma(\theta)\, e^{-i\rho\theta} \cosh\frac{\theta}{4}$$

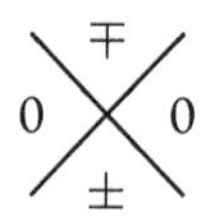

$$= S_{00}^{\pm\mp}(\theta) = -i\gamma(\theta)\, e^{-i\rho\theta} \sinh\frac{\theta}{4}$$

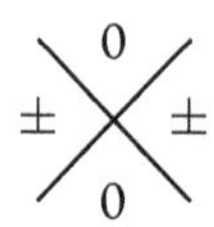

$$= S_{\pm\pm}^{00}(\theta) = \gamma(\theta)\, e^{i\rho\theta} \left(\cosh\frac{\theta}{4} + i\sinh\frac{\theta}{4}\right)$$

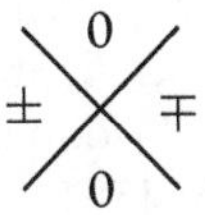

$$= S^{00}_{\mp\pm}(\theta) = \gamma(\theta)\, e^{i\rho\theta}\left(\cosh\frac{\theta}{4} - i\sinh\frac{\theta}{4}\right)$$

where $\rho = (1/2\pi)\log 2$ while the function $\gamma(\theta)$ which implements the unitarity condition reads

$$\gamma(\theta) = \left(\cosh\frac{\theta}{2}\right)^{-1/2} \exp\left[\frac{i}{4}\int_0^\infty \frac{dt}{t}\,\frac{\sin\frac{t\theta}{\pi}}{\cosh^2\frac{t}{2}}\right]. \tag{18.12.31}$$

Notice that, denoting by z_i $(i = 1, 2)$ the SUSY topological charges of each excitation, the two-particle states entering the non-zero scattering amplitudes are all and only those with vanishing total SUSY topological charge $Z = z_1 + z_2$. As a rule of thumb, the role of bosonic and fermionic excitations of this theory are played by those who diagonalize both Q_F and Z

$$|b_+\rangle \simeq (|K_{-0}\rangle + |K_{+0}\rangle), \qquad |f_+\rangle \simeq (|K_{-0}\rangle - |K_{+0}\rangle)$$
$$|b_-\rangle \simeq (|K_{0-}\rangle + |K_{0+}\rangle), \qquad |f_+\rangle \simeq (|K_{0-}\rangle - |K_{0+}\rangle) \tag{18.12.32}$$

even though these assignments could not be taken literally in view of the kink nature of the individual excitation.

REFERENCES

The determination of the scattering theory of the Ising model in a magnetic field is discussed in

Zamolodchikov, A. (1989). 'Integrable Field Theory from Conformal Field Theory', *Advanced Studies in Pure Mathematics*, 19: 641.

The S-matrix of the Yang–Lee model has been computed in

Cardy, J.L. and Mussardo, G. (1989). 'S-matrix of the Yang–Lee Edge Singularity in Two Dimensions', *Physics Letters B*, 225: 275.

The S-matrices of the Toda field theories and their relation with the minimal S-matrices of the statistical models have been studied by several authors. We recommend

Arishtein, A., Fateev, A. and Zamolodchikov, A.B. (1979). 'Quantum S-matrix of the $(1+1)$-Dimensional Toda Chain', *Physics Letters B* 87 (1979), 389.

Braden, H.W., Corrigan, E., Dorey, P.E. and Sasaki, R. (1990). 'Affine Toda Field Theory and Exact S-Matrices', *Nuclear Physics B*, 338: 689.

Braden, H.W., Corrigan, E., Dorey, P.E. and Sasaki, R. (1991). 'Multiple Poles and Other Features of Affine Toda Field Theories', *Nuclear Physics B*, 356: 469.
Christe, P. and Mussardo, G. (1990). 'Elastic S-Matrices in $(1 + 1)$-dimensions and Toda Field Theories', *International Journal of Modern Physics A*, 5: 4581.
Christe, P. and Mussardo, G. (1990). 'Integrable Systems Away from Criticality: The Toda Field Theory and S-matrix of the Tricritical Ising Model', *Nuclear Physics B*, 330: 465
Fateev, V.A. and Zamolodchikov, A.B. (1990). 'Conformal Field Theory and Purely Elastic S-Matrices', *International Journal of Modern Physics A*, 5: 1025.
Sotkov, G. and Zhu, C.J. (1989). 'Bootstrap Fusions and Tricritical Potts Model away from Criticalit', *Phys. Lett. B*, 229: 391.

The scattering theory of geometrical statistical models, as self-avoiding walks or percolation, has been studied in the papers

Zamolodchikov, A.B. (1991). 'Exact S-Matrix Associated with Self-avoiding Polymer Problem in Two Dimensions', *Modern Physics Letters A*, 6: 1807.
Chim L. and Zamolodchikov A.B. (1992). 'Integrable field theory of the q-state Potts model with $0 < q < 4$', *International Journal of Modern Physics A*, 7: 5317.

The S-matrix of the Sine–Gordon model has a long history; we recommend the article in which the exact S-matrix of the model was determined:

Zamolodchikov A.B. and Zamolodchikov Al.B. (1979). 'Factorized S-Matrices in Two Dimensions as the Exact Solutions of Certain Relativistic Quantum Field Theories', *Annals of Physics* 120: 253.

An interesting and simply written article on the properties of the Sine–Gordon S-matrix is

C.J. Goebel, *On the Sine-Gordon S-Matrix*, Prog. Theor. Phys. Suppl. 86 (1986), 261.

The scattering theory for the integrable $\Phi_{1,3}$, $\Phi_{1,2}$ and $\Phi_{2,1}$ deformations of the minimal models admits a general formulation. It requires, though, the sophisticated quantum group tool. The backbone of our understanding of this approach comes mainly from Smirnov and LeClair, and we recommend

LeClair, A. (1989). 'Restricted Sine–Gordon Theory and the Minimal Conformal Series', *Physics Letters B* 230: 103.
Smirnov, F.A. (1990). 'Reductions of the Sine–Gordon model as Perturbation of Minimal Models of Conformal Field Theory', *Nuclear Physics B*, 337: 156.
Smirnov, F.A. (1991). 'Exact S-Matrices for $\Phi_{1,2}$ Perturbed Minimal Models of Conformal Field Theory', *International Journal of Modern Physics A* 6: 1407.

For an introduction to quantum groups see

Majid, S. (1995). *Foundations of Quantum Group Theory*, Cambridge University Press.

The S-matrix for supersymmetric theories are discussed in

Schoutens, K. (1990). 'Supersymmetric and Factorizable Scattering', *Nuclear Physics B* 344: 665.

Shankar R. and Witten, E. (1978). 'S-matrix of the Supersymmetric Non-linear δ Model', *Physics Reviews D*, 17: 2134.

Zamolodchikov, A.B. 'Fractional Spin Integrals of Motion in Perturbed Conformal Field Theory', *Proceeding of the Workshop on Fields, Strings and Quantum Gravity*, 29 May–10 Jun 1989, Beijing, China.

Non-relativistic limit of Toda field theories was proposed and studied in

Bastianello, A., De Luca, A. and Mussardo, G. (2016). 'Non-relativistic Limit of Integrable QFT and Lieb–Liniger Models', *Journal of Statistical Mechanics*, 1612(12): 123104.

Kormos, M., Mussardo G. and Trombettoni, A. (2010). '1D Lieb–Liniger Bose Gas as Non-relativistic Limit of the Sinh–Gordon Model', *Physics Reviews A*, 81: 043606.

PROBLEMS

18.1. Bootstrap equations

Prove that the most general solution of the bootstrap equation relative to a particle bound state of itself

$$S_{AA}(\theta) = S_{AA}\left(\theta - \frac{i\pi}{3}\right) S_{AA}\left(\theta + \frac{i\pi}{3}\right)$$

is given by

$$S_{AA}(\theta) = f_{\frac{2}{3}}(\theta) \prod_i f_{x_i}(\theta) f_{\frac{2}{3}-x_i}(\theta).$$

Study the motion of the poles of the function $S_{AA}(\theta)$ under the shifts induced by the bootstrap equation.

18.2. Analytic structure of the S-matrix of the Bullogh–Dodd model

a. Study the structure of the poles and zeros of the S-matrix of the Bullogh–Dodd model

$$S(\theta) = f_{\frac{2}{3}}(\theta) f_{-\frac{B}{3}}(\theta) f_{\frac{B-2}{3}}(\theta)$$

with

$$B(\lambda) = \frac{\lambda^2}{2\pi} \frac{1}{1+\frac{\lambda^2}{4\pi}}$$

by varying the coupling constant λ.

b. Make the analytic continuation

$$B \to 1 + \frac{3}{i\pi}\beta_0$$

with β_0 real and show that in the limit $\beta_0 \to \infty$ the S-matrix of the Bullogh–Dodd model reduces to the S-matrix of the Yang–Lee model.

18.3. Multiple poles

Prove that the amplitude S_{11} of the fundamental particle cannot have higher-order poles by showing that the resonance angle of two heavier masses is larger than $2\pi/3$. This makes it impossible to draw a diagram the one in Figure 18.2.

18.4. Double poles

Use the values of the resonance angles of the S-matrix of the thermal TIM to explain the double poles that enter the amplitude $S_{1,6}$ in terms of multi-scattering processes.

18.5. Non-relativistic scattering of distinguishable particles

Consider the non-relativistic Hamiltonian

$$H = -\frac{\partial_x^2}{2m_a} - \frac{\partial_y^2}{2m_b} + 2\lambda\delta(x-y) \tag{18.12.33}$$

and let $\chi(x,y)$ be its scattering wavefunction, where the coordinate x is referred to the first particle of mass m_a and y to the second particle of mass m_b.

a. Show that in the centre of mass frame the Hamiltonian can be written as

$$H = -\frac{1}{2M}\partial_X^2 - \frac{1}{2\mu}\partial_Y^2 + 2\lambda\delta(Y)$$

where $X = \frac{m_a x + m_b y}{m_a + m_b}$, $Y = x - y$, M is the total mass $M = m_a + m_b$ and μ the reduced mass $\mu = m_a m_b / M$.

b. Consider now the scattering process in the centre of the mass frame in which the particle of reduced mass μ is approaching the origin with ingoing plane-wave of momentum k. Find the transmission and reflection coefficients T and R, parameterized in terms of the velocity $v = \mu^{-1}k$,

$$R(v) = \frac{-i2\lambda}{v + i2\lambda}, \qquad T(v) = \frac{v}{v + i2\lambda}$$

by solving the eigenstate problem with boundary conditions:

$$\chi_{CM}(Y) = \begin{cases} e^{ikY} + R(v)e^{-ikY} & Y < 0 \\ T(v)e^{ikY} & Y > 0. \end{cases}$$

Notice that for any non-trivial interaction $\lambda \neq 0$ it is *impossible* to have a purely transmissive scattering, since R is always non-zero.

18.6. S-matrix of the Gross–Neveu model

The Gross–Neveu model is a model of n-component neutral Fermi–field $\psi_k(x)$; $k = 1, 2, \ldots, n$ $(n \geq 3)$ with four-fermion interaction

$$\mathcal{L} = \frac{i}{2} \sum_{k=1}^{n} \bar{\psi}_k \gamma^\mu \partial_\mu \psi_k + \frac{g}{8} \left[\sum_{k=1}^{n} \bar{\psi}_k \psi_k \right]^2$$

where $\bar{\psi}_k = \psi_k \gamma^0$ and the 2×2 γ^μ matrices satisfy the anti-commutation relation $\{\gamma^\mu, \gamma^\nu\} = 2g^{\mu\nu}$. Like the bosonic $O(n)$ σ model, the Gross–Neveu model is massive, renormalizable, asymptotically free and explicitly $O(n)$ symmetric. It is also integrable.

With the notation of Section 18.9, the exact S-matrix of the Gross–Neveu model can be obtained solving the unitarity and crossing equations for $S_2(\theta)$

$$S_2(\theta)S_2(-\theta) = \frac{\theta^2}{\theta^2 + \lambda^2}, \qquad S_2(\theta) = S_2(i\pi - \theta)$$

with the initial seed $Q(\theta) = \frac{\theta}{\theta - i\lambda}$ where $\lambda = 2\pi/(n-2)$.

a. With the notation of eqn. (18.9.8), show that in this case we end up in

$$\mathcal{U}^{(-)}(\theta) = \frac{\Gamma\left(-\frac{\lambda}{2\pi} - i\frac{\theta}{2\pi}\right)\Gamma\left(\frac{1}{2} - i\frac{\theta}{2\pi}\right)}{\Gamma\left(\frac{1}{2} - \frac{\lambda}{2\pi} - i\frac{\theta}{2\pi}\right)\Gamma\left(-i\frac{\theta}{2\pi}\right)}.$$

b. Prove that the amplitudes $\mathcal{U}^{(\pm)}$ are related as

$$\mathcal{U}^{(-)}(\theta) = \frac{\sinh\theta + i\sin\lambda}{\sinh\theta - i\sin\lambda}\,\mathcal{U}^{(+)}(\theta).$$

c. Consider the amplitudes with definite isospin channel

$$\begin{aligned} S_{\text{isoscalar}} &= NS_1 + S_2 + S_3, \\ S_{\text{antisym}} &= S_2 - S_3, \\ S_{\text{sym}} &= S_2 + S_3. \end{aligned}$$

Bound states exist only in isoscalar and anti-symmetric isospin channels. Denote these new particles B and B_{ij} and show that their masses are

$$m_B = m_{B_{ij}} = m \frac{\sin\left(\frac{2\pi}{n-2}\right)}{\sin\left(\frac{\pi}{n-2}\right)}.$$

where m is the mass of the elementary fermion.

18.7. Integral representation

Use the expansions

$$\frac{1}{\cosh x} = 2\sum_{k=0}^{\infty}(-1)^k e^{-(2k+1)x}, \qquad \frac{1}{\sinh x} = 2\sum_{k=0}^{\infty} e^{-(2k+1)x}$$

the infinite-product

$$\frac{\Gamma(\alpha)\Gamma(\beta)}{\Gamma(\alpha+\gamma)\Gamma(\beta-\gamma)} = \prod_{k=0}^{\infty}\left[\left(1+\frac{\gamma}{\alpha+k}\right)\left(1-\frac{\gamma}{\beta+k}\right)\right]$$

and the integral

$$\int_0^\infty \frac{dt}{t} e^{-\beta t}\sin(\alpha t) = \frac{1}{2i}\log\left[\frac{1+i\alpha/\beta}{1-i\alpha/\beta}\right]$$

to prove the integral representation of eqns. (18.9.19) and (18.10.12). Isolate a finite number of poles and also recover the mixed representation given in eqn. (18.10.13).

18.8. Sine–Gordon

a. Study the analytic structure of the S-matrix of the solitons of the Sine–Gordon model, identifying all the sequences of the poles in the amplitudes.
b. Using the following definition of the breathers B_n

$$B_n\left(\frac{\theta_1+\theta_2}{2}\right) = \lim_{\theta_1-\theta_2\to in\xi}\left[A(\theta_2)\bar{A}(\theta_1) + (-1)^n \bar{A}(\theta_2)A(\theta_1)\right]$$

compute the S-matrix of these particles by means of the residue of the S-matrix of the solitons.

18.9. Reflectionless points

At $\xi = \pi/n$ $(n = 1, 2, \ldots)$, the amplitude S_R of the Sine–Gordon vanishes and the scattering of the soliton–anti-soliton reduces to a pure transmission. Use the properties of the Γ function to prove that at these values of the coupling constant the transmission amplitude becomes

$$S_T(\theta) = e^{in\pi} \prod_{k=1}^{n-1} \frac{e^{\theta - i(\pi k/n)} + 1}{e^{\theta} + e^{-i(\pi k/n)}}.$$

18.10. Bound states and semi-classical limit

It can be proved that the renormalized coupling constant

$$\xi = \frac{\beta^2}{8} \frac{1}{1 - \frac{\beta^2}{8\pi}}.$$

of the Sine–Gordon model comes from the quantum correction of the classical action. By the same token, it is possible to prove that the exact mass of the soliton/anti-soliton is

$$M = \frac{m}{\xi},$$

where m is the parameter in the Lagrangian. Keeping m fixed, the semi-classical limit $\beta^2 \to 0$ of the Sine–Gordon gives rise to a non-trivial theory.

a. Use the expression above of the mass of the soliton to express differently the mass of the breathers B_n of the Sine–Gordon, given in eqn. (18.10.21).
b. Expand m_n in powers of β^2 and show, that at lowest order, $m_n \simeq n m_1$. So, all these states can be considered as loosely bound states of n 'elementary' bosons B_1.
c. Compute at the order β^4 the binding energy $\Delta E_n \equiv n m_1 - m_n$ of these states.

18.11. Sine–Gordon and non-unitary models

a. Find the value of ξ for which the S-matrix element $S^{(1,1)}(\theta)$ of the Sine–Gordon model coincides with the S-matrix of the Yang–Lee model. Explain why the restriction of the Sine–Gordon model produces a negative residue at the pole $\theta = 2\pi i/3$.
b. Generalize the result above, finding the value of ξ that leads to the equality of the S-matrices $S^{(n,m)}(\theta)$ of the Sine–Gordon model with those of the integrable deformation of the minimal non-unitary models $\mathcal{M}_{2,2n+1}$.

19

Form Factors and Correlation Functions

Elementary, my dear Watson
Arthur Conan Doyle

One of the fundamental problems of statistical mechanics and its QFT formulation is the characterization of the order parameters and the computation of their correlation functions. Beside the intrinsic interest of this problem, the correlation functions are the key quantities in the determination of the universal ratios of the RG and therefore they can have a direct experimental confirmation (Chapter 8). On a general ground, the computation of correlation functions is a difficult task, usually achieved with partial success through perturbative methods.

As mentioned earlier in the book, an exact determination of the operator content and the correlation functions of a two-dimensional theory can be obtained only when the model is at its critical point. In this case, in fact, we have a classification of the order parameters in terms of the irreducible representation of the Virasoro algebra and, moreover, one can get an exact expression of the correlators solving the linear differential equations that they satisfy.

Unfortunately, the simple theoretical scheme of the critical points cannot be generalized once we move away from criticality. In this case, the problem has to be faced with different techniques. This chapter shows significant progress can be made when we deal with integrable theories, characterized by their elastic S-matrix and the spectrum of the asymptotic states. The central quantities are in this case the matrix elements of the various operators on the asymptotic states of the theory, called the *form factors*. The precise definition of these quantities is given below. The general properties related to the unitarity and crossing symmetry lead to a set of functional equations for the form factors that can be explicitly solved in many interesting cases. Once the matrix elements of the operators are known, their correlation functions can be recovered in terms of spectral representation series. It is worth mentioning that these series present remarkable convergence properties.

Hence, the success of the form factor method relies on two points: (a) the possibility to determine exactly the matrix elements of the order parameters on the asymptotic states

Statistical Field Theory: An Introduction to Exactly Solved Models in Statistical Physics. Second Edition. Giuseppe Mussardo, Oxford University Press (2020). © Giuseppe Mussardo. DOI: 10.1093/oso/9780198788102.001.0001

of the theory, identified by the scattering theory; (b) the fast convergence properties of the spectral series. These two steps lead to the determination of the correlation functions away from criticality with a precision that cannot be obtained by other methods.

19.1 General Properties of the Form Factors

An essential quantity for the computation of the matrix elements is the S-matrix of the problem. As shown in the previous chapters, the S-matrix of many two-dimensional systems is particularly simple and can be explicitly found. For the infinite number of conservation laws, the scattering processes of integrable systems are purely elastic and the n-particle S-matrix can be factorized in terms of the $n(n-1)/2$ two-body scattering amplitudes. In the following, for simplicity, we focus our attention on diagonal scattering theories with non-degenerate spectrum. To characterize the kinematic state of the particles we use the rapidities θ_i that enter the dispersion relations

$$p_i^0 = m_i \cosh\theta_i, \quad p_i^1 = m_i \sinh\theta_i. \tag{19.1.1}$$

The two-body S-matrix amplitudes depend on the difference of the rapidities $\theta_{ij} = \theta_i - \theta_j$ and satisfy the unitary and crossing symmetry equations

$$\begin{aligned} S_{ij}(\theta_{ij}) &= S_{ji}(\theta_{ij}) = S_{ij}^{-1}(-\theta_{ij}), \\ S_{i\bar{j}}(\theta_{ij}) &= S_{ij}(i\pi - \theta_{ij}). \end{aligned} \tag{19.1.2}$$

Possible bound states correspond to simple poles (or higher-order odd poles) of these amplitudes, placed at imaginary values of θ_{ij} in the physical strip $0 < \mathrm{Im}\theta < \pi$.

Let us look at how the S-matrix allows us to compute the matrix elements of the (semi)-local operators on the asymptotic states. To this aim, it is useful to introduce an algebraic formalism.

19.1.1 Faddeev–Zamolodchikov Algebra

A key assumption of the form factor theory is that there exist some operators, both of creation and annihilation type, $V^{\dagger}_{\alpha_i}(\theta_i)$, $V_{\alpha_i}(\theta_i)$, that implement a generalization of the usual bosonic and fermionic algebraic relations. We call them *vertex operators*. Denoting by α_i the quantum number that distinguishes the different types of particles of the theory, these operators satisfy the associative algebra in which enters the S-matrix

$$V_{\alpha_i}(\theta_i) V_{\alpha_j}(\theta_j) = S_{ij}(\theta_{ij})\, V_{\alpha_j}(\theta_j) V_{\alpha_i}(\theta_i) \tag{19.1.3}$$

$$V^{\dagger}_{\alpha_i}(\theta_i) V^{\dagger}_{\alpha_j}(\theta_j) = S_{ij}(\theta_{ij})\, V^{\dagger}_{\alpha_j}(\theta_j) V^{\dagger}_{\alpha_i}(\theta_i) \tag{19.1.4}$$

$$V_{\alpha_i}(\theta_i) V^{\dagger}_{\alpha_j}(\theta_j) = S_{ij}(\theta_{ji})\, V^{\dagger}_{\alpha_j}(\theta_j) V_{\alpha_i}(\theta_i) + 2\pi\, \delta_{\alpha_i\alpha_j} \delta(\theta_{ij}). \tag{19.1.5}$$

Any commutation of these operators can be interpreted as a scattering process. The Poincaré group, generated by the Lorentz transformations $L(\epsilon)$ and the translations T_y, acts on the operators as

$$U_L V_\alpha(\theta) U_L^{-1} = V_\alpha(\theta+\epsilon) \tag{19.1.6}$$

$$U_{T_y} V_\alpha(\theta) U_{T_y}^{-1} = e^{ip_\mu(\theta)y^\mu}\, V_\alpha(\theta). \tag{19.1.7}$$

Obviously the explicit form of the creation and annihilation operators crucially depends on the theory in question and their construction is an open problem for most of the models. This difficulty does not stop us from deriving the fundamental equations for the matrix elements starting from the algebraic equations given above.

The vertex operators define the space of the physical states. The vacuum $|0\rangle$ is the state annihilated by $V_\alpha(\theta)$,

$$V_\alpha(\theta)|0\rangle = 0 = \langle 0|V_\alpha^\dagger(\theta),$$

while the Hilbert space is constructed by applying the various vertex operators $V_\alpha^\dagger(\theta)$ on $|0\rangle$

$$|V_{\alpha_1}(\theta_1)\ldots V_{\alpha_n}(\theta_n)\rangle \equiv V_{\alpha_1}^\dagger(\theta_1)\ldots V_{\alpha_n}^\dagger(\theta_n)|0\rangle. \tag{19.1.8}$$

From eqn. (19.1.5), the one-particle states have the normalization

$$\langle V_{\alpha_i}(\theta_i)|V_{\alpha_j}(\theta_j)\rangle = 2\pi\, \delta_{\alpha_i\alpha_j}\delta(\theta_{ij}).$$

The algebra of the vertex operators implies that the vectors (19.1.8) are not all linearly independent. To select a basis of linearly independent vectors we need an additional requirement: for the initial states, the rapidites must be ordered in a decreasing way

$$\theta_1 > \theta_2 > \cdots > \theta_n$$

while, for the final states in an increasing way

$$\theta_1 < \theta_2 < \cdots < \theta_n.$$

These orderings select a set of linearly independent vectors that form a basis in the Hilbert space.

19.1.2 Form Factors

This section exposes the principles of the theory. Unless explicitly stated, in the following we consider the matrix elements between the *in* and *out* states of the particle with the lowest mass of local, scalar and Hermitian operators $\mathcal{O}(x)$

$$
{}_{\rm out}\langle V(\theta_{m+1})\dots V(\theta_n)|\mathcal{O}(x)|V(\theta_1)\dots V(\theta_m)\rangle_{\rm in}. \tag{19.1.9}
$$

We can always place the operator at the origin by using the translation operator, $U_{T_y}\mathcal{O}(x)U_{T_y}^{-1} = \mathcal{O}(x+y)$, and using eqn. (19.1.7), the matrix elements above are given by

$$
\begin{aligned}
&\exp\left[i\left(\sum_{i=m+1}^{n} p_\mu(\theta_i) - \sum_{i=1}^{m} p_\mu(\theta_i)\right)x^\mu\right] \\
&\times {}_{\rm out}\langle V(\theta_{m+1})\dots V(\theta_n)|\mathcal{O}(0)|V(\theta_1)\dots V(\theta_m)\rangle_{\rm in}.
\end{aligned} \tag{19.1.10}
$$

It is convenient to define the functions

$$
F_n^{\mathcal{O}}(\theta_1,\theta_2,\dots,\theta_n) = \langle 0 \mid \mathcal{O}(0) \mid \theta_1,\theta_2,\dots,\theta_n\rangle_{in}. \tag{19.1.11}
$$

called the form factors (FF) (Figure 19.1): they are the matrix elements of an operator placed at the origin between the n-particle state and the vacuum.[1]

For local and scalar operators, the relativistic invariance of the theory implies that the FF are functions of the differences of the rapidities θ_{ij}

$$
F_n^{\mathcal{O}}(\theta_1,\theta_2,\dots,\theta_n) = F_n^{\mathcal{O}}(\theta_{12},\theta_{13},\dots,\theta_{ij},\dots)\ ,i<j. \tag{19.1.12}
$$

The invariance under crossing symmetry permits to recover the most general matrix elements by an analytic continuation of the functions (19.1.11)

$$
F_{n+m}^{\mathcal{O}}(\theta_1,\theta_2,\dots,\theta_m,\theta_{m+1}-i\pi,\dots,\theta_n-i\pi) = F_{n+m}^{\mathcal{O}}(\theta_{ij},i\pi-\theta_{sr},\theta_{kl}) \tag{19.1.13}
$$

where $1 \le i < j \le m$, $1 \le r \le m < s \le n$, and $m < k < l \le n$.

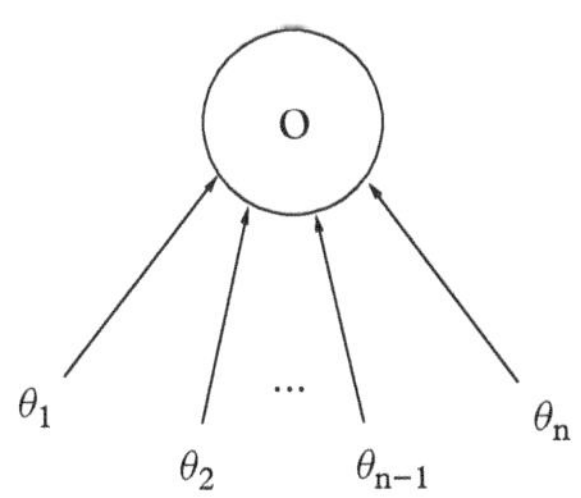

Fig. 19.1 *Form factor of the operator $\mathcal{O}$.*

[1] From now on we use the simplified notation $|\dots V(\theta_n)\dots\rangle \equiv |\dots\theta_n\dots\rangle$ to denote the physical states of the particle with the lowest mass.

Apart from the poles corresponding to the bound states present in all possible channels of this amplitude, the form factors $F_n^{\mathcal{O}}$ are expected to be analytic functions in the strips $0 < \mathrm{Im}\theta_{ij} < 2\pi$.

19.2 Watson's Equations

The FF of a scalar and Hermitian operator $\mathcal{O}$ satisfy a set of equations, known as *Watson equations*, that assume a particularly simple form for the integrable systems

$$F_n^{\mathcal{O}}(\theta_1,\ldots,\theta_i,\theta_{i+1},\ldots,\theta_n) = F_n^{\mathcal{O}}(\theta_1,\ldots,\theta_{i+1},\theta_i,\ldots,\theta_n)S(\theta_i-\theta_{i+1}), \tag{19.2.1}$$
$$F_n^{\mathcal{O}}(\theta_1+2\pi i,\ldots,\theta_{n-1},\theta_n) = e^{2\pi i\gamma}\,F_n^{\mathcal{O}}(\theta_2,\ldots,\theta_n,\theta_1) = \prod_{i=2}^{n} S(\theta_i-\theta_1)F_n^{\mathcal{O}}(\theta_1,\ldots,\theta_n),$$

where γ is the semi-local index of the operator $\mathcal{O}$ with respect to the operator that creates the particles. The first equation is a simple consequence of eqn. (19.1.3), because a commutation of two operators is equivalent to a scattering process. Concerning the second equation, it states the nature of the discontinuity of these functions at the cuts $\theta_{1i} = 2\pi i$. Figure 19.2 shows the graphical representation of these equations. When $n = 2$, eqns. (19.2.1) reduce to

$$\begin{aligned} F_2^{\mathcal{O}}(\theta) &= F_2^{\mathcal{O}}(-\theta)\,S_2(\theta), \\ F_2^{\mathcal{O}}(i\pi-\theta) &= F_2^{\mathcal{O}}(i\pi+\theta). \end{aligned} \tag{19.2.2}$$

The most general solution of the Watson equations (19.2.1) is given by

$$F_n^{\mathcal{O}}(\theta_1,\ldots,\theta_n) = K_n^{\mathcal{O}}(\theta_1,\ldots,\theta_n)\prod_{i<j} F_{\min}(\theta_{ij}). \tag{19.2.3}$$

We now discuss the various terms entering this expression.

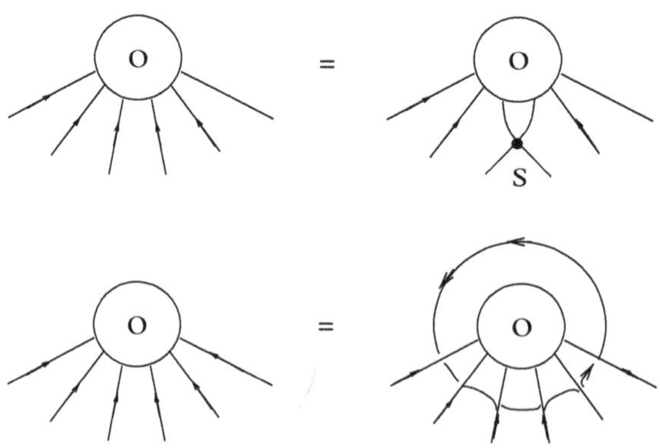

Fig. 19.2 *Graphical form of the Watson equations.*

Minimal Two-particle Form Factor. $F_{\min}(\theta)$ is an analytic function in the region $0 \leq \text{Im}\ \theta \leq \pi$, solution of the two equations (19.2.2), with neither zeros nor poles in the strip $0 < \text{Im}\ \theta < \pi$, and with the mildest behaviour at $|\theta| \to \infty$. These requirements determine uniquely this function, up to a normalization factor $\mathcal{N}$. Its explicit expression can be found by writing the S-matrix as

$$S(\theta) = \exp\left[\int_0^\infty \frac{dt}{t} f(t) \sinh \frac{t\theta}{i\pi}\right].$$

In fact, it is easy to see that $F_{\min}(\theta)$ is given by

$$F_{\min}(\theta) = \mathcal{N} \exp\left[\int_0^\infty \frac{dt}{t} \frac{f(t)}{\sinh t} \sin^2\left(\frac{\pi\hat{\theta}}{2\pi}\right)\right], \quad \hat{\theta} = i\pi - \theta. \tag{19.2.4}$$

Note that for interacting theories, $S(0) = -1$, and therefore the first equation in (19.2.2) forces $F_{\min}(\theta)$ to have a zero at the two-particle threshold

$$F(\theta) \simeq \theta, \quad \theta \to 0 \tag{19.2.5}$$

$K_n^{\mathcal{O}}$ factors. The remaining factors $K_n^{\mathcal{O}}$ in (19.2.3) satisfy the Watson equation but with $S_2 = 1$: this implies that they are completely symmetric functions in the variables θ_{ij}, periodic with period $2\pi i$. Therefore they can be considered as functions of the variables $\cosh\theta_{ij}$. Let us investigate other properties of the functions $K_n^{\mathcal{O}}$. They must have all physical poles expected for the form factors. Remember that, in general, there is a simple pole in the form factors when a cluster made of k particles can reach a kinematical configuration that is equivalent to the one of a single particle (Figure 19.3), with the pole given just by the propagator of the latter particle. If this is the general situation, for the integrable theories there is however an important simplification. In fact, for the factorization property of the S-matrix, it is sufficient to consider only the cases in which the clusters are made of $k = 2$ or $k = 3$: the poles coming from the two-particle clusters are dictated uniquely by the bound states of the S-matrix, while those coming from the three-particle clusters are determined by the crossing processes, although they are also related to the S-matrix (see the discussion in the next section). In conclusion, all the poles of the form factors are determined by the underlying scattering theory and they do not depend on the operator! In the light of this analysis, the functions $K_n^{\mathcal{O}}$ can be parameterized as follows

$$K_n^{\mathcal{O}}(\theta_1, \ldots, \theta_n) = \frac{Q_n^{\mathcal{O}}(\theta_1, \ldots, \theta_n)}{D_n(\theta_1, \ldots, \theta_n)}, \tag{19.2.6}$$

where the denominator D_n is a polynomial in $\cosh\theta_{ij}$ that is fixed only by the pole structure of the S-matrix while the information on the operator $\mathcal{O}$ is enclosed in the

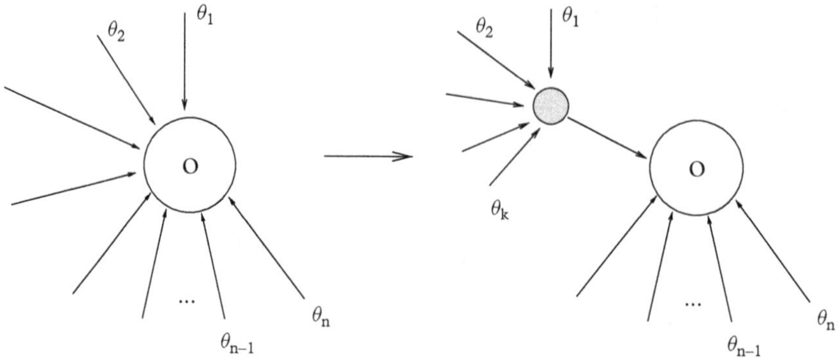

Fig. 19.3 *Kinematic configuration of k-particle responsible for a pole in the form factors.*

polynomial $Q_n^{\mathcal{O}}$ of the variables $\cosh\theta_{ij}$ placed at the *numerator*. We will come back to this important point in the later sections.

Symmetric polynomials. As shown, the functions $K_n^{\mathcal{O}}$ are symmetric under the permutation of the rapidities of the various particles. In many cases it is convenient to change variables, introducing the parameters $x_i \equiv e^{\theta_i}$, so that both numerator and denominator become symmetric polynomials in the x_i variables. A basis in the functional space of the symmetric polynomials in n variables is given by the *elementary symmetric polynomials* $\sigma_k^{(n)}(x_1,\ldots,x_n)$, whose generating function is

$$\prod_{i=1}^{n}(x+x_i) = \sum_{k=0}^{n} x^{n-k}\sigma_k^{(n)}(x_1,x_2,\ldots,x_n). \tag{19.2.7}$$

Conventionally all $\sigma_k^{(n)}$ with $k > n$ and with $n < 0$ are zero. The explicit expressions for the other cases are

$$\begin{aligned}
\sigma_0 &= 1, \\
\sigma_1 &= x_1 + x_2 + \ldots + x_n, \\
\sigma_2 &= x_1x_2 + x_1x_3 + \ldots x_{n-1}x_n, \\
\vdots &\qquad \vdots \\
\sigma_n &= x_1x_2\ldots x_n.
\end{aligned} \tag{19.2.8}$$

The $\sigma_k^{(n)}$ are homogeneous polynomials in x_i, of total degree k but linear in each variable.

Total and partial degrees of the polynomials. The polynomials $Q_n^{\mathcal{O}}(x_1,\ldots,x_n)$ in the numerator of the factor $K_n^{\mathcal{O}}$ satisfy additional conditions coming from the asymptotic behaviour of the form factors. The first condition simply comes from the relativistic invariance: in fact, for a simultaneous translation of all the rapidities, the form factors of a scalar operator[2] satisfy

$$F_n^{\mathcal{O}}(\theta_1+\Lambda,\theta_2+\Lambda,\ldots,\theta_n+\Lambda) = F_n^{\mathcal{O}}(\theta_1,\theta_2,\ldots,\theta_n). \tag{19.2.9}$$

This implies the equality of the total degrees of the polynomials $Q_n^{\mathcal{O}}(x_1,\ldots,x_n)$ and $D_n(x_1,\ldots,x_n)$. Concerning the partial degree with respect to each variable, it is worth anticipating a result discussed in Section 19.8: in order to have a power-law behaviour of the two-point correlation function of the operator $\mathcal{O}(x)$, its form factors must behave for $\theta_i \to \infty$ at most as $\exp(k\theta_i)$, where k is a constant (independent from i), related to the conformal dimension of the operator $\mathcal{O}$.

19.3 Recursive Equations

The poles in the FF induce a set of recursive equations that are crucial for the explicit determination of these functions. As a function of the difference of the rapidities θ_{ij}, the FF have two kinds of simple pole.[3]

Kinematical poles. The first kind of singularity does not depend on whether the model has bound states. It is in fact associated to the kinematical poles at $\theta_{ij} = i\pi$ that come from the one-particle state realized by the three-particle clusters. In turn, these processes correspond to the crossing channels of the S-matrix, as shown in Figure 19.4. The residues at these poles give rise to a recursive equation that links the n-particle and the $(n-2)$-particle form factors

$$-i\lim_{\tilde{\theta}\to\theta}(\tilde{\theta}-\theta)F_{n+2}^{\mathcal{O}}(\tilde{\theta}+i\pi,\theta,\theta_1,\theta_2,\ldots,\theta_n) = \left(1 - e^{2\pi i\gamma}\prod_{i=1}^{n} S(\theta-\theta_i)\right) F_n^{\mathcal{O}}(\theta_1,\ldots,\theta_n). \tag{19.3.1}$$

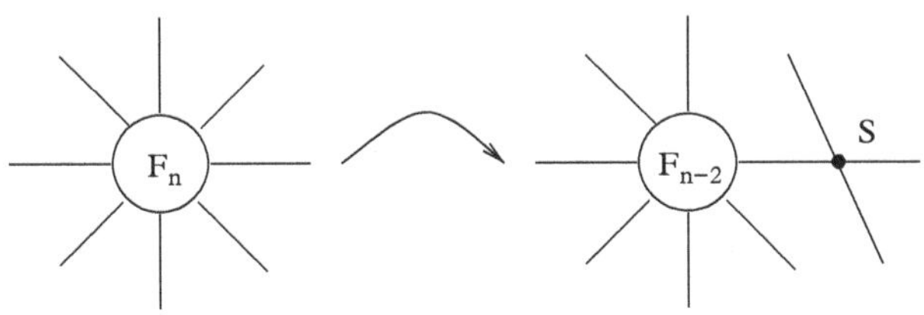

Fig. 19.4 *Recursive equation of the kinematical poles.*

[2] For the form factors of an operator $\mathcal{O}(x)$ of spin s, the equation generalizes as $F_n^{\mathcal{O}}(\theta_1+\Lambda,\theta_2+\Lambda,\ldots,\theta_n+\Lambda) = e^{s\Lambda}F_n^{\mathcal{O}}(\theta_1,\theta_2,\ldots,\theta_n)$.

[3] There could be also higher-order poles, in correspondence with the higher order poles of the S-matrix. Their discussion is however beyond the scope of this book.

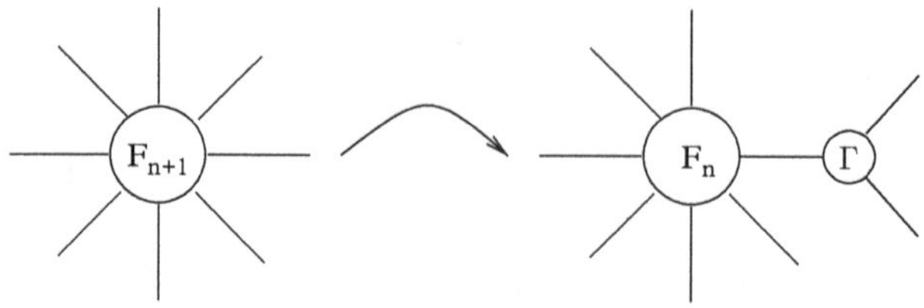

Fig. 19.5 *Recursive equation of the bound state poles.*

Let us denote concisely by $\mathcal{C}$ the map between $F_{n+2}^{\mathcal{O}}$ and $F_n^{\mathcal{O}}$ established by the recursive equation

$$F_{n+2}^{\mathcal{O}} = \mathcal{C}\, F_n^{\mathcal{O}}. \tag{19.3.2}$$

Bound state poles. There is another family of poles in F_n if the S-matrix has simple poles related to the bound states. These poles are at the values of θ_{ij} corresponding to the resonance angles. Let $\theta_{ij} = iu_{ij}^k$ be one of these poles, associated to the bound state A_k present in the channel $A_i \times A_j$. For the S-matrix we have

$$-i \lim_{\theta \to iu_{ij}^k} (\theta - iu_{ij}^k)\, S_{ij}(\theta) = \left(\Gamma_{ij}^k\right)^2 \tag{19.3.3}$$

where Γ_{ij}^k is the on-shell three-particle vertex and for the residue of the form factor F_{n+1} involving the particles A_i and A_j we have

$$-i \lim_{\epsilon \to 0} \epsilon\, F_{n+1}^{\mathcal{O}}(\theta + i\overline{u}_{ik}^j - \epsilon, \theta - i\overline{u}_{jk}^i + \epsilon, \theta_1, \ldots, \theta_{n-1}) = \Gamma_{ij}^k F_n^{\mathcal{O}}(\theta, \theta_1, \ldots, \theta_{n-1}), \tag{19.3.4}$$

where $\overline{u}_{ab}^c \equiv (\pi - u_{ab}^c)$. This equation sets up a recursive structure between the $(n+1)$ and the n particle form factors (Figure 19.5). Let us denote by $\mathcal{B}$ the map between F'_{n+1} and $F_n^{\mathcal{O}}$ set by this recursive equation

$$F_{n+1}^{\mathcal{O}} = \mathcal{B} F_n^{\mathcal{O}}. \tag{19.3.5}$$

When the theory presents bound states, we can show that the two kinds of recursive equation are compatible, so that it is possible to reach the $(n+2)$-particle FF by the n-particle FF either using directly the recursive equation (19.3.1) or applying twice the recursive equation (19.3.4). In terms of the mappings $\mathcal{B}$ and $\mathcal{C}$ we have $\mathcal{C} = \mathcal{B}^2$.

19.4 The Operator Space

At the critical point, we can identify the operator space of a QFT in terms of the irreducible representations of the Virasoro algebra. An extremely interesting point is the characterization of the operator content away from criticality. This can be achieved by means of the form factor theory, although this identification is based on different principles than the conformal theories, it nevertheless allows us to shed light on the classification problem of the operators.

Let's start our discussion with some general considerations. In the form factor approach, an operator $\mathcal{O}$ is defined once all its matrix elements $F_n^{\mathcal{O}}$ are known. Notice the particular nature of all the functional equations—the recursive and Watson's equations—satisfied by the form factors: (i) they are all linear; (ii) they do not refer to any particular operator! This implies that, given a fixed number n of asymptotic particles, the solutions of the form factor equations form a linear space. The classification of the operator content is then obtained by putting in correspondence the vectors of this linear space with the operators.

Kernel solutions. Among the functions of these linear spaces, some belong to the kernel of the operators $\mathcal{B}$ and $\mathcal{C}$: these are the functions $\hat{F}_n^{(i)}$ and $\hat{F}_n^{(j)}$ that satisfy

$$\begin{aligned} \mathcal{B}\hat{F}_n^{(i)} &= 0 \\ \mathcal{C}\hat{F}_n^{(j)} &= 0. \end{aligned} \tag{19.4.1}$$

Their general expression is given in eqn. (19.2.3) but, in this case, the function K_n does not contain poles that give rise to the recursive equations. Hence each of the functions $\hat{F}_n^{(i)}$ and $\hat{F}_n^{(j)}$ is simply a symmetric polynomial in the x_i variables. The vector space of the form factors that belong to the kernels can be further specified by assigning the total and partial degrees of these polynomials.

A non-vanishing kernel of the operators $\mathcal{B}$ and $\mathcal{C}$ has the important consequence that at each level n, if $\tilde{F}_n$ is a reference solution of the recursive equation and $\hat{F}_n$ a function of any of the two kernels, the most general form factor can be written as

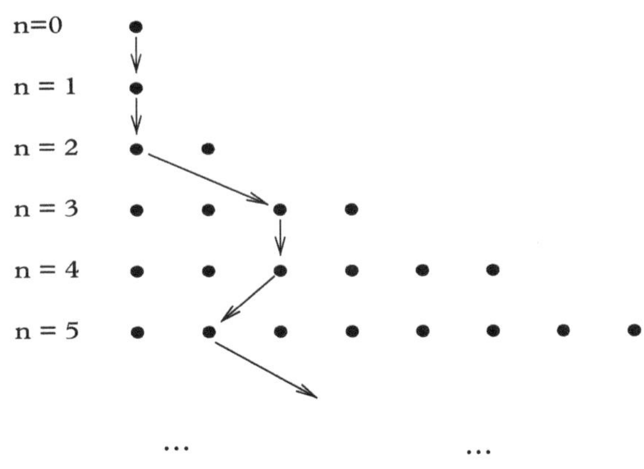

Fig. 19.6 *Vector spaces of the solutions of the form factor equations (the number of dots at each level is purely indicative). An operator is associated to the sequence of its matrix elements F_n.*

$$F_n = \tilde{F}_n + \sum_i \alpha_i \hat{F}_n. \tag{19.4.2}$$

Therefore the identification of each operator is obtained by specifying at each level n the constants α_i. If we graphically represent by dots the linear independent solutions at the level n of the form factor equations, we have Figure 19.6, each operator is associated to a well-defined path on this lattice, with each step $(n+1) \to n$ (or $(n+2) \to n$) ruled by the operator $\mathcal{B}$ (or $\mathcal{C}$). We provide explicit examples of this operator structure when we discuss the form factors of the Ising and the Sinh–Gordon models.

19.5 Correlation Functions

Once we have determined the form factors of a given operator, its correlation functions can be written in terms of the spectral representation series using the completeness relation of the multi-particle states

$$1 = \sum_{n=0}^{\infty} \int \frac{d\theta_1 \dots d\theta_n}{n!(2\pi)^n} \, |\theta_1, \dots, \theta_n\rangle \, \langle \theta_1, \dots, \theta_n|. \tag{19.5.1}$$

For instance, for the two-point correlation function of the operator $\mathcal{O}(x)$ in the Euclidean space, we have

$$\begin{aligned} \langle \mathcal{O}(x)\, \mathcal{O}(0) \rangle &= \sum_{n=0}^{\infty} \int \frac{d\theta_1 \dots d\theta_n}{n!(2\pi)^n} < 0|\mathcal{O}(x)|\theta_1, \dots, \theta_n >_{\text{in in}} < \theta_1, \dots, \theta_n|\mathcal{O}(0)|0 > \\ &= \sum_{n=0}^{\infty} \int \frac{d\theta_1 \dots d\theta_n}{n!(2\pi)^n} \mid F_n(\theta_1 \dots \theta_n) \mid^2 \exp\left(-mr \sum_{i=1}^{n} \cosh\theta_i\right), \end{aligned} \tag{19.5.2}$$

where r is the radial distance $r = \sqrt{x_0^2 + x_1^2}$ (Figure 19.7). Similar expressions, although more complicated, hold for the n-point correlation functions. It is worth making some comments to clarify the nature of these expressions and their advantage.

- The integrals that enter the spectral series are all convergent. This is in sharp contrast with the formalism based on the Feynman diagrams, in which we encounter

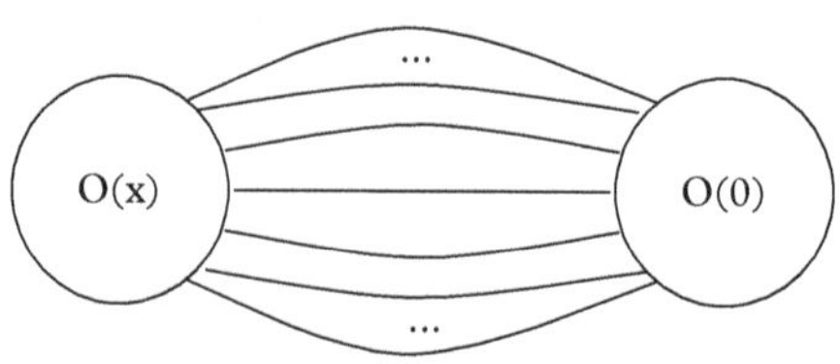

Fig. 19.7 *Spectral representation of the two-point correlation functions.*

the divergences of the various perturbative terms. In a nutshell, the main reason for this difference between the two formalisms can be expressed as follows. The Feynman formalism is based on the quantization of *free* theory and on the *bare* unphysical parameters of the Lagrangian. The renormalization process implements the change from the bare to the physical parameters (such as, the physical value of the mass of the particle). But the form factors employ *ab initio* all the physical parameters of the theory and for this reason the divergences of the perturbative series are absent.

- If the S-matrix depends on a coupling constant, as it happens in the Sinh–Gordon model or in other Toda field theories, each matrix element provides the exact resummation of all terms of perturbation theory.
- If the correlation functions do not have particularly violent ultraviolet singularities (this is the case, for instance, of the correlation functions of the relevant fields), the corresponding spectral series has an extremely fast convergent behaviour for all values of mr. In the infrared region, i.e. for large values of mr, this is evident from the nature of the series, because its natural parameter of expansion is e^{-mr}. The reason of the fast convergent behaviour also in the ultraviolet region $mr \to 0$ is twofold: the peculiar behaviour of the n-particle phase space in two-dimensional theories (see Appendix 17.C) and a further enhancement of the convergence provided by the form factors. To better understand this aspect, consider the Fourier transform of the correlator

$$G(x) = \langle \mathcal{O}(x)\,\mathcal{O}(0)\rangle = \int \frac{d^2p}{(2\pi)^2}\, e^{ip\cdot x}\, \hat{G}(p). \tag{19.5.3}$$

The function $\hat{G}(p)$ can be written as

$$\hat{G}(p) = \int_0^{\infty} d\mu^2\, \rho(\mu^2)\, \frac{1}{p^2+\mu^2}, \tag{19.5.4}$$

where $\rho(k^2)$ is a relativistic invariant function called the *spectral density*

$$\rho(k^2) = 2\pi \sum_{n=0}^{\infty} \int d\Omega_1 \ldots d\Omega_n\, \delta^2(k-P_n)\, |\langle 0\,|\mathcal{O}(0)\,|\theta_1,\ldots,\theta_n\rangle|^2$$

$$d\Omega = \frac{dp}{2\pi E} = \frac{d\theta}{2\pi}, \quad P_n^{(0)} = \sum_{k=0}^{n} \cosh\theta_k, \quad P_n^{(1)} = \sum_{k=0}^{n} \sinh\theta_k.$$

Since $1/(p^2+\mu^2)$ is the two-point correlation function of the Euclidean free theory with mass μ, i.e. the propagator, eqn. (19.5.4) shows that the two-point correlation function can be regarded as a linear superposition of the free propagators weighted with the spectral density $\rho(\mu^2)$. Notice that the contribution given by the single particle state of mass m in the spectral density is given by

$$\rho_{1part}(k^2) = \frac{1}{2\pi}\delta(k^2 - m^2). \tag{19.5.5}$$

To analyse the behaviour of $\rho(k^2)$ by varying k^2, let us make the initial approximation to take equal to 1 all the matrix elements. In this way, each term of the spectral series coincides with the n-particle phase space

$$\Phi_n(k^2) \equiv \int \prod_{k=1}^{n} d\Omega_k \, \delta^2(k - P_n). \tag{19.5.6}$$

As shown in Appendix 17.C, in two dimensions the space goes to zero when $k^2 \to \infty$ as

$$\Phi_n(k^2) \simeq \frac{1}{(2\pi)^{n-2}} \frac{1}{(n-2)!} \frac{1}{k^2} \left(\log \frac{k^2}{m^2}\right)^{n-2}. \tag{19.5.7}$$

whereas for $d > 2$ it diverges as

$$\Phi_n(k^2) \sim k^{\frac{n(d-2)-d}{2}}. \tag{19.5.8}$$

On the other hand, $\Phi_n(k^2) = 0$ if $k^2 < (nm)^2$ and near the threshold values we have

$$\Phi(k^2) \simeq A_n \left(\sqrt{k^2} - nm\right)^{\frac{n-3}{2}}. \tag{19.5.9}$$

Hence, we see that for pure reasons related to the phase space we have two different scenarios for the QFTs in two dimensions and in higher dimensions: while in $d > 2$ surpassing the various thresholds the spectral density receives contributions that are more divergent, in $d = 2$ they are all of the same order and all go to zero at large value of the energy. Hence, for $d > 2$ it is practically impossible to approximate the spectral density for large values of k^2 by using the first terms of the series, relative to the states with few particles, whereas in $d = 2$ this is perfectly plausible. If we now include in the discussion the form factors, we see that the situation is even better in $d = 2$! In fact, from the general expression (19.2.3) and for the vanishing of $F_{\min}(\theta_{ij})$ at the origin (eqn. (19.2.5)), the form factors vanish at the n-particle thresholds as

$$|\langle 0 | \mathcal{O}(0) | \theta_1, \ldots, \theta_n \rangle|^2 \simeq \left(\sqrt{k^2} - nm\right)^{n(n-1)}, \quad \theta_1 \simeq \ldots \simeq \theta_n \simeq 0 \tag{19.5.10}$$

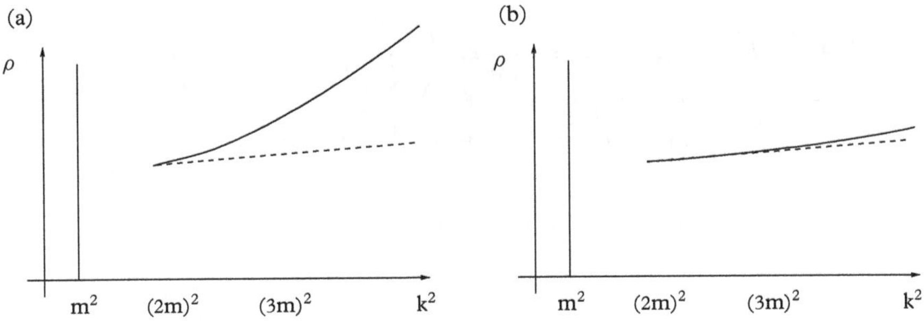

Fig. 19.8 *Plot of the spectral series in a model in d = 4 (a) and in d = 2 (b). The contribution of the two-particle state is given by the dashed line. In d = 4 this do not provide a good approximation of $\rho(k^2)$ for large values of k^2 while in d = 2 it gives very often an excellent approximation of this quantity.*

while, for large values of their rapidities, they typically tend to a constant.[4] This scenario implies that the spectral density of the correlation functions of the two-dimensional integrable models usually flatten more at the thresholds and therefore becomes a very smooth function varying k^2. For all these reasons, the spectral density can be approximated with a great accuracy just taking the first terms of the series, even for large values of k^2, therefore leading to a fast convergent behaviour also in the ultraviolet region.

19.6 Form Factors of the Stress-energy Tensor

The stress-energy tensor is an operator that plays an important role in QFT and its form factors have special properties. For its conservation law $\partial_\mu T^{\mu\nu}(x) = 0$, this operator can be written in terms of an auxiliary scalar field $A(x)$ as

$$T_{\mu\nu}(x) = \left(\partial_\mu \partial_\nu - g_{\mu\nu}\Box\right) A(x). \tag{19.6.1}$$

In the light-cone coordinates, $x^{\pm} = x^0 \pm x^1$, its components are

$$\begin{aligned} T_{++} = \partial_+^2 A, T_{--} = \partial_-^2 A, \\ \Theta = T_\mu^\mu = -\Box A = -4\partial_+\partial_- A. \end{aligned}$$

Introducing the variables $x_j = e^{\theta_j}$ and the elementary symmetric polynomials $\sigma_i^{(n)}$, it is easy to see that

[4] This is what usually happens for the form factors of the strongly relevant operators.

$$
\begin{aligned}
F_n^{T_{++}}(\theta_1,\ldots,\theta_n) &= -\frac{1}{4}m^2\left(\frac{\sigma_{n-1}^{(n)}}{\sigma_n^{(n)}}\right)^2 F_n^A(\theta_1,\ldots,\theta_n),\\
F_n^{T_{--}}(\theta_1,\ldots,\theta_n) &= -\frac{1}{4}m^2\left(\sigma_1^{(n)}\right)^2 F_n^A(\theta_1,\ldots,\theta_n),\\
F_n^{\Theta}(\theta_1,\ldots,\theta_n) &= m^2\,\frac{\sigma_1^{(n)}\sigma_{n-1}^{(n)}}{\sigma_k} F_n^A(\theta_1,\ldots,\theta_n).
\end{aligned}
\tag{19.6.2}
$$

Solving for F_n^A, we have

$$
\begin{aligned}
F_n^{T_{++}}(\theta_1,\ldots,\theta_n) &= -\frac{1}{4}\frac{\sigma_{n-1}^{(n)}}{\sigma_1^{(n)}\sigma_n^{(n)}} F_n^{\Theta}(\theta_1,\ldots,\theta_n),\\
F_n^{T_{--}}(\theta_1,\ldots,\theta_n) &= -\frac{1}{4}\frac{\sigma_1^{(n)}\sigma_n^{(n)}}{\sigma_{n-1}^{(n)}} F_n^{\Theta}(\theta_1,\ldots,\theta_n).
\end{aligned}
\tag{19.6.3}
$$

Hence, the whole set of the form factors of $T_{\mu\nu}$ can be recovered by the form factors of the trace Θ. This is a scalar operator and therefore its form factors depend on the differences of the rapidities $\theta_{ij} = \theta_i - \theta_j$. Moreover, since the form factors of T_{--} and T_{++} must have the same singularities of those of Θ, $F_n^{\Theta}(\theta_1,\ldots,\theta_n)$ (for $n > 2$) has to be proportional to the combination $\sigma_1^{(n)}\sigma_{n-1}^{(n)}$ of the elementary symmetric polynomials. This combination corresponds to the relativistic invariant given by the total energy and momentum of the system.

For the normalization of these matrix elements, the recursive structure reduces the problem of finding the normalization of the form factors of $\Theta(x)$ on the one- and two-particle states, i.e. $F_1^{\Theta}(\theta)$ and $F_2^{\Theta}(\theta_{12})$. The normalization of $F_2^{\Theta}(\theta_{12})$ can be determined by using the total energy of the system

$$
E = \frac{1}{2\pi}\int_{-\infty}^{+\infty} dx^1\, T^{00}(x). \tag{19.6.4}
$$

Computing the matrix element of both terms of this equation on the asymptotic states $< \theta'|$ and $|\theta >$, for the left-hand side we have

$$
\langle\theta'|\,E\,|\theta\rangle = 2\pi\, m\cosh\theta\,\delta(\theta'-\theta).
$$

On the other hand, keeping into account that $T^{00} = \partial_1^2 A$ and using the relation

$$
\langle\theta'|\,\mathcal{O}(x)|\theta\rangle = e^{i(p^\mu(\theta')-p^\mu(\theta))x_\mu}\, F_2^{\mathcal{O}}(\theta,\theta'-i\pi)
$$

that holds for any Hermitian operator $\mathcal{O}$, we obtain

$$F_2^{\partial_1^2 A}(\theta_1,\theta_2) = -m^2 (\sinh\theta_1 + \sinh\theta_2)^2 F_2^A(\theta_{12}).$$

From eqns. (19.6.2) and (19.6.4) it follows that the normalization of F_2^{Θ} is given by

$$F_2^{\Theta}(i\pi) = 2\pi m^2. \tag{19.6.5}$$

However, there is no particular constraint on the one-particle form factor of $\Theta(x)$ coming from general considerations

$$F_1^{\Theta} = < 0 \mid \Theta(0) \mid \theta >. \tag{19.6.6}$$

This is a free parameter of the theory, related to the intrinsic ambiguity of $T^{\mu\nu}(x)$, since this tensor can always be modified by adding a total divergence (see Problem 19.1).

19.7 Vacuum Expectation Values

The recursive equations enable us to determine the form factors $F_n^{\mathcal{O}}$ in terms of the previous $F_{n-1}^{\mathcal{O}}$ or $F_{n-2}^{\mathcal{O}}$. At the bottom of this iterative structure there are, as its initial seeds, the lowest quantities $F_0^{\mathcal{O}}$ (i.e. the vacuum expectation value of the operator $\mathcal{O}$ and F_1 (i.e. its matrix elements on one-particle states). Presently it is not known how to determine in general all the one-particle matrix elements. However, the situation is much better for the vacuum expectation values: they can be computed exactly for several operators both of the Sine–Gordon and Bullogh–Dodd models, as well as of restriction thereof. The theoretical steps that lead to these results are quite technical but well described in the series of papers quoted at the end of the chapter and will not be reviewed here. This section presents the most relevant formulae, in particular, the exact vacuum expectation values of primary fields in integrable perturbed CFT's, with respect to the deformations $\Phi_{1,3}$, $\Phi_{1,2}$ and $\Phi_{2,1}$. In the following, to denote such theories we use the notation

$$\mathcal{S}_m^{(k,l)\pm} = \mathcal{S}_m^{(CFT)} \pm \lambda \int d^2x \, \Phi_{k,l}(x), \tag{19.7.1}$$

where $\mathcal{S}_m$ is the action of the mth conformal minimal model, $\Phi_{r,s}$ is the relevant primary field that leads to an integrable model and $\lambda > 0$ is its dimensional coupling constant setting the scale of the QFT (the sign of the coupling only makes sense after fixing the normalization of the fields $\Phi_{r,s}$). Hereafter

$$x \equiv (m+1)k - ml.$$

Integrable theory $\mathcal{S}_m^{(1,3)-}$. For $\lambda > 0$, $\Phi_{1,3}$ induces a massless flow between the minimal models $\mathcal{M}_m \to \mathcal{M}_{m-1}$ (see Section 15.6). For $\lambda < 0$, $\Phi_{1,3}$ drives instead the conformal model into a massive phase where there are kinks interpolating the $(m-1)$ degenerate vacua labelled as

$$\mathbf{a} = 0, \frac{1}{2}, \ldots, \frac{(m-2)}{2}.$$

For the vacuum expectation values of the primary fields on the various vacua we have

$$\langle \mathbf{a} | \Phi_{k,l} | \mathbf{a} \rangle^{(1,3)-} = \frac{\sin\left(\frac{\pi(2a+1)}{m}((m+1)k - ml)\right)}{\sin\frac{\pi(2a+1)}{m}} F_{k,l}^m(x) \tag{19.7.2}$$

where

$$F_{k,l}^m(x) = \left(M \frac{\sqrt{\pi}\,\Gamma\left(\frac{m+3}{2}\right)}{2\Gamma\left(\frac{m}{2}\right)}\right)^{2\Delta_{k,l}} \mathcal{Q}_{1,3}(x)$$

and

$$\mathcal{Q}_{1,3}(\eta) = \exp\left\{\int_0^\infty \frac{dt}{t}\left[\frac{\cosh(2t)\sinh((\eta-1)t)\sinh((\eta+1)t)}{2\cosh(t)\sinh(mt)\sinh((1+m)t)} - \frac{\eta^2-1}{2m(m+1)}e^{-4t}\right]\right\}.$$

In the formula above

$$M = \frac{2\Gamma\left(\frac{m}{2}\right)}{\sqrt{\pi}\,\Gamma\left(\frac{m+1}{2}\right)}\left[\frac{\pi\lambda(1-m)(2m-1)}{(1+m)^2}\sqrt{\frac{\Gamma\left(\frac{1}{m+1}\right)\Gamma\left(\frac{1-2m}{m+1}\right)}{\Gamma\left(\frac{m}{m+1}\right)\Gamma\left(\frac{3m}{m+1}\right)}}\right]^{\frac{1+m}{4}}$$

is the common mass of the kinks expressed in term of the coupling constant λ.

Integrable theory $\mathcal{S}_m^{(1,2)}$. For the integrable model $\mathcal{S}_m^{(1,2)}$, the vacuum structure of the theory depends on whether m is odd or even.

- **m even.** When m is even, the field $\Phi_{1,2}$ is even under the Z_2 spin symmetry and the two theories $\mathcal{S}_m^{(1,2)\pm}$ are different although related by duality. The number of vacua of $\mathcal{S}_m^{(1,2)+}$ is equal to $(m-2)/2$, while the number of vacua of $\mathcal{S}_m^{(1,2)-}$ is equal to $m/2$. Their label is

$$\mathbf{a} = \frac{1}{2}, \frac{3}{2}, \ldots, \frac{m-3}{2}, \quad \lambda > 0,$$
$$\mathbf{a} = 0, 1, \ldots, \frac{m-2}{2}, \quad \lambda < 0.$$

- ***m* odd.** In this case the field $\Phi_{1,2}$ is odd under the Z_2 symmetry and the two theories $\mathcal{S}_m^{(1,2)\pm}$ are equal. There are $(m-1)/2$ degenerate vacua in both theories that we label as

$$a = \frac{1}{2}, \frac{3}{2}, \ldots, \frac{m-2}{2}, \quad \lambda > 0,$$
$$a = 0, 1, \ldots, \frac{m-3}{2}, \quad \lambda < 0.$$

The vacuum expectation values of the primary fields on the various vacua are:

$$\langle \mathbf{a} | \Phi_{k,l} | \mathbf{a} \rangle^{(1,2)} = \frac{\sin\left(\frac{\pi(2a+1)}{m}((m+1)k - ml)\right)}{\sin \frac{\pi(2a+1)}{m}} G^m_{k,l}(x) \tag{19.7.3}$$

where

$$G^m_{k,l}(x) = \left(M \frac{\pi(m+1)\Gamma\left(\frac{2m+2}{3m+6}\right)}{2^{\frac{2}{3}}\sqrt{3}\Gamma\left(\frac{1}{3}\right)\Gamma\left(\frac{m}{3m+6}\right)} \right)^{2\Delta_{k,l}} \mathcal{Q}_{1,2}(x)$$

and

$$\begin{aligned} \mathcal{Q}_{1,2}(\eta) = \exp\Bigg\{ \int_0^\infty \frac{dt}{t} \Bigg[& \frac{\sinh((m+2)t)\sinh((\eta-1)t)\sinh((\eta+1)t)}{\sinh(3(m+2)t)\sinh(2(m+1)t)\sinh(mt)} \times \\ & (\cosh(3(m+2)t) + \cosh((m+4)t) - \cosh((3m+4)t) + \cosh(mt) + 1) \\ & - \frac{\eta^2 - 1}{2m(m+1)} e^{-4t} \Bigg] \Bigg\} \end{aligned}$$

In the formula above

$$M = \frac{2^{\frac{m+5}{3m+6}}\sqrt{3}\Gamma\left(\frac{1}{3}\right)\Gamma\left(\frac{m}{3m+6}\right)}{\pi\Gamma\left(\frac{2m+2}{3m+6}\right)} \left[\frac{\pi^2\lambda^2\Gamma\left(\frac{3m+4}{4m+4}\right)\Gamma\left(\frac{1}{2}+\frac{1}{m+1}\right)}{\Gamma\left(\frac{m}{4m+4}\right)\Gamma\left(\frac{1}{2}-\frac{1}{m+1}\right)} \right]^{\frac{m+1}{3m+6}}$$

is the mass of the kinks expressed in terms of the coupling constant λ.

Integrable theory $\mathcal{S}_m^{(2,1)}$. For this theory the situation is reverse with respect to the previous one: $\Phi_{2,1}$ is odd under the Z_2 symmetry when m is even (and the theory is independent of the sign of its coupling), while is a Z_2 even field when m is odd (and the theories with $\lambda > 0$ and $\lambda < 0$ are related by duality). For the degenerate vacua, in this case we have the following labels

- when m is even, both for $\lambda > 0$ and $\lambda < 0$, their number is $m/2$ and

$$a = \frac{1}{2}, \frac{3}{2}, \ldots, \frac{m-1}{2}, \quad \lambda > 0,$$
$$a = 0, 1, \ldots, \frac{m-2}{2}, \quad \lambda < 0;$$

- when m is odd, their number is $(m-1)/2$ for $\lambda > 0$, and $(m+1)/2$ for $\lambda < 0$, with

$$a = \frac{1}{2}, \frac{3}{2}, \ldots, \frac{m-2}{2}, \quad \lambda > 0,$$
$$a = 0, 1, \ldots, \frac{m-1}{2}, \quad \lambda < 0.$$

The vacuum expectation values of the primary fields on the various vacua are

$$\langle \mathbf{a}|\Phi_{k,l}|\mathbf{a}\rangle^{(2,1)} = \frac{\sin\left(\frac{\pi(2a+1)}{m+1}((m+1)k - ml)\right)}{\sin\frac{\pi(2a+1)}{m+1}} H_{k,l}^m(x) \qquad (19.7.4)$$

where

$$H_{k,l}^m(x) = \left(M \frac{\pi m \Gamma\left(\frac{2m}{3m-3}\right)}{2^{\frac{2}{3}}\sqrt{3}\Gamma\left(\frac{1}{3}\right)\Gamma\left(\frac{m+1}{3m-3}\right)} \right)^{2\Delta_{k,l}} \mathcal{Q}_{2,1}(x)$$

and

$$\mathcal{Q}_{2,1}(\eta) = \exp\left\{ \int_0^\infty \frac{dt}{t} \left[\frac{\sinh((m-1)t)\sinh((\eta-1)t)\sinh((\eta+1)t)}{\sinh(3(m-1)t)\sinh(2mt)\sinh((m+1)t)} \times \right.\right.$$
$$(\cosh(3(m-1)t) + \cosh((m-3)t) - \cosh((3m-1)t) + \cosh((m+1)t) + 1)$$
$$\left.\left. - \frac{\eta^2 - 1}{2m(m+1)} \mathrm{e}^{-4t} \right] \right\}. \qquad (19.7.5)$$

The mass of the kinks, as a function of the coupling constant λ, is expressed by

$$M = \frac{2^{\frac{m-4}{3m-3}}\sqrt{3}\Gamma\left(\frac{1}{3}\right)\Gamma\left(\frac{m+1}{3m-3}\right)}{\pi\Gamma\left(\frac{2m}{3m-3}\right)}\left[\frac{\pi^2\lambda^2\Gamma\left(\frac{3m-1}{4m}\right)\Gamma\left(\frac{1}{2}-\frac{1}{m}\right)}{\Gamma\left(\frac{m+1}{4m}\right)\Gamma\left(\frac{1}{2}+\frac{1}{m}\right)}\right]^{\frac{m}{3m-3}}.$$

19.8 Ultraviolet Limit

In the ultraviolet limit, the correlation functions of the scaling operators has a power-law behaviour, dictated by the conformal dimension of the operator

$$G(r) = \langle \mathcal{O}(r)\mathcal{O}(0)\rangle \simeq \frac{1}{r^{4\Delta}}, \quad r \to 0. \tag{19.8.1}$$

We may wonder how the spectral series (19.5.2), that is based on the exponential terms e^{-kmr}, is able to reproduce a power law in the limit $r \to 0$. The answer to this question comes from an interesting analogy.

Feynman gas. Note that the formula (19.5.2) is formally similar to the expression of the grand canonical partition function of a fictitious one-dimensional gas

$$\mathcal{Z}(mr) = \sum_{n=0}^{\infty} z^n Z_n. \tag{19.8.2}$$

To set up the vocabulary of this analogy, we identify the coordinates of the gas particles with the rapidities θ_i, while the Boltzmann weight relative to the interactive potential of the gas with the modulus square of the form factors

$$e^{-V(\theta_1,\ldots,\theta_n)} \equiv |\langle 0|\mathcal{O}(0)|\theta_1,\ldots,\theta_n\rangle|^2. \tag{19.8.3}$$

Finally, let us identify the fugacity of the gas with

$$z(\theta) = \frac{1}{2\pi} e^{-mr\cosh\theta}. \tag{19.8.4}$$

We use this to devine the *Feynman gas* analysed at the end of Chapter 2. The only difference with respect to the standard case is the coordinate dependence of the fugacity of this gas. Although the coordinates of the particles of this gas span the infinite real axis, the effective volume of the system is however determined by the region in which the fugacity (19.8.4) is significantly different from zero (Figure 19.9). Note that $z(\theta)$ is a function that rapidly goes to zero outside a finite interval and, in the limit $mr \to 0$, presents a plateau of height $z_c = 1/(2\pi)$ and width

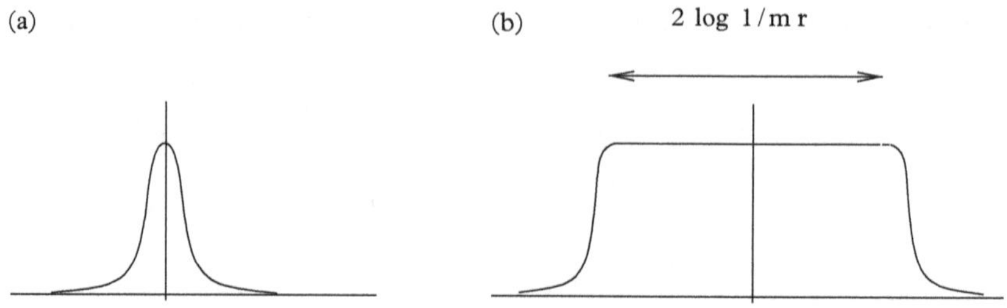

Fig. 19.9 *Plot of the fugacity as a function of θ: (a) for finite values of (mr); (b) in the limit $(mr) \to 0$.*

$$L \simeq 2 \log \frac{1}{mr}.$$

The equation of state of a one-dimensional gas is given by

$$\mathcal{Z} = e^{p(z)L},$$

where $p(z)$ is the pressure as a function of the fugacity. Following this analogy, for the two-point correlation function in the limit $(mr) \to 0$ we have

$$G(r) = \mathcal{Z} = e^{p(z_c)L} \simeq e^{2p(z_c)\log 1/(mr)} = \left(\frac{1}{mr}\right)^{2p(z_c)}, \tag{19.8.5}$$

i.e. a power-law behaviour! Moreover, comparing with the short-distance behaviour of the correlator given in eqn. (19.8.1), there is an interesting result: the conformal dimension can be expressed in terms of the pressure of this fictitious one-dimensional gas, evaluated at the plateaux value of the fugacity

$$2\Delta = p(1/2\pi). \tag{19.8.6}$$

Beside from the thermodynamics of the Feynman gas, the conformal dimension of the operators can be also extracted by applying the sum rule given by the Δ-theorem (see Chapter 15)

$$\Delta = -\frac{1}{2\langle \mathcal{O} \rangle} \int_0^\infty dr\, r\, \langle \Theta(r) \mathcal{O}(0) \rangle. \tag{19.8.7}$$

To compute this quantity, it is necessary to know the form factors of the operator $\mathcal{O}(x)$ and the trace of the stress-energy tensor $\Theta(x)$.

c-theorem sum rule. An additional control of the ultraviolet limit of the theory is provided by the sum-rule of the c-theorem: it gives the central charge of the CFT associated to the ultraviolet limit of the massive theory through the integral

$$c = \frac{3}{2}\int_0^{\infty} dr\, r^3 \,\langle \Theta(r)\Theta(0)\,\rangle_c.$$

Using the spectral representation of this correlator we have

$$c = \sum_{n=1}^{\infty} c_n, \tag{19.8.8}$$

where the n-particle contribution is

$$\begin{aligned} c_n \;=\; & \frac{12}{n!}\int_0^{\infty}\frac{d\mu}{\mu^3}\int_{-\infty}^{\infty}\frac{d\theta_1}{2\pi}\cdots\frac{d\theta_n}{2\pi} \\ & \times\, \delta\Big(\sum_{i=1}^{n}\sinh\theta_i\Big)\,\delta\Big(\sum_{i=1}^{n}\cosh\theta_i-\mu\Big)\,|\langle 0|\Theta(0)|\theta_1,\ldots,\theta_n\rangle|^2. \end{aligned} \tag{19.8.9}$$

Usually this series presents a very fast behaviour. This permits to obtain estimations rather accurate of the central charge c, with an explicit check of the entire formalism of the S-matrix and form factors. It is easy to understand the reason of this fast convergence by studying the integrand, shown in Figure 19.10: the term r^3 kills the singularity of the correlator at short distance (therefore the integrand vanishes at the origin), while it weights more correlator at large distances. But this is just the region where few terms of the spectral series are very efficient in approximating the correlation function with high accuracy.

Asymptotic behaviour. Finally, let us discuss the upper bound on the asymptotic behaviour of the form factors dictated by the ultraviolet behaviour of the correlator (19.8.1). To establish this bound, we start by noting that in a massive theory we have

$$M_p \equiv \int d^2x\,|x|^p\,\langle \mathcal{O}(x)\mathcal{O}(0)\,\rangle_c \qquad < +\infty \qquad \text{if} \qquad p > 4\Delta_{\mathcal{O}} - 2, \tag{19.8.10}$$

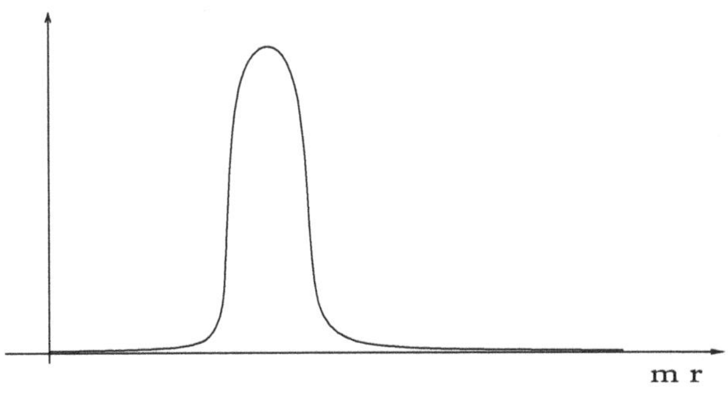

Fig. 19.10 *Plot of the integrand $r^3\langle\Theta(r)\Theta(0)\rangle$ in the C-theorem sum rule.*

Employing now the spectral representation of the correlator (19.5.3) and integrating on p, μ and x, we get

$$M_p \sim \sum_{n=1}^{\infty} \int_{\theta_1 > \ldots > \theta_n} d\theta_1 \ldots d\theta_n \frac{|F_n^{\mathcal{O}}(\theta_1, \ldots, \theta_n)|^2}{\left(\sum_{k=1}^{n} m_k \cosh\theta_k\right)^{p+1}} \, \delta\left(\sum_{k=1}^{n} m_k \sinh\theta_k\right). \tag{19.8.11}$$

Eqn. (19.8.10) can be now used to find an upper limit on the real quantity y_Φ, defined by

$$\lim_{|\theta_i| \to \infty} F_n^{\mathcal{O}}(\theta_1, \ldots, \theta_n) \sim e^{y_\Phi |\theta_i|}. \tag{19.8.12}$$

In fact, taking the limit $\theta_i \to +\infty$ in the integrand of (19.8.11), the delta function forces some other rapidities to move at $-\infty$ as $-\theta_i$. Because the matrix element $F_n^{\mathcal{O}}(\theta_1, \ldots, \theta_n)$ depends on the differences of the rapidities, it contributes to the integrand with the factor $e^{2y_\Phi|\theta_i|}$ in the limit $|\theta_i| \to \infty$. Hence, eqn. (19.8.10) leads to the condition

$$y_{\mathcal{O}} \leq \Delta_{\mathcal{O}}. \tag{19.8.13}$$

This equation provides information on the partial degree of the polynomial $Q_n^{\mathcal{O}}$. Note, however, that this conclusion may not apply for non-unitary theories because not all terms of the expansion on the intermediate states are necessarily positive in this case.

19.9 The Ising Model at $T \neq T_c$

This section presents the form factors and the correlation functions of the relevant operators $\epsilon(x)$, $\sigma(x)$ and $\mu(x)$ of the two-dimensional Ising model when the temperature T is away from its critical value. For the duality of the model, we discuss equivalently the case $T > T_c$ or $T < T_c$. Suppose the high-temperature phase where the scattering theory of the off-critical model involves only one particle with an S-matrix $S = -1$. There are no bound states. The particle A is created by the magnetization operator $\sigma(x)$, so that it is odd under the Z_2 symmetry of the Ising model, with its mass given by $m = |T - T_c|$.

Let us now employ the form factor equations to find the matrix elements of the various operators on the multi-particle states. The first step is the determination of the function $F_{min}(\theta)$ that satisfies

$$\begin{aligned} F_{\min}(\theta) &= -F_{\min}(-\theta) \\ F_{\min}(i\pi - \theta) &= F_{\min}(i\pi + \theta). \end{aligned} \tag{19.9.1}$$

The minimal solution is

$$F_{\min}(\theta) = \sinh\frac{\theta}{2}. \tag{19.9.2}$$

19.9.1 The Energy Operator

Let us discuss initially the form factors of the energy operator $\epsilon(x)$ or, equivalently, those of the trace of the stress-energy tensor, since the two operators are related by

$$\Theta(x) = 2\pi m \epsilon(x). \tag{19.9.3}$$

This is an even operator under the Z_2 symmetry and therefore it has matrix elements only on states with an even number of particles, F^{Θ}_{2n}. The recursive equation of the kinematical poles are particularly simple

$$-i \lim_{\tilde{\theta}\to\theta} (\tilde{\theta}-\theta) F^{\Theta}_{2n+2}(\tilde{\theta}+i\pi,\theta,\theta_1,\theta_2,\ldots,\theta_{2n}) = \left(1-(-1)^{2n}\right) F^{\Theta}_{2n}(\theta_1,\ldots,\theta_{2n}) = 0. \tag{19.9.4}$$

Taking into account the normalization of the trace operator $F^{\Theta}_2(i\pi) = 2\pi m^2$, the simplest solution of all these equations is

$$F^{\Theta}_{2n}(\theta_1,\ldots,\theta_{2n}) = \begin{cases} -2\pi i m^2 \sinh\frac{\theta_1-\theta_2}{2}, & n=2 \\ 0, & \text{otherwise.} \end{cases} \tag{19.9.5}$$

As discussed in Section 19.4, note that the identification of the operator Θ with this specific sequence of form factors is equivalent to put equal to zero all coefficients of the kernel solutions $F^{(i)}_{2n}$ at all the higher levels.

We must verify that (19.9.5) is the correct sequence of the form factors of the trace operator comes from its two-point correlation function and from the c-theorem. For the correlator we get

$$\begin{aligned} G^{\Theta}(r) = \langle \Theta(r)\Theta(0)\rangle &= \frac{1}{2}\int \frac{d\theta_1}{2\pi}\frac{d\theta_2}{2\pi} |F^{\Theta}_2(\theta_{12})|^2 \, e^{-mr(\cosh\theta_1+\cosh\theta_1)} \\ &= \frac{m^4}{2}\int d\theta_1\, d\theta_2 \sinh^2\frac{\theta_1-\theta_2}{2}\, e^{-mr(\cosh\theta_1+\cosh\theta_2)} \\ &= \frac{m^4}{4}\int d\theta_1\, d\theta_2 \left[\cosh(\theta_1-\theta_2)-1\right] e^{-mr(\cosh\theta_1+\cosh\theta_2)} \\ &= m^4\left(\left[\int d\theta \cosh\theta\, e^{-mr\cosh\theta}\right]^2 - \left[\int d\theta\, e^{-mr\cosh\theta}\right]^2\right) \\ &= m^4\left(K_1^2(mr)-K_0^2(mr)\right) \end{aligned} \tag{19.9.6}$$

where, in the last line, we used the integral representation of the modified Bessel functions

$$K_\nu(z) = \int_0^\infty dt \cosh\nu t\, e^{-z\cosh t}.$$

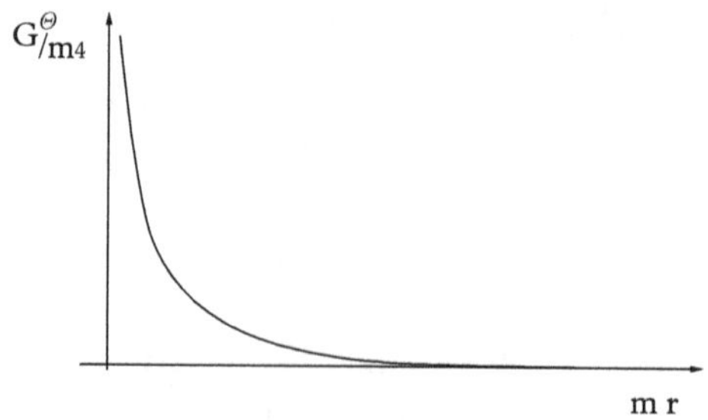

Fig. 19.11 *Plot of the two-point correlation function of the trace of the stress-energy tensor for the thermal Ising model.*

Hence we have

$$G^{\Theta}(r) = \langle \Theta(r)\Theta(0)\rangle = m^4\Big[K_1^2(mr) - K_0^2(mr)\Big]. \tag{19.9.7}$$

whose plot is in Figure 19.11. This function has the correct ultraviolet behaviour associated to the energy operator

$$G^{\Theta}(r) \to \frac{m^2}{|x|^2}, \quad |x| \to 0. \tag{19.9.8}$$

Substituting the expression above in the c-theorem, we get the correct value of the central charge of the Ising model

$$c = \frac{3}{2}\int_0^{\infty} dr\, r^3 \langle \Theta(r)\Theta(0)\rangle = \frac{1}{2}. \tag{19.9.9}$$

19.9.2 Magnetization Operators

In the high-temperature phase, the order parameter $\sigma(x)$ is odd under the Z_2 symmetry while the disorder operator $\mu(x)$ is even. Hence, $\sigma(x)$ has matrix elements on states with an odd number of particles, F^{σ}_{2n+1}, whereas $\mu(x)$ on an even number, F^{μ}_{2n}. In writing down the residue equations relative to the kinematical poles, we keep into account that the operator μ has a semi-local index equal to $1/2$ with respect to the operator $\sigma(x)$ that creates the asymptotic states. Denoting by F_n the form factors of these operators (for n even they refer to $\mu(x)$ while for n odd to $\sigma(x)$), we have the recursive equation

$$-\,i \lim_{\tilde{\theta}\to\theta} (\tilde{\theta} - \theta) F_{n+2}(\tilde{\theta} + i\pi, \theta, \theta_1, \theta_2, \ldots, \theta_{2n}) = 2F_n(\theta_1, \ldots, \theta_{2n}). \tag{19.9.10}$$

As for any form factor equation, these equations admit an infinite number of solutions that can be obtained by adding all possible kernel solutions at each level. The minimal solution is the one chosen to identify the form factors of the order and disorder operators

$$F_n(\theta_1,\ldots,\theta_n) = H_n \prod_{i<j}^{n} \tanh\frac{\theta_i-\theta_j}{2}. \qquad (19.9.11)$$

The normalization coefficients satisfy the recursive equation

$$H_{n+2} = i H_n.$$

The solutions with n even are therefore fixed by choosing $F_0 = H_0$, namely with a non-zero value of the vacuum expectation of the disorder operator

$$F_0 = \langle 0|\mu(0)|0\rangle = \langle \mu \rangle, \qquad (19.9.12)$$

while those with n odd are determined by the real constant F_1 relative to the one-particle matrix element of $\sigma(x)$

$$F_1 = \langle 0|\sigma(0)|A\rangle. \qquad (19.9.13)$$

Adopting the conformal normalization of both operators

$$\langle \sigma(x)\sigma(0)\rangle = \langle \mu(x)\mu(0)\rangle \simeq \frac{1}{|x|^{1/4}}, \quad |x| \to 0 \qquad (19.9.14)$$

it is possible to show that $F_0 = F_1$ and the vacuum expectation value F_0 can be computed using eqn. (19.7.2)

$$F_0 = F_1 = 2^{1/3} e^{-1/4} A^3 m^{1/4}, \qquad (19.9.15)$$

where $A = 1.282427..$ is called the Glasher constant. Vice versa, if we choose $F_0 = F_1 = 1$ (as we do hereafter) for the ultraviolet behaviour of the correlation functions we have

$$\langle \sigma(x)\sigma(0)\rangle = \langle \mu(x)\mu(0)\rangle \simeq \frac{2^{-1/3}e^{1/4}A^{-3}}{|x|^{1/4}} = \frac{0.5423804\ldots}{|x|^{1/4}}, \quad |x| \to 0. \qquad (19.9.16)$$

There are several way to check the correct identification of the form factors of the order/disorder operators. A direct way is to employ the Δ-theorem. In fact, using the matrix elements of $\mu(x)$ and $\Theta(x)$, we can compute their correlator, following the same procedure as in eqn. (19.9.6)

$$\begin{aligned}\langle \Theta(r)\mu(0)\rangle &= \frac{1}{2}\int \frac{d\theta_1}{2\pi}\frac{d\theta_2}{2\pi} F^{\Theta}(\theta_{12})\bar{F}^{\mu}(\theta_{12})\, e^{-mr(\cosh\theta_1+\cosh\theta_2)} \\ &= -m^2\langle \mu \rangle \left[\frac{e^{-2mr}}{2mr} + Ei(-2mr)\right] \qquad (19.9.17)\end{aligned}$$

where

$$Ei(-x) = -\int_x^\infty \frac{dt}{t}\, e^{-t}.$$

Substituting this correlator in the formula of the Δ-theorem, one obtains the correct value of the conformal dimension of the disorder operator

$$\Delta = -\frac{1}{2\langle\mu\rangle}\int_0^\infty dr\, r\langle\Theta(r)\mu(0)\rangle = \frac{1}{4\pi}\int_0^\infty d\theta\, \frac{\sinh^2\theta}{\cosh^3\theta} = \frac{1}{16}. \tag{19.9.18}$$

Another way to determine the conformal dimension of the magnetization operators consists of solving the thermodynamics of the Feynman gas associated to the form factors. Using the nearest-neighbour approximation discussed in Chapter 2, the pressure of this gas satisfies the integral equation (Problem 19.2)

$$z_c^{-1} = 2\pi = \int_0^\infty dx\, \tanh^2\frac{x}{2}\, e^{-px}, \tag{19.9.19}$$

whose numerical solution is

$$p \simeq 0.12529\ldots. \tag{19.9.20}$$

Comparing with the exact value

$$p = 2\Delta = \frac{1}{8} = 0.125, \tag{19.9.21}$$

we see that the relative precision is less than one part in a thousand! This result confirms the validity of the form factor solution for the magnetization operators and also it explicitly shows the convergence property of the spectral series.

19.9.3 The Painlevé Equation

The two-point correlation functions of the magnetization operators are given by

$$\begin{aligned}\langle\mu(r)\mu(0)\rangle &= \textstyle\sum_{n=0}^\infty g_{2n}(r)\\ \langle\sigma(r)\sigma(0)\rangle &= \textstyle\sum_{n=0}^\infty g_{2n+1}(r)\end{aligned}$$

where

$$g_n(r) = \frac{1}{n!}\int\left[\prod_{k=1}^n \frac{d\theta_k}{2\pi}\, e^{-mr\cosh\theta_k}\right]\prod_{i<j}\tanh^2\frac{\theta_{ij}}{2}.$$

These expressions can be further elaborated: posing $u_i = e^{\theta_i}$ and using

$$\tanh^2 \frac{\theta_i - \theta_j}{2} = \left(\frac{u_i - u_j}{u_i + u_j}\right)^2,$$

we get

$$\prod_{i<j} \tanh^2 \frac{\theta_{ij}}{2} = \prod_{i<j} \left(\frac{u_i - u_j}{u_i + u_j}\right)^2 = \det W, \qquad (19.9.22)$$

where the matrix elements of the operator W are

$$W_{ij} = \frac{2\sqrt{u_i u_j}}{u_i + u_j}.$$

Combining the two correlators

$$G^{(\pm)}(r) = \langle \mu(r)\mu(0)\rangle \pm \langle \sigma(r)\sigma(0)\rangle = \sum_{n=0}^{\infty} \lambda^n g_n(r) \qquad (19.9.23)$$

(with $\lambda = \pm 1$) and using (19.9.22) we obtain

$$G^{(\pm)}(r) = \sum_{n=0}^{\infty} \frac{\lambda^n}{n!} \int \left[\prod_{k=1}^{n} \frac{d\theta_k}{2\pi} e^{-mr\cosh\theta_k}\right] \det W. \qquad (19.9.24)$$

The last expression is nothing else but the Fredholm determinant of an integral operator V, whose kernel is

$$V(\theta_i, \theta_j, r) = \frac{E(\theta_i, r)\, E(\theta_j, r)}{u_i + u_j}$$

$$E(\theta_i, r) = (2u_i e^{-mr\cosh\theta_i})^{1/2}.$$

Hence

$$G^{(\pm)}(r) = \mathrm{Det}\,(1 + \lambda V). \qquad (19.9.25)$$

The remarkable circumstance that the correlation functions are expressed in terms of the Fredholm determinant of an integral operator is crucial for studying their properties. The detailed discussion is beyond the scope of this book and here we simply present the main conclusions.

First of all, the expression given in eqn. (19.9.25) permits to solve *exactly* the thermodynamics of the Feynman gas associated to the form factors of the correlation function $G^{(+)}(r)$. The exact expression of the pressure of the Feynman gas is given by

$$p(z) = \frac{1}{4}\int \frac{dp}{2\pi}\log\left[1+\left(\frac{2\pi z}{\sinh \pi p}\right)^2\right]$$
$$= \frac{1}{4\pi}\arcsin(2\pi z) - \frac{1}{4\pi^2}\arcsin^2(2\pi z).$$

Substituting in this formula the plateau value of the fugacity, $z = z_c = 1/(2\pi)$, one obtains the exact value of the conformal dimension of the magnetization operators, $p = 2\Delta = 1/8$.

Secondly, using the Fredholm determinant (19.9.25), it is possible to show that the correlators can be concisely written as

$$\begin{pmatrix} \langle \mu(r)\mu(0)\rangle \\ \langle \sigma(r)\sigma(0)\rangle \end{pmatrix} = \begin{pmatrix} \cosh \frac{\Psi(s)}{2} \\ \sinh \frac{\Psi(s)}{2} \end{pmatrix} \exp\left[-\frac{1}{4}\int_s^{\infty} dt\, t\left[\left(\frac{d\Psi}{dt}\right)^2 - \sinh^2\Psi\right]\right] \tag{19.9.26}$$

($s = mr$), where $\Psi(s)$ is a function solution of the differential equation

$$\frac{d^2\Psi}{ds^2} + \frac{1}{s}\frac{d\Psi}{ds} = 2\sinh(2\Psi), \tag{19.9.27}$$

with boundary conditions

$$\begin{aligned} \Psi(s) &\simeq -\log s + \text{costant}, \quad s \to 0 \\ \Psi(2) &\simeq 2/\pi\, K_0(2s), \quad s \to \infty. \end{aligned} \tag{19.9.28}$$

With the substitution $\eta = e^{-Psi}$, the differential equation becomes the celebrated Painlevé differential equation of the third kind

$$\frac{\eta''}{\eta} = \left(\frac{\eta'}{\eta}\right)^2 - \frac{1}{s}\left(\frac{\eta'}{\eta}\right) + \eta^2 - \frac{1}{\eta^2}. \tag{19.9.29}$$

This equation was originally created by Wu, McCoy, Tracy and Barouch by studying the scaling limit of the lattice Ising model. It has also been derived by Jimbo, Miwa and Ueno by using the monodromy theory of the differential equations.

19.10 Form Factors of the Sinh–Gordon Model

This section studies the form factors of an integrable Lagrangian theory defined by the Sinh–Gordon model. The action is

$$\mathcal{S} = \int d^2x\left[\frac{1}{2}(\partial_\mu\phi)^2 - \frac{m^2}{g^2}\cosh g\phi(x)\right], \tag{19.10.1}$$

and it possesses the Z_2 symmetry $\phi \to -\phi$. The exact S-matrix relative to the particle created by the field $\phi(x)$ is given by

$$S(\theta,B) = \frac{\tanh\frac{1}{2}(\theta - i\frac{\pi B}{2})}{\tanh\frac{1}{2}(\theta + i\frac{\pi B}{2})}, \tag{19.10.2}$$

where B is a function of the coupling constant g

$$B(g) = \frac{2g^2}{8\pi + g^2}. \tag{19.10.3}$$

The theory does not have bound states; therefore the form factors satisfy the recursive equations coming from the kinematic poles only. As discussed in Chapter 18, the S-matrix is invariant under the transformation

$$B \to 2 - B, \tag{19.10.4}$$

namely, under the weak/strong duality

$$g \to \frac{8\pi}{g}. \tag{19.10.5}$$

The Z_2 symmetry implies that the even (odd) operators have form factors different from zero only on asymptotic states with an even (odd) number of particles. The simplest odd field is just $\phi(x)$, with the normalization given by

$$F_1^{\phi}(\theta) = < 0 \mid \phi(0) \mid \theta >_{\text{in}} = \frac{1}{\sqrt{2}}. \tag{19.10.6}$$

One of the most important field is the stress-energy tensor

$$T_{\mu\nu}(x) = 2\pi\,(:\partial_\mu\phi\partial_\nu\phi - g_{\mu\nu}\mathcal{L}(x):) \tag{19.10.7}$$

where :: denotes the normal order of the composite operators. Its trace $T_\mu^\mu(x) = \Theta(x)$ is normalized as

$$F_2^{\Theta}(\theta_{12} = i\pi) = {}_{\text{out}}< \theta_1 \mid \Theta(0) \mid \theta_2 >_{\text{in}} = 2\pi m^2, \tag{19.10.8}$$

while F_1^{Θ} is a free parameter. The following only discusses the case $F_1^{\Theta} = 0$: this is equivalent to regard the Sinh–Gordon model as a deformation of the CFT with central charge $c = 1$ (see Chapter 16 and Problem 19.1).

19.10.1 Minimal Form Factor

The first step to the solution of the form factor equation consists of finding the minimal two-particle form factor. Expressing the S-matrix as

$$S(\theta) = \exp\left[8\int_0^\infty \frac{dx}{x}\sinh\left(\frac{xB}{4}\right)\sinh\left(\frac{x}{2}(1-\frac{B}{2})\right)\sinh\frac{x}{2}\sinh\left(\frac{x\theta}{i\pi}\right)\right].$$

we have

$$F_{\text{min}}(\theta,B) = \mathcal{N}\exp\left[8\int_0^\infty \frac{dx}{x}\frac{\sinh\left(\frac{xB}{4}\right)\sinh\left(\frac{x}{2}(1-\frac{B}{2})\right)\sinh\frac{x}{2}}{\sinh^2 x}\sin^2\left(\frac{x\hat{\theta}}{2\pi}\right)\right], \quad (19.10.9)$$

($\hat{\theta}\equiv i\pi-\theta$), with the normalization given by

$$\mathcal{N} = \exp\left[-4\int_0^\infty \frac{dx}{x}\frac{\sinh\left(\frac{xB}{4}\right)\sinh\left(\frac{x}{2}(1-\frac{B}{2})\right)\sinh\frac{x}{2}}{\sinh^2 x}\right].$$

The analytic structure of this function can be studied using its representation in terms of an infinite product of Γ functions (see Problem 19.3)

$$F_{\text{min}}(\theta,B) = \prod_{k=0}^{\infty}\left|\frac{\Gamma\left(k+\frac{3}{2}+\frac{i\hat{\theta}}{2\pi}\right)\Gamma\left(k+\frac{1}{2}+\frac{B}{4}+\frac{i\hat{\theta}}{2\pi}\right)\Gamma\left(k+1-\frac{B}{4}+\frac{i\hat{\theta}}{2\pi}\right)}{\Gamma\left(k+\frac{1}{2}+\frac{i\hat{\theta}}{2\pi}\right)\Gamma\left(k+\frac{3}{2}-\frac{B}{4}+\frac{i\hat{\theta}}{2\pi}\right)\Gamma\left(k+1+\frac{B}{4}+\frac{i\hat{\theta}}{2\pi}\right)}\right|^2 \quad (19.10.10)$$

$F_{\text{min}}(\theta,B)$ has a simple zero at $\theta=0$ since $S(0)=-1$ and its asymptotic behaviour is

$$\lim_{\theta\to\infty} F_{\text{min}}(\theta,B) = 1.$$

It satisfies the functional equation

$$F_{\text{min}}(i\pi+\theta,B)F_{\text{min}}(\theta,B) = \frac{\sinh\theta}{\sinh\theta+\sinh\frac{i\pi B}{2}} \quad (19.10.11)$$

that can be proved employing its representation (19.10.10). For the numerical evaluation of this function it is useful to use the mixed representation given by

$$F_{\min}(\theta,B) = \mathcal{N}\prod_{k=0}^{N-1}\left[\frac{\left(1+\left(\frac{\hat{\theta}/2\pi}{k+\frac{1}{2}}\right)^2\right)\left(1+\left(\frac{\hat{\theta}/2\pi}{k+\frac{3}{2}-\frac{B}{4}}\right)^2\right)\left(1+\left(\frac{\hat{\theta}/2\pi}{k+1+\frac{B}{4}}\right)^2\right)}{\left(1+\left(\frac{\hat{\theta}/2\pi}{k+\frac{3}{2}}\right)^2\right)\left(1+\left(\frac{\hat{\theta}/2\pi}{k+\frac{1}{2}+\frac{B}{4}}\right)^2\right)\left(1+\left(\frac{\hat{\theta}/2\pi}{k+1-\frac{B}{4}}\right)^2\right)}\right]^{k+1}$$

$$\times \exp\left[8\int_0^\infty \frac{dx}{x}\frac{\sinh\left(\frac{xB}{4}\right)\sinh\left(\frac{x}{2}(1-\frac{B}{2})\right)\sinh\frac{x}{2}}{\sinh^2 x}(N+1-Ne^{-2x})\,e^{-2Nx}\sin^2\left(\frac{x\hat{\theta}}{2\pi}\right)\right].$$

The convergence of the integral in this formula can be improved by increasing the value of N.

19.10.2 Recursive Equations

The Sinh–Gordon does not have bound states. Hence the only recursive equations come from the kinematical poles relative to the three-particle clusters. Using the identity

$$(p_1+p_2+p_3)^2 - m^2 = 8m^2\cosh\frac{1}{2}\theta_{12}\cosh\frac{1}{2}\theta_{13}\cosh\frac{1}{2}\theta_{23},$$

all possible poles in these channels are taken into account using the parameterization

$$F_n(\theta_1,\dots,\theta_n) = H_n\,Q_n(x_1,\dots,x_n)\prod_{i<j}\frac{F_{\min}(\theta_{ij})}{x_i+x_j} \tag{19.10.12}$$

where $x_i = e^{\theta_i}$ and H_n are normalization factors. The expression above has simple poles each time that the difference of two rapidities θ_{ij} is equal to $i\pi$. The functions $Q_n(x_1,\dots,x_n)$ are symmetric polynomials in x_i. For the form factors of the scalar operators, the total degree of these polynomials must be equal to the one of the denominator, given by $n(n-1)/2$. The partial degree of Q_n depends instead on the asymptotic behaviour of the operator $\mathcal{O}$. With the parameterization above, the recursive equations can be expressed as recursive equations for the polynomials Q_n

$$(-)^n\,Q_{n+2}(-x,x,x_1,\dots,x_n) = x\mathcal{C}_n(x,x_1,x_2,\dots,x_n)\,Q_n(x_1,x_2,\dots,x_n) \tag{19.10.13}$$

where we have introduce the function

$$\mathcal{C}_n = \frac{-i}{4\sin(\pi B/2)}\left(\prod_{i=1}^n\left[(x+\omega x_i)(x-\omega^{-1}x_i)\right]-\prod_{i=1}^n\left[(x-\omega x_i)(x+\omega^{-1}x_i)\right]\right)$$

with $\omega = \exp(i\pi B/2)$. The normalization constants H_n in (19.10.12) satisfy the recursive equations

$$H_{2n+1} = H_1 \mu^{2n}, \quad H_{2n} = H_2 \mu^{2n-2},$$

with

$$\mu \equiv \left(\frac{4 \sin(\pi B/2)}{F_{min}(i\pi, B)} \right)^{\frac{1}{2}}$$

where H_1 and H_2 are the initial conditions, fixed by the operator. Using the generating function of the elementary symmetric polynomials, the function C_n can be written as

$$C_n(x, x_1, \ldots, x_n) = \sum_{k=1}^{n} \sum_{m=1,odd}^{k} [m] \, x^{2(n-k)+m} \sigma_k^{(n)} \sigma_{k-m}^{(n)} (-1)^{k+1} \tag{19.10.14}$$

where we have introduced the symbol $[n]$ defined by

$$[n] \equiv \frac{\sin(n\frac{B}{2})}{\sin\frac{B}{2}}.$$

Note that the elementary symmetric polynomials satsify the recursive equation

$$\sigma_k^{(n+2)}(-x, x, x_1, \ldots, x_n) = \sigma_k^{(n)}(x_1, x_2, \ldots, x_n) - x^2 \sigma_{k-2}^{(n)}(x_1, x_2, \ldots, x_n). \tag{19.10.15}$$

19.10.3 General Properties of the Q_n Solutions

The form factors of the derivative operators present a factorized form: for instance, for the operator $\partial\bar{\partial}\phi$ we have $Q_n = \sigma_{n-1}\sigma_1 \tilde{Q}_n$. For this reason, it is convenient to focus the attention on the so-called *irreducible operators*, whose form factors cannot be factorized, and use them as building blocks for the form factors of all other operators. The polynomials Q_n of the irreducible operators satisfy a series of interesting results coming from the recursive equations (19.10.13). Let us initially show that the partial degree of Q_n satisfies the inequality

$$\deg(Q_n) \leq n - 1. \tag{19.10.16}$$

It is easy to see that this result holds for Q_1 and Q_2. To show that it also holds for the higher polynomials, let' consider the two cases (a) $Q_n \neq 0$ and (b) $Q_n = 0$ separately.

- In the case (a) the proof is by induction. Assume $\deg(Q_n) \leq n-1$. Since C_n is bilinear in $\sigma^{(n)}$ (see eqn. (19.10.14)), the partial degree of $Q_{n+2}(-x, x, x_1, \ldots x_n)$ in the variables $x_1, \ldots x_n$ is less or equal to $n+1$. But the partial degree of

$Q_{n+2}(x_1, x_2, \ldots x_{n+2})$ is equal to the partial degree of $Q_{n+2}(-x, x, x_1 \ldots, x_n)$, hence the partial degree of Q_{n+2} must be less or equal to $n+1$.

- In the case (b), the space of the solutions is given by the kernel of the operator $\mathcal{C}$, namely

$$Q_{n+2}(-x, x, \ldots, x_{n+2}) = 0.$$

In the space of the polynomials $\mathcal{P}$ of total degree $\frac{(n+2)(n+1)}{2}$, there is only one solution of this equation, given by

$$Q_{n+2} = \prod_{i<j}^{n+2} (x_i + x_j). \tag{19.10.17}$$

This polynomial has partial degree $n+1$ and coincides with the polynomial of the denominator of eqn. (19.10.12).

We have thus shown that the partial degree of Q_n must be less or equal to $(n-1)$ for any irreducible scalar operator. The first consequence is that the form factors of these operators cannot diverge when $\theta_i \to \infty$. The second consequence is the presence of an additional parameter at each step of the iterative procedure. This comes from a simple argument: the dimension of the space of the polynomials Q_n is given by the dimension of the space of the polynomials Q_{n-2} *plus* the dimension of the kernel. Since the kernel is one dimensional, the dimension of the space of the solutions increases exactly by one at each iterative step. With the initial conditions $\dim(Q_1) = \dim(Q_2) = 1$, we finally get

$$\dim(Q_{2n-1}) = \dim(Q_{2n}) = n. \tag{19.10.18}$$

Hence the most general form factor of an irreducible scalar operator belong to a linear space that can be spanned by a basis Q_n^k

$$Q_{2n}(A_1^{(2n)}, \ldots, A_n^{(2n)}) = \sum_{p=1}^{n} A_p^{(2n)} Q_{2n}^p \tag{19.10.19}$$
$$Q_{2n-1}(A_1^{(2n-1)}, \ldots, A_n^{(2n-1)}) = \sum_{p=1}^{n} A_p^{(2n-1)} Q_{2n-1}^p.$$

Each polynomial above defines a matrix element of an operator of the Sinh–Gordon model. Note that the dimension of this linear space grows exactly as the number of powers ϕ^k $(k < n)$ of the elementary field. This means that the matrix elements of the composite operators ϕ^k can be obtained as linear combination of the above functions.

19.10.4 The Elementary Solutions

A remarkable class of solutions of the recursive equations (19.10.13) is given by[5]

$$Q_n(k) = ||M_{ij}(k)||, \tag{19.10.20}$$

where $M_{ij}(k)$ is the $(n-1) \times (n-1)$ matrix

$$M_{ij}(k) = \sigma_{2i-j}\,[i-j+k]. \tag{19.10.21}$$

and $||M||$ denotes the determinant of the matrix M. These polynomials are called *elementary solutions*: they depends on an arbitrary integer k and satisfy

$$Q_n(k) = (-1)^{n+1} Q_n(-k). \tag{19.10.22}$$

Although all $Q_n(k)$ are solutions of (19.10.13), not all of them are linearly independent. The simplest reason is that the dimension of the space of the solutions at the level $N = 2n$ (or $N = 2n-1$) is at most n. Among the first representatives we have

$$Q_3(k) = \left\| \begin{matrix} [k]\sigma_1 & [k+1]\sigma_3 \\ [\text{k-1}] & [k]\sigma_2 \end{matrix} \right\|.$$

Using the trigonometric identity $[n]^2 - [n-1][n+1] = 1$, it is easy to see that this expression satisfies eqn. (19.10.13) (with $A_0^1 = 1$) for any integer k.

These solutions allow us to express simultaneously all the form factors of the elementary field $\phi(x)$ and the trace $\Theta(x)$ of the stress-energy tensor. In fact, it is possible to prove that the matrix elements of $\phi(x)$ are given by $Q_n(0)$. Note that the form factors relative to an even number of particles are automatically zero, in agreement with the Z_2 symmetry of the model. Those with an odd number of asymptotic particles vanish when $\theta_i \to \infty$, in agreement with the perturbative evaluation of these matrix elements given by the Feynman diagrams. The form factors of $\Theta(x)$ are instead given by the even polynomials $Q_{2n}(1)$, that go to a finite limit when $\theta_i \to \infty$, once again in agreement with their perturbative computation. A further confirmation of the validity of this identification can be obtained by using the c-theorem. Employing just the two-particle form factor, we have the following approximated value of the ultraviolet central charge

$$c^{(2)} = \frac{3}{2F_{\min}^2(i\pi)} \int_0^\infty \frac{d\theta}{\cosh^4\theta}\, |F_{\min}(2\theta)|^2. \tag{19.10.23}$$

[5] For simplicity we have suppressed the dependence of $Q_n(k)$ on the variables x_i.

B	$\frac{g^2}{4\pi}$	$\Delta c^{(2)}$
$\frac{1}{10}$	$\frac{2}{19}$	0.9989538
$\frac{3}{10}$	$\frac{6}{17}$	0.9931954
$\frac{2}{5}$	$\frac{1}{2}$	0.9897087
$\frac{1}{2}$	$\frac{2}{3}$	0.9863354
$\frac{2}{3}$	1	0.9815944
$\frac{7}{10}$	$\frac{14}{13}$	0.9808312
$\frac{4}{5}$	$\frac{4}{3}$	0.9789824
1	2	0.9774634

Table 19.1 *Approximate values of the central charge of the Sinh–Gordon model obtained by using only the two-particle form factor of $\Theta(x)$ in the c-theorem.*

The numerical values for different values of the coupling constant $g^2/4\pi$ are collected in Table 19.1. From this table we can see that the sum rule is saturated by the two-particle form factor even for large values of the coupling constant: this proves once again the fast convergent behaviour of the spectral series.

It is vital to understand which are the operators $\Psi_k(x)$ associated to the elementary solutions $Q_n(k)$ $(k \neq 0)$. For the sequence of the form factors related to $Q_n(k)$, let us choose the normalization as follows

$$H_1^k = \mu[k], \;\; H_2^k = \mu^2\,[k]. \tag{19.10.24}$$

The present conjecture is that the operators Ψ_k correspond to the vertex operators $e^{kg\phi}$. A non-trivial check of this conjecture is provided by the computation of the conformal dimensions $\Delta_k(g)$ that emerge in their ultraviolet limits. These quantities can be computed analysing the limit $x \to 0$ of the correlation function

$$\begin{aligned} G_{k,m}(x) &= \langle \Psi_k(x)\,\Psi_m(0)\rangle = \\ &= \sum_{n=0}^{\infty} \int \frac{d\beta_1 \ldots d\beta_n}{n!(2\pi)^n} F_n^{\Psi_k}(\beta_1 \ldots \beta_n) F_n^{\Psi_m}(\beta_n \ldots \beta_1) \exp\left(-mr\sum_{i=1}^{n} \cosh\beta_i\right). \end{aligned}$$

At the first order in g, we have $\Delta_k(g) = -k^2 g^2/8\pi$ that coincides with the conformal dimension of the vertex operators $e^{kg\phi(x)}$, computed using the Gaussian conformal theory.

19.11 The Ising Model in a Magnetic Field

The Ising model in a magnetic field has quite a rich S-matrix: it has eight massive exitations and 36 elastic scattering amplitudes, some of them with higher order poles. In addition to the functional and recursive equations, the form factors of this theory also satisfy other recursive equations related to the higher poles of the S-matrix. The relative formulae can be found in Delfino and Mussardo and Delfino and Simonetti mentioned at the end of the chapter. Here we only report the main results about the form factors of the the energy operator $\epsilon(x)$ and of magnetization operator $\sigma(x)$. In this theory, the latter operator is proportional to the trace

$$\Theta(x) = 2\pi h(2 - 2\Delta_\sigma)\sigma(x). \tag{19.11.1}$$

Relying on the fast convergence of the spectral series, for the correlation functions of these operators we can focus our attention on the one- and two-particle form factors. To begin with, we need to fix some notation. For the S-matrix of the particles A_a and A_b we have

$$S_{ab}(\theta) = \prod_{\alpha \in \mathcal{A}_{ab}} (f_\alpha(\theta))^{p_\alpha} \tag{19.11.2}$$

where

$$f_\alpha(\theta) \equiv \frac{\tanh\frac{1}{2}(\theta + i\pi\alpha)}{\tanh\frac{1}{2}(\theta - i\pi\alpha)}. \tag{19.11.3}$$

The set of the numbers $\mathcal{A}_{ab}$ and their multiplicity p_α can be found in Tables 18.3 and 18.4. It is convenient to parameterize the two-particle form factors of this theory as

$$F_{ab}^{\mathcal{O}}(\theta) = \frac{Q_{ab}^{\Phi}(\theta)}{D_{ab}(\theta)} F_{ab}^{min}(\theta), \tag{19.11.4}$$

where $D_{ab}(\theta)$ and $Q_{ab}^{\mathcal{O}}(\theta)$ are polynomials in $\cosh\theta$: the latter is fixed by the singularities of the S-matrix, the former depends on the operator $\mathcal{O}(x)$. The minimal form factors can be written as

$$F_{ab}^{min}(\theta) = \left(-i\sinh\frac{\theta}{2}\right)^{\delta_{ab}} \prod_{\alpha \in \mathcal{A}_{ab}} (G_\alpha(\theta))^{p_\alpha}, \tag{19.11.5}$$

where

$$G_\alpha(\theta) = \exp\left\{2\int_0^\infty \frac{dt}{t} \frac{\cosh\left(\alpha - \frac{1}{2}\right)t}{\cosh\frac{t}{2}\sinh t} \sin^2\frac{(i\pi - \theta)t}{2\pi}\right\}. \tag{19.11.6}$$

For large values of the rapidity, we have

$$G_\alpha(\theta) \sim \exp(|\theta|/2), |\theta| \to \infty, \tag{19.11.7}$$

independently from the index α.

From the analysis of the singularities of the form factors, we arrive at the following expression of the denominator

$$D_{ab}(\theta) = \prod_{\alpha \in \mathcal{A}_{ab}} (\mathcal{P}_\alpha(\theta))^{i_\alpha} (\mathcal{P}_{1-\alpha}(\theta))^{j_\alpha}, \tag{19.11.8}$$

where

$$\begin{array}{llll} i_\alpha = n+1, & j_\alpha = n, & \text{se} & p_\alpha = 2n+1; \\ i_\alpha = n, & j_\alpha = n, & \text{se} & p_\alpha = 2n, \end{array} \tag{19.11.9}$$

having introduced the notation

$$\mathcal{P}_\alpha(\theta) = \frac{\cos \pi\alpha - \cosh\theta}{2\cos^2 \frac{\pi\alpha}{2}}. \tag{19.11.10}$$

Both quantities $F_{ab}^{min}(\theta)$ and $D_{ab}(\theta)$ are normalized to be equal to 1 when $\theta = i\pi$.

c_1	0.472038282
c_2	0.019231268
c_3	0.002557246
c_{11}	0.003919717
c_4	0.000700348
c_{12}	0.000974265
c_5	0.000054754
c_{13}	0.000154186
c_{partial}	0.499630066

Table 19.2 *Central charge given by the partial sum of the form factors entering the c-theorem.* $c_{ab..}$ *denotes the contribution of the state* $A_a A_b$. *The exact result is* $c = 1/2$.

	σ	ϵ
Δ_1	0.0507107	0.2932796
Δ_2	0.0054088	0.0546562
Δ_3	0.0010868	0.0138858
Δ_{11}	0.0025274	0.0425125
Δ_4	0.0004351	0.0069134
Δ_{12}	0.0010446	0.0245129
Δ_5	0.0000514	0.0010340
Δ_{13}	0.0002283	0.0065067
Δ_{partial}	0.0614934	0.4433015

Table 19.3 *Conformal dimensions $\Delta_{\mathcal{O}}$ given by the partial sum of the form factors of the correlation functions entering the Δ-theorem. $\Delta_{ab..}$ denotes the contribution of the state $A_a A_b \ldots$. The exact values are $\Delta_\sigma = 1/16 = 0.0625$ and $\Delta_\varepsilon = 1/2$.*

The polynomials of the numerator can be expressed as

$$\mathcal{Q}_{ab}^{\mathcal{O}}(\theta) = \sum_{k=0}^{N_{ab}^{\mathcal{O}}} c_{ab,\mathcal{O}}^{(k)} \cosh^k \theta. \tag{19.11.11}$$

The condition $\left[F_{ab}^{\mathcal{O}}(\theta)\right]^* = F_{ab}^{\mathcal{O}}(-\theta)$ follows from the monodromy condition satisfied by the form factors and from the property $S_{ab}^*(\theta) = S_{ab}(-\theta)$. This means that the coefficients $c_{ab,\mathcal{O}}^{(k)}$ are real numbers and their values identify the different operators.

The degrees of the polynomials are fixed by the conformal dimension of the operators and, both for $\sigma(x)$ and $\epsilon(x)$, we have in particular $N_{11}^{\Phi} \leq 1$. Therefore the initial conditions of the recursive equation for the form factors of the two relevant operators consists of *two* free parameters, i.e. the coefficients $c_{11,\mathcal{O}}^{(0)}$ and $c_{11,\mathcal{O}}^{(1)}$. Furthermore, it the number of free parameters does not increase implementing the bootstrap equations. Consider, for instance the condition $N_{12}^{\mathcal{O}} \leq 2$, which seems to imply three new coefficients $c_{12,\mathcal{O}}^{(k)}$ $(k = 1, 2, 3)$ for $F_{12}^{\mathcal{O}}(\theta)$. However, the amplitudes $S_{11}(\theta)$ and $S_{12}(\theta)$ have three common bound states. This circumstance gives rise to three equations

$$\frac{1}{\Gamma_{11}^c} \text{Res}_{\theta = iu_{11}^c} F_{11}^{\Phi}(\theta) = \frac{1}{\Gamma_{12}^c} \text{Res}_{\theta = iu_{12}^c} F_{12}^{\Phi}(\theta), \qquad c = 1, 2, 3$$

that allow us to fix the three coefficients $c_{12,\mathcal{O}}^{(k)}$ in terms of the two coefficients $c_{11,\mathcal{O}}^{(k)}$.

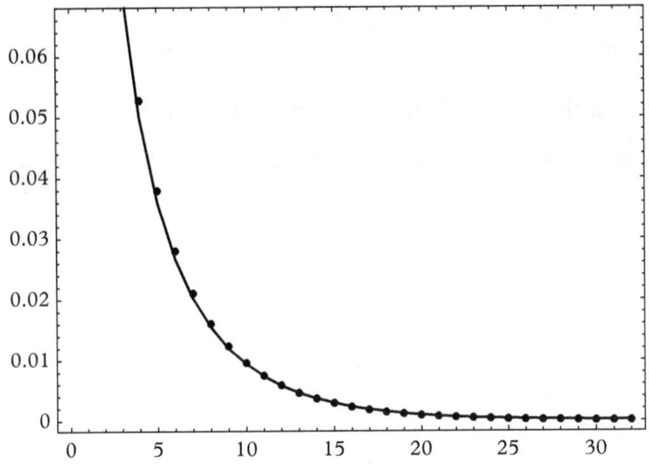

Fig. 19.12 *Plot of the correlation function $\langle\sigma(r)\sigma(0)\rangle$ for the Ising model in a magnetic field. The continuous line is the determination obtained with the first eight form factors, while the dots are the numerical determination of the correlators obtained by a Monte Carlo simulation.*

There is an additional information on the numerator Q_{ab} of the operator $\Theta(x)$. In fact, from the conservation law $\partial_\mu T^{\mu\nu} = 0$ it follows that the polynomials Q_{ab}^{Θ} contain the factor

$$\left(\cosh\theta + \frac{m_a^2 + m_b^2}{2m_a m_b}\right)^{1-\delta_{ab}} \tag{19.11.12}$$

The determination of the coefficients $c_{ab}^{(k)}$ and the one-particle form factors of the two operators $\sigma \sim \Theta$ and ϵ has been done in the papers cited at the end of the chapter and their values can be found there.

By employing these lowest form factors we can compute the correlation functions and perform some non-trivial checks by applying the sum rules of the c-theorem and Δ-theorem. The relative results are given in the Tables 19.2 and 19.3. A successful check of the correlation function $\langle\sigma(r)\sigma(0)\rangle$ has also been done versus the numerical determination of this function (Figure 19.12).

REFERENCES

The FF program has been proposed by

Karowski, M. and Weisz, P. (1978). 'Exact Form-Factors In $(1+1)$-Dimensional Field Theoretic Models with Soliton Behavior', *Nuclear Physics B*, 139: 455.

Smirnov, F. (1992). *Form Factors in Completely Integrable Models of Quantum Field Theory*, World Scientific.

The form factors of the thermal Ising model are studied in detail in

Yurov, V.P. and Zamolodchikov, A.B. (1991). 'Correlation Functions of Integrable 2-D Models of Relativistic Field Theory: Ising Model', *International Journal of Modern Physics A*, 6: 3419.

The determination of the correlators of the Ising model on the lattice and their relation with the Painleve' equation are discussed in

Jimbo, M., Miwa T. and Ueno, K. (1981). 'Monodromy Preserving Deformation of Linear Ordinary Differential Equations with Rational Coefficients: I. General Theory and τ-function, *Physics D*, 2: 306.

McCoy, B., Tracy, C. and Wu, T.T. (1977). 'Painlevé Functions of the Third Kind, *Journal of Mathematical Physics*, 18: 1058.

Wu, T.T., McCoy, B., Tracy, C. and Barouch, E. (1976). 'Spin-spin Correlation Functions for the Two-dimensional Ising model: Exact Theory in the Scaling Region', *Physics Review B*, 13: 316.

The classification of the operators obtained in terms of the different solutions of the form factor equations is proposed in

Cardy, J. and Mussardo G. (1990). 'Form-Factors Of Descendent Operators In Perturbed Conformal Field Theories' *Nuclear Physics B*, 340: 387.

The form factors of the Sinh–Gordon model have been determined in

Fring, A., Mussardo, G. and Simonetti, P. (1993). 'Form-factors for Integrable Lagrangian Field Theories: The Sinh–Gordon Theory', *Nuclear Physics B*, 393: 413.

Koubek, A and Mussardo, G. (1993). 'On the Operator Content of the Sinh–Gordon Model', *Physics Letters B*, 311: 193.

The matrix elements of the Ising model in a magnetic field have been computed in

Delfino, G. (2004). 'Integrable Field Theory and Critical Phenomena. The Ising Model in a Magnetic Field', *Journal of Physics A*, 37: R45.

Delfino, G. and Mussardo, G. (1995). 'The Spin-spin Correlation Function in the Two-dimensional Ising Model in a Magnetic Field at $T = T_c$', *Nuclear Physics B*, 455: 724.

Delfino, G. and Simonetti, P. (1996). 'Correlation Functions in the Two-dimensional Ising Model in a Magnetic Field at $T = T_c$', *Physics Letters B*, 383: 450.

The convergence property of the spectral series has been analysed by several authors.

Cardy, J. and Mussardo, G. (1993). 'Universal Properties of Self-Avoiding Walks from Two-Dimensional Field Theory', *Nuclear Physics B*, 410: 451.

Zamolodchikov, A.B. (1991). 'Two-point Correlation Function in Scaling Lee–Yang Models, *Nuclear Physics B*, 348: 619.

The vacuum expectation values of several operators of integrable field theories have been determined in a remarkable series of papers.

Fateev, V., Lukyanov, S., Zamolodchikov, A.B. and Zamolodchikov, Al.B. (1998). 'Expectation Values of Local Fields in Bullogh–Dodd Model and Integrable Perturbed Conformal Theories, *Nuclear Physics B*, 516: 652.

Fateev, V., Lukyanov, S., Zamolodchikov, A.B. and Zamolodchikov, Al.B. (1999). 'Expectation Values of Descendent Fields in the Sine–Gordon Model', *Nuclear Physics B*, 540: 587.

Lukyanov, S. and Zamolodchikov, A.B. (1997). 'Exact Expectation Values of Local Fields in Quantum Sine–Gordon Model', *Nuclear Physics B*, 493: 571.

PROBLEMS

19.1. Form factors of a free theory

Consider the theory of a free bosonic field $\phi(x)$ associated to a particle A of mass m.

a. Compute the form factors of $\phi(x)$ and prove that $\langle 0|\phi(0)|A\rangle = 1/sqrt2$. Show that the Euclidean correlation function is given by

$$\langle \phi(x)\phi(0)\rangle = \frac{1}{\pi} K_0(mr).$$

b. Show that the arbitrariness of the one-particle form factor of the trace of the stress-energy tensor

$$F_1^{\Theta} = \langle 0|\Theta(0)|A\rangle \equiv -\sqrt{2\pi}\, m^2 Q$$

corresponds to the possibility of redefining the stress-energy tensor by adding a total divergence

$$\Theta(x) = 2\pi \left(m^2\phi^2 + \frac{Q}{\sqrt{\pi}} \Box\phi \right).$$

c. Use the c-theorem and the form factors of $\Theta(x)$ to show that the central charge in the ultraviolet region is given by

$$c = 1 + 12Q^2.$$

19.2. Feynman gas

a. Derive the equation of state of the Feynman gas associated to the form factors of the magnetization operators in the nearest-neighbour approximation. Prove that the pressure $p(z)$ satisfies the integral equation (19.9.19).

b. Justify the accuracy of the approximation of the conformal dimensions computing the average number of particles per unit length by means of the formula

$$\frac{\langle N \rangle}{L} = z \frac{\partial p}{\partial z}$$

and checking the very dilute nature of the gas.

19.3. Infinite products

Using the integral

$$\int \frac{dt}{t} e^{-\beta t} \sin^2 \frac{\alpha t}{2} = \frac{1}{4} \log \frac{\alpha^2 + \beta^2}{\beta^2},$$

and the identity satisfied by the Γ functions

$$\frac{\Gamma(\alpha)\Gamma(\beta)}{\Gamma(\alpha+\gamma)\Gamma(\beta-\gamma)} = \prod_{k=0}^{\infty} \left[\left(1 + \frac{\gamma}{\alpha+k}\right) \left(1 - \frac{\gamma}{\beta+k}\right) \right],$$

to derive the expression of $F_{\text{min}}(\theta)$ of the Sinh–Gordon model.

19.4. Cluster properties

Consider the form factors of a scattering theory based on the functions

$$f_x(\theta) = \frac{\tanh \frac{1}{2}(\theta + i\pi x)}{\tanh \frac{1}{2}(\theta - i\pi x)}$$

that have the property $\lim_{\theta \to \infty} f_x(\theta) = 1$.

a. Using the Watson equation satisfied by the form factors $F_n^{\mathcal{O}_a}(\theta_1, \ldots, \theta_n)$ of an operator $\mathcal{O}_a$, prove that taking the limit

$$\lim_{\Lambda \to \infty} F_n^{\mathcal{O}_a}(\beta_1 + \Delta, \ldots, \beta_m + \Delta, \beta_{m+1}, \ldots, \beta_n) = F_m^{\mathcal{O}_b}(\beta_1, \ldots, \beta_m) F_{n-m}^{\mathcal{O}_c}(\beta_{m+1}, \ldots, \beta_n)$$

the form factor factorizes in terms of two functions both satisfying the Watson equations. Hence they can be considered the form factors of the operators $\mathcal{O}_b$ and $\mathcal{O}_c$. This expresses the cluster property of the form factors.

b. Prove that the form factors of the elementary solutions of the Sinh–Gordon model are self-clustering quantities.

19.5. Correlation functions of the Ising model

Use the fermionic representation of the energy operator of the Ising model, $\epsilon = i\bar{\psi}\,\psi$, and the mode expansion of the fermionic field in terms of the creation and annihilation operators, to compute the matrix elements of $\epsilon(x)$ and its two-point correlation function.

19.6. Form factors of the Yang–Lee model

Using the form factors of the Sinh–Gordon model, obtain the form factors of the Yang–Lee by using the analytic continuation $B \to -\frac{2}{3}$.

Part 5

Finite Size Effects

20

Thermodynamic Bethe Ansatz

In quantum mechanics there are principles that are certain and these are much more important for the world and for us than the uncertainty principle.

Hans A. Bethe

20.1 Introduction

The thermodynamics of a QFT in an infinite volume can be determined by its S-matrix. Originally proposed by Dashen, Ma and Berstein, has been later widely used to study the thermal properties of the integrable field theories in $(1+1)$ dimensions. The reason consists of the particularly simple properties of these scattering matrices and the possibility to generalize to the relativistic case the thermodynamical Bethe ansatz (TBA) techniques successfully applied to the non-relativistic problems by Yang and Yang. In the TBA approach, the derivation of the thermodynamics of a purely elastic scattering theory reduces to find the solution of a set of non-linear integral equations that rule the energies of the particle excitations and their statistical distribution.

The TBA equations for the relativistic models with a diagonal S-matrix have been derived by Al.B. Zamolodchikov. Several applications have been made by Zamolodchikov himself and many other authors. In addition to the generalization of the TBA to the non-diagonal S-matrices, further advances have been accomplished in the computation of the energies of the excited states, in the analysis of systems with generic integrable boundary conditions and also in the discovery of interesting relations with the Schröedinger equation in quantum mechanics. This chapter exposed the main ideas of this approach, and for all the advanced topics of the subject we refer the reader to the articles listed at the end of the chapter.

20.2 Casimir Energy

Consider a $(1+1)$-dimensional Euclidean QFT defined on a cylinder, with periodic boundary conditions in both the R and L directions. There are two equivalent ways to quantize the theory on such a geometry: for the symmetry of the two directions, we can

Statistical Field Theory: An Introduction to Exactly Solved Models in Statistical Physics. Second Edition. Giuseppe Mussardo, Oxford University Press (2020). © Giuseppe Mussardo. DOI: 10.1093/oso/9780198788102.001.0001

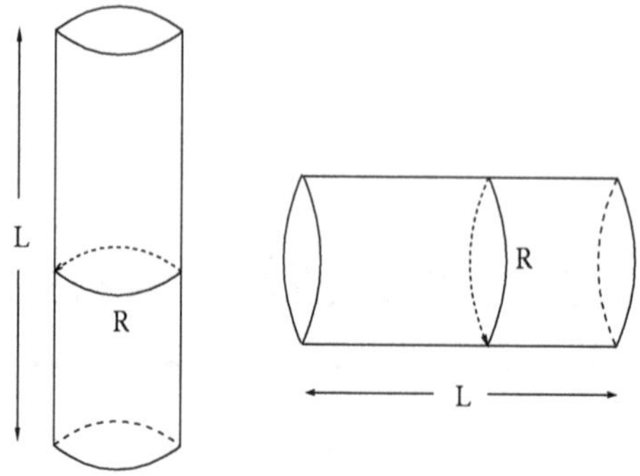

Fig. 20.1 *Cylinder geometry with periodic boundary conditions on both directions, with the two different channels of quantization.*

equivalently choose as time direction one of the two axes and consider the other as space direction.[1] Hence, the partition function can be written either as

$$Z(R,L) = \mathrm{Tr}\, e^{-L\mathcal{H}_R}, \tag{20.2.1}$$

or as

$$Z(R,L) = \mathrm{Tr}\, e^{-R\mathcal{H}_L}, \tag{20.2.2}$$

where $\mathcal{H}_R$ and $\mathcal{H}_L$ are the Hamiltonians of the system quantized along the R and L axes, and the trace is a sum done on their eigenstates. The two Hamiltonians can be expressed in terms of the stress-energy tensor $T_{\mu\nu}$, where x and y denote the coordinates along the R and L axes, respectively. In fact we have

$$\mathcal{H}_R = \frac{1}{2\pi}\int T_{yy}\, dx,$$

while

$$\mathcal{H}_L = \frac{1}{2\pi}\int T_{xx}\, dy.$$

The quantization scheme in which the role of the time direction is played by the L axis will be denoted as the *L-channel*, while the other one as the *R-channel*.

[1] In the context of the CFT this is the basis of the modular invariance, see Section 11.7.

When $L \to \infty$, the expression in (20.2.1) clearly reduces only to the lowest term, given by the ground state energy $E_0(R)$ of $\mathcal{H}_R$

$$Z(R,L) \simeq e^{-LE_0(R)}. \tag{20.2.3}$$

But, taking the limit $L \to \infty$ in the second expression (20.2.2) is equivalent to the thermodynamic limit of an one-dimensional quantum system defined along the L axis at temperature $T \equiv 1/R$. In this case, the limiting form of the partition function can be written as

$$Z(R,L) \simeq e^{-LRf(R)}, \tag{20.2.4}$$

where $f(R)$ is the free energy per unit length of the system at temperature $1/R$. Comparing the two limiting expressions (20.2.3) and (20.2.4) of the partition function, we find

$$E_0(R) = Rf(R). \tag{20.2.5}$$

This equation states the important relation between the Casimir energy $E_0(R)$ of the ground state on a finite volume and the free energy $f(R)$ of the one-dimensional quantum system at infinite volume but at temperature $T = 1/R$.

For the translation invariance of the cylinder geometry along the two axes, the one-point correlation functions are independent on the coordinates. In particular, we have

$$\langle T_{yy}\rangle = 2\pi\frac{E_0(R)}{R}, \quad \langle T_{xx}\rangle = 2\pi\frac{dE_0(R)}{dR},$$

and, for the one-point correlation function of the trace of the stress-energy tensor $\Theta = (T_{xx} + T_{yy})$, we have

$$\langle\Theta\rangle = \frac{2\pi}{R}\frac{d}{dR}[RE_0(R)]. \tag{20.2.6}$$

Furthermore, for theories that are invariant under parity and with a unique ground state, it holds

$$\langle T_{xy}\rangle = \langle T_{yx}\rangle = 0.$$

It is convenient to parameterize the ground state energy as

$$E_0(R) = -\frac{\pi\tilde{c}(r)}{6R}, \tag{20.2.7}$$

where $r = m_1 R$ is a purely dimensionless variable, with m_1 the lowest mass gap of the theory. As we will show later, the scaling function $\tilde{c}(r)$ can be determined for any value of r by the TBA equations based on the scattering data. However, there is a simple limit of this expression: in the ultraviolet limit $r \to 0$, the behaviour of the ground state energy is controlled by the underlying CFT

$$E_0(R) = \frac{2\pi}{R}\left(\Delta_{min} + \overline{\Delta}_{min} - \frac{c}{12}\right), \tag{20.2.8}$$

and, for a theory in which $\Delta_{min} = \overline{\Delta}_{min}$, the function $\tilde{c}(r)$ goes to the effective central charge

$$\lim_{r\to 0} \tilde{c}(r) = c - 24\Delta_{min}. \tag{20.2.9}$$

Notice that this limit establishes an important relation between the scattering theory of a massive QFT and the conformal theory that rules its short-distance behaviour. The confirmation and the validity of many scattering theories proposed to describe the deformations of CFTs can been accomplished thanks to the relation above.

The following sections derive the TBA equations following the original proposals by Yang–Yang and by Al. Zamolodchikov, discussing all their important consequences. The last two sections, using the simple example of a free massive theory, show how to derive the ground state energy $E_0(R)$ at a finite volume by directly quantizing the theory in its L-channel.

20.3 Bethe Relativistic Wave Function

Consider a $(1+1)$-dimensional integrable theory defined on a circumference of length L. Let us assume that the spectrum consists of a set of particles A_a $(a = 1, 2, \cdots n)$ with masses m_a, and that their scattering amplitudes are purely diagonal and characterized by their phase shifts $\delta_{ab}(\theta)$, where $S_{ab}(\theta) = e^{i\delta_{ab}(\theta)}$. The lowest mass determines the correlation length of the system through $\xi = 1/m_1$. The particles can be either bosons or fermions.

The Hilbert space of such a theory is rather simple. In fact, given any N-particle state, the integrability of the theory ensures that the time evolution of this state preserves both the identity of the particles and their momenta. In this case, it makes sense to associate to any state of such a relativistic system a wave function $\Psi(x_1, \ldots, x_N)$. In the configurational space of the N-particle state, we can select $N!$ regions where the particles are well separated each other, i.e. $| x_i - x_{i+1} | \gg \xi$, so that we can neglect all relativistic effects induced by the virtual processes. Each of these domains is identified by the ordering $x_{i_1} \ll x_{i_2} \ll x_{i_3} \cdots \ll x_{i_N}$ of the coordinates of the particles. In these region, the expression of the wave function is particularly simple, since it is given by plane waves

$$\Psi(x_{i_1}, x_{i_2}, \cdots, x_{i_N}) = \prod_{k=1}^{N} e^{i p_{i_k} x_{i_k}}. \tag{20.3.1}$$

Notice that the exchange of two particles maps one domain into another, and each of these transition is equivalent to multiply the wave function by the corresponding scattering amplitude. Imposing the periodic (anti-periodic) boundary condition for the wave function of the bosonic (fermionic) particles, we have the quantization condition[2] for the momenta p_i:

$$e^{i p_i L} \prod_{j \neq i}^{N} S(\theta_i - \theta_j) = \pm 1, \quad i = 1, 2, \cdots N. \tag{20.3.2}$$

Using the rapidity variable to express the momenta and considering the terms in the exponents of this equation, we can write it as

$$m_i L \sinh\theta_i + \sum_{j \neq i}^{N} \delta_{ij}(\theta_i - \theta_j) = 2\pi n_i, \tag{20.3.3}$$

where

$$\tilde{\delta}_{ij}(\theta) = -i \ln S_{ij}(\theta).$$

The numbers $\{n_i\}$ assume integer values for the bosons and half-integers for the fermions. Together with the rapidity values that solve eqn. (20.3.3), they identify the states of the Bethe ansatz

$$| \, n_1, \theta_1; n_2, \theta_2; \ldots; n_N, \theta_N \rangle.$$

The energy and the momentum of these states are

$$E = \sum_{i=1}^{N} m_i \cosh\theta_i, \quad p = \sum_{i=1}^{N} m_i \sinh\theta_i. \tag{20.3.4}$$

Both these quantities have the *same* expression as in the case of N free particles. The difference, though, is that in the free case, the rapidities of the particles can take arbitrary values, whereas in the interacting case, their values are determined by the quantization relation (20.3.3), in which it enters the phase shifts $\delta_{ab}(\theta)$ of the scattering processes.

[2] Notice that, in the absence of interactions, this leads to the usual quantization condition of the momenta in a finite volume, $p_i = \frac{2\pi n_i}{L}$.

20.3.1 Selection Rules

The Bethe wave function must be symmetric (anti-symmetric) under the exchange of two identical bosons (fermions) with the same value of their rapidities. It is then necessary to consider the selection rules coming from the identity of the particles. Since for the diagonal S-matrices the unitarity condition implies $S_{aa}^2(0) = 1$, there could be two different cases:

1. In the first case

$$S_{aa}(0) = -1,$$

and this leads to a wave function that is *anti-symmetric* under the exchange of two particles with the same rapidity. If the two particles are bosons, this is clearly in conflict with their Bose statistics. This implies that two bosons A_a cannot have the same value of the rapidity, namely each value of θ can be assigned at most to one particle only. Hence all integers $n_i^{(a)}$ of the species a in eqn. (20.3.3) must be different. Vice versa, if the identical particles are fermions, the anti-symmetry of the wave function perfectly matches with their Fermi-Dirac statistics and there is no restriction on the integers $n_i^{(a)}$. In the context of the Bethe ansatz, the condition $S = -1$ is called the *fermionic type*, independently from the bosonic or the fermionic nature of the particles A_a.

2. In the second case,

$$S_{aa}(0) = 1, \tag{20.3.5}$$

the situation is opposite to the previous one: this condition gives rise to a *symmetric* wave function under the exchange of two particles of the same species with the same rapidity. Hence, if the two particles are bosons, this is compatible with their Bose statistics and there is no restriction on the integers $n_i^{(a)}$. Vice versa, if the two particles are fermions, each value of the rapidity can be taken only by one particle, i.e. all integers $n_i^{(a)}$ of the species a must be necessarily different. In the context of the Bethe ansatz, the condition $S = +1$ is called the *bosonic type*, independently from the bosonic or the fermionic nature of the particles A_a.

20.4 Derivation of Thermodynamics

The quantization conditions (20.3.3) for the rapidities of the particles form a complicate set of transcendental equations. They simplify in the thermodynamic limit, on which both $L \to \infty$ and the total number of particles $N_a \to \infty$ but keeping their ratio fixed. In such a limit, the spectrum of the rapidities, solutions of the equation (20.3.3), becomes dense and the distance between two adjacent levels is of the order $(\theta_i - \theta_{i+1}) \sim 1/mL$.

It is convenient to introduce the continuous densities $\rho_a^{(r)}(\theta)$ relative to the distribution of the rapidities of the particles, defined as the number of particles A_a with rapidity between θ and $\theta + \Delta\theta$ divided for $L\Delta\theta$. In terms of these densities, the energy per unit length of the system can be written as

$$E[\rho^{(r)}] = \sum_{a=1}^{n} \int_{-\infty}^{+\infty} m_a \cosh\theta \, \rho_a^{(r)}(\theta) \, d\theta. \tag{20.4.1}$$

For $m_1 L = L/\xi \gg 1$, the quantization equation (20.3.3) becomes

$$\frac{m_a}{2\pi} \sinh\theta_i^{(a)} + \sum_{b=1}^{n} (\tilde{\delta}_{ab} * \rho_b^{(r)})(\theta_i) = \frac{n_i^{(a)}}{L}, \tag{20.4.2}$$

where $*$ denotes the convolution of the functions

$$(f * g)(\theta) = \int_{-\infty}^{+\infty} \frac{d\theta'}{2\pi} f(\theta - \theta') g(\theta').$$

Each time that $n_i^{(a)}$ is a set of admissible quantum numbers, the corresponding solution $\theta_i^{(a)}$ of (20.4.2) is said to be a *root* of the species a and the density of these solutions around the value θ is denoted with the function $\rho_a^{(r)}(\theta)$ introduced above. However, these equations admit solutions in $\theta_i^{(a)}$ also for integer values of $\tilde{n}_i^{(a)}$ that are necessarily in relation to the occupied states. Such solutions, associated to the integers $\tilde{n}_i^{(a)}$ that do not correspond to the admissible quantum numbers, are called *holes* of the species a and their density around the value θ is denoted by $\rho_a^{(h)}(\theta)$. The possibility of having these two types of solution is due, in definitive, to two circumstances. The first is that

$$\mathcal{J}_a(\theta) = \frac{m_a}{2\pi} \sinh\theta_i^{(a)} + \sum_{b=1}^{n} (\tilde{\delta}_{ab} * \rho_b^{(r)})(\theta), \tag{20.4.3}$$

are monotonically increasing functions, as we will show later. The second is the absence of certain integers in the sequence of the quantum numbers $n_i^{(a)}$ of the physical states. This derives from the previous discussion on the selection rules: for instance, in the case of bosonic particles but with $S(0) = -1$, choosing an ordering for the variables $\theta_i^{(a)}$, the integers $n_i^{(a)}$ of the physical states must necessarily be a strictly increasing sequence and some integers may be missed in this sequence.

Therefore, in the thermodynamic limit there are the densities both of the roots and the holes. The total density ρ_a of the occupied and empty levels of the particle A_a is equal to the derivative of the functions $\mathcal{J}_a(\theta)$

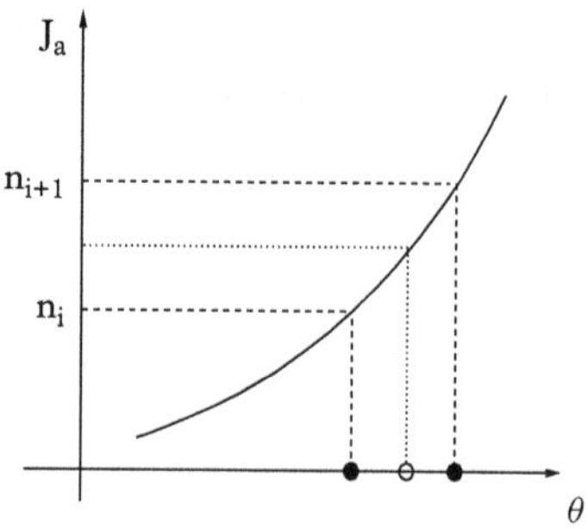

Fig. 20.2 *Plot of the function $\mathcal{J}_a(\theta)$ and graphical solution of eqn. (20.4.16) for three different integers, where two of them, n_i and n_{i+1}, are admissible quantum numbers. The roots are given by • while the roots by ∘.*

$$\rho_a(\theta) = \rho_a^{(r)}(\theta) + \rho_a^{(h)}(\theta) = \frac{d}{d\theta}\mathcal{J}_a(\theta) = \frac{1}{2\pi} m_a \cosh\theta + \sum_{b=1}^{n} (\varphi_{ab} * \rho_a^{(r)})(\theta), \quad (20.4.4)$$

where

$$\varphi_{ab}(\theta) = \frac{d}{d\theta}\tilde{\delta}_{ab}(\theta). \quad (20.4.5)$$

Properties of the functions $\varphi_{ab}(\theta)$. The functions $\varphi_{ab}(\theta)$ satisfy

$$\varphi_{ab}(-\theta) = \varphi_{ab}(\theta),$$

as it can be seen using the unitarity of the amplitudes S_{ab}. For a S-matrix

$$S_{ab}(\theta) = \prod_{\alpha \in \mathcal{A}_{ab}} s_\alpha(\theta)$$

expressed in terms of the functions $s_\alpha(\theta)$

$$s_\alpha(\theta) = \frac{\sinh\theta + i\sin\alpha\pi}{\sinh\theta - i\sin\alpha\pi}$$

we have

$$\varphi_{ab}(\theta) = \sum_{\alpha \in \mathcal{A}_{ab}} \varphi_\alpha(\theta), \quad (20.4.6)$$

where wehave defined

$$\varphi_\alpha(\theta) = -i\frac{d}{d\theta}\log s_\alpha(\theta) = -\frac{\sin\alpha\pi}{\cosh\theta - \cos\alpha\pi}. \tag{20.4.7}$$

It is easy to see that $\varphi_{ab}(\theta)$ are periodic functions with period $2\pi i$. For $\theta \neq 0$, they can be written as

$$\varphi_{ab}(\theta) = -\sum_{s=1}^{\infty}\varphi_{ab}^{(k)} e^{-k|\theta|}, \tag{20.4.8}$$

$$\varphi_{ab}^{(k)} = 2\sum_{\alpha\in A_{ab}}\sin(k\pi\alpha).$$

Notice that, inserting this expansion in the logarithm derivative of the bootstrap equation (17.4.68), we have

$$\varphi_{il}^{(k)} = \varphi_{ij}^{(k)} e^{-ik\bar{u}_{jl}^k} + \varphi_{ik}^{(k)} e^{ik\bar{u}_{lk}^j}. \tag{20.4.9}$$

Comparing now with the consistency equations of the conserved charges, eqn. (17.5.74), we see that the linearly independent columns and the rows of the matrix $\varphi^{(k)} = (\varphi_{ab}^{(k)})$ are solutions (although sometimes trivial) of these equations. The connection between $\varphi_{ab}^{(k)}$ and the eigenvalues $\chi_s^{(a)}$ of the conserved charges is established by

$$\varphi_{ab}^{(s)} = \varphi_{11}^{(s)}\chi_s^{(a)}\chi_s^{(b)}. \tag{20.4.10}$$

The index 1 of this formula refers to the particle of the theory with the lowest mass, with the normalization of the conserved charges set by $\chi_s^{(1)} = 1$. Note that for $s = 1$, eqn. (20.4.10) reduces to ($\hat{m}_a = m_a/m_1$)

$$\varphi_{ab}^{(1)} = \varphi_{11}^{(1)}\hat{m}_a\hat{m}_b. \tag{20.4.11}$$

In the thermodynamic limit, there is a large number $N_a \sim L\rho_a(\theta)\,\Delta\theta$ of levels in each interval $\Delta\theta$ of the rapidities and there are about $n_a \sim L\rho_a^{(r)}\Delta\theta$ particles distributed among them. Since these densities are not strongly influenced by the local redistributions of the particles, the number of different ways of distributing the particles among these levels is given by

$$\Omega_a = \frac{[L\rho_a(\theta)\,\Delta\theta]!}{[L\rho_a^{(r)}(\theta)\Delta\theta]!\,[L\rho_a^{(h)}(\theta)\Delta\theta]!},$$

in the fermionic case and by

$$\Omega_a = \frac{[L(\rho_a(\theta) + \rho_a^{(r)}(\theta) - 1)\,\Delta\theta]!}{[L\rho_a^{(r)}(\theta)\Delta\theta]!\,[L(\rho_a(\theta) - 1)\Delta\theta]!},$$

in the bosonic case. Correspondingly, the entropy per unit length $\mathcal{S} = \ln(\prod_a \Omega_a)$ is expressed by

$$\mathcal{S}_{\text{fermi}}[\rho, \rho^{(r)}] = \sum_{a=1}^{n} \int_{-\infty}^{+\infty} d\theta\, [\rho_a \ln \rho_a - \rho_a^{(r)} \ln \rho_a^{(r)} - (\rho_a - \rho_a^{(r)}) \ln(\rho_a - \rho_a^{(r)})],$$

$$\mathcal{S}_{\text{bose}}[\rho, \rho^{(r)}] = \sum_{a=1}^{n} \int_{-\infty}^{+\infty} d\theta\, [(\rho_a + \rho_a^{(r)}) \ln(\rho_a + \rho_a^{(r)}) - \rho_a \ln \rho_a - \rho_a^{(r)} \ln(\rho_a^{(r)})].$$

In terms of the densities ρ_a and $\rho_a^{(r)}$, the free energy per unit length is given by the functional

$$f[\rho, \rho^{(r)}] = E[\rho^{(r)}] - T\,\mathcal{S}[\rho, \rho^{(r)}]. \tag{20.4.12}$$

To derive the thermodynamics of the system at its thermal equilibrium with temperature $T = 1/R$, it is necessary to minimize the free energy with respect the two densitites ρ_a and $\rho_a^{(r)}$, subjected to the constraint (20.4.4). This minimization problem can be solved by using a Lagrange multiplier and can be elegantly expressed by introducing the *pseudo-energies* $\epsilon_a(\theta)$, defined in the two cases by the formulae

$$\frac{\rho_a^{(r)}}{\rho_a} = \frac{e^{-\epsilon_a}}{1 + e^{-\epsilon_a}}, \quad e^{-\epsilon_a} = \frac{\rho_a^{(r)}}{\rho_a - \rho_a^{(r)}} \quad \text{fermionic case,} \tag{20.4.13}$$

$$\frac{\rho_a^{(r)}}{\rho_a} = \frac{e^{-\epsilon_a}}{1 - e^{-\epsilon_a}}, \quad e^{-\epsilon_a} = \frac{\rho_a^{(r)}}{\rho_a + \rho_a^{(r)}} \quad \text{bosonic case.} \tag{20.4.14}$$

Using these quantities, the extremum condition reduces to the integral equations

$$m_a R \cosh\theta = \epsilon_a(\theta) \pm \sum_{b=1}^{n} \int \varphi_{ab}(\theta - \theta') \log(1 \pm e^{-\epsilon_b(\theta')}) \frac{d\theta'}{2\pi}, \tag{20.4.15}$$

where the upper sign refers to the fermionic case and the lower one to the bosonic case. The free energy at equilibrium is then given by

$$f(R) = \mp \frac{1}{R} \sum_{a=1}^{n} \int_{-\infty}^{+\infty} m_a \cosh\theta \log\left(1 \pm e^{-\epsilon_a(\theta)}\right) \frac{d\theta}{2\pi}, \tag{20.4.16}$$

where $\epsilon_a(\theta)$ is solution of the integral equation (20.4.15). Therefore the partition function is expressed by

$$Z(L,R) = \exp\left[\pm L\sum_{a=1}^{n}\int_{-\infty}^{+\infty} m_a \cosh\theta \log(1\pm e^{-\epsilon_a(\theta)})\frac{d\theta}{2\pi}\right]. \tag{20.4.17}$$

We have thus achieved the complete determination of the thermodynamics of the integrable models with diagonal S-matrix. In the next sections we analyse the behaviour of the free energy in different regimes of r and we study some significant examples.

It is useful to accompany the derivation of thermodynamics given earlier with a series of comments. The first comment concerns the conceptual difference that exists between the energy levels of free theories and interacting integrable theories. For free theories the levels are simply determined by the quantization of the states of one particle and they can be either empty or occupied by one or more of the N particles of the system, while in integrable models the levels are instead determined in a self-consistent way with the statistical distribution of the particles themselves. This is the reason behind the non-linear integral equations (20.4.15) for the pseudo–energies ϵ_a. Notice that these quantities determine the distribution of the particles but, in turn, they are also determined by them, as it can be seen from their definition, eqns. (20.4.13) and (20.4.14).

The second comment concerns some mathematical properties of the pseudo-energies. It is interesting to note that, even though $\epsilon_a(\theta)$ satisfies a non-linear integral equation, the derivatives $\partial_R\epsilon_a$ e $\partial_\theta\epsilon_a$ are instead solutions of linear integral equations. Deriving eqn. (20.4.15) with respect to R, we have in fact

$$\partial_R\epsilon_a = m_a \cosh\theta + \sum_{b=1}^{n}\int_{-\infty}^{+\infty} \varphi_{ab}(\theta-\theta')\frac{e^{-\epsilon_b(\theta')}}{1\pm e^{-\epsilon_b(\theta')}}\partial_R\epsilon_b(\theta')\frac{d\theta'}{2\pi}, \tag{20.4.18}$$

and, analogously

$$\partial_\theta\epsilon_a = m_a R \sinh\theta + \sum_{b=1}^{n}\int_{-\infty}^{+\infty} \varphi_{ab}(\theta-\theta')\frac{e^{-\epsilon_b(\theta')}}{1\pm e^{-\epsilon_b(\theta')}}\partial_{\theta'}\epsilon_b(\theta')\frac{d\theta'}{2\pi}. \tag{20.4.19}$$

These equations can be written in a compact form by defining the integral operators $\hat{K}_a^{\pm}$, whose kernel is

$$\hat{K}_a^{\pm}(\theta,\theta') = \sum_{b=1}^{n}\varphi_{ab}(\theta-\theta')\frac{e^{-\epsilon_b(\theta')}}{1\pm e^{-\epsilon_b(\theta')}}. \tag{20.4.20}$$

Hence

$$\left(1-\hat{K}_a^{\pm}\right)\partial_R\epsilon_a = e_a,$$
$$\left(1-\hat{K}_a^{\pm}\right)\frac{1}{R}\partial_\theta\epsilon_a = k_a, \tag{20.4.21}$$

where

$$(\hat{K}\,\partial_R\epsilon) = \int \hat{K}(\theta,\theta')\,\partial_R\epsilon(\theta')\,\frac{d\theta'}{2\pi},$$

(analogously for $(\hat{K}\partial_\theta\epsilon)$), with the notation $e_a = m_a\cosh\theta,\; k_a = m_a\sinh\theta$. Eqns. (20.4.21) are linear integral equation for the quantities $\partial_R\epsilon_a$ and $\partial_\theta\epsilon_a$ since the functions $\epsilon_b(\theta)$ entering the definition of the kernel $\hat{K}_a^{\pm}$ are regarded as assigned functions, known once the original integral equation (20.4.15) is solved. It is possible to invert eqn. (20.4.21) by introducing the resolvents $\hat{L}_a^{\pm}$, that satisfy

$$(1-\hat{K}_a^{\pm})(1-\hat{L}_a^{+}) = 1.$$

In this way

$$\partial_R\epsilon_a = (1+\hat{L}_a^{\pm})\,e_a,\quad \frac{1}{R}\partial_\theta\epsilon_a = (1+\hat{L}_a^{\pm})\,k_a. \tag{20.4.22}$$

Since $1+\hat{L}_a = \sum_{n=0}^{\infty}\hat{K}^n$, $\partial_R\epsilon_a$ are then expressed by the Fredholm series

$$\partial_R\epsilon_a = e_a + \sum_{b=1}^{n}\varphi_{ab} * \left(\frac{e^{-\epsilon_b}}{1+e^{-\epsilon_b}}\,e_a\right) + \cdots,$$

with an analogous result for $\frac{1}{R}\partial_\theta\epsilon_a$.

The third comment refers to the nature of the system of the TBA equations for the S-matrices of fermionic and bosonic type. Till now we have presented on the same footing the two cases but there is strong reason to believe that the only consistent interacting theories are those of fermionic type, with $S_{aa}(0) = -1$. In other words, the only diagonal bosonic type S-matrix that gives rise to a consistent set of TBA equations is given by the free theory, for which we have identically $S = 1$. From the mathematical point of view, the problem with the bosonic type TBA equations comes from the term $\log(1-e^{-\epsilon_a})$, present in the integral of eqn. (20.4.15) that determined the pseudo-energies. If it happens that, varying r, one of the ϵ_a becomes negative in an interval of θ, the TBA equations give rise to complex solutions that do not have a natural physical interpretation.

In the light of these remarks, we focus our attention on the fermionic TBA systems. In this case, it is easy to prove the statement previously made on the monotonic nature of the functions $\mathcal{J}_a(\theta)$. In the fermionic TBA, the eqn. (20.4.15) satisfied by $\epsilon_a(\theta)$ implies that these are real functions of θ for any value of r. This also implies the positivity of the densities $\rho_a^{(r)}$ and $\rho_a(\theta)$. But these functions are just the derivatives of the functions $\mathcal{J}_a(\theta)$, so that $\frac{d}{d\theta}\mathcal{J}_a(\theta) > 0$ and $\mathcal{J}_a(\theta)$ are strictly increasing functions.

20.5 The Meaning of Pseudo-energy

The pseudo-energies $\epsilon_a(\theta)$ admit an interesting physical interpretation. It is necessary to initially note that the final expression of the partition function given by the TBA, eqn. (20.4.17), is formally identical to the partition function of a gas of *free* quasi-particles,[3] the only difference being that the energy of each of these particles is given by $\epsilon_a(\theta)/R$ rather than $m_a \cosh\theta$. This observation suggests that, in integrable theories, the only effect of the temperature consists of modifying the excitation energies of the particles, which are now measured with respect to the thermal ground state of the system. To better clarify this observation and to understand the nature of the pseudo-energies, it is convenient to analyse the simplest case of a system with only one species of particles. The following demonstrates the one-to-one correspondence between the partition function as obtained by the TBA

$$Z(L,R) = \exp\left[\pm L \int \frac{d\theta}{2\pi} m_a \cosh\theta \, \log(1 \pm e^{-\epsilon(\theta)})\right]. \tag{20.5.1}$$

and the partition function as given by the usual sum on the states

$$Z(L,R) = \sum_{n=0}^{\infty} \frac{1}{n!} \int \frac{d\theta_1}{2\pi} \cdots \frac{d\theta_n}{2\pi} \langle \theta_n \cdots \theta_1 | \theta_1 \cdots \theta_n \rangle \prod_{i=1}^{n} e^{-\epsilon(\theta_i)}, \tag{20.5.2}$$

where the scalar product is computed using the standard rules relative to the free fermionic or bosonic cases (depending on the type of the TBA equations), while the energy of the particles is given by the pseudo-energy $\epsilon(\theta)/R$. Let us study the fermionic case, leaving the derivation of the bosonic case as exercise to the reader. We start by defining

$$F(R) = \int \frac{d\theta}{2\pi} \cosh\theta \log(1 + e^{-\epsilon(\theta)}). \tag{20.5.3}$$

This function can be expanded in series as

$$F(R) = \sum_{n=0}^{\infty} \frac{(-1)^{n+1}}{n} I_n(R), \tag{20.5.4}$$

[3] The dispersion relations of the free quasi-particle differ from those of the usual particle and they are given in eqn. (20.5.7).

where

$$I_n(R) \equiv \int \frac{d\theta}{2\pi} \cosh\theta \, e^{-n\epsilon(\theta)}.$$

The TBA partition function has a power series expansion in powers of (mL)

$$Z(L,R) = 1 + (mL)F(R) + \frac{(mL)^2}{2!}(F(R))^2 + \cdots + \frac{(mL)^n}{n!}(F(R))^n + \cdots. \qquad (20.5.5)$$

Let's now compute the partition function using its alternative definition, given in eqn. (20.5.2). We need to employ a regularization of the square of the δ function (that enters the scalar product $\langle\theta_n\cdots\theta_1|\theta_1\cdots\theta_n\rangle$) provided by the free fermionic theory defined on a sufficiently large volume L

$$[\delta(\theta-\theta')]^2 \equiv \frac{mL}{2\pi} \cosh(\theta)\,\delta(\theta-\theta'). \qquad (20.5.6)$$

We have then

$$Z(L,R) = 1 + Z_1 + Z_2 + \cdots + Z_n + \cdots$$

where the first terms are given by

$$\begin{aligned} Z_1 &= \int \frac{d\theta}{2\pi} < \theta|\theta > e^{-\epsilon(\theta)} = \int \frac{d\theta}{2\pi} d\theta' \delta(\theta-\theta') < \theta'|\theta > e^{-\epsilon(\theta)} \\ &= mL \int \frac{d\theta}{2\pi} \cosh\theta \, e^{-\epsilon(\theta)} = (mL)\, I_1; \end{aligned}$$

$$\begin{aligned} Z_2 = \frac{1}{2}\int \frac{d\theta_1}{2\pi}\frac{d\theta_2}{2\pi} < \theta_2\theta_1|\theta_1\theta_2 > e^{-\epsilon(\theta_1)-\epsilon(\theta_2)} = \frac{1}{2}\int \frac{d\theta_1}{2\pi}\frac{d\theta_2}{2\pi}\Big[(2\pi)^2(\delta(\theta_1-\theta_1)\delta(\theta_2-\theta_2) + \\ -\delta(\theta_1-\theta_2)\delta(\theta_2-\theta_1))\Big]\, e^{-\epsilon(\theta_1)-\epsilon(\theta_2)} = \frac{1}{2}(mL)^2 I_1^2 - \frac{1}{2}(mL) I_2, \end{aligned}$$

Similarly,

$$Z_3 = \frac{(mL)^3}{3!} I_1^4 - \frac{(mL)^2}{2} I_1 I_2 + \frac{(mL)}{3} I_3,$$

$$Z_4 = \frac{(mL)^4}{4!} I_1^4 - (mL)^3 I_1^2 I_2 + \frac{(ML)^2}{2}\left[\frac{2}{3} I_1 I_3 + \left(\frac{I_2}{2}\right)^2\right] - \frac{(mL)}{4} I_4.$$

It is not difficult to extend the computation to a higher order and show that the series (20.5.2) precisely coincides with the one given in eqn. (20.5.5), the only difference being the different arrangement of their terms. In fact, summing all those proportional to (mL) present in each Z_n, we recover the function $F(R)$, while the sum of all the terms proportional to $(mL)^k$ present in each Z_n precisely reproduces the higher powers $(F(R))^k$.

This important result implies that all physical properties of the system depend on the quasi-particle excitations with respect to the ground state of the TBA. These excitations have an *effective energy* $\tilde{e}(\theta) = \epsilon(\theta)/R$ and an *effective momentum* $\tilde{k}(\theta)$ given by

$$\begin{aligned} \tilde{e}(\theta) &= \epsilon(\theta)/R, \\ \tilde{k}(\theta) &= k(\theta) + 2\pi(\delta * \rho_1)(\theta). \end{aligned} \tag{20.5.7}$$

Hence, in the presence of the temperature, we may regard the rapidity θ as the parameter that expresses the dispersion relations of the quasi-particle excitations. This result, derived in the Yang–Yang non-relativistic case, can be easily generalized to the relativistic case, as shown in the box below.

Dressed energy and momentum

Let (n_j, θ_j) and (n'_j, θ'_j) be two Bethe states satisfying eqn. (20.3.3), where $n'_j = n_j$ except for $j = \alpha$. Subtracting the two equations, we have

$$mL(\sinh\theta'_j - \sinh\theta_j) = \sum_i [\tilde{\delta}(\theta_j - \theta_i) - \tilde{\delta}(\theta'_j - \theta'_i)]. \tag{20.5.8}$$

$(j \neq \alpha)$. Since $\theta'_j \approx \theta_j$, we can introduce a function $\chi(\theta)$ and write

$$L(\sinh\theta'_j - \sinh\theta_j) \approx \chi(\theta_j)\cosh\theta_j. \tag{20.5.9}$$

In the thermodynamic limit, eqn. (20.5.8) can be written as

$$2\pi(1 - \hat{K})(\rho\chi) = \tilde{\delta}(\theta - \theta_\alpha) - \tilde{\delta}(\theta - \theta'_\alpha), \tag{20.5.10}$$

where ρ is the density of the levels and $\hat{K}$ is the integral operator defined in eqn. (20.4.20), here restricted to the case in which there is only one type of particle. We also have

$$\delta(\theta - \theta_\alpha) - \delta(\theta - \theta'_\alpha) = \int_{\theta_\alpha}^{\theta'_\alpha} d\theta' \hat{K}(\theta, \theta')(1 + e^{\epsilon(\theta')}). \tag{20.5.11}$$

Using the resolvent $\hat{L}$ of $\hat{K}$, we can invert (20.5.10)

$$\rho\chi(\theta) = \int_{\theta_\alpha}^{\theta'_\alpha} \frac{d\theta'}{2\pi} \hat{L}(\theta,\theta')(1+e^{\epsilon(\theta')}). \tag{20.5.12}$$

Consider now the difference in energy ΔE between the two Bethe states

$$\Delta E = \sum_j m\,[\cosh\theta'_j - \cosh\theta_j] = \tag{20.5.13}$$

$$= m\cosh\theta'_\alpha - m\cosh\theta_\alpha + m\int d\theta\,\sinh\theta\,\frac{\chi(\theta)\rho(\theta)}{1+e^{\epsilon(\theta)}}.$$

Substituting (20.5.12) and using the property $\hat{L}(\theta,\theta')(1+e^{\epsilon(\theta')}) = (1+e^{\epsilon(\theta)})\hat{L}(\theta',\theta)$, together with (20.4.22), we have

$$\Delta E = \frac{1}{R}(\epsilon(\theta'_\alpha) - \epsilon(\theta_\alpha)). \tag{20.5.14}$$

The dressed momentum is obtained in a similar way

$$\Delta P = \sum_j m\,[\sinh\theta'_j - \sinh\theta_j] = \tag{20.5.15}$$

$$= m\sinh\theta'_\alpha - m\sinh\theta_\alpha + m\int d\theta\,\cosh\theta\,\frac{\chi(\theta)\rho(\theta)}{1+e^{\epsilon(\theta)}}.$$

Substituting again (20.5.12) and using eqn. (20.4.22),

$$\Delta P = \tilde{k}(\theta'_\alpha) - \tilde{k}(\theta_\alpha) \tag{20.5.16}$$

where $\tilde{k}$ is defined in eqn. (20.5.7).

The interpretation given above of the pseudo-energy finds interesting application in the computation of the correlation functions of the integrable models at finite temperature.

20.6 Infrared and Ultraviolet Limits

This section studies in more detail the scaling function $\tilde{c}(r)$ for the TBA systems of fermionic type. Using eqns. (20.2.5) and (20.2.7), this function is given by

$$\tilde{c}(r) = \frac{3}{\pi^2} r \sum_{a=1}^{n} \hat{m}_a \int_{-\infty}^{+\infty} L_a(\theta) \cosh\theta \, d\theta, \tag{20.6.1}$$

where we have defined $\hat{m}_a = m_a/m_1$ and

$$L_a(\theta) = \log\left(1 + e^{-\epsilon_a(\theta)}\right).$$

It is easy to find the behaviour of this function for large values of r (the infrared limit): we have in fact

$$\epsilon_a(\theta) \simeq \hat{m}_a r \cosh\theta, \quad L_a(\theta) \simeq e^{-\hat{m}_a r \cosh\theta}, \tag{20.6.2}$$

so that $\tilde{c}(r)$, for $r \to \infty$, behaves as

$$\tilde{c}(r) \simeq \frac{3}{\pi^2} r \sum_{a=1}^{n} \hat{m}_a \int_{0}^{+\infty} d\theta \cosh\theta \, e^{-r\hat{m}_a \cosh\theta} = \frac{3}{\pi^2} r \sum_{a=1}^{n} \hat{m}_a K_1(\hat{m}_a r), \tag{20.6.3}$$

where $K_1(z)$ is the modified Bessel function. For $r \to \infty$ the Bessel function decreases exponentially and the behaviour of the system is the one of a free theory made of n particles of masses m_a (see Section 20.11).

The opposite limit, $r \to 0$, corresponds to the ultraviolet or conformal limit[4] of the massive theory. To compute the value of the scaling function $\tilde{c}(r)$ of this limit, we need to study some properties of the integral equation (20.4.15).

The solutions $\epsilon_a(\theta)$ are even functions of θ. For $r \to 0$, they flatten and become constant in the region $-\ln\frac{2}{r} \ll \theta \ll \ln\frac{2}{r}$, whereas they tend to the free values (20.6.2) outside this interval.[5] The constant values ϵ_a can be found solving the transcendental equation

$$\epsilon_a = \sum_{b=1}^{n} N_{ab} \ln\left(1 + e^{-\epsilon_b}\right), \tag{20.6.4}$$

where N_{ab} is the positive symmetric matrix given by

$$N_{ab} = -\int_{-\infty}^{+\infty} \frac{d\theta}{2\pi} \varphi_{ab}(\theta) = -\frac{1}{2\pi} \left[\delta_{ab}(+\infty) - \delta_{ab}(-\infty)\right]. \tag{20.6.5}$$

[4] Since r is a scaling variable, given by $r = m_1 R = R/\xi$, the limit $r \to 0$ can be equivalently regarded as the limit in which the correlation length ξ diverges, $\xi \to \infty$.

[5] This is actually the situation for the TBA equations relative to the minimal S-matrices, while for the S-matrices associated to Lagrangian models, as for instance the S-matrix of the Sinh–Gordon or Toda models, the functions $\epsilon_a(\theta)$ diverge as $\log(m_a r)$ in the limit $r \to 0$.

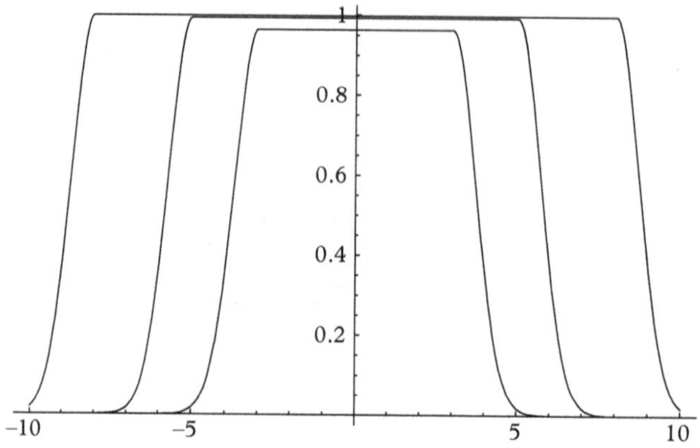

Fig. 20.3 *Behaviour of the function $L_a(\theta)$ when $r \to 0$.*

For $r \to 0$, the plateau of the curve enlarges (Figure 20.3). In this limit, the curve rapidly decreasing outside the plateau assumes a universal shape. This can be determined noting that, for large values of θ, the right-hand side of eqn. (20.4.15) can be written as

$$\hat{m}_a r \cosh\theta \sim \frac{1}{2}\hat{m}_a r e^{\theta} = \hat{m}_a e^{\left(\theta - \ln\frac{2}{r}\right)},$$

and the dependence on r of the functions $\epsilon_a(\theta)$ reduces then to a simple shift[6]

$$\theta \to \theta - \ln\left(\frac{2}{r}\right).$$

Hence, for $r \to 0$, their behaviour at the edges of the interval is universal and dictated by the equation

$$\hat{m}_a e^{\theta} = \tilde{\epsilon}_a(\theta) + \sum_{b=1}^{n} (\varphi_{ab} * \tilde{L}_b)(\theta), \tag{20.6.6}$$

where the functions $\tilde{\epsilon}_a(\theta)$ assume the constant values ϵ_a for $\theta \ll \ln\left(\frac{2}{r}\right)$ and increase exponentially to infinity when $\theta \to \infty$. The corresponding functions $\tilde{L}_a(\theta)$ interpolate bewteen 0 and their limiting value given in eqn. (20.6.4). For this reason, the universal functions $\tilde{\epsilon}_a$ are called the *kink solutions* of the TBA equations. Expressed in terms of these functions, the value of the scaling function $\tilde{c}(r)$ at $r = 0$ assumes the form

[6] We discuss the behaviour of $\epsilon_a(\theta)$ for positive values of θ, since the behaviour for negative values can be recovered by parity.

$$\tilde{c}(0) = \frac{6}{\pi^2} \sum_{a=1}^{n} \int_0^{\infty} d\theta \, \tilde{L}_a(\theta) \, \hat{m}_a e^{\theta}. \tag{20.6.7}$$

Substituting for $\hat{m}_a e^{\theta}$ the derivative of the left-hand side of eqn. (20.6.6)

$$\hat{m}_a e^{\theta} = \frac{d\tilde{\epsilon}_a(\theta)}{d\theta} - \sum_{b=1}^{n} \left(\varphi_{ab} * \frac{e^{-\tilde{\epsilon}_b}}{1+e^{-\tilde{\epsilon}_b}} \frac{d\tilde{\epsilon}_b}{d\theta} \right)(\theta), \tag{20.6.8}$$

we have

$$\tilde{c}(0) = \frac{6}{\pi^2} \sum_{a=1}^{n} \int_0^{\infty} d\theta \, \tilde{L}_a(\theta) \left[\frac{d\tilde{\epsilon}_a(\theta)}{d\theta} - \sum_{b=1}^{n} \left(\varphi_{ab} * \frac{e^{-\tilde{\epsilon}_b}}{1+e^{-\tilde{\epsilon}_b}} \frac{d\tilde{\epsilon}_b}{d\theta} \right)(\theta) \right]. \tag{20.6.9}$$

Since $\tilde{\epsilon}_a$ are monotonically increasing functions, the first term on the right-hand side simply becomes

$$\int_0^{\infty} d\theta \, \tilde{L}(\theta) \frac{d\tilde{\epsilon}_a(\theta)}{d\theta} = \int_{\epsilon_a}^{\infty} d\epsilon \, \tilde{L}(\epsilon). \tag{20.6.10}$$

The convolution term in (20.6.9) can be analogously substituted using the same equation (20.6.6). After an integration by parts, the final result is given by

$$\tilde{c}(0) = \sum_{a=1}^{n} \tilde{c}_a(\epsilon_a), \tag{20.6.11}$$

where

$$\tilde{c}_a(\epsilon_a) = \frac{6}{\pi^2} \left[\int_{\epsilon_a}^{\infty} dx \ln(1+e^{-x}) + \frac{1}{2} \epsilon_a \ln(1+e^{-\epsilon_a}) \right] = \frac{6}{\pi^2} L\left(\frac{1}{1+e^{\epsilon_a}} \right),$$

and $L(x)$ is the dilogarithm function

$$L(x) = -\frac{1}{2} \int_0^x dt \left[\frac{\ln t}{1-t} + \frac{\ln(1-t)}{t} \right].$$

In conclusion, the effective central charge of the conformal limit of the massive theory with purely elastic S-matrix is obtained through the following steps:

1. Solve the transcendental equation (20.6.4)
2. Substitute their constant solutions ϵ_a in eqn. (20.6.11).

The following sections provide some significant examples of this result.

20.7 The Coefficient of Bulk Energy

In a theory with a mass scale, the additivity of the energy requires a linear growing of the energy of the ground state with the dimension R of the system

$$E_0 \sim \mathcal{E}_0 R.$$

$\mathcal{E}_0$, the bulk term, can be interpreted as the singular part of the infinite volume energy due to the fluctuations present in the system. Usually this is not a universal quantity, since it depends on the regularization scheme adopted. However, in a perturbed CFT, the regularization scheme is fixed by the requirement that the off-critical quantities adiabatically go to their conformal values. Hence, in this case, it is possible to extract a universal term $\mathcal{E}_0$ that only depends on the scattering data. Since E_0 is directly related to the scaling function $\tilde{c}(r)$, the bulk term $\mathcal{E}_0$ is given by

$$\mathcal{E}_0 = -\frac{\pi}{12} m_1^2 \left. \frac{1}{r} \frac{d\tilde{c}}{dr} \right|_{r=0} . \tag{20.7.1}$$

For the computation of this limit, let us introduce the functions

$$\psi_a(\theta) = \left(\partial_r + \frac{1}{r} \partial_\theta \right) \epsilon_a(\theta).$$

As discussed at the end of Section 20.4, the functions ψ_a satisfy the linear integral equations

$$\psi_a(\theta) = \hat{m}_a e^\theta + \sum_{b=1}^{n} \left(\varphi_{ab} * \frac{\psi_b}{e^{\epsilon_b} + 1} \right)(\theta).$$

Using eqn. (20.6.1),

$$\frac{1}{r} \frac{d\tilde{c}(r)}{dr} = -\frac{3}{\pi^2} \sum_{a=1}^{n} \int_{-\infty}^{+\infty} d\theta \, \hat{m}_a \, e^{-\theta} \frac{\psi_a(\theta)}{e^{\epsilon(\theta)} + 1}.$$

When $r \to 0$, the integrand is localized near the edge of the flat region and its behaviour is determined by the kink solutions $\tilde{L}_a(\theta)$. Hence we get

$$\left. \frac{1}{r} \frac{d\tilde{c}(r)}{dr} \right|_{r=0} = \frac{3}{\pi^2} \sum_{a=1}^{n} \hat{m}_a \int_{-\infty}^{\infty} d\theta \, e^{-\theta} \, \partial_\theta \tilde{L}_a(\theta) \equiv -\frac{3}{\pi^2} \sum_{a=1}^{n} \hat{m}_a T_a. \tag{20.7.2}$$

To compute the right-hand side of this equation, let us proceed as follows. Considering initially the asymptotic expansion for $\theta \to -\infty$ of the convolution term, we have

$$\sum_{b=1}^{n} (\varphi_{ab} * \tilde{L}_b)(\theta) = -\epsilon_a + \frac{e^{\theta}}{2\pi} \sum_{b=1}^{n} \varphi_{ab}^{(1)} T_b + \cdots$$

where $\varphi_{ab}^{(1)}$ is the first term of the expansion of these functions, as given in eqn. (20.4.8). Comparing with the exponential term in eqn. (20.6.8), we arrive at

$$\sum_{b=1}^{n} \varphi_{ab}^{(1)} T_b = 2\pi .$$

Using now eqn. (20.4.11), we obtain

$$\sum_{b=1}^{n} \varphi_{ab}^{(1)} T_b = \varphi_{11}^{(1)} \hat{m}_a \sum_{b=1}^{n} \hat{m}_b T_b,$$

where $\varphi_{11}^{(1)}$ is the corresponding quantity relative to the lightest particle. Hence the bulk energy term is determined by the S-matrix of the lightest particle as

$$\mathcal{E}_0 = \frac{m_1^2}{2\varphi_{11}^{(1)}}. \tag{20.7.3}$$

A direct measurement of this quantity can be achieved by a numerical diagonalization of the transfer matrix of the theory.

20.8 The General Form of the TBA Equations

For the diagonal scattering theories related to the simply laced Lie algebras, there is an extremely elegant formulation of the TBA equations. Besides the great level of generality, this formulation has also the advantage to highlight the common structure of all these theories. Chapter 14, the number of particles of these theories is equal to the rank r of the algebra $\mathcal{A}$. Another important quantity to keep in mind is the Coxeter number h of these algebras.

In order to give such a formulation, we introduce the notation

$$\nu_a(\theta) = m_a R \cosh\theta, \tag{20.8.1}$$

and note that the original TBA equations of these theories, expressed by

$$-\nu_a + \epsilon_a + \sum_{b=1}^{r} \varphi_{ab} \star \log[1 + \exp(-\epsilon_b)] = 0, \tag{20.8.2}$$

can be rewritten in the universal form

$$-\nu_a + \epsilon_a + 2\sum_{b=1}^{r} I_{ab}\varphi_h \star \{\nu_b - \log[1 + \exp(\epsilon_b)]\}, \tag{20.8.3}$$

where I_{ab} is the incidence matrix of the Dynkin diagram of the corresponding algebra $\mathcal{A}$ and $\varphi_h(\theta)$ is the universal kernel

$$\varphi_h(\theta) = \frac{h}{2\cosh\frac{h\theta}{2}}, \tag{20.8.4}$$

with h the Coxeter number of the relative algebra. The equivalence between the sets of equations (20.8.2) and (20.8.3) is based on the important identity

$$\left(\delta_{ab} - \frac{1}{2\pi}\varphi_{ab}(k)\right)^{-1} = \delta_{ab} - \frac{1}{2\cosh(k/h)} I_{ab} \tag{20.8.5}$$

that holds for the Fourier transforms

$$\varphi_{ab}(k) = \int_{-\infty}^{+\infty} \varphi_{ab}(\theta)\exp(ik\theta)\, d\theta$$

of the original kernels. Another important relation in the derivation of eqn. (20.8.3) is provided by

$$\sum_{b=1}^{r} I_{ab}\, m_b = 2m_a \cos\frac{\pi}{h}. \tag{20.8.6}$$

As a by-product of eqn. (20.8.5), one of its remarkable consequences is the universal expression of the matrix N_{ab} for the scattering theories associated to the Lie algebras

$$N = I(2 - I)^{-1}. \tag{20.8.7}$$

To derive this expression it is sufficient to substitute $k = 0$ in (20.8.5), into account that $N_{ab} = -\frac{1}{2}\varphi_{ab}(0)$.

Eqns. (20.8.3) can be now analytically continued to the values $\theta \pm i\pi/h$. Using the relation

$$\nu_a\left(\theta + i\frac{\pi}{h}\right) + \nu_a\left(\theta - i\frac{\pi}{h}\right) = \sum_{b=1}^{r} I_{ab}\nu_b(\theta), \tag{20.8.8}$$

they can be written in terms of a set of functional equations for the quantities $Y_a(\theta) = \exp[\epsilon_a(\theta)]$

$$Y_a\left(\theta + i\frac{\pi}{h}\right) Y_a\left(\theta + i\frac{\pi}{h}\right) = \prod_{b=1}^{r} [1 + Y_b(\theta)]^{I_{ab}}. \tag{20.8.9}$$

These equations are completely independent on the energy terms $\nu_a(\theta)$ of the particles and involve only the basic information of the algebras encoded in the incidence matrix I_{ab}. Furthermore, they present the periodic properties

$$\begin{aligned} Y_a\left(\theta + i\pi\tfrac{h+2}{h}\right) &= Y_{n-a+1}(\theta), \quad && A_n \text{ series} \\ Y_a\left(\theta + i\pi\tfrac{h+2}{h}\right) &= Y_a(\theta), && D_n \text{ and } E_n \text{ series.} \end{aligned} \tag{20.8.10}$$

For the series A_n, we should keep in mind that the symmetry of these algebras imposes $Y_a(\theta) = Y_{n-a+1}(\theta)$, so that the periodicity condition is also satisfied for this series.

The periodic properties of $Y_a(\theta)$ have important consequences. First of all, they imply that the solutions of the original equations (20.8.2), with $\nu_a(\theta)$ given in (20.8.1), are entire functions of θ and, consequently, they can be expanded in a Laurent series

$$Y_a(\theta) = \sum_{n=-\infty}^{\infty} Y_a^{(n)}\, t^n, \tag{20.8.11}$$

where $t = \exp([(2h/(h+2)]\theta)$. These series are convergent on all the complex plane of θ, except at $t = 0$ and $t = \infty$. In particular, for the solutions of eqn. (20.8.2), the symmetry $\theta \to -\theta$ requires that $Y_a^{(n)} = Y_a^{(-n)}$. In the t-plane, the functional equation (20.8.9) becomes

$$Y_a(\Omega t) Y_a(\Omega^{-1} t) = \prod_{b=1}^{r} [1 + Y_b(t)]^{I_{ab}}. \tag{20.8.12}$$

where $\Omega = \exp[2i\pi/(h+2)]$. These are the most general form of the TBA equations and they may have several classes of solutions. Obviously, among the solutions of eqns. (20.8.12), there are also those that are entire functions in t. The kink solution, for instance, corresponding to an energy term given by $\nu_a(\theta) = m_a R \exp(\theta)$ instead of (20.8.1), is an example of this set of solutions. Notice that posing $t = 0$ in

(20.8.12), one obtains the algebraic transcendental equations (20.6.4) for the quantities $z_a = \exp[\epsilon_a(0)]$, that are crucial quantities to obtain the effective central charge.

The second consequence of the periodicity of the functions $Y_a(\theta)$ concerns the behaviour of the solutions of eqn. (20.8.2) when $R \to 0$. In this limit we saw that the functions $\log[1+\exp[-\epsilon(\theta)]]$ acquire a plateau of height $\log[1+1/z_a]$ in the central interval $-\log(1/m_1 R) \ll \theta \ll \log(1/m_1 R)$ and rapidly tend to zero outside this interval. For $R \to 0$ this plateau enlarges and this implies that the integral equations (20.8.2) or (20.8.3) are somehow local in the central interval in the rapidity space while, for large R, the two edges of the plateau influence each other only through the wave length terms fixed by the periodicity. Hence the function $f(R) = RE_0(R)/2\pi$ admits a regular series expansion with respect the variable $G^2 = (m_1 R)^{4h/(h+2)}$, except for a bulk term energy proportional to R^2

$$f(R) = -\frac{c_{eff}}{12} - \frac{\mathcal{E}_0}{2\pi} R^2 + \sum_{n=1}^{\infty} f_{2n}\, G^{2n}. \tag{20.8.13}$$

This analytic structure of $f(R)$ is in full agreement with the conformal perturbation theory, set in this case by a perturbing operator of conformal dimension $\Delta = 1 - h/(h+2)$ (for the scattering theory based on the Lie algebras, the corresponding perturbative terms are given by even powers of the coupling constants).

The analysis of the TBA done for the Lie algebras can be extended to the most general case. In particular, we can show that, beside the bulk term, the expansion of the free energy dictated by the TBA is a regular function with respect to the variable $G = (m_1 R)^{2-2\Delta}$, where Δ is the conformal dimension of the perturbing field

$$f(R) = -\frac{c_{eff}}{12} - \frac{\mathcal{E}_0}{2\pi} R^2 + \sum_{n=1}^{\infty} f_n\, G^n. \tag{20.8.14}$$

Comparing with the perturbative expression of this quantity we can derive an important relationship between the coupling constant and the lowest mass of the theory.

20.9 The Exact Relation $\lambda(m)$

The TBA allows us to determine the exact relation that holds between the coupling constant of a perturbed CFT and its lowest mass. This section presents the basic idea that leads to this formula, listing afterward the formulae of the various integrable theories.

Consider the action of a perturbed CFT, with the perturbation given by a relevant field of conformal dimension Δ

$$\mathcal{S} = \mathcal{S}_0 + \lambda \int d^2x\, \phi(x). \tag{20.9.1}$$

Let us assume that such a deformation defines a massive integrable field theory, characterized by its S-matrix. The coupling constant λ is a dimensional quantity, expressed in terms of the lowest mass m_1 by the relation

$$\lambda = \mathcal{D} m_1^{2-2\Delta}. \tag{20.9.2}$$

Once the normalization of the operator $\phi(x)$ is fixed, the coefficient $\mathcal{D}$ is a pure number that can be extracted by the comparison between the TBA and the perturbative series. For the normalization of the operator we take the conformal one, given by

$$\langle \phi(x_1)\phi(x_2)\rangle \simeq \frac{1}{|x_{12}|^{4\Delta}}, \quad x_{12} \to 0.$$

As discussed, the free energy of an integrable theory can be computed in terms of the Bethe ansatz equations and this leads to the general expression (20.8.14). On the other hand, this quantity can be computed in conformal perturbation theory, using eqn. (20.9.1). The corresponding series is

$$f_{pert}(R) = -c_{eff} - \frac{R^2}{2\pi}\sum_{n=1}^{\infty}\frac{(-\lambda)^n}{n!}\int \langle \phi(X_1)\ldots\phi(X_n)\rangle_c\, d^2X_1\ldots d^2X_n, \tag{20.9.3}$$

where $X_i = (x_i, y_i)$ are the coordinates on the cylinder and the connected correlation functions are those of the conformal theory on the cylinder. Using the mapping $z = \exp(-2\pi i\zeta/R)$ where $\zeta = x + iy$ is the complex coordinate of the cylinder, the perturbative terms can be written as integrals of the connected correlation functions on the Euclidean plane

$$\begin{aligned} f_{pert}(R) &= -c_{eff} - \frac{R^2}{2\pi}\sum_{n=1}^{\infty}\frac{(-\lambda)^n}{n!}\left(\frac{2\pi}{R}\right)^{2(\Delta-1)n+2} \\ &\times \int \langle V(0)\phi(z_1,\bar{z}_1)\ldots\phi(z_n,\bar{z}_n)V(\infty)\rangle_c \prod_{i=1}^{n}(z_i\bar{z}_i)^{\Delta-1}\, d^2z_1\ldots d^2z_n. \end{aligned} \tag{20.9.4}$$

In this expression, $V(z,\bar{z})$ is the operator that creates the lowest energy state on the cylinder, i.e. the field associated to Δ_{min} (for the unitary theories, $V = \mathbf{1}$). If the field ϕ is odd under a Z_2 symmetry of the original conformal model—a hypothesis that we will make in the following for simplicity we have an even series in λ.

All the integrals of the perturbative series are ultraviolet convergent if $\Delta < 1/2$ and, on the cylinder for its finite size, they are also convergent in the infrared. Using a dimensional argument, it is not difficult to see that the perturbative series is an expansion in the parameter $g^2 \equiv \lambda^2 R^{2(2-2\Delta)}$

$$f_{pert}(R) = -c_{eff} + \mathcal{F}_2 g^2 + \mathcal{F}_4 g^4 + \cdots.$$

Let us assume that this series converges in a finite domain around the origin where it defines a function $\mathcal{F}(g)$. On the other side, for thermodynamic reasons, we know that in the limit $R \to \infty$ we have

$$f_{pert}(R) \sim \frac{\mathcal{E}_0}{2\pi} R^2.$$

This behaviour is related to the analytic continuation of $\mathcal{F}(g)$ outside the domain of convergence of the original series. Since in QFT the normalization of the free energy is chosen in such a way to vanish at infinity, it is necessary to subtract the quantity above from the perturbative series, so that the final expression is

$$f(R) = -c_{eff} - \frac{\mathcal{E}_0}{2\pi} R^2 + \sum_{n=1} \mathcal{F}_{2n} g^{2n}. \tag{20.9.5}$$

The relation between the coupling constant and the lowest mass is obtained by comparing this series with the original series of the TBA, eqn. (20.8.14). Taking the first term of both and simplifying the common factor $R^{2(2-2\Delta)}$ we have

$$\mathcal{F}_2 \lambda^2 = f_2 \, m_1^{2(2-2\Delta)}. \tag{20.9.6}$$

The proportional coefficient $\mathcal{D}$ between the coupling constant λ and the mass m_1 is then

$$\mathcal{D}^2 = f_2 / \mathcal{F}_2. \tag{20.9.7}$$

This coefficient has been determined exactly for many integrable models. For all the theories related to the simply laced Toda models, this has been achieved by Fateev. Let us present the relevant formula. Section 16.7 showed that the Toda field theories for particular imaginary values of the coupling constant given by

$$\beta^2 = \frac{p}{p+1} \qquad p = k+h, k+h+1, \ldots$$

describe the integrable deformation of the field of conformal dimension

$$\Delta = 1 - \frac{h}{k+h+1}.$$

of the coset model

$$\frac{G_k \times G_1}{G_{k+1}}.$$

Let us denote generically the perturbed action of these models as

$$\mathcal{S} = \mathcal{S}_{CFT} + \lambda \int d^2x\, \Phi_\Delta(x).$$

For these theories the relation that links λ to the lowest mass of the system is given by

$$\lambda = \left[\frac{\pi\, m_1\, k(\mathcal{G})\, \Gamma\left(\frac{k+h+1}{h}\right)}{\Gamma\left(\frac{1}{h}\right)\Gamma\left(\frac{k+h}{h}\right)} \right]^{4hu} \times \frac{(1-hu)^{-2}(1-(h+1)u)^2\, \gamma(qu)\, \gamma\left(\frac{h-2q+2}{2}u\right)}{\pi^2\, \gamma(u)\, \gamma\left(\frac{h+2q-2}{2}u\right)\, \gamma((h-q)u))\, \gamma((h+1)u)} \tag{20.9.8}$$

where h is the Coxeter number of the algebra $\mathcal{G}$, $\gamma(x)$ is given by $\gamma(x) = \Gamma(x)/\Gamma(1-x)$ and

$$k(\mathcal{G}) = \left(\prod_{i=1}^{r} q_i^{q_i} \right)^{1/2h}, \qquad u = \frac{1}{k+h+1}.$$

The quantities q_i are the integer numbers that enter the definition of the maximal root of the simply laced algebra, see eqn. (16.6.2), whereas $q = \max q_i$. For the ADE we have $q(A) = 1$, $q(D) = 2$, $q(E_6) = 3$, $q(E_7) = 4$, $q(E_8) = 6$.

20.10 Examples

This section uses the TBA associated to several integrable models to compute the effective central charge that emerges in the ultraviolet limit. As we will see, this gives a strong support to the S-matrix description of the off-critical deformations.

20.10.1 Yang–Lee

Beside the free theories, that will be discussed in Section 20.11, the simplest TBA system is provided by the Yang–Lee S-matrix presented in Chapter 18. In this case the kernel is given by

$$\varphi(\theta) = -\sqrt{3}\left(\frac{1}{2\cosh\theta + 1} + \frac{1}{2\cosh\theta - 1} \right). \tag{20.10.1}$$

The bulk energy term is then

$$\mathcal{E}_0 = \frac{m^2}{2 \sin \frac{2\pi}{3}}.$$

To discuss the conformal limit of this scattering theory we need first to find the plateau value of the pseudo-energy, solution of the transcendental equation

$$\epsilon_0 = \log\left(1 + e^{-\epsilon_0}\right). \tag{20.10.2}$$

Taking the exponential of both terms and posing $x = e^{\epsilon_0}$, it reduces to the algebraic equation

$$x^2 - x - 1 = 0 \qquad \epsilon_0 = \log\left(\frac{\sqrt{5}+1}{2}\right)$$

and, because

$$L\left(\frac{2}{3+\sqrt{5}}\right) = \frac{\pi^2}{15}$$

for the effective central charge we get the value

$$\tilde{c}(0) = \frac{2}{5}. \tag{20.10.3}$$

Notice that for the Yang–Lee conformal model, $c = -22/5$ while $\Delta_{min} = -1/5$. The effective central charge is then $c_{eff} = c - 24\Delta_{min} = 2/5$, in agreement with the value above. The exact relation between the coupling constant λ of the field that perturbs the conformal theory is given by $i\lambda = \mathcal{D} m^{12/5}$ with

$$\mathcal{D} = -\frac{25}{12}\left(\frac{\pi}{12}\right)^{1/5}\left(\frac{\Gamma\left(\frac{5}{6}\right)}{\Gamma\left(\frac{1}{3}\right)}\right)^{12/5}\left(\frac{\Gamma\left(\frac{2}{5}\right)\Gamma\left(\frac{6}{5}\right)}{\Gamma\left(-\frac{1}{5}\right)\Gamma\left(\frac{3}{5}\right)}\right)^{1/2} = 0.09704845...i. \tag{20.10.4}$$

Note the explicit presence of the imaginary number i for the non-unitarity of the model.

20.10.2 The Ising Model in a Magnetic Field

The S-matrix proposed for the Ising model in an external magnetic field involves particles of different masses, and the S-matrix amplitudes of the lowest particle is (with the notation of Chapter 17)

$$S_{11}(\theta) = f_{2/3}(\theta) f_{2/5}(\theta) f_{1/15}(\theta).$$

This amplitude determined the bulk energy term

$$\mathcal{E}_0 = \frac{m_1^2}{2(\sin\frac{2\pi}{3} + \sin\frac{2\pi}{5} + \sin\frac{\pi}{15})}.$$

From the exact expressions of all other amplitudes, given in Chapter 18, we can determine the N matrix

$$N_{ab} = \begin{pmatrix} 3 & 4 & 6 & 6 & 8 & 8 & 10 & 12 \\ 4 & 7 & 7 & 10 & 12 & 14 & 16 & 20 \\ 6 & 8 & 11 & 12 & 16 & 16 & 20 & 24 \\ 6 & 10 & 12 & 15 & 18 & 20 & 24 & 30 \\ 8 & 12 & 16 & 18 & 23 & 24 & 30 & 36 \\ 8 & 14 & 16 & 20 & 24 & 27 & 32 & 40 \\ 10 & 16 & 20 & 24 & 30 & 32 & 39 & 48 \\ 12 & 20 & 24 & 30 & 36 & 40 & 48 & 59 \end{pmatrix}.$$

This permits to derive the plateu values of the pseudo-energy solving eqn. (20.6.4)

$$\begin{array}{lll} e^{\epsilon_1} = 2 + 2\sqrt{2} & e^{\epsilon_2} = 5 + 4\sqrt{2} & e^{\epsilon_3} = 11 + 8\sqrt{2} \\ e^{\epsilon_4} = 16 + 12\sqrt{2} & e^{\epsilon_5} = 42 + 30\sqrt{2} & e^{\epsilon_6} = 56 + 40\sqrt{2} \\ e^{\epsilon_7} = 152 + 108\sqrt{2} & e^{\epsilon_8} = 543 + 384\sqrt{2}. & \end{array} \tag{20.10.5}$$

Computing the dilogarithmic functions associated to these values, we have

$$\begin{array}{lll} \tilde{c}_1 = 0.2100096.. & \tilde{c}_2 = 0.120269.. & \tilde{c}_3 = 0.068324.. \\ \tilde{c}_4 = 0.0500483.. & \tilde{c}_5 = 0.023056.. & \tilde{c}_6 = 0.018087.. \\ \tilde{c}_7 = 0.0076889.. & \tilde{c}_8 = 0.002515.. & \end{array} \tag{20.10.6}$$

whose sum is $\tilde{c}(0) = \frac{1}{2}$. This is a highly non-trivial check of the validity of the S-matrix proposed for the magnetic deformation of the Ising model. The exact relation that links the lowest mass m_1 of this model to the coupling constant, given in this case by the magnetic field, is $m_1 = C\,h^{\frac{8}{15}}$ where

$$C = \frac{4\sin\frac{\pi}{5}\Gamma\left(\frac{1}{5}\right)}{\Gamma\left(\frac{2}{3}\right)\Gamma\left(\frac{8}{15}\right)}\left(\frac{4\pi^2\Gamma\left(\frac{3}{4}\right)\Gamma^2\left(\frac{13}{16}\right)}{\Gamma\left(\frac{1}{4}\right)\Gamma^2\left(\frac{3}{16}\right)}\right)^{\frac{4}{5}} = 4.40490858... \tag{20.10.7}$$

20.10.3 The Tricritical Ising Model

The thermal deformation of the TIM is described by an exact S-matrix based on seven particles, where the amplitude of the fundamental particle is

$$S_{11} = -f_{1/9}(\theta) f_{4/9}(\theta).$$

Hence, the bulk energy term is

$$\mathcal{E}_0 = \frac{m_1^2}{2(\sin\frac{\pi}{9}\,\sin\frac{4\pi}{9})}.$$

From the other scattering amplitudes we can obtain the N matrix of this model

$$N_{ab} = \begin{pmatrix} 2 & 2 & 3 & 4 & 4 & 5 & 6 \\ 2 & 3 & 4 & 4 & 6 & 6 & 8 \\ 3 & 4 & 6 & 6 & 8 & 9 & 12 \\ 4 & 4 & 6 & 7 & 8 & 10 & 12 \\ 4 & 6 & 8 & 8 & 11 & 12 & 16 \\ 5 & 6 & 9 & 10 & 12 & 14 & 18 \\ 6 & 8 & 12 & 12 & 16 & 18 & 23 \end{pmatrix}.$$

The solutions of the plateau equations of the pseudo-energies are

$$\begin{array}{lll} e^{\epsilon_1} = 2+\sqrt{5} & e^{\epsilon_2} = (5+3\sqrt{5})/2 & e^{\epsilon_3} = 6+3\sqrt{5} \\ e^{\epsilon_4} = 8+4\sqrt{5} & e^{\epsilon_5} = (33+15\sqrt{15})/2 & e^{\epsilon_6} = 27+12\sqrt{5} \\ e^{\epsilon_7} = 80+36\sqrt{5}. & & \end{array} \tag{20.10.8}$$

Computing the dilogarithmic functions at these values, we get

$$\begin{array}{lll} \tilde{c}_1 = 0.228828.. & \tilde{c}_2 = 0.184429.. & \tilde{c}_3 = 0.1054611.. \\ \tilde{c}_4 = 0.084686.. & \tilde{c}_5 = 0.049684.. & \tilde{c}_6 = 0.0335404.. \\ \tilde{c}_7 = 0.013369.. & & \end{array} \tag{20.10.9}$$

whose sum gives $\tilde{c}(0) = 7/10$, that is, the central charge of the TIM. This provides an explicit confirmation of the validity of the S-matrix for the thermal deformation of this model. The exact relation between the lowest mass m_1 and the coupling constant $\tau = T - T_c$ is $m_1 = \mathcal{C}\tau^{\frac{5}{9}}$ where

$$\mathcal{C} = \left(\frac{2\Gamma\left(\frac{2}{9}\right)}{\Gamma\left(\frac{2}{3}\right)\Gamma\left(\frac{5}{9}\right)}\right)\left(\frac{4\pi^2\,\Gamma\left(\frac{2}{5}\right)\Gamma^3\left(\frac{4}{5}\right)}{\Gamma^3\left(\frac{1}{5}\right)\Gamma\left(\frac{3}{5}\right)}\right)^{5/18} = 3.745378362... \tag{20.10.10}$$

20.11 Thermodynamics of the Free Field Theories

A particularly simple example of a TBA equation is associated to the theories where there is only one massive excitation with a constant S-matrix $S = \pm 1$. In these theories it is not necessary to solve the integral equations (20.4.15) to derive the thermodynamics since we have identically

$$\epsilon(\theta) = r\cosh\theta, \tag{20.11.1}$$

and, for the central charge

$$c_{\pm}(r) = \mp\frac{6}{\pi^2}\int_0^{\infty} d\theta\, r\cosh\theta\,\log(1 \mp e^{-r\cosh\theta}). \tag{20.11.2}$$

Apart from the prefactor $-\pi/6R^2$, these expressions are precisely the free energies of a relativistic gas with Bose/Fermi statistics at temperature $T = 1/R$ (see Appendix 1.B). discuss their analytic structure. Expanding the logarithm in powers of $\exp[-r\cosh\theta]$ and integrating term by term, we get

$$c_{\pm}(r) = \frac{6r}{\pi^2}\sum_{n=1}^{\infty}\frac{(\pm 1)^{n-1}}{n}K_1(nr), \tag{20.11.3}$$

where $K_1(z)$ is the modified Bessel function. Taking now the limit $r \to 0$, we have

$$c_{\pm}(0) = \frac{6}{\pi^2}\sum_{n=1}^{\infty}\frac{(\pm 1)^{n-1}}{n^2} = \begin{cases} 1 \\ \frac{1}{2} \end{cases}. \tag{20.11.4}$$

Moreover, using

$$\frac{d}{dx}[xK_1(x)] = -xK_0(x),$$

we obtain

$$\frac{1}{r}\frac{d}{dr}c_{\pm}(r) = -\frac{6}{\pi^2}\sum_{n=1}^{\infty}(\pm 1)^{n-1}K_0(nr). \tag{20.11.5}$$

Using the identities

$$\sum_{n=1} K_0(nx)\cos nxt = \frac{1}{2}\left(\gamma_E + \log\frac{x}{4\pi}\right) + \frac{\pi}{2x\sqrt{1+t^2}} +$$
$$\frac{\pi}{2}\sum_{l=1}^{\infty}\left\{\frac{1}{\sqrt{x^2+(2l\pi - tx)^2}} - \frac{1}{2l\pi}\right\} + \frac{\pi}{2}\sum_{l=1}^{\infty}\left\{\frac{1}{\sqrt{x^2+(2l\pi + tx)^2}} - \frac{1}{2l\pi}\right\}$$

$$\sum_{n=1}(-1)^n K_0(nx)\cos nxt = \frac{1}{2}\left(\gamma_E + \log\frac{x}{4\pi}\right) +$$
$$+\frac{\pi}{2}\sum_{l=1}^{\infty}\left\{\frac{1}{\sqrt{x^2+(2l\pi - tx)^2}} - \frac{1}{2l\pi}\right\} + \frac{\pi}{2}\sum_{l=1}^{\infty}\left\{\frac{1}{\sqrt{x^2+(2l\pi + tx)^2}} - \frac{1}{2l\pi}\right\}$$

and integrating eqn. (20.11.5), we arrive at the expressions: for the bosonic case $S = 1$

$$c_+(r) = 1 - \frac{3r}{\pi} + \frac{3r^2}{2\pi^2}\left[\log\frac{1}{r} + \frac{1}{2} + \log 4\pi - \gamma_E\right] + \qquad (20.11.6)$$
$$-\frac{6}{\pi}\sum_{n=1}^{\infty}\left(\sqrt{(2n\pi)^2 + r^2} - 2n\pi - \frac{r^2}{4n\pi}\right)$$

while, for the fermionic case $S = -1$

$$c_-(r) = \frac{1}{2} - \frac{3r^2}{2\pi^2}\left[\log\frac{1}{r} + \frac{1}{2} + \log\pi - \gamma_E\right] + \qquad (20.11.7)$$
$$+\frac{6}{\pi}\sum_{n=1}^{\infty}\left(\sqrt{(2n-1)^2\pi^2 + r^2} - (2n-1)\pi - \frac{r^2}{2(2n-1)\pi}\right).$$

Figure 20.4 shows the plot of these functions.

20.12 *L*-channel Quantization

The formulae obtained by the TBA for the finite volume vacuum energy of free theories can be directly derived by quantizing them in the *L*-channel. This section presents the explicit formula for the bosonic case, since similar expressions can be easily reproduced for the fermionic case. Let $\phi(x,t) = \phi^\dagger(x,t)$ be the real bosonic field defined in the interval $(-\frac{R}{2}, \frac{R}{2})$, with periodic boundary conditions

$$\phi(x+R,t) = \phi(x,t) \qquad (20.12.1)$$

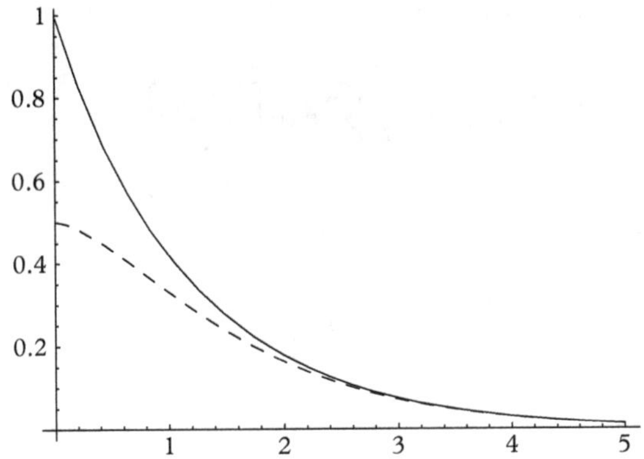

Fig. 20.4 *Plot of the functions $c_+(r)$ (continuous line) and $c_-(r)$ (dashed line).*

at any time. The action is

$$\mathcal{A} = \int dt \int_{-\frac{R}{2}}^{\frac{R}{2}} dx \frac{1}{2}\left[(\partial_\mu \phi)^2 - m^2\phi^2\right].$$

Defining the conjugate momentum of the field

$$\Pi(x,t) = \frac{\partial \phi}{\partial t}(x,t),$$

for the Hamiltonian of the system we have

$$H = \frac{1}{2}\int_{-\frac{R}{2}}^{\frac{R}{2}} dx[\Pi^2 + (\nabla\phi)^2 + m^2\phi^2]. \tag{20.12.2}$$

Let us assume the commutation relations

$$\begin{aligned} [\phi(x,t), \Pi(y,t)] &= i\delta_p(x-y), \\ [\phi(x,t), \phi(y,t)] &= [\Pi(x,t), \Pi(y,t)] = 0, \end{aligned} \tag{20.12.3}$$

where $\delta_p(x-y)$ is the periodic version of the usual Dirac delta function: in addition to the usual properties, in this case it also satisfies

$$\delta_p(x+R) = \delta_p(x).$$

Its explicit representation is given by

$$\delta_p(x) = \frac{1}{R}\sum_{-\infty}^{\infty} \exp\left[\frac{2\pi i n}{R} x\right]. \tag{20.12.4}$$

It is now necessary to solve the equation of motion of the field $\phi(x,t)$

$$\left[\frac{\partial^2}{\partial t^2} - \nabla^2 + m^2\right]\phi(x,t) = 0, \tag{20.12.5}$$

together with the boundary conditions (20.12.1). There is a standard procedure to do so: because the field is periodic along the space direction, it admits a Fourier expansion

$$\phi(x,t) = \sum_{-\infty}^{\infty} c_n(t) \exp\left[\frac{2\pi n i}{R} x\right]. \tag{20.12.6}$$

It is convenient to introduce the notation

$$p_n = \frac{2\pi n}{R}, \quad \omega_n = \sqrt{p_n^2 + m^2}, \quad n = 0, \pm 1, \pm 2 \ldots.$$

Substituting the expansion (20.12.6) in the equation of motion (20.12.5), we obtain

$$\left[\frac{d^2}{dt^2} + \omega_n^2\right] c_n(t) = 0,$$

whose solution is

$$c_n(t) = a_n e^{-i\omega_n t} + a_n^{\dagger} e^{i\omega_n t}.$$

Hence, the field and its conjugate momentum are expressed by

$$\phi(x,t) = \sum_{-\infty}^{+\infty} N_n \left[a_n e^{i(p_n x - \omega_n t)} + a_n^{\dagger} e^{-i(p_n x - \omega_n t)}\right], \tag{20.12.7}$$

$$\Pi(x,t) = \sum_{-\infty}^{+\infty} N_n (-i\omega_n) \left[a_n e^{i(p_n x - \omega_n t)} - a_n^{\dagger} e^{-i(p_n x - \omega_n t)}\right],$$

where N_n is a normalization that can be fixed by imposing the quantization conditions (20.12.3). Choosing

$$N_n = \frac{1}{\sqrt{2\omega_n R}},$$

eqn. (20.12.3) becomes the commutation relation among the a_n modes

$$\left[a_n, a_m^\dagger\right] = \delta_{n,m}, \qquad [a_n, a_m] = \left[a_n^\dagger, a_m^\dagger\right] = 0. \tag{20.12.8}$$

Substituting the expressions of $\phi(x,t)$ and $\Pi(x,t)$ in the Hamiltonian (20.12.2), we have

$$H = \frac{1}{2}\sum_{-\infty}^{+\infty} \omega_n (a_n a_n^\dagger + a_n^\dagger a_n) = \frac{1}{2}\sum_{-\infty}^{+\infty} \omega_n \left[a_n^\dagger a_n + \frac{1}{2}\right], \tag{20.12.9}$$

where we used

$$\int_{-\frac{R}{2}}^{\frac{R}{2}} e^{i(p_n - p_m)x}\, dx = R\delta_{n,m}.$$

The ground state energy of the theory is then

$$E_0(R) = \frac{1}{2}\sum_{-\infty}^{+\infty} \omega_n = \frac{1}{2}\sum_{-\infty}^{+\infty} \sqrt{\left(\frac{2n\pi}{R}\right)^2 + m^2}. \tag{20.12.10}$$

This expression needs however to be regularized by subtracting the term coming from the continuous limit of the infinite volume in order to implement the correct normalization

$$\lim_{R\to\infty} \mathcal{E}_0^{\text{vac}}(R) = 0.$$

Hence, for the finite volume ground state energy we have

$$\mathcal{E}_0^{\text{vac}}(R) = \frac{1}{2}\sum_{n=-\infty}^{\infty} \sqrt{\left(\frac{2\pi n}{R}\right)^2 + m^2} - \frac{1}{2}\int_{-\infty}^{\infty} dn \sqrt{\left(\frac{2\pi n}{R}\right)^2 + m^2}. \tag{20.12.11}$$

Selecting out the zero mode, this expression can be written as

$$\mathcal{E}_0^{\text{vac}}(R) = \frac{m}{2} + \frac{2\pi}{R}\sum_{n=1}^{\infty} \sqrt{n^2 + \left(\frac{r}{2\pi}\right)^2} - \frac{2\pi}{R}\int_0^{\infty} dn \sqrt{n^2 + \left(\frac{r}{2\pi}\right)^2}, \tag{20.12.12}$$

where $r \equiv mR$. Since the divergence of the series is due to the large n behaviour of the first two terms of the expansion

$$\sqrt{n^2 + \left(\frac{r}{2\pi}\right)^2} \simeq n + \frac{1}{2}\left(\frac{r}{2\pi}\right)^2 \frac{1}{n} + \mathcal{O}\left(\frac{1}{n^2}\right),$$

start by subtracting and adding these divergent terms

$$S(r) \equiv \sum_{n=1}^{\infty} \sqrt{n^2 + \left(\frac{r}{2\pi}\right)^2} = \sum_{n=1}^{\infty} \left\{ \sqrt{n^2 + \left(\frac{r}{2\pi}\right)^2} - n - \frac{1}{2}\left(\frac{r}{2\pi}\right)^2 \frac{1}{n} \right\} + \\ + \sum_{n=1}^{\infty} n + \frac{1}{2}\left(\frac{r}{2\pi}\right)^2 \sum_{n=1}^{\infty} \frac{1}{n}. \tag{20.12.13}$$

The first series on the right is now convergent, while the last two terms must be paired with analogous terms coming from the integral, whose divergence has to be treated in a similar way of the series. Hence, subtracting and adding these divergent terms in the integral

$$I(r) \equiv \int_0^{\infty} dn \sqrt{n^2 + \left(\frac{r}{2\pi}\right)^2} = \\ = \int_0^{\infty} dn \left\{ \sqrt{n^2 + \left(\frac{r}{2\pi}\right)^2} - n \right\} + \int_0^{\infty} dn\, n, \tag{20.12.14}$$

we can pair the last term of this expression with the one in (20.12.13) and implement the well-known regularization

$$\sum_{n=0}^{\infty} n - \int_0^{\infty} n\, dn = \lim_{\alpha \to 0} \left[\sum_{n=0}^{\infty} n e^{-\alpha n} - \int_0^{\infty} n e^{-\alpha n}\, dn \right] = -\frac{1}{12}. \tag{20.12.15}$$

However, the first term in (20.12.14) still contains a logarithmic divergence, as it can be seen explicitly computing the integral by means of a *cut-off* Λ, in the limit $\Lambda \to \infty$

$$\int_0^{\Lambda} dn \left\{ \sqrt{n^2 + \left(\frac{r}{2\pi}\right)^2} - n \right\} = \frac{1}{2}\left(\frac{r}{2\pi}\right)^2 \ln 2\Lambda + \frac{1}{4}\left(\frac{r}{2\pi}\right)^2 - \frac{1}{2}\left(\frac{r}{2\pi}\right)^2 \ln \frac{r}{2\pi}. \tag{20.12.16}$$

This divergence is fixed by subtracting and adding the term $\frac{1}{2}\left(\frac{r}{2\pi}\right)^2 \ln \Lambda$. Combining this last term with the analogous one coming from the series, we obtain

$$\lim_{\Lambda\to\infty}\left(\sum_{n=1}^{\Lambda}\frac{1}{n}-\ln\Lambda\right)=\gamma_E,$$

where γ_E is the Euler–Mascheroni constant, whereas the remaining part of (20.12.16), with the subtraction done above, is now finite.

Gathering together all the terms, the finite expression of the finite volume ground state energy is

$$\mathcal{E}_0^{\text{vac}}(R)=\frac{1}{R}\left[-\frac{\pi}{6}+\frac{r}{2}+\frac{r^2}{4\pi}\left(\ln\frac{r}{4\pi}+\gamma_E-\frac{1}{2}\right)+\sum_{n=1}^{\infty}\left(\sqrt{(2\pi n)^2+r^2}-2\pi n-\frac{r^2}{4\pi n}\right)\right]. \tag{20.12.17}$$

It is easy to see that eqn. (20.12.17) satisfies the modular invariance, that imposes its equality with the expression obtained by the TBA

$$\mathcal{E}_0^{\text{vac}}(R)=-\frac{\pi c(r)}{6R},$$

where

$$c(r)=-\frac{6r}{\pi^2}\int_0^{\infty}d\theta\,\cosh\theta\,\ln\left(1-e^{-r\cosh\theta}\right).$$

Moreover, it is easy to verify that the regularization adopted above ensures the perfect agreement between the expressions for the one-point correlation functions $\langle\phi^{2k}\rangle$, done either in the R or the L channels.

It is useful to notice that the result (20.12.17) can be obtained in a simplest way by using a prescription that automatically ensures the subtraction of the various divergent terms. This consists of ignoring completely the divergent part of the integral, keeping its finite part and regularizing the divergent series according to the formulae

$$\left.\sum_{n=1}^{\infty} n\right|_{\text{reg}}=-\frac{1}{12}, \tag{20.12.18}$$

$$\left.\sum_{n=1}^{\infty}\frac{1}{n}\right|_{\text{reg}}=\gamma_E+\ln\frac{r}{2\pi}. \tag{20.12.19}$$

Eqn. (20.12.18) is the standard regularization provided by the Riemann-zeta function $\zeta(-1)$, where $\zeta(s)=\sum_{n=1}^{\infty}\frac{1}{n^s}$. This regularization corresponds to the normal order of the operators in the infinite volume. However, for the logarithmic divergence, the regularization of the second series is intrinsically ambiguous, and it finite value can be determined according to the earlier discussion.

20.13 LeClair–Mussardo formula

Many applications, especially in condensed matter physics, employ correlation functions at finite temperature. These quantities can be viewed as correlation functions on an infinite cylindrical geometry, where the spacial coordinate $-\infty < x < \infty$ runs along the length of the cylinder and the Euclidean time lives on a circle of radius $R = 1/T$, where T is the temperature. This is precisely the picture we used in deriving the thermodynamics of integrable models using the Bethe ansatz equation, where the Hamiltonian evolution is along the circumference of the cylinder while the Hilbert space lives on the infinite line of the coordinate x. Hence, the states of the Hilbert space to sum over in performing thermodynamic averages have the usual multi-particle description in infinite volume: this implies that the matrix elements of operators do not depend on R and therefore they can be expressed by the usual temperature-independent FFs (see Chapter 19). Formally,

$$\langle \mathcal{O} \rangle_R = Z^{-1} \mathrm{Tr}\left(e^{-RH} \mathcal{O}\right) \qquad (\mathrm{R-channel}) \tag{20.13.1}$$

where Z is the partition function. In the case of one-point functions at finite temperature of integrable models, there exists a formula, known as LeClair–Mussardo (LM) formula, which allows us to express these quantities in terms of a special limit of the form factors and the pseudo-energy $\epsilon(\theta)$ entering the TBA. The LM formula reads

$$\langle \mathcal{O} \rangle_R \equiv \sum_{n=0}^{\infty} \frac{1}{n!} \int_{-\infty}^{\infty} \left(\prod_{i=1}^{n} \frac{d\theta_i}{2\pi} f(\theta_i) \right) \langle \overleftarrow{\theta} | \mathcal{O}(0,0) | \overrightarrow{\theta} \rangle_{\mathrm{conn}}, \tag{20.13.2}$$

where $|\overrightarrow{\theta}\rangle \equiv |\theta_1, \ldots, \theta_n\rangle$ ($\langle \overleftarrow{\theta}| \equiv \langle \theta_n, \ldots, \theta_1|$) denotes the asymptotic multi-particle states of the Integrable QFT expressed in terms of the rapidities θ_i, with relativistic dispersion relation $E(\theta) = m\cosh\theta$, $p(\theta) = m\sinh\theta$. The LM formula employs the connected diagonal FF of the operator $\mathcal{O}$, which are *finite* functions of the rapidities defined as

$$\begin{aligned} \langle \overleftarrow{\theta} | \mathcal{O} | \overrightarrow{\theta} \rangle_{\mathrm{conn}} &\equiv F^{\mathcal{O}}_{2n,conn}(\theta_1, \ldots \theta_n) \\ &= FP\left(\lim_{\eta_i \to 0} \langle 0 | \mathcal{O} | \overrightarrow{\theta}, \overleftarrow{\theta} - i\pi + i\overleftarrow{\eta} \rangle \right) \end{aligned} \tag{20.13.3}$$

where $\overleftarrow{\eta} \equiv (\eta_n, \ldots, \eta_1)$ and FP in front of the expression means taking its finite part, i.e. omitting all the terms of the form η_i/η_j and $1/\eta_i^p$ where p is a positive integer, in taking the limit $\overleftarrow{\eta} \to 0$ in the matrix element given above. Formula (20.13.2) also employs the filling factor $f(\theta)$ of the one-particle state given (for fermionic-type S-matrix) by

$$f(\theta_i) = \frac{1}{1+e^{\epsilon(\theta_i)}}, \tag{20.13.4}$$

where the pseudo-energy $\epsilon(\theta)$ is solution of the TBA equation.

Let us sketch the derivation of the LM formula; interested readers are referred to the for further details. First, we evaluate the traces present in eqn. (20.13.1) in terms of the particle basis, so that

$$\langle \mathcal{O}\rangle_R = Z^{-1}\sum_{n=0}\frac{1}{n!}\int\prod_{i=1}^{n}\left(\frac{d\theta_i}{2\pi}\right)\langle\theta_n\ldots\theta_1|\mathcal{O}(0)||\theta_1\ldots\theta_n\rangle. \tag{20.13.5}$$

As it stands, however, this expression is highly problematic: all terms of the sum in the numerator as well as those present in Z are in fact *divergent*. As shown in Section 20.5, in the latter the divergencies come from the normalization of the eigenstates, $\langle\theta'_m,\ldots\theta'_1|\theta_1\ldots\theta_n\rangle = \prod_i\delta(\theta'-\theta)$, which gives rise to $[\delta(0)]^n$ when $\theta'_i=\theta$. In the former, the divergencies come from the FF $\langle\theta'_m\ldots\theta'_1|\mathcal{O}(0)||\theta_1\ldots\theta_n\rangle$, once evaluated at $\theta'_i=\theta_i$. These divergencies are an unavoidable consequence of the kinematical pole structure of the FFs.

The first cure of the divergencies is to define the theory on a large but finite interval L so that the rapidities of the n-particle states entering (20.13.5) are solutions of the Bethe equation (20.3.3). Let us denote the finite volume eigenstates associated with the integers $\{n_i\}$ as $|\theta_1,\ldots,\theta_n\rangle_L$: the corresponding density of states is given by the Jacobian $\mathcal{J}(\theta_1,\ldots,\theta_n)=\det\mathcal{J}_{jk}$, with $\mathcal{J}_{jk}=\frac{\partial \mathcal{J}_j}{\partial\theta_k}$ and the functions $\mathcal{J}_i$ given by the derivative of the right-hand side of (20.3.3)

$$\mathcal{J}_i(\theta_1,\ldots,\theta_n) = mL\cosh\theta_i + \sum_{h\neq i}\varphi(\theta_i-\theta_k). \tag{20.13.6}$$

We now state, without proving, the relation between the diagonal FFs in finite volume and the infinite-volume connected defined in eqn.(20.13.3): this relation reads

$$\begin{aligned}\langle\theta_1,\ldots,\theta_n|\mathcal{O}|\theta_1,\ldots,\theta_n\rangle_{\mathrm{L}} &= \frac{1}{\mathcal{J}_n(\theta_1,\ldots,\theta_n)}\\ \times\sum_{\{\theta_+\}\bigcup\{\theta_-\}} & F^{\mathcal{O}}_{2l,\mathrm{conn}}(\theta_-)\,\overline{\mathcal{J}}_{n-l}(\{\theta_+\},\{\theta_-\}),\end{aligned} \tag{20.13.7}$$

where the sum runs on all possible bipartite partitions of the set of rapidities $\{\theta_1,\ldots,\theta_n\}$ in two disjoint sets made by l and $n-l$ rapidities, and $\overline{\mathcal{J}}_{n-l}(\{\theta_+\},\{\theta_-\})=\det\mathcal{J}_+$ is the restricted determinant of the sub-matrix $\mathcal{J}_+$ corresponding to the particles in the set $\{\theta_+\}$ in the presence of those in $\{\theta_-\}$. Notice that the relation (20.13.7) involves the kernel $\varphi(\theta)$ of the Bethe ansatz eqns. (20.13.6). Equivalently, the relation (20.13.7) can be written as

$$\langle \theta_1, \ldots, \theta_n | \mathcal{O} | \theta_1, \ldots, \theta_n \rangle_{\rm L} = \frac{1}{\mathcal{J}_n(\theta_1, \ldots, \theta_n)} \sum_{\{\theta_+\} \bigcup \{\theta_-\}} F_{2l,\rm sym}(\theta_-)\, \mathcal{J}_{n-l}(\theta_+), \qquad (20.13.8)$$

where, as before, the sum runs on all possible bipartite partitions of the set of rapidities $\{\theta_1, \ldots, \theta_n\}$ in two disjoint sets made by l and $n-l$ rapidities, while $F_{2l,\rm sym}(\theta_1, \ldots, \theta_l)$ are finite functions defined by the *symmetric limit*

$$\begin{aligned} F_{2n,\rm sym}(\theta_1, \ldots, \theta_n) &= \langle \theta_1, \ldots, \theta_n | \mathcal{O} | \theta_1, \ldots, \theta_n \rangle_{\rm sym} \qquad (20.13.9) \\ &= \left(\lim_{\eta \to 0} \langle 0 | \mathcal{O} | \theta_1 + i\pi + i\eta, \ldots, \theta_n + i\pi + i\eta, \theta_1, \ldots, \theta_n \right) \rangle. \end{aligned}$$

Notice that, while eqn.(20.13.7) employs $\overline{\mathcal{J}}_{n-l}(\{\theta_+\}, \{\theta_-\})$ that contains information both on $\{\theta_+\}$ and its complementary set $\{\theta_-\}$, eqn.(20.13.8) instead employs the density of states $\mathcal{J}_{n-l}(\theta_+)$.

Since for any local operator $\mathcal{O}$ its $F_2^{\mathcal{O}}(\theta)$ is a constant, $F_2 = \langle \theta | \mathcal{O}(0,0) | \theta \rangle$ and therefore $F_2 = F_{2,\rm conn} = F_{2,\rm sym}$. Expressing now $F_{2n,\rm sym}$ in terms of $F_{2n,\rm conn}$, for the next few cases we have

$$F_{4,\rm sym}(\theta_1, \theta_2) = F_{4,\rm conn}(\theta_1, \theta_2) + 2\varphi(\theta_1 - \theta_2) F_{2,\rm conn};$$

$$\begin{aligned} F_{6,\rm sym}(\theta_1, \theta_2, \theta_3) &= F_{6,\rm conn}(\theta_1, \theta_2, \theta_3) + \left[F_{4,\rm conn}(\theta_1, \theta_2) (\varphi(\theta_1 - \theta_3) + \varphi(\theta_2 - \theta_3)) + \right. \\ &\text{permutations}] + 3F_{2,\rm conn}\, [\varphi(\theta_1 - \theta_2)\varphi(\theta_1 - \theta_3) + \text{permutations}]. \end{aligned}$$

It is then clear that the finite-volume diagonal matrix elements (20.13.8) are expressed in terms of the infinite-volume connected FFs $F_{2l,\rm conn}(\theta_1, \ldots, \theta_l)$ and the kernel $\varphi(\theta)$ of the Bethe ansatz equations (20.13.6).

With all the information collected above, coming back to the thermal average (20.13.5), with the regularization given by the finite interval L its right-hand side can be written in terms of the LM formula as

$$\begin{aligned} \langle \mathcal{O} \rangle_R &= \lim_{L \to \infty} \frac{\mathrm{Tr}\left(e^{-\mathcal{H}} \mathcal{O}\right)_L}{\left(\mathrm{Tr} e^{-\mathcal{H}}\right)_L} = \\ &= \sum_{n=0}^{\infty} \frac{1}{n!} \int_{-\infty}^{\infty} \left(\prod_{i=1}^{n} \frac{d\theta_i}{2\pi} f(\theta_i) \right) \langle \overleftarrow{\theta} | \mathcal{O}(0,0) | \overrightarrow{\theta} \rangle_{\rm conn}. \qquad (20.13.10) \end{aligned}$$

We examine in detail the low and high–temperature limits of the one-point functions. First notice that if the anomalous dimension of the operator $\mathcal{O}$ is given by $2\Delta_{\mathcal{O}}$, its vacuum expectation value may be expressed as $\langle 0 | \mathcal{O} | 0 \rangle \equiv \langle \mathcal{O} \rangle_0 = Y m^{2\Delta_{\mathcal{O}}}$, with Y a pure number. Hence, the series (20.13.10) relative to its one-point function at finite temperature can be cast in the scaling form

$$\frac{\langle \mathcal{O} \rangle_R}{\langle \mathcal{O} \rangle_0} = H(mR), \tag{20.13.11}$$

where $H(mR)$ is a function of the scaling variable mR.

In the low-temperature limit $R \to \infty$, the pseudo-energy goes as $\epsilon(\theta) \sim mR \cosh\theta$ and therefore the N-th term of the series entering $H(mR)$ vanishes asymptotically as e^{-NmR}, so that the leading correction is given by

$$H(mR) \to 1 + \frac{\langle A|\mathcal{O}|A\rangle}{\pi} K_0(mR) + \cdots \tag{20.13.12}$$

where $K_0(x)$ is the usual Bessel function, and $|A\rangle = |\theta\rangle$.

On the other hand, in the high-temperature limit $R \to 0$, the one-point correlators may become scaling invariant functions[7] i.e.

$$\langle \mathcal{O} \rangle_R \to \frac{1}{R^{\eta}}, \tag{20.13.13}$$

with a power-law exponent η ruled by the underlying CFT. This is the case, for instance, of massive integrable models obtained as deformation of a CFT by a strong relevant field. Under this hypothesis, the limit $R \to 0$ can be controlled by means of a conformal perturbation theory on the cylinder. Let

$$\mathcal{A} = \mathcal{A}_{CFT} + \lambda \int \Phi(x) d^2x \tag{20.13.14}$$

be the action of the off-critical model, where $\Phi(x)$ is the relevant operator that gives rise to the massive integrable theory, with $\lambda \sim m^{2-2\Delta_\Phi}$. The general structure of the perturbation series for the one-point functions is then given by

$$\langle \mathcal{O} \rangle_R = \sum_{n=0}^{n} d_n, \tag{20.13.15}$$

where

$$d_n = \frac{(-\lambda)^n}{n!} \int_{cyl} \langle \xi | \mathcal{O}(0) \Phi(X_1) \cdots \Phi(X_n) | \xi \rangle_{conn} d^2X_1 \cdots d^2X_n, \tag{20.13.16}$$

and X_i are points on the cylinder and the connected correlations functions are calculated in the unperturbed CFT. The field ξ entering eqn. (20.13.16) is relative to the conformal operator of lowest anomalous dimension in the theory that plays the role of the vacuum

[7] This is not always the case, because there are theories, e.g. thermal Ising model, with logarithmic behaviour that may spoil the pure power-law behaviour given in the text.

state on the cylinder (for unitary model, this field is the identity operator). Notice that the first term of the series is given by

$$\langle \xi | \mathcal{O} | \xi \rangle = \left(\frac{2\pi}{R} \right)^{2\Delta_{\mathcal{O}}} C_{\xi \mathcal{O} \xi}, \qquad (20.13.17)$$

and this will be different from zero only if the conformal structure constant $C_{\xi\mathcal{O}\xi}$ does not vanish. In this case, the exponent η in (20.13.13) will coincide with the anomalous dimension of the field $\mathcal{O}$ itself, $\eta = 2\Delta_{\mathcal{O}}$, otherwise the exponent η will be given by $\eta = 2\Delta_{\mathcal{O}} - q(2 - 2\Delta_{\Phi})$, where the integer q is the first non-vanishing term in the series (20.13.15).

The above considerations open the possibility to extract conformal data out of the high-temperature limit of the one-point function. This is easily checked by the analysis of some simple models that have only one massive particle in the spectrum.

REFERENCES

The use of the S-matrix to determine the thermodynamics of a QFT appears in

Dashen, R., Ma, S.K. and Bernstein, H.J. (1969). 'S-matrix Formulation of Statistical Mechanics', *Physics Review*, 187: 345.

The original paper on the TBA for the integral models is

Yang, C.N. and Yang, C.P. (1969). 'Thermodynamics of a One-dimensional System of Bosons with Repulsive Delta-function Interaction', *Journal of Mathematical Physics*, 10: 1115.

The generalization to the relativistic case has been proposed in

Zamolodchikov, Al.B. (1990). 'Thermodynamic Bethe Ansatz in Relativistic Models: Scaling 3-State Potts and Lee–Yang Models', *Nuclear Physics B*, 342: 695.

On the relation between the TBA and the deformed conformal theories we recommend

Klassen, T. and Melzer, E. (1990). 'Purely Elastic Scattering Theories and their Ultraviolet Limits', *Nuclear Physics B*, 338: 485.
Klassen, T. and Melzer, E. (1991). 'The Thermodynamics of Purely Elastic Scattering Theories and Conformal Perturbation Theory', *Nuclear Physics B*, 350: 635.

The relationship between the coupling constant and the lowest mass of the integrable models is examined in

Zamolodchikov, Al.B. (1995). 'Mass scale in the Sine–Gordon Model and its Reductions', *International Journal of Modern Physics A*, 10: 1125.

Fateev, V. (1994). 'The Exact Relations Between the Coupling Constants and the Masses of Particles for the Integrable Perturbed Conformal Field Theories', *Physics Letters B*, 324: 45.

The formalism based on the TBA to compute both the free energy finite temperature correlation functions in integrable models is explored in

LeClair, A. and Mussardo, G. (1999). 'Finite Temperature Correlation Functions in Integrable QFT', *Nuclear Physics B*, 552: 624.

About the LeClair–Mussardo formula, the interested reader is referred to

Kostov, I., Serban D. and Vu, D.L. (2017). 'TBA and Tree Expansion', In: 10th International Symposium on Quantum Theory and Symmetries and 12th International Workshop on Lie Theory and its Applications in Physics, vol. 2. Jun 2017, Varna, Bulgaria. Springer Proceedings in Mathematics & Statistics series, pp. 77–98.

Lukyanov, S. (2001). 'Finite Temperature Expectation Values of Local Fields in the Sinh–Gordon Model', *Nuclear Physics B*, 612: 391.

Pozsgay, B. (2011). 'Mean Values of Local Operators in Highly Excited Bethe States', *Journal of Statistical Mechanics*, 2011: P01011.

Pozsgay B. and Takacs, G. (2008). 'Form Factors in Finite Volume II: Disconnected Terms and Finite Temperature Correlators', *Nuclear Physics B*, 788: 209.

The TBA is deeply related to the monodromy matrix of integrable models. This subject has been investigated in a remarkable series of papers:

Bazhanov, V.Z., Lukyanov, S. and Zamolodchikov, A.B. (1996). 'Integrable Structure of Conformal Field Theory, Quantum KdV Theory and Thermodynamic Bethe Ansatz', *Communications in Mathematical Physics*, 177: 381.

Bazhanov, V.Z., Lukyanov, S. and Zamolodchikov, A.B. (1997). 'Integrable Structure of Conformal Field Theory 2. Q Operator and DDV Equation', *Communications in Mathematical Physics*, 190: 247.

Bazhanov, V.Z., Lukyanov, S. and Zamolodchikov, A.B. (1999). 'Integrable Structure of Conformal Field Theory 3. The Yang–Baxter Relation', *Communications in Mathematical Physics*, 200: 297.

The TBA can be also extended to compute the finite volume energies of the excited states, as discussed in

Bazhanov, V. Z., Lukyanov, S. and Zamolodchikov, A.B. (1997). 'Integrable Quantum Field Theories in Finite Volume: Excited State Energies', *Nuclear Physics B*, 489: 487.

Dorey, P.and Tateo, R. (1996). 'Excited States by Analytic Continuation of TBA Equations', *Nuclear Physics B*, 482: 639.

Fendley, P. (1992). 'Excited State Thermodynamics', *Nuclear Physics B*, 374: 667.

The interesting relationship between the TBA and the spectrum of one-dimensional quantum mechanical systems is discussed in

Bazhanov, V.Z., Lukyanov, S. and Zamolodchikov, A.B. (2001). 'Spectral Determinants for Schrödinger Equation and Q-Operators of Conformal Field Theory', *Journal of Statistical Physics*, 102: 567.

Dorey P. and Tateo, R. (1999). 'Anharmonic Oscillators, the Thermodynamic Bethe Ansatz, and Nonlinear Integral Equations', *Journal of Physics A*, 32: L419.

The TBA equation has been also applied to study the cascade of RG flows $\mathcal{M}_p \to \mathcal{M}_{p-1}$ between the minimal models of CFF.

Zamolodchikov, Al.B. (2006). 'Resonance Factorized Scattering and Roaming Trajectories', *Journal of Physics A*, 39: 12847.

We recommend Generalization of the roaming trajectory model is discussed in

Dorey P. and Ravanini, F. (1993). 'Generalizing the Staircase Models', *Nuclear Physics B*, 406: 708.

PROBLEMS

20.1. Non-relativistic gas

Consider a one-dimensional gas of N non-relativistic bosons on an interval of length L, with two-body repulsive interaction given by a delta-function. The Hamiltonian of such a system is

$$H = -\sum_{i=1}^{N} \frac{\partial^2}{\partial x_i^2} + 2c \sum_{i>j} \delta(x_i - x_j) \quad c > 0.$$

a. Find the phase shift of the two-body scattering process and write the Bethe ansatz equations.
b. Analyse the solutions in the thermodynamical limit $N \to \infty, L \to \infty, N/L = \rho$, ρ finite.

20.2. Simple TBA system

Consider the TBA equations for a relativistic system made of one massive particle and with kernel

$$\varphi(\theta) = \frac{1}{2\pi}\delta(\theta).$$

a. Solve explicitly the equation for the pseudo-energy $\epsilon(\theta)$ and show that it is given by

$$\epsilon(\theta) = \log\left[e^{mR\cosh\theta} - 1\right].$$

b. Plot the scaling function

$$c(R) = \frac{6}{\pi^2} mR \int_0^\infty \cosh\theta \log(1 + e^{-\epsilon(\theta)})\, d\theta$$

and compute its limit at $R = 0$.

20.3. *L*-channel for Majorana fermions

Consider the Dirac action of a Majorana massive fermion on a finite volume

$$\mathcal{S} = \int dt \int_{-\frac{R}{2}}^{\frac{R}{2}} dx\, \bar{\psi}\left(i\gamma^\mu \partial_\mu - m\right)\psi.$$

Quantize this system in the canonical way and show that the finite volume ground state energy $E_0(R)$ can be written as

$$E_0(R) = -\frac{\pi\, c_-(r)}{6R},$$

where the scaling function $c_-(r)$ coincides with the expression given by the TBA.

21

Boundary Field Theory

The only boundary that exists is the one in your own mind.
Anonymous

21.1 Introduction

The QFTs far are defined in an infinite volume and are highly symmetric: CFTs and deformations thereof, for instance, are invariant under translations, rotations, boosts and inversions; CFTs themselves are also scaling invariant. As shown in previous chapters, these symmetries pose severe constraints on correlation functions and permit to study non-perturbative aspects of the corresponding QFT. Homogeneous systems, however, are in many cases just a mathematical idealization of the real physical samples which present instead boundary effects and various types of inhomogeneity or defects. It is an interesting problem in statistical mechanics to estimate the influence of all these inhomogeneities, taking into account boundary effects and boundary conditions. Correlation functions of systems with boundaries are expected to be richer and more intricate: in a CFT with boundary, for example, scalar operators can have non-zero one-point functions while we know that they are generally forbidden in a CFT defined in an infinite volume.

It is worth stressing that boundary field theories is a subject in itself, too large to be covered here in all its aspects: in $(1 + 1)$ dimensions, for instance, boundary field theories are relevant to numerous condensed matter phenomena (the most prominent example is the Kondo effect and other 'impurity' problems), to non-perturbative aspects of of D-branes and other space-time defects in string theory, and also are of great interest from the conceptual point of view of understanding RG flows beyond surface critical behaviour. Interested readers are referred to the literature quoted at the end of this chapter for mastering their details. This chapter guides the reader to the main results of Boundary CFTs (BCFTs), as developed by Cardy, and later to integrable deformations of these theories, a formalism developed by Ghoshal and Zamolodchikov and based on S-matrix amplitudes, which also includes reflection amplitudes. In both cases, a key concept consists of the so-called boundary states: these are particular non-renormalizable states of the Hilbert space of the theory, which encode the boundary conditions imposed on the field theory.

Statistical Field Theory: An Introduction to Exactly Solved Models in Statistical Physics. Second Edition. Giuseppe Mussardo, Oxford University Press (2020). © Giuseppe Mussardo. DOI: 10.1093/oso/9780198788102.001.0001

21.2 Stress-energy Tensor in Boundary CFT

As a prototype situation of a BCFT with boundary, we consider a critical system defined on the semi-infinite geometry of the upper half-plane, with the boundary taken to be along the real axis (Figure 21.1). In such a system, fields are defined only on the upper half-plane $Im\, z \geq 0$ and in order to preserve this geometry under conformal transformations $z \to f(z)$, only real analytic maps are allowed

$$\overline{f(z)} = f(\bar{z}). \tag{21.2.1}$$

As we will see later, this constraint implies that holomorphic and anti-holomorphic sectors of the theory are no longer independent. Here we introduce the correlation functions of the primary fields

$$G(z_1, \bar{z}_1, \ldots z_n, \bar{z}_n) = \langle \varphi_{\Delta_i, \bar{\Delta}_i}(z_1, \bar{z}_1) \ldots \varphi_{\Delta_n, \bar{\Delta}_n}(z_n, \bar{z}_n) \rangle, \tag{21.2.2}$$

and we reconsider the conformal Ward identity

$$\begin{aligned} &\frac{1}{2\pi i} \oint_C dz\, \epsilon(z) \left[\sum_{i=1}^{n} \left[\frac{\Delta_i}{(z - z_i)^2} + \frac{1}{z - z_i} \partial_i \right] \langle \phi_1(z_1, \bar{z}_1) \ldots \rangle - \langle T(z) \phi_1(z_1, \bar{z}_1) \ldots \rangle \right] \\ &- \frac{1}{2\pi i} \oint_C d\bar{z}\, \bar{\epsilon}(\bar{z}) \left[\sum_{i=1}^{n} \left[\frac{\bar{\Delta}_i}{(\bar{z} - \bar{z}_i)^2} + \frac{1}{\bar{z} - \bar{z}_i} \bar{\partial}_i \right] \langle \phi_1(z_1, \bar{z}_1) \ldots \rangle - \langle \bar{T}(\bar{z}) \phi_1(z_1, \bar{z}_1) \ldots \rangle \right] = 0 \end{aligned} \tag{21.2.3}$$

derived in Chapter 10. In the infinite plane, $\bar{\epsilon}(\bar{z})$ and $\epsilon(z)$ are independent functions and this leads to two equations that involve separately $T(z)$ and $\bar{T}(\bar{z})$. But, in the presence of the boundary, the anti-holomorphic coordinates $\bar{z}_i$ of the the fields have to be identified with the holomorphic coordinates $z_i^* = \bar{z}_i$ on the lower half-plane. This allows us to introduce a *mirror image* of the system on the lower half-plane (Figure 21.2), paying attention that this extension fulfils the boundary conditions. A natural requirement is

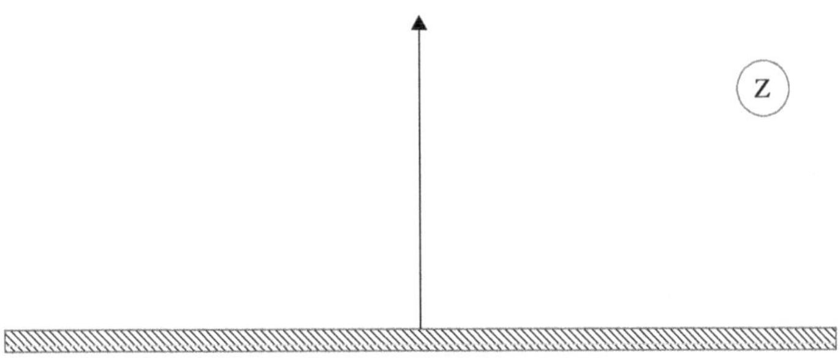

Fig. 21.1 *Upper half-plane with the boundary placed on the horizontal axis.*

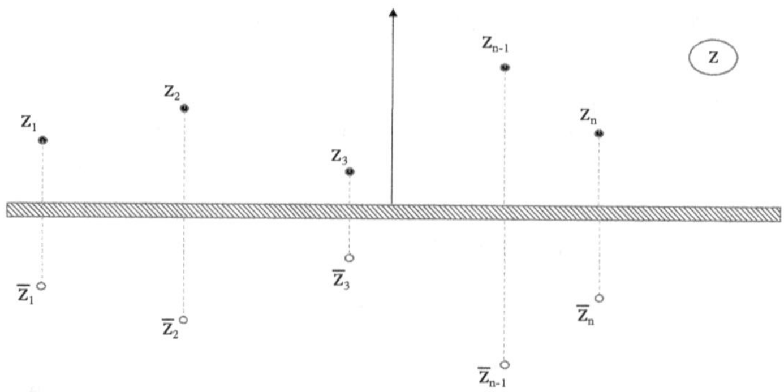

Fig. 21.2 *Original points $z_1, \ldots z_n$ and their mirror images $\bar{z}_1, \ldots, \bar{z}_n$.*

that there is no flow of energy through the boundary, a condition expressed in the original Cartesian coordinates as $T_{xy} = 0$ at the boundary and which translates in the analytic continuation of the stress-energy tensor component $T(z)$ into the lower half-plane according to the equation

$$T(z) = \bar{T}(z), \quad Im z < 0 \tag{21.2.4}$$

so that $T = \bar{T}$ at the boundary. This condition enforces that the correlators of T are those of $\bar{T}$, analytically continued into the lower half-plane. Eqn. (21.2.4) is equivalent to have conformal boundary conditions: for a given CFT in the bulk, there may be several ways of implementing it and one of the goals of BCFT is indeed to classify all possible conformal boundary conditions. We will see later how this program can be implemented.

Coming back to the conformal Ward identity, using eqn. (21.2.4), the second term in eqn. (21.2.3) is then just an integration along the mirror image $\bar{S}$ of the contour S placed on the upper half-plane (Figure 21.3): notice that the direction of the integration along the mirror contour is reversed, according to the relative minus sign present in the original Ward identity (21.2.3). Due to the identity $T = \bar{T}$ on the boundary, the horizontal part of the two contours cancels out and we can put together the two disjoint contours into a single one C which encircles both the original points and their mirror images wrt the horizontal axis. Indeed, the correlation functions (21.2.2), defined on the upper half-plane, satisfy the same differential equation as the correlators $\tilde{G}$ on the entire plane as functions of the $2n$ purely holomorphic variables $z_1, \ldots z_{2n}$, where $z_{n+i} = z_i^*$. In other words, an n-point function in the upper half-plane is equivalent to the purely holomorphic part of a $2n$-point function on the infinite plane. Drawing an analogy with electrostatics, the property just described for the correlation functions of BCFT reminds how one finds the electric potential that fulfils proper boundary conditions by placing fictitious electric charges in unphysical regions of the space, without spoiling the differential equation satisfied by the potential in the region where there are the actual charges: the validity of the method of images in both cases is simply due to the linearity of their differential equations.

21.3 Conformal Boundary Operators

If we adopt a radial quantization, the equal time surfaces will be given by semi-circles centred on a point of the boundary taken as the origin. The Hamiltonian in such a scheme coincides with the dilatation operator D, given by the integral on the contours shown in Figure 21.3

$$D = \frac{1}{2\pi i}\int_S zT(z)\,dz - \frac{1}{2\pi i}\int_{\bar{S}} \bar{z}\bar{T}(\bar{z})\,d\bar{z} = \int_C z\,T(z)\,dz = L_0, \tag{21.3.1}$$

where L_0 is the operator of the Virasoro algebra. Other operators L_n can be defined along the same lines,

$$L_n = \int_C z^{n+1}\,T(z)\,dz. \tag{21.3.2}$$

For the reality condition (21.2.4) of the stress-energy tensor, in the presence of the boundary there is *only* one Virasoro algebra. The eigenstates of L_0 correspond to the *boundary operators* $\varphi_j(0)$ acting on the vacuum state $|0\rangle$. In order to discuss in full generality the properties of these operators, it is convenient to consider an infinitely long strip of width L, related to the upper half-plane by the conformal mapping $w = L/\pi \log z$ (Figure 21.4). Putting $w = t + i\sigma$, the time translation along the t-axis defines the H, given in terms of the generator of scale transformations in the upper half-plane

$$H(L) = \frac{\pi}{L}\left(L_0 - \frac{c}{24}\right). \tag{21.3.3}$$

Since the existence of the Virasoro algebra depends only on the fact that the boundary conditions are conformally invariant at both the edges of the strip, i.e. at $\sigma = 0$ and $\sigma = L$, we can have *two* different boundary conditions on the two sides of the strip, here denoted as (α, β). Correspondingly, we denoted the Hamiltonian with these conformal invariant

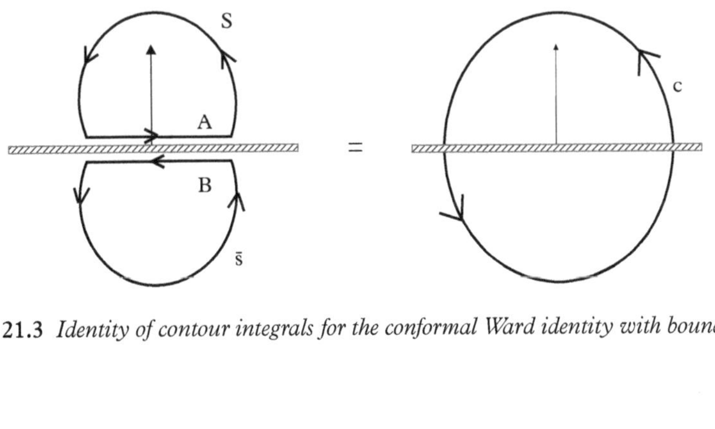

Fig. 21.3 *Identity of contour integrals for the conformal Ward identity with boundary.*

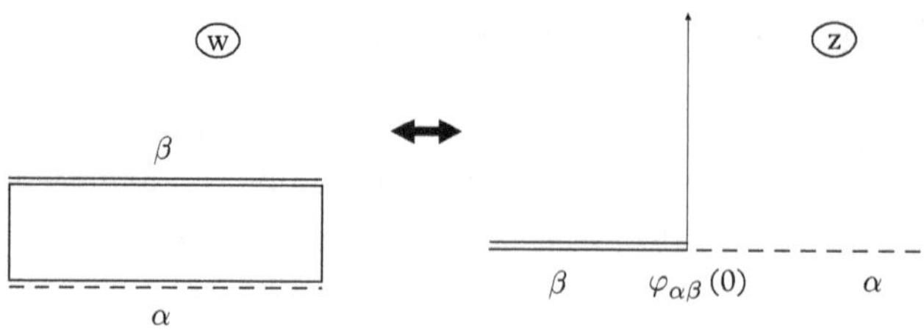

Fig. 21.4 *Left-hand side: strip with boundary conditions α and β. Right-hand side: upper half-plane with insertion of the boundary operator $\phi_{\alpha\beta}(0)$ that swaps the boundary conditions at the origin.*

boundary conditions $H_{\alpha\beta}$. Of course, the eigenstates of this Hamiltonian must fall into irreducible representations of the Virasoro algebra.

To this aim let us consider the geometry of an annulus of width L with boundary conditions (α, β) on its two sides and of length R, with periodic boundary conditions along this direction (see Figure 21.6). For later use, let us define the modular parameters

$$q = e^{2\pi i\tau}, \quad \tilde{q} = e^{-2\pi i/\tau}; \quad \tau = \frac{i}{2}\frac{R}{L}. \tag{21.3.4}$$

One way to compute the partition function is according to

$$Z_{\alpha,\beta}(L, R) = \text{Tr}\, e^{-RH_{\alpha\beta}(L)} = \sum_{c} n^{\gamma}_{\alpha\beta}\, \chi_{\gamma}(q), \tag{21.3.5}$$

where $\chi_{\gamma}(q)$ are the characters of the irreducible Virasoro highest-weight representations R_{γ}

$$\chi_{\gamma}(q) = \text{Tr}_{R_{\gamma}}\left(q^{L_0 - c/24}\right), \tag{21.3.6}$$

while $n^{\gamma}_{\alpha\beta}$ is the number of times that the irreducible representation R_{γ} occurs in the spectrum of $H_{\alpha\beta}(R)$. These integers $n^{\gamma}_{\alpha\beta}$ give the operator content of the theory with boundary conditions (α, β). These are integer numbers normalized as $n^{0}_{\alpha\alpha} = 1$, where 0 here stays for the channel of identity operator. Notice that now the partition function is *linear* in the characters, rather than quadratic as it was with the periodic boundary conditions discussed in Section 11.7. The next section shows an alternative expression for $Z_{\alpha\beta}(L, R)$ that unveils the nature of the boundary states.

Returning back from the strip geometry to the original upper half-plane region, the boundary conditions α and β on the two-sided of the strip give rise to the situation shown in Figure 21.4: on the negative real axis, we have a conformal boundary condition α, on the positive real axis a conformal boundary condition β and a discontinuity at $a = 0$. This implies that in the radial quantization scheme, this boundary state is no longer annihilated by L_{-1} and it can be regarded as created by the action of a *boundary operator*

$\phi_{\alpha\beta}(0)$ acting on the true vacuum $|0\rangle$ (Figure 21.4). This operator is nothing else that the primary field with the lowest conformal dimension Δ_γ for which $n^\gamma_{\alpha\beta} \neq 0$ and we show some explicit examples below. Acting with other local operators on this boundary state, we can obtain the other allowed boundary state for which $n^\delta_{\alpha\beta} \neq 0$. For the two-point function of the boundary operators, we have

$$\langle \phi_{\alpha\beta}(x_1)\,\phi_{\beta\alpha}(x_2)\rangle = \frac{1}{(x_1 - x_2)^{2\Delta_{\alpha\beta}}}, \tag{21.3.7}$$

where $\Delta_{\alpha\beta} = \Delta_{\beta\alpha}$ is the boundary, or surface, scaling dimension. Note that a boundary operator and its bulk counterpart do not necessarily share the same scaling dimensions, as it will become evident next.

Boundary Magnetic Field in the Ising Model. As an example of surface scaling dimension which is different from the bulk scaling dimension, let us consider the Ising model on the upper half-plane with free boundary condition at the boundary. We want to show that the surface scaling dimension of the spin operator is equal to $\Delta_\sigma^{(\mathrm{sur})} = 1/2$ while we know that $\Delta_\sigma^{(\mathrm{bulk})} = 1/16$. We can prove this by taking the two-point function of the spin operator in the presence of the boundary

$$G(x_1,y_1;x_2,y_2) = \langle \sigma(x_1,y_1)\sigma(x_2,y_2)\rangle, \tag{21.3.8}$$

where y_i $(i = 1,2)$ are the distances of the boundary of each point (Figure 21.5) and, at y_1 and y_2 sufficiently small but fixed, see how this correlation function decays for large values of $(x_1 - x_2)$, $G \sim (x_1 - x_2)^{-2\eta}$. This parameter η is the surface scaling dimension of the spin we seek. Using this strategy, we can proceed to the actual computation of the correlation function $G(x_1,y_1;x_2,y_2)$. In the presence of the boundary and with the mirror image method, this correlation function is expressed in terms of the four-point correlation function of the spin operator and can be cast in the form

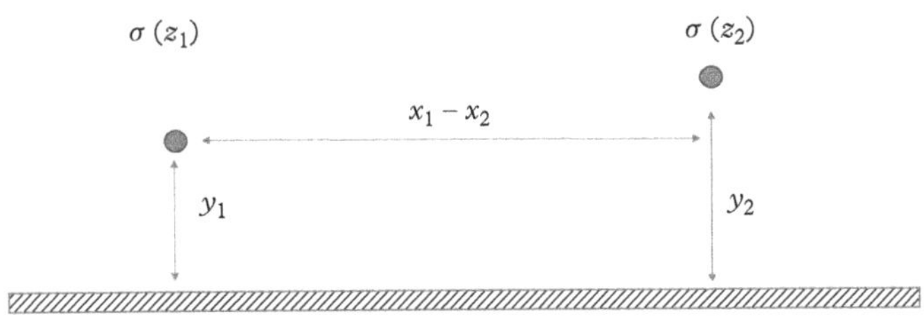

Fig. 21.5 *Two-point function of the spin operators in Ising model with free boundary condition at the boundary.*

$$G(x_1,y_1;x_2,y_2) = \left[\frac{(z_1-\bar{z}_1)(z_2-\bar{z}_2)}{(z_1-z_2)(\bar{z}_1-\bar{z}_2)(z_1-\bar{z}_2)(\bar{z}_1-z_2)}\right]^{1/8} F(\zeta) \qquad (21.3.9)$$

where $z_1 = x_1 + iy_1$, $z_2 = x_2 + iy_2$ and ζ is the harmonic ratio

$$\zeta = \frac{(z_1-z_2)(\bar{z}_1-\bar{z}_2)}{(z_1-\bar{z}_1)(z_2-\bar{z}_2)} = -\frac{(x_1-x_2)^2+(y_1-y_2)^2}{4y_1y_2}. \qquad (21.3.10)$$

The function $F(\zeta)$ is given by a proper linear combination of the hypergeometric functions $\sqrt{\sqrt{1-\zeta}\pm 1}$ relative to the bulk four-point function of the spin operator (Section 14.2). For the free boundary condition we have $\langle\sigma\rangle = 0$ even at the boundary and this uniquely fixes the function $F(\zeta)$

$$F(\zeta) = \sqrt{\sqrt{1-\zeta}+1} - \sqrt{\sqrt{1-\zeta}-1}, \qquad (21.3.11)$$

Going back to the expression (21.3.9), taking the limit $(x_1 - x_2) \to \infty$ at fixed y_1 and y_2, we have

$$G(x_1,y_1;x_2,y_2) \simeq \text{const}\,(y_1y_2)^{3/8}\,(x_1-x_2)^{-1}, \qquad (21.3.12)$$

and from this behaviour in $(x_1 - x_2)$ we can extract that the surface dimension of the spin field is $\Delta_\sigma^{(\text{sur})} = 1/2$.

21.4 Conformal Boundary States

To enlighten the nature of the boundary states let us again consider CFT defined on the geometry of an annulus (Figure 21.6). We have two alternative ways to compute the partition function of this system.

1. Considered previously, the first consists of regarding the direction along R as time axis. In this interpretation, the Hamiltonian is given by $H_{\alpha\beta}$ and therefore the partition function is expressed as in eqn. (21.3.5).
2. Alternatively, we can regard the time direction as the axis along L. In this interpretation, the Hamiltonian is the one relative to periodic boundary conditions and the partition function is given by the matrix element between the so-called *boundary states*

$$Z_{\alpha\beta}(L,R) = \langle\alpha|e^{-LH}|\beta\rangle. \qquad (21.4.1)$$

These boundary states belong to the Hilbert space of the bulk theory made of states in the tensor product $\mathcal{V}_\Delta \otimes \bar{\mathcal{V}}_{\bar{\Delta}}$). However, due to the boundary condition (21.2.4), they are

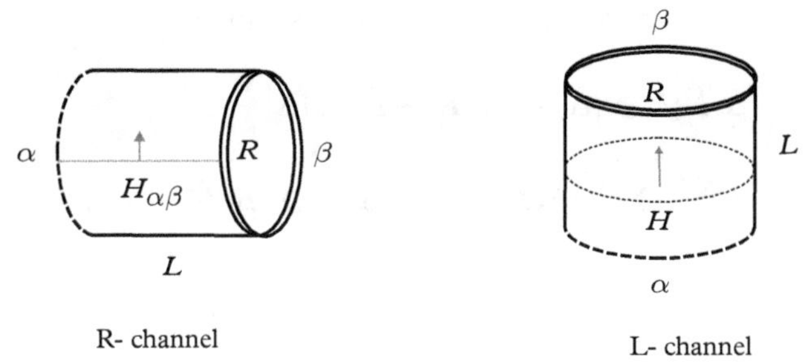

Fig. 21.6 *Partition function of the annulus with boundary conditions α and β at the two sides of the width R and periodic boundary condition along the L direction. Left-hand side: the R channel, with time translation implemented by the Hamiltonian $H_{\alpha\beta}$. Right-hand side: the L channel with time translation implemented by the bulk Hamiltonian H.*

solutions of the equation

$$\left(L_n - \bar{L}_{-n}\right) |B\rangle = 0. \tag{21.4.2}$$

Interestingly enough, one can find the general solutions of these equations given by the so-called *Ishibashi states*. For the minimal models of CFT, they are constructed as follows. First let us consider the sub-space of level N of a primary field of dimension Δ, with dimension $d_\Delta(N)$, and let us choose an orthonormal basis in this space[1] denoted as $|\Delta, N; j\rangle$, with $j = 1, 2, \ldots, d_\Delta(N)$. Therefore, the general element of the basis in the Hilbert space of the bulk theory is given by

$$|\Delta, N; j\rangle \otimes |\bar{\Delta}, M; k\rangle. \tag{21.4.3}$$

It is now easy to see that there are as many solutions of the eqns. (21.4.2) as primary fields of the minimal models of CFT and they are given by

$$|\Delta\rangle\rangle = \sum_{N=0}^{\infty} \sum_{j=1}^{d_\Delta(N)} |\Delta, N; j\rangle \otimes |\Delta, N; j\rangle. \tag{21.4.4}$$

Clearly, these are not normalizable states, since such a sum has no chance to converge in the Hilbert space of our states. Nevertheless, their scalar product with states of definite energy are well defined: it is this set of scalar products that identifies our boundary states. Therefore, it is sufficient to show it holds the infinite identities

$$\langle \Delta, \tilde{N}; j| \otimes \langle \bar{\Delta}, \tilde{M}; k| \left(L_n - \bar{L}_{-n}\right) |\Delta\rangle\rangle = 0. \tag{21.4.5}$$

[1] Of course taking care of the null vector states at that level.

Substituting the expression for $|\Delta\rangle\rangle$, we see that this equation is indeed satisfied

$$
\begin{aligned}
&\langle \Delta, \tilde{N}; l| \otimes \langle \bar{\Delta}, \tilde{M}; k| \left(L_n - \bar{L}_{-n}\right) |\Delta\rangle\rangle \\
&= \sum_{N=0}^{\infty} \sum_{j=1}^{d_{\Delta(N)}} \Big[\langle \Delta, \tilde{N}; l|L_n|\Delta, N; j\rangle \otimes \langle \bar{\Delta}, \tilde{M}; k|\Delta, N; j\rangle \\
&\quad - \langle \Delta, \tilde{N}; l|\Delta, N; j\rangle \otimes \langle \bar{\Delta}, \tilde{M}; k|L_{-n}|\Delta, N; j\rangle \Big] \\
&= \sum_{N=0}^{\infty} \sum_{j=1}^{d_{\Delta(N)}} \Big[\delta_{\tilde{M},N} \delta_{k,j} \langle \Delta, \tilde{N}; l|L_n|\Delta, N; j\rangle \\
&\quad - \delta_{\tilde{N},N} \delta_{l,j} \langle \bar{\Delta}, \tilde{M}; k|L_{-n}|\Delta, N; j\rangle \Big] \\
&= \langle \Delta, \tilde{N}; l|L_n|\Delta, \tilde{M}; k\rangle - \langle \bar{\Delta}, \tilde{M}; k|L_{-n}|\Delta, \tilde{N}; l\rangle = 0,
\end{aligned}
\tag{21.4.6}
$$

having used the orthogonality of the basis and the Hermitian relations $L_n = L_n^\dagger$. Notice that we can easily computed the matrix elements between the Ishibashi states of the translator operator along the cylinder

$$
\begin{aligned}
&\langle\langle \Delta'|e^{-LH}|\Delta\rangle\rangle = \\
&\sum_{N,N'=0} \sum_{j'=1}^{d_{\Delta(N')}} \sum_{j=1}^{d_{\Delta(N)}} \langle \Delta', N', j'| \otimes \langle \Delta', N', j'|e^{-2\pi L/R(L_0+\bar{L}_0-c/12)}|\Delta, N, j\rangle \otimes |\Delta, N, j\rangle \\
&= \delta_{\Delta',\Delta} \sum_{N=0} \sum_{j=1}^{d_{\Delta(N)}} e^{-4\pi L/R(\Delta+N-c/24)} = \chi_\Delta(\tilde{q})\, \delta_{\Delta',\Delta}.
\end{aligned}
\tag{21.4.7}
$$

Remember (Section 11.7) that the characters $\chi_a(\tilde{q})$, expressed in terms of $\tilde{q}$, are related to those expressed in terms of the variable q by the linear transformation S_a^b associated to the modular group

$$
\chi_a(\tilde{q}) = \sum_b S_a^b \chi(q). \tag{21.4.8}
$$

With all these results, we now come back to eqn. (21.4.1) and consider the case where we have a diagonal partition function on the torus, i.e. given by the expression $Z_{\text{torus}} = \sum_j \chi_j(q)\chi_j(\bar{q})$: in this case each conformal family appears once and just once in the spectrum of the Hamiltonian H. We express each boundary state $|\alpha\rangle$ and $|\beta\rangle$ as linear combination of the Ishibashi states

$$
|\alpha\rangle = \sum_\Delta |\Delta\rangle\rangle \, \langle\langle \Delta|\alpha\rangle,
$$

so that the partition function (21.4.1) is given by

$$Z_{\alpha\beta}(L,R) = \langle\alpha|e^{-LH}|\beta\rangle = \sum_{\Delta}\langle\alpha|\Delta\rangle\rangle\,\langle\langle\Delta|\beta\rangle\,\chi_{\Delta}(\tilde{q}). \tag{21.4.9}$$

Using the modular transformation (21.4.8) and comparing this expression with the one given in eqn. (21.3.5), we arrive at the following identities

$$n_{ab}^{\Delta} = \sum_{\Delta'} S_{\Delta'}^{\Delta}\,\langle a|\Delta'\rangle\rangle\langle\langle\Delta'|b\rangle \tag{21.4.10}$$

$$\langle a|\Delta'\rangle\rangle\langle\langle\Delta'|b\rangle = \sum_{\Delta} S_{\Delta}^{\Delta'}\,n_{ab}^{\Delta}. \tag{21.4.11}$$

The solution of these equations for the diagonal partition function of the minimal models goes as follows: since the matrix elements S_0^{Δ} of the moular S-matrix are all non-negative, we can define a boundary state

$$|\tilde{0}\rangle \equiv \sum_{\Delta}(S_0^{\Delta})^{1/2}\,|\Delta\rangle\rangle, \tag{21.4.12}$$

such that $n_{\tilde{0}\tilde{0}}^{\Delta} = \delta_0^{\Delta}$: this means that the only representation entering $H_{\tilde{0}\tilde{0}}$ is the identity representation of the unit operator. In a similar way, we can introduce a boundary state

$$|\tilde{h}\rangle = \sum_{j}\frac{S_h^j}{(S_0^j)^{1/2}}\,|j\rangle, \tag{21.4.13}$$

such that $n_{\tilde{0}\tilde{h}}^{i} = \delta_h^i$, i.e. the only representation present in the spectrum of $H_{\tilde{0}\tilde{h}}$ is the representation h.

It is interesting to see an explicit example of this analysis and the simplest one is offered by the Ising model.

Ising model. In the ordering given by the conformal dimensions $0, \frac{1}{2}, \frac{1}{16}$, the modular S-matrix of this model is given by

$$S = \begin{pmatrix} \frac{1}{2} & \frac{1}{2} & \frac{1}{\sqrt{2}} \\ \frac{1}{2} & \frac{1}{2} & -\frac{1}{\sqrt{2}} \\ \frac{1}{\sqrt{2}} & -\frac{1}{\sqrt{2}} & 0 \end{pmatrix}. \tag{21.4.14}$$

Hence, the boundary states are

$$
\begin{aligned}
|\tilde{0}\rangle \equiv |+\rangle &= \frac{1}{\sqrt{2}}|0\rangle + \frac{1}{\sqrt{2}}|\epsilon\rangle + \frac{1}{\sqrt[4]{2}}|\sigma\rangle \\
|\tilde{\epsilon}\rangle \equiv |-\rangle &= \frac{1}{\sqrt{2}}|0\rangle + \frac{1}{\sqrt{2}}|\epsilon\rangle - \frac{1}{\sqrt[4]{2}}|\sigma\rangle \\
|\tilde{\sigma}\rangle \equiv |f\rangle &= |0\rangle - |\epsilon\rangle .
\end{aligned}
\tag{21.4.15}
$$

The first two states differ just from the sign of the last term, which is a $\mathbb{Z}_2$ odd state. This observation suggests that they should correspond to the boundary states associated to the fixed boundary conditions $|+\rangle$ and $|-\rangle$, where the spins are all up or down at the boundary. The other boundary state which is conformal invariant corresponds to the free boundary condition, where each spin can be equivalently up or down: then, this should correspond to the state $|f\rangle$.

21.4.1 Boundary Entropy

The boundary states may have some internal degrees of freedom and therefore a non-zero entropy. This can be explicitly computed according as follows. Consider the partition function as given in eqn. (21.4.1): in the limit $L/R \gg 1$, such an expression factorizes as

$$
Z_{ab}(L,R) \simeq \langle a|0\rangle \, e^{-LE_0(R)} \, \langle 0|b\rangle \equiv g_a g_b \, e^{\pi c L/(6R)}, \tag{21.4.16}
$$

where $|0\rangle$ is the ground state of the periodic Hamiltonian H, $E_0(R)$ is its ground state energy while the product $g_a g_b$ can be related to a constant, length-independent zero temperature entropy $\mathcal{S}(0) = \log(g_a g_b)$. This becomes evident taking the point of view that the partition function (21.4.1) can be regarded as associated to a two-dimensional classical system of length L and at an inverse temperature $\beta = R$. Using the expression (21.4.16), we can easily obtain the free energy $F_{ab} = -R^{-1} \log Z_{ab}$ and the entropy $\mathcal{S}_{ab} = -R^2 (\partial F_{ab}/\partial R)$ and the result is

$$
\mathcal{S}_{ab} = \frac{\pi c}{3R} L + \mathfrak{f}_a + \mathfrak{f}_b + \mathcal{O}(1/L). \tag{21.4.17}
$$

Each individual term $\mathfrak{f}_a = \log(\langle a|0\rangle)$ and $\mathfrak{f}_b = \log(\langle b|0\rangle)$ is the boundary entropy of the corresponding boundary state. In the case of the Ising model discussed above, the overlap between the boundary state and the vacuum state $|0\rangle$ can be easily computed using eqn. (21.4.15) and therefore we have

$$
\log g_+ = \log g_- = \log \frac{1}{\sqrt{2}} = -0.3465\ldots \tag{21.4.18}
$$

$$
\log g_{\text{free}} = \log 1 = 0. \tag{21.4.19}
$$

21.5 Operator Product Expansion Involving a Boundary Operator

Consider a CFT with a boundary condition **a** which does not break conformal invariance. A primary field $\phi(x,y)$ of such a theory nearby the boundary, admits an OPE

$$\phi(x,y) \simeq \sum_n (2y)^{\Delta_n - 2\Delta_\Phi} C^a_{\Phi,\lambda_n} \lambda_n(x), \quad y \to 0, \tag{21.5.1}$$

which involves the boundary operators, here denoted as $\lambda_n(x)$. The origin of this relation is clear: the original field gives rise to its mirror image and the OPE refers to this pair of operators but, with respect to an analogous OPE in the bulk, the structure constants depends on the boundary condition **a** implemented on the boundary. Taking as example the Ising model, from the usual fusion rules of the fields with themselves of this model

$$\begin{aligned} \sigma\sigma &= 1 + \epsilon, \\ \epsilon\epsilon &= 1, \end{aligned} \tag{21.5.2}$$

we infer that for the boundary OPE of the field σ, the boundary fields which may appear in this expansion are 1 and ϵ, while in the boundary OPE of the field ϵ can only appear the operator 1. The same kind of considerations apply to other minimal conformal models as well with the final result which can be summarized as follows: the boundary OPE of the field Φ resembles the fusion rules of the operator Φ with itself, but the structure constant $C^a_{\Phi\lambda_n}$ are genuine new quantities which depend, behind the field Φ, by the boundary condition a.

Following Cardy and Lewellen, we present how to compute the structure constant relative to the identity channel, i.e. $C^a_{\Phi 1} = A^a_\Phi$. This quantity can be expressed as

$$A^a_\Phi = \frac{\langle \Phi | a \rangle}{\langle 0 | a \rangle}, \tag{21.5.3}$$

a result that comes from comparing two equivalent expressions of the one-point function of the operator Φ on a semi-infinite cylinder (of width R and with coordinate τ along the cylinder) which terminates with a circular boundary, with boundary condition a: the first expression is obtained using the transfer matrix formalism and, in the limit $\tau \to \infty$, this gives rise to

$$\langle \Phi(\tau) \rangle_a \simeq \left(\frac{2\pi}{R} \right)^{2\Delta_\Phi} \frac{\langle \Phi | 0 \rangle}{\langle 0 | a \rangle} \exp(-4\pi \Delta_\Phi \tau / R). \tag{21.5.4}$$

The second expression is obtained, instead, by a conformal map from the upper half-plane to the same cylinder geometry, with the result given by

$$\langle \Phi(\tau) \rangle_a = A_\Phi^a \left(\frac{2\pi}{R}\right)^{2\Delta_\Phi} \frac{\exp(-4\pi \Delta_\Phi \tau / R)}{1 - \exp(-8\pi \Delta_\Phi \tau / R)}. \tag{21.5.5}$$

Comparing these two expressions in the limit $\tau \to \infty$, we get eqn. (21.5.3). To proceed, we require the expression of the boundary state $|a\rangle$ as given by eqn. (21.4.13), so that the final result is

$$A_\Phi^a = \frac{S_a^\Phi}{S_a^0} \left(\frac{S_0^0}{S_0^\Phi}\right)^{1/2}. \tag{21.5.6}$$

For the Ising model, the physical boundary states correspond to the fixed boundary conditions $|+\rangle$ and $|-\rangle$ and to the free boundary condition $|f\rangle$. Using this formula and the matrix element (21.4.14) of the modular matrix of the Ising model, we get

$$(2y)^{1/8} \langle \sigma \rangle_\pm = \pm 2^{1/4}, \quad (2y)^{1/8} \langle \sigma \rangle_f = 0, \tag{21.5.7}$$

and

$$(2y) \langle \epsilon \rangle_\pm = 1, \quad (2y) \langle \epsilon \rangle_f = -1. \tag{21.5.8}$$

Notice that there is *no* coupling to the boundary identity operator of the spin field σ in the free boundary condition. Hence the only boundary field entering its OPE is the field ϵ of dimension $\Delta_\epsilon = 1/2$, which provides a further justification to the surface dimension equal to $1/2$ for the spin field.

21.6 Massive Integrable Boundary Field Theory

We now consider an integrable massive field theory in the bulk, characterized by a set of massive particles $|A_a(\theta)\rangle$ $(a = 1, 2 \ldots r)$ and the S-matrix amplitudes $S_{ab}(\theta_1 - \theta_2)$, where θ_i is a rapidity variable. To keep the analysis as simple as possible, in the following we assume that the particles are all distinguishable, so that the S-matrix is completely diagonal. Suppose now that the theory is confined to the space region $x < 0$ and at $x = 0$ there is a boundary, associated to a boundary condition for the corresponding field theory. For a generic boundary condition, the integrability of the original theory in the bulk is lost but there may exist special boundary conditions that are compatible with the existence of an infinite number of conserved charges: in those cases, we are in the presence of a massive integrable boundary field theory. Free or fixed boundary conditions in general preserve the integrability of the bulk theory, but there may be other boundary conditions that also keep the integrability: see Ghoshal and Zamolodchikov in the chapter references for further details.

If the boundary condition at $x = 0$ preserves the integrability of the model, when a particle $|A_a(\theta)\rangle$ hits the boundary it is reflected with an amplitude $R_a(\theta)$ preserving its own identity (see Figure 21.7(a))

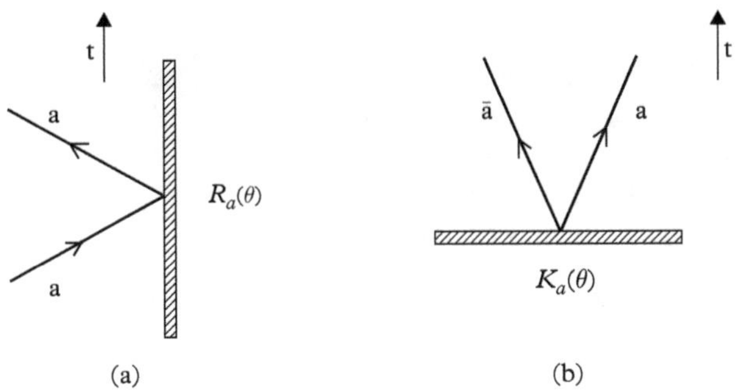

Fig. 21.7 *(a) A particle of type a hits the boundary placed at x = 0 and is reflected with an amplitude $R_a(\theta)$; (b) a pair of particle and anti-particle of type a is emitted with an amplitude $K_a(\theta)$ from the boundary placed at t = 0.*

$$|A_a(\theta)\rangle = R_a(\theta)\,|A_a(-\theta)\rangle. \tag{21.6.1}$$

For distinguishable particles there are as many as reflection amplitudes $R_a(\theta)$ as the number of particles r, one for each particle, and their expression depends of course on the boundary conditions used. In Figure 21.7(a) the boundary is placed at $x = 0$ and time flows vertically. But we can rotate the geometry, interchanging time with space, and therefore place the boundary at $t = 0$ (Figure 21.7(b)): this is the crossed channel of the previous situation; therefore the incoming particle becomes an outgoing anti-particle and the previous scattering process now appears as a virtual emission pair of particle and anti-particle from the boundary, with an amplitude $K_a(\theta)$, related to the previous amplitude through the relation

$$K_a(\theta) = R_a\left(\frac{i\pi}{2} - \theta\right). \tag{21.6.2}$$

The set of reflection amplitudes satisfy a certain number of constraints.

1. **Unitarity**. The first constraint comes from the unitary condition that reads

$$R_a(\theta)\,R_a(-\theta) = 1. \tag{21.6.3}$$

2. **Crossing**. To derive the crossing property of the reflection amplitude it is convenient to go to the crossed channel, where the physical process is a virtual emission of a pair of particle and anti-particle. Once this pair is emitted, it can braid through the S-matrix of the bulk, giving rise to the equation

$$K_a(\theta) = S_{\bar{a}a}(2\theta)\,K_a(-\theta), \tag{21.6.4}$$

whose graphical intepretation is shown in Figure 21.8.

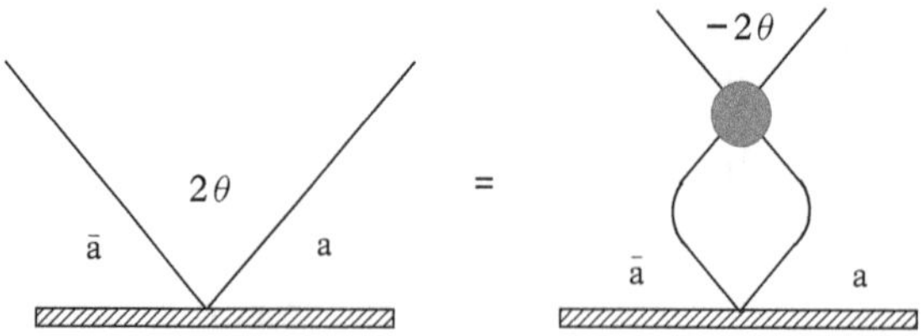

Fig. 21.8 *Boundary cross-unitarity condition.*

In the more general scattering theories there are also boundary Yang–Baxter equations relative to the factorization of the boundary amplitudes, but in our case of diagonal scattering processes they are trivially satisfied.

21.7 Boundary States

In the picture in which the boundary is placed at $t = 0$, the boundary condition is encoded into a boundary state $|B\rangle$ as much as it was in the conformal situation previously analysed. Such a state belongs to the space of states of the bulk theory and therefore expressed as linear superposition of infinitely many multi-particle states

$$|B\rangle = \sum_{N=0} \int [d\theta_i]\, K_N(\theta_1,\theta_2,\ldots,\theta_N) A_{a_1}(\theta_1) A_{a_2}(\theta_2)\ldots A_{a_N}(\theta_N)|0\rangle, \qquad (21.7.1)$$

but with the amplitudes $K_N(\theta_1,\ldots,\theta_N)$ severely constrained by the presence of the infinite conservation laws coming from the integrability of the theory. Let us discuss in more detail. Firstly, the infinite number of integrals of motion are diagonalized by the asymptotic states

$$P_s\,|A_a(\theta)\rangle = \gamma_a^{(s)}\, e^{s\theta}\,|A_a(\theta)\rangle, \quad \bar{P}_s\,|A_a(\theta)\rangle = \gamma_a^{(s)}\, e^{-s\theta}\,|A_a(\theta)\rangle \qquad (21.7.2)$$

where $\gamma_a^{(s)}$ are constants. The analog of eqn. (21.4.2) in this case is expressed by

$$(P_s - \bar{P}_s)\,|B\rangle = 0. \qquad (21.7.3)$$

In order to solve these equations notice that the eigenvalues of the combination $(P_s - \bar{P}_s)$ on a multi-particle states $|A_{a_1}(\theta_1)A_{a_2}(\theta_2)\ldots A_{a_N}(\theta_N)\rangle$ is given by

$$2\sum_{i=1}^{N} \gamma_{a_i}^{(s)}\,\sinh(s\theta), \qquad (21.7.4)$$

and therefore the only multi-particle states which can enter the expansion (21.7.1) are in terms of pairs[2] of equal mass particle/anti-particles $A_{\bar{a}}(\theta)A_a(-\theta)$ of opposite rapidities! So, the first term of the expansion has an amplitude that is given by the quantity $K_a(\theta)$

$$|B\rangle = \left(1 + \frac{1}{2}\int_{-\infty}^{\infty} d\theta\, K_a(\theta)A_{\bar{a}}(\theta)A_a(-\theta) + \ldots\right)|0\rangle. \tag{21.7.5}$$

It is now easy to argue that the multi-particle amplitudes $K_N(\theta_1,\ldots,\theta_N)$ have to be expressed in terms of product of the two-particle amplitudes $K_a(\theta)$: indeed, it would be enough to appeal to the factorization properties of the integrable theories in order to propose an exact expression of the boundary states given by

$$|B\rangle = \exp\left[\frac{1}{2}\int_{-\infty}^{\infty} d\theta\, K_a(\theta)A_{\bar{a}}(\theta)A_a(-\theta)\right]|0\rangle. \tag{21.7.6}$$

21.8 Massive Boundary Ising Model

A very instructive example of the previous discussion is provided by the Ising model at $T \neq T_c$ in zero magnetic field. Chapter 9 shows that this model admits a formulation in terms of a free Majorana fermion, with mode expansion in terms of annihilation and creation operators $A(\theta)$ and $A^{\dagger}(\theta)$ given in eqn. (9.7.19), and an exact S-matrix amplitude $S = -1$. According to the eqns. (21.6.2), (21.6.3) and (21.6.4), for any integrable boundary condition the reflection amplitude $R(\theta)$ satisfies in this case the conditions

$$R(\theta)\,R(-\theta) = 1, \quad R\left(\frac{i\pi}{2} - \theta\right) = -R\left(\frac{i\pi}{2} + \theta\right). \tag{21.8.1}$$

Notice that we can define the boundary scattering amplitude according to the relation

$$A^{\dagger}(\theta)\,|B\rangle = R(\theta)A^{\dagger}(-\theta)\,|B\rangle, \tag{21.8.2}$$

which will become useful soon. Given that the theory is free, i.e. linear in the femionic modes, there are two obvious boundary conditions which preserve the integrability of the bulk theory

$$\left(\Psi + \bar{\Psi}\right)_{x=0} = 0; \tag{21.8.3}$$
$$\left(\Psi - \bar{\Psi}\right)_{x=0} = 0. \tag{21.8.4}$$

[2] There could also be the case of zero rapidity particle $A_a(0)$ entering the expansion of the boundary state (21.7.1), which we discuss later.

We can argue that the first corresponds to the fixed boundary condition while the second to the free boundary conditions. An argument in support of these interpretation comes from expressing the original spin operator of the Ising model in terms of the Majorana fermion and showing that the fixed boundary condition where the spins at the boundary are all equal to $+1$ (or -1) corresponds to the condition (21.8.3), while the free boundary condition where the spins at the boundary can take any value corresponds instead to eqn. (21.8.4). We will see that another justification of these interpretations comes from the exact expressions of the amplitudes given below that correspond to the two cases.

Fixed Boundary Condition. Using the mode expansion of eqn. (9.7.19), the operators A and $A^\dagger$ satisfy the conditions

$$\left(\omega e^{\frac{\theta}{2}} + \bar{\omega} e^{-\frac{\theta}{2}}\right) A(\theta) = -\left(\bar{\omega} e^{\frac{\theta}{2}} + \omega e^{-\frac{\theta}{2}}\right) A(-\theta), \tag{21.8.5}$$

$$\left(\bar{\omega} e^{\frac{\theta}{2}} + \omega e^{-\frac{\theta}{2}}\right) A^\dagger(\theta) = -\left(\omega e^{\frac{\theta}{2}} + \bar{\omega} e^{-\frac{\theta}{2}}\right) A^\dagger(-\theta). \tag{21.8.6}$$

Comparing now with the definition (21.8.2), we see that the solution is

$$R_{\text{fixed}}(\theta) = i \tanh\left(\frac{i\pi}{4} - \frac{\theta}{2}\right). \tag{21.8.7}$$

For the amplitude in the crossing channel we have

$$K(\theta) = i \tanh\frac{\theta}{2}. \tag{21.8.8}$$

Free Boundary Condition. Repeating the same computation for the condition given by $(\Psi - \bar{\Psi})_{x=0} = 0$, the reflection amplitude is given in this case by

$$R_{\text{free}}(\theta) = -i \coth\left(\frac{i\pi}{4} - \frac{\theta}{2}\right) \tag{21.8.9}$$

and correspondingly

$$K_{\text{free}} = -i \coth\frac{\theta}{2}. \tag{21.8.10}$$

It is now important to notice that this amplitude has a pole at $\theta = 0$ which points out the existence of a zero-momentum one-particle state in the expression of the boundary state associated to this case

$$|B_{\text{free}}\rangle = (1 + A^\dagger(0) + \ldots)|0\rangle, \tag{21.8.11}$$

The presence of this extra contribution, which has zero energy, further justifies to identify (21.8.9) as the reflection amplitude relative to the free boundary condition and (21.8.7) to the fixed boundary condition. Indeed, spin configurations associated to the free boundary conditions may be described as containing an infinitely long fluctuating domain wall which separates two domains of opposite magnetizations: this domain wall corresponds to a zero-momentum particle emitted by the boundary.

Boundary magnetic field. Interestingly enough, the presence of a boundary magnetic field preserves the integrability of the theory. This because the boundary magnetic field has surface dimension $\Delta_\sigma^{(\mathrm{sur})} = 1/2$ and therefore the action, expressed in terms of Majoarana fermions, still remains quadratic! In fact, the boundary spin operator can be expressed as

$$\sigma_B(y) = \frac{1}{2}(\Psi + \bar{\Psi})_{x=0}(y)\, a(y), \tag{21.8.12}$$

where $a(y)$ is a fermionic boundary operator that describes the ground state degeneracy of the vacuum in the free boundary condition: it anti-commutes with Ψ, $\bar{\Psi}$ and satisfies $a^2 = 1$, so that for the free boundary condition we have $a|\pm\rangle = |\mp\rangle$. The action of the model in the half-plane *HP* with boundary $\mathcal{D}$ placed at $x = 0$ can be written as

$$\begin{aligned} \mathcal{A} = &\frac{1}{2\pi}\int_{HP} dxdy\left[\Psi\partial_{\bar{z}}\Psi + \bar{\Psi}\partial_z\bar{\Psi} + im\Psi\bar{\Psi}\right] \\ &+ \int_{\mathcal{D}} dy\left[-\frac{i}{4\pi}\Psi\Psi + \frac{1}{2}a\frac{da}{dy} - ih\frac{1}{2}a(\Psi + \Psi)\right]. \end{aligned} \tag{21.8.13}$$

From the variation of the action, in addition to the familiar Dirac equation for the two components of the fermion in the bulk, we also get the boundary condition

$$i\frac{d}{dy}(\Psi - \bar{\Psi})_{x=0} = \frac{h^2}{2}(\Psi + \bar{\Psi})_{x=0}. \tag{21.8.14}$$

Expanding in modes and using eqn. (21.8.2) we get the reflection amplitude

$$R_h(\theta) = i\tanh\left(\frac{i\pi}{4} - \frac{\theta}{2}\right)\frac{\kappa - i\sinh\theta}{\kappa + i\sinh\theta}, \qquad \kappa = 1 - \frac{h^2}{2m}. \tag{21.8.15}$$

Notice that, varying h from 0 to ∞, this amplitude interpolates between the amplitudes relative to the free and fixed boundary conditions, therefore realizing an RG flow on the boundary. For the amplitude of the crossed channel we have

$$K_h(\theta) = i\tanh\frac{\theta}{2}\,\frac{\kappa + \cosh\theta}{\kappa + \cosh\theta}. \tag{21.8.16}$$

For any non-zero value of h this amplitude no longer has the pole at $\theta = 0$, since the presence of a boundary magnetic field splits the degeneracy of the two vacua present at the free boundary condition. However, for $h < h_c = \sqrt{2m}$, we can parameterize κ as $\kappa = \cos v$ and there is a pole at $\theta = i(\pi/2 - v)$ that corresponds to a boundary bound state whose interpretation is in terms of boundary histeresis, present for small enough magnetic field. When $h > h_c$ there is no longer the pole but a pair of resonances localized at $\theta = -i\pi/2 \pm \theta_0$, where we have parameterized κ as $\kappa = -\cosh\theta_0$, with θ_0 real and positive.

21.9 Correlation Functions

The best way to compute correlation functions of an integrable massive QFT with boundary is to take full advantage of the solution of the theory in the bulk and to employ the exact expression of the boundary state given in eqn. (21.7.6). In this approach the boundary is then placed at $t = 0$ and therefore the correlation functions can be expressed as

$$\langle \mathcal{O}_1(x_1,t_1)\ldots\mathcal{O}_n(x_n,t_n)\rangle = \frac{\langle 0|T_t\,[\mathcal{O}_1(x_1,t_1)\ldots\mathcal{O}_n(x_n,t_n)]\,|B\rangle}{\langle 0|B\rangle}. \qquad (21.9.1)$$

In this geometry, the Hilbert space of the theory is as in the bulk and therefore, even in the presence of the boundary line, the local operators $\mathcal{O}_i$ can be completely characterized by the bulk FFs $< \theta_1,\ldots,\theta_n \mid \mathcal{O}_i \mid \theta_{n+1},\ldots,\theta_m >$.

Let u present the simplest example of a correlation function, given by the one-point function of the energy operator of the Ising model with a boundary magnetic field. This correlation function can be computed through the formula

$$\epsilon_0(t) = \sum_{n=0}^{\infty} < 0 \mid \epsilon(x,t) \mid n >< n \mid B >. \qquad (21.9.2)$$

The energy operator couples the vacuum only to the two-particle state and for its matrix element in the Euclidean space we have

$$\begin{aligned} < 0 \mid \epsilon(x,t) \mid \theta_1,\theta_2 > &= -2\pi m i \sinh\frac{\theta_1-\theta_2}{2} \qquad (21.9.3)\\ &\times \exp\left[-mt(\cosh\theta_1+\cosh\theta_2)+imx(\sinh\theta_1+\sinh\theta_2)\right]. \end{aligned}$$

Hence, in this case the sum (21.9.2) consists of only one term and the one-point function of the energy operator can be expressed as

$$\epsilon_0(t) = -im\int_0^{\infty} d\theta\,\sinh\theta\,K_h(\theta)\,e^{-2mt\cosh\theta}. \qquad (21.9.4)$$

The one-point function does not depend on x, as a consequence of the translation invariance along this axis. The above integral reduces to closed expressions in terms of Bessel functions in the case of free and fixed boundary conditions, which are given respectively by

$$\epsilon_0^{(\pm)}(t) = \mp m\,[K_1(2mt) \pm K_0(2mt)]\,. \tag{21.9.5}$$

In the short distance limit $mt \to 0$, they have the scaling form $\epsilon_0^{(\pm)}(t) \sim \mp\frac{1}{2t}$, in agreement with the BCFT prediction (21.5.8). In the large distance limit, they instead decay exponentially with an extra power term, which is different in the two cases, i.e.

$$\begin{aligned} \epsilon_0^{(+)}(t) &\sim -m\sqrt{\tfrac{\pi}{mt}}\,e^{-2mt}, \\ \epsilon_0^{(-)}(t) &\sim \tfrac{m}{8}\sqrt{\tfrac{\pi}{(mt)^3}}\,e^{-2mt}. \end{aligned} \tag{21.9.6}$$

Turning our attention to the one-point function of the energy operator in the presence of a boundary magnetic field, we have

$$\epsilon_0(t,h) = -m\int_0^\infty d\theta(\cosh\theta - 1)\frac{\cosh\theta + \kappa}{\cosh\theta - \kappa}e^{-2mt\cosh\theta}. \tag{21.9.7}$$

By varying h, we may switch between the free and the fixed boundary. However, how can we interpolate between the two functions $\epsilon_0^{(\pm)}(t)$, which present quite a different behaviour? To understand better this aspect, we write down a differential equation satisfied by $\epsilon_0(t,h)$. Observe that eqn. (21.9.7) can be expressed as

$$\epsilon_0(r,h) = -m\int_0^\infty d\theta\int_0^\infty d\alpha\,(\cosh\theta - 1)(\cosh\theta + \kappa)e^{-r\cosh\theta - \alpha(\cosh\theta - \kappa)}, \tag{21.9.8}$$

($r \equiv 2mt$) and the resulting integral can be expressed in terms of modified Bessel function as

$$\epsilon_0(r,h) = -me^{-\kappa r}\int_r^\infty d\eta\, e^{\kappa\eta}\left[\frac{K_1(\eta)}{\eta} + (1-\kappa)(K_0(\eta) - K_1(\eta))\right]. \tag{21.9.9}$$

Taking the derivative with respect to r in both sides of this equation, we obtain the following differential equation satisfied by $\epsilon_0(r,h)$

$$\frac{\partial\epsilon_0(r,h)}{\partial r} + \kappa\,\epsilon_0(r,h) = m\left[\frac{K_1(r)}{r} + (1-\kappa)(K_0(r) - K_1(r))\right]. \tag{21.9.10}$$

This differential equation is quite useful to control the short and large distance behaviour of the one-point function $\epsilon_0(r,h)$. For $r \to 0$, we can parameterize the solution as $\epsilon_0(r,h) \sim Cr^{-\delta}$. Substituting into eqn. (21.9.10), it is easy to see that for *any* finite κ, the term which determines the short distance behaviour in the RHS is $m\frac{K_1(r)}{r}$, therefore $C=-m$ and $\delta=1$, independent of the value of the magnetic field. Hence, as far as h is finite, all the curves $\epsilon_0(r,h)$ follow at a short distance scale the behaviour associated to the free boundary condition. For $r\to\infty$, we look for a solution of the form $\epsilon_0(r,h)\sim e^{-r}r^{-\omega}\sum_{k=0}^{\infty}a_k r^{-k}$. Substituting into (21.9.10), using the large distance expansion of the Bessel functions, and comparing the power series we have then the following cases:

1. For $\kappa=1$, $a_0=-2m\sqrt{\frac{\pi}{2}}$ and $\omega=\frac{1}{2}$, so that $\epsilon_0(r,h=0)\sim a_0 e^{-r}r^{-1/2}$.
2. For $-1<\kappa<1$, we have $a_0=-m\sqrt{\frac{\pi}{2}\frac{1+\kappa}{1-\kappa}}$ and $\omega=\frac{3}{2}$, so that $\epsilon_0^{(h)}(r)\sim a_0 e^{-r}r^{-3/2}$. In this range $a_0<0$ and the function approaches the real axis from below.
3. For $\kappa=-1$ the first leading term of the expansion vanishes and for the next leading term we have $a_1=-m\frac{27}{16}\sqrt{\frac{\pi}{2}}$, $\omega=\frac{3}{2}$ and the corresponding solution goes to zero much faster, $\epsilon_0(r,h=2\sqrt{m})\sim a_1 e^{-r}r^{-5/2}$.
4. For $\kappa<-1$, we have $a_0=-m\sqrt{\frac{\pi}{2}\frac{1+\kappa}{1-\kappa}}$, $\omega=\frac{3}{2}$ and $\epsilon_0^{(h)}\sim a_0 e^{-r}r^{-3/2}$. In this range $a_0>0$ and the function approaches the real axis from above.

The one-point functions relative to some of these cases are shown in Figure 21.9. The result of this analysis is that for the boundary magnetic field smaller than $\tilde{h}=2\sqrt{m}$, the one-point function does not have any zero at finite value of r, whereas for $h>\tilde{h}$, it crosses the horizontal axes at a finite value of r, and its large r behaviour closely follows the behaviour relative to fixed boundary condition.

The above picture is consistent with an RG analysis. In fact, the boundary magnetic field is a relevant operator, which therefore cannot affect the behaviour of the correlation functions near the boundary, as far as it assumes finite values. Hence, sufficiently close to $r=0$ the boundary always appears subject to free boundary conditions. The boundary magnetic field, however, influences the observables at large distance scales and when $h>\tilde{h}$, the boundary always appears as subject to the fixed boundary condition for an observer placed at $r\to\infty$. In an intermediate scale, there is a non-trivial crossover between the two different behaviours, which becomes rather sharp with increasing h. The function $r(h)$, implicitly defined as the zeros of the one-point function $\epsilon_0(r(h),h)=0$, may be interpreted as a phase diagram of the theory, which divides the free and the fixed boundary condition regions.

It is also interesting to study the massless limit of the one-point function (21.9.7). This corresponds to the physical situation of a critical Ising model in the bulk but in the presence of a boundary condition, which breaks the conformal invariance of the model. With the change of variable $y=2mt\cosh\theta$, the integral (21.9.7) becomes

$$\epsilon_0(t,h)=\frac{1}{2t}\int_{2mt}^{\infty}dy e^{-t}\sqrt{\frac{y-2mt}{y+2mt}\frac{y+2mt\kappa}{y-2mt\kappa}}. \tag{21.9.11}$$

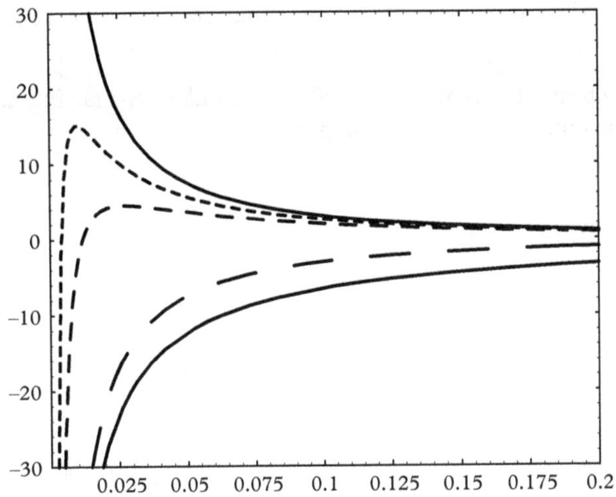

Fig. 21.9 *One-point function of the energy operator versus (mt) for different values of the boundary magnetic field. Upper full line: fixed boundary condition. Lower full line: free boundary condition. Long dashed line: $h < h_c$. Short dashed lines: curves with $h > h_c$.*

In the limit $m \to 0$, we have $2mt\kappa \to -h^2 t \equiv -z$ and the one-point function of the energy operator can be expressed as

$$\epsilon_0(t,h) = \frac{1}{2t}\left[1 + 2ze^{z}Ei(-z)\right], \tag{21.9.12}$$

where $Ei(z)$ is the exponential integral function.

REFERENCES

John Cardy has developed boundary conformal field theories in a series of papers:

Cardy, J.L. (1984). 'Conformal Invariance and Surface Critical Behavior', *Nuclear Physics B*, 240[FS12]: 514.

Cardy, J.L. (1986). 'Effect of Boundary Conditions on the Operator Content of Two-dimensional Conformally Invariant Theories', *Nuclear Physics B*, 275[FS17]: 200.

Cardy, J.L. (1989). 'Boundary Conditions, Fusion Rules and the Verlinde Formula', *Nuclear Physics B*, 324: 581.

Cardy, J.L. (2006). 'Boundary Conformal Field Theory', *Encyclopedia of Mathematical Physics*, Elsevier.

Cardy, J.L. and Lewellen, D.C. (1991). 'Bulk and Boundary Operators in Conformal Field Theory', *Physics Letters B*, 259: 274.

The massive boundary integrable theory and the relative S-matrix approach has been proposed in

Ghoshal, S. and Zamolodchikov, A.B. (1994). 'Boundary S-matrix and Boundary State in Two-dimensional Integrable Quantum Field Theory', *International Journal of Modern Physics A*', 9: 3841.

Other relevant papers include:

Cherednik, I.V. (1984). Untitled. *Teor. Matem. Fiz.*, 61: 55.
Chim, L. (1995). 'Boundary S-matrix for the Integrable q Potts Model', *Journal of Physics A*, 28: 7039.
De Martino, A., Moriconi, M. and Mussardo, G. (1998). 'Reflection Scattering Matrix of the Ising Model in a Random Boundary Magnetic Field', *Nuclear Physics B*, 509: 615.
Fring, A. and Koberle, R. (1994). 'Affine Toda Field Theory in the Presence of Reflecting Boundaries', *Nuclear Physics B*, 419: 647.
Ghoshal, S. (1994). 'Boundary S-matrix of the O(N) Symmetric Nonlinear Sigma Model', *Physics Letters B*, 334: 363.

Several papers address the boundary ground state energies in computed integrable QFTs:

Chatterjee, R. (1996). 'Exact Partition Function and Boundary State of 2D Massive Ising Model with Boundary Magnetic Field', *Nuclear Physics B*, 468: 439.
Dorey, P., Pocklington, A., Tateo, R. and Watts, G. (1998). 'TBA and TCSA with Boundaries and Excited States', *Nuclear Physics B*, 525: 641.
LeClair, A., Mussardo, G., Saleur, H. and Skorik, S. (1995). 'Boundary Energy and Boundary States in Integrable Quantum Field Theories', *Nuclear Physics B*, 453: 581.

A series of papers have discussed the computation of correlation functions of the boundaried Ising model:

Konik, R., LeClair, A. and Mussardo, G. (1996). 'On Ising Correlation Functions with Boundary Magnetic Field', *International Journal of Modern Physics A*, 11: 2765.
Chatterjee, R. and Zamolodchikov, A.B. (1994). 'Local Magnetization in Critical Ising Model with Boundary Magnetic Field', *Modern Physics Letters A*, 9: 2227.

PROBLEMS

21.1. Boundary States for a bosonic field

Consider the two analytic and anti-analytic $U(1)$ currents $j(z) = \sum_n j_n z^{-n-1}$ and $\bar{j}(\bar{z}) = \sum_n \bar{j}_n \bar{z}^{-n-1}$ related to a massless bosonic field $\varphi(z, \bar{z})$ by the relations $j(z) = i\partial_z \varphi$ and $\bar{j}(\bar{z}) = i\partial_{\bar{z}} \varphi$. Let us use the conformal map $z = \exp[\tau + i\sigma]$ to map the upper half-plane into an infinitely long strip of width π.

a. Show that it holds the relations

$$\partial_\sigma \varphi = \sum_{n=-\infty}^{\infty} \left(j_n e^{-n(\tau+i\sigma)} - \bar{j}_n e^{-n(\tau-i\sigma)} \right),$$

$$i\partial_\tau \varphi = \sum_{n=-\infty}^{\infty} \left(j_n e^{-n(\tau+i\sigma)} + \bar{j}_n e^{-n(\tau-i\sigma)} \right).$$

b. The Neumann boundary condition at $\sigma = 0$ is implemented by the condition $\partial_\sigma \varphi|_{\sigma=0} = 0$, while the Dirichlet boundary condition by $\partial_\tau \varphi|_{\sigma=0} = 0$. Show that these imply the following relations between the modes j_n and $\bar{j}_n$

$$\begin{aligned} j_n - \bar{j}_n &= 0 \quad \text{Neumann boundary condition.} \\ j_n + \bar{j}_n &= 0 \quad \text{Dirichlet boundary condition.} \end{aligned}$$

c. Now swap the role of σ and τ and consider the boundary states associated to both Neumann and Dirichlet boundary conditions

$$\begin{aligned} \partial_\tau \varphi|_{\tau=0} |B_N\rangle &= 0 \quad \text{Neumann boundary state.} \\ \partial_\sigma \varphi|_{\tau=0} |B_D\rangle &= 0 \quad \text{Dirichlet boundary state.} \end{aligned}$$

Show that their exact expression (up to normalization) is given by

$$|B_N\rangle = \exp\left[-\sum_{k=1}^{\infty} \frac{1}{k} j_{-k} \bar{j}_{-k} \right] |0\rangle,$$

$$|B_D\rangle = \exp\left[+\sum_{k=1}^{\infty} \frac{1}{k} j_{-k} \bar{j}_{-k} \right] |0\rangle.$$

21.2. Bogoliubov transformation and boundary state

Consider a free massive scalar field $\varphi(x,t)$ in $(1+1)$ dimension with mass

$$m = \begin{cases} m_- & t < 0 \\ m_+ & t \geq 0. \end{cases}$$

Therefore it admits two different mode expansions, with two sets of annihilation and creation operators $(A_-, A_-^\dagger)$ and $(A_+, A_+^\dagger)$, which refer to $t < 0$ and $t \geq 0$ cases respectively

$$\varphi(x,t) = \int \frac{dk}{2\pi\sqrt{2E_a}} \left[A_a(k)\, e^{i(kx - E_a t)} + A_a^\dagger(k)\, e^{-i(kx - E_a t)} \right], \quad a = \pm$$

where $E_a = \sqrt{m^2 + k^2}$.

a. Use the continuity of the field $\varphi(x,t)$ at $t=0$ to show that the two set of modes are related by a Bogoliubov transformation

$$A_+(k) = c(k)A_-(k) + d(k)A_-^\dagger(-k),$$
$$A_+^\dagger(k) = c(k)A_-^\dagger(k) + d(k)A_-(-k).$$

and compute explicitly the coefficients $c(k)$ and $d(k)$.

b. In terms of the modes for $t<0$, the vacuum of the theory is identified by $A_-(k)|0\rangle$. Use the Bogoliubov transformation to show that the boundary state $|B\rangle$ which lives at $t=0$ is expressed in terms of the modes $(A_+, A_+^\dagger)$ as

$$|B\rangle = \exp\left[\int_{-\infty}^{\infty} dk\, K(k) A_+^\dagger(k) A_+^\dagger(-k)\right]|0\rangle$$

where $K(k) = d(k)/c(k)$.

Part 6

Non-Integrable Aspects

22

Form Factor Perturbation Theory

Nobody is perfect!
Billy Wilder

22.1 Breaking Integrability

The integrable QFTs analysed in the previous chapters provide the exact solution of many statistical models away from the critical point. Despite the elegance and the undeniable success of these methods, the generic situation that occurs in statistical physics is that of non-integrable dynamics: many interesting statistical models fall within this class and therefore it would be highly desirable to develop an appropriate formalism to deal with the lack of integrability. This task is notoriously difficult for the rich phenomenology that arises in the corresponding QFT when the dynamics is not integrable: there are decays and production scattering processes, confinement phenomena and nucleation of false vacua, resonance peaks in the cross sections, etc. All these physical aspects are usually accompanied by a great mathematical complexity. To see this, it is sufficient to consider the analytic structure of the S-matrix of such theories: once we give up the integrability condition, the infinite number of thresholds of the production processes gives rise to nested patterns of branch cut singularities, in addition to the pole structure associated to the bound states or the resonances (see Figure 22.1).

It is worth stressing that for non-integrable theories even the most basic question—the spectrum of the physical excitations—poses a problem of considerable difficulty. Determining the spectrum is indeed a dynamical problem which cannot be easily solved by e.g. looking at the fields employed in the Lagrangian of the model: there are indeed systems that have dynamical bound states do not appear in the Lagrangian but there are also others where the fields employed in the Lagrangian do not correspond to any physical excitation! Chapter 18 showed a famous example of such a rich phenomenology is provided by the Sine–Gordon model for a two-dimensional bosonic field $\varphi(x)$ with interaction given by the potential

Statistical Field Theory: An Introduction to Exactly Solved Models In Statistical Physics. Second Edition. Giuseppe Mussardo, Oxford University Press (2020). © Giuseppe Mussardo. DOI: 10.1093/oso/9780198788102.001.0001

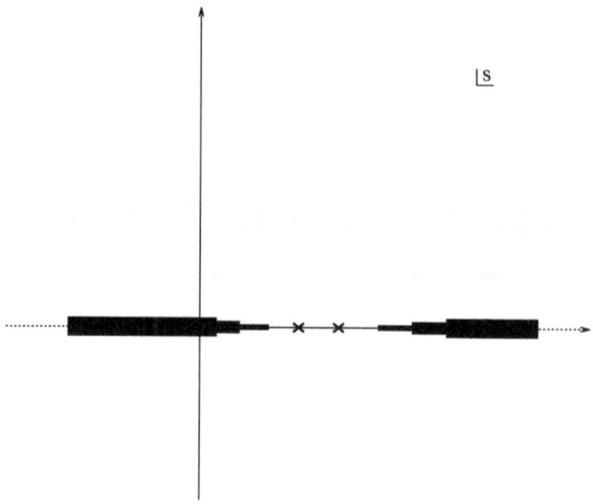

Fig. 22.1 *Analytic structure of the S-matrix of non-integrable models. At the production thresholds, the scattering amplitudes develop new branch cuts.*

$$U_{SG}(\varphi) = \frac{m_0^2}{\beta^2}\,[1 - \cos(\beta\varphi)], \tag{22.1.1}$$

where, in terms of familiar perturbative arguments, we might be tempted to conclude that the theory has always a neutral excitation above it while such a conclusion is wrong if $\beta^2 > 4\pi$: beyond this value, such a particle—the very same particle described by the field $\varphi(x)$ itself—does not exist!

We emphasize this simply to argue that should expect that the determination of the spectrum of an interactive QFT which is not integrable may be quite a difficult problem of non-perturbative nature. Such a problem can be obviously addressed from a numerical point of view, by employing for instance the truncated conformal space approach discussed in Chapter 25. Here, we present instead an analytical method to address the study of two-dimensional non-integrable theories, showing that certain progress can be achieved depending on the degree of '*non-integrability*', so to speak. For instance, if the non-integrable theory in question may be regarded as small deformation of an integrable model, one can apply the so-called form factor perturbation theory (FFPT) to follow the evolution of the spectrum moving away from the integrable direction. With this method a great deal of information can be obtained on several aspects of non-integrable QFTs, e.g. confinement of topological excitations, adiabatic shift of the mass of non-topological excitations, decays of higher mass particles, presence of resonances, etc.

Chapters 23 and 24 present the so-called *semi-classical method*, an approach which has the advantage to rely neither on integrability nor of small breaking thereof. While, in general, the application of this method may be quite involved, it drastically simplifies if applied to the analysis of QFTs with vacua degeneracy and topological kink excitations. In this case, we show that, with relatively small efforts, we can easily get the semi-classical spectrum of neutral bound states of the non-integrable theories. Chapter 23 is dedicated to purely bosonic theories while Chapter 24 to fermionic theories.

Although the detailed analysis of all physical aspects of non-integrable models go beyond the scope of this book, in the following four chapters we present a series of results that are particularly helpful in understanding the class of universality of some important models. Our study starts from a perturbative approach based on the exact FFs of the integrable models discussed in Chapter 19.

22.2 Multiple Deformations of the Conformal Field Theories

Let us focus the attention on a particular class of non-integrable models, defined in terms of a conformal action deformed by two relevant operators, each of them giving rise *individually* to an integrable model

$$\mathcal{A} = \mathcal{A}_{CFT} + \lambda_1 \int d^2x \varphi_1(x) + \lambda_2 \int d^2x \varphi_2(x). \tag{22.2.1}$$

We emphasize that there are indeed many interesting physical systems that belong to this class of non-integrable models. Let us briefly discuss two of them.

- The first is the Ising model at temperature T, different from the critical value T_c, and in an external magnetic field h. Its action is given by

$$\mathcal{A} = \mathcal{A}_{CFT} + \tau \int d^2x \varepsilon(x) + h \int d^2x \sigma(x), \tag{22.2.2}$$

  with $\tau = T - T_c$. When $h = 0$, the above action corresponds to the integrable theory of the thermal deformation: this theory has only one particle excitation and an elastic S-matrix equal to $S = -1$. On the contrary, when $T = T_c$ it is the integrable theory of the Ising model in a magnetic field: its spectrum consists of eight massive particles and the relative S-matrix amplitudes were discussed in Chapter 18.
- The second example is provided by the multi-frequency Sine–Gordon model. The action is given by

$$\mathcal{A} = \int d^2x \left[\frac{1}{2}(\partial_\mu \varphi)^2 + \lambda_1 \cos \beta\varphi + \lambda_2 \cos \alpha\varphi \right]. \tag{22.2.3}$$

  When $\lambda_1 = 0$, this action gives rise to the integrable theory of the Sine–Gordon model with frequency α. In addition to the soliton states, such a theory has a

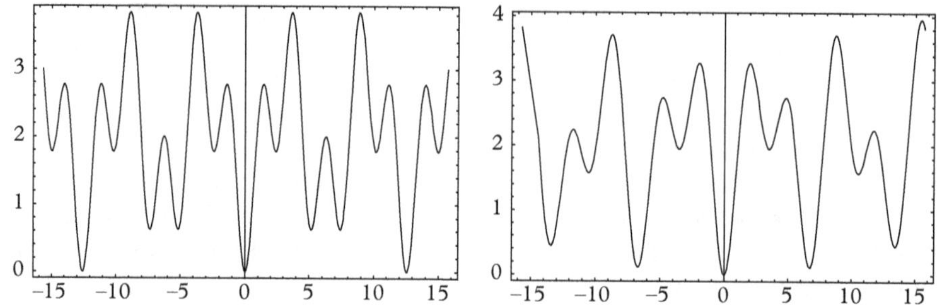

Fig. 22.2 *Potential of the multiple-frequency Sine–Gordon model: rational (a) and irrational (b) ratio of the frequencies.*

number of neutral bound states given by[1] $N_2 = \left[\frac{\pi}{\xi_\alpha}\right]$, where $\xi_\alpha = \frac{\alpha^2}{8}/(1-\alpha^2/8\pi)$. Viceversa, if $\lambda_2 = 0$, we have again a Sine–Gordon model but of frequency β and correspondingly a different number of neutral bound states, $N_1 = \left[\frac{\pi}{\xi_\beta}\right]$. When the ratio of α and β is a rational number, the potential of the theory has an infinite number of periodic and degenerate vacua. On the contrary, when the ratio of the frequencies is an irrational number, the potential has only one vacuum that can always be placed at the origin (Figure 22.2).

To study the field theory associated to an action as in eqn. (22.2.1), it is convenient to regard it as *a deformation of an integrable action* rather than as a multiple deformation of a conformal theory. By taking this point of view and grouping differently the terms, the action (22.2.1) can then be written as

$$\mathcal{A} = \mathcal{A}^i_{\text{int}} + \lambda_j \int d^2x\, \varphi_j(x), \tag{22.2.4}$$

$(i = 1, 2, j \neq i)$. There are several advantages in doing so.

1. The first advantage becomes evident by going to the Minkoswki space. In fact, the non-integrable theory can be analysed starting from the basis of the Hilbert space provided by the particle excitations associated to the integrable model $\mathcal{A}^i_{\text{int}}$. Although the spectra of $\mathcal{A}$ and $\mathcal{A}^i_{int}$ would be different, the massive basis provided by the particles of the integrable model will be certainly more appropriate than the conformal basis, as far as the infrared properties of the non-integrable model are concerned.
2. The second advantage consists of the exact solvability of the integrable models, in particular, the possibility of computing exactly all the matrix elements (i.e. the

[1] $[x]$ denotes the integer part of the real number x.

FFs) of local and non-local operators of such theories. Hence, in complete analogy with ordinary quantum mechanics, we can set up a perturbative approach based on the FFs of the integrable models. As shown later, this perturbative approach will enable us to reach a remarkable series of predictions about the mass correction, the decay processes or the correction to the scattering amplitudes.

3. When each deformation is individually integrable, there is the obvious freedom of using any of them as a starting point. Each choice selects a particular basis of the particles and gives rise to a particular perturbation theory. However, since the actual dynamics of the model should be insensitive to such a choice, there should be a series of mathematical identities that links one perturbative series to the other.

22.3 Form Factor Perturbation Theory

Let us consider a QFT in the Minkowski space, defined by the action

$$\mathcal{A} = \mathcal{A}_0 + \mathcal{A}_I = \mathcal{A}_0 - \lambda \int d^2x\, \Psi(x), \tag{22.3.1}$$

where $\mathcal{A}_0$ denotes here the Minkowskian action of the unperturbed theory and Ψ one of its operators (we choose the minus sign in front of the coupling to simplify the formulae below). We suppose that the QFT associated to the action $\mathcal{A}_0$ is exactly solvable (although not necessarily free), so that the spectrum of its particles, their scattering amplitudes and the matrix elements of the its operators, in particular those of the perturbing field Ψ, are assumed to be all known. For the sake of simplicity, this section considers the case of an isospectral perturbation of a solvable theory the spectrum of the total action $\mathcal{A}$ is made of the same number of particles of the unperturbed one $\mathcal{A}_0$: i.e. the new interaction $\mathcal{A}_I$ changes the values of the masses of the physical particles but not their stability properties.[2]

Let us now describe the properties of the theory (22.3.1). Under the hypothesys that the interaction term is turned off at $t \to \pm\infty$, it is possible to adopt the formalism of the asymptotic '*in*' and '*out*' states. We are interested in computing the scattering amplitude

$$\begin{aligned} S\{q_1,\ldots,q_n \to q'_1,\ldots,q'_m\} &= {}^{\text{out}}\langle q'_1,\ldots,q'_m|q_1,\ldots,q_n\rangle^{\text{in}} = \\ &= {}^{\text{in}}\langle q'_1,\ldots,q'_m|S|q_1,\ldots,q_n\rangle^{\text{in}}, \end{aligned} \tag{22.3.2}$$

where q_i and q'_j label the momenta of the *in*-going and *out*-going set of particles. Since in the remote past $t \to -\infty$ the interaction is not present yet, the asymptotic *in* states coincide with the unperturbed ones

$$|q_1,\ldots,q_n\rangle^{\text{in}} = |q_1,\ldots,q_n\rangle_0^{\text{in}}.$$

[2] We will comment later on the more general case.

As usual, the scattering operator S in eqn. (22.3.2) can be obtained as the limit

$$S = \lim_{t\to+\infty} U(t,-t)$$

of the time evolution operator $U(t,t_0)$, solution of the equation

$$i\frac{d}{dt}U(t,t_0) = H\,U(t,t_0), \qquad U(t_0,t_0) = 1, \tag{22.3.3}$$

where $H = H_0 + H_I$ denotes the Hamiltonian of the theory (22.3.1).

Interactive representation. Following the standard quantum mechanical procedure,[3] the operator U can be factorized as $U = U_0 U_I$, where U_0 and U_I are the solutions of eqn. (22.3.3) with H replaced by H_0 and $\tilde{H}_I(t) = U_0^{-1} H_I U_0$ respectively. Then, we can write the scattering operator of the theory (22.3.1) as $S = S_0 S_I$, where $S_0 = \lim_{t\to+\infty} U_0(t,-t)$ is the unperturbed and exactly known scattering matrix

$$S_0\{q_1,\ldots,q_n \to q'_1,\ldots,q'_m\} = {}_0^{\text{out}}\langle q'_1,\ldots,q'_m|q_1,\ldots,q_n\rangle_0 = \\ {}_0^{\text{in}}\langle q'_1,\ldots,q'_m|S_0|q_1,\ldots,q_n\rangle_0^{\text{in}},$$

while S_I has the usual formal representation

$$S_I = \lim_{t\to+\infty} U_I(t,-t) = T\exp\left(i\mathcal{A}_I[\Psi]\right).$$

The scattering amplitude is therefore given by

$$\begin{aligned} {}^{\text{out}}\langle q'_1,\ldots,q'_m|q_1,\ldots,q_n\rangle^{\text{in}} &= {}_0^{\text{in}}\langle q'_1,\ldots,q'_m|T\exp\left(-i\lambda\int d^2x\Psi(x)\right)|q_1,\ldots,q_n\rangle_0^{\text{in}} = \\ &= \sum_{k=0}^{+\infty}\frac{(-i\lambda)^k}{k!}\int d^2x_1\ldots d^2x_k\, {}_0^{\text{in}}\langle q'_1,\ldots,q'_m|T\left(\Psi(x_1)\ldots\Psi(x_k)\right)|q_1,\ldots q_n\rangle_0^{\text{in}}, \end{aligned} \tag{22.3.4}$$

where (22.3.4) has been used in order for absorbing the factor S_0.

In ordinary Lagrangian perturbation scheme based on free theories, the computation of scattering amplitudes would now proceed through the use of creation/annihilation operators and Wick theorem, finally leading to the diagrammatic expansion that is characteristic of the Feynman covariant perturbation theory. This approach, however, cannot be generally followed here because we might not know, in general, a local Lagrangian formulation of the theory associated to $\mathcal{A}_0$. However, the exact solution of its dynamics, in the form specified at the beginning of this section, naturally suggests computing the scattering amplitudes (22.3.4) within the same framework used to deal

[3] See, for instance, C. Cohen Tannoudji, C., Liu, B. and Laloe, F. (1977). *Quantum Mechanics*, John Wiley and Sons.

with ordinary time-dependent perturbation theory in quantum mechanics. In other words, let us initially insert between the operators $\Psi(x_l)$ and $\Psi(x_{l+1})$ $(l = 1,\dots,k-1)$ in the second line of (22.3.4) a sum over a complete set of asymptotic states of the unperturbed theory

$$\sum_n |n\rangle_{\text{out out}}\langle n| = 1 = \sum_n |n\rangle_{\text{in in}}\langle n|,$$

with $|n\rangle$ denoting an asymptotic state containing n on–shell particles. The integrations on the space coordinates in (22.3.4) can be immediately performed: they lead to delta functions that constraint the total momentum of the intermediate states to coincide with that of the initial and final states. In doing the integrations on the time variables, the time ordering prescription gives rise in this case to the appearence of energy denominators. The final expression is

$$\begin{aligned}
&{}^{\text{out}}\langle q'_1,\dots,q'_m|q_1,\dots,q_n\rangle^{\text{in}} = {}^{\text{in}}_{0}\langle q'_1,\dots,q'_m|q_1,\dots,q_n\rangle^{\text{in}}_{0} + \\
&+(2\pi)^2\delta^{(2)}\left(\sum_{j=1}^{m} q'_j - \sum_{j=1}^{n} q_j\right) \times \\
&\times \Bigg\{-i\lambda\ {}^{\text{in}}_{0}\langle q'_1,\dots,q'_m|\Psi(0)|q_1,\dots,q_n\rangle^{\text{in}}_{0} \\
&+\frac{1}{2\pi i}\sum_{k=2}^{+\infty}(2\pi\lambda)^k \sum_{n_1}\cdots\sum_{n_{k-1}}\Bigg[\frac{\delta(Q-P_1)\dots\delta(Q-P_{k-1})}{(E-E_1+i\epsilon)\dots(E-E_{k-1}+i\epsilon)}\times \\
&\times\ {}^{\text{in}}_{0}\langle q'_1,\dots,q'_m|\Psi(0)|n_1\rangle_0\cdots{}_0\langle n_{k-1}|\Psi(0)|q_1,\dots,q_n\rangle^{\text{in}}_{0}\Bigg]\Bigg\},
\end{aligned} \tag{22.3.5}$$

where E and E_i (Q and P_i) denote the total energy (momentum) of the initial state and of the i–th intermediate state, respectively. Each intermediate sum can be equivalently taken either on the basis of the *in* states or on that of *out* states. Since the matrix elements between asymptotic states of the perturbing operator $\Psi(x)$ are supposed to be known, the scattering amplitudes (22.3.5) are (in principle) computable quantities, order by order in the coupling constant λ. The above expansion over intermediate states must be contrasted with the usual formalism of covariant perturbation theory in which both energy and momentum are conserved in the internal lines of Feynman diagrams but the corresponding particles are off-shell.

Refinements: normalization conditions. The formula (22.3.5) however is not completely correct because the new interaction changes both the vacuum energy density and the mass of the particles. We have to refine the action $\mathcal{A}_I$ with the introduction of some counterterms to take properly into account the correct normalization of the states. We impose the validity of the following conditions for any value of the coupling constant: the normalization of the vacuum state

$$\langle 0|0\rangle = {}_0\langle 0|0\rangle_0 = 1, \tag{22.3.6}$$

and the normalization of the one-particle states

$$ {}^{\text{out}}\langle q'|q\rangle^{\text{in}} = {}^{\text{out}}_{0}\langle q'|q\rangle^{\text{in}}_{0} = 2\pi E\,\delta(q'^{1} - q^{1}). \tag{22.3.7} $$

The two conditions given above should be enforced order by order in perturbation theory when one uses eqn. (22.3.5) to compute the vacuum to vacuum transition and the one-particle amplitudes.

The condition (22.3.6) leads to a subtraction of a constant term $\delta\mathcal{E}_{vac}(\lambda)$ from the interaction density. This extra term obviously measures the variation of the vacuum energy density under the effect of the perturbation. This effect is usually ignored in Lagrangian perturbation theory with the prescription of disregarding the disconnected vacuum bubble diagrams. We keep track of this term here because, for the class of models we are considering, it is a measurable quantity.

To enforce the correct one-particle normalization we need to introduce a 'mass' term operator in the interaction density. This operator, denoted here by $O^{(2)}(x)$, can be defined in terms of its (unperturbed) FFs given by

$$ F_n^{O^{(2)}} = {}_0\langle 0|O^{(2)}(0)|q_1,\ldots,q_n\rangle = \delta_{n,2}. \tag{22.3.8} $$

With this definition, the coefficient in front of the operator $O^{(2)}(x)$ in the interaction density plays the role of a mass counterterm $\delta m^2(\lambda)$ and has to be determined by imposing eqn. (22.3.7) order by order in the coupling λ.

In summary, the correct formula for the scattering amplitude is given by

$$ \begin{aligned} {}^{\text{out}}\langle q'_1,\ldots,q'_m|q_1,\ldots,q_n\rangle^{\text{in}} &= {}^{\text{in}}_{0}\langle q'_1,\ldots,q'_m|q_1,\ldots,q_n\rangle^{\text{in}}_{0} + \\ &-i(2\pi)^2\delta^{(2)}\left(\sum_{j=1}^{m} q'_j - \sum_{j=1}^{n} q_j\right)\times \\ &\times\Bigg\{ {}^{\text{in}}_{0}\langle q'_1,\ldots,q'_m|\left(\lambda\Psi(0) - \frac{1}{2}\delta m^2 O^{(2)}(0) - \delta\mathcal{E}_{\text{vac}}\right)|q_1,\ldots,q_n\rangle^{\text{in}}_{0} + \\ &+\sum_{n_1}\frac{2\pi\,\delta(Q-P_1)}{(E-E_1+i\epsilon)}\,{}^{\text{in}}_{0}\langle q'_1,\ldots,q'_m|\left(\lambda\Psi(0) - \frac{1}{2}\delta m^2 O^{(2)}(0) - \delta\mathcal{E}_{\text{vac}}\right)|n_1\rangle_0 \times \\ &{}_0\langle n_1|\left(\lambda\Psi(0) - \frac{1}{2}\delta m^2 O^{(2)}(0) - \delta\mathcal{E}_{\text{vac}}\right)|q_1,\ldots,q_n\rangle^{\text{in}}_{0} + \ldots\Bigg\}. \end{aligned} \tag{22.3.9} $$

Let us remark that the above expansion appears as the most physical one since it deals with the true physical degrees of freedom of the problem. But, as in any QFT, divergent contributions are expected to emerge when the above formula is applied beyond the first perturbative order. A general discussion on such divergences and on the renormalization procedure that must be adopted to deal with the infinities, seems to be an interesting open problem in the unconventional setting we are considering: notice that the perturbing operator $\Psi(x)$ has in general non-vanishing matrix elements on *all* the asymptotic states and therefore a sum of an infinite number of terms is required at any perturbative order beyond the first. However, the general aspects of this problem beyond the scope of this book and will not be further investigated here. We rather concentrate our attention only on the first-order approximation because, as we show, this term is enough to catch the leading effects induced by a small perturbation that breaks the integrability.

22.4 First-order Perturbation Theory

Let us now apply the results of the previous section to study the action (22.2.1), where the first two terms defined an integrable theory perturbed by the relevant scalar operator $\varphi_2(x)$. Let x_1 and x_2 be the scaling dimensions of the two operators φ_1 and φ_2. The theory depends in this case on the two dimensionful couplings constants[4] λ_1 and λ_2. Since $\lambda_1 \sim M^{2-x_1}$ and $\lambda_2 \sim M^{2-x_2}$ (where M is a mass scale), we can decide to use λ_1 as dimensionful parameter of the theory while the dimensionless combination

$$\chi \equiv \lambda_2 \, \lambda_1^{-\frac{2-x_2}{2-x_1}} \tag{22.4.1}$$

as a label of the different RG trajectories that originate from the fixed point at $\lambda_1 = \lambda_2 = 0$. For example, if $N(\chi)$ denotes the number of stable particles in the spectrum of the theory, their masses can be expressed as

$$m_a(\lambda_1, \chi), = \mathcal{C}_a(\chi) \, \lambda_1^{\frac{1}{2-x_1}}, \qquad a = 1, 2, \ldots, N(\chi), \tag{22.4.2}$$

where $\mathcal{C}_a(\chi)$ is an amplitude that characterizes the whole trajectory. Similarly, the vacuum energy density can be written as

$$\mathcal{E}_{\text{vac}}(\lambda_1, \chi) = \mathcal{E}(\chi) \, \lambda_1^{\frac{2}{2-x_1}}. \tag{22.4.3}$$

Dimensionless quantities, as for instance mass ratios, only depend on χ and therefore they do not vary along the trajectories of the RG.

Once the new interaction $\lambda_2 \int d^2x \, \varphi_2(x)$ is switched on in the action, the integrability of the unperturbed theory is generally lost and the S-matrix amplitudes become complicated quantities. Inelastic processes of particle production are no longer forbidden and, as a consequence, the analytic structure of the scattering amplitudes present additional cuts due to the higher thresholds. In particular, their expression is no longer factorized into the sequence of two-body scattering amplitudes and, even in elastic channels, the only surviving restriction on the final momenta comes from energy-momentum conservation.

[4] For the sake of simplicity of notation, we assume that no other terms are generated by renormalization effects, as it happens for the models that we will discuss later. The first-order corrections do not depend on this assumption.

The knowledge of the matrix elements of the perturbing field $\varphi_2(x)$ ensures the possibility to compute perturbatively both the amplitudes of the inelastic processes and the corrections to the elastic ones. To the first order in λ_2, with an obvious extension of the notation, equation (22.3.9) reads

$$
\begin{aligned}
&{}^{out}\langle b_1(q_1^1)\dots b_m(q_m^1)|a_1(p_1^1)\dots a_n(p_n^1)\rangle^{in} \\
&\simeq \delta_{mn}\,{}^{out}_{\ 0}\langle b_1(q_1^1)\dots b_n(q_n^1)|a_1(p_1^1)\dots a_n(p_n^1)\rangle_0^{in} - i\delta^2\left(\sum_{k=1}^{n} p_k^\mu - \sum_{k=1}^{m} q_k^\mu\right) \qquad (22.4.4)\\
&\times {}^{out}_{\ 0}\langle b_1(q_1^1)\dots b_n(q_m^1)|\left(\lambda_2\varphi_2(0) - \frac{1}{2}\sum_{a,b=1}^{N}\delta M_{ab}^2\, O_{ab}^{(2)}(0) - \delta\mathcal{E}_{\rm vac}\right)|a_1(p_1^1)\dots a_n(p_n^1)\rangle_0^{in}.
\end{aligned}
$$

The 'mass operator' $O_{ab}^{(2)}(x)$ is defined assigning its form factors. With an obvious generalization of eqn.(22.3.8), they are given by

$$
F_{a_1\dots a_n}^{O_{ab}^{(2)}}(\theta 1,\dots,\theta_n) = \delta_{n2}\delta_{aa_1}\delta_{ba_2}.
$$

The first order corrections to the masses of the particles and to the vacuum energy density are obtained imposing the conditions (22.3.7) and (22.3.6). The result is

$$
\delta M_{\bar{b}a}^2 \simeq 2\lambda_2 F_{\bar{b}a}^{\varphi_2}(i\pi,0)\,\delta_{m_a m_b}, \qquad (22.4.5)
$$

$$
\delta\mathcal{E}_{\rm vac} \simeq \lambda_2\ [{}_0\langle 0|\varphi_2|0\rangle_0]\,. \qquad (22.4.6)
$$

Role of the rapidity. We must be careful in using the rapidity parameterization in eqn. (22.4.4). To illustrate this point let us consider the first-order correction to some elastic process $ab \to cd$. In the unperturbed theory, this process is characterized by the scattering amplitude $S_{ab}^{cd}(\theta)$, where $\theta = \theta_1 - \theta_2$ denotes the rapidity difference of the colliding particles. In two dimensions, the momenta of the particles in a two-body elastic collision are individually conserved even in absence of integrability, so that the general elastic amplitude $S_{ab}^{cd}(\theta,\chi)$ can be introduced through the relation

$$
{}^{out}\langle c(\theta_1)d(\theta_2)|a(\theta_3)b(\theta_4)\rangle^{in} = (2\pi)^2\delta(\theta_1-\theta_3)\delta(\theta_2-\theta_4)\,S_{ab}^{cd}(\theta_1-\theta_2,\chi).
$$

Away from the integrable direction (i.e. $\chi = 0$), the scattering amplitude $S_{ab}^{cd}(\theta,\chi)$ is no longer a meromorphic function of θ for the opening of inelastic channels. In computing the correction to $S_{ab}^{cd}(\theta,\chi)$ around $\chi = 0$, we must take into account that the total energy of the colliding system is fixed and therefore the variation in the masses given by eqn. (22.4.5) induces a corresponding change in the rapidity difference, expressed by

$$
\delta\theta \simeq -\frac{m_a\delta m_a + m_b\delta m_b + (m_b\delta m_a + m_a\delta m_b)\cosh\theta}{m_a m_b \sinh\theta}. \qquad (22.4.7)
$$

Then the correction to the amplitude can be decomposed as

$$\delta S_{ab}^{cd}(\theta,\chi) = \frac{\partial S_{ab}^{cd}(\theta)}{\partial \theta}\delta\theta + \left.\frac{\partial S_{ab}^{cd}(\theta,\chi)}{\partial \chi}\right|_{\chi=0} \delta\chi. \tag{22.4.8}$$

The first-order result for this quantity is obtained by using formula (22.4.4). Taking into account the cancellation occurring between the disconnected parts of the FFs and the contributions of the counterterms, we finally obtain

$$\delta S_{ab}^{cd}(\theta,\chi) \simeq -i\lambda_2 \frac{F_{\bar{c}\bar{d}ab}^{\varphi_2}(\theta)}{m_a m_b \sinh\theta}, \tag{22.4.9}$$

where

$$F_{\bar{c}\bar{d}ab}^{\varphi_2}(\theta_1-\theta_2) \equiv F_{\bar{c}\bar{d}ab}^{\varphi_2}(\theta_1+i\pi,\theta_2+i\pi,\theta_1,\theta_2). \tag{22.4.10}$$

Cancellation of $i\pi$ singularities. Notice that the right-hand side of (22.4.9) employes the expression of the FF at very special values of the rapidity variables. According to eqn. (19.3.1), the FFs present pole singularities whenever the rapidities of a particle–anti-particle pair differ by $i\pi$ and these kinematical poles are often explicitly inserted into the denominator of their parameterization. Apart from a term encoding the monodromy properties, Chapter 20 showed that the parameterization of the FFs may be written as $\mathcal{Q}/\mathcal{D}$ where both $\mathcal{Q}$ and $\mathcal{D}$ are polynomials in the variables $\cosh\theta_{ij}$: the denominator is uniquely fixed by the pole structure of the S-matrix whereas the numerator is determined by means of the residue equations, as for instance those of eqn. (19.3.1). From the finiteness of the left-hand side of eqn. (22.4.9), we expect that the '$i\pi$ singularities' of the denominator of the FFs $F_{\bar{a}\bar{b}ab}^{\varphi_2}(\theta_1,\theta_2,\theta_3,\theta_4)$ should be cancelled by the polynomial $\mathcal{Q}$, once evaluated at the specific rapidity configuration of eqn. (22.4.10). This should hold in general whenever the perturbing operator is local with respect to the fields that create the particles in the unperturbed theory.[5]

Universal ratios. Eqns. (22.4.5), (22.4.6) and (22.4.9) are the main results of this Chapter. The best use of these formulae is to get rid of the explicit dependence on the normalization of the perturbing operator by defining universal quantities, as for instance ratios of the mass shifts. Hence, under the validity of the linear approximation, all the universal quantities of non-integrable field theories can be entirely expressed in terms of FFs of the integrable ones. Chapter 25 presents a comparison of the theoretical predictions with their numerical determinations.

Trivial deformation and its consequences. It is particularly instructive to specialize the discussion done so far to the 'trivial' case in which the perturbing operator $\varphi_2(x)$ coincides with the same operator $\varphi_1(x)$ that defines the initial integrable theory. In

[5] We defer to the sections devoted to the Ising model and the multi-frequency Sine–Gordon for the discussion of the case in which this condition is not fulfilled.

this case, of course, the physics should be invariant, since the result of the additional perturbation simply corresponds to a shift of the coupling constant of the original integrable model by an amount $\delta\lambda_1 = \lambda_2$. The variations of the masses of the particles and the vacuum energy density corresponding to such a shift can be directly computed from eqns. (22.4.2) and (22.4.3), respectively. But, at the same time, we can also apply our general formulae (22.4.5) and (22.4.6) to estimate the first-order corrections. The two different routes coincide as long as the following identities are valid

$$\begin{aligned} F_{\bar{a}a}^{\Theta}(i\pi,0) &= 2\pi m_a^2, \\ \mathcal{E}_{\text{vac}} &= \tfrac{1}{4\pi}\langle 0|\Theta|0\rangle, \end{aligned} \tag{22.4.11}$$

where $\Theta(x) = 2\pi\lambda_1(2-x_1)\varphi_1(x)$ is the trace of the energy-momentum tensor for the trajectory $\chi = 0$. The two relationships above are indeed true and can be easily derived by other means (see Chapter 20): it is interesting to notice that, in this context, their validity emerges as consistency equations. By the same token, considering higher multi-particle scattering processes, we can generate an infinite number of identities involving the form factors of the original integrable field theory. For instance, next to (22.4.11), a new identity is obtained by comparing eqn. (22.4.8) with eqn. (22.4.9): since χ is constant in the case we are considering, we have

$$\frac{\partial S_{ab}^{cd}(\theta)}{\partial\theta} = -\frac{1}{2\pi i}\,\frac{F_{\bar{c}\bar{d}ab}^{\Theta}(\theta)}{s_{ab}(\theta)}. \tag{22.4.12}$$

This identity provides a simple and unique way to normalize the four-particle form factors of the stress-energy tensor. It may be then particularly useful in the study of massless field theories where the first relationship in eqn. (22.4.11) cannot be used for this purpose.

It is also obvious that the first-order inelastic amplitudes computable by formula (22.4.4) must vanish identically when we choose $\varphi_2(x) = \varphi_1(x)$. This is ensured by the fact that the form factors of the stress-energy tensor $F_{a_1\ldots a_n}^{\Theta}(\theta_1,\ldots,\theta_n)$ factorize the term $P_\mu P^\mu$, with $P^\mu = \sum_{i=1}^n p_i^\mu$ denoting the total energy-momentum of the set of particles. Since $p_i^\mu \to -p_i^\mu$ when the ith particle is crossed from the initial to the final state, P^μ is zero for a set of particles entering a physical scattering process. Only in the case of elastic scattering, the zeros coming from the factor $P_\mu P^\mu$ cancel the kinematical poles and relations analogous to eqn. (22.4.12) are obtained.

22.5 Non-locality and Confinement of the Excitations

Let us consider in more details the mass correction formula, given by

$$\delta m_a^2 \simeq 2\lambda F_{a\bar{a}}^{\varphi_2}(i\pi), \tag{22.5.1}$$

where the the form factor of the operator $\varphi_2(x)$ is defined by the matrix element

$$F_{a\bar{a}}^{\varphi_2}(\theta) \equiv \langle 0|\varphi_2(0)|a(\theta_1)\bar{a}(\theta_2)\rangle. \qquad (22.5.2)$$

Let us recall that the two-particle FF of an integrable theory satisfies the equations

$$F_{a\bar{a}}^{\mathcal{O}}(\theta) = S_{a\bar{a}}^{b\bar{b}}(\theta) F_{\bar{b}b}^{\mathcal{O}}(-\theta), \qquad (22.5.3)$$

$$F_{a\bar{a}}^{\mathcal{O}}(\theta + 2i\pi) = e^{-2i\pi\gamma_{\mathcal{O},a}} F_{\bar{a}a}^{\mathcal{O}}(-\theta). \qquad (22.5.4)$$

In the second equation the explicit phase factor $e^{-2i\pi\gamma_{\mathcal{O},a}}$ is inserted to take into account a possible semi-locality of the operator which interpolates the particles and the operator $\mathcal{O}(x)$. If $\gamma \neq 0$, the two-particle FF presents a pole at $\theta = \pm i\pi$, with the residue given by

$$-i\,\mathrm{Res}_{\theta=\pm i\pi} F_{a\bar{a}}^{\mathcal{O}}(\theta) = (1 - e^{\mp 2i\pi\gamma_{\mathcal{O},a}})\langle 0|\mathcal{O}|0\rangle. \qquad (22.5.5)$$

According whether the perturbing field is local or non-local with respect to the asymptotic particles, we see that there are two different scenarios:

- If the field that breaks integrability is a local operator, the mass correction of the particles is finite. In other words, the particle will be also present in the perturbed non-integrable model, with an adiabatic shift of the value of its mass.
- Viceversa, if the perturbing field is non-local, the mass correction of the particles is divergent. This implies the confinement of the particles, which occurs as soon as the non-integrable perturbation is turned on. In this case the particle does not belong any longer to the spectrum of the deformed non-integrable theory, for this reason we say that it is subjected to *confinement*, as it happens to quarks in chromodynamics.

There are several ways to show the confinement phenomena, the simplest of computing the propagator $\langle A(p)A(-p)\rangle$ of the particle A in the perturbed theory. At the tree level approximation (Figure 22.3), this corresponds to a geometrical series that can be explicitly summed so that

$$\langle A(p)A(-p)\rangle \simeq \frac{1}{p^2 - m^2 - \delta m^2}. \qquad (22.5.6)$$

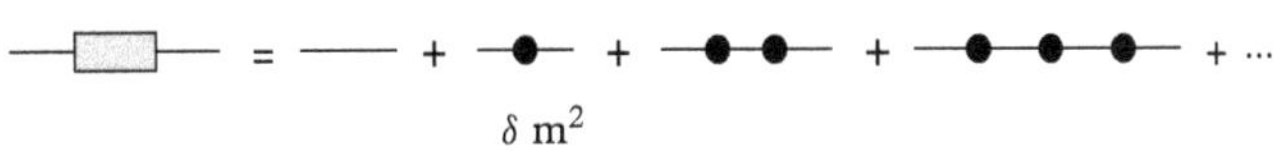

Fig. 22.3 *Perturbative series of the propagator.*

If $\delta m^2 = \infty$, the propagator obviously vanishes, i.e. the particle does not propagate any longer it disappears from the spectrum. A more intuitive explanation of the confinement phenomenon comes from the analysis of the Ising model (which will be discussed in Chapter 25 where it will be compared with its numerical analysis) and the multi-frequency Sine–Gordon model.

22.6 Multi-frequency Sine–Gordon Model

Let us consider the perturbed Sine–Gordon model

$$\mathcal{L} = \frac{1}{2}(\partial_\mu \varphi)^2 + \mu \cos \alpha\varphi + \lambda \cos(\beta\varphi + \delta). \tag{22.6.1}$$

The quantum model depends on μ and λ through the dimensionless variable

$$\eta \equiv \lambda \mu^{-(1-\Delta_\beta)/(1-\Delta_\alpha)} = \lambda \mu^{-(8\pi - \beta^2)/(8\pi - \alpha^2)}.$$

In terms of η, we can identify two perturbative regimes of the model (22.6.1) where we can apply the FFPT: the first regime is obtained in the limit $\eta \to 0$, while the other is reached for $\eta \to \infty$. In the last case, we swap the role played by the two operators. Consider firstly the system in the vicinity of $\eta = 0$: in the unperturbed situation ($\eta = 0$), the field undergoes to the following discontinuity $\varphi \to \varphi + 2\pi/\alpha$ across a soliton configuration. As a consequence, the exponential operator $e^{i\beta\varphi(x)}$ is semi-local with respect to the kink with semi-locality index $\gamma_{\alpha,s} = \beta/\alpha$. According to eqn.(22.5.5), the kink–anti-kink FF of the perturbing operator $\Upsilon = \cos(\beta\varphi + \delta)$ contains annihilation poles at $\theta = \pm i\pi$, with residues given by[6]

$$-i\,\mathrm{Res}_{\theta=\pm i\pi} F^{\Phi}_{a\bar{a}}(\theta) = [\cos\delta - \cos(\delta \mp 2\pi\beta/\alpha)]\, \langle 0|e^{i\beta\varphi}|0\rangle. \tag{22.6.2}$$

Hence, for generic values of the frequency β (and the phase-shift δ) of the perturbing operator $\Upsilon(x)$, the above residues are different from zero. Consequently, the kink and the anti-kink of the original Sine–Gordon model become unstable excitations of the perturbed theory. This result can be easily understood by noting that the perturbed Lagrangian (22.6.1) loses its original $2\pi/\alpha$-periodicity (see plot (a) in Figure 22.4).

The possibility the changes of the spectrum in both perturbative limits ($\eta \to 0$ and $\eta \to \infty$) allows us to deduce interesting information about its evolution also in the intermediate, non-perturbative region. In certain cases, for instance, there may be a topological excitation in one limit that is no longer present in the other. When this happens, the very nature of topological excitations requires that a change in the vacuum

[6] We choose as a vacuum state of the unperturbed theory the one at the origin, characterized by $\langle 0|\varphi|0\rangle = 0$. With this choice the unperturbed theory is invariant under the reflection $\varphi \to -\varphi$ and therefore it holds the equality $\langle 0|e^{i\epsilon\varphi}|0\rangle = \langle 0|e^{-i\epsilon\varphi}|0\rangle$.

structure of the theory shall take place somewhere in the non-perturbative region, namely that there must be a quantum phase transition. Lines of phase transition are then expected to appear in the multi-frequency Sine–Gordon model for particular values of the parameters: as discussed by Delfino and Mussardo in their work on the multi-frequency Sine–Gordon model, this circumstance occurs when $|\delta| = \pi/n$ and $\beta/\alpha = 1/n$. Once these values of the parameters are fixed, by varying η a change of the vacuum state takes place at η_c: at this critical value, the QFT may be regarded as a massless RG flow that interpolates between the CFT with central charge $c = 1$ (in the ultraviolet regime) and the one associated to the Ising model, with $c = 1/2$ (in the infrared regime). Simply note that, at η_c, the effective theory at the vacuum nearby the origin is given by the massless ϕ^4 associated to the class of universality of the Ising model: for η slightly smaller than η_c, there are two degenerate ground states separated by a potential barrier, for η slightly greater than η_c there is only one vacuum with a parabolic shape, while at $\eta = \eta_c$ the quadratic term vanishes, giving rise to an effective ϕ^4 behaviour with massless excitations (plot (b) in Figure 22.4). An interesting statistical mechanics system described by the double Sine–Gordon theory is the Ashkin–Teller model.

22.6.1 The Generalized Ashkin–Teller Model

The Ashkin–Teller model describes two planar Ising models interacting through a local four spin interaction. It is defined by the lattice Hamiltonian

$$H_{AT} = \sum_{(i,j)} [\mathcal{J}(\sigma_1^i \sigma_1^j + \sigma_2^i \sigma_2^j) + K \sigma_1^i \sigma_1^j \sigma_2^i \sigma_2^j], \tag{22.6.3}$$

where $\sigma_{1,2}^i = \pm 1$, the sum is over nearest neighbours and the same coupling $\mathcal{J}$ has been chosen for the two Ising models (isotropic case). A scaling limit description of the model can be formulated in terms of the Euclidean action of a two-layer Ising system (Figure 22.5)

$$\mathcal{A}_{AT} = \mathcal{A}_1^{Ising} + \mathcal{A}_2^{Ising} + \tau \int d^2x\, (\varepsilon_1(x) + \varepsilon_2(x)) + \rho \int d^2x\, \varepsilon_1(x)\varepsilon_2(x), \tag{22.6.4}$$

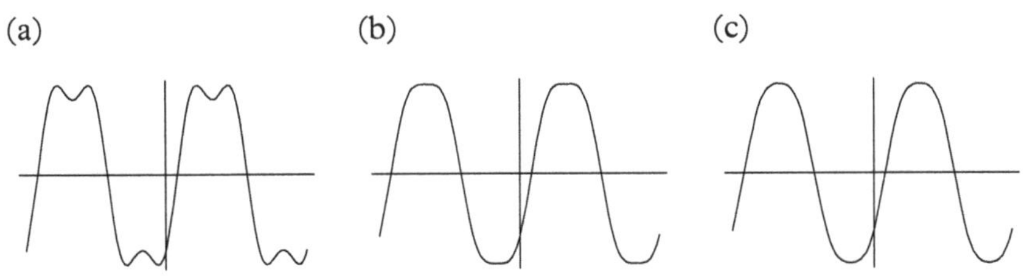

Fig. 22.4 *Potential of the multi-frequency Sine–Gordon model with $\beta/\alpha = 1/3$ and $\delta = \pi/3$ in the vicinity of η_c: (a) $\eta < \eta_c$, (b) $\eta = \eta_c$ and (c) $\eta > \eta_c$.*

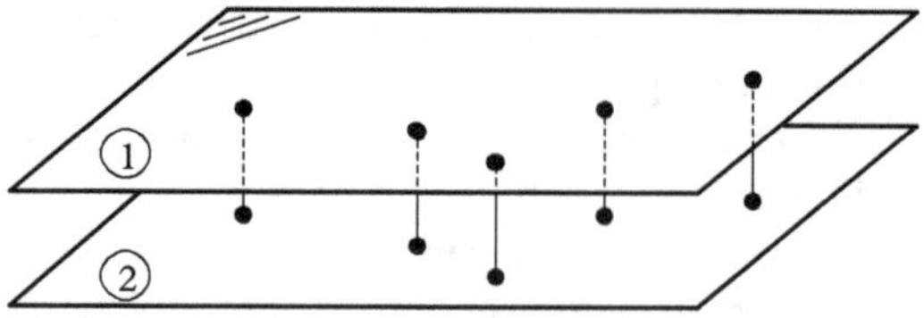

Fig. 22.5 *Two-layer Ising system.*

where $\mathcal{A}_i^{Ising}$ and $\varepsilon_i(x)$ denote the fixed point action and the energy operator of the i-th Ising model, respectively. In this language, the scaling limit of the Ashkin–Teller model appears as a CFT with central charge $C = 1/2 + 1/2 = 1$ perturbed by the two operators $\mathcal{E} \equiv \varepsilon_1 + \varepsilon_2$ and $\varepsilon \equiv \varepsilon_1\varepsilon_2$.

Let us firstly keep the two Ising models at their critical temperature ($\tau = 0$). Since the operator ε is marginal ($\Delta_\varepsilon = 1/2 + 1/2$), its addition to the action of the two critical Ising models does not spoil the criticality of the model, but it leads to a line of $C = 1$ fixed points parameterized by the values of the coupling ρ. It is a well-known property of the model that, while the spin fields σ_1 and σ_2 retain along the critical line the conformal dimensions they have at the decoupling point $\rho = 0$ (i.e. $\Delta_\sigma = 1/16$), the dimensions of the total energy operator $\mathcal{E} = \varepsilon_1 + \varepsilon_2$ and of the so-called 'polarization' operator $P = \sigma_1\sigma_2$ become instead ρ-dependent. Their ratio, however, remains the same as at the decoupling point

$$\frac{\Delta_P(\rho)}{\Delta_{\mathcal{E}}(\rho)} = \frac{\Delta_P(0)}{\Delta_{\mathcal{E}}(0)} = \frac{1/16 + 1/16}{1/2} = \frac{1}{4}. \tag{22.6.5}$$

Thus, the action (22.6.4) with $\tau \neq 0$ can be seen as a $C = 1$ CFT perturbed by the operator $\mathcal{E}$ with ρ-dependent conformal dimension, namely as a pure Sine–Gordon model with the frequency α of its cosine term determined by the relation

$$\Delta_{\mathcal{E}}(\rho) = \alpha^2/8\pi. \tag{22.6.6}$$

Note that, since $\Delta_{\mathcal{E}}(0) = 1/2$, the decoupling point corresponds to $\alpha = \sqrt{4\pi}$. For this value of the coupling, the SG model is equivalent to a free Dirac fermion (see Section 12.4 and the Problem 12.7), as also expected from the well-known equivalence between the thermal Ising model and a free Majorana fermion.

Consider now a generalization of the Ashkin–Teller model defined by the action

$$\mathcal{A}_h = \mathcal{A}_{AT} + h\int d^2x\,\sigma_1(x)\sigma_2(x), \tag{22.6.7}$$

with an additional spin–spin interaction between the two planar Ising models. For $\tau = 0$, it can be shown that the action (22.6.7) defines an integrable massive field theory. For $\tau \neq 0$, there is however the possibility of fine-tuning the two parameters τ and h in such a way that a massless RG flow is obtained. The infrared fixed point can be easily identified by means of the following physical consideration. Let us consider first the strong–coupling limit $h \to \infty$: in this regime, the last term in (22.6.7) forces the spins σ_1 and σ_2 to assume the same value on each site. Hence, the resulting system is effectively reduced to a single Ising model. The reduced system is generally massive but, as any other Ising model, it may become critical provided the temperature of the two original layers is appropriately tuned to its critical value. It can be easily argued that the strong–coupling limit scenario just described should also occur for finite values of h: the coupling h tends to order the system (hence introducing a finite correlation length) but this tendency can be contrasted by a value of the temperature sufficiently high. In conclusion, for each value of ρ, there should be a critical line in the τh-plane along which the system flows from the $C = 1$ fixed point to the $C = 1/2$ Ising fixed point. For the symmetries of the problem, the approach to the Ising fixed point must occur along one of the irrelevant directions of the conformal family of the identity operator: in fact, the massless flow is along the line which separates the high- and the low-temperature phases, hence the conformal families of the magnetic and energy operators of the Ising model are both ruled out since the former is odd under the Z_2 spin symmetry while the latter is odd under the high/low temperature duality of the model. So, only the conformal family of the identity operator remains.

The previous discussion can be rephrased in the bosonic language. Since $\mathcal{E} \sim \cos\alpha\varphi$, eqn. (22.6.5) suggests the identification $P \sim \cos(\alpha\varphi/2 + \delta_P)$, with δ_P to be fixed. The model (22.6.7) then corresponds to the two-frequency SG model (22.6.1), with the identification $\mu = \tau$, $\lambda = h$, $\beta = \alpha/2$, $\delta = \delta_P$, with α determined in terms of ρ through (22.6.6). We found in the previous section that, for $\beta/\alpha = 1/n$, a flow to the Ising fixed point takes place in the $\lambda\mu$-plane if $|\delta| = \pi/n$. This implies $|\delta_P| = \pi/2$, so that $P \sim \sin(\beta\varphi/2)$.

REFERENCES

The FFPT of non-integrable models has been introduced in the articles

G. Delfino, G. Mussardo, P. Simonetti, *Nonintegrable quantum field theories as perturbations of certain integrable models*, Nucl. Phys. B 473 (1996), 469.

G. Delfino, G. Mussardo, *Non-integrable aspects of the multifrequency Sine-Gordon model*, Nucl. Phys. B 516 (1988), 675.

The decay processes of the heavier particles of the Ising model in a magnetic field is discussed in

G. Delfino, P. Grinza, G. Mussardo, *Decay of particles above threshold in the Ising field theory with magnetic field*, Nucl. Phys. B 737 (2006), 291.

Other properties of the spectrum in the Ising model have also been studied in

Rutkevich, S.B. (2005). 'Large n Excitations in the Ferromagnetic Ising Field Theory in a Weak Magnetic Field: Mass Spectrum and Decay Widths', *Physics Review Letters* 95: 250601.

Zamolodchikov, A. and Ziyatdinova, I. (2011). 'Inelastic Scattering and Elastic Amplitude in Ising Field Theory in a Weak Magnetic Field at $T > T_c$. Perturbative Analysis', *Nuclear Physics B*, 849 [FS]: 654.

The infinite tower of bound states in the Ising model has been firstly identified in

McCoy, B.M. and Wu, T.T. (1978). 'Two-dimensional Ising Field Theory in a Magnetic Field: Breakup of the Cut in the Two-point Function', *Physics Review D*, 18: 1259.

The analytic properties of the free energy of the Ising model by varying T and h can be found discussed in

P. Fonseca, A. Zamolodchikov *Ising field theory in a magnetic field: Analytic properties of the free energy*, J. Stat. Phys. 110 (2003), 527.

Other references which may be useful to understand the non-integrable effects are

Controzzi, D. and Mussardo, G. (2004). 'On the Mass Spectrum of the Two–dimensional O(3) Sigma Model with θ Term, *Physics Review Letters*, 92: 021601.
Delfino, G. and Grinza, P. (2008). 'Confinement in the q-state Potts Field Theory', *Nuclear Physics B*, 799: 364.
Lepori, L., Mussardo, G. and Toth, G. (2008). 'The Particle Spectrum of the Tricritical Ising Model with Spin Reversal Symmetric Perturbations', *Journal of Statistics*, 08/09: P09004.
Mussardo, G., Riva, V., Sotkov, G. and Delfino, G. (2006). 'Kink Scaling Functions in 2-D Nonintegrable Quantum Field Theories', *Nuclear Physics B*, 736: 259.
Mussardo G. and Takacs, G. (2009). 'Effective Potentials and Kink Spectra in Non-integrable Perturbed Conformal Field Theories', *Journal of Physics A: Mathematical Theories*, 42: 304022.
Takacs, G. (2010). 'Form Factor Perturbation Theory from Finite Volume', *Nuclear Physics B*, 825: 466.

PROBLEMS

22.1. Tricritical Ising model with even perturbations

Consider the TIM deformed by its even fields $\epsilon(x)$ (energy density field) and $t(x)$ (vacancy density field) with conformal weights given by $\Delta_2 = 1/10$ and $\Delta_4 = 3/5$ respectively. The action of the perturbed model can be formally written as

$$\mathcal{A} = \mathcal{A}_0 + g_2 \int \varepsilon(x)\, d^2x + g_4 \int t(x)\, d^2x,$$

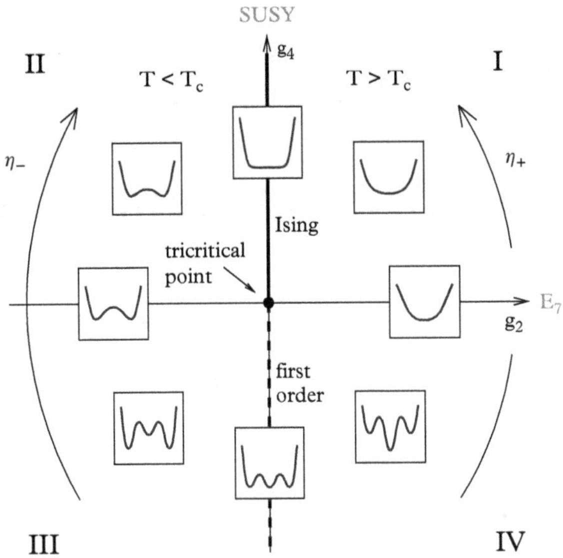

Fig. 22.6 *Landau–Ginzburg potentials associated with the various parts of the phase space.*

where $\mathcal{A}_0$ is the action of the TIM at its conformal point, whereas g_2 and g_4 are the two coupling constants of our theory. The theory depends upon the dimensionless combinations $\eta_\pm = g_4/|g_2|^{\frac{4}{9}}$, where η_+ if $g_2 > 0$ and η_- if $g_2 < 0$.

The model is integrable both when $g_2 = 0$ (this is the supersymmetry line of the model) and $g_4 = 0$ (this is the integrable line associated to the E_7 structure). In more details: the $g_4 > 0$ half line consists of a massless RG flow to the Ising critical point in which the supersymmetry is spontaneously broken, while the $g_4 < 0$ half line consists of first-order phase transition points with unbroken supersymmetry. Along the $g_4 = 0$ line the particle spectrum of the model and its exact S-matrix are related to the E_7 Lie algebra, where the high- and low-temperature phases are related by duality.

a. Use the identification $\varepsilon \sim: \phi^2 :, t \sim: \phi^4 :$ to express the perturbed action (22.6.1) in terms of a Landau–Ginzburg potential and study the evolution of this potential by moving the couplings. Show that the different shapes of the potential match with those shown in Figure 22.6.

b. Use the self-duality of the model to argue that the operator $t(x)$ is *local* with respect to the low-temperature kinks.

c. Using the FFPT nearby the low-temperature axis to show that that the kink excitations do not get confined once the E_7 integrable theory is perturbed by the operator $t(x)$.

d. From the shape of the potential nearby the vertical negative axis of Figure 22.6, argue that the operator $\epsilon(x)$ is non-local with respect to the kinks relative to the first-order phase line of the model.

23

Particle Spectrum by Semi-classical Methods

Life can be simpler if you just take the correct point of view.

Anonymous

23.1 Introduction

The FFPT allows us to control the spectrum of a QFT if this is sufficiently close to an integrable direction. This chapter aims to present another approach for getting the spectrum of a two-dimensional QFT. Such an approach is based on semi-classical methods: in QFT context, it has been promoted in a series of papers by Dashen, Hasslacher and Neveu, and also by Jackiw in separate collaborations with Goldstone and Rebbi. We urge the interested reader to consult these references, together with Rajaraman and the references listed at the end of the chapter for detailed discussions on the many different facets of this semi-classical approach.

The main difference between the FFPT and the semi-classical approach is the following: while the former is based on the S-matrix theory or, more precisely, on the exact scattering amplitudes and FFs of the perturbed integrable models, the latter is based instead on the Lagrangian density of the model, irrespectively whether or not it describes an integrable system. The two approaches sometimes coincide while in general complement each other, i.e. one needs to employ both methods to recover the whole spectrum of the model under scrutiny. It is worth stressing that for the Sine–Gordon model, the semi-classical calculations of the mass spectrum are exact and, for this reason, this model is a benchmark of the method and plays the same role played by the hydrogen atom in quantum mechanics where the semi-classical Bohr–Sommerfeld quantization gives the energy levels exactly.

Rather than making an extended presentation of the semi-classical methods—a task far beyond the scope of this book—this chapter instead focuses on a particular class of two-dimensional QFTs where the application of the semi-classical methods is particularly simple, direct and efficient: those are the models which have several degenerate ground states connected ones to the others by kink configurations. We

Statistical Field Theory: An Introduction to Exactly Solved Models in Statistical Physics. Second Edition. Giuseppe Mussardo, Oxford University Press (2020). © Giuseppe Mussardo. DOI: 10.1093/oso/9780198788102.001.0001

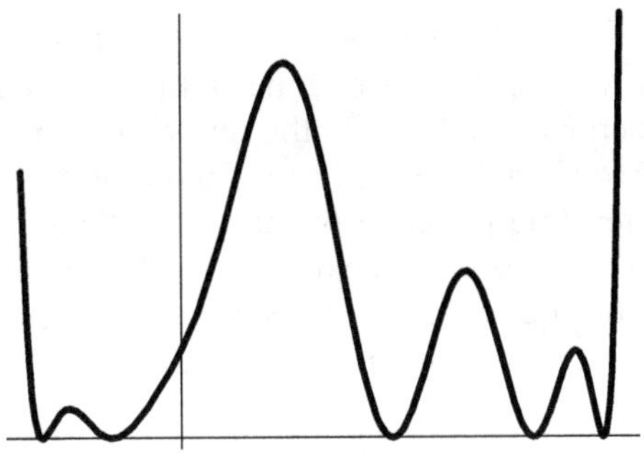

Fig. 23.1 *Potential $U(\varphi)$ of a quantum field theory with kink excitations.*

consider purely bosonic theories described by a scalar real field $\varphi(x)$, with a Lagrangian density

$$\mathcal{L} = \frac{1}{2}(\partial_\mu \varphi)^2 - U(\varphi, \lambda), \tag{23.1.1}$$

where λ is some coupling constant and the potential $U(\varphi)$ possesses several degenerate minima at $\varphi_a^{(0)}$ ($a = 1, 2, \ldots, n$; see Figure 23.1). These minima correspond to the different vacua $| \, \mathbf{a} \rangle$ of the associated QFT.

For such a class of theories there is a very simple formula which expresses the semi-classical mass spectrum of the topological neutral bound states. To this aim, it is useful to discuss first some general features of the topological excitations which exist in two-dimensional Lagrangian theories having several vacua.

23.2 Kinks

For Lagrangian theories with a vacua structure as the one given in eqn. (23.1.1), the most basic excitations are kinks and anti-kinks, i.e. topological configurations which interpolate between two neighbouring vacua. We have already encountered such solutions in Section 16.3 in the contest of the integrable Sine–Gordon model: our aim here twofold, to show that this kind of solutions can appear in a broader context and to underline their important role for the analysis of models which are not necessarily integrable.

Let us remind that the kink configurations correspond to the classical static solutions of the equation of motion, i.e.

$$\partial_x^2 \varphi(x) = U'[\varphi(x), \lambda], \tag{23.2.1}$$

subjected to the boundary conditions

$$\begin{aligned} \varphi(-\infty) &= \varphi_a^{(0)} \\ \varphi(+\infty) &= \varphi_b^{(0)} \end{aligned}, \quad b = a \pm 1. \tag{23.2.2}$$

Let us denote by $\varphi_{ab}(x)$ the classical solution which interpolates between $\varphi_a^{(0)}$ (at $x=-\infty$) and $\varphi_b^{(0)}$ (at $x=+\infty$); conventionally the *kink* corresponds to the field configuration $\varphi_{ab}(x)$ when $b=a+1$, while the *anti-kink* corresponds to the field configuration $\varphi_{ba}(x)$ *anti-kink* (Figure 23.2). In the following, however, it is better to keep in mind that we will often find easier to use the word 'kink' to denote both of them, especially in discussions of general context. As already argued in Section 16.3, these kink configurations can be equivalently obtained as solutions of the simpler first-order differential equation (with the same boundary conditions as above)

$$\frac{d\varphi}{dx} = \pm\sqrt{2U(\varphi)}, \tag{23.2.3}$$

where the $\pm$ signs refer to the kink and the anti-kink respectively. Let us briefly review some basic features of the kink/anti-kink solutions particularly relevant for their physical consequences. In the following, unless differently stated, we take $b=a+1$.

Asymptotic behaviour. As we show here, for finding the spectrum of the topological neutral excitations it is important to know how $\varphi_{ab}(x)$ approaches at $x=\mp\infty$ the two vacua $\varphi_i^{(0)}$ $(i=a,b)$. These asymptotic behaviours are ruled by the curvatures $\omega_i \equiv \sqrt{U''(\varphi_i^{(0)})}$ $(i=a,b)$ at the two vacua: if these quantities are different from zero, the approaches to the two vacua are exponentially fast. To prove this statement, we introduce the variables $\eta_i \equiv \left(\varphi(x)-\varphi_i^{(0)}\right)$ $(i=a,b)$ and expand the right-hand side of eqn. (23.2.3) (for our choice of the indices a and b, the one with positive sign) in power of η_i in the vicinity of the minimum $\varphi_a^{(0)}$ (for $x\to-\infty$) or $\varphi_b^{(0)}$ (for $x\to+\infty$) so that

$$\frac{d\eta_i}{dx} = \omega_i\,\eta_i + \alpha_2^{(i)}\eta_i^2 + \alpha_3^{(i)}\eta_i^3 + \ldots. \tag{23.2.4}$$

The solution of these differential equations provides the sought asymptotic approaches to the two vacua

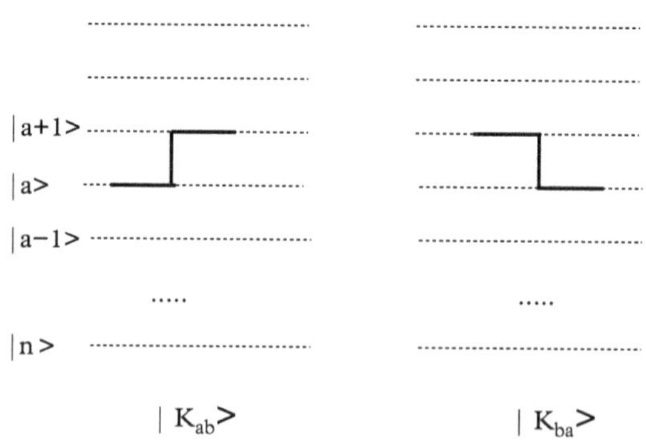

Fig. 23.2 *Schematic kink and anti-kink configurations.*

$$\varphi(x) = \varphi_a^{(0)} + e^{\omega_a x} + \sum_{n=2}^{\infty} \mu_n^{(a)} e^{n\omega_a x}, \quad x \to -\infty \tag{23.2.5}$$

$$\varphi(x) = \varphi_b^{(0)} - e^{-\omega_b x} + \sum_{n=2}^{\infty} \mu_n^{(b)} e^{n\omega_b x}, \quad x \to +\infty \tag{23.2.6}$$

where the coefficients $\mu_n^{(i)}$'s can be iteratively computed in terms of the $\alpha_n^{(i)}$s in (23.2.4). Notice that, in all the exponential terms of these series, the exponents are expressed in terms of *integer multiples* of the curvature ω_i of the corresponding vacuum. When $\omega_a \neq \omega_b$, the exponential approaches to the two vacua are of course different and, in this case, the kink configuration $\varphi_{ab}(x)$ may be not related in a simple way to the anti-kink configuration $\varphi_{ba}(x)$ (by a parity transformation, for instance). As discussed later, this fact has far-reaching consequences on the spectrum of the theory.

Classical mass of the kinks. Associated to each solution $\varphi_{ab}(x)$ there is a classical energy density given by

$$\epsilon_{ab}(x) = \frac{1}{2}\left(\frac{d\varphi_{ab}}{dx}\right)^2 + U(\varphi_{ab}(x)),$$

which, in light of (23.2.3), can be also written as

$$\epsilon_{ab}(x) = 2U(\varphi_{ab}(x)).$$

Integrating this density on the entire real axis, we get the classical value of the mass of this excitation

$$M_{ab} = \int_{-\infty}^{\infty} \epsilon_{ab}(x)\, dx. \tag{23.2.7}$$

Notice that it is not necessary to know the exact solution of the kinks to get their masses M_{ab}. In fact, it can be computed in terms of the potential $U(\varphi)$ alone: using the monotonic behaviour of the kink solution, making a change of variable $t = \varphi_{ab}(x)$ and using once again eqn. (23.2.3), we have in fact

$$M_{ab} = \int_{\varphi_a}^{\varphi_b} \sqrt{2\,U(\varphi)}\, d\varphi. \tag{23.2.8}$$

This equation shows that the mass of the kink $\varphi_{ab}(x)$ is always equal to the mass of the anti-kink $\varphi_{ba}(x)$, even when the two configurations are not related one to the other in a simple way. Generally the masses M_{ab} of the kinks are non-perturbative quantities of the coupling constant λ of the theory, and they diverge when $\lambda \to 0$. This allows us to identify the semi-classical regime as the one where the coupling constant is sufficiently small, so that the masses of the kinks are higher than any other mass scale. Notice that, at the quantum level, the masses M_{ab} of the various kinks will experience a *finite renormalization* with respect to their classical values (23.2.8), an effect that can be found discussed in the original papers by Dashen and colleagues and that we take explicitly into account later.

Dispersion relations. Using the relativistic invariance of Lagrangian theory (16.2.1), we can set the static solutions $\varphi_{ab}(x)$ in motion by making a Lorentz transformation $\varphi_{ab}(x) \rightarrow \varphi_{ab}\left[(x \pm vt)/\sqrt{1-v^2}\right]$: these new configurations correspond to lumps of energy localized at the maximum of the energy density $\epsilon_{ab}(x)$ which fulfill a relativistic dispersion relation. For this reason they are naturally associated to the quantum *kink states* $| K_{ab}(\theta) \rangle$, where θ is the rapidity variable which parameterizes their relativistic dispersion relation

$$E = M_{ab} \cosh\theta, \quad P = M_{ab} \sinh\theta. \tag{23.2.9}$$

Graphical representation. The various kink configurations can be conveniently represented as in Figure 23.3. This graphical representation is obviously a very crude image of the actual exponential approaches to the two vacua (which could be also different), yet it provides a useful rule of thumb to build up and depict the multi-kink states of the quantum theory: these are given by a string of kinks, satisfying the adjacency condition of the consecutive indices for the continuity of the field configuration, as shown in Figure 23.3.

$$| K_{a_1,a_2}(\theta_1) K_{a_2,a_3}(\theta_2) K_{a_3,a_4}(\theta_3) \ldots \rangle, \quad (a_{i+1} = a_i \pm 1).$$

Notice that, associated to the field theory (16.2.1), there is the conserved current $j^{\mu}(x) = \mathcal{N} \epsilon^{\mu\nu} \partial_{\nu} \varphi(x)$, whose charge

$$\mathcal{T} = \mathcal{N} \int_{-\infty}^{\infty} \partial_1 \varphi(x) = \mathcal{N} [\varphi(+\infty) - \varphi(-\infty))], \tag{23.2.10}$$

assumes integer number values provided we properly choose the constant $\mathcal{N}$ in reference to the various values of the vacua $\varphi_n^{(0)}$. These integers $\mathcal{T}$ label the different topological

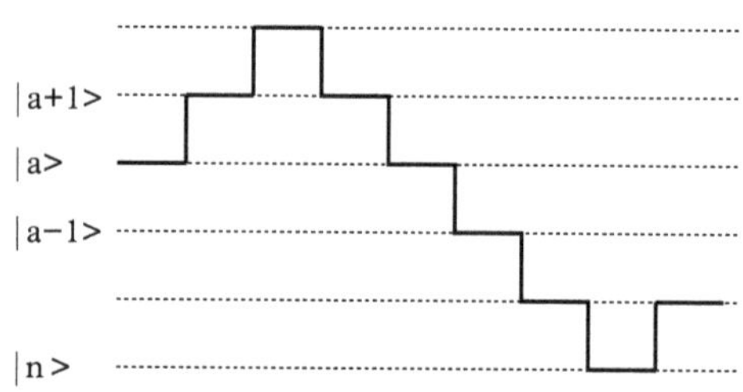

Fig. 23.3 *Multi-kink configurations.*

sectors of the theory, where $\mathcal{T} = 0$ refers to the Hilbert space build upon the various vacua $|\,\mathbf{a}\rangle$, while $\mathcal{T} = \pm 1$ to the sectors of kink and anti-kink. Other values of $\mathcal{T}$ refer to higher multi-kink sectors.

Neutral bound states. In addition to the topological kink states, in the quantum theory there may exist other excitations in the guise of ordinary scalar and neutral particles (breathers) $|\,B_c(\theta)\,\rangle_a$ $(c = 1, 2, \ldots)$ that exist above each of the vacua $|\,\mathbf{a}\rangle$. These excitations belong to the $\mathcal{T} = 0$ sector of the theory. For a theory based on a Lagrangian of a single real field, these states are all non-degenerate: in fact, there are no extra quantities that commute with the Hamiltonian and that can give rise to a multiplicity of them. The only exact (i.e. unbroken) symmetries for a Lagrangian as (16.2.1) may be the discrete ones, like the charge conjugation $\mathcal{C}$, for instance, or the parity transformation P. In this case, however, being neutral excitations, they will be simply either even or odd eigenvectors of $\mathcal{C}$.

These neutral particles must be identified as the bound states of the kink-anti-kink configurations $|\,K_{ab}(\theta_1)\,K_{ba}(\theta_2)\,\rangle$ that start and end at the same vacuum $|\,\mathbf{a}\rangle$, with the 'tooth' shapes shown in Figure 23.4. Notice that if the vacuum $|\,\mathbf{a}\rangle$ is connected to *two* different vacua rather than only one, there are *two* different kink–anti-kink states starting and ending at $|\,\mathbf{a}\rangle$, i.e. $|\,K_{a,a-1}K_{a-1,a}\rangle$ and $|\,K_{a,a+1}K_{a+1,a}\rangle$. However, we argue later that the only two-kink state that matters for the spectrum of the breathers is the one made of the kinks with the *lower* mass between the two values $M_{a,a\pm 1}$ or, in presence of a degeneracy of the two states $|\,K_{a,a\pm 1}K_{a\pm 1,a}\rangle$, a special linear combination of them. Taken this rule as granted, let us then assume that the kink-anti-kink configuration $|\,K_{ab}(\theta_1)K_{ba}(\theta_2)\rangle$ is the one made of the kink with the lowest mass M^*_{ab} originating from the vacuum $|\,\mathbf{a}\rangle$. Hence the breathers $|\,B_c\rangle_a$ correspond to the poles at the imaginary values $i\,u^c_{ab}$ within the physical strip $0 < \mathbf{Im}\theta < \pi$ of the rapidity difference $\theta = \theta_1 - \theta_2$ of this state

$$|\,K_{ab}(\theta_1)\,K_{ba}(\theta_2)\,\rangle \simeq i\,\frac{g^c_{ab}}{\theta - iu^c_{ab}}\;|\,B_c\,\rangle_a, \tag{23.2.11}$$

where g^c_{ab} is the on-shell three-particle coupling between the kinks and the neutral particle. Knowing the resonance value $i\,u^c_{ab}$, the mass of the bound states are simply obtained by substituting this quantity into the expression of the Mandelstam variable s of the two-kink channel

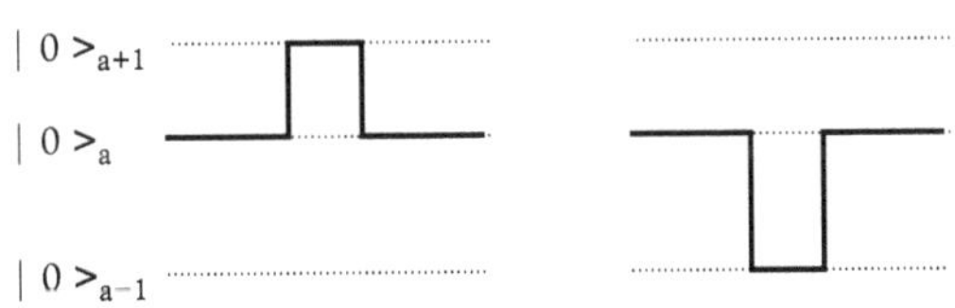

Fig. 23.4 *Kink–anti-kink configurations that may give rise to a bound state nearby the vacuum* $|\,0\,\rangle_a$.

$$s = 4M_{ab}^{*2} \cosh^2 \frac{\theta}{2} \quad \longrightarrow \quad m_c = 2M_{ab}^* \cos \frac{u_{ab}^c}{2}. \tag{23.2.12}$$

The determination of these resonance values u_{ab}^c and the semi-classical computation of the spectrum of these bosonic neutral excitations are discussed next.

23.3 A Semi-classical Formula for the Kink Matrix Elements

The starting point for getting the semi-classical spectrum of the neutral bound states is a remarkably simple formula for the matrix element of the field $\varphi(x)$ on the kink states shown in Figure 23.5. Such a formula reads[1]

$$f_{ab}^{\varphi}(\theta) = \langle K_{ab}(\theta_1) \mid \varphi(0) \mid K_{ab}(\theta_2) \rangle \simeq \int_{-\infty}^{\infty} dx \, e^{iM_{ab}\theta x} \, \varphi_{ab}(x), \tag{23.3.1}$$

where $\theta = \theta_1 - \theta_2$. Notice that, by substituting in the above formula $\theta \to i\pi - \theta$, the corresponding expression may be interpreted as the following FF

$$F_{ab}^{\varphi}(\theta) = f(i\pi - \theta) = \langle a \mid \varphi(0) \mid K_{ab}(\theta_1) K_{ba}(\theta_2) \rangle, \tag{23.3.2}$$

where it enters the neutral kink–anti-kink states around the vacuum $\mid \mathbf{a} \rangle$ that we are interested in.

Let us present the derivation of the formula (23.3.1) into account that, if λ is the dimensionless coupling constant of the theory, the semi-classical regime is characterized by the limit $\lambda \to 0$, where the mass of the kink becomes arbitrarily larger than any

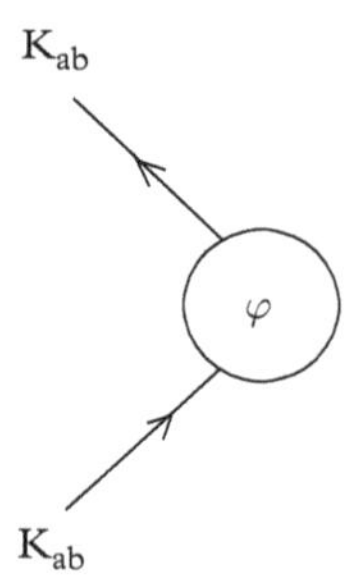

Fig. 23.5 *Matrix element between kink states.*

[1] The matrix element of the field $\varphi(y)$ is easily obtained using $\varphi(y) = e^{-iP_\mu y^\mu} \varphi(0) e^{iP_\mu y^\mu}$ and acting with the conserved energy-momentum operator P_μ on the kink state. Moreover, for the semiclassical matrix element $F_{ab}^{\mathcal{G}}(\theta)$ of the operator $G[\varphi(0)]$, we should employ $G[\varphi_{ab}(x)]$. For instance, the matrix element of $\varphi^2(0)$ is given by the Fourier transform of $\varphi_{ab}^2(x)$.

other energy scales. Consider initially the Heisenberg equation of motion satisfied by the field $\varphi(x)$

$$\left(\partial_t^2 - \partial_x^2\right) \varphi(x,t) = -U'[\varphi(x,t)],$$

and then sandwich it between the kink states of momentum p_1 and p_2. By using

$$\langle K_{ab}(p_1) \mid \varphi(x,t) \mid K_{ab}(p_2)\rangle = e^{-i(p_1-p_2)_\mu x^\mu} \langle K_{ab}(p_1) \mid \varphi(0) \mid K_{ab}(p_2)\rangle \qquad (23.3.3)$$

we have

$$\begin{aligned} &\left[-(p_1-p_2)_\mu (p_1-p_2)^\mu\right] \langle K_{ab}(p_1) \mid \varphi(0) \mid K_{ab}(p_2)\rangle \\ &= -\langle K_{ab}(p_1) \mid U'[\varphi(0)] \mid K_{ab}(p_2)\rangle. \end{aligned} \qquad (23.3.4)$$

Once we have extracted the x^μ-dependence of the matrix element (23.3.3), the remaining expression $\langle K_{ab}(p_1) \mid \varphi(0) \mid K_{ab}(p_2)\rangle$ can only depend on the relativistic invariants of the channel of the two kinks expressed by the Mandelstam variables. These can be expressed in terms of the difference of the rapidities of the two kinks, a fact that suggests adopting the rapidity variables and writing eqn. (23.3.4) as

$$\begin{aligned} &2M_{ab}^2(\cosh\theta - 1) \langle K_{ab}(\theta_1) \mid \varphi(0) \mid K_{ab}(\theta_2)\rangle \\ &= -\langle K_{ab}(\theta_1) \mid U'[\varphi(0)] \mid K_{ab}(\theta_2)\rangle, \end{aligned} \qquad (23.3.5)$$

where $\theta = \theta_1 - \theta_2$. Assuming that the kinks are sufficiently slow, their dispersion relations can be approximated by the non-relativistic expressions

$$E = M_{ab}\cosh\theta \simeq M_{ab}\left(1+\frac{\theta^2}{2}\right), \quad P = M_{ab}\sinh\theta \simeq M_{ab}\,\theta \ll M_{ab}.$$

In this quasi-static regime, we can define the matrix element of the field $\varphi(0)$ between the kink states as the Fourier transform with respect to the Lorentz invariant difference $\theta = \theta_1 - \theta_2$

$$f_{ab}^{\varphi}(\theta) = \langle K_{ab}(\theta_1) \mid \varphi(0) \mid K_{ab}(\theta_2)\rangle \simeq \int_{-\infty}^{\infty} dx\, e^{iM_{ab}\theta x}\, \hat{f}(x), \qquad (23.3.6)$$

with the inverse Fourier transform given by

$$\hat{f}(x) = \int \frac{d\theta}{2\pi}\, e^{-iM_{ab}\theta x} f_{ab}^{\varphi}(\theta).$$

In the quasi-static limit, the left-hand side of equation (23.3.5) becomes $M_{ab}^2 \theta^2 f_{ab}^{\varphi}(\theta)$ which, in real space, corresponds to $-\frac{d^2}{dx^2}\hat{f}_{ab}(x)$. For what concerns the right-hand side, assuming that $U'[\phi(0)]$ can be expressed in terms of powers of the field $\varphi(x)$ (either as a finite sum or an infinite series)

$$U'[\varphi] = \sum_{n=1} \alpha_n \varphi^n.$$

Let us consider the matrix elements of the generic term of this expression on the kink states $\langle K_{ab}(\theta_1) \mid \varphi^n(0) \mid K_{ab}(\theta_2)\rangle$. Inserting $(n-1)$ times a complete set of states, for these matrix elements we have

$$\sum_{m_1,\ldots m_{n-1}} \langle K_{ab}(\theta_1) \mid \varphi(0) \mid m_1\rangle \langle m_1 \mid \varphi(0) \mid m_2\rangle \ldots \langle m_{n-1} \mid \varphi(0) \mid K_{ab}(\theta_2)\rangle. \quad (23.3.7)$$

The only states that contribute to the above sums are those having the same topological charge of the kink K_{ab}, with the lowest mass states given precisely by the kinks $K_{ab}(\theta_i)$ themselves. By truncating the sums, keeping just these states and using the definition (23.3.6), we have then

$$\langle K_{ab}(\theta_1) \mid \varphi^n(0) \mid K_{ab}(\theta_2)\rangle \simeq \int_{-\infty}^{\infty} dx\, e^{iM_{ab}\theta x} \left(\hat{f}(x)\right)^n.$$

Hence, at the leading order $1/\lambda$, the function $\hat{f}(x)$ satisfies the same differential equation (23.2.1) satisfied by the static kink solution, i.e.

$$\frac{d^2}{dx^2}\hat{f}_{ab}(x) = U'[f_{ab}(x)],$$

in this way arriving then to the result (23.3.1).

23.4 Universal Mass Formula

The appealing aspect of the formula (23.3.1) is that it unveils an interesting relation between the Fourier transform of the *classical* configuration of the kink—i.e. the solution $\varphi_{ab}(x)$ of the differential equation (23.2.3)—and the *quantum* matrix element of the field $\varphi(0)$ between the vacuum $\mid \mathbf{a}\rangle$ and the two-particle kink state $\mid K_{ab}(\theta_1) K_{ba}(\theta_2)\rangle$. Notice that, once the solution of eqn. (23.2.3) has been found and its Fourier transform has been taken, the poles of $F_{ab}(\theta)$ within the physical strip of θ identify the neutral bound states which couple to φ: their mass can be extracted by using eqn. (23.2.12), while the on-shell three-particle coupling g_{ab}^c can be obtained from the residue at these poles (Figure 23.6)

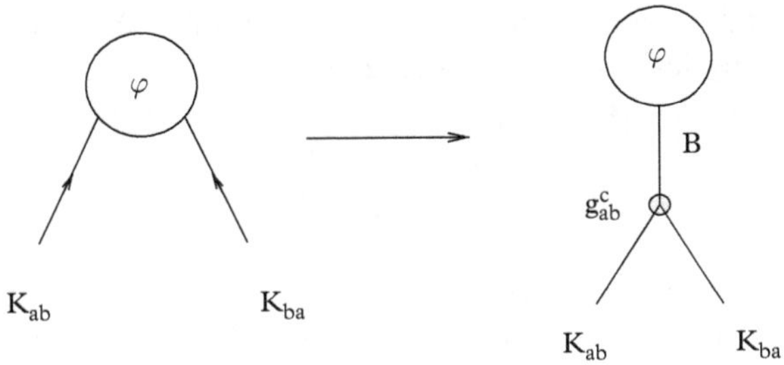

Fig. 23.6 *Residue equation for the matrix element on the kink states.*

$$\lim_{\theta \to i u^c_{ab}} (\theta - i u^c_{ab}) F_{ab}(\theta) = i g^c_{ab} \langle a \mid \varphi(0) \mid B_c \rangle. \tag{23.4.1}$$

For a generic theory we already commented that the classical kink configuration $\varphi_{ab}(x)$ is not related in a simple way to the anti-kink configuration $\varphi_{ba}(x)$. It is precisely for this reason that neighbouring vacua may have different spectra of neutral excitations, as discussed below. It is also worth noting that extracting the masses of the bound states through the poles of the Fourier transform of the kink solutions is a procedure much simpler than carrying out directly the semi-classical quantization of the breather solutions, as it was done originally by Dashen, Hasslacher and Neveu. The reason is that the classical breather configurations depend also on time and they have, in general, a more complicated structure than the kink ones. Moreover, in non-integrable theories, these configurations do not exist as exact solutions of the partial differential equations of the field theory. On the contrary, observe that in order to apply eqn. (23.3.1), it is simply needed the solution of an *ordinary* differential equation, the one given by (23.2.3), solution that can be found either exactly or numerically. Moreover, for the purpose of locating the poles of the Fourier transform it is sufficient to look at the exponential behaviour of the kink solutions at $x \to \pm\infty$, expressed in eqns. (23.2.5) and (23.2.6). Given that they consist of a series made of multiples of the same exponential (expressed by the curvature ω_a at that minimum), we can rapidly conclude that the poles of $F^{\varphi}_{ab}(\theta)$ are regularly spaced by

$$\xi_a \equiv \frac{\omega_a}{\pi M_{ab}}. \tag{23.4.2}$$

Focusing our attention on the kink of lower mass starting from the vacuum $\mid \mathbf{a} \rangle$, the semi-classical mass spectrum of the bosonic neutral bound states above the vacuum $\mid \mathbf{a} \rangle$ assumes then the universal form

$$m^B_a(n) = 2M^*_{ab} \sin\left(n \frac{\pi\, \xi_a}{2}\right) \tag{23.4.3}$$
$$n = 1, 2, \ldots N^B_a$$

where N_a^B is the total number of bound states on top of it: this is the number of poles that fall within the physical strip and it is given by

$$N_a^B = \left[\frac{1}{\xi_a}\right], \tag{23.4.4}$$

where $[x]$ expresses the integer part of the number x. Notice that the mass formula for the breathers of the Sine–Gordon model (Section 18.10), matches perfectly the general mass formula (23.4.3). Posing

$$m_1^B \equiv m_a^B(1) = 2M_{ab}\sin\left(\frac{\pi\,\xi_a}{2}\right),$$

the formula (23.4.3) can be equivalently written as

$$m_a^B(n) = m_1^B\,\frac{\sin\left(n\,\frac{\pi\,\xi_a}{2}\right)}{\sin\left(\frac{\pi\,\xi_a}{2}\right)} \tag{23.4.5}$$

$$n = 1, 2, \ldots N_a^B.$$

Main conclusions. It is possible to draw some main conclusions of our analysis:

1. If $\xi_a > 1$ the poles of $F_{ab}^{\varphi}(\theta)$ are outside the physical strip and therefore the vacuum $|\,\mathbf{a}\,\rangle$ does not support any bound state.
2. If $\frac{1}{2} < \xi_a < 1$ on the vacuum $|\,\mathbf{a}\,\rangle$ there is only one bound state.
3. If $\frac{1}{3} < \xi_a < \frac{1}{2}$, on the vacuum $|\,\mathbf{a}\,\rangle$ there are instead two bound states.
4. If $\xi_a < \frac{1}{3}$ and the theory is non-integrable, out of the possible $\left[\frac{1}{\xi_a}\right]$ bound states only the lowest two will be stable, the others being resonances. Indeed, for a simple geometrical property of the trigonometric sine function of multiple angles, we have $m_n > 2m_1$ (for $n > 3$) and therefore these higher particles are generically expected to decay.

Hence, as a general conclusion, the semi-classical analysis of a generic non-integrable bosonic theory predicts that such a theory cannot have more than two stable bound states in its spectrum.

An important issue concerns the range of validity of the formula (23.3.1) and the consequent mass formula (23.4.5). Formula (23.3.1) was derived assuming that the kink momentum was small compared to its mass together with the possibility of neglecting intermediate higher particle contributions. These two assumptions usually end up in the combined condition $\theta \simeq \mathcal{O}(\lambda) \ll 1$, where λ is the adimensional coupling constant of the theory. This allows us to substitute in the result of the Fourier transform, $\theta \to \sinh\theta$

(since θ is infinitesimal), but keeping untouched all expressions containing θ/λ. This constraint may result in a different level of accuracy on various physical quantities, with the precision that may also depend on the model under investigation. However, properly handled, the semi-classical formalism remains one of the most powerful methods for extracting the spectrum of the bound states in kink-like theories. To show its efficiency, the next sections analyse some important examples, starting first from QFTs with two symmetric vacua, then a model with two asymmetric vacua and finishing with the discussion of the double Sine–Gordon model.

23.5 Symmetric Wells

A prototype example of a potential with two symmetric wells is the φ^4 theory in its broken phase. The potential is given in this case by

$$U(\varphi) = \frac{\lambda}{4}\left(\varphi^2 - \frac{m^2}{\lambda}\right)^2 .$$

Let us denote with $|\pm 1\rangle$ the vacua corresponding to the classical minima $\varphi_{\pm}^{(0)} = \pm\frac{m}{\sqrt{\lambda}}$. By expanding around them, $\varphi = \varphi_{\pm}^{(0)} + \eta$, we have

$$U(\varphi_{\pm}^{(0)} + \eta) = m^2\eta^2 \pm m\sqrt{\lambda}\,\eta^3 + \frac{\lambda}{4}\eta^4. \tag{23.5.1}$$

Hence ordinary perturbation theory predicts the existence of a neutral particle above each of the two vacua, with a bare mass given by $m_b = \sqrt{2}m$, irrespectively of the value of the coupling λ. Instead, what is the result of the semi-classical analysis.

The kink solutions are given in this case by[2]

$$\varphi_{-a,a}(x) = a\frac{m}{\sqrt{\lambda}}\tanh\left[\frac{mx}{\sqrt{2}}\right], \quad a = \pm 1 \tag{23.5.2}$$

and their classical mass is

$$M_0 = \int_{-\infty}^{\infty} \epsilon(x)\,dx = \frac{2\sqrt{2}}{3}\frac{m^3}{\lambda}.$$

The classical value of the mass of the kink receives corrections from the small oscillations around the classical static configurations, as shown in the original articles by Dashen and colleagues. In this model, the correction changes the kink mass as

[2] In the following we will discard the integration constant x_0 on which is localized the kink solution.

$$M = \frac{2\sqrt{2}}{3}\frac{m^3}{\lambda} - m\left(\frac{3}{\pi\sqrt{2}} - \frac{1}{2\sqrt{6}}\right) + \mathcal{O}(\lambda). \tag{23.5.3}$$

It is convenient to define

$$c = \left(\frac{3}{2\pi} - \frac{1}{4\sqrt{3}}\right) > 0,$$

and also the adimensional quantities

$$g = \frac{3\lambda}{2\pi m^2}; \quad \xi = \frac{g}{1 - \pi c g}. \tag{23.5.4}$$

In terms of them, the mass of the kink can be expressed as

$$M = \frac{\sqrt{2}m}{\pi\,\xi} = \frac{m_b}{\pi\,\xi}. \tag{23.5.5}$$

Since the kink and the anti-kink solutions are equal functions (up to a sign), their Fourier transforms have the same poles. Hence, the spectrum of the neutral particles will be the same on both vacua, in agreement with the Z_2 symmetry of the model. To be specific, it is more convenient to consider the Fourier transform of the derivative of the kink configuration

$$\left(\frac{d\varphi}{dx}\right)_{-a,a} = a\,\frac{m^2}{\sqrt{2\lambda}}\,\frac{1}{\cosh^2\frac{mx}{\sqrt{2}}}. \tag{23.5.6}$$

In this way we can get rid of the constant, albeit singular, term of the Fourier transform (23.3.1) leaving however untouched the pole structure. As a result, we have

$$f_{-a,a}(\theta) = \int_{-\infty}^{\infty} dx\, e^{iM\theta\, x} \varphi_{-a,a}(x) = i\,a\sqrt{\frac{2}{\lambda}}\,\frac{1}{\sinh\left(\frac{\pi M}{\sqrt{2}m}\theta\right)}.$$

By making now the analytical continuation $\theta \to i\pi - \theta$ and using the above definitions (23.5.4), we arrive to

$$F_{-a,a}(\theta) = \langle a \mid \varphi(0) \mid K_{-a,a}(\theta_1)K_{a,-a}(\theta_2)\rangle = A_a\,\frac{1}{\sinh\left(\frac{(i\pi - \theta)}{\xi}\right)}, \tag{23.5.7}$$

where A_a is a constant, given by

$$A_a = i\left[am\left(\frac{3(1+\pi c\xi)}{(\pi\xi)}\right)^{1/2}\right].$$

The poles of this expression are located at

$$\theta_n = i\pi(1-\xi n), \quad n = 0, \pm 1, \pm 2, \ldots \tag{23.5.8}$$

and, for $\xi \geq 1$, none of them is in the physical strip $0 < \text{Im}\,\theta < \pi$. Consequently, in the range of the coupling constant

$$\frac{\lambda}{m^2} \geq \frac{2\pi}{3}\frac{1}{1+\pi c} = 1.02338\ldots \tag{23.5.9}$$

the theory does not have any neutral bound states above both vacua. Vice versa, if $\xi < 1$, there are $n = \left[\frac{1}{\xi}\right]$ neutral bound states, where $[x]$ denotes the integer part of the number x. Their semi-classical masses are given by

$$m_b^{(n)} = 2M\sin\left[n\frac{\pi\xi}{2}\right] = n\,m_b\left[1 - \frac{3}{32}\frac{\lambda^2}{m^4}n^2 + \ldots\right]. \tag{23.5.10}$$

Note that the leading term is given by multiples of the mass of the elementary boson $|\,B_1\rangle$. Therefore the nth breather may be considered as a loosely bound state of n of it, with the binding energy provided by the remaining terms of the above expansion. However, for the non-integrability of the theory, all particles with mass $m_n > 2m_1$ will eventually decay. It is easy to see that, if there are at most two particles in the spectrum, it always holds the inequality $m_2 < 2m_1$. However, as previously discussed, if $\xi < \frac{1}{3}$, for the higher particles we always have

$$m_k > 2m_1, \quad \text{for } k = 3, 4, \ldots n.$$

Important remarks. Let us comment on the general scenario emerging from the semi-classical analysis of the Φ^4 Landau–Ginzburg theory. An obvious question concerns the conclusion on the absence of the bound states for $\lambda > \lambda_c$, with the critical value λ_c given in eqn. (23.5.9): after all, this value is not infinitesimal and it might be quite outside the range of validity of the semi-classical approximation we are using. In order to support the scenario coming from the semi-classical analysis let us focus on the following arguments:

1. Concerning λ_c, in other models the critical value of the coupling is definitely not small either. In Sine–Gordon for instance, the critical value beyond which there are non-bound states is given by $\beta_c^2 = 4\pi$.

2. There exists an exact mapping between the kinks of φ^4 and those of the Sine–Gordon model and this may help in arguing about the critical values of these models. by rescaling the coordinate $x \to mx$ and the field $\varphi \to \beta\varphi$ of the Sine–Gordon theory, its Lagrangian can be written as

$$\mathcal{L} = \frac{m^2}{\beta^2}\left[\frac{1}{2}(\partial\varphi)^2 + (\cos\phi - 1)\right], \qquad (23.5.11)$$

and under the substitution

$$\varphi_\pm(x,t) = \pi \pm 4 \arctan \Phi(x,t),$$

it becomes

$$\mathcal{L} = \frac{m^2}{\beta^2}\frac{1}{(1+\Phi^2)^2}\left[\frac{1}{2}(\partial\Phi)^2 - \frac{1}{8}(\Phi^2-1)^2\right].$$

The static solutions of the Sine–Gordon model coincide with those of φ^4, once the coupling constants of the two theories are related as

$$\frac{\lambda}{m^2} = \frac{\beta}{2}.$$

By inserting in this relation the critical value of the coupling of the Sine–Gordon, $\beta_c = \sqrt{4\pi}$, we may get an estimate of the critical value of φ^4 theory

$$\frac{\lambda_c}{m^2} = \sqrt{\pi} = 1.77245$$

Although different, this value is of the same order of magnitude of the one (23.5.9) previously obtained, and it would have been too ambitious to obtain a better agreement between these two critical values of λ. After all, even though the static solutions of the Sine–Gordon and the Φ^4 LG theory are similar, their time–dependent solutions are instead rather different (the Sine–Gordon has an integrable dynamics, whereas φ^3 does not).

3. The possibility of having bound states in the Φ^4 theory in its broken phase is related to the cubic interaction present in the potential (23.5.1). This term, in fact, provides an effective attractive interaction between the particles which, for low values of λ/m^2, overcomes the repulsive interaction given by the φ^4 term. However, the coefficient in front of φ^3 scales as $\sqrt{\lambda}$, whereas the one in front of φ^4 as λ. Therefore, it seems plausible that there should exist a sufficient large value of λ/m^2 where the repulsive force prevails on the attractive one and to have a bound state.
4. There is an additional piece of information encoded in the semi-classical formula of the poles (23.5.8) that can be further exploited. Namely, suppose that the

dimensionless coupling constant g is very small, so that both the formula of the poles and the corresponding one of the masses (23.5.10) can be trusted. These expressions show that, by increasing g, the common tendency of all particles is to move *forward* the threshold of the two-kink state and to decay afterward. If nothing occurs to stop this motion of the poles by increasing the coupling constant, the actual scenario of the theory will be then the one predicted by the semi-classical analysis, i.e. the existence of a critical value g_c (surely different from the one given by the semi-classical approximation) but with no neutral particles beyond it.

5. An independent check of the semi-classical scenario is provided by the numerical analysis of the model done in terms of the truncated conformal space approach discussed in Chapter 25.

23.6 Asymmetric Wells

Let us now see how the situation changes if we consider a potential with two asymmetric wells. Using only power interactions, it is easy to see that to get such a potential we must necessarily employ higher powers than φ^4, the simplest example being the Φ^6 Landau–Ginzburg theory. Remember that in Section 11.6 we showed that such a theory is relevant for the class of universality of the TIM and we will see that the information available on this model provide a nice confirmation of the scenario emerging from the semi-classical analysis.

We can parameterize the class of potentials with two asymmetric wells as

$$U(\varphi) = \frac{\lambda}{2}\left(\varphi + a\frac{m}{\sqrt{\lambda}}\right)^2\left(\varphi - b\frac{m}{\sqrt{\lambda}}\right)^2\left(\varphi^2 + c\frac{m^2}{\lambda}\right), \tag{23.6.1}$$

with a, b, c all positive numbers. To simplify the notation, it is convenient to use the dimensionless quantities obtained by rescaling the coordinate as $x \to mx$ and the field as $\varphi(x) \to \sqrt{\lambda}/m\varphi(x)$. In this way the Lagrangian of the model becomes

$$\mathcal{L} = \frac{m^6}{\lambda^2}\left[\frac{1}{2}(\partial\varphi)^2 - \frac{1}{2}(\varphi + a)^2(\varphi - b)^2(\varphi^2 + c)\right]. \tag{23.6.2}$$

The minima of this potential are localized at $\varphi_0^{(0)} = -a$ and $\varphi_1^{(0)} = b$ and the corresponding ground states are denoted by $|\,0\rangle$ and $|\,1\rangle$. The curvatures of the potential at these vacua are given by

$$\begin{aligned} U''(-a) &\equiv \omega_0^2 = (a+b)^2(a^2+c); \\ U''(b) &\equiv \omega_1^2 = (a+b)^2(b^2+c). \end{aligned} \tag{23.6.3}$$

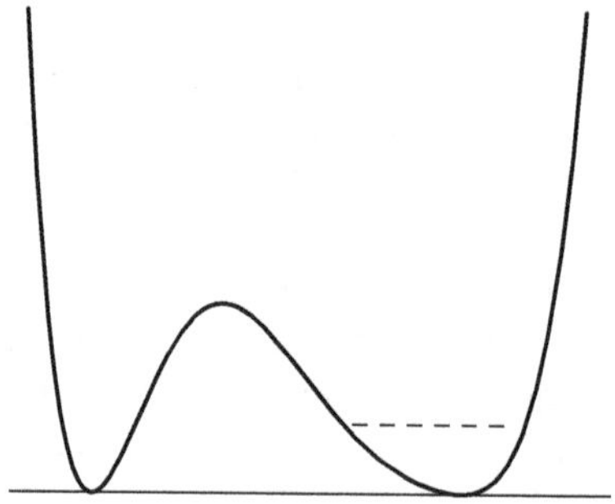

Fig. 23.7 *Example of φ^6 potential with two asymmetric wells, with a bound state only above one of the two vacua.*

Hence, for $a \neq b$, we have two asymmetric wells (Figure 23.7). To be definite, let us assume that the curvature at the vacuum $| \, 0 \rangle$ is greater than the one at the vacuum $| \, 1 \rangle$, i.e. $a > b$.

The problem we address is whether the spectrum of the neutral particles $| \, B \rangle_s$ $(s = 0, 1)$ could be different at the two vacua, in particular, whether it would be possible that one of them (say $| \, 0 \rangle$) has no neutral excitations, whereas the other has just one neutral particle. Ordinary perturbation theory shows that both vacua have neutral excitations, although with different value of their mass:

$$m^{(0)} = (a+b)\sqrt{2\,(a^2+c)}, \quad m^{(1)} = (a+b)\sqrt{2\,(b^2+c)}. \tag{23.6.4}$$

Let's see, instead, what is the conclusion coming from the semi-classical analysis. The kink equation is given in this case by

$$\frac{d\varphi}{dx} = \pm(\varphi + a)(\varphi - b)\sqrt{\varphi^2 + c}. \tag{23.6.5}$$

We will not attempt to solve exactly this equation but we can present nevertheless its main features: the kink solution interpolates between the values $-a$ (at $x = -\infty$) and b (at $x = +\infty$) while the anti-kink solution does the, but with the important difference that its behaviour at $x = -\infty$ is different from that of the kink. As a matter of fact, the behaviour at $x = -\infty$ of the kink is always equal to the behaviour at $x = +\infty$ of the anti-kink (and vice versa), but in this theory the two vacua are approached differently, as explicitly shown in Figure 23.8. This follows from the general analysis encoded in the equations (23.2.5) and (23.2.6). Hence, we have the following exponential approach to the vacuum $| \, 0 \rangle$

$$\varphi_{0,1}(x) \simeq -a + A\exp(\omega_0 x), \quad x \to -\infty$$

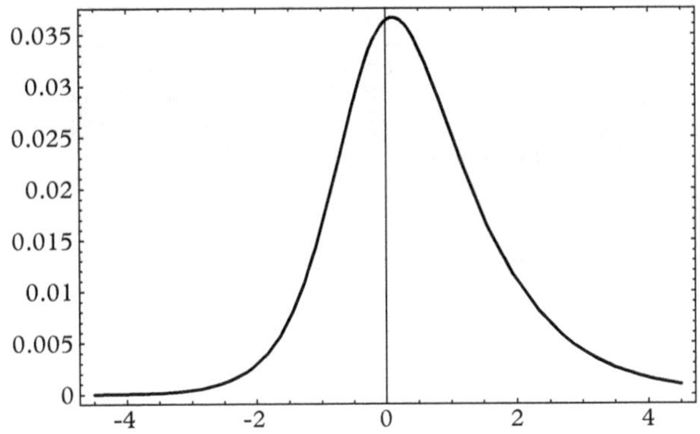

Fig. 23.8 *Typical shape of* $\left(\frac{d\varphi}{dx}\right)_{01}$, *obtained by a numerical solution of eqn. (23.6.5).*

where $A > 0$ is a arbitrary costant (its actual value can be fixed by properly solving the non-linear differential equation). For the exponential approach to the vacuum $\mid 1\rangle$ of the anti-kink we have instead

$$\varphi_{1,0}(x) \simeq b - B\exp(\omega_1 x), \quad x \to -\infty$$

where $B > 0$ is another constant. Since $\omega_0 \neq \omega_1$, the asymptotic behaviour of the two solutions gives rise to the following poles in their Fourier transform

$$\begin{aligned} \mathcal{F}(\varphi_{0,1}) &\to \frac{A}{\omega_0 + ik} \\ \mathcal{F}(\varphi_{1,0}) &\to \frac{-B}{\omega_1 + ik}. \end{aligned} \tag{23.6.6}$$

In order to locate the pole in terms of the variable θ, we reintroduce the correct units. Assuming we have solved the differential equation (23.6.5), the integral of its energy density gives the common mass of the kink and the anti-kink. In terms of the constants in front of the Lagrangian (23.6.2), its value will be given by

$$M = \frac{m^5}{\lambda^2}\alpha,$$

where α is a number (typically of order 1), coming from the integral of the adimensional energy density (23.2.7). Hence, the first pole[3] of the Fourier transform of the kink and the anti-kink solution are localized at

[3] In order to determine the others, one should look for the sub-leading exponential terms of the solutions.

$$\theta^{(0)} \simeq i\pi\left(1-\omega_0\,\frac{m}{\pi M}\right)=i\pi\left(1-\omega_0\,\frac{\lambda^2}{\alpha m^4}\right)$$
$$\theta^{(1)} \simeq i\pi\left(1-\omega_1\,\frac{m}{\pi M}\right)=i\pi\left(1-\omega_1\,\frac{\lambda^2}{\alpha m^4}\right) \tag{23.6.7}$$

It is now easy to see that choosing the coupling constant in the range

$$\frac{1}{\omega_0}<\frac{\lambda^2}{m^4}<\frac{1}{\omega_1}, \tag{23.6.8}$$

the first pole will be out of the physical sheet whereas the second will still remain inside it! Hence, in this regime the theory will have only one neutral bound state, localized at the vacuum $|\,1\,\rangle$. This result may be expressed by saying that the appearance of a bound state depends on the order in which the topological excitations are arranged: an anti-kink–kink configuration gives rise to a bound state whereas a kink–anti-kink does not.

Finally, notice that the value of the adimensional coupling constant can be chosen such that the mass of the bound state around the vacuum $|\,1\,\rangle$ becomes equal to mass of the kink. This happens when

$$\frac{\lambda^2}{m^4}=\frac{\alpha}{3\omega_1}.$$

For this value of the coupling the Φ^6 Landau–Ginzburg (23.6.2) shares the same features of the TIM perturbed by its sub-leading magnetization (the exact scattering matrix was discussed in Section 18.11.3). A simple way to check the above scenario would be to study the finite volume energy spectrum with periodic boundary conditions for the field

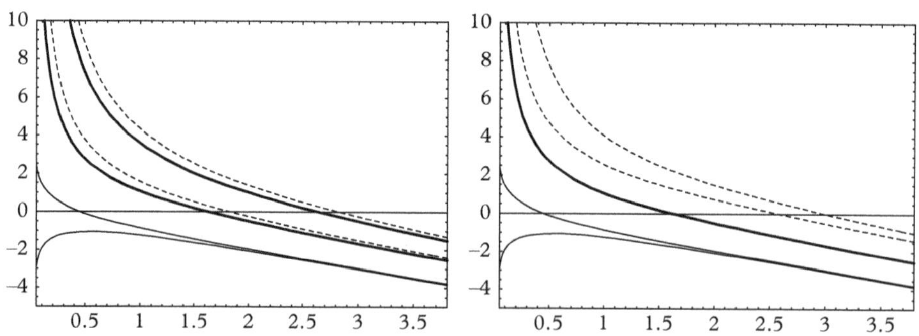

Fig. 23.9 *Finite-size energy levels of Landau–Ginzburg potentials versus the width R of an infinite cylinder. Left-hand side: symmetric well potential, with an asymptotic double degeneracy of all states. Right-hand side: asymmetric well potential, which displays an individual energy line corresponding to the bound state above only one vacuum.*

$\varphi(x, \tau)$ at the edge of the cylinder of width R, as discussed in Chapter 25. We expect that, at a finite R, the energies of the two vacua will be exponentially split for the tunnelling process of the kinks but this will not necessarily be the case for their excitation. Where only one bound state is above one of the two vacua, the energy level of this case will correspond to a single, isolated curve placed between the vacua energy and the threshold lines: this situation should be contrasted with the one relative to a potential with two symmetric wells (Figure 23.9).

23.7 Double Sine–Gordon Model

The semi-classical analysis of the Sine–Gordon model was carried out in seminal papers by Dashen and, for what concerns the spectrum, the final output coincides with the exact result coming from the S-matrix theory (Section 16.3). The ultimate reason of this agreement stays in the perfect equivalence between the kinks connecting the vacua placed at the right and left sides of each vacuum, together with the elastic scattering experienced by the kinks of the Sine–Gordon model: in particular, the elasticity of their S-matrix allows justification of the truncation of the matrix elements (23.3.7) to the kink states alone and therefore to obtain a closed differential equation for the kink matrix element that finally leads to the formula (23.3.1) and to the universal mass formula (23.4.3).

This section considers the interesting situation of a vacuum state in communication with two neighbouring ones through kinks of different masses. A prototype of this situation is given by the double Sine–Gordon model, with potential given by

$$V(\varphi) = -\frac{\mu}{\beta^2} \cos\beta\varphi - \frac{\lambda}{\beta^2} \cos\left(\frac{\beta}{2}\varphi + \delta\right) + C.$$

By choosing

$$\delta = \frac{\pi}{2}, \quad C = \frac{1}{\beta^2}\left(\mu + \frac{\lambda^2}{8\mu}\right),$$

and by varying λ, the shape of the potential changes (Figure 23.10). In the parameter

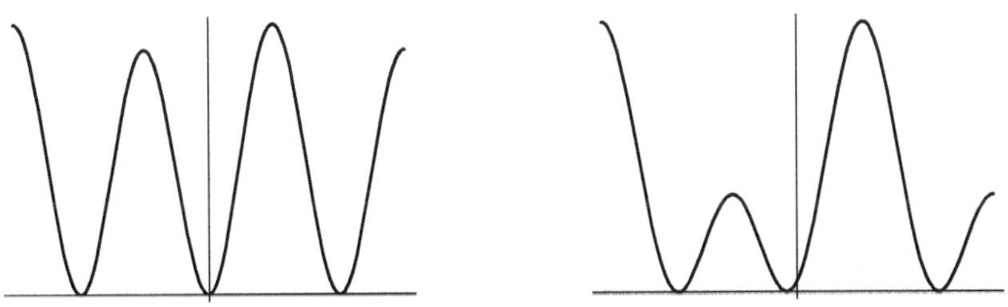

Fig. 23.10 *Shapes of the double Sine–Gordon potential by varying λ.*

space, there are two regions that are qualitatively different: the first given by $0 < \lambda < 4\mu$, the second given by $\lambda > 4\mu$. They are separated by the value $\lambda = 4\mu$ (where the curvature at the minima vanishes) which can be identified as a phase transition point of the double Sine–Gordon to the Ising model.[4]

Let us focus our analysis in the coupling constant region where $\lambda < 4\mu$. Starting from the pure Sine–Gordon model, once we switch on the coupling constant λ in the potential (23.7), there are several effects: the original minima of the Sine–Gordon, located at $\varphi_{\min} = 0, \frac{2\pi}{\beta}$ (mod $\frac{4\pi}{\beta}$), remain degenerate but they move to $\varphi_{\min} = -\varphi_0$, $\frac{2\pi}{\beta} + \varphi_0$ (mod $\frac{4\pi}{\beta}$), with $\varphi_0 = \frac{2}{\beta} \arcsin \frac{\lambda}{4\mu}$. The different minima maintain, however, the same curvature, given by

$$m^2 = \mu - \frac{1}{16} \frac{\lambda^2}{\mu}. \tag{23.7.1}$$

In correspondence with the above shifts, there are two different types of kinks, one called 'large kink' and interpolating through the higher barrier between $-\phi_0$ and $\frac{2\pi}{\beta} + \phi_0$, the other called 'small kink' and interpolating through the lower barrier between $-\frac{2\pi}{\beta} + \phi_0$ and $-\varphi_0$. Their classical expressions are given by

$$\varphi_L(x) = \frac{\pi}{\beta} + \frac{4}{\beta} \arctan\left[\sqrt{\frac{4\mu + \lambda}{4\mu - \lambda}} \tanh\left(\frac{m}{2} x\right)\right] \quad (\text{mod}\ \frac{4\pi}{\beta}), \tag{23.7.2}$$

$$\varphi_S(x) = -\frac{\pi}{\beta} + \frac{4}{\beta} \arctan\left[\sqrt{\frac{4\mu - \lambda}{4\mu + \lambda}} \tanh\left(\frac{m}{2} x\right)\right] \quad (\text{mod}\ \frac{4\pi}{\beta}). \tag{23.7.3}$$

For the, we can neglect the periodic structure of this potential[5] and concentrate our attention only on the three vacua around the origin, here denoted by $|\,0\,\rangle$ (the one near the origin) and $|\pm 1\,\rangle$ (the other two). Around the vacuum $|\,0\,\rangle$, the admitted quantum kink states are

$$|\,L\rangle = |\,K_{0,1}\,\rangle \ \text{ and } \ |\,\overline{S}\rangle = |\,K_{0,-1}\,\rangle,$$

together with the corresponding anti-kink states $|\,\overline{L}\,\rangle = |\,K_{1,0}\,\rangle$ and $|\,S\,\rangle = |\,K_{-1,0}\,\rangle$. The topological charges of these kinks are different, and given by

[4] Interested readers are referred to Delfino and Mussardo on the multi-frequency Sine–Gordon model quoted in Chapter 22 for the of this phase transition.

[5] The periodicity of the potential obviously implies that the following analysis applied as well to any other vacuum of this theory.

$$\begin{aligned} \mathcal{T}_L = -\mathcal{T}_{\bar{L}} = 1 + \tfrac{\beta\phi_0}{\pi}, \\ \mathcal{T}_S = -\mathcal{T}_{\bar{S}} = 1 - \tfrac{\beta\phi_0}{\pi}. \end{aligned} \tag{23.7.4}$$

The classical masses of the large and small kink get split when we switch on λ, and their exact value is easily computed

$$M_{L,S} = \frac{8m}{\beta^2}\left\{1 \pm \frac{\lambda}{\sqrt{16\mu^2 - \lambda^2}}\left(\frac{\pi}{2} \pm \arcsin\frac{\lambda}{4\mu}\right)\right\}. \tag{23.7.5}$$

The expansion of this formula for small λ is given by

$$M_{L,S} \underset{\lambda \to 0}{\longrightarrow} \frac{8\sqrt{\mu}}{\beta^2} \pm \frac{\lambda}{\beta^2}\frac{\pi}{\sqrt{\mu}} + O(\lambda^2), \tag{23.7.6}$$

where the first term is the classical mass of the unperturbed Sine–Gordon kink.

23.7.1 Dynamics of Long and Short Kinks

Let us now apply the semi-classical formula (23.3.1) for obtaining the spectrum of the neutral particles at the vacuum $|\,0\rangle$. For the form factor $F^{\varphi}_{L\bar{L}}(\theta) = f^{\varphi}_{LL}(i\pi - \theta)$ of the large kink (23.7.2) (with $\lambda < 4\mu$) we have

$$F^{\varphi}_{L\bar{L}}(\theta) = i\frac{4\pi}{\beta}\frac{1}{i\pi - \theta}\frac{\sinh\left[\alpha\frac{M_L}{m}(i\pi - \theta)\right]}{\sinh\left[\pi\frac{M_L}{m}(i\pi - \theta)\right]},$$

where

$$\alpha = 2\arctan\sqrt{\frac{4\mu + \lambda}{4\mu - \lambda}},$$

while m and M_L are given by (23.7.1) and (23.7.5), respectively. For the form factor $F^{\varphi}_{S\bar{S}}(\theta) = f^{\varphi}_{SS}(i\pi - \theta)$ of the small kink (23.7.3) (with $\lambda < 4\mu$) we have instead

$$F^{\varphi}_{\bar{S}S}(\theta) = -i\frac{4\pi}{\beta}\frac{1}{i\pi - \theta}\frac{\sinh\left[\alpha\frac{M_S}{m}(i\pi - \theta)\right]}{\sinh\left[\pi\frac{M_S}{m}(i\pi - \theta)\right]},$$

where

$$\alpha = 2\arctan\sqrt{\frac{4\mu - \lambda}{4\mu + \lambda}},$$

while m and M_S are given by (23.7.1) and (23.7.5), respectively.

By looking at the poles of these expressions within the physical strip, we have then two towers of neutral particles above the vacuum $| \, 0 \rangle$: the one coming from the bound states of the $| \, L\bar{L} \rangle$

$$m_{(L)}^{(n)} = 2M_L \sin\left(n_L \frac{m}{2M_L}\right), \quad 0 < n_L < \left[\pi \frac{M_L}{m}\right], \tag{23.7.7}$$

the other coming from the bound states of $| \, \bar{S}S \rangle$

$$m_{(S)}^{(n)} = 2M_S \sin\left(n_S \frac{m}{2M_S}\right), \quad 0 < n_S < \left[\pi \frac{M_S}{m}\right]. \tag{23.7.8}$$

As a matter of fact, this situation is not peculiar of the double Sine–Gordon model but it occurs each time that there are kinks of different masses originating from the same vacuum. Here, it is instructive to look at Problem 2. Notice that when $M_s = M_L$ the masses of the two towers of bound states' masses coincide, inducing an apparent degeneracy in the spectrum. However, as discussed in the introduction, the corresponding state cannot be degenerate. Hence, one of the two spectra (23.7.7) or (23.7.8) has to be spurious. It is then interesting to understand what went wrong with the semi-classical formula if applied to the case of asymmetric kinks.

First of all, let us notice that each kink solution knows only *half* of the shape of the vacuum state from which it originates: for instance, the long kink explores the shape of the potential only on the right of the origin. For what this kink solution is concerned, the shape of the potential to the left of the origin could be arbitrarily changed without effecting the behaviour of this solution. The same considerations also apply to the short kink, which are determined only by the shape of the potential to the left of the origin.

This implies that when we employ the long-kink solution to extract the mass spectrum we refer to a potential that is not the *actual* one but a potential $U_L(\varphi)$, whose values for $\varphi < 0$ are obtained by the specular image of those for $\varphi > 0$ of the original potential. Viceversa, when we employ the short-kink solution, it is as we are referring to a potential $U_R(\varphi)$, whose values for $\varphi > 0$ are the specular image of those for $\varphi < 0$, which determine in this case the short-kink solution. Hence it is not surprising that, applying eqn. (23.3.1) in the case of kinks with different mass, each of them gives rise to a different spectrum of bound states above the same vacuum.

Moreover, at the leading order in the coupling constant—the same for which it holds the semi-classical form factor (23.3.1)—the short- and long-kink states are invisible to each other. existence only at the next leading order in the coupling constant. For instance, in the expression (23.3.7) involving the long kink, the first sub-leading terms are given by the matrix elements with a couple of short-kink and anti-kink state, as

$$\langle L(\theta_1) \mid \varphi(0) \mid L\bar{S}S \rangle \, \langle L\bar{S}S \mid \varphi(0) \mid L \rangle \ldots \langle L \mid \varphi(0) \mid L(\theta_2) \rangle . \tag{23.7.9}$$

With $M_L > M_S$, these terms (as well as all the others obtained by inserting more times the couples $\bar{S}S$) are always present, no matter what are the values of the external rapidities

θ_1 and θ_2 of the long kink. Hence, it becomes impossible to pin down their presence by appealing to the perturbation expansion in the coupling constant. But this spoils the possibility of obtaining a closed differential equation for the form factors of the kink, as the one that has led to the semi-classical expression (23.3.1).

23.7.2 The importance of small kinks

Once the origin of the discrepancy of the spectra coming from the two-kink solutions has been clarified, what remains is understanding the correct spectrum of bound states. Although the exact expression of the mass formula is unknown, the spectrum can be studied in a relatively simple way in two different cases: (i) when the asymmetric kinks have approximately the same mass; (ii) when the mass of one of them is much smaller than the mass of the other.

Let us consider the first case. This situation can be realized starting by a symmetric configuration of the potential (which we assume to be an even function $V(\varphi) = V(-\varphi)$) and slightly deforming it by an infinitesimal deformation $\lambda\,\delta V(\varphi)$, with $\delta V(\varphi)$ odd under $\varphi \to -\varphi$. Its effect is to decrease one the maxima of the potential and to increase the other. This is, for instance, what happens in the double Sine–Gordon for small value of λ. Under this deformation, the masses of the kinks change as $M_{L,S} \simeq M \pm \lambda \mathcal{M}$. Denote by $\mid b\rangle_L$ and by $\mid b\rangle_R$ the bound state of the long and the short kinks which, in the unperturbed theory, have equal mass. The actual breather of the unperturbed theory is a linear combination of the two 'half-breathers' $\mid b\rangle_{L,S}$, which can be determined by imposing that it is an eigenvector of the parity transformation P ($P^2 = 1$). Let us assume, for instance, that this state is given by

$$\mid B\rangle = \frac{\mid b\rangle_L - \mid b\rangle_R}{\sqrt{2}}, \tag{23.7.10}$$

In the unperturbed potential, its mass can be equivalently written as

$$m_B = \frac{1}{2}(m_L + m_S). \tag{23.7.11}$$

Switching now δV, at the first order in λ the state (23.7.10) does not change. The masses m_L and m_R receive, instead, a linear correction of opposite sign,

$$\delta m_{L,S} \simeq \pm\left(2\sin\left(\frac{m}{2M}\right) - \frac{m}{M}\cos\left(\frac{m}{2M}\right)\right). \tag{23.7.12}$$

Plugging into (23.7.11), the mass of the breather remains then unchanged. Notice that this result matches with the first-order FFPT, as it can be seen by employing the parity operator P

$$\lambda\,\langle B|\delta V(\varphi)\,|B\rangle = \lambda\,\langle B|P(P\delta V(\varphi)P)\,P|B\rangle = -\lambda\,\langle B|\delta V(\varphi)\,|B\rangle = 0.$$

We now consider the second case, i.e. when one of the kinks is much heavier than the other. In this case the semi-classical spectrum is *essentially* determined by the short-kink solution, i.e.

$$m_{(S)}^{(n)} \simeq 2M_S \sin\left(n_S \frac{m}{2M_S}\right), \quad 0 < n_S < \left[\pi \frac{M_S}{m}\right], \tag{23.7.13}$$

with the stable part of the spectrum obtained only for $n_S \leq 2$. There are some arguments that support this scenario.

1. The actual mass of the bound state obviously depends on both M_S and M_L but the correction induced by the long kink is expected to be suppressed with respect to the one of the short kink by the ratio $\left(\frac{M_S}{M_L}\right)^2$. Therefore, if the mass of the long kink becomes heavier, its influence on the mass of the neutral particle should become increasingly less relevant. In particular, when $M_L \to \infty$, the long kink decouples from the theory and the masses of the breathers are only determined by the dynamics of the remaining short kink.
2. Another argument in favour of the formula (23.7.13) comes directly from the Watson equations satisfied by the form factors of the short- and long-kink solutions. First we consider the scattering process of the two-body small kink–anti-kink state $|\,\bar{S}(\hat{\theta}_1)S(\hat{\theta}_2)\,\rangle$. As far as in the centre of mass (defined by $\frac{\theta}{2} \equiv \theta_1 = -\theta_2$ and $\frac{\hat{\theta}}{2} \equiv \hat{\theta}_1 = -\hat{\theta}_2$), we have the inequality

$$M_S \cosh\hat{\theta} < M_L \cosh\theta, \tag{23.7.14}$$

there is no possibility of converting the above state into a $|\,L(\theta_1)\bar{L}(\theta_2)\,\rangle$. Therefore the scattering process of this short kink–anti-kink state can only be elastic

$$|\,\bar{S}(\hat{\theta}_1)\,S(\hat{\theta}_2)\,\rangle \longrightarrow |\,\bar{S}(\hat{\theta}_1)\,S(\hat{\theta}_2)\,\rangle. \tag{23.7.15}$$

The elastic region obviously extends further by making the mass of the long kink heavier. In this region, the Watson equation satisfied by any form factor of the short kink–anti-kink state is then

$$F_{\bar{S}S}^{\mathcal{O}}(\hat{\theta}) = S_{SS}(\hat{\theta})\,F_{\bar{S}S}^{\mathcal{O}}(-\hat{\theta}), \tag{23.7.16}$$

where $S_{SS}(\hat{\theta})$ is the elastic S-matrix of the process (23.7.15). Assuming that this elastic process proceeds through the exchange of the scalar particles, we have the diagram of Figure 23.11. Eqn. (23.7.16) implies that the ratio $F_{\bar{S}S}^{\mathcal{O}}(\hat{\theta})/F_{\bar{S}S}^{\mathcal{O}}(-\hat{\theta})$ is a pure phase for the real values of $\hat{\theta}$ below the inelastic threshold given by eqn. (23.7.14), in agreement with the semi-classical result. Trusting the semi-classical result of the form factors also for complex values of the rapidity

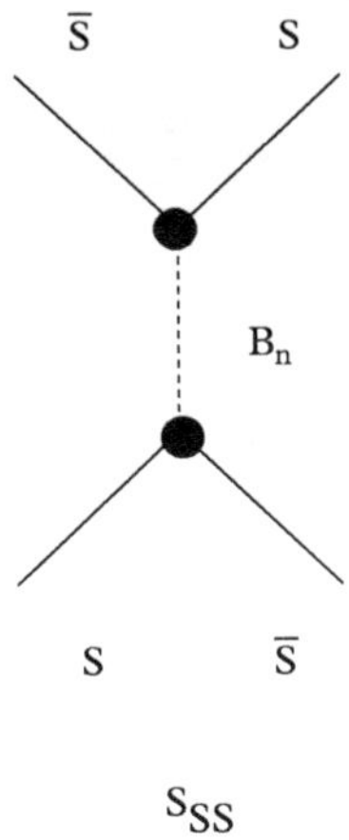

Fig. 23.11 *Elastic scattering amplitude of the short kink–anti-kink state.*

(in particular for those concerning the location of their poles), we arrive at the mass spectrum (23.7.13).

But why is the situation different for the long kink–anti-kink scattering? Notice that for the non-integrability of the theory, the scattering processes of the state $|\, L(\theta_1)\, \bar{L}(\theta_2)\,\rangle$ in this case *always* involves, as a final state, $|\, \bar{S}(\hat{\theta}_1)\, S(\hat{\theta}_2),\rangle$, namely the production process

$$|\, L(\theta_1)\, \bar{L}(\theta_2)\,\rangle \longrightarrow |\, \bar{S}(\hat{\theta}_1)\, S(\hat{\theta}_2)\,\rangle. \tag{23.7.17}$$

is always present. Going in the centre of mass, the rapidity of the outcoming small kink–anti-kink state is determined by

$$M_S \cosh\hat{\theta} = M_L \cosh\theta.$$

In addition to this production process, in the scattering of the long kink–anti-kink state there is also its elastic part

$$|\, L(\theta_1)\, \bar{L}(\theta_2)\,\rangle \longrightarrow |\, L(\theta_1)\, \bar{L}(\theta_2)\,\rangle. \tag{23.7.18}$$

For values of θ below other inelastic thresholds, the Watson equation satisfied by any form factors of the long kink–anti-kink state is then

$$F^{\mathcal{O}}_{L\bar{L}}(\theta) = S_{LL}(\theta)\, F^{\mathcal{O}}_{L\bar{L}}(-\theta) + S_{LR}(\theta)\, F^{\mathcal{O}}_{\bar{S}S}(-\hat{\theta}), \tag{23.7.19}$$

where $S_{LL}(\theta)$ is the S-matrix relative to the process (23.7.18), whereas $S_{LR}(\theta)$ is the one of (23.7.17). Assuming that, also in this case, these scattering processes are dominated by the exchange of the scalar particles, we have the diagrams shown in Figure 23.12. In contrast with the short-kink case, this time the ratio

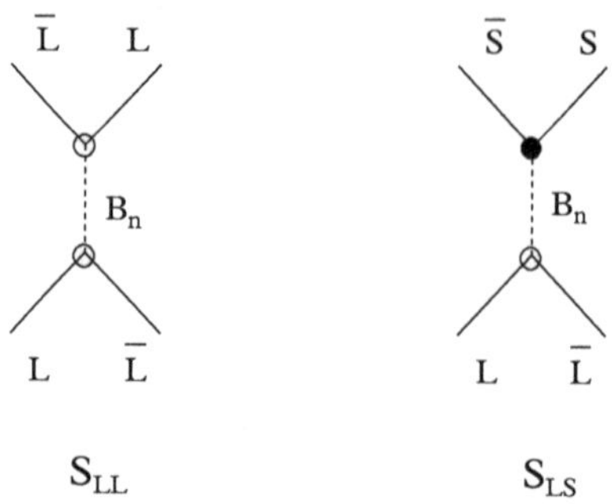

Fig. 23.12 *Elastic and production amplitudes of the long kink–anti-kink scatterings.*

$F^{\mathcal{O}}_{L\bar{L}}(\theta)/F^{\mathcal{O}}_{L\bar{L}}(-\theta)$ can never be a pure phase, not even for real values of θ. Hence the semi-classical result relative to the long kink cannot be the correct one since from the semi-classical analysis the ratio of the long-kink form factors is always a pure phase expression.

Notice that in the production diagram $| L\bar{L}\rangle \to | \bar{S}S\rangle$ it appears the same three-particle coupling of the elastic scattering $| \bar{S}S\rangle \to | \bar{S}S\rangle$. This implies that the intermediate particles of the two processes are the same. Assuming that their masses are those extracted by the short kink–anti-kink bound states, we can predict where the correct position of the poles of the long kink–anti-kink form factors should be: denoting the position of these poles by $iu^n_{\bar{S}S} = i\pi(1-\xi^n_{\bar{S}S})$, with $\xi^n_{\bar{S}S} = n\frac{m}{M_S}$, and $iu^n_{L\bar{L}} = i\pi(1-\xi^n_{L\bar{L}})$, the resonance value $\xi^n_{L\bar{L}}$ is determined by the relation

$$M_L \sin\left(\frac{u^n_{L\bar{L}}}{2}\right) = M_R \sin\left(\frac{u^n_{\bar{S}S}}{2}\right),$$

shown in Figure (23.13). This relation is similar to the refraction law of the light between two media of refraction indices M_L and M_R.

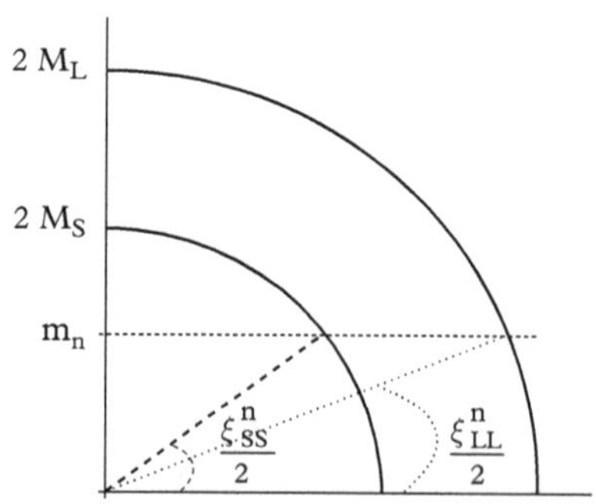

Fig. 23.13 *Relation between the resonance angles of the short and long kink–anti-kink state.*

REFERENCES

Dashen, Hasslacher and Neveu promoted the semi-classical methods in quantum field theories in a series of papers:

Dashen, R.F., Hasslacher, B. and Neveu, A. (1974). 'Nonperturbative Methods and Extended Hadron Models in Field Theory: 1. Semi-classical Functional Methods', *Physics Review D*, 10: 4114.

Dashen, R.F., Hasslacher, B. and Neveu, A. (1974). 'Nonperturbative Methods and Extended Hadron Models in Field Theory: 2. Two-Dimensional Models and Extended Hadrons', *Physics Review D*, 10: 4130.

Dashen, R.F., Hasslacher, B. and Neveu, A. (1974). 'Nonperturbative Methods and Extended Hadron Models in Field Theory: 3. Four-dimensional Nonabelian Models', *Physics Review D*, 10: 4138.

Dashen, R.F., Hasslacher, B. and Neveu, A. (1975). 'The Particle Spectrum in Model Field Theories from Semi-classical Functional Integral Techniques', *Physics Review D*, 11: 3424.

Dashen, R.F., Hasslacher, B. and Neveu, A. (1975). 'Semi-classical Bound States in an Asymptotically Free Theory', *Physics Review D*, 12: 2443.

The semi-classical formula of the matrix elements on the kink states and its relation to the classical kink configuration is discussed in

Goldstone J. and Jackiw, R. (1975). 'Quantization of Nonlinear Waves', *Physics Review D*, 11: 1486.

while its refined version, in terms of the relativistic invariants of the two-body channel was derived in

G. Mussardo, V. Riva and G. Sotkov, *Finite volume form-factors in semiclassical approximation*, Nucl. Phys.B 670 (2003), 464.

A thorough analysis of the non-integrable bosonic models in terms of the semi-classical methods, including the symmetric and asymmetric well potential and the double Sine–Gordon model, together with the general conclusions about the number of possible bound states, can be found in

Mussardo, G. (2007). 'Neutral Bound States in Kink-like Theories', *Nuclear Physics B*, 779: 101.

For other interesting references on the double Sine–Gordon model, see

Bajnok, Z., Palla, L., Takacs G. and Wagner, F. (2001). 'Nonperturbative Study of the Two- frequency Sine–Gordon Model', *Nuclear Physics B*, 601: 503.

Campbell, D.K., Peyrard, M. and Sodano, P. (1986). 'Kink–anti-kink Interactions in the Double Sine–Gordon Equation', *Physica* 19: 165.

Fabrizio, M., Gogolin, A.O. and Nersesyan, A.A. (2000). 'Critical Properties of the Double-frequency Sine–Gordon Model with Applications', *Nuclear Physics B*, 580: 647.

Takacs, G. and Wagner, F. (2006). 'Double Sine–Gordon model revisited, *Nuclear Physics B*, 741: 353.

For a general introduction to classical solutions of field theories and their quantum interpretation the standard reference is

R.Rajaraman, *Solitons and instantons*, Amsterdam, North Holland, 1982.

....

PROBLEMS

23.1. Decay processes

Consider the decay processes of the higher particles in the spectrum of Φ^4 theory,

$$B_k \to r B_1 + s B_2$$

where r and s are all those integers which satisfy

$$m_k \geq r m_1 + s m_2, \quad r + s = n.$$

Take as an explicit example $\xi < \frac{1}{5}$. The decay width of these processes is given by the Fermi golden rule

$$\Gamma = \int (2\pi)^2 \delta^2(P - p_1 - \cdots - p_n) \mid T_{fi} \mid^2 \frac{1}{2E} \prod_{i=1}^{n} \frac{dp_i}{(2\pi)2E_i},$$

where P denotes the 2-momentum of the decay particle whereas the amplitude T_{fi} is given by the matrix element $T_{fi} = \langle B_k(P) \mid B_1(p_1) \ldots B_1(p_s) B_2(p_{s+1}) \ldots B_2(p_n) \rangle$.

a. Assuming T_{fi} to be of the same order of magnitude for the various processes, use the phase space expressions discussed in Appendix 17.C to estimate the decay rates

$$\begin{array}{lcl} \Gamma^5_{11} & \cdots\cdots & B_5 \to B_1 + B_1 \\ \Gamma^5_{12} & \cdots\cdots & B_5 \to B_1 + B_2 \\ \Gamma^5_{22} & \cdots\cdots & B_5 \to B_2 + B_2 \end{array}$$

with respect to the decay rate Γ^3_{11} of the process $B_3 \to B_1 + B_1$.

b. Show that there exists a critical value ξ_c where the particle B_5 has its mass exactly equal to $2m_2$ and discuss the consequences of this critical value for the decay processes.

23.2. Short and long kinks

Consider a simplified version of a potential with three vacua configurations, realized by the potential shown in the Figure below

$$V(\varphi) = \frac{m^2}{2}\begin{cases}(\varphi+2b)^2, & \varphi \leq -b;\\ \varphi^2, & -b \leq \varphi \leq a;\\ (\varphi-2a)^2, & \varphi > a.\end{cases}$$

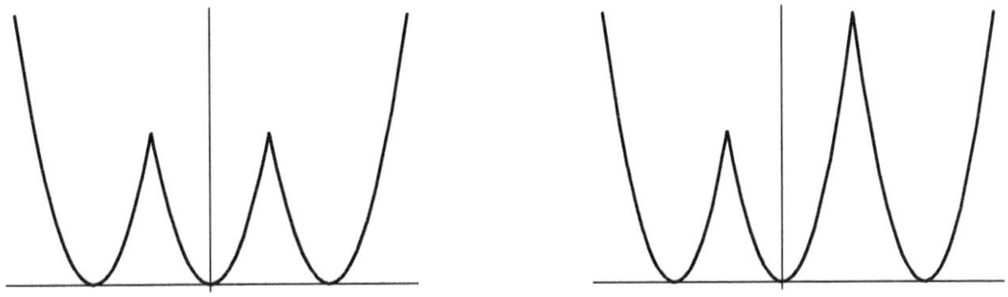

Shape of V (φ) with a = b (left hand side) and with a > b (right hand side).

a. Prove that the explicit configurations of the long and short kink of this potential are given by

$$\varphi_L(x) = \begin{cases} a\,e^{mx}, & x \leq 0,\\ 2a - a\,e^{-mx}, & x \geq 0,\end{cases} \qquad \varphi_{\bar{S}}(x) = \begin{cases} -b\,e^{mx}, & x < 0,\\ -2b + b\,e^{-mx}, & x \geq 0\end{cases}$$

b. Show that their classical masses are

$$M_L = ma^2, \quad M_S = mb^2.$$

c. Compute the form factors involving the long and the short kinks and show that they are given by

$$F^{\varphi}_{L\bar{L}}(\theta) = \frac{i}{M_L(i\pi-\theta)}\left[\frac{1}{\theta - i\pi(1-\xi_L)} - \frac{1}{\theta - i\pi(1+\xi_L)}\right],$$

$$F^{\varphi}_{\bar{S}S}(\theta) = -\frac{i}{M_S(i\pi-\theta)}\left[\frac{1}{\theta - i\pi(1-\xi_S)} - \frac{1}{\theta - i\pi(1+\xi_S)}\right].$$

Determine ξ_L and ξ_S.

d. Show that there are apparently *two* towers of particles, of different masses, above the vacuum $|\,0\rangle$, expressed by the formulae

$$m_L = 2m a^2 \sin\frac{\pi}{2a^2}, \quad m_S = 2m b^2 \sin\frac{\pi}{2b^2}.$$

e. Argue that one of these two spectra has to be spurious.

23.3. Instantons

Instantons are *finite* action classical solutions of Euclidean equations of motion and are closely related to tunnelling phenomena among degenerate vacua. Consider a unit-mass particle in one dimension, under the potential $V(q) = 1 - \cos q$.

a. Write down the Euclidean action of this problem and the corresponding equation of motion.

b. Show that the finite-action solutions of the Euclidean equation of motion are the same as the finite-energy static solitons of the Sine–Gordon model.

24

Interacting Fermions and Supersymmetric Models

Bosons, fermions, all those damn particles you cannot see. That's what drove me to drink. But now I can see them all.

Anonymous

24.1 Introduction

This chapter presents some interesting applications of both the semi-classical formalism and the FFPT to models that involve fermions and bosons. Free fermions are subjected to Dirac equation but a profound change of the Dirac equation is achieved when the constant mass term is replaced by an inhomogeneous term that varies with the position as dictated by an underlying bosonic field. Only profiles that are significantly different from the homogenous case are interesting because small variations give rise to small, insignificant changes in the mass of the field. In (1+1) systems, such profiles are those given by kink configurations encountered, particularly those in the previous chapter. This chapter shows that, under very general conditions, the semi-classical formalism allows us to find, in addition to the spectrum of bosonic bound states, also the spectrum of the fermionic bound states: indeed, for models involving fermions interacting with bosons which possess topologically non-trivial configurations, labelled by the integer $\mathcal{T}$, there is a remarkable formula that expresses the matrix element of the fermion field on the kink states in terms of the Fourier transform of the so-called zero-mode of the fermion field itself. The existence of this peculiar zero mode solution of the fermion field is related both to the presence of kink configurations of the bosonic field and a condition on the Yukawa interaction of the fermion field.

A closer look at the mass formulae for bosonic and fermionic bound states reveals that sometimes these two spectra become degenerate for particular values of the coupling constants of the theory. This remarkable fact signals the presence of a supersymmetry, namely a symmetry that links bosons and fermions, which we have already encountered in Chapter 18. In light of this circumstance, this chapter studies the evolution of the spectrum of supersymmetric theories using the FFPT, finally arriving at a notable result: contrary to the purely bosonic theories with degenerate vacua, where essentially any deformation that breaks the integrability of the original theory also gives rise to

Statistical Field Theory: An Introduction to Exactly Solved Models in Statistical Physics. Second Edition. Giuseppe Mussardo, Oxford University Press (2020). © Giuseppe Mussardo. DOI: 10.1093/oso/9780198788102.001.0001

the confinement of the kinks, in supersymmetric theories this is no longer true: in these theories, the kinks remain physical excitations of the theory also if the original integrability is broken. The absence of confinement of the kink excitations is a simple result of the unbroken realization of supersymmetry.

24.2 Fermion in a Bosonic Background

Let us now consider a two-dimensional theory involving a scalar field $\varphi(x)$ and a real Majorana fermion field

$$\psi(x) = \begin{pmatrix} \psi_1 \\ \psi_2 \end{pmatrix} \tag{24.2.1}$$

with Lagrangian density

$$\mathcal{L} = \frac{1}{2}(\partial_\mu \varphi)^2 - U(\varphi, \lambda) \; + i\bar{\psi}\gamma^\mu \partial_\mu \psi - V(\varphi, g)\bar{\psi}\psi. \tag{24.2.2}$$

As coupling constants we have now both λ and g. In addition to the bosonic potential $U(\varphi,\lambda)$ (taken as in Chapter 23 as a potential with a degenerate set of minima connected by the kinks $\varphi_{ab}(x)$), we have also a Yukawa interaction $V(\varphi)$ for the fermion. For any kink configuration $\varphi_{ab}(x)$, this interaction satisfies the conditions

$$V(\varphi_{ab}(x), g) = \begin{cases} v_a(g), & x \to -\infty \\ v_b(g), & x \to +\infty \end{cases} \tag{24.2.3}$$

where both limits are finite, with their product always strictly negative

$$v_a(g)\, v_b(g) < 1. \tag{24.2.4}$$

Hereafter, we denote by $V_{ab}(x)$ the potential $V(\varphi, g)$ computed on the kink configuration $\varphi_{ab}(x)$, also skipping for simplicity the presence of the coupling constant g in this expression. $\mathcal{T}$ will denote the topological charge.

For the γ matrices we adopt the Weyl representation given by

$$\gamma^0 = \sigma_2 = \begin{pmatrix} 0 & -i \\ i & 0 \end{pmatrix}; \;\; \gamma^1 = i\sigma_1 = \begin{pmatrix} 0 & i \\ i & 0 \end{pmatrix}, \;\; \gamma^5 = \gamma^0\gamma^1 = \sigma_3 = \begin{pmatrix} 1 & 0 \\ 0 & -1 \end{pmatrix}.$$

The charge-conjugation matrix C, satisfying

$$(C\gamma^0)(\gamma^\mu)^*(C\gamma^0)^{-1} = -\gamma^\mu,$$

is given by $C = \gamma^0$ and maps the fermion ψ to its conjugate particle ψ_c according to

$$\psi_c = (C\gamma^0)\psi^*. \tag{24.2.5}$$

In this representation a Majorana fermion (which satisfies the neutrality condition $\psi_c = \psi$) has then both components real

$$\psi_1^* = \psi_1, \quad \psi_2^* = \psi_2.$$

Notice that the Lagrangian (24.2.2) has no explicit fermion mass term and therefore the fermion appears to be massless, although this is not the case. First of all, the interaction term $V(\varphi(x))$ plays the role of an effective mass for the fermion field $\psi(x)$: this is particularly true when the bosonic field $\varphi(x)$ has been prepared in a $\mathcal{T} = 0$ trivial state, such as one of its vacua, since in these cases the effective mass of the fermion at the vacuum $|\,\mathbf{a}\rangle$ is given by the constant value $V(\varphi_a^{(0)}) = v_a$. However, as for the case of the bosonic spectrum, to unveil the actual spectrum of the fermion in the $\mathcal{T} = 0$ sector (and see, in particular, whether the particle associated to $\psi(x)$ also belongs or not to the physical spectrum!) we need to study first the higher topological sectors of the theory.

Let us consider then the $\mathcal{T} = 1$ sector (analogous considerations hold for $\mathcal{T} = -1$), with the bosonic field prepared in the one kink static configurations $\varphi_{ab}(x)$. Although, strictly speaking, such a classical configuration refers to the equation of motion (23.2.1) where the fermion field is absent, it is natural to assume that the latter will have a negligible effect on the former as far as the mass of the kink itself exceeds the 'mass energy' of the fermion for both vacua, i.e. if

$$M_{ab} \gg |\, v_i \,|, \quad i = a, b. \tag{24.2.6}$$

This is the condition that expresses the semi-classical limit of the Lagrangian (24.2.2) and that allows us to show, in the $\mathcal{T} = 1$ sector, fermion has a new qualitative non-perturbative feature: it possesses localized zero-energy mode, namely a normalizable solution of zero energy of the Dirac equation. Indeed, if $\varphi_{ab}(x)$ is a *static* kink configuration, we can also look for a *static* solution of the Dirac equations, which take the form

$$\begin{aligned} \partial_x \psi_2^{(0)} + V_{ab}(x)\,\psi_1^{(0)} &= 0, \\ \partial_x \psi_1^{(0)} + V_{ab}(x)\,\psi_2^{(0)} &= 0. \end{aligned} \tag{24.2.7}$$

These are nothing else but the 0-eigenvalue condition $H_D\,\psi^{(0)} = 0$ for the Dirac Hamiltonian

$$H_D = -i\alpha\,\partial_x + V_{ab}(x)\,\beta, \tag{24.2.8}$$

where $\alpha = \gamma^0\gamma^1 = \sigma_3$ and $\beta = \gamma^0$. Solutions of the two equations (24.2.7) can be found in terms of the linear combinations of the two components of the fermion field

$$\psi_{\pm}^{(0)}(x) \equiv \left(\psi_1^{(0)} \pm \psi_2^{(0)}\right) = A_{\pm} \exp\left[\mp \int_{x_0}^{x} V_{ab}(t)\, dt\right]. \qquad (24.2.9)$$

In light of the asymptotic behaviours (24.2.4) of the potential $V_{ab}(t)$ and the condition (24.2.4) on their signs, it is clear that only one of the two solutions $\psi_{\pm}^{(0)}(x)$ is normalizable (the other is divergent at infinite). Therefore, let us assume that $v_a < 0$ and $v_b > 0$: in this case, the normalizable solution is $\psi_{+}(x)$ and correspondingly we have to impose $A_{-} = 0$, so that the final result for wave function of the zero mode is

$$\psi_{ab}^{(0)}(x) = A\,\hat{\psi}\, \exp\left[-\int_{x_0}^{x} V_{ab}(t)\, dt\right], \quad \hat{\psi} = \begin{pmatrix} 1 \\ -1 \end{pmatrix} \qquad (24.2.10)$$

where A is a normalization constant. Notice that this wave function is localized at the position where the kink solution jumps between the two vacua (Figure 24.1).

In the $\mathcal{T} = 1$ sector, the quantization of the fermion field can be achieved in terms of its eigenmode expansion

$$\psi(x,t) = a_0\, \psi_{ab}^{(0)}(x) + \sum_{p} \left[b_p\, e^{-iE_p t}\, \psi_{E_p}(x) + b_p^{\dagger}\, e^{iE_p t}\, \psi_{-E_p}(x)\right], \qquad (24.2.11)$$

where the operator a_0 is relative to the zero-mode solution, while b_p and $b_p^{\dagger}$ are the operators associated to the positive/negative energy solutions $\psi_{\pm E}(x)$ of the Dirac Hamiltonian in the kink background $\varphi_{ab}(x)$

$$H_D\, \psi_{\pm E}(x) = \pm E\, \psi_{\pm E}(x). \qquad (24.2.12)$$

The operators b_p and $b_p^{\dagger}$ satisfy the anti-commutation relations

$$\{b_p, b_k^{\dagger}\} = \delta(p-k), \qquad \{b_p, a_0\}, = \{b_p^{\dagger}, a_0\} = 0.$$

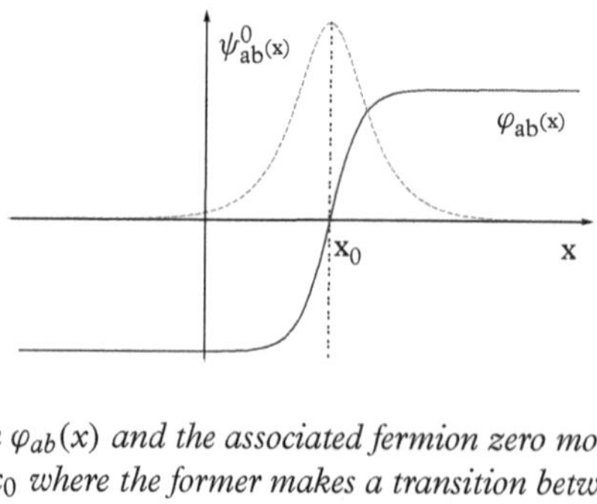

Fig. 24.1 *Kink configuration $\varphi_{ab}(x)$ and the associated fermion zero mode $\psi_{ab}^{(0)}(x)$. The latter is always localized at the point x_0 where the former makes a transition between the two vacua.*

The operator a_0 is instead a Majorana self-conjugated zero mode operator, $a_0 = a_0^{\dagger}$ that, properly normalized, fulfills the condition

$$a_0^2 = 1. \tag{24.2.13}$$

The a_0 mode carries zero energy and therefore the original quantum state $| K_{ab}(\theta)\rangle$ has to be actually doubly degenerate! We will use the notation $| K_{ab}^{\pm}(\theta)\rangle$ to identify these two degenerate states, related one to the other by

$$\begin{aligned} | K_{ab}^{+}(\theta)\rangle &= a_0 \, | K_{ab}^{-}(\theta)\rangle, \\ | K_{ab}^{-}(\theta)\rangle &= a_0 \, | K_{ab}^{+}(\theta)\rangle. \end{aligned} \tag{24.2.14}$$

Notice that we can consider $| K_{ab}^{+}\rangle$ to be the charge-conjugated state of $| K_{ab}^{-}\rangle$, and vice-versa. Given their symmetry in sharing the Majorana fermion, there is a certain arbitrariness in assigning the fermion parity of these states: our choice is to regard $| K_{ab}^{\pm}(\theta)\rangle$ as kinks that have Majorana fermion parity $\mathcal{P} = \pm 1$. The same considerations apply to $\mathcal{T} = -1$ sector, i.e. the sector of the anti-kinks $| K_{ba}\rangle$, which now become doublets $| K_{ba}^{\pm}(\theta)\rangle$. We will graphically distinguish these new kinks by putting on their previous graphical representation a yellow up arrow on the those of fermion parity $+1$ while a green down arrow on those of fermion number -1, as shown in Figure 24.2.

We can now use this double multiplicity of kinks and anti-kinks to *unfold* the original vacuum structure. Let us consider two bosonic neighbourhood vacua $| \mathbf{a}\rangle$ and $| \mathbf{b}\rangle$ connected by the kinks and anti-kinks of different fermion number: the adjacency properties of the initial vacua, carried by the kinks $| K_{ab}^{\pm}\rangle$, can be equivalently expressed

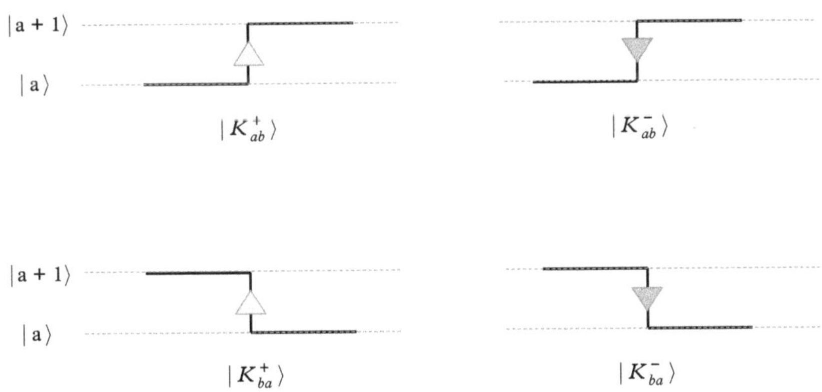

Fig. 24.2 *Graphical representation of the fermionic kinks $| K_{ab}^{\pm}\rangle$ and anti-kinks $| K_{ba}^{\pm}\rangle$. The yellow up arrow denotes those of fermion parity $\mathcal{P} = +1$, while the down green arrow those of fermion parity $\mathcal{P} = -1$.*

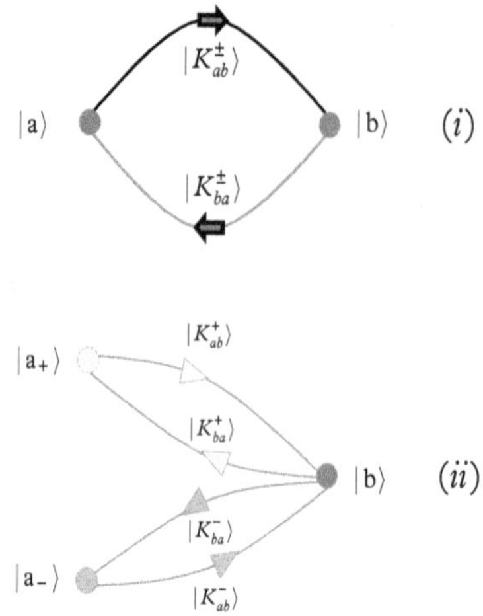

Fig. 24.3 *(i) Adjacency diagrams of original vacua connected by fermionic kinks; (ii) splitting the original vacuum* $|\,\mathbf{a}\rangle$ *into two vacua* $|\,\mathbf{a}_{\pm}\rangle$*, where each of them is related to the vacuum* $|\,\mathbf{b}\rangle$ *by kinks of the same Majorana fermion parity.*

as the adjacency conditions of a new set of vacua, in which we *split* one of the original vacua—say $|\,\mathbf{a}\rangle$—into two new ones $|\,\mathbf{a}_{\pm}\rangle$, while we leave the other $|\,\mathbf{b}\rangle$ untouched, and we connect this new set of vacua as follows: $|\,\mathbf{a}_{+}\rangle$ is linked to $|\,\mathbf{b}\rangle$ only by kink/anti-kink of fermion parity $+1$, while $|\,\mathbf{a}_{-}\rangle$ is connected to $|\,\mathbf{b}\rangle$ only by those of fermion parity -1 (Figure 24.3). This construction can be repeated for any two neighbourhood vacua and will be useful later.

24.3 The Fermionic Bound States in $\mathcal{T}=0$ Sector

Let

$$g^{\pm}_{ab}(\theta) = \langle K^{\mp}_{ab}(\theta_1)\,|\,\psi(0)\,|\,K^{\pm}_{ab}(\theta_2)\rangle$$

be matrix elements of the fermion field $\psi(x)$ on the kink states $|\,K^{\pm}_{ab}(\theta)\rangle$. In the semi-classical approximation we have

$$g^{+}_{ab}(\theta) = g^{-}_{ab}(\theta) \simeq \int_{-\infty}^{\infty} dx e^{iM_{ab}\theta x}\,\psi^{(0)}_{ab}(x), \qquad (24.3.1)$$

where $\theta = (\theta_1 - \theta_2)$ and $\psi^{(0)}_{ab}(x)$ is the zero-mode configuration of the fermion in the presence of the kink background $\varphi_{ab}(x)$. The identity of the two matrix elements $g^{\pm}_{ab}(\theta)$ is simply due to the real condition on the Majorana fermion $\psi(x)$: in presence of a complex

Dirac fermion, these matrix elements would be instead complex conjugate one to the other and eqn. (24.3.1) would change as

$$\langle K^{\mp}_{ab}(\theta_1) \mid \psi(0) \mid K^{\pm}_{ab}(\theta_2)\rangle = \langle K^{\mp}_{ab}(\theta_2) \mid \psi^{\dagger}(0) \mid K^{\pm}_{ab}(\theta_1)\rangle^{*}. \tag{24.3.2}$$

The derivation of the formula (24.3.1) follows exactly from the same steps used to derive the analogous formula (23.3.1) for the purely bosonic case, namely the equation of motion of the fermion field sandwiched between the kink states of momentum p_1 and p_2, with a proper truncation of the insertion of intermediate states according to the semi-classical approximation.

We can now study the fermionic bound states spectrum in the $\mathcal{T} = 0$ sector, making the analytic continuation $\theta \to i\pi - \theta$ of (24.3.1) and going to the crossed channel

$$G^{\pm}_{ab}(\theta) = g^{\pm}_{ab}(i\pi - \theta) = \langle a \mid \psi(0) \mid K^{\pm}_{ab}(\theta_1) K^{\mp}_{ba}(\theta_2)\rangle.$$

To identify the $\mathcal{T} = 0$ bound states with fermion quantum number above the vacuum $\mid \mathbf{a}\rangle$, we have to look at the poles of $G^{\pm}_{ab}(\theta)$ in the physical strip $0 < \text{Im}\theta < \pi$. Notice that there are *two* states which share the same poles, i.e. $\mid K^{+}_{ab} K^{-}_{ba}\rangle$ and $\mid K^{-}_{ab} K^{+}_{ba}\rangle$: these states are charge-conjugated one to the other, in agreement with the general result that any solution relative to $+E$ is paired with another solution relative to $-E$. Notice that, unfolding the original vacua, these states $\mid K^{\pm}_{ab} K^{\mp}_{ba}\rangle$ have the graphical representation shown in Figure 24.4.

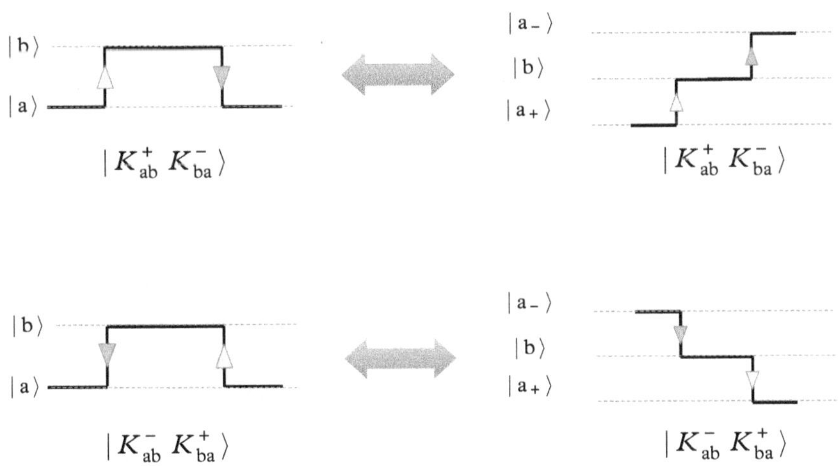

Fig. 24.4 *Two-particle fermionic kinks in the graphical representation of the original vacua (figures on the left) and of the unfolded vacua (figures on the right).*

In light of the conditions (24.2.3) and (24.2.4), the zero-mode goes exponentially to zero at $x \to \pm\infty$ as

$$\psi_{ab}^{(0)}(x) = A\hat{\psi} \times \begin{cases} \exp\left[v_a(g)x\right], & x \to -\infty, \\ \exp\left[-v_b(g)x\right], & x \to +\infty. \end{cases} \tag{24.3.3}$$

For what concerns the bound states relative to the vacuum $|\,\mathbf{a}\rangle$, as in the bosonic case, the only relevant poles are those coming from the behaviour of $\psi_{ab}^{(0)}(x)$ at $x \to -\infty$: in particular, the first pole of $G_{ab}^{\pm}(\theta)$ is simply fixed by the asymptotic value v_a

$$\theta_1 = i\pi\left(1 - \frac{v_a}{\pi M_{ab}^*}\right).$$

If

$$\eta_a(g) \equiv \frac{v_a(g)}{\pi M_{ab}^*} < 1 \tag{24.3.4}$$

this pole is in the physical strip and there are two $\mathcal{T} = 0$ fermionic bound states of mass

$$m_1^F(g) = \pm 2M_{ab}^* \sin\left(\frac{\pi\,\eta_a(g)}{2}\right). \tag{24.3.5}$$

Let's make few comments about this formula.

1. The $\pm$ signs refer to the paired solutions of energy $\pm E$. From now on for simplicity we concentrate on one of them the positive energy bound states.
2. When $\eta_a < 1$, we have that $|\,m_1^F(g)\,| < 2M_{ab}^*$, i.e. we are in presence of genuine $\mathcal{T} = 0$ fermionic bound states made of the kink–anti-kink of lowest mass carrying Majorana fermion number.
3. In the very deep semi-classical limit, when $M_{ab}^* \to \infty$, expanding in series the formula (24.3.5), we have that $|\,m_1\,| \simeq v_a$, which is the value of the mass usually associated to the elementary fermion ψ in the $\mathcal{T} = 0$ sector at the vacuum $|\,\mathbf{a}\rangle$.
4. If, on the contrary, moving the coupling constant g, we reach a value g_c for which $\eta_a(g_c) \geq 1$, then $|\,m_1\,|$ overpasses the threshold value $2M_{ab}^*$ and leaves the physical spectrum. Therefore, quite similarly to the purely bosonic case analysed in Chapter 23, the scenario is as follows: the particle associated with the fermion present in the Lagrangian (24.2.2) has to be considered, on a general ground, not an elementary particle but a bound state of the two kinks with different fermion parity, and it belongs to the physical spectrum of the theory if and only if the condition (24.3.4) is satisfied.

In order to get the other poles eventually present in the Fourier transform of $\psi_{ab}^{(0)}(x)$, we need to analyse its exponential approach at $x \to -\infty$. As shown in one of the problems at the end of the chapter, under the assumption that the interaction term $V_{ab}(\varphi)$ is analytic around the vacuum configurations $\varphi_a^{(0)}$, the fermionic zero mode has an asymptotic expansion for $x \to -\infty$ given by

$$\psi_{ab}^{(0)}(x) = A\,\hat{\psi}\, e^{v_a x} \left(1 + \sum_{n=1} t_n\, e^{n\omega_a x}\right), \tag{24.3.6}$$

where t_n are all computable quantities. The presence of these exponential terms induces a series of equidistant poles in the function $G_{ab}^{\psi}(\theta)$ localized at

$$\theta_n = i\pi(1 - \eta_a - n\xi_a). \tag{24.3.7}$$

They are on the physical strip as far as $n \leq (1 - \eta_a)/\xi_a$, whose integer part is therefore the total number of fermionic bound state in the $\mathcal{T} = 0$ sector

$$N_a^F = \left[\frac{1 - \eta_a}{\xi_a}\right]. \tag{24.3.8}$$

The general semi-classical expression of the masses of the fermionic bound states is thus given by

$$m_n^F = 2M_{ab}^* \sin\left(\frac{\pi(\eta_a + (n-1)\xi_a)}{2}\right) \tag{24.3.9}$$

$$n = 1, 2, \ldots, N_a^F.$$

This is the main result of the semi-classical formula for the fermionic bound states. Comparing this mass formula with the one of the bosonic bound states given in eqn. (23.4.3), we see that the only significant difference is the shift induced by the quantity η_a. The next section revisits the case of a symmetric well potential, leaving the study of the spectrum of the asymmetric well potential to one of the exercises of this chapter.

24.4 Symmetric Wells

Let us discuss and compare the bosonic and fermionic bound states for the familiar symmetric double well bosonic potential here written as

$$U(\varphi) = \frac{\lambda^2}{2}\left(\varphi^2 - \frac{m^2}{2\lambda^2}\right)^2. \tag{24.4.1}$$

The spectrum of the neutral particles, discussed in Section 23.5, is given by

$$m^B_{\pm,n} = 2M \sin\left[n\frac{\pi\xi}{2}\right], \qquad n = 1,2,\ldots N^B_{\pm}. \tag{24.4.2}$$

Let us now add the fermion and compute the spectrum of its bound states. Since $V(\varphi)$—evaluated on the kink solution given in eqn. (23.5.2)—must assume opposite values at $x \to \pm\infty$, the simplest way to satisfy this requirement consists of a potential simply proportional to the bosonic field $\varphi(x)$ itself

$$V(\varphi) = g\varphi(x),$$

where g is another coupling constant. The asymptotic values of V at $x \to \pm\infty$ are given by

$$v_a = -v_{-a} = g\hat{\varphi}.$$

Let us now briefly analyze the fermionic spectrum in the $\mathcal{T} = 0$ sector.

Fermionic bound states in $\mathcal{T} = 0$ sector. With the potential $V(\varphi)$ given above, the fermionic zero mode can be computed exactly and it is given by

$$\psi^{(0)}_{-a,a}(x) = A\hat{\psi}\left(\frac{1}{\cosh\frac{mx}{\sqrt{2}}}\right)^r,$$

where r is the ratio of the couplings

$$r = \frac{g}{\lambda}.$$

Posed $\omega_a = \sqrt{2}m$, its behaviour at $x \to -\infty$ is given by

$$\psi^{(0)}_{a,a}(x) \simeq 2^r A\hat{\psi}\, e^{v_a x}\left[1 - r e^{\omega_a x} + r(r+1)\, e^{2\omega_a x} + \cdots\right],$$

and therefore the poles of the matrix element $G_{-aa}(\theta)$ are located at

$$\theta_n = i\pi\,(1 - \eta - (n-1)\xi), \quad n = 1,2,\ldots$$

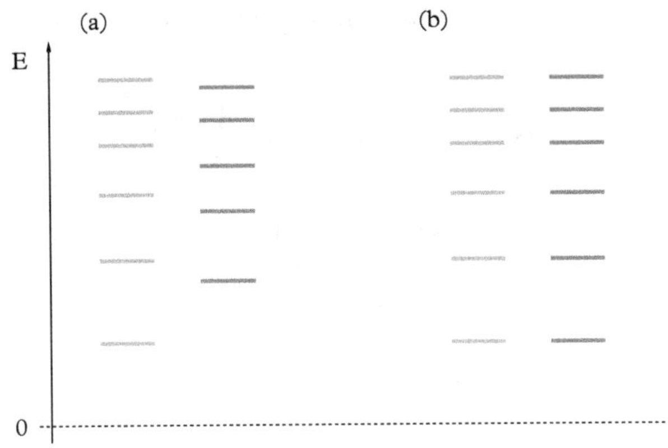

Fig. 24.5 *Spectrum of boson and fermion bound states in the $\mathcal{T} = 0$ sector for: (a) generic values of the couplings g and λ; (b) for g = λ.*

with

$$\eta = \frac{v_a}{\pi M}, \qquad \xi = \frac{\omega_a}{\pi M}.$$

Hence, the masses of the fermionic bound states at both vacua $|\pm\rangle$ are given by

$$m^F_{\pm,n} = 2M \sin\left(\frac{\pi(\eta + (n-1)\xi)}{2}\right), \qquad n = 1, 2, \ldots, N^F_{\pm}. \tag{24.4.3}$$

Comparing this expression with the spectrum of the bosons, eqn. (24.4.2), we see that the masses of bosons and fermions, as well as their number, are in general different (see Figure 24.5). There is however a special case where the two spectra match exactly: it is when $\eta = \xi$, namely $g = \lambda$. The next section shows that this is not at all a coincidence, but the condition under which the theory is invariant under an interchange of bosons with fermions, i.e. a supersymmetric transformation.

24.5 Supersymmetric Theory

Chapter 18 showed that it is possible to set up exact S-matrices which take into account degenerate bosonic and fermionic excitations. Here we integrate those considerations with a Lagrangian formulation of a QFT, which is invariant under a transformation which changes a fermion into a boson, and vice versa. Hereafter we discuss the case when only one supersymmetric fermionic charge $Q = (Q_+, Q_-)$ is present, a case known in the literature as $N = 1$ supersymmetric field theories. The best way to formulate a theory that is supersymmetric under an $N = 1$ supersymmetry transformation is to adopt the *superfield formalism*, where the bosonic and fermionic fields, together with a real auxiliar field $F(x)$, can be conveniently organized into a real superfield $\Phi(x,\theta)$ which admits the expansion

$$\Phi(x,\theta) = \varphi(x) + \bar{\theta}\,\psi(x) + \frac{1}{2}\bar{\theta}\,\theta F(x). \tag{24.5.1}$$

The space coordinates $x^\mu = (x^0, x^1)$ and the two real Grassmann coordinates $\theta_\alpha = (\theta_1, \theta_2)$ describe the $N=1$ superspace.[1] Using this geometrical picture, a supersymmetric transformation is equivalent to a translation in superspace

$$x^\mu \to x^\mu + i\bar{\xi}\gamma^\mu\theta, \quad \theta_\alpha \to \theta_\alpha + \xi_\alpha \tag{24.5.2}$$

under which the variation of the superfield is given by

$$\delta\Phi(x,\theta) = \bar{\xi}_\alpha\, \mathcal{Q}_\alpha\, \Phi(x,\theta), \tag{24.5.3}$$

with $\mathcal{Q}_\alpha = \partial/\partial\bar{\theta}_\alpha + i(\gamma^\mu\theta)_\alpha\,\partial_\mu$. The most general action invariant under the supersymmetric transformation (24.5.3) is given by

$$\mathcal{A} = \int d^2x\, d^2\theta \left[\frac{1}{4}(\bar{D}_\alpha\Phi)\, D_\alpha\Phi + W(\Phi)\right], \tag{24.5.4}$$

where $\int d^2\theta\bar{\theta}\theta = 2$, with the covariant derivative D_α given by

$$D_\alpha \equiv \frac{\partial}{\partial\bar{\theta}_\alpha} - (i\partial_\mu\gamma^\mu\theta)_\alpha.$$

$W(\Phi)$ is the so-called superpotential, assumed to be an analytic function of Φ. Integrating on the Grassmann variables, the action (24.5.4) becomes

$$\mathcal{A} = \int d^2x \left\{\frac{1}{2}\left[(\partial_\mu\varphi)^2 + i\bar{\psi}\gamma^\mu\partial_\mu\psi + F^2\right] + F\,W'(\varphi) - \frac{1}{2}W''(\varphi)\bar{\psi}\psi\right\},$$

where $W'(\varphi) = dW(\varphi)/d\varphi$, etc. Finally, eliminating the auxiliary field F from its algebraic equation of motion, i.e. substituting $F \to -W'(\varphi)$ in the expression above, and rescaling for convenience the fermion field as $\psi \to \sqrt{2}\psi$, we arrive at the general form of the Lagrangian density for a supersymmetric theory given by

$$\mathcal{L} = \frac{1}{2}\left[(\partial_\mu\varphi)^2 - [W'(\varphi)]^2\right] + + i\bar{\psi}\gamma^\mu\partial_\mu\psi - \frac{1}{2}W''(\varphi)\bar{\psi}\psi. \tag{24.5.5}$$

[1] Notice that, for the fermionic nature of the variables θ_i, the expansion (24.5.1) truncates at the second order in θ.

The conserved supercurrent associated to the transformation (24.5.2) is expressed by

$$\mathcal{J}^{\mu}_{\alpha}(x) = (\partial_{\nu}\varphi)(\gamma^{\nu}\gamma^{\mu}\psi)_{\alpha} - iF(\gamma^{\mu}\psi)_{\alpha}, \tag{24.5.6}$$

and the conserved supercharges is therefore

$$Q_{\alpha} = \int dx^1\, \mathcal{J}^0_{\alpha}. \tag{24.5.7}$$

Together with the stress-energy tensor

$$\mathcal{T}^{\mu\nu}(x) = i\bar{\psi}\gamma^{\mu}\partial^{\nu}\psi + \partial^{\mu}\varphi\,\partial^{\nu}\varphi - \frac{1}{2}g^{\mu\nu}\left[(\partial_{\alpha}\varphi)^2 - F^2\right], \tag{24.5.8}$$

and the topological current

$$\xi^{\mu}(x) = -\epsilon^{\mu\nu}F\,\partial_{\nu}\varphi = \epsilon^{\mu\nu}\partial_{\nu}W(\varphi), \tag{24.5.9}$$

these charges close the supersymmetry algebra

$$\{Q_{\alpha}, \bar{Q}_{\beta}\} = 2(\gamma_{\lambda})_{\alpha\beta}P^{\lambda} + 2i(\gamma_5)_{\alpha\beta}\,\mathcal{Z}, \tag{24.5.10}$$

where $P^{\lambda} = \int \mathcal{T}^{0\lambda}(x)\,dx^1$ is the total energy and momentum, whereas $\mathcal{Z}$ plays now the role of the topological charge

$$\mathcal{Z}_{ab} = \int \xi^0(x)\,dx^1 = [W(\varphi)]^{+\infty}_{-\infty} \equiv W(\varphi_b) - W(\varphi_a). \tag{24.5.11}$$

Although closely related to the conventional kink topological charge $\mathcal{T}$, the charge $\mathcal{Z}$ may however differ from it, as we show below. A convenient and explicit form way of the supersymmetry algebra (24.5.10) is the following

$$Q_+^2 = P_+, \quad Q_-^2 = P_-, \quad Q_+Q_- + Q_-Q_+ = 2\mathcal{Z}, \tag{24.5.12}$$

where we have used the notation $(Q_1, Q_2) \equiv (Q_-, Q_+)$ and $P_{\pm} = P_0 \pm P_1$.

24.6 General Results in SUSY Theories

The existence of an algebra that relates the bosonic and fermionic sectors implies important consequences. First of all, comparing the Lagrangian eqn. (24.5.5) with the original one (24.2.2), we see that in the $N = 1$ SUSY model both potentials $U(\varphi)$ and $V(\varphi)$ come from the same function $W(\varphi)$, i.e.

$$U(\varphi) = \tfrac{1}{2}(W'(\varphi))^2,$$
$$V(\varphi) = W''(\varphi). \tag{24.6.1}$$

$U(\varphi)$ is then intrinsicaly positive and the relationship of this potential with the spontaneous breaking of supersymmetry: in fact, when $U(\varphi)$ has zeros, these zeros are the true vacua of the theory and supersymmetry is unbroken; vice-versa, if $U(\varphi)$ has local minima that are not zeros of this function, supersymmetry will be spontaneously broken at these minima. The local minima may be regarded as meta-stable vacua, as long as there exists a true vacuum somewhere in the landscape of $U(\varphi)$. We will return to these issues later.

Focusing on the bosonic and fermionic bound states in $\mathcal{T} = 0$ sector, let us show that the general pattern of these spectra is fixed by the following two results.

1. **An identity concerning the potentials.** For all SUSY models it is easy to prove that for the fermionic and bosonic potentials we have

$$v_a = \pm\omega_a. \tag{24.6.2}$$

   Such an identity is easy to prove since

$$v_a = V(a) = W''(a),$$

   while

$$\omega_a^2 = \frac{d^2U}{d\varphi^2} = (W''(a))^2 + W'(a)\,W'''(a) = (W''(a))^2,$$

   where the last step comes from $W'(a) = 0$, since φ_a is one of the vacua of $W'(\varphi)$.
2. **Exact value of the classical mass of kinks.** In supersymmetric theories the classical mass of the kinks can be computed entirely in terms of the algebra. Indeed the minima of $U(\varphi)$ are connected by supersymmetric kinks, solutions of the so-called BPS equations

$$(Q_+ \pm Q_-) \mid K_{ab}\rangle = 0, \tag{24.6.3}$$

   where the $\pm$ refers to the kink $\mid K_{ab}\rangle$ and the anti-kink $\mid K_{ba}\rangle$, respectively. The bosonic component of these equations is just the familiar condition (23.2.3)

$$\frac{d\varphi_{ab}}{dx} = \pm F(\varphi_{ab}) = \pm W'(\varphi_{ab}). \tag{24.6.4}$$

Thanks to the identity

$$P_+ + P_- = (Q_+ \pm Q_-)^2 \mp 2Z_{ab} \tag{24.6.5}$$

that follows from the supersymmetry algebra (24.5.12), the classical mass M_{ab} of the kinks is then simply expressed by their topological charge: for the kink (anit-kink) at rest, the left-hand side is given by $P_+ + P_- = 2M$, whereas the right-hand side, using eqn. (24.6.3), is equal to $\mp 2Z_{ab}$, therefore

$$M_{ab} = |Z_{ab}|. \tag{24.6.6}$$

Notice that this expression is a particular case of the general formula (23.2.8), given that in supersymmetry $\sqrt{2U(\varphi)}$ is equal to $W(\varphi)$ and the topological charge Z_{ab} is expressed by eqn. (24.5.11).

Degeneracy of the bosonic and fermionic bound states. We can now use the two results given above to reach a very general conclusion for the spectra of supersymmetry theories. Namely, given the identity $\omega_a = | v_a |$ and the expression (24.6.6) of the mass of the kinks, for SUSY theories we have

$$\eta_a = \xi_a,$$

and therefore the spectra (23.4.3) and (24.3.9) of bosonic and fermionic bound states at each true ground state $| \mathbf{a} \rangle$ of zero energy are always necessarily paired.

24.7 Integrable SUSY Models

There is a set of theories where we can test the semi-classical results: those are the integrable exactly solvable models. Here we focus the attention on a paradigmatic example of these theories, i.e. the SUSY Sine–Gordon model with superpotential $W(\Phi)$ given by

$$W(\Phi) = \frac{m}{\lambda^2} \cos(\lambda \Phi).$$

This gives rise to the SUSY Sine–Gordon Lagrangian

$$\mathcal{L} = \frac{1}{2}(\partial_\mu \varphi)^2 + \frac{m^2}{2\lambda^2} \sin^2 \lambda \varphi + i\bar{\psi}\gamma^\mu \partial_\mu \psi - m\bar{\psi}\psi \cos \lambda \varphi. \tag{24.7.1}$$

The exact S-matrix of this theory can be found generalizing the results discussed in Chapter 18: this leads to the exact spectrum of the $Q = 0$ sector of this model

$$m^B(n) = m^F(n) = 2M \sin\left(\frac{n\hat{\lambda}^2}{2}\right), \qquad n = 1, 2, \ldots \left[\frac{1}{\hat{\lambda}^2}\right], \tag{24.7.2}$$

where M is the mass of the kink while $\hat{\lambda}^2$, the so-called renormalized coupling constant, is given by

$$\hat{\lambda}^2 = \frac{\lambda^2}{2\pi} \frac{1}{1 - \frac{\lambda^2}{4\pi}}. \tag{24.7.3}$$

This is the exact $\mathcal{T} = 0$ spectrum of the theory. Let us see how this spectrum can be recovered using the semi-classical approach.

The minima of the scalar potential $U(\varphi)$ are localized at $\varphi_k^{(0)} = k\pi/\lambda$. Since they are also the zeros of $U(\varphi)$, they are supersymmetric vacua of the quantum theory. The theory (24.7.1) is invariant under the Z_2 parity $\varphi \to -\varphi$ and under the shift

$$\varphi \to \varphi + \frac{2\pi\, n}{\lambda}, \tag{24.7.4}$$

so we can restrict our attention to the interval $(0, 2\pi/\lambda)$ of the field φ. The elementary kink and the anti-kink solutions of (23.2.3) of the SSG model are explicitly given by

$$\varphi_{\pm}^{cl}(x) = \frac{2}{\lambda} \arctan(e^{\pm mx}), \tag{24.7.5}$$

and connect the adjacent vacua $\varphi_0^{(0)} = 0$ and $\varphi_1^{(0)} = \pi/\lambda$, all other kinks or anti-kinks of the model being equivalent to them. In this model $\mathcal{Z}$ differs from the usual topological charge $\mathcal{T}$, since the latter counts the kink number while the former counts the kink number modulo 2.

In addition to the explicit expression of the kinks, there is also a closed formula also the fermionic zero mode, since

$$\int_{x_0}^{x} V(\varphi_+^{cl}(x)dx = \int_{\varphi_+^{cl}(x_0)}^{\varphi_+^{cl}(x)} V(\varphi) \left(\frac{d\varphi}{dx}\right)^{-1} dx = \int_{\varphi_+^{cl}(x_0)}^{\varphi_+^{cl}(x)} \frac{\cos t}{\sin t}\, dt = \log\left[\sin(\varphi_+^{cl}(x))\right],$$

where we have used eqn.(24.6.4) for the derivative of the kink solution and choose x_0 such that $\varphi_+^{cl}(x_0) = \pi/(2\lambda)$. In this way, the zero mode can be exactly expressed as

$$\psi_{01}^{(0)}(x) = A\,\hat{\psi}_0 \sin\left[2\arctan(e^{mx})\right] = A\,\hat{\psi}_0 \frac{1}{\cosh(mx)},$$

and, as a matter of fact, is equal to the zero-mode of the double well potential at the supersymmetric point $g = \lambda$. About the nature of the SUSY vacua, we can take every

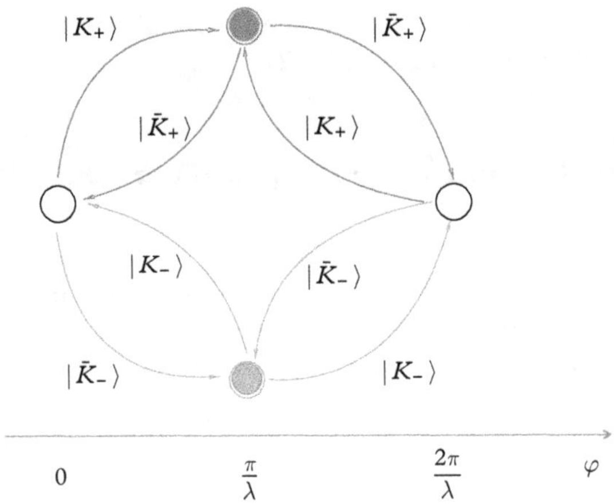

Fig. 24.6 *Structure and adjacency diagram of the vacua in the SUSY Sine–Gordon in the periodic interval of the model. This structure has to be periodically repeated for the other vacua.*

vacuum $\varphi_{2n}^{(0)} = 2n\pi/\lambda$ to be non-degenerate (these are the vacua where $V(\varphi_{2n}^{(0)}) > 0$), while we can unfold those located at $\varphi_{2n+1}^{(0)} = (2n+1)\pi/\lambda$ (where $V(\varphi_{2n+1}^{(0)}) < 0$): the adjacency properties of the new vacua are shown in Figure 24.6. This way of unfolding the vacua is similar to what has been done in the supersymmetry deformation of the TIM (Chapter 18). The classical mass of these kinks, according to the formula (24.6.6), is given by

$$M_{cl} = \frac{2m}{\lambda^2}.$$

while its finite quantum correction by

$$M_{cl} \to M = 2m\left[\frac{1}{\lambda^2} - \frac{1}{4\pi}\right],$$

The final expression that can be written as

$$M \equiv \frac{m}{\pi\hat{\lambda}^2},$$

where $\hat{\lambda}$ is the effective coupling constant of the SUSY Sine–Gordon model given in (24.7.3). Thanks to this formula, all the minima of $U(\varphi)$ have a curvature equal to $\omega_a = m$ and therefore for the semi-classical parameter η_a and ξ_a we have

$$\xi_a = \eta_a = \frac{m}{\pi M} = \hat{\lambda}^2.$$

Hence, for this exactly solvable model, the semi-classical spectra (23.4.3) and (24.3.9) precisely coincide with the exact ones (24.7.2).

24.8 Non-integrable Multi-frequency Super Sine-Gordon Models

The SUSY Sine–Gordon model has provided an important check of our formulae for the mass spectrum of bosons and fermions. However, the actual usefulness of the semi-classical formulae is of course in the absence of an exact solution of the model. This is, for instance, the case of the double well theory previously analysed in Section 24.4: even though this model can be elegantly put in a SUSY form when $g = \lambda$, with the superpotential given by

$$W(\Phi) = \lambda\Phi^3 - \frac{m^2}{2\lambda}\Phi, \tag{24.8.1}$$

it remains nevertheless non-integrable and therefore the discussion in Section 24.4 becomes particularly valuable.

Here, however, we are interested in a class of non-integrable SUSY models that interpolate between two integrable theories. A representative of this type of models is the multi-frequency SUSY Sine–Gordon model with action

$$\mathcal{A} = \int d^2x d^2\theta \left[\frac{1}{4}(\bar{D}_\alpha\Phi)D_\alpha\Phi - \frac{\mu}{\alpha}\cos(\alpha\Phi) - \frac{\lambda}{\beta}\cos(\beta\Phi)\right], \tag{24.8.2}$$

We restrict our attention to the case in which the ratio ω of the two frequencies is a rational number

$$\omega \equiv \frac{\beta}{\alpha} = \frac{p}{q}, \tag{24.8.3}$$

with p and q two co-prime natural numbers, $p < q$. Since any irrational number can be approximated with arbitrary precision by the sequence of rational approximants given by its continued fraction expressions, the case when ω is irrational may be regarded as a particular limit of the rational situation.

The two trigonometric interactions enter the action (24.8.2) in a symmetric way. Hence, there are two natural perturbative regimes of the theory:[2] $\lambda \to 0$ (with μ fixed) or $\mu \to 0$ (with λ fixed). Below we the first regime, since the other can be simply recovered by swapping the role of the two operators. The question to address concerns the fate of the kinks of the original SSG: are they going to be confined once the new interactions

[2] As in the bosonic case, the quantum theory actually depends on the appropriate dimensionless combination of the two couplings.

are switched on? Or, on the contrary, are they going to remain stable excitations of the perturbed action (24.8.2)? To make any progress, let us first eliminate the auxiliary field F and focus on the resulting bosonic interaction $U(\varphi.\lambda)$ associated to the action (24.8.2)

$$U(\varphi,\lambda) = \frac{1}{2}\left[\mu \sin(\alpha\varphi) + \lambda \sin(\beta\varphi)\right]^2. \tag{24.8.4}$$

The formulae simplify if we rescale the field $\varphi \to \varphi/\alpha$ and the coupling constant $\lambda \to \lambda/\mu$ in such a way that, up to an overall constant, $U(\varphi,\lambda)$ assumes the form

$$U(\varphi,\lambda) = \left[\sin\varphi + \lambda \sin(\omega\varphi)\right]^2 = \left[\sin\varphi + \lambda \sin\left(\frac{p}{q}\varphi\right)\right]^2. \tag{24.8.5}$$

Notice that, no matter how we vary λ, the origin $\varphi^{(0)} = 0$ is always a zero of this expression, i.e. this model always possesses a true supersymmetric vacuum. We comment later on the possibility of having meta-stable vacua, i.e. vacua where the supersymmetry is spontaneously broken, by varying λ. Let us also notice that, at lowest order in λ, all the vacua of the original SSG potential continue to remain degenerate, i.e. the original kinks are always stable at weak coupling! In other words, for supersymmetric theory there is no confinement of the kink once we perturb the integrable theory, contrary to what happens in purely bosonic theories (Chapter 22).

In order to show the stability of the kinks, let us follow the evolution of the zeros of (24.8.5) when we switch on λ. At $\lambda = 0$, they are placed at $\varphi_n^{(0)} = n\pi$. Switching on λ, we can look for the new location of the zeros in power-series in λ, i.e.

$$\varphi_n^{(0)} \simeq n\pi + \lambda\epsilon_n^{(1)} + \cdots. \tag{24.8.6}$$

Substituting this expression into the equation $U(\varphi,\lambda) = 0$, it is easy to prove that, at the first order in λ, $U(\varphi,\lambda)$ has the same zeros of $U(\varphi,0)$. This can be seen as a simple consequence of the fact that U is given, in supersymmetric theory, by the square of a function: namely, if $U(x) = [f(x)]^2$ and $f(x)$ has zeros at $x = x_1, x_2, \ldots$ in the interval I, perturbing this function by $\lambda g(x)$ $U(x,\lambda) = [f(x) + \lambda g(x)]^2 \simeq [f(x)]^2 + 2\lambda f(x)g(x)$. Therefore, if $g(x)$ does not have zeros in the interval I, at the lowest order in λ, $U(x,\lambda)$ has the same zeros of $U(x)$ in I. The actual shift of the zeros can be computed by imposing that $U(\varphi,\lambda)$ has zeros also at the second order in λ: the consistency of this request comes from the result there is always a solution in terms of $\epsilon_n^{(1)}$, given by

$$\epsilon_n^{(1)} = (-1)^n \sin\frac{np\pi}{q}. \tag{24.8.7}$$

Hence, at weak coupling, the zeros $\varphi_n^{(0)}$ shift their position by $\epsilon_n^{(1)}$ but they do not disappear.

The stability of the kinks at the lowest order in λ and the role played by supersymmetry becomes even more evident if we apply the FFPT. For doing so, the first thing to do is to identify the operator Υ that, at lowest order in λ, spoils the integrability of the original model. This is given by

$$\Upsilon = \sin\varphi \sin(\omega\varphi) + \omega\bar{\psi}\psi \cos(\omega\varphi). \tag{24.8.8}$$

To compute its semi-local index $\gamma_{\Upsilon,a}$ on the kink state, we can use the fact that the fermionic and the bosonic field are components of the same superfield and that the exponential operators $e^{i\eta\varphi}$ has a semi-local index $\gamma = \frac{\eta}{2}$ with respect to the kink of this theory.[3] If we quantize the theory by choosing as vacuum the zero at the origin, for the residue of the FF of Υ that controls the confinement of the kinks we have

$$\text{Res}_{\theta=\pm i\pi} F^{\Upsilon}_{a\bar{a}}(\theta) = [1+\cos(\pi\omega)]\ {}_0\langle 0|\Upsilon(0)|0\rangle_0. \tag{24.8.9}$$

The above quantity consistently vanishes if $\omega = 1$, i.e. if we perturb the action by the same trigonometric term that enters its original definition: in this case the deformation just redefines the original coupling constant μ and this cannot have consequences on the confinement of the kink. For all other values of ω, the term $[1+\cos(\pi\omega)]$ is different from zero but the residue is nevertheless zero, because it is the vacuum expectation value of Υ that vanishes in this case! This is a consequence of the unbroken supersymmetry of the theory: in fact, Υ is nothing else that an additional term in the trace $\Theta(x)$ of the stress-energy tensor of theory and, since the supersymmetry is exact, from the condition $Q_{\pm}\mid 0\rangle_0 = 0$ simply follows ${}_0\langle 0|\Theta(0)|0\rangle_0 = 0$. It is interesting to notice that the absence of the confinement of the kinks is also independent from the rational or irrational nature of ω.

The vanishing of the vacuum expectation value of the perturbing operator Υ is due to the simultaneous presence of a bosonic and a fermion term in its expression, where the vacuum expectation value of one term cancels the vacuum expectation of the other. The exact cancellation obviously happens for an opportune fine-tuning of the relative couplings of these terms—a fact that simply expresses the supersymmetry invariance of the theory. In the bosonic multi-frequency Sine–Gordon, there is no such cancellation and, as a result, in the perturbed theory the kinks are confined.

The vanishing of the residue (24.8.9) implies the absence of confinement of the kinks at weak coupling but not that their mass remains untouched. As a matter of fact, the masses vary and, in particular, the kinks are no longer degenerate when λ is switched on. Namely, the original kink excitations split into families of *long kinks* and *short kinks*: the former overpass higher barriers, the latter lower barriers. The actual computation of the new masses of the kinks can be done by employing their topological charges, i.e. by using eqns. (23.7.4) and (24.6.6), together with the shift of the zeros given by (24.8.7).

[3] Notice that this is half the semi-local index of the bosonic Sine–Gordon case, since the supersymmetric bosonic potential $\sin^2\varphi$ has a frequency that is twice the frequency of the usual Sine–Gordon model.

24.9 Phase Transition and Meta-stable States

The analysis done so far applies to the weak coupling regime $\lambda \to 0$. We may wonder, however, about the evolution of the topological part of the spectrum by varying λ, either at finite values of this coupling or its strong coupling regime $\lambda \to \infty$. The answer is quite interesting.

Consider the bosonic and the fermionic potential terms of the theory in their rescaled form

$$\hat{U} = \left[\sin\varphi + \lambda \sin\left(\frac{p}{q}\varphi\right)\right]^2. \tag{24.9.1}$$

$$\hat{V} = \left[\cos\varphi + \lambda \frac{q}{p} \cos\left(\frac{p}{q}\varphi\right)\right] \bar{\psi}\psi.$$

In the presence of λ, the periodicity of the theory is no longer equal to 2π but has become $2\pi q$. Therefore it is convenient to consider these interactions, in particular $\hat{U}$, on this extended interval.

The evolution of the spectrum by varying λ is constrained by a simple argument. At $\lambda = 0$, the sequence of the zeros comes from the first trigonometric term, and they become slightly displaced from their original position, without changing their number, when λ is sufficiently small. Denoting by $N_z(\lambda)$ the number of zeros of $\hat{U}$ at a given λ in the enlarged interval $(0, 2\pi q)$, we have

$$N_z(0) = 2q + 1,$$

where we have also included the zero at the origin. On the other hand, when $\lambda \to \infty$, the number of zeros becomes

$$\lim_{\lambda \to \infty} N_z(\lambda) = 2p + 1,$$

i.e. the number of zeros in the interval $(0, 2\pi q)$ that come from the second trigonometric term in $\hat{U}$. Hence, by varying λ from 0 to ∞ there is a variation of the number of zeros of $\hat{U}$ equal to

$$\Delta N_z = 2(q - p).$$

Since the kinks owe their existence to the zeros, a variation of their number implies that certain kinks will disappear by moving λ from $\lambda = 0$ to $\lambda = \infty$. Since they are topological excitations, their disappearance signals that certain phase transitions will take place in the model at some critical values of the coupling, $\lambda = \lambda_c^{(n)}$. At these values, the system has massless excitations that rule its long-distance behaviour. As discussed below, $\lambda_c^{(n)}$ are identified as the values where $N_z(\lambda)$ jumps by a step of two units (Figure 24.7).

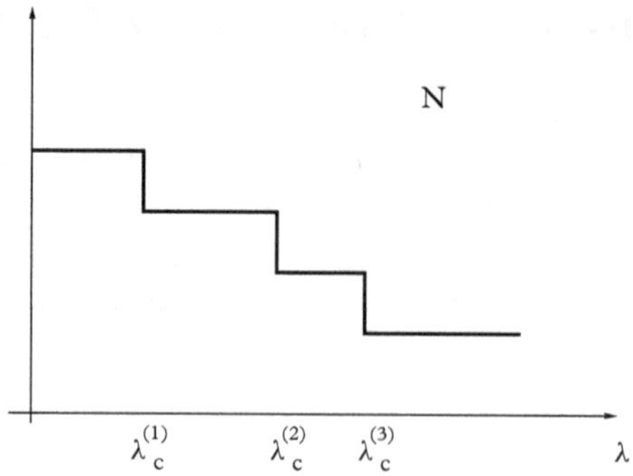

Fig. 24.7 *$N_z(\lambda)$ vs λ, with the sequence of the critical values $\lambda_c^{(n)}$.*

The way in which the number of zeros changes, by varying with continuity λ, is that pairs of zeros collide and then move on immaginary values. When this happens, the barrier between the two colliding vacua vanishes and, correspondingly, the kinks that connect them become massless. Notice that, in the evolution of the zeros, those placed at $\varphi^{(0)} = 0$, $\varphi^{(0)} = \pi q$ and $\varphi^{(0)} = 2\pi q$ will never disappear. There is, however, a different way in which the disappearance of the zeros is implemented: whether $(p-q)$ is an even or an odd number. In fact, when $(q-p)$ is an odd number, among the pairs of colliding zeros there will always be a couple of them placed just on the right and on the left position of $\varphi^{(0)} = \pi q$: at some critical value $\hat{\lambda}_c$, these zeros will collide, but in the meantime, strangle the zero at πq that is in between. When $(q-p)$ is an even number, this will not happen.

To understand in more detail what is going on, it is useful to consider two examples: the first refers to $\omega = \frac{1}{3}$, i.e. $(q-p) = 2$, the second instead to $\omega = \frac{2}{3}$, with $(q-p) = 1$.

First we consider the case $\omega = \frac{1}{3}$. By varying λ, the landscape of $\hat{U}$ changes, as shown in the sequence of the pictures of Figure 24.8. Observe that, just before reaching λ_c, there are two zeros in the interval $(0, 3\pi)$ that are going to merge together. We call these vacua $\varphi_1^{(0)}$ and $\varphi_2^{(0)}$. At $\lambda = \lambda_c$ the barrier between these two zeros vanishes and, soon after λ_c, a meta-stable vacuum is created at their middle position $\varphi = \varphi_{mv}$. For λ close to λ_c, the shape of $\hat{U}$ at φ_{mv} can be parameterized as

$$\hat{U} \sim \left[(\varphi - \varphi_{mv})^2 + (\lambda - \lambda_c)\right]^2.$$

In terms of the superfield formalism, such a form of the bosonic potential comes from a superpotential as

$$W(\hat{\Phi}) = \frac{1}{3}\hat{\Phi}^3 + (\lambda - \lambda_c)\hat{\Phi}, \tag{24.9.2}$$

where $\hat{\Phi} = \Phi - \Phi_{mv}$. Such a superpotential is locally the same as the one of the TIM perturbed by the superfield $\hat{\Phi}$ (see Section 13.2, with $p = 3$): in terms of the Virasoro

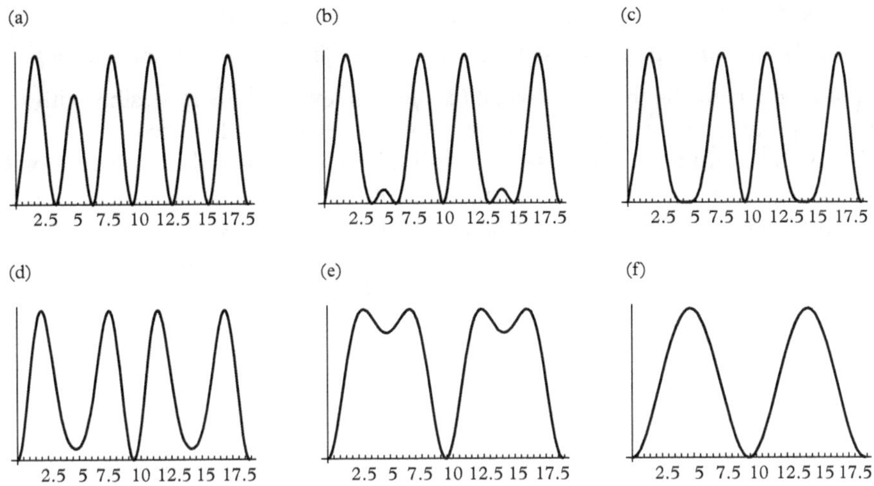

Fig. 24.8 *Plot of* $\hat{U}$ *in the interval* $(0, 6\pi)$ *for* $\omega = 1/3$ *by varying* λ*: (a) small value of* λ*; (b)* $\lambda < \lambda_c$*; (c)* $\lambda = \lambda_c$*; (d)* $\lambda > \lambda_c$*; (e)* $\lambda \gg \lambda_c$*; (f)* $\lambda \to \infty$.

fields, this deformation corresponds to a $\Phi_{1,3}$ deformation of the model with a positive coupling which induces a massless flow from the TIM to the Ising model (Section 15.6). Let us see how this information fits into our general discussion. Since at $\lambda = \lambda_c$ some topological excitations disappear, a phase transition takes place in the model. If we had originally quantized the theory either at $\varphi_1^{(0)}$ or at $\varphi_2^{(0)}$, the effective dynamics at any of these two vacua appears as a cascade of massless flows: the first, at $\lambda = \lambda_c$, from $c = 3/2$ (the value of the central charge associated to the bosonic and the Majorana fermionic fields) to $c = 7/10$ (the central charge of the TIM) and soon after overpassed λ_c, from $c = 7/10$ to $c = 1/2$, the last value being the central charge of the Ising model. These jumps in the central charge are consistent with the c-theorem and the unitarity of the theory. Although the zero at the origin always leaves the supersymmetry of the model exact, the quantization around the metastable vacuum state at φ_{mv} realizes however a spontaneously supersymmetric breaking situation, with a spectrum given by a massive scalar and a massless fermion. This can be specifically checked by studying the potential V associated to the superpotential $W(\hat{\Phi})$ above

$$V(\varphi)\overline{\psi}\,\psi = W''(\varphi)\overline{\psi}\,\psi = 2(\varphi - \varphi_{mv})\,\overline{\psi}\,\psi.$$

For $\lambda > \lambda_c$ (and for all the range of values of λ that allow us to consider the lifetime of this metastable vacuum sufficiently long), the term that multiplies $\bar{\psi}\psi$ is indeed zero at $\varphi = \varphi_{vm}$. The massless fermion is nothing else but the usual Majorana fermion of the critical Ising model. Increasing λ further and reaching values much larger than λ_c, the lifetime of the metastable vacuum shortens and the effective theory based on this pseudo-vacuum loses, at certain point, its validity. As a matter of fact, at strong coupling this pseudo-vacuum is going to be finally absorbed into the maximum of the second term present in $\hat{U}$, i.e. $\sin\left(\frac{1}{3}\varphi\right)$. It is easy to check that the situation just discussed for the case $\omega = \frac{1}{3}$ also occurs in all other cases when $(q-p)$ is an even number, possibly with several critical values of λ where phase transitions of the type of the TIM occur.

We now discuss the second case, $\omega = \frac{2}{3}$. The evolutation of the landscape of $\hat{U}$ by varying λ is shown in the sequence of pictures of Figure 24.9. When λ is just switched on, the number of zeros does not change but the barriers that separate them start to lower. Concentrating the attention on the barriers placed in the middle of the plots, their evolution near the critical value $\hat{\lambda}_c$ of the coupling follows the patter encoded into this superpotential

$$W(\hat{\Phi}) = \frac{1}{4}\hat{\Phi}^4 + \frac{1}{2}(\lambda - \hat{\lambda}_c)\hat{\Phi}^2, \tag{24.9.3}$$

so that

$$\hat{U} \sim (\varphi - \varphi_{vm})^2 \left[(\varphi - \varphi_{mv})^2 + (\lambda - \hat{\lambda}_c)\right]^2,$$

with $\varphi_{mv} = 3\pi$. Comparing with the Landau–Ginzburg description of the superconformal minimal models discussed in Section 13.2, the superpotential (24.9.3) is locally the same as the one of the Gaussian model (with $p = 4$ and central charge $c = 1$) perturbed by the superfield $\hat{\Phi}^2$. In this case, overpassing the critical value $\hat{\lambda}_c$, supersymmetry still remains exact at the vacuum $\varphi^{(0)} = \varphi_{mv}$. It is easy to see that a similar situation occurs

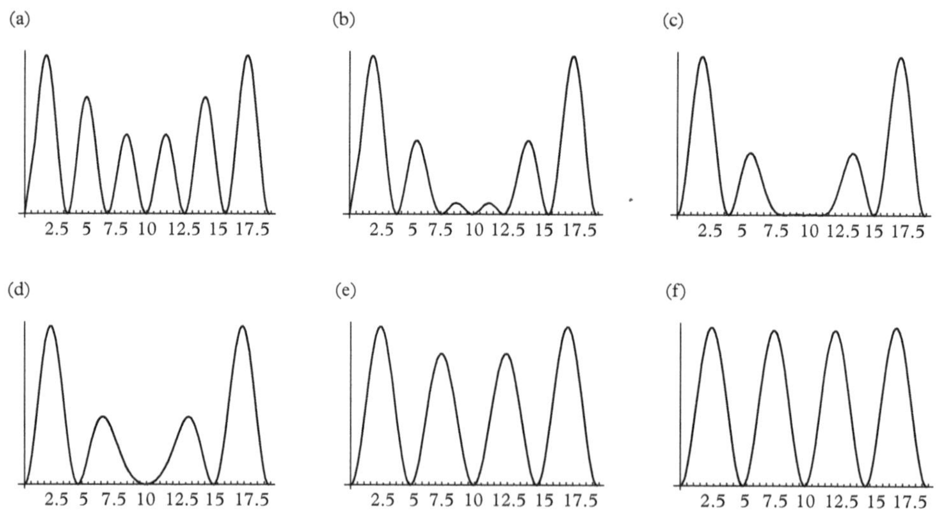

Fig. 24.9 *$\hat{U}$ for $\omega = 2/3$ by varying λ: (a) small value of λ (b) $\lambda < \hat{\lambda}_c$; (c) $\lambda = \hat{\lambda}_c$; (d) $\lambda > \hat{\lambda}_c$; (e) $\lambda \gg \hat{\lambda}_c$; (f) $\lambda \to \infty$.*

in all other case when $(q-p)$ is an odd number, the only difference possibly being the existence of a sequence of other critical values $\lambda_c^{(n)}$ (in addition to $\hat{\lambda}_c$) where TIM-like phase transitions occur at other vacua. This happens if $(q-p) > 1$. If we had quantized the theory by choosing as the vacuum the one placed at $\varphi^{(0)} = \pi q$, the dynamics at this vacuum state consists, at $\lambda = \hat{\lambda}_c$, in a massless flow from $c = 3/2$ to $c = 1$, and then from $c = 1$ to $c = 0$, without breaking supersymmetry.

24.10 Summary

This chapter has derived the semi-classical formula (24.3.9) for the fermionic spectrum of an interacting QFT of boson and fermion. Together with the formula (23.4.3) for the bosons, these expressions provide interesting information on those QFTs made of degenerate vacua. Notice that the formula (24.3.9) remains valid also when the fermion is of Dirac type rather than Majorana, modifying correspondingly the expansion of the fermion in the presence of the kink, eqn. (24.2.11), and using the more general relation between matrix elements given in eqn. (24.3.2). So, these semi-classical formulae allow us to identify the particle excitations and estimate one of their most important characteristics, i.e. their mass. Of course, in absence of integrability, not all the particles entering the semi-classical formulae are stable: to determine which of them will decay simply involves identifying the proper threshold of the decay channel. The main difference with respect to the bosonic case is the presence of the shift η_a in the fermionic formula (24.3.9) with respect to the bosonic one (23.4.3), which may influence the value of the thresholds. However it is natural to think that, as in the bosonic case, only few excitations will be generally stable on each vacua, while all the others become resonances.

We also analysed the breaking of integrability of a supersymmetric theory: contrarily to the purely bosonic theories, this does not lead to a confinement of the kink exitations. The stability of the kink excitations is due to the unbroken supersymmetry invariance of the theory although there is a variation of the spectrum by varying the coupling constant. As a particular example of the pattern that emerges by breaking integrability in supersymmetric models, we studied the multi-frequency super Sine–Gordon model. In this theory, at weak value of the coupling that breaks integrability, there is no longer the degeneracy of the original kinks, since the kinks split into a sequence of *long* and *short* kinks, whose mass can be computed in terms of their topological charge $\mathcal{Z}$. At finite values of the coupling, there exist critical points where phase transitions take place in the system, namely pairs of kinks become massless and then disappear from the set of asymptotic states. Hence, the evolution of the spectrum by varying the coupling can be seen as an RG trajectory which passes by several critical points, finally ending up in the phase described by the strong coupling regime of the theory.

REFERENCES

The semi-classical methods in QFTs with fermions are discussed in

Dashen, R.F., Hasslacher B. and Neveu A. (1974). 'Nonperturbative Methods and Extended Hadron Models in Field Theory 2. Two-Dimensional Models and Extended Hadrons', *Physics Review D*, 10: 4130.

Jackiw R. and Rebbi C. (1976). 'Solitons with Fermion Number 1/2', *Physics Review D*, 13: 3398.

The semi-classical mass formula for the bound states of fermions was derived in

Mussardo, G. (2015). 'Bound States of Majorana Fermions in Semi-classical Approximation', *Journal of Statistical Mechanics*, P12003.

For general references on supersymmetry and two-dimensional supersymmetric models, see

Ahn, C. (1991). 'Complete S-Matrices of Supersymmetric Sine-Gordon Theory and Perturbed Superconformal Minimal Models', *Nuclear Physics B*, 354: 57.

Bajnok, Z., Dunning, C., Palla, L., Takacs, G. and Wagner, F. (2004). 'SUSY Sine–Gordon Theory as a Perturbed Conformal Field Theory and Finite Size Effects', *Nuclear Physics B*, 679: 521.

Cooper, F., Khare, A. and Sukhatme, U. (1995). 'Supersymmetry and Quantum Mechanics', *Physics Reports*, 251: 267.

Freund, P. (1986). *Introduction to Supersymmetry*, Cambridge University Press, Ch. 13.

Kastor, D.A., Martinec E.J. and Shenker, S.H. (1989). 'RG Flow in $N = 1$ Discrete Series', *Nuclear Physics B*, 316: 590.

Shankar, S. and Witten, E. (1978). 'S-matrix of Supersymmetric σ Model', *Physics Review D*, 17: 2134.

Shifman M. and Yung, A. (2009). *Supersymmetric Solitons*, Cambridge University Press.

Witten, E. (1981). 'Dynamical Breaking of Supersymmetry', *Nuclear Physics B*, 185: 513.

Witten, E. and Olive, D. (1978). 'Supersymmetry Algebras that include Topological Charges', *Physics Letters B* 76: 97.

For the absence of kink confinement in the breaking of integrability in supersymmetric models, see

Mussardo, G. (2007). 'Kink Confinement and Supersymmetry', *Journal of High Energy Physics*, 07/08: 003.

PROBLEMS

24.1. Supersymmetric quantum mechanics

Consider two quantum mechanics Hamiltonians

$$\begin{aligned} H_+ &= -\frac{\hbar^2}{2m}\frac{d^2\psi}{dx^2} + V_+(x), \\ H_- &= -\frac{\hbar^2}{2m}\frac{d^2\psi}{dx^2} + V_-(x). \end{aligned}$$

where the potential terms are obtained in terms of a *superpotential* $W(x)$ as

$$V_\pm(x) = W^2(x) \mp \frac{\hbar}{\sqrt{2m}} W'(x).$$

a. Show that $H_\pm$ can be written as

$$H_+ = A^\dagger A, \quad H_- = AA^\dagger.$$

in terms of some differential operators A and $A^\dagger$ of the first order in the derivative. Identify these operators.

b. Assume we have chosen $W(x)$ such that the ground state energy of H_+ is identically zero. Show that the eigenvalues of the two Hamiltonians $H_\pm$ are related as

$$E_n^{(-)} = E_{n+1}^{(+)}, \quad E_0^{(+)} = 0.$$

Find the relation between the relative eigenfunctions of the two Hamiltonians.

c. Show that, introducing the operators

$$Q = \begin{pmatrix} 0 & 0 \\ A & 0 \end{pmatrix}, \quad Q^\dagger = \begin{pmatrix} 0 & A^\dagger \\ 0 & 0 \end{pmatrix},$$

and grouping the two Hamiltonians as

$$H = \begin{pmatrix} H_+ & 0 \\ 0 & H_- \end{pmatrix}$$

we have the following relations

$$\begin{aligned} &\{Q, Q^\dagger\} = H, \quad \{Q, Q\} = \{Q^\dagger, Q^\dagger\} = 0, \\ &[H, Q] = [H, Q^\dagger] = 0. \end{aligned}$$

d. Suppose that the potentials $V_{\pm}(x)$ are finite as $x \to -\infty$ or as $x \to +\infty$ or both and let us define $W_1 = \lim_{x\to+\infty} W(x)$ and $W_2 = \lim_{x\to-\infty} W(x)$. Show that the reflection and transmission coefficients of the two potentials are related as

$$R_+(k) = \left(\frac{W_2 + ik}{W_2 - ik}\right) R_-(k), \quad T_+(k) = \left(\frac{W_1 - ik'}{W_2 - ik'}\right) T_-(k),$$

where k and k' are given by

$$k = \sqrt{E - W_2^2}, \quad k' = \sqrt{E - W_1^2}.$$

Analyse the implication of these relations and their analytic structure.

24.2. Exponential behaviour of the fermionic zero mode

The poles present in the Fourier transform of $\psi_{ab}^{(0)}(x)$ are determined by the exponential behaviour at $x \to -\infty$ of this function. This behaviour depends on the interaction term $V(\varphi(x))$, which we assume can be expanded nearby the vacuum values $\varphi_a^{(0)}$ as

$$V_{ab}(\varphi(x)) = v_a + V'_{ab}(\varphi(x) - \varphi_a^{(0)}) + \frac{1}{2} V''_{ab}(\varphi(x) - \varphi_a^{(0)})^2 + \cdots .$$

a. Substitute in $V_{ab}(x)$ the asymptotic expression of the kink configuration $\varphi(x)$ and show that around $x \to -\infty$ it can be generally written as

$$V_{ab}(\varphi(x)) = v_a + \sum_{n=1}^{\infty} d_n e^{n\omega_a x}, \quad x \to -\infty,$$

where ω_a is the curvature of the *bosonic potential* $U(\varphi)$ at $\varphi = \varphi_a^{(0)}$, while d_n are coefficients determined by the various derivatives V_{ab}^n and the expansion coefficients $\mu_n^{(a)}$ in eqn. (23.2.5).

b. Show that, for $x \to -\infty$, it holds

$$\int_{x_0}^{x} V_{ab}(t)\, dt = -v_a x + \sum_{n=1} \hat{d}_n e^{n\omega_a x}, \quad x \to -\infty,$$

where all terms but the first one are exponentially small.

c. Use the results above to show that for $x \to -\infty$ the fermionic zero mode has the asymptotic expansion

$$\psi_{ab}^{(0)}(x) = A\hat{\psi}\,\exp\left[-\int_{x_0}^{x} V_{ab}(t)\,dt\right] = A\hat{\psi}\,\exp\left[-v_a x + \sum_{n=1} \hat{d}_n e^{n\omega_a x}\right]$$

$$= A\hat{\psi}\,e^{v_a x}\left(1 + \sum_{n=1} t_n\, e^{n\omega_a x}\right),$$

where t_n are the final coefficients resulting from the series expansion of the exponential function and the coefficients $\hat{d}_n$.

24.3. Asymmetric well potential

It is interesting to see what happens when the bosonic potential has two asymmetric wells $|\,\mathbf{a}\rangle$ and $|\,\mathbf{b}\rangle$: in this case the asymptotic behaviours of the kink at $x \pm \infty$ are different and therefore we should expect to find two different spectra piling up on the two vacua. The purely bosonic model was discussed in Section 23.6, which showed that there could exist a range of the coupling constant where one vacuum (say $|\,\mathbf{a}\rangle$) has no bound states on top of it, while the other vacuum $|\,\mathbf{b}\rangle$ can have instead one or more. Let us now consider how the presence of the fermion may change this picture. In opportune units, as bosonic potential $U(\varphi)$ of this model we choose

$$U(\varphi) = \frac{1}{2}(\varphi + a)^2(\varphi - b)^2(\varphi^2 + c),$$

while as potential term $V(\varphi)$ we take

$$V(\varphi) = g\varphi(x).$$

a. Compute the curvature at $\varphi_0^{(0)} = -a$ and $\varphi_0^{(0)} = b$ and show that $a \neq b$, the two wells are asymmetric. From now on choose $a > b$.

b. Show that in this model there is a zero-mode $\psi_{ab}^{(0)}$ corresponding to the kink configuration $\varphi_{ab}(x)$ and another one $\psi_{ba}^{(0)}$ relative to the anti-kink $\varphi_{ba}(x)$, with the behaviour at $x \to \mp\infty$ of one equal to that one of the other at $x \to \pm\infty$, so that

$$\psi_{ab}^{(0)}(x) = A\hat{\psi}\,\exp\left[-\int_{x_0}^{x} V_{ab}(t)\,dt\right] \to A\hat{\psi}\,e^{v_a x}\left(1 + \sum_{n=1} t_n\, e^{n\omega_a x}\right), \quad x \to -\infty$$

$$\psi_{ba}^{(0)}(x) = A\hat{\psi}\,\exp\left[-\int_{x_0}^{x} V_{ba}(t)\,dt\right] \to A\hat{\psi}\,e^{v_b x}\left(1 + \sum_{n=1} t_n\, e^{n\omega_b x}\right), \quad x \to -\infty$$

where

$$v_a \equiv -V(-a) = g\,a\,\tfrac{m}{\lambda},$$

$$v_b \equiv \;\; V(b) \quad = g\,b\tfrac{m}{\lambda}.$$

c. Posing $\eta_k = v_k/(\pi M)$ and $\xi_k = \omega_k/(\pi M)$ $(k = a, b)$, where M is the mass of the kink, show that there are two different fermion spectra on the two vacua

$$m_k^F(n_k) = 2M \sin\left(\frac{\pi}{2}(\eta_k + (n_k - 1)\xi_k)\right), \quad n_k = 1, 2, \ldots \left[\frac{1-\eta_k}{\xi_k}\right].$$

d. Show that varying the coupling constants λ and g results in many curious situations at the two vacua as, for instance, no bosonic bound state on the vacuum $|\,\mathbf{a}\rangle$ and only one on the vacuum $|\,\mathbf{b}\rangle$ but a complete reverse situation for what concerns the fermionic spectrum!

25

Truncated Hilbert Space Approach

Young man, everything in physics boils down to numbers!
Statement of an old physicist

A very important aspect of strongly coupled QFTs is how to make predictions on their basic quantities, for instance how to extract the value of the masses of their excitations or of the various coupling constants. If a QFT is integrable, those questions find an answer in the bootstrap solution of the theory, as previously discussed. If, instead, the theory is not-integrable but admits a semi-classical approach, an estimate of several physical quantities can be reached using semi-classical quantization methods. But, what to do when all these approaches fail? Here we present a quite efficient numerical method that allows us to obtain a numerical estimate of a fair number of non-perturbative data of two dimensional QFTs, such as ratios of the various masses, two-particle S-matrices (at least below the inelastic thresolds), on-shell three-point couplings and so on. The method does not reply on integrability and therefore it is equally suited to both integrable or non-integrable theories, providing in the first case a non-trivial confirmation of several quantities extracted from the exact analytic solution of the various models, while in the second case a powerful tool to extract values of highly non-trivial quantities of generically non-integrable models. The method goes under the name of truncated Hilbert space approach (THSA) and it is quite different from more familiar numerical methods used to study QFTs as, for instance, lattice formulations and related Monte Carlo simulations. This chapter shows that, we can study the spectrum of the excitation of a QFT along all the relevant directions of its class of universality and follow their evolution by varying the coupling constants. In a nutshell, the THSA consists of the following steps:

1. Define initially the theory on a finite volume, hereafter taken to be a circumference of radius R of an infinitely long cylinder along the orthogonal direction considered as time-axis. In such a finite geometry, the spectrum becomes discrete.
2. Truncate the Hilbert space of the theory to finite number of states up to a given energy E_t which fixes the level of truncation. In the following the Hilbert space will be given in terms of the conformal states organized in Verma modules (simply because we will be mainly interested in described models obtained as deformation

Statistical Field Theory: An Introduction to Exactly Solved Models in Statistical Physics. Second Edition. Giuseppe Mussardo, Oxford University Press (2020). © Giuseppe Mussardo. DOI: 10.1093/oso/9780198788102.001.0001

of CFTs) but, changing the basis, the approach can be easily generalized to other cases as well.

3. Diagonalize the truncated Hamiltonian matrix numerically and extract eigenvalues and eigenvectors as functions of R and the coupling constants.

The truncated Hilbert space approach has its root in ordinary quantum mechanics. For this reason, and to present its three step implementation (given above) in a simple way, we the numerical determination of the spectrum of simple but quite instructive examples of quantum mechanics.

25.1 Truncated Hamiltonians of Quantum Mechanics

Let us consider the one-dimensional quantum Hamiltonian of a particle of mass m in presence of a potential $V(x)$

$$H = \frac{p^2}{2m} + V(x). \tag{25.1.1}$$

For simplicity, we focus our attention on the class of power-law potentials

$$V(x) = \lambda\, |x|^a, \tag{25.1.2}$$

where both λ and a are positive real quantities. Given the nature of each of these potentials, the associated spectrum is discrete, with energy levels all non-degenerate. Moreover, the sequence E_n of eigenvalues is monotonic increasing and goes to infinity for $n \to \infty$. This Hamiltonian gives rise to the Schrödinger equation

$$\left(-\frac{\hbar^2}{2m}\frac{d^2}{dx^2} + \lambda |x|^a!\right) \psi_n(x) = E_n\, \psi_n(x). \tag{25.1.3}$$

Since H commutes with the parity operator P which sends $x \to -x$, the wave functions are also eigenfunctions of P, such that $\psi(-x) = (-1)^n \psi_n(x)$. It is useful to write the Schrödinger equation in dimensionless variables as follows. Let us initially introduce a length scale ξ, which we will fix later, and define the dimensionless quantity

$$\eta = \frac{x}{\xi}. \tag{25.1.4}$$

Hence, in terms of η, eqn. (25.1.3) becomes

$$\left(-\frac{\hbar^2}{2m\xi^2}\frac{d^2}{d\eta^2}+\lambda\,\eta^a|\eta|^a\right)\psi_n(\eta)=E_n\,\psi_n(\eta). \tag{25.1.5}$$

If we now define

$$\mathcal{E}\equiv\frac{\hbar^2}{m\xi^2},\quad b\equiv\frac{m\lambda}{\hbar^2}\xi^{a+2}, \tag{25.1.6}$$

we have

$$\left[-\frac{1}{2}\frac{d^2}{d\eta^2}+b\,|\eta|^a\right]\psi_n(\eta)=\frac{E_n}{\mathcal{E}}\,\psi_n(\eta). \tag{25.1.7}$$

Notice that $\mathcal{E}$ has the same dimension of an energy scale, while b is a purely dimensionless quantity, which therefore can be regarded as the strength of the coupling constant λ once expressed in unity of $\hbar^2/m\xi^{a+2}$. Expressing the energies E_n in unity of $\mathcal{E}$, the Schrödinger equation (25.1.7) is then written in terms of purely dimensionless quantities. As for η, we also introduce the dimensionless expression $\hat{p}$ of the momentum operator

$$\hat{p}=\frac{\xi}{\hbar}p,\quad \hat{p}\rightarrow -i\frac{d}{d\eta}. \tag{25.1.8}$$

Notice that, given the relations in eqn. (25.1.6), the two quantities $\mathcal{E}$ and b scale differently by varying ξ. Two observers, using two different unities to measure distances, could however compare the values they extract for the energy levels by using the law (25.1.6). For later convenience, it is useful to fix the reference units in such as way that it holds the corresponding principle, namely that the energy levels E_n, when $n\rightarrow\infty$, coincide with the values obtained by a semi-classical approximation. Their expression is given by

$$E_n^{sc}=E_0^{sc}\left(n+\frac{1}{2}\right)^{\frac{2a}{a+2}}, \tag{25.1.9}$$

where

$$E_0^{sc}=\left[\sqrt{\frac{\pi}{2m}}\frac{\Gamma\left(\frac{3}{2}+\frac{1}{a}\right)}{\Gamma\left(1+\frac{1}{a}\right)}\hbar\lambda^{1/a}\right]^{\frac{2a}{a+2}}. \tag{25.1.10}$$

Let us now suggest that the two energy scales $\mathcal{E}$ and E_0^{sc} coincide. Extracting initially ξ from the second equation in (25.1.6)

$$\xi = \left(\frac{\hbar^2 b}{m\lambda}\right)^{\frac{1}{a+2}}, \tag{25.1.11}$$

and substituting into the first equation, we have

$$\mathcal{E} = \left[\left(\frac{\lambda}{b}\right)^{\frac{1}{a}} \frac{\hbar}{\sqrt{m}}\right]^{\frac{2a}{a+2}}. \tag{25.1.12}$$

Imposing this quantity equal to E_0^{sc}, we get how the dimensionless quantity b depends on the exponent a of the potential

$$b(a) = \left[\sqrt{\frac{2}{\pi}} \frac{\Gamma\left(1+\frac{1}{a}\right)}{\Gamma\left(\frac{3}{2}+\frac{1}{a}\right)}\right]^a, \tag{25.1.13}$$

In terms of $b(a)$, the length scale lξ is then fixed from eqn. (25.1.11).

25.1.1 Harmonic Oscillator

A significant case of polynomial potential is the harmonic oscillator, usually written as

$$V(x) = \frac{1}{2} m\omega^2 x^2. \tag{25.1.14}$$

The spectrum of this Hamiltonian can be obtained on the basis of a purely algebraic method. Indeed, let us introduce the annihilation and creation operators A and $A^\dagger$ defined by

$$x = \frac{\xi_{osc}}{\sqrt{2}}\left(A + A^\dagger\right); \quad p = i\frac{\hbar}{\xi_{osc}} \frac{1}{\sqrt{2}}\left(A^\dagger - A\right), \tag{25.1.15}$$

where we introduce the quantity ξ_{osc}, which is the typical scale of the harmonic oscillator.[1]

$$\xi_{osc} = \sqrt{\frac{\hbar}{m\omega}}. \tag{25.1.16}$$

[1] It is easy to see that it coincides with the expression (25.1.11) for $a = 2$.

Hence the Hamiltonian of this system can be written as

$$H = \frac{p^2}{2m} + \frac{1}{2}m\omega^2 x^2 = \hbar\omega\left(A^\dagger A + \frac{1}{2}\right), \tag{25.1.17}$$

with

$$[H,A] = -\hbar\omega A, \quad [H,A^\dagger] = \hbar\omega A^\dagger, \tag{25.1.18}$$

and

$$[A,A^\dagger] = 1. \tag{25.1.19}$$

Applying A to an eigenstate of H with eigenvalue E_n, we get another eigenstate of H but with eigenvalue $E_n - \hbar\omega$, while applying $A^\dagger$ we get instead another eigenstate of H with eigenvalue $E_n + \hbar\omega$. However, the chain of decreasing eigenstates must stop (since the Hamiltonian is a positive operator) and therefore a lowest eigenstate must exist, hereafter denoted as $\mid 0\rangle$, identified by the condition

$$A \mid 0\rangle = 0. \tag{25.1.20}$$

Using eqn. (25.1.17), its energy is $E_0 = \frac{\hbar\omega}{2}$ and the entire sequence of eigenvectors $\mid n\rangle$ and eigenvalues of H

$$E_n = \left(n + \frac{1}{2}\right)\hbar\omega, \tag{25.1.21}$$

can be obtained by an iterative use of the creation operator $A^\dagger$

$$\mid n\rangle = \frac{1}{\sqrt{n!}}(A^\dagger)^n \mid 0\rangle, \quad \langle n \mid n\rangle = 1. \tag{25.1.22}$$

The annihilation and creation operators act on these states as

$$A \mid n\rangle = \sqrt{n} \mid n-1\rangle; \quad A^\dagger \mid n\rangle = \sqrt{n+1} \mid n+1\rangle. \tag{25.1.23}$$

and therefore in the base of the eigenstates of H their matrix representation is given by the infinite-dimensional matrices

$$A = \begin{pmatrix} 0 & \sqrt{1} & 0 & 0 & 0 & \dots \\ 0 & 0 & \sqrt{2} & 0 & 0 & \dots \\ 0 & 0 & 0 & \sqrt{3} & 0 & \dots \\ 0 & 0 & 0 & 0 & \sqrt{4} & \dots \\ \vdots & \vdots & \vdots & \vdots & \vdots & \vdots \end{pmatrix}, \quad A^{\dagger} = \begin{pmatrix} 0 & 0 & 0 & 0 & 0 & \dots \\ \sqrt{1} & 0 & 0 & 0 & 0 & \dots \\ 0 & \sqrt{2} & 0 & 0 & 0 & \dots \\ 0 & 0 & \sqrt{3} & 0 & 0 & \dots \\ \vdots & \vdots & \vdots & \vdots & \vdots & \vdots \end{pmatrix}. \tag{25.1.24}$$

In the Schrödinger representation A e $A^{\dagger}$ are expressed by the differential operators

$$A = \sqrt{\frac{m\omega}{2\hbar}}\left(x + \frac{\hbar}{m\omega}\frac{d}{dx}\right); \quad A^{\dagger} = \sqrt{\frac{m\omega}{2\hbar}}\left(x - \frac{\hbar}{m\omega}\frac{d}{dx}\right),$$

and therefore the wave function of the ground state, coming from eqn. (25.1.20), satisfies the differential equation

$$\left(x + \frac{\hbar}{m\omega}\frac{d}{dx}\right)\psi_0(x) = 0, \tag{25.1.25}$$

whose solution is given by

$$\psi_0(x) = \left(\frac{m\omega}{\hbar\pi}\right)^{1/4} e^{-\frac{m\omega}{2\hbar}x^2} = \left(\frac{1}{\pi\xi^2}\right)^{1/4} \exp^{-\frac{x^2}{2\xi^2}}, \tag{25.1.26}$$

with normalization given by

$$\int_{-\infty}^{+\infty} \psi_0^2\, dx = 1.$$

25.1.2 Basis of the Truncated Hamiltonian

In order to study the spectrum of the Hamiltonian given by

$$H = \frac{p^2}{2m} + \lambda \mid x \mid^a. \tag{25.1.27}$$

we can choose a basis $\mid \phi_n\rangle$ in the Hilbert space and evaluate all matrix elements of the Hamiltonian (25.1.27) on this basis, $H_{n,m} = \langle \phi_n \mid H \mid \phi_m\rangle$: truncating the Hilbert space up to the first N vectors of the basis, the Hamiltonian can be approximated by the $N \times N$ matrix

$$H_{(N)} = \begin{pmatrix} H_{11} & H_{12} & \cdot & \cdot & \cdot & H_{1N} \\ \cdot & \cdot & \cdot & \cdot & \cdot & \cdot \\ \cdot & \cdot & \cdot & \cdot & \cdot & \cdot \\ \cdot & \cdot & H_{nm} & \cdot & \cdot & \cdot \\ \cdot & \cdot & \cdot & \cdot & \cdot & \cdot \\ H_{N1} & H_{N2} & \cdot & \cdot & \cdot & H_{NN} \end{pmatrix}, \tag{25.1.28}$$

that can be numerically diagonalized. The corresponding eigenvalues $\hat{E}_1, \ldots, \hat{E}_N$ provide a numerical estimate of the first N energy levels of the Hamiltonian (25.1.27). For a given value N of the truncation, the accuracy depends of course on the basis that has been chosen: if this basis is closer to the actual set of eigenstates of the Hamiltonian, one observes a better rate of accuracy. Moreover, the precision varies by varying the level, being better for the lower levels but worst for those which are closer to the level N of truncation. The rate of convergence of the eigenvalues can be studied by enlarging systematically the dimension N of the matrix and studying the difference of the energy levels $\Delta E_k^{(N)} = (\hat{E}_k^{(N+1)} - \hat{E}_k^{(N)})$ as functions of N.

We now discuss a very efficient method to obtain the sequence of truncated Hamiltonians $H_{(N)}$ using as a basis the one associated to an harmonic oscillator with a proper frequency ω. The optimized value of this quantity can be obtained using the variational principle, which states that for any normalized quantum states $|\,\psi\rangle$ the ground state energy E_0 of a system with Hamiltonian H satisfies the inequality

$$E_0 \leq \frac{\langle \psi \,|\, H \,|\, \psi \rangle}{\langle \psi \,|\, \psi \rangle}. \tag{25.1.29}$$

This equation can be used to put an upper bound to the ground state energy but also to select the best computational basis for diagonalizing an Hamiltonian as in eqn. (25.1.27). In fact, let us take as wave function of the state $|\,\psi\rangle$ the normalized wave function of a harmonic potential with frequency ω, here used as variational parameter

$$\psi_\omega(x) = \left(\frac{m\omega}{\hbar\pi}\right)^{1/4} e^{-\frac{m\omega}{2\hbar}x^2}.$$

We need to compute the matrix element of H, on this state as function of ω

$$\langle \psi_\omega \,|\, \left(\frac{p^2}{2m} + V(x)\right) |\, \psi_\omega \rangle. \tag{25.1.30}$$

This matrix element is given by

$$E(\omega) = \langle \psi_\omega \,|\, H \,|\, \psi_\omega \rangle = \frac{\hbar\omega}{4} + \frac{\lambda}{\sqrt{\pi}} \Gamma\left(\frac{a+1}{2}\right) \left(\frac{\hbar}{m\omega}\right)^{a/2}. \tag{25.1.31}$$

The minimal value of this expression is determined by the condition

$$\frac{dE(\omega)}{d\omega} = 0,$$

whose solution is

$$\omega_0 = \left[\frac{2a\lambda\Gamma\left(\frac{a+1}{2}\right)\hbar^{(n-2)/2}}{\sqrt{\pi}\, m^{a/2}} \right]^{2/(a+2)}. \tag{25.1.32}$$

Substituting this value in eqn. (25.1.31), we have an estimate of the ground state energy of the system which, in unit of $\mathcal{E} = E_0^{sc}$ is given by

$$E_0 \leq \mathcal{E}\left(\frac{a+2}{4a}\right)\left[\frac{2a\,b(a)}{\sqrt{\pi}}\Gamma\left(\frac{a+1}{2}\right)\right]^{2/(a+2)}. \tag{25.1.33}$$

Once we have identified the best choice of ω for the ground state energy, we can also use this value to set up an entire basis in the Hilbert space given by the eigenfunctions $|\,\psi_n\rangle$ of the harmonic oscillator with such a frequency. This harmonic oscillator basis has the advantage of allowing an exact determination of the matrix elements $H_{n,m}$ on purely algebraic methods. To set such an algebraic method, we see that the matrix elements of the kinetic term $p^2/(2m)$ of the Hamiltonian can be computed using the expression of p in terms of the creation and annihilation operators A and $A^\dagger$ and their action on the eigenstates of the harmonic oscillator. There is however a subtlety: in fact, the expression (25.1.15) of the momentum p involves the length scale ξ_{osc} of the harmonic oscillator (eqn. (25.1.16)), while the Hamiltonian (25.1.27) involves a length scale

$$\xi(a) = \left(\frac{\hbar^2}{m\lambda}b(a)\right)^{1/(a+2)}.$$

Therefore, to express correctly the operator p in terms of A and $A^\dagger$, we need the dimensionless ratio of the two length scales

$$R(a) = \frac{\xi_{osc}}{\xi(a)} = \left[\frac{\sqrt{\pi}}{2ab(a)\Gamma\left(\frac{a+1}{2}\right)}\right]^{1/(a+2)}. \tag{25.1.34}$$

Hence, the dimensionless representation of the momentum operator on the basis of the harmonic oscillator with the optimized frequency ω is

$$\hat{p} = \frac{i}{\sqrt{2}R(a)}(A^\dagger - A). \tag{25.1.35}$$

Let us now discuss how to obtain the matrix elements of the potential term in the favourable cases when $a = 2n$, with n be an integer. In these cases, we can neglect the presence of the absolute value of x in V and therefore we can implement the representation (25.1.15) of this operator in terms of the operators A and $A^\dagger$. However, here we must also take into account the presence of the two different length scales ξ_{osc} and $\xi(a)$ and employ the ratio $R(a)$, arriving to the operatorial representation of the Hamiltonian (25.1.27) given by

$$H = \mathcal{E}\left[-\frac{1}{4(R(2n))^2}(A^\dagger - A)^2 + \frac{b(2n)[R(2n)]^{2n}}{2^n}(A^\dagger + A)^{2n}\right]. \tag{25.1.36}$$

The matrix elements of this Hamiltonian can be performed on a purely algebraic method, using the commutation relation (25.1.19) of A and $A^\dagger$ and the expression (25.1.22) of the eigenstates of the harmonic oscillator. To find the truncated expression of the Hamiltonian (25.1.36) at the level N, a way to speed up such a computation consists of using the truncated expression at level N of infinite-dimensional representations of A and $A^\dagger$ given in eqn. (25.1.24): doing so, it remains only to perform multiplications and sums of finite-dimensional matrices. Notice, however, that this procedure introduces an error in the matrix elements of the higher states: indeed, any finite dimensional representation of A and $A^\dagger$ cannot satisfy the commutation relations (25.1.19), since the trace of a commutator of finite-dimensional matrices always vanishes. Therefore, denoting with A_N and $A_N^\dagger$ the truncation at level N of the matrices (25.1.24), they satisfy the commutation relation

$$[A_N, A_N^\dagger] = \mathbf{1}_N - (N)\,\delta_{N,N}. \tag{25.1.37}$$

Hence any finite-dimensional representation of the operators A and $A^\dagger$ gives rise to an explicit error in the matrix elements of the higher state. But this drawback is amply compensated by the convenient implementation of all other matrix elements, done using only algebraic rules of finite matrices.

25.2 Truncated Hamiltonian of the Deformed Conformal Models

Let us now apply the THSA to study the spectrum of a CFT perturbed by one or more relevant scaling fields ϕ_k with angular momentum $\Delta_k - \overline{\Delta}_k = 0$ and scaling dimension $\Delta_k + \overline{\Delta}_k = \eta_k$. The Euclidean action for such a theory is

$$\mathcal{A}_\lambda = \mathcal{A}_0 + \sum_k \lambda_k \int \phi_k(z, \overline{z})\, d^2z. \tag{25.2.1}$$

The coupling constants λ_k have dimension $y_k = 2 - \eta_k > 0$. In order to study such a theory on an infinite cylinder with periodic boundary condition along its circumference of length R, we can use the logarithmic mapping

$$w = u + iv = \frac{R}{2\pi} \log z$$

so that the action (25.2.1) defines a Euclidean QFT on such a geometry, with the Hamiltonian (i.e. the logarithm of the transfer matrix) given by

$$H_{\{\lambda_k\}} = H_0 + \sum_k \lambda_k V_k. \qquad (25.2.2)$$

The conformal part H_0 of the Hamiltonian can be expressed in terms of the Virasoro generators L_0, L_0 , and the central charge c,

$$H_0 = \frac{2\pi}{R}\left(L_0 + \overline{L}_0 - \frac{c}{12}\right), \qquad (25.2.3)$$

while the interaction terms V_k are given by

$$V_k = \int_0^R \phi_k(z,\overline{z})(\omega,\overline{\omega})\, dv. \qquad (25.2.4)$$

Since $\phi_k(z,\overline{z})$ are scalar fields in the plane, both H_0 and all V_k commute with the momentum operator on the cylinder

$$K = \frac{2\pi}{R}\left(L_0 - \overline{L}_0\right). \qquad (25.2.5)$$

The following give a matrix representation of the Hamiltonian $H_{\{\lambda_k\}}$ and, truncating it to a sufficient large number N of elements, we find the energy levels of the perturbed CFT varying the circumference R and also as functions of the various couplings λ_k. In order to implement this program, we choose as basis of our Hilbert space the one spanned by the eigenstates of H_0: these are the conformal states discussed in Chapter 10, labelled by their energy, their momentum and additional quantum numbers. They organize into conformal families, i.e. products of holomorphic and anti-holomorphic Verma modules (irreducible representations of the Virasoro algebra), where each Verma module contains one primary state $\mid \Delta_a\rangle$ of dimension Δ_a and descendant states of dimension $\Delta_a + n$. The number of linearly independent states at level n is given by the character of the corresponding Verma module

$$\chi_{\Delta_a}(q) = q^{\Delta_a - c/24} \sum_{n=0}^{\infty} d(n,\Delta_a)\, q^n. \qquad (25.2.6)$$

The generic expression of these descendant states was given in eqn. (10.8.17), here repeated for convenience

$$\begin{gathered} \mid \Delta_a; n_1, n_2, \dots n_k\rangle \equiv L_{-n_1} L_{-n_2} \dots L_{-n_k} \mid \Delta_a\rangle \\ n_1 \leq n_2 \leq \dots \leq n_k. \end{gathered} \tag{25.2.7}$$

On the basis of the conformal states, the unperturbed H_0 is then a diagonal matrix

$$H_0 = \frac{2\pi}{R} \begin{pmatrix} * & & \ddots & \dots \\ & * & \cdot & \dots \\ & & & \\ \dots & & & * \end{pmatrix} \tag{25.2.8}$$

with the diagonal matrix elements generically expressed in terms of the central charge and the conformal dimensions of the fields of the various Verma modules

$$(H_0)_{e,f} = \frac{2\pi}{R}(2\Delta_a + 2n - c/24)\,\delta_{e,f}. \tag{25.2.9}$$

There are several ways of ordering these diagonal elements: better captures the effect of interactions V_k is ordering the eigenstates in terms of increasing dimensions within each conformal family. This means the following: once we fix a real number Λ and an energy scale E_t in terms of the relation

$$E_t = \frac{2\pi}{R}\Lambda, \tag{25.2.10}$$

we truncate the states up to those which satisfy the condition

$$\Delta_a + n_a \leq \Lambda + c/24. \tag{25.2.11}$$

For any given conformal dimension Δ_a of the primary fields, this condition selects the corresponding integer n_a and correspondingly the number of descendent fields associated to each primary state $\mid \Delta_a\rangle$ to include in the truncated Hamiltonian. We can then order the conformal states such that H_0 is expressed as

$$H_0 = \frac{2\pi}{R} \begin{pmatrix} D_1 & 0 & 0 & 0 & 0 \\ 0 & D_2 & 0 & 0 & 0 \\ & & & & \\ 0 & 0 & 0 & \ddots & 0 \\ 0 & 0 & 0 & 0 & D_l \end{pmatrix} \tag{25.2.12}$$

where the diagonal block matrices D_a refer to the ath conformal families (ordered in the increasing values of Δ_a) and are made of n_a diagonal elements.

Let us now discuss the form of the interaction terms. Using translation invariance along the v direction of the cylinder, the matrix elements of each V_k between the generic conformal states can be written as

$$\langle\phi_a|V_k|\phi_b\rangle = \frac{R}{2\pi}\langle\phi_a|\phi_k(0,0)|\phi_b\rangle\,\delta_{K_i,K_j}. \tag{25.2.13}$$

In turn, these quantities can be expressed in terms of the conformal three-point functions in the plane: in fact, if both conformal states $|\phi_a\rangle$ and $|\phi_b\rangle$ correspond to primary fields, we simply have

$$\langle\phi_a|\phi_k(0,0)|\phi_b\rangle = \left(\frac{2\pi}{R}\right)^{2\Delta_k} C_{\phi_a,\phi_k,\phi_b} \tag{25.2.14}$$

where C_{ϕ_a,ϕ_k,ϕ_b} is the structure constant of the corresponding three-point function (see eqn. (10.4.7)). If, on the other hand, one or both conformal states correspond instead to descendent states, we can make a repeated use of the commutation relations

$$[L_m,\phi_{\Delta,\overline{\Delta}}(z,\overline{z})] = z^m(m\Delta + z\partial/\partial z)\phi_{\Delta,\overline{\Delta}}(z,\overline{z}),$$

to compute the relative matrix elements, with the final result given by an expression proportional to the same structure constant C_{ϕ_a,ϕ,ϕ_b} times a rational functions of the central charge c and the conformal dimensions Δ_{ϕ_i}, Δ_{ϕ_j} and Δ_k. For instance, for descendent fields which are pure derivatives (i.e. made of powers of L_{-1} applied to the primary field) for the analytic part of the matrix element we have

$$\langle\Delta_a|L_1^{r_a}\phi_{\Delta_k}(0)L_{-1}^{r_b}|\Delta_b\rangle = C_{\phi_a,\phi_k,\phi_b}\, r_a!\, r_b$$
$$\sum_{l=0}^{\min(r_a,r_b)} \frac{\Delta_a+\Delta_b-\Delta_k)_l\,(\Delta_k+\Delta_a-\Delta_b)_{r_a-l}(\Delta_k+\Delta_b-\Delta_a)_{r_b-l}}{l!(r_a-l)!(r_b-l)!}$$

where $(a)_m = \Gamma(a+m)/\Gamma(a)$.

Notice that each V_k corresponds to a block form matrix

$$V_k = \left(\frac{R}{2\pi}\right)^{1-2\Delta_k} \begin{pmatrix} 0 & \blacksquare & 0 & .. & \blacksquare \\ \blacksquare & 0 & \blacksquare & .. & 0 \\ 0 & \blacksquare & \blacksquare & .. & \blacksquare \\ . & . & & & . \\ \blacksquare & 0 & \blacksquare & .. & \blacksquare \end{pmatrix} \tag{25.2.15}$$

where the black boxes $\blacksquare$ stay for the non-zero blocks of the matrix, corresponding to the conformal families a and b for which $C_{\phi_a,\phi_k,\phi_b} \neq 0$.

25.2.1 General features of the finite-size energy levels

Figure 25.1 show a typical behaviour of the energy level as a function of R. To understand such a behaviour let us focus, for simplicity, on the case when it is present only one relevant perturbation of the CFT

$$H(R,\lambda) = H_0 + \lambda V = \frac{2\pi}{R}\mathbf{D} + \lambda\,(2\pi)^x R^{1-x}\,\mathbf{C}, \tag{25.2.16}$$

with $\mathbf{D}$ a diagonal matrix while $\mathbf{C}$ a block form matrix according to the non-zero structure constants. Notice the different powers in R of the pre-factors of the two matrices. We assume that this Hamiltonian corresponds to a massive QFT. In this case, the coupling constant λ is a dimensionful quantity related to the mass m_1 of the lightest particle of the theory as

$$\lambda = \mathcal{D} m_1^{2-x}, \tag{25.2.17}$$

where $\mathcal{D}$ is a pure number. Using this relation and taking into account that the natural scale in which measuring the energies is $1/R$, we can rewrite the expression above as

$$H(R,c) = \frac{2\pi}{R}\left[\mathbf{D} + 2\pi\,\mathcal{D}\left(\frac{m_1 R}{2\pi}\right)^{2-x}\mathbf{C},\right]. \tag{25.2.18}$$

This formula clearly shows that the energy levels differ from the pure $1/R$ conformal behaviour due do the presence of a non-zero mass term. Moreover, this expression shows that the spectrum depends on the dimensionless parameter $\mathcal{D}$, analogously to the case

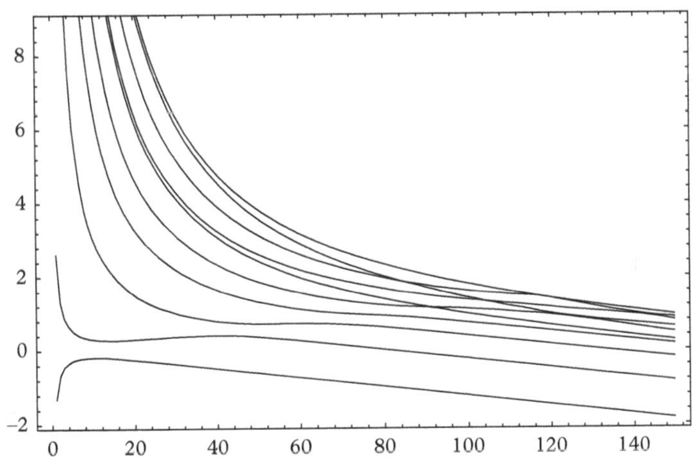

Fig. 25.1 *Energy level $E_a(R)$ as functions of the circumference R.*

of quantum mechanics discussed previously: different choices of $\mathcal{D}$ give rise to energy outputs related by simple scale transformations. In fact, under a transformation of length scales $R \to \beta R$ we have,

$$H(R,\lambda) \to \frac{1}{\beta} H(\beta R, \lambda \beta^{-2+x}), \tag{25.2.19}$$

which we can write as

$$H(R,\lambda) = \lambda^{1/(2-x)} \, H(\lambda^{1/(2-x)} R, 1). \tag{25.2.20}$$

Different values of $\mathcal{D}$ correspond then to a change of scale for the length R and for the energies E_i. As in quantum mechanics, it is however useful to fix the convenient value of $\mathcal{D}$ for which the mass of the lightest particle is fixed to be equal to 1. In the integrable cases, this choice can be fixed in terms of the TBA (Section 20.9).

Let us now discuss the general features of the energy levels as functions of R. Without any truncation, the finite size energy levels E_i ($i = 0, 1, 2, \ldots$) will have the scaling form

$$E_i(R,\lambda) = \frac{2\pi}{R} f_i \left(\frac{R}{\xi} \right), \tag{25.2.21}$$

where $\xi(\lambda) = m_1^{-1}$ is the finite correlation length. Since the CFT is deformed by a relevant operator, the ultraviolet behaviour $R \to 0$ of these eigenvalues is dictated purely by H_0

$$f_i \simeq x_i - \frac{c}{12} \qquad (R \ll \xi), \tag{25.2.22}$$

where $\eta_i = \Delta_i + \bar{\Delta}_i$ are the conformal dimensions of the various conformal states. Let us consider now the infrared regime $R \to \infty$: with a finite correlation length ξ, the system behaves as if it is made of a number R/ξ of independent sub-systems and, given that the energies are extensive quantities, their behaviour for $R \to \infty$ is then fixed to be

$$E_i(R) \simeq \frac{\epsilon_0}{\xi^2} R + m_i \qquad (R \gg \xi), \tag{25.2.23}$$

where m_i is the mass scale associated to the ith level while the dimensionless constant ϵ_0 is the density of energy per volume. If the theory is integrable, such a constant ϵ_0 can be computed in terms of the TBA (Section 20.7). This implies that the scaling functions f_i behave as

$$f_i \simeq \frac{1}{2\pi}\left[\epsilon_0\left(\frac{R}{\xi}\right)^2 + \frac{m_i}{m_1}\frac{R}{\xi}\right] \qquad (R \gg \xi). \tag{25.2.24}$$

It is pretty remarkable that there is a linear behaviour in R of all energy level by the sum of two terms, as those in eqn. (25.2.16), with completely different power laws. Such a linear behaviour in R of the energy eigenvalues is evident by looking at Figure 25.1.

It is worth emphasizing that the levels actually present in the spectrum depend on the boundary conditions chosen on the finite size geometry, as we will show in some of the examples discussed later. For instance, if the theory has kink/anti-kink excitations $|\, K(\theta)$ and $|\, \overline{K}(\theta)\rangle$ which interpolate between different vacua, as in the Sine–Gordon model for instance, with periodic boundary conditions it is impossible to see energy levels corresponding directly to these excitations. But, of course, we could observe energy levels corresponding to neutral two-particle states as $|\, K(\theta_1)\overline{K}(\theta_2)\rangle$.

An interesting situation occurs when the deformation of CFT gives rise to an integrable field theory. In this case, in addition to the Hamiltonian, there are other (local) conserved and commuting quantities $\mathcal{Q}_i$

$$[\mathcal{Q}_i, \mathcal{Q}_j] = 0. \tag{25.2.25}$$

Therefore they can be simultaneously diagonalized together with H. The existence of these conserved charges implies the occurrence of intersections of the eigenvalue curves for those which belong to different representations of these symmetries. Indeed, even if they have the same energy, eigenvalues belonging to different representations can be always be distinguished in terms of their different quantum numbers with respect to these conserved quantities. The crossing of the eigenvalues are then the analogous of the elastic scattering processes of the particles in integrable field theories. Vice versa, if the system is not integrable, when two eigenvalues become very close, the corresponding eigenvectors mix one to the other, with the result that there is a net repulsion between the two levels, with a gap proportional to the matrix element of the Hamiltonian between these states. The two situations are illustrated in Figure 25.2.

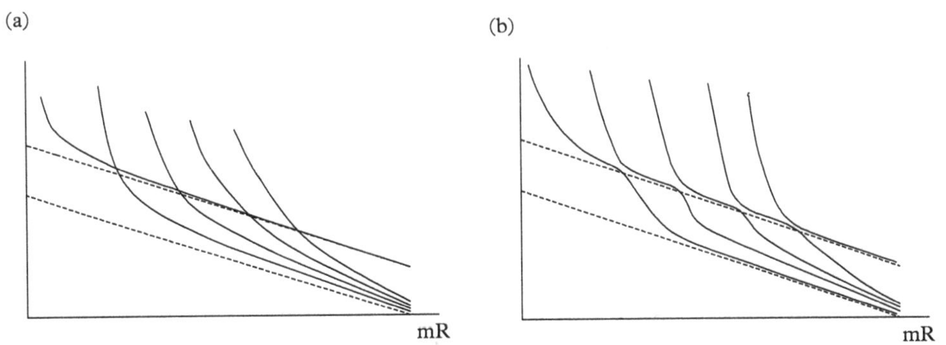

Fig. 25.2 *Crossing of energy levels when the theory is integrable (a); repulsion of energy levels when the theory is not integrable (b).*

25.2.2 Effects of truncation

The previous considerations need to be refined in the presence of a truncation of the Hilbert space. In the infinite dimensional Hilbert space, the matrix elements of H_0 are unbounded and therefore for any value of R, no matter how large, the eigenstates of the full Hamiltonian $H(R,\lambda)$ are always given in terms of linear combination of the conformal basis. However, the truncation fixed by the cut-off Λ (see eqn. (25.2.10)), introduces an additional length scale ρ into the problem which for the energy levels leads to a scaling form as

$$E_i(R) = \frac{2\pi}{R} f_i^{trunc}\left(\frac{R}{\xi}, \frac{R}{\rho}\right). \tag{25.2.26}$$

In this case, for $R \gg \rho$, the matrix elements of H_0 become negligible with respect to those of λV and therefore, for large R the eigenstates are actually those of the matrix V alone! This is clearly a non-physical effect of the truncation of the Hilbert space and correspondingly, instead of a linear scaling in R, the energy levels scale as

$$E_i \sim \lambda R^{1-x} \qquad\qquad (R \gg \rho). \tag{25.2.27}$$

While the correlation length ξ characterizes the *crossover* from the ultraviolet to the infrared behaviours, the scale ρ rules instead the truncation effects. Therefore, in order to extract physical predictions relative to the system, we must include a sufficient number of states such that $\rho \ll \xi$. In this case, for the energy levels we have the three different scaling regimes (25.2.22), (25.2.24) and (25.2.23), separated by two *crossover* points, the first around $R \sim \xi$ and the second $R \sim \rho$. To identify these three regions it is useful to define an *effective scale exponent* according to the formula

$$\zeta(R) = \frac{d}{d\ln R} \ln E_0 . \tag{25.2.28}$$

The ultraviolet regime is when $\zeta \simeq -1$, the infrared regime when $\zeta \simeq 1$, while the unphysical truncation regime is identified when ζ assumes the value $\simeq (1-x)$.

Let us now analyse the role of the truncation scale Λ from a different point of view, namely what is the effect of neglecting higher energy states in the computation of the lowest levels? This question can be approached along the same lines of the RG ideas in Chapter 8. To this aim, let us split our Hilbert space in two sets

$$\mathcal{H} = \mathcal{H}_l + \mathcal{H}_h, \tag{25.2.29}$$

where $\mathcal{H}_l$ is made of states with energies lower that the truncation energy E_t while $\mathcal{H}_h$ of states with energies higher than E_t. Any vector $\mid v\rangle$ of the Hilbert space can be decomposed in terms of its low- and high-energy components, $\mid v\rangle = \mid v_l\rangle + \mid v_h\rangle$. The original eigenvalue problem

$$H v = E v, \tag{25.2.30}$$

can be then written as

$$\begin{aligned} H_{ll} v_l + H_{lh} v_h &= E v_l \\ H_{hl} v_l + H_{hh} v_h &= E v_h \end{aligned} \tag{25.2.31}$$

where H_{ab} is the part of the Hamiltonian acting on the sectors a and b. Eliminating v_h from the second equation and substituting into the first one, we have

$$\left[H_{ll} - H_{lh}(H_{hh} - E)^{-1} H_{hl}\right] v_l = E v_l, \tag{25.2.32}$$

which can be also written as

$$[H_{trunc} + \Delta H]\, v_l = E v_l, \tag{25.2.33}$$

where

$$\Delta H = -V_{lh}(H_0 + V_{hh} - E)^{-1} V_{hl}. \tag{25.2.34}$$

In writing this last expression we have used the fact that H_0 is diagonal on the employed basis. Hence, the mixing between high- and low-energy states comes only from the interaction term V. If we imagine V is accompanied by a (small) coupling constant in front, we can expand this expression in series, with the first term given by

$$\Delta H \simeq -V_{lh}(H_0 - E)^{-1} V_{hl}, \tag{25.2.35}$$

The evaluation of this term is complicated by the fact that it depends on the energy E we are looking for. Replacing it with some reference energy E^*, we see that the corrections to the truncated energy levels are given by

$$(\Delta H)_{ab} = -\sum_{E_k > E_t} \frac{V_{ak} V_{kb}}{E_k - E^*}. \tag{25.2.36}$$

Handling this term requires an infinite summation on the higher energy states which are in $\mathcal{H}_h$. If we succeed to express this correction as

$$\Delta H \simeq \sum_c V_c \tag{25.2.37}$$

where V_c are operators which act on the Hilbert space $\mathcal{H}_l$ of the lowest energy states and expressed in the same way as the original deformation V, i.e. as integral of a local density, then the net effect of this correction to the energy levels may be simply regarded as a renormalization of the original couplings. In other words, the new Hamiltonian that also includes this correction term may be thought as an improved THSA Hamiltonian. Since ΔH employs high energy states, we expect that it can be related to the short-distance behaviour of the perturbing operator and the OPE thereof. This is indeed the case: let us initially introduce the (imaginary) time dependence of the operators in the interaction picture as

$$\mathcal{O}(\tau) = e^{H_0\tau}\,\mathcal{O}(0)\,e^{-H_0\tau} \tag{25.2.38}$$

so that we can express eqn. (25.2.35) as

$$(\Delta H) = -\sum_{c\in\mathcal{H}_h}\int_0^{\infty} e^{(E-H_0)\tau}\,V(\tau)|c\rangle\langle c|V(0)\,d\tau. \tag{25.2.39}$$

Using eqn. (25.2.4), it is clear that to evaluate the matrix elements $(\Delta H)_{ab}$ we need to consider the following quantities

$$\begin{aligned} D_{ab}(\tau) &= \int_0^R dx_1\,dx_2 \sum_{c\in\mathcal{H}_h} \langle a|\phi_k(x_1,\tau)|c\rangle\,\langle c|\phi_k(x_2,0)|b\rangle \\ &= \sum_{c\in\mathcal{H}_h} e^{-(E_c-E_a)\tau}\int_0^R dx_1\,dx_2\,(\phi_k)_{ac}(x_1,0)\,(\phi_k)_{cb}(x_2,0), \end{aligned}$$

where $(\phi_k)_{ac} = \langle a|\phi_k|c\rangle$. Observe that $D_{ab}(\tau)$ involves terms as $e^{-E_c\tau}$ relative to states with energies E_c higher than the energy scale of truncation, i.e. $E_c > E_t$: to evaluate these quantities the strategy consists of using the OPE of the perturbing field ϕ_k but disregarding the power terms of $e^{-2\pi\tau/R}$ smaller than the integer Λ defined in eqn. (25.2.10), since $D_{ab}(\tau)$ involves a sum on the high-energy states c. The identification of these power terms can be done using the expansion

$$(1-z)^{-a} = \sum_{p=0}^{\infty}\frac{1}{p!}\frac{\Gamma(a+p)}{\Gamma(a)}\,z^p \equiv \sum_{p=0} S(p,a)\,z^p.$$

Therefore, using the OPE

$$\begin{aligned} \phi_k(x_1,\tau)\phi_k(x_2,0) &= \sum_{\varphi} C_{\varphi,k,k}\left(\frac{R}{2\pi}\right)^{2\Delta_\varphi-4\Delta_k} \\ &\qquad |z_1-z_2|^{-4\Delta_k+2\Delta_\varphi}\,|z_1|^{2\Delta_k}\,|z_2|^{2\Delta_k-2\Delta_\varphi}\,\varphi(x_2,0)+\cdots \end{aligned}$$

with

$$z_1 = e^{-2\pi/R(\tau+ix_1)}, \quad z_2 = e^{-2\pi/Rix_2},$$

we have

$$D_{ab}(\tau) = \sum_{\varphi}\left(\frac{R}{2\pi}\right)^{2\Delta_\varphi-4\Delta_k} C_{\varphi,k,k}R\sum_{2n>\Lambda} S^2(n,2\Delta_k-\Delta_\varphi)e^{-\frac{2\pi\tau(2n+2\Delta_k)}{R}}$$
$$\int_0^R dx_2\,\langle a|\varphi(x_2,0)|b\rangle.$$

Together with eqn. (25.2.39), this expression allows us to express $(\Delta H)_{ab}$ as

$$(\Delta H)_{ab} = -\; R^2\sum_{\varphi} C_{\varphi,k,k}\left(\frac{R}{2\pi}\right)^{2\Delta_\varphi-4\Delta_k} \tag{25.2.40}$$
$$\times\sum_{2n>\Lambda} S^2(n,2\Delta_k-\Delta_\varphi)\left(\frac{1}{E_a-E+\frac{2\pi}{R}(2n+2\Delta_k)}\right)\langle a|\varphi(0,0)|b\rangle\,\delta_{P_a,P_b}$$

where δ_{P_a,P_b} enforces the equality of the momentum of the states $|a\rangle$ and $|b\rangle$. Notice that that matrix elements $(\Delta H)_{ab}$ are not symmetric in a,b. This is a consequence of the choice made in eqn. (25.2.39) where the Hermiticity of this Hamiltonian term was lost. If we had instead represented ΔH as

$$\Delta H = -\sum_{c\in\mathcal{H}_h}\int_0^\infty d\tau\, e^{E\tau}\,V(\tau)|c\rangle\langle c|V(0)e^{-H_0\tau}, \tag{25.2.41}$$

and had repeated the same steps as before, we would end up in the same expression as in eqn. (25.2.41) but with E_a replaced by E_b (we mark $(\Delta H)_{ab}$ with a prime to indicate this swap). Hence, we can take as fully Hermitian Hamiltonian $H_0+\lambda V_{ll}+\Delta\tilde{H}$, where $\Delta\tilde{H} = 1/2(\Delta H+\Delta H')$, as a model that is free of cut-off effects up to second order in perturbation theory in V.

In summary, we have succeeded in writing ΔH as a sum over the single local fields $\varphi(x)$ that arise in the OPE of the original perturbing field. As written, we can compute the correction to the eigenstate $\sum_{a\in\mathcal{H}_h} c_a|a\rangle$ with energy E via

$$\delta E = \sum_{a,b\in\mathcal{H}_l} c_a c_b(\Delta\tilde{H})_{ab}. \tag{25.2.42}$$

However if we approximate $\Delta\tilde{H}$ by dropping the dependence on $(E-E_a)$ and $(E-E_b)$ (which is weak provided $E_t \gg E$), we can use $\Delta\tilde{H}$ to study the theory $H_0+\lambda V_{ll}+\Delta\tilde{H}$ equipped with the cut-off E_t and in this way we can find the entire low-lying spectrum of the theory, $H_0 + \lambda V$ without cut-off, making relative errors of $O(\lambda(\frac{R}{N})^{2-4\Delta_k})$, $O(\frac{E}{E_t})$. To see in more detail the effect of $\Delta\tilde{H}$, let us focus on the sum, denoted by W, that appears in $\Delta\tilde{H}$ once we drop the dependence on $(E - E_a)$ and $(E - E_b)$

$$W \equiv \sum_{2n>\Lambda} S^2\left(n, 2\Delta_k - \Delta_\varphi\right) \frac{1}{2n + 2\Delta_k}. \tag{25.2.43}$$

Since for large n, $S(n,a)$ goes as

$$S(n,a) \sim n^{a-1}, \tag{25.2.44}$$

W has a dependence on the cut-off Λ given by

$$W \sim \Lambda^{-2+4\Delta_k - 2\Delta_\varphi}. \tag{25.2.45}$$

Surely in the OPE of the perturbing field ϕ_k with itself there is the channel of the identity, for which $\Delta_\varphi = 0$. In this channel eqn. (25.2.45) predicts that the dependence on the cut-off Λ (i.e. on the energy E_t associated to the truncation of the Hilbert space) is mild as far as $4\Delta_k - 2 < 0$, i.e. $\Delta_k < 1/2$. When $\Delta_k \geq 1/2$, there will be a divergence term in *all* the truncated energy levels. This effect can be easily cured by studying *energy differences* $(E_n - E_0)$ with respect to the ground state energy E_0 because the divergent term drops in these differences. Most often this is the only divergence that matters. Other channels give rise to convergent corrections if it is satisfied the inequality

$$2\Delta_k < \Delta_\varphi - 1, \tag{25.2.46}$$

and divergent corrections otherwise. For their proper treatment we refer to the specialized literature quoted at the end of the chapter.

25.3 Finite-size Mass Corrections

Diagonalizing the truncated Hamiltonian H, on top of the ground state energy $E_0(R)$ there is the first excited energy $E_1(R)$, which corresponds to the zero momentum one-particle state $|\,A_a\rangle$ of the lowest excitation of the system. At large R, the gap between these two energy levels identifies the mass of this excitation which, choosing properly the coupling constant λ, can be taken to be $m_a = 1$. Figure 25.1 shows at large R the level $E_1(R)$ indeed has the same straight line of $E_0(r)$ and it is shifted by 1. The approach to the asymptotic value is exponentially in R and shows a very interesting interplay between finite size effects and QFT data.

The corrections come from processes that only exist on a finite geometry and therefore vanish when $R = \infty$. They take place for virtual production and annihilation of particles 'around the world'. The leading finite-size correction, which is present only in theories

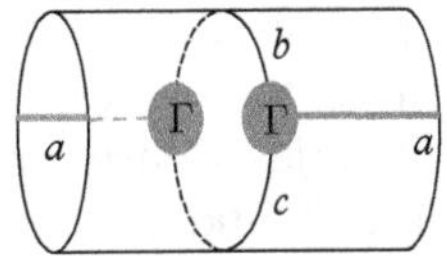

Fig. 25.3 *Leading finite size correction to one-particle energy, where Γ is the on-shell three-particle coupling.*

with appropriate cubic couplings, comes from the process in Figure 25.3. This can can be understood as due to virtual processes in which the given particle A_a itself disintegrates into two constituents A_b and A_c, which travel around the cylinder before recombining to give back the original particle A_a. Notice that the intermediate lines are given by the free propagator of the relative particles and therefore contribute with exponentially small terms for large R. Working out the details, this virtual process gives rise to the following correction (known as μ term) to the asymptotic mass

$$\Delta m_a^{(\mu)}(R) = -\frac{1}{8m^2}\sum_{b,c}\theta(m_a^2 - |m_b^2 - m_c^2|)\,\frac{\Gamma_{abc}^2}{\mu_{abc}}\,e^{-\mu_{abc}R}, \tag{25.3.1}$$

where Γ_{abc} is the on-shell three-particle vertex of the three particles while

$$\mu_{abc} = \frac{m_b m_c}{m_a}\,\sin u_{bc}^a, \tag{25.3.2}$$

where u_{bc}^a is defined by the mass triangle formula $m_a^2 = m_b^2 + m_c^2 + 2m_b m_c \cos u_{bc}^a$, see eqn. (17.4.59). This formula applies for all one-particle states of an integrable QFT, while it applies only to particles with mass below the threshold of production in all other non-integrable model. Notice that for unitary theories, where the on-shell three-particle vertices are real, this is a *negative* correction to the asymptotic value of the mass.

What we have described, however, is not the only correction here is in fact another that exists in any theory (known as F term) and that involves the scattering amplitudes of the given particle with other particles in the theory (Figure 25.4). As in the previous case, it can be also interpreted as arising from virtual particles 'travelling around the world' once, before annihilating again, and gives rise to the mass correction

$$\Delta m_a^{(F)} = -\frac{1}{8\pi m_a}\sum_b \mathcal{P}\int_{-\infty}^{\infty} d\theta\, e^{-m_b R\cosh\theta}\left[S_{ab}\left(\theta + i\frac{\pi}{2}\right) - 1\right], \tag{25.3.3}$$

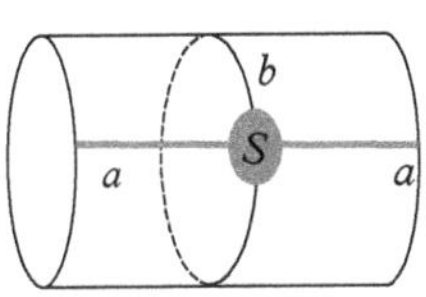

Fig. 25.4 *Next to leading finite size correction to one-particle energy, where S is the elastic S-matrix.*

where $\mathcal{P}$ stays for the principal part of the integral. The F-term can have either sign, even if we restrict the attention to unitary theories. Therefore, if the μ-term dominates, as it happens in most theories with bound states, the mass correction will be negative in a unitary theory for large R and the curve approaches its asymptotic value from below.

25.4 The Scaling Region of the Ising Model

Let us now use the THSA, together with the FFPT presented in Chapter 22, to discuss the scaling region of the Ising model and, in particular, the evolution of the mass spectrum of this model when we move its couplings along the path C (Figure 25.5) in the plane ($\tau = T - T_c, h$): this path starts from the low-temperature phase of the model and ends at its high-temperature phase, here represented by the points (a) and (f) respectively. The action of the model is given by

$$\mathcal{A} = \mathcal{A}_{CFT} + \tau \int d^2x \varepsilon(x) + h \int d^2x \sigma(x). \tag{25.4.1}$$

As discussed previously, such an action defines a family of field theories identified by $\chi \equiv \tau |h|^{-8/15} \in (-\infty, +\infty)$, a dimensionless RG invariant quantity. The spectrum of the theory changes in a very significant way moving χ: indeed, in the low-temperature phase (corresponding to $\chi = -\infty$ and to the point (a) of the curve C), the model has two degenerate vacua and therefore its excitations consists of the topological kink and anti-kink that interpolate between the two ground states. Along the magnetic axes ($\chi = 0$, corresponding to the point (d) of the curve C), the spectrum of the model consists instead of eight particles, with value of the masses given in Chapter 18. Finally, in the high-temperature phase (i.e. $\chi = +\infty$), the system has a unique vacuum and only a massive excitation above it. Let us see how this scenario can be recovered by employing the FFPT and confirmed by using the THSA.

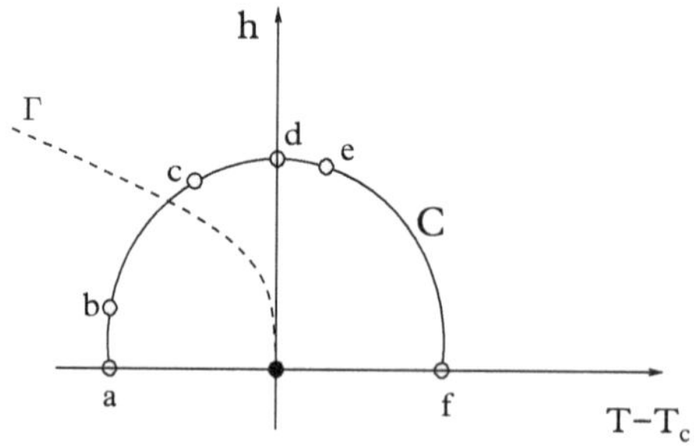

Fig. 25.5 *Interpolation curve in the plane (τ, h) between the low and high-temperature phases of the Ising model. Γ is a renormalization group trajectory, identified by the dimensionless parameter χ.*

25.4.1 Analysis of the Ising Model through FFPT

Let's start our analysis from the point (a), where the massive excitations are the kink/anti-kinks that interpolate between the two degenerate vacua. By switching on the magnetic field, the model moves to the point (b) of the curve C. The FF of the perturbing field σ on the two-particle kink/anti-kink state is given by

$$F^{\sigma}(\theta_1 - \theta_2) = \langle 0|\sigma(0)|A(\theta_1)A(\theta_2)\rangle = \tanh\frac{\theta_1 - \theta_2}{2}.$$

Therefore, using the first-order term of the FFPT given in eqn. (22.4.5), it is easy to see that this FF leads to an infinite correction to the mass of the kinks. In other words, the kinks get immediately confined as soon as the magnetic field is switched on. Looking at the effective potential of the theory, it is easy to see that this is indeed the correct conclusion: no matter how small the magnetic field may be, it lifts the degeneracy of the two vacua (Figure 25.6) and consequently there is no longer the possibility of having topological configurations.

Consider now the effect of the magnetic field on a state made of a kink and an anti-kink separated by a distance R. When the magnetic field is absent, the energy of this state is essentially equal to $2M$, i.e. the sum of the masses of the kink and the anti-kink. The energy of this state depends very weakly on the distance R because this field configuration takes values on the zeros of the effective potential and, no matter how large the distance R could be, the energy of this state remains unchanged. This situation changes by switching on the magnetic field since, in this case, at every point of the space there is an energy gap equal to $2h$ and the energy U of this state becomes a linear function of R, $U(R) = 2M + 2hR$. This attractive interaction between the kink and the anti-kink gives rise to a discrete spectrum of bound states (Figure 25.7). Regarding the kinks as very massive and quasi-static objects, the energy of the bound states can be obtained by solving the quantum mechanical problem of the bound states for a linear potential, well-known in quantum mechanics. The result is simply

$$E_k \equiv m_k = (2 + h^{2/3}\gamma_k^{2/3})M, \tag{25.4.2}$$

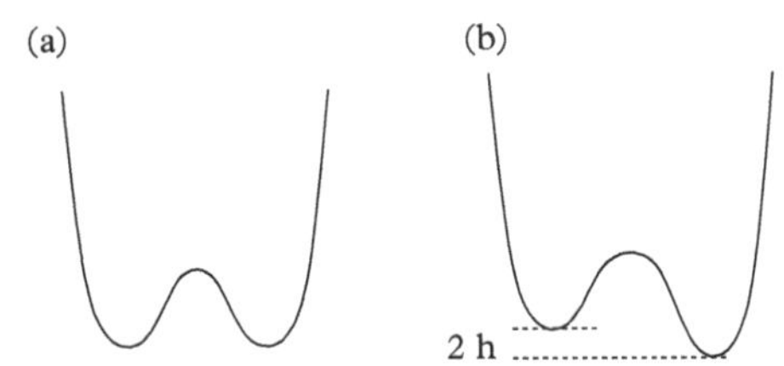

Fig. 25.6 *Effective potential in the low-temperature phase (a) and in the presence of an infinitesimal magnetic field (b). In the last case, the two minima are no longer degenerate and the kink/anti-kink disappear from the spectrum of the asymptotic states.*

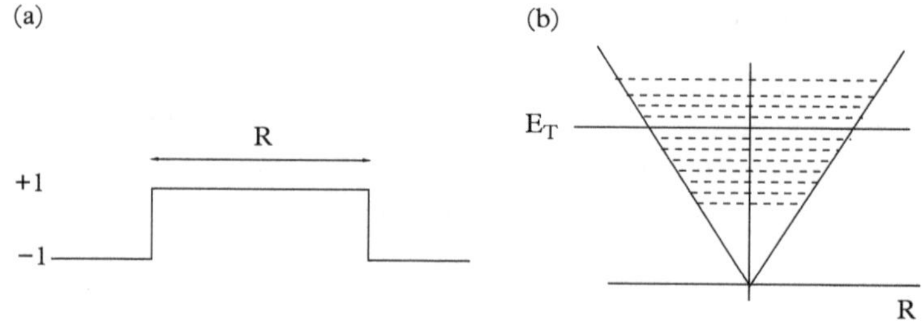

Fig. 25.7 *(a) kink–anti-kink state separated by a distance R; (b) kink–anti-kink potential in the presence of a magnetic field h and its bound states. The stable bound states are identified by the condition* $E_n < E_T$.

where γ_k are the positive roots of the equation

$$\mathcal{J}(\gamma_k) = \mathcal{J}_{\frac{1}{3}}\left(\frac{1}{3}\gamma_k\right) + \mathcal{J}_{-\frac{1}{3}}\left(\frac{1}{3}\gamma_k\right) = 0,$$

($\mathcal{J}_\nu(x)$ is the Bessel function of order ν) (Figure 25.8). Obviously not all the bound states shown in Figure 25.7 are stable: those that are stable are identified by the condition $m_n < 2m_1$, while all particles with a mass higher than the threshold $2m_1$ decay into particles of lower masses. When χ increases, i.e. when the system moves clockwise along the curve C of Figure 25.5, the number of bound states monotonically decreases. At the point (d) of Figure 25.5, there are the eight stable particles of the Zamolodchikov's solution of the Ising model in a magnetic field, and the value of their mass can be found in Section 18.4.2.

Moving away from the magnetic axes by means of the operator $\varepsilon(x)$, the lowest three particles change the value of their masses, while the remaining higher five particles decay into the low-energy channels. To estimate both effects, we need the FFs of the energy

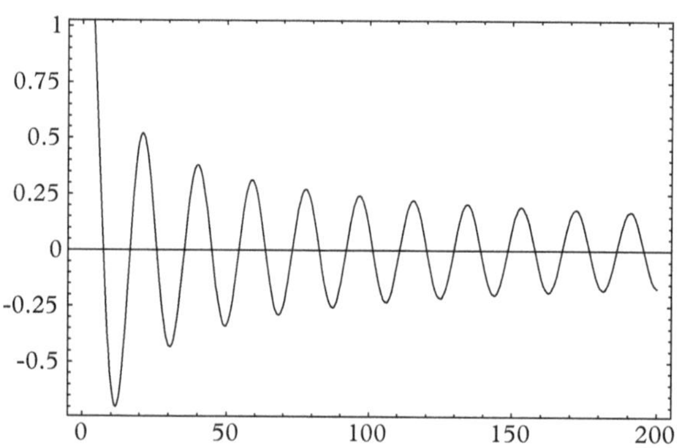

Fig. 25.8 *Plot of the function* $\mathcal{J}(x)$. *The zeros of this function determine the energies of the bound states of the Ising model coming from the original kink–anti-kink state.*

operator but in the integrable theory of the Ising model in a magnetic field. Here we simply report their expressions

$$
\begin{aligned}
\langle 0|\varepsilon(0)|0\rangle &= m_1, \\
F_{11}^{\varepsilon}(i\pi) &= \langle 0|\varepsilon(0)|A_1(\theta+i\pi)A_1(\theta)\rangle = -17.8933..m_1, \\
F_{22}^{\varepsilon}(i\pi) &= \langle 0|\varepsilon(0)|A_2(\theta+i\pi)A_2(\theta)\rangle = -24.9467..m_1, \\
F_{33}^{\varepsilon}(i\pi) &= \langle 0|\varepsilon(0)|A_3(\theta+i\pi)A_3(\theta)\rangle = -53.6799..m_1, \\
F_{44}^{\varepsilon}(i\pi) &= \langle 0|\varepsilon(0)|A_4(\theta+i\pi)A_4(\theta)\rangle = -49.3169..m_1.
\end{aligned}
$$

The first equation may be regarded as the normalization condition of the energy operator $\varepsilon(x)$. The corrections of the universal ratios are given by

$$
\begin{aligned}
\frac{\delta\mathcal{E}_{vac}}{\delta m_1} &= \frac{\langle 0|\varepsilon|0\rangle}{F_{11}^{\varepsilon}(i\pi)} m_1^0 = -0.0558..m_1^0, \\
\frac{\delta m_2}{\delta m_1} &= \frac{F_{22}^{\varepsilon}(i\pi)}{F_{11}^{\varepsilon}(i\pi)} \frac{m_1^0}{m_2^0} = 0.8616.., \\
\frac{\delta m_3}{\delta m_1} &= \frac{F_{33}^{\varepsilon}(i\pi)}{F_{11}^{\varepsilon}(i\pi)} \frac{m_1^0}{m_3^0} = 1.5082...
\end{aligned}
\tag{25.4.3}
$$

In turn, as discussed below, these quantities can be independently determined by a numerical analysis of the model in terms of the THSA and the numerical values determined by this approach are

$$
\begin{aligned}
\frac{\delta\mathcal{E}_{vac}}{\delta m_1} &\simeq -0.05\, m_1^0, \\
\frac{\delta m_2}{\delta m_1} &\simeq 0.87, \\
\frac{\delta m_3}{\delta m_1} &\simeq 1.50.
\end{aligned}
\tag{25.4.4}
$$

As shown in the expressions above, there is a satisfactory agreement between the theoretical and the numerical estimates.

Breaking the integrability of the Ising model in a magnetic field has a more dramatic effect on the five particles with a mass above threshold. Their stability is indeed only due to integrability and, once this is broken, they decay. In the perturbative approach, the decay process is associated to the presence of a negative imaginary part in the mass that is a second-order perturbative effect in τ (Figure 25.9).

$$
\mathrm{Im}\, m_c^2 = -\sum_{a\leq b,\, m_a+m_b\leq m_c} m_c \Gamma_{c\to ab} \simeq -\tau^2 \sum_{a\leq b,\, m_a+m_b\leq m_c} 2^{1-\delta_{ab}} \frac{|f_{cab}|^2}{m_c m_a \left|\sinh\theta_a^{(cab)}\right|},
$$

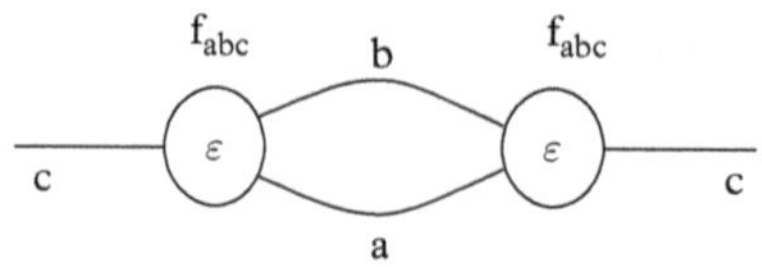

Fig. 25.9 *Perturbative diagram at the second order in τ relative to the imaginary part of the mass of the particle c. The intermediate particles a and b satisfy the on-shell conditions $p_a^2 = m_a^2$ e $p_b^2 = m_b^2$. When $c > 5$, there are additional diagrams with more intermediate particles.*

where $\Gamma_{c \to ab}$ is the decay amplitude of the particle A_c into the two particles $A_a A_b$, whereas

$$f_{cab} = F^{\varepsilon}_{cab}(i\pi, \theta_a^{(cab)}, \theta_b^{(cab)})\Big|_{\tau=0}.$$

The rapidities $\theta_a^{(cab)}$ and $\theta_b^{(cab)}$ are fixed by the conservation of the energy and the momentum in the decay process $A_c \to A_a A_b$ in the rest frame of the particle A_c. In the above equations all masses are the unperturbed values at $\tau = 0$. When $c > 5$ the sum must be completed including the contribution of the decay channels with more than two particles in the final state. Once the decay amplitudes $\Gamma_{c \to ab}$ are known, one can determine the lifetime t_c of the unstable particle A_c given by

$$t_c = \frac{1}{\Gamma_c}, \qquad \Gamma_c = \sum_{a \leq b} \Gamma_{c \to ab}. \tag{25.4.5}$$

For the Ising model, the relevant matrix elements are

$$|f_{411}| = (36.73044..)\,|\langle\varepsilon\rangle|_{\tau=0}$$
$$|f_{511}| = (19.16275..)\,|\langle\varepsilon\rangle|_{\tau=0}$$
$$|f_{512}| = (11.2183..)\,|\langle\varepsilon\rangle|_{\tau=0}.$$

where the normalization of the operator ε is fixed by its vacuum expectation value

$$\langle\varepsilon\rangle_{\tau=0} = (2.00314..)\,|h|^{8/15}.$$

The imaginary part of the mass of the first two particles, which are over threshold, is given by

$$\mathrm{Im}\, m_4^2 \simeq (-840.172..)\left(\frac{\tau\langle\varepsilon\rangle_{\tau=0}}{m_1}\right)^2 = (-173.747..)\,\tau^2$$
$$\mathrm{Im}\, m_5^2 \simeq (-240.918..)\left(\frac{\tau\langle\varepsilon\rangle_{\tau=0}}{m_1}\right)^2 = (-49.8217..)\,\tau^2.$$

The ratio of their lifetime is universal

$$\lim_{\tau \to 0} \frac{t_4}{t_5} = \lim_{\tau \to 0} \frac{m_4 \, \text{Im} \, m_5^2}{m_5 \, \text{Im} \, m_4^2} = 0.23326.. \tag{25.4.6}$$

While the particle A_4 can only decay into A_1A_1, the particle A_5 can also decay into the channel A_1A_2. The ratios of the amplitudes of these decays

$$b_{c \to ab} = \frac{m_c|_{\tau=0} \, \Gamma_{c \to ab}}{|\text{Im} \, m_c^2|}$$

are given by

$$\lim_{\tau \to 0} b_{5 \to 11} = 0.47364.., \qquad \lim_{\tau \to 0} b_{5 \to 12} = 0.52635...$$

Notice that eqn. (25.4.6) predicts that the lifetime of the particle A_5 is almost four time longer than the lifetime of the particle A_4. This paradoxical result, in contradiction with the intuitive idea that a heavy particle should decay faster than a light one, finds its explanation once again in the peculiar behaviour of the phase space in two dimensions. For the decay process $A_c \to A_a A_b$ the phase space in d-dimension is given by

$$\int \frac{d^{d-1}\vec{p}_a}{p_a^0} \frac{d^{d-1}\vec{p}_b}{p_b^0} \, \delta^d(p_a - p_b) \sim \frac{p^{d-3}}{m_c}, \tag{25.4.7}$$

where $p = |\vec{p}_a| = |\vec{p}_b|$ is the value in the rest frame. For fixed decay products, p grows with m_c: in $d = 2$, this term joins the factor m_c in the denominator and leads to a suppression of the phase space. In the Ising model, eqns. (25.4.6)–(25.4.6) show that this suppression is further enhanced by the dynamics (i.e. by the vaues of the matrix elements) in a way that is not compensated by the additional decay channels.

If we keep moving along the curve C, we first meets a value χ_1 at which the mass of the particle A_3 becomes larger than $2m_1$ and, later on, a second value χ_2 at which also the mass of the particle A_2 becomes larger than $2m_1$. When $\chi > \chi_2$ the spectrum of the stable particles of the theory consists of only one excitation. In the limit $\chi \to +\infty$, this is nothing but the particle of the integrable theory of the high-temperature phase of the model.

25.4.2 Analysis of the Ising Model through THSA

It is now very instructive to study numerically the evolution of the spectrum of the Ising model by implementing the THSA and to confirm the picture coming from the FFPT. In the conformal basis ordered as (I, ϵ, σ), where I stays for the conformal family of the identity operator, ϵ and σ the conformal families of the energy and spin operators respectively, according to the fusion rules of the CFT of this mode the two matrices associated to the two deformations of the conformal point of the model have the following block form

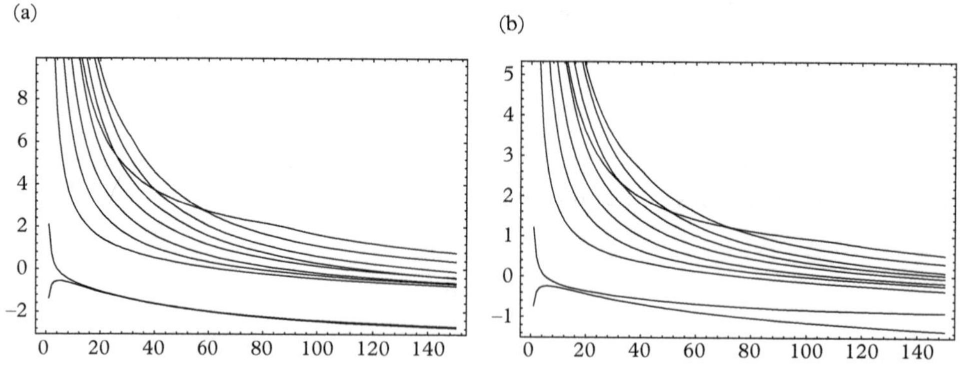

Fig. 25.10 *Low-energy lines of the Ising model defined on a cylinder with periodic boundary conditions, as functions of the cylinder width R, at the points (a) and (b) of the curve C in Figure 25.5.*

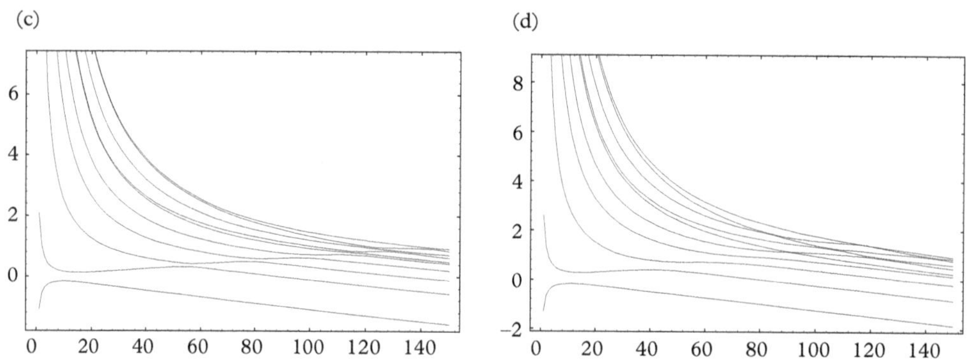

Fig. 25.11 *Low-energy lines of the Ising model defined on a cylinder with periodic boundary conditions, as functions of the cylinder width R, at the points (c) and (d) of the curve C in Figure 25.5.*

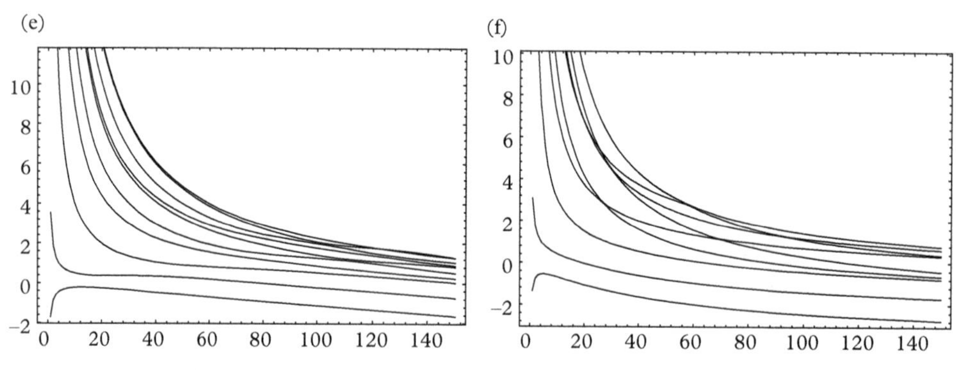

Fig. 25.12 *Low-energy lines of the Ising model defined on a cylinder with periodic boundary conditions, as functions of the cylinder width R, at the points (e) and (f) of the curve C in Figure 25.5.*

$$V_\epsilon = \begin{array}{c} \\ I \\ \epsilon \\ \sigma \end{array}\!\!\begin{array}{c} \begin{array}{ccc} I & \epsilon & \sigma \end{array} \\ \begin{bmatrix} 0 & \blacksquare & 0 \\ \blacksquare & 0 & 0 \\ 0 & 0 & \blacksquare \end{bmatrix} \end{array}, \quad V_\sigma = \left(\frac{R}{2\pi}\right)^{7/8} \begin{array}{c} \\ I \\ \epsilon \\ \sigma \end{array}\!\!\begin{array}{c} \begin{array}{ccc} I & \epsilon & \sigma \end{array} \\ \begin{bmatrix} 0 & 0 & \blacksquare \\ 0 & 0 & \blacksquare \\ \blacksquare & \blacksquare & 0 \end{bmatrix} \end{array}. \tag{25.4.8}$$

The lowest energy levels of the model defined on an infinite cylinder of width R (with periodic boundary conditions) are shown in Figure 25.11, labelled with the same letters used to identify the points in the phase diagram of Figure 25.5.

- In plot (a) of Figure 25.5, the model has two degenerate ground states. On a cylinder with a finite width R, these two ground states are exponentially split $\Delta E \sim e^{-MR}$, where M is the mass of the kink interpolating between the two vacua, and they correspond to the two lowest energy lines of plot (a) in Figure 25.10. With periodic boundary conditions, the next excited line corresponds to the two-particle states made of neutral configurations of kink–anti-kink. The gap between the ground state energy and the first exited state is then $2M$. Notice that the conformal dimension of the perturbing operator is in this case $\Delta = 1/2$ and, in agreement with the discussion which followed eqn. (25.2.45), the energy present a logarithm divergence with respect to the energy of truncation: this is evident from the bending of the energy lines shown in Figure 25.5.
- Plot (b) shows the energy lines once we the low-temperature phase by adding a magnetic field h: the degeneracy of the lowest as well as of the higher eigenvalues is immediately removed. The two originally degenerate ground-state levels have now been split into: (i) a unique ground-state energy line and (ii) an excited state whose mass gap diverges in the large volume limit $R \to \infty$. The degeneracy of the lowest threshold line has also been lifted, giving rise to a sequence of one-particle energy levels, the lowest values of which can be checked to be in reasonable agreement with eqn. (25.4.2).
- By increasing the magnetic field h, these features of the spectrum are further enhanced, as shown in plot (c) of Figure 25.11: the divergent energy line of the initial degenerate ground states meets all other lines at smaller values of R (and therefore it decouples faster from the remaining spectrum) whereas the other eigenvalues start to assume the structure and the values predicted by the Zamolodchikov solution for $T = T_c$, here shown in plot (d). Observe that, while the eigenvalues have the typical repulsive behaviour of a non-integrable situation all along the path from the point (a) to (c), on the contrary they cross each other once we reach the integrable situation of the pure magnetic axis in plot (d). Moreover, the fact that in the pure magnetic case the approach to the asymptotic value of the masses is reached from below can be simply interpreted as the echo of the divergent line coming from the degenerate ground-state energies of the low-temperature phase.
- Moving away from the integrable magnetic axis and toward the high-temperature phase, the upper five particles decay while the number of the particles below

threshold changes moving along curve C of Figure 25.5: in plot (e) of Figure 25.12, there are two particles below threshold, while in plot (f) there is only one particle, relative to the integrable high-temperature phase of the model.

REFERENCES

For quantum field theory in (1+1)-dimensions, the truncated Hilbert space approach

Yurov, V.P., and Zamolodchikov, A.B., *Truncated Conformal Space Approach to Scaling Lee-Yang Model*, Int. J. Mod. Phys. A6 (1991) 4557-4578.

Important applications for several other classes of universality are covered in

Bajnok, Z., Palla, L., Takacs, G. and Wagner, F. (2000). 'The k-folded Sine–Gordon Model in Finite Volume', *Nuclear Physics B*, 587: 585.

Beria, M., Brandino, G.P., Lepori, L., Konik, R.M. and Sierra, G. (2013). 'Truncated Conformal Space Approach for Perturbed Wess–Zumino–Witten SU(2)k Models', *Nuclear Physics B*, 877: 457.

Coser, A., Beria, M., Brandino, G.P., Konik, R.M. and Mussardo, G. (2014). 'Truncated Conformal Space Approach for 2D Landau–Ginzburg Theories', *Journal of Statistical Mechanics*, 1412: P12010.

Lässig, M., Mussardo G. and Cardy, J.L. (1991). 'The Scaling Region of the Tricitical Ising Model in Two Dimensions', *Nuclear Physics B*, 348: 591.

Rychkov, S. and Vitale, L.G. (2015). Hamiltonian Truncation Study of the Φ4 Theory in Two Dimensions I', *Physics Review D*, 91: 085011.

Rychkov, S. and Vitale, L.G. (2016). 'Hamiltonian Truncation Study of the Φ4 Theory in Two Dimensions II. The Z2-broken Phase and the Chang Duality', *Physics Review D*, 93: 065014.

Yurov, V.P. and Zamolodchikov, Al.B. (1991). 'Truncated Fermionic Space Approach to the Critical 2-D Ising Model with Magnetic Field', *International Journal of Modern Physics A*, 6: 4557.

The effect of truncation and the RG improvement of the THSA are covered in

Hogervorst, M., Rychkov, S. and van Rees, B.C. (2015). 'Truncated Conformal Approach in d-dimensions: A Cheap Alternative to Lattice Field Theory?', *Physics Review D*, 91: 025005.

Konik R.M. and Adamov, Y. (2009). 'Renormalization Group for Treating 2D Coupled Arrays of Continuum 1D Systems', *Physics Review Letters*, 102: 097203.

The THSA has been also used to study FFs at finite volume: we recommend

Feher, G. and Takacs, G. (2011). 'Sine–Gordon Form Factors in Finite Volume, *Nuclear Physics B*, 852: 441.

The THSA has been applied to analyse non-integrable deformations of CFT: we recommend

Delfino, G., Mussardo, G., and Simonetti, P. (1996). 'Nonintegrable Quantum Field Theories as Perturbations of Certain Integrable Models', *Nuclear Physics B*, 473: 469.

Fonseca, P. and Zamolodchikov, A.B. (2003). 'Ising Field Theory in a Magnetic Field: Analytic Properties of the Free Energy', *Journal of Statistical Physics*, 110: 527.

Lepori, L., Mussardo, G. and Tóth, G.Z. (2008). 'The Particle Spectrum of the Tricritical Ising Model with Spin Reversal Symmetric Perturbations', *Journal of Statistical Mechanics*, 2008: P09004.

PROBLEMS

1. Semi-classical energy spectrum

Consider the semi-classical quantization condition

$$\oint p(x)\,dx = \left(n + \frac{1}{2}\right) h,$$

where

$$p(x) = \sqrt{2m(E - V(x))},$$

is the momentum of the classical particle with energy E equal to the energetic level E_n. The integral is along the periodic orbit of the particle.

Compute the semi-classical energy levels for a potential $V(x) = \lambda |x|^a$.

2. Variational principle

Using the completeness relation

$$\sum_{n=0}^{\infty} |n\rangle \langle n| = 1,$$

show that for any quantum Hamiltonian the energy of the ground state satisfies the inequality

$$E_0 \leq \frac{\langle \psi \mid H \mid \psi \rangle}{\langle \psi \mid \psi \rangle}.$$

3. Expectation values in the harmonic oscillator

Consider the wave function $\psi_\omega(x)$ of the ground state of an harmonic oscillator with frequency ω, here used as variational parameter

$$\psi_\omega(x) = \left(\frac{m\omega}{\hbar\pi}\right)^{1/4} e^{-\frac{m\omega}{2\hbar}x^2}.$$

Show that it holds We need to compute the matrix element of H, on this state as function of ω

$$\langle\psi_\omega \mid \frac{p^2}{2m} \mid \psi_\omega\rangle = \frac{\hbar\omega}{4},$$

$$\langle\psi_\omega \mid |x|^a \mid \psi_\omega\rangle = \frac{\lambda}{\sqrt{\pi}}\Gamma\left(\frac{a+1}{2}\right)\left(\frac{\hbar}{m\omega}\right)^{a/2}.$$

Index

Ablowitz, M. 394
action 239, 240, 241, 247, 260, 269, 283, 288, 304–310
action-angle variables 575
Adamov, Y. 972
ADE classification 435, 436
adjacency 886, 917, 918, 929
adjoint representation 492, 502, 506, 513, 514, 533
affine Toda theory 577, 589, 596, 597, 603, 604, 653, 676, 682, 687, 689, 694, 700, 701, 704
Aharony, A. 102
Ahn, C. 938
Akutsu, Y. 233
Alvarez-Gaume, L. 438
Amit, D. 285, 295, 316
analytic continuation 87, 88, 276, 286, 421, 450, 478, 582, 604, 642, 644, 645, 685, 715, 747, 816, 838, 918
analytic theory of S-matrix 623, 671, 683
Andrew, G.E. 232
anti-commutation relations 57, 322, 337, 338, 339, 520, 734, 741, 853, 916
anti-ferromagnetic systems 42, 53, 72, 102, 154
anti-particles 273, 288, 630, 644, 649, 673, 697, 849, 851, 873
Appel, K. 101
approximation 9, 89, 316, 757, 780, 786
  mean field 18, 106, 104–109, 111, 116, 153
  of Bethe-Peierls 116, 120
Arishtein, A. 737
Askhin-Teller model 487, 877, 878
associativity 346, 410, 411, 484, 489, 646, 673
asymptotic states 548, 622, 623, 630, 632, 634, 636, 642, 647, 650, 651, 653, 656, 722, 728, 744, 756, 768, 773, 778, 828, 850, 869, 871
attractive fixed point 297, 301, 303, 561

Backlund transformation 619
Bagnuls, C. 316
Bajnok, Z. 909, 938, 972
Baker-Campbell-Hausdorff formula 322
Balian, R. 186
Bander, M. 339
bare coupling constant 267, 708, 755, 893
bare mass 577, 708, 755, 893
Barouch, E. 20, 40, 772, 784
Barton, R. 285
Bastianello, A. 739
Bateman, P.T. 439
Baxter, R.J. 20, 40, 211, 518, 541
Bazhanov, V.V. 232, 833, 834
Belavin, A. 394
Beria, M. 972
Berlin, T.H. 130, 154
Bernard, D. 671
Bernstein, H.J. 832
β function 313–315
Bervillier, C. 316
Bessel functions 74, 75, 76, 78, 80, 84, 86, 90, 92, 135, 137, 248, 249, 570, 667, 668, 807, 821, 831, 855, 856, 966
Bethe Ansatz 231–234, 555, 678, 691, 791–834
Bethe-Peierls approximation 116–120, 153
Bethe wave function 794–796
Binney, J.J. 285
binomial coefficients 141–143
block variables 292, 293, 297, 302–304
Blume-Capel model 261, 262, 530
Bogoliubov, N.N. 285
Bogoliubov trasformation 319, 332, 334, 340, 859, 860
Boltzmann
  constant 5, 24, 158
  factor 122, 172–174, 294
  weight 301, 763
bootstrap
  conformal 346, 394, 622
  chains 655
  principle 650, 651
  scattering theory 619, 646, 647–656, 672, 674, 676–682, 687, 688, 691–699, 714, 724–727, 738, 739, 782, 799, 943
Bose-Einstein condensation 6, 32, 36, 41, 45, 136, 275
boson
  statistics 32, 45, 796
  compactified 453
  partition function 451, 452, 455
  propagator 146, 246–250, 249, 250, 254, 264, 276, 278, 287, 306, 412, 468, 634, 875
bosonization 31, 443, 468–471, 475, 517, 523–525, 582
  of Ising model 523–525
bound states 590, 601, 603, 622, 632, 637, 646, 647, 650–655, 662, 663, 674, 677, 679, 681, 683, 685, 686, 691–695, 698, 707, 711–715, 716, 721, 724, 728, 729, 730, 742, 743, 745, 748–752, 766, 782, 854, 863, 865, 866, 880, 883, 884, 887–890, 900–927, 938, 941, 942, 966

boundary
  bound state 854
  conditions 34, 37, 48, 55, 59, 62–69, 72, 123, 124, 144, 156, 178, 179, 204, 206, 212, 213, 217, 229, 230, 320, 329, 340, 385, 399, 429, 445, 451, 453, 455, 458, 459, 460, 461, 463–468, 478, 498, 520, 527, 529, 620, 629, 741, 772, 791, 792, 795, 822, 824, 836, 837, 840–860, 883, 884, 900, 952, 957, 970, 971
  entropy 846
  magnetic field 841, 853–858
  operator 839, 840, 841, 847
  reflection amplitude 849, 853
  state 841, 843, 844, 845, 846, 848, 850, 851, 854, 858, 859, 860
Braden, H.W. 737, 738
braid group 31, 46, 47
Brandino, G.P. 972
brownian motion 141–157
Brush, S. 41
bulk energy, *see* energy
Bullogh-Dodd model 588, 589, 590, 593, 619, 621, 654, 676–679, 716, 723, 724, 739, 740, 759, 785

c-theorem 547, 562, 563, 565, 566, 570, 574, 578, 764, 765, 767, 768, 778, 779, 781, 783, 935
Calabrese, P. 394
Callan, C. 315
Callan-Symanzik equation 558
Campbell, D.K. 909
canonical ensemble 22–29, 41, 42, 43, 44
Cappelli, A. 435, 439, 482, 515
Capra, F. 672
Cardy, J.L. 316, 394, 429, 439, 542, 574, 737, 784, 836, 857, 972
Casimir energy 365, 378, 379, 381, 397, 791, 793
Cauchy formula 138, 353, 361, 364, 396, 468, 672
causality 240, 282, 287, 623, 629, 672, 675
Cayley, A. 97
CDD 649, 672
central charge 362, 363, 364, 365, 381, 385, 443, 447, 547, 562–567, 572, 577–580, 583, 588, 589, 597, 621, 764, 765, 768, 773, 778, 779, 781, 785, 794, 809, 814, 817, 818, 820, 821, 877, 878, 935, 936, 952
  coset models 492, 502, 503, 605
  minimal models 399, 401, 402, 403, 404, 405, 530, 541, 621, 676, 680, 696, 708
  parafermionic models 504, 517
  superconformal models 480, 482
Ceva, H. 339
characters 433, 434, 435, 437, 466, 527, 528, 529, 952
charge
  at infinity 579, 580, 588, 589, 596
  conserved 582, 584, 590, 607, 608, 609, 611, 614, 634, 636–639, 642, 648, 650, 651–654, 671, 677, 679, 686, 687, 692, 697, 734, 737, 799, 848
charge conjugation 483, 487, 637, 643, 711, 714, 887, 914
Chatterjee, R. 858
Chebyshev, polynomial 599, 560
chemical potential 33, 36, 261, 298, 530
Cherednik, J.V. 858
Chim, L. 738, 858
Christe, P. 738
chromatic polynomial 72, 101, 102
Chu, F. 619
class of universality 3, 12, 15, 260, 261, 311, 368, 406, 518, 519, 530, 531, 541, 545, 607, 609, 696, 865, 877, 897, 972
classification
  ADE, *see* ADE classification 435, 436
  of modular invariant partition functions 436
  of simple Lie algebras 435, 436
Clebsh-Gordon coefficients 717
cluster 717, 749, 751, 775, 786, 787
Coleman, S. 468, 471, 710
Coleman-Mandula theorem 638, 639, 671
combinatorial factors 252, 254
combinatoric approach 190–209
commutator 211, 366, 367, 370, 472, 505, 610, 731, 951
confinement 186, 863, 864, 874, 875, 876, 880, 914, 931, 932, 937, 938
conformal algebra 365, 367, 368, 373
conformal anomaly, *see* central charge
conformal blocks 522
conformal bootstrap see *boostrap*
conformal dimension 401, 405, 407, 409, 414, 427
  minimal models 405
  parafermionic models 485
  superconformal models 481
  WZW models 494
conformal families 375, 399, 433, 495, 549, 943, 952, 953
conformal generators 358, 362, 369, 370, 372, 433
conformal mapping/transformation 12, 348, 353, 357, 358, 359, 365, 370, 384, 385
  primary and quasi-primary fields 351, 352, 358, 359, 360, 361, 362, 370, 371, 372, 375, 376, 378, 380, 381, 397, 400, 407, 408, 409, 412, 416, 418, 420, 424, 436, 440, 477, 478–482, 485, 490, 491, 494–496, 531, 610–612, 676, 708, 759, 760–762, 837, 841, 843, 847, 954
  stress-energy tensor 272, 284, 287, 350, 355, 360–368, 371, 376, 378

383, 384, 397, 403, 414, 415, 447, 457, 561, 462, 479, 484, 491, 502–505, 557–559, 562–565, 572, 573, 576, 577, 583, 597, 607, 611, 757, 764, 767, 773, 778, 785, 792, 793, 837–839, 874, 925, 932
conformal Ward identity 358, 359, 360, 361, 362, 364, 376, 377, 384, 397, 429, 460, 491, 496, 556, 573, 606, 607, 837, 838, 839
connected correlation functions 62, 255, 257, 258, 259, 278, 831
continuum limit 156, 197, 231, 305, 307, 313, 319, 334, 335, 337, 338, 437, 482, 489, 518, 519, 520, 534, 708
Controzzi, D. 880
Cooper, F. 515, 938
correlation functions 9, 20, 48, 76, 104, 240, 246, 247, 276, 277, 278, 286, 338, 340, 342, 344, 349, 351, 360, 362
connected 253, 376, 381, 395, 399, 408, 413, 414, 417, 418, 420, 426, 480, 495, 505, 517, 520, 522, 524, 531, 542, 546, 548, 550, 556, 558, 575, 744, 746, 748, 750, 763–784, 787, 793, 806, 807, 815, 827, 833–838, 854, 855, 858
four-point 345, 441
spectral representation 628, 757, 758
three-point 314, 352, 378, 381, 423, 554, 954
two-point 21, 53, 306, 352, 363, 378, 456, 468
correlation length 3, 6, 10, 11, 12, 16, 18, 51, 62, 66, 76, 78, 79, 80, 81, 84, 103, 199, 237, 239, 249, 262, 289, 290, 291, 292, 298, 299, 301, 302, 304–309, 324, 341, 344, 385, 501, 541, 546, 559, 793, 794, 807, 828, 879, 956, 958
conserved charges and currents, *see* integrals of motion
consistency equations 346, 417, 622, 651, 653, 654, 655, 680, 787, 692, 698, 799, 974
coordination number 107, 112, 116, 175, 176, 245, 539, 540
Cornwell, J.F. 516
Corrigan, E. 737, 738
Coser, A. 972
coset construction 503, 516, 526
Coulomb gas 186, 399, 402, 412–423, 438, 443, 450, 481, 521, 523, 531, 539, 542, 579
coupling constants 5, 50, 55, 58, 59, 112, 124, 147, 155, 164, 169–175, 179, 180, 181, 189, 211–223, 229, 237, 240–245, 250, 251, 260–267, 288, 289, 291, 298, 301–318, 322–324, 413, 425, 426, 448, 475, 499, 500, 501, 530, 538, 545–558, 559–568, 570, 583, 586, 589, 601–606, 615, 621, 655, 657, 676, 677, 679, 689, 691, 696, 707, 718, 732, 743, 763, 779, 814, 815
exact relation with the mass 816, 818–820, 832, 833
counting method 612, 613, 677, 686, 727
Coxeter, H.S.M. 439
Coxeter
numbers 604, 614
exponents 494, 510, 514, 515, 526, 532, 589, 596, 598, 616, 617, 654, 691, 692, 700, 811, 812, 817
critical behavior 15, 183, 290, 307, 527, 676
critical dimensions 18, 137, 242
lower 242
upper 261, 263, 266
critical exponents 3, 13, 16, 18, 20, 66, 67, 136, 161, 200, 243, 261, 289, 307, 309, 312, 471
in mean field theory 109
of two-dimensional Ising model 519
of two-dimensional Tricritical Ising model 530, 534
of Potts model 536
of spherical model 137, 156
critical opalescence 9
critical phenomena 12, 17, 18, 40, 41, 154, 161, 237289, 299, 313, 316, 319, 368, 394, 399, 515, 539, 541, 574
critical surface 298, 299, 341
critical temperature 3, 4, 11, 14, 15, 19, 44, 48, 50, 112, 114, 120, 129, 135, 155, 175, 196, 262, 487, 519, 533, 696, 727, 766
cross section 627, 629, 656, 657, 663, 664
crossing symmetry 631, 632, 633, 641, 643–650, 657, 671, 687, 705, 709, 711, 724, 730, 732, 741, 745, 749, 751, 852
Curie temperature 4, 11, 15
cut-off parameter 241, 248, 249, 252, 254, 256, 266, 267, 303, 328, 412, 413, 548, 551, 552, 556–564, 578, 708, 826, 958, 961, 962
cyclic matrix 123, 124, 132, 204
cylinder
theory defined on a 368, 369, 370, 378–382, 385, 389, 429, 430, 433, 444, 447, 448, 462–464, 517, 527, 528, 791–793, 815, 828, 831, 832, 844, 847, 900, 901, 943, 952, 954, 963, 971

Dashen, R. 791, 832, 882, 886, 891, 893, 901, 909, 938
Das Sarma, S. 439
decay process 601, 626, 627, 628, 664, 673, 688, 863, 867, 879, 910, 911, 967, 968, 968
de Cloizeaux, J. 102

Dedekind η function 434, 453, 473
deformation
  of conformal theory 400, 545, 547, 548, 549, 550, 562, 565, 570, 576, 579, 613, 759, 864, 865
  integrable theory 619, 676, 678, 680, 681, 684, 685, 686, 692, 715, 716, 719, 721, 723–729, 815, 866, 867
  magnetic 605
  multiple 615, 616, 617
  thermal 606, 694, 695, 696, 699, 820
de Gennes, P.G. 77, 102
Deguchi, T. 233
Delfino, G. 21, 40, 316, 572, 574, 780, 784, 877, 879, 880, 901, 973, 976
Δ-theorem 572
De Luca, A. 739
De Martino, A. 858
de Morgan, A. 96–101
Derrick theorem 620
descendant states 374, 375, 376, 377, 378, 381, 400, 401, 404, 407, 408, 429, 576, 606, 612, 613, 784, 785, 952, 953, 954
diagrams
  connected 258
  Feynman rules for 252–256, 291, 603, 677, 754
differential equations
  Bessel see *Bessel functions*
  hypergeometric 437, 438
  for conformal correlation functions 376, 407, 408, 412, 418, 420, 421, 425, 495, 496, 521, 523, 539
  Painleve' 20, 770, 772, 784
Di Francesco, P. 394
dilogarithm function 809, 819, 820
dimension
  engineering 305
  anomalous 10, 306, 341, 299, 403, 427, 428, 449, 455, 536, 541, 546, 559, 566, 567, 569, 571, 606, 830, 832
dimension of space 17, 45, 281, 343
dimensional analysis 398
dimensional regularization 45
dimer method 189, 200–210
Dirac
  delta 120, 823
  equation 161, 319, 336, 337, 340, 853, 913, 915, 916, 919
  fermion 443, 455, 467, 523, 710, 937
disorder operator 319, 326, 329, 339, 459, 460, 461, 482, 486, 520, 522, 525, 532, 768, 769, 770
divergence
  infrared 548, 549
  of correlation length 6, 12, 16, 66, 976
  superficial 264, 265, 266
  ultraviolet 497, 548, 549, 556, 558, 578, 597, 655, 755, 827, 870
Dodd, R.K. 618
Dolan, L. 394
domain, magnetic 10, 162, 163, 167, 853
Dorey, P.E. 591, 619, 737, 738, 833, 834, 858
Dotsenko, V. 412, 438, 542
Double Sine-Gordon model 877, 893, 901, 903, 907, 909, 910
Dowrick, N.J. 285
Drazin, P.G. 618
Drouffe, J.M. 154, 186
dual Coxeter number see *Coxeter*
dual lattice 166, 168, 171, 177, 180, 183, 325, 451
duality
  curl and divergence 177
  of two-dimensional Ising model 161, 162, 163, 170–172, 340, 346, 521, 524, 684, 766, 879
  Wannier-Warnier 3, 327, 481, 520
  two-dimensional systems 181, 186
  self-duality 187, 330, 331, 528, 685
  s-t channel 657, 658
  strong-weak 676, 679, 691, 773
  tricritical Ising model 531, 532, 692, 881
  Z_n two dimensional models 487
Dunning, C. 938
Dynkin diagrams 437, 439, 510, 511, 593–604, 654, 688, 693, 694, 699, 812

Eden, R.J. 671
ϵ-expansion 18
Efetov, K. 515
effective action
effective Hamiltonian 290, 291, 294, 295
effective potential 260, 262, 736, 880
effective theories 267, 609
Eguchi, T. 619
Eilbeck, J.C. 618
eigenvalues
  of transfer matrix 60, 62, 69, 70, 78, 125–133, 211, 214, 222–227, 231, 233, 441
  of Renormalization Group 291, 297, 8, 658, 311–314
elastic scattering processes 277, 278, 608, 626, 628, 631–639, 641–643, 646, 703, 704, 730, 731, 733, 735, 736, 738, 744, 745, 780, 791, 809, 832, 865, 871, 880, 901, 906, 908
El-Showk, S. 394
energy
  bulk 814, 818, 819, 820
  Casimir see *Casimir*
  ground state 379, 464, 794, 810, 827, 835, 846, 858, 950
  internal 78, 115, 119
ensemble 6, 22–37, 43, 45, 85, 153, 320, 664
entanglement 381–386, 398
entropy 4–8, 24, 29, 45, 54, 112, 113, 163, 800, 846
equation of motion 44, 270, 271, 272, 283, 284, 337, 338, 363, 429, 453, 456, 457, 581, 585, 587, 619, 620, 824, 889, 912, 915
equation of state 30, 42, 84, 104, 110, 131, 134, 156

ergodic hypothesis 23, 24, 28, 44
Euler 90, 96, 98, 99, 101, 271, 283, 827
exceptional algebra 511, 526, 532, 595, 600, 684, 687, 694, 700
expectation value 7, 24, 28, 51, 61, 78, 104, 256, 260, 261, 276, 295, 303, 309, 329, 330, 331, 342, 344, 272, 383, 397, 419, 460, 461, 462, 548, 550, 552, 555, 656, 759, 760, 761, 769, 785, 830, 833, 932, 968
exponents see *critical exponents*
extensive quantities 7, 8, 26, 195, 385, 956

Fabrizio, M. 910
Faddev, L. 232, 618
Faddev-Zamolodchikov algebra 646, 673
Fateev, V. 412, 438, 482, 489, 516, 737, 738, 785, 816, 833
Feher, G. 972
Fendley, P. 209, 833
fermion 30–34
  di Dirac 443, 444, 446–470, 475, 919, 924–942
  di Majorana 161, 261, 320–340, 363, 483, 523, 527, 528, 735, 787, 835, 851, 852, 853, 913, 916, 972
ferromagnet 4, 5,6, 19, 21, 22, 40, 41, 53, 72, 102, 113, 155, 162, 186, 339, 704, 879
Feynman, R.P. 39
Feynman
  diagram 251–257, 264, 269, 270, 276, 278, 287, 291, 589, 592, 597, 603, 604, 619, 631, 632, 657, 677, 679, 683, 685, 755, 778, 869
  gas 48, 84, 86, 763, 764, 770, 771, 786
  path integral 284
finite-size 900, 955, 962, 963
Fioravanti, D. 316
first order perturbation theory
  conformal 606, 607
  form factor 871–874, 881, 965, 968, 969
first order phase transition 5, 6, 10, 36, 44, 114, 262, 263, 736
Fisher, A.J. 285
Fisher, M. 206, 207, 208, 209, 538
fixed point, *see* renormalization group
fluctuations 9, 10, 11, 12, 18, 22, 25, 26, 27, 39, 106, 112, 157, 242, 243, 248, 260, 263, 289, 290, 292, 341, 347, 358, 530, 538, 707, 810
fluctuation-dissipation theorem 16, 52, 310
Fock space 275, 448, 548, 678
Fokas, A. 394
Fonseca, P. 21, 40, 880, 973
form factors 744–787
  Ising in a magnetic field 779–783
  minimal 749
  bound state recursive equations 752
  kinematical recursive equations 751
  thermal Ising model 767, 768, 769–772
  Sinh-Gordon model 773–778
  stress-energy tensor 757
form factor perturbation theory 863–880
Forrester, P.J. 232, 518, 541
Fourier transform 44, 124, 132, 142, 144, 145, 180, 181, 182, 184, 185, 194, 195, 204, 241, 244–252, 257–259, 332, 444, 672, 707, 755, 812, 824, 888, 889, 890–894, 913, 921, 940
four color problem 73, 95, 96, 97, 98, 99, 101, 102, 166
four-point correlation function 278, 345, 346, 352, 394, 418, 423, 441, 485, 521, 522, 523, 539
Fradkin, E. 339
Freedman, M. 439
free energy 8, 15, 17, 30, 33, 34, 40, 253, 301, 308, 309, 312, 537, 793
  of one-dimensional Ising model 51, 60, 67
  of two-dimensional Ising model 227
  of one-dimensional Potts model 69, 71
  of Thermodynamic Bethe Ansatz 800
free field theory
  bosonic 247, 249, 272, 343, 364, 395, 443, 444, 445, 447, 453, 482, 504, 578, 708, 821, 859, 868
  fermionic 456–470, 475, 520, 523, 851, 878
Freund, P. 938
Friedan, D. 404, 438, 515, 541
Fring, A. 784, 858
fugacity 26, 85, 763, 764, 772
fusion rules 399, 408, 410, 411, 412, 428, 439, 440, 441, 476, 481, 482, 518, 519, 531, 535, 536, 538, 654, 691, 694, 718, 847, 857, 969, 975, 977

Γ function 86
Gaudin, M. 232
Gaussian free model 305, 450, 453, 482, 517, 578, 779, 936
Gaussian integral 84, 131, 141, 143, 243, 248, 251, 277, 304
Gaussian model 120–129, 204, 303
generating function 92, 145–150, 202, 203, 205, 207, 210, 374, 750, 776
Ghoshal, S. 836, 848, 858
Giamarchi, T. 471
Gibbons, J.D. 618
Ginsparg, P. 394, 471
Goddard, P. 516
Goebel, C.J. 738
Gogolin, A. 471, 910
Goldenfeld, N. 316
Goldstone, J. 909
Gomez, C. 438
Gram matrix 375, 401, 402, 404

graph 187, 190, 191, 192, 206, 207, 252, 254, 255, 256, 265, 286, 577, 589, 592, 632, 681, 703, 704
Green function 249, 413, 629
Green, B. 672
Green, M. 672
Grinza, P. 879, 880
Gross-Neveu model 741

Haken, W. 101
harmonic ratios 350, 352, 353, 357, 388, 485, 539, 842
Hasslacher, B. 882, 891, 909, 938
highest-weight representation 400, 407, 438, 840
Hibbs, A.R. 284
Hogervorst, M. 972
hydrodynamics 396
hypergeometric functions 420–422, 437–439, 486, 539, 842
Huang, K. 39, 316
Hughes, B.D. 154
Huse, D. 518, 541

incidence matrix 688, 700, 812, 813
instanton 619, 910, 912
integrable models 547, 575, 577, 588, 590, 593, 596, 612, 613, 615, 616, 619, 622, 641, 650, 678, 680, 681, 710, 715, 716, 722, 738, 739, 743, 744, 757, 759, 784, 785, 791, 801, 806, 816, 817, 828, 831, 832, 836, 863–867, 871, 879, 882, 883, 896, 909, 927, 929, 930, 943, 972
integral representation 667, 710, 712, 742
integrals of motion 24, 575, 576, 606, 609, 613, 651, 739, 850
invariant functions 286
irrelevant operators 242, 245, 291, 298, 299, 304, 309, 318, 426, 429, 531, 561, 565, 567, 568, 600, 879
Ising, Ernst 19, 22
Ising model 3, 18, 19, 40
  boundary 851–856, 858
  *E*_8 coset description
  *E*_8 masses 687–691, 737, 780, 783, 787, 818, 819
  fermionic formulation 319–338
  as minimal model 519–526
  characters in 527
  mean field 109, 111, 112, 116
  one-dimensional 48–55, 62, 64, 301, 303
  transfer matrix 55, 212, 214–230
  two-dimensional 20, 155, 161, 163, 164, 166–170, 180, 182, 187, 190
Istrail, S. 40
Itzykson, C. 154, 186, 339, 435, 439, 471

Jackiw, R. 882, 909, 938
Jacobi identity 474, 489, 505, 506
Jannick, G. 102
Jimbo, M. 20, 40, 772, 784
Johnson, R.S. 618
Jordan-Wigner transformation 319, 331

Kac determinant 399, 400, 401, 402, 403, 406, 438
Kac, M. 154, 189
Kac, V. 438
Kac table 407, 410, 414, 417, 429, 440, 481, 518, 521, 526, 527, 534, 537, 567, 568, 680, 697
Kac-Moody algebra 485, 489, 490, 493, 496, 497, 501–505, 514–517
Kadanoff, L. 102, 154, 186, 315, 339
Kainen, P.C. 102
Karowski, M. 783
Kasteleyn, P.W. 203, 209
Kastor, D.A. 938
Kaufman, B. 40
Kent, A. 516
kernel
  of Bethe Ansatz 801, 802, 812, 817, 829, 830, 834
  solutions of form factors 753, 767, 768, 777
Khare, A. 515, 938
kinematical poles 751, 767, 768, 775, 829, 873, 874
kink 326, 330, 331, 684, 722, 726, 728, 736, 760, 865, 881, 884, 885, 887, 891, 893, 898, 901, 911, 928, 930, 933, 938, 957, 964, 966
  multi-kink 718, 886
  S-matrix 725, 730, 907, 908
kink solution of Thermodynamic Bethe Ansatz 808, 810, 813
Klassen, T. 832
Knizhnik, V.G. 516
Knizhnik-Zamolodchikov equation 496
Koberle, R. 858
Kogut, J. 316
Konik, R. 858, 972
Kormos, M. 739
Kostov, I. 833
Koubek, A. 784
Kramers-Wannier 3, 20, 163, 169, 186, 327, 481, 520
Kubo, R. 39

lagrangian 239, 240, 243, 250, 262, 269, 271, 283, 284, 285, 570, 576, 578, 588, 593, 620, 651, 656, 686, 730, 755, 772, 868, 870, 882, 883, 887
  renormalizable 267
  Toda 596, 654, 676, 679, 685, 689, 691, 694, 701, 807, 896
Landau, L.D. 39, 671
Landau-Ginzburg theory 264, 268, 696
  for minimal models 427, 520, 530
  for superconformal models 481, 482
  for Yang-Lee model 676, 678, 723
Landshoff, P.V. 671
Laplace equation 139, 395, 396
Laplace transform 666, 668, 669
Lässig, M. 972
lattice 3, 5, 6, 9, 12, 19, 20, 51, 56, 64, 65, 68, 71, 72, 73, 106, 111, 112, 116, 117, 132, 135, 144, 148, 154,

161, 162, 164, 166, 167, 171–198, 202, 206, 207, 209, 215, 220, 228, 238, 240, 320, 325, 326, 337
lattice models 41, 76, 103, 107, 137, 141, 262, 291, 293, 296, 303, 304, 437, 482, 486, 487, 488, 516, 518, 519, 520, 533, 536, 537, 541, 716, 772
Lawrie, I. 541
Lax pair 620
LeClair, A. 671, 738, 833, 858
LeClair-Mussardo formula 828, 829, 831, 833
Legendre transformation 256, 257, 269, 271, 444
Lepori, L. 880, 972, 973
Lie algebra 435, 436, 438, 490, 493, 496, 499, 505–516, 593, 596, 599, 603
Lieb, E.H. 339, 468, 471
Lieb-Liniger model 702, 704, 739
Lifsits, E.M. 39, 671, 717
Liouville theorem 24, 90, 358
Liouville theory 579, 580, 583, 588, 589, 596, 605, 621, 688, 692
Lipkin, H. 671
locality 350, 479, 483, 486, 651, 716, 874, 875, 876
logistic map 296, 317
Lorentz
  invariance 287, 395, 586, 624, 625, 627, 641, 889
  transformation 280, 281, 282, 634, 637, 642, 746
Ludwig, A.W.W. 574
Lukyanov, S. 785, 883, 834
Luther, A. 471

Ma, S.K. 832
magnetization 4, 5, 8, 9, 13, 14, 17, 20, 43, 51, 52, 60, 67, 107, 108, 109, 112, 118, 119, 120, 136, 155, 162, 197, 199, 209, 311, 312, 481, 519, 531, 532, 537, 538, 614, 615, 616, 618, 696, 727, 735, 766, 768, 770, 772, 780, 786, 853, 858, 900
Majid, S. 738
Majorana fermion 161, 319, 336, 337, 363, 443, 456, 457, 464, 483, 520, 523, 684, 735, 835, 851, 852, 878, 914, 915, 917, 918, 920, 935, 937, 938
majority rule 293
Manakov, S. 618
Mandula, J. 638, 639, 671
marginal operator 298, 305, 318, 548, 561, 566, 878
Markov processes 157, 193
Martinec, E.J. 938
mass formula 648, 763, 816, 874, 890, 892, 913, 938
mass spectrum 21, 597, 598, 599, 600, 603, 622, 648, 656, 686, 693, 701, 704, 722, 744, 794, 865, 867, 882, 885, 886, 888, 892, 894, 898, 905, 913, 922, 928, 935, 948
mass triangle 648
massless flow 546, 567, 569, 735, 760, 877, 879, 881, 937
Mathieu, P. 394
Mattis, D.C. 339, 471
maximal root 596, 605, 817
maximal weights 493
mean field theory 107, 108–119, 161, 243, 245, 263, 315, 318, 530
McCoy, B. 20, 40, 772, 784, 880
McLaughlin, D. 619
Melzer, E. 832
microscopic scale 3, 12, 237, 241, 306, 384, 539
Miwa, T. 20, 40, 772, 784
Moessner, R. 209
modular group 386, 431, 432, 441, 453, 454, 467, 473, 528, 840, 844, 845, 848
modular invariance 399, 429, 430, 433–439, 455, 482, 515, 518, 534, 535, 792, 827
Moebius transformations 356, 357, 359, 365, 386–393, 397, 419, 432, 521
monodromy invariance 421, 422, 425, 429, 458, 496, 772, 782, 784, 833
Montroll, E.W. 154
Moriconi, M. 858
Morris, H.C. 618
multicritical points 263, 438, 541
Mussardo, G. 21, 40, 316, 515, 518, 618, 672, 737, 738, 739, 780, 784, 828, 829, 833, 858, 863, 877, 879, 901

Nayak, C. 439
Nersesyan, A. 471, 910
Neveu, A. 458, 463, 466, 478, 479, 481, 532, 741, 882, 891, 909, 938
Newman, M.E.J. 285
Nienhuis, B. 541, 542
Niss, M. 41
Noether theorem 283, 288, 363, 364, 415, 577
non-unitary models 440, 538, 818
non-integrable models 40, 861, 863–866, 873, 875, 879, 880, 891, 892, 930, 931, 943, 963, 971, 973, 978
non-locality 486, 874, 875, 978
non-relativistic limit 629, 632, 662, 674, 701, 702, 703, 717, 739, 740, 791, 805, 834, 889
normal order 274
Novikov, S. 618
null vector 400, 402, 403, 407, 408, 412, 418, 440, 495, 611, 612, 615, 617, 618, 621, 843

Olive, D.I. 516, 671, 938
O(n) model 73, 539, 704
one-particle irreducible 257, 258
Onsager, L. 3, 20, 40
operator algebra 339, 343, 344, 346
OPE, Operator Product Expansion 342, 362, 367, 369, 372, 414, 415, 448, 458, 477, 478, 480, 483, 496, 501, 560
  off-critical 549, 554, 555
operator space 752
optical theorem 628, 629

order parameter 8, 11, 12, 16, 17, 18, 43, 117, 118, 186, 239, 240, 241, 243, 248, 263, 304, 338, 341, 342, 399, 483, 518, 530, 735, 744, 768
Osborn, H. 394, 395

p-state model 178
Painlevè equation 20, 770, 772, 784
Palla, L. 619, 909, 938, 972
parafermions 476, 482, 483, 484, 485, 487, 488, 489, 504, 505
Parisi, G. 285
parity symmetry 362, 483, 520, 638, 643, 659, 661, 663, 731, 793, 808, 885, 887, 905, 917, 918, 920, 928, 944
partition function 3, 6, 7, 8, 16, 20, 25, 26, 29, 30, 32, 40, 41, 42, 43, 48, 59, 62, 64, 66, 68, 70, 71, 73, 77, 82, 83, 85, 103, 106, 120, 125, 133, 137, 161, 164, 166, 170, 174, 177, 179, 183, 207, 213, 230, 241, 243, 250, 254, 277, 295, 303, 305, 321
  of boson on a torus 452, 453, 455, 473
  of fermion 455, 462, 465, 466, 467, 471
  of minimal models 430, 433, 434, 436, 437, 439, 482, 527, 528, 534, 535
  of Thermodynamical Bethe Ansatz 793, 803, 804, 828, 929
Pasquier, V. 437, 439
Patasinskij, A.Z. 39
Pauli matrices 47, 231, 321, 322, 340, 456, 471, 499, 620
Paulos, M.F. 516, 517, 536
Peierls, R.E. 22, 40, 116–120, 153, 154
Peirls argument 19, 161, 162, 163, 166, 186
Pethick, C. 41
Petkou, A.C. 395
Peyrard, M. 909
percolation 72, 102, 406, 536
permutation group 47, 48, 67, 113, 200, 201, 202, 203, 521, 609, 697, 718, 723
Perron-Frobenius theorem 214, 233, 598, 688, 693
perturbation theory
  conformal 548–554, 574, 814, 815, 831, 832
  Feynman 250, 251, 253, 255, 314, 577, 597, 655, 657, 755
  Form Factor 863–867
Peschel, I. 468, 471
Peskin, M.E. 285
Pfaffian 200–205, 208
Pfeuty, P. 316, 339
phase (high-low temperature, see *temperature*)
phase space 23, 24, 75, 755, 756
phase transitions 5–12, 15, 17, 39, 40, 41, 43, 44, 48, 51, 60, 66, 106, 111, 112, 123, 129, 134, 136, 147, 161, 188, 196, 237, 242, 249, 262, 536, 538, 877
physical amplitude 625, 634
physical sheet 631, 663, 713, 715, 724, 745
physical regions 633, 645, 673
Pitaevski, L. 39, 41, 618
Poisson resummation formula 184–187, 435, 472
Pokrovskij, V.I. 39
Poland, D. 394
Polchinski, J. 672
Polkinghorne, J.C. 671
poles 358, 631, 650, 652, 656, 657, 658, 663, 715, 724, 729, 854, 873, 876, 890, 891, 894, 899, 904, 920, 922
  double 362, 468, 682
  multiple 683, 684, 688, 689, 738, 740, 780
  simple 224, 632, 646, 647, 649, 662, 677, 678, 681, 687, 692, 693, 698, 713, 745, 749, 750, 752, 768
Polyakov, A. 341, 347, 394, 516
Polyakov theorem 350, 358
Polyakov-Wiegmann identity 517
polymers 17, 77, 102, 709, 710, 738
Potts model 41, 42, 48, 70, 73, 81, 83, 98, 102, 104, 243, 487, 518, 533, 609, 613, 832, 858, 880
  fusion rules 533–536, 541, 738
  mean field theory 112–116
  one-dimensional 67–71
  S-matrix 696–699
Pozsgay, B. 833
primary operators 351, 352, 359, 360, 361, 362, 370, 371, 372, 375, 378, 380, 397, 400, 401, 407, 408, 409, 411, 412, 418, 420, 424, 436, 440, 447, 478–485, 490, 491, 494, 495, 532, 550, 561, 572, 606, 610, 611, 612, 676, 708, 759, 760, 761, 837, 841, 843, 847, 952, 953, 954
probability 7, 22, 24, 25, 26, 28, 29, 117, 118, 121, 122, 141–144, 147–149, 152, 153, 156–158, 193, 198, 240, 268, 278, 294, 623–628, 662–664, 678
propagator 146, 246–250, 254–256, 264, 276, 278, 287, 306, 307, 344, 346, 395, 412, 413, 468, 569, 578, 603, 604, 634, 683, 749, 755, 875, 876, 963
pseudo-energy 800, 801, 803, 805, 806, 818, 819, 828, 829, 831, 835

Qiu, Z. 404, 438, 515, 541
quantum anomaly 358, 364
quantum dimension 441, 442
quantum field theory 235–282, 285, 304, 305, 574, 576, 578, 582–616, 658, 738, 784, 785, 858, 879
quantum group 102, 212, 228, 231, 232, 233, 411, 438, 716, 718, 722, 723, 733
quantum hamiltonian 58, 319, 322, 325, 331, 340, 702
quantum mechanics of a particle 55, 125, 284, 548, 674

quantum Pitagora's
theorem 475
quantum statistics 27–31, 33, 35, 37
quasi-primary operators 351, 352, 359, 371, 610, 611, 612, 676

R-matrix 227–233, 718
radial quantization 368, 369, 839, 840
Rajaraman, R. 619, 882, 910
Ramond fields 458, 463, 466, 480, 481, 532, 541, 735, 541, 735
rapidity variable 338, 620, 634, 635, 639, 642, 643, 645, 646, 675, 701, 703, 723, 781, 795, 796, 797, 805, 814, 851, 872, 873, 886, 887, 889, 906, 907
Ravanini, F. 834
Rattazzi, R. 395
recursive method
one-dimensional Ising model 48–53
Rees, B.C. 972
Reichl, R.P. 39, 153
reflection amplitude 648, 660, 661, 673, 674, 697, 698, 705, 711, 740, 743, 836, 849, 851, 852, 853, 858
reflection formula 89
renormalization 12, 154, 264, 265, 267, 286, 496, 548, 549, 551, 552, 564, 573, 578, 590, 597, 755, 871, 886, 960
Renormalization Group 3, 13, 17, 59, 238, 242, 289, 290–314, 501, 547, 556, 615
β functions 313–315, 557, 558, 559, 561
block spins 292–294, 297, 300, 301, 302, 315
Callan-Symanzik equation 558
decimation 300
eigenvalues 291, 297, 312, 313
fixed points 170, 175, 296, 298, 299–318, 341, 347, 389, 390–392, 477, 545, 548, 559, 561, 562, 563, 566, 567, 568, 569, 574, 577
majority rule 293
scaling operators 345, 364, 446, 447, 534, 763
Reshetikin, N.Y. 232
resonances 552, 574, 647, 652, 653, 654, 655, 657, 662, 663, 664, 674, 675, 680, 681, 683, 688, 687, 688, 692, 740, 752, 834, 854, 863, 864, 887, 888, 892
Riemann function 38, 44, 447, 452, 827
Riemann sphere 356, 357, 388, 414, 498
Riemann surface 382, 382, 415, 631, 643, 645, 648
Riva, V. 880, 909
RSOS model 232, 518, 682, 719, 720–728
Rutkevich, S.B. 879
Rychkov, V.S. 394, 395, 972

S-matrix 547, 575, 588, 622–675
Ising model in a magnetic field 686
Ising model in a thermal field 684
O(n) models 704
RSOS models 722
thermal 3-state Potts model 696
thermal tricritical Ising model 692
thermal tricritical 3-state Potts model 697
Toda field theory 700
tricritical Ising model sub-leading magnetization 727
Sine-Gordon 710
SUSY 730
Yang-Lee 676
Saary, T.L. 102
Saddle point method 88, 133, 134, 136, 137, 139, 140, 141, 270, 639, 707
Saleur, H. 102, 858
Sasaki, R. 737, 738
Sato, M. 40
Savit, R. 186
scaling invariance 12
scaling law 15, 16, 17, 43, 66, 111, 137, 200, 299, 308, 313, 315, 351
scaling operators 345, 364, 446, 609, 763
scaling region 40, 545, 676
scattering process 267, 277, 278, 575, 588, 591, 599, 601, 603, 604, 629, 630, 688, 749, 791, 795, 818, 849, 850
elasticity 587, 608, 622, 626, 634, 635–664, 672, 676, 699, 702–711, 714, 715, 719, 720–724, 735, 738, 740–743, 746, 748
Schoeder, D.V. 285
Schultz, T.D. 339
Schwarz, J.H. 672
Schwarz derivative 365
Scott, A. 619
screening operators 416–418, 424
Seiberg, N. 186
self-avoiding walks 518, 540, 709, 738, 784
semi-classical
limit 270, 743, 885, 915, 920
formula 889, 937
methods 864, 909, 913, 938
spectrum 865, 882, 884, 890, 892, 894–906, 928, 973
Senechal, D. 394
Serban, D. 833
series expansion 64–66, 88, 90, 91, 103, 164, 166, 183, 184, 187, 197, 258, 289, 314, 329, 343, 464, 589, 669
Shankar, S. 739, 938
Shenker, S. 438, 515, 541, 938
Shifman, M. 574, 938
Shirkov, D.V. 285
Skorik, S. 858
Sierra, G. 438, 972
Simonetti, P. 21, 572, 574, 780, 784, 879, 973
Simon, P. 316, 672
Simons, S.H. 439
Simmons-Duffin, D. 394
Sinh-Gordon theory 576–583, 587, 588, 590, 593, 619, 684, 685, 686, 715, 739,

Sinh-Gordon theory (*cont.*) 755, 779, 784, 786, 807, 833
Sine-Gordon theory 231, 468, 471, 475, 565, 570, 572, 582–587, 590, 591, 593, 604, 612, 617, 619, 621, 636, 676, 682, 685, 710, 711, 715, 718, 719, 720–725, 732, 733, 734, 743, 759, 785, 833, 882, 883
  double 893, 902, 903, 904, 905, 907, 909, 910
  multi-frequency 866, 873, 877, 879
  supersymmetric 927, 929, 930, 931, 938
Slodowy, P. 439
Smith, H. 41
solid angle 73, 74, 75, 95, 248, 665, 667, 668
solitons 250, 287, 585–587, 590, 591, 608, 609, 618–620, 711, 714, 715, 718–722, 733, 734, 742, 743, 783, 865, 876, 910, 912, 938
Sonhius, M.F. 515
Sondhi, S.L. 209
SOS model 181, 183, 232, 518, 541
Sotkov, G. 515, 541, 738, 880, 909
Sodano, P. 909
specific heat 14, 26, 44, 53, 54, 56, 111, 196, 311
spectrum 21, 28, 44, 45, 125, 127, 223, 225, 275, 289, 319, 325, 332, 334, 340, 341, 345, 439, 455, 497, 512, 517, 547, 575, 582, 584, 587, 590, 597, 598, 599, 600, 603, 614, 616, 617, 622, 623, 637, 648, 650, 652, 654, 655, 659, 660, 680, 686, 687, 692, 693, 701, 704, 714, 715, 722, 744, 794, 832, 844, 863, 865, 879, 883, 885, 891–915, 919, 921, 927, 930, 933, 943, 951, 955
spherical model 130, 131, 133–137, 141, 147, 154, 156, 204, 245, 315, 318
spontaneous magnetization 4, 5, 9, 13, 14, 20, 51, 67, 108, 109, 120, 136, 155, 162, 163, 199, 209, 259, 260
spontaneous symmetry breaking 261, 330, 516, 590, 684, 728, 735, 881, 926, 931, 935
star-triangle transformation 161, 172, 173, 216
Stanley, H.E. 40, 130, 154
Staniskov, M. 515, 541
statistics
  Bose-Einstein 32, 36, 41, 45, 136, 275, 702, 796
  Fermi-Dirac 32–35
Stauffer, D. 102
Stern, A. 439
stress-energy tensor 272, 284, 287, 350, 355, 360, 362–368, 371, 376, 378, 384, 414, 415, 447, 457, 461, 479, 484, 491, 502, 505, 514, 557, 559, 564, 565, 572, 575, 577, 583, 607, 611, 757, 767, 778, 785, 793, 837, 838, 874, 925, 932
string theory 446, 489, 658, 672
Stringari, S. 41
structure constants 345, 346, 381, 422, 423, 428, 484, 550, 553, 832, 847, 954, 955
Sugawara formula 491, 492, 495, 502
sum rule
  c-theorem 563, 565, 570, 571
  Δ-theorem 573
supersymmetry 476, 482, 515, 516, 532, 541, 608, 730, 735, 881, 913, 914, 923, 926, 927, 931, 935, 937, 938
susceptibility 14, 16, 20, 52, 53, 66, 109, 155, 184, 310, 311
Sutherland, B. 232
Suzuki, M. 232
Symanzik, K. 315
symmetric polynomials 750, 753, 757, 775, 776
Takacs, G. 833, 880, 909, 910, 938, 972
Takahashi, M. 232
Takhtajan, L. 618
Tarasov, V.O. 233
Tateo, R. 833, 834
Taylor, J.R. 671
temperature 6, 8, 18, 20, 26, 29, 30, 34, 36, 37, 42, 43, 45, 48, 51, 53, 54, 60, 64, 66, 72, 79, 81, 109, 158, 245, 382, 537, 538, 569, 609, 800, 803, 805, 828, 833
  critical 3, 11, 44, 50, 111, 112, 129, 135, 136, 155, 162, 175, 176, 181, 188, 196, 262, 307, 519, 696, 727, 878
  Curie 4
  high-temperature expansion 64, 71, 165, 166, 184, 187, 189, 200, 206, 207, 209, 708
  high-temperature phase 18, 20, 71, 124, 161, 163, 168, 174, 176, 263, 301, 303, 330, 331, 538, 684, 685, 708, 766, 768, 831, 832, 964, 969, 971, 972, 977
  low-temperature expansion 163, 166, 171, 187
  low-temperature phase 20, 124, 130, 161, 168, 169, 174, 176, 181, 263, 301, 303, 331, 684, 692, 831, 879, 881, 965
tensor product representation 28, 227
Tetradis, N. 316
Thermodynamics Bethe Ansatz 791–830
  dressed energy and momentum 805, 806
  holes and roots 797
  selection rules 796, 797
theta functions 472
Thirring model 468, 471, 475, 710, 711
threshold 630, 631, 633, 668, 688, 749, 756, 863, 864, 871, 879, 897, 901, 906, 907, 920, 937, 943, 963, 966, 967, 968, 971

time ordering 624
time reversal 643
Tonni, E. 383, 394, 395
Toda field theories
imaginary coupling constant
real coupling constant 577, 589, 593, 595–606, 619, 653, 655, 676, 680, 682, 684, 687, 688, 689, 691, 692, 694, 700–704, 737–739, 755, 807, 816, 858
topological excitations 250, 287, 584, 585, 587, 590, 734, 864, 876, 883, 884, 900, 913, 935
Toth, G. 880
Toulouse, G. 316
torus geometry 399, 406, 429, 431, 432, 433, 443, 450, 451, 462, 465, 471, 482, 518, 528, 529, 844
Tsvelik, A. 471
Tracy, C. 20, 40, 772, 784
trace anomaly 358, 364, 365
transfer matrix 55, 56, 57, 61, 62, 63, 65, 68, 69, 103, 124, 125, 129, 154, 161, 172, 189, 211–234, 238, 319, 320, 321, 323, 380, 716, 811, 847
transmission amplitude 660, 661, 705, 711, 740, 743, 940
triangle of masses 601, 647, 648, 652, 653, 963
tricritical Ising model 261–263, 316, 399, 406, 482, 515, 517, 518, 530, 531, 541, 617, 692, 693, 695, 697, 727, 728, 735, 738, 820, 880, 881
tricritical Potts model 606, 609, 613
truncated Hilbert Space approach 943–972
twist field 383, 384
two-point function 9, 21, 53, 62, 76, 77, 247, 253, 254, 257, 258, 276, 306, 351, 352, 362, 363, 364, 378, 380, 413–416, 425, 450, 456, 459, 461, 468, 521, 524, 554, 555, 576, 578, 754, 755, 764, 767, 768, 770, 785, 787, 841, 880

Ueno, K. 772, 784
ultraviolet divergences (*see* divergence)
ultraviolet limit 763–765, 779, 794, 806–809, 817, 877
unitarity condition
conformal
representations 404, 438, 440, 518, 538, 563
S-matrix 628, 641–650, 675, 697, 698, 705, 706, 709, 712, 722, 724, 729, 732, 736, 737, 741, 744, 796, 798, 818, 849, 850, 935
unitarity paradox 677, 678
universal ratios 15, 110, 308, 309, 310, 311, 313, 316, 318
universality see *class of universality*

vacuum expectation values see *expectation values*
vacuum state 238, 275, 276, 329, 330, 370, 372, 448, 459, 471, 490, 493, 494, 495, 517, 548, 549, 678, 685, 721, 726, 727, 735, 746, 832, 846, 853, 854, 870, 898, 900, 901, 902, 926, 935, 941, 942
Verlinde algebra 411, 412, 439, 441, 653, 857
Verma module see *conformal family*
vertex functions 257, 258, 259, 264, 265, 266, 267
vertex operator 413–419, 424, 447, 448, 450, 468, 471, 504, 517, 524, 605, 621, 746, 779
Vichi, A. 394, 395
vicinity of the critical point 17, 110, 161, 196, 305, 341, 344, 545, 548–572
Virasoro algebra
generator 362, 363, 367, 368, 370, 462, 476, 479, 502, 503, 504, 516, 744, 752, 839, 840
characters 433, 434, 466, 527, 528
minimal models 399–433, 480, 482
Vitale, L.G. 972
Vdovichenko, N.V. 189, 190, 192, 209
Vu, D.L. 833

Wadati, M. 233
Wagner, F. 909, 910, 938, 972
Wannier, G.H. see *Kramers-Wannier*
Ward identity 358–364, 384, 397, 460, 496, 556, 573, 606, 607, 838, 839
Ward, J.C. 189, 209
Watson equation 748, 749, 753, 786, 907
Wegner, F. 186
Weinberg, S. 285, 671
Weisz, P. 783
wells
asymmetric 730, 897, 898, 899, 941
symmetric 893, 895, 901, 921
Wess-Zumino-Witten model 476, 496, 498, 499, 501, 516, 517, 972
West, B.J. 154
Wetterich, C. 316
White, A.R. 316, 671
Wick's theorem 251, 252, 276, 343, 363, 364, 413, 419, 468, 520, 521, 570, 571, 868
Wilczek, F. 41, 394
Wilson, K.G. 316
Wilson, R. 102
Witten, E. 186, 516, 672, 739, 938
Wu, T.T. 20, 40, 772, 784, 880
Wybourne, B. 516

XXZ model 233, 340
XY model 6, 181, 183

Yang, C.N. 199, 209, 794, 805, 832
Yang, C.P. 794, 805, 832
Yang, S.K. 619
Yang-Baxter equation 46, 47, 189, 212, 227, 228, 229, 230–233, 399, 640, 641, 643, 646, 648, 673, 705, 712, 718, 723, 732, 833, 850
Yang-Lee model 406, 518, 537, 542, 554, 555, 574, 654, 676, 677, 678, 679, 680

Yang-Lee model (*cont.*) 737, 740, 743, 785, 787, 817, 818, 972
Yung, A. 938
Yurov, V.P. 784, 972

Z_N models 81–83, 482–487
Zakhorov, V. 574, 618
Zamolodchikov, A.B. 3, 21, 394, 438, 482, 489, 516, 612, 619, 646, 671, 737, 738, 739, 745, 785, 791, 794, 833, 834, 836, 848, 858, 880, 966, 971, 973
Zamolodchikov, Al.B. 394, 574, 671, 738, 784, 785, 832, 833, 834, 972
Zee, T. 285
zero mode 444, 445, 446, 448, 451, 455, 459, 460, 467, 490, 528, 825, 915, 917, 918, 920, 922, 941
Zhu, C.J. 738
Zinn-Justin, J. 285, 574
Zuber, J.B. 339, 435, 439, 471

The manufacturer's authorised representative in the EU for product safety is Oxford University Press España S.A. of el Parque Empresarial San Fernando de Henares, Avenida de Castilla, 2 – 28830 Madrid (www.oup.es/en or product.safety@oup.com). OUP España S.A. also acts as importer into Spain of products made by the manufacturer.

www.ingramcontent.com/pod-product-compliance
Lightning Source LLC
LaVergne TN
LVHW062201200926
848484LV00006B/1295
* 9 7 8 0 1 9 8 7 8 8 1 0 2 *